KIRK-OTHMER

ENCYCLOPEDIA
OF CHEMICAL
TECHNOLOGY

Third Edition

VOLUME 6

**Chocolate and Cocoa
to
Copper**

ERRATA

VOLUME 6

Page	Line	For:	Read:
87	7	$CrSO_4.H_2O$ [*19812-12-0*]	$CrSO_4.7H_2O$ [*7789-05-1*]
90	between 36–37		**Chromium(VI) Compounds**
99	6 from bottom	1.8 to 5 g Cr	$(1.8–5) \times 10^{-11}$ g Cr
330	3	$\tan E_y/E_x$	$\cot^{-1} E_y/E_x$
330	3	$\tan K/\beta(1 - K)$	$\cot^{-1} K/\beta(1 - K)$
333	Fig. 7	Compressor efficiency, η_c	Conversion efficiency, η_e
		Locus max η_c	Locus max η_e
340	Fig. 10 caption	B, coal(seed/wt, 0.5% Cs)	B, coal(seed/wt, 0.7% K)
		C, coal(seed/wt, 0.7% K)	C, oil(seed/wt, 0.5% Cs)
		D, coal(seed/wt, 0.5% K)	D, nat. gas(seed/wt, 0.5% Cs)
341	Fig. 11 caption	$\theta = 0.95$	$\phi = 0.95$
343	Fig. 12 vertical scale	wt %	wt
354	3 from bottom	516 MJ/kg	516 g/GJ
361	eq. 74	$\alpha = \tan (E_y/E_x = \tan K/[\beta(1 - K)]$	$\alpha = \cot^{-1} (E_y/E_x) = \cot^{-1} K/[\beta(1 - K)]$
786	bottom of page		(structure missing as shown)

(2)

KIRK-OTHMER

ENCYCLOPEDIA OF CHEMICAL TECHNOLOGY

THIRD EDITION

VOLUME 6

CHOCOLATE AND COCOA

TO

COPPER

A WILEY-INTERSCIENCE PUBLICATION

John Wiley & Sons

NEW YORK • CHICHESTER • BRISBANE • TORONTO

Library of Congress Cataloging in Publication Data:

Main entry under title:
    Encyclopedia of chemical technology.

    At head of title: Kirk-Othmer.
    "A Wiley-Interscience publication."
    Includes bibliographies.
    1. Chemistry, Technical—Dictionaries. I. Kirk,
Raymond Eller, 1890–1957. II. Othmer, Donald Frederick,
1904–   III. Grayson, Martin. IV. Eckroth, David.
V. Title: Kirk-Othmer encyclopedia of chemical technology.

TP9.E685   1978   660'.03   77-15820
ISBN 0-471-02042-7

Printed in the United States of America

CONTENTS

EDITORIAL STAFF
FOR VOLUME 6

Executive Editor: **Martin Grayson**
Associate Editor: **David Eckroth**
Production Supervisor: **Michalina Bickford**
Editors: **Galen J. Bushey**   **Caroline L. Eastman**   **Anna Klingsberg**
                                      Lorraine van Nes

CONTRIBUTORS
TO VOLUME 6

Roger J. Araujo, *Corning Glass Works,*
    mogenic materials
Ronald D. Archer, *University of Mas*
    compounds
Fred W. Billmeyer, Jr., *Rensselaer Po*
Edward F. Bouchard, *Pfizer Inc., N*
L. J. Chinn, *G. D. Searle & Co., Ch*
W. P. Clinton, *General Foods Co*
F. B. Colton, *G. D. Searle & Co*
Jesse H. Day, *Ohio University*
    Chromogenic materials
R. W. Drisko, *U. S. Naval Ci*
    marine
P. A. Dougall, *Kennecott*
E. J. Finnegan, *Hershey F*
J. E. Gearien, *Universit*
    inhibitors

Stewart W

Thomas G. Webb
J. H. Westbrook, Gen
    conversio

Stanley C. Zink, The Black Claw
B. L. Zoumas, Hershey Foods Corp
Samuel Zuckerman, Hershey
    chromium alloys

Daan M. Zwick, Eastman Kodak Company, Rod
    drugs, and cosmetics
    H. Kohnstamm & Co

Ralph E. Grim, *University of Illinois, Urbana, Illinois,* Uses under Clays

M. Hamell, *General Foods Corp., White Plains, New York,* Coffee

Thomas J. Harrison, *International Business Machines Corporation, Boca Raton, Florida,* Computers

Winslow H. Hartford, *Belmont Abbey College, Belmont Abbey, North Carolina,* Chromium compounds

W. B. Hillig, *General Electric Co., Schenectady, New York,* Composite materials

Seymore Hochberg, *E. I. du Pont de Nemours & Co., Inc., Philadelphia, Pennsylvania,* Coatings, industrial

Thomas H. Jukes, *University of California, Berkeley, Berkeley, California,* Choline

W. D. Keller, *University of Missouri, Columbia, Missouri,* Survey under Clays

Earl C. Makin, *Monsanto Chemical Intermediates Co., Texas City, Texas,* Clathration

Edward G. Merritt, *Pfizer Inc., New York, New York,* Citric acid

F. R. Morral, *Consultant, Columbus, Ohio,* Cobalt compounds

Edward E. Mueller, *Alfred University, Alfred, New York,* Colorants for ceramics

Charles G. Munger, *Consultant, Fallbrook, California,* Coatings, resistant

John B. Newkirk, *University of Denver, Denver, Colorado,* Cobalt and cobalt alloys

A. H. Nishikawa, *Hoffmann LaRoche Inc., Nutley, New Jersey,* Chromatography, affinity

Michael Perch, *Koppers Co., Inc., Verona, Pennsylvania,* Carbonization under Coal conversion processes

F. Planinsek, *University of Denver, Denver, Colorado,* Cobalt and cobalt alloys

Miguel F. Refojo, *Eye Research Institute of Retina Foundation, Boston, Massachusetts,* Contact lenses

W. F. Ringk, *Consultant, Westfield, New Jersey,* Cinnamic acid, cinnamaldehyde, and cinnamyl alcohol

D. N. Schulz, *Firestone Tire & Rubber Co., Akron, Ohio,* Copolymers

Joseph Senackerib, *H. Kohnstamm & Co., Inc., New York, New York,* Colorants for foods, drugs, and cosmetics

Edgel Stambaugh, *Battelle Memorial Institute, Columbus, Ohio,* Desulfurization under Coal conversion processes

A. Stefanucci, *General Foods Corp., White Plains, New York,* Coffee

D. P. Tate, *Firestone Tire & Rubber Co., Akron, Ohio,* Copolymers

Fred N. Teumac, *Uniroyal, Inc., Mishaka, Indiana,* Coated fabrics

John R. Thirtle, *Eastman Kodak Company, Rochester, New York,* Color photography

W. M. Tuddenham, *Kennecott Copper Corporation, Salt Lake City, Utah,* Copper

Karl S. Vorres, *Institute of Gas Technology, Chicago, Illinois,* Coal

Vivian K. Walworth, *Polaroid Corporation, Cambridge, Massachusetts,* Color photography, instant

_ _ay, *MHD Consultant, Columbia, Maryland,* Magnetohydrodynamics under Coal _ _n processes

_ _er, *Consultant, Vienna, West Virginia,* Colorants for plastics

_ _eral Electric Company, Schenectady, New York, Chromium and

_ _son Co., Fulton, New York, Coating processes

_ _ration, Hershey, Pennsylvania, Chocolate and cocoa

_ _o., Inc., New York, New York, Colorants for foods,

_ _hester, New York, Color photography

NOTE ON CHEMICAL ABSTRACTS
SERVICE REGISTRY NUMBERS
AND NOMENCLATURE

Chemical Abstracts Service (CAS) Registry Numbers are unique numerical identifiers assigned to substances recorded in the CAS Registry System. They appear in brackets in the *Chemical Abstracts* (CA) substance and formula indexes following the names of compounds. A single compound may have many synonyms in the chemical literature. A simple compound like phenethylamine can be named β-phenylethylamine or, as in *Chemical Abstracts*, benzeneethanamine. The usefulness of the Encyclopedia depends on accessibility through the most common correct name of a substance. Because of this diversity in nomenclature careful attention has been given the problem in order to assist the reader as much as possible, especially in locating the systematic CA index name by means of the Registry Number. For this purpose, the reader may refer to the CAS Registry Handbook-Number Section which lists in numerical order the Registry Number with the Chemical Abstracts index name and the molecular formula; eg, **458-88-8,** Piperidine, 2-propyl-, (*S*)-, $C_8H_{17}N$; in the Encyclopedia this compound would be found under its common name, coniine [*458-88-8*]. The Registry Number is a valuable link for the reader in retrieving additional published information on substances and also as a point of access for such on-line data bases as Chemline, Medline, and Toxline.

In all cases, the CAS Registry Numbers have been given for title compounds in articles and for all compounds in the index. All specific substances indexed in *Chemical Abstracts* since 1965 are included in the CAS Registry System as are a large number of substances derived from a variety of reference works. The CAS Registry System identifies a substance on the basis of an unambiguous computer-language description of its molecular structure including stereochemical detail. The Registry Number is a machine-checkable number (like a Social Security number) assigned in sequential order to each substance as it enters the registry system. The value of the number lies in the fact that it is a concise and unique means of substance identification, which is independent of, and therefore bridges, many systems of chemical nomenclature. For polymers, one Registry Number is used for the entire family; eg, polyoxyethylene (20)sorbitan monolaurate has the same number as all of its polyoxyethylene homologues.

Registry numbers for each substance will be provided in the third edition index (eg, Alkaloids will show the Registry Number of all alkaloids (title compounds) in a table in the article as well, but the intermediates will have their Registry Numbers shown only in the index). Articles such as Absorption, Adsorptive separation, Air conditioning, Air pollution, Air pollution control methods will have no Registry Numbers in the text.

Cross-references have been inserted in the index for many common names and for some systematic names. Trademark names appear in the index. Names that are incorrect, misleading or ambiguous are avoided. Formulas are given very frequently in the text to help in identifying compounds. The spelling and form used, even for industrial names, follow American chemical usage, but not always the usage of *Chemical Abstracts* (eg, *coniine* is used instead of *(S)-2-propylpiperidine, aniline* instead of *benzenamine,* and *acrylic acid* instead of *2-propenoic acid*).

There are variations in representation of rings in different disciplines. The dye industry does not designate aromaticity or double bonds in rings. All double bonds and aromaticity will be shown in the *Encyclopedia* as a matter of course. For example, tetralin has an aromatic ring and a saturated ring and its structure will appear in the

Encyclopedia with its common name, Registry Number enclosed in brackets, and parenthetical CA index name, ie, tetralin, *[119-64-2]* (1,2,3,4-tetrahydronaphthalene). With names and structural formulas, and especially with CAS Registry Numbers, the aim is to help the reader have a concise means of substance identification.

CONVERSION FACTORS, ABBREVIATIONS, AND UNIT SYMBOLS

SI Units (Adopted 1960)

A new system of measurement, the International System of Units (abbreviated SI), is being implemented throughout the world. This system is a modernized version of the MKSA (meter, kilogram, second, ampere) system, and its details are published and controlled by an international treaty organization (The International Bureau of Weights and Measures) (1).

SI units are divided into three classes:

BASE UNITS

length	meter[†] (m)
mass[‡]	kilogram (kg)
time	second (s)
electric current	ampere (A)
thermodynamic temperature[§]	kelvin (K)
amount of substance	mole (mol)
luminous intensity	candela (cd)

[†] The spellings "metre" and "litre" are preferred by ASTM; however "-er" will be used in the Encyclopedia.

[‡] "Weight" is the commonly used term for "mass".

[§] Wide use is made of "Celsius temperature" (t) defined by

$$t = T - T_0$$

where T is the thermodynamic temperature, expressed in kelvins, and $T_0 = 273.15$ K by definition. A temperature interval may be expressed in degrees Celsius as well as in kelvins.

SUPPLEMENTARY UNITS

plane angle	radian (rad)
solid angle	steradian (sr)

DERIVED UNITS AND OTHER ACCEPTABLE UNITS

These units are formed by combining base units, supplementary units, and other derived units (2–4). Those derived units having special names and symbols are marked with an asterisk in the list below:

Quantity	Unit	Symbol	Acceptable equivalent
*absorbed dose	gray	Gy	J/kg
acceleration	meter per second squared	m/s^2	
*activity (of ionizing radiation source)	becquerel	Bq	1/s
area	square kilometer	km^2	
	square hectometer	hm^2	ha (hectare)
	square meter	m^2	
*capacitance	farad	F	C/V
concentration (of amount of substance)	mole per cubic meter	mol/m^3	
*conductance	siemens	S	A/V
current density	ampere per square meter	A/m^2	
density, mass density	kilogram per cubic meter	kg/m^3	g/L; mg/cm^3
dipole moment (quantity)	coulomb meter	C·m	
*electric charge, quantity of electricity	coulomb	C	A·s
electric charge density	coulomb per cubic meter	C/m^3	
electric field strength	volt per meter	V/m	
electric flux density	coulomb per square meter	C/m^2	
*electric potential, potential difference, electromotive force	volt	V	W/A
*electric resistance	ohm	Ω	V/A
*energy, work, quantity of heat	megajoule	MJ	
	kilojoule	kJ	
	joule	J	N·m
	electron volt[†]	eV[†]	
	kilowatt-hour[†]	kW·h[†]	

[†] This non-SI unit is recognized by the CIPM as having to be retained because of practical importance or use in specialized fields (1).

Quantity	Unit	Symbol	Acceptable equivalent
energy density	joule per cubic meter	J/m^3	
*force	kilonewton	kN	
	newton	N	$kg \cdot m/s^2$
*frequency	megahertz	MHz	
	hertz	Hz	$1/s$
heat capacity, entropy	joule per kelvin	J/K	
heat capacity (specific), specific entropy	joule per kilogram kelvin	$J/(kg \cdot K)$	
heat transfer coefficient	watt per square meter kelvin	$W/(m^2 \cdot K)$	
*illuminance	lux	lx	lm/m^2
*inductance	henry	H	Wb/A
linear density	kilogram per meter	kg/m	
luminance	candela per square meter	cd/m^2	
*luminous flux	lumen	lm	$cd \cdot sr$
magnetic field strength	ampere per meter	A/m	
*magnetic flux	weber	Wb	$V \cdot s$
*magnetic flux density	tesla	T	Wb/m^2
molar energy	joule per mole	J/mol	
molar entropy, molar heat capacity	joule per mole kelvin	$J/(mol \cdot K)$	
moment of force, torque	newton meter	$N \cdot m$	
momentum	kilogram meter per second	$kg \cdot m/s$	
permeability	henry per meter	H/m	
permittivity	farad per meter	F/m	
*power, heat flow rate, radiant flux	kilowatt	kW	
	watt	W	J/s
power density, heat flux density, irradiance	watt per square meter	W/m^2	
*pressure, stress	megapascal	MPa	
	kilopascal	kPa	
	pascal	Pa	N/m^2
sound level	decibel	dB	
specific energy	joule per kilgram	J/kg	
specific volume	cubic meter per kilogram	m^3/kg	
surface tension	newton per meter	N/m	
thermal conductivity	watt per meter kelvin	$W/(m \cdot K)$	
velocity	meter per second	m/s	
	kilometer per hour	km/h	
viscosity, dynamic	pascal second	$Pa \cdot s$	
	millipascal second	$mPa \cdot s$	
viscosity, kinematic	square meter per second	m^2/s	

Quantity	Unit	Symbol	Acceptable equivalent
	square millimeter per second	mm^2/s	
volume	cubic meter	m^3	
	cubic decimeter	dm^3	L(liter) (5)
	cubic centimeter	cm^3	mL
wave number	1 per meter	m^{-1}	
	1 per centimeter	cm^{-1}	

In addition, there are 16 prefixes used to indicate order of magnitude, as follows:

Multiplication factor	Prefix	Symbol	Note
10^{18}	exa	E	
10^{15}	peta	P	
10^{12}	tera	T	
10^{9}	giga	G	
10^{6}	mega	M	
10^{3}	kilo	k	
10^{2}	hecto	h[a]	[a] Although hecto, deka, deci, and centi are
10	deka	da[a]	SI prefixes, their use should be avoided
10^{-1}	deci	d[a]	except for SI unit-multiples for area
10^{-2}	centi	c[a]	and volume and nontechnical use of
10^{-3}	milli	m	centimeter, as for body and clothing
10^{-6}	micro	μ	measurement.
10^{-9}	nano	n	
10^{-12}	pico	p	
10^{-15}	femto	f	
10^{-18}	atto	a	

For a complete description of SI and its use the reader is referred to ASTM E 380 (4) and the article Units and Conversion Factors which will appear in a later volume of the *Encyclopedia*.

A representative list of conversion factors from non-SI to SI units is presented herewith. Factors are given to four significant figures. Exact relationships are followed by a dagger. A more complete list is given in ASTM E 380-76(4) and ANSI Z210.1-1976 (6).

Conversion Factors to SI Units

To convert from	To	Multiply by
acre	square meter (m^2)	4.047×10^3
angstrom	meter (m)	1.0×10^{-10}†
are	square meter (m^2)	1.0×10^{2}†
astronomical unit	meter (m)	1.496×10^{11}
atmosphere	pascal (Pa)	1.013×10^5
bar	pascal (Pa)	1.0×10^{5}†
barrel (42 U.S. liquid gallons)	cubic meter (m^3)	0.1590
Bohr magneton μ_β	J/T	9.274×10^{-24}
Btu (International Table)	joule (J)	1.055×10^3

† Exact.

To convert from	*To*	*Multiply by*
Btu (mean)	joule (J)	1.056×10^3
Btu (thermochemical)	joule (J)	1.054×10^3
bushel	cubic meter (m^3)	3.524×10^{-2}
calorie (International Table)	joule (J)	4.187
calorie (mean)	joule (J)	4.190
calorie (thermochemical)	joule (J)	4.184[†]
centipoise	pascal second (Pa·s)	1.0×10^{-3}[†]
centistoke	square millimeter per second (mm^2/s)	1.0[†]
cfm (cubic foot per minute)	cubic meter per second (m^3/s)	4.72×10^{-4}
cubic inch	cubic meter (m^3)	1.639×10^{-5}
cubic foot	cubic meter (m^3)	2.832×10^{-2}
cubic yard	cubic meter (m^3)	0.7646
curie	becquerel (Bq)	3.70×10^{10}[†]
debye	coulomb·meter (C·m)	3.336×10^{-30}
degree (angle)	radian (rad)	1.745×10^{-2}
denier (international)	kilogram per meter (kg/m)	1.111×10^{-7}
	tex[‡]	0.1111
dram (apothecaries')	kilogram (kg)	3.888×10^{-3}
dram (avoirdupois)	kilogram (kg)	1.772×10^{-3}
dram (U.S. fluid)	cubic meter (m^3)	3.697×10^{-6}
dyne	newton (N)	1.0×10^{-5}[†]
dyne/cm	newton per meter (N/m)	1.00×10^{-3}[†]
electron volt	joule (J)	1.602×10^{-19}
erg	joule (J)	1.0×10^{-7}[†]
fathom	meter (m)	1.829
fluid ounce (U.S.)	cubic meter (m^3)	2.957×10^{-5}
foot	meter (m)	0.3048[†]
footcandle	lux (lx)	10.76
furlong	meter (m)	2.012×10^{-2}
gal	meter per second squared (m/s^2)	1.0×10^{-2}[†]
gallon (U.S. dry)	cubic meter (m^3)	4.405×10^{-3}
gallon (U.S. liquid)	cubic meter (m^3)	3.785×10^{-3}
gallon per minute (gpm)	cubic meter per second (m^3/s)	6.308×10^{-5}
	cubic meter per hour (m^3/h)	0.2271
gauss	tesla (T)	1.0×10^{-4}
gilbert	ampere (A)	0.7958
gill (U.S.)	cubic meter (m^3)	1.183×10^{-4}
grad	radian	1.571×10^{-2}
grain	kilogram (kg)	6.480×10^{-5}
gram force per denier	newton per tex (N/tex)	8.826×10^{-2}
hectare	square meter (m^2)	1.0×10^4[†]
horsepower (550 ft·lbf/s)	watt (W)	7.457×10^2
horsepower (boiler)	watt (W)	9.810×10^3
horsepower (electric)	watt (W)	7.46×10^2[†]
hundredweight (long)	kilogram (kg)	50.80
hundredweight (short)	kilogram (kg)	45.36
inch	meter (m)	2.54×10^{-2}[†]
inch of mercury (32°F)	pascal (Pa)	3.386×10^3

[†] Exact.

[‡] See footnote on p. xii

To convert from	To	Multiply by
inch of water (39.2°F)	pascal (Pa)	2.491×10^2
kilogram force	newton (N)	9.807
kilowatt hour	megajoule (MJ)	3.6†
kip	newton (N)	4.48×10^3
knot (international)	meter per second (m/s)	0.5144
lambert	candela per square meter (cd/m²)	3.183×10^3
league (British nautical)	meter (m)	5.559×10^3
league (statute)	meter (m)	4.828×10^3
light year	meter (m)	9.461×10^{15}
liter (for fluids only)	cubic meter (m³)	1.0×10^{-3}†
maxwell	weber (Wb)	1.0×10^{-8}†
micron	meter (m)	1.0×10^{-6}†
mil	meter (m)	2.54×10^{-5}†
mile (U.S. nautical)	meter (m)	1.852×10^3†
mile (statute)	meter (m)	1.609×10^3
mile per hour	meter per second (m/s)	0.4470
millibar	pascal (Pa)	1.0×10^2
millimeter of mercury (0°C)	pascal (Pa)	1.333×10^2†
minute (angular)	radian	2.909×10^{-4}
myriagram	kilogram (kg)	10
myriameter	kilometer (km)	10
oersted	ampere per meter (A/m)	79.58
ounce (avoirdupois)	kilogram (kg)	2.835×10^{-2}
ounce (troy)	kilogram (kg)	3.110×10^{-2}
ounce (U.S. fluid)	cubic meter (m³)	2.957×10^{-5}
ounce-force	newton (N)	0.2780
peck (U.S.)	cubic meter (m³)	8.810×10^{-3}
pennyweight	kilogram (kg)	1.555×10^{-3}
pint (U.S. dry)	cubic meter (m³)	5.506×10^{-4}
pint (U.S. liquid)	cubic meter (m³)	4.732×10^{-4}
poise (absolute viscosity)	pascal second (Pa·s)	0.10†
pound (avoirdupois)	kilogram (kg)	0.4536
pound (troy)	kilogram (kg)	0.3732
poundal	newton (N)	0.1383
pound-force	newton (N)	4.448
pound per square inch (psi)	pascal (Pa)	6.895×10^3
quart (U.S. dry)	cubic meter (m³)	1.101×10^{-3}
quart (U.S. liquid)	cubic meter (m³)	9.464×10^{-4}
quintal	kilogram (kg)	1.0×10^2†
rad	gray (Gy)	1.0×10^{-2}†
rod	meter (m)	5.029
roentgen	coulomb per kilogram (C/kg)	2.58×10^{-4}
second (angle)	radian (rad)	4.848×10^{-6}
section	square meter (m²)	2.590×10^6
slug	kilogram (kg)	14.59

† Exact.

To convert from	To	Multiply by
spherical candle power	lumen (lm)	12.57
square inch	square meter (m²)	6.452×10^{-4}
square foot	square meter (m²)	9.290×10^{-2}
square mile	square meter (m²)	2.590×10^{6}
square yard	square meter (m²)	0.8361
stere	cubic meter (m³)	$1.0^{\dagger}$
stokes (kinematic viscosity)	square meter per second (m²/s)	$1.0 \times 10^{-4\dagger}$
tex	kilogram per meter (kg/m)	$1.0 \times 10^{-6\dagger}$
ton (long, 2240 pounds)	kilogram (kg)	1.016×10^{3}
ton (metric)	kilogram (kg)	$1.0 \times 10^{3\dagger}$
ton (short, 2000 pounds)	kilogram (kg)	9.072×10^{2}
torr	pascal (Pa)	1.333×10^{2}
unit pole	weber (Wb)	1.257×10^{-7}
yard	meter (m)	$0.9144^{\dagger}$

† Exact.

Abbreviations and Unit Symbols

Following is a list of commonly used abbreviations and unit symbols appropriate for use in the *Encyclopedia*. In general they agree with those listed in *American National Standard Abbreviations for Use on Drawings and in Text (ANSI Y1.1)* (6) and *American National Standard Letter Symbols for Units in Science and Technology (ANSI Y10)* (6). Also included is a list of acronyms for a number of private and government organizations as well as common industrial solvents, polymers, and other chemicals.

Rules for Writing Unit Symbols (4):

1. Unit symbols should be printed in upright letters (roman) regardless of the type style used in the surrounding text.

2. Unit symbols are unaltered in the plural.

3. Unit symbols are not followed by a period except when used as the end of a sentence.

4. Letter unit symbols are generally written in lower-case (eg, cd for candela) unless the unit name has been derived from a proper name, in which case the first letter of the symbol is capitalized (W,Pa). Prefix and unit symbols retain their prescribed form regardless of the surrounding typography.

5. In the complete expression for a quantity, a space should be left between the numerical value and the unit symbol. For example, write 2.37 lm, *not* 2.37lm, and 35 mm, *not* 35mm. When the quantity is used in an adjectival sense, a hyphen is often used, for example, 35-mm film. *Exception:* No space is left between the numerical value and the symbols for degree, minute, and second of plane angle, and degree Celsius.

6. No space is used between the prefix and unit symbols (eg, kg).

7. Symbols, not abbreviations, should be used for units. For example, use "A," not "amp," for ampere.

8. When multiplying unit symbols, use a raised dot:

N·m for newton meter

In the case of W·h, the dot may be omitted, thus:

$$Wh$$

An exception to this practice is made for computer printouts, automatic typewriter work, etc, where the raised dot is not possible, and a dot on the line may be used.

9. When dividing unit symbols use one of the following forms:

$$m/s \text{ } or \text{ } m \cdot s^{-1} \text{ } or \text{ } \frac{m}{s}$$

In no case should more than one slash be used in the same expression unless parentheses are inserted to avoid ambiguity. For example, write:

$$J/(mol \cdot K) \text{ } or \text{ } J \cdot mol^{-1} \cdot K^{-1} \text{ } or \text{ } (J/mol)/K$$

but *not*

$$J/mol/K$$

10. Do not mix symbols and unit names in the same expression. Write:

$$joules \text{ } per \text{ } kilogram \text{ } or \text{ } J/kg \text{ } or \text{ } J \cdot kg^{-1}$$

but *not*

$$joules/kilogram \text{ } nor \text{ } joules/kg \text{ } nor \text{ } joules \cdot kg^{-1}$$

ABBREVIATIONS AND UNITS

A	ampere	amt	amount
A	anion (eg, H*A*)	amu	atomic mass unit
a	atto (prefix for 10^{-18})	ANSI	American National Standards Institute
AATCC	American Association of Textile Chemists and Colorists	AO	atomic orbital
ABS	acrylonitrile–butadiene–styrene	APHA	American Public Health Association
		API	American Petroleum Institute
abs	absolute		
ac	alternating current, *n*.	aq	aqueous
a-c	alternating current, *adj*.	Ar	aryl
ac-	alicyclic	*ar-*	aromatic
ACGIH	American Conference of Governmental Industrial Hygienists	*as-*	asymmetric(al)
		ASH-RAE	American Society of Heating, Refrigerating, and Air Conditioning Engineers
ACS	American Chemical Society		
AGA	American Gas Association		
Ah	ampere hour	ASM	American Society for Metals
AIChE	American Institute of Chemical Engineers	ASME	American Society of Mechanical Engineers
AIP	American Institute of Physics	ASTM	American Society for Testing and Materials
alc	alcohol(ic)		
Alk	alkyl	at no.	atomic number
alk	alkaline (not alkali)	at wt	atomic weight

av(g)	average
bbl	barrel
bcc	body-centered cubic
Bé	Baumé
bid	twice daily
BOD	biochemical (biological) oxygen demand
bp	boiling point
Bq	becquerel
C	coulomb
°C	degree Celsius
C-	denoting attachment to carbon
c	centi (prefix for 10^{-2})
ca	circa (approximately)
cd	candela; current density; circular dichroism
CFR	Code of Federal Regulations
cgs	centimeter–gram–second
CI	Color Index
cis-	isomer in which substituted groups are on same side of double bond between C atoms
cl	carload
cm	centimeter
cmil	circular mil
cmpd	compound
COA	coenzyme A
COD	chemical oxygen demand
coml	commercial(ly)
cp	chemically pure
cph	close-packed hexagonal
CPSC	Consumer Product Safety Commission
D-	denoting configurational relationship
d	differential operator
d-	dextro-, dextrorotatory
da	deka (prefix for 10^1)
dB	decibel
dc	direct current, n.
d-c	direct current, adj.
dec	decompose
detd	determined
detn	determination
dia	diameter
dil	dilute
dl-; DL-	racemic
DMF	dimethylformamide
DMG	dimethyl glyoxime
DOE	Department of Energy

DOT	Department of Transportation
dp	dew point; degree of polymerization
dstl(d)	distill(ed)
dta	differential thermal analysis
(E)-	entgegen; opposed
ϵ	dielectric constant (unitless number)
e	electron
ECU	electrochemical unit
ed.	edited, edition, editor
ED	effective dose
emf	electromotive force
emu	electromagnetic unit
eng	engineering
EPA	Environmental Protection Agency
epr	electron paramagnetic resonance
eq.	equation
esp	especially
esr	electron-spin resonance
est(d)	estimate(d)
estn	estimation
esu	electrostatic unit
exp	experiment, experimental
ext(d)	extract(ed)
F	farad (capacitance)
f	femto (prefix for 10^{-15})
FAO	Food and Agriculture Organization (United Nations)
fcc	face-centered cubic
FDA	Food and Drug Administration
FEA	Federal Energy Administration
fob	free on board
FPC	Federal Power Commission
fp	freezing point
frz	freezing
G	giga (prefix for 10^9)
g	gram
(g)	gas, only as in H_2O(g)
g	gravitational acceleration
gem-	geminal
glc	gas-liquid chromatography
g-mol wt; gmw	gram-molecular weight
grd	ground
Gy	gray

H	henry	LPG	liquefied petroleum gas
h	hour; hecto (prefix for 10^2)	ltl	less than truckload lots
ha	hectare	lx	lux
HB	Brinell hardness number	M	mega (prefix for 10^6); metal (as in MA)
Hb	hemoglobin		
HK	Knoop hardness number	M	molar
HRC	Rockwell hardness (C scale)	m	meter; milli (prefix for 10^{-3})
HV	Vickers hardness number	m	molal
hyd	hydrated, hydrous	m-	meta
hyg	hygroscopic	max	maximum
Hz	hertz	MCA	Manufacturing Chemists' Association
^{i}(eg, Pr^i)	iso (eg, isopropyl)		
i-	inactive (eg, i-methionine)	MEK	methyl ethyl ketone
IACS	International Annealed Copper Standard	meq	milliequivalent
		mfd	manufactured
ibp	initial boiling point	mfg	manufacturing
ICC	Interstate Commerce Commission	mfr	manufacturer
		MIBC	methylisobutyl carbinol
ICT	International Critical Table	MIBK	methyl isobutyl ketone
ID	inside diameter; infective dose	MIC	minimum inhibiting concentration
IPS	iron pipe size	min	minute; minimum
IPT	Institute of Petroleum Technologists	mL	milliliter
		MLD	minimum lethal dose
ir	infrared	MO	molecular orbital
ISO	International Organization for Standardization	mo	month
		mol	mole
IUPAC	International Union of Pure and Applied Chemistry	mol wt	molecular weight
		mom	momentum
IV	iodine value	mp	melting point
J	joule	MR	molar refraction
K	kelvin	ms	mass spectrum
k	kilo (prefix for 10^3)	mxt	mixture
kg	kilogram	μ	micro (prefix for 10^{-6})
L	denoting configurational relationship	N	newton (force)
		N	normal (concentration)
L	liter (for fluids only) (5)	N-	denoting attachment to nitrogen
l-	*levo*-, levorotatory	n (as n_D^{20})	index of refraction (for 20°C and sodium light)
(l)	liquid, only as in NH_3(l)		
LC$_{50}$	conc lethal to 50% of the animals tested		
LCAO	linear combination of atomic orbitals	^{n} (as Bu^n), n-	normal (straight-chain structure)
lcl	less than carload lots		
LD$_{50}$	dose lethal to 50% of the animals tested	n	nano (prefix for 10^{-9})
		na	not available
liq	liquid	NAS	National Academy of Sciences
lm	lumen		
ln	logarithm (natural)	NASA	National Aeronautics and Space Administration
LNG	liquefied natural gas		
log	logarithm (common)	nat	natural

NBS	National Bureau of Standards
neg	negative
NF	*National Formulary*
NIH	National Institutes of Health
NIOSH	National Institute of Occupational Safety and Health
nmr	nuclear magnetic resonance
NND	New and Nonofficial Drugs (AMA)
no.	number
NOI- (BN)	not otherwise indexed (by name)
NOS	not otherwise specified
nqr	nuclear quadrople resonance
NRC	Nuclear Regulatory Commission; National Research Council
NRI	New Ring Index
NSF	National Science Foundation
NTSB	National Transportation Safety Board
O-	denoting attachment to oxygen
o-	ortho
OD	outside diameter
OPEC	Organization of Petroleum Exporting Countries
OSHA	Occupational Safety and Health Administration
owf	on weight of fiber
Ω	ohm
P	peta (prefix for 10^{15})
p	pico (prefix for 10^{-12})
p-	para
p.	page
Pa	pascal (pressure)
pd	potential difference
pH	negative logarithm of the effective hydrogen ion concentration
pmr	proton magnetic resonance
pos	positive
pp.	pages
ppb	parts per billion
ppm	parts per million
ppt(d)	precipitate(d)
pptn	precipitation
Pr (no.)	foreign prototype (number)
pt	point; part

PVC	poly(vinyl chloride)
pwd	powder
qv	quod vide (which see)
R	univalent hydrocarbon radical
(*R*)-	rectus (clockwise configuration)
rad	radian; radius
rds	rate determining step
ref.	reference
rf	radio frequency, *n*.
r-f	radio frequency, *adj.*
rh	relative humidity
RI	Ring Index
RT	room temperature
^{s} (eg, Bu^s); *sec*-	secondary (eg, secondary butyl)
S	siemens
(*S*)-	sinister (counterclockwise configuration)
S-	denoting attachment to sulfur
s-	symmetric(al)
s	second
(s)	solid, only as in H_2O(s)
SAE	Society of Automotive Engineers
SAN	styrene–acrylonitrile
sat(d)	saturate(d)
satn	saturation
SCF	self-consistent field
Sch	Schultz number
SFs	Saybolt Furol seconds
SI	Le Système International d'Unités (International System of Units)
sl sol	slightly soluble
sol	soluble
soln	solution
soly	solubility
sp	specific; species
sp gr	specific gravity
sr	steradian
std	standard
STP	standard temperature and pressure (0°C and 101.3 kPa)
SUs	Saybolt Universal seconds
syn	synthetic

^{t} (eg, Bu^t), t-, *tert*-	tertiary (eg, tertiary butyl)	Twad	Twaddell
T	tera (prefix for 10^{12}); tesla (magnetic flux density)	UL	Underwriters' Laboratory
		USDA	United States Department of Agriculture
t	metric ton (tonne) temperature	USP	*United States Pharmacopeia*
TAPPI	Technical Association of the Pulp and Paper Industry	uv	ultraviolet
tex	tex (linear density)	V	volt (emf)
THF	tetrahydrofuran	var	variable
tlc	thin layer chromatography	*vic*-	vicinal
TLV	threshold limit value	vol	volume (not volatile)
trans-	isomer in which substituted groups are on opposite sides of double bond between C atoms	vs	versus
		v sol	very soluble
		W	watt
		Wb	Weber
		Wh	watt hour
TSCA	Toxic Substance Control Act	WHO	World Health Organization (United Nations)
		wk	week
		yr	year
		(Z)-	zusammen; together

Non-SI (Unacceptable and Obsolete) Units   *Use*

Å	angstrom	nm
at	atmosphere, technical	Pa
atm	atmosphere, standard	Pa
b	barn	cm^2
bar†	bar	Pa
bhp	brake horsepower	W
Btu	British thermal unit	J
bu	bushel	m^3; L
cal	calorie	J
cfm	cubic foot per minute	m^3/s
Ci	curie	Bq
cSt	centistokes	mm^2/s
c/s	cycle per second	Hz
cu	cubic	exponential form
D	debye	C·m
den	denier	tex
dr	dram	kg
dyn	dyne	N
erg	erg	J
eu	entropy unit	J/K
°F	degree Fahrenheit	°C; K
fc	footcandle	lx
fl	footlambert	lx
fl oz	fluid ounce	m^3; L
ft	foot	m
ft·lbf	foot pound-force	J
gf den	gram-force per denier	N/tex
G	gauss	T
Gal	gal	m/s^2
gal	gallon	m^3; L

† Do not use bar (10^5Pa) or millibar (10^2Pa) because they are not SI units, and are accepted internationally only for a limited time in special fields because of existing usage.

Non-SI. (Unacceptable and Obsolete) Units		Use
Gb	gilbert	A
gpm	gallon per minute	(m^3/s); (m^3/h)
gr	grain	kg
hp	horsepower	W
ihp	indicated horsepower	W
in.	inch	m
in. Hg	inch of mercury	Pa
in. H_2O	inch of water	Pa
in.-lbf	inch pound-force	J
kcal	kilogram-calorie	J
kgf	kilogram-force	N
kilo	for kilogram	kg
L	lambert	lx
lb	pound	kg
lbf	pound-force	N
mho	mho	S
mi	mile	m
MM	million	M
mm Hg	millimeter of mercury	Pa
mμ	millimicron	nm
mph	miles per hour	km/h
μ	micron	μm
Oe	oersted	A/m
oz	ounce	kg
ozf	ounce-force	N
η	poise	Pa·s
P	poise	Pa·s
ph	phot	lx
psi	pounds-force per square inch	Pa
psia	pounds-force per square inch absolute	Pa
psig	pounds-force per square inch gage	Pa
qt	quart	m^3; L
°R	degree Rankine	K
rd	rad	Gy
sb	stilb	lx
SCF	standard cubic foot	m^3
sq	square	exponential form
thm	therm	J
yd	yard	m

BIBLIOGRAPHY

1. The International Bureau of Weights and Measures, BIPM, (Parc de Saint-Cloud, France) is described on page 22 of Ref. 4. This bureau operates under the exclusive supervision of the International Committee of Weights and Measures (CIPM).
2. *Metric Editorial Guide (ANMC-75-1)*, American National Metric Council, 1625 Massachusetts Ave. N.W., Washington, D.C. 20036, 1975.
3. *SI Units and Recommendations for the Use of Their Multiples and of Certain Other Units (ISO 1000-1973)*, American National Standards Institute, 1430 Broadway, New York, N. Y. 10018, 1973.
4. Based on *ASTM E 380-76 (Standard for Metric Practice)*, American Society for Testing and Materials, 1916 Race Street, Philadelphia, Pa. 19103, 1976.
5. *Fed. Regist.*, Dec. 10, 1976 (41 FR 36414).
6. For ANSI address, see Ref. 3.

R. P. LUKENS
American Society for Testing and Materials

C Continued

CHOCOLATE AND COCOA

The name *Theobroma cacao*, "food of the gods," indicating both the legendary origin and nourishing qualities of chocolate, was bestowed upon the cacao tree by Linnaeus in 1720. All cocoa and chocolate products are derived from the cocoa bean, the seed of the fruit of this tree. Davila Garibi, a contemporary Mexican scholar, has traced the derivation of the word from basic root words of the Mayan language to its adoption as chocolate in Spanish (1).

The terms cocoa and cacao are often used interchangeably in the literature. Both terms describe various products from harvest through processing. In this article, the term cocoa will be used to describe products in general and the term cacao reserved for botanical contexts. Cocoa traders and brokers frequently use the term raw cocoa to distinguish unroasted cocoa beans from finished products and this term is used to report statistics for cocoa bean production and consumption.

Standards for Cocoa and Chocolate

In the United States, chocolate and cocoa are foods standardized by the U.S. Food and Drug Administration under the Federal Food, Drug and Cosmetic Act. The current definitions and standards resulted from prolonged discussions between the U.S. chocolate industry and the Food and Drug Administration (FDA). The definitions and standards originally published in the *Federal Register* of December 6, 1944 have since been revised only slightly.

The Codex Alimentarius Committee on Cocoa Products, established by the Food and Agricultural Organization (FAO) of the United Nations in 1963, has also drafted product standards. The draft standards for several products, including sweetened

1

chocolate and milk chocolate, have been sent to member governments for acceptance. A draft standard becomes an official Codex standard when it has been accepted by a sufficient number of countries.

There are several major differences between the current U.S. standards for chocolate and cocoa products and those proposed by Codex, including a difference in the use of the word chocolate. According to U.S. standards, chocolate refers only to the material prepared from grinding cocoa beans (see below), also called chocolate liquor. According to Codex standards, chocolate designates a category of products as well as specific products. The FDA has already stated its intention to adopt the Codex food standards (2).

Cocoa

Production. Worldwide cocoa bean production has increased from about 75,000 metric tons in 1895 to well over 1,000,000 t today. Ghana is by far the largest single producer, followed by Nigeria and Brazil. Table 1 lists production statistics for these three countries, as well as for other countries with a long history of production.

Recently, government assistance programs have enabled Brazilian farmers to make significant progress in cocoa production (3). Because this assistance has emphasized research to increase cocoa bean yields and prevent disease, many experts predict that Brazil will be the leading producer of beans in the future.

Consumption. Consumption of cocoa beans has, for the most part, paralleled supply. The United States has been the largest single consumer of cocoa beans for many years. Table 2 gives the annual tonnage of cocoa beans imported into the United States and other leading consumer countries.

Prior to World War II, less than 50,000 t/yr (8%) of the world's annual usage of beans was processed in the country of origin. At present, however, about one half of the world supply undergoes some processing in the producing country. Though some of this cocoa is consumed, most is converted to cocoa butter, cocoa cake, cocoa powder, and unsweetened chocolate (chocolate liquor) for export. Brazil is leading the trend by processing as much as 40% of its annual crop.

Table 1.   Production of Raw Cocoa, Thousands of Metric Tons [a]

	1966–1967	1976–1977 (estimated)	1977–1978 (forecast)
Ghana	382	320	320
Nigeria	267	165	220
Brazil	175	234	249
Ivory Coast	150	230	255
Cameroon	86	82	90
Dominican Republic	28	30	30
other	263	278	300
Total	*1351*	*1339*	*1464*

[a] Ref. 3.

Table 2.   Consumption of Raw Cocoa, Thousands of Metric Tons[a]

	1960	1965	1970	1975	1977[b]	1978[c]
United States	213	285	266	208	186	180
United Kingdom	75	102	82	73	77	72
FRG	116	157	126	139	140	132
Netherlands	85	118	115	119	127	127
France	53	63	40	34	35	35
Eastern Europe and USSR	74	147	190	277	204	215
producing countries	155	216	286	368	395	396
Total	*931*	*1335*	*1357*	*1459*	*1373*	*1360*

[a] Ref. 3.
[b] Estimated.
[c] Forecast.

Marketing.   An International Cocoa Agreement, signed in 1972, governs the marketing of much of the world's cocoa. The International Cocoa Organization (London, England), which does not include the United States, is authorized by this agreement to purchase cocoa when prices fall below a predetermined figure. The purpose of the agreement was to guarantee producing countries a minimum price and thus prevent large fluctuations in the market. However, in recent years, cocoa has been in short supply and the Agreement has never been invoked. To date, over $86 million has been collected to purchase surplus cocoa.

Most chocolate and cocoa products imported into the United States are marketed through the New York Cocoa Exchange, Inc. The Exchange was established in 1925 in the hope of forming a futures market that would permit hedging and provide a reliable barometer of the value of cocoa. Reference 5 provides additional information on the functions of the New York Cocoa Exchange and a thorough review of the worldwide marketing of cocoa.

Cocoa Beans

Significant amounts of cocoa beans are produced in about 30 different localities. These areas are confined to latitudes 20° north or south of the equator. Although cocoa trees thrive in this very hot climate, young trees require the shade of larger trees such as banana, coconut and palm for protection.

A cocoa tree produces its first crop in three to four years and a full crop after six to seven years. A full grown tree can reach a height of 12 to 15 m but is normally trimmed to 5 to 6 m to permit easy harvest. Cocoa pods are harvested twice a year, once in May, and again in October or November.

Fermentation (Curing).   Prior to shipment from producing countries most cocoa beans undergo a process known as curing, fermenting, or sweating. These terms are used rather loosely to describe a procedure in which seeds are removed from the pods, fermented, and dried. Unfermented beans, particularly from Haiti and the Dominican Republic, are used in the United States.

The age-old process of preparing cocoa beans for market involves specific steps that allegedly promote the activities of certain enzymes. Various methods of fermentation are used to the same end.

Fermentation (qv) plays a major role in flavor development of beans by mechanisms that are not well understood (6). Because freshly harvested cocoa beans are covered with a white pulp, rich in sugars, fermentation begins almost immediately upon exposure to air. The sugars are converted to alcohol, and finally, to acetic acid which drains off, freeing the cotyledon from the pulpy mass. The acetic acid and heat formed during fermentation penetrate the skin or shell, killing the germ and initiating chemical changes within the bean that play a significant role in the development of flavor and color. During this initial stage of fermentation, the beans acquire the ability to absorb moisture which is necessary for many of the chemical reactions that follow.

Commercial Grades. Most cocoa beans imported into the United States are one of about a dozen commercial varieties that can be generally classified as Criollo and Forastero. Criollo beans have a light color, a mild, nutty flavor, and an odor somewhat like sour wine. Forastero beans have a strong, somewhat bitter flavor and various degrees of astringency. The Forastero varieties are more abundant and provide the basis for most chocolate and cocoa formulations. Table 3 shows the main varieties of cocoa beans imported into the United States. The varieties are usually named for the country or port of origin.

Bean Specifications. Cocoa beans vary widely in quality necessitating a system of inspection and grading to ensure uniformity. Producing countries have always inspected beans for proper curing and drying as well as for insect and mold damage. Only recently, however, has a procedure for grading beans been established at an international level. This ordinance, reached primarily through the efforts of FAO, has been adopted by Codex as the model ordinance for inspection and grading of beans. It classifies beans in two major categories according to the fraction of moldy, slaty, flat, germinated, and insect-damaged beans (7).

Cocoa beans are sometimes evaluated in the laboratory to distinguish and characterize flavors. The procedure usually consists of the following steps: beans are roasted at a standardized temperature for a specific period of time, shelled, usually by hand, and ground or heated slightly to obtain chocolate liquor. The liquor's taste is evaluated by a panel of experts who characterize and record the particular flavor profile. The Chocolate Manufacturers' Association of Americas recently formed a committee to standardize this laboratory evaluation (7).

Blending. Most chocolate and cocoa products consist of blends of beans chosen for flavor and color characteristics. Cocoa beans may be blended before or after roasting, or nibs may be blended before grinding. In some cases finished liquors are blended. Common, or basic beans are usually African or Brazilian, and constitute the bulk of most blends. More expensive flavor beans from Venezuela, Trinidad, Ecuador, etc, are added to impart specific characteristics. The blend is determined by the end use or type of product desired.

Table 3.   Main Varieties of Cocoa Beans Imported into the United States

Africa	South America	West Indies	Other
Accra (Ghana)	Arriba (Ecuador)	Trinidad	New Guinea
Ivory Coast	Venezuelean	Grenada	Malaysia
Lagos	Bahia (Brazil)	Sanchez (Dominican Rep.)	Samoa
Nigeria			
Fernando Po			
Sierra Leone			

Manufacture of Cocoa and Chocolate Products

The cocoa bean is the basic raw ingredient in the manufacture of all cocoa products. The beans are converted to chocolate liquor, the primary ingredient from which all chocolate and cocoa products are made. Figure 1 depicts the conversion of cocoa beans to chocolate liquor, and in turn, to cocoa powder, cocoa butter, and sweet and milk chocolate, the chief chocolate and cocoa products manufactured in the United States.

CHOCOLATE LIQUOR

Chocolate liquor is the solid or semiplastic food prepared by finely grinding the kernel or nib of the cocoa bean. It is also commonly called chocolate, baking chocolate, or cooking chocolate. In Europe, chocolate liquor is often called chocolate mass.

Cleaning. Cocoa beans are imported in the United States in 70-kg bags. The beans can be processed almost immediately or stored for later use. They are usually fumigated prior to storage.

The first step in the processing of cocoa beans is cleaning. Stones, metals, twigs,

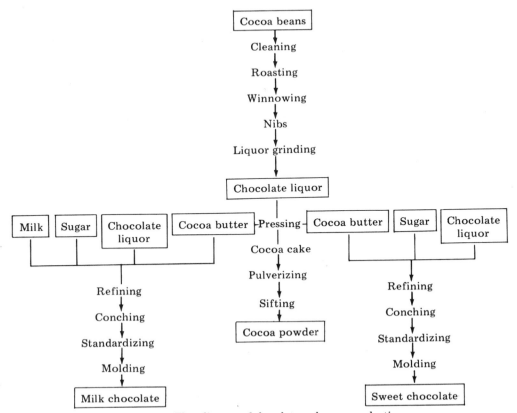

Figure 1. Flow diagram of chocolate and cocoa production.

twine and other foreign matter are usually removed by passing beans in a large thin layer over a vibrating screen cleaner. Large objects are retained as the beans fall through to a lower screen. The second screen removes sand and dirt that have adhered to the beans. Strategically placed magnets are commonly used to remove small pieces of metal.

Roasting.   The chocolate flavor familiar to us is primarily developed during roasting which promotes reactions among the latent flavor precursors in the bean. Good flavor depends upon the variety of bean and the curing process used. The bacterial or enzymatic changes that occur during fermentation presumably set the stage for the production of good flavor precursors.

Although flavor precursors in the unroasted cocoa bean have no significant chocolate flavor themselves, they react to form highly flavored compounds. These flavor precursors include various chemical compounds such as: proteins, amino acids, reducing sugars, tannins, organic acids, and many unidentified compounds.

The cocoa bean's natural moisture combined with the heat of roasting cause many chemical reactions other than flavor changes. Some of these reactions remove unpleasant volatile acids and astringent compounds; partially break down sugars; modify tannins and other nonvolatile compounds with a reduction in bitterness; and convert proteins to amino acids which react with sugars to form flavor compounds, particularly pyrazines (8). To date, over 300 different compounds, many of them formed during roasting, have been identified in the chocolate flavor (5).

Roasting conditions are usually adjusted to produce different types of flavors. Low, medium, full, and high roasts can be developed by varying time and temperature. Low roasts produce mild flavors and light color; high roasts produce strong flavors and dark color (9).

Roasters have evolved from the gas heated rotary drum roasters to continuous feed roasters. Some newer roasters are designed to subject each bean to the same temperature and environmental conditions, minimizing transfer of cocoa butter from the nib to the shell and producing a constant roast. Normal roasting temperature varies from 100 to 140°C, and roasting time from 45 to 90 minutes.

Winnowing.   Winnowing, often called cracking and fanning, is one of the most important operations in cocoa processing. It is a simple process that involves separating the nib, or kernel, from the inedible shell. Failure to remove shell results in lower quality cocoa and chocolate products, more wear on nib grinding machines, and lower efficiency in all subsequent operations.

Because complete separation of shells and nibs is virtually impossible, various countries have established maximum allowable limits of shell in finished products. The maximum in the United States is 1.75% of shell by weight. However, U.S. manufacturers average from 0.05 to 1%.

The analysis of cocoa shell as determined by Knapp (10) is given in Table 4. In the United States, shells are often used as mulch or fertilizer for ornamental and edible plants, or for animal feed. Recent studies demonstrate that cocoa shells incorporated in their diets can improve the appetites of ruminants (see Pet and other livestock feeds). This phenomenon is attributed to the theobromine content of shells.

Grinding.   The final step in chocolate liquor production is the grinding of the kernel or nib of the cocoa bean. The nib is a cellular mass containing about 50 to 55% cocoa fat (cocoa butter). Grinding liberates the fat locked within the cell wall while producing temperatures as high as 110°C.

Table 4. Analyses of Cocoa Shell from Roasted Cocoa Beans

Component	Shell percentage
water	3.8
fat	3.4
ash	
total	8.1
water-soluble	3.5
water-insoluble	4.6
silica, etc	1.1
alkalinity (as K_2O)	2.6
chlorides	0.07
iron (as Fe_2O_3)	0.03
phosphoric acid (as P_2O_5)	0.8
copper	0.004
nitrogen	
total nitrogen	2.8
protein nitrogen	2.1
ammonia nitrogen	0.04
amide nitrogen	0.1
theobromine	1.3
caffeine	0.1
carbohydrates	
glucose	0.1
sucrose	0.0
starch (Taka-diastase method)	2.8
pectins	8.0
fiber	18.6
cellulose	13.7
pentosans	7.1
mucilage and gums	9.0
tannins	
tannic acid	
(Lowenthal's method)	1.3
cocoa-purple and cocoa-brown	2.0
acids	
acetic (free)	0.1
citric	0.7
oxalic	0.32
extracts	
cold water	20.0
alcohol (85%)	10.0

Nibs are usually ground while they are still warm after roasting. The original horizontal three tier stone mills and vertical disk mills have been replaced by modern horizontal disk mills, which have much higher outputs and are capable of grinding nibs to much greater fineness. Two modern machines in particular account for a large percentage of liquor grinding. One uses a pin mill mounted over a roller refiner. The pin mill grinds the nibs to a coarse but fluid liquor. The liquor is delivered to a roll refiner that reduces the particle size to a very fine limit. The second type is a vertical ball mill. Coarsely ground nib is fed to the base of a vertical cylinder which contains small balls in separate compartments. A central spindle causes the balls to rotate at very high speeds, grinding the liquor between them and against the internal wall of the cylinder (11).

COCOA POWDER

Cocoa powder (cocoa) is the food prepared by pulverizing the material remaining after part of the fat (cocoa butter) is removed from chocolate liquor. The U.S. chocolate standards define three types of cocoas based on their fat content. These are: (*1*) breakfast, or high-fat cocoa containing not less than 22% fat; (*2*) cocoa, or medium-fat cocoa, containing less than 22% fat but more than 10%; and, (*3*) low-fat cocoa, containing less than 10% fat.

Cocoa powder production today is an important part of the cocoa and chocolate industry because of increased consumption of chocolate-flavored products. Cocoa powder is the basic flavoring ingredient in most chocolate-flavored cookies, biscuits, cakes, and ice cream. It is also used extensively in the production of confectionery coatings for candy bars.

Manufacture.   When chocolate liquor is exposed to pressures of 34–41 MPa (5000–6000 psig) in a hydraulic press, and part of the fat (cocoa butter) is removed, cocoa cake (powder) is produced. The original pot presses used in cocoa production had a series of pots mounted vertically one above the other. These have been supplanted by horizontal presses which have four to twenty-four pots mounted in a horizontal frame. The newer presses are capable of complete automation, and by careful selection of pressure, temperature, and time of pressing, cocoa cake of a specified fat content can be produced.

Cocoa powder is produced by grinding cocoa cake. Cocoa cake warm from the press breaks easily into large chunks, but is difficult to grind into a fine powder. Cold, dry air removes the heat generated during most grinding operations in order to ensure that the cocoa butter remains in a solid, stable phase within the cocoa powder. Because cocoa powder is a relatively high fat product, great care must also be taken to prevent the absorption of undesirable odors and flavors.

Commerical cocoa powders are produced for various specific uses and many cocoas are alkali treated or "Dutched" to produce distinctive colors and flavors. The alkali process can involve the treatment of nibs, chocolate liquor, or cocoa with a wide variety of alkalizing agents (12).

Cocoa powders not treated with alkali are known as cocoa, natural cocoa or American processed cocoa. Natural cocoa has a pH of about 5.4 to 5.8 depending upon the type of cocoa beans used. Alkali processed cocoa ranges in pH from about 6 to as high as 8.5.

COCOA BUTTER

Cocoa butter is the common name given to the fat obtained by subjecting chocolate liquor to hydraulic pressure. It is the main carrier and suspending medium for cocoa particles in chocolate liquor and for sugar and other ingredients in sweet and milk chocolate.

The FDA has not legally defined cocoa butter, and no standard exists for this product under the U.S. Chocolate Standards. For the purpose of enforcement, the FDA defines cocoa butter as "the edible fat obtained from sound cocoa beans either before or after roasting." Cocoa butter, as a pharmaceutical compound, is refined to theobroma oil and is so defined in *The United States Pharmacopeia*.

The Codex Committee on Cocoa and Chocolate Products defines cocoa butter as "the fat produced from one or more of the following: cocoa beans, cocoa nibs, cocoa mass (chocolate liquor), cocoa cake, expeller cake or cocoa dust (fines) by a mechanical process and/or with the aid of permissible solvents" (13). It further states that, "cocoa butter shall not contain shell fat or germ fat in excess of the proportion in which they occur in the whole bean."

Codex has also defined the various types of cocoa butter in commercial trade (13). Press cocoa butter is defined as fat obtained by pressure from cocoa nib or chocolate liquor. In the United States, this is often referred to as prime pure cocoa butter. Expeller cocoa butter is defined as the fat prepared by the expeller process. In this process, cocoa butter is obtained directly from whole beans by pressing in a cage press. Expeller butter usually has a stronger flavor and darker color than prime cocoa butter, and is filtered with carbon or otherwise treated prior to use. Solvent extracted cocoa butter is cocoa butter obtained from beans, nibs, liquor, cake, or fines by solvent extraction (qv), usually with hexane. Refined cocoa butter is any of the above cocoa butters that has been treated to remove impurities or undesirable odors and flavors.

Composition and Properties. Cocoa butter is a unique fat with specific melting characteristics. It is a solid at room temperature (20°C), starts to soften around 30°C and melts completely just below body temperature. Its distinct melting characteristic makes cocoa butter the preferred fat for chocolate products.

Cocoa butter is composed mainly of glycerides of stearic, palmitic and oleic fatty acids. The triglyceride structure of cocoa butter has been worked out by Hilditch and Stainsby (14) and by Meara (15) and is as follows: tri-saturated, 3%; mono-unsaturated—oleo-distearin, 22%; oleo-palmitostearin, 57%; oleo-dipalmitin, 4%; di-unsaturated—stearo-diolein, 6%; palmito-diolein, 7%; tri-unsaturated, tri-olein, 1%.

Four crystalline forms of cocoa butter are generally recognized: α, β, β', and γ. The γ form, the least stable, has a melting point of 17°C. It changes rapidly to the α form which melts at 21–24°C. At normal room temperature the α form changes to the β' form, melting at 27–29°C, and finally, the β form is reached. It is the most stable form with a melting point of 34–35°C (16).

Since cocoa butter is a natural fat, derived from different varieties of cocoa beans, no single set of specifications or chemical characteristics can apply. Codex has attempted to define the physical and chemical parameters of the various types of cocoa butter (17) (see Table 5).

Cocoa Butter Substitutes and Equivalents. In the past 25 years, many fats have been developed to replace part or all of the added cocoa butter in chocolate-flavored products. These fats fall into two basic categories commonly known as cocoa butter substitutes and cocoa butter equivalents. Neither can be used in the United States in standardized chocolate products, but they are used in small amounts, usually up to 5% of the total weight of the product, in some European countries.

Cocoa butter substitutes do not chemically resemble cocoa butter and are compatible with cocoa butter only within specified limits. Cocoa butter equivalents are chemically similar to cocoa butter and can replace cocoa butter in any proportion without deleterious physical effects (18–19).

Cocoa butter substitutes and equivalents differ greatly with respect to their method of manufacture, source of fats, and functionality, and are produced by several physical and chemical processes (20–21).

For example, cocoa butter substitutes are produced from lauric acid fats such as coconut, palm, and palm kernel oils by fractionation and hydrogenation; from do-

Table 5. Properties and Composition of Cocoa Butter[a]

Characteristics	Press cocoa butter	Expeller cocoa butter	Refined cocoa butter
refractive index, n_D 40°C	1.456–1.458	1.453–1.459	1.453–1.462
melting behavior			
slip point, °C	30–40	30–34	30–34
clear melting point, °C	31–35	31–35	31–35
free fatty acids			
(mol % oleic acid)	0.5–1.75	0.5–1.75	0–1.75
saponification value			
(mg KOH/g fat)	192–196	192–196	192–196
iodine value (Wijs)	33.8–39.5	35.6–44.6	35.7–41.0
unsaponifiable matter	not more than	not more than	not more than
(petroleum ether % m/m)	0.35%	0.40%	0.50%

[a] Contaminants not to exceed 0.5 mg/kg of arsenic, 0.4 mg/kg of copper, 0.5 mg/kg of lead and 2.0 mg/kg of iron. From ref. 17.

mestic fats such as soy, corn, and cotton seed oils by selective hydrogenation; or from palm kernel stearines by fractionation. Cocoa butter equivalents can be produced from palm kernel oil by fractional crystallization; from glycerol and selected fatty acids by direct chemical synthesis; or from edible beef tallow by acetone crystallization.

Cocoa butter substitutes of all types enjoy widespread use in the United States chiefly as ingredients in chocolate-flavored products. Cocoa butter equivalents are not widely used because of their higher price and limited supply. At present the most frequently used cocoa butter equivalent in the United States is that derived from palm kernel oil but a synthesized product is expected to be available in the near future.

SWEET AND MILK CHOCOLATE

Most chocolate consumed in the United States is consumed in the form of sweet chocolate or milk chocolate. Sweet chocolate is chocolate liquor to which sugar and cocoa butter have been added. Milk chocolate contains these same three ingredients and milk or milk solids.

The U.S. definitions and standards for sweet chocolate are quite specific (22). Sweet chocolate must contain at least 15% chocolate liquor by weight and must be sweetened with sucrose, dextrose, or corn syrup solids in specific ratios. Semi-sweet chocolate and bittersweet chocolate, though often referred to as sweet chocolate, must contain a minimum of 35% chocolate liquor. The three products, sweet chocolate, semi-sweet chocolate, and bittersweet chocolate are often called simply chocolate or dark chocolate to distinguish them from milk chocolate. Table 6 gives some typical formulations for sweet chocolates (5).

Sweet chocolate can contain milk or milk solids, nuts, coffee, honey, malt, salt, and other spices and flavors as well as emulsifiers. Many different kinds of chocolate can be produced by careful selection of bean blends, controlled roasting temperatures, and varying amounts of ingredients and flavors (23).

The most popular chocolate in the United States is milk chocolate. The U.S. Chocolate Standards state that milk chocolate shall contain no less than 3.66 wt % of milk fat and not less than 12 wt % of milk solids. In addition, the ratio of nonfat milk

Table 6.   Typical Formulations for Sweet (Dark) Chocolates

Ingredients	% in formulation		
	No. 1	No. 2	No. 3
chocolate liquor	15.0	35.0	70.0
sugar	60.0	50.4	29.9
added cocoa butter	23.8	14.2	
lecithin	0.3	0.3	
vanillin	0.9	0.1	0.1
Total fat	*32.0*	*33.0*	*37.1*

solids to milk fat must not exceed 2.43:1 and the chocolate liquor content must be not less than 10 percent by weight. Some typical formulations of milk chocolate and some compositional values are shown in Table 7 (5).

Table 7.   Typical Formulations for Milk Chocolate

Ingredients	% in formulation		
	No. 1	No. 2	No. 3
chocolate liquor	11.0	12.0	12.0
dry whole milk	13.0	15.0	20.0
sugar	54.6	51.0	45.0
added cocoa butter	21.0	21.6	22.6
lecithin	0.3	0.3	0.3
vanillin	0.1	0.1	0.1

Production.   The only major difference in the production of sweet and milk chocolate is that in the production of milk chocolate, water must be removed from the milk. Many milk chocolate producers in the United States use dry milk powder. Others condense fresh whole milk with sugar, and either dry it, producing milk crumb, or blend it with chocolate liquor and then dry it, producing milk chocolate crumb. These crumbs are mixed with additional chocolate liquor, sugar, and cocoa butter later in the process (24).

Mixing.   The first step in chocolate processing is the weighing and mixing of ingredients. Today, this is for the most part a fully automated process carried out in a batch or continuous processing system. In batch processing, all the ingredients for one batch are automatically weighed into a mixer and mixed for a specific period of time. The mixture is conveyed to storage hoppers directly above the refiners. In the continuous method, ingredients are metered into a continuous kneader which produces a constant supply to the refiners (25). The continuous process requires very accurate metering and rigid quality control procedures for all raw materials.

Refining.   The next stage in chocolate processing is refining. This is essentially a fine grinding operation in which the coarse paste from the mixer is passed through steel rollers and converted to a drier powdery mass. This process breaks up crystalline sugar, fibrous cocoa matter, and milk solids.

Tremendous advances have been made in the design and efficiency of roll refiners. The methods currently used for casting the rolls have resulted in machines capable of very high output and consistent performance. The efficiency of the newer refiners has also been improved by hydraulic control of the pressure between the rolls and

Table 8.   Variations in Theobromine and Caffeine Content of Various Chocolate Liquors

Country of origin	Theobromine, %	Caffeine, %	Total, %	Theobromine to caffeine ratio
New Guinea	0.818	0.329	1.147	2.49:1
New Guinea	0.926	0.330	1.256	2.8:1
Malaysia	1.05	0.252	1.302	4.17:1
Malaysia	1.01	0.228	1.238	4.45:1
Brazil (Bahia)	1.21	0.183	1.393	6.61:1
Nigeria (Main Lagos)	1.73	0.159	1.889	14.9:1
Nigeria (Light Lagos)	1.23	0.137	1.367	8.99:1
Dominican Republic (Sanchez)	1.57	0.177	1.757	8.93:1
Dominican Republic (Sanchez small)	1.25	0.261	1.511	4.77:1
Africa (Fernando Po)	1.47	0.064	1.534	23.2:1
Mexico (Tabascan)	1.41	0.113	1.523	12.4:1
Trinidad	1.24	0.233	1.473	5.30:1
maximum	1.73	0.330	1.889	23.2:1
minimum	0.818	0.064	1.147	2.49:1

Table 9.   Theobromine and Caffeine Content of Finished Chocolate Products

Product	Theobromine, %	Caffeine, %
baking chocolate	1.38	0.092
chocolate flavored syrup	0.24	0.014
cocoa, 15% fat	1.46	0.250
dark sweet chocolate	0.41	0.078
milk chocolate	0.19	0.018

thermostatic control of cooling water to the rolls. Modern refiners can process 900 kg of paste per hour.

Particle size is extremely important to the overall quality of sweet and milk chocolate. Hence the refining process, which controls particle size, is critical. Fine chocolates usually have no particles larger than 25 or 30 μm. This is normally accomplished by passing the paste through refiners more than once. However, smooth chocolates can be produced with only a single pass through the refiners if the ingredients are ground prior to mixing.

Conching.   After refining, chocolate is subjected to conching, a step critical to the flavor development of high quality chocolates. Conching is a kneading process in which chocolate is slowly mixed, allowing moisture and volatile acids to escape while smoothing the remaining chocolate paste.

The earliest type of conche consisted of a tank with a granite bed on which the chocolate paste from the refiners was slowly pushed back and forth by a granite roller. This longitudinal conche is still used, and many experts consider it best for developing subtle flavors.

Conching temperatures for sweet chocolate range from 55–85°C, and from 45–55°C for milk chocolate. Higher temperatures are sometimes used for milk chocolate if caramel or butterscotch flavors are desired (26).

Table 10. Composition of Cocoa Beans and Products Made Therefrom, Whole Weight Basis in %

	No. of sample	Total solids	Total protein[a]	Cocoa protein[b]	Milk protein[c]	Fat	Ash	Total carbo- hydrates[d]
Whole cocoa beans								
Ghana	1	92.9	10.1	10.1		47.8	2.7	30.3
	2	94.0	9.8	9.8		51.6	2.6	28.0
	3	94.5	10.2	10.2		46.4	2.9	33.0
	4	94.7	10.3	10.3		46.3	3.1	33.0
Bahia	1	94.0	10.0	10.0		49.3	2.7	30.0
	2	94.1	10.2	10.2		48.6	2.7	30.6
	3	95.1	10.2	10.2		48.2	2.7	32.0
	4	94.9	10.2	10.2		48.4	2.7	31.6
Chocolate liquor								
natural	1	98.4	9.4	9.4		56.2	2.4	28.5
	2	98.5	9.5	9.5		54.1	2.6	30.5
	3	98.9	10.1	10.1		57.0	2.4	27.4
	4	98.5	10.2	10.2		55.1	2.6	28.6
Dutch	1	98.6	9.2	9.2		55.4	3.8	28.5
	4	99.2	9.4	9.4		56.0	3.8	28.1
Cocoa								
natural	1	96.3	18.4	18.4		12.8	4.6	56.9
	2	96.2	18.4	18.4		16.4	4.8	52.9
	3	97.4	19.8	19.8		12.7	4.5	56.5
Dutch	1	97.1	17.5	17.5		12.0	8.3	55.9
	2	97.4	18.3	18.3		14.3	7.4	53.7
Sweet chocolate								
	1	99.6	3.4	3.4		35.1	1.0	59.4
	2	99.3	3.8	3.8		36.5	1.0	57.3
	4	99.5	3.6	3.6		35.0	1.0	59.2
Milk chocolate								
12% whole milk solids	1	99.2	4.2	1.0	3.2	34.7	0.9	59.2
	2	99.5	4.3	1.1	3.1	30.2	1.0	63.8
	3	99.6	4.5	1.1	3.4	32.3	0.9	61.6
	4	99.5	4.0	1.4	2.6	29.6	1.0	64.6
20% whole milk solids	1	98.8	6.6	1.3	5.2	34.4	1.5	56.1
	2	99.5	6.5	1.2	5.2	33.1	1.4	58.3
	3	99.4	6.8	1.4	5.4	30.5	1.5	60.4

[a] Total protein = milk protein + cocoa protein.
[b] Cocoa protein = (total nitrogen − milk nitrogen) × 4.7.
[c] Milk protein = milk nitrogen × 6.38.
[d] Total carbohydrate by difference using cocoa N × 5.63.

Several other kinds of conches are also used today. The popular rotary conche can handle chocolate paste in a dry stage direct from the refiners (27). The recently developed continuous conche actually liquifies and conches in several stages and can produce up to 3000 kg of chocolate per hour in a flow area of only 34 m^2.

The time of conching varies from a few hours to many days. Many chocolates receive no conching. Non-conched chocolate is usually reserved for inexpensive candies, cookies, and ice cream. In most operations, high quality chocolate receives extensive conching for as long as 120 hours.

Flavors, emulsifiers, or cocoa butter are often added during conching. The fla-

Table 11. Amino Acid Content of Cocoa and Chocolate Products

	Whole beans[a]	Chocolate liquor[b] Natural	Dutch	Cocoa[c] Natural	Dutch	Sweet choc.[d]	Milk chocolate[e] 12% MS	20% MS
				mg/g				
tryptophan	1.2	1.3				0.6		
threonine	3.5	3.9	3.6	7.7	8.0	1.5	1.8	2.8
isoleucine	3.3	3.8	4.0	7.0	7.4	1.4	2.2	3.5
leucine	5.3	6.0	6.3	11.5	11.3	2.3	3.8	6.1
lysine	4.8	5.1	5.1	8.7	8.3	1.9	2.4	3.9
methionine	0.7	1.1	0.9	2.0	1.7	0.4	0.9	1.4
cystine	1.4	1.1	1.0	2.1	2.1	0.4	0.3	0.4
phenylalanine	4.1	4.9	5.3	9.9	9.7	1.7	2.2	3.6
tyrosine	2.6	3.5	3.6	7.8	8.0	1.2	1.9	3.0
valine	5.1	5.8	6.3	11.1	10.9	2.1	2.7	4.3
arginine	5.0	5.3	5.1	11.3	11.3	1.9	1.2	1.9
histidine	1.6	1.7	1.7	3.4	3.0	0.6	0.7	1.0
alanine	3.8	4.3	4.1	8.7	8.4	1.5	1.4	2.3
aspartic acid	9.2	10.0	9.8	19.1	18.3	3.9	3.5	5.5
glutamic acid	12.8	14.1	14.1	28.0	26.2	5.7	8.5	13.7
glycine	4.0	4.4	4.5	8.3	8.5	1.6	1.0	1.6
proline	3.4	3.7	3.9	7.6	7.5	1.4	3.4	5.5
serine	3.7	4.1	4.0	6.8	8.2	1.5	1.9	3.0
Total AA								
Recovered[f]	*75.5*	*84.1*	*83.3*	*162.0*	*158.8*	*31.6*	*39.8*	*63.4*

[a] Whole beans = 48% fat, 5% moisture, 10% shell.
[b] Chocolate liquor = 55% fat.
[c] Cocoa = 13% fat.
[d] Sweet chocolate = 35% chocolate liquor, 35% total fat.
[e] 12% MS milk chocolate = 12% whole milk solids, 10% liquor, 32% total fat; 20% MS milk chocolate = 20% whole milk solids, 13% liquor, 33% total fat.
[f] Total AA recovered = sum of individual amino acids.

voring materials most commonly added in the United States are vanillin, a vanilla-like artificial flavor, and natural vanilla (28) (see Flavors). Cocoa butter is added to adjust viscosity for subsequent processing.

Several chemical changes occur during conching including a rise in pH and a decline in moisture as volatile acids (acetic) and water are driven off. These chemical changes have a mellowing effect on the chocolate (29).

Standardizing. In standardizing or finishing, emulsifiers and cocoa butter are added to the chocolate to adjust viscosity to final specifications.

Lecithin (qv) is by far the most common emulsifier in the chocolate industry (5). It is a natural product, a phospholipid, possessing both hydrophilic and hydrophobic properties. The hydrophilic groups of the lecithin molecules attach themselves to the water, sugar, and cocoa solids present in chocolate. The hydrophobic groups attach themselves to cocoa butter. This reduces both the surface tension between cocoa butter and the other materials present and the viscosity. Less cocoa butter is then needed to adjust the final viscosity of the chocolate.

The amount of lecithin required falls within a narrow range of about 0.2–0.6% (30). It can have a substantial effect on the amount of cocoa butter used, reducing the final fat content of chocolate by as much as 5%. Because cocoa butter is usually the

Table 12. Fatty Acid Composition of Raw Cocoa Beans and Cocoa Butter

	No. of sample	Fatty acid[a]					
		14:0	16:0	18:0	18:1	18:2	20:0
Cocoa Beans							
Ghana	1	0.16	28.31	34.30	34.68	2.55	
	2	0.53	30.20	31.88	33.55	3.84	
	3	0.19	31.72	32.57	32.82	2.70	
	4	0.23	31.50	32.39	33.06	2.82	
Bahia	1	0.15	29.29	31.70	35.24	3.62	
	2	0.12	26.68	32.06	37.90	3.24	
	3	0.25	33.99	28.80	33.62	3.34	
	4	0.19	30.91	30.37	35.22	3.31	
Natural cocoa butter							
	1	0.15	27.08	32.64	35.61	3.63	0.89
	2	0.19	27.68	32.64	35.03	3.63	0.83
	3	0.14	28.42	32.55	34.71	3.23	0.95
	4	0.14	27.29	32.41	35.36	3.70	1.10
Dutch cocoa butter							
	1	0.16	27.23	32.69	35.54	3.31	1.07
	2	0.15	26.63	34.24	34.68	3.52	0.78
	4	0.15	26.47	33.53	35.45	3.40	1.00

[a] Expressed as mole percent and calculated from peak areas of the gas chromatograms.

most costly ingredient in the formulation, the savings to a large manufacturer can be substantial.

Lecithin is usually introduced in the standardizing stage, but can be used earlier in the process. Some lecithin is often added during mixing or in the later stages of conching. The addition at this point has the added advantage of reducing the energy necessary to pump the product to subsequent operations since the product viscosity is reduced.

Viscosity control of chocolate is quite complicated because chocolate does not behave as a true liquid owing to the presence of cocoa particles. This non-Newtonian behavior is best described by Steiner (31) using a flow relation discovered by Casson (32). Steiner shows that when the square root of rate of shear is plotted against the square root of shear stress for chocolate, a straight line is produced. With this Casson relationship method two values are obtained, Casson viscosity and Casson yield value, which describe the flow of chocolate. The chocolate industry has been slow to adopt the Casson relationship, however, and the simpler MacMichael viscometer method is still preferred.

Tempering. Tempering follows conching and standardizing in the processing of chocolate. The state or physical structure of the fat base in which sugar, cocoa, and milk solids are suspended is critical to the overall quality and stability of chocolate. Production of a stable fat base is somewhat complicated because the cocoa butter in chocolate exists in several polymorphic forms. Tempering eases the transition from the liquid state to a stable solid form.

A stable crystalline form for chocolate depends primarily upon the method of cooling the fat present in liquid chocolate. To avoid the grainy texture and poor color and appearance of improperly cooled chocolate, the chocolate must be tempered or cooled down so as to form cocoa butter seed crystals (33). This is usually accomplished

Table 13.   Vitamin Content of Cocoa Beans and Chocolate Products,[a] Whole Weight Basis

	No. of sample	B_1	B_2	Pantothenic acid	Niacin	B_6
				mg/100 g		
Whole cocoa beans						
Ghana	1	0.21	0.16	0.24	0.19	0.22
	2	0.17	0.18	0.35	1.07	0.21
	3	0.19	0.18	0.57	0.91	0.18
	4	0.16	0.15	0.32	0.52	0.01
Bahia	1	0.14	0.18	0.34	0.46	0.61
	2	0.17	0.18	0.35	1.13	0.16
	3	0.13	0.27	0.61	1.00	0.16
	4	0.16	0.16	0.38	0.81	0.09
Chocolate liquor						
	1	0.08	0.17	0.20	0.88	0.09
	2	0.11	0.16	0.27	1.02	0.20
	3	0.08	0.15	0.17	1.01	0.16
	4	0.05	0.11	0.15	0.29	0.02
Cocoa						
	1	0.05	0.19	0.33	1.34	0.17
	2	0.13	0.23	0.35	1.53	0.17
	3	0.15	0.22	0.32	1.37	0.24
Milk chocolate						
	1	0.07	0.10	0.37	0.14	0.02
	2	0.11	0.24	0.37	0.38	0.02
	3	0.07	0.16	0.45	0.21	0.07
	3a	0.10	0.25	0.61	0.24	0.08
	4	0.15	0.33	0.32	1.11	0.20

[a] Vitamins A and C—negligible amounts present.

Table 14.   Tocopherols of Chocolate of Cocoa Beans and Chocolate Products

	Total tocopherol	Alpha tocopherol
	mg/100 g	
Bahia–Ghana beans	10.3	1.0
liquor, natural	10.9	1.1
liquor, Dutch	10.0	0.8
cocoa butter, natural	19.2	1.2
cocoa butter, Dutch	18.7	1.1
cocoa, natural	2.3	0.2
cocoa, Dutch	2.2	0.2
dark chocolate	6.0	0.7
milk chocolate, 12% milk	5.6	0.7
milk chocolate, 20% milk	6.3	0.7

by cooling chocolate in a water-jacketed tempering kettle which has an internal scraper or mixer. Warm chocolate contacts the cool surface of the kettle (15°C) causing the cocoa butter to solidify or seed. This solid is immediately mixed into the still liquid chocolate, and the mixture allowed to cool to around 29°C where the seeding process continues. The viscosity of the chocolate increases rapidly during this time and it becomes too thick if not handled quickly.

Table 15.   Mineral Element Content of Cocoa and Chocolate Products (by Atomic Absorption Spectrophotometry) [a]

Product	Ca	Fe	Mg	P [b]	K	Na	Zn	Cu	Mn
				mg/100 g					
raw Accra nibs	59.56	2.50	232.16	385.33	626.70	11.98	3.543	1.93	1.60
raw Bahia nibs	52.73	2.45	229.11	383.33	622.55	13.55	3.423	1.94	2.06
natural cocoa	115.93	11.34	488.51	7716.66	1448.56	20.12	6.306	3.62	3.77
Dutch cocoa	111.41	15.52	475.98	7276	2508.58	81.14	6.37	3.61	3.75
chocolate liquor	59.39	5.61	265.23	3996.66	679.61	18.89	3.53	2.05	1.85
12% milk chocolate	106.41	1.23	45.56	159	156.64	80.09	0.773	1.02	0.282
20% milk chocolate	174	1.40	52.26	207.96	346.33	115.4	1.24	0.126	0.139
dark chocolate	26.33	2.34	93.7	142.9	302.53	18.63	1.50	0.432	0.345

[a] Data from duplicate analyses of each of three samples. Mean values.
[b] Total phosphorus—ash below 550°C (AOAC Procedure).

In another method of tempering, solid chocolate shavings are added as seed crystals to liquid chocolate at 32–33°C. This is a particularly good technique for a small confectionery manufacturer who does not produce his own chocolate. However, the shavings are sometimes difficult to disperse and may cause lumps in the finished product (23).

Molding.   The final stage in the processing of chocolate is molding. The three basic methods of molding are block, shell, and hollow molding (23). Block molding predominates. Chocolate, either plain or mixed with nuts, raisins or other ingredients, is deposited in molds, allowed to cool, and removed from the molds as solid pieces. Shell molding is a very complicated process. Chocolate is deposited into metal molds and, by a reversal process, a layer of liquid chocolate remains clinging to the mold's inner surface. After this layer of chocolate has cooled, it is filled with a confection, such as caramel, or more liquid chocolate. The molds used in hollow molding are divided in two halves and connected by a hinge. Chocolate is deposited in one half. The mold is then closed and rotated so that the entire mold is coated. Easter eggs and other hollow chocolate products are produced by this process.

Theobromine and Caffeine

Chocolate and cocoa products, like coffee, tea and cola beverages, contain alkaloids (1). The predominant alkaloid in cocoa and chocolate products is theobromine, although significant amounts of caffeine may be present, depending upon the origin of the beans. Published values for theobromine and caffeine content of chocolate vary widely, mainly because of natural differences in various beans and differences in methodology. This latter problem has been alleviated by the recent introduction of high pressure liquid chromatography (hplc) which has greatly improved the accuracy of analyses. Kreiser and Martin (34) have published hplc values for theobromine and caffeine in a number of chocolate liquor samples (Table 8). Of the 12 varieties tested, the ratio of theobromine to caffeine varied widely from 2.5:1 for New Guinea liquor to 23.2:1 for that obtained from Fernando Po. Total alkaloid content, however, remained fairly constant, ranging from 1.5 to 1.89%.

The theobromine and caffeine contents of several finished chocolate products as determined by hplc in the authors' laboratories are presented in Table 9.

Nutritional Properties of Chocolate Products

Chocolate and cocoa products supply proteins, fats, carbohydrates, vitamins, and minerals. The Chocolate Manufacturers' Association of the United States, McLean, Virginia, recently completed a nutritional analysis of a wide variety of chocolate and cocoa products representative of those generally consumed in the United States. The analyses were conducted by Philip Keeney's laboratory at the Pennsylvania State University and complete nutritional data for the various products analyzed are given in Tables 10 to 15. Where possible, data on more than one sample of a given variety or type product are presented.

BIBLIOGRAPHY

"Chocolate and Cocoa" in *ECT* 1st ed., Vol. 3, pp. 889–918, by W. Tresper Clarke, Rockwood & Co.; "Chocolate and Cocoa" in *ECT* 2nd ed., Vol. 5, pp. 363–402, by B. D. Powell and T. L. Harris, Cadbury Brothers Limited.

1. W. T. Clarke, *The Literature of Cacao,* American Chemical Society, Washington, D.C., 1954.
2. B. Siebers, *Manuf. Confect.* **57**(8), 52 (1977).
3. Gill and Duffus, *Cocoa Market Report No. 278,* December 1977.
4. B. Bartley and P. de T Alvim, *Manuf. Confect.* **56**(6), 44 (1976).
5. L. R. Cook, *Chocolate Production and Use,* Magazines for Industry, Inc., New York, 1972.
6. *Report of the Cocoa Conference,* Cocoa, Chocolate and Confectionery Alliance, London, 1957.
7. C. E. Taneri, *Manuf. Confect.* **52**(6), 45 (1972).
8. G. A. Reineccius, P. G. Keeney, and W. Weissberger, *J. Agric. Food Chem.* **20**(2), 202 (1972).
9. H. R. Riedl, *Confect. Prod.* **40**(5), 193 (1974).
10. A. W. Knapp and A. Churchman, *J. Soc. Chem. Ind. (London)* **56**, 29 (1937).
11. A. Szegvaridi, *Manuf. Confect.* **50**, 34 (1970).
12. H. J. Schemkel, *Manuf. Confect.* **53**(8), 26 (1973).
13. *Report of Codex Committee on Cocoa Products and Chocolate,* Codex Alimentarius Commission, 10th Session, Geneva, 1974.
14. T. P. Hilditch and W. J. Stainsby, *J. Soc. Chem. Ind.* **55**, 95T (1936).
15. M. L. Meara, *J. Chem. Soc.,* 2154 (1949).
16. S. J. Vaeck, *Manuf. Confect.* **40**(6), 35 (1960).
17. *Draft Standard for Cocoa Butter,* Codex Committee on Cocoa Products and Chocolate.
18. K. Wolf, *Manuf. Confect.* **57**(4), 53 (1977).
19. J. Robert Ryberg, *Cereal Sci. Today* **15**(1), 16 (1970).
20. P. Kalustian, *Candy Snack Industry* **141**(3), (1976).
21. B. O. M. Tonnesmann, *Manuf. Confect.* **57**(5), 38 (1977).
22. *Code of Federal Regulations,* No. 21, Part 14—Cacao Products, April 1, 1974.
23. B. W. Minifie, *Chocolate, Cocoa and Confectionery: Science and Technology,* AVI, Westport, Conn., 1970.
24. B. Christiansen, *Manuf. Confect.* **56**(5), 69 (1976).
25. H. R. Riedl, *Confect. Prod.,* **42**(41), 165 (1976).
26. L. R. Cook, *Manuf. Confect.* **56**(5), 75 (1976).
27. E. M. Chatt, "Cocoa," in Z. I. Kertesz, ed., *Economic Crops,* Vol. 3, Interscience Publishers, Inc., New York, 1953, p. 185.
28. H. C. J. Wijnougst, *The Enormous Development in Cocoa and Chocolate Marketing Since 1955,* H. C. J. Wijnougst, Mannheim, Germany, 1957, p. 161.
29. J. Kleinert, *Manuf. Confect.* **44**(4), 37 (1964).
30. R. Heiss, *Twenty Years of Confectionery and Chocolate Progress,* AVI, Westport, Conn., 1970, p. 89.

31. E. H. Steiner, *Inter. Choc. Rev.* **13**, 290 (1958).
32. N. Casson, *Brit. Soc. Rheo. Bull.* ns **52**, (Sept. 1957).
33. W. N. Duck, ref. 5, p. 22.
34. W. R. Kreiser and R. A. Martin, *J. Assoc. Off. Analy. Chem.* **61**(6), (1978).

B. L. ZOUMAS
E. J. FINNEGAN
Hershey Foods Corporation

CHOLINE

Choline [*123-41-1*] (trimethyl(2-hydroxyethyl)ammonium hydroxide), $[(CH_3)_3NCH_2CH_2OH]^+OH^-$, derives its name from bile (Greek *cholē*), from which it was first obtained. Choline is a colorless, hygroscopic liquid with an odor of trimethylamine. Choline or a precursor is needed in the diet as a constituent of certain phospholipids universally present in protoplasm. This makes choline an important nutritional substance. It is also of great physiological interest because one of its esters, acetylcholine (ACh), appears to be responsible for the mediation of parasympathetic nerve impulses and has been postulated to be essential to the transmission of impulses of all nerves. ACh and other more stable compounds that simulate its action are pharmacologically important because of their powerful effect on the heart and on smooth muscle. Choline is used clinically in liver disorders and as a constituent in animal feeds.

Choline was isolated from ox bile in 1849 by Strecker. During 1900 to 1920, the observations of Reid Hunt led to interest in the vasodepressor properties of the esters of choline, and in the 1920s, Loewi showed that acetylcholine was presumably the "vagus-substance." The nutritional importance of choline was recognized in the 1930s, when Best and his colleagues found that choline would prevent fatty infiltration of the liver in rats. Subsequent observations showed that choline deficiency could produce cirrhosis (1) or hemorrhagic kidneys (2) in experimental animals under various conditions.

Physical and Chemical Properties

Choline is a strong base ($pK_B = 5.06$ for 0.0065–0.0403 M solutions) (3). It crystallizes with difficulty and is usually known as a colorless deliquescent syrupy liquid, which absorbs carbon dioxide from the atmosphere. Choline is very soluble in water and in absolute alcohol but insoluble in ether. It is stable in dilute solutions but in concentrated solutions tends to decompose at 100°C, giving ethylene glycol, polyethylene glycol, and trimethylamine (4).

Choline is not usually encountered as the free base but as a salt; the one most commonly used is the chloride, $[(CH_3)_3N(CH_2CH_2OH)]^+Cl^-$. As a quaternary ammonium hydroxide, choline reacts with hydrochloric acid to form the chloride and water, whereas primary, secondary, and tertiary amines combine with hydrochloric acid to form so-called hydrochlorides without the elimination of water (5) (see Amines;

Ammonium compounds). Choline chloride is a crystalline deliquescent salt, usually with a slight odor of trimethylamine, and with a strongly brackish taste somewhat resembling that of ammonium chloride (6). It is very soluble in water, freely soluble in alcohol, slightly soluble in acetone and chloroform, and practically insoluble in ether, benzene, and ligroin. Its aqueous solutions are neutral to litmus and are stable. The specific gravity of these solutions is a straight-line function between pure water and the value of 1.10 for the 80% solution, which represents the approximate limit of solubility. Choline chloride absorbs moisture from the atmosphere at relative humidities greater than 20% at 25.5°C. Choline dihydrogen citrate $(CH_3)_3N(CH_2CH_2OH)C_6H_7O_7$, is prepared by analogous methods. It has the same pharmacological action as the chloride, but contains a lower proportion of choline. It is not as deliquescent as the chloride, absorbs moisture from the atmosphere only at relative humidities greater than 56% at 25.5°C. It is more palatable than the chloride.

Biological Functions

In nutrition, the most important function of choline appears to be the formation of lecithin (qv) and other choline-containing phospholipids. Lecithin (qv) is most simply formulated as:

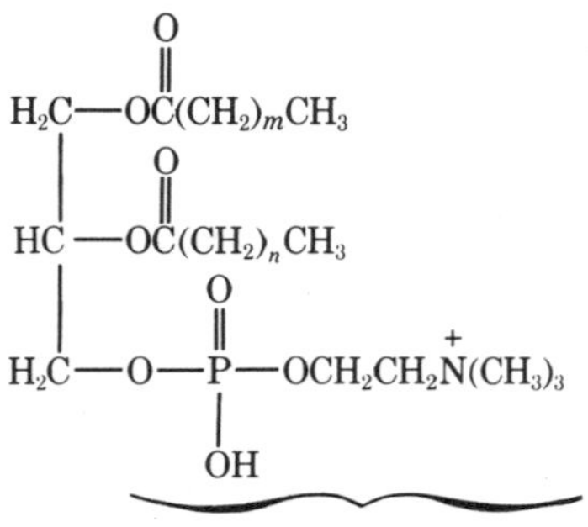

$$
\begin{array}{l}
\quad\quad\quad\quad O \\
\quad\quad\quad\quad \| \\
H_2C\!-\!OC(CH_2)_mCH_3 \\
\quad\quad\quad\quad\quad O \\
\quad\quad\quad\quad\quad \| \\
HC\ \!-\!OC(CH_2)_nCH_3 \\
\quad\quad\quad\quad\quad\quad O \\
\quad\quad\quad\quad\quad\quad \| \quad\quad\quad\quad + \\
H_2C\!-\!O\!-\!P\!-\!OCH_2CH_2N(CH_3)_3 \\
\quad\quad\quad\quad\quad | \\
\quad\quad\quad\quad\quad OH
\end{array}
$$

phosphorylcholine portion; $m, n = 10$–16

It is regarded as a fat in which one of the fatty acid molecules has been replaced by a phosphoric acid derivative of choline (phosphorylcholine) via cytidine diphosphate choline. The replacement changes the physical properties of the fat so that lecithin is readily dispersible with water. This property is important in the transport of fats in the blood. In choline deficiency, fats tend to accumulate in the liver presumably because they are not transformed into lecithin and hence are not carried away from the liver by the circulating blood. Owing to this effect, choline is said to have a lipotropic (fat-moving) action. Other lipotropic substances are known, such as methionine, $CH_3SCH_2CH_2CH(NH_2)COOH$, and betaine $(CH_3)_3\overset{+}{N}CH_2COO^-$, which furnish labile methyl groups that unite with 2-aminoethanol to form choline in the body. Choline itself can also yield labile methyl groups for the methylation of other organic compounds (2,7–17).

Fatty infiltration of the liver has been observed to precede cirrhosis in experimental animals receiving diets low in choline and other substances that can furnish labile methyl groups, and can thus serve as precursors of choline.

Figure 1 shows some of the biological reactions involving labile methyl groups.

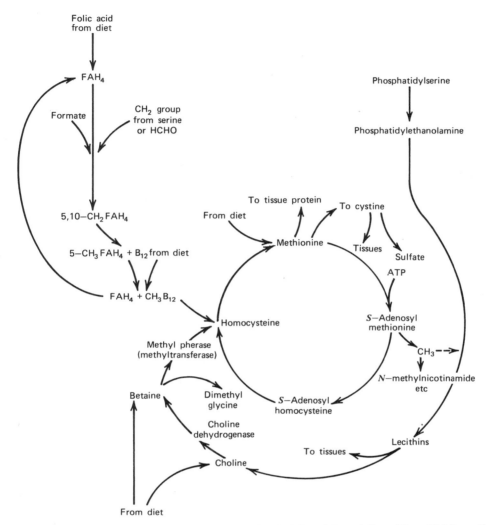

Figure 1. The choline and methionine cycles showing the origin and disposition of labile methyl groups. FAH_4 = tetrahydrofolic acid; CH_3B_{12} = methylated vitamin B_{12}; and ATP = adenosine triphosphate.

The groups can originate from serine, formaldehyde, or formate by enzymatic reactions involving tetrahydrofolic acid (FAH_4) so that the compound N^5,N^{10}-methylenetetrahydrofolic acid (5,10-CH_2FAH_4) is formed. This undergoes hydrogenation to form 5-methyltetrahydrofolic acid (5-CH_3FAH_4) from which the methyl group is transferred to a vitamin B_{12} compound, shown in the diagram as CH_3B_{12}. This compound methylates homocysteine to produce methionine, which may become activated by adenosine triphosphate (ATP) with the formation of S-adenosylmethionine. The methyl group attached to sulfur in this compound can be transferred to various receptor molecules. One of these is 2-aminoethanol which is thereby converted to choline. Betaine is formed by dehydrogenation of choline, and can furnish a methyl group to homocysteine in an alternative pathway, catalyzed by methylpherase, for methionine biosynthesis.

Occurrence

Choline occurs widely in nature and, prepared synthetically, it is available as an article of commerce. Soybean lecithin and egg-yolk lecithin have been used as natural sources of choline for supplementing the diet. Other important natural-food sources include liver and certain legumes (18–19).

Preparation

An earlier procedure for the production of choline from natural sources, such as the hydrolysis of lecithin (20), has no present-day application.

Choline is produced and used for medicinal and nutritional purposes as the chloride, dihydrogen citrate, and bitartrate and as tricholine citrate.

Preparation of Choline Salts

The chlorohydrin process (21) (eq. 1) has been used for the preparation of ace-tyl-β-alkylcholine chloride (see Chlorohydrins) (22). Choline bromide has been produced by allowing ethylene bromide to react with trimethylamine at 110–120°C and heating the resulting bromoethyltrimethylammonium bromide with water at 160°C (23).

$$N(CH_3)_3 + ClCH_2CH_2OH \rightarrow (CH_3)_3\overset{+}{N}CH_2CH_2OH\ Cl^- \tag{1}$$

The preparation of salts may be carried out more economically by the neutralization of choline produced by the chlorohydrin synthesis. A modification produces choline carbonate as an intermediate that is converted to the desired salt (24). The most practical production procedure is that of Ulrich and Ploetz, in which 300 parts of a 20% solution of trimethylamine is neutralized with 100 parts of concentrated hydrochloric acid, and the solution is treated for 3 h with 50 parts of ethylene oxide under pressure at 60°C (35).

Specifications and Standards

Choline Chloride [67-48-1].   The USP reagent standards are as follows (6): "White crystals or crystalline powder. Very soluble in water and freely soluble in alcohol. It is hygroscopic and usually has a slight amine-like odor. Its solution is neutral to litmus paper." The USP also lists tests and an assay procedure.

Choline Dihydrogen Citrate [77-91-8].   Choline dihydrogen citrate is a white, crystalline, granular substance possessing an acid taste, mp 105–107.5°C, and is freely soluble in water, very slightly soluble in alcohol, and practically insoluble in benzene, chloroform, and ether. The pH of a 25% solution is about 4.25.

Tricholine Citrate Concentrate [546-63-4].   This compound is a clear, faintly yellow to light-green syrupy aqueous liquid containing 65.0 ± 2.0% tricholine citrate. It usually has a slight amine odor. It should have a pH of 9.0–10.0 and should contain not more than 0.2% trimethylamine, 0.5% ethylene glycol, 10 ppm of formaldehyde, and 0.1% residue on ignition. Its limit for heavy metals is 20 ppm and it should contain more than 0.2% chlorides or sulfates.

Choline Bitartrate [87-67-2]. This substance is a white crystalline material possessing an acid taste. It melts at 149–153°C. Analysis by cobaltous chloride shows more than 99% as the bitartrate. Free ethylene glycol is less than 0.25%, with free alkali at 0.0%.

Analysis

In biological materials, various nonspecific precipitants have been used in the gravimetric determination of choline, including potassium triiodide, platinum chloride, gold chloride, and phosphotungstic acid (26). Choline may also be determined spectrophotometrically and by microbiological and physiological assay methods.

Reinecke salt, used in the spectrophotometric determination of choline, forms an insoluble complex with choline and with other organic bases. Compounds with carboxyl groups, such as betaine or carnitine, which form reineckates soluble in alkali, may be separated from choline reineckate when the precipitation occurs at an elevated pH (27).

The use of mutant 34486 of *Neurospora crassa* for the microbiological assay of choline has been described (28). Choline obtained by hydrolysis from a sample of biological material is adsorbed on a zeolite column. The growth-promoting effect of this choline on *N. crassa* is a measure of the choline content of the original sample. Mutant 34486 also responds to certain precursors, derivatives, and analogues of choline including methylaminoethanol, dimethylaminoethanol, acetylcholine, phosphorylcholine, arsenocholine, and ethyl-substituted cholines. Methionine in comparatively high concentrations also produces growth, but is not adsorbed on zeolite (28).

A physiological method has also been used in which the choline is extracted after hydrolysis from a sample of biological material and acetylated. The acetylcholine is then assayed by a kymographic procedure, in which its effect in causing contraction of a piece of isolated rabbit intestine is measured (29).

Uses

Choline has a low toxicity (30–31) and is used clinically and as a dietary supplement for poultry.

The Committee on Dietary Allowances, Food and Nutrition Board, National Research Council, states (32): "Choline ... is usually considered to be a vitamin. The most prominent signs of choline deficiency in mammals are fatty infiltration of the liver and hemorrhagic kidney disease. ... There is little evidence to suggest that, in man, administration of choline alleviates fatty liver, cirrhosis, chronic liver disease, or other defects that, at least superficially, resemble those associated with choline deficiency." Choline deficiency has not been demonstrated in man, and it seems unlikely that it could occur, except possibly in instances where young children have severe protein deficient diets, rich in highly refined products (32). However, choline has been used for treatment in patients with cirrhosis of the liver. The recommended dosage is 1–6 g daily, in terms of choline chloride. The addition of cystine or methionine to the treatment is sometimes recommended (1,7,10,33). There have been several clinical reports describing megaloblastic anemias that were refractory to liver but responded to choline (34–35). These cases appeared to have been complicated by fatty infiltration of the liver or bone marrow, which responded to choline, so that the antianemic effect of choline was presumably secondary.

Choline is reported as beneficial to patients with tardive dyskinesia (36), a motor disorder, characterized by involuntary twitches, that typically occurs in susceptible persons after chronic ingestion of antipsychotic drugs. The 20 patients, in a double-blind study received 0.15–0.20 g of choline chloride per kg of body wt/d for 2 weeks. Choreic movements decreased in nine patients, worsened in one, and were unchanged in 10 (see also Lecithin).

An independent investigation of the relationship between dietary choline and acetylcholine formation in the brain led to the same conclusion as that cited above (37).

Administration of choline to patients receiving the anticancer drug methotrexate was suggested (38) as a means of protecting the liver against fatty metamorphosis and other lesions.

As a therapeutic agent, choline is administered orally in the form of syrups or elixirs containing the chloride, citrates or bitartrate, or in the form of compressed tablets or capsules of the dihydrogen citrate. Choline is also given in small doses as a nutritional supplement in combination with a variety of other materials. In dry pharmaceutical-dosage forms, the dihydrogen citrate is usually preferred because of its lower tendency to absorb atmospheric moisture. Both salts have been used parenterally.

In the feeding of animals, choline is often added to chicken and turkey feeds as a dietary supplement. This use has resulted from observations that young chickens and turkeys are unable to utilize methionine efficiently as a precursor of choline and that they have a high dietary requirement for choline. According to the Committee on Animal Nutrition of the U.S. National Research Council, young chickens require 0.15% choline in the diet, and young turkeys require 0.2%. Choline deficiency may cause a deformity of the bones, known as perosis. Examination of Table 1 shows that many commonly used poultry feedstuffs, especially the cereal grains, do not supply sufficient choline, and therefore choline chloride is sometimes added to chicken and turkey starting mashes. Choline has been found to be of importance in the nutrition of swine, but its addition to rations for these animals is not practiced extensively (39–42) (see Pet and livestock feeds).

In 1977 the price of choline chloride, feed grade, 70%, was $0.66/kg, choline dihydrogen citrate $4.60/kg, choline bitartrate $6.10/kg, and tricholine citrate, 65%, $2.97/kg.

Derivatives

The most important derivatives of choline are acetylcholine, acetyl-β-methylcholine, and carbamylcholine (43–45). More than 200 other choline derivatives have been synthesized and studied, but are not satisfactory for clinical use (44).

Acetylcholine (ACh), $CH_3CO_2CH_2CH_2N(CH_3)_3OH$, the "vagus substance," is produced in the body in small quantities by the stimulation of the cholinergic (parasympathetic) nerves; it is produced at the nerve endings, and it is the chemical substance that mediates the impulses of this group of nerves in the body. Its formation in the body is apparently caused by the action of an enzyme, choline acetylase, which transforms choline into acetylcholine. The powerful pharmacological activity of acetylcholine is evident from the data presented in Table 1. Its action, however, is transient because there exists in tissues another enzyme, cholinesterase, which hydrolyzes

Table 1. Comparative Properties of Acetylcholine Chloride, Acetyl-β-methylcholine Chloride, and Carbamylcholine Chloride[a]

Property	Acetylcholine chloride [51-84-3]	Acetyl-β-methylcholine chloride [62-51-1]	Carbamylcholine chloride [51-83-2]
mp, °C	149–152	171–173	202–203 (dec)
soly	v sol, H_2O and ROH; insol, ethers	v sol, H_2O and ROH	v sol, H_2O; mod sol, ROH
stability			
in air	hygroscopic	hygroscopic	nonhygroscopic
in water	unstable	unstable	stable
susceptibility to cholinesterase	++	+	−
nicotinelike action	+	−	++
LD_{50} in rats, mg/kg			
intravenous	22	20	0.1
subcutaneous	250	75	4
oral	2500	750	40
equally effective doses on blood pressure of rabbits, mg/kg			
intravenous	0.002	0.002	0.002
subcutaneous	1.0	0.2	0.1
oral	1000	50	2
min effective cathartic dose in dogs, mg/kg			
subcutaneous	0.8	0.05	0.01
oral	40	25	0.2

[a] Most properties are presented in ref. 45.

acetylcholine into choline and acetate. Extensive discussions concerning choline acetylase and cholinesterase appear in references 46–49 (see Cholinesterase inhibitors).

Acetylcholine occurs in nature only in small amounts because it is rapidly transformed into choline. Acetylcholine is prepared from choline chloride and acetic anhydride. It is available as the chloride or the bromide. The chloride is an odorless, crystalline, hygroscopic salt, decomposed by hot water or alkali. Acetylcholine salts are assayed by measuring their effect on isolated rabbit intestine (29).

Acetylcholine is used in small quantities in the form of salts as a vasodilator and to slow the rate of the heart beat. The salts are given intramuscularly, subcutaneously, or by iontophoresis, but are ineffective when given by mouth, presumably because of hydrolysis of acetylcholine in the gastrointestinal tract. The salts have been used in various diseases such as varicose ulcers, Raynaud's disease, thromboangiitis obliterans, and paroxysmal tachycardia, but in most cases the results have not been encouraging and other parasympathomimetic drugs with less transient effects are preferred (see Psychopharmacological agents).

Acetyl-β-methylcholine chloride (methacholine chloride (USP XIX)),

$CH_3CO_2CH(CH_3)CH_2N(CH_3)_3Cl$, is used as a parasympathetic stimulant and as an antiepinephrine substance. Its activity is compared with that of acetylcholine in Table 1. It is used in paroxysmal tachycardia. Acetyl-β-methylcholine chloride in therapeutic doses frequently causes nausea and vomiting, and atropine sulfate for parenteral use should always be on hand as an antidote. In overdosage, the salts of acetyl-β-methylcholine are quite toxic (44).

Carbamoylcholine chloride (carbachol (USP XIX)), $H_2NCO_2CH_2CH_2N(CH_3)_3Cl$, is used to increase peripheral circulation in cases of vasospasm in peripheral vascular disease. This treatment may be given to avert threatened gangrene. Carbamylcholine chloride is also used in a subcutaneous dosage of 0.25 mg to treat urinary retention of neurogenic origin. The salts of carbamylcholine are quite toxic in overdosage (44).

Chlorocholine chloride [999-81-5] (2-chloroethyltrimethylammonium chloride), $((CH_3)_3NCH_2CH_2Cl)Cl$, mp 245°C (dec) is a colorless, crystalline, deliquescent salt, very soluble in cold water and alcohols, stable in aqueous solution. It has an ammoniacal odor.

Chlorocholine chloride (CCC) is a plant growth regulant that is used to produce compactness and sturdiness in plants. The treated plant specifically has shorter internodes, dark green leaves, and shorter petioles. The plant usually remains dwarfed when irrigated and fertilized adequately. CCC is rapidly decomposed by soil microorganisms and does not persist in the soil for more than one growing season. Its effects on green plants are described in reference 50. It is marketed, for use in food crop plants, as Cycocel.

The growth regulating effect of CCC starts to appear within one week of treatment. One of its most useful applications is the prevention of lodging and flattening of wheat by wind. This increases crop yields, especially in conjunction with use of high levels of nitrogenous fertilizers. Favorable results have also been obtained with barley, rice, oats, and rye, but the principal use in cereal grains is for wheat. CCC is applied usually at rates of 1–5 kg/ha (see Plant growth substances). CCC and gibberellic acid produce opposite growth effects in plants and it has been postulated that there is a specific competition between gibberellin and CCC for one or several active sites in plants (50). It retards stem elongation by competing with the gibberellin system of plants (51). CCC inhibits the production of gibberellin by *Fusarium moniliforme* (52).

Chlorocholine chloride is also used in the United States as a growth regulant for ornamental plants, especially poinsettas and azaleas. It produces more compact plants with more abundant and darker foilage, and induces the formation of multiple buds on azaleas.

Related compounds and CCC appear to give similar plant growth effects, differing only quantitatively. In general, treated plants have shortened, thickened internodes and slightly small, thickened leaves. Subsequent research has been concentrated on CCC as the most desirable member of the chemical series. The compound is effective when applied to either the foliage or roots of sensitive plants.

BIBLIOGRAPHY

"Choline" in *ECT* 1st ed., Vol. 3, pp. 919–927, by T. H. Jukes, American Cyanamid Co., Lederle Laboratories Div. and G. H. Schneller, American Cyanamid Co., Calco Chemical Div.; "Choline" in *ECT* 2nd ed., Vol. 5, pp. 403–413, by T. H. Jukes, University of California, Berkeley.

1. C. L. Connor and I. L. Chaikoff, *Proc. Soc. Exptl. Biol. Med.* **39,** 356 (1938).
2. W. H. Griffith and D. J. Mulford, *J. Am. Chem. Soc.* **63,** 929 (1941).
3. C. W. Prince and W. C. M. Lewis, *Trans. Faraday Soc.* **29,** 775 (1933).
4. E. Kahane and J. Lévy, *Biochimie de la Choline et de ses Dérivés,* Hermann, Paris, 1938.
5. W. M. Latimer and W. H. Rodebush, *J. Am. Chem. Soc.* **42,** 1419 (1920).
6. *The Pharmacopeia of the United States of America,* 19th rev. ed. (USP XIX), Mack Publishing Co., Easton, Pa., 1974, p. 731.
7. A. J. Beams, *J. Am. Med. Assoc.* **130,** 190 (1946).
8. C. H. Best and C. C. Lucas, "Choline—Chemistry and Significance as a Dietary Factor," in R. S. Harris and K. V. Thimann, eds., *Vitamins and Hormones: Advances in Research and Applications,* Vol. 1, Academic Press, Inc., New York, 1943, pp. 1–58.
9. H. Blumberg and E. V. McCollum, *Science* **93,** 598 (1941).
10. G. O. Broun and R. O. Muether, *J. Am. Med. Assoc.* **118,** 1403 (1942).
11. V. du Vigneaud, *Harvey Lectures, Ser. 38,* 39 (1942–1943).
12. C. Entenman and I. L. Chaikoff, *J. Biol. Chem.* **138,** 477 (1941).
13. C. Entenman, I. L. Chaikoff, and M. Montgomery, *J. Biol. Chem.* **155,** 573 (1944).
14. J. V. Lowry, L. L. Ashburn, and W. H. Sebrell, *Quart. J. Studies Alc.* **6,** 271 (1945).
15. P. J. C. Mann and J. H. Quastel, *Biochem. J.* **31,** 869 (1937).
16. A. W. Moyer and V. du Vigneaud, *J. Biol. Chem.* **143,** 373 (1942).
17. D. Stetten, Jr., *J. Biol. Chem.* **140,** 143 (1941).
18. A. Z. Hodson, *J. Nutrition* **29,** 137 (1945).
19. J. M. McIntire, B. S. Schweigert, and C. A. Elvehjem, *J. Nutrition* **28,** 219 (1944).
20. Ger. Pat. 193,449 (Dec. 22, 1906), J. D. Reidel (Aktiengesellschaft Berlin).
21. U.S. Pat. 2,774,759 (1956), Blackett and Soliday (to American Cyanamid).
22. U.S. Pat. 2,198,629 (Apr. 30, 1940), R. T. Major and H. T. Bonnett (to Merck & Co., Inc.).
23. M. Kruger and P. Bergell, *Ber. Deut. Chem. Ges.* **36,** 2901 (1903).
24. Fr. Pat. 736,107 (Apr. 29, 1932), F. Korner.
25. U.S. Pat. 2,137,314 (Nov. 22, 1938), H. Ulrich and E. Ploetz (to I. G. Farbenindustrie A.G.).
26. I. Sakakibaia and T. Yoshinaga, *J. Biochem. (Japan)* **23,** 211 (1936).
27. D. Glick, *J. Biol. Chem.* **156,** 643 (1945).
28. N. H. Horowitz and G. W. Beadle, *J. Biol. Chem.* **150,** 325 (1943).
29. J. P. Fletcher, C. H. Best, and O. M. Solandt, *Biochem. J.* **29,** 2278 (1935).
30. H. C. Hodge, *Proc. Soc. Exptl. Biol. Med.* **58,** 212 (1945).
31. M. W. Neumann and H. C. Hodge, *Proc. Soc. Exptl. Biol. Med.* **58,** 87 (1945).
32. Committee on Dietary Allowances, *Recommended Dietary Allowances,* Food and Nutrition Board, National Research Council, National Academy of Sciences, 8th ed., 1974, p. 64.
33. D. H. Copeland and W. D. Salmon, *Am. J. Pathol.* **22,** 1059 (1946).
34. L. J. Davis and A. Brown, *Blood* **2,** 407 (1947).
35. F. B. Moosnick, E. M. Schleicher, and W. E. Peterson, *J. Clin. Invest.* **24,** 278 (1945).
36. J. H. Growdon and co-workers, *New Engl. J. Med.* **297,** 524 (1977); R. J. Wurtman and J. H. Growdon, *Hospital Practice* (3), 71 (1978).
37. D. R. Haubrich, *J. Neurochem.* **27,** 1305 (1976).
38. M. Freeman-Narrod, S. A. Narrod, and R. P. Custer, *J. Natl. Cancer Inst.* **59,** 1013 (1977).
39. R. J. Evans, *Poultry Sci.* **22,** 266 (1943).
40. T. H. Jukes, *J. Nutrition* **22,** 315 (1941).
41. T. H. Jukes, J. J. Oleson, and A. C. Dornbush, *J. Nutrition* **30,** 219 (1945).
42. T. H. Jukes and A. D. Welch, *J. Biol. Chem.* **146,** 19 (1942).
43. G. Alles, *Physiol. Revs.* **14,** 276 (1934).
44. L. Goodman and A. Gilman, *The Pharmacological Basis of Therapeutics,* The Macmillan Co., New York, 1941, pp. 349–375.
45. H. A. Molitor, *J. Pharmacol. Exptl. Therap.* **58,** 337 (1936).
46. R. Hawkins and B. Mendel, *Brit. J. Pharmacol.* **2,** 173 (1947).
47. F. Lipmann and co-workers, *J. Biol. Chem.* **167,** 869 (1947).
48. D. Nachmansohn, "The Role of Acetylcholine in the Mechanism of Nerve Activity," in ref. 8, Vol. 3, 1945, pp. 337–360.
49. D. Nachmansohn, *Chemical and Molecular Basis of Nerve Activity,* Academic Press, Inc., New York, 1959.
50. N. E. Tolbert, *J. Biol. Chem.* **235,** 475 (1960).

51. J. A. Lockhart, *Plant Physiol.* **37**, 759 (1962).
52. H. Kende, H. Ninnemann, and A. Lang, *Die Naturwissenschaften,* **50**, 599 (1963).

THOMAS H. JUKES
University of California, Berkeley

CHOLINESTERASE INHIBITORS

The cholinesterases are a group of hydrolytic enzymes present in both vertebrates and insects. Of these, acetylcholinesterase (AChE) is specifically used for the hydrolysis of the ion acetylcholine (1);

$$\underset{\text{(1)}}{CH_3\overset{\overset{\displaystyle O}{\|}}{C}OCH_2CH_2\overset{+}{N}(CH_3)_3}$$

a neurotransmitter at synapses and at neuroeffector junctions of the cholinergic nervous system. Inhibition of AChE is a major cause of increased stimulation of the cholinergic nervous system through increased concentrations of (1). Cholinesterase (ChE) inhibitors are useful in medicine and as insecticides (see Insect control technology). They have potential application as nerve gases in chemical warfare (see Chemicals in war).

Biochemistry of Cholinergic Nerve Transmission

Nerve impulses from higher centers are transported within cholinergic nerves by the action potential. The ions of greatest importance involved in nerve transmission are Na^+ and K^+. Upon receipt of the nerve impulse, acetylcholine (ACh) is liberated from storage vesicles at cholinergic nerve endings (1). It then interacts with a structure (receptor) on the membrane surrounding the next neuron (synapse), or the motor endplate, or the neuromuscular junction. This effects transmission of impulse by altering the permeability of the postjunctional membrane to permit the flow of ions necessary for excitation.

The interaction of acetylcholine with its receptor is reversible. The receptor–acetylcholine complex dissociates and the liberated ACh is rapidly hydrolyzed. This extremely rapid hydrolysis is catalyzed AChE (see below).

$$(1) + H_2O \xrightarrow{\text{AChE}} CH_3CO_2H + HOCH_2CH_2\overset{+}{N}(CH_3)_3$$

Since this hydrolytic reaction proceeds at an extremely rapid rate, the liberated ACh is rapidly destroyed after reacting with its receptor. Thus, the ACh liberated by a single nerve impulse essentially transmits that, and only that nerve impulse to the effector cell.

Chemical compounds that inhibit AChE reduce the hydrolysis of ACh and increase its concentration in the vicinity of the receptors. Thus, such compounds mimic stimulation of cholinergic nerves and the structures under their control.

The cholinergic nervous system is part of the autonomic nervous system which is responsible for the contraction of muscles not under voluntary control. In this role, the cholinergic nervous system is responsible for contraction of smooth muscle in the gastrointestinal and urinary tracts. It also increases the discharge of exocrine glands thereby increasing salivation, lacrimation, and the flow of gastric juices. The cholinergic nervous system maintains the tone of striated muscle. Certain medical problems require activation of the cholinergic nervous system, which theoretically can be achieved by the administration of ACh or analogous compounds. However, their action is fleeting because of rapid hydrolysis by the cholinesterases; a more practical approach is often the administration of AChE inhibitors (2). Some of the therapeutic uses of ACh are described in the article Choline. AChE inhibitors are also used in the treatment of myasthenia gravis, a disease characterized by muscular weakness; in this case, its action results from the increased tone of striated muscle. Inhibitors are also used to relieve ocular pressure in glaucoma; this results because ACh promotes draining from the canal of Schlemm (2).

Since insects have nervous systems mediated by ACh, some inhibitors of this enzyme are effective insecticides. In attempts to prepare specifically safer and more effective insecticides, a large number of ChE inhibitors have been prepared and examined for their effect on insects.

The toxicity of overdoses of these compounds, as well as their debilitating effects on humans, has made some of them potential nerve gases for use in chemical warfare.

Structure and Mechanism of Action of AChE

Owing to its biological importance, AChE has been the subject of intense study. Difficulty in enzyme purification has hampered studies of both its structure and the mechanism of its hydrolytic reaction. It has been established, however, that its active site contains two or more dicarboxylic amino acid residues as well as a serine and a histidine residue. It is postulated that these play a role in the hydrolysis of ACh.

Since most AChE inhibitors act by serving as slowly hydrolyzed substrates for the enzyme, a discussion of the binding and hydrolysis of ACh by the enzyme is in order. ACh is attracted to the active site through the formation of an ionic bond between its positively charged quaternary ammonium group and an ionized carboxyl group in the active site. Hydrophobic binding accounts for the attachment of the remaining portion of the molecule to the receptor with the possible aid of a hydrogen or dipole–dipole bond between the ester group and an appropriate group on the enzyme surface. This places the hydrolyzable ester group in the vicinity of a serine and a histidine residue in the active site.

Two mechanisms have been advanced to account for the enzymatic hydrolysis of ACh. One of these resembles a mechanism for the acid hydrolysis of esters (3), and the other the basic hydrolytic mechanism (4). The acid hydrolysis mechanism embodies protonation of the carbonyl oxygen of the ester group by a hydrogen from the imidazole ring of histidine. This results in an increased positive character of the carbon atom, which increases the ease of nucleophilic attack by the hydroxyl group of serine on the carboxy carbon.

An alternative mechanism involves the transfer of the hydrogen on the serine hydroxyl group to the closely situated imidazole ring via an ionized carboxyl group. This provides alkoxide character to the serine residue which increases its nucleophilicity sufficiently to attack the carbon atom of the ester group.

Both mechanisms include the attack of the serine hydroxyl on the carbonyl carbon of ACh with formation of the transition state shown below.

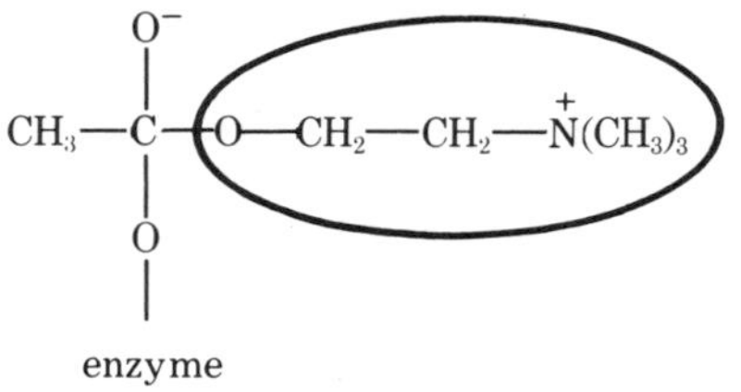

A reaction proceeds with the encircled choline fragment serving as the leaving group, and the acetyl group becomes esterified to the serine hydroxyl group of the enzyme.

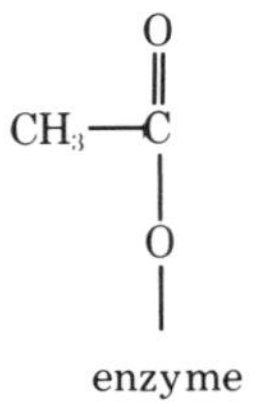

The regeneration of the enzyme is necessary in a final step for its continued role in the metabolism of the neurotransmitter. This involves the hydrolysis of the acetylated enzyme which occurs at a rate measurable in microseconds.

Compounds That Inhibit ChE

A number of compounds inhibit the hydrolysis of ACh by AChE (5–6). Most of these are esters that serve as substrates for AChE and are hydrolyzed by it via the mechanism previously discussed. In all cases they act as inhibitors since their hydrolyses proceed at rates appreciably lower than that of ACh.

Two types of compounds are used as inhibitors of AChE: (1) esters of carbamic acid, and (2) derivatives of the acids of phosphorus. In both cases the rate of enzymatic hydrolysis for each inhibitor is considerably less than that for ACh. Thus, the enzyme is occupied in their hydrolyses for a prolonged period of time, and is unavailable for continued hydrolysis of ACh. The reduced rate of hydrolysis of carbamate and phosphate esters by AChE results from the final step of the mechanism described above—the regeneration of the esterified enzyme (6). In the case of the carbamate esters, the enzyme is carbamoylated (see below)

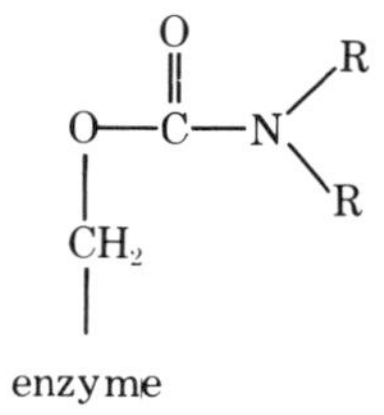

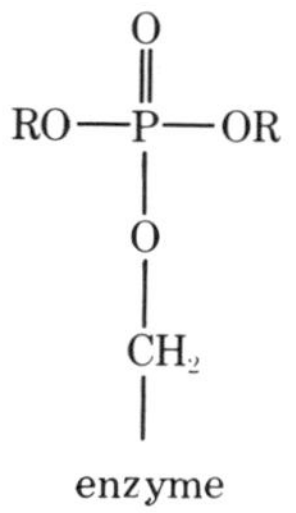

and in the case of the esters of phosphorus, it is phosphorylated. The rate required for the regeneration of the carbamoylated enzyme can be expressed in minutes, whereas that for the acetylated enzyme is measured in microseconds. The rate of hydrolysis of the phosphorylated enzyme is measured in hours. Untreated phosphorylated enzyme may, through a process known as aging, form an irreversible complex from which the enzyme is not regenerated. This aging process is assumed to involve the loss of an alkoxide group of the phosphate.

Because of their prolonged action, the phosphate esters are used only for topical application in medicine but have been widely employed as insecticides. Carbamate esters are employed both in medicine and as insecticides.

Carbamate Esters That Inhibit AChE. Physostigmine [57-47-6] (**2**), an alkaloid from *Physostigma venenosum* (see Alkaloids), was one of the first compounds to be used in medicine for its action as a cholinesterase inhibitor (7–9). Its structural relationship to ACh, which is probably responsible for its affinity to the active site, should be noted.

Attempts to prepare more easily synthesized analogues of physostigmine with useful anticholinesterase activity led to the synthesis of the trimethylammonium carbamates (10–12). The most widely medically used of these is neostigmine bromide [*114-80-7*] (**3**).

Neostigmine bromide inhibits AChE at concentrations as low as $10^{-6}\,M$. Other structurally related inhibitors useful in medicine are pyrodostigmine bromide [*101-26-8*] (**4**) and demecarium bromide [*56-94-0*] (**5**).

A variety of other carbamates have been synthesized (13) and evaluated. The

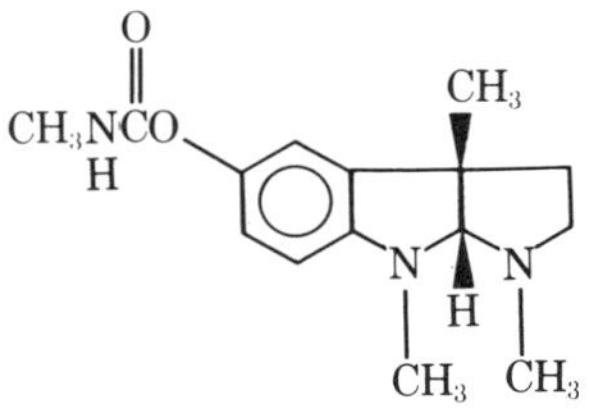

(**2**)

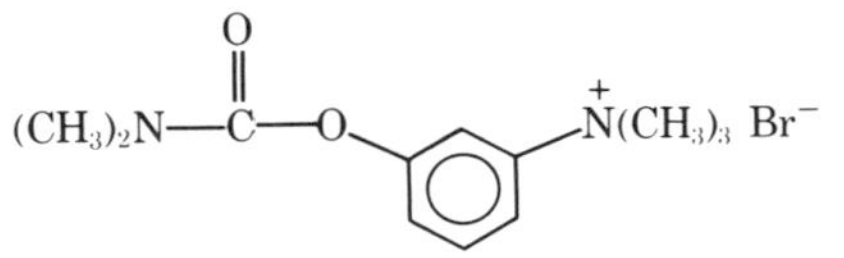

(**3**)

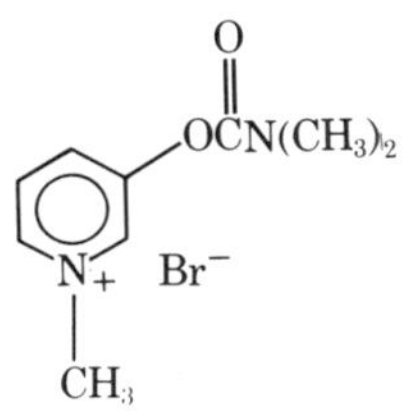

(**4**)

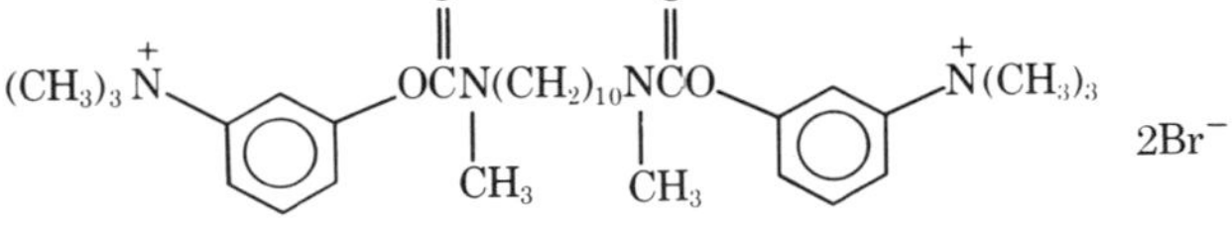

(**5**)

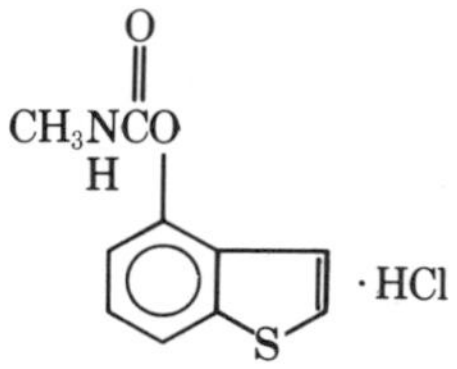

(6)

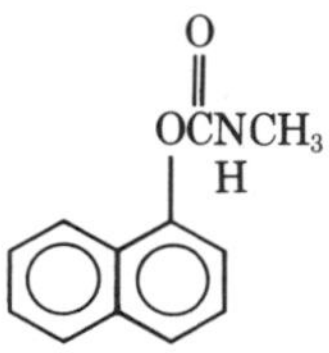

(7)

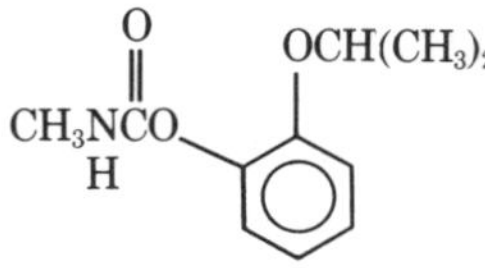

(8)

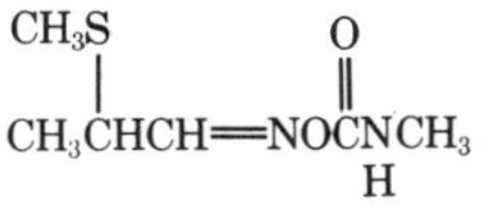

(9)

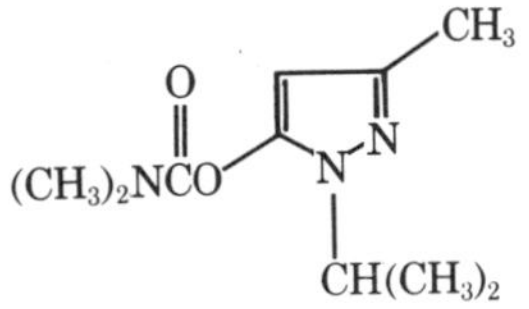

(10)

compounds mobam [1079-33-0] (6), carbaryl (7), baygon (8), adicarb (9), and isolan (10) are examples of carbamates that have proven useful as insecticides (see Insect control technology).

The structural differences between the medically useful cholinesterase inhibitors and those used as insecticides are apparent. There are a number of reasons for this. The primary reason is probably the difference in structure required to ensure passage through the cuticle of an insect when applied as a contact spray.

Most of the carbamates (qv) can be prepared by one of three methods shown in the following set of equations (14).

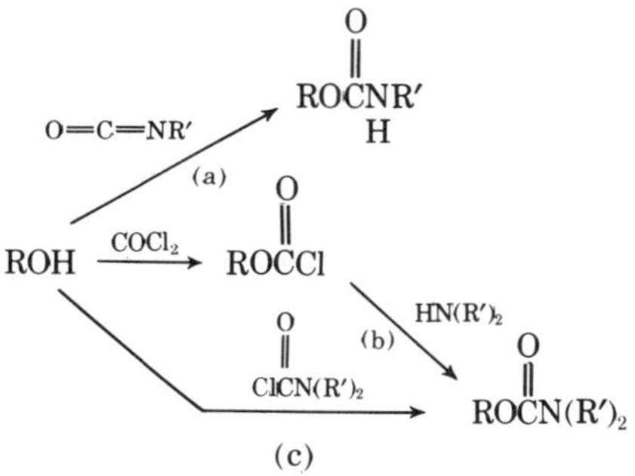

Reaction (a) provides an excellent route to the desired compounds when the necessary isocyanates are available. Reaction (b) is the most general approach.

Derivatives of Phosphorus Acids. Derivatives of phosphoric acid, pyrophosphoric acid, phosphonic acid, and their thio-derivatives are inhibitors of cholinesterases (see Phosphoric acid; Phosphorus compounds). In general, derivatives of the phosphorus acids have a longer duration of action than do the carbamates. Thus, they have a limited use in medicine, but wider application as insecticides.

Diisopropyl fluorophosphate [55-91-4] (11) (DFP), and echothiopate iodide [513-10-0] (12), are applied topically to relieve intraocular tension of glaucoma.

Among the compounds used as insecticides are dicapthon (13), tetraethyl pyrophosphate (TEPP) [107-49-3] (14), malathion (15), and ethyl p-nitrophenylphenyl phosphonothioate (EPN) [2104-64-5] (16).

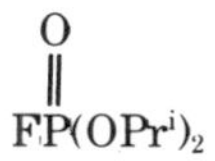

$$\overset{\text{O}}{\overset{\|}{\text{FP(OPr}^i)_2}}$$

(11)

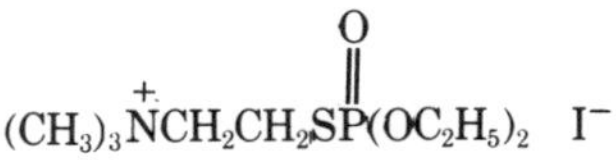

$$(CH_3)_3\overset{+}{N}CH_2CH_2\overset{\overset{\text{O}}{\|}}{S}P(OC_2H_5)_2 \quad I^-$$

(12)

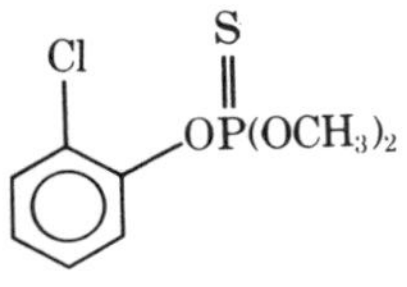

(13)

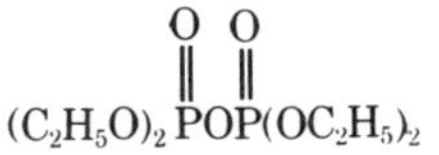

$$(C_2H_5O)_2\overset{\overset{\text{O}\ \ \text{O}}{\|\ \ \|}}{POP}(OC_2H_5)_2$$

(14)

$$(CH_3O)_2\overset{\overset{\text{S}}{\|}}{P}S\underset{\underset{CH_2CO_2C_2H_5}{|}}{CH}CO_2C_2H_5$$

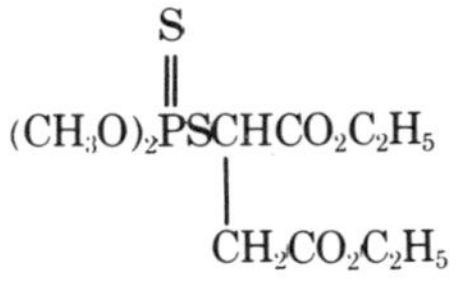

(15)

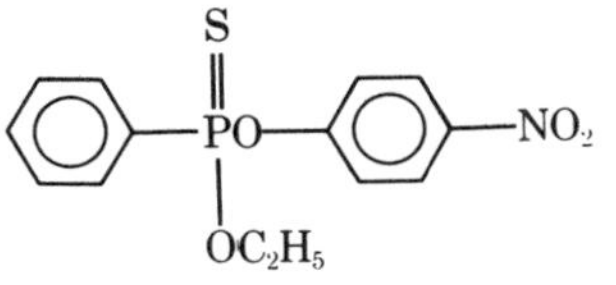

(16)

The previously discussed compounds hold considerable promise in the control of insects. Their rate of hydrolysis can be altered by structural modification to provide compounds sufficiently stable for efficient application, but with rates of hydrolysis that decrease residue problems. Differences in structural requirements for penetration of the insect's cuticle when applied by contact, and possible species differences in the structures of AChE, suggests an opportunity to develop more species-specific insecticides.

The phosphate derivatives are usually prepared from halogen derivatives of the corresponding acids as shown in the preparation of parathion (17) (15).

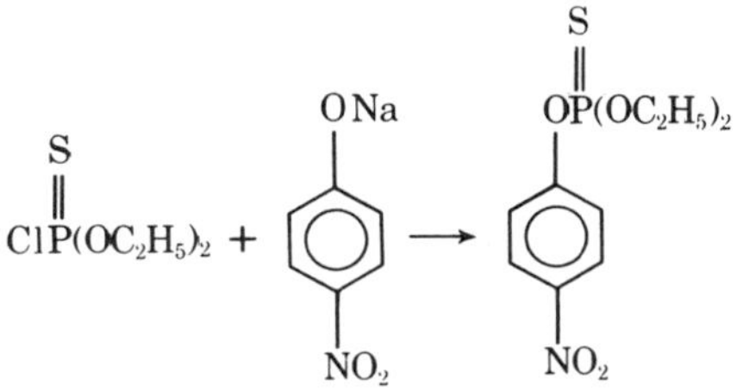

(17)

In some cases the esterification of the alcohol or phenol can be catalyzed by pyridine or other suitable tertiary amines.

Pyrophosphates, eg, TEPP (14), can be formed by condensation of the substituted acid halides:

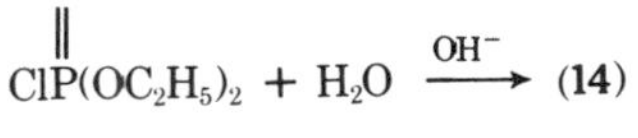

$$\overset{\overset{\text{O}}{\|}}{ClP}(OC_2H_5)_2 + H_2O \overset{OH^-}{\longrightarrow} (14)$$

or by condensation of two molecules of the acid derivative using a condensing agent such as dicyclohexylcarbodiimide (DCC).

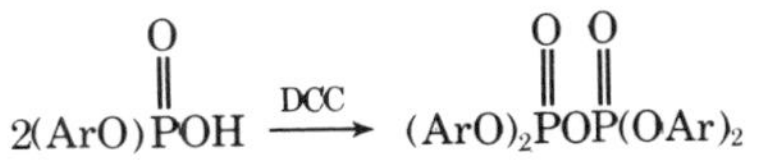

$$2(ArO)\overset{\overset{\text{O}}{\|}}{P}OH \overset{DCC}{\longrightarrow} (ArO)_2\overset{\overset{\text{O}\ \ \text{O}}{\|\ \ \|}}{POP}(OAr)_2$$

Table 1. LD$_{50}$ of Some Insecticides in Rats by Oral Administration[a]

Compound and structure no.	CAS Registry No.	LD$_{50}$, mg/kg
carbaryl (7)	[63-25-2]	850
baygon (8)	[114-26-1]	0.83
adicarb (9)	[116-06-3]	0.8
isolan (10)	[119-38-0]	23
dicapthon (13)	[2463-84-5]	400
malathion (15)	[121-75-5]	1500
parathion (17)	[56-38-2]	6

[a] Refs. 13 and 16.

Toxicity of AChE Insecticides

The toxicity of cholinesterase inhibitors varies widely. Table 1 lists the LD$_{50}$ in male rats for compounds (7–10,13,15, and 17).

The relatively high toxicity of some of these compounds coupled with the long duration of action of the phosphorus derivatives poses a hazard to those involved in their application and transportation. This can be reduced by limiting their utilization to those trained properly and by the prompt administration of an effective antidote in case of poisoning.

The toxic effects of the carbamates can be counteracted by administration of atropine or similar anticholinergic drugs (see Alkaloids). Additional treatment may be necessary for counteracting the toxicity of the phosphorus derivatives. Hydrolysis of phosphorylated AChE is increased by the administration of pralidoxime iodide (18) (2-PAM).

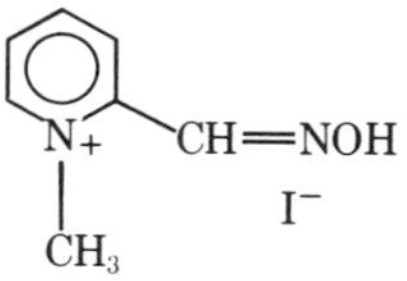

(18)

This compound is attracted by its positive charge to the anionic site of the phosphorylated enzyme, which places the nucleophilic hydroxyl group close to the phosphorus atom. After nucleophilic attack, the enzyme becomes the leaving group and is regenerated. Since regeneration of the enzyme requires appreciable time, atropine should be administered with (18).

BIBLIOGRAPHY

1. W. O. Foye, *Principles of Medicinal Chemistry,* Lea and Febiger, Philadelphia, Pa., 1975, pp. 322–323.
2. G. B. Koelle in L. S. Goodman and A. Gilman, eds., *The Pharmacological Basis of Therapeutics,* 5th ed., MacMillan Publishing Co., New York, 1975, pp. 445–447.
3. Ref. 1, p. 335.
4. A. Burger, ed., *Medicinal Chemistry,* 3rd ed., Part 47, Wiley-Interscience, New York, 1970, p. 1350.

5. C. O. Wilson, O. Giswald, and R. F. Doerge, *Textbook of Organic Medicinal Chemistry and Pharmaceutical Chemistry,* 7th ed., J. B. Lippincott Co., Philadelphia, Pa., 1977.
6. Ref 4, p. 1307.
7. J. Jobst and O. Hesse, *Ann. Chem. Liebigs* **129,** 247 (1864).
8. E. Stedman and G. Barger, *J. Chem. Soc.* **127,** 247 (1925).
9. P. L. Julian and J. Pilel, *J. Am. Chem. Soc.* **57,** 755 (1935).
10. E. Stedman, *Biochem. J.* **20,** 719 (1926).
11. *Ibid.,* **23,** 17 (1929).
12. E. Stedman and E. Stedman, *J. Chem. Soc.,* 609 (1929).
13. R. J. Kehr and H. W. Dorough, *Carbamate Insecticides: Chemistry and Biochemistry,* CRC Press, Cleveland, Ohio, 1976, pp. 88, 90.
14. Ref. 8, pp. 21.
15. P. Alexander and Z. M. Bacq, *International Series of Monographs on Pure and Applied Biology,* Vol. 13, Pergamon Press, New York, 1961, p. 29.
16. R. White, *Pesticides in the Environment,* Part I, Marcel Dekker Inc., New York, 1971, pp. 95, 98.

J. E. Gearien
University of Illinois at the Medical Center

CHROMATOGRAPHY. See Analytical methods; Chromatography, affinity.

CHROMATOGRAPHY, AFFINITY

When compared to the more familiar forms of chromatography (see Analytical methods), affinity chromatography is somewhat of a misnomer in that the process is more of an extraction process using a rather selective adsorbent (1). This process developed by biochemists (including immunochemists) enables the efficient isolation of biological macromolecules or biopolymers (qv) by making use of a feature unique to these substances. Biopolymers such as enzymes and antibodies are capable of recognizing certain chemical structures with a high degree of selectivity and then binding to them. To emphasize this point, some scientists have preferred the term biospecific affinity chromatography (2).

The process of affinity chromatography is quite simple. As shown in Figure 1, it involves a selective adsorbent which is placed in contact with a solution containing several kinds of substances including the desired species, the ligate. The ligate is selectively adsorbed to the ligand, which is attached via a leash (or tether) to the insoluble support or matrix. The nonbinding species are removed by washing. The ligate is then recovered by eluting with a specific desorbing agent.

Although it has been only infrequently attained, it is theoretically possible that a one-step purification process for the ligate can be achieved by proper design of the sorbent and careful selection of the desorbing agent. This possibility is one of the great attractions of affinity chromatography. This is further emphasized by a comparison, as shown in Figure 2, to classical methods of biopolymer isolation which generally rely on gross molecular properties such as solubility, molecular weight, or isoelectric point. In addition to the saving of labor, implicit in fewer steps, higher product yields can

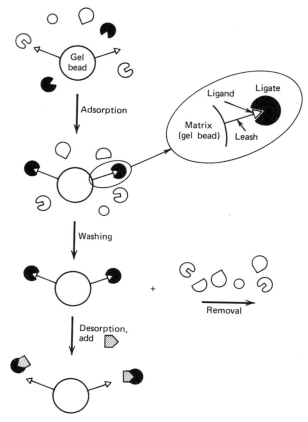

Figure 1. Affinity purification process.

often be obtained with the shorter scheme. If the yield in an average process step is 80%, a scheme involving seven steps would afford only a 21% yield, overall.

The affinity principle has been recognized for several decades. Immunologists starting in the 1930s took advantage of it to purify antibody molecules (immunoglobulins, in particular IgG). The first synthetic immunosorbents were prepared in the early 1950s by attaching haptens (the chemical entity recognized by immunoglobulins) to cellulose (3).

In the early twentieth century, enzymologists debated over the protein nature of enzymes (qv) even after 1926 when Sumner crystallized urease (4). In the mid 1930s Northrup performed what today can be called an affinity adsorption experiment to prove that the enzyme pepsin was a crystallizable protein (5). Thus, the protein nature of enzymes was established. This was followed by the development of a variety of techniques for enzyme purification based on their protein properties.

Early applications of the affinity principle to enzyme isolation depended on the use of naturally occurring adsorbents. Therefore, Holmbergh was able to adsorb malt α-amylase on starch in the presence of a cold 50% ethanol solution of maltose and separate it from β-amylase (6). In the mid 1940s a more concerted effort was reported by Hockenhull and Herbert (7) who purified, by 300-fold, the amylase from the bac-

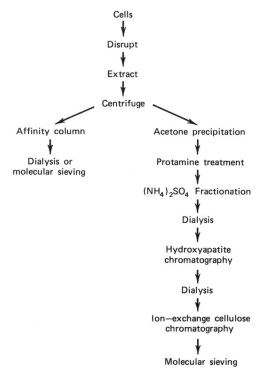

Figure 2. Comparison of an affinity purification scheme, using conventional methods. Courtesy of *Chem. Tech.* (2a).

terium *Clostridium butylicum* by adsorption on potato starch in the presence of ethanol.

In 1953 the first synthetic enzyme-specific adsorbent was reported. Lerman had succeeded in immobilizing phenolic compounds to cellulose in order to adsorb and purify mushroom tyrosinase (8). A decade passed before Arsenis and McCormick revived the notion of deliberately designing affinity sorbents, this time for flavin-binding enzymes (9). By 1967 a review by Baker appeared on the concept of enzyme-specific adsorbents (10). In the following year, the term affinity chromatography was introduced and became widely used (11). The convenience of using agarose gel and cyanogen bromide coupling chemistry, developed by Porath and co-workers (12), facilitated the preparation of a wide variety of affinity sorbents, and many examples soon came forth.

Biological Specificity or Affinity

Ligand–Ligate Combinations. In addition to enzymes and antibody molecules there are several other classes of biopolymers that, as ligates, exhibit significant affinities to appropriate ligands. The known permutations and combinations are listed in Table 1.

Accordingly, almost all significant classes of biopolymers are represented because, in varying degrees, the high structure recognition property is a distinctive feature of

Table 1. Affinity Systems

Ligand, immobile entity	Ligate, soluble entity
inhibitor, cofactor, prosthetic group, polymeric substrate	enzymes; apoenzymes
enzyme	polymeric inhibitors
nucleic acid, single strand	nucleic acid, complementary strand
hapten; antigen	antibody
antibody (IgG)	proteins; polysaccharides
monosaccharide; polysaccharide	lectins; receptors
lectin	glycoproteins; receptors
small target compounds	binding proteins
binding protein	small target compounds

all biological macromolecules. To a much lower extent this recognition ability may be found in certain small molecules, eg, those with optical activity (chirality). An inquiry into these possibilities has been presented elsewhere (13).

It is evident from Table 1 that affinity interactions are reversible. The insolubilization of one partner of an interacting pair allows the isolation of the other. A circular process pathway has been occasionally reported (14). For example, a protein, X, brought to high purity by laborious classical purification methods is used (as an antigen) to immunize test animals to produce specific (anti-X) antibodies. A sample of protein X is immobilized to obtain a selective sorbent which is used to isolate anti-X from the hyperimmune animal serum. The highly purified antibody so obtained is in turn immobilized. This second sorbent comprised of anti-X ligands can now be used to harvest protein X from its crude source, possibly in a single step. The complex way of obtaining this selective immunosorbent is justified by the simplification achieved for the further isolation of more protein X, as well as by its reusability in any repeat isolations of this ligate. Thus, with repetitive use the immunosorbent route may reduce labor significantly.

Degrees of Selectivity.   To prepare an affinity sorbent, the first task is to find a suitable ligand. The choice is usually made from likely candidates suggested by the appropriate literature, eg, inhibitors (or substrate analogues) for enzymes, or lectins for glycoproteins, etc. The preferred ligand compound shows a high structure selectivity for the desired ligate and also possesses a second functional site where immobilization may be effected without adversely affecting ligate binding.

What is structure selectivity? Perhaps the best known model is the lock and key concept for explaining enzyme specificity. The structural details of the proper substrate molecule (the key) enable it to bind to the enzyme-active site (the lock) and undergo transformation (ie, turn). The fit need not always be absolute. As there are master keys that can fit the same lock, there are alternative substrates or inhibitors, which may be substrate analogues, that can also bind the enzyme, although not with the same efficiency or ability to turn over. Affinity sorbents are vastly more selective than absorbents such as ion-exchange resins, but they are generally not absolutely selective for a single species of ligate molecule. Varying degrees of specificity are observed. Generally, the less complex ligands exhibit poorer (less stringent) selectivity and the more complex ones show stronger (more stringent) selectivity. Thus an indole structure may bind lysozyme and chymotrypsin, which are functionally rather unrelated enzymes. But L-tryptophan, which contains a chiral center as well as the indole group as a ligand, may bind chymotrypsin well but not lysozyme.

This concern for selectivity is a practical one. Sorbents containing less stringent ligands may be useful in the isolation of several types of ligate molecules; however, the extent of purification of any one of the desired ligate types may be low. Conversely, sorbents bearing very stringent ligands may afford an extensive single-step purification of a ligate species, but separate and substantial effort may be required for the isolation of different ligates. In practice, both kinds of examples are found.

General Ligands. Many ligands are selective for a group or class-type of ligate species. For example, immobilized aminobenzamidine has been used to isolate trypsin, a pancreatic protease; thrombin, a blood enzyme that promotes clotting (15); and urokinase, a protease from urine that promotes dissolution of blood clots (16). These enzymes are catalytically related but differ in their source and biological function. Similarly, various NAD (nicotine adenine dinucleotide)-dependent dehydrogenases can be separated on a gel containing immobilized 5'-AMP (5'-adenosine monophosphate) which comprises one-half of the NAD molecule and is sufficient for binding these enzymes. Such group-selective ligands have been referred to as general affinity ligands (17–18).

Cibacron Blue F 3GA (Ciba) is a compound of related interest of synthetic origin; it is presently gaining widespread use as a general affinity ligand. This blue dye, when bonded to agarose, is assumed to have the following structure (see Dyes, reactive).

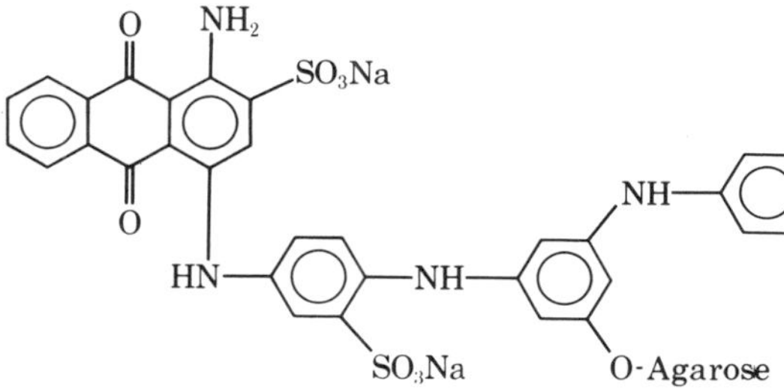

R = H or SO_3Na
R' = SO_3Na or H

There are apparently sufficient similarities between portions of the dye molecule and NAD or NADP that permit recognition and binding by enzymes, although they are somewhat different in structure than naturally occurring cofactors. There is a dinucleotide fold in the binding-site of enzymes (mainly kinases and dehydrogenases) that binds the dye (19–20). Thus a rationale has developed for using this dye as a general affinity ligand. The great practical potential of sorbents bearing this dye has been shown (21). Although not as well studied, a related group of dyes called Procion (Imperial Chemical Industries) has also been investigated as general affinity ligands (22).

Although the previous discussion mainly draws upon examples from enzymology, similar conceptual reasoning can be applied when considering isolation of other types of biopolymers. For example, the lectin, concanavalin A, is frequently used as a general ligand for polysaccharides and glycoproteins (23). Concanavalin A is known to selectively bind glucose and mannose residues in polymers. Thus when incorporated into a sorbent, it can be used to isolate a wide variety (ie, with respect to biological function) of biopolymers as long as they contain one of the two sugar groups.

Size of Ligands. Ligands come in all sizes—from a mol wt of a few hundred (as with aminobenzamidine) to a mol wt of several tens of thousands (as with concanavalin A), or more. The observed intrinsic selectivity is not usually determined by ligand size; however, practical considerations of sorbent preparation are affected by ligand size. Large, macromolecular ligands take up proportionately large amounts of space in the carrier support, thus allowing lower concentrations of functional entity per unit volume. Conversely, small molecule ligands can be attached in higher functional density in the support. Avidity or strength of binding is proportional to the ligand density in the carrier gel. Consequently, ligand size determines avidity, ie, if a carrier is saturated with ligands and other factors are held equal, large ones will yield sorbents of lower avidity and small ones will yield high avidity sorbents.

Among the reported macromolecular ligands, the cases where the gel or support material itself exhibits selectivity and binding towards the ligate population deserve separate mention. A number of such examples were reported before the recent era where most affinity sorbents are products of chemical synthesis. Although their numbers are somewhat limited to naturally occurring polymers, the success of their use has nevertheless been great. Some examples have been: chitin for isolation of lysozyme (24), dextran (Sephadex) for glucoamylase (25), elastin for elastase (26), collagen for collagenase (27), and agarose for agarase (28).

Theoretical Considerations

The theoretical underpinnings of affinity adsorption or chromatography are not yet fully established. However considerable progress has been made and several attempts have been reported (29–31). The main points of concern are avidity or the strength of ligate binding to ligand, and elution behavior or the prediction of solute (ligate) migration through the adsorbent bed (see also Adsorptive separation, liquids).

Avidity is essentially described by the association constant for ligate (E) and ligand (L) (eqs 1–2):

$$E = L \underset{k_b}{\overset{k_a}{\rightleftharpoons}} EL \tag{1}$$

where

$$K_{assoc} = \frac{k_a}{k_b} = \frac{[EL]}{[E].[L]} \tag{2}$$

The association process (or adsorption, since L is part of the insoluble phase) can be described approximately by a Langmuir-type isotherm (eq. 3):

$$[EL] = \frac{K_{assoc}\,[E]\,[L_0]}{1 + K_{assoc}\,[E]} \tag{3}$$

which can be derived from equation 2 (eq. 4) (32):

$$[L_0] = [L] + [EL] \tag{4}$$

The practical importance of avidity resides in the fact that the ligand capacity of the support matrix is finite. Therefore, the limiting ligand concentrations in agarose gels are about 5 μeq/mL, 15–20 μeq/mL, and 50–60 μeq/mL for 2% gel, 4% gel, and 6%

gel, respectively (33). For a fixed upper limit of ligand density one can calculate degree of binding with the aid of a rewritten form of equation 3 (eq. 5):

$$\frac{[EL]}{[E_o]} = \frac{K_{\text{assoc}} [L]}{1 + K_{\text{assoc}} [L]} = \theta \tag{5}$$

If $K_{\text{assoc}} = 1000$, then for $[L] = 50$ mM (ie, 50 μeq/mL), $\theta = 0.98$ or about 98% of the protein in contact with the sorbent would be bound. However, if $[L] = 1$ mM, then $\theta = 0.5$, and the protein may be expected to wash through the sorbent bed. Therefore, the concentration limits for ligand density in agarose gels suggest the need for association constants $\geqslant 1000$.

With a sorbent that successfully binds the desired ligate, the next task is to determine a means of efficient recovery. In many instances the approach has been to simply alter the solvent medium in contact with the sorbent so that the association constant is diminished to the point where the ligate falls off. This can be achieved by changing pH, ionic strength, or solvent composition, such as by adding water-miscible organic solvents. The approach is successful as long as the sorbent selectivity precludes binding of unwanted contaminants. Usually, however, other solutes are nonspecifically adsorbed; thus a more selective desorption process is in order.

The most simple approach to selective recovery of a ligate is to use a soluble form of the ligand as the desorbing agent. The process is illustrated with the following example of a small enzyme (mol wt: 10^4–2×10^4) adsorbed to an affinity sorbent made from 4% agarose gel. When the soluble desorbing agent, C, is introduced to the system, it will compete with the ligand in binding by enzyme and cause the latter to become soluble (eqs. 6–9).

$$E + L \rightleftharpoons EL \text{ (insoluble)} \tag{6}$$

$$\text{and association constant is } K_L = \frac{[EL]}{[E] [L]} \tag{7}$$

also

$$E + C \rightleftharpoons EC \text{ (soluble)} \tag{8}$$

$$\text{and association constant is } K_C = \frac{[EC]}{[E] [C]} \tag{9}$$

Substance C therefore shifts the enzyme partitioning from insoluble to soluble phase. From adsorption chromatography analysis (eq. 10) (34):

$$V_e = V_t + K_p(V_t) \tag{10}$$

where V_e is the observed elution volume of the ligate when it interacts with the sorbent; V_t is the ligate elution volume when no interaction with sorbent occurs; and K_p is the partition coefficient of the ligate between insoluble and soluble phases.

The total amount of ligate molecules involved in the partitioning is (eq. 11):

$$[E_t] = [E] + [EC] + [EL] \tag{11}$$

By combining equations 7, 9, 10, and 11 (see refs. 35–36) one obtains (eq. 12):

$$V_e = V_t + \frac{K_L[L]}{1 + K_C[C]} \cdot V_t \tag{12}$$

When no C is present, and E is bound, V_e will be very large. When a suitably high concentration of C is contacted with the sorbent, the absorbed E will be displaced and emerge at a volume governed by the ratio: $K_L L/L + K_C C$.

Several other approaches to quantitative description of the affinity process have been reported (37–39).

Materials and Methods

Preparation of sorbent requires proper selection of ligand and then an appropriate coupling chemistry to attach it to the carrier support or matrix.

Supports. In research applications, agarose gels and cross-linked agarose gels have been the most widely used support materials. They are convenient to handle and fairly easy to form chemically. Their hydrophilicity makes them relatively free of nonspecific binding by proteins. However, their compressibility, which leads to low throughput, and relatively high cost make them less attractive as carriers in large-scale processing, as in manufacturing. Nevertheless, these drawbacks may not be obstacles where the isolation of very precious substances are concerned.

A rigidly stable column packing may be found in controlled-pore glass (CPG) beads. This material is commercially available with surface coatings comprised of glycerol- and glycol-like structures to eliminate nonspecific protein adsorption. Although high throughputs can be obtained with columns packed with CPG, this carrier is even more expensive than agarose gel beads.

Cellulose particles have been long used by immunochemists since the earliest days of synthetic affinity sorbents. It appears to have been adequate in the isolation of immunoglobulins, which typically exhibit high affinity constants ($K_a \geqslant 10^8$). However, compared to agarose gels, cellulose particles are formed with more difficulty and therefore, have received less attention in the preparation of affinity sorbents for enzymes, which often exhibit ligand affinities of ca 10^4–10^5. Cellulose is perhaps the cheapest of the support matrices.

Two lesser used support materials are polyacrylamide gel beads and Sephadex a proprietary (Pharmacia Fine Chemicals) gel bead made from dextran and epichlorohydrin (see Acrylamide polymers; Chlorohydrins). Polyacrylamide can be substituted by replacing the amide nitrogen. Although convenient methods have been developed for doing this, the softness of these beads allows for poor column packings and the low molecular porosity yields a sorbent with poor ligand availability to the ligate. Surveys are available that include detailed properties and uses of the various support materials (40–42).

Coupling. A variety of covalent coupling methods for attaching ligands and their leashes to supports have been developed; their choice and use, however, depends on the chemical nature of the support itself. For carbohydrates such as agarose or cellulose, the reaction with cyanogen bromide in aqueous alkali has been the most common. Yet this coupling method has been subject to criticism since the resulting chemical linkage (an isourea, among others) is susceptible to degradation and may contain positively charged groups which can cause nonspecific adsorption (43). Thus the use of bisoxiranes (see Epoxy resins) and epihalohydrins has been advocated and developed (44–45) and found to avoid the drawbacks associated with cyanogen bromide chemistry.

Suitable derivatives of polyacrylamide gels can be obtained by treatment with hydrazine (46), after which a variety of acylating and alkylating reagents may be employed for further chemical conversion. Controlled-pore glass can be treated with γ-aminopropyltriethoxysilane to cover silanol groups that give rise to nonspecific adsorption by proteins. The γ-aminopropyl moiety provides loci to which ligands may be attached. Derivations of the glyco- or glycerol-phase coated CPG beads may be obtained by treatment with sodium periodate (47).

Once connecting points have been introduced to the support, an assortment of reactions may be employed to couple with the leash or ligand. The choice of a coupling is dictated by the chemical nature of the ligand or leash. Reviews on these reactions may be consulted for details (42,48).

Operational Modes

The introduction suggested that the term affinity chromatography was somewhat misleading in view of how it is most frequently practiced. This is more evident when the various modes of application are considered. Thus, a more generic and logical term is affinity purification or affinity separation, which has been used by a number of authors.

Column Operation. The most commonly reported mode of using affinity sorbents is in column packing. The chief advantages of such a mode are: freedom from handling or attritional losses of packing, high efficiency of solute (ligate) contact with sorbent, and ease of automating the process. Limitations may be the low throughput rates owing to resistance of the packed bed and clogging of the bed by insoluble (sometimes soluble) materials in the sample solution. A schematic arrangement of an affinity sorbent column system is shown in Figure 3. The sample mixture is pumped onto the column, followed by a washing buffer, after which a valve is switched to admit the desorbing agent. A valve on the column exit can direct the effluent stream to waste or to sample collection. The effluent containing the ligate may be passed through a molecular sieve column or a flow-through dialyzer to remove the desorbing agent before sample collection (see Molecular sieves). A monitor on the flow stream is useful in locating sample zones and facilitating the collection of sample fractions.

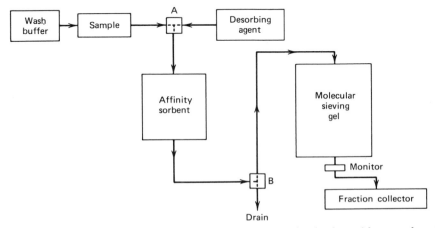

Figure 3. Affinity purification carried out in columns, where valve A selects either sample and wash solutions or the desorbing solution to be directed to sorbent column, and valve B can direct the column effluent to waste or to molecular sieve column and eventual collection.

An example of a separation obtained with the column system is shown in Figure 4. The molecular sieve column packed with Sephadex G-25 gel allows the high molecular weight (39,000) thrombin to elute first and be separated from the low molecular weight (156) benzamidine, the selective desorbing agent (15).

The column operation can be automated so that the functional cycle is repeated reproducibly. This is particularly useful in any production situation where labor costs need to be minimized. An example of an automated recycling system applied to immunosorbents is shown in Figure 5. This system, which has been named Cyclum by its developers (49), has a timer–programmer that controls a set of pumps and valves involved in the various sub operations: loading, washing, recovery, and regeneration.

Multi-Sequence Columns.   To conserve precious raw materials, which may be the source of several desired substances, it would be logical to employ several affinity columns in sequence. This would be particularly attractive in production situations. At present, examples from the research literature are few. One of the earlier examples used a rat liver extract that was first passed through a column containing the following structure (50):

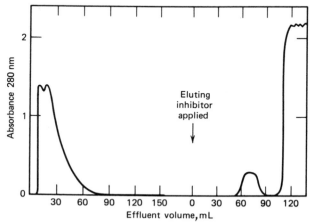

The ligand was an inhibitor of dihydrofolate reductase which caused the enzyme to be removed from the extract. Subsequently, the nonabsorbing proteins from this

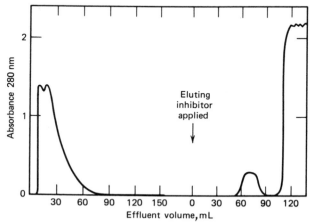

Figure 4.   Example of affinity purification of enzyme. Note that impure thrombin is applied to a sorbent containing *m*-aminobenzamidine as a ligand. On the left-hand side, the monitor tracing (absorbance) indicates nonactive (and nonbinding) protein washing through the columns. On the right-hand side, addition of benzamidine solution (indicated by arrow) results in thrombin displacement. Owing to molecular sieve column, thrombin emerges first (small peak) and is separated from benzamidine (large peak). Courtesy of *Arch. Biochem. Biophys.* (15).

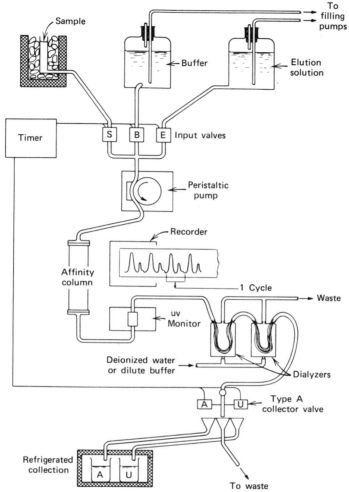

Figure 5. The Cyclum automated affinity purification system. Where the master timer controls overall process: sample loading, washing, etc; the desired ligate is accumulated by repeating process cycle; and the monitor recorder makes a tracing of each batch cycle processed. S, sample; B, buffer; E, eluant; A, absorbed material; U, unbound material. Courtesy of *Anal. Biochem.* (49).

column were passed into a second column that contained an inhibitor for guanine deaminase:

$$\text{H}_2\text{N} \quad \text{HN} \quad \text{O} \quad \text{N} \quad \text{N} \quad \text{N} \quad \text{—O(CH}_2)_2\text{NH——Agarose}$$

The enzyme adsorbed here was subsequently recovered with a solution of 0.5 m*M* guanine, the substrate.

Sequential Isolation of Ligates. As discussed in the section on the selectivity of ligands, a general ligand, or one to which several functionally related biopolymers might bind, is an attractive entity because of its wider range of utility. The strategy for using a sorbent containing a general ligand is shown in Figure 6. The general affinity sorbent is first contacted with a crude mixture from which several ligate species are adsorbed. After washing out nonadsorbing components, the sorbent is eluted with a solution of desorbing agent at low concentration, whereupon the weakly bound ligates are preferentially removed. Next, a solution containing a moderate concentration of desorbing agent is applied, and the more tightly bonded ligates are recovered. Finally, a high concentration of the desorbing agent is applied to recover the most tightly bound ligates. Thus, ligate molecules differing in avidity for the same affinity sorbent are separated. In addition to the stepwise elutions previously described, ligate recovery may be effected by gradient elution. A variety of examples involving both techniques is found in practice (17–18).

It is also possible to selectively recover different functionally unrelated ligates bound simultaneously to a general affinity sorbent. This may be accomplished by successive elution with different desorbing agents each of which is able to remove only one of the bound ligates. The mechanism of this technique may be explained as follows:

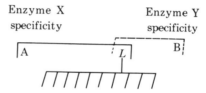

The specificity range of enzyme X could include compound A in addition to the ligand *L*, whereas enzyme Y is able to recognize compound B and the ligand, but not A. In the absence of compounds A or B, both enzymes are able to bind to ligand *L*. Now when compound A, at sufficiently high concentration, is contacted with the sorbent carrying both enzymes, enzyme X can be displaced owing to its avidity for this substance. Similarly, when compound B is applied, enzyme Y will preferentially shift from the ligand to the soluble phase. An example of this has been recently reported (51).

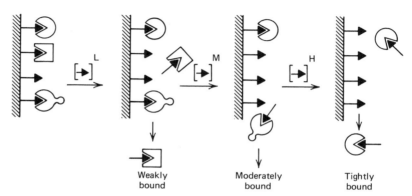

Figure 6. Stepwise (or gradient) elution from general ligand sorbent.

Batch Operation. This mode has been employed frequently when column arrangements have suffered flow rate or clogging problems, especially during the sample loading step. It is also convenient to work in batch mode when long contact times for ligate and sorbent are needed to obtain complete binding (52–53). A simplified scheme for carrying out the process is shown in Figure 7. The affinity sorbent particles are merely suspended in the sample mixture and agitated for the desired length of time, after which they are collected by filtration, or alternatively by sedimentation and fluid aspiration. After washing to remove nonbound or unwanted substances, the particles are suspended in a desorbing solution and the ligate is finally recovered in the liquid phase. An alternative recovery procedure is to pack the washed ligate-bound particles into a column tube and percolate with the desorbing solution.

Tea-Bag Batch. By containing affinity sorbent beads inside of porous fabric bags, one can immerse the sorbent into a ligate solution and, after an appropriate contact time, easily withdraw it, like a tea bag. By having several bags, each containing a different sorbent, it is possible to simultaneously extract different ligate species from the same starting mixture. A process is shown schematically in Figure 8 (54). Simultaneous extraction offers advantages in saving time, in decreasing losses of labile components, and of conserving precious starting materials, eg, human blood plasma, which is the source of several clinically useful proteins (see Blood, fractionation).

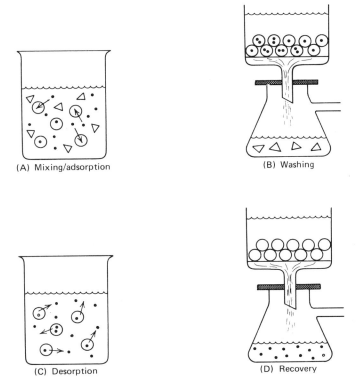

(A) Mixing/adsorption

(B) Washing

(C) Desorption

(D) Recovery

Figure 7. Batch processing of biopolymers on affinity sorbent. Note that O = affinity gel bead; ● = ligate molecule; and △ = contaminant.

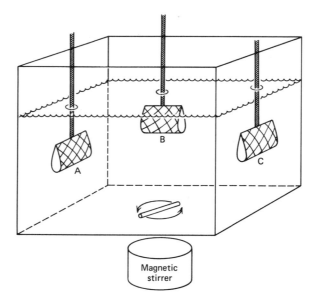

Figure 8. Tea-bag processing of several biopolymers on three different sorbents, A, B, and C, simultaneously. Courtesy of *Biochem. Biophys. Acta* (54).

Continuous Belt. This method is, in part, a variation on the tea bag in that the affinity sorbent is also enclosed in a porous, fine-meshed fabric, except that it is shaped into a continuous belt. As with the tea bag arrangement, very high affinities must prevail between ligate and sorbent for the continuous belt sorbent to work. As shown schematically in Figure 9, the belt is constructed in an endless loop and transported through a sequence of tanks containing sample solution, rinse bath, recovery solution, and finally, regeneration bath. The continuous feature of the system, of course, is at-

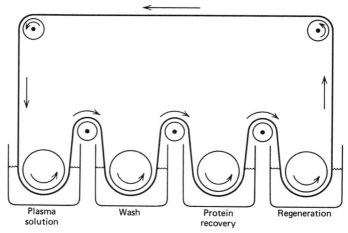

Figure 9. Continuous belt process for affinity adsorption of biopolymers. Courtesy of B. L. Wong and S. E. Charm (55).

tractive for manufacturing applications. An apparatus has been developed by Charm and colleagues to demonstrate the concept (55). However, much work is needed to determine longevity, efficiency, and selectivity of the sorbent belt as well as the cost–effectiveness of the whole system.

Large-Scale. At present only a few reports exist that deal with large-scale applications of affinity sorbents where sorbent volumes are >1 L. In one detailed study β-galactosidase from *Escherichia coli* was isolated on a 1.8-L column of agarose containing immobilized *p*-aminophenyl-β-D-thiogalactopyranoside (56). An electromechanical control system for automating the isolation process was included in this report. Unfortunately, the adsorbent system, adopted from earlier work (57), has been found to be nonbiospecific and the successful purification of the enzyme from *E. coli* was somewhat fortuitous (58). Scientists from a commercial enzyme supply house (Worthington Biochemical Corp.) have reported on the affinity purification of plasminogen, the inactive precursor of plasmin using a 1-L bed of agarose gel containing lysine (59). The adsorbent was originated by Deutsch and Mertz (60). The isolation of hepatitis B surface antigen using a 3.5-L column of concanavalin A linked to agarose gel has been reported (61). In spite of the paucity of examples here, it is expected that as the applications of the affinity method increase, there will be more occasions for isolating biopolymers at large-scale.

Analytical Applications

Until now, in both scale and variety the bulk of affinity principle applications have been in isolation of biopolymers for research purposes. The hundreds of reported examples cover a considerable variety of enzymes, inhibitors, antibodies, receptors, and binding proteins. For details, the reader is directed to some of the general subject reviews listed in the reference section.

Another area of application, although less frequent, has been the study of enzyme reaction mechanisms with immobilized affinity ligands. Still another is that of medical diagnostics applications of immunosorbents that use the affinity principle in a most rapidly growing technique called radioimmunoassay (RIA). This methodology is being used to detect and quantitate a wide variety of substances of clinical interest from metabolites and cancer antigens to drugs being used in therapy (see Radiochemical technology). One example is reviewed here to provide a glimpse of the possibilities.

RAST (Radioallergosorbent Test). This diagnostic system is perhaps one of the earliest commercial successes exploiting the affinity concept. As a test for the determination of immunoglobulin E (IgE) titer in atopic (allergic) individuals (62), it has been very helpful to the physician guiding the course of treatment. Because of its *in vitro* nature, it is safer for the patient and spares him from discomfort as well. The test scheme is outlined in Figure 10. The test vehicle is a paper disk containing bonded allergen of a known type. This disk is contacted with a sample of the atopic patient's serum. During incubation, IgE molecules (if present in the sample) bind selectively to the allergen-coated disk. After washing, the exposed disk is contacted with isotope-labeled anti-IgE, which will bind to any IgE molecules already selectively bound to the disk. The disk is again rinsed free of any unbound material, then placed in a gamma counter to record the amount of ^{125}I on the disk. By comparison with calibration standards, the amount of IgE found in the patient's serum is quantitated. The method described here is referred to as the sandwich technique and two affinity ad-

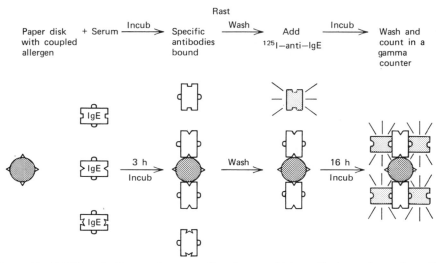

Figure 10. RAST, radioallergosorbent test. Note that O = IgE specific determinant phase. Courtesy of U. Lundkvist (62).

sorption steps are involved. This strategy affords greater detection sensitivity and freedom from background interference.

Biomedical Applications

In contrast to the many examples of affinity purification of biopolymers, those representing a more practical application of the affinity principle are few. However, because of their heuristic value to ongoing and future research, those selected examples are reviewed here in some detail. They are mostly attempts to solve biomedical problems and represent only beginnings of some novel approaches to treat afflictions that have largely defied more conventional modes of treatment.

Jaundice Treatment.   This disease is characterized by the abnormal accumulation of bilirubin, a breakdown product of hemoglobin metabolism, in the circulatory system. Owing to its high avidity for albumin, bilirubin is tightly bound to this plasma protein and is unable to clear through the kidney and be eliminated. With protracted accumulation of bilirubin in the circulation, the subject takes on an abnormal yellow skin coloration, long associated with this disease. A more serious consequence than coloration is the possibility of permanent brain damage (kernicterus), especially in neonates, which can result if high titers of bilirubin are sustained for extended periods. To treat neonatal jaundice as well as other situations where severe liver dysfunction has occurred (bilirubin is normally degraded further in the liver), a program has been undertaken to develop a method of selectively removing excess bilirubin from blood by means of a specially designed extracorporeal filtration device (63).

The filter consists of an immobilized albumin gel packed into a column. The extracorporeal circuit including the column is shown in Figure 11 (64). By having the ligand albumin immobilized in the gel at sufficiently high concentration, the ligate bilirubin carried by the plasma is adsorbed. Thus the harmfully high concentrations of bilirubin in the subject's circulation can be brought down to a safer, acceptable level.

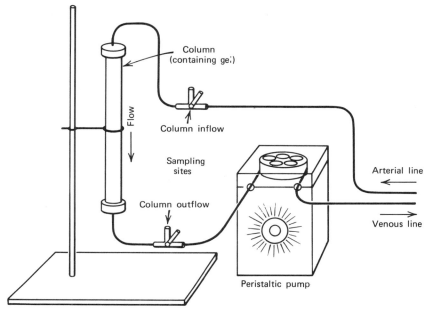

Figure 11. Experimental extracorporeal circuit for bilirubin removal. The column contains agarose gel bonded with serum albumin. Courtesy of *J. Clin. Invest.* (64).

Owing to the complexity of blood composition, as well as the lability of many of its components, much developmental work is still required before treatment of human patients by this technique becomes routine practice (65–66).

Hypercholesterolemia. Coronary atherosclerotic disease has been associated for some time with high levels of cholesterol in the circulation (hypercholesterolemia). Individuals suffering from this problem are those with a familial syndrome and who do not readily respond to a cholesterol-free diet (see Cardiovascular agents).

Studies have been initiated to treat familial hypercholesterolemics by an affinity adsorption process to remove cholesterol directly from their blood (67). There is selective removal from the blood plasma of the low density lipoproteins (LDL) and very low density lipoproteins (VLDL) which are the carriers of cholesterol. Development of the sorbent was based on earlier studies (68) in which it was shown that LDL and VLDL could be selectively adsorbed by heparin–agarose gel beads. Instead of an extracorporeal shunt device, blood was collected from subjects into blood packs filled with heparin–agarose gel (67). After mixing with gentle agitation, the blood was reinfused to the subject through a transfusion filter. No doubt because of its well-known anticoagulant properties, heparin in its immobilized form here posed few problems during trials on human subjects. More studies are needed, however, to determine the long term prognosis for this type of treatment.

Systemic Lupus Erythmatosus. This bewildering malady is apparently partially mediated by antibodies formed by the subject to his own DNA (an example of an auto-immune disorder). As an alternative to immunosuppression (ie, procedures to prevent more antibody formation), anti-DNA immunoglobulins are selectively removed from the circulation as a means of treating the disease (69).

Single-stranded DNA can be adsorbed on cellulose particles, which are then entrapped in agar to form the immunosorbent. The sorbent is packed into a column which is then arranged in an extracorporeal circuit drawing blood from the femoral artery of a rabbit and fed back into its femoral vein. As a model test, a rabbit was preimmunized with calf thymus DNA to develop a high anti-DNA titer. On passing the rabbit blood through the extracorporeal device over a 4-h period, the anti-DNA titer was found to decrease by 65%. Moving on from this initial immunochemical model, studies of a biological animal model that resembles the actual disease more closely are underway. This is a necessary prelude to human testing.

Removing Hepatitis Contaminants.   The widespread use of whole blood, as well as blood fractions in transfusions, is still attended with a high risk of hepatitis infection. This virus is transmitted by unsuspecting carriers and diseased individuals through their blood donations to the transfusion recipients. The development of a radioimmunoassay to check each unit of blood for the presence of the viral antigen has helped reduce, but not eliminate incidence of transmission (see Blood fractionation). Because of the procedures used in isolating human plasma albumin, this product is free from hepatitis. However, other clinically useful plasma protein fractions cannot, at present, be prepared so that they are rigorously free from contamination. Therefore, it is of great interest to be able to treat each collected unit of blood in a way to remove the danger of hepatitis transmission. One approach has been developed using immunosorbents for the selective removal of viral antigen (55,70). Hepatitis-contaminated plasma was treated for appropriate lengths of time with antihepatitis B immunoglobulin immobilized on agarose gel. The plasma thus obtained was tested by inoculation into gibbons and rhesus monkeys, which showed no disease symptoms even after 8 mo of observation. These same animals when inoculated with untreated plasma (or plasma fractions) showed disease by 16 weeks after inoculation (70).

BIBLIOGRAPHY

 1. J. Porath and L. Sundberg in M. L. Hair, ed., *The Chemistry of Biosurfaces*, Vol. 2, Marcel Dekker, New York, 1972, pp. 633–661.
 2. J. Porath, *Biochimie* **55,** 943 (1973).
2a. A. H. Nishikawa, *Chem. Tech.* **5,** 564 (1975).
 3. D. H. Campbell, E. Luescher, and L. S. Lerman, *Proc. Nat. Acad. Sci. USA* **37,** 575 (1951).
 4. J. B. Sumner, *J. Biol. Chem.* **69,** 435 (1926).
 5. J. H. Northrup, *J. Gen. Physiol.* **17,** 165 (1934).
 6. O. Holmbergh, *Biochem. Z.* **258,** 134 (1933).
 7. D. J. D. Hockenhull and D. Herbert, *Biochem. J.* **39,** 102 (1945).
 8. L. S. Lerman, *Proc. Nat. Acad. Sci. USA* **39,** 232 (1953).
 9. C. Arsenis and D. B. McCormick, *J. Biol. Chem.* **239,** 3093 (1964).
10. B. R. Baker, *Design of Active-Site-Directed Irreversible Enzyme Inhibitors*, John Wiley & Sons, Inc., New York, 1967, Chapt. 13.
11. P. Cuatrecasas, M. Wilchek, and C. B. Anfinsen, *Proc. Nat. Acad. Sci. USA* **61,** 636 (1968).
12. R. Axen, J. Porath, and S. Ernback, *Nature* **214,** 1302 (1967).
13. G. M. Whitesides and A. H. Nishikawa in J. B. Jones, C. J. Sih, and D. Perlman, eds., *Applications of Biochemical Systems in Organic Chemistry*, Part II, Wiley-Interscience, New York, 1976, Chapt. 8.
14. R. Cornell and S. E. Charm, *Biotech. Bioeng.* **18,** 1171 (1976).
15. H. F. Hixson and A. H. Nishikawa, *Arch. Biochem. Biophys.* **154,** 501 (1973).
16. U.S. Pat. 3,746,622 (July 17, 1973), A. H. Nishikawa and H. F. Hixson (to Xerox Corp.).
17. S. Barry, P. Brodelius, and K. Mosbach, *FEBS Lett.* **70,** 261 (1976).
18. C. R. Lowe, M. J. Harvey, and P. D. G. Dean, *Eur. J. Biochem.* **42,** 1 (1974).
19. S. T. Thompson, K. H. Cass, and E. Stellwagen, *Proc. Nat. Acad. Sci. USA* **72,** 669 (1975).

20. E. Stellwagen, *Acct. Chem. Res.* **10,** 92 (1977).
21. R. L. Easterday and I. M. Easterday in R. B. Dunlap, ed., *Adv. Exp. Med. Biol.* **42,** 123 (1974).
22. J. K. Baird and co-workers in ref. 17, p. 61.
23. H. Lis, R. Lotan, and N. Sharon, *Ann. N. Y. Acad. Sci.* **234,** 232 (1974).
24. T. Imoto and K. Yagishita, *Agr. Biol. Chem.* **37,** 465 (1973).
25. S. Sivakami and A. N. Radhakrishnan, *Indian J. Biochem. Biophys.* **10,** 283 (1973).
26. N. H. Grant and K. C. Robbins, *Arch. Biochem. Biophys.* **66,** 396 (1957).
27. P. M. Gallop, S. Seifter, and E. Meilman, *J. Biol. Chem.* **227,** 891 (1957).
28. A. I. Usov and L. I. Miroshnikova, *Carbohyd. Res.* **43,** 204 (1975).
29. P. Andrews, B. J. Kitchen, and D. J. Winzor, *Biochem. J.* **135,** 897 (1973).
30. P. C. Wankat, *Anal. Chem.* **46,** 1400 (1974).
31. D. J. Graves and Y-T. Wu in W. B. Jacoby and M. Wilchek, eds., *Methods in Enzymology,* Vol. 84, Academic Press, Inc., New York, 1974, p. 140.
32. A. H. Nishikawa, P. Bailon, and A. H. Ramel in ref. 21, p. 33.
33. A. H. Nishikawa, P. Bailon, and A. H. Ramel, *J. Macromol. Sci. Chem.* **A10,** 149 (1976).
34. L. R. Snyder, *Principles of Adsorption Chromatography,* Marcel Dekker, New York, 1968, Chapt. 2.
35. B. M. Dunn and I. M. Chaiken, *Proc. Nat. Acad. Sci. USA* **71,** 2382 (1974).
36. T. C. J. Gribnau, *Ph.D. dissertation,* Catholic University, Nijmegen, The Netherlands, 1977.
37. R. J. Brinkworth, C. J. Masters, and D. J. Winzer, *Biochem. J.* **151,** 613 (1975).
38. R. C. Bottomley, A. C. Storer, and I. P. Trayer, *Biochem. J.* **159,** 667 (1976).
39. K-I. Kasai and S-I. Ishii, *J. Biochem.* **77,** 261 (1975).
40. G. P. Royer, *Chem. Tech.* **4,** 694 (1974).
41. W. H. Scouten, *Am. Lab.* **6**(8), 23 (1974).
42. N. Weliky and H. H. Weetall, *Immunochem.* **2,** 293 (1965).
43. A. H. Nishikawa and P. Bailon, *Arch. Biochem. Biophys.* **168,** 576 (1975).
44. J. Porath, J-C. Janson, and T. Laas, *J. Chromatogr.* **60,** 167 (1971).
45. A. H. Nishikawa and P. Bailon, *J. Solid-Phase Biochem.* **1,** 33 (1976).
46. J. K. Inman and H. M. Dintzis, *Biochemistry* **8,** 4074 (1969).
47. C. J. Sanderson and D. V. Wilson, *Immunochem.* **8,** 163 (1971).
48. H. Guilford, *Chem. Soc. Rev.* **2,** 249 (1973).
49. N. G. Anderson and co-workers, *Anal. Biochem.* **66,** 159 (1975).
50. B. R. Baker and H.-U. Siebenieck, *J. Med. Chem.* **14,** 799 (1971).
51. P. Bailon and A. H. Nishikawa, *Prep. Biochem.* **7,** 61 (1977).
52. L. Gyenes and A. H. Sehon, *Can. J. Biochem.* **38,** 1235 (1960).
53. T. Ternynck and S. Avrameas, *FEBS Lett.* **23,** 24 (1972).
54. L. Sundberg, J. Porath, and K. Aspberg, *Biochim. Biophys. Acta* **221,** 394 (1970).
55. S. E. Charm and B. L. Wong, *J. Macromol. Sci. Chem.* **A10,** 53 (1976).
56. P. J. Robinson and co-workers, *Biotech. Bioeng.* **16,** 1103 (1974).
57. E. Steers, P. Cuatrecasas, and H. B. Pollard, *J. Biol. Chem.* **246,** 196 (1971).
58. A. H. Nishikawa and P. Bailon, *Arch. Biochem. Biophys.* **168,** 576 (1975).
59. L. H. Hsu and H. R. Bungay, *168th National Meeting of ACS,* Atlantic City, N.J., Sept. 11, 1974.
60. D. G. Deutsch and E. T. Mertz, *Science* **170,** 1095 (1970).
61. A. R. Neurath, A. M. Prince, and J. Giacolone, *Experientia,* **34,** 414 (1978).
62. U. Lundkvist in R. Evans, ed., *Advances in Diagnosis of Allergy: RAST,* Vol. 85, Symposia Specialists, Miami, Fla., 1975.
63. P. D. Berk, P. H. Plotz, and B. F. Scharschmidt in T. M. S. Chang, ed., *Biomedical Applications of Immobilized Enzymes and Proteins,* Vol. 1, Plenum Press, New York, 1977, p. 297.
64. B. F. Scharschmidt and co-workers, *J. Clin. Invest.* **53,** 786 (1974).
65. B. F. Scharschmidt and co-workers, *J. Lab. Clin. Med.* **89,** 101 (1977).
66. R. D. Hughes and co-workers, *Biomater. Med. Devices. Artif. Organs.* **5,** 205 (1977).
67. P.-J. Lupien, S. Moorjani, and J. Awad, *Lancet i,* 1261 (1976).
68. P. H. Iverius, *J. Biol. Chem.* **247,** 2607 (1972).
69. D. S. Terman and co-workers, *Clin. Exp. Immunol.* **24,** 231 (1976).
70. B. L. Wong and S. E. Charm in T. M. S. Chang, ed., *Biomedical Applications of Immobilized Enzymes and Proteins,* Vol. 2, Plenum Press, New York, 1977, p. 131.

General References

R. B. Dunlap, ed., *Immobilized Biochemicals and Affinity Chromatography, Adv. Exp. Med. Biol.,* **42** Plenum Press, New York, (1974).
H. F. Hixson and E. P. Goldberg, eds., *Polymer Grafts in Biochemistry,* Marcel Dekker, New York, 1976.
W. B. Jakoby and M. Wilchek, eds., *Affinity Techniques, Methods in Enzymology,* **34,** Academic Press, Inc., New York, 1974.
C. R. Lowe and P. D. G. Dean, *Affinity Chromatography,* Wiley-Interscience, New York, 1974.

A. H. NISHIKAWA
Hoffmann LaRoche Inc.

CHROME DYES.   See Azo dyes; Dyes—Application and evaluation.

CHROMIUM AND CHROMIUM ALLOYS

Chromium [7440-47-3] is the 21st element in the earth's crust in relative abundance, ranking with V, Zn, Ni, Cu, and W. With atomic number 24, it belongs to Group VIB of the periodic table whose other members are molybdenum and tungsten. Its neighbors are vanadium and manganese.

Chromium was first isolated and indentified as a metal by the French chemist, Vauquelin, in 1798 working with a rare mineral, Siberian red lead (crocoite, $PbCrO_4$). He chose to name it chromium, from the Greek word chroma meaning color, because of the wide variety of brilliant colors displayed by compounds of the new metal. An early application of chromium compounds was as pigments, particularly chrome yellow, $PbCrO_4$; basic chromium sulfate was used for tanning of hides where the reaction of chromium with collagen raises the hydrothermal stability of the leather and also renders it resistant to bacterial attack (see Leather). The most important application of chromium, namely its use as an alloying element, was gradually developed during the 19th century, mainly in France, and led to chromium steels (1–2). Their first major structural application was in the famous Eads Bridge across the Mississippi (1867–1874). Further technological developments included improved oxidation resistance and hardenability, and the superior corrosion resistance of a ferritic 12.8% chromium–iron and of austenitic alloys of 18 Cr–8 Ni variety. Chromite was employed as a furnace refractory as early as 1879 (see Chromium compounds; Pigments, inorganic; Steel).

Occurrence and Mining

The only commercial ore, chromite, has the ideal composition $FeO.Cr_2O_3$, ie 68% Cr_2O_3, 32% FeO or ca 46% chromium. Actually the Cr/Fe ratio varies considerably and the ores are better represented as $(Fe,Mg)O.(Cr,Fe,Al)_2O_3$. Table 1 gives the classification of chromite ores.

Chromite deposits occur in olivine and pyroxene type rocks and their derivatives. Geologically they appear in stratiform deposits several feet thick covering a very wide area and are usually mined by underground methods. Podiform deposits, ie, isolated lenticular, tabular, or pod-shaped bodies ranging in size from a kilogram to several million tons are mined by both surface and underground methods, depending on size and occurrence. Most chrome ores are rich enough for hand sorting. However, fines or lower-grade ores can be effectively concentrated by gravity separation methods yielding products as high as 50% Cr_2O_3 with the Cr/Fe ratio of the original ore usually unchanged (see Gravity concentration). Decreasing world supplies (3) of high-grade lumpy ore and increasing availability of high-grade fines and concentrates has increased the use of three agglomeration methods, (*a*) briquetting with a binder, (*b*) production of an oxide pellet by kiln firing, and (*c*) production of a prereduced pellet by furnace treatment. (For analysis, see Chromium compounds.)

Properties

The valence states of chromium are +2, +3, and +6, the latter two being the most common. The +2 and +3 states are basic, whereas the +6 is acidic, forming ions of the type $(CrO_4)^{2-}$ (chromates) and $(Cr_2O_7)^{2-}$ (dichromates). The blue-white metal is refractory and very hard. Its properties are listed in Table 2 (4–9).

Oxidation tests in oxygen at atmospheric pressure on a chromium specimen containing 0.04% carbon showed the formation of an oxide film 0.15 μm thick in 2 h at 700°C and 2.4 μm thick in 1 h at 900°C (7).

Chromium is highly acid resistant and is only attacked by hydrochloric, hydrofluoric, and sulfuric acids.

Mechanical Properties. Perhaps more so than any other common metal, the mechanical properties of chromium (6,7,9) depend on purity, history, grain size, strain rate, and surface condition. Most reported mechanical properties for chromium are thus those of an ill-defined dilute alloy of unique history and metallurgical condition and hence of little value for handbook quotation. More meaningful data are those reported for swaged iodide chromium as shown in Table 3. Not only is the ductile-

Table 1. Composition of Chromite Ores [a]

Grade	Composition	Ratio Cr:Fe
metallurgical, high Cr	46% Cr_2O_3 min	>2:1[b]
chemical, high Fe	40 to 46% Cr_2O_3	1.5 to 2:1
refractory, high Al	>20% Al_2O_3	
	>60% $Al_2O_3 + Cr_2O_3$	

[a] Ref. 2.

[b] Some South African ore of a ratio ca 1.5:1 is also used for ferrochromium production.

Table 2. Physical Properties of Chromium

Property	Value
atomic weight	51.996
isotopes, %	
50	4.31
52	83.76
53	9.55
54	2.38
crystal structure	bcc
a_0, nm	0.2844–0.2848
density at 20°C, g/cm^3	7.19
melting point, °C	1875
boiling point, °C	2680
vapor pressure, 130 Pa[a], °C	1610
heat of fusion, kJ/mol[b]	13.4–14.6
latent heat of vaporization at bp, kJ/mol[b]	320.6
specific heat at 25°C, kJ/(mol·K)[b]	23.9 (0.46 kJ/(kg·K))
linear coefficient of thermal expansion at 20°C	6.2×10^{-6}
thermal conductivity at 20°C, W/(m·K)	91
electrical resistivity at 20°C, $\mu\Omega$·m	0.129
specific magnetic susceptibility at 20°C	3.6×10^{-6}
total emissivity at 100°C, nonoxidizing atm	0.08
reflectivity, R	
λ, nm	300 500 1000 4000
%	67 70 63 88
refractive index	
α	1.64–3.28
λ	2,570–6,080
standard electrode potential, valence 0 to 3+, V	0.71
ionization potential, V	
1st	6.74
2nd	16.6
half-life of ^{51}Cr isotope, days	27.8
thermal neutron scattering cross section, m^2	6.1×10^{-28}
elastic modulus, GPa[c]	250
compressibility[a,d] at 10–60 TPa	70×10^{-3}

[a] To convert Pa to mm Hg, multiply by 0.0075.
[b] To convert J to cal, divide by 4.184.
[c] To convert GPa to psi, multiply by 145,000.
[d] 99% Cr; to convert TPa to megabars, multiply by 10.

to-brittle transition temperature (DBTT) dependent on the several variables cited above, but the potential utility of the metal is impaired by the fact that the ductility below this transition is essentially nil. To achieve measurable ductility, impurity contents should be below the following limits: O, 2000 ppm; N, 100 ppm; C, 100 ppm; H, 20 ppm; Si, 1500 ppm; S, 150 ppm.

Production

An overview of the production processes (2,4–8,10) leading to chromium metal and the various chromium compounds is shown in the simplified flow sheet of Figure 1. Very little chromite, of course, is processed all the way to ductile chromium, since

Table 3. Mechanical Properties of Room Temperature Swaged Iodide Chromium[a]

Condition	Yield strength, 0.2% offset, MPa[b]	Ultimate strength, MPa[b]	Elongation, %[c]	Reduction of area, %
wrought	362	413	44	78
recrystallized		282	0	0

[a] Pure chromium made by the iodide process.
[b] To convert MPa to psi, multiply by 145.
[c] Percent elongation in a 6 mm gage length.

most can be used in the various intermediate forms. For example, the chromite ore itself, mixed with small amounts of lime or magnesia, is made into refractory brick; ferrochrome is used directly in steel making; chrome alum is used as a mordant and in tanning leather; pigments, metal finishing agents, wood preservatives, etc, are made from sodium dichromate; and chromic acid, in addition to being the main source of chromium for electroplating, is also used for metal finishing, organic syntheses, and in the manufacture of catalysts (see Catalysis; Electroplating; Metal surface treatment).

Ferrochrome. Ferrochrome is usually made by reduction of chromite with coke in a three-phase electric submerged arc furnace. This process leads inevitably to a high-carbon ferrochrome [11114-46-8] whose use was formerly restricted to high-carbon steels. With increasing use of argon-oxygen decarburization (AOD) and similar processing of alloy and stainless steels, this limitation is less severe (10). Care is taken to keep sulfur low as it embrittles both Cr metal and the Fe–Ni and Ni-base alloys to which Cr is added.

Low-carbon ferrochromes cannot be made by carbonaceous reduction unless accompanied by top blowing with oxygen. Aluminum, or especially silicon, is frequently used as the reducing agent. When silicon is employed, high-silicon ferrochrome, practically carbon-free, is first produced in a submerged arc furnace, and then treated in an open arc-type furnace with a synthetic slag containing Cr_2O_3. A ferrochromium of very low carbon content (0.01%) is produced in a solid-state process by heating high-carbon ferrochrome with oxidized ferrochrome in a high vacuum with the carbon removed as carbon monoxide (Simplex process). In the other smelting processes the molten ferrochrome is tapped from the furnace, cast into chills, broken into lumps, and graded.

The compositions of several different grades of ferrochrome are given in Table 4.

Chromium Metal by Pyrometallurgical Reduction. The principal pyrometallurgical process for commercial chromium metal is the reduction of Cr_2O_3 by aluminum.

$$Cr_2O_3 + 2\,Al \rightarrow 2\,Cr + Al_2O_3$$

The chromium oxide is mixed with aluminum powder and placed in a refractory-lined vessel and ignited with barium peroxide and magnesium powder. The reaction is exothermic and self-sustaining. Chromium metal of 97–99% purity is obtained, the chief impurities being aluminum, iron, and silicon (carbon, sulfur, and nitrogen are about 0.03, 0.02, and 0.045%, respectively).

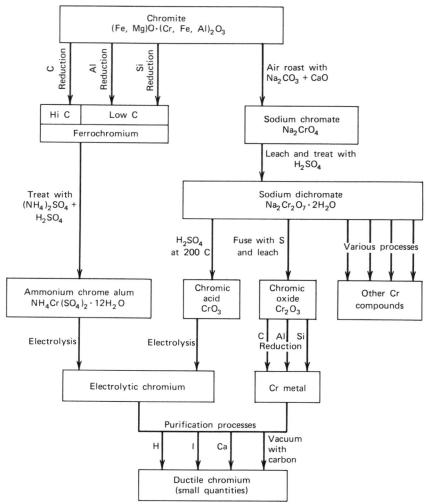

Figure 1.   Simplified flow chart for the production of metallic chromium and chromium compounds from chromite.

Commercial chromium metal may also be produced from the oxide by reduction with silicon in an electric arc furnace.

$$2\,Cr_2O_3 + 3\,Si \rightarrow 4\,Cr + 3\,SiO_2$$

The product is similar to that obtained by the aluminothermic process; however, the aluminum content is lower and silicon may run as high as 0.8%.

The oxide may also be reduced with carbon at low pressure.

$$Cr_2O_3 + 3\,C \rightarrow 2\,Cr + 3\,CO$$

Briquets of mixed, finely divided oxide and carbon are heated to 1275–1400°C in a refractory container. The minimum pressure is about 40 Pa (0.3 mm Hg) for reduction at 1400°C. Lower pressures or higher temperatures cause excessive volatilization of

Table 4. Composition of Typical Ferrochromium Alloys and Chromium Metal,[a] Percent of Principal Constituent[b]

Grade	Chromium	Silicon	Carbon	Sulfur[c]	Phosphorus[c]	Other
ferrochromium						
high-carbon	66–70	1–2	5–6.5	0.04	0.03	
high-carbon, high-silicon						
blocking chrome	55–63	8–12	4–6	0.03		
exothermic ferrochrome	41–51	9–14	3.6–6.4	0.03		
foundry ferrochrome	55–63	8–12	4–6			
refined chrome	53–63	2.5[c]	3–5	0.03		
SM ferrochrome	60–65	4–6	4–6			4–6 manganese
charge chromium						
50–55 percent chromium	50–56	3–6	6–8	0.04	0.03	
66–70 percent chromium	66–70	3[c]	5–6.5	0.04	0.03	
low-carbon:						
0.025 percent carbon	67–75	1[c]	0.025[c]	0.025	0.03	
0.05 percent carbon	67–75	1[c]	0.05[c]	0.025	0.03	
Simplex	63–71	2.0[c]	0.01 or 0.025			
ferrochromium–silicon:						
36/40 grade	35–37	39–41	0.05[c]			
40/43 grade	39–41	42–45	0.05[c]			
chromium metal						
electrolytic	99.3[d]	0.01[c]	0.02[c]	0.03		0.5 oxygen[c] 0.05 nitrogen[c]
aluminothermic	99.3[d]	0.15[c]	0.05[c]	0.015	0.01	0.2 oxygen[c] 0.3 aluminum[c]

[a] Ref. 2.
[b] Difference between sum of percentages shown and 100 percent is chiefly iron content.
[c] Maximum.
[d] Minimum.

chromium. The product contains about 0.02% silicon, less than 0.03% iron, and 0.015% carbon, 0.001% nitrogen, and 0.04% oxygen.

Electrowinning of Chromium.   *Chrome Alum Electrolysis.*   The Union Carbide Corporation's Metals Division Plant at Marietta, Ohio, is a typical chrome–alum plant with a capacity of 2000 net metric tons per year.

In this process (see Fig. 2), high-carbon ferrochromium is leached with a hot solution of reduced anolyte plus chrome alum mother liquor and makeup sulfuric acid. The slurry is then cooled to 80°C by the addition of cold mother liquor from the ferrous ammonium sulfate circuit, and the undissolved solids, mostly silica, are separated by filtration. The chromium in the filtrate is then converted to the nonalum form by several hours' conditioning treatment at elevated temperature.

Ammonium chrome alum, $NH_4Cr(SO_4)_2.12H_2O$, exists in a violet and a green modification. Their properties are quite different with regard to conductivity, solubility, and ionization. Above 50°C, the green complex is more stable; at room temperature it changes slowly to the violet form with a change in pH. In solutions of ammonium chromium alum, the chromium (III) ion can exist in a variety of forms, depending on time, temperature and past conditions. At higher temperatures, a variety of green nonalum ions such as $[Cr(H_2O)_5-SO_4]^+$, $[Cr(H_2O)_5(OH)]^{2+}$, and $[(SO_4)-(H_2O)_4Cr-O-Cr(H_2O)_5]^{2+}$ form, whereas the violet hexaquo ion $[Cr(H_2O)_6]^{3+}$ predominates in cool, dilute solutions of moderate acidity. It is only the latter which permits crystallization of the desired ammonium chromium alum.

Once the green, "non-alum" forms of the chromium ion have formed, their revision to the hexaquo form on cooling is sufficiently slow that on chilling to 5°C, a crude ferrous ammonium sulfate can be crystallized, removing nearly all the iron from the system. This crude iron salt is treated with makeup ammonium sulfate, heated again to retain the chromium impurities in the green noncrystallizable form, and then cooled to separate the bulk of the iron as a technical ferrous ammonium sulfate which is sold for fertilizer and other purposes. The mother liquor from this crystallization is returned to the filtration step.

The mother liquor from the crude ferrous sulfate crystallization contains nearly all the chromium. It is clarified and aged with agitation at 30°C for a considerable period to reverse the reactions of the conditioning step. Hydrolysis reactions are being reversed; therefore, the pH increases. Also, sulfate ions are released from complexes and the chromium is converted largely to the hexaquo ion, $[Cr(H_2O)_6]^{3+}$. Ammonium chrome alum precipitates as a fine crystal slurry. It is filtered and washed and the filtrate sent to the leach circuit; the chrome alum is dissolved in hot water, and the solution is used as cell feed.

The principal electrolytic cell reactions are shown in Figure 3. A diaphragm-type cell prevents the sulfuric acid and chromic acid formed at the anode from mixing with the catholyte and oxidizing the divalent chromium. Electrolyte is continuously fed to the cells to maintain the proper chromium concentration. The catholyte pH of each is controlled by adjusting the amount that flows through the diaphragms into the anolyte compartments. Control of the pH between narrow limits governs the successful electrodeposition of chromium as well as the preservation of divalent chromium at the cathode.

The analyses of the solutions in the electrolytic circuit and cell operating data are given in Tables 5 and 6, respectively. The current efficiency of 45% shown in Table 6 includes low efficiencies that always prevail during the startup of a reconditioned

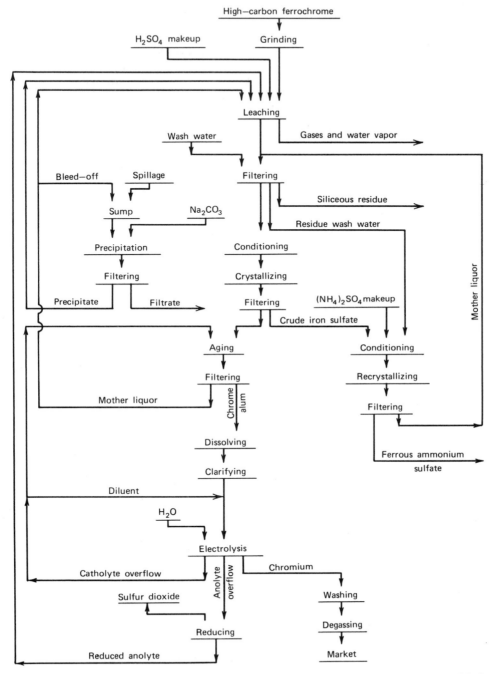

Figure 2. Flow sheet for production of electrolytic chromium by the chrome alum process, Marietta Plant, Union Carbide Corporation, Marietta, Ohio.

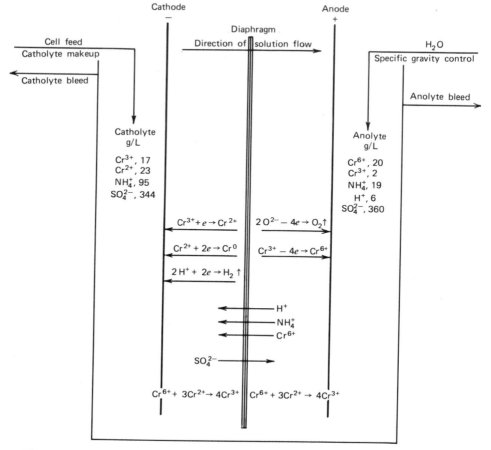

Figure 3. Principal electrolytic cell reactions in chromium production by the chrome alum process.

Table 5. Typical Analyses of Solutions in Electrolysis of Chrome Alum, g/L

	Total Cr	Cr^{6+}	Cr^{3+}	Cr^{2+}	Fe	NH_3	H_2SO_4
cell feed	130	0	130	0	0.2	43	3
circulating mixture	65	0	63	2	0.1	68	1
catholyte	24	0	11.5	12.5	0.035	84	
anolyte	15	13	2	0	0.023	24	280

cell. The 2.1–2.4 pH range used in the plant also results in a somewhat lower current efficiency but provides a safe operating latitude.

At the end of the 72-h cycle, the cathodes are removed from the cells, washed in hot water, and the brittle deposit (3–6 mm thick) is stripped by a series of air hammers. The metal is then crushed by rolls to 50 mm size and again washed in hot water. The metal contains about 0.034% hydrogen and, after drying, is dehydrogenated by heating

Table 6.   Operating Data for Electrowinning of Chromium from Chrome Alum

cathode current density, A/m^2	753
cell potential, V	4.2
current efficiency, %	45
electrical consumption, MJ/kg^a	67
pH of catholyte	2.1–2.4
catholyte temperature, °C	53 ± 1
deposition time, h	72
cathode material	Type 316 stainless steel
anode material, wt %	1–99 Ag–Pb

^{a} To convert J to cal, divide by 4.184.

to at least 400°C in stainless steel cans. A typical product composition is shown below:

chromium	99.8%	sulfur	0.025%	oxygen	0.50%	silicon	0
iron	0.14%	copper	0.01%	hydrogen	0.004%	phosphorus	0
carbon	0.01%	lead	0.002%	nitrogen	0.01%	manganese	0
						aluminum	0

Chromic Acid Electrolysis.   The chromic acid, CrO_3, is obtained from sodium dichromate (see Chromium compounds). Small amounts of an ionic catalyst, specifically sulfate, chloride, or fluoride, are essential. Fluoride and complex fluoride catalyzed baths have become especially important in recent years. The cell conditions for the chromic acid process are given in Table 7.

The low current efficiency of this process is due to the evolution of hydrogen at the cathode; this occurs because the hydrogen deposition overvoltage on chromium is significantly more positive than that at which chromous ion deposition would be expected to commence. This hydrogen evolution at the cathode surface also increases the pH of the catholyte beyond 4 which may result in the precipitation of $Cr(OH)_3$ and $Cr(OH)_2$, causing a partial passivation of the cathode and a reduction in current efficiency. The latter is also inherently low, as six electrons are required to reduce hexavalent ions to chromium metal.

Plating variables for this process may be summarized as follows: Higher operating temperatures (87°C) enable the oxygen content of the metal to be reduced to 0.01%; The ratio CrO_3:SO_4 should be below 100 to obtain low-oxygen metal; Current efficiencies greater than 8% are associated with high oxygen contents; Better current efficiencies are obtained at low current densities.

The metal obtained by this process contains less iron and oxygen than that from

Table 7.   Cell Conditions for Chromic Acid Process for Electrowinning of Chromium^a

bath composition, g/L	300 CrO_3, 4 sulfate ion
temperature, °C	84–87
current density, A/m^2	9500
current efficiency, %	6–7
plating time, h	80–90
production rate, g/week	1000

^{a} From Ref. 6.

the chrome alum electrolyte. The gas content is 0.02% O, 0.0025% N, and 0.009% H. If desired, the hydrogen content can be still further lowered by a dehydrogenation treatment.

Purification. The metal obtained from both electrolytic processes contains considerable oxygen, which is believed to cause brittleness at room temperature. For most purposes the metal as plated is satisfactory. However, if ductile metal is desired, the oxygen can be removed by hydrogen reduction, the iodide process, calcium refining, or melting in a vacuum in the presence of a small amount of carbon.

In the hydrogen reduction process the electrolytic metal is heated to about 1500°C in a closed-circuit stream of dry, pure hydrogen. Although this process reduces the oxygen content considerably and the nitrogen content somewhat less so, it has little effect on the other impurities. A typical product from this process might contain 0.005 wt % oxygen and 0.001 wt % nitrogen.

In the iodide purification process (Van Arkel process) the impure chromium and iodine are sealed in an evacuated bulb containing an electrically heated wire. The bulb is heated to the temperature of formation of chromium(III) iodide. This chromic iodide reacts with more of the impure metal to form chromium(II) iodide which diffuses to the hot wire where it decomposes and deposits chromium. The freed iodine forms additional chromous iodide, and the process continues (cyclic process). Iodination at 900°C and decomposition at 1000–1300°C may also be carried out separately (straight-flow process), but with lower efficiency. The analyses of metal treated by iodide purification showed 0.0008% oxygen, 0.003% nitrogen, 0.00008% hydrogen, and 0.001% carbon; the other impurities were very low. This metal is produced by the Chromalloy Corporation under the trade name Iochrome.

In the calcium refining process the chromium reacts with calcium vapor at about 1000°C in a titanium-lined bomb which is first evacuated and then heated to the proper temperature. A pressure of about 2.7 Pa (20 μm Hg) is maintained during heating until the calcium vapor reaches the cold end of the bomb and condenses. This allows the calcium vapor to pass up through the chromium metal where it reacts with the oxygen. Metal obtained by this process contained 0.027% oxygen, 0.0018% nitrogen, 0.008% carbon, 0.012% sulfur, and 0.015% iron.

Carbon purification is accomplished by addition to the melt or to the solid charge before melting. Melting is carried out in vacuum and completion of the deoxidation process is observed by a rapid pressure drop indicating when the evolution of CO is complete.

The vacuum melting process can upgrade chromium at a modest cost; the other purification processes are very expensive. Thus iodide chromium is about 100 times as expensive as the electrolytic chromium and, therefore, is used only for laboratory purposes or special biomedical applications.

Consolidation and Fabrication. Chromium metal may be consolidated by powder metallurgy techniques or by arc melting in an inert atmosphere (5,7,11–12).

For powder metal consolidation the metal is first ball-milled using chromium-plated balls. The powder is then pressed at 300–500 MPa (3000–5000 atm) with or without a binder such as a wax; the binder is then removed by heating to about 300°C before sintering. The sintering operation which follows is carried out at high temperatures, 1450–1500°C, in a slow stream of purified hydrogen, helium, or argon. Sometimes sintering and purification can be combined, and in this case a large volume of purified hydrogen is needed to ensure adequate oxygen removal.

In a chemical vapor deposition (CVD) variant of conventional powder metallurgy processing, fine chromium powder is obtained by hydrogen reduction of CrI_2 and simultaneously combined with fine ThO_2 particles (see under Chromium-based alloys). This product is isostatically pressed to 70 MPa (700 atm) and 1100°C for 2 h. Compacts are steel clad and hot rolled to sheets (11).

Vacuum melting of chromium must take place in highly refractory crucibles such as pure zirconia, beryllia, or alumina lined with thoria. Under proper conditions, the nitrogen content of chromium-base alloys can be lowered to 0.01% or less, and carbon can be used as a deoxidizer to obtain alloys with 0.01% carbon and oxygen. Volatilization of some chromium is a drawback in this process. Arc melting into a water-cooled copper mold in an inert atmosphere has been successfully used for chromium and has the advantage that no refractory material comes in contact with the molten metal. The chromium must be given an oxygen purification before melting if low-oxygen material is required. It is claimed that addition of a small amount of yttrium or other rare earth element during melting to act as a scavenger improves the workability and mechanical properties.

The initial cast structure of arc melted ingots must be carefully broken down by hot working in order to permit subsequent warm working. Forging, swaging, and extrusion are all possible, with ease of working increasing from forging to extrusion, and particularly with hydrostatic extrusion. Hot working usually takes place in a steel or stainless sheath to protect the metal from contamination and this sheath is later removed by acid dissolution on completion of the working. The working process seems to make the material more ductile over and above benefits resulting from break-up of the as-cast structure (12). Electropolishing is often used to further improve the apparent ductility of sheet and wire samples. Elimination of the surface layer in this way removes any air-contaminated layer of chromium and also minimizes the effects of any notches.

Electroplating, Chromizing, and other Chromium-Surfacing Processes

Electroplating of chromium (13,14–17) on various substrates is practiced in order to realize a more decorative and corrosion or wear-resistant surface. About 80% of the chromium employed in metal treatment is used for chromium plating; over 50% is for decorative chromium plating. Hard chromium plating differs from decorative plating mostly in terms of thickness. Hard chromium plate may be 10 to several 100 μm thick, whereas the chromium layers in a decorative plate may be as little as 0.25 μm or about 2 g Cr/m² of surface. Hard plating is noted for its excellent hardness, wear resistance, and low coefficient of friction. Decorative chromium plating retains its brilliance because air exposure immediately forms a thin, invisible protective oxide film. The chromium is not applied directly to the surface of the metal but rather over a nickel plate which in turn is laid over a copper plate. Since the chromium plate is not free of cracks, pores, and similar imperfections, the intermediate nickel layer must provide the basic protection. Indeed, optimum performance is obtained when a controlled but high density of microcracks in the chromium is achieved (40–80 microcrack intersections per linear millimeter) leading to reduced local galvanic current density at the imperfections and increased cathode polarization. Modern practice uses a duplex nickel layer containing a small amount of sulfur. In addition to the familar applications of chromium plate in the automotive and plumbing fields, it has recently been introduced as a substitute for tin plate on steel for canning purposes. Since 1926, commercial chromium plating has used a chromic acid bath (Cr^{6+}) to which various catalyzing anions have been added. Currently much attention is being given the development of trivalent plating baths (see Electroplating).

Chromizing (16) is the other principal method of obtaining a chromium-rich surface on steel. The material to be treated is embedded in a mixture of ferrochrome powder, a chromium halide, alumina, and sometimes NH_4Cl. The chromium is diffused in by a furnace treatment at about 1100°C to produce an effective stainless steel surface whose mean composition is about 18% Cr and whose thickness is controlled by the time of treatment. This is an economical process improving the corrosion resistance of steel parts where cut edges and appearance are not important considerations, eg, automotive exhausts, heat exchangers, and silos.

Other surface processes (16) using chromium include sputtering, ion implantation, chemical vapor deposition, metal spraying, cladding, and weld overlayment. Only the latter two have commercial significance. Stainless clad steel (single- or double-faced) has been prepared, for more than 40 years, by hot rolling a duplex or triplex metal sandwich. It is an attractive means of conserving expensive stainless steel. Weld overlayment is used where clad is not available or unreliable or where wear resistance is an important consideration. Flame and arc welding are both practiced. The most popular alloys deposited are the Co–Cr based Stellites and similar compositions, but chromium carbides and oxides can also be deposited (see Metal surface treatments).

Economic Aspects

During much of the 19th century, the U.S. was the principal world producer of chromite when deposits in Maryland, Pennsylvania, and Virginia were exploited. However, today the United States is completely dependent on imports from the USSR, South Africa, Turkey, and Rhodesia.

The general features of the supply and demand relationships for the U.S. are shown in Figure 4. It does not, however, show that in the past 5 to 10 years the import to the U.S. (as well as to other highly developed countries) has shifted from chromite to ferrochrome as countries with low-cost energy and natural resources have installed the furnace capacity to convert the ore to ferrochrome. Since supply routes from the eastern hemisphere sources are long and world political problems pose a constant threat of interruption of supply, chromium consuming industries have been forced to maintain large stocks.

The distribution of chromium consumption in the U.S. by physical form is shown in Table 8 and detailed breakdown of consumption areas in Table 9. Growth has been modest over the past two decades with few changes in the consumption pattern. A recent study (2) has projected a 3.4% growth in U.S. chromium consumption leading to a total U.S. primary chromium demand in the year 2000 of 1,000,000 metric tons. Stainless steels dominate all other classes of chromium use since chromium is absolutely essential to the production of stainless steel. It is unlikely that research will lead to the development of a chromium-free stainless steel. In view of the importance of stainless steel to today's industrial economy and the fact that substitute materials increase cost or impair performance, U.S. vulnerability resulting from the near monopoly control of chromite by Soviet and African sources is extreme. The price of Soviet

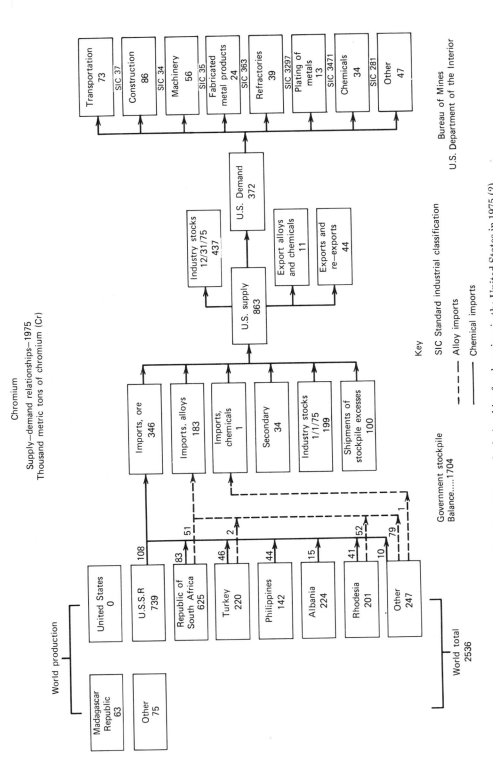

Figure 4. Supply and demand relationships for chromium in the United States in 1975 (2).

Table 8.   Chromium Consumption in the United States[a]

	Contained chromium, 1000 metric tons		% of total[b]	
	1964	1976	1964	1976
metal production				
high-carbon ferrochrome	75	144	21	44
low-carbon ferrochrome	94	49	27	15
ferrochrome silicon	29	24	8	7
other	9	8	3	2
Total	*207*	*225*	*59*	*68*
refractories	103	44	29	13
chemicals	41	58	12	18
Grand total	*351*	*327*	*100*	*100*

[a] From *Minerals Yearbook,* 1964 and 1976.
[b] Totals may not add due to rounding.

chromite approximately tripled between 1974 and 1977 after having been virtually constant for many years.

Prices for ferrochromium depend on both size and chromium and carbon content. In the United States requirements for air pollution controls have increased costs. It has been estimated that during the 1970–1974 period, 60% of new investment dollars in the domestic ferroalloy industry was spent on pollution control equipment rather than for increased capacity of new furnaces (2,10).

CHROMIUM ALLOYS

In addition to inorganic compounds of chromium, important as pigments and tanning agents, and certain organic compounds used in greases, catalysts and plastic compounding agents, there are a number of metallic compounds of chromium significant either in their own right or as metallurgical constituents in Cr-bearing alloys. The carbide Cr_3C_2 is important as a wear-resistant gage material, CrB for oil-well drilling, and $Cr_xMn_{2x}Sb$ as a magnetic material with unique characteristics. The intermetallic compounds Cr_3Al, Cr_3Si, and Cr_2Ti are encountered in developmental oxidation-resistant coatings. $Cr_{23}C_6$, Cr_7C_3, CrFe (σ phase), and $Cr_{12}Fe_{36}Mo_{10}$ (χ phase) are found as constituents in many alloy steels and $CrAl_7$ and CoCr in aluminum and cobalt-based alloys. The chromium-rich interstitial compounds (Cr_2H, Cr_2N, $Cr_{23}C_6$) play an important role in the effect of trace impurities on the properties of unalloyed chromium. The intermetallic and the interstitial compounds of chromium are stabilized by electronic and/or spatial factors and are not to be regarded as simple ionic or covalent compounds.

Chromium-Based Alloys

Alloying has not solved the problem of resistance to gaseous embrittlement. Alloying with yttrium improves the resistance to embrittlement by high-temperature exposure to oxygen but not to nitrogen-bearing atmospheres, and a barrier coating approach must be used. Furthermore, although solid solution additions can improve high temperature strength by three- to four-fold over unalloyed chromium, these have

Table 9.   U.S. Chromium Consumption Pattern in 1977

	Quantity consumed[a], (thousand metric tons)	Fraction of U.S. consumption[b], %
metallurgical		
wrought stainless and heat resisting steels	210	51.3
tool steels	5	1.3
wrought alloy steels	39	9.5
cast alloy steels	13	3.1
alloy cast irons	7	1.8
nonferrous alloys	13	3.1
other	5	1.3
Total	292	*71.5*
refractories		
chrome and chrome-magnesite	9	2.2
magnesite-chrome brick	13	3.1
granular chrome-bearing	24	6.0
granular chromite	9	2.2
Total	55	*13.5*
chemicals		
pigments	16	4.0
metal finishing	14	3.3
leather tanning	10	2.4
drilling muds	3	0.7
wood treatment	4	0.9
water treatment	4	0.9
chemical manufacture	5	1.1
textiles	2	0.4
catalysts	<1	<0.3
other	5	1.1
Total	*64*	*15.0*
Grand total	*411*	*100*

[a] Exclusive of scrap.
[b] Totals may not add due to rounding.

also resulted in increases in the ductile-to-brittle transition temperature (DBTT). A better combination of properties has been achieved through precipitation or dispersion hardening. The second phases may be oxides, carbides, or borides.

It appears that nitrogen embrittlement is the result of dislocation pinning by the presence of small Cr_2N particles on certain crystallographic planes rather than by elemental nitrogen retained in solid solution. On the other hand, the improvement of ductility experienced with oxide, carbide, or boride particles is apparently achieved through the generation and multiplication of free dislocations by one of several hypothesized mechanisms. Chromium–thorium oxide dispersions prepared by CVD (see under Consolidation and Fabrication) have been found not only to lower the DBTT relative to unalloyed chromium but largely to preserve this improvement even after a 1 hour anneal at 1200°C (11). Silicides (9), rare earths (Y and Y + La) (9), and a Cu–30Pd alloy (18) seem to improve protection against nitrogen embrittlement (100–200h at 1150°C). The leading developmental chromium-based alloys are shown in Table 10. Creep rupture results indicate a temperature advantage of 110–140°C

Table 10. Leading Developmental Chromium Based Alloys

Country	Common name	Composition, wt %
United States	C-207	Cr–7.5W–0.8Zr–0.2Ti–0.1C–0.15Y
	CI-41	Cr–7.1Mo–2Ta–0.09C–0.1(Y + La)
	IM-15	Cr–1.7Ta–0.1B–0.1Y
	Chrome-30	Cr–6MgO–0.5Ti
	Chrome-90	Cr–3MgO–2.5V–0.5Si
	Chrome-90S	Cr–3MgO–2.5V–1Si–0.5Ti–2Ta–0.5C
	CVD-Cr	Cr–3ThO$_2$
Australia	Alloy E	Cr–2Ta–0.5Si–0.1Ti
	Alloy H	Cr–2Ta–0.5Si–0.5RE
	Alloy J	Cr–2Ta–0.5Si
Soviet Union	VKh-1I	Cr–(0.3–1.0Y)–0.02C
	VKh-2	Cr–(0.1–0.35)V–0.02C
	VKh-2I	Cr–(0.3–1.0)Y–(0.1–0.35)V–(0.1–0.2Ti)–0.02C
	VKh-4	Cr–32Ni–1.5W–0.3V–0.2Ti–0.08C

over the strongest superalloys with 100-h stress-rupture strengths as high as 140 MPa (20,300 psi) at 1150°C. Further improvements in low-temperature toughness and high-temperature nitridation resistance are still needed.

Stainless Steels (1,5,20–21,23–27)

The stainless quality is conferred on steels if they contain enough chromium to form a protective surface film. About 12 wt % chromium is required for protection in mild atmospheres or in steam. With 18–20 wt % chromium, sufficient protection is achieved for satisfactory performance in a wide variety of more destructive environments, including those occurring in the chemical, petrochemical, and the power-generating industries. Stainless grades with 25 wt % chromium or more and containing other alloying elements such as molybdenum provide even higher corrosion resistance. In certain stainless steels, chromium depresses martensite transformation below room temperature; by thus stabilizing austenite it permits achievement of desired mechanical properties without loss of corrosion resistance (see Steel).

Stainless steels are classified in terms of their microstructures as austenitic, martensitic, ferritic, duplex (austenite + ferrite), and precipitation hardening (PH). The microstructure type is determined by base composition and heat treatment and, in turn, dictates the properties, especially strength, toughness, and corrosion resistance. The compositions of the leading stainless steels are shown in Tables 11–17 and the major groupings are illustrated diagramatically in Figure 5. Since most stainless steels are based on the Fe–Cr–Ni system, their relationships are conveniently interpreted in reference to the metastable phase diagram for such alloys with a carbon content of 0.1%, rapidly cooled from 1000°C as shown in Figure 6. Table 18 shows the relative production of the various stainless grades (19).

The *austenitic* stainless steels have won wide favor because of their outstanding corrosion and oxidation resistance coupled with good formability. They cannot, however, be hardened by heat treatment which, in certain types, may impair corrosion resistance by allowing precipitation of chromium carbide near the grain boundaries (sensitization) and hence local impoverishment in chromium. Variants of the basic

Table 11. Austenitic Grades of Stainless Steels

UNS[a] no.	AISI[b] type	Max C, %	Max Mn, %	Max Si, %	Cr, %	Ni, %	Other[c], %
S20100	201	0.15	7.50	1.00	16.0–18.0	3.5–5.5	0.25 max N
S20200	202	0.15	10.00	1.00	17.0–19.0	4.0–6.0	0.25 max N
S30100	301	0.15	2.00	1.00	16.0–18.0	6.0–8.0	
S30200	302	0.15	2.00	1.00	17–19	8–10	
S30215	302B	0.45	2.00	3.00	17–19	8–10	
S30300	303	0.15	2.00	1.00	17–19	8–10	0.15 min S
S30323	303Se	0.15	2.00	1.00	17–19	8–10	0.15 min Se, 0.020 max P, 0.060 max S
S30400	304	0.08	2.00	1.00	18–20	8–12	
S30403	304L	0.03	2.00	1.00	18–20	8–12	
S30500	305	0.12	2.00	1.00	17–19	10–13	
S30800	308	0.08	2.00	1.00	19–21	10–12	
S30900	309	0.20	2.00	1.00	22–24	12–15	
S30908	309S	0.08	2.00	1.00	22–24	12–15	
S31000	310	0.25	2.00	1.50	24–26	19–22	
S31008	310S	0.08	2.00	1.50	24–26	19–22	
S31400	314	0.25	2.00	3.00	23–26	19–22	
S31600	316	0.08	2.00	1.00	16–18	10–14	2.0–3.0 Mo
S31603	316L	0.03	2.00	1.00	16–18	10–14	2.0–3.0 Mo
S31700	317	0.08	2.00	1.00	18–20	11–15	3.0–4.0 Mo
S32100	321	0.08	2.00	1.00	17–19	9–12	5XC min Ti
S34700	347	0.08	2.00	1.00	17–19	9–13	10XC min Nb + Ta
S34800	348	0.08	2.00	1.00	17–19	9–13	10XC min Nb + Ta; 0.10 max Ta

[a] Unified Numbering System (22).
[b] American Iron and Steel Institute.
[c] Except as noted all have 0.045 max P; 0.030 max S.

Table 12. New Proprietary Mn-N-Rich Austenitic Stainless Steels

UNS no	Common name (former)	Typical Composition, %					
		C	Mn	Cr	Ni	N	Other
S24000	Nitronic 33 (18-3 Mn)	0.05	12.0	18.0	3.2	0.32	
S21904	Nitronic 40 (21-6-9)	0.03	9.0	21.0	7.0	0.30	
S20910	Nitronic 50 (22-13-5)	0.04	5.0	21.2	12.5	0.30	2.2 Mo, 0.2 Cb, 0.2 V
S21400	USS Tenelon (XM-31)	0.10	15.0	18.0		0.40	
S21460	Cryogenic Tenelon (XM-14)	0.10	15.1	17.5	5.5	0.42	
S21600	AL 216 (XM-17)	0.06	8.3	20.0	6.0	0.37	2.5 Mo

Table 13. Martensitic or Hardenable Grades of Stainless Steel

UNS no.	AISI type	C, %	Max Mn, %	Max Si, %	Cr, %	Ni, %	Other[a], %
S40300	403	0.15 max	1.00	0.50	11.5–13.0		
S40500	405	0.08 max	1.00	1.00	11.5–14.5		0.10 to 0.30 Al
S41000	410	0.15 max	1.00	1.00	11.5–13.5		
S41400	414	0.15 max	1.00	1.00	11.5–13.5	1.25–2.5	
S41600	416	0.15 max	1.25	1.00	12.0–14.0		0.15 min S, 0.060 P max
S42000	420	0.15 min	1.00	1.00	12.0–14.0		
S42200	422	0.20–0.25	1.00	0.75	11.0–13.0	0.50–1.0	0.75–1.25 Mo 0.15–0.30 V, 0.75–1.25 W 0.025 P max, 0.025 S max
S43100	431	0.20 max	1.00	1.00	15.0–17.0	1.25–2.50	
S44002	440A	0.60–0.75	1.00	1.00	16.0–18.0		0.75 max Mo
S44003	440B	0.75–0.95	1.00	1.00	16.0–18.0		0.75 max Mo
S44004	440C	0.95–1.25	1.00	1.00	16.0–18.0		0.75 max Mo

[a] 0.040 max P and 0.030 max S unless otherwise noted.

Table 14. Ferritic or Nonhardenable Grades of Stainless Steel

UNS no.	AISI type	Max C, %	Max Mn, %	Max Si, %	Cr, %	Mo, %	Ni, %	Other[a], %
S40900	409	0.08	1.00	1.00	10.5–11.75		0.5 max	(6 × C, .75 max) Ti; 0.045 P max, 0.045 S max
S42900	429	0.12	1.00	1.00	14.0–16.0			
S43000	430	0.12	1.00	1.00	14.0–18.0			
S43020	430F	0.12	1.25	1.00	14.0–18.0			0.15 min S, 0.060 max P, 0.60 Mo (optional)
S43400	434	0.12	1.00	1.00	14.0–18.0	1.0		
S43600	436	0.12	1.00	1.00	16.0–18.0	0.75–1.25		(Cb + Ti) = 5 × C min −0.70 max
S44200	442	0.20	1.00	1.00	18.0–23.0			
S44600	446	0.20	1.50	1.00	23.0–27.0			0.24 max N

[a] 0.040 max P, 0.030 max S unless otherwise noted.

Table 15. Super-ferritic Grades

UNS no.	Common name	Max C, %	Max Mn, %	Max Si, %	Cr, %	Mo, %	Ni, %	Other[a], %
S44400	18-2-Ti	0.025	1.00	1.00	17.5–19.5	1.75–2.50	1.0 max	0.025 max N, C + N <0.030 desirable; Ti + Cb = 0.20 + 0.4 (C + N) min = 0.80 max
S44626	26-1-Ti	0.060	0.75	0.75	25.0–27.0	0.75–1.50	0.50 max	0.040 max N, 0.2 to 1.0 Ti typical [7 × (C + N) min] C + N = 0.050 typical, Cu = 0.20 max, S = 0.020 max
	26-1-Cb	0.005	1.00	1.00	25.5–27.5	0.6–1.4		Cb = 13–29 × N %
	28-2-4-Cb	0.015	1.00	1.00	26.0–30.0	1.75–2.25	3.6–4.4	0.035 max N, C + N ≤0.040 with Cb ≥0.2 + 12 (C + N)%
S44700	29-4	0.010	0.30	0.20	28.0–30.0	3.5–4.2		0.020 max N, C + N ≤0.025, 0.15 max Cu[b]
S44800	29-4-2	0.010	0.30	0.20	28.0–30.0	3.5–4.2		0.020 max N, C + N ≤0.025, 0.15 max Cu[b]

[a] 0.040 max P, 0.030 max S unless otherwise noted.
[b] 0.025 max P, 0.020 max S.

Table 16. Duplex (Ferrite + Austenite) Grades of Stainless Steel

UNS no.	AISI type	Max C, %	Max Mn, %	Max Si, %	Cr, %	Ni, %	Other[a], %
S31500		0.030	1.20–2.00	1.4–2.0	18.0–19.0	4.25–5.25	2.5–3.0 Mo, 0.030 max P
	327	0.10	1.00	1.00	25.0–26.0	4.8	
S32900	329	0.10	2.00	0.75	25.0–30.0	3.0–6.0	1.0–2.0 Mo

[a] 0.040 max P, 0.030 max S unless otherwise noted.

Table 17. Precipitation Hardenable Stainless Steels

UNS no.	Common name	C	Mn	Si	Typical composition, %				Other[a]
					Cr	Ni	Mo	Al	
S13800	PH 13–8 Mo	0.05 max	0.10 max	0.10 max	12.25–13.25	7.5–8.5	2.0–2.5	0.90–1.35	0.010 N, 0.01 max P, 0.008 max S
S15500	15–5 PH	0.07 max	1.0 max	1.0 max	14.5–15.5	3.5–5.5			2.4–4.5 Cu, Cb + Ta = 0.15–0.45
S17400	17–4 PH	0.07 max	1.0 max	1.0 max	15.5–17.5	3.0–5.0			3.0–5.0 Cu, Cb + Ta = 0.15–0.45
S17700	17–7 PH	0.09 max	1.0 max	1.0 max	16.0–18.0	6.5–7.75		0.75–1.5	0.040 max S
	PH 15–7 Mo	0.09 max	1.0 max	1.0 max	14.0–16.0	6.5–7.75	2.0–3.0	0.75–1.5	

[a] 0.040 max P, 0.030 max S unless otherwise noted.

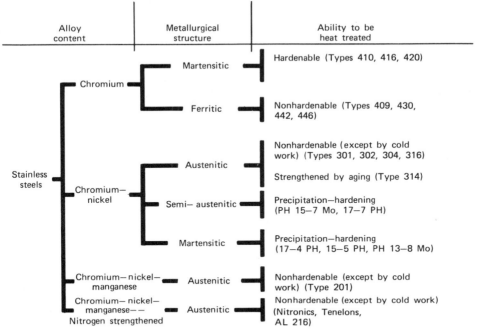

Figure 5. Diagramatic illustration of major classes of stainless steels (20).

Table 18.   Relative Production of Major Stainless and Heat-Resistant Steel Grades, 1975

Grades	Percent
304	27.4
301	8.7
316	5.2
410	5.1
304L	4.6
316L	4.0
430	4.0
201[a], 202[a]	2.8
300 series[a], all other	7.3
400 series[b], all other	6.7
other chromium nickel stainless steels	5.9
501[b], 502[b], and all other high chromium heat resistant steels	3.3
all others	15.0

[a] The principal alloy elements are chromium and nickel.

[b] The principal alloy element is chromium.

18 Cr–8 Ni composition contain Mo for improved resistance to sulfuric acid or Ti or Cb to prevent the undesirable precipitation of chromium carbide at grain boundaries. Sulfur or selenium additions improve machinability. Recent additions to the austenitic class are the nitrogen strengthened Cr–Mn proprietary alloys, leading grades of which are shown in Table 12. They have higher strength at both room and elevated tem-

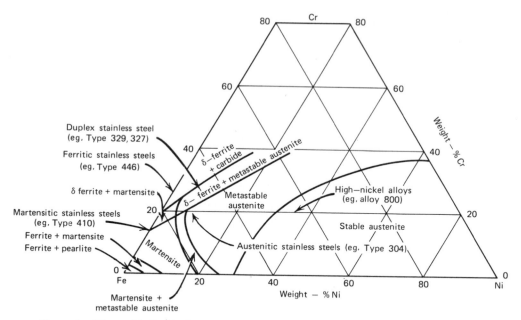

Figure 6.   Metastable Fe–Ni–Cr "ternary" phase diagram with C = 0.1% and for alloys cooled rapidly from 1000°C showing the locations of austenitic, duplex, ferritic and martensitic stainless steels with respect to the metastable phase boundaries. For higher carbon contents than 0.1% martensite lines occur at lower alloy contents (21).

peratures than standard 300 grades as well as improved corrosion resistance and low temperature toughness.

Ferritic stainless steels with low carbon and high chromium can be heated to the melting point without transformation to the austenitic structure as shown in Figure 7. This fact ensures their freedom from quench hardening, thereby facilitating fabrication, but eliminates the possibility of grain refinement by heat treatment. Such steels are widely used for architectural work, automotive trim, and equipment in the chemical and food industries. Types 430 and 446 are especially resistant to oxidation and hence are favored for furnace parts, heat exchangers, and other high temperature equipment. Recently some new grades in this class have appeared, the so-called super-ferritics, (see Table 15). They offer superior toughness, weldability, and stress corrosion cracking resistance at strength levels comparable to those of conventional ferritics.

Martensitic stainless steels have somewhat higher carbon contents than the ferritic grades for the equivalent chromium level. They are therefore subject to the austenite–martensite transformation on heating and quenching and can be hardened significantly. The higher-carbon martensitic types, eg 420 and 440 are typical cutlery compositions, whereas the lower-carbon grades are used for special tools, dies, and machine parts and equipment subject to combined abrasion and mild corrosion.

Leading grades of the so-called precipitation hardening (PH) types are shown in Table 17. Hardened by precipitating a nickel-rich intermetallic compound in a martensite matrix, they are shipped either in the martensitic or austenitic condition. The austenite is transformed to martensite before the precipitation-aging heat treatment. Compared to standard stainless steels, especially the martensitic grades,

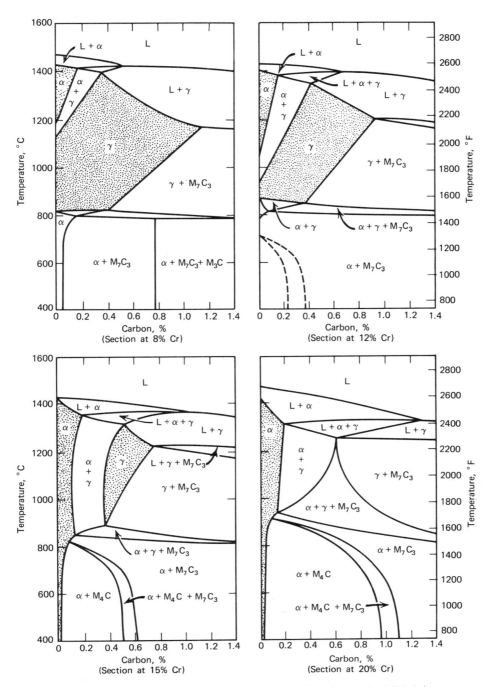

Figure 7. Fe–Cr–C vertical sections at constant chromium content (23–24).

77

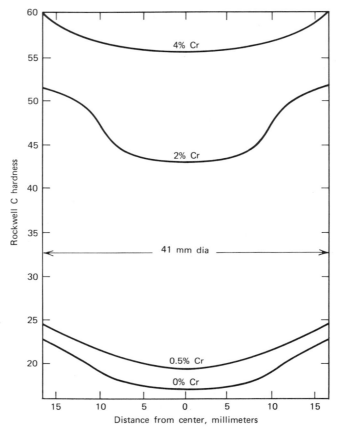

Figure 8.   Effect of chromium on hardenability of steel as indicated by hardness distribution across 41 mm rounds of oil-quenched 0.35% C steel (25).

the PH stainless steels offer high yield strength—some even above 1400 MPa (203,000 psi),—good ductility, and corrosion resistance.

Other Alloy Steels (5,25)

In low-alloy steels chromium contributes more to hardenability, tempering resistance, and toughness than to solid-solution hardening or oxidation resistance. In the high-chromium tool steel compositions chromium carbides improve the high hot hardness. The marked effect of small additions of chromium on hardenability is shown in Figure 8. Other alloying elements show a similar or greater effect, but chromium is one of the cheapest. The effect of chromium on resistance to tempering is shown in Figure 9.

Wrought alloy steels, alloyed cast irons and steels, and tool steels account for 20–25% of the annual U.S. consumption of chromium. The largest category are the wrought alloy steels classified as shown in Table 19. In terms of chromium consumption, the most important classes are the CrMo, NiCrMo, and Cr steels (in that order). They may be further subdivided into carburizing and through-hardening grades. Such

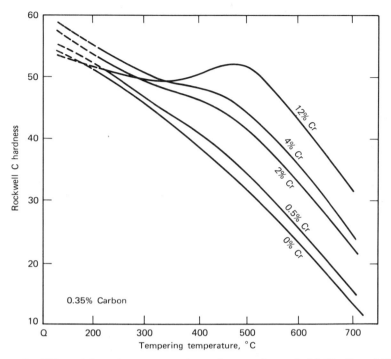

Figure 9. Effect of chromium on tempering resistance of quenched 0.35% C steel (25).

steels are extensively used in machinery, construction, and other structural work, and in machine parts such as bearings, gears, rolls, springs, and shafting. The AISI 500 series steels shown in Table 20 with their higher chromium content find application in the oil industry where their resistance to corrosion and oxidation is especially important.

Nonferrous Alloys

Nonferrous alloys account for only about 2% of the total chromium used in the United States. Nonetheless, some of these applications are unique and constitute a vital role for chromium. For example, chromium confers corrosion and oxidation resistance on the nickel-base superalloys used in jet engines; the familiar electrical resistance heating elements are made of a Ni–Cr alloy; and a variety of Fe–Ni and Ni-based alloys used in a diverse array of applications (especially in the nuclear reactor field) depend on chromium for oxidation and corrosion resistance.

Recovery and Reuse

At the present time only about 15% of chromium consumed in the United States is recycled and this largely from stainless steel scrap. Today because of environmental considerations much of the chromium formerly lost from plating operations is recovered. However this does not amount to a significant tonnage. Furthermore, steps are now being taken to improve recovery of the substantial chromium losses incurred

Table 19.   Basic Numbering System for Chromium-Bearing Low-Alloy Steels

Alloy	AISI/SAE designation[a]	UNS[b] no.
nickel–chromium steels		
Ni 1.25; Cr 0.65	31XX	G31XXZ
Ni 3.50; Cr 1.57	33XX	G33XXZ
chromium–molybdenum steels		
Cr 0.50 and 0.95; Mo 0.25, 0.20, and 0.12	41XX	G41XXZ
nickel–chromium–molybdenum steels		
Ni 1.82; Cr 0.50 and 0.80; Mo 0.25	43XX	G43XXZ
Ni 1.05; Cr 0.45; Mo 0.20	47XX	G47XXZ
Ni 0.55; Cr 0.50 and 0.65; Mo 0.20	86XX	G86XXZ
Ni 0.55; Cr 0.50; Mo 0.25	87XX	G87XXZ
Ni 3.25; Cr 1.20; Mo 0.12	93XX	G93XXZ
Ni 1.00; Cr 0.80; Mo 0.25	98XX	G98XXZ
chromium steels		
Cr 0.27, 0.40, and 0.50	50XX	G50XXZ
Cr 0.80, 0.87, 0.92, 0.95, 1.00, and 1.05	51XX	G51XXZ
Cr 0.50	501XX	G509ZZ
Cr 1.02	511XX	G519ZZ
Cr 1.45	521XX	G529ZZ
chromium–vanadium steels		
Cr 0.80 and 0.95; V 0.10 and 0.15 (min)	61XX	G61XXZ
boron-treated chromium steels	XXBXX[c]	various

[a] The XX after the designation is left open for carbon content; thus, 3110 equals a 0.10% carbon, nickel–chromium steel, and in this case XX = 0.10.

[b] The XX after the designation follows the AISI/SAE system and hence indicates carbon content; the Z is either an arbitrary number or 0. For a fuller description and tabulations of this new numbering system see Ref 22.

[c] The XX before the B is for the type numbers and after the B for carbon content.

Table 20.   Composition of Heat-Resisting Chromium Steels

UNS no.	AISI type	C, %	Mn[a], %	Si[a], %	P[a], %	S[a], %	Cr, %	Mo, %
S50100	501	over 0.10	1.00	1.00	0.04	0.03	4.0–6.0	0.40–0.65
S50200	502	0.10 (max)	1.00	1.00	0.04	0.03	4.0–6.0	0.40–0.65
S50300	503	0.15 (max)	1.00	1.00	0.04	0.04	6.0–8.0	0.45–0.65
S50400	504	0.15 (max)	1.00	1.00	0.04	0.04	8.0–10.0	0.90–1.10

[a] Maximum.

in the past from refractory and foundry applications of chromite grain (see Recycling).

Future Trends

World political problems and the unique and vital industrial role of chromium are stimulating research and development to create new economically feasible sources of chromium and to achieve a more conservative use of chromium. Thus we can expect to see process developments permitting the recovery of chromium from lower-grade ores, and the replacement of chromium-containing materials by other materials, development of low chromium grades, especially in stainless steels, and extended and improved recycling. Further research may lead to practical chromium plating processes based on Cr^{3+} rather than Cr^{6+} and to Cr-based alloys of acceptable ductility that can effectively exploit chromium's potential as a high-temperature material.

BIBLIOGRAPHY

"Chromium and Chromium Alloys" in *ECT* 1st ed., Vol. 3, pp. 935–940, by J. J. Vetter, Diamond Alkali Company; "Chromium and Chromium Alloys" in *ECT* 2nd ed., Vol. 5, pp. 451–472, F. E. Bacon, Union Carbide Corporation.

1. M. A. Streicher, "Stainless Steels: Past, Present and Future," in *Stainless Steel '77*, Climax Molybdenum Corporation, in press.
2. J. L. Morning, *Chromium—1977, U.S. Bur. Mines Rpt. MCP-1*, May 1977.
3. P. R. Grabfield, "Chrome Ore Sources and Availability with Emphasis on Metallurgical Use," *Iron and Steel Metallurgy*, Oct. 1975, pp. 16–26; Nov. 1975, pp. 27–35.
4. M. J. Udy, *Chromium, Chemistry of Chromium and Its Compounds*, Vol. 1, Reinhold Publ. Corp., New York, 1956.
5. M. J. Udy, *Chromium, Metallurgy of Chromium and Its Alloys*, Vol. 2, Reinhold Publ. Corp., New York, 1956.
6. *Ductile Chromium*, American Society for Metals, Cleveland, Ohio, 1957.
7. A. H. Sully and E. A. Brandes, *Chromium*, Plenum Publ. Corp., New York, 2nd ed., 1967.
8. C. L. Rollinson, *The Chemistry of Chromium, Molybdenum & Tungsten*, Vol. 21, Pergamon Texts in Inorganic Chemistry, 1975.
9. W. D. Klopp, "Chromium-Base Alloys," in C. Sims and W. Hagel, eds., *The Superalloys*, John Wiley & Sons, Inc., New York, 1972, pp. 175–196.
10. R. Urquhart, *Miner. Sci. Eng.* **4**, 48 (1972).
11. N. D. Veigel and co-workers, *Development of a Chromium-Thoria Alloy, NASA Rpt. CR-72901*, March 10, 1971.
12. D. P. Shashkov "The Effect of Plastic Deformation on the Brittleness of Chromium," in J. L. Walter, J. H. Westbrook, and D. A. Woodford, eds., *Grain Boundaries in Engineering Materials*, Claitor's Publ. Div., Baton Rouge, La., 1975, pp. 657–668.
13. G. Dubpernell, *Electrodeposition of Chromium from Chromic Acid Solutions*, Pergamon Press, Oxford, 1977, 148 pp.
14. G. Dubpernell and F. A. Lowenheim, *Modern Electroplating*, 3rd ed., John Wiley & Sons, Inc., New York, 1974, pp. 87–151.
15. B. L. McKinney and C. L. Faust, *J. Electrochem. Soc.*, **124**, 379C (1977).
16. R. M. Burns and W. W. Bradley, *Protective Coatings for Metals*, 3rd ed., Reinhold, New York, 1967, pp. 243–255.
17. R. Weiner, *Electrolytic Chromium Plating*, Finishing Publications Ltd., 1978.
18. A. Ankara, *Met. Technol.* **4**, 279 (1977).
19. *Metal Statistics*, Fairchild Publishers, Inc., 1976, p. 207.
20. *ARMCO Stainless Steels*, Armco Steel Corp., Middletown, Ohio, 1968.
21. M. O. Speidel in R. W. Staehle and M. O. Speidel, eds., *Stress Corrosion of Austenitic Stainless Steels*, ARPA report, in press.
22. *Unified Numbering System for Metals and Alloys*, SAE HS 1086a or ASTM DS-56A, 2nd ed., 1977.
23. W. Tofaute, A. Sponheuer, and H. Bennek, *Arch. Eisenhuttenwes.* **8**, 499 (1934).
24. W. Tofaute, C. Küttner, and A. Büttinghaus, *Arch. Eisenhuttenwes.* **9**, 606 (1936).
25. E. D. Bain and H. W. Paxton, *Alloying Elements in Steel*, 2nd ed., ASM Cleveland, Ohio, 1966, 201 pp.
26. *Source Book on Stainless Steels*, American Society for Metals, Cleveland, Ohio, 1976, 408 pp.
27. D. Peckner and I. M. Bernstein, *Handbook of Stainless Steels*, McGraw-Hill, New York, 1977, ca 1200 pp.

General References

U.S. Bureau of Mines, *Minerals Yearbook,* Vol. I, 1964, 1976.

T. N. Irvine, ed., *Chromium: Its Physicochemical Behaviour and Petrologic Significance,* Carnegie Institution of Washington Conference, Geophysical Laboratory, Pergamon, 1977.

M. J. Udy, "Chromium," *Chemistry of Chromium and Its Compounds,* Vol. 1, Reinhold Publ. Corp., New York, 1956.

M. J. Udy, "Chromium," *Metallurgy of Chromium and Its Alloys,* Vol. 2, Reinhold Publ. Corp., New York, 1956.

Ductile Chromium, American Society for Metals, Cleveland, Ohio, 1957.

A. H. Sully and E. A. Brandes, *Chromium,* 2nd ed., Plenum Publ. Corp., New York, 1967.

C. L. Rollinson, *The Chemistry of Chromium, Molybdenum & Tungsten,* Vol. 21, Pergamon Texts in Inorganic Chemistry, 1975.

G. Dubpernell, *Electrodeposition of Chromium from Chromic Acid Solutions,* Pergamon Press, Oxford, 1977.

W. D. Klopp "Chromium-Base Alloys," in C. Sims and W. Hagel eds., *The Superalloys,* John Wiley & Sons, Inc., 1972.

Source Book on Stainless Steels, American Society for Metals, Cleveland, 1976.

ARMCO Stainless Steels, 1968.

E. D. Bain and H. W. Paxton, *Alloying Elements in Steel,* 2nd ed., ASM Cleveland, 1966.

E. R. Parker and co-workers, *Contingency Plan for Chromium Utilization,* NMAB Report 335, 1977.

D. Peckner and I. M. Bernstein, *Handbook of Stainless Steels,* McGraw-Hill, New York, 1977, ca 1200 pp.

J. H. WESTBROOK
General Electric Company

CHROMIUM COMPOUNDS

When chromium was discovered at the end of the 18th century (1–2), it was found to be a principal constituent of chromite, the sole commercial source of chromium and its compounds. Attack of the ore by alkaline oxidation was soon discovered, and by 1800 the roasting of chromite with lime and potash, the basis of most present technologies, was in use. The principal early use of chromium compounds was in the manufacture of lead chromate (chrome yellow) followed by the development of mordant dyeing. Other milestones in the chemistry and technology of chromium compounds were the discovery of chromyl compounds in 1824, of chromous compounds in 1844, of chrome tanning in 1858 (commercialization attained about 1884), of organochromium compounds in 1919 (although not characterized as "sandwich" compounds until 1957 (3)), and of chromium plating as a practical process (4–5) about 1926 (see Chromium and chromium alloys).

The United States is completely dependent on imports for its chromium ore, most of which is imported from the Union of South Africa. Grade B ore is generally used, although concentrates are available. Typical analyses of ores used in American plants during the period from 1934 to the present are given in Table 1. Compared to ores from other sources, Transvaal ores are very uniform in analysis as received. They are high in iron and contain appreciable vanadium and titanium. The iron and vanadium contents are of importance in the extractive technology.

Table 1. Typical Analyses of Chrome Ores Used for Manufacture of Chromium Compounds[a]

Constituent	% by weight of constituent in ore						
	1	2	3	4	5	6	7
Cr_2O_3	47.72	48.77	50.19	53.57	44.51	46.55	40.99
total iron as Fe	9.23	14.35	15.18	10.10	19.28	19.72	21.40
SiO_2	7.81	4.95	2.33	6.41	3.24	1.20	0.99
Al_2O_3	6.43	9.46	12.27	9.67	15.36	16.00	18.18
MgO	21.36	10.97	13.03	13.08	10.68	9.98	10.21
MnO	0.30				0.26		0.19
CaO	0.46	1.90	0.36	0.15		0.06	0.02
TiO_2					0.45	0.50	0.73
other				0.25	0.30	0.34	
				NiO	V_2O_3	V_2O_3	
H_2O+	4.25	1.76	1.54	0.67			1.07

[a] Refs. 16 and 17.
KEY: 1. Fantoche mine, New Caledonia; 2. Great Dyke, Rhodesia; 3. Acoje, Philippines; 4. U.S.S.R.; 5. Transvaal Grade B; 6. Transvaal concentrates; 7. Stillwater complex concentrates, Montana; used only in pilot-plant studies.

Physical Properties

Some of the properties of a group of important chromium compounds are given in Table 2. Many of these are important commercial compounds (6–9). More detailed data on solubility are given in references 10–12 and 13; on specific gravity and viscosity of solutions in references 10, 11 and 13–15; and on eutectics with ice in references 10, 15 and 13. Data on specific heats, boiling and freezing points of solutions, and pH may also be found in these references.

Chemical Properties

The important physical and chemical properties of the compounds are given in detail in standard reference books and advanced texts (18–23).

In its compounds, chromium may use any of its six $3d$ and $4s$ electrons and may show any oxidation state from −2 to +6. Those of +2, +3, and +6 are the most important, whereas the −1 and −2 oxidation states are of little significance. Commercial applications center about the +6 state, with some interest in the +3 state. Research interest at the present centers around kinetic studies involving the +3 state and structural studies of the π-bonded complexes of the lower oxidation states.

The various oxidation states and stereochemistry are summarized in Table 3. Compounds of commercial interest are discussed under Manufacture, and compounds used as pigments or dyes are discussed under Uses.

Chromium(0) Compounds. The chemical compounds of Cr(0) include most of the π-bonded complexes of the element. Very active research in the past few years has led to the preparation of large numbers of these compounds. In general, the fundamental basis of these compounds is a variety of neutral π-electron-rich donors such as CO and C_6H_6, leading to $Cr(CO)_6$ and $Cr(C_6H_6)_2$ with a sandwich structure, and the half-sandwich $Cr(C_6H_6)(CO)_3$ [12082-05-5]. Diphenyl, bipyridyl, as well as toluene, cumene, and mesitylene may substitute for benzene.

Table 2. Physical Properties of Typical Chromium Compounds

Compound	CAS Registry No.	Formula	Appearance	Crystal system and space group	Density, g/cm³	mp, °C	bp, °C	Solubility
Oxidation state 0								
chromium carbonyl	[13007-92-6]	$Cr(CO)_6$	colorless crystals	orthorhombic, C_{2r}^9	1.77_{18}	150 (dec) (sealed tube)	151 (dec)	sl sol CCl_4; insol H_2O, $(C_2H_5)_2O$, C_2H_5OH, C_6H_6
dibenzene chromium(0)	[1271-54-1]	$(C_6H_6)_2Cr$	brown crystals	cubic, Pa_3	1.519	284-285	subl 150 (vacuum)	insol H_2O; sol C_6H_6
Oxidation state +1								
bis(biphenyl)chromium(I) iodide	[12099-17-1]	$(C_6H_5C_6H_5)_2CrI$	orange plates		1.617_{16}	178	dec	sol C_2H_5OH, C_5H_5N
Oxidation state +2								
chromous acetate	[628-52-4]	$Cr_2(C_2H_3O_2)_4\cdot2H_2O$	red crystals	monoclinic, $C2c$	1.79			sl sol H_2O; sol acids
chromous chloride	[10049-05-5]	$CrCl_2$	white crystals	tetragonal, D_{4b}^{14}	2.93	815	1120	sol H_2O to blue sol, absorbs O_2
chromous ammonium sulfate	[25638-51-1]	$CrSO_4(NH_4)_2SO_4\cdot6H_2O$	blue crystals	monoclinic, C_{2h}^5				sol H_2O, absorbs O_2
Oxidation state +3								
chromic chloride	[10025-73-7]	$CrCl_3$	bright purple plates	hexagonal, $D_3^{3\ or\ 5}$	2.87_{25}	subl	885	insol H_2O, sol presence Cr^{2+}
chromic acetylacetonate	[13681-82-8]	$Cr(CH_3COCHCOCH_3)_3$	red-violet crystals	monoclinic	1.34	208	345	insol H_2O; sol C_6H_6
chromic potassium sulfate (chrome alum)	[7788-99-0]	$KCr(SO_4)_2\cdot12H_2O$	deep purple crystals	cubic, A_n^6	1.826_{15}	89 (incongruent)		sol H_2O
chromic chloride hexahydrate	[10060-12-5]	$(Cr(H_2O)_4Cl_2)Cl\cdot2H_2O$	bright green crystals	triclinic or monoclinic	1.835_{25}	95		sol H_2O, green soln turning green-violet
chromic chloride hexahydrate		$(Cr(H_2O)_6)Cl_3$	violet crystals	rhombohedral, D_{3d}^6		90		sol H_2O, violet soln turning green-violet
chromic oxide	[1308-38-9]	Cr_2O_3	green powder or crystals	rhombohedral, D_{3d}^6	5.22_{25}	2435	ca 3000	insol

			Appearance	Crystal system	Density	mp, °C	bp, °C	Solubility
Oxidation state +4								
chromium(IV) oxide	[12018-01-8]	CrO_2	dark-brown or black powder	tetragonal, D_{4h}^{14}	4.98 (calcd)		dec to Cr_2O_3	sol acids to Cr^{3+} and Cr^{6+}
chromium(IV) chloride	[15597-88-3]	$CrCl_4$	stable only at high temp				830	
Oxidation state +5								
barium chromate(V)	[12345-14-1]	$Ba_3(CrO_4)_2$	black-green crystals	same as $Ca_3(PO_4)_2$				sl dec H_2O; sol dil acids to Cr^{3+} and Cr^{6+}
Oxidation state +6								
chromium(VI) oxide	[1333-82-0]	CrO_3	ruby-red crystals	orthorhombic, C_{20}^{16}	2.7_{25}	197	dec	v sol H_2O; sol CH_3COOH, $(CH_3CO)_2O$
chromyl chloride	[14977-61-8]	CrO_2Cl_2	cherry-red liquid		1.9145_{25}	−96.5	115.8	insol H_2O, hydrolyzes; sol CS_2, CCl_4
ammonium dichromate	[7789-09-5]	$(NH_4)_2Cr_2O_7$	red-orange crystals	monoclinic	2.155_{25}	dec 180		sol H_2O
potassium dichromate	[7778-50-9]	$K_2Cr_2O_7$	orange-red crystals	triclinic	2.676_{25}	398	dec	sol H_2O
sodium dichromate	[7789-12-0]	$Na_2Cr_2O_7\cdot 2H_2O$	orange-red crystals	monoclinic	2.348_{25}	84.6 incongruent	dec	v sol H_2O
potassium chromate	[7789-00-6]	K_2CrO_4	yellow crystals	orthorhombic	2.732_{18}	971		sol H_2O
sodium chromate	[7775-11-3]	Na_2CrO_4	yellow crystals	orthorhombic, D_{2h}^{17}	2.723_{25}	792		sol H_2O
potassium chlorochromate		$KCrO_3Cl$	orange crystals	monoclinic	2.497_{39}	dec		sol H_2O, hydrolyzes
silver chromate	[7784-01-2]	Ag_2CrO_4	maroon crystals	monoclinic	5.625_{25}			v sl sol H_2O; sol dil acids
barium chromate	[10294-40-3]	$BaCrO_4$	pale yellow solid	orthorhombic	4.498_{25}	dec		v sl sol H_2O; sol strong acids
strontium chromate	[7789-06-2]	$SrCrO_4$	yellow solid	monoclinic, C_{2h}^5	3.895_{15}	dec		sl sol H_2O; sol dil acids
lead chromate	[7758-97-6]	$PbCrO_4$	yellow solid; orange solid; red solid	orthorhombic; monoclinic, C_{2h}^5; tetragonal	6.12_{15}	844		practically insol H_2O; sol strong acids

Table 3. Oxidation States and Stereochemistry of Chromium [a]

Oxidation state	Coordination number	Geometry	Example	
			Formula	CAS Registry No.
Cr(−11)	?	?	$Na_2[Cr(CO)_5]$	[51233-19-3]
Cr(−1)	6?	octahedral	$Na_2[Cr_2(CO)_{10}]$	[15616-67-8]
Cr(0)	6	octahedral	$Cr(CO)_6$	[13007-92-6]
			$Cr(C_6H_6)_2$	[12001-92-7]
Cr(I)	6	octahedral	$[Cr(C_6H_5)_2]^+[Cr(bipy)_3]^+$	
Cr(II) [b]	4	distorted tetrahedral	$CrCl_2(CH_3CN)_2$	[56890-91-6]
	6	distorted octahedral	$CrCl_2$	[10049-05-5]
			$Cr(H_2O)_6^{2+}$	
Cr(III) [c]	4	distorted tetrahedral	$CrCl_4^-$	[15597-88-3]
			probably $CrCl_4$	
	6 [d]	octahedral	$[Cr(NH_3)_6]^{3+}$	
			$Cr(acac)_3$	[21679-31-2]
			$[Cr(CN)_6]^{3-}$	
Cr(IV)	4	tetrahedral	$Cr(OC_4H_9)_4$	[68438-63-1]
			Ba_2CrO_4	[68900-59-4]
	6	octahedral	CrF_6^{2-}	
			CrO_2	[12018-01-8]
Cr(V)	4	tetrahedral	CrO_4^{3-}	
	5	?	CrF_5	[14884-42-5]
			$CrOCl_4^-$	
	6	octahedral	$[CrOCl_5]^{2-}$	
	8	quasi-dodecahedral	CrO_8^{3-}	
Cr(VI)	4	tetrahedral	CrO_4^{2-}	
			CrO_3	[1333-82-0]
			CrO_2Cl_2	[14977-61-8]

[a] Adapted from reference 20.
[b] Coordination numbers of 5 and 7 are also known.
[c] Coordination numbers of 3 and 5 are also known.
[d] Practically all Cr^{3+} compounds have this coordination number.

A typical preparation is that of chromium hexacarbonyl (24):

$$Al + CrCl_3 + 6\ CO \xrightarrow[\substack{140°C,\ 15-20\ MPa \\ (150-200\ atm)}]{AlCl_3,\ C_6H_6} AlCl_3 + Cr(CO)_6$$

Chromium hexacarbonyl is the only colorless chromium compound. It is volatile and dissociates on further heating, giving a deposit of metal contaminated with carbide. Under different conditions, CrO [12018-00-7] may be formed. The diarene compounds have also been investigated as possible sources of vapor-deposited chromium. However, they tend to be unstable. The more stable ones, such as dimesitylenechromium [1274-07-3], give carbides on heating. Dicumenechromium was at one time introduced commercially for vapor deposition (25).

Chromium(I) Compounds. The Cr(0) compounds are readily oxidized to Cr(I) compounds. The dibenzenechromium cation $[Cr(C_6H_6)_2]^+$ is orange-yellow, forms a strong base, and difficultly soluble salts with large anions such as I^- and $[B(C_6H_5)_4]^-$. The Cr(I) ion is formed by the following reaction:

$$AlCl_3 + 3\ CrCl_3 + 2\ Al + 6\ C_6H_6 \xrightarrow{AlCl_3\ +\ C_6H_6} 3\ Cr(C_6H_6)_2^+ + 3\ AlCl_4^-$$

The $[Cr(C_6H_6)_2]^+$ ion can then be reduced to $Cr(C_6H_6)_2$ with hypophosphite (26).

Chromium(II) Compounds. The aqueous chemistry of chromium in water solution concerns the oxidation states 2, 3, and 6 almost exclusively. Standard electrode potentials are listed in Table 4 and show that the Cr^{2+} ion is a powerful reducing agent and it is so used in analytical procedures. It is prepared as needed by dissolving electrolytic chromium in the appropriate acid (27). The chromous ion in aqueous solution $[Cr(H_2O)_6]^{2+}$ is blue, and blue hydrated salts, isomorphous with those of Fe^{2+} and Mg^{2+}, are formed. Examples are $CrSO_4.H_2O$ [19812-12-0], $CrCl_2.4H_2O$ [13931-94-7], and $(NH_4)_2Cr(SO_4)_2.6H_2O$.

Divalent chromium also forms dimeric red hydrated salts with aliphatic acids, an example being the acetate, $(Cr(OCOCH_3)_2)_2.2H_2O$ [628-52-4]. These salts are stable to oxidation in dry air, although the moist powder oxidizes fast enough to ignite. In the dimeric crystal, the Cr-Cr distance is so short that the four d electrons of each Cr are paired and the compounds are diamagnetic. In this respect Cr^{2+} resembles Cu^{2+}.

Transition states, known as ligand-bridged complexes, involved in the oxidation of Cr(II), have been studied (28).

The anhydrous halides are prepared by the action of hydrogen halides on the metal at 600–700°C or by reduction of the trihalides with hydrogen or aluminum at 500–600°C (29). They are stable in air as long as the relative humidity is sufficiently low to prevent formation of the readily oxidized hydrates. Chromium(II) sulfide [12018-06-3] (1:1), CrS, and chromium(II) oxide (1:1), CrO, are also known.

In this oxidation state a number of coordination compounds form with, for example, hydrazine, dipyridyl, ethylenediamine, cyanide, and thiocyanate. The hydrazine complex, $Cr(N_2H_4)_2Cl_2$ [56890-91-6], is quite insoluble and stable toward oxidation.

Chromium(III) Compounds. In the important oxidation state of +3, chromium's extensive chemistry is dominated by the formation of stable, kinetically inert d^2sp^3 complexes in which chromium is octahedrally six-coordinate. The oxidation potentials (see Table 4) also serve to emphasize the thermodynamic stability of the +3 state. The compounds are referred to as a "chromic compound" in most technical applications.

Chromic trioxide [1308-38-9], chromium sesquioxide, chromium(III) oxide, Cr_2O_3, is an important green pigment (see under Chromium oxide greens). It has the rhombohedral structure of α-Al_2O_3 and α-Fe_2O_3. However, when Cr_2O_3 is introduced in small quantities into the α-Al_2O_3 lattice, the color is red rather than green; larger amounts produce a green color. The color difference relates to the energies of the ligand field splitting (30). Other minerals are also colored by Cr(III) (31).

Table 4. Standard Electrode Potentials of Chromium(II), Chromium(III), and Chromium(IV)

Reaction	E^0, V
$Cr^{2+} + 2\,e^- \rightarrow Cr$	-0.91
$Cr^{3+} + 3\,e^- \rightarrow Cr$	-0.74
$Cr^{3+} + e^- \rightarrow Cr^{2+}$	-0.41
$CrO_4^{2-} + 4\,H_2O + 3\,e^- \rightarrow Cr(OH)_3 + 5\,OH^-$	-0.13
$Cr(OH)_3 + 3\,e^- \rightarrow Cr + 3\,OH^-$	-1.3
$[Cr(OH)_4(H_2O)_2]^- + 3\,e^- \rightarrow Cr + 4\,OH^- + 2\,H_2O$	-1.2
$Cr_2O_7^{2-} + 14\,H^+ + 5\,H_2O + 6\,e^- \rightarrow 2\,Cr(H_2O)_6^{3+}$	1.33

The precipitate formed by the addition of a base to a Cr(III) salt solution is a hydrous oxide of indefinite composition, $Cr_2O_3 \cdot x\,H_2O$, although it is frequently referred to as chromic hydroxide, $Cr(OH)_3$ [1308-14-1]. On addition of more base to freshly precipitated hydrous oxide, the precipitate redissolves. It is not clear whether this action is due to peptization or the formation of ions of the type $[Cr(OH)_6]^{3-}$. Some support is lent to the latter hypothesis by the existence of crystalline salts such as $Ba_3[Cr(OH)_6]_2$ [12009-86-8]. Such salts are the only true chromites. Compounds of the type $M_2^{1+}Cr_2O_4$ and $M^{2+}Cr_2O_4$ are mixed oxides, with the chromium occupying the octahedral holes in a closed-packed lattice of oxide ions. The M^{2+} oxides are largely spinels and the ore, chromite, is an important example of the class.

Chromic trichloride, chromium(III) chloride, $CrCl_3$, is the most important of the anhydrous halides of Cr(III). Its deep-purple plates reflect an unusual crystal lattice, consisting of a close-packed cubic array of chlorine atoms, in which two-thirds of the octahedral holes between every other plane of chlorine atoms are occupied by chromium (32). The chromic tribromide [10031-25-1] lattice is similar. Another curious property of these layer-structure chromium halides is their very slow hydration and solution in water, although they dissolve readily and exothermically in the presence of Cr(II). The finely divided halides are very hygroscopic.

The solution chemistry of Cr(III) is coordination chemistry. Equilibria between various ion species are reached slowly and thousands of Cr^{+3} complex ions are known. The kinetics and mechanisms whereby one ligand substitutes for another have been a major field of inorganic research for the past 25 years and the literature is most extensive (18–20).

The simple hydrated ion $[Cr(H_2O)_6]^{3+}$ exists at room temperature in solutions of the violet chloride and sulfate and the chromium alums. It is violet and shows strong dichroism. On warming such solutions the color changes to green, the result of hydrolysis, polymerization, and displacement of coordinated water by other ions. Stable complexes, which resist precipitation by hydroxyl ion, are formed by many organic anions such as oxalate and those of hydroxy acids. Trivalent chromium is known to coordinate with virtually all electron-pair donors, with chelating agents and strong Lewis bases forming particularly stable complexes.

With ligands such as 2,4-diketones, which combine the functions of a monoprotic acid and bidentate ligand, neutral complexes are formed. Chromic tri(acetylacetonate) is a nonelectrolyte, nearly insoluble in water, and soluble in organic solvents. It may be melted and distilled without appreciable decomposition.

Complexes of $CrCl_3$ with Lewis-base organic solvents can be very conveniently prepared by refluxing $CrCl_3$ in the presence of a small amount of zinc in a Soxhlet extractor. The acetone, THF, and pyridine complexes can be prepared in this way (33).

When hydroxyl ions are added to Cr(III) solutions in quantities insufficient for precipitation, basic ions are formed, which then polymerize and condense, with formation of *ol* and *oxo* bridges. Such ions are of the proper size to cross-link protein fibers and play an important part in the chemistry of tanning. Some typical reactions involving OH^-, a weakly coordinated ion such as SO_4^{2-}, and a strongly coordinated ion such as $C_2O_4^{2-}$ are shown in Figure 1 (16).

Chromium(III) compounds are reduced to chromium(II) compounds by hypophosphites, electrolysis, or active metals such as zinc, magnesium, and aluminum in acid solution. In basic solution Cr(III) is readily oxidized to CrO_4^{2-} by hypochlorite,

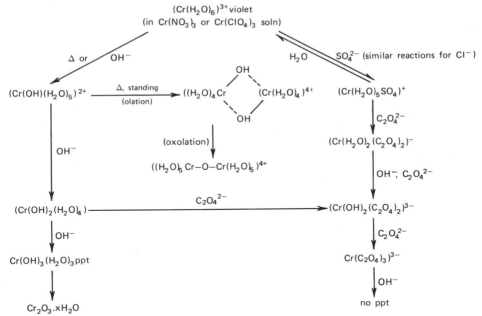

Figure 1. Some typical ion relationships of chromium (III).

hypobromite, peroxide, and oxygen under pressure at high temperatures. Heating of chromium compounds or minerals in air in the presence of alkalies also yields chromate. In acid solution, as can be seen from the oxidation potentials in Table 4, Cr(III) is harder to oxidize. Chlorate, concentrated perchloric acid, lead dioxide, sodium bismuthate, permanganate, peroxydisulfate–silver ion, and anodic oxidation typify the oxidants required.

The action of peroxides is unusual. In basic solution, the final product is CrO_4^{2-}. However, a transitory brownish-red color is often observed, and if the reaction is carried out at low temperatures, red unstable salts of the formula M_3CrO_8 are obtained, in which the CrO_8^{3-} ion, a Cr(V) species, is present. In acid solution an unstable blue compound extractable with ether, of the formula CrO_5, and blue salts in which the $Cr_2O_{12}^{2-}$ ion is present are formed. They are probably Cr(VI) compounds; they decompose rapidly into Cr(III) and oxygen.

Chromium(IV) Compounds. The importance of the +4 and +5 oxidation states, like that of the 0 and +1 states, has increased in recent years and there are now commercial applications.

Chromium(IV) oxide, chromium dioxide, CrO_2, is obtained by hydrothermal reaction of Cr(VI) or Cr(III) compounds in the presence of oxygen. According to a U.S. patent (34), an oxide having an average oxidation state of 4 to 6 is heated in the presence of water at 300–350°C. The stability of chromium dioxide has been studied (35). It has a tetragonal lattice similar to rutile, TiO_2.

Chromium tetrafluoride, [*10049-11-3*], CrF_4, a volatile halide, is formed by the action of fluorine on CrF_3. Both $CrCl_4$ and chromium tetrabromide [*51159-56-9*], $CrBr_4$ exist at elevated temperatures in the systems $CrCl_3$–Cl_2 and $CrBr_3$–Br_2, but have not been isolated.

Although Cr(IV) does not form anions in aqueous solution, compounds of the alkaline earths, such as dibarium chromate (1:1) Ba_2CrO_4, containing discrete CrO_4^{4-} groups, have been prepared.

Chromium(IV) also forms organic compounds. When $Cr(C_6H_6)_2$ is treated with di-t-butyl peroxide, a deep blue sublimable compound, tetrabutoxychromium [53221-57-1], $Cr(OC_4H_9)_4$, is obtained in which the oxidation state of +4 has been confirmed by magnetic measurements. Chromium(IV) compounds function as intermediates in the oxidation of organic compounds by CrO_3 and can be isolated in the Etard reaction, that is oxidation by chromyl compounds (36).

The peroxy compound, $CrO_4 \cdot 3NH_3$ [7168-85-3], is obtained by warming an ammoniacal ammonium perchromate solution followed by cooling. Structural data show that three ammonia molecules and two peroxide ions are coordinated to Cr(IV).

Chromium(V) Compounds. The chemistry of chromium in oxidation state +5 is quite similar to that of the +4 state. There are more ionic chromium V compounds, but disproportionation in aqueous systems prevents their isolation from water solution.

Chromium pentoxide, chromium(V) oxide [12218-36-9] (2:5), Cr_2O_5, like CrO_2 can be prepared by the decomposition of CrO_3. It is frequently deficient in oxygen, accounting for such early formulas as Cr_5O_{12}; a definite compound of Cr(III) and Cr(VI) of this composition exists (37). Of the halides, only chromium pentafluoride [14884-42-5], CrF_5, exists. It was formerly considered the highest fluoride of chromium. Oxyhalide salts like $KCrOF_4$ [68258-60-6] are known but are readily hydrolyzed. The preferred 4-coordination of Cr(V) lends stability to $CrOF_3$ [43997-25-7] and $CrOCl_3$.

A nearly complete series of salts containing the CrO_4^{3-} ion has been prepared. They resemble the corresponding phosphates structurally and are generally green in color. The alkali metal salts M_3CrO_4 are hygroscopic and unstable in moist air (38). The barium, strontium, and calcium chromates(V) are also known, and the calcium chromite-chromate [12139-05-8], $9CaO \cdot 4CrO_3 \cdot Cr_2O_3$ reported from the roasting of chromite with lime and soda ash is more properly formulated as $Ca_3(CrO_4)_2$ [12205-18-4]. This compound is readily prepared by heating CaO and $CaCrO_4$ to about 1150°C. Hydroxy compounds, such as $Ba_5(CrO_4)_3(OH)$ [12009-86-8] analogous to apatite exist, and quite probably fluoro compounds (39). Furthermore, rare-earth chromates of the formula $(RE)CrO_4$ are known, where (RE) may be La, Pr, Nd, Y, and the series Sm through Lu (40). Chromium(V) compounds also act as intermediates in organic oxidations (36).

In this state chromium has its greatest industrial applications as a consequence of its acidic and oxidant properties, and its ability to form strongly colored and slightly soluble salts.

Chromium(VI) bears a strong structural resemblance to hexavalent sulfur. The chromates, the oxide CrO_3, and the chromyl compounds CrO_2X_2, all resemble the corresponding sulfur compounds in stoichiometry, crystal structure, and solubility.

The principal differences are to be found in the oxidizing nature of chromium(VI), the weaker acidity of H_2CrO_4, the nonexistence of solid dichromates containing the $HCrO_4^-$ ion, the existence of polychromates—particularly dichromates—and much less energetic hydration of the oxide CrO_3, and the softer nature, eg, more polarizable of the CrO_4^{2-} ion.

The dissociation equilibria in chromic acid solutions are given as (20):

$$\text{H}_2\text{CrO}_4 \rightleftharpoons \text{H}^+ + \text{HCrO}_4^- \qquad\qquad -0.613 \;(\text{p}K_a)$$
$$\text{HCrO}_4^- \rightleftharpoons \text{H}^+ + \text{CrO}_4^{2-} \qquad\qquad 5.9$$
$$\text{Cr}_2\text{O}_7^{2-} + \text{H}_2\text{O} \rightleftharpoons 2\,\text{HCrO}_4^- \qquad\qquad 2.2$$

The dissociation $\text{H}_2\text{Cr}_2\text{O}_7 \rightleftharpoons \text{H}^+ + \text{HCr}_2\text{O}_7^-$ appears to be that of a strong acid. In the presence of chloride and sulfate ions at low pH, other equilibria exist:

$$\text{HCrO}_4^- + \text{H}^+ + \text{Cl}^- \rightleftharpoons \text{CrO}_3\text{Cl}^- + \text{H}_2\text{O}$$
$$\text{HCrO}_4^- + \text{HSO}_4^- \rightleftharpoons \text{CrO}_3(\text{SO}_4)^{2-} + \text{H}_2\text{O}$$

In solution, at pH 6, the tetrahedral yellow CrO_4^{2-} ion is the principal species; at pH 2–6, the orange HCrO_4^- and $\text{Cr}_2\text{O}_7^{2-}$ ions predominate at moderate concentrations. At very low pH, the principal species is $\text{HCr}_2\text{O}_7^{2-}$. In very concentrated solutions, $\text{Cr}_3\text{O}_{10}^{2-}$ and $\text{Cr}_4\text{O}_{13}^{2-}$ ions exist; salts of larger cations such as $(\text{NH}_4)_2\text{Cr}_3\text{O}_{10}$ [34390-97-1] and $\text{K}_2\text{Cr}_4\text{O}_{13}$ [12422-53-6] are known.

Chromium trioxide, chromic anhydride, chromium(VI) oxide, CrO_3, is the chromic acid of commerce. It is a red solid, melting at 197°C with slow decomposition. The decomposition becomes rapid and exothermic at about 220°C, forming largely impure Cr_2O_5 [12218-36-9]. Chromium trioxide is deliquescent and very soluble in water.

In solution in organic solvents, such as glacial acetic acid or pyridine, mixtures of compounds are formed that are useful and powerful oxidants in organic syntheses; however, they must be handled with care, as they readily decompose explosively.

Of the halides, only chromium hexafluoride [13843-28-2], CrF_6, is known, a yellow compound made from Cr and F_2 at 20–35 MPa (200–350 atm) and 200°C. At atmospheric pressure, it is stable only below −100°C (41). Oxyhalides include CrOF_4 [23276-90-6], and the important chromyl compounds, CrO_2X_2, of which the chloride is the best-known. Other chromyl compounds include the fluoride, nitrate, acetate, and perchlorate.

Halochromates, such as KCrO_3Cl [16037-50-6] and $\text{NH}_4\text{CrO}_3\text{F}$ [58501-09-0], are prepared from acid solutions of chromates containing halide ion. The alkali salts are orange, resembling dichromates. The corresponding bromo and iodo compounds are unstable, but can be isolated because of the slow rate of oxidation of halide ions by chromium(VI) at low temperatures.

The chromates, containing the tetrahedral CrO_4^{2-} ion in the crystal lattice and in solution, are an important class of salts. Because d orbitals enter into the bonding, H_2CrO_4 [7738-94-5] is a softer acid than H_2SO_4. Chromates of hard bases resemble the sulfates closely; those of soft bases are less soluble and deeper colored than would be expected, thus Ag_2CrO_4 [7784-01-2] is maroon in color. Lead chromate [15804-54-3] is of particular interest, being trimorphic. The stable modification is orange-yellow and monoclinic; an unstable orthorhombic form is yellow, isomorphous with lead sulfate, PbSO_4, and stabilized by it. An orange-red tetragonal form is similar, isomorphous with lead molybdate(VI), PbMoO_4, and stabilized by it. On these properties depends the versatility of PbCrO_4 as a pigment.

Chromates are converted to dichromates by acids. The dichromates are generally soluble, the solubility tending to increase with decreasing size and increasing charge of the cation. With the exception of the heavier alkali metal dichromates, the salts are hydrated. Hard bases, forming yellow chromates, form orange-red dichromates. The dichromates of soft bases resemble the corresponding chromates in color.

Many chromate esters and salts of organic bases are known; the esters, particularly of primary alcohols, are generally very unstable. All organic chromates are photosensitive and decompose on exposure to light.

Oxidation by chromium(VI) compounds usually takes place in acid solution:

$$Cr_2O_7^{2-} + 14\,H^+ + 5\,H_2O + 6\,e^- \rightleftharpoons 2\,Cr(H_2O)_6^{3+}$$

Many important organic intermediates were formerly made by chromic acid oxidation. For example, benzoic acid from toluene, anthraquinone from anthracene, phthalic anhydride from naphthalene, terephthalic acid from p-xylene, and saccharin from o-toluenesulfonamide. Today, catalytic oxidation is used, but chromic acid oxidation and the Etard reaction using chromyl chloride remain important research tools.

Another important property of the chromates is their ability to passivate metal surfaces and inhibit corrosion (see under Uses; see Corrosion and corrosion inhibitors).

The thermodynamic properties of the principal chromium compounds have recently been critically examined (42).

Manufacture

The two primary industrial compounds of chromium made directly from chrome ore are sodium chromate and sodium dichromate. Secondary chromium compounds produced in substantial quantity include potassium chromate and dichromate, ammonium dichromate, chromic acid (chromium(VI) oxide), and various formulations of basic chromic sulfate used principally for leather tanning. These are made by the manufacturers listed under Economic Aspects. For information on the manufacturing technology of other commercial compounds see under Uses.

Sodium Chromate and Dichromate. Sodium chromate and dichromate are made by treatment of the calcine obtained by roasting chrome ore with soda ash to which lime and/or leached calcine may be added. In present practice, chemical grade ore from the Transvaal is currently used. Ores from Montana, Oregon, and Alaska are of poorer grade, limited supply, and processing and transportation costs are higher. For procedures and equipment formerly used, see references 43–45. Modern processes have to meet today's safety and environmental standards.

The chrome ore is crushed if necessary, dried, and ground in ball mills to ca 74 μm (200 mesh). It is then mixed with soda ash and, optionally, with lime and leached residue from a previous roasting operation, if needed. In American and European practice, a variety of kiln mixes have been used containing up to 57 parts of lime per 100 parts of ore; the roasting may be carried out in one, two, or three stages; and there may be as much as three parts of leached residue per part of ore. These adaptations are responses to the variations in kiln roast and the capabilities of the furnaces used. A typical mix for the first stage of a double roast using lime is: ore, 24 parts; soda ash, 15 parts; lime, 12 parts; residue, 49 parts.

After thorough mixing, the mixture is roasted in a mechanical furnace, usually a rotary kiln. An oxidizing atmosphere is essential, and the basic reaction for a theoretical chromite is:

$$4\,FeCr_2O_4 + 8\,Na_2CO_3 + 7\,O_2 \rightarrow 2\,Fe_2O_3 + 8\,Na_2CrO_4 + 8\,CO_2$$

The temperature in the hottest part of the kiln is closely controlled using automatic equipment and a radiation pyrometer, and generally kept at about 1100–1150°C. Time of passage is about four hours, varying with the kiln mix being used. The rate of oxidation increases with temperature. However, the maximum temperature is limited by the tendency of the calcine to become sticky and form rings or balls in the kiln, and by factors such as loss of Na_2O by volatilization, and increased rate of attack on the refractory lining.

A gas-fired furnace with a revolving annular hearth also has been used to roast chrome ore (44). The mix is charged continuously at the outer edge of the hearth; a water-cooled helical screw moves it toward the inner edge where it is discharged. Mixes containing a much higher soda ash content (28% Na_2CO_3) can be handled in these furnaces.

Lime has two functions. First, it increases the roasting rate appreciably. Second, it converts the alumina and silica in the ore to insoluble calcium aluminates and silicates. The leach solutions then require no additional processing.

The roast from the kilns is discharged through a cooler and then goes to the leaching operation. If lime is used, the hydraulic calcium compounds, plus the calcium chromates in various oxidation states present, delay the leaching process markedly. Continuous countercurrent leaching procedures have been reported (43) but the usual procedure has been to use a battery of large false-bottomed leaching boxes called filters. Liquid is pumped onto the charge, percolates downward, and is drawn off as a strong solution of sodium chromate containing some excess alkali. If no lime is present, or only very small amounts, the leaching is more rapid, but the leach solutions contain sodium aluminate as a major impurity in addition to sodium silicate and vanadate.

The leached calcine combined from all intermediate stages is dried, ground to ca 150 μm (100 mesh), and recycled as diluent to the kiln mixes.

The process to this point is essentially a metallurgical operation. For environmental control, all stacks and vents must be protected. Electrostatic precipitators are desirable on kiln and residue dryer stacks. Leaching operations should be hooded and stacks equipped with scrubbers. Recovered chromate values can be returned to the leaching-water cycle.

Technical developments in the roasting and leaching area, as recorded in the patent literature, include refinements in pelletizing the mix fed to the kilns (46–48) and in the preoxidation of the ore prior to roasting (49). Both of these variants aim to increase the kiln capacity, the first through increasing the permissible fraction of soda ash in the mix, the second through increasing the effective rate of oxidation.

If the sodium chromate solution contains appreciable amounts of sodium aluminate, the alumina is precipitated at this point and may be recovered. The solution is run through polishing filters and into a battery of batch hydrolyzing tanks. Sodium dichromate solution from subsequent processing is slowly added, maintaining the pH at about 9. The precipitate is largely gibbsite, $Al_2O_3.3H_2O$. A variant, based on raising the pH to 10 and then lowering it to 6.5, is claimed to give very readily filterable Al_2O_3 (50).

The neutralized, alumina-free sodium chromate solution may be marketed as solution of sp gr 1.38 (40° Bé) or evaporated to dryness or crystallized to give a technical grade of sodium chromate or sodium chromate tetrahydrate [*10034-82-9*]. If the fuel for the kilns contains sulfur, the product contains sodium sulfate as an impurity. This is isomorphous with sodium chromate and hence difficult to separate. High-purity sodium chromate must be made from purified sodium dichromate.

Sodium chromate is usually converted to the dichromate by a continuous process of treatment with sulfuric acid, evaporation of sodium dichromate and precipitation of sodium sulfate, and finally crystallization of sodium dichromate. The recovered sodium sulfate may be used for other purposes. The dichromate mother liquor may be returned to the evaporators or marketed as 69% sodium dichromate solution.

A simplified flow sheet of the sodium dichromate operation is shown in Figure 2. Carbon dioxide under pressure can be substituted for part of the sulfuric acid in which case sodium bicarbonate is precipitated. This can be calcined to soda ash for use in the kilns (43–45).

Other Chromates and Dichromates.   Potassium and ammonium dichromates are generally made from sodium dichromate by a crystallization process involving equivalent amounts of potassium chloride or ammonium sulfate. In each case the solubility relationships are favorable so that the desired dichromate can be separated on cooling, whereas the sodium chloride or sulfate crystallizes out on boiling. For certain uses, ammonium dichromate low in alkali salts is required; this may be prepared by the reaction of ammonia with chromic acid. Care must be taken when drying ammonium dichromate, because decomposition starts at 185°C and becomes violent and self-sustaining at slightly higher temperatures.

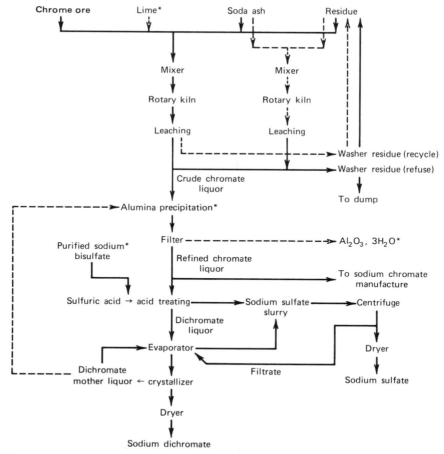

Figure 2.   Flow sheet for the production of sodium chromate and dichromate. The asterisks and dashed lines indicate optional steps.

Potassium chromate is prepared by the reaction of potassium dichromate and potassium hydroxide. Sulfates are the most difficult impurity to remove, because of the isomorphism of potassium sulfate and potassium chromate.

The wet operations employed in the modern manufacture of the chromates and dichromates are completely enclosed and all stacks and vents are equipped with scrubbers and entrainment traps to prevent contamination of the plant and its environment. The continuous equipment used greatly facilitates this task. The material trapped may be cycled.

Chromic Acid. Chromic acid is produced by the reaction of sulfuric acid with sodium dichromate:

$$Na_2Cr_2O_7 + 2\,H_2SO_4 \rightarrow 2\,CrO_3 + 2\,NaHSO_4 + H_2O$$

Traditionally, sodium dichromate dihydrate is mixed with 66° Bé (sp gr 1.84) sulfuric acid in a heavy-walled steel or cast-iron reactor heated externally and provided with a sweep agitator. Water is driven off and the hydrous bisulfate melts at about 160°C. As the temperature is gradually increased, the molten bisulfate provides an excellent heat transfer medium for melting the chromic acid at 197°C without appreciable decomposition. As soon as the chromic acid melts, the agitator is stopped and the mixture separates into a heavy layer of molten chromic acid and a light layer of molten bisulfate. The chromic acid is tapped and flaked on water-cooled rolls to produce the customary commercial form. The bisulfate contains dissolved CrO_3 and soluble and insoluble chromic sulfates. It was formerly discarded, but environmental considerations now dictate purification and return of the bisulfate to the treating operation.

Molten chromic acid decomposes at its melting point at a significant rate; the lower oxides formed impart darkness and turbidity to the water solution. Accordingly, both temperature and time are important in obtaining a quality product.

Instead of the dihydrate and sulfuric acid, 20% oleum and anhydrous sodium dichromate may be used. In this case, the reaction requires little if any external heat, and liquid chromic acid is spontaneously produced. This procedure is the basis for a continuous process (51).

Another process depends on the addition of a large excess of sulfuric acid to a concentrated solution or slurry of sodium dichromate. A crude chromic acid, containing bisulfate and some sulfuric acid is precipitated and separated. The solution contains little chromium(III) and may be used for converting chromate to dichromate. The crude chromic acid is then melted in a small amount of bisulfate as before, with bichromate added to convert any excess sulfuric acid (52,53). The small amount of bisulfate may be treated for chromium values in the plant's recovery system.

These newer processes (51–53) lend themselves well to effective environmental controls.

Chromic Sulfate, Basic. Basic chromic sulfate is manufactured as a proprietary product under various trade names for use in leather tanning. It is generally made by reduction of sodium dichromate in the presence of sulfuric acid, and contains sodium sulfate, small amounts of organic acids if carbohydrate reducing agents are used, plus various additives. The compounds are sold on a specification of chromic oxide content and basicity, the latter value referring to the degree to which the chromium is converted to the theoretical hydroxide. Thus, $Cr(OH)SO_4$ [12336-95-7] is 33.3% basic. The Cr_2O_3 content of solid chrome tanning compounds ranges from 20.5 to 25%, and basicity ranges from 33 to 58%. Solutions are also available. For specifications, see manufacturer's catalogues (13–14). The two reducing agents commonly used are sugar and sulfur dioxide, although other organic materials have been used during sugar shortages.

Sugar reduction. To sodium dichromate solution in an agitated acid resistant tank is added the sugar required for reduction and the stoichiometric amount of sulfuric acid. The reaction is highly exothermic and must be carried out with care to prevent foaming and boiling over. The reaction is completed by boiling.

Theoretically, glucose is oxidized to CO_2 and water:

$$4\ Na_2Cr_2O_7 + 12\ H_2SO_4 + C_6H_{12}O_6 + 26\ H_2O \rightarrow 8\ [Cr(OH)(H_2O)_5]SO_4 + 4\ Na_2SO_4 + 6\ CO_2$$

However, not all the glucose is oxidized, nor does the formula given represent the exact state of the chromium complex because hydrolysis, polymerization, and entrance of sulfate into the complex occur.

Sulfur dioxide reduction. With sulfur dioxide, a 33.3% basic chromic salt is automatically obtained:

$$Na_2Cr_2O_7 + 3\ SO_2 + 11\ H_2O \rightarrow 2\ [Cr(OH)(H_2O)_5]SO_4 + Na_2SO_4$$

Pure sulfur dioxide is bubbled through the sodium dichromate solution in an acid-resistant tank, whereas sulfur burner gas is passed through a ceramic-packed tower countercurrent to descending dichromate solution. After reduction is complete, steam is bubbled through the solution to decompose any dithionate that may have formed, and to remove excess sulfur dioxide.

After reduction any desired additives, such as aluminum sulfate, are incorporated, the solution is aged. It is then spray dried. Careful temperature control during drying is necessary to obtain a highly water-soluble solid product.

Economic Aspects

The total chrome ore consumption in the United States in 1975, including chemical, metallurgical (the principal use), and refractory grades, was 881,000 metric tons, as compared to 1,447,000 t in 1974 (54). The largest supplier was the Transvaal (Union of South Africa) with 150,600 t (as compared to 228,000 t in 1974) of an average content of 44.9% Cr_2O_3. Estimated world production is about 220,000 t sodium dichromate ($Na_2Cr_2O_7 \cdot 2H_2O$) or equivalent per year (excluding the U.S.S.R. and the Eastern European states). Prices of the most important chromium chemicals are given in Table 5, production and shipment data in Table 6.

In July 1976, U.S. capacity for sodium dichromate production was estimated at 155,600 t/yr, distributed as follows:

Allied Chemical Corp., Baltimore, Md.	59,000 t/yr
Diamond Shamrock, Castle Hayne, N. C.	69,400 t/yr
PPG Industries, Corpus Christi, Texas	27,200 t/yr

However, expansion in progress at Castle Hayne should increase Diamond Shamrock's capacity to 82,600 t/yr by the end of 1979 (55).

Table 5. Prices of Important Chromium Chemicals[a], September 1978

Compound	CAS Registry No.	$/kg
sodium dichromate, dihydrate	[7789-12-0]	0.816
sodium chromate, anhydrous	[7775-11-3]	0.865
potassium dichromate	[7789-50-9]	1.058
potassium chromate	[7789-00-6]	1.257
ammonium dichromate	[7789-09-5]	1.720
chromic acid	[1333-82-0]	1.720

[a] For prices of pigments, see under Uses. From *Chemical Marketing Reporter*.

Table 6. Production and Shipments of Chromium Chemicals[a]

Chemical	Production, metric tons	Shipments Metric tons	Shipments $1000
Sodium dichromate and chromate			
1977	143,649	na	na
1976	142,776	95,281	51,124
1973	143,936	95,316	28,639
1967	122,737	85,251	23,080
Chromic acid[b]			
1977	26,000(E)		
1976	25,900(E)		
1973	26,100(E)		
1970	19,960	19,743	11,919

[a] Ref. 56.

[b] Production and shipment data for chromic acid have not been published since 1972; therefore, the figures for 1973 and 1976 were estimated by assuming that 27% of dichromate production was used for chromic acid. E = estimated.

Specifications and Packaging

Chromates and dichromates are sold in both technical and reagent (57–58) grades. Chlorides and sulfates are the principal impurities. Both manufacturers' and U.S. GSA specifications exist for the technical grades (59–61,62). Provisions of these specifications are shown in Table 7.

Packaging is generally in multiwall paper bags, fiber drums, or steel drums. Chromic acid is packed only in steel drums. Both chromic acid and ammonium dichromate require ICC yellow labels for interstate shipment.

Sodium dichromate, sodium chromate, and mixtures thereof are shipped as concentrated solutions in tank cars and trucks. The chloride and sulfate contents are usually somewhat higher than in the crystalline product. Sodium dichromate is customarily shipped at a concentration of 69% $Na_2Cr_2O_7 \cdot 2H_2O$, which is close to the eutectic composition freezing at $-48.2°C$.

Analytical and Test Methods

The industrial analysis (14,62–66) of chromium chemicals and chromium-containing materials may be divided into several categories: analysis of (*a*) ores, refrac-

Table 7. U.S. Government Specifications for Technical Chromium Chemicals[a]

Specification	Na$_2$Cr$_2$O$_7$·2H$_2$O	CrO$_3$	Na$_2$CrO$_4$	K$_2$Cr$_2$O$_7$	(NH$_4$)$_2$Cr$_2$O$_7$
number	0-S-595	0-C-303	0-S-588	0-P-559	0-A-498
assay, % min	99.0[b]	99.5	98.5	99.0	99.7
chloride, % max	0.1	0.10	0.1	0.1	0.005
sulfate, % max	0.2	0.20	1.0	0.1	0.06
water insol, % max	0.2	0.10		0.1	0.02
particle size, U.S. sieve		10% min passes through −30 mesh[c]		passes completely through 10 mesh[c]; less than 25% passes through −100 mesh	
pH at 25°C, min	12.5				3.2[d]
moisture and volatiles at 120°C, max		0.2	0.5		
other[e]					13.25% min NH$_3$[f]

[a] Ref. 59.

[b] Assay as % Na$_2$Cr$_2$O$_7$ after drying at 120°C.

[c] −30 mesh is < 590 μm; 10 mesh is ~ 2000 μm; −100 mesh is < 149 μm.

[d] 20% w/v.

[e] A limit on vanadium is set by some commercial users.

[f] Some users specify a maximum fixed alkali content.

tories, and other insoluble materials for major amounts of chromium; (b) commercial chromium compounds for chromium content (assay); (c) materials involved in processes using chromium compounds; (d) traces of chromium in effluents, air samples, biological materials, etc, usually in regard to health and environmental considerations; and (e) important impurities in chromium compounds.

Ores and similar insoluble materials are analyzed by fusing a pulverized sample with sodium peroxide in an ingot iron or other suitable crucible. The sample is then leached, acidified with sulfuric acid, and reoxidized with a small amount of permanganate. The excess permanganate is destroyed with hydrochloric acid, and Cr(VI) titrated with standardized ferrous sulfate, recording the end point either potentiometrically or with a redox indicator such as o-phenanthroline ferrous complex.

X-ray spectroscopy is available as a rapid control method when the range of samples being tested is small and similar standards are run concurrently.

Chromium(VI) compounds are usually assayed by direct titration with ferrous sulfate in sulfuric acid solution, as described above. Chromium(III) compounds, or compounds in other oxidation states, are first oxidized with sodium peroxide in alkaline solution. The solution is boiled to destroy the peroxide, acidified, and titrated. If organic matter is present, various acid oxidants are usually employed.

As many chromium compounds are deliquescent, proper sampling and handling procedures are essential.

Solutions containing chromium compounds can be usually analyzed by the same methods as chromium compounds. It is occasionally necessary to distinguish between oxidation states of chromium, usually (III) and (VI). In this case, one sample is titrated directly for Cr(VI) and another is oxidized as described above. Insoluble organic materials such as wood and leather processed with chromium are dry-ashed and fused as described for ores or wet-ashed with oxidizing acids. Pigments are fused or dissolved in acid and titrated. Instrumental methods such as spectrophotometry, x-ray spectroscopy, and atomic absorption spectroscopy are being used more and more. Procedures of this type are described in the standard methods of such organizations as ASTM, the American Leather Chemists' Association, and the American Wood Preservers' Association.

Trace amounts of chromium are determined by converting the chromium quantitatively to Cr(VI), adjusting the pH, and adding s-diphenylcarbazide solution. An intense red-violet color, sensitive to 0.003 ppm Cr, appears. The absorbance is measured at 540 nm and compared to known standards. The application of this method to wastewaters, air samples, and biological materials (67) are numerous.

The quantitative oxidation of chromium to Cr(VI) is difficult to achieve at very low concentrations, therefore, atomic absorption spectroscopy with improved vaporization methods is becoming the method of choice for complex samples (67–69). Whereas flame vaporization is sensitive to 0.1 ppm using the 359.3 nm line, atomic fluorescence is sensitive to 0.05 ppm. With the Lvov furnace and Mini-Mossman atomizer 0.001 ppm, or an absolute amount of 1.8 to 5 g Cr can be detected.

Impurities in industrial chromium compounds (14,65) include chlorides, sulfates, insoluble matter, vanadium (in $Na_2Cr_2O_7.2H_2O$), and fixed alkalies (in $(NH_4)_2Cr_2O_7$). In general, the Mohr titration is used for chloride, barium precipitation for sulfate, a variety of methods for vanadium, and the flame photometer for alkalies. A wider variety of tests are required for reagent chemicals (57–58).

Health, Safety, and Environmental Considerations

Several aspects must be considered in regard to the effects of chromium compounds: acute and chronic toxicity; chromium compounds in nutrition; their natural occurrence in rural and urban environment; and technology of control. An excellent recent survey of the medical and biological effects of chromium and its compounds is reference 69a. References 69b and 69c are also useful.

Acute and Chronic Toxic Effects.   From the practical standpoint, the only chromium compounds encountered are those in oxidation states +3 and +6. Compounds in the +3 state have no established toxicity. Thus the chief health problems associated with chromium compounds are related to chromium(VI) compounds.

Acute systemic poisoning is rare. Fatalities in humans have resulted from the accidental use of potassium chromate in an ointment. Kidney damage has resulted from nonfatal ingestion. Continued ingestion of small quantities, as in drinking water, of up to 25 ppm Cr(VI) appears to be without toxic effect (70). Federal drinking water standards are at 0.05 ppm, but a recent report suggests that higher concentrations may be without adverse effect (71). No well-defined fatal dose has been established; however, it seems to be on the order of 3–5 g for adult humans.

Acute effects of chromates are mainly on the skin and mucous membranes, and have been observed in workers for 150 years. The breathing of dusts or mists containing chromium(VI) compounds leads to ulceration and eventual perforation of the cartilaginous portion of the nasal septum. As a result of a 1928 study of exposures and nasal perforations of workers in chromium plating plants (72), exposure standards have been set at 0.1 mg CrO_3/m^3, or 0.05 mg $Cr(VI)/m^3$ (73–75). Standards for soluble chromic and chromous salts are set at 0.5 mg Cr/m^3, and for chromium metal and insoluble chromic compounds at 1.0 mg Cr/m^3 (74–75).

In addition to these effects, chromates may produce ulcers when a cut or abrasion in the skin is contaminated. If sufficient chromate is brought into contact with skin lesions, fatalities may result.

Dermatitis and allergic reactions may result from prolonged exposure of the skin to chromate. In the majority of workers dermatitis results only from prolonged exposure, or under conditions such as exposure to solvents where natural skin oils are removed. However, a few individuals exhibit a characteristic allergic reaction on exposure to very small amounts of chromates.

Lung cancer as a result of long-term exposure to chromate dusts was observed in German chromate plants as early as 1912. Various studies demonstrate beyond doubt that a significant incidence of lung cancer cases results from exposure to dusts in manufacturing plants (76–80). Despite the wide use of chromium compounds, lack of reported lung cancer cases in the consuming industries suggests that the hazard is confined largely to the manufacturing industries.

Animal experiments have implicated only calcium, zinc, lead, and chromic chromates as being potentially carcinogenic. Calcium and crhomic chromates (the latter possibly a Cr(IV) or Cr(V) oxide) are encountered in manufacturing plants, as is also $Ca_3(CrO_4)_2$. Zinc and lead chromates are commercial pigments and cases of lung cancer have been reported due to exposure in plants manufacturing these products.

The rate of incidence appears to be decreasing in recent years (81–83), principally because of reduced working hours and improved conditions.

Chromium in Nutrition. Although chromium is widely distributed in plant and animal tissues, it has only been recently recognized as an essential element in both plant and animal nutrition (see Mineral nutrients).

The case for plant nutrition is only poorly documented. Workers at Southwest Research Institute found that sugar-producing crops had a high Cr_2O_3 content and that the Cr_2O_3 taken from the soil on harvesting needed to be returned, for which very finely ground chromite ore was used (84). Other workers reported increased yields of cucurbits from increased feeding of chromium. Circumstantial evidence is strong here, for Hawaii and Cuba, both producers of sugar-rich crops, have basaltic and lateritic soils that are high in chromium.

Certain evidence for chromium as an essential element in animal and human nutrition was reported in 1969 (85). Further work has established improvement in glucose tolerance, especially in the elderly, with supplemental chromium intake of the order of 150 $\mu g/d$. A definite recommended intake has not been established. The U.S. daily average of 80 $\mu g/d$ is probably marginal, as chromium is poorly assimilated from many foods.

Chromium may play a role in sugar production in plants and sugar metabolism in animals.

Chromium Compounds in the Environment. Chromium is widely distributed in the earth's crust, but is concentrated in the basic, ultramafic rocks. At an overall crustal concentration of 125 ppm Cr, it is the twentieth most abundant element. Chromium is very insoluble and leaches very little into natural waters. A survey of chromium in 15 North American rivers showed 0.7–84 ppb, with most samples in the 1–10 ppb range. Public water supplies range from no detectable to 36 ppb Cr, the median being 0.43 ppb. Nearly all foodstuffs contain chromium in the range of 20 to 590 ppb, resulting in a daily intake for humans between 10 and 400 μg; the average is about 80 μg. About 20 U.S. cities, many of them industrial, show measurable amounts of chromium, generally about 0.01–0.02 μg Cr/m^3 air. Baltimore, which has both a chromium chemicals and stainless steel plant, has shown significantly higher values in most years, recording 0.301 μg Cr/m^3 in 1960 and 0.10 $\mu g/m^3$ in 1969. The general trend of Cr values is downward as plants install more effective dust control equipment. Rural air contains no detectable chromium (see Air pollution).

Environmental Control. Plants have to be designed and operated so that emissions are kept to a practical minimum. However, it is recognized that wastes are unavoidable. Waste chromium can, however, be converted to insoluble, inert chromium(III) compounds.

Stacks on kilns, leaching tanks, open boiling vessels, plating tanks, and other installations emitting dusts and mists containing chromium should be equipped with suitable precipitators (for insoluble dusts) or wet scrubbers (for mists or soluble dust) (see Air pollution control methods). Material entrapped in scrubbers can frequently be recycled. It is generally not feasible to further process any recovered dry material to remove Cr(VI), although it may be returned to roasting or leaching.

Wastes containing chromium compounds in solution may result from a variety of operations. Drainage from roofs, inadvertent leakage, multiple effect evaporator entrainment in barometric condenser water represent dilute or intermittent wastes that are extremely dilute (5–500 ppm Cr) and hence not readily processed unless concentrated. It is not generally appreciated that such wastes can be combined with municipal sewage or other waters having a high COD. Activated sludge plants can handle up to 50 ppm chromate without effect (86). It has also been observed that Cr(VI) discharged into a heavily polluted harbor was completely removed at a distance of about 300 m (87).

If the waste containing in most cases both Cr(III) and Cr(VI) cannot be disposed of directly, the chromium is usually reduced to the +3 state, followed by addition of lime to raise the pH (88). The slurry is settled and the clear decantate returned to the watershed. The accumulated slurry is eventually used as fill.

Other processes employ ferrous sulfate, scrap iron (89) or sulfide wastes (90) as reducing agents.

Alternatively concentration can be effected by ion exchange, by which Cr(VI) can be reduced to 0.05 ppm.

The direct precipitation of Cr(VI) wastes with barium and recovery of the resulting chromate values have been patented (91).

Several companies (88,92) and the EPA have issued bulletins for wastewater treatments in chromium-consuming industries (see Water).

As long as solid chromium wastes contain no Cr(VI), no environmental problem exists. They may be used as fill, and if the chromium is precipitated in sewage sludge, the use of the sludge is not impaired (see Wastes).

The waste resulting from the roasting and leaching step in the chromate manufacturing process traditionally contains lime, and residual chromium(VI) bleeds slowly through desorption and disproportionation. If used as fill, it must be covered with organic wastes, such as sewage sludge, to prevent phytotoxicity. Elimination of lime from the kiln mixes solves this problem.

The adoption of standards in 1971 (73–75) established design parameters for plants. Since 1950, all major U.S. producers have built new plants, and details of methods used to reduce air pollution have been published (93–95). Complete enclosure of operations and maintenance of negative pressure within equipment to prevent escape of dust or mist are now required.

As standards for effluents and control of the work area have tightened, newer plants have improved techniques further. Consequently, investment in designs satisfying environmental requirements has increased sharply. All dry stacks (kilns, dryers, etc) must be equipped with high-efficiency electrostatic precipitators to prevent particulate emissions. The leaching operation is totally enclosed and equipped with wet scrubbers to return chromate mists to the process. Evaporators must be equipped with high efficiency entrainment separators, and all wastes must be processed to either recover or precipitate chromate before discharge. Employees are monitored by wearing personal samplers in order to detect and correct any malfunction of the control equipment (96).

The Manufacturing Chemists' Association has published chemical safety data sheets now under revision describing methods for safe handling of chromates, dichromates, and chromic acid (97). Compilations listing fire and explosion hazards also exist. Ammonium dichromate also decomposes in a self-sustaining reaction when heated. Numerous reducing agents, mostly organic compounds, react violently with chromic acid (98–99).

Uses

Chromium compounds are essential to many industries. The distribution of major uses is given in Table 8.

Metal Finishing and Corrosion Control. The most important use of chromium compounds in metal finishing is that of chromic acid in chromium plating (100–105) which consumes most of the chromic acid produced (see Electroplating; Metal surface treatment).

Unlike most metals, chromium is best plated from solutions in which it is present as an anion in a high oxidation state. The use of the lower oxidation states is confined to the electrolytic production of chromium metal (see Chromium and chromium alloys).

The deposition of chromium from chromic acid solutions also requires the presence of a catalyst anion, usually sulfate, although fluoride and fluosilicate have been extensively used. The amount of the catalyst anion must be carefully regulated. Neither pure chromic acid or solutions with excess catalysts will produce a satisfactory plate. Even with carefully controlled temperature and current density, and bath compositions, chromium plating is one of the most difficult electroplating operations. Throwing power and current efficiency are notably poor, making good racking procedures and electrical practices essential.

The deposition process is still not fully understood; however, the following facts are established:

Plating takes place under a cathode film, which must have rather exactly defined thickness and permeability controlled by the catalyst.

In a bath operating at maximum efficiency Cr(III) is present, but the plate is deposited from a Cr(VI) species.

Although the Cr(VI) species involved is unknown, concentration and pH considerations suggest ion dipoles such as CrO_3H^+.

Decorative chromium plating, in which the chromium plate has a thickness of 0.2–0.5 μm, is widely used for automobile body parts, appliances, plumbing fixtures, and many other products. It is customarily applied over a nonferrous base in the plating of steel plates. To obtain the necessary corrosion resistance, the nature of the undercoat and the porosity and stresses of the chromium are all controlled carefully. Thus microcracked, microporous, crack-free or conventional chromium may be plated over duplex and triplex nickel undercoats (see also Electroless plating).

Hard chromium plating (101) is a successful way of protecting a variety of in-

Table 8. **Distribution of Uses of Chromium Chemicals**[a]

Use	Percent
metal finishing and corrosion control	37
pigments and allied products	26
leather tanning and textiles	20
wood preservation	5
drilling muds	4
other uses (catalysts, intermediates)	8

[a] Ref. 84. From *Chemical Marketing Reporter.*

dustrial applications from wear and friction. The most important examples are cylinder liners and piston rings for internal combustion engines. Hard chromium plating must be applied over a hard substrate such as steel, and is applied in a wide variety of thicknesses ranging from 1 to over 300 μm.

Both decorative and hard chromium plate use essentially the same bath compositions, but operating conditions vary. Some of these are summarized in Table 9.

Thin chromium plate or TFS for cans was developed in 1963 (106,107). A deposit 0.05 μm thick is obtained by electrolysis from a dilute (50 g/L) CrO_3 solution. The deposit contains substantial quantities of hydrous oxides.

Black and colored plates are also largely oxide (108). Sulfate is absent from the bath, and small amounts of fluoride or fluosilicate are used as catalyst.

The higher current yields and reduced environmental effects continues to give impetus to plating research based on chromium(III). In general such baths contain a chromium(III) salt, a complexing agent, a conducting salt, and a surfactant (109). The plate, although bright, has a grayish appearance rather than the bluish hue obtained from a chromic acid electrolyte.

Chromium compounds are used in etching and bright-dipping of copper and its alloys. A typical composition for the removal of scale after heat-treating contains 30 g $Na_2Cr_2O_7.2H_2O$ and 240 mL conc H_2SO_4/L. It is used at 50–60°C.

Still another use of chromium compounds in metal finishing is the production of chemical conversion coatings. These are produced on nonferrous metals, mainly magnesium, aluminum, zinc, and cadmium, by immersion in a chromium(VI) solution containing an activating ion at a controlled pH. For color modification, other ingredients are added. The formulations are largely proprietary, and the largest use today is for colorless coatings on zinc; olive-drab colors were extensively used during and after World War II (111–113).

Oxide films on aluminum are produced by anodizing in a chromic acid solution. The films are heavier than those produced by chemical conversion and thinner and

Table 9.   Typical Bath Composition and Operating Conditions for Chromium Plating [a]

Composition [b]	1	2	3	4
CrO_3, g/L	250	250	400	300
SO_4^{2-}, g/L	2.5	2.5		0.6
$SrSO_4$			satd	
K_2SiF_6			satd	
Cr^{3+}, g/L	equil.	equil.	equil.	equil.
NaOH, g/L				50
ethanol, mL/L				1.0
type of plate	hard	decorative	decorative	dull
temp, °C	50	40	50	19
current density, A/m^2	10,000	1,000	8,000	6,000
current efficiency, %	18	18	24	30

[a] Ref. 101.
[b] Numerous proprietary formulations are available.

Note: Baths 1 and 2. Conventional SO_4^{2-} catalyst baths. Bath 3. So-called SRHS bath, in which a mixed catalyst is maintained by difficulty soluble salts (110). Bath 4. Bornhauser bath, used to some extent in Europe. The plate must be buffed.

more impervious than those produced by the more common sulfuric acid anodizing. They impart exceptional corrosion resistance and paint adherence, and were widely used on military aircraft assemblies during World War II. The films may be dyed. The usual procedure is to anodize at 35°C for 30 min in a solution of 100 g total CrO_3/L at a pH of 0.8–0.9 (114).

Dichromates or chromic acid are used as sealers or after-dips to improve the corrosion resistance of various coatings on metals. They are so used in sealing phosphate coatings on galvanized iron or steel, sulfuric acid anodic coatings on aluminum, and TFS plate on steel (see Corrosion and corrosion inhibitors).

Chromates are used to inhibit metal corrosion in recirculating water systems. When methanol was extensively used as an antifreeze, chromates could be successfully used as corrosion inhibitor for cooling systems in locomotive diesels and automobiles (115). Steel immersed in dilute chromate solutions does not rust, and corrosion of other metals is similarly prevented (116). The exact mechanism of the inhibition is not known, although it is agreed that polarization of the local anodes which serve as corrosion foci is important. In the inhibition of iron and steel corrosion a film of γ-Fe_2O_3, in which some Cr is present, appears to form.

The concentration of chromate required to inhibit corrosion may range from 50 to 20,000 ppm, depending on conditions, a pH of 8–9 is usually optimum. The inclusion of chromium compounds in formulations permits the use of such corrosive salts as zinc chloride and copper sulfate in steel cylinders.

Pigments. Recent figures on the consumption of chromium chemicals for the production of pigments are shown in Table 10.

Chromium pigments can be further classified into chromate color pigments based on lead chromate, chromium oxide greens, and corrosion inhibiting pigments based on difficultly soluble chromate. An excellent discussion of these pigments is given in a recent encyclopedia (117); an older reference is also useful (118).

Chromate Pigments Based on Lead. These pigments can be further subdivided into primrose, lemon, and medium yellows, and chrome orange [1344-38-3], molybdate orange [12656-85-8], and normal lead silicochromates. Although earlier emphasis was on pure lead compounds, modern pigments contain additives to improve working

Table 10. Production of Chromium Pigments and Corresponding Consumption of Sodium Dichromate, 1976–1977, Metric Tons [a]

Pigment	Production		Sodium dichromate consumption [b]	
	1976	*1977*	*1976*	*1977*
Chrome Yellow and Orange	32,055	28,154	11,800	10,400
Molybdate Orange	14,340	11,473	6,000	4,800
Chrome Oxide Green [c]	3,700	2,700	7,400	5,400
Zinc Yellow [c]	2,500	1,700	2,500	1,700
Chrome Green + other	(d)	(d)	600	500
Total	*52,595*	*44,027*	*28,300*	*22,800*

[a] Ref. 56.
[b] Estimated, with allowance for process losses
[c] Estimated on basis of 1974–1975 data.
[d] Not listed separately, cannot be estimated.

properties, hue, light fastness, and crystal size and shape and to maintain metastable structures (119).

The chemical composition of these pigments is shown in Table 11. Details for present commercial procedures are not disclosed. The pigments are characterized as follows:

Medium yellows, an orange-yellow, are essentially pure monoclinic lead chromate.

Light lemon or primrose yellows containing up to 40% lead sulfate, contain some or all of the lead chromate in the metastable orthorhombic form, which is stabilized by lead sulfate and other additives. The higher the orthorhombic content, the greener the shade.

Chrome oranges, which are basic lead chromate [18454-12-1], $PbCrO_4.PbO$.

Molybdate oranges, which are tetragonal solid solutions of lead sulfate, lead chromate, and lead molybdate. An aging step is required in precipitation to permit development of the orange tetragonal form.

Lead silicochromate, essentially medium chrome yellow [1344-37-2] precipitated on silica, has been developed for use in traffic paints where the silica gives better abrasion resistance.

Table 11. Chemical Composition and Analytical Specifications for Chromate Color Pigments [a]

	$PbCrO_4$, %	$PbSO_4$, %	Other components, %
primrose chrome yellow			
theoretical	77.3	22.7	
ASTM spec, min	50		
actual,			
min	52.0	4.2	
max	82.7	25.9	
lemon chrome yellow			
theoretical	72.7	27.3	
ASTM spec, min	65		
actual,			
min	52.4	17.4	
max	68.8	39.0	
medium chrome yellow			
theoretical	100.0		
ASTM spec, min	87		
actual,			
min	82.4	nd[b]	
max	98.2	nd[b]	
chrome orange			
theoretical	59.2		PbO, 40.8
ASTM spec, min	55		
actual	58.2		PbO, 39.4
molybdate orange			
theoretical	82.3	2.8	$PbMoO_4$, 14.9
ASTM spec, min	70		
lead silicochromate			
theoretical	50.0		SiO_2, 50.0

[a] Refs. 119, 120.

[b] nd = not determined.

Chrome green [7758-97-6], not to be confused with chromic oxide green [1308-39-9], is a mixture of a light chrome yellow (lemon or primrose) with a blue, usually iron blue. The pigment may be produced by grinding, mixing in suspension, or precipitating the yellow on the blue. The latter method is the preferred; the former is hazardous because the pigment, containing both oxidizing (chromate) and reducing (ferrocyanide) components, may undergo spontaneous combustion. In recent years, phthalocyanine blues have replaced iron blues to some extent. Furthermore, organic greens and chromic oxide have displaced chrome green, which has poor acid and alkali resistance.

The 1978 prices of some of these pigments are listed below (121):

chrome yellow (cp)	$1.83/kg
chrome orange	$1.83/kg
molybdate orange	$2.40/kg
chrome green (cp)	$3.00/kg

Chromium Oxide Greens. These pigments comprise both the pure anhydrous oxide Cr_2O_3, and hydrated oxide, or Guignet's green (122). The three following manufacturing processes appear to be in use:

An alkali dichromate is reduced in a self-sustaining dry reaction by a reducing agent such as sulfur, carbon, starch, wood flour, or ammonium chloride. For pigment use, the reducing agent is generally sulfur. When a low-sulfur grade is needed in the manufacture of aluminothermic chromium, a carbonaceous reducing agent is employed:

$$Na_2Cr_2O_7 + S \rightarrow Na_2SO_4 + Cr_2O_3$$
$$Na_2Cr_2O_7 + 2\,C \rightarrow Na_2CO_3 + Cr_2O_3 + CO$$

The mixture is ignited with an excess of reducing agent in a reverberatory furnace or small kiln, transferred to leaching tanks, filtered, washed, dried, and pulverized. The product is +99% Cr_2O_3; metallurgical grades contain less than 0.005% of sulfur.

Chromate–dichromate solutions are reduced by sulfur in a boiling alkaline suspension (123–124).

$$2\,Na_2CrO_4 + Na_2Cr_2O_7 + 6\,S + 2\,x\,H_2O \rightarrow 2\,Cr_2O_3.x\,H_2O + 3\,Na_2S_2O_3$$

Excess NaOH is used to start the reaction and not over 35% of the chromium is added as dichromate. At the end of the reaction, the thiosulfate is removed by filtration and recovered. The hydrous oxide slurry is then acidified to pH 3–4 and washed free of sodium salts. On calcination at 1200–1300°C, a fluffy pigment oxide is obtained, which may be densified and strengthened by grinding. The shade can be varied by changes in the chromate–dichromate ratio, and by additives.

A dichromate or chromate solution is reduced under pressure to produce a hydrous oxide, which is filtered, washed, and calcined at 1000°C. The calcined oxide is washed to remove sodium chromate, dried, and ground. Sulfur, glucose, sulfite, and reducing gases may be used as reducing agent, and temperatures may reach 210°C and pressures 4–5 MPa (40–50 atm).

Chromic oxide green is the most stable green pigment known. It is used where chemical and heat resistance are required and is a valuable ceramic color (see Colorants for ceramics). It is used in coloring cement and granulated rock for asphalt roofing. An interesting application is in camouflage paints, as the infrared reflectance of chromic oxide resembles green foliage. A minor use is in the coloring of synthetic gem stones. Ruby, emerald, and the dichroic alexandrite all owe their color to chromic oxide (31) (see Gems).

A substantial amount of chromic oxide is used in the manufacture of chromium metal and aluminum-chromium master alloys (see Chromium and chromium alloys).

Guignet's green [12001-99-9], or hydrated chromic oxide green, is not a true hydrate, but a hydrous oxide, $Cr_2O_3 \cdot x\,H_2O$, in which x is about 2. It is obtained from the production of hydrous oxide at elevated temperature (and sometimes pressure) in a borax or boric acid melt.

Although Guignet's green is permanent, it does not withstand use in ceramics. It has poor tinting strength but is a very clean, transparent, bluish green. It is used mainly in metallic automotive finishes.

The 1978 prices are: chromic oxide green, $1.720/kg and Guignet's green $4.630/kg (121).

Corrosion Inhibiting Pigments.   These pigments derive their effectiveness from the low solubility of chromate. The major pigment of this group is zinc chromate (see below) or zinc yellow; others include zinc tetroxychromate, basic lead silicochromate, strontium chromate, and barium potassium chromate (125).

The chemical composition of some of these pigments is shown in Table 12.

Zinc yellow [37300-23-5] became an important corrosion-inhibiting pigment for aircraft during World War II; however, the war production rate of 11,000 metric tons per year has not since been reached. Now it is widely used for corrosion inhibition on auto bodies, light metals, and steel, and in combination with red lead and ferric oxide for structural steel painting.

Zinc yellow is not a normal zinc chromate, having the empirical formula $K_2O \cdot 4ZnO \cdot 4CrO_3 \cdot 3H_2O$ [12433-50-0]. It belongs to the group of salts having the general formula $M(I)_2O \cdot 4M(II)O \cdot 4CrO_3 \cdot 3H_2O$ (126). The sodium zinc salt has occasionally been used as a pigment; the sodium copper salt has been tested as an antifouling marine pigment and is an ingredient of dips for auto bodies.

Zinc yellow is made by a variety of processes, all based on the reaction of zinc compounds, chromates, and potassium salts in aqueous solution. If chloride- and especially sulfate-free products are desired, they are exluded from the system. In one process, for example, zinc oxide is "swollen" with potassium hydroxide and the chromates are added as a solution of potassium tetrachromate [12422-53-6] (127):

$$4\,ZnO + K_2Cr_4O_{13} + 3\,H_2O \rightarrow K_2O \cdot 4\,ZnO \cdot 4CrO_3 \cdot 3H_2O$$

The final pH is 6.0–6.6. Care must be taken in washing to avoid hydrolysis and loss of chromate.

The 1978 price of zinc yellow was $1.90/kg (121).

Zinc tetroxychromate [13530-65-9] (approximately $4ZnO \cdot ZnCrO_4 \cdot x\,H_2O$) has a somewhat lower chromate solubility than zinc yellow and has been used in wash primers.

Strontium chromate, $SrCrO_4$, is used increasingly despite its high cost. It works well on light metals, and is compatible with some latex emulsions where zinc compounds cause coagulation. It is also an ingredient of some proprietary formulations for chrome plating (see above).

Basic lead silicochromate [11113-70-5] (National Lead Co. designation Pigment

Table 12. Analytical Specifications and Compositions (Percent by Weight) of Corrosion Inhibiting Pigments[a]

	CrO_2	ZnO	Other metal oxides	H_2O combined	SO_3	Other
zinc chromate						
type I, low sulfate						
theoretical	45.8	37.2	10.8 K_2O	6.2		
ASTM spec	41 min	35–40	13 max K_2O		0.2 max	0.1 max Cl
typical	45.0	36.0	10.0	6.0	0.05	
type II, regular						
ASTM spec	41 min	35–40	13 max K_2O		3.0 max	0.8 max Cl
typical	44.0	38.0	10.0	6.0	1.0	
zinc tetroxychromate						
typical	17.0	71.0		10.0	0	
strontium chromate						
theoretical			47 min SrO			
ASTM spec	44 min		49.0		0.2 max	0.1 max Cl
typical	47.0				0.2	
basic lead silicochromate						
ASTM spec	5.1–5.7		46–49 PbO	0.2 max.		45.5–48.5 SiO_2
typical	5.4		47.0			47.0

[a] Refs. 120,125.

109

M-50) is a composite in which basic lead chromate (chrome orange) is precipitated onto a lead silicate–silica base. It does not have an appreciable chromate solubility and depends on lead oxide for its effectiveness.

Barium potassium chromate, $K_2Ba(CrO_4)_2$, (National Lead Co. designation Pigment E) has been sold commercially, but is not believed to be now in use.

Leather Tanning and Textiles. Although compounds of chromium(VI) are the most important commercially, the bulk of the applications in the textile and tanning industries depend on the ability of chromium(III) to form stable complexes with proteins, cellulosic materials, dyestuffs, and various synthetic polymers (see Leather). The chemistry is complex and still imperfectly understood in many cases. The common denominator is the coordinating ability of chromium(III).

The chrome tanning of leather is one step in a complicated series of operations leading from the raw hide to the finished product. Chrome tanning is the most important tannage for all hides except heavy cattle hides, which are usually vegetable tanned. In heavy shoe uppers and soles, a chrome tanned leather is frequently given a vegetable retan to produce chrome retan leather. The annual consumption of hides by the leather industry appears to be decreasing (131). Recent technical bulletins (132) and proprietary formulations of chromic sulfate are available (14).

Sodium dichromate and various chromic salts are employed in the textile industry (133–134). The former is used as oxidant and as a source of chromium; for example, to dye wool and synthetics with mordant acid dyes, oxidize vat dyes and indigosol dyes on wool, after-treat direct dyes and sulfur dyes on cotton to improve washfastness and oxidize dyed wool (see Dyes, application). Premetallized dyes are also employed. These are hydroxyazo or azomethine dyes in which chromium or other metals are combined in the dye (see Azo dyes).

A typical premetallized dye is designated Chromolan Black NWA [5610-64-0] (CI 15711). The commercial product also contains some of the 1:1 chelate.

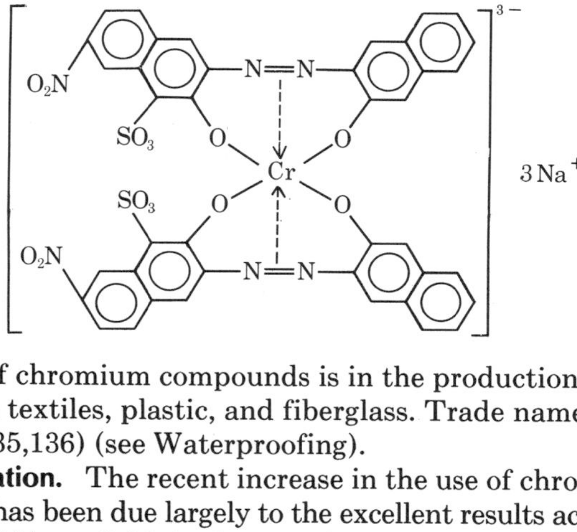

Another use of chromium compounds is in the production of water- and oil resistant coatings on textiles, plastic, and fiberglass. Trade names are Quilon, Volan, and Scotchgard (135,136) (see Waterproofing).

Wood Preservation. The recent increase in the use of chromium compounds in wood preservation has been due largely to the excellent results achieved by chromated copper arsenate (CCA), available in three modifications under a variety of trade names. The treated wood is free from bleeding, is paintable and of an attractive olive-green color. Thus CCA is widely used, especially in treating utility poles, building lumber, and wood foundations.

Chromium compounds are also used in fire-retardant formulations where their function is to prevent leaching of the fire retardant from the wood and corrosion of the equipment employed (see Flame retardants, inorganic).

Chromium-containing wood preservatives and fire-retardants are listed in Table 13 (137).

Chromium compounds have a triple function in wood preservation (138). Most importantly, after impregnation of the wood the chromium(VI) compounds used in the formulations react with the wood extractives and the other preservative salts to produce difficultly soluble complexes from which preservative leaches only very slowly. This mechanism has been studied in the laboratory (139–144) and the field (145). Finally, although most of the chromium is reduced to chromium(III), there is probably some slight contribution of the chromium to the preservative value (146).

Drilling Muds in the Petroleum and Natural Gas Industry. Since 1941, chromium chemicals have been used in the drilling of wells to combat fatigue corrosion cracking of drill strings, with about 1 t of sodium chromate being used per year for an average West Texas well. Other early uses were in gas-condensate wells in Louisiana and East Texas.

More recently, however, the industry has turned to proprietary drilling-mud formulations, specially designed to suit the aqueous environment and rock strata in which the well is located. In addition to heavy minerals such as barite, and both soluble and difficultly soluble chromates for corrosion control, many of these formulations contain chromium lignosulfonates. These are prepared like a tanning compound from sodium dichromate, but using lignosulfonate waste from sulfite pulp mills as the reducing agent. This use amounts to about 4% of the total chromium compound consumption (147–148) (see Petroleum, drilling fluids).

Miscellaneous Uses. A large number of chromium compounds have been sold in small quantities for a variety of uses (6–9). These, with their uses, are described in Table 14.

Catalysts. A more important minor use of chromium compounds is in the manufacture of catalysts, consuming about 1500 metric tons of sodium dichromate equivalent annually (see Table 14).

Chromium catalysts are used in a great variety of reactions, including hydrogenations, oxidations, and polymerizations (149–151). Most of the details are proprietary and many patents are available (see Catalysis).

Chromia-alumina catalysts are prepared by impregnating γ-alumina shapes with a solution of chromic acid, ammonium dichromate, or chromic nitrate, followed by gentle calcination. Zinc and copper chromites are prepared by coprecipitation and ignition, or by thermal decomposition of zinc or copper chromates, or organic amine complexes thereof. Many catalysts have spinel-like structures (152).

Photosensitive reactions. The reduction of chromium(VI) by organic compounds is highly photosensitive. This property is used in photosensitive dichromate-colloid systems (see Photoreactive polymers; Printing processes).

The basic principle is simple (153). A dichromate-colloid system is applied to a metal printing plate. This soluble material is exposed to an image, and, where light strikes, the photochemical reaction reduces the dichromate. The chromium(III) produced forms an insoluble complex with the colloid in a reaction similar to that of dye mordanting or leather tanning. The unreacted colloid is washed off, exposing bare metal that can be etched. Some of the colloids used are shellac, glue, albumin, casein,

Table 13. Chromium Compounds in Wood Preservatives and Fire Retardants, 1976[a]

Chromium compound	Consumption metric tons	Cr content, % CrO_3	$Na_2Cr_2O_7 \cdot 2H_2O$ equiv, metric tons
wood preservative			
fluor-chrome-arsenic phenol	111	37	61
chromated copper arsenate			
Type A	1115	65.5	1088
Type B	2082	35.3	1095
Type C	4554	47.5	3223
acid copper chromate	321	68.2	326
chromated zinc chloride	233	20	69
fire retardant			
chromated zinc chloride (FR)	222	14.4	47
pyresote	598	3.3	29
Totals	*9236*		*5938*

[a] Ref. 137.

112

Table 14. Properties and Uses of Miscellaneous Chromium Compounds

Name[a]	Formula	CAS Registry Number	Physical properties	Uses	Refs.
Oxidation state +6					
aluminum chromate	variable		yellow, amorphous	in ceramics	
ammonium chromate	$(NH_4)CrO_4$		yellow crystals, sp gr density 1.90, soly 27.00% at 25%	flexible printing, photosensitization	154–155, 157
barium chromate	$BaCrO_4$	[10294-40-3]	see Table 2	pyrotechnics, high-temp, batteries	
barium dichromate	$BaCr_2O_7 \cdot 2H_2O$	[10031-16-0]	orange, rhombic, dec in H_2O	in ceramics	158
barium potassium chromate	$K_2Ba(CrO_4)_2$	[27133-66-0]	yellow, sp gr 3.65	corrosion-inhibiting pigment	
cadmium chromate	$CdCrO_4$	[14312-00-6]	yellow, insol	in catalysts, pigment	
cadmium dichromate	$CdCr_2O_7 \cdot H_2O$	[69239-51-6]	orange, sol H_2O	metal finishing	
calcium chromate	$CaCrO_4$	[13765-19-0]	yellow, tetragonal soly 2.23% at 20°C	metal primers, corrosion inhibitor, high temp batteries	154–155
calcium dichromate	$CaCr_2O_7 \cdot 4.5H_2O$	[14307-33-6]	orange crystals, sp gr 2.136, soly (anhyd.) 59.56% at 30°C; tr to 3 H_2O at 42°C	metal finishing	159
cesium chromate	Cs_2CrO_4	[13454-78-9]	yellow, hexagonal or orthorhombic 4.23, soly 48.8% at 20°C	electronics	
chromic chromate	variable	[11056-30-7]	brown, amorphous, hydrous	in catalysts, mordants	160
chromyl chloride	CrO_2Cl_2	[14977-61-8]	see Table 2	organic oxidation, Etard reaction	
cobalt chromate	$CoCrO_4$	[13455-25-9]	gray-black, orthorhombic	in ceramics	161
copper chromate, basic	$4CuO \cdot CrO_3 \cdot xH_2O$	[12433-14-6]	brown, amorphous	fungicides, in catalysts	
copper dichromate	$CuCr_2O_7 \cdot 2H_2O$	[13675-47-3]	red-brown triclinic, sp gr 2.286, vs H_2O	in catalysts, wood preservatives	
copper sodium chromate	$Na_2O \cdot 4CuO \cdot 4CrO_2 \cdot 3H_2O$	[68399-60-0]	maroon, triclinic, sp gr 3.57, v sl sol	antifouling pigment	162
lithium chromate	$Li_2CrO_4 \cdot 2H_2O$	[7789-01-7]	yellow, orthorhombic, sp gr 2.149 soly 49.6 at 30°C, tr. to anhyd. 74.6°C	corrosion inhibitor, especially in air-conditioning and nuclear reactors	163
lithium dichromate	$Li_2Cr_2O_7 \cdot 2H_2O$	[10022-48-7]	orange-red crystals, sp gr 2.34, soly 66.08 at *40°C	corrosion inhibitor	164
magnesium chromate	$MgCrO_4 \cdot 5H_2O$	[16569-85-0]	yellow triclinic, sp gr 1.954, soly 35.39% at 25°C tr, to 7 H_2O at 17.2°C	corrosion inhibitor for gas turbines; refractories	
magnesium dichromate	$MgCr_2O_7 \cdot 6H_2O$	[16569-85-0]	orange-red orthorhombic, sp gr 2.002, sol. (as anhyd.) 58.52% at 30°C, tr to 5 H_2O at 48.5°C	in catalysts, refractories	165

Table 14. (continued)

Name[a]	Formula	CAS Registry Number	Physical properties	Uses	Refs.
mercuric chromate	$HgCrO_4$	[13444-75-2]	red, orthorhombic, sl sol	antifouling formulations	
mercurous chromate	Hg_2CrO_4		red, v sl sol	antifouling formulations	
morpholine chromate	$(OC_4H_8NH_2)_2CrO_4$	[36969-05-8]	yellow oily material	vapor phase corrosion inhibitor	
nickel chromate	$NiCrO_4$	[14721-18-7]	maroon, v sl sol dark red	in catalysts	
*pyridine–chromic acid	$CrO_3 \cdot 2C_5H_5N$		crystals, expl. on warming	research oxidant	
pyridine dichromate	$(C_5H_5NH)_2Cr_2O_7$	[20039-37-6]	orange crystals	photosensitizer in photoengraving in ceramics	
nickel potassium chromate	$K_2Ni(CrO_4)_2 \cdot 6H_2O$	[15275-09-9]			
silver chromate	Ag_2CrO_4	[7784-01-2]	see Table 2	in catalysts	
strontium chromate	$SrCrO_4$	[7789-06-2]	see Table 2	corrosion-inhibiting pigment, plating additive	
tetramino copper chromate	$Cu(NH_3)_4CrO_4$	[13870-96-7]	dark green needles	catalysts, gas absorbent	
zinc sodium chromate	$Na_2O.4ZnO.4CrO_3 \cdot 3H_2O$	[68399-59-7]	yellow, triclinic; sl sol sp gr 3.24	corrosion-inhibitive pigment	
Oxidation state +3					
ammonium tetrathiocyanato diamino chromate(III)	$NH_4(NH_3)_2Cr(SCN)_4$	[13573-16-5]	Reinecke's salt, red cubic	identification amines and alkaloids	
basic chromic chloride	$Cr(OH)Cl(H_2O)_4\ Cl_2$	[12506-63-7]	many compositions available and isopropanol soln	mfg Quilon, Volan, Scotchgard, all trademarks	
chromic acetate	$Cr(OCOCH_3)_3 \cdot xH_2O$	[1066-30-4]	usually sold as solution, basic salts used	printing and dyeing textiles	15
chromic acetylacetonate	$Cr(CH_3COCHCOCH_3)_3$	[13681-82-8]	see Table 2	preparation of Cr complexes, in catalysts, antiknock compounds	
chromic ammonium sulfate	$NH_4Cr(SO_4)_2 \cdot 12H_2O$	[10022-47-6]	violet octahedral, sp gr 1.72, mp 94°C, sol	electrolyte for mfg Cr metal	166
chromic chloride	$CrCl_3$	[10025-73-7]	see Table 2	chromizing, Cr metal, organochromium compounds	167–168
*chromic chloride (hydrated)	$CrCl_3 \cdot 6H_2O$		three isomers, see Table 2	mordant, tanning, chromium complexes	169
chromic fluoborate	$Cr(BF_4)_3$	[27519-39-7]	usually available as solution	chromium plating, in catalysts	
chromic fluoride	$CrF_3 \cdot 9(4)H_2O$	[16671-27-5]	usually sold as solution	mordants, in catalysts	
chromic fluoride	CrF_3	[7788-97-8]	green rhombohedral, sp gr 3.78, mp 1100°C, insol H_2O	chromizing	
chromic formate, basic	$Cr(OH)(HCOO)(H_2O)_4(HCOO)$		one of few crystalline basic Cr salts; green needles, sol H_2O	skein printing of cotton tanning	
chromic hydroxide	$Cr_2O_3 \cdot xH_2O$	[11056-30-7]	gray-green amorphous powder	see Guignet's Green under Uses	

Name	Formula	CAS Registry Number	Properties	Uses	References
chromic lactate	$Cr(CH_3CHOHCOO)_3 \cdot xH_2O$	[19751-95-2]	sold as solution	mordant ingredient of Universal mordant 9333	34, 170
chromic naphthenate	no def. formula		sold as soln in petr. solvents	textile preservative	
chromic nitrate	$Cr(NO_3)_3 \cdot 9H_2O$	[7789-02-8]	violet crystals, sp gr 1.80, mp 66.3°C, soly 74% at 25°C other hydrates known	in catalysts, textiles, mfg CrO_2	
chromic phosphate	$CrPO_4$	[27096-04-4]	green powder, also soln in H_3PO_4	pigments, phosphate coating, wash primers	
chromic potassium oxalate	$K_3[(Cr(C_2O_4)_3] \cdot 3H_2O$		violet crystals	mordant Eriochromal Mordant, and Chromosol	
chromic potassium sulfate	$KCr(SO_4)_2 \cdot 12H_2O$	[7788-99-0]	chromic alum; see Table 2	hardening photographic emulsions insolubilizing gelatin	
chromic sulfate	$Cr_2(SO_4)_3 \cdot xH_2O$	[15005-90-0]	green, amorph., also violet $Cr_2(SO_4)_3 \cdot 18H_2O$	pigment, ceramics, in catalysts	
cobalt chromite	$CoCr_2O_4$	[12016-69-2]	turquoise blue cubic, spinel structure	in catalysts, especially for automobile exhaust	149
copper chromite	$CuCr_2O_4$	[12018-10-9]	black, tetragonal, distorted spinel		54
ferrous chromite	$FeCr_2O_4$	[12068-77-8]	black, cubic, spinel	impure (Fe,Mg) $(Cr,AlFe)_2O_4$ is chrome ore of commerce	
magnesium chromite	$MgCr_2O_4$	[12053-26-8]	brown, cubic spinel, sp gr 4.415	refractory	171
triphenyl chromium tetrahydrofuranate	$(C_6H_5)_3Cr \cdot 3THF$		only V-bonded organo chromium	unstable	172
zinc chromite	$ZnCr_2O_4$	[12018-19-8]	green, cubic spinel, sp gr 5.30	in catalyst	
Other oxidation states					
Oxidation state 0					
chromium carbonyl	$Cr(CO)_6$	[13007-92-6]	see Table 2, vp 0.1 MPa (1 atm) at 48°C dec. 130°C, expl. 210°C	synthesis of "sandwich" compounds	24, 173
dicumene chromium	$[(CH_3)_2CHC_6H_5]_2Cr$	[12001-89-7]	est bp 300°C, exp 210°C	prepn of Cr carbides by thermal vapor deposition	25, 174–175
Oxidation state +2					
chromous chloride	$CrCl_2$	[10049-05-5]	see Table 2, stable in dry air	chromizing, prepn of Cr metal	29
Oxidation state +4					
chromium(IV) oxide	CrO_2	[12018-01-8]	see Table 2	mfg magnetic tapes	16, 156
manganese(III) chromate(IV)	$Mn_2CrO_5 \cdot xH_2O$		black-green powder	catalysts, in ceramics	
Oxidation state +5					
calcium chromate(V)	$Ca_3(CrO_4)_2$	[12205-18-4]	green crystals, sim. $Be_3(CrO_4)$ (Table 2) isomorphous with $Ca_2(PO_4)_2$	corrosion inhibitive pigment, may be carcinogenic	176

ᵃ * Indicates a different degree of hydration.

gum arabic, and gelatin. During the past twenty years, the industry has generally shifted to more readily controlled synthetic materials, such as poly(vinyl alcohol).

Batteries.   The shelf life of dry batteries is increased from 50 to 80% by the use of a few grams of dichromate or zinc chromate near the zinc anode. This polarizes the anode on open circuit but does not interfere with current delivery.

Since World War II, the U.S. space program and the military have used small amounts of insoluble chromates, largely barium and calcium chromates, as activators and depolarizers in fused-salt batteries. The technology of these batteries has only recently been declassified and general descriptions are available (154–155) (see Batteries).

Magnetic tapes.   Chromium dioxide, CrO_2, is used as a ferromagnetic material in high-fidelity magnetic tapes. Chromium dioxide has several technical advantages over the magnetic iron oxides generally used (34,156) (see Magnetic tape).

Reagent-grade chemicals.   Potassium dichromate is an important analytical standard, and other chromium chemicals, in reagent grades, find considerable laboratory use (57–58). This use, though small, is most important in wet analyses (see Fine chemicals).

BIBLIOGRAPHY

"Chromium Compounds" in *ECT* 1st ed., Vol. 3, pp. 941–995, by J. J. Vetter, Diamond Alkali Company, and C. Mueller, General Aniline & Film Corp.; and under "Chromium Compounds" in *ECT* 2nd ed., Vol. 5, pp. 473–516, by W. H. Hartford and R. L. Copson, Allied Chemical Corporation.

1. M. E. Weeks, *Discovery of Elements*, 7th ed., Journal of Chemical Education, Easton, Pa., 1968, pp. 271–281.
2. *Chromium Chemicals, Their Discovery, Development, and Use,* Mutual Chemical Company, New York, 1941.
3. H. H. Zeiss and M. Tsutsui, *J. Am. Chem. Soc.* **79,** 3062 (1957).
4. U.S. Pat. 1,591,188 (April 20, 1926), C. G. Fink (to United Chromium).
5. G. J. Sargent, *Trans. Am. Electrochem. Soc.* **37,** 479 (1920).
6. M. J. Udy, ed., *Chromium,* Vol. 1, Reinhold Publishing Co., New York, 1956, pp. 283–422.
7. *Chemical Sources USA,* Directories Publishing Co., Flemington, N. J., 1977.
8. *Chemical Week 1979 Buyers' Guide Issue,* McGraw-Hill, New York, 1978.
9. *1976–1977 OPD Chemical Buyers' Directory,* Schnell Publishing Co., New York, 1976.
10. W. H. Hartford, *Ind. Eng. Chem.* **41,** 1993 (1949).
11. R. A. Kearly, *J. Chem. Eng. Data* **9,** 548 (1964).
12. Seidell, *Solubilities—Inorganic and Metal-Organic Compounds,* Vol. 1, 4th ed., D. Van Nostrand Co., Inc., New York, 1956, p. 871.
13. *Chromium Chemicals,* Diamond Shamrock Chemical Co., Cleveland, Ohio 1972.
14. *Chromium Chemicals, Bulletin 52,* Industrial Chemicals Division, Allied Chemical Corporation, Syracuse, New York, (1960).
15. Ref. 6, pp. 219–221.
16. W. H. Hartford, "Chromium", in I. M. Kolthoff and P. J. Elving eds., *Analytical Chemistry of the Elements,* Vol. 8, part II, Interscience, New York, 1963, p. 278.
17. Mutual Chemical Company of America, 1934–1957, unpublished analyses.
18. "Chrom" in *Gmelins Handbuch der Anorganischen Chemie,* 8th ed., System—no. 52, 1963–1965.
19. C. L. Rollinson, "Chromium, Molybdenum, and Tungsten" in J. C. Bailar, Jr., and co-workers, eds., *Comprehensive Inorganic Chemistry,* Vol. 3, Pergamon, Oxford, 1973, pp. 623–700.
20. F. A. Cotton and G. Wilkinson, *Advanced Inorganic Chemistry,* 3rd ed., Interscience, New York, 1972, pp. 528–922.
21. J. W. Mellor, *A Comprehensive Treatise on Inorganic and Theoretical Chemistry,* Vol 11, Longmans, Green & Co., New York, 1931, pp. 122–483.
22. P. Pascal, *Nouveau Traite de Chimie Minerale,* Vol 14, Masson et Cle, Paris, 1959, pp. 33–551.

23. Ref. 6, pp. 113–251.
24. Ger. Pat. 1,007,305 (May 2, 1957), E. O. Fischer and W. Hafner (to Badische Anilin-und Soda-Fabrik A.G.).
25. *Dicumenechromium,* Data Sheet, Union Carbide Corp., New York, 1960.
26. E. O. Fischer and W. Hafner, *Z. für Naturforsch.* **106,** 665 (1955).
27. D. G. Holah and J. P. Fackler, Jr. in E. L. Mutterties, ed., *Inorganic Syntheses,* Vol. 10, McGraw-Hill, New York, 1967, pp. 26–35.
28. Ref. 20, pp. 675–678, 834.
29. U.S. Pats. 3,414,428 (Dec. 3, 1968), and 3,497,316 (Feb. 24, 1970), W. R. Kelly and W. B. Lauder (to Allied Chemical).
30. B. M. Loeffler and R. G. Burns, *Am. Sci.* **64,** 636 (1977).
31. W. H. Hartford, *Rocks and Miner.* **52,** 169 (1977).
32. A. F. Wells, *Structural Inorganic Chemistry,* 3d ed., Oxford University Press, London, 1975, p. 355.
33. J. P. Collman and E. P. Kittleman in H. F. Holtzclaw, ed., *Inorganic Syntheses,* Vol 8., McGraw-Hill, New York, 1966, pp. 150–151.
34. U.S. Pat. 3,117,093 (Jan. 7, 1964), P. Arthur, Jr. and J. N. Ingraham (to E. I. du Pont de Nemours & Co.).
35. O. Fukunaga and S. Saito, *J. Am. Ceram. Soc.* **51**(7) 362 (1968).
36. K. B. Wiberg and R. Eisenthal, *Tetrahedron* **20,** 1151 (1964).
37. Ref. 32, pp. 945, 947.
38. R. Scholder and H. Schwarz, *Z. Anorg. Allgem. Chem.* **363,** 10 (1968).
39. *Ibid.,* **326,** 11 (1963).
40. H. Schwarz, *Z. Anorg. Allgem. Chem.* **323,** 275 (1963).
41. O. Glemser, *Science* **143,** 1058 (1964).
42. I. Dellien, F. M. Hall and L. G. Hepler, *Chem. Revs.* **76**(3), 283 (1976).
43. F. Ullmann, *Enzyklopädie der Technischen Chemie,* Vol 3, Urban and Schwarzenberg, Berlin, 1929, pp. 400–433.
44. F. McBerty and B. H. Wilcoxon, *FIAT Rev. Ger. Sci., PB22627, Final Report no. 796,* 1946.
45. Ref. 6, pp. 262–282.
46. U.S. Pat. 3,095,266 (June 25, 1963), W. B. Lauder and W. H. Hartford (to Allied Chemical Corp.).
47. U.S. Pat. 3,853,059 (Dec. 3, 1974), C. P. Bruen, W. W. Low, and E. W. Smalley (to Allied Chemical Corp.).
48. S. African Pat. 74-03,604 (April 28, 1975), C. P. Bruen and co-workers (to Allied Chemical Corp.).
49. U.S. Pat. 3,816,095 (June 11, 1974), C. P. Bruen, W. W. Low, and E. W. Smalley (to Allied Chemical Corp.).
50. U.S. Pat. 3,899,568 (Aug. 12, 1975), D. G. Frick, T. R. Morgan, and T. L. Streeter (to Allied Chemical Corp.).
51. U.S. Pat. 1,873,589 (Aug. 23, 1932), P. R. Hines (to Harshaw Chemical Co.).
52. U.S. Pat. 3,065,055 (Nov. 20, 1962), T. S. Perrin and R. E. Banner (to Diamond Alkali Co.).
53. U.S. Pat. 3,607,026 (Sept. 21, 1971), T. S. Perrin, R. E. Banner, and J. O. Brandstaetter (to Diamond-Shamrock Corp.).
54. *1975 Minerals Yearbook,* Vol. 1, U.S. Bureau of Mines, Washington, D.C., 1977, pp. 343–354.
55. E. F. Foley, Jr., Diamond-Shamrock Corp., Painesville, O., private communication, 1977.
56. *Inorganic Chemicals,* Report M28A (76)-14, U.S. Bureau of the Census, Washington, D.C., 1977.
57. *Reagent Chemicals,* 5th ed., American Chemical Society, Washington, D.C., 1974.
58. J. Rosin, *Reagent Chemicals and Standards,* 5th ed., D. Van Nostrand Reinhold Co., New York, 1967.
59. General Services Administration, *Federal Specifications* O-S-5956, April 23, 1970; O-C-303c, March 12, 1970; O-S-588b, Nov. 12, 1970; O-P-559, May 26, 1952; O-A-498b, Sept. 7, 1966 (cancelled 1976); General Services Administration, Washington, D.C.
60. *Federal Specification O-C-303c (1),* March 12, 1970.
61. *Federal Specification O-S-588b (1),* Nov. 12, 1970.
62. W. H. Hartford in F. D. Snell and L. C. Ettre, eds., *Encyclopedia of Industrial Chemical Analysis,* Vol. 9, John Wiley & Sons, Inc., New York, 1970, pp. 680–709.
63. E. D. Olsen and C. C. Foreback in F. D. Snell and L. C. Ettre, eds., *Encyclopedia of Industrial Chemical Analysis,* Vol. 9, John Wiley & Sons, Inc., New York, 1070, pp. 632–680.

64. W. H. Hartford in F. D. Snell and L. C. Ettre, eds., *Encyclopedia of Industrial Chemical Analysis*, Vol. 9, John Wiley & Sons, Inc., New York, 1970, pp. 176–213.

65. Ref. 54, pp. 300–377.

66. N. H. Furman, *Standard Methods of Chemical Analysis*, Vol. 1, 6th ed., Van Nostrand-Reinhold, New York, pp. 350–377.

67. *Standard Methods for Examination of Water and Wastewater*, 13th ed. American Public Health Association, Washington, D.C., 1971, pp. 155–159.

68. J. W. Robinson, *Atomic Absorption Spectroscopy*, Marcel Dekker, New York, 1975, pp. 88–89, 154, 160–172.

69. J. A. Rawa and E. L. Henn, *Am. Lab.* **9**(8) 31 (1977).

69a. Division of Medical Sciences, National Research Council, *Chromium*, Report of Panel, A. M. Baetjer, Chairman, National Academy of Sciences, Washington, D.C., 1974.

69b. J. E. McKee and H. W. Wolf, *Water Quality Criteria, Publication 3-A*, 2nd ed., The Resources Agency of California, State Water Quality Control Board, Sacramento, Calif., 1963, pp. 163–166.

69c. J. W. Jaworski, ed., *Effects of Chromium in the Canadian Environment, Publication 15107 of the Environmental Secretariat*, Associate Committee on Scientific Criteria for Environmental Quality, National Research Council of Canada, Ottawa, Ont. 1976.

70. H. W. Davids and M. Lieber *Wat. Sew. Works* **98**, 528 (1951).

71. *Drinking Water and Health, Fed. Regis.* **42**(132) 35770 (1977).

72. J. J. Bloomfield and W. Blum, *Public Health Rep.* **43**, 2330 (1928).

73. American National Standards Institute, *USAS Z37.7-1943. (rev. 1971)* New York, 1971.

74. American Conference of Governmental Industrial Hygienists, *Threshold Limit Values of Airborne Contaminants and Physical Agents with Intended Changes*, Cincinnati, Ohio, 1971.

75. U.S. Department of Labor, *Occupational Safety and Health Standards*, Ch. XVII, Part 1910.93. Tables G1 and G2, *Publ. Fed. Register* **36**(157) 15101 (1971).

76. Federal Security Agency, *Health of Workers in Chromate-Producing Industry*, U.S. Public Health Service Publication No. 192 Washington, D.C., 1953.

77. A. M. Baetjer, *A.M.A. Arch. Hyg. Occup. Med.* **2**, 487 (1950).

78. Ref. 6, pp. 76–104.

79. W. Machle and F. Gregorius. *Public Health Rep.* **63**(35) 1114 (1928).

80. F. H. Taylor, *Am. J. Public Health* **56**, 218 (1966).

81. A. M. Baetjer, personal communication, 1976.

82. F. H. Westheimer, *Chem. Rev.* **45**, 419 (1949).

83. J. Cairns, *Sci. Am.* **233**, 64 (1975).

84. Southwest Research Institute, private communication, 1956.

85. W. Mertz, *Physiol. Rev.* **49**, 163 (1969).

86. W. A. Moore and co-workers, *Purdue University Engineering Bull. Ext. Ser. No. 106 (158–82)*, 1960.

87. W. H. Hartford, personal observation at Baltimore, Md., and Greenville, S. C. 1950–1958, 1972.

88. *Practical Guide to Treatment of Chromium Waste Liquors*, Allied Chemical Corporation, Industrial Chemicals Division, Syracuse, New York, 1975.

89. U.S. Pat. 3,027,321 (March 27, 1962), R. P. Selm and B. T. Hulse (to Wilson & Co.).

90. U.S. Pat. 3,294,680 (Dec. 29, 1966), L. E. Lancy (to Lancy Laboratories).

91. U.S. Pat. 3,552,917 (Jan. 5, 1971), C. O. Weiss (to M-T Chemicals, Inc.); U.S. Pat. 3,728,273 (April 17, 1973), C. P. Bruen and C. A. Wamser (to Allied Chemical Corp.).

92. *Treatment of Chromium Bearing Wastes*, Diamond Shamrock Corporation, Soda Products Division, Cleveland, Ohio.

93. L. C. Palmer and G. E. Best, *Am. Ind. Hyg. Assoc. Quart.* **14**, 294 (1953).

94. H. G. Bourne, Jr., and W. R. Rushin, *Ind. Med. Surg.* **19**, 568 (1950).

95. H. G. Bourne, Jr., P. M. Frazier, and H. T. Yee, *Ind. Med. Surg.* **20**, 498 (1951).

96. C. W. Rudolf, private communication, 1977.

97. Manufacturing Chemists Association, *Chemical Safety Data Sheets SD-44, SD-45, SD-46*, Washington, D.C., 1952.

98. I. Bretherick, ed., *Handbook of Reactive Chemical Hazards*, CRC Press, Cleveland, Ohio 1975, pp. 711–714.

99. National Fire Protection Association, *Manual of Hazardous Chemical Reactions*, Publication 491, M, 4th ed., Boston, 1971, pp. 79–82.

100. W. H. Hartford, *Chromium,* in A. J. Bard, ed., *Applied Electrochemistry of the Elements,* Marcel Dekker, New York, Oct. 1977.
101. G. Dubpernell in F. A. Lowenheim, ed., *Modern Electroplating,* 3rd ed., Wiley-Interscience, New York, 1974, pp. 87–151.
102. P. Morriset and co-workers, *Chromium Plating,* Robert Draper, Teddington, Middlesex, England, 1954.
103. J. M. Hosdowich in M. J. Udy, ed., *Chromium,* Vol II, Reinhold, New York, 1958, pp. 65–91.
104. Allied Chemical Corporation, Industrial Chemicals Division, *Practical Guide to Chromium Plating.* Technical Service Report 17.60R, Syracuse, New York, 1971.
105. F. A. Loewenheim and M. R. Moran, *Faith, Keyes, and Clark's Industrial Chemicals,* 4th ed., Wiley-Interscience, New York, 1975, pp. 716–721.
106. H. Uchida and A. Horiguchi, *Met. Prog.* **83,** 113 (1963).
107. U.S. Pat. 3,113,845 (Dec. 10, 1963), H. Uchida and A. Yanabu (to Fuji Iron and Steel Co., Ltd.).
108. P. Caokan, G. Barnafoldi, and I. Royik, *Bull. Doc. Cent. Inform. Chrome Dur.,* Paris, June 1971, pp. 11–A50.
109. B. Chalkley, paper presented at American Electroplaters Society meeting, Denver, Col., June 1977.
110. U.S. Pat. 2,640,022 (May 26, 1953), J. E. Stareck (to United Chromium, Inc.).
111. C. W. Ostrander in K. W. Graham, ed., *Electroplating Engineering Handbook,* Reinhold, New York, 1955, pp. 366–376.
112. F. W. Eppensteiner and M. R. Jenkins in *Metal Finishing Guidebook and Direcotry,* Metals and Plastics Publications, Inc., Hackensack, N. J. 1977, pp. 540–557.
113. C. W. Ostrander, private communication, 1976.
114. Allied Chemical Corporation, Industrial Chemicals Division, *Chromic Acid Anodizing of Aluminum,* Technical Service Applications Bulletin 103, Syracuse, 115. Ref. 26, pp. 406–422.
115. Ref. 6, pp. 406–422.
116. Allied Chemical Corporation, Industrial Chemicals Division, *Chromates for Corrosion Inhibition,* Technical Service Applications Bulletin No. 101. Morristown, N. J.
117. T. C. Patton, *Pigment Handbook,* Vol. 1, Wiley-Interscience, New York, 1973.
118. C. H. Love, *Important Inorganic Pigments,* Hobart, Washington, D.C., 1947.
119. Ref. 118, pp. 357–389.
120. W. H. Hartford in F. D. Snell and L. C. Ettre, eds., *Encyclopedia of Industrial Chemical Analysis.* Vol. 17, John Wiley & Sons, New York, 1973, pp. 197–201.
121. *Chem. Mark. Rep.* **213,** (1978).
122. Ref. 118, pp. 351–357.
123. U.S. Pat. 2,246,907 (July 30, 1940), O. F. Tarr and L. G. Tubbs (to Mutual Chem. Co. of America).
125. Ref. 118, pp. 843–861.
126. W. H. Hartford, *J. Am. Chem. Soc.* **72,** 1286 (1950).
127. U.S. Pat. 2,415,394 (Feb. 4, 1947), O. F. Tarr and M. Darrin (to Mutual Chemical Co. of America).
128. G. D. McLaughlin and E. R. Theis, *Chemistry of Leather Manufacture,* Reinhold, New York, 1945, Chapts. 14–16.
129. F. O'Flaherty, W. T. Roddy and P. M. Lollar, *The Chemistry and Technology of Leather,* Vol. 2, Reinhold, New York, 1958, pp. 221–323; Vol. 3, pp. 184–460.
130. Ref. 6, pp. 302–314.
131. R. A. Smith, *Quo Vadis Chromium,* Producers Council of Canada, Toronto, 1976 (copies available from Allied Chemical Corp.) Syracuse, New York.
132. Allied Chemical Corporation, Industrial Chemicals Division, *Leather Group Technical Bulletins* 77-I-77-IV, 77-IX, Syracuse, N. Y., 1977.
133. Ref. 6, pp. 283–301.
134. H. A. Lubs, *The Chemistry of Synthetic Dyes and Pigments,* Robert E. Krieger Publishing Co., Huntington, N. Y., 1972, pp. 153, 160, 161, 247, 258, 261, 284, 426.
135. U.S. Pat. 2,683,156 (July 6, 1954), R. F. Iler (to E. I. du Pont de Nemours & Co., Inc.).
136. U.S. Pat. 2,662,835 (Dec. 16, 1953), T. S. Reid (to Minnesota Mining & Manufacturing Co.).
137. Ernst and Ernst, *Proc. Am. Wood Preservers Assoc.* **73,** 186 (1977).
138. W. H. Hartford in D. D. Nicholas, ed., *Wood Deterioration and its Prevention by Preservative Treatments,* Vol. 2, Syracuse Univ. Press, Syracuse, N. Y., 1973, pp 1–120.
139. S. E. Dahlgren and W. H. Hartford, *Holzforschung,* **26**(2) 62 (1972).
140. *Ibid.* **26**(3) 105 (1972).
141. *Ibid.* **26**(4) 142 (1972).

142. S. E. Dahlgren and W. H. Hartford, *Holzforschung,* **28**(2) 58 (1974).
143. *Ibid.* **29**(3) 84 (1975).
144. *Ibid.* **29**(4) 130 (1975).
145. R. D. Arsenault, *Proc. Am. Wood Preservers' Assoc.* **71,** 126 (1975).
146. W. H. Hartford, *Proc. Am. Wood Preservers' Assoc.* **72**(172) (1976).
147. W. F. Rogers, *Composition and Properties of Oil Well Drilling Fluids,* 3rd ed., Gulf Publishing Co., Houston, 1963, pp. 420–422.
148. W. G. Skelly and D. E. Dieball, *J. Soc. Petrol. Engineers,* **10**(2) 140 (1970).
149. U.S. Pat. 3,532,457 (Oct. 6, 1970), K. H. Koepernik (to Kali-Chemie A.G.).
150. O. F. Joklik, *Chem. Eng.* **80**(23) 49 (1973).
151. U.S. Pat. 3,007,905 (Nov. 7, 1961), G. C. Bailey (to Phillips Petroleum Co.).
152. Ref. 32, pp. 489–498.
153. Ref. 6, pp. 385–405.
154. R. H. van Domelyn and R. D. Wehrle, *Proc. 9th Intersociety Energy Conservation Engineering Conf.* Am. Soc. Mech. Engineers, New York, 1975, pp. 665–670.
155. F. Tepper, in ref. 154, pp. 671–677.
156. U.S. Pat. 2,885,365 (May 5, 1959), A. L. Oppegard (to E. I. du Pont de Nemours & Co., Inc.).
157. R. H. Comyn, M. L. Couch, and R. E. McIntyre, *Report TR-635,* Diamond Ordnance Fuze Laboratories, 1958.
158. M. L. Kastens and M. J. Prigotsky, *Ind. Eng. Chem.* **41,** 2376 (1949).
159. W. H. Hartford, K. A. Lane, and W. A. Meyer, Jr., *J. Am. Chem. Soc.* **72,** 3353 (1950).
160. W. H. Hartford and M. Darrin, *Chem. Rev.* **58,** 1 (1958).
161. U.S. Pat. 3,824,160 (July 30, 1974), W. H. Hartford (to Allied Chemical Corp.).
162. W. H. Hartford, K. A. Lane, and W. A. Meyer, Jr., *J. Am. Chem. Soc.* **72,** 1286 (1950).
163. U.S. Pat. 2,764,553 (Sept. 25, 1956), W. H. Hartford (to Allied Chemical Corp.).
164. W. H. Hartford and K. A. Lane, *J. Am. Chem. Soc.* **70,** 647 (1948).
165. R. L. Costa and W. H. Hartford, *J. Am. Chem. Soc.* **80,** 1809 (1958).
166. F. E. Bacon, *ECT* 2nd ed., Vol 5, Wiley-Interscience, New York, 1964, pp. 451–472.
167. U.S. Pat. 3,305,303 (Feb. 21, 1967), W. H. Hartford and E. B. Hoyt (to Allied Chemical Corp.).
168. U.S. Pat. 3,309,172 (March 14, 1967), W. H. Hartford and E. B. Hoyt (to Allied Chemical Corp.).
169. B. K. Preobrazhenskii and O. M. Lilova, *Zh. Neorgan. Khim.* **8,** 779 (1963).
170. W. H. Hartford and R. W. McQuaid, unpublished data.
171. A. M. Alper and co-workers, *J. Am. Ceram. Soc.* **47,** 30 (1964).
172. W. Herwig and H. Zeiss, *J. Am. Chem. Soc.* **81,** 4798 (1959).
173. Ref. 32, p. 76.
174. W. H. Metzger, Jr., *Plating* **49,** 1176 (1962).
175. E. W. Smalley, Jr., Allied Chemical Corp., Syracuse, N. Y., private communication, 1968.
176. K. P. Yorks, Allied Chemical Corp., Syracuse, N. Y., private communication, 1968.

WINSLOW H. HARTFORD
Belmont Abbey College

CHROMOGENIC MATERIALS

PHOTOCHROMIC

Photochromism is the phenomenon in which absorption of electromagnetic energy by a system has resulted in a change in its color. It can also be defined as the phenomenon in which a single chemical species is caused, by the absorption of electromagnetic radiation, to reversibly change to a state having an absorption spectrum different than the first. The definition says nothing about the rates of the forward or reverse reaction. The word reversibly does not exclude examples having immeasurably slow reverse rates, merely those in which the state resulting from the photolysis is thermodynamically much more stable than the initial state. The photodecomposition of silver nitrate [7761-88-8] is an example of an irreversible photochemical reaction that results in a change of color. The single chemical species may be a molecule, an ion, or even a trapped electron or hole. The definition of photochromism, in terms of a single chemical species, is intended to exclude reversible processes operating through a cyclic series of reactions, eg, the photoreduction of methylene blue [61-73-4] in the presence of ferrous ions and the subsequent oxidation by air.

Figure 1 illustrates the time dependence of the absorbance of a photochromic material. A species in some initial state A, with an absorption coefficient α, is irradiated beginning at time t_1, and undergoes transitions to some other state B with an absorption coefficient β. The absorbance of the system changes at a finite rate until a steady-state value is reached when the exciting and reverse reactions are exactly balanced. Upon cessation of irradiation at time t_2, the absorbance of the system reverts to its original value.

Specification of the rates of the forward and reverse reactions is an important part of the characterization of photochromic materials. The forward reaction (and

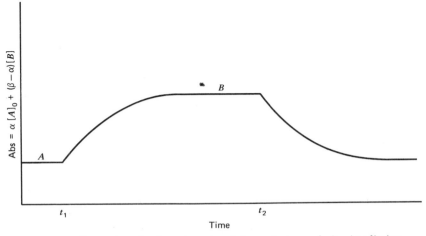

Figure 1. Absorption of a photochromic substance during and after irradiation.

sometimes the reverse reaction as well) is initiated by the system absorbing light sufficiently energetic to cause transitions from state A to state B. The rate at which this occurs depends on the intensity and wavelength distribution of the exciting light, the extinction coefficient of the photochromic compound and other components (eg, solvent), concentration, and temperature. A quantum yield of less than one is obtained if not all the A species that are activated are converted to B. The quantum loss can be attributed to deactivating processes such as fluorescence, conversion to heat, or permanent chemical change. The last is related to fatigue, ie, a gradual diminishing of photochromic response as the material is repeatedly cycled between states A and B. In many cases the reverse reaction also involves crossing a potential energy barrier, the energy being supplied either thermally or by optical excitation.

For many systems the activating radiation is in the near ultraviolet or blue part of the visible spectrum. The photochromic material is normally, but not necessarily, more highly colored in the activated state than in the initial state. In such a case, the reverse reaction is called bleaching. If the energy to cross the barrier in the reverse reaction is supplied by radiation, the process is called optical bleaching, otherwise it is referred to as thermal bleaching.

Photochromic materials, which may be either organic or inorganic, derive their character from a large number of dissimilar mechanisms, and they exhibit properties that vary enormously.

Organic Photochromic Materials

Organic photochromic materials have not yet been proven as being useful although they have been studied extensively. They are all plagued by small spectral shifts between the excited and unexcited species, slow reversal times, fatigue, or any combination of the three. Mechanical constraints imposed for certain applications also eliminates many organic materials from consideration.

The number of organic molecules that exhibit photochromism is much too large for individual discussion. Only a few are needed to illustrate the most important mechanisms leading to photochromism.

Heterolytic Cleavage. The photoactivation of an organic molecule may result in the cleavage of a single bond. The resulting segments may or may not be connected by other bonds. In heterolytic or polar cleavage the charged species formed may be fairly stable and unreactive; simple thermal rebonding to form the original molecule may be the most important reaction. As is the case for all reactions that involve broken bonds, the rate of these reactions is highly dependent on the geometry of the system. Side reactions, as already mentioned, can cause fatigue of the system.

Spiropyrans illustrate heterolytic cleavage without complete fragmentation of the molecule (1–2) (see Fig. 2). They are perhaps the most extensively studied of the systems exhibiting this type of photochromism. The spiropyrans are not usually photochromic in the solid state, but often show this phenomenon in solution. The term solution includes not only dilute fluid solutions but also viscous or even rigid systems such as gels and plastics. Since many applications require a rigid material, solutions in plastics have received considerable attention. In some cases spiropyrans may be incorporated in the polymer chains themselves by copolymerization with a suitable side chain monomer (3). Little quantitative data on the fatigue in plastic films has been published but it is known that fatigue does occur faster with increasing concentration (4).

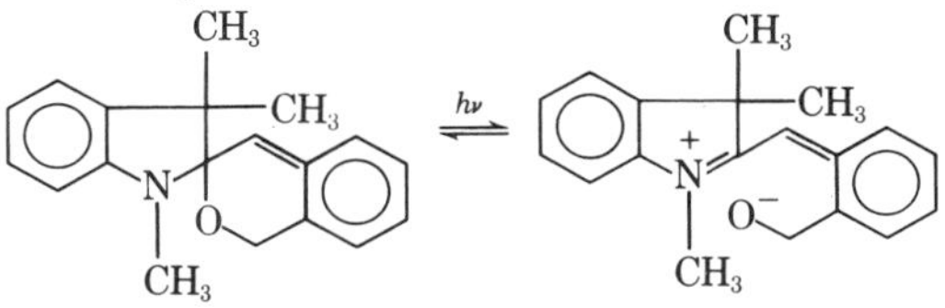

Figure 2. Heterolytic cleavage.

In compounds that undergo heterolytic cleavage into fragments not connected by other bonds, the reverse reaction is bimolecular and the thermal fade rate is even more highly dependent on the environment than it is in spiropyrans.

Homolytic Cleavage. The absorption of light can lead to the rupture of a single bond in such a way that it produces radicals. In many cases the homolytic rupture of C–C, N–N, S–S, C–N, or C–Cl bonds gives rise to photochromism. At present, little or nothing is known about systems in which a C–O, C–S, N–O, N–S, or N–Cl bond undergoes a reversible photolytic scission.

The bis(imidazoles) are believed to undergo a C–C cleavage (5) when irradiated with ultraviolet, and absorption in the visible results (see Fig. 3). These compounds show photochromism not only in liquid solutions, but also in solid solution in KBr pellets and even in the pure crystalline state. In the crystalline state, the decay of the photochromic darkening is very slow at room temperature.

In frozen benzene solutions of the imidazoles an unusual form of light storage occurs. In the temperature range between −70° and −20°C, irradiation does not cause darkening. However, if the material once irradiated is allowed to warm to 15°C in the dark, the coloration slowly appears.

The potential for incorporating these compounds directly in a polymer structure (6) makes them attractive for the many applications requiring rigid materials.

Cis–Trans **Isomerization.** Photochromism has been observed in compounds that can isomerize about a single ethylenic group such as the stilbenes as well as in compounds that isomerize about a nitrogen–nitrogen double bond or a carbon–nitrogen double bond (7). All of these photochromic materials share a number of properties. In principle, both the *trans* → *cis* and the reverse process can be brought about by irradiation. In nearly all the cases studied, the *trans* isomer is more stable; hence, only the *cis* → *trans* process occurs thermally. The rate of the thermal process depends not only on the compound but also on the solvent, pH value, and sometimes on specific catalysts. Thermal fade times as short as milliseconds and as long as months have been observed.

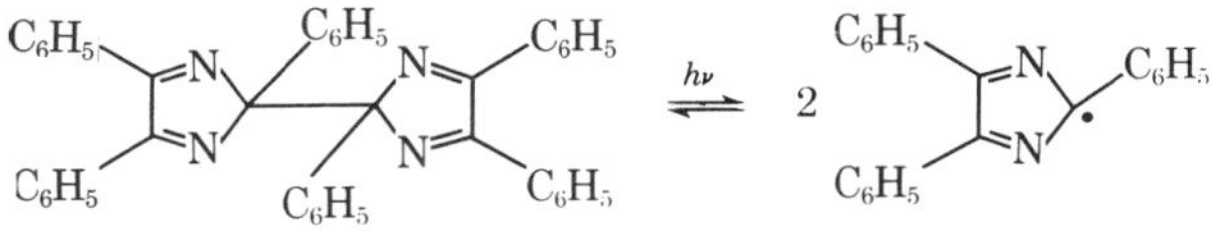

[*39915-24-7; 52603-38-0*]   [*35398-30-0*]

Figure 3. Homolytic cleavage.

In general, the quantum yield for the *trans* → *cis* isomerization increases with temperature and is sensitive to solvent effects. By contrast, the *cis* → *trans* isomerization quantum yield is rather insensitive to temperature and environment.

The absorption bands of both forms are normally broad enough so that there is considerable spectral overlap. Consequently, it is seldom possible to obtain 100% of either isomer by photolyzing a mixture of the two.

The compounds undergoing *cis–trans* isomerization have only limited applicability as photochromic materials not only because of the small spectral shifts normally encountered but also because of the very serious fatigue problems encountered.

Photoinduced Tautomerism.   The two major forms of tautomerism are hydrogen transfer and valence tautomerism. The latter overlaps with other categories already discussed. For example, heterolytic cleavage in the spiropyrans is also an example of valence tautomerism.

Valence tautomerism is also illustrated by the fulgides (see Fig. 4). Coloration occurs upon irradiation of the crystal at room temperature. Reversal requires heating or irradiation with visible light. Some of the fulgides exhibit minimal fatigue but their long reversal time limits their usefulness to information storage applications.

Reversible photoenolizations, ie, H transfer tautomerism (9), have been observed in a number of *ortho*-substituted phenyl ketones. In this rearrangement a hydrogen migrates to the carbonyl oxygen. The lifetimes of the enols are usually very short, making photochromism difficult to observe at room temperatures. Unfortunately, these reactions generally have poor photochromic reversibility. The hydrogen-transfer step often has directly competing photochemical side reactions.

Triplet States.   Absorption of light in the uv region by an organic molecule leads to the formation of excited molecular states. If the molecule reaches a triplet state, that is a state in which unpaired spins are parallel, the lifetime of the excited state can be long enough so that a significant change in the color of the material occurs. For several reasons, molecules with reasonably long-lived triplet states seem promising as photochromic materials. No bonds are broken in the formation of a triplet state so the reverse reaction is not slowed by a viscous or even rigid matrix and is virtually independent of temperature. As a further consequence of the fact that no bonds are broken, photodegradation can be expected to be minimized. The triplet-state absorption is fairly broad so that absorption over appreciable regions of the visible can be anticipated.

A large number of polycyclic hydrocarbons have been investigated for photochromic behavior (10). Fatigue, although not as bad as that exhibited in other kinds of organic photochromics, is still a serious problem. Both the action of oxygen and photodegradation of the host can contribute to the fatigue. Rather strong luminescence

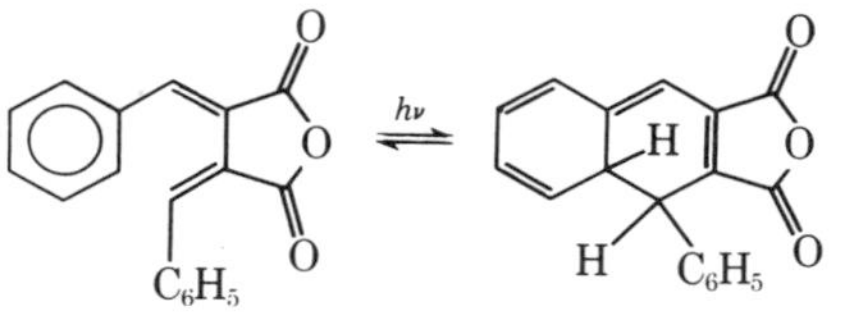

[53244-00-1]

Figure 4.   Valence tautomerism.

characterizes almost all these materials and is undesirable for some applications. In many cases, because the lifetimes of the excited state are not long enough, the materials do not darken sufficiently when exposed to sunlight to be useful for sunglass applications. Deuteration increases the triplet lifetime of polycyclic hydrocarbons and the amount of darkening they show (11).

Inorganic Photochromic Systems

Most inorganic photochromic materials are solids that have a band gap in excess of three eV and consequently do not absorb in the visible in their unactivated state. Absorption of light having energy equal to or greater than the band gap creates electron-hole pairs. Color centers are formed by the trapping of the electrons and holes at defects in the structure. Photochromism, when it is observed in inorganic solids, is determined by the nature of the defects characteristic of the solid.

The alkali halides are the most studied of the crystals containing Schottky defects. When an alkali halide is irradiated with ultraviolet at low temperatures ($\sim$77 K), electrons are trapped at anion vacancies where they absorb visible light (12). The position of the absorption band, called an F band, varies with the lattice constant of the alkali halide. In spite of trapping depths estimated (13) to be in excess of 1.7 eV, these centers slowly, thermally bleach at room temperature. Irradiation in the F center band leads to optical bleaching.

Unlike many other photochromic materials, these systems do not return to their original state when optically bleached. Instead, new color centers are formed. The interaction of vacancies gives rise to a plethora of different defects and a corresponding number of different color centers. Sodium azide [26628-22-8] shows many of the same phenomena as the alkali halides (14). It is, perhaps, because of the complexity of these systems that most authors do not consider the darkening reversible and choose not to use the name photochromism to describe the coloring and bleaching of these systems.

A study (15) of calcium fluoride doped with La, Ce, Gd, and Tb showed that irradiation of these materials produces absorption bands characteristic of the rare earth impurities. Reversal may be stimulated optically as well as thermally. Photochromism has been observed in doped titania crystals and in doped titanates (16–19).

Doped anatase, TiO_2, is not photochromic, whereas mixtures of rutile, TiO_2, and anatase are. When doped with Fe, a mixture containing 5% rutile showed the strongest photochromism. In the case of Cr and Ni doping, darkening was stronger with increasing rutile content illustrating again the structure-sensitive nature of photochromism. Magnesium titanate, $MgTiO_3$, which exists in the close-packed ilmenite structure, is not photochromic (20), whereas strontium titanate [12060-59-2] ($SrTiO_3$), calcium titanate [12049-50-2] ($CaTiO_3$), and barium titanate [12047-27-7] ($BaTiO_3$), which all have a perovskite structure, are photochromic when doped with Fe^{3+}, Sb^{5+}, or V^{5+}.

Hackmanite [1302-90-5], $Na_2O.Al_2O_3.2\,SiO_2.\frac{1}{2}NaCl$, sodalite, is a naturally occurring mineral which shows photochromism because of the sulfur impurities. The color center responsible for the induced absorption is an F center, ie, an electron trapped at a halogen ion vacancy (21). This was confirmed by esr measurements (22). The initial step in the photolysis of hackmanite is believed to be the absorption of light by S_2^{2-} to produce S_2^{-} and an electron, which is trapped.

The existence of Frenkel defects (23) is important to photochromism in silver halides (see under Photochromic glass).

Photochromism has been observed in a large number of complex mercury salts (24). The mechanisms of darkening have not been clearly elucidated. Photochromism in copper chloride was ascribed to a disproportion reaction (25). The spectrum observed in a darkened aqueous solution of copper chloride can be explained on the basis of light scattering by slightly elongated colloids of copper (19). Until now, no application has been reported for either the mercury salts or the copper chloride except for the use of the latter in photochromic glasses.

Photochromic Glass

Photochromic glass has several advantages over other photochromic materials simply because it is glass. By standard glass melting and forming techniques, it can be made in any desired size or shape. Its transparency and durability under chemical attack, scratching, or moderate heating, offers advantages for some applications (see Glass). Most important is that photochromic glass usually shows no fatigue. Photochromism is occasionally observed in homogeneous glass as well as in several varieties containing a suspension of crystallites (26). In the latter case, photochromic glass has the additional advantage that its physical properties can often be varied independently of its photochromic properties.

Homogeneous Glass. There are several families of homogeneous glasses that darken reversibly in response to ultraviolet excitation. Certain strongly reduced alkali–silicate glasses are photochromic (27). Doping with cerium, europium, or zirconia considerably strengthens the induced darkening. The dopants do not, however, appreciably influence the position of the peak absorption. This depends only on the specific alkali used in the glass, illustrating the importance of structure for glasses as well as for inorganic crystals.

The darkening in these glasses is probably caused by the trapping of an electron in an oxygen-deficient site and thus is analogous to an F-center in a crystal. The fading is essentially complete in five minutes after cessation of the irradiation. Moreover, the fading time actually decreases with each darkening and bleaching cycle with an attendant diminishing of darkening level. Heating the glass to the annealing point restores the original photochromic properties. Optical bleaching is also noted in these glasses.

The tendency of reduced alkali silicates to fatigue is unique among photochromic glasses and limits their usefulness. Furthermore, the reducing conditions required during melting make it difficult to make samples of these photochromic glasses larger than a few hundred grams.

In a family of homogeneous glasses based on the $CdO-B_2O_3-SiO_2$ system, the darkening depended primarily on the level of cadmium oxide, the optimal level being about 50 mol % (28). Doping with copper or silver increased the darkening sensitivity slightly. These glasses fade much more slowly than the reduced alkali silicates; however, they do not fatigue and they are easily made in large quantities.

Many glasses having a stoichiometry similar to nepheline, $Na_2O.Al_2O_3.2SiO_2$, would, if doped with sodium chloride and sodium sulfide, darken when irradiated with short wavelength ultraviolet (253.7 nm) (29). Appropriate heat treatment causes the precipitation of a phase identified by x-ray diffraction as hackmanite. Unfortunately, high contrast between darkened and undarkened state is not obtained in either the glass–ceramic form or the glassy form of these materials.

Suspension of Crystallites in Glasses. Photochromic materials have been made by precipitating silver molybdate [*13765-74-7*] or silver tungstate [*13465-93-5*] from a sodium aluminoborosilicate-base glass (30). These materials were characterized by low darkening efficiencies.

Thallium halide can be precipitated by heat treatment from a potassium barium aluminophosphate glass and the resulting material can be darkened by exposure to near uv (31). Darkening is considerably enhanced by doping with copper or silver. Thermal fading is quite slow. Half-fading times are in tens of hours but optical bleaching is quite fast.

Transparent photochromic glasses containing crystallites of mixed copper and cadmium halides (32) can darken in sunlight to ca 20% in less than 10 min, and the half-fade times are about a half hour or less. These glasses show no optical bleaching. The green color that characterizes the darkened state makes them an attractive candidate for use as automatic sunglasses.

Silver halides can be precipitated to produce transparent photochromic glasses (33). The precipitated crystallites range in size from about 8–15 nm. When the particles are smaller than 8 nm, the glasses do not darken well; when larger than 15 nm, the glass is hazy. The average spacing between particles is of the order of 100 nm.

A small amount of copper increases the darkening by several orders of magnitude by trapping the holes created during photolysis. Electrons are trapped by interstitial silver ions (Frenkel defects) that aggregate to form light-absorbing specks. Optical bleaching is the result of the ejection of an electron from the silver speck upon the absorption of visible light. Thermal bleaching occurs when an electron tunnels from the silver speck to a cupric ion. Diffusion of the copper ions replenishes the supply of cupric ions close enough to the speck for tunneling to occur.

When copper-doped silver halides are precipitated in certain lanthanum borate-base glasses, a dark state is the stable state but the coloration can be optically bleached (34).

If a piece of photochromic glass is elongated under a sufficiently high stress, it polarizes light in the darkened state (35) because the silver halide particles become elongated and the material therefore exhibits anisotropy in its refractive index.

Applications

Many proposed applications of photochromic materials are related to their ability to act as a light valve; other applications are based on the change in the material itself. Examples of the latter include use of photochromic materials as uv or visible dosimeters, pigment for suntanning dolls, and self-developing photographic film (see Color photography, instant). Transparency is not important for these applications. Information storage is perhaps the most studied application of this type (34).

Examples of the use of photochromics as light valves include suggestions for an automatic photochromic camera iris, photochromic windows, sunroofs, and protective devices. The use of photochromics for protection against sudden high intensity flashes such as nuclear explosions or lasers has been proposed (10). The use of photochromic glasses as automatic sunglasses is already a commercial reality.

Many photochromic materials have been carefully studied and many applications for photochromics have been proposed. In most cases, the proposed applications required a combination of properties not exhibited by any single photochromic material. Use of photochromics in the field of information storage and display in particular has received much attention but has not yet been commercially successful. Only in the ophthalmic field have photochromics had commercial success and thus far only photochromic glasses containing silver halides have been successfully used in that application.

BIBLIOGRAPHY

1. D. A. Reeves and F. W. Wilkinson, *J. Chem. Soc. Faraday Trans. II* **69**, 1381 (1973).
2. G. Smets, *Pure Appl. Chem.* **30**, 1 (1972).
3. G. Smets, *Polym. Prep. Am. Chem. Soc.* **9**, 211 (1968).
4. Jpn. Pat. 28,289 (1970), H. Ono and co-workers.
5. T. Hayashi and K. Maeda, *Bull. Chem. Soc. Jpn.* **38**, 2202 (1968).
6. K. Akagane and co-workers, *Shikizai Kyokaishi*, **45**, 159 (1972).
7. G. M. Wyman, *Chem. Rev.* **55**, 625 (1965).
8. A. Santiago and R. S. Becker, *J. Am. Chem. Soc.* **90**, 3654 (1968).
9. J. D. Margerum and L. J. Miller in G. H. Brown, ed., *Photochromism*, Wiley-Interscience, New York, 1971, Chapt. VI, p. 557.
10. W. R. Dawson and M. W. Windsor, *Appl. Optics* **8**, 1045 (1969).
11. U.S. Pat. 3,475,507 (Oct. 28, 1969), J. A. Sedlak (to American Cyanamid Co.).
12. J. H. Schulman and W. D. Compton, *Color Centers in Solids,* Pergamon Press, London, Eng., 1962.
13. N. F. Mott and R. W. Gurney, *Electronic Processes in Ionic Crystals,* Dover, New York, 1964.
14. G. J. King and co-workers, *J. Chem. Phys.* **35**, 1442 (1961).
15. D. L. Staebler and Z. J. Kiss, *Bull. Am. Phys. Soc.* **12**, 670 (1967).
16. B. W. Faughnam and Z. J. Kiss, *Phys. Rev. Lett.* **21**, 1331 (1968).
17. Z. J. Kiss and W. Phillips, *Phys. Rev.* **180**, 924 (1969).
18. W. A. Weyl and T. Forland, *Ind. Eng. Chem.* **42**, 257 (1950).
19. J. Baer and F. K. McTaggert, *J. Appl. Chem.* **5**, 643 (1955).
20. S. M. Neuder, *NASA-TND2258,* (1964).
21. D. B. Medved, *Am. Mineral.* **21**, 764 (1936).
22. W. G. Hodgson, J. S. Brinen, and E. F. Williams, *J. Chem. Phys.* **47**, 3719 (1967).
23. J. Frenkel, *Z. Physik* **35**, 52 (1926); ref. 13, p. 26.
24. G. H. Brown and W. G. Shaw, *Rev. Pure Appl. Chem.* **11**, 1 (1961).
25. G. H. Hecht and G. Miller, *Z. Physik. Chem.* **202**, 403 (1953).
26. R. J. Araujo in M. Tomozawa and R. H. Doremus, eds., *Treatise on Materials Science and Technology,* *Vol. 12,* Academic Press, Inc., New York, 1977, p. 91.
27. A. J. Cohen and H. L. Smith, *Science U.S.A.* **137**, 981 (1962).
28. U.S. Pat. 3,615,771 (Oct. 26, 1971), G. S. Meiling (to Corning Glass Works).
29. U.S. Pat. 3,923,529 (Dec. 2, 1975), R. J. Araujo, G. H. Beall, and L. G. Sawchuk (to Corning Glass Works).
30. U.S. Pat. 3,293,052 (Dec. 20, 1966), L. G. Sawchuk and S. D. Stookey (to Corning Glass Works).
31. J. D. MacKenzie and S. Sakka, *Bull. Chem. Soc. Jpn.* **46**, 848 (1973).
32. U.S. Pat. 3,325,299 (June 13, 1967), R. J. Araujo (to Corning Glass Works).
33. W. H. Armistead and S. D. Stookey, *Science* **144**, 150 (1964).
34. T. P. Seward, III, *J. Appl. Phys.* **46**, 689 (1975).
35. U.S. Pat. 3,540,793 (Nov. 17, 1970), R. J. Araujo, W. H. Cramer, and S. D. Stookey (to Corning Glass Works).

General References

G. H. Brown, ed., *Photochromism,* Wiley-Interscience, New York, 1971.
G. H. Dorion and A. F. Wiebe, *Photochromism,* Focal Press, London, Eng., 1970.

ROGER J. ARAUJO
Corning Glass Works

ELECTROCHROMIC AND THERMOCHROMIC

Electrochromic Materials

Electrochromism (1) is a color change caused by an electric current. It has found some use in determination of dipole moments of excited molecules and in the study of photosynthesis. Many uses are proposed in the patent literature for display devices and duplication of images (see also Chemiluminescence; Digital displays; Electrophotography).

Solid Electrochromic Materials. An electrochromic cell consists of the electrolyte solution and a pair of transparent electrodes, one of which bears the electrochromic material.

The electrodes usually consist of a thin layer of stannic oxide (NESA glass), SnO_2, indium (2), or cadmium stannate, Cd_2SnO_4, on glass (3). The electrochromic material is most often tungsten trioxide [1314-35-8], WO_3, although the whole range of transition metal oxides appears in patents (4–9). The electrolyte layer may be sulfuric acid with gelling agents and metal salts added to improve conductivity and reduce leakage (10–15). The completed cell may be encapsulated to preserve its integrity.

When about 1 V is applied, the WO_3 layer turns dark blue; reversing the current restores the original color.

The mechanism of color formation in the WO_3 cells probably involves the simultaneous migration of positive ions, such as hydronium or alkali metal (16) from the electrolyte layer, and electrons from the cathode into the WO_3 to form a tungsten bronze (17). Films prepared under different conditions, such as by evaporation, sputtering, or spraying, differ by orders of magnitude in their electrochromic sensitivity (18–19). The use of alkali and other metals as dopants gives a variety of color effects (20–21). It has been shown that a chemical reaction occurs, since the colored material is soluble in water whereas the uncolored material is not. The presence of some water in the WO_3 is necessary (22–23). In the case of vanadium pentoxide [1314-62-1], V_2O_5, immersed in distilled water, a negative potential changes the color from yellow through gold, azure, dark blue, and black, but the effect is stated not to be due to electroreduction or the formation of a bronze (24).

Potential uses appearing in the patent literature include a variety of display devices such as instrument faces and signs, the formation of optical images (25–26), writing with light (27–28), and facsimile transmission (29).

Liquid Cells. Electrochemical cells that reduce a soluble organic material to a colored form in cells of similar construction to those discussed above have been investigated using a variety of color-forming materials. These include salts of aromatic substituted pyridines and bipyridines (30–39), pyrazolines (40), and picrylhydrazyl compounds (41), as well as solid complexes of diphthalocyanine with lanthanide metals as the negative electrode. A full range of colors is generated in lutetium diphthalocyanine [58692-72-1] by variation of the applied voltage (42–43). The salts of 1,1'-dialkyl-4,4'-bipyridinium ion (viologens) have been studied extensively (44–48). In a different kind of cell, a pair of metal electrodes is placed in a transparent case with the electrolyte; one of the electrodes is anodizable and its color is a function of applied voltage (46).

Photosynthesis. In photosynthesis a burst of light causes a shift in the spectrum of the photosensitive pigment molecule whose orientation is fixed by the membrane in which it resides. A study of thin layers of chlorophyll and carotenoids in oriented films has been made using the electrochromic effect. Correspondence of the two phenomena demonstrate that this part of the photosynthetic reaction is the impression of an electric field rather than a conformation change in the membrane. Thus electrochromic measurements are useful for the calculation of electric field changes in biological membranes. For these experiments the electrochromic cell is in effect a capacitor of very thin, semitransparent aluminum plates, with the electrochromic material molecularly oriented in a thin film (49–56).

Dipole Measurement. Platt's electrochromic effect is a shift of absorption without any chemical reaction. The effect is described by the following equation:

$$\Delta E - \Delta E_o = h\,\Delta\nu_o = \Delta\mu \cdot F - \tfrac{1}{2}\,\Delta\alpha F^2$$

where μ is the component of the dipole moment and α is the polarizability, both in the direction of the applied field. $\Delta\mu$ is the change in dipole moment between electronic states and $\Delta\alpha$ is the difference in polarizabilities. Since the change in dipole moment between ground and excited states is proportional to the impressed field strength, then measurement of wavelength shifts at different field strengths allows calculation of $\Delta\mu$ and $\Delta\alpha$. This measurement of change in dipole moment gives also the dipole moment of the excited state when that of the ground state is known. It is similar to the polarizability, whose variance is a quadratic function of field strength (57–60). The theory and quantum mechanics have been discussed at length (61). The electrochromic material can be studied as a solid matrix in a polymer which provides orientation of the molecules and prevents rotational motion (62–65). Gas-phase electrochromism has been studied for aniline, yielding dipole moment data for the transition from ground to excited state in good agreement with previous work (66).

Thermochromic Materials

Thermochromic compounds change color reversibly with temperature, gradually in solution, sharply on melting. Potential uses are many but the best established is the irreversible system that leaves a record of heat distribution on an engine part or a reaction vessel (67). In reversible thermochromism the number of color change cycles may be very great but practical uses for continuous repetitive cycles have been limited by irreversible side reactions. Thermochromism has also found a place in chemical analysis. About 2000 compounds have been reported as thermochromic, although the number of compound types is small. Some representative organic compounds are shown in Table 1. The mechanism of the thermochromic transition varies with compound type. Much of the literature is more concerned with mechanisms than with documenting colors, temperatures, and possible uses. The literature has been reviewed to 1963 for organic compounds (80) and to 1968 for inorganic compounds (81).

Reversible Color Changes of Organic Compounds. Many compounds that can be designated as overcrowded ethylenes are thermochromic. Structural features in common are at least one ethylene group, a multiplicity of aromatic rings, and a hetero atom, usually nitrogen or oxygen. The ethylene bond provides a route for extension of conjugation and places a restriction on possible molecular shapes, whereas the hetero atom provides a source or sink for excess electron charge. The exact mechanism is not worked out in all cases and more than one point of view has been expressed (71,82–84).

Table 1. Some Representative Types of Organic Thermochromic Compounds

Name	CAS Registry Number	Structure	Ref.
tetrakis(*p*-methoxyphenyl)ethylene	[10019-24-6]	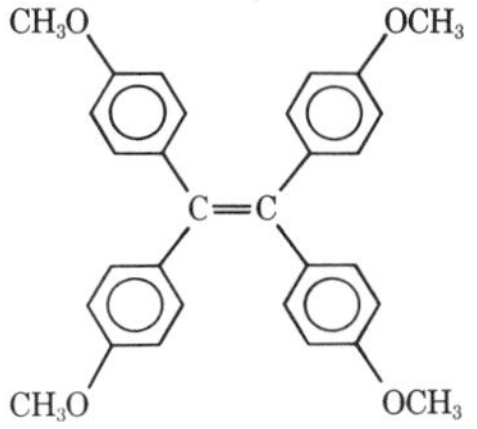	68
3-[2-(2-hydroxy-1-naphthyl)vinyl]-5,5-dimethyl-2-cyclohexene-1-one	[28785-69-5]	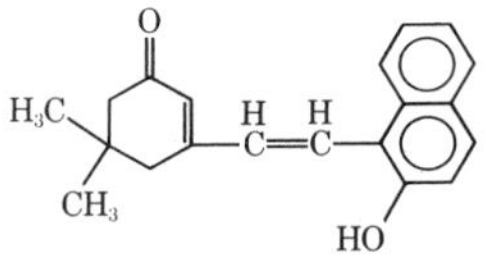	69
2-phenyl-4-(2-phenyl-4*H*-1-benzopyran-4-ylidene)-4*H*-1-benzopyran	[4388-05-0]	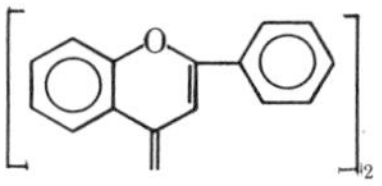	70
10-(10-oxo-9(10*H*)-anthracenylidene)9(10*H*)-anthracenone	[434-85-5]	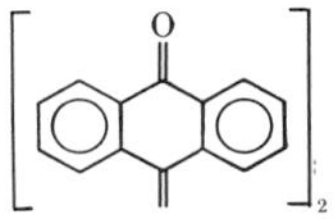	71–72
4-[[3-(4-methoxyphenyl-4-methylbenzo[*f*]quinolin-1(4*H*)-ylidene]ethylidene]-2,5-cyclohexadien-1-one	[56046-07-2]	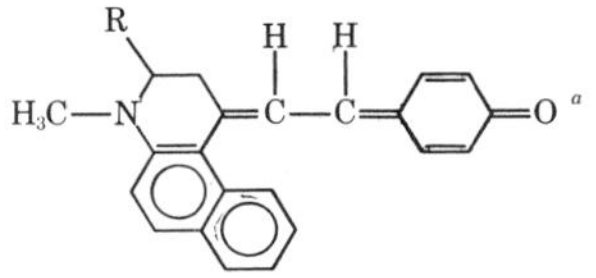 a	73
4-[2-(2,3-dihydro-9,9-dimethyloxazolo[3,2-α]indol-9a(9*H*)-yl)ethenyl]-*N,N*-dimethylbenzeneamine	[59335-14-7]	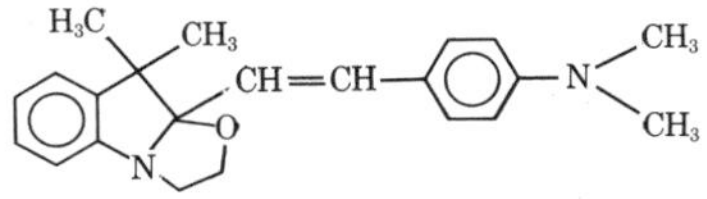	74
4,4'-(2*H*-naphtho[1,8-*bc*]]furan-2-ylidene)bis[*N,N*-dimethylbenzeneamine]	[56935-83-2]	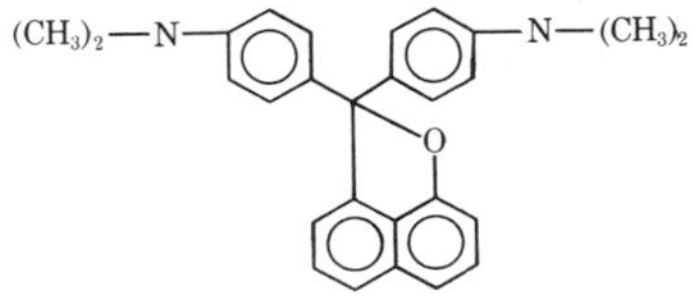	75
oxidized dimer of 2,4,5-triphenylimidazole	[484-47-9]	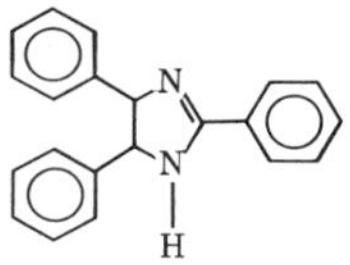	76–77
1-(4-hydroxyphenyl)-2,4,6-triphenyl pyridinium perchlorate	[56524-90-4]	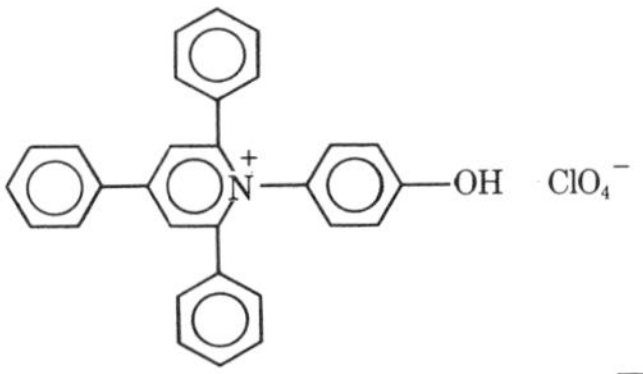	78
1-(2-benzothiazolyl)-3-(1-methylethyl)-5-(nitrophenyl) formazan	[52644-55-0]	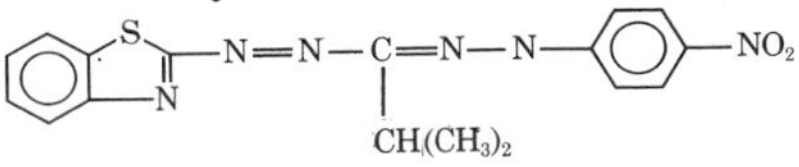	79

a R = 4-methoxyphenyl

131

In solution the color change is reversible for a large number of cycles; in the solid state the reversibility depends on the length of time at higher temperatures.

Spiropyrans become deeply colored on heating; eg, colorless di-β-naphthospiro-pyran [178-10-9] (1) gives a blue–violet melt. The spiro molecule itself (1) is colorless, or has color conferred by substituents, whereas the colored form (2) is one in which the spiro ring has opened to give a polar structure stabilized by a number of resonance structures.

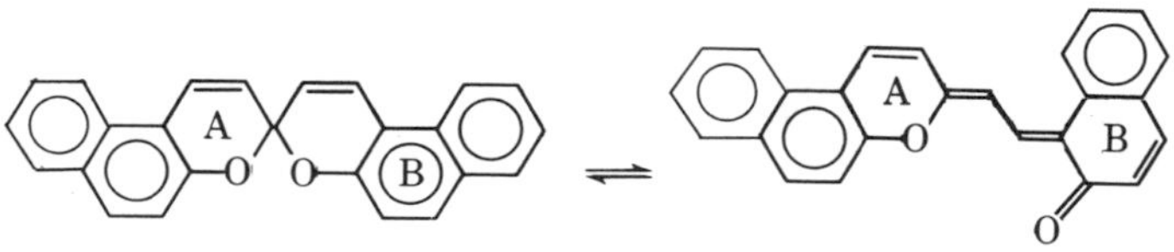

colorless spiropyran   colored form
(1)   (2)

Color formation is favored by heat, polar solvents, or adsorption on silica, magnesia, or alumina. The open form is present as a salt in acid solution. The essential units of thermochromic structure are (1) the pyran ring containing the spiro carbon, (2) a heterocyclic ring A which is usually fused to one or more aromatic rings, and (3) the ring B which is a system of one or more fused aromatic rings. The indolinobenzospirans have been of particular interest. The chemistry of the spiropyrans, their preparation, and spectra have been reviewed (85–91). Although most of the spiropyrans have a naphtho- or substituted benzopyran ring, a number of A rings have been used, such as thiazolidine, xantho, and thiaxantho, as well as a acridine and pyran in an iso–spiro configuration. Spirofurans are also reported as thermochromic (74–75).

Anils (3) (92), pyridine or quinoline quaternary amines (93), and naphthothiazoles (94) are generally yellow, turning orange to red on heating. The thermochromic transition is due to an equilibrium between the hydroxy form and the quinone (95–98).

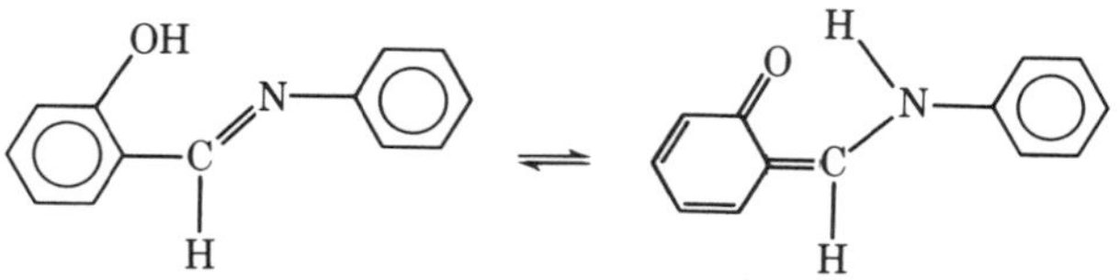

anil   colored form
(3)   (4)
[779-84-0]   [52828-01-0]

Low-molecular-weight polymers with two to six salicylidene units (4) are made by the reaction of dialdehydes with diamines. The brightly colored polymers, from yellow to red, precipitate almost immediately on mixing the reactants. Some of these polymers are reversibly thermochromic (99). Dimers of triphenylimidazole formed by oxidative degradation are thermochromic because of dissociation of the dimer into free radicals. The color change is from pale yellow to reddish purple (76,100). Dimers of tetraphenylpyrrole also give piezo- and photochromic forms, both of which are thermochromic (77,101). Quinones that bear a —NHCH$_2$CH$_2$OH substituent are light in color but become deeply colored when their solutions are heated, owing to a quinol–quinone equilibrium. A patent was issued on their use in the paper or ink of official documents so that they can be easily tested for genuineness (102–105).

Pyridinium-N-phenol betaines are ionic compounds whose solutions are markedly thermochromic. The thermochromic mechanism in this case is due to the lowering of the ground-state energy of the betaine by polar attraction to the solvent molecule. The thermal effect is to decrease this orientation (78). Triphenylmethane dyes such as malachite green [569-64-2] and pararosaniline [569-61-9] are thermochromic in bisulfite solution, reversibly if the SO_2 formed is not allowed to escape (106–108).

Thermochromic materials that are a combination of an electron-donating compound that forms a weak association with an electron-accepting material may form a brightly colored complex in the dissolved or melted state, which vanishes on cooling or freezing. These behave essentially as mixtures in the solid state, but in the liquid state form colored complexes. Electron donors include aromatic amines and substituted hindered aromatics; the acceptors include nitro-substituted aliphatic or aromatic compounds and heterocyclics. At least two color changes at different temperatures may be obtained from a mixture of materials (109–110). Dyes in admixture with acidic reagents have been incorporated into polymers yielding thermochromic properties. Typically, a dye such as crystal violet [548-62-9] or malachite green may be mixed with a phenol, organic acid, or HI and the mixture incorporated into a polymer film (111–112). Poly(vinyl acetal) [27790-26-7] resins with some methanol present are transparent at room temperature but turn milky and opaque on heating. Calcium chloride intensifies the opacity and lowers the transition temperature (113). Complex formazans in aqueous-organic media are colored in the aci-form (79). Nitrobenzylidenepyridinecarboxylic acid hydrazides have been called photothermochromic; the pale yellow compound turns red in sunlight and the color can be bleached by gentle heating (114). The polymer of at least one diacetylene monomer is green at temperatures below about 135°C and red above that temperature, because of the establishment of resonance-stabilized domains of extended conjugation (115).

Reversible Color Changes of Inorganic Compounds. A few organic complexes of metal ions are reversibly thermochromic. The nitrogen bases complexed with Cu(II) (116–128) or Ni(II) (129–134) are most numerous, their color change being caused by a change from tetrahedral to square-planar geometry or by a change in the number and kinds of ligands present. Very important to the color change are the anions, most often ClO_4^- or BF_4^-, and the choice of solvent. β-Diketone complexes in pyridine shift colors to longer wavelength or more intense absorption (122). The copper complexes are listed in Table 2 and miscellaneous compounds in Table 3. Complexes with Ni(II) behave similarly to the copper complexes listed in Table 2, except that thermochromism occurs with a greater variety of anions present and at higher temperatures (118–119,130–133).

The double salts Ag_2HgI_4 and Cu_2HgI_4 have been extensively studied (see Table 4). The thermochromic transition is an order–disorder transition in which the metal ions may exchange places in the lattice. Fast heating rates make the color change of the silver salt take place over a larger temperature range; a sharply defined change is found if the material is held just below the transition temperature so that a rise of a degree would be effective. If the material is held for extended time above the transition temperature a slow decomposition takes place (144–152). The copper salt has been investigated as an infrared sensitive surface for recording images (149–150), as

Table 2. Thermochromism of Solid Copper Complexes

Compound or ion	CAS Registry Number	Color At room temp	At °C	Ref.
$CuX_n(ONO)_{2-n}(NH_3)_2$				
$n = 0$	[21710-47-4]	purple	green, 30	124
X = Cl, $n = 0.15$		green	purple, −5	124
X = Cl, $n = 0.30$		green	purple, −50	124
X = Br, $n = 0.55$		green	purple, ±10	124
X = Br, $n = 0.23$		green	purple, −40	124
$[(NH_3CH_2CH_2)_2NH_2]Cl(CuCl_4)$	[56508-39-5]	yellow	pale green, low temp, orange–brown, 120	125
$[(C_2H_5)_2NH_2]_2CuCl_4$	[52003-08-4]	green	yellow, 43	116
			yellow, 67	126
$[(CH_3)_2CHNH_3]_2CuCl_4$	[52003-02-8]	green	yellow, 60	116
$\left[\begin{matrix} NH_2 \\ \parallel + \\ CH_3C\!=\!\!=\!NH_2 \end{matrix}\right]_2 CuCl_4{}^{2-}$	[22142-86-5]	bright yellow green on cooling		127 128
$[(C_2H_5)_2NCH_2CH_2NH_2]_2CuX_2$				
X = BF_4^-	[52646-61-4]	red	purple, 15	118–120
X = ClO_4^-	[52646-62-5]	red	purple, 65	118
			violet, 35	117
X = NO_3^-	[52646-63-6]	red	purple, 160	120
Cu(II)–picoline	[57358-74-4]	violet–blue	yellow	121

a temperature indicator (153), and for calendering rolls (154). Silver iodide–lead iodide salts in various proportions turn from yellow to orange or red at 97–137°C (134,144).

Aqueous solutions of some simple metal salts are thermochromic when heated to 230°C. These include $FeCl_3$ [7705-08-0], $CuCl_2$ [7447-39-4], $CoCl_2$ [7646-79-9], $NiCl_2$ [7718-54-9], UO_2Cl_2 [7791-26-6], hexachloroplatinic acid [16941-12-1], and tetrachloroauric acid [16903-35-8] (155–156). The formation of weak complexes between organic electron-donor molecules, such as benzene and toluene, with inorganic electron acceptor molecules include $VOCl_3$, $TiCl_4$, UF_6, OsO_4, and the like. Acceptor–donor bonding is responsible for the color; the loss of color on freezing is due to phase separation (157). Thermochromic ceramic coatings may be based on lead (158) or cadmium salts (159). Some molybdenum salts are thermochromic (160–161).

Fluorescence Thermochromism. Monovalent copper halides complexed with nitrogen bases show strong ultraviolet fluorescence. The fluorescence color depends on the temperature and is called fluorescence thermochromism. Some compounds go through several changes in fluorescence color on cooling from room temperature to −190°C. The color changes are reversible and the mechanism is not clear. The number of cycles possible has not yet been checked. Possible applications have scarcely been explored, though patents have been issued on the use of impregnated paper whose genuineness can be checked by its color under uv light at different temperatures. Possible analytical applications are the identification of nitrogen bases by use of a cuprous iodide-impregnated paper; even isomers are distinguishable. It is remarkable that so far as is known only copper compounds show the property of fluorescence

Table 3.   Thermochromism of Miscellaneous Metal Complexes

Compound	CAS Registry Number	Color		Ref.
		At room temp	At °C	
$Sn[(CH_3)_2NCS_2]_2{}^a$	[30051-58-2]	orange	gold, -55	135
$Co\ (CH_3(CH_2)_3CH(CH_2CH_3)\text{-}CH_2OPO_3)_2$	[69070-53-7]	pink	deep blue, 50	136
$CoCl_2 + (NH_2)_2C{=}NH.HCl^b$		blue	pink, above 30	137
$[Co(CO_4)]_2\ Hg^c$	[13964-88-0]	pink	blue, 70–140	138
$[(C_2H_5)_4N]VCl_4.2CH_3CN^d$	[19817-98-2]	red–brown	yellow, -196	139
$Cd_2P_3Cl^d$	[37204-84-5]		light orange, -196; black, 300	140
$[(SiCl)_n + CaCl_2{}^a]^e$		yellow–brown	red, above 200	141
$(CH_3)_2Se_2{}^f$	[7101-31-7]		yellow, -196; red, 23	142
$SrTiO_3{}^d$	[12060-59-2]		function of annealing	143

^{a} Solution.
^{b} In sealed tube.
^{c} In molten imidazole.
^{d} Solid.
^{e} In nitrogen atmosphere.
^{f} On melting or in hydrogen atmosphere.

Table 4.   Thermochromic Iodide Double Salts

Compound	CAS Registry Number	Formula	Color change at °C	Ref.
silver mercury iodide (2:1:4)	[12344-40-0]	Ag_2HgI_4	yellow to orange, 51	144
copper mercury iodide (2:1:4)	[13876-85-2]	Cu_2HgI_4	red to black, 70; becomes red, 160; deep red, 220	144
thallium mercury iodide (2:1:4)	[31298-28-9]	Tl_2HgI_4	yellow to red, 116	144
lead mercury iodide (1:1:4)	[31298-29-0]	$PbHgI_4$	orange to darker red, 134	144
mercury iodide (1:2)	[7774-29-0]	HgI_2	red to yellow, 127; 150	144–145
silver iodide	[7783-96-2]	AgI	yellow to brown, 147	144–145

thermochromism (see Table 5). Possible applications must include all those suggested for thermochromism itself, with possible advantages particular to fluorescence thermochromism (162–166). A very sensitive method of separation and analysis of N-heterocyclic compounds uses chromatographic or electrophoretic separation followed by identification by means of the fluorescence thermochromism of the CuI complexes on the paper (167).

Irreversible Thermochromism, Temperature-Indicating Paints. Commercially available temperature-indicating paints and crayons change color between 40 and 1300°C, with an accuracy of about ±5%. Manufacturers' literature indicates which of these are suitable for use in the presence of steam, under water, and in various environments. Most have a single color change, but several have up to four distinct color changes at well-defined temperatures. Color changes may occur notably lower if the

Table 5. Fluorescence Thermochromism of Cu(I) Halide Complexes with Nitrogen Bases[a]

Ligand	Halogen	Color		
		At room temp	At −196°C	At −180°C
ammonia	I	orange		red
methylamine	I	rose		red
dimethylamine	I	orange		red
trimethylamine	I	yellow		red
diethylamine	I	orange		blue–violet
triethylamine	I			red
isopropylamine	I			deep red
aniline[b]	I	orange		red–violet
pyrrolidine	I	orange		red–violet
piperidine	I	yellow		red–orange
pyridine	I	yellow		red
2-picoline	Br	bright red	dark blue	
2-picoline	I	orange	blue	
3-picoline	I	deep yellow	sky blue	
4-picoline	Cl	red	turquoise	
4-picoline	I	blue	blue	
2,4-lutidine	Br	orange		blue
2,4-lutidine	I	ocher		blue
2-ethylpyridine	I	yellow	blue	
3-ethylpyridine	I	rose	blue	
4-ethylpyridine	I	orange	orange	
4-aminopyridine	Cl	yellow	turquoise	
(4-aminopyridine)$_3$	Cl	yellow	blue	
4-aminopyridine	Br	rose	blue	
4-aminopyridine	I	rose	blue	
nicotinic acid ethyl ester	I	yellow		blue
nicotine[b]	I	red		blue–green
quinoline	I	orange		orange
morpholine	I	orange		deep red
hexamethylenetetramine	I	yellow		red[c], violet[d]

[a] Compounds of the formula CuLX, where L = ligand, X = halogen.
[b] In benzene.
[c] In methanol.
[d] In ethanol.

indicator is in a moderate vacuum of about 650–1300 Pa (5–10 mm Hg). The heating rate is an important factor since a rapidly heated object may exceed the temperature of the color indicator because of lag in heat flow or may change color if held too long at a temperature somewhat lower than the expected transition temperature. The manufacturers supply data relating temperature of color change to heating rate. The color change for some materials in the lower temperature ranges (up to ca 150°C) is caused by loss of water of hydration so that these indicators may revert to original color after a certain time. Heat-indicating paints and crayons are used to provide warning of the development of hot spots. A permanent record of the heat history of the object is given, which can be photographed and filed. Difficult-to-reach places in complex electrical equipment or moving parts of engines, such as pistons, turbines, or shafts, are easily monitored. Areas of heat flow or failures in insulation in furnaces of all kinds are made visible. Reactor vessels, indoor or out, display the entire area in which a heat

Table 6.   Irreversible Color Changes of Cobalt Salts

Co(II) salt	CAS Registry Number	Formula	Color at 25°C	Temp 2, °C	Temp 3, °C	Ref.
carbonate	[513-79-1]	$CoCO_3$	violet	black, 330		168
fluoride[a]	[10026-17-2]	CoF_2	orange	pink, 84		169
chloride[a]	[31277-41-5]	$CoCl_2$	pink	dark blue, 75		168
chloride hexahydrate	[7791-13-1]	$CoCl_2.6H_2O$	wine red	blue, 50		170
pentaammine chlorodi- chloride	[13859-51-3]	$Co(NH_3)_5ClCl_2$	rose	violet, 120	turquoise, 170; black, 230	170
iodide	[30262-15-8]	CoI_2	pink	green, 50		169
sulfate[a]	[10124-43-3]	$CoSO_4$	dark pink	violet, 110		168
sulfate heptahy- drate	[10026-24-1]	$CoSO_4.7H_2O$	dark reds	lilac, 60	blue, 170; red violet, 235	170
phosphate	[18773-90-5]	$Co_3(PO_4)_2$	pink	blue, 112		169
ammonium phosphate hydrate	[16827-96-6]	$CoNH_4PO_4.H_2O$	red	blue, 140	blue–gray, 500	171
hexaamine phosphate hexahydrate	[16674-72-9]	$Cp(NH_3)_6PO_4.6H_2O$	red	violet, 185	blue, 285	170
formate	[544-18-3]	$Co(HCO_2)_2$	pink	deep lavender, 116		169
acetate tetrahydrate	[19621-58-0]	$Co(C_2H_3O)_2.4H_2O$	violet	blue–violet, 95	dark green, 170	170
oxalate	[814-89-1]	CoC_2O_4	pink	black, 400		168
oxalate dihy- drate	[5965-38-8]	$CoC_2O_4.2H_2O$	rose	lilac, 150	black, 230	170

[a] Complex with hexamethylenetetramine.

limit has been reached, and conventional heat sensors can be dispensed with (67). The use of a multiple-change paint, or a set of stripes or dots of several single-transition paints, can show the heat history of an entire piece of equipment. Pellets, paste-on labels, and printing inks are also available.

Applications are made in electrical, manufacturing, and chemical industries, in aeronautics and space flight, and in the development of various types of engines. Salts and organic complexes of cobalt, chromium, nickel, manganese, iron, vanadium, and copper are used most. The color change is due to loss of water of hydration, the splitting out of some other small molecule, or decomposition. Actual color transition temperature depends not only on the chemical compound but also on the materials and binders with which it is mixed. Considerable variation can be made by adding one or more isomorphous compounds, whether or not they are thermochromic. Table 6 lists a number of typical cobalt compounds. Reported transition temperature values in some cases differ considerably (168–174). Calibration is advisable before using these indicators or making new formulations.

Detailed information is available from the following manufacturers: Tempil Div., Big Three Industries, Inc., Hamilton Boulevard, South Plainfield, New Jersey, 07080; Telatemp Corp., Fullerton, California 92635; and A. W. Faber-Castell, Stein bei Nürnberg, Federal Republic of Germany.

BIBLIOGRAPHY

1. J. R. Platt, *J. Chem. Phys.* **34,** 862 (1961).
2. U.S. Pat. 3,940,205 (Feb. 24, 1976), R. S. Crandall and B. W. Faughnan (to RCA Corp.).
3. U.S. Pat. 3,970,365 (July 20, 1976), R. D. Giglia (to American Cyanamid Co.).
4. Jpn. Pats. 74 121,797, 74 121,798 (Nov. 21, 1974), E. Inoue and A. Izawa (to Dainippon Printing Co., Ltd.).
5. U.S. Pat. 4,006,966 (Feb. 8, 1977), M. D. Meyers and H. P. Landi (to American Cyanamid Co.).
6. U.S. Pat. 3,978,007 (Aug. 31, 1976), R. D. Giglia and R. H. Clasen (to American Cyanamid Co.).
7. Can. Pat. 984,605 (Mar. 2, 1976), M. D. Meyers and H. P. Landi (to American Cyanamid Co.).
8. U.S. Pat. 4,009,935 (Mar. 1, 1977), B. W. Faughnan and R. S. Crandall (to RCA Corp.).
9. Y. Imai and K. Nakamura, *Jpn. J. Appl. Phys.* **16**(8), 1471 (1977).
10. U.S. Pat. 3,995,943 (Dec. 7, 1976), R. J. Jasinski (to Texas Instruments, Inc.).
11. U.S. Pat. 3,971,624 (July 27, 1976), P. Bruesch and co-workers (to Brown, Boveri und Cie. A.G.).
12. U.S. Pat. 3,708,220 (Jan. 2, 1973), M. D. Meyers and T. A. Augurt (to American Cyanamid Co.).
13. Jpn. Pat. 77 05,684 (Jan. 17, 1977), K. Inoue (to Suwa Seikosha Co., Ltd.).
14. Neth. Appl. 74 11,935 (Mar. 25, 1975), (to Matsushita Elec. Industrial Co., Ltd.).
15. U.S. Pat. 3,968,639 (July 13, 1976), D. J. Berets, G. A. Castellion, and G. Haacke (to American Cyanamid Co.).
16. M. Green, W. C. Smith, and J. W. Weiner, *Thin Solid Films* **38**(1), 89 (1976).
17. H. N. Hersh, W. E. Kramer, and J. H. McGee, *Appl. Phys. Lett.* **27**(12), 646 (1975).
18. O. F. Schirmer and co-workers, *J. Electrochem. Soc.* **124**(5), 749 (1977).
19. H. R. Zeller and H. U. Beyeler, *Appl. Phys.* **13**(3), 231 (1977).
20. I. Lefkowitz and G. W. Taylor, *Optics Commun.* **15**(3), 340 (1975).
21. E. K. Sichel, J. I. Gittleman, and J. Zelez, *Appl. Phys. Lett.* **31**(2), 109 (1977).
22. T. C. Arnoldussen, *J. Electrochem. Soc.* **123**(4), 527 (1976).
23. R. Hurditch, *Electron. Lett.* **11**(7), 142 (1975).
24. A. I. Gavrilyuk and F. A. Chudnovskii, *Pis'ma Zh. Tekh. Fiz.* **3**(4), 174 (1977).
25. Jpn. Pat. 76 126,142 (Nov. 4, 1976), T. Hara and co-workers (to Canon K.K.).
26. Jpn. Pat. 77 04,236 (Jan. 13, 1977), M. Matsushima and co-workers (to Canon K.K.).
27. Jpn. Pat. 76 76,185 (July 1, 1976), N. Oode, M. Tobita, and Y. Arai (to Nippon Electric Co., Ltd.).
28. U.S. Pat. 3,589,896 (June 29, 1971), D. E. Wilcox (to U.S. Dept. of the Air Force).
29. U.S. Pat. 3,943,528 (Mar. 9, 1976), D. L. Camphausen (to Xerox Corp.).
30. Ger. Pat. 2,527,638 (May 6, 1976), J. G. Allen (to Imperial Chemical Industries Ltd.).
31. U.S. Pat. 4,018,508 (Apr. 19, 1977), M. J. McDermott and co-workers (Imperial Chemical Industries Ltd.).
32. Jpn. Pat. 76 149,185 (Dec. 21, 1976), H. Takeshita (to Suwa Seikosha Co., Ltd.).
33. U.S. Pat. 3,317,266 (May 2, 1967), W. R. Heller and co-workers (to International Business Machines Corp.).
34. V. I. Steblin, R. I. Marinchenko, and E. V. Steblina, *Opt. Spektrosk.,* **42**(2), 405 (1977); *Chem. Abstr.* **86,** 163015y (1977).
35. Jpn. Pat. 76 140,892 (Dec. 4, 1976), M. Yamamoto, T. Kawata, and M. Yamana (to Tokyo Electrical Engineering College).
36. Jpn. Pat. 77 66,886 (June 2, 1977), K. Iwasa and Y. Katagiri.
37. U.S.S.R. Pat 566,863 (July 30, 1977), I. V. Shelepin, N. I. Karpova, and V. A. Barachevskii; *Chem. Abstr.* **87,** 143993h (1977).
38. K. Murao, M. Sato, and N. Shito, *Proc. Conf. Solid State Devices* **8,** 329 (1977).
39. Jpn. Pat. 77 24,169 (Feb. 23, 1977), N. Yoshike, S. Kondo, and M. Fukai.
40. R. V. Pole, G. T. Sincerbox, and M. D. Shattuck, *Appl. Phys. Lett.* **28**(9), 494 (1976).
41. Jpn. Pat. 76 134,391, (Nov. 20, 1976), H. Mori (to Sony Corp.).
42. P. N. Moskalev and I. S. Kirin, *Russ. J. Phys. Chem.* **46**(7), 1019 (1972).

43. M. M. Nicholson and R. V. Galiardi, *U.S. NTIS, AD Rep. AD-A039596*, National Technical Information Service, Springfield, Va., 1977, 67 pp.

44. T. Kawata and co-workers, *Jpn. J. Appl. Phys.* **14**(5), 725 (1975).

45. I. V. Shelepin and co-workers, *Electrokhimiya* **13**(1), 32 (1977).

46. J. Bruinink and P. van Zanten, *J. Electrochem. Soc.* **124**(8), 1232 (1977).

47. U.S. Pat. 3,961,842 (June 8, 1976), R. J. Jasinski (to Texas Instruments Inc.).

48. M. Yamana, *Jpn. J. Appl. Phys.* **15**(12), 2469 (1976).

49. S. Schmidt and R. Reich, *Ber. Bunsenges. Phys. Chem.* **76**(11), 1202 (1972); *ibid.* **76**(7), 589, 599 (1972); R. Reich, R. Scheerer, and K.-U. Sewe, *ibid.* **80**(3), 245 (1976); R. Reich and R. Scheerer, *ibid.* **80**(6), 542 (1976).

50. K.-U. Sewe and R. Reich, *Z. Naturforsch. Teil C* **32C**(3–4), 161 (1977).

51. R. Reich and co-workers, *Biochim. Biophys. Acta* **449**(2), (1976).

52. G. P. Borisevich and co-workers, *Biofizika* **20**(2), 250 (1975).

53. P. Borisevich, A. A. Kononenko, and A. B. Rubin, *Photosynthetica* **11**(1), 81 (1977).

54. M. Kluge and C. B. Osmund, *Naturwissenschaften* **58**(8), 414 (1971).

55. W. K. Cheng, *Biomembranes* **7,** 56 (1975).

56. K.-U. Sewe and R. Reich, *FEBS Lett.* **80**(1), 30 (1977).

57. H. Buecher and co-workers, *Chem. Phys. Lett.* **3**(7), 508 (1969).

58. H. Buecher and H. Kuhn, *Z. Naturforsch. Teil B,* **25**(12), 1323 (1970).

59. J. L. Stevenson, S. Ayers, and M. M. Faktor, *J. Phys. Chem. Solids* **34**(2), 235 (1973).

60. H. Labhart, *Ber. Bunsenges. Phys. Chem.* **80**(3), 240 (1976).

61. W. Liptay, *Ber. Bunsenges. Phys. Chem.* **80**(3), 207 (1976); *Angew. Chem. Int. Ed.* **8**(3), 177 (1969); E. C. Lim, ed., *Advances on Electronic Excitation and Relaxation,* Academic Press, Inc., New York, 1974, p. 129.

62. F. P. Chernyakovskii and V. A. Gribanov, *Russ. J. Structural Chem.* **9**(3), 387 (1968).

63. I. M. Kanevskii and co-workers, *Russ. J. Phys. Chem.* **45**(2), 232 (1971); *ibid.* **45**(4), 483 (1971).

64. A. R. Hill and M. M. Malley, *J. Mol. Spectrosc.* **40**(2), 428 (1971).

65. B. P. Bespalov and L. M. Blinov, *Fiz. Tverd. Tela* **11**(4), 1032 (1969).

66. G. C. Causley, J. D. Scott, and B. R. Russell, *Rev. Sci. Instrum.* **48**(3), 264 (1976).

67. *Chem. Eng. News* **50**(3), 92 (Jan. 16, 1978).

68. N. Latif, I. Zeid, and N. Mishriky, *Chem. Ind.* (*London*) (23), 754 (1969).

69. R. Lemke, *Chem. Ber.* **103**(9), 3003 (1970).

70. G. Kortum and P. Kreig, *Chem. Ber.* **102**(9), 3033 (1969).

71. R. Kornstein, K. A. Muszkat, and S. Sharafy-Ozeri, *J. Am. Chem. Soc.* **95**(19), 6177 (1973).

72. H. Kuroda and R. Sudo, *Bull. Chem. Soc. Jpn.* **39**(5), 1059 (1966).

73. N. S. Kozlov, O. D. Zhikhareva, and I. P. Stremok, *Dokl. Akad. Nauk B* **19**(4), 349 (1975).

74. Ger. Pat. 2,541,666 (Apr. 1, 1976), M. Hayami and S. Torikoshi.

75. Ger. Pat. 2,454,695 (May, 28, 1975), H. Balli.

76. B. D. West, *Organic Thermochromic Materials, SC-CR-67-2555, Supt. Docs. Y3-A47 22SC-CR,* U.S. Atomic Energy Commission, Washington, D.C., 1967, 9 pp.

77. H. Maeda, A. Chinone, and T. Hayashi, *Bull. Chem. Soc. Jpn.* **43**(5), 1431 (1970).

78. K. Dimroth, C. Reichardt, and A. Schweig, *Ann.* **669,** 95 (1963); *ibid.* **661,** 1 (1963).

79. G. N. Lipunova and co-workers, *Khim. Geterotsikl. Soedin.* (4), 493 (1974); N. N. Gulemina and co-workers, *Sh. Prikl. Spektrosk.* **22**(5), 941 (1975); U.S.S.R. Pat. 536,184 (Nov. 25, 1976), G. M. Petrova, L. V. Rodnenko, and N. P. Bednyagina; N. Gulemina and co-workers, *Zh. Org. Khim.* **13**(5), 1108 (1977).

80. J. H. Day, *Chem. Rev.* **63,** 65 (1963).

81. J. H. Day, *Chem. Rev.* **68**(6), 649 (1968).

82. J. F. D. Mills and S. C. Nyburg, *J. Chem. Soc.,* 308, 927 (1963).

83. G. Kortum, *Ber. Bunsenges. Phys. Chem.* **78**(4), 391 (1974).

84. A. Schoenberg and E. Singer, *Tetrahedron Lett.* (24), 1925 (1975).

85. A. Mustafa, *Chem. Rev.* **43,** 509 (1948).

86. C. Schiele and H. O. Kalinowski, *Angew. Chem. Int. Ed.* **5**(4), 416 (1966).

87. C. Schiele and M. Ruch, *Tetrahedron Lett.,* (37), 4413 (1966); C. Schiele, D. Hendriks, and M. Ruch, *Tetrahedron Lett.* (37), 4409 (1966).

88. G. Arnold, *Z. Naturforsch. Teil B,* **21**(3), 291 (1966).

89. C. Schiele, A. Wilhelm, and G. Paal, *Justus Liebigs Ann. Chem.* **722,** 162 (1969).

90. Y. N. Malkin and co-workers, *Izv. Akad. Nauk. SSSR Ser. Khim.* (3), 555 (1976).

91. A. Bertoluzza, C. Concilio, and P. Finelli, *J. Appl. Chem. Biotechnol.* **27,** 225 (1977).
92. E. Hadjoudis and D. Geogiou-Hadjoudis, *U.S. Contract DA-ERO-124-74-G-0073, AD A 013507,* NTIS, Springfield, Va., June 1975.
93. U.S. Pat. 3,763,151 (Oct. 2, 1973), J. Kazan, Jr.
94. S. Avramovici and co-workers, *An. Stiint Univ. Al. I. Cuza Iasi* 18(1), 61 (1972).
95. E. Hadjoudis and F. Milia, *Adv. Nucl. Quadrupole Reson.* **7,** 133 (1974). E. Hadjoudis and co-workers, *Solid State Commun.* 21(6), 541 (1977).
96. M. D. Cohen and co-workers, *J. Chem. Soc.,* 2041, 2060 (1964); *ibid.,* 329, 334, 373 (1967).
97. V. I. Minkin and co-workers, *Zh. Org. Khim.* 6(2) 348 (1970).
98. M. I. Knyazhanskii and co-workers, *Zh. Fiz. Chim.* 46(1), 178 (1972).
99. J. J. Laverty and Z. G. Garland, *J. Polym. Sci. B,* 7(2), 161 (1969).
100. T. Hayashi and co-workers, *Bull. Chem. Soc. Jpn.* 37(10), 1563 (1964); *ibid.* 38(5), 857 (1965); *ibid.* 43(2), 429 (1970).
101. K. Tomita and N. Yoshida, *Tetrahedron Lett.* (17), 1169 (1971); *Bull. Chem. Soc. Jpn.* 45(12), 3584 (1972).
102. K. H. Konig, *Chem. Ber.* **92,** 257, 1789 (1959); Ger. Pat. 1,228,972 (Nov. 17, 1966).
103. G. Loeber, *Z. Chem.* 3(9), 359 (1963).
104. J. H. Day and A. Joachim, *J. Org. Chem.* **30,** 4107 (1965).
105. H. Jancke and co-workers, *Z. Chem.* 17(3), 105 (1977).
106. R. N. McNair, *J. Org. Chem.* 33(5), 1947 (1968).
107. R. Simoni and M. Toschi, *Z. Naturforsch. B* 25(5), 461 (1970).
108. S. Hamai and H. Kokubun, *Bull. Chem. Soc. Jpn.* 47(9), 2085 (1974).
109. U.S. Pat. 3,874,240 (Apr. 1, 1975), A. Rembaum (to U.S. National Aeronautic and Space Administration).
110. P. R. Hammond and L. A. Burkart, *J. Phys. Chem.* 74(3), 639 (1970).
111. Ger. Pat. 2,327,723 (Dec. 20, 1973), Jpn. Pats., 75 75,640 (June 20, 1975), 75 75,991 (June 21, 1975), 75 107,040 (Aug. 23, 1975), N. Nakasuji and co-workers (to Pilot Ink Co.).
112. Jpn. Pat. 73 93,404 (Dec. 3, 1973), T. Teranishi and T. Ishiki (to Pilot Ink Co.).
113. G. V. Okatova, A. Y. Kuznetsov, and L. N. Shabanova, *Sov. J. Opt. Technol.* 40(4), 250 (1973); *ibid.* 42(1), 48 (1975); *ibid.* 40(5), 313 (1973).
114. R. M. Ellam and co-workers, *Chem. Ind.* (*London*) (2), 74 (1974).
115. G. J. Exharos, W. M. Risen, Jr., and R. H. Baughman, *J. Am. Chem. Soc.* 98(2), 481 (1976).
116. R. D. Willett and co-workers *Inorg. Chem.* 13(10), 2510 (1974).
117. H. Yokoi, M. Sai, and I. Isobe, *Bull. Chem. Soc. Jpn.* **42,** 2232 (1969).
118. L. Fabbrizzi, M. Micheloni, and P. Paoletti, *Inorg. Chem.* 13(12), 3019 (1974).
119. J. R. Ferraro and co-workers, *Inorg. Chem.* 15(10), 2342 (1976).
120. A. B. P. Lever, E. Mantovani, and J. C. Donini, *Inorg. Chem.* 10(11), 2424 (1971).
121. N. Iwasaki, K. Sone, and H. Ojima, *Bull. Chem. Soc. Jpn.* 48(8), 2279 (1975).
122. H. Yokoi, M. Sai, and I. Isobe, *Bull. Chem. Soc. Jpn.* 45(4), 1100 (1972).
123. R. D. Gillard and G. Wilkinson, *J. Chem. Soc.,* 5885 (1963).
124. Y. Mori, H. Inoue, and M. Mori, *Inorg. Chem.* 14(5), 1002 (1975).
125. G. L. Ferguson and B. Zaslow, *Acta Crystallogr. Sect. B.* 27(4), 849 (1971).
126. D. R. Hill and D. W. Smith, *J. Inorg. Nucl. Chem.* 36(2), 466 (1974).
127. L. A. Bares, K. Emerson, and J. E. Drumheller, *Inorg. Chem.* 8(1), 131 (1969).
128. M. Vacatello, *Ann. Chem.* (*Rome*) 64(1–2), 13 (1974).
129. N. Iwasaki, K. Sone, and Y. Fukuda, *Z. Anorg. Allg. Chem.* 412(2), 170 (1975).
130. A. Ouchi, T. Takeuchi, and I. Taminaga, *Bull. Chem. Soc. Jpn.* 43(8), 2609 (1970).
131. N. F. Curtis and D. A. House, *J. Chem. Soc.,* 6194 (Nov. 1965).
132. T. Matsumoto, M. Sato, and M. Nakamura, *Nippon Kagaku Zashi* 92(5), 472 (1971).
133. R. D. Gillard and M. V. Twigg, *Inorg. Chim. Acta* 6(1), 150.
134. D. M. L. Goodgame and L. M. Venanzi, *J. Chem. Soc.,* 616 (1963).
135. D. L. Perry, J. L. Margrave, and W. H. Waddell, *Inorg. Chim Acta* 18, 129 (1976).
136. J. E. Barnes, J. H. Setchfield, and G. O. R. Williams, *J. Inorg. Nucl. Chem.* **38,** 1065 (1976).
137. P. V. Gogorishvili, D. A. Gogorishvili, and E. N. Zedelashvili, *Soobshch. Akad. Nauk Gruz. SSR* 57(1), 61 (1970).
138. F. Seel and J. Rodrian, *J. Organometal. Chem.* 16(3), 479 (1969).
139. R. D. Bereman and C. H. Brubaker, Jr., *J. Inorg. Nucl. Chem.* 32(8), 2557 (1970).
140. U.S. Pat. 3,725,310 (Apr. 3, 1973), P. C. Donohue (to E.I. du Pont de Nemours & Co., Inc.).

141. E. Hengge and G. Scheffler, *Monatsh. Chem.* **95**(6), 1450 (1964).
142. J. E. Kuder and M. A. Lardon, *Ann. N.Y. Acad. Sci.* **192,** 147 (1972).
143. R. L. Wild, E. M. Rockar, and J. C. Smith, *Phys. Rev. B* **8**(8), 3828 (1973).
144. W. W. Wendlandt, *Science* **149**(3571), 1085 (1963).
145. Z. Halmos and W. W. Wendlandt, *Thermochimica Acta* **7**(2), 113 (1973).
146. D. Grafstein and co-workers, *NASA Contract NSA 12-89, NASA Symposium Sept., 1969, NSA 1.21;159.*
147. W. L. Flint, D. Grafstein, and J. Menczel, *NASA Contract Rep. 1969, NASA-CR-86136, NTIS Doc. No. N6921546,* 86 pp.
148. W. W. Wendlandt and S. S. Bradley, *Thermochimica Acta* **1**(6), 529 (1970).
149. J. S. Chivian, R. N. Claytor, and D. D. Eden., *Appl. Phys. Lett.* **15**(4), 123 (1969).
150. J. S. Chivian and co-workers, *Appl. Opt.* **11**(11), 2649 (1972).
151. J. S. Chivian, *Mat. Res. Bull.* **8**(7), 795 (1973).
152. U.S. Pat. 3,516,185 (June 23, 1970), E. H. Hilborn and D. Grafstein (to U.S. National Aeronautics and Space Administration).
153. Jpn. Pat. 76 102,049 (Sept. 9, 1976), Y. Tamai.
154. U.S.S.R. Pat. 352,922 (Sept. 29, 1972), D. Vijuma and G. Bike.
155. T. Katsurai and Y. Makide, *Sci. Pap. Inst. Phys. Chem. Res.* **68**(3), 100 (1974).
156. T. Katsurai and K. Sone, *Bull. Chem. Soc. Jpn.* **41**(2), 519 (1968).
157. P. R. Hammond and co-workers, *J. Chem. Soc. A,* 3800 (1971).
158. U.S.S.R. Pat. 514,945 (June 5, 1976), W. Spaethling and H. Seidel.
159. Ger. Pat. 2,314,722 (Oct. 11, 1973), M. P. Borom and R. C. DeVries.
160. Y. Yoshino, I. Taminaga, and S. Uchida, *Bull. Chem. Soc. Jpn.* **44**(5), 1435 (1971).
161. T. Yamase and I. Ikawa, *Bull. Chem. Soc. Jpn.* **50**(3), 746 (1977).
162. H. D. Hardt, *Naturwissenschaften* **61**(3), 107 (1974); *Chem. Exp. Didakt* **2**(2–8), 265 (1976).
163. H. D. Hardt and H. D. DeAhna, *Naturwissenschaften* **57**(5), 244 (1970); H. D. DeAhna and H. D. Hardt, *Z. Anorg. Allg. Chem.* **387**(1), 61 (1972).
164. H. D. Hardt and H. Gechnizdjani, *Z. Anorg. Allg. Chem.* **397**(1), 16, 23 (1973); *Inorg. Chim. Acta* **15**(1), 47 (1975).
165. H. D. Hardt, H. Gechnizdjani, and A. Pierre, *Naturwissenschaften* **59**(8), 363 (1972); Ger. Pat. 2,226,994 (Dec. 20, 1973).
166. H. D. Hardt and A. Pierre, *Z. Anorg. Allg. Chem.* **402**(1), 107 (1973); *Fresenius' Z. Anal. Chem.* **265**(5), 337 (1973).
167. H. Wagner and H. Lehmann, *Fresenius' Z. Anal. Chem.* **283**(2), 115 (1977).
168. Ger. Pat. 2,515,474 (Oct. 30, 1975), L. Glover, Jr., and E. F. Lopez (to Raychem Corp.).
169. Von K. Th. Wilke and W. Opfermann, *Z. Phys. Chem.* (*Leipzig*) **224,** 237 (1963).
170. S. P. Gvozdov and A. A. Erunova, *Izvest. Vysshykh Ucheb. Zavedenif Khim. Khim. Tekhnol.* (5), 154 (1958).
171. K. Nagase, H. Yokobayashi, and K. Sone, *Bull. Chem. Soc. Jpn.* **49**(6), 1563 (1976).
172. J. E. Cowling, P. King, and A. L. Alexander, *Ind. Eng. Chem.* **45**(10), 2317 (1953).
173. W. H. Duffey, *Soc. Auto. Engineers, Proceedings Combined Power Plant and Transportation Meeting, Cleveland, Ohio,* Pct. 18–21, 1965, paper 650705, 6 pp.
174. L. C. Tyte, *Inst. Mech. Eng. Proc.* (*London*) **152,** 226 (1945).

General References

J. A. Castellano, *Opt. Laser Technol.* **7**(6), 259 (1975), this is a review of display devices, including electrochromism.
G. Elliott, *Radio Electron. Eng.* **46**(6), 281 (1976), a review of displays including electrochromism.
B. W. Faughnan, R. S. Crandall, and P. M. Heyman, *RCA Review* **36,** 177 (1975), an overall view of WO_3 amorphous films as electrochromic materials.
J. Bruinik, "Electrochromic Display Devices" in A. R. Kmetz and F. K. van Willisen, eds., *Nonemissive Electrooptic Displays,* Plenum Publishing, New York, 1976.
H. Labhart, *Adv. Chem. Phys.* **13,** 179 (1967), theoretical treatment of spectral shifts in an electric field to find dipole moments and polarizabilities.
J. K. Fischer, D. M. van Bruening, and H. Labhart, *Appl. Opt.* **15**(11), 2812 (1967), discusses light modulation by electrochromic cells.

K.-U. Sewe and R. Reich, *FEBS Lett.* **80**(1), 30 (1977), influence of the chlorophylls on the electrochromism of carotenoids in the membranes of photosynthesis.
L. C. Tyte, "Temperature-Indicating Paints," *Inst. Mech. Eng. (London), Proc.* **152**, 226 (1945).
W. H. Duffey, "Temperature Indicating Paints as they Assist in Gas Turbine Design Processes," *Soc. Auto. Eng. Proc. Oct. 1965.*

JESSE H. DAY
Ohio University

CINNAMIC ACID, CINNAMALDEHYDE, AND CINNAMYL ALCOHOL

These unsaturated aromatic compounds and their derivatives, characterized by the grouping $C_6H_5CH{=}CHCH_2{-}$, are important in the flavors, perfumes, cosmetics, pharmaceuticals, graphic arts, plastics, and polymers industries. Many derivatives of cinnamic acid and the corresponding aldehyde and alcohol have been prepared (1).

CINNAMIC ACID

Cinnamic acid [*621-82-9*], 3-phenyl-2-propenoic acid, $C_6H_5CH{=}CHCOOH$, exists partly in the free state and partly in the form of esters in ethereal oils, resins, storax (styrax), balsams of Tolu and Peru, and coca leaves. Commercial cinnamic acid occurs as pale yellow to off-white crystals with a characteristic aromatic odor. Its solubility in water is only 0.546 g/L at 25°C. It is more soluble in absolute alcohol and very soluble in ether.

Physical and Chemical Properties

Cinnamic acid occurs in cis (1) and trans (2) forms.

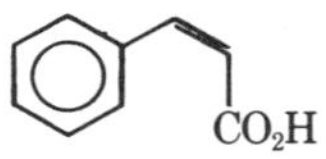

(1) 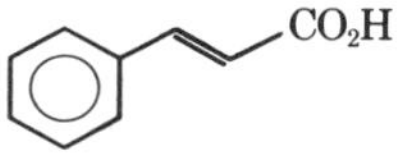   (2)

Ordinary cinnamic acid possesses the trans configuration [*140-10-3*]; mp, 133°C; bp, 300°C; d_4^4 1.2475; heat of combustion, 4477 kJ/mol (1040 kcal/mol); K_{25} 3.5×10^{-5}. The cis form presents an unusual case of physical isomerism (polymorphism), as shown by Biilmann (2). There are three crystalline *cis*-cinnamic acids [*102-94-3*], known as Liebermann's allocinnamic acid , mp 68°C, Liebermann's isocinnamic acid , mp 58°C, and Erlenmeyer's isocinnamic acid , mp 42°C.

Cinnamic acid undergoes reactions typical of a carboxyl group, an olefinic double bond, and the benzene nucleus. As an acid, it forms salts and esters. It also reacts with phosphorus tri- or pentachloride or with thionyl chloride to form cinnamoyl chloride, mp 35°C, which is soluble in carbon tetrachloride and petroleum ether. This derivative reacts with isobutyl salicylate to form isobutyl salicyl cinnamate, a patented sunscreening agent (3). The *cis*-cinnamic acids may be prepared from the mixture of acids resulting from hydrolyses of cocaine alkaloids in the manufacture of cocaine.

Cinnamic acid nitrated with fuming nitric acid yields a mixture of *o*- and *p*-nitrocinnamic acid. Sulfonation of cinnamic acid with 20% oleum at ordinary temperatures yields chiefly the para and some meta sulfonic acid.

Manufacture

trans-Cinnamic acid is synthesized by the following methods.

The Perkin reaction, used commercially in the United States, utilizes benzaldehyde (qv), acetic anhydride, and anhydrous sodium or potassium acetate.

$$C_6H_5CHO + (CH_3CO)_2O \xrightarrow{CH_3COONa} C_6H_5CH{=}CHCOOH + CH_3COOH$$

Potassium acetate substituted for the sodium salt decreases the reaction time for comparable yields (4–6). A trace of pyridine used as a catalyst (7) when 1.5 moles of anhydride and 0.65 mole of sodium acetate are heated with 1 mole of benzaldehyde for 8 h at 180°C supposedly increases the yield of cinnamic acid from 50 to 85%. This yield could not be duplicated, however (8). Johnson gives an excellent review of the Perkin reaction in which reaction mechanisms, rates, temperatures, and catalysts are discussed (9).

In the commercial process, anhydrous sodium acetate, acetic anhydride, and benzaldehyde in a 1.0:1.5:1.0 mole ratio are loaded into a 1900 L (500 gal), glass-lined agitated, jacketed still. The still is equipped with a fractionating column, partial condenser, and thermometer wells in the condenser, still, and jacket. The jacket is heated by oil to 190–195°C. The condenser is regulated so that at 170°C the acetic acid distills and the anhydride is returned to the still. The reaction is complete after 8–10 h of heating. A solution of sodium carbonate is then run into the reaction mass, with agitation, until the mixture is alkaline. Live steam blown through the reaction solution removes any unreacted benzaldehyde in about 2 h, after which the hot reaction mass is treated with decolorizing carbon and stirred for ½ h. The hot liquor is run off through the bottom outlet to a filter press and the clarified liquid is pumped into crocks. Concentrated hydrochloric acid is added to the sodium cinnamate solution in the crocks. Chopped ice is then added to rapidly cool the precipitated cinnamic acid. The acid suspension is pumped to a rubber-lined centrifuge, washed with cold water, and dried. The cinnamic acid obtained melts at 131–133°C, and the yields average 85% of theory based on benzaldehyde consumed. No pollution problem arises in this process.

In another commercial method of producing cinnamic acid, benzal chloride and anhydrous sodium acetate are heated to 180–200°C.

$$C_6H_5CHCl_2 + 2\ CH_3COONa \rightarrow C_6H_5CH{=}CHCOOH + 2\ NaCl + CH_3COOH$$

Because benzal chloride is cheaper than benzaldehyde, this process is used by manufacturers who obtain by-product benzal chloride from their benzyl chloride plants. The yields are comparable to those obtained through the Perkin synthesis.

Cinnamic acid esters may be made conveniently through the Claisen condensation. For example, ethyl cinnamate is obtained from benzaldehyde and ethyl acetate in the presence of sodium ethylate.

$$C_6H_5CHO + CH_3COOC_2H_5 \xrightarrow{C_2H_5ONa} C_6H_5CH{=}CHCOOC_2H_5 + H_2O$$

The ester may be hydrolyzed to cinnamic acid by the usual methods.

Economic Aspects

The Tarriff Commission has not reported production or sales figures for cinnamic acid since 1949. The price of cinnamic acid has risen from the average $4.86/kg in 1964 to $13.20/kg in 1977.

Shipping and Handling Information

Details about toxicity are unknown. Chronic local toxicity indicates that cinnamic acid is slightly allergenic. There are no DOT regulations for cinnamic acid. Its freight classification is: Chemicals—NOIBN.

Specification and Standards

Cinnamic acid is available commercially in a technical grade that meets the following specifications: mp 132–134°C; moisture, 0.2% maximum; ash, 0.25% maximum; assay (titration), 98.5% minimum; odor, aromatic; appearance, pale yellow to off-white crystals. A refined grade is also available. The technical grade is sold in 11-, 22-, 45- and 56-kg fiber drums: the refined grade in 500-g and 2.5-kg bottles.

Uses

Use of cinnamic acid derivatives has increased significantly in the past decade. Poly(vinyl cinnamate) is used as a light-sensitive, film-forming coating for photoresists (10–11); in a color photographic process (12); for overhead projection copy (13); and in xerography (see Photoreactive polymers; Electrophotography) (14). Phenoxycinnamate polymers are used in photolithography and a photomechanical process (15) (see Reprography). Other applications of cinnamic acid and its derivatives include: general use in cosmetics (16) and as sun-screening agents (3, 17); in the plastics and polymer field, as acrylate co-polymers (18), poly(vinyl cinnamate) preparation (19), glassy polymers of Pb, Cd, and Zn cinnamates (20), and fire retardants (see Flame retardants) (21–22); in agriculture as a fungicide (qv) (23–24), insecticides (25), growth control (26), and preservation of fruits, vegetables, and flowers (27); in electroplating (qv) of nickel, copper, tin, and zinc (28–29); in corrosion inhibition of iron in neutral and alkaline chloride media (30); and in the medical and pharmaceutical fields, as a relaxant and analgesic (31–33).

CINNAMALDEHYDE

Cinnamaldehyde [*104-55-2*], 3-phenyl-2-propenal, $C_6H_5CH=CHCHO$, occurs naturally in Chinese cinnamon oil from the leaves and twigs of *Cinnamonum cassia*. It is obtained by steam distillation and further rectified by vacuum distillation. It also occurs in Sri Lanka (Ceylon) cinnamon oil from *C. zeylanicum*. Cinnamaldehyde is a yellow mobile liquid which turns to a dark brown viscous liquid on exposure to light and air. It has the characteristic odor of cinnamon oil and a burning aromatic taste. Cinnamaldehyde is slightly soluble in water, infinitely soluble in ether, and insoluble in petroleum ether; its solubility in 50% alcohol is 1 part in 25 parts, and in 70% alcohol, 1 part in 2–3 parts. It is volatile with steam.

Physical and Chemical Properties

Cinnamaldehyde has the following properties: mp − 7.5°C; bp 252°C at 101 kPa (760 mm Hg), with partial decomposition, bp 128–130°C at 2.7 kPa (20 mm Hg); d_{20}^{20} 1.1102; n_D^{20} 1.61949. It reacts typically as an aldehyde. With sodium bisulfite solutions, cinnamaldehyde forms addition compounds that serve for its quantitative determination. With hydroxylamine and phenylhydrazine, cinnamaldehyde gives the corresponding aldoximes (*syn,* mp 139°C, and *anti,* mp 76°C) and phenylhydrazone (mp 168°C), respectively. At room temperature it reacts slowly with a 1% solution of hydrogen chloride in methanol to form cinnamaldehyde dimethyl acetal, $C_6H_5CH=CHCH(OCH_2)_3$, bp 126°C at 1.5 kPa (11 mm Hg). With acetic anhydride it forms a diacetate (mp 85°C) and with acetic anhydride and sodium acetate it undergoes the Perkin reaction to form cinnamylacetic acid, 5-phenyl-2,4-pentadienoic acid, $C_6H_5CH=CHCH=CHCOOH$. A modified Perkin reaction leads to diphenylpolyenes (34).

$$C_6H_5CH=CHCHO + C_6H_5CH_2COOH \xrightarrow[\text{(CH}_3\text{CO)}_2\text{O}]{\text{PbO}} \underset{\substack{\text{1,4-diphenyl-1,3-butadiene}\\ \text{(bistyryl).}}}{C_6H_5CH=CHCH=CHC_6H_5}$$

Grignard reaction between cinnamaldehyde and methyl magnesium iodide or bromide forms 1-methyl-3-phenyl-2-propen-1-ol (α-methylcinnamyl alcohol), $C_6H_5CH=CHCH(CH_3)OH$.

Cinnamaldehyde is easily oxidized at the aldehyde group, the double bond, or both: in air, to cinnamic acid; in hot nitric acid, to benzoic acid and benzaldehyde; in hot calcium hypochlorite, to benzoic acid; in chromic acid, to benzoic and acetic acids; and with oxygen to all of these products and others (35). Bromine and chlorine add readily to the double bond at ordinary temperatures to form the dihalides, $C_6H_5CHXCHXCHO$, which rearrange to α-halocinnamaldehydes, $C_6H_5CH=CX-CHO$. Hydrogenation of cinnamaldehyde occurs at either the carbonyl group, yielding cinnamyl alcohol, or the double bond, yielding hydrocinnamaldehyde, $C_6H_5CH_2CH_2CHO$, mp 47°C. The latter is best prepared from cinnamaldehyde dimethyl acetal by reaction with sodium ethylate, or with nickel or palladium and hydrogen, followed by hydrolysis (with 3% sulfuric acid) of the acetal produced.

Nitration of cinnamaldehyde with potassium nitrate and sulfuric acid (conc) produces *ortho-* and *para*-nitrocinnamaldehyde. The most significant alkyl-substituted derivative of cinnamaldehyde is α-pentylcinnamaldehyde, C_6H_5-$CH{=}C(C_5H_{11})CHO$, bp 140°C at 0.7 kPa (5 mm Hg), made by condensing benzaldehyde and heptaldehyde in the presence of dilute sodium hydroxide in an aqueous alcohol medium at room temperature. It is an important perfume with a jasmine odor; α-methyl- and α-hexylcinnamaldehyde are also used in perfumes.

Manufacture

Cinnamaldehyde is made commercially by the condensation of benzaldehyde and acetaldehyde in the presence of dilute sodium hydroxide (36), hydrochloric acid, or sodium ethylate.

$$C_6H_5CHO + CH_3CHO \xrightarrow{NaOH} C_6H_5CH{=}CHCHO + H_2O$$

It may be purified through the bisulfite addition compound.

Other methods of synthesizing cinnamaldehyde include hydrolysis of cinnamyl dichloride by cold water; oxidation of cinnamyl alcohol, eg, by passing the alcohol vapor and air over platinum black; treatment of (1-chloroallyl)benzene, $C_6H_5CH_2CH{=}CHCl$, with phosphorus pentachloride and conversion of the resulting ((2,3,3-trichloropropyl)benzene, $C_6H_5CH_2CHClCHCl_2$, to cinnamaldehyde with hydrochloric acid (37); or condensation of styrene, $C_6H_5CH{=}CH_2$, with formylmethylaniline, $C_6H_5N(CH_3)CHO$, in the presence of phosphorus oxychloride (38).

Economic Aspects

In 1974 the U.S. Tariff Commission reported the production of cinnamaldehyde at 796 metric tons, sales at 567 t, and an average price of $1.46/kg. The market price in December 1977 was $3.30/kg.

Specifications and Standards

Cinnamaldehyde used to flavor food, beverage, or medicinal products must meet the following NF IX specifications or standards (39): assay, not less than 98% (by oximation); specific gravity, not less than 1.048 and not more than 1.052 at 25°C; refractive index, not less than 1.618 and not more than 1.623 at 20°C; chlorinated compounds, any opalescence produced after 5 min by 1 mL of cinnamaldehyde treated with isopropyl alcohol, nitric acid, and silver nitrate solution and heated to boiling should not be greater than that produced by 0.1 mL of 0.01 N hydrochloric acid treated in the same manner; hydrocarbons, no oil should separate when the sodium bisulfite compound is formed with cinnamaldehyde.

Cinnamaldehyde is sold in 500-g and 2-kg cans and 44-kg carboys and should be kept in well filled, light-resistant containers and protected from excessive heat.

Uses

Cinnamaldehyde is primarily used in the flavor industry for imparting a cinnamon flavor to all kinds of food, beverage, and medicinal products, and in the liquor industry for liqueurs and cordials (see Flavors). It gives a spicy and oriental note to perfumes (qv), and was once used in soaps. In electroplating (qv), cinnamaldehyde is used in bright tin-plating baths (40) and in Ni, acid Cu, Co, Mn, Fe, Zn, and Sn plating baths (41); in medicine, in an inclusion compound with β-dextrin (see Clathration) (42); and in polymers for introduction of organic groups into ethylenically-unsaturated hydrocarbons (43). Several miscellaneous applications of cinnamaldehyde mentioned in the literature includes as an attractant for termites, as a corrosion inhibitor for sulfamic acid baths in cleaning galvanized iron and zinc, as an emulsion fog inhibitor for photographic film and in a photographic hardening bleach bath in color processes. These uses have not, however, materially increased cinnamaldehyde production.

CINNAMYL ALCOHOL

Cinnamyl alcohol [104-54-1], 3-phenyl-2-propen-1-ol, styrylcarbinol, cinnamic alcohol, $C_6H_5CH=CHCH_2OH$, does not occur free in nature, but is usually found as an ester, eg, cinnamyl cinnamate in storax and balsam of Peru. It forms colorless needles and has an odor resembling hyacinths. Cinnamyl alcohol is slightly soluble in water and very soluble in alcohol and ether.

Physical and Chemical Properties

Cinnamyl alcohol has the following properties: mp 33°C; bp 257°C (cor) at 101 kPa (760 mm Hg) bp 142–145°C at 1.9 kPa (14 mm Hg); d_{35}^{35} 1.0397, n_D^{33} 1.57580.

With organic acids, cinnamyl alcohol forms esters used mainly in the perfume industry (see Perfumes). The action of hydrochloric acid on cinnamyl alcohol gives cinnamyl chloride, and a glacial acetic acid solution of hydrobromic acid produces cinnamyl bromide. Bromine adds at the double bond to give 2,3-dibromo-3-phenyl-1-propanol, $C_6H_5CHBrCHBrCH_2OH$. Cinnamyl alcohol is oxidized in air in the presence of platinum black to cinnamaldehyde; with chromic acid to cinnamic acid; with hot potassium hydroxide and lead dioxide to cinnamic acid and benzaldehyde; and with hot nitric acid to benzaldehyde.

Manufacture

Cinnamyl alcohol is made commercially by the Meerwein-Ponndorf reduction of cinnamaldehyde using aluminum isopropylate or ethylate. Yields are 85% of theory (44). Good yields of cinnamyl alcohol have been obtained by the catalytic reduction of cinnamaldehyde with platinum black catalyst promoted with ferrous chloride and zinc acetate under a hydrogen pressure of 100–200 kPa (2–3 atm). The alcohol produced is free from cinnamaldehyde (45). Limited amounts of cinnamaldehyde are produced by the alkaline hydrolysis of storax, of which one constituent is cinnamyl cinnamate (styracin). Cinnamyl diacetate, $C_6H_5CH=CHCH(OOCCH_3)_2$, produced by the action of acetic anhydride on cinnamaldehyde, is reduced to cinnamyl alcohol with acetic acid and iron filings.

Economic Aspects

In 1974 the U.S. Tariff Commission reported the production of 148 metric tons and sales of 148 t with an average price of $3.64/kg. The market price of cinnamyl alcohol in December 1977 was $7.16–7.38/kg.

Specifications and Standards

Because most cinnamyl alcohol is produced for the perfume industry, only the perfume grades are commercially available. The two perfume grades manufactured depend upon the source of the alcohol. Cinnamyl alcohol produced by the hydrolysis of storax is preferred by many perfumers and commands a higher price than that produced by the reduction of cinnamaldehyde. Each manufacturer has his own set of specifications. The following list is typical: white to pale yellow crystals; mp 32–33°C; bp 118–120°C at 0.7 kPa (5 mm Hg); d_{35}^{35} 1.039–1.041. The specifications and standards set up by the Essential Oil Association are recommendations only, and have no official status in the industry. Cinnamyl alcohol is sold in fractional, 500-g and 2.2-kg bottles and 22-kg cans.

Uses

Cinnamyl alcohol is used in the perfume industry for its high fixation value in blends containing hyacinth, jasmine, lilac, lily of the valley, narcissus, or rose perfumes, for its balsamic aroma in many compositions, and in the productions of esters which are even more valuable than the alcohol as fixatives. Cinnamyl alcohol has also been used to produce photosensitive polymers (46), lithographic plates (47), and chemically resistant coatings. It possesses ovicidal action on mosquitoes, and is used in Ni plating baths and as a plant growth inhibitor (see Insect control technology; Plant growth substances).

BIBLIOGRAPHY

"Cinnamic Acid, Cinnamaldehyde, and Cinnamyl Alcohol" in *ECT* 1st ed., Vol. 4, pp. 1–8, by W. F. Ringk, Benzol Products Company; "Cinnamic Acid, Cinnamaldehyde, and Cinnamyl Alcohol" *ECT* 2nd ed., Vol. 5, pp. 517–523, by W. F. Ringk, Benzol Products Company.

1. F. K. Beilstein, *Handbuch der Organischen Chemie,* Vol. 6, 4th ed., Springer, Berlin, pp. 570–571; 1st Suppl., Vol. 6, p. 281; 2nd Suppl., Vol. 6, pp. 525–528; 3rd Suppl., Vol. 6, pp. 2670–2680; Vol. 7, pp. 348–359; 1st Suppl., Vol. 7, pp. 187–190; 2nd Suppl., Vol. 7, pp. 273–283; 3rd Suppl., Vol. 7, pp. 1364–1371; Vol. 9, pp. 572–609, 1st Suppl., Vol. 9, pp. 225–251; 2nd Suppl., Vol. 9, pp. 377–407; 3rd Suppl., Vol. 9, pp. 2401–2403.
2. E. Biilmann, *Ber. Deut. Chem. Ges.* **42,** 182 (1909); **43,** 568 (1910).
3. U.S. Pat. 2,974,089 (Mar. 7, 1961), E. Alexander and co-workers (to Revlon Inc.).
4. F. L. Chappell, Jr., *Studies of the Perkin Reaction,* PhD Thesis, Cornell University, 1933.
5. P. Kalnin, *Helv. Chim. Acta* **11,** 977 (1928).
6. H. Meyer and R. Beer, *Monatsh.* **34,** 649 (1913).
7. G. Bacharach and F. Brogan, *J. Am. Chem. Soc.* **50,** 3333 (1928).
8. J. R. Johnson, "The Perkin Reaction and Related Reactions" in R. Adams, ed., *Organic Reactions,* Vol. 1, John Wiley & Sons, Inc., New York, 1942, p. 218.
9. Ref. 8, pp. 210–265.
10. U.S. Pats. 2,670,285, 2,670,286, 2,670,287 (Feb. 23, 1954), L. M. Minsk and co-workers (to Eastman Kodak Co.).

11. U.S. Pat. 3,148,064 (Sept. 8, 1964), F. J. Rauner and J. J. Murray (to Eastman Kodak Co.).
12. U.S. Pat. 3,255,002 (June 7, 1966), H. G. Rogers (to Polaroid Corp.).
13. U.S. Pat. 3,218,168 (Nov. 16, 1965), W. R. Workman (to 3M Co.).
14. U.S. Pat. 3,307,941 (Mar. 7, 1967), R. W. Gundlach (to Xerox Corp.).
15. U.S. Pat. 3,387,976 (June 11, 1968), J. L. Sorkin (to Harris Intertype Corp.).
16. H. Tanabe, *Am. Perfumer Cosmet.* **77**, 25 (1962); *Takeda Kenkyusho Nempo* **20**, 249 (1961).
17. U.S. Pat. 3,390,051 (June 25, 1968), J. A. Baker and co-workers (to Nicholas Ltd.).
18. U.S. Pat. 3,388,189 (June 11, 1968), C. Mazzolini and S. LoMonaco (to Monsanto Co.).
19. U.S. Pat. 3,329,664 (July 4, 1967), M. Tsuda (to Japan Bureau of Industrial Technics).
20. U.S. Pat. 3,024,222 (Mar. 6, 1962), M. L. Freedman and S. B. Elliott (to Ferro Chemical Corp.).
21. U.S. Pat. 3,080,406 (Mar. 5, 1963), B. S. Marks and B. O. Shoepfle (to Hooker Chemical Co.).
22. U.S. Pat. 2,996,528 (Aug. 15, 1961), B. S. Marks and B. O. Shoepfle (to Hooker Chemical Co.).
23. U.S. Pat. 3,175,941 (Mar. 30, 1965), J. Dekker (to North American Philips Co.).
24. U.S. Pat. 3,149,031 (Sept. 15, 1964), P. J. Stoffel and B. J. Beaver (to Monsanto Co.).
25. U.S. Pat. 3,259,648 (July 5, 1966), H. E. Hennis (to The Dow Chemical Co.).
26. U.S. Pat. 3,183,075 (May 11, 1965), B. L. Walworth (to American Cyanamid Co.).
27. U.S. Pats. 2,819,972, 2,819,973 (Jan. 14, 1958), A. A. Robbins.
28. V. P. Seliverstov and co-workers, *Zashch. Met.* **12**, 206 (1976).
29. U.S. Pat. 3,977,949 (Aug. 31, 1976), W. E. Rosenberg (to Columbia Chemical Co.).
30. H. C. Gatos, *Ann. Univ. Ferrara Sez.* **5**, Suppl. (1960), 257 (1961).
31. U.S. Pat. 2,954,373 (Sept. 27, 1960), S. L. Shapiro (to U.S. Vitamin & Pharm. Corp.).
32. U.S. Pat. 3,012,030 (Dec. 5, 1961), P. A. J. Janssen.
33. U.S. Pat. 3,268,408 (Aug. 23, 1966), T. Naito (to Bristol-Banyu Res. Inst.).
34. R. Kuhn, *Helv. Chim. Acta* **11**, 103 (1928).
35. A. W. Pound and J. R. Pound, *J. Phys. Chem.* **38**, 1045 (1934).
36. G. Peine, *Ber.* **17**, 2117 (1884).
37. L. Bert and R. Annequin, *Compt. Rendu* **192**, 1315 (1931).
38. Brit. Pat. 504,125 (Apr. 20, 1939) (to I. G. Farbenindustrie A.G.).
39. *The National Formulary*, 9th ed., Committee on National Formulary, American Pharmaceutical Association, Washington, D.C., 1950.
40. Neth. Appl. 6,501,841 (Aug. 15, 1966) (to N. V. Phillips Gloeilampenfabrieken).
41. Fr. Pat. 518,960 (Mar. 29, 1968) (to Vikers, S.A. Ltd.).
42. Ger. Pat. 895,769 (Nov. 5, 1953), K. Freudenberg and co-workers (to Knoll A.G. Chemische Fabrieken).
43. U.S. Pat. 3,574,777 (Apr. 13, 1971), R. F. Heck (to Hercules Inc.).
44. H. Meerwein and R. Schmidt, *Ann. Chem.* **444**, 221 (1925).
45. W. Tuley and R. Adams, *J. Am. Chem. Soc.* **47**, 3061 (1925).
46. Fr. Demande 2,009,112 (Jan. 30, 1970), L. Katz and co-workers (to G.A.F. Corp.).
47. S. African Pat. 69,03, 590 (Nov. 27, 1969), J. F. Houle (to Eastman Kodak Co.).

W. F. RINGK
Consultant

CINNAMON. See Flavors and spices.

CINNAMYL ALCOHOL, $C_6H_5CH{=}CHCH_2OH$. See Cinnamic acid, cinnamaldehyde, and cinnamyl alcohol.

CITRAL, $(CH_3)_2C{=}CH(CH_2)_2C(CH_3){=}CHCHO$. See Flavors and spices; Perfumes; Terpenoids.

CITRIC ACID

Citric acid [77-92-9] (2-hydroxy-1,2,3-propanetricarboxylic acid), $C_6H_8O_7$, a natural constituent and common metabolite of plants and animals, is the most versatile and widely used organic acid in the field of foods and pharmaceuticals.

Citric acid is also used in many industrial applications to sequester ions, neutralize bases, and act as a buffer. In cosmetics it is used as a buffer to control pH in shampoos, hair rinses, and setting lotions. Many citrates, especially the neutral sodium salt, are used extensively in food and pharmaceutical products and in detergents. Esters of citric acid are used commercially as plasticizers in the preparation of polymer compositions, protective coatings, and adhesives.

Citric acid was first isolated in the crystalline form by Scheele in 1784 from the juice of lemons. Much earlier the alchemist Vincentius Bellovacensis had recognized that lemon and lime juices contained a special acid substance and referred, in his *Speculum Naturale,* written about 1200, to the use of lemon juice as an acid solvent. Liebig in 1838 recognized citric acid as a hydroxy tribasic acid and Grimoux and Adam synthesized citric acid in 1880 from glycerol. Wehmer indicated in 1893 that certain fungi produce citric acid when grown on sugar solutions. Currently, commercial production of citric acid is derived almost exclusively from fermentation processes.

Physical Properties

Anhydrous citric acid, mol wt 192.13, crystallizes from hot concentrated aqueous solutions. The crystals are translucent and colorless; they belong to the holohedral class of the monoclinic system. The melting point of the anhydrous form is 153°C; the density is 1.665. Citric acid is optically inactive and manifests no piezoelectric effect.

Citric acid monohydrate [5949-29-1] mol wt 210.14, crystallizes from cold aqueous solutions; the mean transition temperature from monohydrate to anhydrous is 36.6 ± 0.15°C. The monohydrate crystals, which are colorless and translucent, belong to the orthorhombic system. Citric acid monohydrate is stable in air of normal humidity, but loses water in dry air or in a vacuum over sulfuric acid. On gentle heating, the monohydrate crystals soften at about 70–75°C with the loss of water, and finally melt completely in the range 135–152°C. On rapid heating, the crystals melt at 100°C, solidify as they become anhydrous, and melt sharply at 153°C to a liquid of density 1.542.

Citric acid is a relatively strong organic acid as indicated by the first dissociation constant, 8.2×10^{-4} at 18°C. Second and third dissociation constants are 1.77×10^{-5} and 3.9×10^{-7}, respectively. pK values at 25°C are pK_1 3.128, pK_2 4.761, and pK_3 6.396.

The molecular refractivity of the monohydrate crystal is 67.11; the indexes of refraction, n_D^{20}, are reported as 1.493, 1.498, and 1.509. Heats of combustion at 25°C are 1.96 MJ/mol (468.5 kcal/mol) (1) and 1.952 MJ/mol (466.6 kcal/mol) (2) for citric acid anhydrous and monohydrate, respectively.

The surface tension of a 0.167 molal solution in contact with air is 69.51 mN/m (dyn/cm) at 30°C. The equivalent conductivity is 8.0×10^{-4} S/cm^2 at 25°C. The densities of aqueous solutions are given in Table 1. The freezing point depression and

Table 1.   Densities of Aqueous Citric Acid Solutions at 15°C[a]

Citric acid, monohydrate, wt %	d_{15}^{15}	Concn of soln		Citric acid, monohydrate, wt %	d_{15}^{15}	Concn of soln	
		g/L	lb/gal			g/L	lb/gal
6	1.0227	61.36	0.5121	40	1.1709	468.4	3.909
10	1.0392	103.9	0.8673	46	1.1998	551.9	4.606
16	1.0632	170.1	1.420	50	1.2204	610.2	5.092
20	1.0805	216.1	1.803	56	1.2514	700.8	5.848
26	1.1060	287.6	2.400	60	1.2738	764.3	6.378
30	1.1244	337.3	2.815	66	1.3071	862.7	7.199
36	1.1515	414.5	3.460				

[a] Ref. 4.

Table 2.   Freezing Point Depression and Boiling Point Elevation of Aqueous Citric Acid Solutions[a]

Concn, mol/1000 g H_2O	fp depression, °C	bp elevation, °C	Concn, mol/1000 g H_2O	fp depression, °C	bp elevation, °C
0.01	0.023		2.00	1.00	1.214
0.05	0.042		5.00		3.512
0.10	0.203		10.00		8.39
0.50	0.965	0.284	20.00		16.6
1.00	1.94	0.577			

[a] Ref. 5.

Table 3.   Solubility of Anhydrous Citric Acid in Water[a]

Temperature, °C	Citric acid, wt %	Solid phase	Temperature, °C	Citric acid, wt %	Solid phase
10	54.0	$C_6H_8O_7 \cdot H_2O$	60	73.5	$C_6H_8O_7$
20	59.2	$C_6H_8O_7 \cdot H_2O$	70	76.2	$C_6H_8O_7$
30	64.3	$C_6H_8O_7 \cdot H_2O$	80	78.8	$C_6H_8O_7$
36.6[b]	67.3	$C_6H_8O_7 \cdot H_2O + C_6H_8O_7$	90	81.4	$C_6H_8O_7$
40	68.6	$C_6H_8O_7$	100	84.0	$C_6H_8O_7$
50	70.9	$C_6H_8O_7$			

[a] Ref. 5.
[b] Transition point.

boiling point elevation of aqueous solutions are given in Table 2. Additional data on properties of concentrated aqueous solutions, 0.5–30.0 wt %, have been summarized (3). Citric acid is readily soluble in water (Table 3), moderately soluble in alcohol (Table 4), but only sparingly soluble in diethyl ether (5). The solubility in various organic solvents is listed in Table 5. Anhydrous citric acid is insoluble in chloroform, benzene, carbon disulfide, carbon tetrachloride, and toluene.

Table 4. Solubility of Hydrated and Anhydrous Citric Acid in Aqueous Solutions of Ethyl Alcohol at 25°C[a]

Ethyl alcohol, wt %	Monohydrate		Anhydrous		
	d_{25}, satd soln	Soly, g/100 g satd soln	Ethyl alcohol, wt %	d_{25}, satd soln	Soly, g/100 g satd soln
20	1.286	66.0	20	1.297	62.3
40	1.257	64.3	40	1.246	59.0
50	1.237	63.3	60	1.190	54.8
60	1.216	62.0	80	1.120	48.5
80	1.163	58.1	100	1.010	38.3
100	1.068	49.8			

[a] Ref. 6.

Table 5. Solubility of Hydrated and Anhydrous Citric Acid in Some Organic Solvents at 25°C[a,b]

Solvent	d_{25}, satd soln	Soly, g/100 g satd soln
	Monohydrate	
amyl acetate	0.8917	5.980
amyl alcohol	0.8774	15.430
ethyl acetate	0.9175	5.276
diethyl ether	0.7228	2.174
chloroform	1.4850	0.007
	Anhydrous	
amyl acetate	0.8861	4.22
diethyl ether (abs)	0.7160	1.05

[a] 100 g methanol dissolves 197 g citric acid monohydrate at 19°C; 100 g propyl alcohol dissolves 62.8 g citric acid monohydrate at 19°C.
[b] Ref. 6.

Chemical Properties

When heated to 175°C, citric acid (1) (Fig. 1) is partially converted to aconitic acid (2) by elimination of water, and to acetonedicarboxylic acid (3) by the loss of carbon dioxide and water. Acetonedicarboxylic acid in turn decomposes to form acetone (4) and carbon dioxide. Rapidly heated to a higher temperature, aconitic acid (2) loses water to give its anhydride acid (5), which changes to itaconic anhydride (6) by the elimination of carbon dioxide. A part of the itaconic anhydride becomes citraconic anhydride (7). Above 175°C citric acid yields an oily distillate, which crystallizes as itaconic acid (8). Further heating yields an uncrystallizable oil, which is citraconic anhydride (7). Citraconic acid (9) is formed by the addition of water to its anhydride (7) and also by the molecular rearrangement of itaconic acid (8). Mesaconic acid (10) can also be formed by the molecular rearrangement of itaconic acid (8) or by heating citraconic acid (9) in a small quantity of water at 200°C. The hydrogenation of citric acid (1) or aconitic acid (2) yields tricarballylic acid (11). The reduction of itaconic (8), citraconic (9), or mesaconic (10) acids yields methylsuccinic acid (12).

Digestion of citric acid with fuming sulfuric acid or oxidation with potassium

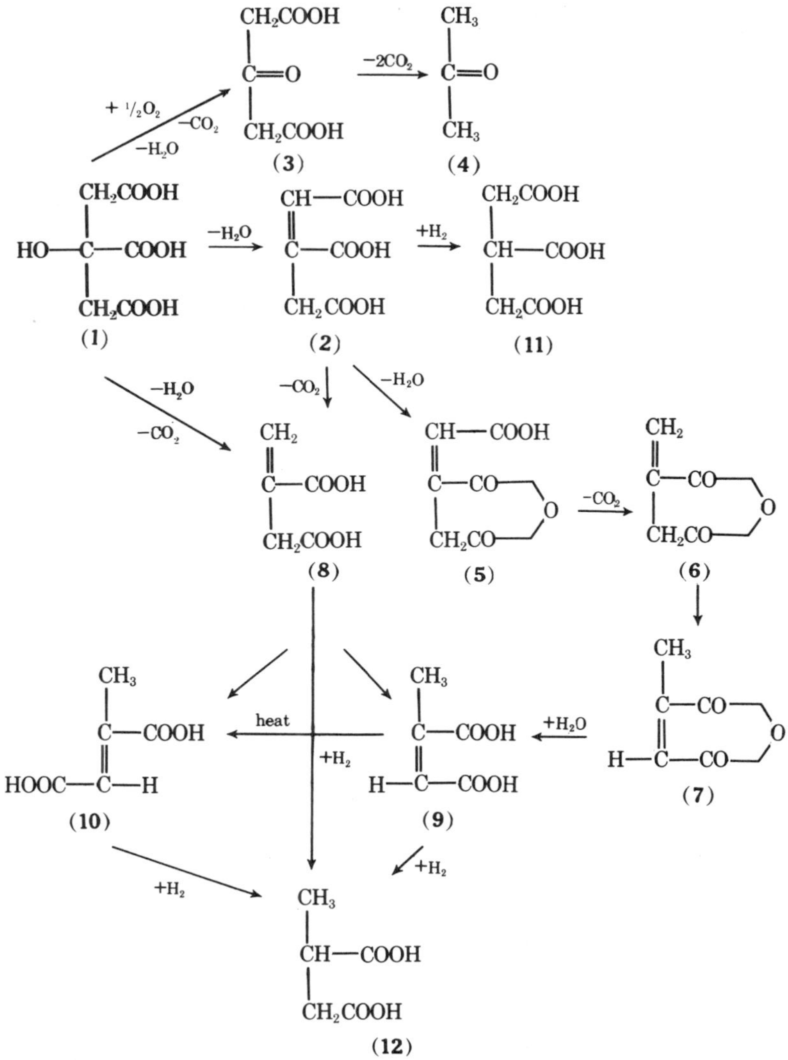

Figure 1. Thermal decomposition of citric acid.

permanganate solution yields acetonedicarboxylic acid. Above 35°C, oxidation with potassium permanganate produces oxalic acid. Citric acid decomposes to form oxalic acid and acetic acid when fused with potassium hydroxide or oxidized with nitric acid. Acetonedicarboxylic acid is useful in various syntheses since the two methylene groups between carbonyl groups possess considerable reactivity like that of acetoacetic ester or malonic ester; acetonedicarboxylic acid enters into similar condensation reactions.

As a tribasic acid, citric acid manifests the usual properties of a polybasic acid. It forms a variety of salts including those of the alkali metal and alkaline earth metal families (see under Salts). In addition, citric acid forms a variety of esters, amides, and acyl chlorides. Mixed compounds such as the salts of the acid esters can also be formed. The anhydride itself cannot be formed, but acyl derivatives of the acid can be dehydrated to form acyl citric anhydrides. The hydroxyl group may form acyl derivatives, ethers, etc. A wide variety of such mixed compounds is possible, and many have been prepared and studied (7–9).

In aqueous solution, citric acid can be mildly corrosive toward carbon steels and, therefore, should be used with an appropriate inhibitor. It is not corrosive to stainless steels which are most often employed as the material of construction for processes involving citric acid.

At proper pH in aqueous media, citric acid's hydroxyl and carboxylic acid groups act as multidentate ligands forming complexes or chelates with metal ions (10). These chelating reactions are the basis for many of today's industrial processes, including elimination or control of metal-ion catalysis, lowering of metal oxidation potentials, removal of corrosion products (ie, Fe^{3+}), regeneration of ion-exchange resins, recovery of valuable metals by precipitation of insoluble chelates, decontamination of radioactive materials, quenching reactions, and driving reactions to completion (see Chelating agents).

Occurrence

Citric acid occurs widely in both the plant and animal kingdoms (11). It is found most abundantly in the fruits of the citrus species such as lemons (4.0–8.0%), grapefruit (1.2–2.1%), tangerines (0.9–1.2%), oranges (0.6–1.0%), and limes. It also is found in currants (black, 1.5–3.0%; red, 0.7–1.3%), raspberries (1.0–1.3%), gooseberries (1.0%), strawberries (0.6–0.8%), and apples (0.008%), and in pineapples, sloeberries, cranberries, whortleberries, cherries, tamarind, and berries of the mountain ash. In vegetables, citric acid is found in potatoes (0.3–0.5%), tomatoes (0.25%), asparagus (0.08–0.2%), turnips (0.05–1.1%), peas (0.05%), butternut squash (0.007–0.025%), corn (kernels, 0.02%), lettuce (0.016%), eggplant (0.01%), and in beets. It is a constituent of wine (0.4 g/L). As the free acid or a salt, it is found in the seeds and juices of a wide variety of flowers and plants.

The citrate ion occurs in all animal tissues and fluids (11), eg, in human whole blood (15 ppm), blood plasma (25 ppm), red blood cells (10 ppm), milk (500–1250 ppm), urine (100–750 ppm), semen (2000–4000 ppm), cerebrospinal fluid (25–50 ppm), mammary gland (3000 ppm), thyroid gland (750–900 ppm), kidney (20 ppm), bone (7500 ppm), amniotic fluid (17–100 ppm), saliva (4–24 ppm), sweat (1–2 ppm), tears (5–7 ppm), and skeletal muscle, liver, and brain (2–100 ppm); in cow's milk (0.08–0.23%); and in chicken's liver (108 ppm), kidney and heart (142 ppm), brain (472 ppm), pancreas (667 ppm) and other living tissues (507 ppm). The total circulating citric acid in the serum of man is approximately 1 mg/kg of body weight. Normal daily excretion in the urine of humans is 0.2–1.0 g.

Citric acid is produced by many yeasts, molds and bacteria. Wehmer, in about 1893, reported that two species of mold, which he called *Citromyces pfefferianus* and *C. glaber,* could produce citric acid by fermentation of sugar solutions.

Physiological Role of Citric Acid.   Citric acid occurs in the terminal oxidative metabolic system of all but a very few organisms. This system (Fig. 2), variously referred to as the Krebs cycle (for its discoverer, H. A. Krebs), the tricarboxylic acid cycle, or the citric acid cycle, is a metabolic intermediate cycle involving the terminal steps

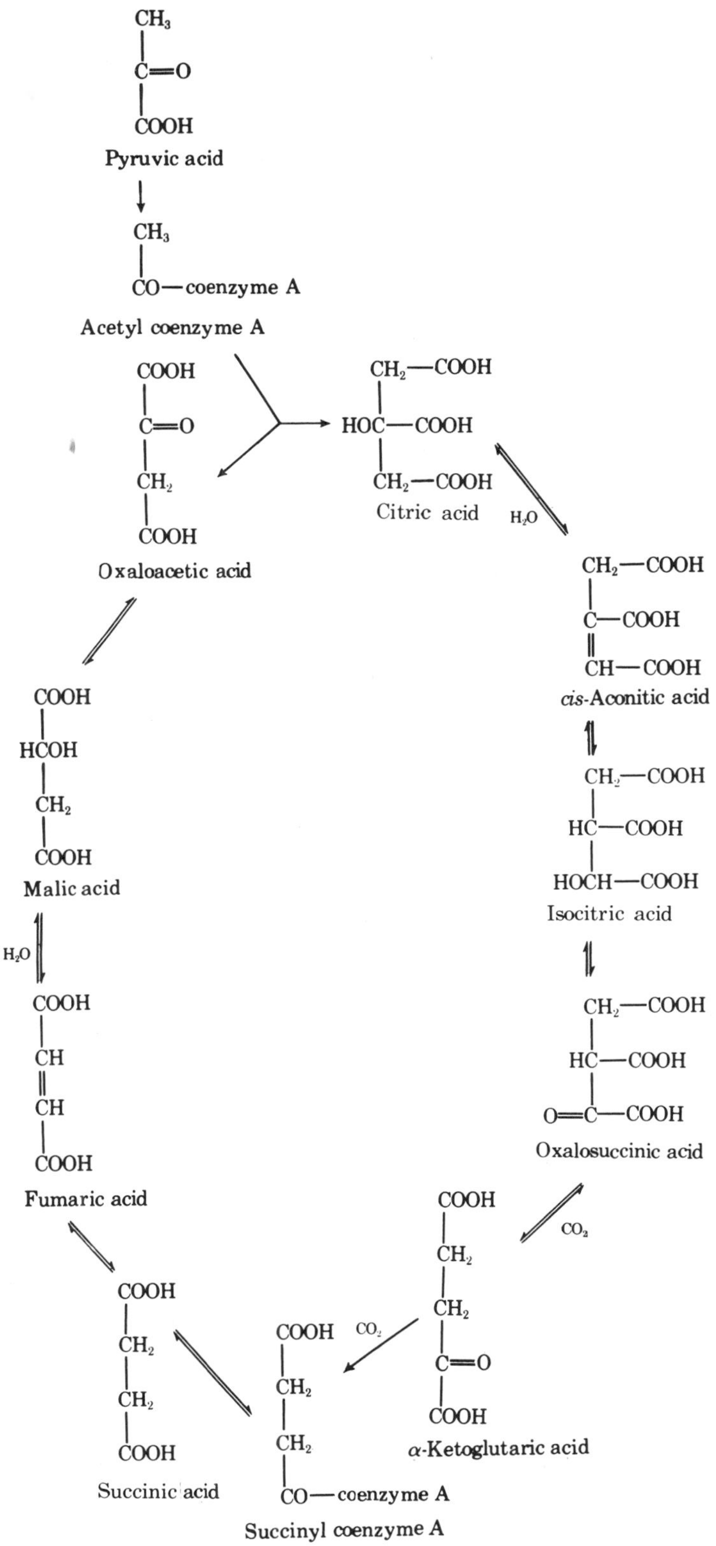

CH$_3$

C=O

COOH

Pyruvic acid

↓

CH$_3$

CO—coenzyme A

Acetyl coenzyme A

COOH

C=O

CH$_2$

COOH

Oxaloacetic acid

CH$_2$—COOH

HOC—COOH

CH$_2$—COOH

Citric acid   H$_2$O

CH$_2$—COOH

C—COOH

CH—COOH

cis-Aconitic acid

CH$_2$—COOH

HC—COOH

HOCH—COOH

Isocitric acid

COOH

HCOH

CH$_2$

COOH

Malic acid

CH$_2$—COOH

HC—COOH

O=C—COOH

Oxalosuccinic acid

H$_2$O

COOH

CH

CH

COOH

Fumaric acid

COOH

CH$_2$

CH$_2$

C=O

COOH

α-Ketoglutaric acid

CO$_2$

COOH

CH$_2$

CH$_2$

COOH

Succinic acid

CO$_2$

COOH

CH$_2$

CH$_2$

CO—coenzyme A

Succinyl coenzyme A

Figure 2. Krebs (citric acid) cycle.

155

in the conversion of carbohydrates, fats, or proteins to carbon dioxide and water with concomitant release of energy necessary for growth, movement, luminescence, chemosynthesis, and reproduction. This cycle also provides the carbonaceous materials from which amino acids and fats are synthesized by the cell.

Manufacture and Processing

The traditional method of preparing citric acid was by extraction from the juice of certain citrus fruits, such as lemons and limes, and later from pineapple wastes. With the development of fermentation technology these processes became obsolete (see Fruit juices).

Fermentation. A comprehensive review of fermentation technology has been published elsewhere (12) (see Fermentation). A summary of the various fermentation processes follows.

Surface Process. The microbial production of citric acid on a commercial scale was begun by Pfizer Inc. in 1923 based primarily on the work of Currie (13) who found that certain strains of *Aspergillus niger,* when grown on the surface of a sucrose-and-salts solution, produced significant amounts of citric acid. Variations of this surface culture technique still account for a substantial portion of the world's production of this acid.

The surface methodology consists of inoculating with spores of *A. niger* shallow aluminum or stainless steel pans containing sugar solution along with sources of assimilable nitrogen, phosphate, magnesium, and various trace minerals. Growth of the mold occurs on the solution surface forming a rubbery, convoluted mycelial mass. Air passed over the surface provides oxygen and controls temperature by evaporative cooling.

Because of its relatively low cost, molasses has proven to be a preferred source of sugar for microbial production. Since it is a by-product of sugar refining, molasses varies considerably and not all types are suitable for citric acid production. Beet molasses is preferred to cane, but there are considerable yield variations within each type (see Sugar). Beet molasses must be treated prior to use to improve yield. Treatment may consist of ion-exchange or chemical precipitation or chelation of metallic ions.

Submerged Process Using Aspergillus Niger. In this process the microorganism, *A. niger,* is grown dispersed through a liquid medium. The fermentation vessel usually consists of a sterilizable tank of several hundred cubic meters (thousands of gallons) capacity equipped with a mechanical agitator and a means of introducing sterile air.

The origins of the submerged process date back to the work of Amelung (14). Perquin (15), using a small flask to simulate deep culture conditions, established that phosphate limitation plays an important role in the submerged production of citric acid by *A. niger.* Martin and Waters (16) related the influence of *A. niger* morphology to citric acid production from beet molasses. Subsequently, Clark (17) found that specific quantities of ferrocyanide leads to formation of an optimum mycelial form.

In practice the mold spores are produced under controlled aseptic conditions. When harvested they are used in specific quantities to seed the inoculum fermenter, a stage prior to the production fermenter containing a medium designed to develop cellular mass and to control morphology rather than to produce acid. The inoculum is then transferred aseptically to the production fermenter. Medium constituents in this stage are chosen to foster acid production rather than growth. Progress is monitored by periodically determining the acid and sugar content of the vessel. In addition, pH, dissolved oxygen, and solids content are recorded. Initial sugar concentrations and acid yield compare favorably with those of the surface process. As in the surface process, molasses requires treatment prior to use. Selection of raw materials for the submerged process suffers from the same constraints as the surface method except that it permits a wider choice of materials.

A patent was issued in 1951 (18) for the use of *A. niger* in submerged culture with a medium consisting of pure sucrose or a solution of cation-depleted invert molasses. Subsequent patents (19) cover production by *A. niger* on a substrate produced from starch treated with amylase and amyloglucosidase. More recently, glucose solutions derived directly from corn have been used. Yields from substrates of this type have been reported to exceed 90% (20).

Submerged Process Using Yeast. Until about 1969 or 1970 *A. niger* was considered to be the only organism that could be used to produce citric acid in commercial quantities. A patent issued in 1970 (21) challenged this prevailing point of view by demonstrating the production of citric acid by species of yeast (eg, *Candida guilliermondii*) grown submerged in a medium containing either glucose or blackstrap molasses with an equivalent amount of sugar. Fermentation time was shorter than with *A. niger*. A subsequent improvement patent (22) quotes citric acid concentrations of 110 g/L (see Yeast).

Candida strains are also used in a novel process that permits production of citric acid from hydrocarbons. In 1970 a patent was issued for the conversion of C_9 to C_{20} normal paraffins by *Candida lipolytica* (23). Citric acid weight yields in excess of 100% were claimed. Other patents issued in 1972 (24) describe a procedure for selecting Candida mutants. One example shows a 138% citric acid weight yield from a C_{13} to C_{15} *n*-paraffin mixture. In 1974 Pfizer patented a continuous process (25) for fermentation by *Candida lipolytica* using a single vessel to which paraffin is continuously added and fermented broth continuously withdrawn (see also Foods, nonconventional).

Energy Requirements. The energy requirement of the surface culture method is low. Small electric motors are used to drive the fans that provide the air necessary to maintain the optimum fermentation temperature. Energy requirements of 1.3–2.6 MJ/m^3 (4.8–9.5 Btu/gal) of fermenting medium are sufficient for both cooling and aeration (26). Submerged culture techniques must provide agitation, aeration, and cooling. Energy demands range from 8–16 MJ/m^3 (28.5–57.0 Btu/gal) depending on the type and size of the fermenter (26).

Recovery. Citric acid is generally recovered from a fermented aqueous solution by first separating the microorganism (using rotary filtration or centrifugation) and then precipitating the citrate ion as the insoluble calcium salt. Classically, the calcium citrate salt has been used to separate fermentation by-products and other impurities from the citrate ion. If fermented solutions are of sufficiently high purity, it may be possible to recover crude citric acid by direct crystallization; however, commercial success with this method has not been reported.

Recovery of citric acid via calcium salt precipitation is a highly complex process. Although the chemistry is straight-forward, the engineering principles, separation techniques, and unit operations employed dictate a sophisticated commercial process. The fermented solution, which has been separated from the microorganism, may be treated with a calcium hydroxide slurry in two stages. The first removes oxalic acid when present as a by-product, and the second precipitates the citrate. The citrate is formed by adding a lime slurry to obtain a neutral pH. After sufficient reaction time, the slurry is filtered and the precipitate washed free of soluble impurities. The resulting calcium citrate is acidified with sulfuric acid. This reaction converts the calcium citrate to calcium sulfate and citric acid in the presence of free sulfuric acid. Calcium sulfate is filtered and washed free of citric acid solution. Both the calcium citrate and calcium sulfate reactions are generally performed in agitated reaction vessels and filtered on commercially available filtration equipment. Control of the reactions is accomplished using conventional pH instrumentation.

The aqueous citric acid solution, which is saturated with calcium sulfate, is concentrated to a designated specific gravity. The calcium sulfate that forms during this concentration step is separated by filtration. In some cases, solutions are deionized at this stage to improve subsequent crystallization steps and reduce the amount of recycled material. The clear citric acid solution undergoes a series of crystallization steps to achieve the physical separation of citric acid from the remaining trace impurities. This is accomplished in evaporation, crystallization, and centrifugation equipment. A conventional crystallization scheme consists of a batch vacuum-pan evaporator or a forced circulating evaporator coupled with auxiliary tankage and appropriate centrifuge equipment. Within these systems the crystals formed are separated from the mother liquor and advanced to the next crystallization step. Both batch and continuous units have been employed in this operation. The choice of batch or continuous equipment, or a combination, is generally made on the basis of process adaptability and economics.

In some cases trace-impurity removal and decolorization with adsorptive carbon must be accomplished between successive crystallization steps to meet the specifications for USP/FCC citric acid. Various salts of citric acid can be formed at this stage utilizing commercial neutralizing alkalis and dissolved citric acid crystals.

The finished citric acid is dried and sifted in conventional rotary drying (qv) and material handling equipment. The dryer is generally one employing a countercurrent hot air source to remove residual moisture. Because anhydrous citric acid is hygroscopic, care must be taken to reach the final moisture specification during drying and to avoid storage in areas of high temperature and high humidity.

An alternative recovery process (27) involves the extraction of citric acid from fermentation broths using a mixture of trilaurylamine, n-octanol, and a C_{10} or C_{11} isoparaffin. This is followed by reextraction of the citric acid from the solvent phase into water with the aid of heat. Efficient citric acid extraction is achieved through a series of countercurrent steps which ensure intimate contact of the aqueous and nonaqueous phases. When transfer of the citric acid to the solvent phase is complete, the citric acid is reextracted into water, again using a multistage countercurrent system. The two steps differ only in the temperature at which they are performed.

This recovery process, therefore, consists of selectively transferring citric acid via a solvent from an aqueous solution containing various by-products to another aqueous solution in which the citric acid is more concentrated and contains substantially less by-products. The final processing steps begin with a diluent wash of the aqueous solution by the hydrocarbon solvent, followed by passage of the acid solution through granular activated-carbon columns. Effluent from the carbon columns is processed through a conventional sequence of evaporator-crystallizer steps to complete the manufacturing process.

Waste Utilization. The residual solubles and biomass from the fermentation process contain nutrients that can be recycled usefully as animal feeds and for other agricultural purposes. The Association of American Feed Control Officials (28) lists solids and fermentation solubles from citric acid production as acceptable ingredients in animal feeds. The value of citric acid mycelia as a fertilizer and soil conditioner also has been reported (29) (see Pet and other livestock feeds).

Chemical Syntheses. A number of syntheses of citric acid have appeared in the chemical and patent literature since the 1880 report by Grimoux and Adam of a route based on the reaction of glycerol-derived 1,3-dichloroacetone with cyanide (30). These include a spate of recent syntheses aimed at discovering routes competitive with fermentation. None of these has achieved commercial status.

The early syntheses are of the following six types: conversion of acetonedicarboxylate and derivatives to the cyanohydrins followed by hydrolysis (31); Reformatsky and Grignard reactions on oxaloacetic and glyoxylic esters (32); photolysis of a mixture of glycolic and malic acids (33); benzilic acid rearrangement of 3,4-diketoadipic acid (34); oxidative degradation of quinic acid (35); aldol condensation of pyruvate with oxaloacetic acid to form citroylformic acid and lactone followed by oxidative decarboxylation (36). Other syntheses include: base-catalyzed condensation of acetonedicarboxylate with formate (37) conversion of γ-chloroacetoacetic ester to the cyanohydrin followed by hydrolysis, displacement of chloride by cyanide, and hydrolysis (38); electrolytic oxidation and then reduction of succinic acid to form a mixture of acids including citric acid (39).

More recently, diverse synthetic approaches to citric acid from relatively inexpensive starting materials have been described in the patent literature: oxidative conversion of maleic and fumaric derivatives to the corresponding oxaloacetates followed by cycloaddition with ketene to form β-lactones and then hydrolysis (40); bimolecular decarboxylative condensation of oxaloacetic acid to form citroylformic acid followed by oxidative decarboxylation (41); carboxylation of acetone using an alkali metal phenolate catalyst in solvents such as dimethylformamide or glyme (dimethyl ether of ethylene glycol) to form acetonedicarboxylate, followed by reaction with cyanide to form the cyanohydrin (42) and then acid or base hydrolysis (43); condensation of ketene and phosgene to form acetonedicarboxylic acid chloride (44) or its cyclic precursor (45), reaction with alcohol to form the corresponding acetonedicarboxylate, followed by cyanide reaction to form cyanohydrin, and then acid hydrolysis (46); chlorination of diketene and hydrolysis to form 4-chloroacetoacetic acid, conversion to cyanohydrin and hydrolysis to the 3-carbamoyl derivative, reaction with base and then cyanide to form the corresponding 4-cyano derivative followed by hydrolysis (47); epoxidation of itaconate, reaction with cyanide, and hydrolysis (48); condensation of isobutylene with formaldehyde to form 3-methylene-1,5-pentanediol followed by oxidation with nitrogen dioxide and nitric acid (49); oxidation of 1-hydroxy-3-cyclopentenecarboxylic acid derivatives (50); hydration of *cis*- and *trans*-aconitic acids and rearrangement of isocitric and alloisocitric acids by heating in alkali (51–52).

Economic Aspects

Citric acid manufacturing plants are found in over 30 countries throughout the world. Actual production figures are not available; however, it can be estimated conservatively that 1976 United States production was more than double the 25–35 thousand metric tons estimated for 1963. Prices of citric acid and its major salts as of January 1978 are as follows (53).

citric acid anhydrous	$1.29/kg
citric acid hydrous	$1.565/kg
ammonium citrate, dibasic	$2.73/kg
calcium citrate	$4.365/kg
potassium citrate	$1.455/kg
sodium citrate, dihydrate	$1.15/kg

Specifications, Standards, and Quality Control

Citric acid for pharmaceutical use in the United States must meet the following USP specifications (54): conform to the pyridine–acetic anhydride test for identity and contain not less than 99.5% of citric acid calculated on the anhydrous basis; the anhydrous acid must contain no more than 0.5% water and the hydrous form no more than 8.8%; ash may not exceed 0.05%; the limit on heavy metals is 0.001% max, and within the heavy metals group, the maximum level of arsenic tolerated is 0.0003%. The USP monograph also prescribes tests for turbidity limits for oxalate (calcium precipitation) and sulfate (barium precipitation). Substances readily carbonizable in the presence of hot concentrated H_2SO_4 (90°C for 60 min) must be such that the visual appearance of the specimen solution under test is not darker than a matching fluid containing one part of cobaltous chloride colorimetric solution (C.S.) to nine parts ferric chloride (C.S.).

The same forms of citric acid are commercially available for food (55) and feed (56) applications, and both must meet the specifications of the Food Chemicals Codex (FCC) (57). With two exceptions, the FCC requirements are essentially equivalent to the USP specifications. The FCC lists no limit for sulfate. On the other hand, the FCC does limit the abundance of trace substances having uv absorbance properties in the range of 280–400 nm (58). The limit also is mandated for food use by FDA regulations (59). Specifications are also available for the 50% aqueous technical solution forms of citric acid for industrial applications (60).

Particle-Size Specifications.   Citric acid USP/FCC is available in a variety of United States standard sieve sizes. The three major forms for the anhydrous (61) are:

granular	max 2% on 16-mesh (1.19 mm)
	max 10% through 50-mesh (0.30 mm)
fine-granular	max 3% on 30-mesh (0.59 mm)
	max 5% through 100-mesh (0.15 mm)
powder	max 2% on 60-mesh (0.25 mm)
	min 50% through 200-mesh (0.07 mm)

For the hydrous acid, the three major particle-size specifications (62) are:

granular	max 10% on 16-mesh
	max 10% through 50-mesh
fine-granular	max 3% on 30-mesh

	max 10% through 100-mesh
fine-granular XX	max 20% on 60-mesh
	max 20% through 100-mesh

Analytical and Test Methods

Aqueous titration with $1N$ NaOH remains the official method (54,57) for assaying the pure chemical entity. Although not specific, the procedure is satisfactory in the absence of interfering substances and when conducted in conjunction with an appropriate test for identity. The determination of citric acid in foods, blood, and other biological materials is readily accomplished by its conversion to pentabromoacetone which is measured either gravimetrically (63) or colorimetrically (64). Trace quantities of citric acid can be determined by a spectrophotometric method based on the widely used Furth and Herrmann reaction with pyridine and acetic anhydride (65). An enzymatic method (66) which is species-specific for the citrate moiety is used as a combined assay–identification test for the pure chemical entity, as a trace method for monitoring biodegradability of citric acid and its salts in complex fluids, and as a procedure for measuring the level of citrate in such commercial formulations as dishwashing detergents.

The test for readily carbonizable substances (RCS), generally trace carbohydrates, is a judgment method that requires considerable experience on the part of the analyst. The purpose of the test is not to detect RCS materials *per se* but rather to serve as a measure of the effectiveness of impurities removal.

The test for uv-absorbing substances (58) is straightforward in principle but difficult to carry out in practice. The limit of detection is of the order of one nanogram per gram. This is in many cases less than the background, particularly in urban areas. Special facilities, reagents, and ultra-clean glassware are required to avoid false positive responses.

Citric acid and sodium citrate are widely used as buffering reagents in biological assay systems. The chelating properties of these compounds, particularly their affinity for calcium and magnesium ions, must be accounted for in designing biological assays (67).

Shipment and Storage

Citric acid, USP, FCC, anhydrous granular and fine granular is shipped in 110-kg drums and the powder in 90.72-kg drums. All granulations are also shipped in 45.36-kg and 22.68-kg multiwall paper bags which may be palletized and shrink-wrapped with polyethylene film. Citric acid is also shipped in semibulk corrugated boxes. Citric acid, USP, FCC, hydrous granular, fine granular and fine granular XX is shipped in 113.4-kg drums and the powder in 90.72-kg drums. A technical-grade 50% solution of citric acid may be shipped in tank cars or tank trucks.

Citric acid, USP, FCC, anhydrous and hydrous, can be stored in dry form without difficulty; however, high humidity conditions and elevated temperatures should be avoided to prevent caking. Materials packaged with desiccants are commercially available (68).

Health and Safety Factors (Toxicology)

Health Aspects. Citric acid is universally accepted as a safe food ingredient. The U.S. Food and Drug Administration lists citric acid as a multiple purpose generally-recognized-as-safe (GRAS) food substance and as a GRAS sequestrant limited only by good manufacturing practice (69). Citric acid is also approved by the Joint FAO/WHO Expert Committee on Food Additives for use in foods without limitation (70).

The use of citric acid and certain of its salts and esters was recently evaluated by a Special Committee On GRAS Substances (SCOGS) of the Federation of American Societies for Experimental Biology under contract with the FDA (71). The evaluation was based largely on two scientific literature reviews prepared for the FDA that summarize the world's applicable scientific literature from 1920 through 1973 (11). Factors considered were citric acid's wide distribution as a natural component of man's diet, its well-established metabolic pathway, the fact that citrate ingested by infants and adults is considered to be completely metabolized, and the general absence of any adverse effects disclosed in the scientific literature. The amount of citrate added to foods, estimated at 500 mg per person per day, does not constitute a significant addition to the total body load from natural sources. The committee concluded that: "There is no evidence in the available information on citric acid, sodium citrate, potassium citrate, calcium citrate, ammonium citrate, isopropyl citrate, stearyl citrate, and triethyl citrate that demonstrates, or suggests reasonable ground to suspect, a hazard to the public when used at levels that are now current or that might reasonably be expected in the future."

Environmental Aspects. Citric acid, part of the metabolic cycle in living organisms, is found throughout the ecosphere. Citrates are leached from rotting vegetation and produced by soil microorganisms. As a result, they occur naturally in soil and water and have been detected at low levels throughout the ecosystem. Citric acid has been found in sea water at levels of 35–145 ppb (72). Similar levels have been found in brackish water, river water, and dirty snow. Citrates degrade readily when in contact with a variety of microorganisms that are found in soil, natural waters or sewage treatment systems. Biodegradability studies, conducted in Swisher columns and in draw-and-fill columns, showed that citric acid and sodium citrate are 99% degraded in 2 h (73). Metal chelate degradation studies indicate that calcium citrate degrades as rapidly as sodium citrate or citric acid. Degradation rates for some other chelates are:

Chelate	CAS Registry No.	Degradation, %
aluminum citrate	[813-92-3]	90 <8 h
ferric citrate	[2338-05-8]	90 <8 h
cadmium citrate	[49707-39-3]	99 in 24 h
copper citrate	[866-82-0]	90 in 24 h

Thus it is seen that metals do not prevent biodegradation of citrate. The slower degradation rate for copper is believed to be caused by copper-ion toxicity and not the chelation of citrate.

Sewage treatment is not adversely affected by citrate (73–74). As might be expected for a compound so widespread in living systems, aquatic toxicity is very low (75).

Uses

The food and pharmaceutical industries utilize citric acid extensively because of its general recognition of safety, pleasant acid taste, high water solubility, and chelating and buffering properties. It is also used in cosmetics and toiletries as a buffer, and in a wide variety of industrial applications as a buffering and chelating agent. Citric acid is also a reactive intermediate in chemical synthesis.

Food Uses. *Beverages.* Citric acid is used extensively in carbonated beverages (qv) to provide tartness and complement fruit and berry flavors. It also increases the effectiveness of antimicrobial preservatives. The high solubility of citric acid is important to this application because the acid is dissolved in the beverage syrup before addition of the carbonated water. The amount of acid used depends on the flavor of the product (76), as well as taste evaluation and customer preference. The acid level of most fruit-flavored carbonated beverages falls in the range of 0.1–0.25%.

Uncarbonated fruit drinks are a mixture of water, fruit juice, sugar, acid, color, and flavor. Citric acid is used to adjust the pH to provide uniform acidity. It is a natural component of fruit juices and its character blends well with the flavor systems. Total titratable acidity of these products generally is in the range of 0.25–0.4%.

Soft drink mixes have made great gains in popularity in recent years (77). The basic ingredients are sugar, acid, flavor, and color but many of the more sophisticated formulations also include buffers, vitamins, fruit pulp, cloud, and bodying and free-flow agents. Citric acid is the acidulant of choice for this application because of its natural tang and its rapid solubility. Acid levels may vary from 1.5 to 5%, depending on flavor.

Jellies, Jams, and Preserves. Citric acid is used in jams and jellies to provide tartness and to adjust the pH of the product. The pH must be adjusted within very narrow limits for optimum gelation (78). Depending on the type of pectin used, the pH may range from 3.0 to 3.4. In the manufacture of jams, jellies, and preserves, all the ingredients other than the acid are cooked to the proper soluble solids content. The citric acid then is added as a 50% solution to ensure good distribution throughout the batch.

Candy. Citric acid is added to candy primarily for tartness. Most of the acid purchased by the confectionery industry is used in flavors containing 0.5–1.0% acid but sours may contain as much as 2% citric acid. To minimize sucrose inversion the acid is added with the color and flavor to the molten candy glass after the cook. Compressed candy tablets, agar jellies, and starch-based jellies also use citric acid as a flavor adjunct. In pectin jellies, citric acid also functions to adjust pH in order to attain maximum gel strength.

Frozen Food. The chelating and pH adjustment features of citric acid enable it to optimize the stability of frozen food products by enhancing the action of antioxidants and inactivating enzymes.

Citric acid, alone or in combination with erythorbic acid, helps prolong the shelflife of frozen fish (79) and shellfish. Citric acid inactivates some enzymes and chelates trace metals that cause rancidity in fish and off-color development in shellfish. The seafood product is generally dipped in a solution of 0.25% citric acid and 0.25% erythorbic acid just before freezing.

Citric acid also inhibits color and flavor deterioration in frozen fruit (80). Here again the function is to inhibit enzymatic and trace metal-catalyzed oxidation. Levels of 0.1–0.3% citric acid in combination with 100–200 ppm of erythorbic acid have been found effective.

Fats and Oils. The ability of citric acid to chelate trace metals that act as prooxidants enables it to be used as an antioxidant synergist in fats, oils, and fat-containing foods. Levels of 0.005–0.02% are added during the refining process (see Fats and fatty oils).

Other Food Uses. Citric acid is used in meat products (qv) to improve the peelability of frankfurters, as an antioxidant synergist in sausage, and as an anticoagulant. Where Federal Standards of Identity permit, citric acid may be used in canned vegetables to lower pH, to reduce heating requirements, and to inhibit the growth of microorganisms. As a flavor adjunct, citric acid is used in sherbets and water ices. In gelatin desserts it is used for flavor and to control pH. Citric acid is used in wine (qv) to adjust acidity (81) (see Flavors; Food additives).

Pharmaceutical, Cosmetic, and Toiletry Uses. *Pharmaceuticals.* Effervescence is a popular delivery system for oral dosage of some medications (82–83). The reaction of citric acid with a bicarbonate–carbonate source in water produces carbon dioxide (effervescence) and a salt of the acid. It provides rapid dissolution of active ingredients, improved palatability and solubilizing action for cathartics and analgesics (84–86).

Citric acid is used in oral pharmaceutical liquids, elixirs, and suspensions to buffer to pH 3.5–4.5 and thus maintain stability of active ingredients, and to enhance the activity of preservatives. In conjunction with fruit flavors, citric acid provides a desirable tart taste that assists in masking the bitter medicinal flavor (87). Addition of 0.02% citric acid to liquid dosage forms complexes trace iron and copper ions and retards degradation of active ingredients (88).

In chewable tablets citric acid (0.1–0.2%) is used to provide tartness to citrus, berry, and grape flavors. It serves the same purpose in low-residue diet preparations for post-operative patients and geriatric care.

Blood clotting is dependent upon the presence of ionic calcium. Removal of the calcium to prevent clotting is necessary in the preparation of fluid whole blood for further processing or for transfusion purposes. Citric acid and sodium citrate are used in the preparation of anticoagulant solutions (89) (see below under Salts).

Cosmetics and Toiletries. Citric acid is a standard ingredient in cosmetic formulations for pH adjustment, and in antioxidant systems as a metallic-ion chelator. It is also included in acid hair rinses and permanent wave neutralizer solutions (see Hair preparations). Citric acid and sodium citrate combinations, to buffer pH between 4.5 and 5.0, enhance the action of methyl parabens in a preservative system for cosmetics (90). Improved luster and resilience of hair is produced by low pH (4.0–6.0) shampoos and antidandruff shampoos through the use of 0.25–1.86% citric acid (91–92).

A propellant for aerosol-type dispensers has been developed from a combination of citric acid with a carbonate–bicarbonate source to provide carbon dioxide (93) (see Aerosols). Effervescent denture cleanser tablets containing citric acid and a carbonate–bicarbonate source are commonly used. The effervescence achieves rapid and even dissolution of the active cleansers in the solution (94) (see Dentifrices).

Industrial Applications. Citric acid is used in many industrial applications to sequester metal ions, neutralize bases, and act as a buffer to achieve desired acid levels in aqueous media. In nonaqueous applications, the metal sequestering properties, as well as a variety of reactions of citric's carboxylic acid groups, have proven useful. Citric acid effectively improves the performance of a wide range of nontoxic, noncorrosive and biodegradable products in processes that meet current ecological and safety standards (73).

Metal Cleaning. Citric acid-based metal cleaning formulations efficiently remove metal oxidation products from the surface of ferrous and nonferrous metals. As a weak acid, citric acid causes little damage to substrate metals (95–101), yet as a metal ion sequestrant, it accelerates removal of the metal oxides. Faster acting, but more corrosive and hazardous strong acids have been replaced by citric acid in many steel pickling systems in order to eliminate chloride-stress cracking and to minimize hydrogen embrittlement of high strength steels (102–103).

Preoperational and operational cleaning of iron and copper oxides from boilers, reheater and superheater tubing, nuclear reactors, and fabricated stainless steel components are performed with citric acid, especially where chloride cannot be tolerated (102,104–110). The Citrosolv Process (105), one of the earliest and most representative of the citrate-based metal cleaning formulas, combines citric acid and nitrogen-containing bases such as ammonia with appropriate inhibitors and an oxidizing agent. It cleans and passivates the water sides of such installations. Citric formulations perform best at temperatures greater than 37.8°C but lower temperatures can be employed with longer cleaning times. Best results are obtained by estimating the amount of deposit to be removed and adjusting the level of citric acid to meet the conditions. Deposits of magnetite in boiler tubes, eg, are effectively removed by adding 2.5 kg of citric acid in the cleaning solution per kg of magnetite in the plant being cleaned.

Electrolytic techniques have been combined with simple dip-tank methods of pickling to accelerate the removal of metal oxides from metal surfaces by citric acid solutions (111–112). Solutions of 0.1% citric acid are very effective in the electropickling of copper and its alloys. This solution generally also contains iron (Fe^{3+}) salts. Citric acid in combination with these salts removes scale, produces a bright, smooth, natural finish, and prevents discoloration. Citric acid is used also in alkaline anodic pickling and, when combined with sulfuric acid in pickling processes, contributes to the stability of the pickling solution (113).

High pressure spraying equipment and steam-cleaning apparatus have been employed when dip-tank techniques are not possible. Citric acid solutions, ammoniated to pH 5, have been applied with steam-cleaning apparatus to remove imbedded iron and scale that remain on stainless steel after machining and forming processes.

Citric acid has become more attractive in recent years for metal-cleaning applications because it is suited to either chemical treatment or a nonpolluting incineration of spent solvents (114–115). The citric acid molecule incinerates to carbon dioxide and water leaving the metallic contents of the solutions in the ash. Metals that are valuable can be recovered from the ash; toxic or radioactive metals can be disposed of in a compact form (see Metal surface treatments).

Detergents. The detergent-building properties of citrate enable it to be used as a rapidly biodegradable, environmentally acceptable phosphate substitute in a variety of household products (73,116). These include heavy-duty laundry powders (117–123) and liquids (121,124), hard surface cleaners, dishwashing powders (125) and liquids (126), presoak and light-duty powders. Citrate is also effective as a co-builder with aluminum silicates in nonphosphate detergent powders (122). Neutralizing the technical-grade 50% citric acid solution is the most economical way of introducing the citrate in these products (127). Citric acid is also used for its buffering action, and its ability to sequester trace-metal ions in unbuilt dishwashing liquids, hot-water-extraction rug cleaners, drain cleaners (128), bubble baths, and fabric softeners (129) (see Surfactants and detersive systems).

Agriculture.   Citric acid chelates trace elements in micronutrient solutions used by farmers to remedy zinc, iron (130), copper (131), magnesium, and manganese (132) soil deficiencies. Since citric acid renders phosphates soluble, it also is effective in maximizing phosphorus uptake by crops (133–134) (see Mineral nutrients).

Mineral and Pigment Slurries.   Citric acid's dispersing effect has made it the material of choice to reduce viscosity in several important slurry systems. In the mining and hydraulic transport of finely divided phosphate rock, viscosity problems are often caused by the swelling of clay components. Treatment with citric acid or one of its salts controls this swelling and allows the pumping of a high-solid, concentrated slurry (135). Similarly, citric acid retards the settling of concentrated slurries of titanium dioxide pigments, facilitating their handling and shipment in this bulk form (136). In the paper, paint, and textile industries citric acid is used in paper sizing (137) and coating dispersions (138–140); for treatment of paint fillers and pigments (141); and for textile treatment with titanium dioxide dispersions (see Dispersants; Pigments).

Electrodeposition of Metals.   The quality of plated metals often is determined by factors related directly to characteristics of the metal complexes in the plating solution that delivers metal ions for reduction at the surface of the substrate metal. Among these factors are plate thickness, adhesion to the substrate metal, porosity of the plate, luster, and hardness. Citrates have been reported to assist in the plating of copper (142), nickel (143–144), chromium, lead (145), silver (146), tungsten (147–150), antimony (151), manganese (152), molybdenum (153–154), palladium (155), rhenium (156–157), and iron (158) on various substrate metals. Colored electrodeposits of molybdenum–oxygen compounds can be obtained in baths containing citric acid complexes (see Electroplating).

Nonelectrolytic Deposition of Metals.   The first practical method for deposition of nickel on catalytic metal or activated substrates without electricity was developed by Brenner and Riddell (159). With their method, citrate provides a bright plate in alkaline baths and acts as a sequestering agent to prevent undesirable precipitation of basic nickel salts. Nickel plate deposited from acid baths gives semibright surfaces. More recently, citric acid has been reported to buffer (160) and to stabilize (161) electroless nickel plating baths more efficiently than other available acids (see Electroless plating).

Sulfur Dioxide Absorption.   The citrate process for the recovery of elemental sulfur from sulfur dioxide emissions in waste gas was conceived and reported by U.S. Bureau of Mines investigators at Salt Lake City (162). This work led to pilot plant studies at two metal smelters (163–164). The feasibility of the citrate process for controlling the relatively low-level emissions of power plants was first studied on a Pfizer Inc. power plant in Terre Haute, Indiana.

The chemistry of the process has been studied through its principal functional steps (absorption, regeneration, and sulfur melting) and important nonfunctional aspects (oxidation and purging). The citrate molecule acts primarily as an efficient buffering agent allowing better than 95% sulfur dioxide removal from industrial stack gases, high sulfur dioxide-solution loading and the recovery of high quality precipitated sulfur (165). Variations on the process have been reported (166–168). A unit currently under construction, sponsored jointly by the Bureau of Mines, the EPA, and private industry, will scrub the stack gas emitted from a 65-MW power plant.

Citric acid greatly increases the sulfur dioxide absorption capacity of aqueous calcium carbonate by improving the latter's solubility (169) (see Sulfur recovery).

Photography. Citric acid, its salts, and esters are used by the photographic industry to adjust the pH of sizing solutions for papers (170), as a component of printing plate emulsions (171), to form metallic complexes in various bleaches (172), fixers (173–175) and stabilizers (176), and in polymeric esters useful for photographic image formation (177) (see Photography).

Oil Well Treatments and Cements. Millions of barrels of oil that heretofore remained in the reservoir sands are now recovered with water-flooding methods. Citric acid is used in a variety of surfactant systems. It functions primarily as an iron sequestrant, helping to unplug deposits and prevent replugging of cleaned wells (178–181) (see Petroleum, drilling fluids).

Citric acid has been used to lengthen the setting period of cements for deep wells. Citric acid contributes to plugging-agent uniformity and distribution by controlling the precipitation time of the gelatinous hydroxide in a selective process for plugging wells (182).

Textiles. Citric acid acts as a buffer in the manufacture of the glyoxal resins used to give textiles a high-quality durable-press finish. It has been reported to increase the soil-release property of cotton with wrinkle-resistant finishes (183) and to be a component of the catalyst for curing fabrics impregnated with ketone–aldehyde precondensates (184). Citric acid also may be cross-linked with cellulose (185). Applying titanium, chrome, or zirconium mordants to wool in the dyeing process improves the visual and wear properties of fibers and increases the flame resistance of wool shag carpets (186–188). Citric acid serves as the complexing agent in this process, stabilizing the metal ion in the aqueous solution, allowing for its hydrolysis in the wool. Citric acid provides both chelation and pH control below pH 5 during disperse dyeing of polyesters and during acid dyeing of nylon and polypropylene carpeting (189). Citric acid is effective in improving the brightness of peroxide-bleached nylon (190) and is a component in a noncorrosive chlorite nylon bleach (191). Citric acid can replace phosphates as the buffer for the slightly acid pH range necessary in acid, basic, and multiple dying of nylon carpet (192).

Concrete, Mortar, and Plaster. Hydroxy carboxylic acids, their salts, and derivatives act as set-retarding, water-reducing admixtures for concrete (193). The tricarboxylic acid structure makes it particularly effective in lengthening setting times. As such it is often used to offset the rapid-setting effects of heat and certain accelerating components such as gypsum. In addition, citric acid serves to reduce the amount of water required, resulting in greater workability without loss of strength (194). Citric acid is often combined with accelerating or air-entraining admixtures to gain benefits such as high early strength (195), or improved frost resistance (196). The mechanism of citric's action both on concrete and mortars (197–198), and on the setting of gypsum plaster (199) has been reported.

Citric acid has been found useful for removing spilled cement from work areas and equipment. Set cement is loosened and easily flushed away without harming metal surfaces or paint finishes (200). A method of protecting concrete from attack by organic lubricants by treatment with an aqueous citric acid solution has been reported (201).

Refractories and Molds. Citric acid is a binder for refractory cements, imparting volume stability as well as increased green, dry, and fired strength (202–203). It is used in a polyesterification reaction to produce a ceramic material for electrical condensers (204), and in foundry and glass-making molds (205), and in sand molds for metal castings (206) (see Refractories).

Adhesives. Citric acid acts as a gelation retarder in several types of adhesive formulations (207–211), as a hardener component in others (212–213), and as a primer for polyamide adhesives (214–215). It controls pH to provide optimum conditions for good tack properties (216) and reacts to form esters in formulations for bonding steel to steel (217–218) (see Adhesives).

Paper. Citric acid is used in processing paper (qv) and packaging materials (219–221), and to stabilize diazo papers by preventing precoupling of diazo and coupler.

Polymers. The preferred nucleating or blowing agent in polymeric foams for food and beverage use is citric acid (222–228) or its sodium (229) or diammonium salt (230). Citric acid also functions in many other polymeric processes, eg, as a chelator in production of rubber-modified vinyl polymer (231), as part of the initiator mix for a methyl methacrylate polymer (232), and as the inhibitor for stabilizing an acrylamide (233). Citric acid is the starting material in the manufacture of many esters used in the plastics industry (234).

Tobacco. Citric acid, a natural constituent of the tobacco leaf, increases during the curing process (235). Additional citric acid is used to enhance the flavor and to effect a more complete combustion of certain tobaccos (see Tobacco substitutes).

Waste Treatment. In the treatment of industrial waste waters by reverse osmosis (qv), citric acid is effective in preventing scale and slime formation in the module (236) and in cleaning modules that have been fouled (237).

Salts

The salts of citric acid of commercial significance include dibasic ammonium citrate, calcium citrate, ferric ammonium citrate, potassium citrate, and sodium citrate. Important physical and chemical properties of these salts are described in Table 6. Sodium citrate, USP, FCC, granular and fine granular, is shipped in 124.74-kg drums and 45.36-kg multiwall paper bags; the powder is shipped in 90.72-kg bags. Dibasic ammonium citrate is shipped in 113.4-kg drums, calcium citrate, FCC, and potassium citrate, NF, FCC, in 90.72-kg drums, and ferric ammonium citrate in 45.36-kg drums.

Food Uses. Sodium citrate is the citrate salt of major importance in the food and beverage industries. It is used as an emulsifier, stabilizer, and acidity and flavor modifier. The FDA lists sodium citrate as a multiple-purpose, generally-recognized-as-safe (GRAS) food substance and as a GRAS sequestrant limited only by good manfacturing practice (238).

Dairy Products. Federal Standards of Identity permit up to 3% sodium citrate in process cheese, process cheese food and process cheese spread (239). Sodium citrate prevents fat separation and imparts flexibility to slices and uniform melt-down qualities.

Sodium citrate is an important stabilizer used in whipping cream and vegetable-based dairy substitutes. Addition of sodium citrate to ice cream, ice milk, and

Table 6. Physical and Chemical Properties of Citrate Salts

Name	CAS Registry No.	Chemical formula	Formula weight	Physical form	Solubility, g/100 mL [a]			Characteristics in air/ on heating
					Water	Alcohol	Other solvent	
ammonium citrate dibasic	[3012-65-5]	$(NH_4)_2HC_6H_5O_7$	226.19	white granules or powder	100^{25}	SS	ether I	stable in air
calcium citrate tetrahydrate	[5785-44-4]	$Ca_3(C_6H_5O_7)_2 \cdot 4H_2O$	570.51	white needles or powder	0.085^{18} 0.096^{23}	0.0065^{18}	ether I	loses most water at 100°C becomes anhydrous at 120°C
anhydrous	[813-94-5]	$Ca_3(C_6H_5O_7)_2$	498.44					
ferric ammonium citrate brown	[1185-57-5] or [7050-19-3]	Fe 16.5–18.5% NH_3 ca 9% $H_3C_6H_5O_7 \cdot H_2O$ ca 65%		reddish brown granules, garnet-red scales or brownish yellow powder	$>50^{25}$	I	ether I	deliquesces in air, reduced to ferrous salt by light
green	[1185-57-5] or [7050-19-3]	Fe 14.5–16% NH_3 ca 7.5% $H_3C_6H_5O_7 \cdot H_2O$ ca 75%		green transparent scales, granules or powder	$>50^{25}$	I	ether I	deliquesces in air, reduced readily to ferrous salt by light
potassium citrate monohydrate	[6100-05-6]	$K_3C_6H_5O_7 \cdot H_2O$	324.42	white granules or powder	167^{15} 199^{31}	I	ether I glycerine S	deliquesces in moist air; becomes anhydrous at 180°C; decomposes at 230°C
anhydrous	[866-84-2]	$K_3C_6H_5O_7$	306.40					
sodium citrate[b] dihydrate	[6132-04-3]	$Na_3C_6H_5O_7 \cdot 2H_2O$	294.10	white crystals or powder	71^{25} 167^{100}	I	ether I	stable in air; becomes anhydrous at 150°C with decomposition
pentahydrate	[6858-44-2]	$Na_3C_6H_5O_7 \cdot 5\frac{1}{2}H_2O$	357.15	white crystals or granules	92.6^{25} 250^{100}	VSS	ether I	effloresces in air
anhydrous	[68-04-2]	$Na_3C_6H_5O_7$	258.07	white granules or powder	57^{25}	I	ether I	decomposes at melting point

[a] Temperature (in °C) is indicated by superscript; I = insoluble; S = soluble; SS = slightly soluble; VSS = very slightly soluble.
[b] Water solubility data (3).

169

frozen custard before pasteurization and homogenization reduces the viscosity of the mix and renders it easier to whip. Low levels of sodium citrate are highly effective in reducing solids precipitation in evaporated milk during processing and storage (see Milk products).

Other Foods.   Sodium citrate reduces the sharpness of the acid taste in some beverages if the quantity of acid is relatively large. The salt tends to mellow the flavor in lemon-lime carbonated beverages, and in club soda it imparts a cool, saline taste and aids retention of carbonation.

In products requiring close regulation of pH, such as jams, jellies, preserves, pectin candies, and gelatin desserts, acidity is often controlled by addition of sodium citrate.

The other salts of citric acid are used in the food industry either as a nutrient source in fortified foods or as a source of metal ion for a technical function. Calcium citrate can be used as a firming agent for tomatoes or to assist the gel formation of low methoxyl pectin. Potassium citrate assists the formation of certain carrageenin gels. It is also used where sodium ion is undesirable. Ferric ammonium citrate, a source of nutrient iron, is particularly useful in milk fortification. Calcium citrate and potassium citrate are listed by the FDA as multiple purpose GRAS food substances and as GRAS sequestrants, and calcium citrate and manganese citrate [5968-88-7] are listed as GRAS nutrients and/or dietary supplements (240).

Pharmaceutical Uses.   Sodium citrate is used in a broad range of pharmaceuticals as a buffer to maintain optimum pH for maximum stability of active ingredients. Sodium or potassium citrate is used in pharmaceutical preparations as a blood and urinary alkalizer. In larger doses sodium citrate is used as a saline cathartic.

The anticoagulant property of sodium citrate is employed in sodium citrate solutions for plasma and blood fractionation (qv) (241). Sodium citrate–citric acid combinations are also used for this purpose.

Ferric ammonium citrates, brown and green, are mild tasting, very soluble deliquescent salts used in syrups and elixirs as iron sources for treatment of anemia. Calcium citrate is used in prenatal products as a calcium source.

Cosmetics and Toiletry Uses.   Studies have reported that a 5% solution of zinc citrate [546-46-3] is effective as a mouth rinse in reducing plaque concentrations on human teeth (242). Patents have been issued on the manufacture of zinc citrate with reduced astringency for use in dental creams (243), and on the use of zinc citrate in toothpastes at levels of 0.25–10% as an antiplaque and anticalculus agent (244–245).

Industrial Uses.   Sodium citrate is an effective, ecologically-acceptable phosphate replacement in laundry detergents and household cleaners (73,116–129). Other industrial uses of sodium citrate include water conditioning and sodium chloride replacement in the regeneration of ion-exchange (qv) resins. Sodium citrate is used to retard cementing of oil wells, and in certain concrete, mortar, and gypsum formulations (193–200), in electroless plating (qv) of nickel (159), and as buffer and complexing agent in various metal-electroplating baths (142–158).

Dibasic ammonium citrate is particularly useful in two metal cleaning applications: in the removal of rust and mill scale from newly fabricated stainless steel, and in the cleaning of oxidized coatings from aluminum surfaces (see Metal surface treatments). Like sodium citrate, ammonium citrate is an effective iron chelant in ion-exchange resin regeneration solutions (246). Dibasic ammonium citrate is also

Table 7. Physical Properties of Citric Acid Esters[a]

Name	CAS Registry No.	Formula weight	Density d_{25}^{25}	Density lb/gal at 25°C	bp at 133 Pa[b], °C	n_D^c	Solubility in water, g/100 mL soln[c]	Evaporation at 105°C, g/(cm²·h)
triethyl citrate	[77-93-0]	276.29	1.136	9.47	126–127	$1.4405^{24.5}$	6.5	0.000107
tri-n-butyl citrate	[77-94-1]	360.43	1.042	8.70	169–170	$1.4429^{24.5}$	0.002	0.000011
tricyclohexyl citrate	[4132-10-9]	438.57	1.7_{20}	(solid at 25°C)	57^d	nd	nd	none
acetyl triethyl citrate	[77-89-4]	318.31	1.135^{23}	9.47^{23}	131–132	$1.4386^{23.0}$	0.72	0.000096
acetyl tri-n-butyl citrate	[77-90-7]	402.46	1.046	8.7	172–174	$1.4408^{25.5}$	0.002	0.000009
acetyl tri-2-ethylhexyl citrate	[144-15-0]	570.81	0.983	8.21	225	$1.441^{25.0}$	I	nd

[a] Ref. 5.
[b] To convert Pa to mm Hg, multiply by 0.0075.
[c] nd = not determined; I = insoluble.
[d] Melting point.

Table 8. Compatibility of Citric Acid Esters with Certain Resins[a,b]

Resins	Triethyl citrate	Tri-n-butyl citrate	Acetyl triethyl citrate	Acetyl tri-n-butyl citrate	Tricyclohexyl citrate	Acetyl tri-2-ethylhexyl citrate
cellulose acetate	C	P	C	P	I	P
cellulose acetate butyrate	C	P	C	P	nd	P
cellulose nitrate	C	C	C	C	C	C
ethyl cellulose	C	C	C	C	nd	P
polystyrene	P	C	C	C	C	nd
poly(vinyl acetate)	C	C	C	C	C	I
poly(vinyl chloride)	nd	C	nd	C	C	C
poly(vinyl chloride acetate)	C	C	C	C	nd	C
poly(vinyl butyral)	C	C	C	C	C	I

[a] Ref. 5.
[b] C = compatible; P = partially compatible; I = incompatible; nd = not determined.

recommended for removing surface discoloration from concrete flatwork (247). Ferric ammonium citrate is used in the production of blueprint papers (see Printing processes). Sodium and potassium citrates have been used as burning rate modifiers in papers (248).

Esters

A large number of citric acid esters have been prepared. Their properties are detailed in Table 7. Most liquid citrate esters are available in tank-car or tank-truck quantities and in drums and pails; tricyclohexyl citrate, a solid at room temperature, is available in 90.72-kg drums.

Triethyl, acetyl triethyl, tri-*n*-butyl and acetyl tri-*n*-butyl citrates are used commercially as plasticizers in the preparation of polymer compositions, protective coatings, adhesives, and similar substances. The low toxicity of these plasticizers makes them especially suitable for products that come in contact with food. These esters have been permitted by the FDA as indirect food additives for use in a variety of products that come in contact with food (249). In addition, triethyl citrate, FCC, is permitted by the FDA for use in egg whites as a GRAS substance (250). Triethyl citrate and acetyl tri-*n*-butyl citrate are considered to be GRAS as flavor ingredients by the Flavor and Extract Manufacturers Association of the United States (FEMA) (251).

Triethyl citrate has been used as an ingredient in cosmetic and fragrance formulations. Tricyclohexyl citrate is a solid plasticizer used as a component of thermosensitive adhesives. Acetyl tri-*n*-butyl citrate is used commercially as a surface lubricant in the manufacture of tin-plated steel.

Triallyl citrate [6299-73-6] and acetyl triallyl citrate [115-72-0] are polyunsaturated monomers for making specialty resins. With peroxide catalysts and relatively modest heat, these nearly colorless liquids undergo an easily controlled homopolymerization to a clear, hard, thermoset polymer.

Isopropyl citrate [1587-21-9] at a level not to exceed 0.02% and monoisopropyl citrate [25168-59-6] in an amount limited only by good manufacturing practice are listed as GRAS sequestrants by the FDA (252). A variety of partial and complete fatty alcohol esters of citric acid, most notably tristearyl citrate [7775-50-0], have been used commercially as surfactants, as lubricating agents, and as emollients. Stearyl citrate is listed by the FDA as a GRAS sequestrant for food use at a level not to exceed 0.15% (253).

Calcium monostearyl citrate [25134-28-5] has been used in a formulated heat stabilizer for processing vinyl polymers. The compatibility of a number of citrate esters with certain resins may be found in Table 8.

BIBLIOGRAPHY

"Citric Acid" in *ECT* 1st ed., Vol. 4, 8–23, by G. B. Stone, Chas. Pfizer & Co., Inc.; "Citric Acid" in *ECT* 2nd ed., Vol. 5, pp. 524–540, by Lewis B. Lockwood and William E. Irwin, Miles Chemical Company.

1. R. C. Wilhoit and D. Shiao, *J. Chem. Eng. Data* **9,** 595 (1964).
2. F. P. Chappel and F. E. Hoare, *Trans. Faraday Soc.* **54,** 367 (1958).
3. A. V. Wolf, M. G. Brown, and P. G. Prentiss, *Handbook of Chemistry and Physics,* CRC Press Inc., Cleveland, Ohio, 1975, pp. 225.
4. Z. Gerlach, *Ann.* **8,** 295 (1869).

5. Pfizer Inc., New York, technical files.
6. A. Seidell, *Solubilities of Inorganic and Organic Compounds,* 3rd ed., Vol. 2, D. Van Nostrand Co., Inc., New York, 1941, pp. 427–429.
7. U.S. Pat. 3,997,596 (Dec. 14, 1976), F. Smeets (to Citrex S.A., Belgium).
8. Cyclo Chemicals Corp., Miami, Fla., *Citrest-Citric Acid Fatty Esters,* technical information.
9. C. J. Knuth and A. Bavley, *Plast. Technol.* **3**, 555 (1957).
10. *Pfizer Organic Chelating Agents, Technical Bulletin No. 32,* Pfizer Chemicals Division, New York, 1972.
11. *Scientific Literature Review on GRAS Food Ingredients—Citrates, PB-223 850,* National Technical Information Service, Springfield, Va., April 1973; *Scientific Literature Review on GRAS Food Ingredients—Citric Acid, PB-241 967,* National Technical Information Service, Springfield, Va., Oct. 1974.
12. L. M. Miall in A. H. Rose, ed., *Primary Products of Metabolism,* Academic Press, Inc., New York, 1978.
13. J. N. Currie, *J. Biol. Chem.* **31**, 15 (1917).
14. H. Amelung, *Chem. Ztg.* **54**, 118 (1930).
15. L. H. C. Perquin, *Bijdrage Tot De Kennis Der Oxydative Dissimilatic Van Aspergillus niger van Tiegham,* Meinema, Delft, Ger., 1938.
16. S. M. Martin and W. R. Waters, *Ind. Eng. Chem.* **44**, 2229 (1952).
17. D. S. Clark, *Can. J. Microbiol.* **8**, 133, 1962.
18. Brit. Pat. 653,808 (Mar. 23, 1951), R. L. Snell and L. B. Schweiger (to Miles Laboratories Inc.).
19. U.S. Pat. 3,285,831 (Nov. 15, 1966), E. J. Swarthout (to Miles Laboratories Inc.).
20. Brit. Pat. 1,145,520 (Mar. 19, 1969), M. A. Batti (to Miles Laboratories Inc.).
21. Brit. Pat. 1,182,983 (Mar. 4, 1970), F. Roberts, Jr. (to Pfizer Inc.).
22. Brit. Pat. 1,293,786 (Oct. 25, 1972), J. H. Fried (to Pfizer Inc.).
23. Brit. Pat. 1,211,246 (Nov. 4, 1970) (to Takeda Chemical Industries Ltd.).
24. Brit. Pat. 1,297,243 (Nov. 22, 1972), H. Fukuda and co-workers (to Takeda Chemical Industries Ltd.).
25. Brit. Pat. 1,369,295 (Oct. 2, 1974), L. M. Miall and G. F. Parker (to Pfizer Inc.).
26. *Ullmanns Encyklopädie der Technischen Chemie,* 4th ed., Vol. 9, Urban & Schwarzenberg, Munich, Berlin, Ger., 1975, pp. 624–636.
27. *Code of Federal Regulations, Title 21, §173.280,* U.S. Government Printing Office, Washington, D.C.. 1977.
28. *The Association of American Feed Control Officials, Official Publication,* 1977, Section 36, pp. 87–88.
29. H. C. DeRoo, *The Connecticut Agricultural Experimental Station Bulletin* (750), (1975).
30. E. Grimoux and P. Adam, *C. R. Acad. Sci. Paris* **90**, 1252 (1880); *Bull. Soc. Chim. Fr.* **36**, 18 (1881).
31. H. V. Pechmann and M. Dunschmann, *Ann. Chem.* **261**, 162 (1891); A. Haller and A. Held, *Ann. Chim. Phys.* **23**, 175 (1891).
32. W. T. Lawrence, *J. Chem. Soc.* **71**, 457 (1897); E. Ferrario, *Gazz. Chim. Ital.* **38**, 99 (1908).
33. E. Baur, *Chem. Ber.* **46**, 852 (1913).
34. H. Franzen and F. Schmitt, *Chem. Ber.* **58**, 222 (1925).
35. H. O. L. Fischer and G. Dangschat, *Helv. Chim. Acta* **17**, 1196 (1934); E. Baer, J. M. Grosheintz, and H. O. L. Fischer, *J. Am. Chem. Soc.* **61**, 2607 (1939).
36. F. Knopp and C. Martius, *Hoppe-Seyler's Z. Physiol. Chem.* **242**, 204 (1936); C. Martius, *ibid.,* **279**, 96, (1943).
37. A. M. Gakhokidze and A. P. Guntsadze, *J. Gen. Chem. U.S.S.R.* **17**, 1642 (1947).
38. P. E. Wilcox, C. Heidelberger, and V. R. Potter, *J. Am. Chem. Soc.* **62**, 5019 (1950).
39. M. Taniyama, *Toho-Reiyon Kenkyu Hokoku* **1**, 40 (1954).
40. U.S. Pat. 3,356,721 (Dec. 5, 1967), R. H. Wiley (to Miles Laboratories, Inc.).
41. U.S. Pat. 3,751,458 (Aug. 7, 1973); 3,755,436 (Aug. 28, 1973); 3,783,154 (Jan. 1, 1974), R. H. Wiley (to Miles Laboratories, Inc.); R. H. Wiley and K. S. Kim, *J. Org. Chem.* **38**, 3582 (1973).
42. U.S. Pat. 3,770,796 (Nov. 6, 1973), W. W. Lawrence, T. W. McKay, and K. E. Wiegand (to Ethyl Corporation).
43. U.S. Pats. 3,798,266 (Mar. 19, 1974), G. Bottaccio and co-workers (to Montecatini Edison S.p.A.); 3,912,778 (Dec. 14, 1975), E. Alneri and co-workers (to Montecatini Edison S.p.A.).
44. U.S. Pat. 3,773,821 (Nov. 20, 1973), F. Broussard (to Lonza Ltd.).

45. U.S. Pat. 3,963,775 (June 15, 1976), N. Heyboer (to Akzona Incorporated).
46. U.S. Pat. 3,950,397 (Apr. 13, 1976), J. G. Batelaan (to Akzona Incorporated); U.S. Pat. 3,917,686 (Nov. 4, 1975), H. A. Bruson (to Bjorksten Research Laboratories, Inc.).
47. U.S. Pats. 3,769,337 (Oct. 30, 1973), K. E. Wiegand (to Ethyl Corporation); 3,769,338 (Oct. 30, 1973), M. J. Dagani and T. H. Pearson (to Ethyl Corporation); 3,843,692 (Oct. 22, 1974), 3,853,711 (Oct. 22, 1974), 3,911,002 (Oct. 7, 1975), 3,960,941 (June 2, 1976), 3,962,287 (June 8, 1976), 3,965,168 (June 22, 1976), K. E. Wiegand (to Ethyl Corporation).
48. Brit. Pat. 1,421,310 (Jan. 14, 1976), U. Romano and M. M. Mauri (to Snamprogetti S.p.A.).
49. U.S. Pat. 4,022,823 (May 10, 1977), J. B. Wilkes and R. G. Wall (to Chevron Research Company).
50. U.S. Pats. 3,852,322 (Dec. 3, 1974); 3,950,390 (Apr. 13, 1976), H. Faubl (to Pfizer Inc.).
51. U.S. Pat, 4,022,803 (May 10, 1977), E. N. Gutierrez and V. Lamberti (to Lever Brothers Company).
52. U.S. Pat. 4,056,567 (Nov. 1, 1977), V. Lamberti and E. N. Gutierrez (to Lever Brothers Company).
53. *Chem. Mark. Rep.* **213**(3), 30 (Jan. 16, 1978).
54. The United States Pharmacopeial Convention, Inc., *The United States Pharmacopeia,* 19th ed., Mack Publishing Co., Easton, Pa., 1975.
55. *Code of Federal Regulations, Title 21, Food and Drugs, §170.30(h) (1),* U.S. Government Printing Office, Washington, D.C., 1977.
56. *Code of Federal Regulations, Title 21, Food and Drugs, §570.30(d),* U.S. Government Printing Office, Washington, D.C., 1977.
57. The National Research Council, *Food Chemicals Codex,* 2nd ed., National Academy of Sciences, Washington, D.C., 1972.
58. The National Research Council, *Food Chemicals Codex,* 2nd ed., Suppl., National Academy of Sciences, Washington, D.C., 1975.
59. *Code of Federal Regulations, Title 21, Food and Drugs, §173.165,* U.S. Government Printing Office, Washington, D.C., 1977.
60. *Citrosol® 50T and 50W,* Specifications FG H2300 (Sept. 14, 1976); FG H2310 (Sept. 14, 1976), Pfizer Inc., New York, 1976.
61. *Citric Acid, USP, FCC Anhydrous,* Specifications FG-H2041 (Mar. 14, 1975); FG-H2042 and FG-H2044 (Feb. 28, 1975), Pfizer Inc., New York, 1975.
62. *Citric Acid, USP, FCC, Hydrous,* Specifications FG-H1041, FG-H1042, and FG-H1043 (Mar. 13, 1975), Pfizer Inc., New York, 1975.
63. G. W. Pucker, C. C. Sherman, and H. B. Vickery, *J. Biol. Chem.* **113,** 235 (1936).
64. D. A. Perlman, H. A. Lardy, and J. J. Johnson, *Ind. Eng. Chem. Anal. Ed.* **16,** 515 (1944).
65. C. G. Hartford, *Anal. Chem.* **34,** 426 (1962).
66. J. A. Taraborelli and R. P. Upton, *J. Am. Oil Chem. Soc.* **52,** 248 (1975).
67. R. T. Wood, D. M. Cohen, and K. P. Munnelly, *Antimicrob. Agents Chemother.* **8,** 30 (1975).
68. *Food Acidulants, Technical Bulletin,* Pfizer Inc., New York, 1977.
69. *Code of Federal Regulations, Title 21, §182.1033, 182.6033,* U.S. Government Printing Office, Washington, D.C., 1977.
70. *FAO Nutrition Meetings Report Series No. 40 A,B,C,* Food and Agriculture Organization of the United Nations World Health Organization, New York, 1967, p. 134.
71. *Tentative Evaluation of the Health Aspects of Citric Acid, Sodium Citrate, Potassium Citrate, Calcium Citrate, Triethyl Citrate, Isopropyl Citrate, and Stearyl Citrate as Food Ingredients, PB 280 954,* National Technical Information Service, Springfield, Va., 1977.
72. P. Creach, *C.R. Acad. Sci. Paris* **240,** 2551 (1955).
73. *Environmental Impact of Citrates, Information Sheet No. 2030,* Pfizer Chemicals Division, New York, 1974.
74. E. E. Shannon and L. J. Kamp, *Environmental Report Series EPS 4-WP-73-3,* Canadian Water Pollution Control Directorate, Ontario, Can., 1972.
75. E. E. Shannon, N. W. Schmidtke, and P. J. A. Fowlie, *Environmental Protection Service Report Project No. 73-3-7,* National Research Council of Canada, Ontario, Can., 1977.
76. *Beverage Industry Annual Manual 1976–1977,* Magazines for Industry, New York, 1977, p. 208.
77. J. D. Stacey, *Beverage World,* (May 1977).
78. *Preservers Handbook,* Sunkist Growers, Inc., Ontario, Calif., 1964.
79. K. H. Moledina and co-workers, *J. Food Sci.* **42,** 759 (1977).
80. C. E. Wells, D. C. Martin and D. A. Tichenor, *J. Am. Dietetic Assoc.* **61,** 665 (1972).

81. M. A. Amerine, H. W. Berg, W. V. Cruess, *Technology of Wine Making,* 2nd ed., The AVI Publishing Co., Inc., Westport, Conn., 1967.
82. U.S. Pat. 2,999,293 (Sept. 12, 1961), J. White and R. Kolb (to Warner Lambert Pharmaceutical Company).
83. *Remingtons' Pharmaceutical Sciences,* 15th ed., Mack Publishing Company, Inc., Easton, Pa., 1975, p. 1574.
84. *The National Formulary,* 14th ed., Am. Pharm. Assoc., Washington, D.C., 1975, pp. 389–390.
85. Ref. 83, p. 742.
86. *Handbook of Non-Prescription Drugs,* 5th ed., Am. Pharm. Assoc., Washington, D.C., 1977, pp. 3–17.
87. D. Entriken and C. Becker, *J. Am. Pharm. Assoc.* **43,** 693 (1954).
88. S. Bhattacharya and co-workers, *J. Indian Chem. Soc.* **31,** 231 (1954).
89. *USP XIX,* The United States Pharmacopeial Convention Inc., Rockville, Md., 1975, pp. 33–35.
90. L. MacDonald, *Am. Perfum.* **76**(7), 22 (1961).
91. M. Ash and I. Ash, *Formulary of Cosmetic Preparations,* Chemical Publishing Company, Inc., New York, 1977.
92. L. Smith and M. Weinstein, *Household Pers. Prod. Ind.* **14**(10), 54 (1977).
93. U.S. Pat. 3,718,236 (Feb. 27, 1973), E. M. Reyner and M. E. Reyner.
94. Ref. 87, p. 260.
95. D. C. Horner, *Electroplat. Met. Finish.,* 75 (Mar. 1968).
96. *Ibid.,* 113 (Apr. 1968).
97. Czech Pat. 145, 781 (Oct. 15, 1972) J. Trefry.
98. Brit. Pat. 1,270,185 (Apr. 12, 1972), D. B. Cofer and co-workers (to Southwire Co.).
99. Jpn. Kokai 72 25,078 (Oct. 19, 1972), M. Miyazaki (to Tokyo Shibaura Electric Co., Ltd.).
100. Ger. Offen. 2,041,871 (Mar. 11, 1971), P. F. Heller (to Amchem Products, Inc.).
101. U.S. Pat. 3,492,238 (Jan. 27, 1970), C. Wohlberg (to U.S. Atomic Energy Commission).
102. W. J. Blume, *Mater. Perform.* **16**(3), 15 (1977).
103. U.S. Pat. 3,806,366 (Apr. 23, 1974), D. B. Cofer and co-workers (to Southwire Co.).
104. U.S. Pat. 3,664,870 (May 23, 1972), A. W. Oberhofer and co-workers (to Nalco Chemical Co.).
105. U.S. Pat. 3,072,502 (Jan. 8, 1963), S. Alfano (to Pfizer Inc.); 3,248,269 (Apr. 26, 1966), W. E. Bell (to Pfizer Inc.).
106. L. D. Martin and W. P. Banks, "Electrochemical Investigation of Passivating Systems," *Proceedings of the 35th International Water Conference, Pittsburgh, Pa., Oct. 29–31, 1974.*
107. G. W. Bradley and co-workers, "Investigation of Ammonium Citrate Cleaning Solvents," *Proceedings of the 36th International Water Conference, Pittsburgh, Pa., Nov. 4–6, 1975.*
108. U.S. Pat. 3,003,898 (Oct. 10, 1961), C. F. Reich (to The Dow Chemical Co.).
109. U.S. Pat. 3,496,017 (Feb. 17, 1970), R. D. Weed (to U.S. Atomic Energy Commission).
110. U.S. Pat. 3,013,909 (Dec. 19, 1961), G. P. Pancer and J. L. Zegger (to U.S. Atomic Energy Commission).
111. McCollum and Logan, "Electrolytic Corrosion of Iron in Soils," *Technical Paper No. 25,* Bureau of Standards, Washington, D.C., 1913, p. 7.
112. Hall, Nathaniel, and Hogaboom, "Metal Finishing," *21st Annual Guidebook Directory,* Westwood, N. J. 1953.
113. U.S. Pat. 2,558,167 (June 26, 1951), A. J. Beghin, P. F. Hamberg, and H. E. Smith (to Insl-X Corp.).
114. E. C. Wackenhuth, L. W. Lamb, and J. P. Engle, *Power Eng.,* 68 (Nov. 1973).
115. Jpn. Kokai 75 47,457 (Apr. 26, 1975), Y. Kudo and co-workers (to Mitsubishi Heavy Industries, Ltd.).
116. D. L. Muck and H. L. Gewanter, "The Detergent Building Properties of Trisodium Citrate," *paper presented at the American Oil Chemist Soc., Apr. 1972.*
117. U.S. Pat. 4,028,262 (June 7, 1977), B.-D. Cheng (to Colgate-Palmolive Co.).
118. U.S. Pat. 4,013,577 (Mar. 22, 1977) (to Colgate-Palmolive Co.).
119. U.S. Pat. 4,009,114 (Feb. 22, 1977), J. A. Yurke (to Colgate-Palmolive Co.).
120. Brit. Pat. 1,477,775 (June 26, 1977) (to Hoechst A.G.).
121. Brit. Pat. 1,427,071 (Mar. 3, 1976), M. Filcek and co-workers (to Benckiser G.m.b.H.).
122. U.S. Pat. 3,985,669 (Oct. 12, 1976), H. K. Krummel and T. W. Gault (to the Procter & Gamble Co.).

123. U.S. Pat. 3,956,156 (May 11, 1976), A. N. Osband, F. W. Gray, and J. C. Jervert (to Colgate-Palmolive Co.).
124. U.S. Pat. 4,021,377 (May 3, 1977), P. J. Borchert and J. L. Neff (to Miles Laboratories, Inc.).
125. Belg. Pat. 848,533 (May, 20, 1977) (to Henkel & Cie, G.m.b.H.).
126. U.S. Pat. 4,024,078 (May 7, 1977), A. Gilbert and J. W. Schuette (to The Procter & Gamble Co.).
127. *Citrosol-50 T, W and E, Information Sheet No. 626,* Pfizer Chemicals Division, New York, 1974.
128. U.S. Pat. 3,968,048 (July 6, 1976), J. A. Bolan (to The Drackett Co.).
129. U.S. Pat. 3,920,564 (Nov. 18, 1975), J. J. Greecsek (to Colgate-Palmolive Co.).
130. U.S. Pat. 2,813,014 (Nov. 12, 1957), J. R. Allison and C. A. Hewitt (to Leffingwell Chem. Co.).
131. Brit. Pat. 827,521 (Feb. 3, 1960), I. S. Perold (to Union of South Africa, Dir. of Tech. Service, Dept. of Agriculture).
132. U.S. Pat. 3,869,272 (Mar. 4, 1975), R. J. Windgasen (to Standard Oil Co.).
133. A. Marchesini, P. Sequi, and G. A. Lanzani, *Agrochimica* 10(2), 183 (1966).
134. F. L. Daniel, P. O. Ramaswani, and T. P. Mahadevan, *Madras Agr. J.* 55(1), 31 (1968).
135. U.S. Pat. 4,042,666 (Aug. 16, 1977), H. L. Rice and R. A. Wilkins (to Petrochemicals Co., Inc.).
136. U.S. Pat. 3,663,284 (May 16, 1972), D. J. Stanicoff and H. J. Witt (to Marine Colloids, Inc.).
137. U.S. Pat. 2,952,580 (Sept. 13, 1960), J. Frasch.
138. U.S. Pat. 3,029,153 (Apr. 10, 1962), K. L. Hackley (to Champion Papers, Inc.).
139. U.S. Pat. 3,245,816 (Apr. 12, 1966), H. C. Schwaibe (to Mead Corp.).
140. V. Laskova, *Pap. Cellul.* 30(7–8), 173 (1975).
141. U.S. Pat. 2,336,728 (Dec. 14, 1943), H. W. Hall (to the Dicalite Co.).
142. C. W. Smith and C. B. Munton, *Met. Finish.* 39, 415 (1941).
143. A. W. Hothersall, "The Adhesion of Electrodeposited Nickel to Brass," *paper presented to Electroplaters' and Depositors' Technical Society at the Northampton Polytechnic Institute, London, Eng., May 18, 1932.*
144. Fr. Pat. 813,548 (June 3, 1937) (to The Mound Co. Ltd.).
145. U.S. Pat. 2,474,092 (June 21, 1949), A. W. Liger (to Battelle Development Corp.).
146. C. W. Fleetwood and L. F. Yntema, *Ind. Eng. Chem.* 27, 340 (1935).
147. J. Kashima and F. Fuhushima, *J. Electrochem. Assoc. Jpn.* 15, 33 (1947).
148. W. E. Clark and M. L. Hold, *J. Electrochem. Soc.* 94, 244 (1948).
149. N. N. S. Siddhanta, *J. Indian Chem. Soc. Ind News Ed.* 14, 6 (1951).
150. C. N. Shen and H. P. Chung, *Chin. Sci.* 2, 329 (1951).
151. K. G. Sodeberg and H. L. Pinkerton, *Plating* 37, 254 (1930).
152. W. E. Brodt and L. R. Taylor, *Trans. Electrochem. Soc.* 73, (1938).
153. L. F. Yntema, *J. Am. Chem. Soc.* 54, 3775 (1932).
154. U.S. Pat. 2,599,178 (June 3, 1952), M. L. Holt and H. J. Seim (to Wisconsin Aluminum Research Found).
155. Can. Pat. 443,256 (July 29, 1947), E. M. Wise and R. F. Vines (to The International Nickel Company of Canada, Ltd.).
156. L. E. Netherton and M. L. Holt, *J. Electrochem. Soc.* 95, 324 (1949).
157. *Ibid.,* 98, 106 (1951).
158. Ital. Pat. 444,078 (Jan. 12, 1949), A. Sacco and M. Gandusi.
159. U.S. Pat. 2,532,283 (Dec. 5, 1950), A. Brenner and G. E. Riddell.
160. V. I. Latatuev and N. A. Solokha, *Zh. Prikl. Khim. (Leningrad)* 45(5), 1116 (1972).
161. Jpn. Kokai 74 04,633 (Jan. 16, 1974), M. Tsuchitani (to Song Corp.).
162. D. R. George, L. Crocker, and J. B. Rosenbaum, *Min. Eng.* 22(1), 75 (1970).
163. J. B. Rosenbaum, D. R. George, L. Crocker, "The Citrate Process for Removing SO$_2$ and Recovering Sulfur from Waste Gases," *paper presented at the AIME Environmental Quality Conference, Washington, D.C., June 7–9, 1971.*
164. J. B. Rosenbaum and co-workers, *Sulfur Dioxide Emission Control by Hydrogen Sulfide Reaction in Aqueous Solution—The Citrate System, PB 221914/5,* National Technical Information Service, Springfield, Va., 1973.
165. L. Korosy and co-workers, *Adv. Chem. Ser.* 139, 192 (1975).
166. U.S. Pat. 3,757,488 (Sept. 11, 1973), R. R. Austin and A. L. Vincent (to International Telephone and Telegraph Corp.).
167. U.S. Pat. 3,933,994 (Jan. 20, 1976), G. L. Rounds (to Kaiser Steel Corp.).
168. Ger. Pat. 2,432,749 (Jan. 30, 1975), W. J. Balfánz, R. M. DePirro, and L. P. Van Brocklin (to Stauffer Chemical Co.).

169. T. Wasag, J. Galka, and M. Fraczak, *Ochr. Powietrza* **9**(3), 72 (1975).
170. U.S. Pat. 2,514,689 (July 11, 1950), R. N. Woodward (to Eastman Kodak Co.).
171. Ger. Offen. 2,163,902 (July 13, 1972), Y. Yabuta, T. Tomotsu, and E. Mizuki (to Fuji Photo Film Co., Ltd.).
172. Jpn. Pat. 75 113,233 (Sept. 5, 1975), S. Yamaguchi (to Fuji Photo Film Co., Ltd.).
173. Ger. Offen. 2,416,744 (Oct. 24, 1974), H. Shibaoka (to Fuji Photo Film Co., Ltd.).
174. M. Kawasaki and K. Harada, *Nippon Shashin Gakkaishi* **38**(1), 37 (1975).
175. Jpn. Pat. 75 38,538 (Apr. 10, 1975), M. Kawasaki and K. Harada (to Agency of Industrial Science and Technology).
176. Ger. Offen. 2,361,668 (June 20, 1974), K. Shirasu, H. Iwano, and T. Hatano (to Fuji Photo Film Co., Ltd.).
177. Jpn. Pat. 75 133,316 (Dec. 21, 1974), C. Osada and H. Ono (to Fuji Photo Film Co., Ltd.).
178. R. T. Johansen, J. P. Powell, and H. N. Dunning, *U.S. Bur. Mines Inform. Circ.* **7797**, (1957).
179. *Pfizer Products for Petroleum Production, Technical Bulletin No. 97,* Pfizer Inc., New York, 1961.
180. U.S. Pat. 3,335,793 (Aug. 15, 1967), J. W. Biles and J. A. King (to Cities Service Oil Co.).
181. U.S. Pat. 3,402,137 (Sept. 17, 1968), P. W. Fischer and J. P. Gallus (to Union Oil Co. of California).
182. U.S. Pat. 3,732,927 (May 15, 1973), E. A. Richardson (to Shell Oil Co.).
183. U.S. Pat. 3,754,860 (Aug. 28, 1973), J. G. Frick, Jr., and co-workers (to U.S. Secretary of Agriculture).
184. U.S. Pat. 3,212,928 (Oct. 19, 1965), H. R. Hushebeck (to Joseph Bancroft and Sons Co.).
185. D. D. Gagliardi and F. B. Shippee, *Am. Dyestuff Rep.* **52**, 300 (1963).
186. L. Benisek, *J. Soc. Dyers Colour.*, 277 (Aug. 1971).
187. "Mordanting of Wool: A Dyeing Technique to Increase the Flame Resistance of Wool Shag Carpets," *Wool Facts,* Vol. 1, No. 1, Wool Bureau, Inc., New York, privately presented, Wool Bureau Tech. Center, Woodbury, N. Y., 1971.
188. "Mordanting Wool with Zirconium," *Wool Facts,* Vol. 1, No. 5, Wool Bureau, Inc., New York.
189. R. R. Haynes, J. H. Mathews, and G. A. Heath, *Text. Chem. Color.* **1**(3), 16/74 (1969).
190. U.S. Pat. 2,720,441 (Oct. 11, 1955), J. G. Wallace (to E. I. du Pont de Nemours & Co., Inc.).
191. U.S. Pat. 2,898,179 (Aug. 4, 1959), C. J. Rogers (to E. I. du Pont de Nemours & Co., Inc.).
192. *Replacement of Phosphates with Citric Acid in Nylon Carpet Dyeing, Information Sheet No. 2025,* Pfizer Chemicals Division, New York, 1973.
193. *J. Am. Concr. Inst.* **60**(11), (Nov. 1963).
194. U.S. Pat. 2,174,051 (Sept. 26, 1939), K. Winkler.
195. U.S. Pat. 3,656,985 (Apr. 18, 1972), B. Bonnel and C. Hovasse (to Progil, France).
196. U.S. Pat. 2,542,364 (Feb. 20, 1951), F. A. Schenker and A. Ammann (to Kaspar Winkler Cie).
197. F. Tamas, *N.A.S. N.R.C. Publ.* **1389**, 392 (1966).
198. F. Tamas and G. Liptay, *Proc. Conf. Silicate Ind.* **8**, 299 (1965).
199. E. C. Combe and D. C. Smith, *J. Appl. Chem. (London)* **16**(3), 73 (1966).
200. *Pfizer Organic Chelating Agents, Technical Bulletin No. 32,* Pfizer Inc., Chemicals Division, New York, 1972.
201. U.S. Pat. 4,004,066 (Jan. 18, 1977), A. J. DeArdo (to Aluminum Company of America).
202. U.S. Pat. 3,333,972 (Aug. 1, 1967), J. T. Elmer and B. G. Atlman (to Kaiser Aluminum & Chemical Corp.).
203. Aust. Pat. 259,442 (Jan. 10, 1968), T. Chvatal.
204. Neth. Appl. 6,409,880 (Mar. 1, 1965) (to Sprague Electric Co.).
205. Fr. Appl. 2,172,160 (Nov. 2, 1973), F. Sembera.
206. Jpn. Pat. 75 03,020 (May 15, 1973), S. Sugiyama, Y. Heima, and M. Sugi (to Toshiba Mach. Co. Ltd.).
207. Jpn. Pat. 75 64,388 (May 31, 1975), S. Sakurada and co-workers (to Kuraray Co., Ltd.).
208. Jpn. Pat. 75 64,335 (May 31, 1975), T. Nishida and co-workers (to Kuraray Co., Ltd.).
209. Jpn. Pat. 75 64,342 (May 31, 1975), T. Nishida and co-workers (to Kuraray Co., Ltd.).
210. Jpn. Pat. 75 64,341 (May 31, 1975), T. Nishida and co-workers (to Kuraray Co., Ltd.).
211. Jpn. Pat. 75 13,305 (May 19, 1975), H. Takahashi and A. Nagao (to Nippon Gakki Co., Ltd.).
212. Norw. Pat. 131,891 (May 12, 1975), K. Bergsund.
213. Jpn. Pat. 75 130,829 (Oct. 16, 1975), H. Fukuzaki, T. Kamura, and M. Niki (to Kao Soap Co., Ltd.).
214. Ger. Pat. 1,102,322 (Mar. 16, 1961), D. Aelony (to General Mills, Inc.).

215. Jpn. Pat. 75 33,823 (Nov. 4, 1975), T. Hirose, T. Matsubara, and T. Kato (to Toa Gosei Chemical Industry Co., Ltd.).
216. U.S. Pat. 3,135,648 (June 2, 1964), R. L. Hawkins (to Air Reduction Company, New York).
217. Jpn. Pat. 72 42,127 (Oct. 24, 1973), M. Natsume (to Toho Chemical Industry Co., Ltd.).
218. Jpn. Pat. 74 99,729 (Sept. 20, 1974), I. Sugiyama and M. Kameyama (to Matsumoto Seiyaku Kogyo Co., Ltd.).
219. U.S. Pat. 3,674,619 (July 4, 1972), H. I. Scher and I. S. Ungar (to Esso Research Engineering Co.).
220. Ger. Pat. 1,269,874 (June 6, 1968), R. H. McKillip and R. Henderson (to Olin Mathieson Chemical Corp.).
221. Brit. Pat. 1,079, 762 (June 4, 1969), I. R. Horne and R. N. Lewis (to Bakelite Ltd.).
222. Neth. Appl. 6,605,358 (Oct. 24, 1966) (to Koppers Co.).
223. U.S. Pat. 2,950,263 (Aug. 23, 1960), W. Abbotson, R. Hurd, and H. Jackson (to Imperial Chemical Ind., Ltd.).
224. U.S. Pat. 3,185,588 (May 25, 1965), J. Y. Resnick (to Int Res. & Dev. Co., New York).
225. U.S. Pat. 4,016,110 (Apr. 5, 1977), W. E. Cohrs and co-workers (to The Dow Chemical Co.).
226. V. G. Kharakhash, *Plast. Massy* **6,** 40 (1971).
227. U.S. Pat. 3,523,988 (Aug. 11, 1970), Z. M. Roehr, R. Berger, and P. A. Plasse (to Roehr Metals & Plastics Co.).
228. U.S. Pat. 3,482,006 (Dec. 2, 1969), F. A. Carlson (to Mobil Oil Corp.).
229. U.S. Pat. 3,069,367 (Dec. 18, 1962), R. D. Beaulieu, P. N. Speros, and D. A. Popielski (to Monsanto Chemical Co.).
230. Ger. Pat. 11,144,911 (Mar. 7, 1963), F. Stastny, B. Ikert, and E. F. v. Behr (to Badische Anilin-und Soda-Fabrik).
231. U.S. Pat. 3,660,534 (May 25, 1972), F. E. Carrock and K. W. Ackerman (to Dart Indus.).
232. B. C. Mitra and S. R. Palit, *J. Indian Chem. Soc.* **50**(2), 141 (1973).
233. Jpn. Pat. 72 28,766 (July 29, 1972), K. Okuno, K. Itagaki, and H. Takashi (to Mitsubishi Chemical Industries Co.).
234. J. Jarusek and J. Mleziva, *Chem. Prumyst* **16**(11), 671 (1966).
235. C. O. Jensen, *Chemical Changes During the Curing of Cigar Leaf Tobacco,* The Pennsylvania State College, State College, Pa., American Chemical Society, Sept. 1951.
236. Jpn. Pat. 76 18,280 (Feb. 13, 1976), T. Mizumoto and co-workers (to Ebara-Infilco Co., Ltd.).
237. Jpn. Pat. 75 153,778 (Dec. 11, 1975), T. Mizumoto and co-workers (to Ebara-Infilco Co., Ltd.).
238. *Code of Federal Regulations, Title 21, §182.1751, 182.6751,* U.S. Government Printing Office, Washington, D.C., 1977.
239. *Code of Federal Regulations, Title 21, Part 133,* U.S. Government Printing Office, Washington, D.C., 1977.
240. *Code of Federal Regulations, Title 21, §182.1195, 182.5195, 182.6195 (calcium citrate); 182.1625, 182.6625 (potassium citrate), 182.5449 (manganese citrate),* U.S. Government Printing Office, Washington, D.C., 1977.
241. *The National Formulary* 14th ed., Am. Pharm. Assoc., Washington, D.C., 1975, pp. 650–651.
242. S. Fischman and co-workers, *J. Periodontol.* **44,** 100 (1973).
243. Brit. Pat. 1,373,002 (Nov. 6, 1974), R. Hoyles and D. Macpherson (to Unilever Ltd.).
244. Brit. Pat. 1,373,001 (Nov. 6, 1974), M. Poder, R. Hoyles, and C. Watson (to Unilever Ltd.).
245. Brit. Pat. 1,373,003 (Nov. 6, 1974), R. Hoyles and C. Watson (to Unilever Ltd.).
246. *Pfizer Products for Chemical Cleaning, Technical Bulletin No. 102,* Pfizer Chemicals Division, New York, 1960.
247. N. R. Greening and R. Landgren, *Surface Discoloration of Concrete Flatwork, Bulletin 203,* Res. and Dev. Labs of the Portland Cement Assoc., Skokie, Ill., 1966.
248. *Ind. Eng. Chem.* **53**(1), 28A (1961).
249. *Code of Federal Regulations, Title 21, §175.105, 175.300, 175.320, 175.380, 175.390, 176.170, 177.1210, 178.3910, 181.27 (acetyl triethyl citrate and acetyl tri-n-butyl citrate); 175.105, 175.300, 175.320, 175.380, 175.390, 176.170, 177.1210, 181.27 (triethyl citrate), 175.105 (tri-n-butyl citrate),* U.S. Government Printing Office, Washington, D.C., 1977.
250. *Code of Federal Regulations, Title 21, §182.1911 (triethyl citrate),* U.S. Government Printing Office, Washington, D.C., 1977.
251. R. L. Hall and B. L. Oser, *Food Technol.* **19**(2), 151 (1965).
252. *Code of Federal Regulations, Title 21, §182.6386 (isopropyl citrate), 182.6511 (monoisopropyl citrate),* U.S. Government Printing Office, Washington, D.C., 1977.

253. *Code of Federal Regulations, Title 21, §182.6851,* U.S. Government Printing Office, Washington, D.C., 1977.

EDWARD F. BOUCHARD
EDWARD G. MERRITT
Pfizer Inc.

CITRONELLA OIL.   See Oils, essential.

CITRONELLAL, $(CH_3)_2C{=}CH(CH_2)_2CH(CH_3)CH_2CHO$.   See Perfumes; Terpenoids.

CLATHRATION

Clathrates or cage compounds are organic inclusion species. Definitions of terms frequently used to characterize clathrates include:

Clathrate: a cage structure capable of including another compound within its own structure;

Host: clathrating agent;

Guest: compound in the spaces or cavity of the clathrate structure. Guest molecules have limited ranges in size and shape to fit the geometry of the cavities in the clathrate structure; and

Geometry: the cavities are usually cages or tunnels; layered compounds or combinations of these structures may be considered as clathrates.

Clathrates have been known for over 100 years; however, not until the pioneering work of Powell and associates (1–2) in the late 1940s, using x-ray techniques, did these compounds emerge as definitive structures (2–3).

Clathrates exhibit no strong bonding between host and guest. In these compounds there is no dependence on chemical properties as evidenced by the formation of a hydroquinone clathrate with an inert gas such as argon.

Clathrates also exist where two different components, differing in molecular size, are accommodated by appropriately sized holes in the lattice structure, such as gas hydrates, where one component may stabilize the lattice structure to form second cavities accommodating a different species (see Gas, natural). Other classical types are described elsewhere (see Molecular sieves; Adsorptive separation; Urea and urea derivatives).

Cavities in the clathrate structure include tunnel types, such as in urea and thiourea adducts, and cage compounds, such as in hydroquinone and other phenolic structures, water/ice clathrates (hydrates), nickel cyanide–ammonia structures, and similar metal–organic species.

Layered structures such as graphite (see Carbon), graphitic oxide, and clay minerals have also been termed clathrates. Although they accommodate specific size guests by insertion between the layers of the host material, ionic to covalent and very weak to strong bonding is apparently associated with some structures.

The effect of the clathrate on the encapsulated species is frequently dramatic as in the case of the hydroquinone clathrate of argon. This clathrate is stable at atmospheric pressure and room temperature even though argon has a boiling point of $-185.7°C$.

Furthermore, sulfur dioxide can be encapsulated in hydroquinone and stored at atmospheric pressure in a bottle with essentially no gas odor at room temperature. However, the sulfur dioxide is released when the crystalline clathrate is ground in a mortar and pestle (see Microencapsulation).

Table 1 gives a comparison of the cage dimensions of several clathrates with those of zeolites or molecular sieves (4).

Recently, investigators at the University of Glasgow (5) have developed an approach for synthesizing inclusion hosts that are not directly related to any known known host. This is a radical departure from the present systems developed by chance and will undoubtedly be investigated further.

The design of complexes between synthetic hosts and organic guests and their possible relationship to molecular evolution of biological systems has also been proposed (6). The biological counterpart of a host could be an enzyme, a nucleic acid, or an antibody, whereas a guest might be a substrate, inhibitor, cofactor, or antigen.

Crown ethers and their complexes were used as models to develop predictive effects of structure on the binding potential of hosts for specific guests. Macropolycyclic inclusion complexes have also been investigated (7) (see Chelating agents).

Naturally occurring hosts include condensation polymers of α-amino acids, phosphoric acid, or glucose.

Table 1.   Free Diameters of Some Cage Systems[a], nm

System	Cage	Cage openings
clathrate		
hydroquinone–SO_2	0.52	small
[1786-27-2]		
hydroquinone–Ar	0.42	small
[14343-01-2]		
adducts		
urea-n-alkane	0.52	0.52
thiourea–hydrocarbon adducts	0.61	0.61
water clathrate		
Type I, Cage 1	0.52	small
Cage 2	0.59	
Type II, Cage 1	0.48	small
Cage 2	0.69	small
zeolite 4A		
Cage 1	0.7	0.32
Cage 2	1.18	0.49

[a] Ref. 4.

Hydroquinone Clathrates

Reference 2 summarizes the data available in 1959 on hydroquinone clathrates, as well as others. Attention is directed toward the α and β crystalline forms of hydroquinone and their properties. The α form was obtained by recrystallization of hydroquinone (HQ) from all common solvents that did not form clathrates except methanol and acetonitrile. Methanol as well as other encapsulated species gave the β form containing one molecule of methanol for every 3 molecules of HQ. Figure 1 illustrates the structural unit.

Some observed compositions for β-hydroquinone complexes are shown in Table 2.

The data in Table 2 indicate that nearly all cavities are occupied in the case of methanol and acetonitrile because hydroquinone was recrystallized from these solvents. The other substances listed were prepared from a solvent that dissolves both hydroquinone and the guest component; the solvent is itself not enclosed in the hy-

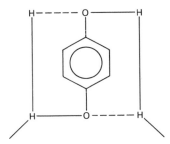

Figure 1. Hydroquinone structural unit in clathrates (8).

Table 2. Observed Compositions of Some β-Hydroquinone Clathrates of Ideal Molecular Ratios M/HQ [a]

M [b]	Molecular ratio M:3HQ	CAS Registry No. of clathrate	Heat of formation, kJ/mol [c]
HCl	0.85	[14342-96-2]	38.5
HBr	0.36	[14342-95-1]	42.7
H_2S	0.64	[14342-95-8]	
CH_3OH	0.97	[16500-84-8]	46
HCO_2H	0.82	[1911-79-1]	51
SO_2	0.88	d	
CO_2	0.74	[14343-02-3]	
CH_3CN	0.99	[1786-26-1]	
argon	0.3–0.85	d	24.5
krypton	0.74	[18932-77-9]	
xenon	0.84	[18932-78-0]	
oxygen		[20471-57-2]	23
nitrogen		[15558-66-4]	24

[a] Reprinted by permission from *Non-Stoichiometric Compounds* (3).
[b] Guest component.
[c] To convert J to cal, divide by 4.184.
[d] See Table 1 for CAS Registry No.

droquinone structure (3). An enclosable molecule is not always available at the instant these cages are closed.

Hydroquinone complexes are formed (9) by linking through hydrogen bonds to form infinite three-dimensional complexes of trigonal symmetry of units shown in Figure 1.

Mixed compounds have also been prepared with two different kinds of molecules in the enclosures. Physical characteristics of the pure and mixed species have been described (10).

The hydroquinone clathrates are very stable at atmospheric pressure and room temperature. They have no smell of the encapsulated species but decompose when heated close to the melting point or when dissolved in water.

The structure of the methanol–hydroquinone clathrate and the effect of molecular shape on cage characteristics have been described in detail (1,11), and the infrared and Raman spectra of β-quinol (hydroquinone) complexes have been reported (12).

A recent Russian publication describes clathrates of krypton and xenon with p-fluorophenol (13) similar to the hydroquinone cage compounds.

Clathrates of hydroquinone, such as the methane clathrate, prepared without a solvent have been described (14). The energy of formation is approximately 25 kJ/mol (6 kcal/mol) compared to 24.7 kJ/mol (5.9 kcal/mol) reported for solution systems.

Thermodynamic Properties. The only significant forces between guest molecules in hydroquinone clathrates are those caused by interaction of a guest molecule with the cage compound (15). Heats of formation are quite low; they are listed in Table 2. Higher values would indicate strong interactions of the guest with the host cage.

For an equation of vapor pressure for a nonpolar gas clathrate of hydroquinone, see reference 16.

Cyclodextrins

Carbohydrates provide examples of two types of inclusion complexes. Both amylose and cyclodextrins form channel structures; cyclodextrins also form cagelike structures (3).

Amylose which consists of glucose units, varies in mol wt from 300,000–1 million depending upon the starch source from which it is derived.

A variety of organic compounds precipitate from solution with amylose as inclusion compounds. Amylose–iodine complexes are well known and can be prepared by adding iodine–potassium iodide solution to starch dispersions or solutions of amylose (see Fig. 2). Table 3 summarizes the molecular dimensions of some of these clathrates that are apparently channel complexes.

The three cyclodextrins have the following characteristics (17):

Cyclodextrin	Cage diameter, nm
α-cyclodextrin [10016-20-3] (6 glucose units)	0.6 (ID)
β-cyclodextrin [7585-39-9] (7 glucose units)	0.75
γ-cyclodextrin [17465-86-0] (8 glucose units)	0.9–1.0

Amylose forms clathrates with long-chain fatty acids such as lauric, palmitic, stearic, and oleic. Fatty acids can be introduced into dried butanol-precipitated am-

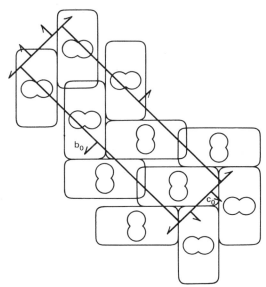

Figure 2. Projection of the structure of cyclohexaamylose–iodine complex [*61507-82-2*] (C₆H₁₀O₅)₆·
I₂·14H₂O, showing the cage structure formed by noncoaxial packing of the cyclodextrin molecules. Iodine
molecules lie on the axis of the dextrin rings. Reprinted by permission from *Non-Stoichiometric Compounds*
(3).

Table 3.   Unit Cell ^{a} Dimensions of Amylose Complexes ^{b}, nm

Complex	a	b	c
butanol			
wet	13.7	25.6	7.8
hydrated	13.7	23.8	8.05
anhydrous	12.97	22.46	7.91
iodine			
hydrated	13.7	23.8	8.05
anhydrous	12.97	22.46	7.91
fatty acid			
hydrated	13.7	23.8	8.05
anhydrous	13.0	23.0	8.05

^{a} Orthorhombic cell.
^{b} Reprinted by permission from *Non-Stoichiometric Compounds* (3).

ylose by suspension in alcoholic or carbon tetrachloride solutions of fatty acids. Most
of the fatty acid in the complex can be displaced by iodine vapor.

Amylose is also precipitated from solution by *n*-butyric, *n*-valeric, and α-bro-
mopropionic acids in a form similar to that of the long-chain fatty acids.

Cyclodextrins (cycloamyloses) provide space in the interior of a single molecule
for the inclusion of guest molecules. Models of β and γ cyclodextrins indicate internal
diameters of about 0.8 and 1.0 nm, respectively, as noted above. Greatest stability
appears to be a guest molecule orientation in the cyclodextrin which allows maximum
interaction with the groups lining the interior. Complexes of iodine and I₂.KI have
been used to illustrate cage-size differences and their behavior (3).

Benzoic and substituted benzoic acids have been used to provide examples of the size effect of guest molecules and reactivity with cyclodextrin types (18).

Proton magnetic resonance studies have provided direct evidence for the inclusion characteristics of cycloamylose (19).

Thermodynamics and kinetics of α-cyclodextrin inclusion compounds have been reported (17) in a study of azo dyes (qv); a mechanism was suggested by which the dyes are enclosed in the cyclodextrin ring.

The binding of azo dyes with α-cyclodextrin is highly stereospecific. Substituent position has little effect on the equilibrium constant. However, the position of the substituent can change the reaction rate by several orders of magnitude.

Cyclodextrins have been used to study complexation behavior of nitroxide radicals (20). Electron spin resonance spectroscopy indicates that this technique is well suited for the study of cyclodextrin's radical association in solution. Furthermore, the inclusion of nitroxide radicals in cyclodextrins is strongly influenced by steric factors and functional group polarity.

A recent study (21) of a Cu^{2+} complex of substituted cyclodextrin revealed accelerated catalytic oxidation activity because of two cyclodextrin cavities in the catalyst molecule instead of one.

Tailoring a structure for specific chemical activity appears possible, particularly in biochemical reactions.

Cyclotriphosphazene Clathrates

Trimers of phosphonitrilic compounds are a recent addition to the list of clathrating agents available to the investigator for separating, purifying, or analyzing organic compounds with a high degree of specificity (22). Structure (1) is the parent compound for a spectrum of clathrating compounds that are channel complexes (23).

Tris-1,2-phenylenedioxycyclotriphosphazene (1), forms clathrates by contact with many organic compounds in liquid or vapor form. The ratio of host to guest depends on the molecular dimensions of the included species. Table 4 illustrates the wide variety of inclusion compounds possible with this interesting structure and the effect of steric size on adduct ratios. Table 5 summarizes observations of inclusion from mixed liquids. The data in Table 5 show the preferential inclusion of less bulky, noncyclic, and almost total exclusion of cyclic components. Replacement has been demonstrated. Included benzene can be totally displaced by xylene or CS_2. The inclusion reaction is indicated by rapid swelling and disintegration of crystallites. X-ray, ir, and mass spectrometry have been used to study these compounds.

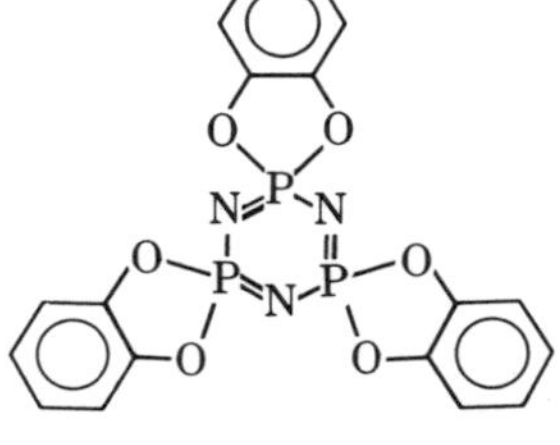

tris 1,2-phenylenedioxycyclotriphosphazene

(1)

Table 4. Inclusion Compounds of Tris-1,2-phenylenedioxycyclotriphosphazene (1)[a,b]

Included compound, M	Moles M: moles (1)	CAS Registry No.	Melting range, °C
tetralin	0.5	[60535-97-9]	220–230
norbornadiene	0.51	[68367-54-4]	236–275
isoctane	0.48–0.50[c]	[68367-47-5]	231–245
trans-decalin	0.47–0.66[c]	[60535-97-9]	237–245
o-xylene	0.46	[60535-88-8]	235–259
o-, m-, and p-xylene	(0.50)[d,e]	[60535-88-8] [68367-55-5] [60535-89-9]	241–256
cumene	0.45	[68367-56-6]	228–248
cyclohexane	0.41–0.42[c] (0.50)[d]	[60535-95-7]	233–245
n-heptane	0.38–0.5[c]	[60535-94-6]	247–253
styrene	0.38 (0.50)[d]	[68367-57-7]	239–256 244–257
ethyl acetate	0.37	[68367-58-8]	233–243
carbon tetrachloride	0.37	[68367-59-9]	233–244
tetrahydrofuran	0.35	[68367-60-2]	233–244
chloroform	0.34–0.42[c] (0.48[f]–0.66[d])	[60535-91-3]	224–245 240–244
methyl methacrylate	0.33	[68367-61-3]	230–267
diethyl ether	0.31	[68367-62-4]	226–243
benzene	0.20[c]–0.25 (0.44[f]–0.50[d,e])	[16469-00-4]	222–245 244–255
acetone	0.20	[68367-63-5]	231–250
carbon disulfide	0.20[g]	[68367-64-6]	226–239
acrylonitrile	0.19	[68367-65-7]	230–257
ethanol	0.11[g]	[68367-66-8]	231–243
(methanol)[h]	0.03[i]	[68367-67-9]	236–243

[a] Unless specified otherwise, the adducts were prepared by spontaneous inclusion and the ratios were determined by weight increase after vacuum drying.
[b] Reproduced with permission from *J. Am. Chem. Soc.* (22).
[c] By mass spectrometry of direct addition adducts.
[d] By mass spectrometry of recrystallized material.
[e] By microanalysis of recrystallized material.
[f] By weight loss of heating of recrystallized material.
[g] Mixture of triclinic and hexagonal forms.
[h] With methanol there was some evidence that the adduct was labile.
[i] Triclinic form.

There appears to be no correlation between inclusion behavior and polarity, π character, or hydrogen bonding capacity (22). Crystal dimension measurements show a hexagonal unit cell about 1.173 by 0.594 nm. In the case of benzene, a density value of 1.39 g/cm^3 indicates two host molecules and approximately one guest molecule per unit cell. Groups of inclusion compounds show similar powder diffraction patterns for different guest molecules.

Naphthalene and biphenylene analogues of the parent structure (see Table 6) have also been studied (24).

Studies of the competitive behavior of the 1,2-phenylene- and 2,3- and 1,8-naphthalenedioxy compounds, with model pairs containing hexadecane as a reference

Table 5.   Inclusion of Mixed Liquids[a]

| Guest, | Molar ratio of guest compounds in | |
organic liquid	Mixture	Adduct[b]
heptane–cyclohexane	0.74	100
hexane–benzene	0.73	20
hexane–cyclohexane	0.91	33
carbon tetrachloride–benzene	0.91	20

[a] Reproduced with permission from *J. Am. Chem. Soc.* (22).
[b] By mass spectrometry. Guest components are noncyclic.

Table 6.   Cage Dimensions of Cyclotriphosphazenes

Cyclotriphosphazene	CAS Registry No. and structure no.	Tunnel dimensions, nm
1,2-phenylenedioxy	[311-03-5] (1)	0.5
2,3-naphthalenedioxy	[6800-70-0]	ca 0.9–1.0
1,8-naphthalenedioxy	[27197-58-6]	0.52–0.7
2,2′-biphenylenedioxy	[92-52-4]	does not form clathrates

guest compound, show that hexadecane is trapped in a tunnel system already stabilized by smaller guest molecules but cannot initiate growth of a hexagonal channel by itself. Coiling of the hexadecane chain would increase the effective diameter to at least 1.0–1.2 nm (24).

The 1,8-naphthalenedioxy structure forms a *p*-xylene clathrate where *p*-xylene molecules are physically trapped (25). The latter *p*-xylene adduct is stable in the atmosphere for a long time and *p*-xylene loss probably occurs at 223–240°C (differential thermal analysis), well below the melting point of the host (mp ca 360°C). Host–host interactions stabilize the crystal structure of the 1,8-naphthalenedioxy complex (25).

Dianin's Compound

Dianin (26) observed the formation of adducts with several common solvents crystallized from structure (2).

The crystal structure of this host molecule and the size, shape of the cavity, and features of included molecules were established over 50 years after the discovery of the clathrating capabilities (27).

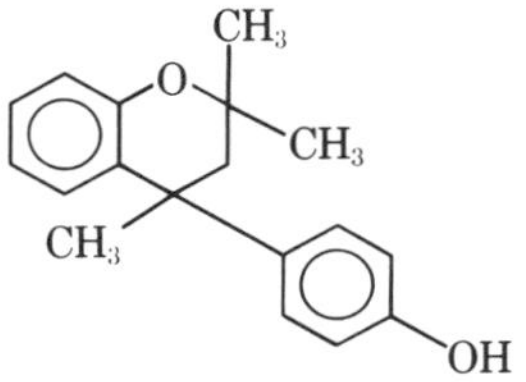

Dianin's compound [472-41-3]
(2)

Studies show a complex containing from 2–9 molecules of Dianin's compound (2), geometrically aligned by the hydrogen bonding of the hydroxyl group to form the cage, similar to the behavior of hydroquinone in its cage-forming action. Most clathrates contain a 6:1 host:guest ratio.

Although the cavities in quinol are roughly spherical, ca 0.4 nm in diameter, Dianin's compound has only one OH group available for hydrogen bonding, forming columns of independent cages with an hour-glass structure about 1.1 nm long and 0.43 nm wide at maximum extension.

Dielectric relaxation studies of clathrates formed from Dianin's compound suggest guest molecules may have considerable freedom of rotational movement (28).

More than 50 crystalline adducts of Dianin's compound have been reported (29). The mole ratio of Dianin's compound to guest component ranges from a low of 2:1 for methanol up to 3:1 for C_2–C_4 alcohols and carbon tetrachloride. Most other adducts formed by Dianin's compound, including aromatic hydrocarbons and derivatives, contained six host molecules per mole of guest; several host:guest ratios of 7:1–9:1 were also reported. The higher ratios involved pyridines, some esters, and 1-methylnaphthalene.

Analogues of Dianin's compound have been studied in which the cage structure was altered without collapsing (30) by removing one of the geminal methyl groups of Dianin's compound.

The length of the cage remains the same at 1.1 nm for the analogue of Dianin's compound with the methyl group removed, but the maximum diameter of the cage increased from 0.42 to 0.71 nm.

Only the carbon tetrachloride clathrate of the analogue was reported. The host:guest ratio was 6:1 for the compound with the methyl group removed, compared to a 3:1 host guest ratio for Dianin's compound.

Thymotides

A number of adducts of alcohols, alkyl halides, and aliphatic ethers with the compound, tri-*o*-thymotide, (3) (31) form channel and clathrate type complexes depending upon the guest dimensions.

Adducts with very long-chain alcohols are extended channels, as shown by x-ray, with an essentially constant enclosing structure. Cavity-type occlusions are formed by bulkier molecules such as chloroform or benzene. However, the length of occluded molecules is a limit of adduct configuration. Channel-type adducts form with alkyl halides, esters, ketones, and ethers where molecular length is less than 0.95 nm.

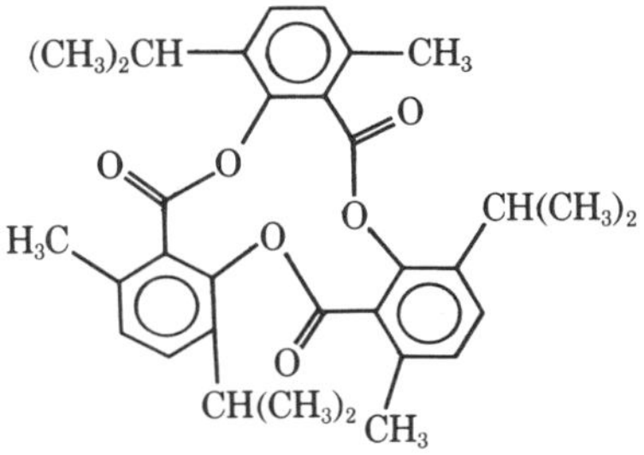

tri-*o*-thymotide [*2281-45-0*]

(3)

The wide variety of tri-*o*-thymotide adducts is attributed to possible expanding molecular dimensions plus the capability of forming trigonal spirals or hexagonal configurations where molecular length is more than 0.95 nm.

All of the molecules forming adducts vary in chemical character and length but have restricted width.

Smaller molecules, such as methyl bromide, form adducts but are less stable thermally. Smaller gas molecules such as argon and carbon dioxide have not been successfully clathrated thus far with tri-*o*-thymotide.

Other Clathrates

Deoxycholic acid has been known for some time to form cholic acids by combining with hydrocarbons, fatty acids, and similar substances (3).

Molecular complexes can form with 4,4-dihydroxytriphenylmethane including channel complexes with *n*-alkanes, *n*-alkenes as well as 2,2,4-trimethylpentane and diisobutene. Halogen-substituted triphenylmethanes have also been studied; however, stabilities of these inclusion complexes were low. Here again, hydrogen bonding across the hydroxyl groups analogous to hydroquinone complexes may be envisioned.

Recently (5) a new approach has been reported for synthesizing inclusion hosts not directly related to any known host. Hexa-substituted benzenes such as hexakis-(phenoxymethyl)benzene and its sulfur analogue retain toluene, dioxan, and xylene. Other hexa-substituted benzenes occlude bromoform, phenyl iodide, cyclohexane, and acetone. The most interesting aspect of this recent work is the synthesis of structures based on the behavior of well-known clathrating agents. This approach obviously requires more study but successes thus far offer encouragement for more exploratory work in the field.

Uses

Clathrate compounds have been investigated (32–33) in partition and adsorption chromatography. Proper selection of the solvent composition for the mobile phase is important particularly in the use of Werner compounds such as tetra(4-methylpyridino)nickel dithiocyanate for separating aromatic isomers, isomeric nitrophenols, nitroanilines, chloronitrobenzenes, and nitrotoluenes (see Coordination compounds). In all cases, quantitative separation of the ortho isomer was obtained. Other Werner complexes of the type MPy_4X_2 have been studied where M is a transition metal ion, Py is a pyridine derivative, and X in most cases is thiocyanate.

Use of these compounds is limited to temperatures below 90°C, above this temperature serious decomposition may occur.

Spirochromans have also been used in adsorption chromatography to separate hydrocarbons (34).

Clathrates may serve for storage of volatile materials.

Separations of mixtures can be achieved by inclusion complexes in systems not readily resolved by classical techniques such as extraction, distillation, etc, particularly where structural differences occur with small differences in boiling points or where the compounds are thermally sensitive. Thus, clathrates are useful analytical tools (35).

Optical resolution of tri-*o*-thymotide clathrates has been reported (36).

Clathrates may be used to dehydrate potable solutions such as fruit juices (qv), coffee (qv) extracts, and solutions such as those of latex that are concentrated in order to reduce shipping costs. The clathrate dehydration process is particularly applicable to products or systems sensitive to thermal degradation (37).

BIBLIOGRAPHY

1. H. M. Powell, *J. Chem. Soc.*, 61 (1948); 2658 (1954).
2. L. Mandelcorn, *Chem. Rev.* **59,** 827 (1959); refs. 33–48 on H. M. Powell and associates work on clathrates.
3. L. Mandelcorn, ed., *Non-Stoichiometric Compounds,* Academic Press, Inc., New York, 1964.
4. R. M. Barrer, *Nature* **178,** 1410 (1956).
5. D. D. MacNicol and D. R. Wilson, *J. Chem. Soc. Chem. Commun.,* 494 (July 7, 1976).
6. D. J. Cram and J. M. Cram, *Acc. Chem. Res.,* 8 (1978).
7. J.-M. Lelm, *Acc. Chem. Res.* **11**(2), 49 (1978).
8. J. F. Brown, Jr., *Sci. Am.* **207**(1), 84 (1962).
9. D. E. Palin and H. M. Powell, *J. Chem. Soc.* 208 (1947).
10. *Ibid.,* 815 (1948).
11. D. E. Palin and H. M. Powell, *J. Chem. Soc.,* 571 (1948).
12. J. Eric, D. Davies, and W. J. Wood, *J. Chem. Soc. Dalton Trans.,* 674 (1975); J. E. D. Davies, *Chem. Lett. Chem. Soc. Jpn.,* 263 (1974).
13. F. I. Kazankina and co-workers, *J. Gen. Chem. USSR* **46**(6), Pt. 2, 1307 (June 1976).
14. R. W. Coutant, *J. Org. Chem.* **39,** 1593 (1974).
15. D. F. Evans and R. E. Richards, *Proc. R. Soc. London* **A223,** 238 (1954).
16. J. H. Van der Waals, *Trans. Faraday Soc.* **52,** 184 (1956).
17. F. Cramer, W. Saenger, and H. Ch. Spatz, *J. Am. Chem. Soc.* **89,** 14 (1967).
18. H. Schlenk and D. M. Sand, *J. Am. Chem. Soc.* **83** 2312 (1961).
19. P. V. Demarco and A. L. Thakkar, *Chem. Commun.* **7,** 2 (1970).
20. J. Martinie, J. Michon, and A. Rassat, *J. Am. Chem. Soc.* **97,** 1818 (1975).
21. Y. Matsui, T. Yokoi, and K. Mochida, *Chem. Lett Tokyo* (10), 1937 (1976).
22. H. R. Allcock and L. A. Siegel, *J. Am. Chem. Soc.* **86,** 5140 (1964).
23. H. R. Allock, *Acc. Chem. Res.* **11**(3), 81 (1978).
24. H. R. Allcock and co-workers, *J. Am. Chem. Soc.* **98,** 5120 (1976).
25. H. R. Allcock, M. T. Stein, and E. C. Bissell, *J. Am. Chem. Soc.* **96,** 4795 (1974).
26. A. P. Dianin, *J. Soc. Phys. Chim. Russe* **46,** 1310 (1914).
27. J. L. Flippin, J. Karle, and I. L. Karle, *J. Am. Chem. Soc.* **92,** 3749 (1970).
28. J. S. Cook, R. G. Heydon, and H. K. Welsh, *J. Chem. Soc. Faraday Trans.,* **2,** 1591 (1974).
29. W. Baker and co-workers, *J. Chem. Soc.,* 2010 (1956).
30. A. D. V. Hardy, J. J. McKendrick, and D. D. MacNicol, *J. Chem. Soc. Commun.,* 355 (1976).
31. D. Lawton and H. M. Powell, *J. Chem. Soc.,* 2339 (1958).
32. W. Kemula and D. Sybilska, *Nature,* 237 (Jan. 30, 1960).
33. A. C. Bhattacharyya and A. Bhattacharyya, *J. Chromatogr.* **41,** 446 (1969); *Anal. Chem.* **41,** 2055 (1969).
34. U.S. Pat. 2,851,500 (Sept. 9, 1958), E. W. Geiser (to Universal Oil Products).
35. E. Grushka, *New Developments in Separation Methods,* Marcel Dekker Inc., New York, 1976, pp. 49–64.
36. A. C. D. Newman and H. M. J. Powell, *J. Chem. Soc.,* 3747 (1952).
37. G. N. Werezak, *Chem. Eng. Prog. Symp. Ser.* **65,** 648 (1969).

EARL C. MAKIN
Monsanto Chemical Intermediates Co.

CLAYS

SURVEY

Clays as they occur in nature are rocks that may be consolidated or unconsolidated. They are distinctive in at least two properties which render them technologically useful.

(1) Plasticity signifies the property of the clay when wetted that permits deformation by application of relatively slight pressure and retention of the deformed shape after release of the pressure. This property distinguishes clays from hard rocks.

(2) Clays are composed of extremely fine crystals or particles, often colloidal in size and usually platy in shape, of clay minerals with or without other rock or mineral particles. The clay minerals, mostly phyllosilicates, are hydrous silicates of Al, Mg, Fe, and other less abundant elements.

The very fine particles yield very large specific-surface areas that are physically sorptive and chemically surface-reactive. Many clay mineral crystals carry an excess negative electric charge owing to internal substitution by lower valent cations, and thereby increase internal reactivity in chemical combination and ion exchange. Clays may have served as substrates that selectively absorbed and catalyzed amino acids in the origin of life. They apparently catalyze petroleum formation in rocks (see Petroleum).

Because clays (rocks) usually contain more than one mineral and the various clay minerals differ in chemical and physical properties, the term clay may signify entirely different things to different technologists.

For geologists, clay refers to sediments or sedimentary rock particles with a diameter of 3.9 μm or less. Soil scientists define clays as disperse systems of the colloidal products of weathering in which secondary mineral particles of dimensions smaller than 2 μm predominate.

Ceramists, who probably process the greatest quantity of clay, usually emphasize aluminosilicate content and plasticity as in the following standard definition (1): Clay is " . . . an earthy or stony mineral aggregate consisting essentially of hydrous silicates and alumina, plastic when sufficiently pulverized and wetted, rigid when dry, and vitreous when fired at a sufficiently high temperature." Ceramists recognize that this definition is open to qualification for their definition (1) of flint fire clay as "practically devoid of natural plasticity," conflicts with their standard definition. The source of the plasticity is defined as (2): " . . . with more or less plasticity due to colloids of organic or mineral nature," but it may arise from other causes (3). Even the origin of the clay has been included in a definition (4): "clays are the weathered products of the silicate rocks . . .," but it has been recognized as a product of deep-seated alteration of silicate minerals by hydrothermal solutions rising from an igneous source as well as a product of surface weathering (5–6). Although clay minerals are usually considered breakdown products of silicates, largely by hydrolysis, they may be built up from hydrates of silica and alumina (7).

A clay deposit usually contains nonclaylike minerals as impurities, although these impurities may actually be essential in determining the unique and specially desired properties of the clay. Both crystalline and amorphous minerals and compounds may be present in a clay deposit (8).

The geologist views clay as a raw material for shale; the pedologist as a dynamic system to support plant life; the ceramist as a body to be processed in preparation for vitrification; the chemist and technologist as a catalyst, adsorbent, filler, coater, or source of aluminum or lithium compounds, etc. A broad definition includes the following properties of clays (rocks):

(1) The predominant content of clay minerals, which are hydrated silicates of aluminum, iron, or magnesium, both crystalline and amorphous. These range from kaolins, which are relatively uniform in chemical composition, to smectites, which vary widely in their base exchange properties, expanding crystal lattice, and proxying of elements. The illite group, typically a component of sediment (9), includes micas, although illite [1273-60-3] is considered a variation of muscovite [1318-94-1] because illite contains less potassium and more water. The chlorite clay minerals resemble metamorphic chlorite [1318-59-8] (10), but aluminous, or at least aluminum-rich, chlorites have also been found in soils (4,11). Vermiculite [1318-00-9], characterized by a highly expanding crystal lattice, and sepiolite [15501-74-3], and palygorskite [12174-11-7], which possess chain or fiber structures, must also be included. Moreover, clay minerals are excellent examples of mixed layering, both random and regular, in layer-structure silicates (Fig. 1).

(2) The possible content of hydrated alumina and iron. Hydrated alumina minerals like gibbsite [14762-49-3], boehmite [1318-23-6], and diaspore [14457-84-2] occur in bauxitic clays. Bauxites grade chemically into hydrated ferruginous and manganiferous laterites. Hence, finely divided M_2O_3, usually hydrated, may be a significant constituent of a clay. Hydrated colloidal silica may play a role in the slippery and sticky properties of certain clays.

(3) The extreme fineness of individual clay particles, which may be of colloidal size in at least one dimension. Clay minerals are usually platy in shape, and less often lathlike and tubular or scroll shaped (13). Because of their fineness they exhibit surface chemistry properties of colloids (14). Some clays possess relatively open crystal lattices and show internal surface colloidal effects. Other minerals and rock particles, which are not hydrous aluminosilicates but which also show colloidal dimensions and characteristics, may occur intimately intermixed with the clay minerals and play an essential role.

(4) The property of thixotropy in various degrees of complexity (3). This includes, in addition to technological applications, the loss of stability as shown by the sometimes catastrophic flow of quick clays, especially in Norway, Sweden, and Canada (15).

(5) The possible content of quartz [14808-60-7] sand and silt, feldspars, mica [12001-26-2], chlorite, opal, volcanic dust, fossil fragments, high-density so-called heavy minerals, sulfates, sulfides, carbonate minerals, zeolites, and many other rock and mineral particles ranging upward in size from colloids to pebbles. An extreme example is that of a clay from western Texas composed of 98.5% dolomite [16389-88-1], $CaMg(CO_3)_2$, and 1.5% iron oxide and alumina; it occurs in rhombic particles averaging 0.008 mm in diameter, and possesses "sufficient plasticity to be molded into bricklets" (16).

The synthesis of clay minerals has been extensively studied (17–21). Many experiments were performed at high temperature, some using synthetic chemicals, others

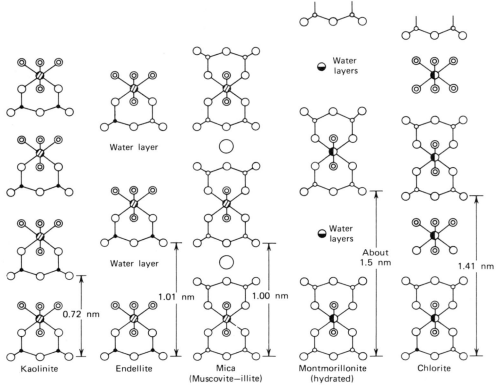

Figure 1. Diagrammatic representation of the succession of layers in some layer lattice silicates (12). O, oxygen; ⊚, (OH); ●, silicon; ○, Si–Al; ⊘, aluminum; ◑, Al–Mg; ○, potassium; ◗, Na–Ca.

using in part naturally occurring minerals. Organic compounds facilitated the synthesis of kaolinite at low temperature by condensing aluminum hydroxide into octahedrally coordinated sheets (18).

Geology and Occurrence

Clays may originate through several processes: (*1*) hydrolysis and hydration of a silicate, ie, alkali silicate + water → hydrated aluminosilicate clay + alkali hydroxide; (*2*) solution of a limestone or other soluble rock containing relatively insoluble clay impurities that are left behind; (*3*) slaking and weathering of shales (clay-rich sedimentary rocks); (*4*) replacement of a preexisting host rock by invading guest clay whose constituents are carried in part or wholly by solution; (*5*) deposition of clay in cavities or veins from solution; (*6*) bacterial and other organic activity, including the extraction of metal cations as nutrients by plants; (*7*) action of acid clays, humus, and inorganic acids on primary silicates; (*8*) alteration of parent material or diagenetic processes following sedimentation in marine and freshwater environments (22–23); and (*9*) resilication of high alumina minerals.

Every state in the United States has within its boundaries clays or shales that may be utilized in the manufacture of bricks, tiles, and other heavy clay products. Some blending of materials is often necessary to control shrinkage of the product, and the economics of manufacture are governed by the demands of fuel, labor, transportation costs, and the market.

Glacial clays, as unassorted glacial till or secondarily deposited melt water, are abundant in the United States north of the Missouri and Ohio Rivers. Quartz-rich sand, silt, or pebbles, especially limestone, may occur mixed with the clay.

Kaolins are plentiful in North Carolina, South Carolina, Georgia (24), Florida, and Vermont. Certain flint clays and other clays with a kaolinitic composition which can replace kaolins in some uses (25) occur in Missouri, Arkansas, Colorado, Texas, Ohio, Indiana, Oregon, and Pennsylvania.

Ball clays, ie, clays with high plasticity and strong bonding power, are obtained primarily from western Tennessee and Kentucky, though some are found in New Jersey. Other plastic clays, especially plastic fire clay, are extensively produced in Missouri, Illinois, Ohio, Kentucky, Mississippi, Alabama, and Arkansas.

Fire clays are those that resist fusion at a relatively high temperature, usually around 1600°C. Missouri, Pennsylvania, Ohio, Kentucky, Georgia, Colorado, New Jersey, Texas, Arkansas, Illinois, and Maryland are major producers of fire clays. Deposits also occur in South Carolina, Indiana, Alabama, California, West Virginia, and Oregon.

Loess is a quartz-rich, clayey silt, windblown in origin but, in some cases, reworked by water. It is prevalent along the Missouri, Mississippi, and Ohio Rivers and their tributaries. Loess has been used primarily for brick making.

Adobe is a calcareous, sandy to silty clay used extensively for making sundried brick for local use in the more arid southwestern states.

Slip clay for glazing pottery is produced near Albany, New York.

Bentonite [1302-78-9] is widely distributed geographically and geologically, and also varies widely in properties. Swelling bentonite occurs in Wyoming, South Dakota, Montana, Utah, Nevada, and California. Bentonite that swells little or not at all occurs in large quantities in Texas, Arkansas, Mississippi, Kentucky, and Tennessee.

Fuller's earth and bleaching clays, though found chiefly in Georgia and Florida, also occur in Arizona, Arkansas, California, Colorado, Illinois, Mississippi, Missouri, Oklahoma, Texas, and Utah.

High alumina clays refer in the ceramic industries to nodular clays, burley-flint clay, burley and diaspore, gibbsitic or bauxitic kaolins (clays), abrasives clays, and others. Since the depletion of diaspore varieties in Missouri and Pennsylvania, most bauxitic kaolin and clay is produced in Alabama and Arkansas.

Though each continent has clays of almost every type, certain deposits are outstanding (26–27). There are tremendous reserves of white kaolin on the Jari and Capim Rivers of Brazil, and deposits of bauxitic clay shared with countries across Brazil's northern borders. England has the famous kaolins of the Cornwall district. Refractory clays in Scotland provide the raw materials for a refractories industry. Deposits similar to those in Cornwall are found in Brittany and neighboring areas in France. Czechoslovakia, the Federal Republic of Germany, and the German Democratic Republic have large reserves of kaolin accompanied by quartz and mica. There are large deposits of bauxitic clays (with bauxite) in Hungary and Yugoslavia. Both flint clay and white kaolin are found in South Africa. Kaolins are found in the People's Republic of China. Japan has notable hydrothermal kaolins, sedimentary kaolin, smectite, and flint clay. Hydrothermal kaolins are also widely distributed in Mexico. Australia has large deposits of flint clay and kaolin clay naturally calcined by the burning of coal beds. New Zealand has hydrothermal clays. India has flint and lateritic clays. Sepiolite is found near Madrid, Spain. Localities producing bauxite almost always have a potential for producing associated high alumina clays.

The commercial value of a clay deposit depends upon market trends, competitive materials, transportation facilities, new machinery and processes, and labor and fuel costs.

Naturally exposed outcrops, geological area and structure maps, aerial photographs, hand and power auger drills, core drills, earth resistivity, and shallow seismic methods are used in exploration for clays (28). Clays are mined primarily by open-pit operation, including hydraulic extraction; however, underground mining is also practiced.

Specific information concerning the geology or occurrence of a particular deposit or variety of clay may be obtained from a state geological survey; general information may be obtained from the U.S. Geological Survey or the U.S. Bureau of Mines. Similar agencies may be contacted in foreign countries, or references may be consulted (8,29–31, and the general references).

Technological Classification

The technological classification of a clay (rock, deposit) should take into account the following factors: (1) The dominant clay mineral type including breakdown into its polymorphs, the sites and amount of charge on it, and shape of clay crystal and particle. (2) The clay minerals present in minor quantities, but perhaps coating the surface of the major constituent. (3) The particle-size distribution of the clay and other minerals. (4) Ion-exchange capacity (cation, anion) and neutral molecule sorption. (5) The type of exchangeable ions present on the clay and degree of saturation of exchange sites. (6) Hygroscopicity of the clay. (7) Reactivity of the clay with organic compounds. (8) Expansion potentialities of the clay mineral lattice. (9) Electrolytes and solutions in association with the clay deposit. (10) The accessory minerals, or mineral impurities, their sizes, homogeneity of mixture, and ion-exchange capacity. (11) Content of organic matter and especially its occurrence, size and discreteness of particles, its adsorption on and/or within the clay crystal units, and protective colloidal action. (12) Presence or absence of bacteria or other living organisms. The pH and other properties of a clay deposit may vary notably within a short time where bacteria are growing. (13) Content of hydrated alumina and/or silica, which are relatively soluble in ground water or in dilute acid or alkali. (14) The structure and texture of the clay deposit, such as lamination, orientation of mineral particles, and other gross features. (15) The rheological properties of both the natural and processed clay. (16) The engineering strength, and sensitivity to moisture, desalination, and shock, eg, the quick clays.

Mineralogy

The development of apparatus and techniques, particularly x-rays, contributed greatly to research on clay minerals. Today crystalline clay minerals are identified and classified (32) primarily on the basis of crystal structure and the amount and locations of charge (deficit or excess) with respect to the basic lattice. Amorphous (to x-ray) clay minerals are poorly organized analogues of crystalline counterparts.

Various techniques are used to study crystalline clay minerals including the polarizing microscope (33), chemical analysis and computation of the mineral formula taking into full account the substitution of atoms (34), staining (35–37), density, possible electrical double refraction (38), dehydration, base exchange (39–40), electron micrographs, transmission electron microscopy (tem) (41) and scanning electron microscopy (sem) (42–44), x-ray or electron powder diffraction patterns (32), differential thermal analysis and thermal balance, imbibition (45), infrared absorption (39,41), and field appearance, especially responses to weathering (46 and the general references) (see Analytical methods).

Clay minerals are divided into crystalline and paracrystalline groups and a group amorphous to x-rays (47).

Although the clays of the different groups are similar, they show vastly different mineralogical, physical, thermal, and technological properties (3). Chemical analysis alone may have limited value in revealing the clay's identity or usefulness. The mineral composition, which reveals the organization of the constituent elements is most important.

CRYSTALLINE AND PARACRYSTALLINE GROUPS

Kaolins. The kaolin minerals include kaolinite [1318-74-7], dickite [1318-45-2], and nacrite [12279-65-1] (all $Al_2O_3.2SiO_2.2H_2O$), and halloysite–endellite ($Al_2O_3.2SiO_2.2H_2O$ and $Al_2O_3.2SiO_2.4H_2O$, respectively) (47–50). The structural formulas for kaolinite and endellite [12244-16-5] are $Al_4Si_4O_{10}(OH)_8$ and $Al_4Si_4O_{10}(OH)_8.4H_2O$, respectively. The kaolinite lattice (12,51) consists of one sheet of tetrahedrally coordinated Si (with O) and one sheet of octahedrally coordinated Al (with O and OH), hence a 1:1 or two-layer structure. A layer of OH completes the charge requirements of the octahedral sheet. The so-called fire clay mineral is a b-axis disordered kaolinite (12); halloysite [12244-16-5] and endellite are disordered along both the a and b axes. Indeed, most variations in the kaolin group originate as structural polymorphs. Representative analyses (47–49) of the kaolin minerals are given in Table 1.

As can be seen from the optical constants given in Table 2, kaolinite and dickite are easily distinguished where they occur in recognizable crystals. Nacrite is relatively rare. Halloysite is usually exceedingly fine grained, showing a mean index of refraction of about 1.546. The index of refraction for endellite varies somewhat with the immersion liquid used; it ranges from 1.540 to 1.552 (48).

X-ray studies show that kaolin minerals have two-layer crystal lattices: a sheet of silica tetrahedra and an alumina–gibbsite sheet. Adjacent cells are spaced about 0.71 nm across the (001) plane. The interplanar spacings normal to the (001) cleavage are the most significant criteria used in x-ray differentiation between the clay mineral groups. Within the kaolin group other x-ray structural differences are used to distinguish the members (12).

Endellite, ie, the hydrated form of halloysite (48), expands to 1.0 nm along the c axis when solvated in ethylene glycol. Halloysite may be differentiated from kaolinite and dickite by treatment with potassium acetate and ethylene glycol (54). Halloysite–endellite minerals are usually tubular, or rolled or scroll-shaped in morphology; this was formerly interpreted as the result of a misfit between the octahedral and tetrahedral sheets (13). An alternative interpretation is that certain crystals have an inherent roundish, elongate morphology (55).

Table 1.   Chemical Analyses of the Kaolin Minerals, %

Component	1[a]	2[b]	3[c]	4[d]	5[e]	6[f]
SiO_2	45.44	40.26	46.5	45.78	42.68	44.90
Al_2O_3	38.52	37.95	39.5	36.46	38.49	38.35
Fe_2O_3	0.80	0.30		0.28	1.55	0.43
FeO				1.08		
MgO	0.08			0.04	0.08	trace
CaO	0.08	0.22		0.50		trace
K_2O	0.14 ⎱	0.74		⎱ 0.25	0.49	0.28
Na_2O	0.66 ⎰			⎰	0.28	0.14
TiO_2	0.16				2.90	1.80
H_2O removed						
at 105°C	0.60	4.45		2.05		
above 105°C	13.60	15.94	14.0	13.40	14.07	14.20
Total	*100.08*	*99.86*	*100.0*	*99.84*	*100.54*	*100.10*

[a] Kaolinite, Roseland, Va. (49).
[b] Halloysite, Huron Co., Ind. (47).
[c] Theoretical kaolinite.
[d] Washed kaolin, Webster, N. C. (6).
[e] Flint fire clay, near Owensville, Mo. (52).
[f] Typical sedimentary kaolin, S. C., Ga., Ala. Courtesy S. C. Lyons.

Table 2.   Optical Constants of Kaolinite, Dickite, and Glauconite [a]

Optical constant	Kaolinite	Dickite	Glauconite
index of refraction			
α	1.561	1.561	1.597
β	1.565	1.563	1.618
γ	1.565	1.567	1.619
optical character	neg	pos	neg
extinction angle,°	3.5	ca 16	
dispersion	$r > v$		
axial angle, 2V	10–57°		20°
X	nearly perpendicular to (001)		dark bluish green [b]
Z, Y			yellow [b]

[a] Ref. 53.
[b] Pleochroic.

     Staining kaolin minerals with aniline dyes produces varied artificial pleochroism which may be sufficiently selective to identify the mineral species (36). In differential thermal analysis, kaolinite shows a strong endothermic peak at about 620°C and a strong exothermic peak at about 980°C, which sharply differentiates it from the other clay mineral groups. Electron micrographs show kaolinite in roughly equidimensional, pseudohexagonal plates and halloysite in lath-shaped crystals. Kaolinite, the most abundant, and halloysite, the second most abundant of the kaolin group of minerals are those most used by industry. Problems of clay mineral differentiation within the kaolin group arise rarely; however, adequate means for identifying and distinguishing the various mineral species are available (30,36). The cationic base-exchange capacity of kaolinite is low, less than 10 meq/100 g of dry clay.

Kaolin most commonly originates by the alteration of feldspar or other aluminum silicate via an intermediate solution phase (56–57) usually by surface weathering (22,58) or by rising warm (hydrothermal) waters. A mica, or hydrated alumina solid phase may intervene between parent and kaolin minerals.

Large deposits of relatively pure kaolinite have developed from parent, feldspar-rich pegmatites, whereas others are secondarily deposited in sedimentary beds after transportation. Colloidal fractions of geologically ancient soils were presumably concentrated in old swamps and leached to develop kaolinitic clay deposits (6,25). Today kaolinite is formed by weathering in an oxidizing environment under acid conditions, and in a reducing environment where the bases such as calcium, magnesium, alkalies, and iron(II) are removed. Removal of the bases is essential in kaolin formation (59). With more intense leaching of silica from the clay an aluminous hydrate remains and the clay becomes bauxitic. Kaolinite may develop from the silication of gibbsite (7). Halloysite–endellite may be formed either by hydrothermal or weathering processes. Allophane [12172-71-3] may have led to halloysite and thence to kaolinite as crystallization became more highly ordered following weathering (60); however, this sequence does not always prevail. Some indications point toward a possible association of acid sulfate waters and mobile potassium with the origin of endellite–halloysite (58).

The textures of kaolin (rock) include varieties similar to examples observed in igneous and metamorphic as well as sedimentary rocks (57). Kaolin grains and crystals may be straight or curved, sheaves, flakes, face-to-face or edge-to-face floccules, interlocking crystals, tubes, scrolls, fibers, or spheres (57).

The kaolin group is transformed at high temperatures to a silica–alumina spinel structure (61) previously interpreted as gamma alumina, thence to mullite [1302-93-8] with or without accompanying cristobalite [14464-46-1] (62–63). The alkali metal, flux, and content of kaolin clay strongly influence the phases formed upon heating (62).

Serpentines.   Substituting 3 Mg in the kaolin structure, $Al_2Si_2O_5(OH)_4$, results in the serpentine minerals, $Mg_3Si_2O_5(OH)_4$. In serpentines all three possible octahedral cation sites are filled, yielding a trioctahedral group carrying a charge of +6, whereas in kaolinite only two thirds of the sites are occupied by Al yielding a dioctahedral group also with a charge of +6. Most serpentine minerals are tubular to fibrous in structure presumably because of misfit, the reverse geometry of halloysite–endellite, between Mg octahedral and tetrahedral layers. Thus, structurally, serpentines are analogues of kaolin minerals, although they depart from aluminous clays in certain other properties.

Chrysotile [12001-27-5] (serpentine) occurs in both clino and ortho structures. Both one-layer ortho and clino, and six-layer ortho (as in nacrite) structures have been observed. Chrysotile transforms at high temperature to forsterite [15118-03-3] and silica.

Amesite [12413-43-3] approximates $(Mg_2Al)(SiAl)O_5(OH)_4$ in composition, cronstedite [61104-63-0], $(Fe_2^{2+}Fe^{3+})(SiFe^{3+})O_5(OH)_4$, and chamosite [12173-07-2], $(Fe^{2+},Mg)_{2.3}(Fe^{2+}Al)_{0.7}(Si_{1.14}Al_{0.86})O_5(OH)_4$. Garnierite [12198-10-6] is possibly a nickel serpentine; a cobalt serpentine has been synthesized (12).

Smectites (Montmorillonites).   Smectites are the 2:1 clay minerals that carry a lattice charge and characteristically expand when solvated with water and alcohols, notably ethylene glycol and glycerol. In earlier literature, the term montmorillonite was used for both the group (now smectite) and the particular member of the group in which Mg is a significant substituent for Al in the octahedral layer. Typical formulas are shown in Table 3. Additional, less common smectites include volkhonskoite [12286-87-2] which contains Cr^{2+}; medmontite [12419-74-8], Cu^{2+}; and pimelite [12420-74-5], Ni^{2+} (12).

Smectites are derived structurally from pyrophyllite [12269-78-2], $Si_8Al_4O_{20}(OH)_4$, or talc [14807-96-6], $Si_8Mg_6O_{20}(OH)_4$, by substitutions mainly in the octahedral layers. Some substitution may occur for Si in the tetrahedral layer, and by F for OH in the structure. When substitutions occur between elements (ions) of unlike charge, deficit or excess charge develops on corresponding parts of the structure. Deficit charges in smectite are compensated by cations (usually Na, Ca, K) sorbed between the three-layer (two tetrahedral and one octahedral, hence 2:1) clay mineral sandwiches. These are held relatively loosely, although stoichiometrically, and give rise to major cation exchange properties of the smectite. Representative analyses of smectite minerals are given in Table 4.

The determination of a complete set of optical constants of the smectite group is usually not possible because the individual crystals are too small. However, suspensions of the clays evaporated on a glass slide usually deposit a pseudocrystal film from which α and γ component values may be determined. The clay films should be soaked in acetone to displace the air and then left to stand in the index immersion oil several hours before measuring indexes of refraction. Representative optical measurements are given in Table 5.

In the montmorillonite–nontronite series, as the iron(III) content increases from 0 to 28%, the α index ranges from 1.523 to 1.590, and the γ from 1.548 to 1.632.

X-ray diffraction patterns yield typical 1.2–1.2 nm basal spacings for smectite partially hydrated in an ordinary laboratory atmosphere. Solvating smectite in ethylene glycol expands the spacing to 1.7 nm, and heating to 550°C collapses it to 1.0 nm. Certain micaceous clay minerals from which part of the smectites' metallic interlayer cations has been stripped or degraded and replaced by H^+ or H_3O^+ expand similarly. Treatment with strong solutions of potassium salts may permit differentiation of these expanding clays (65).

Smectite [12199-37-0] from an oxidized outcrop is stained light blue by a dilute solution of benzidine hydrochloride. The color does not arise from smectite specifically, but from reaction with a high oxidation state of elements such as Fe^{3+} or Mn^{4+} (37).

Table 3. Typical Formulas of Smectite Minerals[a]

Mineral	CAS Registry No.	Formula[b]
montmorillonite	[1318-93-01]	$[Al_{1.67}Mg_{0.33}(Na_{0.33})]Si_4O_{10}(OH)_2$
beidellite	[12172-85-9]	$Al_{2.17}[Al_{0.33}(Na_{0.33})Si_{3.17}]O_{10}(OH)_2$
nontronite	[12174-06-0]	$Fe(III)[Al_{0.33}(Na_{0.33})Si_{3.67}]O_{10}(OH)_2$
hectorite	[12173-47-6]	$[Mg_{2.67}Li_{0.33}(Na_{0.33})]Si_4O_{10}(OH,F)_2$
saponite	[1319-41-1]	$Mg_{3.00}[Al_{0.33}(Na_{0.33})Si_{3.67}]O_{10}(OH)_2$
sauconite	[12424-32-7]	$[Zn_{1.48}Mg_{0.14}Al_{0.74}Fe(III)_{0.40}][Al_{0.99}Si_{3.01}]O_{10}(OH)_2X_{0.33}$

[a] Ref. 34.

[b] More substitution takes place than shown; $Na_{0.33}$ or $X_{0.33}$ refers to the exchangeable base (cation) of which 0.33 equivalent is a typical value.

Table 4. Chemical Analyses of the Smectite Minerals, %

Component	1[a]	2[b]	3[c]	4[d]	5[e]	6[f]
SiO_2	51.14	47.28	43.54	55.86	42.99	34.46
Al_2O_3	19.76	20.27	2.94	0.13	6.26	16.95
Fe_2O_3	0.83	8.68	28.62	0.03	1.83	6.21
FeO			0.99		2.57	
MnO	trace			none	0.11	
ZnO	0.10					23.10
MgO	3.22	0.70	0.05	25.03	22.96	1.11
CaO	1.62	2.75	2.22	trace	2.03	
K_2O	0.11	trace		0.10	trace	0.49
Na_2O	0.04	0.97		2.68	1.04	
Li_2O				1.05		
TiO_2	none			none		0.24
P_2O_6						
F				5.96		
H₂O removed						
at 150°C	14.81	19.72	14.05	9.90	13.65	6.72
above 150°C	7.99		6.62	2.24	6.85	10.67
Total	*99.75*	*100.37*	*100.02*	*102.98*	*100.29*	*99.95*
				100.470–F		

[a] Montmorillonite, Montmorillon, France (34).
[b] Beidellite, Beidell, Colo. (34).
[c] Nontronite, Woody, Calif. (34).
[d] Hectorite, Hector, Calif. (34).
[e] Saponite, Ahmeek Mine, Mich. (64).
[f] Sauconite, Friendsville, Pa. (64).

Table 5. Optical Properties of the Smectite Minerals[a]

Mineral	Source	Index of refraction			Optical character	Axial angle, 2V	Physical form
		α	β	γ			
beidellite	Beidell, Colo.	1.502		1.533			grains
nontronite	Woody, Calif.	1.560	1.585	1.585	negative	small	grains
hectorite	Hector, Calif.	1.485		1.516			film
saponite	Ahmeek Mine, Mich.	1.490	1.525	1.527	negative	moderate	grains
sauconite	Friendsville, Pa.	1.575		1.615	negative		film

[a] Refs. 34, 64.

Transmission electron micrographs show hectorite and nontronite as elongated, lath-shaped units, whereas the other smectite clays appear more nearly equidimensional. A broken surface of smectite clays typically shows a "corn flakes" or "oak leaf" surface texture (42).

The differential thermal analysis curves for smectite commonly show three endothermic peaks and one exothermic peak, which may fall within the ranges 150–320, 695–730, 870–920, and 925–1050°C, respectively. Beidellite exhibits endothermic peaks at about 200 and 575–590°C and an exothermic peak at 905–925°C.

High temperature minerals formed upon heating smectites vary considerably with the compositions of the clays. Spinels commonly appear at 800–1000°C, and dissolve at higher temperatures. Quartz, especially cristobalite, appears and mullite forms if the content of aluminum is adequate (51).

The cation-exchange capacity of smectite minerals is notably high, 80–90 meq or higher per 100 g of air-dried clay, and affords a diagnostic criterion of the group. The crystal lattice is obviously weakly bonded. Moreover, the lattice of smectite is expandable between the silicate layers so that when the clay is soaked in water it may swell to several times its dry volume (eg, bentonite clays). Soil colloids with high cation-exchange capacity facilitate the transfer of plant nutrients to absorbing plant rootlets.

The minerals of the smectite group have been formed by surface weathering, low temperature hydrothermal processes, alteration of volcanic dust in stratified beds (66), action of circulating water of uncertain source along fractures and in veins, and laboratory synthesis. The optimum weathering environment is one in which calcium, iron(II), and especially magnesium are present in significantly high concentrations. Potassium should be low or low in relation to magnesium, calcium, and ferrous iron. Organic matter that exerts reducing action is a usual concomitant, and a neutral to slightly alkaline medium generally prevails under conditions where the alkali and alkaline earth metals are not readily removed. The weathering environment for smectite is different from that in which kaolinite is formed (59). If the system permits effective leaching and H^+ ions become available in sufficient quantity to cause the metallic cations to be easily leached away, kaolinite tends to form. The reverse reaction rarely, if ever, takes place.

Bentonite is a rock rich in montmorillonite that has usually resulted from the alteration of volcanic dust (ash) of the intermediate (latitic) siliceous types. In general, relicts of partially unaltered feldspar, quartz, or volcanic glass shards offer evidence of the parent rock. Most adsorbent clays, bleaching clays, and many clay catalysts are smectites, although some are attapulgite [1337-76-4].

Illites or Micas. Illite is a general term for the clay mineral constituents of argillaceous sediments belonging to the mica group (9); it is not a single pure mineral (67). Mica minerals possess a 2:1 sheet structure similar to montmorillonite except that the maximum charge deficit in mica is typically in the tetrahedral layers and contains potassium held tenaciously in the interlayer space, which contributes to a 1.0 nm basal spacing. Because the micas in argillaceous sediments may be widely diverse in origin, considerable variation exists in the composition and polymorphism of the illite minerals.

The formula of illite can be expressed as $2K_2O.3MO.8R_2O_3.24SiO_2.12H_2O$ (9), and the crystal structure (68) by the formula $K_y(Al_4Fe_4.Mg_4Mg_6)(Si_{8-y})O_{20}(OH)_4$, where y refers to the K^+ ions that satisfy the excess charges resulting when about 15% of the Si^{4+} positions are replaced by Al^{3+}. For representative chemical analysis of illite (fine colloid fraction, Pennsylvania underclay, near Fithian, Vermilion County, Illinois), see Table 6.

Optical constants of illite minerals are difficult to obtain because of the small size of the available crystals. The highest (γ) index of refraction ranges from about 1.588 to 1.610, the birefringence is about 0.033, the optical character is negative, and the axial angle, $2V$, is small, on the order of 5°.

A 1.0 nm basal spacing exhibited in a diffractogram peak that is somewhat broad and diffuse and skewed toward wider spacings characterizes the x-ray diffraction pattern of illite. Polymorphs 1 Md, 1 M, 2 M_1, and 2 M_2 may be present (69); 1 Md and 1 M are most commonly reported. Muscovite derivatives are typically dioctahedral; phlogopite derivatives are trioctahedral.

Table 6. Chemical Analysis of Illite, Glauconite, and Attapulgite, %

Component	Illite	Glauconite	Attapulgite
SiO_2	51.22	48.66	55.03
Al_2O_3	25.91	8.46	10.24
Fe_2O_3	4.59	18.8	3.53
FeO	1.70	3.98	
MgO	2.84	3.56	10.49
CaO	0.16	0.62	
K_2O	6.09	8.31	0.47
Na_2O	0.17		
TiO_2	0.53		
H_2O removed			
at 110°C	7.49		
at 150°C			9.73
above 150°C			10.13
Total	*100.7*	*99.8*	*99.62*

Differential thermal and analysis curves of illite show three endothermic peaks in the ranges 100–150, 500–650, and at about 900°C, and an exothermic peak at about 940°C, or immediately following the highest endothermic peak. Minerals formed from illite at high temperature vary somewhat with the composition of the clay, but usually a spinel-structure mineral followed by mullite at still higher temperatures is observed (34).

The cation-exchange capacity of illite is 20–30 meq/100 g of dry clay. The interlayer potassium exerts a strong bond between adjacent clay structures. Illite that has lost part of its original potassium by weathering processes may be reconstituted with the sorption and incorporation of transient dissolved potassium (65).

Illite was defined as the most abundant clay mineral in Paleozoic shale and is widespread in many other sedimentary rocks; it is common in soils, slates, certain alteration products of igneous rocks, and recent sediments. Its origin has been attributed to alteration of silicate minerals by weathering and hydrothermal solutions, reconstitution, wetting and drying of soil clays, and diagenesis involving other three-layer minerals and potassium during geologic time and pressure under deep burial (23,70–71).

Glauconite. Glauconite [1317-57-3] (72–75) is a green, dioctahedral, micaceous clay rich in ferric iron and potassium. It has many characteristics common to illite, but much glauconite contains some randomly mixed expanding layers, interpreted as montmorillonitic. Glauconite occurs abundantly in sand size pellets or bigger, or in pellets within fossils, notably foraminifera, giving it an organic connotation (12). Occurrences as replacements, matrix, and flakes in sandstone and as a product of diagenesis (70,76) indicate other possible origins. Glauconite is typically formed in a marine environment (77), but glauconitic mica has been reported from nonmarine rocks (46,78).

The chemical analysis of glauconite (Bonneterre, Missouri), is given in Table 6, and the optical constants in Table 2. Powder x-ray diffraction patterns resemble those of illite in which intensities of even-numbered basal spacings are minimal.

The glauconitic green sands of New Jersey have been used in ion-exchange, water-softening installations (see Ion exchange), and as a source of slowly released potassium in soil amendments.

Chlorites and Vermiculites. Chlorite was identified as the mineral yielding a 1.4-nm basal spacing in clays some years after the clay mineral groups previously discussed were fairly well characterized. Chloritic clay minerals are therefore less well explored than the other clay minerals or the chlorite of igneous and metamorphic occurrences (10,78a). Chlorite is widespread in argillaceous sedimentary rocks and in certain soils. Structurally, it has been viewed as three-layer phyllosilicates separated by a brucite [1317-43-7], $Mg(OH)_2$, interlayer, or alternatively, as a four-layer group, or 242, of alternating silica tetrahedral and Mg octahedral sheets (79–81). Both 0.7- and 1.4-nm (basal spacing) polymorphs exist.

Aluminum chlorite [25410-05-3], in which gibbsite, $Al(OH)_3$, proxies in part for brucite, is being discovered in increasing occurrences and abundance (11,82). Chlorite-like structures have been synthesized by precipitating Mg and Al between montmorillonite sheets (83).

The interlayer sheet in chlorite is interpreted as brucitic, $Mg(OH)_2$, but in vermiculite it is thought to be octahedrally coordinated, 6 H_2O about Mg^{2+} (84–86). The basal spacing of vermiculite varies from 1.4 to 1.5 nm with the nature of the interlayer cation and its hydration. The cation-exchange capacity of vermiculite is relatively high and may exceed that of montmorillonite.

Regularly interstratified (1:1) chlorite and vermiculite has been attributed to the mineral corrensite (87). Corrensite [12173-14-7] is being discovered in greater abundance, and its distribution as sedimentary rocks, especially carbonate rocks, is being studied in greater detail (88–89).

Cookeite [1302-92-7], an aluminous chlorite containing lithium, has been found in high-alumina refractory clays and bauxites (90).

Attapulgite and Sepiolite. Attapulgite, named from its occurrence at Attapulgus, Georgia, and sepiolite (meerschaum), named from the Greek word for cuttle fish (whose bones are light and porous) possess chainlike structures, or combination chain–sheet structures (12). The attapulgite structure is similar to palygorskite minerals resembling cardboard, paper, leather, cork, or even fossil skin.

These clays have distinctive uses and properties not shown by platy clay minerals. The Georgia–Florida deposits originated from evaporating sea water (91). Attapulgite and other palygorskites sorb both cations and neutral molecules. Typical cation-exchange capacities are in the order of 20 meq/100 g dry clay. For chemical analysis of attapulgite see Table 6.

Sepiolite is used in drilling muds where resistance to flocculation in briny water is desired (see Petroleum). Sepiolite and attapulgite are best identified by their 110 reflections, 1.21 and 1.05 nm, respectively, in x-ray powder diffraction (92–93).

Mixed-Layer Clay Minerals. In addition to polymorphism due to the disordering and proxying of one element for another, clay minerals exhibit ordered and random intercalation of sandwiches with one another (12). For example, in mixed-layer clay minerals sheets of illite may be interspersed with montmorillonite, or chlorite with one of the others, either randomly or regularly. Corrensite [12173-14-7] has already been cited as an example of a 1:1, ie, regular, alternation of chlorite and vermiculite. Random mixing may consist of two types: (1) where there is little deviation from a mean ratio between two participants intercalated at random which yields a relatively sharply defined intermediate basal spacing between those of the two end members, and (2) a wide ratio between the two that are intercalated, yielding a wide band of spacings. Mixed layering originates, presumably, by either the degradation in random layers of one species, such as the random stripping of potassium from illite by H^+ ions, or by the random precipitation, reconstitution, or growth of layers of a different guest mineral within a layered host. Mixed layering is not restricted to clay minerals where it has been widely observed, but probably also occurs in numerous other mineral groups.

AMORPHOUS AND MISCELLANEOUS GROUPS

Allophane and Imogolite. Allophane is an amorphous clay that is essentially an amorphous solid solution of silica, alumina, and water (47). It may be associated with halloysite or it may occur as a homogeneous mixture with evansite, an amorphous solid solution of phosphorus, alumina, and water. Its composition, hydration, and properties vary. Chemical analyses of two allophanes are given in Table 7.

The index of refraction of allophane ranges from below 1.470 to over 1.510, with a modal value about 1.485. The lack of characteristic lines given by crystals in x-ray diffraction patterns and the gradual loss of water during heating confirm the amorphous character of allophane.

Allophane has been found most abundantly in soils and altered volcanic ash (60,94–95), but may be considerably more widespread than has been recognized. Quantitative estimation in mixtures depends upon a much higher rate of solubility than of its crystalline analogues. It usually occurs in spherical form but has been also observed in fibers.

Imogolite [*12263-43-3*] is an uncommon, thread-shaped paracrystalline clay mineral assigned a formula $1.1\ SiO_2.Al_2O_3.2.3$–$2.8H_2O$ (96). It can be classified as intermediate between allophane and kaolinite.

Table 7. Chemical Analyses of Allophane, %

Component	1^a	2^b
SiO_2	32.30	4.34
Al_2O_3	30.41	41.41
Fe_2O_3	0.23	0.86
MgO	0.29	0.22
CaO	0.02	0.20
$K_2O + Na_2O$	0.10	0.10
TiO_2	none	none
CuO	1.60	1.80
ZnO	4.06	4.30
CO_2	0.65	2.07
P_2O_5	0.02	9.23
SO_3	0.21	0.08
H_2O removed		
at 105°C	16.38	20.92
above 105°C	14.43	14.43
Total	*100.70*	*99.96*

[a] Allophane, Monte Vecchio, Sardinia (47).
[b] Allophane–evansite, Freienstein, Styria (47).

High-Alumina Clay Minerals. Several hydrated alumina minerals should be grouped with the clay minerals because the two types may occur so intimately associated as to be almost inseparable. Diaspore and boehmite, both $Al_2O_3.H_2O$ (Al_2O_3, 85%; H_2O, 15%) are the chief constituents of diaspore clay, which may contain over 75% Al_2O_3 on the raw basis (23). Gibbsite, $Al_2O_3.3H_2O$ (Al_2O_3, 65.4%; H_2O, 34.6%), and cliachite [12197-64-7], the so-called amorphous alumina hydrate (much cliachite is probably cryptocrystalline), as well as the monohydrates, occur in bauxite [1318-16-7] (29,31,97), bauxitic kaolin, and bauxitic clays (98–99).

The hydrated alumina minerals usually occur in oolitic structures (small spherical to ellipsoidal bodies the size of BB shot, about 2 mm in diameter) and also in larger and smaller structures. They impart harshness and resist fusion or fuse with difficulty in sodium carbonate, and may be suspected if the raw clay analyzes at more than 40% Al_2O_3. Their optical properties are radically different from those of common clay minerals, and their x-ray diffraction patterns and differential thermal analysis curves are distinctive.

High alumina minerals are found where intense weathering and leaching has dissolved the silica. It is generally believed that a very humid, subtropical climate is required for this (lateritic) stage of weathering.

BIBLIOGRAPHY

"General Survey" under "Clays" in *ECT* 1st ed., Vol. 4, pp. 24–38, by W. D. Keller, University of Missouri; "Survey" under "Clays" in *ECT* 2nd ed., Vol. 5, pp. 541–560, by W. D. Keller, University of Missouri.

1. ASTM Committee on Standards *J. Am. Ceram. Soc.* **11,** 347 (1928).
2. F. H. Norton, *Refractories*, 2nd ed., McGraw-Hill Book Co., Inc., New York, 1962.
3. R. E. Grim, *Applied Clay Mineralogy*, McGraw-Hill Book Co., Inc., New York, 1962.
4. H. Wilson, *Ceramics—Clay Technology*, McGraw-Hill Book Co., Inc., New York, 1927.
5. H. Ries, *Clays, Their Occurrences, Properties, and Uses,* John Wiley & Sons, Inc., New York, 1927, p. 1.
6. H. Ries, "Clay," in *Industrial Minerals and Rocks,* American Institute of Mining and Metallurgical Engineers, 1937, pp. 207–242.
7. M. Goldman and J. I. Tracey, Jr., *Econ. Geol.* **41,** 567 (1946).
8. T. Sudo, *Mineralogical Study on Clays of Japan,* Maruzen Co., Ltd., Tokyo, 1959.
9. R. E. Grim, R. H. Bray, and W. F. Bradley, *Am. Mineral.* **22,** 813 (1937).
10. M. D. Foster, *U.S. Geol. Surv. Prof. Pap.* 414-A, (1962).
11. J. E. Brydon, J. S. Clark, and V. Osborne, *Can. Mineral.* **6,** 595 (1961).
12. G. Brown, ed., *The X-ray Identification and Crystal Structures of Clay Minerals,* Mineralogical Society, London, 1961.
13. T. F. Bates, F. A. Hildebrand, and A. Swineford, *Am. Mineral.* **35,** 463 (1959).
14. C. E. Marshall, *The Colloid Chemistry of the Silicate Minerals,* Academic Press, Inc., New York, 1949.
15. I. T. Rosenqvist, "Marine Clays and Quick Clay Slides in South and Central Norway," *Guide to Exc. No. C-13, 21st International Geological Congress, Oslo, 1960.*
16. H. Ries, *Am. J. Sci.* **44**(4), 316 (1917).
17. C. DeKimpe, M. C. Gastuche, and G. W. Brindley, *Am. Mineral.* **46,** 1370 (1961).
18. J. Linares and F. Huertas, *Science* **171,** 896 (1971).
19. R. Roy, *C.N.R.S. Groupe Fr. Argiles C.R. Reun. Etud.* **105,** 83 (1962).
20. B. Velde, *Clays and Clay Minerals in Natural and Synthetic Systems,* Elsevier, New York, 1977.
21. C. E. Weaver and L. D. Pollard, *The Chemistry of Clay Minerals,* Elsevier, New York, 1973.
22. W. D. Keller, *Principles of Chemical Weathering,* Lucas Bros., Columbia, Mo., 1957.
23. W. D. Keller, "Processes of Origin of the Clay Minerals," *Proceedings of the Soil Clay Mineral Institute,* Virginia Polytechnic Institute, Blacksburg, Va., 1962.

24. S. H. Patterson and B. F. Buie, "Field Conference on Kaolin and Fuller's Earth, Nov. 14–16, 1974" *Guidebook 14, Georgia Dept. Natl. Resources,* Atlanta, Ga., 1974.
25. W. D. Keller, J. F. Westcott, and A. O. Bledsoe, *Proceedings of the 2nd Conference of Clays and Clay Minerals, National Academy of Science-National Research Council Publication 327,* 1954, pp. 7–46.
26. S. H. Patterson and H. W. Murray, "Clays," *Industrial Minerals and Rocks,* 4th ed., AIME, 1975, pp. 519–585.
27. J. Vachtl, *Proc. XXIII Int. Geol. Cong.* **15,** 13 (1968).
28. M. Kuzvart and M. Bohmer, *Prospecting and Exploration of Mineral Deposits* Academia Press, Prague, Czechoslovakia, 1978.
29. Gy. Bardossy, *Acta Geol. Acad. Sci. Hung.* **6**(1–2), 1 (1959).
30. R. C. Mackenzie, ed., *The Differential Thermal Investigation of Clays,* Mineralogical Society, London, 1957.
31. S. H. Patterson, *U.S. Geol. Surv. Bull.* **1228** (1967).
32. C. M. Warshaw and R. Roy, *Bull. Geol. Soc. Am.* **72,** 1455 (1961).
33. T. R. P. Gibb, Jr., *Optical Methods of Chemical Analyses,* McGraw-Hill Book Co., Inc., New York, 1942, pp. 243–319.
34. C. S. Ross and S. B. Hendricks, *U.S. Geol. Surv. Prof. Pap.* **205-B,** 23 (1945).
35. G. T. Faust, *U.S. Bur. Mines Rep. Invest.* **3522,** (1942).
36. E. A. Hauser and M. B. Leggett, *J. Am. Chem. Soc.* **62,** 1811 (1940).
37. J. B. Page, *Soil Sci.* **51,** 133 (1941).
38. C. E. Marshall, *Z. Kristallogr. Mineral.* **90,** 8 (1935).
39. P. F. Kerr, ed., "Reference Clay Minerals," *American Petroleum Institute Research Project 49,* 1950.
40. C. S. Piper, *Soil and Plant Analysis,* Interscience Publishers, Inc., New York, 1944.
41. H. Beutelspacher and H. Van der Marel, *Atlas of Electron Microscopy of Clay Minerals and Their Admixtures,* Elsevier, New York, 1968.
42. R. L. Borst and W. D. Keller, *Proc. Int. Clay Conf. Tokyo* 871 (1969).
43. W. D. Keller and R. F. Hanson, *Clays Clay Miner.* **23,** 201 (1975).
44. W. D. Keller, *Clays Clay Miner.* **25,** 311 (1977).
45. J. Konta, *Am. Mineral.* **46,** 289 (1961).
46. W. D. Keller, *U.S. Geol. Surv. Bull.* **1150,** (1962).
47. C. S. Ross and P. F. Kerr, *U.S. Geol. Surv. Prof. Pap.* **185-G,** 135 (1934).
48. L. T. Alexander and co-workers, *Am. Mineral.* **28,** 1 (1943).
49. C. S. Ross and P. F. Kerr, *U.S. Geol. Surv. Prof. Pap.* **165-E,** 151 (1930).
50. G. T. Faust, *Am. Mineral.* **40,** 1110 (1955).
51. R. E. Grim, *Clay Mineralogy,* McGraw-Hill Book Co., Inc., New York, 1968.
52. M. H. Thornberry, *Mo. Univ. Sch. Mines Metall. Bull.* **8**(2), 34 (1925).
53. C. S. Ross, *Proc. U.S. Nat. Mus.* **69,** 1 (1926).
54. W. D. Miller and W. D. Keller, "Differentiation between Endellite–Halloysite and Kaolinite by Treatment with Potassium Acetate and Ethylene Glycol," *Proceedings of the 10th National Conference of Clays and Clay Minerals,* 1961, 244–256.
55. F. V. Chukhrov, and B. B. Zvyagin, *Proc. Int. Clay Conf., Jerusalem,* **1,** 11 (1966).
56. L. B. Sand, *Am. Mineral.* **41,** 28 (1956).
57. W. D. Keller, *Clays Clay Miner.* **26,** 1 (1978).
58. W. D. Keller, *Tenth National Conference of Clays and Clay Minerals,* Pergamon Press, New York, 1963, pp. 333–343.
59. W. D. Keller, *Bull. Am. Assoc. Petrol. Geologists* **40,** 2689 (1956).
60. M. Fieldes, *N.Z. J. Sci. Technol.* **37,** 336 (1955).
61. G. W. Brindley and M. Nakahira, *J. Am. Ceram. Soc.* **42,** 311 (1959).
62. M. Slaughter and W. D. Keller, *Am. Ceram. Soc. Bull.* **38,** 703 (1959).
63. F. M. Wahl, R. E. Grim, and R. B. Graf, *Am. Mineral.* **46,** 1064 (1961).
64. C. S. Ross, *Am. Mineral.* **31,** 411 (1946).
65. C. E. Weaver, *Am. Mineral.* **43,** 839 (1958).
66. M. Slaughter and J. W. Earley, *Geol. Soc. Am. Spec. Pap.* **83,** (1965), 95 pp.
67. H. S. Yoder and H. P. Eugster, *Geochim. Cosmochim. Acta* **8,** 225 (1955).
68. R. E. Grim. *Bull. Am. Assoc. Petrol. Geologists* **31,** 1491 (1947).
69. A. A. Levinson, *Am. Mineral.* **40,** 41 (1955).

70. J. F. Burst, Jr., *Proceedings of the 6th National Conference of Clays and Clay Minerals,* Pergamon Press, Inc., New York, 1959, pp. 327–341.

71. W. D. Keller, Diagenesis of Clay Minerals—A Review," *Proceedings of the 11th National Conference of Clays and Clay Minerals, 1962,* Pergamon Press, Inc., New York, 1963.

72. J. F. Burst, Jr., *Bull. Am. Assoc. Petrol. Geologists* **42,** 310 (1958).

73. J. F. Burst, Jr., *Am. Mineral.* **43,** 481 (1958).

74. J. Hower, *Am. Mineral.* **46,** 313 (1961).

75. E. G. Wermund, *Bull. Am. Assoc. Petrol. Geologists* **45,** 1667 (1961).

76. P. M. Hurley and co-workers, *Bull. Am. Assoc. Petrol. Geologists* **44,** 1793 (1960).

77. P. E. Cloud, *Bull. Am. Assoc. Petrol. Geologists* **39,** 484 (1955).

78. W. D. Keller, *Fifth National Conference of Clays and Clay Minerals, National Academy of Science-National Research Council Publication 566,* 1958, pp. 120–129.

78a. A. L. Albee, *Am. Mineral.* **47,** 851 (1962).

79. S. W. Bailey and B. E. Brown, *Am. Mineral.* **47,** 819 (1962).

80. W. F. Bradley, *Second National Conference on Clays and Clay Minerals, National Academy of Science-National Research Council Publication 327,* 1954, pp. 324–334.

81. G. W. Brindley and F. H. Gillery, *Am. Mineral.* **41,** 169 (1956).

82. M. J. Shen and C. I. Rich, *Soil Sci. Soc. Am. Proc.* **26,** 33 (1962).

83. M. Slaughter and I. Milne, eds., *Proceedings of the 7th National Conference of Clays and Clay Minerals,* Pergamon Press, Inc., New York, 1960, pp. 114–124.

84. W. A. Bassett, *Am. Mineral.* **44,** 282 (1959).

85. A. M. Mathieson, *Am. Mineral.* **43,** 216 (1958).

86. G. F. Walker, *Clay Miner. Bull.* **3,** 154 (1957).

87. W. F. Bradley and C. E. Weaver, *Am. Mineral.* **41,** 497 (1956).

88. M. N. A. Peterson, *Am. Mineral.* **46,** 1245 (1961).

89. M. N. A. Peterson, *J. Geol.* **70,** 1 (1962).

90. H. A. Tourtelot and E. F. Brenner-Tourtelot, "Lithium in Flint Clay, Bauxite, Related High-Alumina Materials and Associated Sedimentary Rocks in the United States—a Preliminary Report," *U.S. Geol. Survey Open File Report 77-786,* 1977.

91. S. H. Patterson, *U.S. Geol. Surv. Prof. Pap.* **828,** (1974).

92. S. Caillere and S. Henin, "The X-ray Identification and Crystal Structures of Clay Minerals," *Mineralogical Society Great Britain Monograph, 325–342,* 1961, Chapt. VIII.

93. W. F. Bradley, *Am. Mineral.* **25,** 405 (1940).

94. K. S. Birrell and M. Fieldes, *N.Z. J. Soil Sci.* **3,** 156 (1952).

95. W. A. White, *Am. Mineral.* **38,** 634 (1953).

96. N. Yoshinga and S. Aomine, *Soil Sci. Plant Nutr. Tokyo* **8,** 22 (1962).

97. I. Valeton, *Bauxites,* Elsevier, 1972.

98. A. F. Frederickson, ed., "Problems of Clay and Laterite Genesis," *AIME Symposium Volume,* American Institute Mining and Mechanical Engineers, 1952.

99. M. Gordon, Jr., J. I. Tracey, Jr., and M. W. Ellis, *U. S. Geol. Surv. Prof. Pap.* **299,** (1958).

General References

References 3, 12, 14, 21–22, 24, 26, 30–31, 41, 51, 91, and 97–98 of the numbered bibliography may also be considered general reference works.

W. D. KELLER
University of Missouri, Columbia

USES

Clay materials are composed of extremely small particles of clay minerals. These minerals are generally crystalline, but in some cases their organization is so poor that diffraction indicates them to be amorphous. Clay minerals are essentially hydrous aluminum silicates, with iron or magnesium proxying wholly or in part for the aluminum in some, and with alkalies or alkaline earths present as essential constituents in others. Furthermore, clays may contain varying amounts of so-called nonclay minerals, such as quartz, calcite, feldspar, and pyrites.

The following factors control the properties of clay materials: the identity and relative abundance of the clay mineral components; the identity of the nonclay minerals and their shape, relative abundance, and particle-size distribution; the kind and amount of organic material; the kind and amount of exchangeable ions and soluble salts; and the texture, which refers to the particle-size distribution of the constituent particles, their shape, their orientation in space with respect to each other, and the forces binding them together.

Ceramic Products

Ceramic is defined as "relating to the art of making earthenware or to the manufacture of any or all products made from earth by the agency of fire, as glass, enamels, cements." To this list should be added brick, tile, heat-resisting refractory materials, porcelain, pottery, chinaware, and earthenware. In general, ceramic ware is produced by plasticizing the clay by the addition of water so that it may be shaped or formed by some means into the desired object. Ceramic products may also be formed by dispersing the clay in water to form a slip which is then cast into a plaster mold. In the case of porcelain enamel the slip is sprayed on a metal surface and then fired. After being shaped, the object is dried to increase its strength so that it may be handled, and is then fired at elevated temperatures (frequently in the range of 1090°C) until there has been some vitrification or fusion of the components to develop a glassy bond that makes the shape permanent and strong so that the object does not disintegrate in water (see Ceramics).

Properties. *Plasticity.* Plasticity may be defined as the property of a material that permits it to be deformed under stress without rupturing and to retain the shape produced after the stress is removed. When water is added to dry clay in successive increments, the clay tends to become workable, that is, readily shaped without rupturing. The workability and retention of shape develop within a very narrow moisture range.

Plasticity can be measured by determination of (a) the water of plasticity, ie, the amount of water necessary to develop optimum plasticity (judged subjectively by the operator) or the range of water content in which plasticity is demonstrated (Atterberg limits); (b) the amount of penetration of an object, frequently a needle or some type of plunger, into a plastic mass of clay under a given load or rate of loading and at varying moisture contents; and (c) the stress necessary to deform the clay and the maximum deformation the clay will undergo before rupture at different moisture contents and with varying rates of stress application.

In ceramics, plasticity is usually evaluated by means of the water of plasticity. Ranges of values for common clay minerals are given in Table 1.

Table 1.   Ceramic Properties of Clay Minerals

Property	Kaolinite [1318-74-7]	Illite [12173-60-3]	Halloysite [12244-16-5]	Montmorillonite [1318-93-0]	Attapulgite [1337-76-4]	Allophane [12172-71-3]
water of plasticity[a], %	8.9–56.3	17–38.5	33–50	83–250	93	
green strength[b], kg/cm^2	0.34–3.2	3.2	5	5[c]		
dry strength[a]	69–4840	1490–7420	1965	1896–5723	4482	
linear shrinkage[a], %						
drying[d]	3–10	4–11	7–15	12–23	15	
firing	2–17	9–15	±20	±11	±23	+50

[a] Ref. 1.

[b] Ref. 2.

[c] Calcium montmorillonite.

[d] Values computed as % of dry length after drying test pieces of 6.45 cm^2 cross section in an oven at 105°C for 5 h.

Each clay mineral group can be expected to show a range of values, since particle size, exchangeable cation composition, and crystallinity of the clay mineral also exert an influence. Nonclay mineral component, soluble salts, organic compounds, and texture can also affect the water of plasticity.

In general, a relatively low value for water of plasticity is desired in ceramics and hence kaolinite, illite, and chlorite clays have better plasticity characteristics than attapulgite or montmorillonite. The plasticity values of the first group are changed only slightly by variations in the exchangeable cation composition. However, sodium gives lower values than calcium, magnesium, potassium, and hydrogen. In the case of montmorillonite, the water of plasticity varies considerably with the nature of the exchangeable cation, with sodium giving higher values than the other common exchangeable cations.

Clays composed only of clay minerals may have higher water of plasticity values than desired. Consequently, the presence of nonclay minerals in substantial amounts or the addition of such material serving to reduce the water of plasticity may improve the working characteristics of a clay.

Plasticity in clay–water systems is caused by a bonding force between the particles and water, a lubricant, which permits some movement between the particles under the application of a deforming force. The bonding force is in part a result of charges on the particles. It is now generally believed that the water immediately adsorbed on basal clay mineral surfaces is composed of water molecules in a definite orientation; that is, it is not fluid water (3). This water, composed of water molecules in a definite configuration, also serves as a bond between clay mineral particles. The orientation of the water would decrease outward from the clay mineral surface so that with little adsorbed water the bonding would be strong, whereas with much adsorbed water the bond would be weak or nonexistent. It has been suggested (4–5) that optimum plasticity develops when all the requirements of particle surfaces for oriented water are satisfied and there is a little additional water less rigidly fixed which can act as a lubricant when a deforming force is applied. Furthermore, influence of adsorbed cations on plasticity is to a considerable degree the result of their effect on the extent and perfection of the development of crystalline water, rather than directly on the bonding force between the particles.

The *green strength* of a clay body is the strength measured as transverse breaking strength that prevails while the plasticizing water is still present. As water is continuously added to a dry clay, the strength increases up to a maximum and then decreases. The strength at water of plasticity is, in general, lower than the maximum strength. Values for the common clay minerals are given in Table 1.

As in the case of plasticity, green strength values would be expected to vary with exchangeable cation composition to only a slight degree for kaolinite, illite, and chlorite clays, and to a considerable degree for montmorillonite clays. In the latter, sodium would be expected to provide higher maximum green strength than other common cations. Poorly crystallized varieties of kaolinite and illite yield higher green strength than well-crystallized varieties. The presence of considerable quantities of nonclay minerals reduces the green strength, whereas small amounts may actually increase the strength because they permit the development of a more uniform clay body. Green strength is also related to the particle size with smaller particle size providing the higher strength. If the clay mineral particles develop preferred orientation in certain directions during formation of the ware, the breaking strength will be somewhat greater in the directions transverse to such preferred orientation.

Drying Properties. *Drying shrinkage* is the reduction in size measured either in length or volume that takes place when a mass is dried in order to drive off the pore water and the adsorbed water. Values are given in Table 1.

In general, the drying shrinkage increases as the water of plasticity increases, and for a particular clay mineral it increases as the particle size decreases. In addition, drying shrinkage varies with the degree of crystallinity; thus ball clay (6), which contains relatively poorly ordered kaolinite, shows values at the high end of the range shown in Table 1. Similarly, mixed-layer clays would be expected to have relatively higher drying shrinkage than clays of similar composition, but with the clay mineral mixing being that of discrete particles. The nature of the adsorbed cation causes variations in the amount of drying shrinkage only as it affects the water of plasticity.

The presence of nonclay minerals tends to reduce drying shrinkage depending on their shape, particle-size distribution, and abundance. Granular particles with considerable particle-size range are most effective. The presence of nonclay minerals on the order of about 25% is generally desirable in ceramic bodies to improve their shrinking characteristic. Drying shrinkage is also related to texture; for example, if the clay mass shows parallel orientation of the basal plane surfaces of the clay minerals, shrinkage in the direction at right angles to the basal planes would be substantially greater than in the direction parallel to them (7).

In the initial drying phase of a clay body the volume shrinkage is about equal to the volume of water lost. Beyond a given moisture content there is either no further shrinkage or only a very small amount as the water is lost. The water lost during the shrinkage interval is called shrinkage water and seems to separate the component particles. The critical point at which shrinkage stops is reached when the moisture film around the particle becomes so thin that the particles touch one another sufficiently and shrinkage can go no farther. The water loss following the shrinkage period is called pore water.

In the production of ceramic ware the shape of the ware has to be retained after drying and should be free from cracks and other defects. Excessively slow drying and excessive control of humidity, airflow, etc, should be avoided. In general, clays containing moderate amounts of nonclay minerals are easier to dry than those composed wholly of clay minerals. Furthermore, clays composed of illite, chlorite, and kaolinite are relatively easier to dry than those composed of montmorillonite.

Dry strength is measured as the transverse breaking strength of a test piece after drying long enough, usually at 105°C, to remove almost all the pore and adsorbed water. Values are given in Table 1 and usually show a large range because of variations in particle-size distribution, perfection of crystallinity, and, especially for montmorillonite, the nature of the exchangeable ions.

Large amounts of nonclay mineral components, especially if the particles are well sorted, tend to reduce the dry strength. In general, the dry strength is higher when sodium is the adsorbed cation. The presence of organic material in some clays increases their dry strength and this appears to be partly the explanation for the high dry strength for some ball clays. A major factor in determining dry strength is the particle size of the clay mineral component; the maximum strength increases rapidly as the particle size decreases.

Firing Properties. Heating clay materials to successively higher temperatures results in the fusion of the material. In the 100–150°C range, the shrinkage and pore waters are lost with the attendant dimensional changes. In general, the rate of oxidation increases with increasing temperature. The oxidation of sulfides, which are present in many clays, frequently in the form of pyrite, FeS_2, begins between 400 and 500°C.

Beginning at about 500°C and in some cases continuing to 900°C, the hydroxyl water is driven from the clay minerals. The exact temperature, rate, and abruptness of loss of hydroxyls depend on the nature of the clay minerals and their particle size. Reduction of particle size, particularly if accompanied by poor crystallinity, tends to reduce the temperature interval. In general, kaolinite and halloysite minerals lose their hydroxyls abruptly at 450–600°C. The loss of hydroxyls from the three-layer clay minerals varies greatly with structure and composition, but is generally slower and more gradual than that for kaolinite and halloysite.

The loss of hydroxyls from the clay minerals is usually accompanied by a modification of the structure, but not by its complete destruction. In three-layer clay minerals it is not accompanied by shrinkage, whereas in kaolinite and halloysite loss of hydroxyl water is accompanied by shrinkage, which continues up into the vitrification range.

In the range of 800 to 950°C, the structure of the clay mineral is destroyed, and major firing shrinkage develops. Values for the firing shrinkage are given in Table 1.

The range in shrinkage values is due to variations in the size and shape of the clay mineral particles, the degree of crystallinity, and in the case of the three-layered clay minerals, variations in composition.

At temperatures above about 900°C new crystalline phases develop from all the clay minerals except those containing large amounts of iron, alkalies, or alkaline earths, in which case fusion may result after the loss of structure without any intervening crystalline phase. Frequently, there is a series of new high-temperature phases developing in an overlapping sequence as the mineral is heated to successively higher temperatures. This is followed by complete fusion of the mineral, which takes place in the case of kaolinite at 1650–1775°C. For the three-layered clay minerals, the fusion temperature varies from about 1000 to 1500°C, the lower values being found in materials relatively rich in iron, alkalies, and alkaline earths.

The initial high-temperature phases are frequently related to the structure of the original clay mineral, whereas the later phases developing at higher temperatures are related to its overall composition. In the development of high-temperature phases, nucleation of the new lattice configuration takes place first, followed by a slow gradual growth of the new structure and an increase in its perfection as the temperature is raised above that required for nucleation. Traces of various elements cause substantial changes in the temperature and rate of formation of high-temperature phases.

Miscellaneous. Other important properties are resistance to thermal shock, attack by slag, and thermal expansion in the case of refractories. For whiteware, translucency, acceptance of glazes, etc, may be extremely important. These properties depend on the clay mineral composition and the method of manufacture, such as the forming procedure and intensity of firing.

Raw Material. *Brick.* Almost any clay composition is satisfactory for the manufacture of brick unless it contains a large percentage of coarse stony material that cannot be eliminated or ground to adequate fineness. A high concentration of nonclay material in a silt-size range may cause difficulties by reducing greatly the green and firing strength of the ware. Montmorillonite should be absent or present only in very small amounts; otherwise the shrinkage may be excessive. Clays composed of mixtures of clay minerals with 20–50% of unsorted fine-grain nonclay materials are most satisfactory. Large amounts of iron, alkalies, and alkaline earths, either in the clay minerals or as other constituents, cause too much shrinkage and greatly reduce the vitrification range; thus, a clay with a substantial amount of calcareous material is not desirable. Face bricks, which are superior, can be made from similar material; it is even more desirable to avoid the detrimental components mentioned above. For a light buff or gray face brick, kaolinite clay is preferred.

Tile. Roofing and structural tiles are usually made from the same material as face brick. Drain tiles have a high porosity, which is frequently attained by firing at relatively low temperatures. Frequently, drain tiles are made from clays with about 75% of fine-grained nonclay mineral material, in addition to components that provide high green and dry strength and a low fusion point. Wall and floor tiles are frequently made of mixtures with talc and kaolin as major components.

Terracotta, Stoneware, Sewerpipe, Paving Brick. Clays composed of mixtures of clay minerals containing 25–50% fine-grained unsorted quartz are well suited for the manufacture of terracotta, stoneware, sewerpipe, and paving brick. A small amount of montmorillonite can be tolerated, but a large amount gives undesirable shrinkage and drying properties. In general, clays with low shrinkage, good plastic properties, and a long vitrification range should be used.

Whiteware. Porcelain and dinnerware are made up of about equal amounts of kaolin, ball clay, flint (ground quartz), feldspar, or some other white-burning fluxing material such as talc and nepheline. The kaolin clay is composed of well-crystallized particles of kaolinite. Ball clays are white-burning, highly plastic, and easily dispersible. They provide the plasticity necessary in the forming of the ware and adequate green and dry strength for handling. The chief component of most ball clays is extremely fine-grained and relatively poorly organized kaolinite. However, some ball clays are known, for example, in South Devonshire in Great Britain, that contain remarkably well-ordered kaolinite. Some ball clays also contain small amounts of illite and/or small amounts of montmorillonite, which may add to their desired properties. Also, many ball clays contain a small but appreciable amount of organic material that appears to enhance the desired properties. Small amounts of bentonite are also used in whiteware bodies as replacements of ball clay to increase dry strength.

Porcelain Enamel. The slurry used in enameling is commonly composed of ball clay, frits, and coloring pigments. The frits are finely ground particles of prepared glass with a low fusion point.

Refractories. Refractory products are prepared from a wide variety of naturally occurring materials such as chromite and magnesite or from clays predominantly composed of kaolinite. For many refractory uses a somewhat lower fusion point than that provided by pure kaolinite may be adequate, so that clay materials with a moderate amount of other components as, for example, illite, may be satisfactory. In some cases this may be desirable, as in the case of brick used in lining steel ladles where it seems to provide the necessary reheat expansion.

High-alumina clays are also used extensively for the manufacture of special types of refractories (see Refractories). Such clays may be bauxites, composed essentially of aluminum hydrate minerals (gibbsite, boehmite) and kaolinite, or they may be diaspore clays.

An interesting type of clay used widely in the manufacture of refractories is so-called flint clay. As its name suggests, this clay is very hard and has very slight plasticity even when finely ground. Flint clays are essentially pure, extremely fine-grained kaolinite clays. In some cases, the hardness appears to be due to the presence of a small amount of free silica acting as a cement, whereas in other cases it is the result of an intergrowth of extremely small kaolinite particles.

Molding Sands

Molding sands, which are composed essentially of sand and clay, are used extensively in the metallurgical industry for the shaping of metal by the casting process. Using a pattern, a cavity of the desired shape is formed in the sand, and into this molten metal is poured and then allowed to cool.

The molding sand may be a natural sand containing clay or a synthetically prepared mixture of clean quartz sand and clay. Synthetic sands are widely used because they can be prepared to meet property specifications and their properties are more easily controlled as they are used. Granular particles other than quartz sand, for example, calcined clay, olivine, zircon, or chromite, may be used in rare special instances. Ground bituminous coal (sea coal) and cereal binders may be added to the sand–clay mixture to develop certain properties.

A small amount of water (tempering water) must be added to the molding sand to impart plasticity to develop cohesive strength so that the sand can be molded around the pattern, and to give it sufficient strength to maintain the cavity after the pattern is removed and while the metal is poured into it. These properties vary greatly with the amount of tempering water as well as with the nature of the clay bond.

In foundry practice the same molding sand is used over and over again. The high temperature of the metal dehydrates and vitrifies some of the clay, and fresh clay must be added continuously as the sand is used.

The only adequate test for the satisfactory use of a clay in bonding molding sands is the result obtained by actual use in foundry practice. Laboratory tests only eliminate worthless clays.

The best model of the bonding action of the clays in molding sand is that of a wedge-and-block bond at the interface of the sand grains. The bonding action of clay and water is not that of a glue or adhesive causing the grains to adhere to each other.

The value of a clay for bonding molding sand is usually determined by the green and dry compression strengths of mixtures with varying amounts of the clay and to which varying amounts of tempering water have been added. Other properties such as the bulk density, flowability, permeability, and hot strength may be important. The American Foundrymen's Society has published standard procedures for determining the properties of bonding clays (8).

Raw Materials.   The bentonites, composed essentially of montmorillonite and used extensively in bonding molding sands, are of two types. The type carrying sodium as a principal exchangeable cation is produced largely in Wyoming. The calcium-carrying type is produced in Mississippi and in many countries outside the United States, such as England, Germany, Switzerland, Italy, U.S.S.R., South Africa, India, and Japan. The natural calcium montmorillonite bentonites are occasionally treated with various sodium compounds so that their properties are similar to the properties of the natural sodium bentonites produced in Wyoming.

Plastic clays composed largely of poorly crystalline kaolinite but with small amounts of illite, and at times montmorillonite, are widely used in bonding molding sands especially in the United States. These clays are called fireclays because of the relatively high refractoriness. In the United States such clays are mined extensively in Illinois and Ohio from beds of carboniferous age, where they occur directly beneath beds of coal (underclays).

The third type of clay used in foundries is composed essentially of illite. Most illite clays have a bonding strength and plasticity too low for bonding use, but there are some varieties that have such properties approaching those of montmorillonite. The illite in such clays is fine-grained, poorly organized, and frequently associated with mixed-layer assemblages containing montmorillonite. Illite bonding clays are produced extensively in Illinois.

Properties.   *Compression Strengths.*   *Green compression strengths* in the range from about 35 to 75 kPa (5–11 psi) are desired in actual practice.

Figure 1 shows the maximum green strength for varying amounts of each type of clay up to 15%. This diagram refers to many uses of clays.

Calcium montmorillonite clay gives the highest green compression strength, whereas kaolinite and illite have about the same strength, which is considerably lower than that of either montmorillonite clay. In sands bonded with the calcium montmorillonite, the strength increases approximately in proportion to the amount of clay up to about 8%. Additional clay up to 10% causes only a very slight increase in strength, and clay in excess of this amount causes no further increase.

Sands bonded with the sodium montmorillonite show a similar relation between maximum green compression strength and the amount of clay, except that the reduction in strength per unit of added clay above about 8% is less than that for the calcium montmorillonite. The kaolinite- and illite-bonded sands show a continuing increase in maximum green compression strength up to 15% of clay, but less above 10% for kaolinite and 12% for illite.

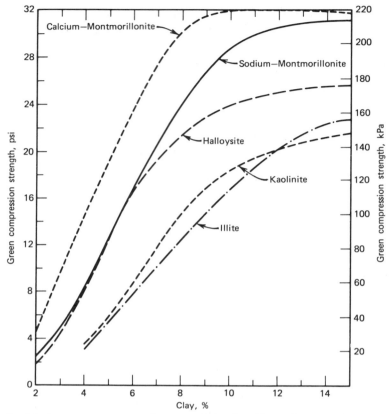

Figure 1.   Maximum green compression strength developed by each type of clay in relation to the amount of clay in the sand clay mixture (9).

Considerably lower dry compression strength is developed by calcium montmorillonite (10) than by sodium montmorillonite clays. Furthermore, calcium montmorillonite clays require a certain minimum amount of tempering water to develop any dry strength which increases with the amount of tempering water up to a maximum value, which is reached abruptly. A striking feature of this clay is that maximum dry compression strength is about the same for all mixtures with more than about 6% clay.

Illite clays develop relatively high dry compression strength (10). Maximum strength is developed with relatively small amounts of tempering water.

The maximum dry compression strength obtained with kaolinite clays is reached abruptly and increases with the amount of clay in the mixture (10). In general, kaolinites yield less dry strength than illites and slightly more water is required to develop equivalent strength in a given mixture.

Drilling Fluids

In the drilling operation a fluid is pumped down through the hollow drill stem, emerging at the bottom of the hole through "eyes" in the bit. During actual drilling it circulates continuously to remove cuttings and rises to the surface in the annular space between the drill stem and the walls of the hole and then flows into a pit for removal of cuttings and entrained gas.

The removal of cuttings is readily facilitated by a fluid with a viscosity higher than that of water. Furthermore, it must be possible to pump the drilling mud. Its viscosity, on the order of 15 mPa·s(= cP), is controlled and it should be markedly thioxotropic; that is, should have gel strength. This is necessary so that the cuttings do not settle to the bottom of the hole and freeze the drill stem when pumping or agitation of the drill stem ceases temporarily.

The drilling fluid serves to confine the formation fluids. Normally such fluids are under pressure at least equal to that of a column of water of equal depth so that densities in excess of water alone are required. However, much higher densities frequently are required necessitating muds of high weight per unit volume which are attained by the addition of such materials as barites. The drilling mud must have sufficient viscosity and gel strength to maintain such weighting material in suspension. An impervious thin coating must be built up on the wall of the hole in order to impede the penetration of water from the drilling fluid into the formations. High impermeability is particularly important in drilling through formations that are permeable and contain fluids under relatively high pressure; it is also important when drilling water may soften formations and cause them to cave or heave into the hole. Argillaceous material added to the drilling fluid from the penetrated formations should be kept to a minimum because such material tends to increase viscosity.

Viscosity and gel strength of the drilling fluid should not be affected by variations in electrolyte concentration and should be subject to some control by chemical treatment. In some cases, high-lime muds with pH values in the range of 12–13 are used to maintain viscosity and gel strength in the presence of clay added from penetrated formations and of large amounts of electrolytes. These muds also contain a protective colloid that may be a tannin material, humic acid derivative, or gelatinized starch; otherwise the system would be completely flocculated.

The suitability of a clay for use in drilling mud is measured primarily by (a) the number of cubic meters (barrels) of mud with a given viscosity (usually 15 mPa·s or cP) obtained from one ton of clay in fresh water and in salt water; (b) the difference in gel strength taken immediately after agitation and 10 min later; (c) the wall-building properties, as measured by the water loss through a filter paper when a 15-mPa·s (= cP) clay is subjected to a pressure of 690 kPa (100 psi) for 30 min; and (d) the thickness of the filter cake produced in the standard water-loss test.

Clays and soils with more than a very small amount of nonclay minerals, particularly in silt and sand sizes, are not suitable for drilling muds because such materials dilute the desired properties and have abrasive action on pumps and other drilling equipment. Local soils and clays of low nonclay mineral content are frequently used to start a well, with special clays being added with increasing depth to control the properties of the drilling fluid.

The most widely used clay for drilling fluids is a bentonite from Wyoming composed of montmorillonite carrying sodium as the major exchangeable cation. This clay gives petroleum yields in excess of 15.9 m^3 per metric ton of clay (100 bbl/t) so that only about 5% clay is adequate to produce the desired viscosity. It also has very high gel strength and low permeability. This bentonite is used worldwide because of its valuable property of producing an impervious clay layer, even when very thin, on the wall of the drilling hole.

Recently, clays with yields up to 47.7 m³/t of mud (300 bbl/t) have been developed by the addition of certain polymers to bentonites.

Muds containing clays composed of attapulgite and sepiolite show small variations in viscosity gel strength with large variations in electrolyte content, and considerable tonnage of these clays are consumed annually. In some deep drilling operations, relatively high temperatures, 370°C, are encountered to which sepiolite muds are especially resistant (11).

In oil-based muds, where oil rather than water is the fluid, clay mineral–organic complexes are used. In such complexes the clay materials are coated with various organic compounds that change them from hydrophilic to oleophilic (see Petroleum, drilling fluids).

Catalysis

Catalysts made from various clay minerals are extensively used (12) in the cracking of heavy petroleum fractions (see Catalysis; Petroleum). These catalysts are produced from halloysites, kaolinites, and bentonites composed of montmorillonite. In the United States, montmorillonite clays from Arizona and Mississippi, halloysite from Utah, and kaolinite clays from sedimentary formations in Georgia are used for this purpose. These clays must be low in iron and substantially free from various elements such as heavy metals which would favor either catalyst poisoning or an unfavorable product distribution.

Preparation.   The process details are proprietary. In one process the clays are first treated with an acid solution, usually sulfuric, at moderately elevated temperatures. Subsequently, the clay is washed to remove alkalies and alkaline earths and partially reduce the alumina content. In another process, a mixture of kaolin and caustic soda is heated under controlled conditions to form a zeolite-type component. In both processes, the clay is prepared in the form of pellets or of "fluid" powder, and calcined at moderately elevated temperatures.

Oil Refining and Decolorization

Clay materials are used widely to decolorize, deodorize, dehydrate, and neutralize mineral, vegetable, and animal oils. Decolorization is generally the major objective of such processes. The oil is filtered through a granular product of 250–2000 μm (10–60 mesh) particles, or placed into contact with finely ground clay of approximately 741 μm (−200 mesh); it is then separated from the clay by a filter pressing operation. The percolation is essentially a low-temperature process, whereas the contact process takes place in the range of 150–300°C.

A wide range of clay materials have been used for decolorization, ranging from fine-grained silts to clays composed of almost pure clay minerals. The materials may be substantially crude clay, such as fuller's earth, or clay that has been prepared by chemical and physical treatment. The name fuller's earth comes from the use of these clays in cleaning or fulling wool with a water slurry of earth, whereby oil and dirt particles were removed from the wool. At the present time, this term is applied to any clay that has adequate decolorizing and purifying capacity to be used commercially in oil refining without chemical treatment. It does not indicate composition or origin.

Clays composed of attapulgite and some montmorillonites possess superior decolorizing powers. For preparation, these clays are dried at 200–315°C and ground to various sizes. The activity of the attapulgite clay is substantially enhanced by extrusion under high pressure at low moisture contents. Montmorillonite clays are not improved by such extrusion. Since only some montmorillonite clays possess substantial decolorizing properties, they have to be evaluated by field tests. Acid activation enhances the decolorizing power of some montmorillonite clays several-fold. Halloysite clays may have high decolorizing power (13–14), as does an unusual illitic clay from northern Illinois (15). Some sepiolite clays from Spain are excellent decolorizing materials (16). At the present time however, only attapulgite and montmorillonite clays are used for commercial decolorization.

Properties. A relatively small amount of clay must reduce color substantially. Furthermore, oil retention must be low; that is, only a small amount of oil is retained by the clay in the course of the decolorizing process. The latter property is particularly important if the oil cannot be reclaimed from the spent clay by a solvent or distillation. The clay must have good filtration characteristics; that is, the oil must pass through fairly rapidly, but the clay must not unduly bind the filters. In the case of edible oils, the earth must not impart an obnoxious odor or taste to the oil.

Paper

Paper is a thin uniform sheet of finely intermeshed and felted cellulose fibers. The cellulose for the highest-quality paper is obtained from cotton and linen, but for most purposes fibers obtained from a large variety of woods are satisfactory (see Paper).

A sheet of cellulose fibers is not well suited to high-fidelity printing because of transparency and irregularities of the surface. These deficiencies are corrected by the addition of binding agents such as starch and resin, and by the incorporation into the fiber stock of mineral fillers such as calcium carbonate or sulfate, and especially pure white clay. Ordinary filled paper still lacks the perfection of surface smoothness required for accurate production of the tiny dots in halftone printing. The quality of the filled sheet is often enhanced by coating its surface with a thin film of finely divided mineral pigment suspended in an adhesive mixture such as starch and casein. Depending on the desired effect, the coating pigment may be calcium carbonate, "satin white" (coprecipitated calcium sulfate and hydrated alumina), titanium dioxide, and especially relatively pure white clay with a specific particle-size distribution. The amount of coating ranges up to 25% of the weight of the paper, the amount of filler up to 35%.

Pure kaolinite clays are used for filling and coating paper, except when special properties are desired. The clays have to be easily dispersible in water and should have, in the crude state, a fairly wide particle-size distribution so that a range of products can be prepared commercially. The clay should be as nearly white as possible or easily bleached to a high degree of whiteness. For coating purposes, low viscosity in clay–water systems is essential; that is, a slurry of high clay concentration with the lowest possible viscosity must be obtained.

In general, the crude clays are washed to remove any grit that may be present, chemically bleached to increase their whiteness, fractionated by settling or centrifugation to develop products with a given particle-size distribution, and then dried.

The clay must be easily freed of grit content if there is any. The grit causes wear in the foundrinier wires of the papermaking machine, and in coating clays causes an abrasive surface on the paper which wears the press plates in printing. Many kaolins could be utilized for paper clay except that the grit cannot be removed economically. Ultraflotation and high-intensity magnetic separation are used in the kaolin industry to increase brightness by removing extremely fine pigment particles, generally iron and titanium minerals (see Flotation; Magnetic separation).

In some paper-coating operations, clays have been found useful that were calcined at temperatures in excess of that required for complete dehydration (540°C) and below that at which new high-temperature phases form when the clay begins to vitrify because calcination tends to increase brightness. After calcination the clay must be ground to an extreme fineness.

In recent years, the relation has been studied between variation in the properties of kaolins, eg, particle-size distribution, composition, particle shape, etc, and the properties of coated paper such as opacity, printability, gloss, etc (17,18). Kaolin producers now offer a variety of products with different compositions and therefore different coating properties. The paper industry now consumes about 75% of all kaolin production.

Attapulgite clay has been used in the manufacture of NCR (no carbon required) paper (19). The clay is used to coat the upper surface of a sheet of paper; the lower surface is coated with minute capsules of certain organic compounds. The lower surface of one sheet rests on the upper surface of the second sheet and the pressure of writing serves to break the capsules of the organic compound, which then penetrates the attapulgite clay causing a color reaction and the writing appears on the second sheet (see Microencapsulation). Recently other materials have largely replaced attapulgite in NCR paper.

At the present time, much research is underway to produce a chemically processed bentonite, kaolin, or sepiolite clay or their combinations for use in a variety of duplicating papers.

Bentonite clays composed of montmorillonite are used for several purposes in papermaking. Pitch, tars, waxes, and resinous material from the wood and from waste cuttings and mixed paper may tend to agglomerate and stick to screens and machine wires, brush rolls, etc, and cause defects in the paper. Addition of small amounts of bentonite prevents the agglomeration of such particles. The addition of 1–2% of such clay at the beater or pulper stage increases retention of pigments by the paper stock, and thus also the uniformity of their distribution throughout the paper. White, or at least very light, highly colloidal bentonites are preferred.

Miscellaneous

Clays are used as ingredients in a vast number of products, for example, asphalt, paints, and rubber. They are also used as agents in many processes, such as water purification, and as a source of raw material, eg, aluminum.

Adhesives. Clays, generally kaolinites, are used in a wide variety of adhesives, such as those prepared with lignins, silicate of soda, starch, latex, and asphalt. They are used extensively in adhesives for paper products and in cements for floor coverings, including linoleum, rubber, and asphalt tile.

In adhesives, clays are not merely inert diluents, but may improve the properties.

Thus, clays tend to reduce the amount of penetration of the adhesives into the members to be joined, cause a faster setting rate and superior bond strength, increase the solid content of the adhesive and the amount of bonded surface, and permit the control of the suspension and viscosity characteristics for more satisfactory application.

Aluminum Ore. Many clays contain as much as 30–40% alumina. The main sources of aluminum are bauxite ores, which are composed of aluminum hydrates, either gibbsite or boehmite (see Aluminum).

The United States has only limited bauxite deposits and imports its requirements from the West Indies, South America, and Australia. Recently, large new deposits have been found in Brazil, Australia, and West Africa.

During World War II, the bauxite sources for the aluminum industry in the United States were threatened and a large research effort was made toward extraction of alumina from clays. Several processes were investigated and several large pilot plants were put in operation. After the end of the war, it was more economical to import bauxite to the United States. However, the economic advantage of using bauxite over clay is relatively slight and improvement in the clay-extraction process or increases in the cost of imported ore may in the future make clay-extraction economical.

Alumina is extracted from clays (20–22) by (a) the acid process, where aluminum is dissolved (chiefly with sulfuric acid), usually after the clay has been heated to elevated temperatures (600°C) to increase the solubility of the aluminum; (b) the aluminum sulfate process, where the clay is baked with ammonium sulfate at approximately 400°C, followed by leaching with water to extract the aluminum sulfate and ammonium sulfate; (d) the lime sinter process, where the clay is treated with ground limestone and calcined at 1375°C to give a mixture of dicalcium silicate ($2CaO.SiO_2$) and pentacalcium trialuminate ($5CaO.3Al_2O_3$), followed by leaching with a dilute alkali carbonate solution to remove the aluminum; (d) the lime-soda sinter process, which is similar to the lime sinter process except that the clay is treated with calcium carbonate and sodium carbonate, calcined to give a mixture of dicalcium silicate and sodium aluminate ($NaAlO_2$), followed by dissolution of the aluminum in a dilute alkali carbonate solution; and (e) several high-temperature electrochemical processes in which there is a direct reaction of aluminum in the clay to form alumina or aluminum nitride, carbide, chloride, sulfide, etc; this reaction is carried out at very high temperatures, generally in an electric furnace. Kaolinite clays are used for these processes because of their high aluminum content and because they contain small amounts of iron, alkalies, and alkaline earths that may cause serious problems in the extraction processes.

In recent years particular attention has been given to the acid extraction process as being the most attractive financially and technically.

Asphalt Products. A large variety of asphalt products, eg, for use in highway construction, contain clay materials as fillers, extenders, and to control viscosity and penetration (see Asphalt).

Radioactive Waste Disposal. The use of clays has been suggested for the disposal of atomic wastes by adsorbing the active elements and then fixing them against leaching by calcination at temperatures adequate to vitrify the clay and thereby to bind the radioactive material in insoluble form (see Nuclear reactors). The disposal of the isotopes of cesium and strontium is particularly important because of their relatively long half-life, potency, and abundance. The problem for strontium can be solved fairly well by its fixation as a carbonate. Certain illite clays have unusual power for the fixation of cesium and chemical pretreatment or heating at moderately low temperatures may enhance the cesium fixation of some three-layer minerals.

Portland Cement.   Portland cement, a mixture of calcium silicate and calcium aluminate minerals, is produced by the calcination of argillaceous limestones or mixtures of limestone and clay. Moderate amounts of iron, alkalies, and alkaline earths, except magnesia which should not be in excess of 5%, apparently are not detrimental to the properties of portland cement. Therefore, any of the clay minerals, with the exception of attapulgite, sepiolite, and some montmorillonites, could be used as an ingredient. However, kaolinite is preferred since it contributes only alumina and silica and is especially desirable in the manufacture of white portland cement and other special types requiring careful control of chemical composition (see Cement).

Pozzolanas.   Pozzolanas are siliceous or siliceous and aluminous materials, natural or artificial, processed or unprocessed, which though not cementitious in themselves, contain constituents that combine with lime in the presence of water at ordinary temperatures to form compounds of low solubility with cementing properties (see Cement).

Pozzolanas are classified, depending upon the substances responsible for the pozzolanic action, in decreasing order of activity as follows: volcanic glass; opal; clay minerals (kaolinites, montmorillonites, illites, mixed vermiculite–chlorite clays, and palygorskite); zeolite; and hydrated aluminum oxide.

In the natural condition, the clay minerals are nonpozzolanic or only weakly so. With calcination, however, particularly in the range of 650–980°C, partial dehydration and structural changes result in significant reactivity for lime.

Medicinals and Cosmetics.   Clays have been used for centuries in therapeutic intestinal adsorbent preparations against intestinal irritation. The clays are believed to function by adsorbing toxins and bacteria responsible for intestinal disorders and by coating the inflamed mucous membrane of the digestive tract (23). Kaolinite clays have been used widely. However, attapulgite clays that have been activated by moderate heating are five to eight times more active than kaolinite clays as adsorbents for alkaloids, bacteria, and toxins (24). By its highly effective adsorption of the aqueous part of inflammatory secretions, attapulgite aids in stool formation.

Clays of various kinds, but particularly those composed of montmorillonite, kaolinite, or attapulgite, have been used for a long time in the preparation of pastes, ointments, and lotions for external use; the best known of these uses is in the preparation of antiphlogistine for checking inflammation. Many cosmetic formulations contain montmorillonite, attapulgite, and kaolinite clays, taking advantage of the clay's softness, dispersion, gelling, emulsifying, adsorption, or other properties. No general specifications are reported for clays for these uses except that they be free from grit, very fine-grained, and dispersible in liquid (see Cosmetics).

Paint.   Paint is essentially a fluid system in which solid bodies, usually identified as pigments, are suspended.

Various types of clays have been used for a long time as paint ingredients, at first as inert fillers. However, clays add valuable properties to paints and are essential components of some types of paint.

Kaolinite clays are used very widely in amounts varying from 2–5% in some enamels, to as much as 45% in some interior wall paints. Calcined kaolins are used as extenders for titania pigment in a variety of paints.

Kaolinite clays impart desirable and controllable surface characteristics, permit high pigment loading, add to the hiding power, and also have good oil-absorption and suspension properties (see Paint).

Bentonite clays, composed essentially of montmorillonite, are used extensively in both oil- and water-base paints. In the latter, the montmorillonite clays act as a suspending and thickening agent.

Montmorillonite is hydrophilic and thus may be difficult to disperse in some oil vehicles. Organic-clad montmorillonite clays, which are oleophilic, are now available that are tailor-made with a variety of organic compounds to meet the requirements of different vehicles, such as cellulose nitrate lacquers, epoxy resins, and vinyl resins.

Pelletizing Ores, Fluxes, Fuels. In the beneficiation of some ores, it is frequently necessary to pulverize the ore to accomplish separations and concentrations. The pulverized ore has to be pelletized or agglomerated into larger units. Bentonites of high dry strength are best suited for pelletizing. The largest market for bentonites is now pelletizing of iron ore. A process for pelletizing finely ground fluorspar flux at 90–200°C with about 1% of Wyoming bentonite and sulfite waste liquor has been patented (25), as has the pelletizing of powdered fuel (26) using about 5% bentonite and quebracho extract (see Pelletizing and briquetting).

Pesticides. Many pesticides are highly concentrated and are in a physical form requiring further preparation to permit effective and economical application. A number of diluents are available in solid or liquid form to bring the pesticide chemicals to field strength. Although carriers of diluents generally are considered to be inert, they have a vital bearing on the potency and efficiency of the pesticide dust or spray, since most dusts and sprays contain only 1–20% active ingredients and 80–99% carrier diluent. The physical properties of the carrier diluents are of great importance in the uniform dispersion, the retention of the pesticide by the plant, and in the preservation of the toxicity of the pesticide. The carrier must not, for example, serve as a catalyst for any reaction of the pesticide that would substantially alter its potency.

Clays composed of attapulgite, montmorillonite, and kaolinite are used for this purpose in finely pulverized as well as granular form. Granular formulations are reportedly less expensive, more easily handled, reduce loss due to wind drift, and produce a most effective coverage (see Insect control technology; Poisons, economic).

Plastics. Mineral filler used in reinforced polyester resins and plastics offers the following advantages (27): it produces a smooth surface finish; reduces cracking and shrinkage during curing; aids in obscuring the fiber pattern of the glass reinforcement; contributes to high dielectric strength, low water absorption, and high wet strength; aids resistance to chemical action and weathering; and controls flow properties. Clay fillers are used extensively in polyester resins and polyvinyl compounds in amounts as high as 60%. Examples of the latter usage are found in electrical insulations, phonograph records, and floor coverings. Clays may also be used effectively in rigid and flexible urethane foams to provide cost reduction and improve deflection values, uniformity, and cell size.

Kaolinite clays are used extensively in the filling of plastics because they are easily dispersible in the resin as compared with other mineral fillers (see Plastics). All fillers increase the viscosity of the mixture, but in the case of clays this property can be controlled by the selection of the proper particle-size distribution. The low specific gravity of kaolinite and its white color are desirable characteristics in plastic compounds. Kaolins clad with certain organic compounds have improved dispersion and flow properties in polyvinyl resin systems (27).

Rubber

Kaolins were first used in rubber as diluents (28), but they also provide desirable reinforcing and stiffening properties as well as increased resistance to abrasion. Kaolins are used with certain types of synthetic rubber for heels and soles, tubing, extruded stock of all types, wire insulation, gloves, adhesives, tire treads, and inner tubes. In general, kaolins of extremely small particle size are desired; thus, so-called hard kaolins with 90% of the particles finer than 2 μm are commonly required. Manganese content must be extremely low because of its deleterious catalytic action (see Rubber).

Montmorillonite clays are used as thickening and stabilizing additives to latex and as emulsion stabilizers in rubber-base paints and in rubber adhesives.

Organic-clad clays of varying types are also used in the rubber industry. Thus clays in which the inorganic cation has been replaced by a substituted organic onium base are reinforcing agents in natural and some synthetic rubbers (29). Coating the clay particles with a strongly adhering organic layer improves the properties as a rubber filler (30).

Water Clarification and Impedance

In the clarification of potable water, colloidal matter is removed by filtration and/or sedimentation processes. Alum is commonly used to flocculate the colloidal material to enhance its settling and filtration rate. A highly colloidal, easily dispersible clay added to the water before the alum is desirable. The alum flocculates the clay, which serves to gather up and collect the colloidal material that would otherwise not settle (31). Bentonites and attapulgite clays have been used for this purpose (see Water). Wyoming-type bentonites are used extensively to impede the movement of water through earthen structures and to retard or stop similar movement through cracks and fissures in rocks and in rock and concrete structures. For example, bentonite is used to stop the seepage of water from ponds and irrigation ditches and to waterproof the outside basement walls of homes (32).

The bentonite is used as a grout, in which a bentonite suspension is injected under pressure into a porous strata, a fissure or crack, or along the foundation of a building, or in a placement method in which the bentonite may be placed to form an unbroken blanket between the earth and the water or mixed with the surface soil of the structure to render the soil impervious (33). In the case of placing a bentonite blanket to seal a pond or lake, the pond need not always be drained as granular bentonite may be added to the water surface. It will settle to the bottom of the pond, forming an impervious blanket.

BIBLIOGRAPHY

"Clays" in *ECT* 1st ed., Vol. 4: "Ceramic Clays," pp. 38–49, by W. W. Kriegel, North Carolina State College; "Fuller's Earth," pp. 49–53, by W. A. Johnston, Attapulgus Clay Co.; "Activated Clays," pp. 53–57, by G. A. Mickelson and R. B. Secor, Filtrol Corporation; "Papermaking, Paint, and Filler Clays," pp. 57–71, by S. C. Lyons, Georgia Kaolin Company; "Rubbermaking Clays," pp. 71–80, by C. A. Carlton, J. M. Huber Corporation; and "Clays (Uses)" in *ECT* 2nd ed., Vol. 5, pp. 560–586, by Ralph E. Grim, University of Illinois.

1. W. A. White, *The Properties of Clays,* Master's Thesis, University of Illinois, Urbana, Illinois, 1947.
2. U. Hofmann, "Fullstoffe und keramische Rohmaterialien," *Rapport Europees Congres Electronen-microscopic,* Ghent, 1954, pp. 161–172.
3. P. F. Low, The Viscosity of Water in Clay Systems, *Proceedings of the 8th National Clay Conference,* Pergamon Press, New York, 1960, pp. 170–182.
4. R. E. Grim and F. L. Cuthbert, *J. Am. Ceram. Soc.* **28,** 90 (1945).
5. R. E. Grim, "Some Fundamental Properties Influencing the Properties of Soil Materials," *Proceedings of the International Conference of Soil Mech. Foundation Eng. 2nd, Rotterdam 1948,* **III,** pp. 8–12.
6. D. A. Holderidge, *Trans. Brit. Ceram. Soc.* **55,** 369 (1956).
7. W. O. Williamson, *Trans. Brit. Ceram. Soc.* **40,** 225 (1941).
8. *Testing and Grading of Foundry Clays,* 6th ed., Am. Foun. Soc., 1952.
9. R. E. Grim and F. L. Cuthbert, *Ill. State Geol. Surv. Rep. Invest. 102* (1945).
10. R. E. Grim and F. L. Cuthbert, *Ill. State Geol. Surv. Rep. Invest. 110* (1946).
11. L. L. Carney and R. L. Meyer, *Am. Inst. Min. Met. Eng. Soc. Petrol. Eng. Paper 6025* (1976).
12. T. H. Mulliken, A. G. Oblad, and G. A. Mills, *Calif. Dept. Nat. Resources Div. Mines Bull. 169* (1955).
13. V. Charrin, *Genie Civ.* **123,** 146 (1946).
14. F. J. Zvanut, *J. Am. Ceram. Soc.* **20,** 251 (1937).
15. R. E. Grim and W. F. Bradley, *J. Am. Ceram. Soc.* **22,** 157 (1939).
16. G. P. C. Chambers, *Silic. Ind.* **24,** 3 (April, 1959).
17. W. M. Bundy, W. D. Johns, and H. H. Murray, *Jour. Tech. Assoc., Pulp and Paper Ind.* **48,** 688 (1965).
18. W. M. Bundy and H. H. Murray, *Clays Clay Miner.* **21,** 295 (1973).
19. U.S. Pat. 2,550,469 (April 24, 1951), K. Barrett, R. Green, and R. W. Sandberg.
20. F. A. Peters, P. W. Johnson, and R. C. Kirby, *U.S. Bur. Mines Rept. Invest. No. 6229* (1963).
21. F. A. Peters, P. W. Johnson, and R. C. Kirby, *U.S. Bur. Mines Rept. Invest. No. 6133* (1963).
22. G. Thomas and T. R. Ingram, *Can. Dept. Mines Tech. Surv., Mines Branch Res. Rept. R45* (1959).
23. L. S. Goodman and A. Gilman, *The Pharmacological Basis of Therapeutics,* 2nd ed., The Macmillan Co., New York, 1955.
24. M. Barr, *J. Am. Pharm. Assoc., Pract. Pharm. Ed.* **19,** 85 (1958).
25. U.S. Pat. 2,220,385 (Mar. 5, 1940), F. C. Abbot and C. O. Anderson.
26. U.S. Pat. 2,217,994 (Oct. 15, 1940), G. G. Rick and C. E. Loetel.
27. J. R. Wilcox, "Controlling Flow Properties with Fillers," *Proc. Soc. Plastics Inc. 9, Sect. 27,* February, 1954.
28. C. C. Davis and J. T. Blake, "Chemistry and Technology of Rubber," *Am. Chem. Soc. Monograph Series No. 74,* American Chemical Society, Washington, D.C. (1937).
29. U.S. Pat. 2,531,396 (1950), L. W. Carter, J. C. Hendricks, and D. S. Bolley.
30. Brit. Pat. 630,418 (1949), N. O. Clark and T. W. Parker.
31. E. Nordell, *Water Treatment for Industrial and Other Uses,* Reinhold Publishing Corp., New York, 1951.
32. R. C. Mielenz and M. E. King, *Calif. Dept. Nat. Resources Div. Mines Bull.* **169,** 196 (1955).
33. C. D. Weaver, *Proc. Soil Sci. Soc. Am.* **11,** 196 (1946).

General Reference

R. E. Grim, *Applied Clay Mineralogy,* McGraw-Hill Book Co., New York, 1962.

RALPH E. GRIM
University of Illinois

CLUTCH FACINGS. See Brake linings and clutch facings.

CMC. See Cellulose derivatives—Esters (sodium carboxymethylcellulose).

COAGULANTS AND ANTICOAGULANTS. See Blood, coagulants and anti-coagulants.

COAL

Coal was used in China, Greece, and Italy more than 2000 years ago; before 400 A.D. it was used by the Romans in Britain. Coal mining began in Germany around the 10th century. English deposits provided coal for shipment during the 13th century. The depletion of wood supplies led to rapidly increased production in the 16th century. In the United States coal mining began around 1700.

Coal is a dark burnable solid, usually layered, that resulted from the accumulation and burial of partially decayed plant matter over earlier geological ages. These deposits were converted to coal through biological changes and later effects of pressure and temperature. Composition as well as chemical and physical properties vary widely for the different kinds of coal. Differences in type are caused by variations in the amounts of different plant parts exemplified by common-banded, splint, cannel, and boghead coals. In Europe the first two are broadly termed ulmic or humic coals. The degree of coalification is referred to as *rank*. Brown coal and lignite, subbituminous coal, bituminous coal, and anthracite make up a natural series with increasing carbon content. The impurities in coal cause differences in grade (see Lignite and brown coal).

Coal is black or brownish-black and is sold in a range of sizes. The color, luster, texture, and fracture vary with the type, rank, and grade. Coal is composed chiefly of carbon, hydrogen, and oxygen with small amounts of nitrogen and sulfur, and varying amounts of moisture and ash or minerals impurities.

The formation of coal, the variation in its composition, its microstructure, and its chemical reactions indicate that coal is a mixture of compounds. The organic composition depends on the degree of biochemical change of the original plant material, on the later coalification by pressure and temperature effects in the deposit, and on mineral matter deposited with the original vegetable material or later. The major organic compounds have resulted from the formation and condensation of polynuclear carbocyclic and heterocyclic ring compounds containing carbon, hydrogen, oxygen, nitrogen, and sulfur. The amount of carbon in aromatic ring structures increases with rank (see Constitution).

Nearly all coal is utilized in combustion and coking. Over 80% is burned directly for generation of electricity (see Coal conversion processes, magnetohydrodynamics), or steam generation for industrial uses, transportation (see Power generation), or residential heating, and metallurgical processes, in the firing of ceramic products, etc. The balance is mostly carbonized to produce coke, coal gas, ammonia, coal tar, and light oil products from which many chemicals are derived (see Coal conversion processes, carbonization). Combustible gases can also be produced by the gasification of coal (see Fuels, synthetic). Carbon products are produced by heat treatment (see Carbon). A relatively small amount of coal is used for miscellaneous purposes such as fillers (qv), pigments (qv), foundry material, and water filtration.

Most organic chemicals can be synthesized from coal carbonization products or by different processes utilizing gases derived from coal. Some aliphatic and aromatic carbocyclic acids can be produced by the oxidation of coal. Processes to produce gaseous and liquid fuels, as well as chemicals have been extensively developed and are being improved in the United States, Europe, and South Africa. Development has been limited by the ability of these products to compete economically with petroleum and natural gas (see Feedstocks).

Coal is one of the world's major consumed commodities. Current annual production averages ca 540×10^6 metric tons in the United States and 3.3×10^9 metric tons for the entire world (2.4×10^9 t coal, and 0.9×10^9 t brown coal and lignite). Reserves of coal are far greater than the known reserves of all other mineral fuels (petroleum, natural gas, oil shale, and tar sands) combined. This is true for the entire world. The reserves of petroleum and natural gas are being depleted more rapidly than coal. Coal consumption has been cyclic and, as energy consumption increases, coal will be used in continually larger amounts. Many new processes for coal conversion and utilization are being developed and will be of significant interest to chemists and chemical engineers (see also Gas, natural; Oil shale; Petroleum; Tar and pitch; Tar sands).

Origin of Coal

Coal evolved from partially decomposed plants in a shallow-water environment. Various chemical and physical changes have altered the plant composition. The changes occurred in two distinct stages, one is biochemical and the other is physicochemical (geochemical) (1–4). Since some parts of plant material are more resistant to biochemical degradation than others, optical variations in petrologically distinguishable coals are observed. The terms vitrain and clarain refer to bright coals, durain is a dull coal, and fusain is a structured fossil charcoal. Subsequent exposure to pressure and heat during the geochemical stage caused the differences in degree of coalification or rank that are observable in the continuous series—peat, brown coal and lignite, subbituminous coal, bituminous coal, and anthracite. The carbon containing deposits in which the inorganic material predominates, such as in oil shale and bituminous shale, are not classified as coal.

Complete decay of plant material is prevented only in environments such as swamps in regions where there is rapid and plentiful plant growth in waterlogged conditions. Peat is formed in such swamps from plant debris such as branches and twigs, bark, leaves, spores and pollen, and even tree trunks which are submerged in the swamp. A series of coal seams have been formed from peat swamps growing in an area that has undergone repeated subsidence. Periods during which vegetation flourished and peat accumulated were followed by rapid subsidence resulting in submergence of the peat swamp and covering of the deposit with silt and sand. Thiessen (5) suggested that the Dismal Swamp of Virginia and North Carolina, which is gradually being flooded by Lake Drummond, is such an area undergoing active subsidence. The autochthonous (*in situ*) theory of coal formation is supported by the existence of peat beds formed in this way from the accumulation of plants and plant debris in place. In the allochthonous (not *in situ*) theory, the coal-producing peat bogs were formed from plant debris that had been transported into deep enclosed basins.

The large coal fields, which include a series of relatively thin coal seams (0.6–3 m), cover large areas which may have been formed from *in situ* material. However, those fields in which the coal seams are thick, cover a limited area, and show a high ash content may have formed from drift material. Accepting that the thinner seams were primarily formed from *in situ* material, most authorities believe these were at least partly formed from transported drift material.

Biochemical Stage. The initial biochemical decomposition of plant matter depends on two factors, the ability of the different plant parts to resist attack and the existing conditions of the swamp water. Fungi and bacteria can cause complete decay of the plant matter if it is exposed to aerated water or to the atmosphere. The decay is less complete if the vegetation is immersed in water containing anaerobic bacteria. Under these conditions the plant protoplasm, proteins, and starches, and to a lesser extent the cellulose are easily digested. Lignin (qv) is more resistant. The most decay-resistant plant parts (for both anaerobic and aerobic decomposition) include the waxy protective layers, ie, cuticles, spore, and pollen walls, as well as resins.

Vitrain (anthraxylon) results from the partial decay of lignin in stagnant water. The relict cell structure of the parent plant tissue can be recognized in many samples.

The clarain of Stopes (6) and bright attritus of Thiessen (5) are finely banded bright parts of coal that evolved from the residues of fine woody material such as branches, twigs, leaves, spores, bark, and pollen. In aerated waters the plant parts were more decomposed and show a higher concentration of resins, spores, and cuticles. Dull coal, called durain by Stopes, was formed under these conditions, and occurs commonly in Europe. It is not as widely recognized in the United States where it is called splint or block coal. More selective chemical and biochemical activity, probably in a drier environment, led to the formation of soft, charcoal-like fusain from woody plant material. The conversion was rapid and probably complete by the end of the peat formation stage. Cannel coal is believed to have formed in aerated water which decomposed all but the spores and pollen. The name is derived from its quality of burning in splints with a candle-like flame. Boghead coal closely resembles cannel coal, but was derived from algae instead of plant spores.

Geochemical Stage. The conversion of peat to bituminous coal is the result of the cumulative effect of pressure and temperature over a long time. The sediment covering the peat provides the pressure and insulation so that internal heat can be applied. This change is termed normal coalification. Conversion to anthracite appears to require the additional intensity of heat and pressure that accompanies mountain building processes.

The chemical composition of the coal changes during coalification. Oxygen and hydrogen decrease, and carbon increases. Volatile matter decreases and calorific value increases. Physical properties changed as a consequence of the effects of pressure. The fixed carbon, or extent of devolatilization, and the calorific content of coal represent the main criteria determining coal rank or degree of coalification.

The change in rank to anthracite involves the application of significantly higher pressures and temperatures. A local increase in temperature generated by igneous activity can alter the included lower rank coal to anthracite or a natural coke. Coal more distant from the disruption was altered proportionately less. Tectonic plate movements involved in mountain building provide pressure for some changes to anthracite.

As a general rule, the older the coal deposit, the more complete the coalification. The most significant bituminous coal fields were deposited in the Carboniferous (ca 300 million years ago) and Upper Cretaceous ages (early Tertiary—ca 100 million years ago).

The lower rank coals came primarily from the Mesozoic or early Tertiary age, and the peat deposits are relatively recent, less than one million years old. However, age alone does not determine rank. The brown coal of the Moscow basin is not buried deeply and although it was deposited in the Lower Carboniferous age, it did not have enough heat and pressure to be converted further.

Coal Petrography

Microscopic examination of coal has been carried out using transmitted light through thin sections, and reflected light from polished samples (1–4). The first studies were made with transmitted light and thin sections. This technique permits the identification of the plant parts that became coal components. Thiessen identified three components termed anthraxylon, attritus, and fusain (7) as a result of his examination of thin sections of bituminous coals from the United States. The anthraxylon occurs in thin bands from <0.02 mm to several mm thick. The color observed in the thin sections ranges from orange to dark red depending on the coal rank. This component evolved from the relict woody tissues which often retain the original plant cell structure.

Thin bands of attrital coal occur along with the anthraxylon and are composed of finely divided plant residues. The attritus consists of translucent plant remains similar to anthraxylon, but smaller than 0.02 mm; opaque material either in the form of discrete granules or black amorphous material with an origin similar to that of fusain; semitranslucent brown matter from plant parts that have imcompletely carbonized; and yellow translucent constituents-resins, spores, pollen articles, and algae that can be identified from their form. Attritus is subdivided into translucent and opaque depending on the proportion of opaque matter in the sample.

Fusain occurs in several forms, ie, thin bands, fibrous particles, and irregularly shaped masses. Usually it is opaque but may be translucent if the section is very thin. The plant cell structure can always be seen and the open spaces are sometimes filled with mineral matter.

In Europe the usual technique used in coal petrography is the examination of polished coal surfaces with reflected light (8–9). This technique has been widely adopted in the United States. The terminology is different from that developed by Thiessen although the petrographic constituents are the same under different viewing conditions. The terminology considers coal as an organic rock consisting of different lithotypes, which are identified visually, and macerals (by analogy with the minerals in inorganic rock) and microlithotypes which are observed microscopically. The four lithotypes in coal originally described by Stopes (6) are vitrain, clarain, durain, and fusain. The principal maceral groups are identified by their reflectivity and form. They include vitrinite, exinite, (including sporinite, cutinite, alginite, and resinite), and inertinite (including micrinite, semifusinite, fusinite, and sclerotinite). Microlithotypes are the typical samples classified according to the relative amount of macerals present in the coal. Eight microlithotypes are recognized including vitrite—mostly vitrinite clarite–vitrinite and exinite; durite–inertinite and exinite; vitrinertinite–vitrinite and inertinite; duroclarite–vitrinite, exinite, and inertinite; and clarodurite–inertinite, exinite, and vitrinite.

The two systems, the Thiessen and Stopes-Heerlen, cannot be directly correlated. Table 1 gives the approximate correlation developed by the International Committee for Coal Petrography.

Mineral Matter in Coal.   The mineral matter (4,10) in coal is the result of several separate processes. Part is caused by the inherent material present in all living matter; another type is caused by the detrital minerals that were deposited during and after the time of peat formation; and a third type is caused by secondary minerals that have crystallized from water which has percolated through the coal seams. Clay is the most common detrital mineral; however, other common ones include quartz, feldspar, garnet, apatite, zircon, muscovite, epidote, biotite, augite, kyanite, rutile, staurolite, topaz, and tourmaline. The secondary minerals are generally kaolinite, calcite, and pyrite. Analyses have shown the presence of almost all elements in at least three trace quantities in the mineral matter (11). Certain elements, ie, germanium, beryllium, boron, and antimony are found primarily with the organic matter in coal. Zinc, cadmium, manganese, arsenic, molybdenum, and iron are found with the inorganic material in coal.

The primary mineral constituents in coal are aluminum, silicon, iron, calcium, magnesium, sodium, and sulfur. The relative concentrations will depend primarily on the geographical location of the coal seam, and will vary from place to place within a given field (see Constitution).

Application of Coal Petrology.   Coal petrology is used to identify the macerals and microlithotypes present in coals. The behavior of coal depends on the properties of the individual constituents, the relative amounts in the coal, and the rank of the coal (1–4). For some purposes, a blend of coals can be selected to achieve desired coking properties. The difference in resistance of macerals to breakage together with other properties is used in the Longwy-Burstlein or Sovaco process (12) in the selection of coals for blending in the optimum proportions for producing high-quality coke. The maceral groups behave differently on heating. Vitrinite from most medium rank coal (9–33% volatile matter) has good plasticity and swelling properties and produces an excellent coke. Inertinite is almost inert and does not soften on heating. Exinite becomes extremely plastic and is almost completely distilled as tar. By careful control of the petrological composition and the rank of a coal blend, its behavior during carbonization can be controlled. The petrological composition of a coal is quantitatively determined on polished surfaces by measuring the volume percentage of the different macerals with a point count method. The rank of the coal is determined microscopically by measuring reflectance of vitrinite with a microphotometer. The coking behavior of a coal or a blend of coals of different ranks can be reliably predicted with these two measurements.

Table 1. Correlation of Thiessen-Bureau of Mines Nomenclature (Thin Section Method) with Stopes-Heerlen Nomenclature (Polished Block Method) of Hard Coal[a]

Transmitted light, Thiessen-Bureau of Mines system			Reflected light, Stopes-Heerlen system	
Banded components	Constituents of attritus[b]		Macerals	Groups of macerals
anthraxylon, translucent			vitrinite more than 14 μm in width	vitrinite
	translucent attritus	translucent humic matter	vitrinite less than 14 μm in width	
		spores, pollen cuticles, algae	sporinite, cutinite, alginite	exinite
		resinous and waxy substance	resinite	
attritus		brown matter, semitrans-lucent	weakly reflecting semifusinite / weakly reflecting massive micrinite / weakly reflecting sclerotinite	
	opaque attritus	granular opaque matter	granular micrinite	inertinite
		amorphous, massive opaque matter	fusinite less than 37 μm in width / strongly reflecting massive micrinite	
		finely divided fusain, sclerotia	strongly reflecting sclerotinite	
fusain, opaque			fusinite and semifusinite more than 37 μm in width	

Transmitted light		Reflected light
Types of coal	Quantitative statements	Microlithotypes
Banded coals		
bright coal	more than 5% anthraxylon	vitrite
	less than 20% opaque attritus	clarite
	more than 5% anthraxylon	duroclarite
semisplint coal	20–30% opaque attritus	vitrinertite
	more than 5% anthraxylon	
splint coal		
	more than 30% opaque attritus	clarodurite
Nonbanded coals		
cannel coal	less than 5% anthraxylon	durite[c]
boghead coal	less than 5% anthraxylon	

[a] An exact correlation of the two systems is not possible. An updated version will be in the new edition of ref 4. Courtesy of Internationale Kommission für Kohlenpetrologie.

[b] For the recognizable botanical entities (for instance "spores," "cuticles") the term "phyteral" may be employed.

[c] Durite is part of banded coals with less than 5% vitrinite that would be placed by definition in the nonbanded coals in the Thiessen-Bureau of Mines system.

Classification Systems

Prior to the 19th century, classifications according to appearance as bright coal, black coal, or brown coal were used. A more detailed classification was needed as coal use increased during the industrial revolution. A number of classification systems have been developed to meet particular needs or to apply to use in certain areas.

Two types of classification may be identified, ie, scientific and commercial. In the scientific category, the Seyler chart has found the widest acceptance (Fig. 1). The commercial classification uses scientific measurements to make the distinctions that occur in the classifications. Both methods are complementary and are used in research. In industry the commercial classification is essential.

Seyler's Classification. Seyler's classification is based on the observation that the amounts of two elements (C and H) present in coals is useful in correlating properties (13). A plot of wt % H vs wt % C, on a dry mineral-matter-free basis is used. Points

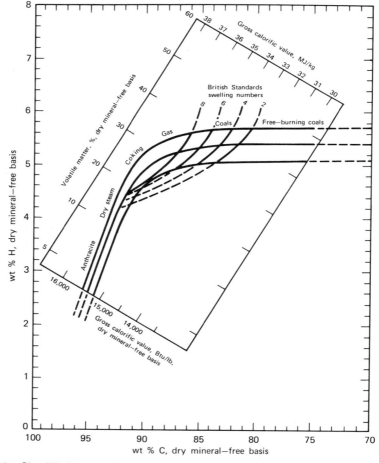

Figure 1.   Simplified form of Seyler's coal classification chart (13). An updated version of Seyler's coal classification is described in reference 4. Note that ASTM uses the "free-swelling index" (14).

representing different coal samples lie along a band. The center band on the chart shows the properties of coals rich in vitrinite. The location of the band indicates the range of the four properties and the interrelationship for a given property. Coals above the band are termed perhydrous (richer in hydrogen as cannel and boghead coals, and the maceral exinite in the usual coals). Coals below the band are termed subhydrous as represented by the maceral inertinite (see p. 238 for the differences between the perhydrous and the usual coals). Other properties may also be correlated with this chart.

Correlations are also made with volatile matter and gross calorific value. A systematic method of nomenclature was proposed in the first version in 1899. This was not adopted because of the complexity of the terms.

Since the chart uses carbon and hydrogen content (Fig. 1), but may also use volatile matter and calorific value, all on a dry, mineral-matter-free basis, coals may be plotted from a variety of data. Other properties also fit into specific areas on the chart, including moisture and swelling indexes. The curve in the band represents a composition range where the properties are changing rapidly. Swelling indexes, coking power, and calorific values are maximized and moisture is minimized. Bright coals, which are mostly vitrain, can be expected to fall within the limits of the band; however, durains may occur above or below since the proportion of oxygen in the coal differs for these.

Seyler's chart is an outstanding scientific classification based on the ultimate analysis of coal. It can contribute to the understanding of the other systems of classification.

ASTM Classification. The development of this classification system results from an extended joint project of the American Standards Association and the ASTM between 1927 and 1938, when it was adopted as a standard means of specification. The higher rank coals are specified by fixed carbon (for volatile matter ≤31%) on a dry, mineral-matter-free basis and lower rank coals are classified by calorific value on the moist, mineral-matter-free basis (Table 2). The latter parameter depends on two properties, the moisture absorbing capacity and the calorific value of the pure coal matter. Some overlap between bituminous and subbituminous coals occurs, and is resolved on the basis of the agglomerating properties.

National Coal Board Classification for British Coals. The classification proposed by the Department of Scientific and Industrial Research (U.K.) in 1946 led to the classification used by the National Coal Board in the United Kingdom with only minor changes (Fig. 2). The two parameters used are the volatile matter on a dry, mineral-matter-free basis and the Gray-King coke-type assay, a measure of coking power. The Gray-King coke type is used as a primary means of classification for lower rank coals. Classes are subdivided at 36% volatile matter. The classification applied to coals with less than 10% ash. High ash coals are cleaned before analysis by a float–sink separation to reduce the ash content below 10%.

U.S.S.R. Coal Classification. The name of a coal field is an adequate description for many coals; however, in the Donets Basin there is a wide range of coals and the volatile matter and coking properties must be specified (3–4). The British Coke Research Association has correlated the U.S.S.R. classification system with the international classification code numbers (16). A description is given of the method for determination of the Sapozhnikov plastometric index used to assess coking properties.

Table 2. ASTM Specifications for Solid Fuels, Excluding Peat[a,b]

Class	Group Name	Symbol	Fixed carbon Dry, %	Fixed carbon Moist, %	Volatile Matter dry, %	Volatile Matter moist, %	Natural moisture, %	Heating values[c] Dry basis (MJ/kg)	Dry basis (kcal/kg)	Moist basis (MJ/kg)	Moist basis (kcal/kg)
I anthracite	*meta*-anthracite	ma	>98	>92	<2	<2	6	32.4	7740	31.4	7500
	anthracite	an	92–98	89–95	2–8	2–8	3	35.5	8000	35.5	8000
	semianthracite	sa	86–92	81–89	8–14	8–15	3	34.7	8300	34.6	8275
II bituminous	low-volatile	lvb	78–86	73–81	14–22	13–21	5	36.6	8741	35.8	8550
	medium volatile	mvb	69–78	65–73	22–31	21–29	7	36.2	8640	34.6	8275
	high-volatile A	hvAb	<69	58–65	>31	>30	5	34.2	8160	>32.5	>7775
	high-volatile B	hvBb	57	53	57	40	7	28.3–34.2	6750–8160	30.2–32.5	7220–7775
	high-volatile C	hvCb	54	45	54	40	16	31.0–35.1	7410–8375	26.7–30.2	6387–7220
								28.3–31.0	6765–7410	24.4–26.7	5832–6387
III subbituminous	subbituminous A	subA	55	45	55	38	18	28.8–31.6	6880–7540	24.4–26.7	5832–6837
	subbituminous B	subB	56	43	56	35	24	27.4–30.3	6540–7230	22.1–24.4	5276–5832
	subbituminous C	subC	53	37	53	36	30	25.1–28.7	5990–6860	19.3–22.1	4610–5276
IV lignite	lignite A	ligA	52	32	52	35	38	20.2–26.6	4830–6360	14.7–19.3	3500–4610
	lignite B	ligB	52	26	52	32	50	<22.0	<5250	<14.7	<3500

[a] Ref. 15.
[b] High volatile C Bituminous coal with heating values of 24.4–26.7 MJ/kg (5830–6390 kcal/kg) is always agglomerating. All other bituminous coals are commonly agglomerating. Anthracite and subbituminous coal and lignite are nonagglomerating.
[c] To convert MJ/kg to Btu/lb, multiply by 430.2.

232

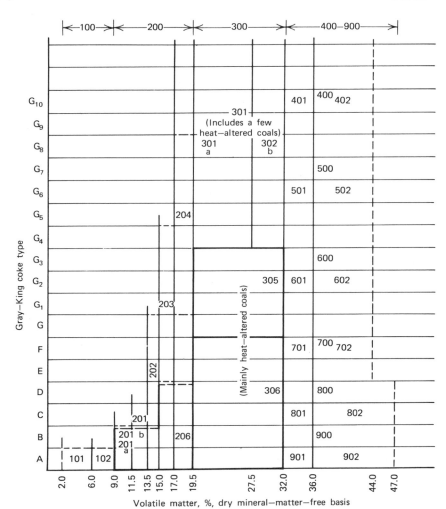

Figure 2. Coal classification system (1956) of National Coal Board (U.K.). Note that - - - defines a limit as found in practice although not a boundary for classification purposes; ——— defines a classification boundary; and 206, 305, and 306 are classes consisting mainly of coals that have been affected by igneous intrusion. (Courtesy of National Coal Board, U.K.).

International Classification. *Hard Coal.* The increasing amount of coal in international commerce since ca 1945 indicated a need for an international system of coal classification. The different terms used to describe similar or identical coals led to confusion. Table 3 compares the classes of the International System with a number of National Systems. The Coal Committee of the European Economic Community considered different national classifications and agreed on a system designated International Classification of Hard Coal by Type, which was published in 1956 (2).

Similar to the other methods, volatile matter and gross calorific value on a moist, ash-free basis are the parameters used. Coals with volatile matter up to 33% are divided into classes 1–5. Coals with volatile matter greater than 33% are divided into classes 6–9. The calorific values are given for a moisture content obtained after equilibrating at 30°C and 96% rh.

Table 3. Classes of the International System Compared with Classes of the National Systems[a]

Classes of the international system			Classes of national systems							
Class no.	Volatile matter, %	Calorific value (calculated to standard moisture content), kJ/g[b]	Belgium	Germany	France	Italy	Netherlands	Poland	United Kingdom	United States
0	0–3					antraciti speciali		*meta*-antracyt		*meta*-anthracite
1A	3–6.5		maigre	Anthrazit	anthracite	antraciti communi	anthraciet	antracyt	anthracite	anthracite
1B	6.5–10									
2	10–14		¼ gras	Mager-kohle	maigre	carboni magri	mager	polantracyt chudy	dry steam	semianthracite
3	14–20		½ gras	Esskohle	demigras	carboni semi-grassi	esskool	polkoksowy *meta*koksowy	coking steam	low-volatile bituminous
4	20–28		¾ gras	Fettkohle	gras à courte flamme	carboni grassi corta fiamma	vetkool	*orto*koksowy	medium-volatile coking	medium-volatile bituminous
5	28–33		gras	Gaskohle	gras propre-ment dit	carboni grassi media fiamma		gazowo koksowy		high-volatile bituminous A
6	>33 (33–40)	32.4–35.4				carboni da gas	gaskool			
7	>33 (32–44)	30.1–32.4			flambant gras	carboni grassi da vapore	gasvlam-kool	gazowy	high-volatile	high-volatile bituminous B
8	>33 (34–46)	25.6–30.1		Gas flamm-kohle	flambant sec	carboni secchi	vlamkool	gazowo-plomienny		high-volatile bituminous C
9	>33 (36–48)	<25.6						plomienny		subbituminous

[a] Ref. 2. Courtesy D. W. van Krevelen and Elsevier Scientific Publishing Co.
[b] To convert kJ/g to Btu/lb, multiply by 430.2.

The nine classes are divided into four groups determined by caking properties as measured through either the free swelling index (14) or the Roga index. These tests indicate properties observed when the coal is heated rapidly.

Brown Coal and Lignite. The brown coals and lignites have been classified separately and have been defined as those coals with heating values less than 23,850 kJ/kg (10,260 Btu/lb, 5700 kcal/kg) (see Lignite and brown coal).

Constitution

Quantitative Chemical Measurements. The functional groups of interest contain O, N, or S (1,2,4,17). The significant oxygen-containing groups found in coals are those of carbonyl, hydroxyl, carboxylic acid, and methoxy. The nitrogen-containing groups include aromatic nitriles, pyridines, and pyrroles. The sulfur is largely found in dialkyl, aryl alkyl thioether and thiophene groups (see Properties and Reactions below).

The relative and absolute amounts of these groups vary with coal rank and maceral type. The major oxygen-containing functional groups in vitrinites of mature coals are phenolic hydroxyl and conjugated carbonyl as in quinones. Evidence exists for hydrogen bonding of hydroxyl and carbonyl groups. There are unconjugated carbonyl groups such as ketones in exinites. The infrared absorption bands are displaced from the normal range for simple ketones by the conjugation in vitrinites. The interactions between the carbonyl and hydroxyl groups affects the normal reactions.

A range of quantitative organic analytical techniques may be used to determine the functional group concentrations. Acetylation is used to determine hydroxyl groups.

Carbonyl groups are difficult to quantify with simple procedures. An electrochemical method has been used rather satisfactorily. Coal extracts are reduced at an appropriate potential in nonaqueous solution. The current passed and hydroxyl content of the product indicate the carbonyl content. Uncertainty about the type of combination of the carbonyl gives an uncertainty factor of 2 in quantitative interpretations of the carbonyl content.

Treatment with mild selective oxidizing agents like benzoquinone (4,18) causes coals to lose much of their hydrogen content. Similarly, a palladium catalyst can cause the evolution of molecular hydrogen (19–20). These methods given an indication (if the approach is valid) of the minimum amount of carbon in the coal that is involved in hydroaromatic rings, and since this amount is close to the total nonaromatic hydrogen for lower rank coals, it is believed that this minimum approximates the true value. Other methods involve dehydrogenation with sulfur (21) and with halogens (22). The results are somewhat less than with benzoquinone.

Hydrogen can be added to the aromatic structures converting them to hydroaromatic rings. The hydrogen addition and removal is generally reversible, but not entirely (20). Coal extracts may be reduced chemically (1,4,23) at ambient or moderate temperature and pressure, either by electrochemical methods using a higher potential than for carbonyl groups, or by reducing agents such as lithium in ethylenediamine.

The complete distribution of carbon and hydrogen in coal requires an additional measurement from a group of interrelated possibilities. The determination of methyl groups is one of the simplest of these. Development of the Kuhn-Roth procedure (24) for selective oxidation has given reasonably consistent methyl group contents over a range of coals (4). Some methyl groups are not determined with this method so that a comparison with results of studies with model compounds and other work is needed (25) to evaluate the significance of the results.

Chemical Constitution of Coals. A summary of the chemical structure of coal as perceived by most of those working in this area is given below (1–2,4). The conclusions are based on experimental measurements referred to earlier. The description refers to vitrinites; differences for other macerals are indicated separately.

Several requirements must be met in developing a structure. The elementary analysis and other physical measurements must be consistent with the structure. Limitations of structural organic chemistry and stereochemistry must be satisfied. Mathematical expressions have been developed that can test the consistency of a set of parameters that describe the molecular structure of coal. A number of postulated structures may be eliminated and the probability that those being considered are correct is increased. Mathematical analyses of this type have been reported (2,4,26–27).

The evidence suggests that the structure for vitrinites in bituminous coals and anthracite has the following characteristics:

(1) The molecule includes a number of small aromatic nuclei or clusters, each usually containing one to four fused benzene rings. The average number of clusters in the molecule is characteristic of the coal rank. This average increases slightly to 90% carbon and then increases rapidly.

(2) The aromatic clusters are partly linked together by hydroaromatic, including alicyclic, ring structures. These rings also contain six carbon atoms thus, upon loss of hydrogen they can become part of the cluster, increasing their average size.

(3) Other linkages between clusters involve short groupings such as methylene, ethylene, ether oxygen, and methoxy.

(4) A significant amount of hydrogen sites on the aromatic and hydroaromatic rings have been substituted by methyl and sometimes larger aliphatic groups.

(5) The oxygen in vitrinites usually occurs in phenolic hydroxyl groups, substituting for hydrogen on the aromatic structures.

(6) A small fraction of the rings, both aromatic and hydroaromatic, have oxygen substituted for a carbon atom. Some of these heterocyclic rings may be five-membered.

(7) A lesser amount of oxygen than that occurring in phenolic groups appears in the carbonyl groups, ie, as quinone forms of aromatic rings or as a ketone attached to a hydroaromatic ring. In both cases the oxygen is hydrogen bonded or chelated to adjacent hydroxyl groups. The reactivity and peak location for the characteristic infrared peaks are different from those in typical quinones and ketones.

(8) The species after oxygen in abundance in vitrinites is nitrogen. It is usually present as a heteroatom in a ring structure or nitrile.

(9) Most of the sulfur in coal, especially if the content exceeds 2%, is associated with the inorganic content. For very low sulfur coals the organic sulfur exceeds the inorganic. The sulfur occurs as a heteroatom in rings in a variety of substituted thiophenes, and in thioethers.

(10) A given piece of coal has a variety of molecules in it with different proportions of the structural features and varying molecular weight. They are composed of planar fused ring clusters. These are linked to nonplanar hydroaromatic structures. The overall structure is irregular, open, and complex. Entanglement between molecules

is possible, leading to difficulties with molecular weight determinations for extracts and to changing properties on heating. The different shapes and sizes of molecules in a piece of coal leads to irregular packing and both the amorphous nature of coal and the extensive ultrafine porosity.

(*11*) Some of the evidence suggests that the molecular weights of extracts representing 5–50% of the coal average 1000–3000. Transient units larger than this will exist owing to aggregation. It is probable that the unextracted coal contains molecular units even larger than this.

Heating above 400°C or mild chemical oxidation changes some hydroaromatic-into aromatic structures. It seems that hydroaromatic structures are involved with tar formation (28); however, the tar probably contains some of the same smaller aromatic structures also found in the coal. Tar and gas production, including its change with rank, is consistent (1–2) with the model described above; most of the larger units remain in the char or coke produced by heating. The average size of the units is increased by polymerization and conversion of hydroaromatic links to aromatic rings.

Some representative quantitative values for the features mentioned, including the variation with rank or maceral type, are given below.

Although agreement on the content of functional groups may have been reached, there has been significant divergence of opinion on the carbon–hydrogen skeleton of coal. Two authoritative groups have favored different ranges of carbon aromaticity, ie, the fraction of carbon in the aromatic form. The British school has favored values in the range of 0.65–0.9 based on nmr and other studies, and the Dutch school, using values obtained from coal density measurements, favors values in the range of 0.8–0.9. There is no direct measurement, all information is indirect; however, the least indirect methods appear to be nmr and ir (see Analytical methods). However, the assumptions needed for the factors used in relating the measurements to structural parameters are significantly different. Furthermore, the measurements for coals of the same rank have a range of values. Since many of the skeletal parameters depend on carbon aromaticity, a small change in aromaticity can significantly change the others.

Table 4 gives some characteristics of coals that are fairly widely accepted.

More recent nmr, ir, and chemical studies suggest a higher methyl concentration than earlier work. For coals of 80–90 wt % carbon (dry, mineral-free) the sum of aromatic and hydroaromatic fractions is usually constant (92 ± 2%) (30). The balance is believed to be aliphatic of which 4.0–4.6% is methyl.

Proton nmr was used by Heredy (31) to determine hydrogen types in soluble fractions obtained from coal of 70–91% carbon under mild conditions. These data were combined with others (32) to obtain the average condensed structure size suggested by nmr. The results are given in Table 5 (17).

For Tables 4 and 5, it is understood that the variation with rank is continuous. Typical vitrinite-rich coals were selected for Table 5. An indication is given where the difference was known that would be expected in a corresponding micrinite and exinite, ie, whether the parameter is greater (>v) or less (<v) than that for vitrinite (1–2,4,29).

Constitution of Brown Coals and Lignites. Studies on lignites and brown coals are more limited than those made on the more mature coals. Their major use is for combustion (1). Chemical study of brown coals has been carried out in Europe (18–19) and Russia, and extensively in Australia (20) for the brown coals of Victoria. The brown

Table 4. Average Characteristics of Coals Rich in Vitrinite[a]

Parameter	Vitrinite					Exinite[b]	Micrinite[b]
C[c]	82.5	85.0	87.5	90.0	92.5	$\sim$v	>v
H[c]	5.3	5.3	5.2	4.9	4.15	>v	<v
O[c]	10.4	7.9	5.5	3.4	2.1	<v	$\gtrsim$v
VM[c] (volatile matter)	39	35.5	31	24.5	13.5	>v	<v
H/C (atomic)	0.765	0.74	0.71	0.65	0.53	>v	<v
CH$_2$ bridges	absent						
type of aromatic nuclei	biphenyl, naphthalene, triphenylene,						
	and phenanthrene may predominate						
average number of rings in	>3	$\sim$3	$\sim$3.5	$\sim$4	$\sim$5	$\gtrsim$v	$\gtrsim$v
aromatic nuclei (R_a)							
hydrogen substitution on							
periphery of aromatic							
rings, %	50+	45+	40	30	10−		
average stacking number							
of parallel aromatic							
clusters	1.3	1.35	1.5	1.8	2.15		
O as hydroxyl, % of coal	6.0	5.0	3.5	2.0	1.0	<v	
O as carbonyl, % of coal	$\sim$2.0	$\sim$1.6	$\sim$1.2	$\sim$0.9			
carbon atoms per free radical	$\sim$8000	$\sim$6000	$\sim$4000	$\sim$3000	$\sim$1500	<v	>v

[a] Refs. 1–2, 4, 29.

[b] An indication is given, where known, of the difference to be expected in the corresponding exinite and micrinite, that is, whether the parameter for these is greater (>v) or less (<v) than for the vitrinite.

[c] Wt % dry, mineral matter-free basis.

coals have a relatively high oxygen content. About two-thirds of their oxygen is bonded carboxyl, acetylatable hydroxyl, and methoxy groups. Some alcoholic hydroxyl groups are believed to exist in brown coals, unlike bituminous coals (see Lignite and brown coal).

Constitution of Anthracites. The anthracites approach graphite in composition (see Carbon, graphite). As such they have higher rank than bituminous coals and have less oxygen and hydrogen. They are less reactive and insoluble in organic solvents. These characteristics are more pronounced as rank increases in the anthracite group. The aromatic carbon fraction of anthracites is at least 0.9, and the number of aromatic rings per cluster is greater than that for the coal with 92.5 wt % C in Table 3, with a value of about 10 for anthracite of 95 wt % C. There is x-ray diffraction evidence (33) to indicate that the aromatic rings are more loosely and variably assembled than those in bituminous coal clusters. This would not be surprising if the anthracite clusters resulted from condensation of smaller clusters during coalification with all the difficulties of motion within the solid coal network.

The anthracites have greater optical and mechanical anisotropy than lower rank coals. The internal pore volume and surface increase with rank after the minimum below about 90 wt % C.

Properties and Reactions

Pieces of coal consist of a mixture of materials somewhat randomly distributed in differing amounts. The mineral matter can be readily distinguished from the organic part. The organic material is a mixture of constituents that have undergone a range

Table 5. The Structural Analysis of Coal, Nmr Analysis of Data[a]

		Coal rank				
	Lignite	Subbit. A	Bit. HVA	Bit. HVA	Bit. HVA	Bit. LV
A, g C/100 g sample[b]	70.6	76.7	82.4	85.1	85.8	90.7
B, g H/100 g sample[c]	4.74 ± 0.13	5.19 ± 0.14	5.57 ± 0.15	5.54 ± 0.18	5.04 ± 0.14	4.84 ± 0.11
			wt % H[a]			
C, aromatic, cond ring	14.3	15.3	8.4	12.9	7.4	18.0
D, aromatic, monarom ring	16.4	7.3	11.2	19.2	16.6	13.0
E, hydroxyl	4.6	5.2	4.7	3.8	3.5	3.9
F, α bridge methylene	14.0	4.5	5.3	2.5	2.4	3.2
G, αCH_2 cond + αCH + ethylene bridges[d]	12.6	7.2	11.1	6.2	5.6	4.7
H, αCH_2 mono + αCH_3[e]	13.1	15.8	19.6	17.4	16.1	16.3
I, β and β^+ [f]	20.8	37.5	29.3	29.2	36.7	32.3
J, βCH_3[g]	2.9	7.2	8.5	9.0	11.5	6.2
			wt % C			
K, aromatic, cond ring	11.5 ± 0.3	12.3 ± 0.3	6.8 ± 0.2	10.0 ± 0.3	5.1 ± 0.1	11.4 ± 0.2
L, aromatic, monarom ring	13.2 ± 0.4	5.9 ± 0.2	9.0 ± 0.3	14.9 ± 0.5	11.6 ± 0.3	8.2 ± 0.2
M, hydroxyl[h]	3.7 ± 0.2	4.2 ± 0.1	3.8 ± 0.1	2.9 ± 0.1	2.4 ± 0.1	2.5 ± 0.0
N, α bridge methylene	16.9 ± 0.5	5.4 ± 0.2	6.4 ± 0.2	2.9 ± 0.1	2.5 ± 0.1	3.1 ± 0.1
O, αCH_2 cond + αCH + ethylene bridges[d,h]	15.2 ± 5.0	8.7 ± 2.9	13.4 ± 4.5	7.2 ± 2.4	5.9 ± 2.0	4.5 ± 1.5
P, αCH_2 mono + αCH_3[e]	8.8 ± 3.5	10.6 ± 4.2	13.1 ± 5.4	11.2 ± 4.5	9.4 ± 3.8	8.6 ± 3.4
Q, $\beta + \beta^+ CH_2$[f]	8.4 ± 0.3	15.1 ± 0.4	11.8 ± 0.3	11.3 ± 0.3	12.8 ± 0.4	10.3 ± 0.4
R, βCH_3[g]	0.8 ± 0.0	1.9 ± 0.1	2.3 ± 0.1	2 .3 ± 0.1	2.7 ± 0.1	1.3 ± 0.0
S, interior carbons	21.5 ± 6.0	35.9 ± 5.0	33.4 ± 0.7	37.3 ± 0.5	47.6 ± 4.3	50.2 ± 3.8
T, peripheral ring carbons	51.7 ± 3.1	35.6 ± 2.6	37.1 ± 3.5	38.9 ± 2.6	28.4 ± 5.7	30.7 ± 4.4
U, ratio of interior to peripheral ring carbons	0.42 ± 0.28	1.0 ± 0.15	0.9 ± 0.2	1.0 ± 0.14	1.7 ± 0.37	1.6 ± 0.26
V, average ring size	2–4	6–8	5–8	6–8	11–23	12–20
W, aliphatic carbons	26.8 ± 3.1	28.5 ± 2.5	29.5 ± 3.5	23.8 ± 2.6	24.0 ± 2.2	19.2 ± 1.9
X, ratio aliphatic to ring carbons	0.37 ± 0.12	0.40 ± 0.05	0.42 ± 0.06	0.31 ± 0.06	0.32 ± 0.04	0.24 ± 0.03

[a] This is the first use of nmr data to calculate the chemical structures in coal.

[b] Ref. 19.

[c] Ref. 20.

[d] Wt % H or wt % C as methylene groups attached to a condensed ring plus methyne groups attached to monoaromatic nuclei.

[e] Wt % H or wt % C as methylene groups attached to monoaromatic nuclei plus methyl groups attached to condensed aromatic nuclei.

[f] Wt % H or wt % C on methylene carbons once or further removed from aromatic nuclei.

[g] Wt % H or wt % C on methyl carbons once or more removed from aromatic nuclei.

[h] These values have high uncertainties because it is not possible to distinguish between the two types of carbon in each case. An average with an uncertainty to cover the possible range of error is used.

of changes under coalification conditions. The properties of the coal reflect the individual constituents and their relative proportions. The emphasis of this section is on the properties of the organic matter, and particularly the macerals. By analogy with geology, these are the constituents that correspond to minerals that make up individual rocks. For coals, these macerals represent particular classes of plant parts that have been transformed into their present state in coal (34). Macerals tend to be consistent in their properties. For this reason most detailed chemical and physical studies have been made on macerals or samples rich in a particular maceral, since separation of macerals is so time consuming. The most predominant maceral is vitrinite, corresponding to the American anthraxylon. The other important macerals include inertinite consisting of micrinite, a dull black amorphous material, and fusinite a dull fibrous material similar to charcoal, and exinite which is relatively fusible and volatile.

The differences in macerals are evident over the range of coals from brown coal or lignite to anthracite. These coals cover the full range of coalification or rank. For the purposes of this section, the definition of rank is that generally accepted as the wt % C, on a mineral-free basis, in the vitrinite associated with the given coal in the seam. The range of rank in which differences between macerals are most significant is 75–92 wt % C content of the vitrinite. These coals are bituminous.

The commercial classification of coals is based on the fixed carbon (or volatile matter) content and the moist heating value in the United States. One scientific correlation is made by plotting the hydrogen content, on a mineral-free basis, against the corresponding carbon content. A similar plot can be made using the commercial criteria of volatile matter and heating value if these alternative parameters are placed on axes at an appropriate angle to the C–H axes. This important observation was first used by C. A. Seyler to develop his coal classification chart (13) (Fig. 2, p. 233).

Table 6 indicates the usual range of composition of commercial coals of increasing rank.

Physical Methods of Examination.   Most physical methods are used to obtain information about the chemical nature of coal rather than the particular properties that are measured. Table 7 lists the methods used and types of information obtained from them.

These methods can be divided into two classes which, in the one case, yield information of a structural nature such as the size of the aromatic nuclei (ie, x-ray diffraction, molar refraction, and calorific value as a function of composition), and in the other class indicate the fraction of the carbon in the aromatic form (ie, ir spectroscopy, nuclear magnetic resonance, and density as a function of composition) (see Analytical methods).

Table 6.   Composition of Main Types of Humic Coals

Type of coal	C,[a] wt %	H,[a] wt %	O,[a] wt %	N,[a] wt %	Moisture as found, wt %	Volatile matter,[a] wt %	Calorific value,[a] kJ/g
peat	45–60	3.5–6.8	20–45	0.75–3.0	70–90	45–75	17–22
brown coals and lignites	60–75	4.5–5.5	17–35	0.75–2.1	30–50	45–60	28–30
bituminous coals	75–92	4.0–5.6	3.0–20	0.75–2.0	1.0–20	11–50	29–37
anthracites	92–95	2.9–4.0	2.0–3.0	0.5–2.0	1.5–3.5	3.5–10	36–37

[a] Dry, mineral matter free basis; to convert kJ/g to Btu/lb, multiply by 430.2.

Table 7. Physical Tests for Structure Analysis [a]

Measurement of	Yields information on
x-ray diffraction	size distribution of aromatic ring systems
	average diameter of aromatic lamellae, nuclei, or clusters (related to R_a)
	mean C–C bond length
	average thickness of the packets of lamellae
uv and visible absorption	aromaticity (fraction of carbon in aromatic structures)
	average number of rings in aromatic nuclei, R_a
reflectance	aromatic surface area (related to R_a)
optical refractive index (molar refraction)	optical anisotropy
ir absorption	characteristic groups such as OH, CH_{ar}, CH_{al}, $(C{=}C)_{ar}$, H_{ar}/H_{al} ratio
^{1}H nmr	H_{ar}/H_{al} ratio
electron spin resonance	free radical content
electrical conductivity	average number of rings in aromatic nuclei, R_a
diamagnetic susceptibility (molar diamagnetic susceptibility)	average number of rings in aromatic nuclei, R_a
dielectric constant	dipole moment
sound velocity	aromaticity
density (molar volume)	aromaticity
	ring condensation index, $2(R_a - 1)/C$ (related to R_a)

[a] Note that $_{ar}$ = aromatic and $_{al}$ = aliphatic.

The scattering of x-rays (4,33) gives information on the average distances between the carbon atoms in coal or insight into the bonding between these atoms. X-ray scattering depends on the number of protons in the nucleus, therefore, carbon is much more effective than hydrogen. The method of analysis is quite complex. The results have been used with information from other methods in the preceding discussion of constitution.

The ultraviolet and visible spectra (4) of coal and solvent extracts show decreasing absorption with increasing wavelength and lack features to aid in interpreting structure except for one peak around 270 nm. This is believed to result from superposition of effects from many species present in these samples. In studies of specific features comparisons are usually made between coal or coal-derived samples and pure compounds, usually aromatic, of interest. The comparisons indicate the probable presence of certain structures or functional groups. Similar statements can be made concerning reflectance and refractive index (1–2). The derived optical anisotropy is especially evident in coals with carbon contents exceeding 80–85 wt %. Measurements perpendicular and parallel to the bedding plane will give different results for optical and some other characteristics.

A significantly greater amount of information concerning the functional groups, such as hydroxyl, present in the coal derived samples can be obtained from infrared absorption (1–2,4); however, this is less specific than the information obtained from an individual organic compound (see Fig. 3). An estimate of the relative amount of hydrogen attached to aromatic and nonaromatic structures can be made by using this method.

Studies may be carried out on raw coal or products derived from the coal. Physical

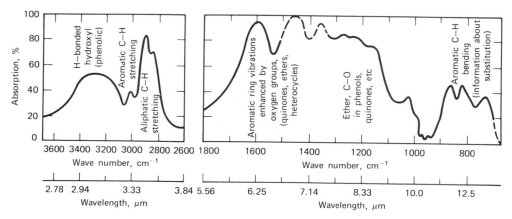

Figure 3.   Typical ir spectrum of a medium-rank coal.

separation is used to separate fractions of extract and aid in the deduction of the parent coal structure. A method of characterizing coal liquids in terms of ten fractions of different functionality was recently described (35).

Nmr spectra (^{1}H) (4,31,36–37) also yield information on bonding for hydrogen and carbon, including the distribution between aromatic and nonaromatic structures. These estimates are somewhat more quantitative than those from ir measurements. Additional estimates may be made of hydrogen in methylene (CH_2) groups. The best data require that the coal is dissolved or degraded, and these materials do not fully represent the original coal sample. Combinations of this information with broad line spectra give the maximum amount of information.

Esr (4) has had more limited use in coal studies. A rough estimate of the free radical concentration or unsatisfied chemical bonds in the coal structure has been obtained as a function of coal rank and heat treatment. For example, the concentration increases from 2×10^{18} radicals/g at 80 wt % carbon to a sharp peak of about 50×10^{18} radicals/g at 95 wt % carbon content, and drops almost to zero at 97 wt % carbon. The concentration of these radicals is less than that of the common functional groups such as hydroxyl however, their existence seems to be intrinsic to the coal molecule and may affect the reactivity of the coal as well as its absorption of ultraviolet radiation.

The other physical measurements (2,4) used in Table 6, except for diamagnetic susceptibility (2) and possibly density (2), are primarily of interest for deriving physical properties of coal. The conclusions concerning the nature of the coal molecule from these physical measurements are inferences based on analogy with properties of pure compounds and are less meaningful than more direct measurements.

Physical Properties.   Most of the physical properties discussed here depend on the direction of measurement compared to the bedding plane of the coal. Additionally, the properties vary according to the history of the piece of coal. They also vary between pieces because of the brittle nature of coal and of the crack and pore structure. One example is electrical conductivity. Absolute values of specific conductivity of coal samples are not easy to determine. A more characteristic value is the energy gap for transfer of electrons between molecules. This is determined by a series of measurements over a range of temperature and is unaffected by the presence of cracks. The velocity of sound is also dependent on continuity in the coal.

The specific electrical conductivity of dry coals is very low (specific resistance 10^{10}–10^{14} $\Omega\cdot$cm), although it increases with rank. Coal has semiconducting properties. The conductivity tends to increase exponentially with increasing temperatures (2,4). As coals are heated to above ca 600°C the conductivity rises especially rapidly owing to rearrangements in the carbon structure, although thermal decomposition contributes somewhat below this temperature. Moisture increases conductivity of coal samples through the water film.

The dielectric constant (ϵ) is also affected by structural changes on strong heating. Also it is very rank dependent, exhibiting a minimum at about 88 wt % C, and rises rapidly for carbon contents over 90 wt % (2,4). The dielectric constant (in debye units) equals the square of the refractive index only at the minimum value. This relationship is theoretically valid for nonconducting and nonpolar substances. For both higher and lower ranks the dielectric constant exceeds the square of the refractive index (the values of the former are 5 D (16.7×10^{-30} C·m) at 70 wt % carbon, 3.5 D (11.7×10^{-30} C·m) at the minimum, and ca 10 D (33.4×10^{-30} C·m) at 91 wt % carbon). The presence of polar functional groups is responsible for this with lower ranks, and for higher ranks it is caused by the increase in electrical conductivity. The dipole moments of lower rank coals, calculated from the difference, are 0–1.8 D (0–6×10^{-30} C·m to 0 C·m). The value of ϵ above 88 wt % C is mainly determined by the polarizability of the π electrons in the aromatic clusters and it is not appropriate to calculate dipole moments. The refractive index values (1–2,4) are not significantly affected by anisotropy, and mean values around 500 nm increase from 1.7 at 70 wt % carbon to 1.8 at 85 wt % carbon to almost 2.0 at 95 wt % carbon.

Density values (2,4) of coals differ considerably, even after correcting for the mineral matter, depending on the method of determination. The true density of coal matter is most accurately obtained from measurement of the displacement of helium after the absorbed gases have been removed from the coal sample. Density values increase with carbon content or rank for vitrinites. They are 1.4–1.6 g/cm³ above 85 wt % carbon where there is a shallow minimum. A plot of density versus hydrogen content gives almost a straight-line relationship. If the reciprocal of density is plotted, the relationship is improved. Values for different macerals as well as for a given maceral of different ranks are almost on the same straight line.

The thermal conductivity and thermal diffusivity are also dependent on the pore and crack structure. Thermal conductivities for coals of different ranks at room temperature are in the range of 0.23–0.35 W/(m·K). The range includes the spread owing to crack variations and thermal diffusivities of (1–2) $\times 10^{-3}$ cm²/s. At 800°C these ranges increase to 1–2 W/(m·K) and (1–5) $\times 10^{-2}$ cm²/s. The increase is mainly caused by radiation across pores and cracks.

The specific heat of coal can be determined by direct measurement or from the ratio of separate measurements of thermal conductivity and thermal diffusivity. The latter method gives values decreasing from 1.25 J/(g·K) (0.3 cal/(g·K)) at 20°C to $\leq$0.4 J/(g·K) ($\leq$0.1 cal/(g·K)) at 800°C.

Ultrafine Structure. Coal contains an extensive network of ultrafine capillaries (1–2,4) that pass in all directions through any particle. The smallest and most extensive passages are caused by the voids from imperfect packing of the large organic molecules. Vapors pass through these passages during adsorption, chemical reaction, or thermal decomposition. The rates of these processes depend on the diameters of the capillaries and any restrictions in them. Most of the inherent moisture in the coal is contained in these capillaries. The porous structure of the coal and products derived from it will have a significant effect on the absorptive properties of these materials.

A range of approaches has been developed for studying the pore structure. For example, heat of wetting by organic liquids is one measure of the accessible surface. The use of liquids with different molecular sizes gives information about restrictions in the pores. Measurements of the apparent density in these liquids give corresponding information about the volume of capillaries. Measurement of the adsorption of gases and vapors provide information about internal volume and surface area. Pores have been classified into three size ranges: (1) micropores (0.4–1.2 nm) measured by CO_2 adsorption at 298 K; (2) transitional pores (1.2–30 nm) from N_2 adsorption at 77 K; and (3) macropores (30–2960 nm) from mercury porosimetry. For coals with less than 75 wt % C, macropores primarily determine porosity. For 76–84 wt % C about 80% of the pore volume is caused by micro and transitional pores. For higher rank coals porosity is caused primarily by micropores (38).

It is generally agreed that bituminous coals have specific internal surfaces in the range of 30–100 m²/g almost entirely from ultrafine capillaries with diameters less than 4 nm. The surface owing to the very fine capillaries can be measured accurately by using methods not too far below room temperature, depending on the gas or vapor used, since diffusion into the particle is very slow at low temperatures. Therefore, measurements at liquid air temperature relate to the external surfaces and macropores and may yield areas that are low by factors of 100. Sorption by neon or krypton near room temperature, and heat of wetting in methanol have been favored methods of obtaining surface area values. The methanol method is affected (factor up to 4) by polar groups but is faster. Total porosity volumes of bituminous and anthracite coal particles are about 10–20% with about 3–10% in the micro range. There are shallow minima in plots of internal area (Fig. 4) or internal porosity against the rank in the range of vitrinite carbon content of 86–90%.

It is possible to use low-angle scattering of x-rays to obtain a value of internal

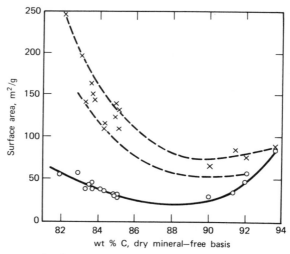

Figure 4.   Surface area of coals as estimated from neon sorption at 25°C and methanol heat of wetting. Note that x = methanol and O = neon.

surface, but this does not distinguish between accessible capillaries and closed pores. This method may give the highest values of all.

The electron microscope has been applied to different coals. Two pore-size ranges appear to have been observed, one of >20 nm and the other <10 nm (39). Fine pores 1–10 nm across have been observed with a lead impregnation procedure (40).

Effectiveness of coal conversion processes depends on rapid contact of gases with the surface. Large internal surfaces will be required for satisfactory rates (41).

Mechanical Properties. Mechanical properties (2,4,42) are important for a number of steps in coal preparation from mining through handling, crushing, and grinding. The properties include elasticity and strength as measured by standard laboratory tests and empirical tests for grindability and friability; and indirect measurements based on particle size distributions. A number of efforts to correlate grinding energy input to size reduction have been made.

Deformation Under Load. The mechanical behavior of coal is strongly affected by the presence of cracks, as shown in the lack of proportionality between stress and strain in compression tests or in attempts to relate strength and rank. However, tests in triaxial compression indicate that, as the confining pressure is increased, different coals tend to exhibit similar values of compressive strength perpendicular to the directions of confining pressure.

Different coals exhibit small amounts of recoverable and irrecoverable strain under load, although this is not observed with anthracites. Dynamic tests have been used to measure the variation of elastic properties with rank. One series of tests with vitrains suggested that coals were mechanically isotropic up to 92 wt % C, with anisotropy increasing above this value. Dynamic tests were used to measure internal energy losses in vibration and to study the fluidity changes of coking coals on oxidation.

The Young's modulus for medium rank coals has been found to be ca 4 GPa (4 $\times 10^{10}$ dyn/cm^2) by several workers. Sharp increases in the Young's and shear moduli have been found in vitrinites with increase in carbon content over 92 wt %.

Strength. The strength of a coal as measured in the usual laboratory tests may not be relevant to mining or size reduction problems where the applied forces are much more complex. There are indications that compressive strength (measured by compression of a disk) may give useful correlations with the ease of cutting with different kinds of equipment. Studies of the probability of survival of pieces of different size suggest that the breaking stress, S, should be most closely related to the linear size, x, rather than the area or volume of the piece. The results of a number of studies are of the form:

$$S \propto x^{-r}$$

where r frequently has the value $\frac{1}{2}$.

The effects of rank on both compressive and impact strength have been studied, and usual minima were found at 20–25% dry, ash-free volatile matter (88–90 wt % carbon). Accordingly, the Hardgrove grindability index exhibits maximum values in this area.

Theoretical Derivations of Size Distribution Relationships. Different models have been used to derive relationships describing the particle size distribution of particles experiencing single and multiple fractures. A model based on fracture at the site of the weakest link and a distribution of weakest links in the system gave results that could be described as well by the Rosin-Rammler relation (43). The latter is based on the concept that fracture takes place at preexisting flaws that are distributed randomly throughout the particle.

   Comminution. A relationship between energy consumption and size reduction is needed for comminution processes. None of the many attempts to develop this have been broadly applicable. One reason is that generation of new surface is only one of many phenomena in the size reduction process. The energy requirements of a comminution system may be estimated from laboratory tests for given amounts of size reduction. For pulverized coal-fired boilers 70% < 74 μm size (-200 mesh) is frequently used. Product size distributions of reproducible forms are claimed to be obtained from a range of graded coal input sizes with careful control of crushing conditions in the laboratory. The efficiency of pulverizers can be calculated from the energy requirements for each product size from a series of tests (see Size reduction).

   The relationship between a particle size and energy consumption obtained from plant data is frequently expressed in terms of Kick's or Rittinger's laws (44) or some modification such as Bond's law (45). These empirical relationships do not provide much insight into the mechanism of the grinding process. The development of a continuous grinding index has been the purpose of more recent work (46). The laboratory test equipment is similar to that used for the Hardgrove test but permits classifying the product and recycling the oversize material. An improved correlation is obtained but may need to be corrected for the effect of relative sizes of the test grinding balls and those used in commercial-scale equipment. This method is especially useful for lower rank coals.

   Empirical Measures of Resistance to Breakage. The size reduction of coal during handling or comminution is the result of the effects of many ill-defined forces. Grindability and friability tests are useful indicators of the size reduction for a given coal with a specific energy consumption. The Hardgrove test gives an index that varies with coal rank, moisture content, and ash and maceral distribution. The higher indexes are indicative of lower energy requirements to achieve a given size reduction and are useful in establishing capacity factors for pulverizers. Grinding is easiest for coals with 75–80% dry, ash-free fixed carbon. Optimum moisture contents have been observed for the younger coals. ASTM tests for grindability and friability exist and the effects of several parameters such as ash content have been studied. There is a need for tests such as the continuous grinding index to more closely simulate grinding system behavior and permit prediction of performance.

   Solvent Extraction. A wide range of organic solvents can dissolve part of coal samples (4,47) but dissolution is never complete and usually requires heating to temperatures sufficient for some thermal degradation or reaction with the solvent to take place (ca 400°C). At room temperature the best solvents are primary aliphatic amines, pyridine, and some higher ketones, especially when used with dimethylformamide. Above 300°C large amounts can be dissolved with phenanthrene, 1-naphthol, and some coal-derived high boiling fractions. Dissolution of up to 40 wt % can be achieved near room temperature, and up to 90% near 400°C. Coals with 80–85% carbon in the vitrinite give the largest yields of extract. Very little coal above 90 wt % C dissolves.

   When the concentration of dissolved coal exceeds about 5% of the solution by weight, the extracted material resembles the parent coal in composition and some properties. The extract consists of the smaller molecules within the range of the parent coal. The recovered extract is relatively nonvolatile and high melting.

A kinetic study of coal dissolution indicated increasing heats of activation for increasing amounts of dissolved coal. These results indicate that weaker bonds were broken initially, and the stronger bonds later (48).

The Pott-Broche process (49) was best known as an early industrial use of solvent extraction of coal, but was ended owing to war damage. The coal was extracted at about 400°C for 1–1.5 h under a pressure of 10–15 MPa (100–150 atm) with a coal-derived solvent. Plant capacity was only 5 t/h with an 80% yield of extract. The product contained less than 0.05% mineral matter and had limited use, mainly in electrodes.

Solvent extraction work was carried out by a number of organizations in the United States. Pilot plants were built in Wilsonville, Alabama, in 1973 and near Tacoma, Washington, in 1974 with capacities of 5 and 45 t/d of coal input, respectively. Heating value of the product solvent refined coal (SRC) is ca 37 MJ/kg (ca 16,000 Btu/lb). Sulfur contents have been reduced from 2–7% initially to 0.9% and possibly less. Ash contents have been reduced from 8–20% to 0.17% (50). These properties permit compliance with 1976 EPA requirements for SO_2 and particulate emissions. The SRC is primarily intended to be used as a boiler fuel in either a solid or molten form (heated to ca 315°C). The solid has a Hardgrove index of 150 (51). Boiler tests have been successfully carried out with a utility boiler.

In the SRC process, crushed coal is slurried with a coal-derived anthracene oil. Hydrogen is added under pressure. The slurry is heated at 400–455°C at 12.4 MPa (1800 psi), and allowed to react up to ca 30 min. The product stream contains some hydrocarbon gases, H_2S, filtration or other separation is used to remove unreacted coal and ash. The residue is used to generate hydrogen for the process. The remaining filtrate is separated into solvent, which is recycled, and SRC.

Properties Involving Utilization. Coal rank is the most important single property for almost every application of coal (see Classification). The rank sets limits on many properties such as volatile matter, calorific value, and swelling and coking characteristics. Other properties of significance include grindability, ash content and composition, sulfur content, and size.

Combustion. Most of the mined coal is burned in boilers to produce steam for electric power generation. The calorific value determines the amount of steam that can be generated. The design and operation of a boiler requires consideration of a number of other properties (see Furnaces, fuel-fired; Burner technology).

In general, high rank coals are more difficult to ignite, requiring supplemental oil firing and slower burning, with large furnaces to complete combustion. The greater reactivity of lower rank coals makes them better suited for cyclone burners which carry out rapid, intense combustion to maximize carbon utilization and minimize smoke emission.

Ash content is important. Ash discharge at high temperature, as molten ash from a slagging boiler, involves substantial amounts of sensible heat. The higher cost of washed coal of lower ash content will not always merit its use. Ash disposal and extra freight costs for high ash coals enter the selection of coal. The current use of continuous mining equipment produces coal with about 25% ash content. The average ash content of steam coal is about 15% in the United States. For some applications, such as chain-grate stokers, a minimum ash content of about 7–10% is needed to protect the metal parts.

Ash fusion characteristics are important in ash deposition with boilers. Ash deposition occurring on the furnace walls is termed slagging, while accumulation on the superheater and other tubes is termed fouling. A variety of empirical indexes have been developed (52–53) to relate fouling and slagging to the ash chemical composition through parameters including acid and base content, sodium, calcium and magnesium, and sulfur.

A related property is viscosity of coal ash; this affects the rate at which ash deposits may flow from the walls, and affects the requirements for ash removal equipment such as wall blowers and soot blowers. The preferred coal ash will have a narrow temperature range through which it passes the plastic range (ca 25–1000 Pa·s or 250–10,000 P) (54).

For pulverized coal firing, a high Hardgrove index or grindability index is desired. This implies a relatively low energy cost for pulverizing since the coal is easier to grind. The abrasiveness of the coal is also important since this determines the wear rate on pulverizer elements.

Moisture content affects handling characteristics. It is most important for fines smaller than 0.5 mm. The lower rank coals have higher moisture contents that affect the freight costs, especially when the coal fields are considerable distances from consumption. The moisture acts as a diluent, lowering flame temperatures and carrying sensible heat out with the flue gases. For pulverized coal firing the moisture content must be low to maximize grindability and avoid clogging. For this reason it may be more desirable to buy dry run of mine coal with up to 30% ash than to clean·and dry the coal. Moisture is sometimes desirable. About 8% is necessary for prevention of combustible loss from a chain-grate stoker.

The moisture content of peat or brown coal that is briquetted for fuel must be reduced to about 15% for satisfactory briquetting. Mechanical or natural means are used because of the cost of thermal drying.

Some minor constituents can cause trouble in firing. High chlorine (>0.6%) is associated with high sodium and complex sulfate deposits that appear to be required to initiate deposition on superheater tubes. Phosphorus (>0.03% of the coal) contributes to phosphate deposits where high firing temperatures are used. Sulfur is also involved in the complex sulfates; however, its most damaging effect is corrosion of the boiler's coolest parts through condensation as sulfuric acid. Control is achieved by setting flue gas temperatures above the acid dew point in the boiler areas of concern.

The sulfur content is important in meeting air quality standards. In the United States, the EPA has set a limit of 516 g SO_2 emission per million kJ (490 g SO_2 10^6 Btu) of coal burned. To meet this, steam coals have to contain less than 1% sulfur. This requirement has forced the addition of SO_2 scrubbing equipment to many boilers, and legislation is contemplated to require scrubbers for all units.

Volatile matter is important for ease of ignition. High rank coals have low volatile matter and burn more slowly. They burn slowly with a short flame and are used for domestic heating, where heat is transferred primarily from the fuel bed. For kilns, long hot flames are needed and the coal should have medium to high volatile matter. The amount of the heating value released with the volatile matter is given in Table 8.

The swelling and caking properties of coal are not important for most boiler firing. Some units, however, such as retort stokers, form coke in their normal operation. The smaller domestic heating units require noncaking coal for satisfactory operation.

The advent of fluidized bed boilers will require coal particles that are about 2–3 mm dia. This technology is being developed with a 30-MW pilot plant near Rivesville, West Virginia (see Reactor technology).

Table 8. Rank and Heating Value in Volatile Matter

Rank (ASTM)	Volatile matter, %	Total heat energy liberated in volatile matter, %
anthracite	<8	5–14
semianthracite	8–14	14–21
low-volatile bituminous	14–22	21–28
medium-volatile bituminous	22–31	28–36
high-volatile A bituminous	>31	36–47

Coke Production. Coking coals are mainly selected on the basis of the quality and amount of coke that they will produce. Gas yield is also considered. About 65–70% of the coal charged is produced as coke. The gas quality depends on the coal rank and is a maximum (in energy in gas per mass of coal) for coals of about 89% carbon (dry, mineral matter-free) or 30% volatile matter.

Coals with 18–32% volatile matter are used to produce hard metallurgical coke. Methods have been developed to blend coals with properties outside this range to produce coke. Several coals are frequently blended to improve the quality of the coke (55). Blending contributes to the quality of shrinkage required to remove the coke from the ovens after initial swelling.

Lower rank coals with up to 40% volatile matter may be used alone or in blends at a gas-making plant. This coke need not be as strong as metallurgical coke and is more reactive and used in the domestic market.

The coal is cleaned so that the coke ash content is not over 10%. An upper limit of 1–2 wt % sulfur is recommended for blast furnace coke. A high sulfur content causes steel to be brittle and difficult to roll. There are seams whose coking properties are suitable for metallurgical coke, but whose sulfur content prevents that application.

Small amounts of phosphorus also make steel brittle. Low phosphorus coals are needed for coke production, especially if the iron ore contains phosphorus.

A Clean Coke process is being developed by U.S. Steel Corp., with partial Department of Energy support, to convert coals not normally used for coke production, including high sulfur coal, to a low ash, low sulfur metallurgical coke, chemical feedstocks, and other fuels. The coal is initially carbonized, then pelletized using a binder derived from another part of the process before heating in a continuous coking kiln.

Gasification. A number of gasifiers are either available commercially or in various stages of development. The range of coals that may be used vary from one gasifier type to another.

The Lurgi fixed bed gasifier operates with lump coal of a noncaking type with an ash composition chosen to avoid a sticky, partly fused ash in the reactor. A slagging version of this gasifier has been tested in Westfield, Scotland, and is planned for a demonstration plant in Noble County, Ohio. Other fixed-bed gasifiers have similar coal requirements.

The Koppers-Totzek gasifier is an entrained bed type. It can gasify lignite and subbituminous or bituminous coal. The coal is fed as a pulverized fuel, usually ground to 70% <74 μm (−200 mesh) as used for pulverized coal fired boilers. Residence times are only a few seconds, therefore coal reactivity is important. The gasifier operates at >1650°C, so that coal ash flows out of the gasifier as a molten slag. Coal ash composition must permit continuing molten ash flow.

Fluidized-bed gasifiers require a coal feed of particles near 2–3 mm dia. Caking coals are to be avoided since they usually agglomerate in the bed. This can be avoided with a pretreatment consisting of a surface oxidation with air in a fluidized bed. A useful fuel gas is produced. Examples of this type include the commercially available Winkler, and the Hygas and U-Gas reactors being developed by the Institute of Gas Technology in Chicago under joint sponsorship of the U.S. Dept. of Energy and the American Gas Association.

Many of the coal selection criteria for combustion apply to gasification, which is a form of partial oxidation.

The larger gasification plants, intended to produce ca 7×10^6 m^3 standard (250 million SCF) of methane per day will be sited near a coal field with an adequate water supply. It is cheaper to transport energy in the form of gas through a pipeline than coal by either rail or pipeline. The process chosen will utilize the available coal in the most economical manner (see Fuels, synthetic-gaseous).

Reactions.   Mature coals (>75% C) tend to be built of assemblages of polynuclear ring systems connected by a variety of functional groups (1–2,4,17). The ring systems themselves contain many functional groups. These so-called molecules differ from each other to some extent in the coal matter. For bituminous coal, a tar-like material fills some of the interstices between the molecules. Generally coals are nonvolatile except for some moisture and light hydrocarbons. The volatile matter produced on carbonization reflects decomposition of parts of the molecule and the release of moisture. Rate of heating affects the volatile matter content. Faster rates give higher volatile matter yields. Coal is not carbon or a hydrocarbon. The composition will depend on the rank. The empirical composition of a vitrinite of ca 84 wt % carbon content is approx $C_{13}H_{10}O$ with small quantities of N and S.

The surface of coal particles undergoes air oxidation. This process may initiate spontaneous combustion in storage piles or weathering with a loss of heating value and coking value during storage. Combustion produces carbon dioxide and water vapor as well as oxides of sulfur and nitrogen. The SO_x results from oxidation of both organic sulfur and inorganic forms such as pyrite. Nitrogen oxides are formed primarily from the nitrogen in the coal, rather than from the air used for combustion. Partial oxidation as carried out in gasification produces carbon monoxide, hydrogen, carbon dioxide, and water vapor. The carbon dioxide will react with hot carbon from the coal to produce carbon monoxide, and steam will react with carbon to produce carbon monoxide and hydrogen. The hydrogen can react with carbon through direct hydrogen gasification:

$$C + 2\,H_2 \rightarrow CH_4$$

at high hydrogen pressure, frequently 6.9 MPa (1000 psi) and moderate temperatures of 650–700°C. The methane may also be produced:

$$CO + 3\,H_2 \rightarrow CH_4 + H_2O$$

in a nickel-catalyzed reactor. The latter reaction is highly exothermic and is used to provide steam for the process. The correct 3:1 ratio of hydrogen and carbon monoxide is accomplished using the water gas shift reaction:

$$CO + H_2O \rightarrow H_2 + CO_2$$

A mixture of CO and H_2, called synthesis gas, may also be used in other catalytic reactors to make methyl alcohol or hydrocarbons:

$$CO + 2\,H_2 \rightarrow CH_3OH$$

$$n\,CO + 2\,n\,H_2 \rightarrow (CH_2)_n + n\,H_2O$$

Surface oxidation short of combustion, or with nitric acid or potassium permanganate solutions, produces regenerated humic acids similar to those extracted from peat or soil. Further oxidation will produce aromatic and oxalic acids, but at least half of the carbon forms carbon dioxide.

Treatment with hydrogen at 400°C and ⩾12.4 MPa (1800 psi) increases the coking power of some coal and produces a change that resembles an increase in rank. Hydrogenation with an appropriate solvent liquefies coal. Noncatalyzed processes primarily produce a tar-like solvent refined coal used as a boiler fuel. Catalysts and additional hydrogen are used in the H-Coal process developed by Hydrocarbon Research, Inc. to produce more liquid products. A 500 t/d plant was built in Catlettsburg, Kentucky, to demonstrate this process by making a coal-derived refinery feedstock. Hydrogen reactions over short times (0.1–2 s) with very rapid heating will produce a range of liquids (benzene, toluene, xylene, phenol). A less rapid heating with lower maximum temperatures will permit removal of some sulfur and nitrogen from the coal (56). These effects have not been commercialized (see Fuels, synthetic-liquid).

Treatment of coal with chlorine or bromine results in addition and substitution of the halogens on the coal. At temperatures up to 600°C chlorinolysis produces carbon tetrachloride, phosgene, and thionyl chloride (57). Treatment with fluorine or chlorine trifluoride at atmospheric pressure and 300°C can produce large yields of liquid products.

Hydrolysis with aqueous alkali has been found to remove ash material including pyrite. A small pilot plant for studying this process was built at Battelle Memorial Institute in Columbus, Ohio. Other studies have produced a variety of gases and organic compounds such as phenols, nitrogen bases, liquid hydrocarbons, and fatty acids totaling as much as 13 wt % of the coal. The products indicate that oxidation and other reactions as well as hydrolysis take place.

The pyritic sulfur in coal can undergo reaction with sulfate solutions to release elemental sulfur. The TRW Co. is developing a process with EPA support to reduce the sulfur content of coal (58). The reaction of coal with sulfuric acid has been used to produce cation exchangers, but was not very efficient and is no longer used.

Many of the products made by hydrogenation, oxidation, hydrolysis, or fluorination are of industrial interest. The increasing cost of petroleum and natural gas is increasing interest in some of these materials primarily as upgraded fuels to meet air pollution control requirements as well as to take advantage of the greater ease of handling of the liquid or gaseous material, and to utilize existing facilities such as pipelines and furnaces. Demonstration plants are planned for conversion of coal to methane (substitute natural gas or SNG), industrial fuel gas, ammonia, coal-derived refinery feedstock, and boiler fuel (see Gas natural; Hydrocarbons). A chemistry based on synthesis gas conversion to other materials is being developed and may find more extensive application in another decade. At present the major production of chemicals from coal involves the by-products of coke manufacturing.

Plasticity of Heated Coals.   *Cause and Nature.*   Coals with a certain range of composition associated with the bend in the Seyler diagram (see Fig. 1) and having 88–90 wt % carbon will soften to a liquid condition when heated (2,4). These are known as prime coking coals. The soft condition is somewhat reversible for a time, but does not persist for many hours at 400°C, and will not be observed above ca 550°C if the sample is continuously heated as in a coking process. This is caused by degradation of coal matter, releasing vapors and resulting in polymerization of the remaining material. The coal does not behave like a Newtonian fluid and only permits empirical measurements of plasticity to be made (see Rheological measurements). Studies indicate that about 10–30% of the coal becomes liquid with a melting point below 200°C. The molten material plasticizes the remaining solid matrix. The molten part of a vitrinite is similar to the gross maceral. A part of the maceral is converted to a form that can be melted after heating to 300–400°C. The molten material is unstable and forms a solid product above 350°C with rates increasing with temperature. The decomposition of the liquid phase is rapid with lower rank noncoking coals, and less rapid with prime coking coals. The material that melts resembles coal rather than tar and, depending on rank, only a slight or moderate amount is volatile.

The fluidity of coal will increase and then decrease at a given temperature. This has been interpreted in terms of the sequence of reactions: coal → fluid coal → semicoke. In the initial step, a part of the coal is decomposed to add to that which normally becomes fluid. In the second step, the fluid phase decomposes to volatile matter and a solid semicoke. The semicoke later fuses with evolution of additional volatile matter to form a high temperature coke.

Formation of a true coke requires that the fluid phase persist long enough during heating for the coal pieces to form a compact mass before solidification occurs from the decomposition. Too much fluidity leads to an expanded froth owing to formation of dispersed bubbles from gas evolution in the fluid coal. Excess bubble formation results in a weak coke. The porous nature of true coke is caused by the bubble formation during the fluid phase. The strength of semicoke is set by the degree of fusion during the fluid stage and the thickness of the bubble walls formed during the frothing. In the final conversion to a hard high temperature coke, additional gas evolution occurs while the solid shrinks and is subjected to thermal stresses. The strength of the resultant coke and the size of the coke pieces are strongly affected by the crack structure produced as a result of the thermal stresses. Strong large pieces of coke are desired to support the ore burden in blast furnaces.

Laboratory Tests.   Several laboratory tests (1,4) are used to determine the desirability of a coal or blend of coals for making coke. These are empirical and are carried out under conditions that approach the coking process. The three properties that have been studied are swelling, plasticity, and agglomeration.

Several dilatometers have been developed to determine the swelling characteristics of coals. The sample is placed in a cylindrical chamber with a piston resting on the coal surface. The piston motion reflects coal volume changes and is recorded as a function of temperature with a constant heating rate. When the coal first softens, contraction is caused by the weight of the piston on softened coal particles that deform to fill void spaces. Swelling then takes place when the particles are fused sufficiently to resist the flow of the evolving gases. The degree of swelling depends on the rate of release of volatile matter and the plasticity of the coal. The mass stabilizes at 450–500°C as the semicoke hardens. The shapes of the curves vary with different dilatometers. A typical curve is shown in Figure 5. Curves obtained with the Hofmann apparatus have been classified into four main Hofmann types (see Fig. 6) which permit distinguishing coals having the optimum softening and swelling properties for producing a strong coke (59).

Free swelling tests are commonly used to measure a coal's caking characteristics. A sample of coal is placed in a crucible or tube without compaction, and heated at a fixed rate to about 800°C. Infusible coals distill without changing appearance or their state of agglomeration. The fusible coals soften, fuse, and usually swell. The profile of the resultant cake is compared with a series of reference profiles so that a swelling index can be assigned. The profiles represent indexes between 0 and 9. The best cokes come from coals with indexes between 4 and 9.

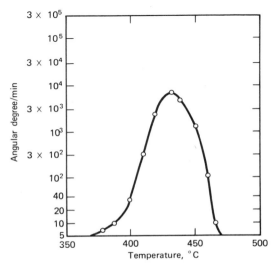

Figure 5. Plasticity curve obtained with the Gieseler plastometer. Heating rate is 2°C/min. (Courtesy of Centre d'Etudes et Recherches des Charbonnages de France).

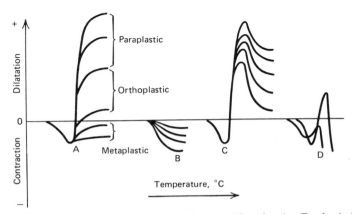

Figure 6. Coal classification system according to Hofmann. Note that A = Eu plastic; B = Sub plastic; C = Per plastic; and D = Fluido plastic. (Courtesy of Centre d'Etudes et Recherches des Charbonnages de France and Brennstoff-Chemie).

The Gray-King assay is obtained from a similar test. The coal is heated to 600°C in a horizontal tube. Standard photographs are used in a comparison of general appearance, profile, and size of the coke mass. Before testing, the more fusible coals are mixed with varying amounts of a standard electrode carbon of carefully selected size. A nonuniform scale termed A-F and G-G9 has been developed from the coke appearance for low swelling coals or from the amount of carbon required to give a standard appearance for the high swelling coals. The U.K. National Coal Board Rank Code Numbers are partly assigned on the Gray-King assay and partly on the volatile matter. The Gray-King assay procedure can also permit evaluation of yields of tar, gas, and liquor.

Plasticity can be studied with a small device known as the Gieseler plastometer. In this device, coal is stirred by paddles driven by a weight suspended from a pulley wheel. The rate of rotation indicates the fluidity and is plotted as a function of the coal temperature. These curves (see Fig. 5) have a well-defined peak for coking coals, usually near 450°C. Softening occurs at 350–400°C. At a normal heating rate of 2°C/min, the fluid hardening may be complete by 500°C.

Several agglutinating and agglomerating tests are used. These indicate the bonding ability of the fusible components and depend on the crushing strength of a coke button produced in some cases with addition of inert material.

The Roga agglutinating test, developed in Poland, provides one of the criteria of the Geneva International Classification System. The coal sample is mixed with carefully sized anthracite, compacted, and heated to 850°C in 15 min. The part of the product that passes through a 1 mm (ca 18 mesh) screen is weighed. A rotating drum further degrades the product. Roga indexes from 0–70 have been determined.

A coherent plastic layer from a few mm to 2–3 cm thick separates the semicoke and coke from the unfused coal in the coke oven. Coking properties are assessed in the U.S.S.R. and some other countries by a measurement of the thickness of this plastic layer. The standardized test by Sapozhnikov (3), widely used in eastern Europe, is the best known of this type. A penetrometer test is used to measure the thickness of the plastic layer in a column of coal heated from the bottom. This test is used for classifying coking coals in the U.S.S.R.

The different standard tests give results that are similar but do not give close correlations with each other.

The behavior of different polymerizing and gas-relating materials has been used to relate the plastic behavior of coal with known kinds of chemical change, notably by van Krevelen (2).

The plastic nature of coal matter is determined by the competition between the reactions that generate the liquid phase, and those that convert it to semicoke. In general, the greater the heating rate, the greater the fluidity or plasticity and the dilatation. The development of the Clean Coke process by U.S. Steel involves fluidized carbonization with rapid heating.

This discussion applies primarily to vitrinites. Inertinite essentially does not contribute to the plastic properties of the coal. Exinite becomes fluid when heated, but also rapidly devolatilizes instead of forming semicoke, has little value as a binder, and can increase the fluidity to an undesirable extent.

Pyrolysis of Coal. Most coals decompose below temperatures of about 400°C (2,4), characteristic of the onset of plasticity. Moisture is released near 100°C, and traces of oil and gases appear between 100–400°C depending on the coal rank. As the temperature is raised in an inert atmosphere at a rate of 1–2°C/min, the evolution of decomposition products reaches a maximum rate near 450°C, and most of the tar is produced in the range of 400–500°C. Gas evolution begins in the same range but most evolves above 500°C. If the coal temperature in a single reactor exceeds 900°C, the tars can be cracked and the yields are reduced and the products are more aromatic. Heating beyond 900°C results in minor additional weight losses but the solid matter changes its structure. The tests for volatile matter indicate the loss in weight at a specified temperature in the range of 875–1050°C from a covered crucible. The weight loss represents the loss of volatile decomposition products rather than volatile components.

The loss of fusible material can be accelerated by reducing the pressure, and can also reduce caking properties. Further reactions of the volatile products are also reduced. Increasing the pressure has little effect. Mild oxidation is used to destroy the caking properties and can eliminate the production of tars.

Nature and Origin of Products. Volatile matter yields decrease with increasing coal rank. For slow heating the final weight loss depends on the maximum temperature (see Fig. 7). A variety of reactions take place. Increasing temperatures provide thermal energy to break the stronger chemical bonds. Much of the decomposition takes place in a short time (apparently <1 s) but is limited by the rate of diffusion of the volatile products through the solid. The liquids result from initial decomposition and gases result from decomposition of liquid material. Very rapid heating rates produce weight losses as high as 72% at 1900°C, suggesting that the intrinsic volatile matter is limited

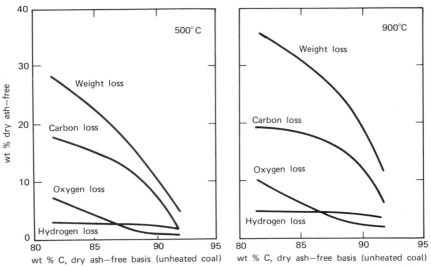

Figure 7. Composition of volatile matter at two temperatures as a function of rank (bright coals). (Courtesy of Institute of Fuel.)

only by the vapor pressure of the initial pyrolysis fragments, and would be expected to increase with temperature and decreasing coal rank (60).

The semicoke or residual char heated to 500°C contains 3–3.5 wt % H and up to 5 wt % O. On further heating to 900°C the char or coke contains only 0.8 wt % H and up to 0.3 wt % O. An aqueous liquor is produced that comes from the moisture in the coal as well as hydroxyl and possibly other oxygen-containing groups. Phenols in the tar are probably derived from hydroxyl aromatic groups in the coal. The total tar yield appears to be proportional to the fraction of aromatic carbon in the coal. Coke oven gas is obtained from a variety of reactions that include cracking some of the tar. The hydrogen in the gas is generated after the char is heated to 400°C but most is evolved in the conversion of the fluid coal to semicoke or coke at 550–900°C. The steam in the ovens can also produce hydrogen on reaction with hot coke.

Chemical Reaction Mechanisms.   An overall picture of the process is generally accepted but the detailed mechanism is still controversial. Information has been obtained from: the sequence of volatile material appearing in a coke plant as determined by gas chromatography; laboratory work simulating coking and minimizing secondary reactions by working in vacuum or sweeping with inert gas (61); laboratory studies with model organic compounds to determine the mechanism by which they are converted to coke, liquid, and gaseous products; and laboratory work with more complex materials including specially synthesized polymers to better provide a model of coal (2,62). Radioactive tracers have been used in the last two studies to follow the transformation to known materials in the products (2). In the last study, gas generating materials were added to aid in simulating the swelling process. Further observations such as the dehydrogenation of coal, which can alter the distribution of products, provides additional information for formation mechanism (63).

Several authors have added to the understanding of the mechanism of coal pyrolysis (61,64–65). A table summarizing the changes was prepared by van Krevelen (65). The early stages involve formation of a fluid through depolymerization and decomposition of coal organic matter containing hydrogen. Around 400–550°C aromatic and nonaromatic groups may condense after releasing hydroxyl groups. The highest yields of methane and hydrogen come from coals with 89–92 wt % C. Light hydrocarbons other than methane are released most readily below 500°C; methane is released at 500°C. The highest rate for hydrogen occurs above 700°C (61).

Reactions of Coal Ash.   The coal ash will pass through many reactors without significant chemical change. High temperature reactions that exceed the ash-softening temperature for the coal permit reactions of the simpler ash constituents to form more complex species. The molten ash behavior affects the slagging or ash removal of these reactors. Correlations of viscosity have been made with a variety of chemical parameters. Iron may be interconverted between the Fe(II) and Fe(III) states with significant reduction in viscosity occuring with increases in ferrous concentrations. Utilization of acid–base concepts appears feasible to correlate the observed effects (66).

Corrosion effects on boiler tubes appear to be initiated in some cases with the formation of a "white layer" of general composition $(Na, K)_3Al(SO_4)_3$. Conditions for initiation of the deposit are favored by coals with high alkali and sulfur contents. The white layer bonds to the tubes and permits growth of ash deposits that insulate the layer and permit further corrosion.

Resources

World Reserves. The first inventory of world coal resources was made during the Twelfth International Geological Congress in Toronto in 1913. An example of the changes since 1913 can be seen from an examination of the coal reserves for Canada. These were estimated to have been 1217×10^9 million metric tons in 1913, based on a few observations and statistical allowance for all possible coalbeds to a minimum thickness of 0.3 m and to a maximum depth of 1220 m below the surface; in 1974, however, the estimate of solid fossil fuel resources (excluding peat) from the World Energy Conference (15) gave the total resources as only 109×10^9 million metric tons, less than 9% of the earlier figures.

The most recent comprehensive review of energy sources is that published in the Survey of Energy Resources 1974, by the World Energy Conference, formerly the World Power Conference (15). These reviews appear at 6-yr intervals. The survey includes reserves as either recoverable or total, and also gives total resources. The total measured reserves of recoverable solid fuels are given as 6×10^{11} metric tons. The total reserves and total resources are 1.4×10^{12} and 10.7×10^{12} metric tons, respectively. These figures are about double the 1913 estimates, primarily because significantly increased reserves have been indicated for the U.S.S.R. Total estimated reserves for the U.S.S.R., United States, and the People's Republic of China together represent almost 90% of the world's solid fossil fuels.

The part of the reserves that is economically recoverable varies by country. The estimates made in the survey show that one-third of total reserves would last 1475 years at the 1974 annual rate of consumption and that the recoverable reserves represent almost 200 times present consumption.

In Table 9, taken from the current U.S. Geological Survey by Averitt (67), a somewhat different basis is used. This gives the estimated total original coal resources of the world. This includes beds ≥ 30 cm thick, and generally <1220 m below surface but includes small amounts between 1220 m and 1830 m. The data from column 1 are

Table 9. Estimated Total Original Coal Resources of the World, By Continents (In 10^9 Metric Tons)[a]

Continent	Identified resources[a]	Estimated hypothetical[b] resources[c]	Estimated total resources[d]
Asia[c]	3,635[d]	6,362	9,997[e]
North America	1,727	2,272	3,999
Europe[f]	273	454	727
Africa	82	145	227
Oceania[g]	64	55	118
Central and South America	27	9	36
Totals	5,808	9,297	15,104

[a] Ref. 67.

[b] Original resources in the ground in beds 30 cm or more thick, and generally less than 1,200 m below surface but includes small amount between 1,200 m and 1,800 m.

[c] Includes European U.S.S.R.

[d] Includes about $2,090 \times 10^9$ metric tons in the U.S.S.R.

[e] Includes about $8,600 \times 10^9$ metric tons in the U.S.S.R.

[f] Includes Turkey.

[g] Australia, New Zealand, and New Caledonia.

from earlier World Power Conference Surveys. "The figures for hypothetical resources (col. 2) and total estimated resources (col. 3) are less reliable but are based on opinions of competent observers and on extrapolations from the figures in column 1." This estimate represents about one third more than the World Energy Conference Survey.

Reserves in the United States. Coal is widely distributed and abundant in the United States (see Fig. 8). A large part of the coal fields consist of lignite and subbituminous coal. Another part of the fields is contained in thin or deep beds that can be mined only with difficulty or great cost.

The current reserve estimates as given by the U.S. Geological Survey (67) in their fifth progress report are indicated in Tables 10 and 11 and are subject to further change as further mapping, prospecting and development is done. These are useful for showing the quantitative distribution of reserves, selecting appropriate areas for further exploration of development, and in planning coal-based industrial activity.

The reserves of 21 states have been classified by overburden thickness, reliability of estimates, and bed thickness. This coal represents about 60% of the total identified tonnage. Of this, 91% is less than 305 meters from the surface, 43% is bituminous, and 58% is in beds thick enough to be mined economically.

On a uniform kJ (Btu) basis, coal constitutes 69% of the total estimated recoverable resources of fossil fuel, whereas petroleum and natural gas are about 7% and oil in oil shale, which is not currently used as a fuel, is about 23%. The smaller reserves of oil and gas are emphasized by a combined production and consumption rate of oil and gas in the United States that is three times that of coal. The total recoverable reserves of coal as of January 1, 1974 are about 360 times the 1974 annual production, and total resources down to 914 meters, excluding thin beds and allowing 50% recovery in mining, is about 1700 times 1974 production.

World Coal Production. The Chinese probably used small amounts of coal for several thousand years before Marco Polo reported their use of "black rocks" for fuel. It was discovered in the 12th century in Britain that certain black rocks, "sea coles," found along the seashore of northeast England could be burned. Thus, coal mining began in Britain and was followed in what is now Belgium, France, and Germany, and later of course extended to all coal producing areas. By 1660 British production reached ca 2 million metric tons a year, and ca 7 million t/yr by 1750. By 1860 world production reached 122 million t/yr.

Statistical data on world coal production from 1860 to 1960 is given in reference 68. A number of fluctuations and intermediate peaks have occurred in coal production since 1860. World production increased to 1140 million metric tons in 1913, giving a 4.2%/yr average rate of increase. The rate slowed and has been erratic since that time. Coal and lignite production rose to 3069 million metric tons in 1971 (15).

United States Coal Production. Coal production in the United States dates back to 1702, and started in earnest about 1820. It has increased with fluctuations to ca 600 million metric tons in 1976. In 1971 the production was 510 million t, which was 16.6% of the world production of coal and lignite. The production of bituminous coal and lignite in 1962–1973 is given in Table 12 (69). Bituminous coal shipments in 1975 totaled 535 million metric tons to United States destinations. Of these, electric utilities received 400×10^6 t (74.4%), coke and gas plants 84×10^6 t (15.7%), retail dealers 4.6×10^6 t (0.9%), and all others ca 50×10^6 t (9.0%) (70).

Anthracite production takes place in Pennsylvania. It fell from 90 million metric

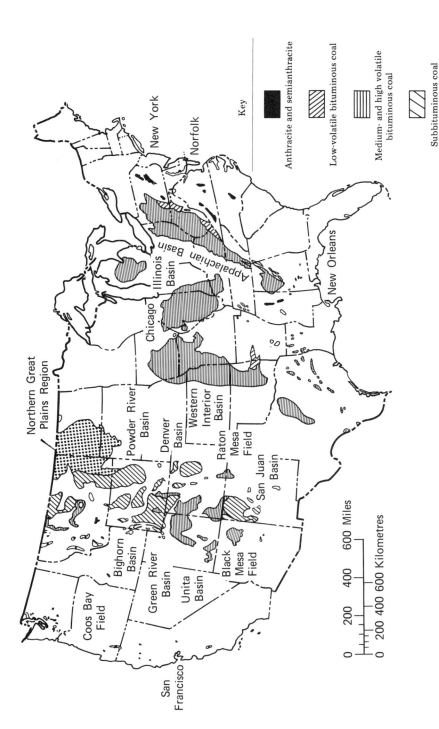

Figure 8a. Coal fields of the conterminous United States.

Key

Anthracite and semianthracite

Low-volatile bituminous coal

Medium- and high volatile bituminous coal

Subbituminous coal

Lignite

New York

Norfolk

New Orleans

Appalachian Basin

Chicago   Illinois
          Basin

Northern Great
Plains Region

Powder River
Basin

Denver
Basin

Western
Interior
Basin

Raton
Mesa
Field

San Juan
Basin

Bighorn
Basin

Green River
Basin

Unita
Basin

Black
Mesa
Field

San
Francisco

Coos Bay
Field

0   200   400   600 Miles

0 200 400 600 Kilometres

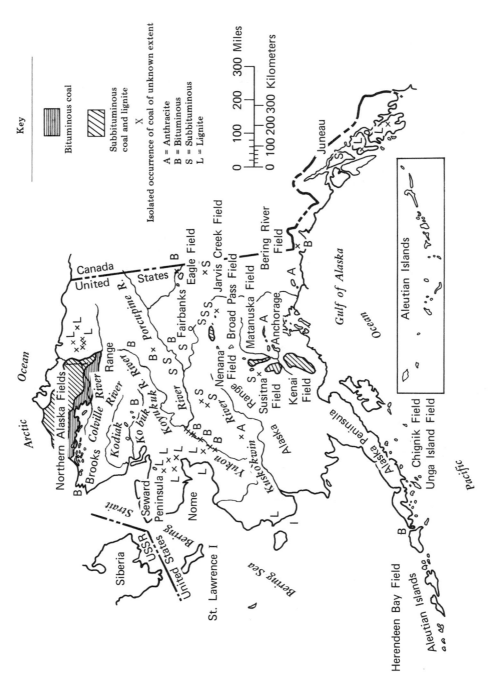

Figure 8b. Coal fields of Alaska.

Table 10. Identified Coal Resources of the United States, January 1, 1974 (In Millions of Metric Tons)[a]

State	Estimated original or remaining identified resources	Resources depleted to Jan. 1, 1974 Production	Resources depleted to Jan. 1, 1974 Production plus loss in mining	Remaining identified resources Jan. 1, 1974
		Bituminous coal		
Alabama	12,477	233	446	12,031
Alaska	17,626	7	14	17,612
Arizona	19,278	7	14	19,264
Arkansas	1,647	81	162	1,485
Colorado	99,793	402	864	98,929
Georgia	22			22
Illinois	132,680	115	230	132,450
Indiana	30,153	168	336	29,817
Iowa	6,565	332	664	5,901
Kansas	16,970	417	34	16,936
Kentucky				
Eastern	30,336	2,365	4,730	25,606
Western	35,270	1,251	2,502	32,768
Maryland	1,089	422	44	1,045
Michigan	270	1/2	84	186
Missouri	28,903	307	614	28,289
Montana	2,144	29	58	2,086
New Mexico	9,932	91	182	9,750
North Carolina	102	1	2	100
Ohio	42,173	2,414	4,828	37,345
Oklahoma	6,456	0	0	6,456
Oregon	45			45
Pennsylvania	58,561	278	556	58,005
Tennessee	2,493	99	198	2,295
Texas	5,534	24	48	5,486
Utah	21,061	15	28	21,033
Virginia	10,610	1,125	2,250	8,360
Washington	1,696	1	2	1,694
West Virginia	105,794	7,470	14,940	90,854
Wyoming	12,007	241	482	11,525
other states[b]	557			554
Total	*712,244*	*17,124*	*34,248*	*678,047*
		Subbituminous coal		
Alaska	100,423	14	28	100,395
Colorado	18,130	114	228	17,902
Montana	160,709	151	302	160,407
New Mexico	46,086	73	146	45,940
Oregon	263	3	6	257
Utah	157			157
Washington	3,805	6	12	3,793
Wyoming	112,179	189	378	111,801
other states[c]	36	4	8	28
Total	*441,788*	*554*	*1,108*	*440,680*
		Lignite		
Alabama	2,000			1,816
Arkansas	318			318
Colorado	18			18
Montana	102,088	5	10	102,078
North Dakota	318,340	140	280	318,060
South Dakota	1,984	1	2	1,982
Texas	9,458	77	121	9,337

Table 10. (continued)

State	Estimated original or remaining identified resources	Resources depleted to Jan. 1, 1974		Remaining identified resources Jan. 1, 1974
		Production	Production plus loss in mining	
Washington	106			106
other states[d]	45	2	4	4
Total	*434,357*	*225*	*417*	*433,756*
		Anthracite and semianthracite		
Arkansas	414	13	26	388
Colorado	82	5	10	72
New Mexico	5	1	2	3
Pennsylvania	17,236	85	170	17,066
Virginia	322	9	18	304
Washington	5			5
Total	*18,064*	*113*	*226*	*17,838*
Total all ranks	*1,606,453*	*18,016*	*35,998*	*1,570,263*

[a] Estimates include beds of bituminous coal and anthracite generally ≥36 cm thick, and beds of subbituminous coal and lignite generally ≥76 cm thick, to an overburden depth of 914 m. Figures are for resources in the ground. Of the reported tonnage, 91% is less than 305 m below the surface.

[b] California, Idaho, Nebraska, and Nevada.

[c] California and Idaho.

[d] California, Idaho, Louisiana, Mississippi, and Nevada.

tons in 1918 to 42 million in 1938; increased in 1944 to 58 million but declined to 23 million in 1957, 15 million in 1961, and 6 million in 1974. Of the 6 million tons produced in 1974, half was mined and the other half was recovered from culm banks.

Probable Trends. The demand for energy is continually increasing. Diminishing reserves of oil and natural gas will require significantly increased amounts of coal in various forms to meet the demand (see Fuels, survey).

The highest energy consumption in the world occurs in the United States. In 1975 it was 11.0 metric tons of coal-equivalent per capita (71).

Table 13 shows that the world average resource to demand ratio (R/D) for 1971 for all solid fuels was 197 years for recoverable reserves, 473 years for total reserves. Estimated coal consumption will reduce the present recoverable reserves at about $1/2\%$/yr however improved knowledge of world reserves may maintain the R/D ratio at about the same level over the next few decades.

The use of bituminous coal and other energy sources in the United States (excluding uses where coal is noncompetitive) is indicated in Table 14 (% on an energy basis).

An estimate of U.S. energy consumption in the recent past and future is shown in Table 15. The trend away from natural gas in the United States has begun since domestic production peaked. Oil consumption continues to grow; however, imports have been growing faster. Production from the North Slope (Alaska) oil fields will have a minor effect on the trend. Natural gas delivery in the United States has been curtailed for electric power generation and industrial use. To compensate, boilers utilizing coal and nuclear fuel will be built. Industrial gas users will switch largely to oil in the near term, with coal gasification to provide low- and intermediate-Btu (5.6–18.6 MJ/m³, 150–500 Btu/SCF) industrial fuel gas over the longer term. The demand for substitute natural gas or high-Btu (37.2 MJ/m³, 1000 Btu/SCF) gas should increase as the more favorable economics of delivering energy as "coal by pipeline" are compared with electricity as "coal by wire."

Table 11. Total Estimated Remaining Coal Resources of the United States, January 1, 1974 (In Millions of Metric Tons)[a]

States	Overburden 914 m				
	Bituminous coal	Subbituminous coal	Lignite	Anthracite and semi-anthracite	Total
Alabama	12,031		1,814		13,845
Alaska	17,611	100,395			118,006
Arizona	19,263				19,263
Arkansas	1,486		318	389	2,193
Colorado	98,991	17,901	18	71	116,981
Georgia	22				22
Illinois	132,452				132,452
Indiana	29,818				29,818
Iowa	5,901				5,901
Kansas	16,935				16,935
Kentucky					
Eastern	25,606				25,606
Western	32,768				32,768
Maryland	1,045				1,045
Michigan	186				186
Missouri	28,290				28,290
Montana	2,086	160,407	102,077		264,570
New Mexico	9,750	45,939		4	55,693
North Carolina	100				100
North Dakota			318,061		318,061
Ohio	37,345				37,345
Oklahoma	6,456				6,456
Oregon	45	258			303
Pennsylvania	58,005			17,066	75,071
South Dakota			1,982		1,982
Tennessee	2,295				2,295
Texas	5,487		9,337		14,824
Utah	21,034	157			21,191
Virginia	8,361			304	8,665
Washington	1,694	3,792	106	5	5,596
West Virginia	90,855				90,855
Wyoming	11,524	111,801			123,325
other states[a]	553	29	42		624
Total	677,991	440,679	433,756	17,837	1,570,260

[a] Estimates include beds of bituminous coal and anthracite generally ≥36 cm thick, and beds of subbituminous coal lignite generally ≥76 cm thick. Figures are for resources in the ground.

[b] California, Idaho, Nebraska, and Nevada.

Mining and Preparation

Mining. Coal is obtained by either surface mining of outcrops or seams near the surface, or by underground mining. The method chosen depends on the geological conditions that may vary from thick, flat seams to thin, inclined seams that are folded and need special methods of mining.

The need for increasing productivity has caused coal mining to change from a labor intensive activity to one that has become highly mechanized. However, after the 1969 Coal Mine Health and Safety Act, underground production had decreased from 15.6 metric tons per man-day to 8.5 metric tons per man-day in 1976.

Table 12.   Production of Bituminous Coal, by States[a] 1962–1973—All Types of Mines (Thousands of Metric Tons)

State, yr of max production	Maximum production quantity	Production by years			Total from earliest record to end of 1973
		1962	1968	1973	
Alabama, 1962	19,051	11,684	14,914	17,955	1,061,200
Arkansas, 1907	2,422	232	191	415	92,278
Colorado, 1917	11,324	3,065	5,042	5,653	521,516
Illinois, 1918	81,003	43,986	56,645	55,836	4,007,342
Indiana, 1918	27,831	14,250	16,770	22,921	1,279,850
Iowa, 1917	8,133	1,025	794	775	331,329
Kansas, 1918	6,860	830	1,150	1,038	268,000
Kentucky, 1973	115,211	62,787	91,766	115,211	3,616,231
Maryland, 1907	5,019	744	1,312	1,591	257,675
Missouri, 1917	5,144	2,627	2,907	4,517	305,635
Montana, 1973	9,026	346	470	9,026	184,935
New Mexico, 1973	8,473	614	3,110	8,473	164,979
N. Dakota, 1973	6,713	2,479	4,070	6,713	137,442
Ohio, 1970	50,213	30,957	43,837	41,136	2,416,205
Oklahoma, 1920	4,398	950	987	2,381	181,806
Pennsylvania, 1918	161,977	59,252	69,127	69,531	8,411,517
Tennessee, 1972	10,214	5,673	7,391	8,158	445,978
Utah, 1947	6,739	3,898	3,915	4,662	293,454
Virginia, 1968	33,534	26,738	33,534	30,725	1,134,135
Washington, 1918	3,703	213	161	2,912	142,069
W. Virginia, 1947	159,806	107,499	132,376	104,542	7,454,660
Wyoming, 1973	12,337	2,330	3,473	12,337	429,734
other states		812	680	9,625	193,958
Total, 1947	*572,089*	*382,965*	*494,635*	*536,143*	*3,333,241*

[a] Source, U.S. Bureau of Mines.

Table 13.   Resource/Demand Ratio R/D (Years)[a]

	Millions of metric tons	High ranking solid fuels	Low ranking solid fuels	All solid fuels
world reserves	1,420,000	494	415	473
recoverable reserves	591,000	198	194	197

[a] Ref. 15.

Strip or open pit mining involves removal of overburden from shallow seams, breaking of coal by blasting or mechanical means, and loading of the coal. The development of very large excavating equipment, including draglines, bulldozers, front-end loaders, and trucks, has been responsible for significantly increased production from strip mining. Strong demand for this equipment has resulted in delivery delays of up to 8 years.

The two methods of underground mining commonly used are room-and-pillar and longwall. In room-and-pillar mining the coal is removed from two sets of corridors that advance through the mine at right angles to each other. Regularly spaced pillars,

Table 14.   Use of Bituminous Coal and Other Energy Sources in the United States[a]

Source	1942	1952	1962	1972
coal	68.7	45.2	32.5	25.6
petroleum	17.1	25.4	23.6	22.8
natural gas	9.0	23.3	38.3	44.4
water power	5.2	6.1	5.5	6.0
nuclear			0.1	1.2

[a] Ref. 71.

Table 15.   U.S. Energy Consumption, 10^{15} J/d (10^6 bbl oil/d-equiv)[a,b]

Source	1950	1960	1970	1975	1980	1990
oil, domestic	34.2 (5.6)	47.7 (7.8)	63.6 (10.4)	57.5 (9.4)	67.9 (11.1)	66.6 (10.9)
imports	5.5 (0.9)	11.6 (1.9)	21.4 (3.5)	36.7 (6.0)	41.0 (6.7)	48.3 (7.9)
shale oil					1.2 (0.2)	4.9 (0.8)
coal	39.7 (6.5)	32.4 (5.3)	45.2 (7.4)	43.4 (7.1)	57.5 (9.4)	74.0 (12.1)
natural gas, domestic	17.7 (2.9)	35.5 (5.8)	63.0 (10.3)	55.6 (9.1)	61.1 (10.0)	52.0 (8.5)
imports		0.6 (0.1)	2.4 (0.4)	2.4 (0.4)	7.3 (1.2)	8.6 (1.4)
nuclear			0.6 (0.1)	4.9 (0.8)	17.1 (2.8)	59.9 (9.8)
hydroelectric	1.2 (0.2)	1.8 (0.3)	2.4 (0.4)	3.1 (0.5)	3.7 (0.6)	3.7 (0.6)
geothermal				0.1 (0.02)	0.9 (0.14)	4.9 (0.8)

[a] Ref. 72.
[b] 6.11×10^9 J/d = 5.8×10^6 Btu/d = one barrel of oil per day-equivalent.

constituting about half of the coal seam, are left behind to support the overhead layers in the mined areas. This method is used extensively in the United States and other nations with large reserves. The pillars may later be removed, leading to possible subsidence of the surface. Longwall mining is used to permit recovery of as much of the coal as possible (73). Two parallel headings are made 100–200 m apart and at right angles to the main heading. The longwall between the two headings is then mined away from the main heading. The equipment provides a movable roof support that advances as the coal is mined and allows the roof to collapse in a controlled manner behind it.

Another method used in Europe for steeply inclined seams is called horizon mining. Horizontal shafts are cut through rock below the coal seams. Vertical connections are made to the seam to permit coal removal.

The mechanical equipment used in underground mining may be either conventional, involving a series of specific operations, or continuous mining equipment. Conventional equipment involves several types of machines in a cycle of operations that include undercutting the seam, drilling blasting holes, blasting the coal from the face of the seam, and loading and transporting the coal from the seam by conveyor belt or shuttle car. Continuous miners use rotating heads equipped with bits to pick or cut through the coal without blasting and load it into a conveying system in a single operation. Preparation for operation limits use of this equipment in most mines to less than half of the time.

Preparation.   Coal preparation is of significant importance to the coal industry and consumers (4,42,74). Preparation normally involves some size reduction of coal particles and the systematic removal of some ash forming material and very fine coal.

The percentage of mined coal that is mechanically cleaned in major coal producing countries has risen during the past 30 years. There are a number of reasons for this. The most important in the United States is the increased use of continuous mining equipment. The nature of this operation tends to include inorganic foreign matter from the floor and ceiling of the seams, and run-of-mine coal currently includes about 25% ash.

The average coal particle is smaller when produced with continuous mining equipment. The depletion of the better quality seams, which are low in ash and sulfur, in many coal fields necessitates cleaning of the remainder.

The economic need to recover the maximum amount of salable coal has led to cleaning of the finer sizes which had previously been discarded. Stringent customer demands for coal meeting definite specifications, regulations requiring the removal of pyrites to control air pollution, increased freight rates, and ash disposal costs all contribute to the upsurge in coal cleaning.

Earlier, the coal was hand-picked for removal of larger pieces of refuse, but higher labor costs have made this method uneconomical. Mechanical separation methods are used, most of which depend on the difference in density between the coal and refuse.

The washability characteristics of a coal determine the extent to which the refuse may be removed, and other problems in cleaning. The laboratory float-and-sink analysis gives information on the percentages and quality of the coal material occupying different density ranges. From this information, graphs are constructed showing the composite quantity and coal quality that will be obtained by cleaning at different specific gravities. This information is considered with the economic factors involved in the sale of the washed coal to choose an optimum method of cleaning. Cleaning plants are usually designed to handle the output of specific mines and to clean for a specific market. The plant will include various types of cleaning methods designed to move the different fractions through several cleaning circuits to maximize optimum recovery.

In many areas, run-of-mine coal is separated into three products: a low gravity, premium priced coal for metallurgical or other special use, a "middlings" product for possible boiler firing, and a high ash refuse.

The complete preparation of coal will usually require several processes. Some of the methods used are outlined below.

Cleaning Methods. *Jigs.* Jig washing is the most widely used of all methods. A bed of particles is subjected to alternate upward and downward currents of water. This causes a moving bed of particles to stratify, with the lighter clean coal particles at the top and the heavier refuse particles at the bottom (see Gravity concentration).

Heavy Medium. This is a simple float-and-sink process that is widely used for coarse coal cleaning. The medium is usually a suspension of pulverized magnetite which is mixed to the desired specific gravity. This method is also used in cyclones for a wide range of coal cleaning.

Trough Washers. The coal is fed to a trough in a stream of water that carries the coal particles forward but allows the heavier particles to sink and be removed.

Washing Tables. These are used for cleaning fine coal. A coal suspension in water flows across a slanted table that oscillates at right angles to the direction of flow. The heavier refuse particles settle onto the table and are trapped by riffles or bars, while the lighter coal particles are carried over the riffles in the current.

Dry Cleaning. The coal passes over a perforated, oscillating table through which air is blown. This method creates a dust problem, although it eliminates the need for drying of the coal. In countries using hydraulic mining or underground dust suppression with water, there is limited opportunity for dry cleaning.

Froth Flotation. This is the most important method for cleaning fine coal. Since very small particles cannot be separated by settling methods, a different approach is used. Air is passed through a suspension of coal in water to which conditioning reagents, usually special oils, have been added. The oils are selected so that the coal particles preferentially attach themselves to the bubbles and separate from the refuse which remains in suspension (see Flotation).

Dewatering. The coal leaving the cleaning plant is very wet and must be at least partially dried to reduce freight charges, meet customer requirements, and avoid freezing. Draining on screens will remove substantial amounts of water from larger coal, but other methods are required for smaller sizes with larger surface areas.

Vibrating screens and centrifuges are used for dewatering. For very fine coal, such as that obtained from flotation, a vacuum filtration with a disk or drum type filter may be used. Flocculants (qv) may be added to aid filtration. They are also used for cleaning wastewater and pollution control. If very low moisture contents are required (as low as 2–3%), thermal drying must be used. A number of dryer types are available including fluidized bed; suspension; rotary- and cascade-dryers, but all are expensive to operate (see Drying).

Storage. Storage of coal may be necessary at any of the major steps in production or consumption, ie, at the mine, preparation plant, or consumer location. Electric utilities have the largest amounts of coal in storage, with stockpiles frequently able to meet 70–100 d normal demand to protect against delays, shortages, price changes, or seasonal demands.

For utilities, two types of storage are used. A small amount of coal in storage meets daily needs and will be continually turned over. This coal is loaded into storage bins or bunkers. Long-term reserves are carefully piled and left undisturbed except as necessary to sustain production.

Coal storage results in some deterioration of the fuel owing to air oxidation. If inadequate care is taken, spontaneous heating and combustion will result. As the rank of coal decreases it oxidizes more easily and must be piled more carefully. Anthracite does not usually present a problem.

The surface of the coal particles oxidize or weather resulting in cracks, finer particles, and reduced agglomeration which may destroy coking properties. If spontaneous heating takes place, the calorific value of the coal is reduced. Hot spots must be carefully dug out and used as quickly as possible. Without spontaneous heating and with good compaction, calorific value losses below 1%/yr have been recorded.

Coal piles are carefully constructed to exclude air, or allow adequate ventilation. The latter requires larger sizes, graded as 4 cm+ without fines, for avoiding heating by ventilation. For exclusion of air mixed sizes provide fines to fill the gaps between larger pieces. Pockets of large sizes must not be allowed since they provide access for air. The coal should be compacted to maximize the bulk density of the coal pile.

Several approaches have been effective for storage:

Large compacted layered piles with sides and top sealed with an oil or asphalt emulsion. Four liters of oil seals one square meter.

Large compacted layered piles with sides and top covered with fines to seal the pile and coarse coal to protect fines from wind and weather. The sides may slope at angles $\leq 30°$.

Piles of compacted layers in open pits with tight sides so that the air has access only at the top.

Sealed bins or bunkers. Airtight storage can be provided for smaller amounts of coal for long times.

Underwater storage in concrete pits. This is expensive and rarely used, but effectively prevents deterioration although it introduces other problems related to handling wet coal.

Large compacted storage piles should be located on hard surfaces, not subject to flooding. A layer of fines may be put down first to facilitate recovery. Each layer of coal should be compacted after it is deposited. The top of the pile should have a slight crown to avoid water accumulation. Excessive heights should be avoided to prevent air infiltration caused by wind. Coal removal should be done in layers followed by compacting and smoothing the surface. Piles should be limited to the same rank of coal depending on the intended use.

Transport.   The usual means of transporting coal are railroad, water, truck, and in some instances by conveyor belt from mine to plant (4) (see Transportation). In 1975 bituminous coal transported to United States destinations was 535 million metric tons; of this 54.3% was shipped by railroad, 24.9% by water, 11.8% by truck, and 9.0% by conveyor and private railroad. The average rail car load of bituminous coal was 76.5 t (18). Electric utilities consumed 74.3% of the bituminous coal transported in the United States in 1975 (70).

For shipment in cold climates, a freeze proofing treatment of inorganic chemicals or oil spray is used. Oil treatment has also been used for dustproofing; wind loss can be prevented by use of an asphalt emulsion on the top of rail cars.

Hydraulic transport has been used to move coal slurries through pipelines (qv) over significant distances (75). A pipeline is under construction to carry coal from Gillette, Wyoming, to White Bluff, Arkansas, a distance of ca 1700 km. This is expected to be completed in 1979 at a cost of $750 million with a capacity of 23×10^6 t/yr, a 97 cm dia and water consumption of 18.5 km^3/yr. Transportation cost is expected to be ca 0.34¢/(t·km). Unit train costs over the same distance are expected to be ca 0.68¢/ (t·km). The longest operational coal pipeline in the United States is the 439-km Black Mesa pipeline between the Black Mesa in Arizona and a power plant in Southern Nevada (76).

Coal pipelines have been built in other countries—France (8.8 km), U.S.S.R. (61 km)—and pipelines are used for transporting limestone, copper concentrates, magnetite, and gilsonite in other parts of the world.

The first coal pipeline, built in Ohio, led to freight rate reductions. The pipeline stopped operation after introduction of the unit train, used exclusively to transport coal from the mine to an electric power generation station.

The slurry is made of pulverized coal (up to 60%) and water. The slurry is dewatered with centrifuges before combustion of the coal.

Hydraulic transport is used in mines and for lifting of coals to the surface in the U.S.S.R., Poland, and France. Pneumatic transport of coal is used over short distances in power plants and steel mills.

The longest (14.6 km) single flight conveyor belt in the world near Uniontown, Kentucky, has a capacity of 1360 t/h.

Economic Aspects

World Trade. Table 16 gives the estimated end use from 1971–1976 of United States bituminous coal exports. About one-third of the bituminous coal exported from the United States in 1976 went to Europe, another third to Asia, and the remainder to North and South America (77).

Table 16.   U.S. Bituminous Coal Exports, 1971 and 1976 (In Thousands of Metric Tons)[a]

Country of destination	1971	1976
North America:		
Canada	15,935	14,966
Mexico	259	228
other	2	
Total N.A.	*16,196*	*15,194*
South America:		
Argentina	490	477
Brazil	1,695	2,033
Chile	188	132
other	23	
Total S.A.	*2,396*	*2,642*
Europe:		
Belgium/Luxembourg	694	1,998
France	2,818	3,109
FRG	2,641	902
Ireland	16	
Italy	2,431	3,820
Netherlands	1,474	3,166
United Kingdom	1,514	765
Total EEC	*11,588*	*13,760*
Greece		422
GDR	70	
Norway	75	112
Portugal	11	234
Romania		192
Spain	2,319	2,280
Sweden	561	740
Switzerland	87	13
Yugoslavia	169	167
Total Europe	*14,880*	*17,920*
Asia:		
Japan	17,867	17,058
Korea		425
Turkey	1	217
Total Asia	*17,868*	*17,700*
Africa:		
Egypt		292
South Africa		114
other	1	
Total Africa	*1*	*406*
Grand total	*51,341*	*53,862*

[a] Source: compiled by International Coal Staff, East Europe Area Office, IDA, Bureau of Mines from official data published by importing countries, and information reported by the Bureau of Census, coal exporters, railways, and coal producers.

United States coke exports totaled 1,190,000 metric tons, mainly to Canada; anthracite exports were 558,000 t, mainly to Canada.

Canada's major source of supply is the United States. Denmark imports most of its coal from Poland, the U.S.S.R., and the United Kingdom. The Federal Republic of Germany imports most of its coal from Poland and the United States. Italy and Australia import most of their coal from the United States, Poland, the FRG, and the U.S.S.R. The United Kingdom imports are mainly from the United States and Australia. A selected list of exports and imports is shown in Table 17.

Coal availability to meet a national (United States) goal of doubling the output to 1130 million metric tons per year by 1985 is unlikely. Time lags of about 4 years for development of an underground mine or to obtain delivery on large draglines used in Western surface mining, put obvious restrictions on expansion. Additional obstacles for implementing plans to open 7 mines in the Rocky Mountain and Northern Plains states include land use, water allocations, community perturbations, manpower problems, transportation expansion, federal leasing policies, states' prerogatives, and Indian land issues. These mines would add 68 million metric tons of annual output. The lower heat content of Western coal would increase the total coal mining requirements to over 1.2 billion metric tons to meet the goal of keeping imports of oil down to 7 million barrels per day by 1985.

The EPA is contemplating ca 50% reduction of the allowable emission of 0.516 kg SO_2 per billion joules (1.2 lbs of SO_2 per million Btu) of coal burned for electric power generation. This would make the use of scrubbing equipment mandatory and probably rule out the use of most solvent refined coal or other processed coals (see Sulfur recovery).

Analysis

There is an official national organization in most countries that is responsible for developing and maintaining standard methods for testing and analysis. In the United States it is the ASTM and in the United Kingdom, the British Standards Institution (BSI). The International Organization for Standardization, in Geneva, formed committee T.C. 27 which is responsible for developing agreed-upon international standard methods.

The following are brief descriptions of several significant tests or analyses. Details of the procedure and equipment may be obtained from the individual national standards.

Table 17.   Coal Exports and Imports, 1976 (Metric Tons)

Country	Exports	Imports
United States	53,900,000	
Austria		2,606,960
FRG	12,804,510	5,970,316
(coke)	7,098,984	1,265,785
Ireland		618,578
S. Africa	5,440,000	
Belgium	706,682	7,909,040

Sampling. The procedures for taking a sample, reducing the size of that sample, and preparation for later analysis are given in ASTM D 2234-76 and D2013-72 (14) and BS 1017 (78). The procedure indicates the minimum weight of total sample required to achieve a chosen degree of accuracy.

Size Analysis. ASTM (14) and BS (79) have a number of standards dealing with size specifications and size analysis procedures including D-197-30 (reapproved 1971) D-410-38 (1976), D-311-30 (1976), D-310-69 (1976), and D-431-44 (1976).

Moisture-Holding Capacity. The bed moisture (equilibrium moisture) is the moisture content of coal after it is equilibrated at 30°C in an atmosphere of 96–97% rh (D1412-74) (14,80). Total moisture is determined (D2961-74) by heating at 107°C to constant weight.

Proximate Analysis. This analysis gives the volatile matter, ash, and moisture of a coal sample (14,16). Fixed carbon is determined by difference. Volatile matter is determined from an empirical weight loss test in which coal is heated in a covered crucible at either 950°C (ASTM, D3172-73) or 900°C (BS).

Ultimate Analysis. This analysis gives the elemental composition of coal in terms of C, H, N, S, and O (D3176-74) (14,81).

Chlorine Determination. The coal is blended with Escka's mixture and burned to convert the chlorine to chloride (14,81). Volhard's method is used to determine chloride by titration (D2361-66 (1972)) (14).

Phosphorus Determination. Phosphorus is converted to soluble phosphate by digesting the coal ash with a mixture of sulfuric, nitric, and hydrofluoric acids (14,81). Phosphate is precipitated as ammonium phosphomolybdate. This may be reduced to give a blue solution that is determined colorimetrically or volumetrically (D2795-69 (1974)) (14).

Calorific Value. Coal is placed in a bomb, pressurized with oxygen, and burned. The temperature rise in the water bath of the calorimeter (see Calorimetry) surrounding the bomb is used to calculate the calorific value (D2015-66 (1976) and D3286-73) (14,81).

Ash Fusibility. A molded cone of ash is heated in a mildly reducing atmosphere and observed with an optical pyrometer. The initial deformation temperature is reached when the cone tip becomes rounded, hemispherical temperature when the cone becomes a hemispherical lump, and fluid temperature when no lump remains (D1857-68 (1974)) (14,81).

Swelling Tests. For the free swelling index (ASTM D720-67 (1977)) (crucible swelling number), a coal sample is rapidly heated to 820°C in a crucible. The profile of the resulting char is compared with a series of standard numbered profiles (14,80–81). For the Roga index weighed amounts of coal and standard anthracite are mixed and carbonized. The product coke is tested in a Roga drum for its resistance to abrasion (80).

Coking Tests. For the Gray-King coke-type assay test (80–81) coal is heated in a retort tube to 600°C and the product coke is compared to a series of standard cokes. For a strongly swelling coal, enough anthracite or electrode carbon is added to the coal to suppress the swelling.

In the Audibert-Arnu dilatometer test (80), a thin cylinder of compressed powdered coal contacting a steel piston is heated at a rate not over 5°C/min. The piston movement is used to calculate the percent dilation.

Hardgrove Grindability Index. A specially sized coal sample is ground in a specifically designed ball and race grinding mill (D409-71). The index is determined from the amount of coal remaining on a 74 μm (200 mesh) screen (14).

Strength Tests. The drop shatter test indicates the resistance of a coal or coke to breakage on impact (see D440-49 (1969)). A sample is dropped in a standard way a number of times from a specified height. For the tumbler test (abrasion index, ASTM D441-75 (1975)) the coal or coke is rotated in a drum to determine the resistance to breakage by abrasion (14,81).

Health and Safety Factors

Coal mining has been a relatively dangerous occupation (82–84). During the period of 1961–1967 the average fatality rate in the United States per million man hours worked was 1.05. In the seven years after the passage of the Federal Coal Mine Health and Safety Act of 1969, the average fatality rate decreased to 0.58, a reduction of 44.8%.

The rates of disabling injuries over three year periods before and after passage of the act for underground mining were 48.60% and 40.07%. The major causes of fatalities are falling rock from mine roofs and faces, haulage, surface accidents, machinery, and explosions. For disabling injuries the major causes are slips and falls, handling of materials, use of hand tools, lifting and pulling, falls of roof, and haulage and machinery.

Gases. Methane is of greatest concern, although others including carbon monoxide and hydrogen sulfide may be found in some mines. Methane must be detected and controlled since mixtures with air of 5–15% are explosive.

The U.S. Mine Health and Safety Act of 1969 provides that a mine will be closed if there is $\geq 1\frac{1}{2}$% methane in the air. The use of an electrical methane detection device is required. High capacity ventilation systems are designed to sweep gases from the cutting face and out of the mine. These systems remove all gases before they become harmful.

Coal Dust Explosions. The explosion from methane tends to be localized, but may start coal dust explosions with more widespread injury and loss of life. All breaking operations result in formation of fine coal particles; some are controlled with water during the mining operation.

Breakage associated with hauling disperses dust. Dust accumulations can be made safe by rockdusting. Powdered limestone is spread over the mine surfaces to cover the dust.

Drainage. Some mines are located beneath subsurface streams, or the coal seams may be aquifers. These mines become flooded and must be continually pumped. In Pennsylvania anthracite mines as much as 30 t of water may be pumped for each ton of coal that is mined (82).

Air oxidation of pyrite leads to sulfate formation and dilute sulfuric acid in the mine drainage. This pollutes streams and water supplies into which the mine water is drained. The U.S. Bureau of Mines and other organizations are studying means of controlling this problem.

Other Hazards. Rocks falling from the roofs of mines cause the largest number of accidents. Roof bolts are placed in holes drilled into the roofs of tunnels to tie the layers of rock together to prevent rock falls.

A disease called pneumoconiosis, also called black lung, results from breathing coal dust over prolonged periods of time. The coal particles coat the lungs and prevent proper breathing.

Governmental Regulations. The U.S. Bureau of Mines (MESA, Mining Enforcement and Safety Administration) studies hazards and advises on accident prevention. MESA also administers laws dealing with safety in mines. Individual states may also have departments of mines to administer their own standards.

The Federal Coal Mine Health and Safety Act set standards for mine ventilation, roof support, coal dust concentration levels, mine inspections, and equipment. As a part of this comprehensive act, miners must receive medical examinations at the employers' expense, and payments are made from the U.S. government to miners who cannot work because of black lung disease.

Uses

Coal as a Fuel. Coal uses as a fuel include electric power generation, industrial heating and steam generation, domestic heating, and railroads and coal processing. About 87% of the world's coal production is burned to produce heat and derived forms ·of energy. The balance is practically all processed thermally to make coke, fuel gas, and liquid by-products (see Fuels, synthetic). The further use of the coke and fuel gas also contribute to the coal consumption for heat. Table 18 gives the consumption of bituminous coal by consumer class and retail deliveries in the United States (see Fuels, survey).

Electric Power Generation. Coal remains the major fuel for thermal electric power generation. Since 1940 the quantity of bituminous coal consumed by electric utilities has almost doubled in each succeeding decade, and growth is expected to continue for many years. In the United States during 1963–1973 the quantity of coal consumed by electric utilities increased by 85%. In 1975, 48% of the total electricity generated in utility power plants in the United States came from coal, and it is predicted that this will decrease to 46% by 1980 and 30% by 1990 (72). Coal consumption for electric power generation is expected to increase by 34% in 1980 and 29% in 1990 according to these predictions. However, government encouragement of coal utilization may increase these amounts. Current goals call for a doubling of coal production by 1985 although achievement of this goal is considered unlikely. The reasons for increased demand for coal include availability of this fuel, relative stability of coal prices, and lack of problems with spent fuel disposal (as contrasted to nuclear power plants) (see Power generation).

The efficiency of coal-fired electric power plants has progressively increased. The overall efficiency of power plants has reached about 39% and is not expected to exceed this with the present configuration of boilers. The addition of pollutant control equipment has put requirements for internal power use on the stations that lower the effective efficiency of the plant. The increased efficiencies have been achieved through use of larger units (up to 1300 MW) and higher pressures to 24.1 MPa (3500 psi) and reheat. Maximum temperatures have not been increased due to difficulties with corrosion owing to coal ash constituents, materials properties, and costs of better alloys. Future increases in efficiency will depend on development of new systems of power generation including fluid-bed boilers, gasification of coal to power a gas turbine with hot exhaust directed to a waste heat boiler in a combined cycle (gas turbine and steam

Table 18. Consumption of Bituminous Coal by Consumer Class and With Retail Deliveries in the United States[a]

Year	Electric power utilities	Railroads Class I	Coking coal	Steel and rolling mills	Cement mills	Other manufacturing and mining industries	Retail deliveries to other consumers	Total U.S. consumption
1933	24,574	65,814	36,368	12,818	2,504	75,909	70,212	288,199
1943	67,164	118,191	92,950	14,392	5,303	131,715	108,972	538,684
1953	101,861	25,161	102,398	7,951	7,409	87,996	54,409	387,184
1958	138,734	3,379	69,472	6,593	7,490	74,686	32,313	332,667
1963	189,636		70,427	6,714	7,383	75,720	21,362	371,243
1968	267,383		82,286	5,132	8,519	75,345	13,811	452,531
1973	350,971		84,691	5,766		55,296	7,439	504,415

[a] Ref. 71.

274

turbine), and magnetohydrodynamics (qv) (see also Furnaces, fuel-fired; Burner technology).

Almost all modern large coal-fired boilers for electric power generation use pulverized coal. The cyclone furnace, built mainly in Germany and the United States, uses larger size coal. The ash is primarily removed as a molten slag from the combustor. Apparently this design is no longer offered in the United States. This method of firing has not been accepted in the United Kingdom because of the higher softening temperature of the ash of the British coals. Stoker firing is generally limited to the smaller obsolete stand-by utility plants and generation plants used by industrial companies.

One significant advantage of pulverized coal boilers is their ability to use any kind of coal, including run-of-mine or uncleaned coals. With the advent of continuous mining equipment, the ash content frequently is ca 25%, and some cleaning is frequently practiced.

A major concern in coal-fired power generation is release of air pollutants. The EPA has placed limits of output of 0.52 g SO_2/MJ (1.2 lb SO_2 per million Btu) of coal input to the plant. For a bituminous coal of 27.9 MJ/kg (12,000 Btu/lb) this places an upper limit of 0.72% on the sulfur content of the coal fuel. Relatively few coals can meet this requirement. The U.S. Bureau of Mines indicated recoverable reserves of 396 billion metric tons in January 1974 (67). Of these 350 billion were categorized by sulfur content; 52% had 1% or less, 24% had between 1.0% and 3.0%, and 24% had more than 3.0% (see Coal conversion processes, desulfurization). Of the low sulfur coal, 88% is found west of the Mississippi (mainly in Montana and Wyoming); this is quite distant from the electric power demand centers in the east. A recent trend to utilization of the western coals has developed. Legislation requiring flue gas desulfurization equipment on all coal-fired boilers may reduce that trend.

Another important factor for future coal use in the United States is the current government policy of no longer allowing oil and gas to be used by new plants for electric power generation. New plants must be either coal-fired or nuclear.

Industrial Heating and Steam Generation. The major users of coal in this category include the iron and steel industry and the food, chemicals, paper, engineering, bricks and other clay products, and cement industries, and a group of miscellaneous consumers (mainly for space heating) including federal and local government installations, armed services, and small industrial concerns. Most of the coal is burned directly for process heat, ie, for drying and firing kilns and furnaces, or indirectly for steam generation for process needs or space heating, and for a small amount of electric power generation.

The use of coal by industry in the United States has diminished significantly in the last decade, especially among small users, because of the greater convenience in storing and handling oil and gas fuels and the higher initial cost of coal-fired equipment. Uncertainties regarding cost and availability of these fuels will probably bring a change back to coal. The coal may be burned in the form of industrial fuel gas because the desire to utilize existing gas consuming equipment, growing availability of gasifiers, and need for pollution control.

Several developments are being pursued to utilize coal directly, ie, automation of controls, coal and ash handling equipment for smaller stoker and pulverized coal-fired units, design of packaged boiler units, and pollution control equipment. In the cement industry coal firing has been used, since the sulfur oxides react with some of the lime to make calcium sulfate in an acceptable amount.

Coal Processing. The major approaches to coal processing or coal conversion are thermal decomposition, including pyrolysis or carbonization (3–4), gasification (see Fuels, synthetic, gases) (4), and hydrogenation (4). The hydrogenation of coal is not currently practiced commercially.

In the United States, the U.S. Office of Coal Research and portions of the successor ERDA and DOE were authorized to contract for research on coal to use it more efficiently and convert it to more desirable energy forms. The budget appropriation for the fiscal year 1978 was over $500 million. Activities range from basic research and establishing integrated operation of new processes in pilot plants through demonstration with commercial-scale equipment.

High-Temperature Carbonization. High temperatures and long processing times are used in carbonizing coking coals in coke ovens or gas retorts. Besides metallurgical or gas coke the products include fuel gas, crude tar, light oils (benzene, toluene, and xylene, referred to as BTX, and solvent naphtha), and ammonia (see Coal conversion processes, carbonization).

Most coal chemicals are obtained from high temperature tar with an average yield over 5% of the coal which is carbonized. The yields in coking are about 70% of the weight of feed coal. Tars obtained from vertical gas retorts have a much more uniform chemical composition than those from coke ovens. Two or more coals are usually blended, and conditions of carbonization will vary depending on the coals used, and will affect the tar composition. The major coal–tar chemicals include phenols, cresols, xylenols, benzene, toluene, naphthalene, and anthracene.

The largest consumer of coke is easily the iron and steel industry. In the United States, ca 600 kg of coke is used to produce a metric ton of steel. Japanese equipment and practice reduce the requirement to 400–450 kg. Coke is used to make calcium carbide (see Carbides), from which acetylene (qv) is made. Synthesis gas for methanol and ammonia production is also made from gasification of coke.

Considerable research has been carried out to produce metallurgical-grade coke from low-rank bituminous and subbituminous coal. This is especially true in areas where coking coal reserves are becoming significantly depleted or are unavailable. The leading countries in this area are the United States (FMC Formcoke and U.S. Steel Clean Coke Processes), Japan (Itoh process), and Germany (BFL process). These processes generally involve carbonization of crushed coal in fluidized beds, agglomerating the semicoke into conveniently sized balls with a binder, and calcining. The advantages of this technology include better heat transfer, shorter carbonizing time, continuous operation, and utilization of a much broader range of coals.

Low-Temperature Carbonization. Lower temperature carbonization of lump coal to ca 700°C, primarily used for production of solid smokeless fuel, gives a different (quantitatively and qualitatively) yield of solid, liquid, and gaseous products than high temperature processes.

Although a number of low temperature processes have been studied, only a few have been used commercially. These have been limited in the types of coal that are acceptable, and the by-products are less valuable than those obtained from high temperature processing.

The Disco process is used in the United States to supply a limited amount of fuel to meet requirements of smoke ordinances. The British Coalite and Rexco processes produced substantial amounts of domestic smokeless fuel.

Development of a fluid-bed method of carbonizing finer coal at ca 400°C has been studied in the United Kingdom. A reactive char would be briquetted without a binder to produce a premium open-fire smokeless fuel.

Gasification. Gasification of coal for fuel gas production has been developed commercially, and improvements in technology are being studied in a number of facilities. In the United States, the purpose of a number of efforts is to produce, a substitute natural gas (methane) with a heating value of about 37.2 MJ/m^3 (1000 Btu/ft^3) for transmission through pipelines (see Fuels, synthetic). In the United Kingdom, the purpose of the program was to develop a process to completely gasify the coal (produce no char) to make town gas with heating value of about 16.8–18.6 MJ/m^3 (450–500 Btu/ft^3) at a cost similar to processes using oil feedstock.

The Lurgi process (4) is the most successful complete gasification process for converting weakly caking coals as well as noncaking ones. The gasification takes place with steam and oxygen at 2–3 MPa (20–30 atm) to produce a 13.0–14.9 MJ/m^3 (350–400 Btu/ft^3) gas. This may be enriched with hydrocarbons to meet town gas specifications. The reactor is a slowly moving bed and is fed lump coal.

The first commercial operation of the Lurgi process was in Germany in 1936 with brown coal (see Lignite and brown coal). It was modified to stir the bed to permit utilization of bituminous coal. One plant was built at the Dorsten Works of Steinkohlengas A.G., and the Sasol plant was built in South Africa to provide synthesis gas for liquid fuels. A large expansion of the Sasol plant is under way.

Brown coal briquettes are used as feed in Australia with the process to make town gas. The British Gas Council has developed a slagging version of the gasifier which is under consideration for a demonstration plant in Noble County, Ohio.

Pulverized coal is used in several gasifiers and was studied in Germany before World War II. The Koppers-Totzek gasifier has been used commercially in different parts of the world. It uses multiple (2 or 4) heads to feed coal, air or oxygen, and steam into an entrained atmospheric pressure reactor. Molten slag is discharged. The Babcock and Wilcox Company also built an entrained bed gasifier for the DuPont Company at Belle, West Virginia, for chemical feedstock (see Feedstocks).

A combined Federal government–American Gas Association program is developing second generation processes for making pipeline quality gas. In these processes coal is prepared, gasified, the gas is cooled, shifted if necessary to adjust the H$_2$/CO ratio to about 3:1, the acid gases (H$_2$S and CO$_2$) are removed and catalytic conversion to methane is carried out. Under this program the Institute of Gas Technology in Chicago has developed the Hygas process in a 68 t/d pilot plant in which the gasifier at 6.9 MPa (1000 psi) accepts a coal slurry, dries it, goes through first stage hydrogenation at 650–730° C, and second stage at 815–930°C before steam–oxygen gasification of the char to obtain high carbon utilization. The process also produces some benzene, toluene, and xylene which are used in the pilot plant to make up the slurry. This process has been operated successfully with lignite, subbituminous, and bituminous coals. The CO$_2$ Acceptor Process was also developed under this program by Consolidation Coal Co. in a 36 t/d pilot plant at Rapid City, South Dakota. Heat to drive the gasification process is provided by the reaction of calcined dolomite (MgO·CaO) with CO$_2$ produced in gasification of lignite or subbituminous coal with steam at 1 MPa (10 atm) and 815°C. The spent dolomite is regenerated at 1010°C in a separate vessel and returned to the gasifier. The process has operated successfully with lignite and subbituminous coal.

Another process under the program is called BI-GAS, developed by Bituminous

Coal Research in a 73 t/d pilot plant at Homer City, Pennsylvania. In this entrained-bed process, pulverized coal slurry is dried and blown into the second stage of the gasifier to contact 1205°C gases at ca 6.9 MPa (1000 psi) for a few seconds residence time. Unreacted char is separated and recycled to the first stage to react with oxygen and steam at ca 1650°C to produce hot gas and molten slag which is tapped. This pilot plant is beginning operation.

The Synthane process is being developed by the DOE at the Pittsburgh Energy Research Center. This fluidized-bed process operates at ca 6.9 MPa (1000 psi) and 980°C to gasify coal and produce some char. It uses subbituminous coal.

A third-generation process called Steam–Iron is also being developed under this program. It is being developed by the Institute of Gas Technology at a pilot plant in Chicago. This plant generates hydrogen from char produced in any gasification process. A gas producer uses air to make a reducing gas from the char in one vessel. The reducing gas converts iron oxide to iron in the upper two stages of a second vessel. Steam is converted to hydrogen and reoxidizes the iron in two stages in the lower half of the vessel.

Two demonstration plants are under consideration and in the design stage. One, using the slagging Lurgi process mentioned above, is sponsored by the Conoco Development Co. The other, using the COGAS process, is sponsored by the Illinois Coal Gasification group and is planned for Perry County, Illinois. This process is similar to the COED process developed by the FMC Corp. in Princeton, New Jersey. Coal is heated in successive stages to devolatilize and then gasify without agglomerating. Product gases are used for the heating. Products include fuel gas, liquids, and char (which may be used in the process). Design capacities of these demonstration plants will be 0.57–1.7 million m³ (20–60 million SCF) of methane per day.

A large commercial plant is planned by a consortium of American Natural Gas and Peoples Gas, Light and Coke, and others for Mercer County, North Dakota. This plant has a design capacity of 3.54 million m³ (125 million SCF) of methane per day and should be complete in about 1981. It will use Lurgi gasifiers.

Intermediate-Btu gas (9.3–18.6 MJ/m³ or 250–500 Btu/ft³), or synthesis gas production is also being developed under government sponsorship. Two demonstration plants are under consideration and in the design stage. One will use the U-Gas process being developed by the Institute of Gas Technology. In this 2540 t/d plant to be located in Memphis, Tennessee, crushed coal will be fed into a fluidized-bed gasifier. Steam and oxygen enter the bed. A part of these gases carry unreacted fines into a hot spout which accelerates gasification and permits the ash to soften and particles to agglomerate. Ash agglomerates discharge below the spout. Product gases will be cleaned and pipelined as industrial fuel gas near 11.2 MJ/m³ (300 Btu/ft³) to industrial consumers up to 32 km away. The second demonstration plant will use the Texaco partial oxidation gasifier developed as a modification of Texaco's oil consuming partial oxidation process. The 1090 t/d plant will be built for W. R. Grace Company to make synthesis gas for an ammonia plant near Baskett, Kentucky. Pulverized coal falls through the reactor at high pressure and temperature to produce the gas which goes through a shift conversion to make hydrogen for the process:

$$CO + H_2O \rightarrow H_2 + CO_2$$

Coal gasification may also be used to provide fuel gas for boilers for electric power generation or for gas turbines for combined cycle power generation. A facility was planned at the Powerton power plant near Pekin, Illinois, to test gasifiers for these purposes.

Underground Gasification. This concept is intended to gasify a coal seam *in situ,* converting the coal into gas and leaving the ash underground. It is presently considered most interesting for deep coal, steeply sloping seams, or thin seams that are not economical to mine by conventional means. This approach involves drilling holes to provide air or oxygen for gasification and removal of product gases and liquids (85).

Depending on whether air or oxygen was used, a low-Btu gas could be produced for boiler or turbine use in electric power production, or an intermediate-Btu gas for an industrial fuel gas, a synthesis gas for chemical or methane production could be provided. This approach has been studied and was considered uneconomic 10–20 years ago. Increasing fuel needs have led to further studies in a number of locations.

In the United States a program is being carried out near Hanna, Wyoming for the Department of Energy. This is intended to examine different approaches to gasification, including use of air and oxygen and obtain data for economic evaluation. Other programs under government sponsorship include a longwall generator concept at the Morgantown, West Virginia, Energy Research Center, and a program managed by the Lawrence Livermore Laboratory.

Industrial testing programs have been carried out by Gulf Research and Development Company in Western Kentucky (86) on a seam at a depth of 32.6 m, and 2.7 m thick. The coal seam was excavated for study after the gasification program. Another program using Russian technology is being carried out by Texas Utilities Services in an East Texas lignite deposit.

A joint Belgian–West German program is aimed at gasifying seams $\geq$1000 m underground and to use the gases for combined cycle electrical generating plants. Later, if initial efforts are effective, hydrogen will be pumped into hot coal seams to make substitute natural gas from the exothermic carbon–hydrogen reaction.

Work with underground coal gasification has been most extensive in the U.S.S.R. They have applied the technology to produce gas for four or five electrical generation stations. An institute for study of the process was established in 1933 and has primarily studied air blown gasification which produced a gas of about 3.35–4.20 MJ/m^3 (90–113 Btu/f^3) heat content. Other work has produced synthesis gas suitable for chemical production.

Liquid Fuels and Chemicals by Gasification of Coal. Coal may be gasified to a mixture of carbon monoxide and hydrogen by using oxygen and steam (87). The product gases may be passed over iron-based catalysts to produce hydrocarbons in the Fischer-Tropsch process, or over zinc or copper catalysts to make methyl alcohol.

The Fischer-Tropsch process has not been economical. The South African Sasol plant (88) has operated successfully with the Kellogg and German Arge (Ruhr Chemie Lurgi) modifications of the Fischer-Tropsch process. The plant was designed to produce 227,000 t/yr of gasoline, diesel oil, solvents, and chemicals from 907,000 metric tons of noncaking high ash bituminous coal. The Lurgi gasification process is used to make the synthesis gas. The capacity of this plant is being expanded substantially.

The success of the Sasol project is attributed to the availability of cheap coal. Plants using Lurgi or Koppers-Totzek gasifiers for making chemicals are located in Australia, Turkey, Greece, India, and Yugoslavia, among others.

A variety of pilot plants using fluid-bed gasifiers have been built in the United States, Germany, and elsewhere. The Winkler process is the only one that has been used on a large scale. It was developed in Germany in the 1920s to make synthesis gas at atmospheric pressure. Plans are being made to develop a pressurized version. Plants using bituminous coal have been built in Spain and Japan with this gasifier (see Fuels, synthetic, liquid).

In the United States, processes under development to produce synthesis gas under pressure include the Texaco partial oxidation process and the U-Gas process by the Institute of Gas Technology. Demonstration plants are being designed to incorporate each of these.

The Koppers-Totzek entrained gasifier is being developed for high pressure by Shell and Krupp-Koppers near a Shell Refinery in Hamburg, Germany.

Gasification and Metallurgy. Some interesting combinations of these technologies include direct reduction of iron ore, and direct injection of coal into the blast furnace (see Iron). In direct reduction, a mixture of carbon monoxide and hydrogen reduces iron ore pellets to elemental iron. These pellets may later be used to feed steel-making processes. Direct reduction has been carried out with synthesis gas from natural gas, but apparently not yet from coal owing to unfavorable economics. Pulverized coal has been successfully injected into the tuyeres of a blast furnace of the Armco Company in Middletown, Ohio, to supplement coke (see Iron by direct reduction).

Hydrogenation to Produce Liquids. Liquefaction of coal to oil was first accomplished by Bergius in 1914. Hydrogen was used with a paste of coal, heavy oil, and a small amount of iron oxide catalyst at 450°C and 20 MPa (200 atm) in stirred autoclaves. The process was developed by the I.G. Farbenindustrie to give commercial quality gasoline as the major product. Twelve hydrogenation plants were operated during World War II to make liquid fuels.

Imperial Chemical Industries in Great Britain hydrogenated coal to produce gasoline until the start of World War II. The process then operated on creosote middle oil until 1958. None of these plants is currently being used to make liquid fuels for economic reasons. The present prices of coal and hydrogen from coal have not made liquid fuels competitive. In those cases where availability is important there is incentive for development of processes.

In the United States, solvent refining of coal is being developed with Department of Energy sponsorship at a 45 t/d pilot plant near Tacoma, Washington. In this process, coal is slurried with a process-derived anthracene oil and heated to ca 425°C at 1.38 MPa (2000 psi) of hydrogen for ca 1 h. After filtration the major product is a low ash, low sulfur tar-like boiler fuel. Another plant at Wilsonville, Alabama, sponsored by the Southern Company and Electric Power Research Institute is also developing this technology with a 5.5 t/d plant. A variation of this process, called SRC-II, which involves further treatment to give a distillate fuel is also being developed.

The H-Coal process developed by Hydrocarbon Research, Inc. is a 545 t/d plant being demonstrated at Catlettsburg, Kentucky, with DOE support. This process uses a slurry of crushed coal and recycle-oil with hydrogen in an ebullated-bed reactor at 20 MPa (200 atm) at ca 455°C with a catalyst. The products range from a light distillate oil to residual fuel.

A contract has been signed for development of the Exxon Donor Solvent Liquefaction process, with partial support by the DOE. A 225 t/d demonstration plant near Baytown, Texas, will be built. The process involves steps similar to solvent refining. In addition, the recycle oil is given a catalytic hydrogenation prior to slurry preparation.

Processes for hydrogen gasification of coal will produce liquid coproducts. The HYGAS process described above produces about 6% liquids as benzene, toluene, and xylene.

Several processes use product gas or synthesis gas to pyrolyze the coal and produce a mixture of products; the Cogas process described above is one of these.

The heavy oils produced by coal hydrogenation need further processing to render them desirable for transportation fuels. The Conoco Coal Development Company, with major support for the DOE, is developing a zinc halide hydrocracking process to produce gasoline from coal extract. A process development unit is starting operation.

Flash hydropyrolysis of coal involves rapid heating in hydrogen to maximize the yield of lighter oils. Processes are in the early stages of development by Rocketdyne, Cities Services, Occidental Research, and the Institute of Gas Technology with DOE support.

Other Uses.   The quantity of coal used for purposes other than combustion or processing described above is relatively insignificant (4,89).

Coal, especially anthracite, has established markets for purifying and filtering agents, in either the natural form or converted to activated carbon (see Carbon). The latter can be prepared from bituminous coal or coke, and is used in sewage treatment, water purification, respirator absorbers, solvent recovery, and in the food industry. Some of these small markets are quite profitable and new uses are continually being sought for this material.

Carbon black (qv) from oil is the main competiton for the product from coal, which is used in filters. Carbon for electrodes is primarily made from petroleum coke, although pitch coke is used in Germany for this product. The pitch binder used for electrodes and other carbon products is almost always a selected coal tar pitch (see Carbon, baked and graphitized products).

The great variety of uses for by-products of coal carbonization is discussed in other articles (see Coal conversion processes, carbonization; Tar and pitch). The use of coal in metallurgical operations has already been indicated. The change to pelletized iron ore represents a substantial market for coke and anthracite for sintering. Direct injection into the blast furnace of an auxiliary fuel, coal, or oil is now practiced to provide heat for the reduction and some of the reducing agent in place of the more expensive coke that serves these purposes.

Some minor uses that may grow in the future include the use of fly ash, cinders, or even coal as a building material; soil conditioners from coal by oxidizing it to humates; and a variety of carbon and graphite products for the electrical industry, and possibly the nuclear energy program.

The growth of synthetic fuels from coal will also provide substantial quantities of by-products including elemental sulfur, fertilizer as ammonia or its salts, and a range of liquid products. The availability of ammonia and straight chain paraffins may permit production of food from fossil fuels.

BIBLIOGRAPHY

"Coal" in *ECT* 1st ed., Vol. 4, pp. 86–134, by H. J. Rose, Bituminous Coal Research Inc.; "Coal" in *ECT* 2nd ed., Vol. 5, pp. 606–678, by I. G. C. Dryden, British Coal Utilisation Research Association.

1. W. Francis, *Coal,* 2nd ed., Edward Arnold & Co., London, 1961.
2. D. W. van Krevelen, *Coal,* Elsevier Scientific Publishing Co., Amsterdam, 1961.
3. H. H. Lowry, ed., *Chemistry of Coal Utilization,* Vols. 1 and 2, John Wiley & Sons, Inc., New York, 1945.
4. *Ibid.,* Suppl. Vol. 1963. An updated version, edited by M. Elliott will be published in late 1979.
5. R. Thiessen, *U.S. Bur. Mines Inform. Circ.,* 7397 (1947).
6. M. C. Stopes, *Proc. R. Soc. London Ser. B* **90,** 470 (1919); *Fuel* **14,** 4 (1935).
7. B. C. Parks and H. J. O'Donnell, *U.S. Bur. Mines Bull.,* 550 (1956).
8. *International Handbook of Coal Petrography,* Centre National de la Recherche Scientifique, Paris, 1963.
9. *Atlas für Angewandte Steinkohlenpetrographie,* Gluckauf, Essen, West Germany, 1951.
10. J. B. Nelson, *Brit. Coal Util. Res. Assoc. Mon. Bull.* **17,** 41 (1953).
11. H. J. Gluskoter and co-workers, *Ill. State Geol. Surv. Circ.,* 499 (1977).
12. R. Loison and P. Foch, *Blast Furn. Coke Oven Raw Mater. Commun. Proc.* **17,** 170 (1958).
13. C. A. Seyler, *Fuel* **3,** 15, 41, 79 (1924); *Proc. S. Wales Inst. Eng.* **53,** 254, 396 (1938).
14. *Annual Book of ASTM Standards, 1977—Part 26, Gaseous Fuels; Coal and Coke; Atmospheric Analysis,* ASTM, Philadelphia, 1977.
15. *Survey of Energy Resources 1974,* World Energy Conference, Central Office, London, 1974.
16. *The Russian Coking Industry 1962,* The British Coke Research Assoc., Special Publication, May 7, 1963.
17. G. L. Tingey and J. R. Morrey, *Coal Structure and Reactivity,* Battelle Energy Program Report, Battelle Pacific Northwest Laboratories, Richland, Washington, Dec. 1973.
18. M. E. Peover, *J. Chem. Soc.,* 5020 (1960).
19. R. Raymond, I. Wender, and L. Reggel, *Science* **137,** 681 (1962).
20. L. Reggel, I. Wender, and R. Raymond, *Fuel* **43,** 75 (1964).
21. B. K. Mazumdar and co-workers, *Fuel* **41,** 121 (1962).
22. B. K. Mazumdar, S. S. Choudhury, and A. Lahiri, *Fuel* **39,** 179 (1960).
23. H. W. Sternberg and co-workers, *Coal Science, Advances in Chemistry Series,* Vol. 55, American Chemical Society, Washington, D.C., 1966, p. 503.
24. R. Kuhn and R. Roth, *Ber. Deut. Chem. Ges.* **66B,** 1274 (1933).
25. R. Bent, W. K. Joy, and W. R. Ladner in ref. 20, p. 75.
26. I. G. C. Dryden, *Fuel* **37,** 444 (1958).
27. I. G. C. Dryden in ref. 21, pp. 55 and 301.
28. A. C. Battacharya, B. K. Mazumdar, and A. Lahiri in ref. 20, p. 181; S. Ganguly and B. K. Mazumdar in ref. 20, p. 281.
29. J. K. Brown, *Brit. Coal Util. Res. Assoc. Mon. Bull.* **23,** 1 (1959).
30. B. K. Mazumdar and co-workers in ref. 23, p. 475.
31. L. A. Heredy, A. E. Kostyo, II, and M. B. Neuworth in ref. 23, p. 493.
32. B. K. Mazumdar, *Fuel* **51,** 284 (1972).
33. L. Cartz and P. B. Hirsch, *Phil. Trans. R. Soc. London Ser. A.* **252,** 557 (1960).
34. W. Spackman, "What is Coal?", *Short Course on Coal Characteristics and Coal Conversion Processes,* Penn. State Univ., Univ. Park, Pa., Oct. 1973.
35. M. Farcasiu, *Fuel* **56,** 9 (1977).
36. H. L. Retcofsky and R. A. Friedel in ref. 23.
37. W. R. Ladner and A. E. Stacey in ref. 20, p. 13.
38. H. Gan, S. P. Nandi, and P. L. Walker, Jr. in ref. 32, p. 272.
39. J. T. McCartney, H. J. O'Donnell, and S. Ergun in ref. 23, p. 261.
40. G. H. Taylor in ref. 23, p. 274.
41. W. H. Wiser, "Some Chemical Aspects of Coal Liquefaction" in ref. 34.
42. H. F. Yancey and M. R. Geer in J. W. Leonard and D. R. Mitchell, eds, *Coal Preparation,* 3rd ed., American Institute of Mining, Metallurgical and Petroleum Engineers, Inc., New York, N.Y. 1968, pp. 3–56.
43. P. Rosin and E. Rammler, *J. Inst. Fuel* **7,** 29 (1933); J. G. Bennett, *J. Inst. Fuel* **10,** 22 (1936).
44. F. Kick, *Dinglers Polytech. J.* **247,** 1 (1883); P. von Rittinger, *Lehrbuch der Aufbereitungskunde,* Ernst and Korn, Berlin, 1867, pp. 595.

45. F. C. Bond, *Min. Eng.* **4,** 484 (1952).
46. S. J. Vecci and G. F. Moore, *Power,* 74 (1978).
47. I. G. C. Dryden, *Fuel* **29,** 197, 221 (1950).
48. G. R. Hill and co-workers in ref. 23, p. 427.
49. A. Pott and co-workers, *Fuel* **13,** 91, 125, 154 (1934).
50. *Environ. Sci. Technol.* **8,** 510 (1974).
51. W. Downs, C. L. Wagoner and R. C. Carr, *Preparation and Burning of Solvent Refined Coal,* presented at American Power Conference, Chicago, Ill., April 1977.
52. R. C. Attig and A. F. Duzy, *Coal Ash Deposition Studies and Application to Boiler Design,* American Power Conference, April 1969.
53. E. C. Winegartner and B. T. Rhodes, *J. Eng. Power* **97,** 395 (1975).
54. *Steam, Its Generation and Use,* The Babcock & Wilcox Co., New York, 1972, p. 15-4.
55. J. A. Harrison, H. W. Jackman, and J. A. Simon, *Ill. State Geol. Surv. Circ.,* 366 (1964).
56. D. K. Fleming, R. D. Smith, and M. R. Y. Aquino, *Am. Chem. Soc. Div. Fuel Chem. Prepr.* **22**(2), 45 (Mar. 1977).
57. S. C. Spalding, Jr., J. O. Burckle, and W. L. Teiser in ref. 23, p. 677.
58. J. W. Hamersma, M. L. Kraft, and R. A. Meyers in ref. 56, pp. 73, 84.
59. H. Hofmann and K. Hoehne, *Brennstoff-Chem.* **35,** 202, 236, 269, 298 (1954).
60. M. D. Kimber and M. D. Gray, *Combust. Flame* **11,** 360 (1967).
61. D. Fitzgerald and D. W. van Krevelen, *Fuel* **38,** 17 (1959).
62. K. Ouchi and H. Honda in ref. 61, p. 429.
63. B. K. Mazumdar, S. K. Chakrabartty, and A. Lahiri, *Proceedings of the Symposium on the Nature of Coal,* Central Fuel Res. Inst., Jealgora, India, 1959, p. 253; S. C. Biswas and co-workers, *Proceedings of the Symposium on the Nature of Coal,* Central Fuel Res. Inst., Jealgora, India, 1959, p. 261.
64. A. Lahiri and co-workers, *J. Mines Metals Fuels* **7,** 13 (1959).
65. Ref. 2, p. 472.
66. K. S. Vorres in ref. 56, **22**(4), 118 (May 1977).
67. P. Averitt, "Coal Resources of the United States, January 1, 1974," *U.S. Geological Survey Bulletin,* 1975, p. 1412.
68. M. K. Hubbert, *Natl. Acad. Sci. Natl. Res. C.* **1000-D,** (1962).
69. *Minerals Year Book 1974,* U.S. Dept. of Interior, Bureau of Mines U.S. Gov't. Printing Office, Washington, D.C., 1974.
70. *Coal Traffic Annual,* 1976 ed., National Coal Association, Washington, D.C., 1976.
71. *Coal Facts, 1974–1975,* National Coal Association, Washington, D.C., 1975.
72. F. X. Murray, *Energy, A National Issue,* Center for Strategic & Intl. Studies, Georgetown Univ., Washington, D.C., 1976.
73. *Coal Age* **82,** 59 (1977).
74. *Coal Age* **68,** 226 (1963).
75. H. S. Ellis, P. J. Redburger, and L. H. Bolt, *Ind. Eng. Chem.* **55,** 18, 29 (1963).
76. *Chem. Eng. News* **55,** 20 (June 27, 1977).
77. *International Coal Trade,* U.S. Dept. of Interior, Bureau of Mines, Vol. 46, No. 6, July, 1977.
78. "The Sampling of Coal and Coke," *Brit. Standards* 1017, Parts 1 and 2, 1960.
79. "Specification for Test Sieves," *Brit. Standards,* 1962, p. 410; Screen Analysis of Coal," *Brit. Standards,* 1946, p. 1293; "Methods for the Use of S. S. Fine Mesh Test Sieves," *Brit. Standards,* 1952, p. 1796; "Size Analysis of Coke," *Brit. Standards,* 1954, p. 2074.
80. *International Classification of Hard Coals by Type,* United Nations Publ. No. 1956 II, E. 4, E/ECE/247; E/ECE/Coal/110 (1956).
81. "Analysis and Testing of Coal and Coke," *Brit. Standards,* Parts 1–16, 1957–1964, p. 1016.
82. *World Book Encyclopedia,* Field Enterprises Educational Corp., Chicago, Ill., 1975, p. 566.
83. Ref. 73, p. 62.
84. S. M. Cassidy, ed., *Elements of Practical Coal Mining,* Port City Press, Inc., Baltimore, Md., 1973.
85. D. R. Stephens, *A.G.A. Mon.* **57,** 6 (1975).
86. D. Raemondi, P. L. Terwilliger, and L. A. Wilson, Jr., *J. Petrol. Tech.* **27,** 35 (1975).
87. C. C. Hall, *J. Inst. Fuel* **23,** 148 (1950); *Research London* **1,** 7 (1956).
88. J. A. Linton and G. C. Tisdall, *Coke Gas* **19,** 402, 442 (1957).
89. R. A. Glenn and H. J. Rose, *The Metallurgical, Chemical and Other Process Uses of Coal,* Bituminous Coal Research, Inc., Pittsburgh, Pa., 1958.

KARL S. VORRES
Institute of Gas Technology

COAL—GASIFICATION OF COAL. See Fuels, synthetic.

COAL—POWER FROM COAL BY GASIFICATION AND MAGNETOHY-DRODYNAMICS. See Coal—Magnetohydrodynamics.

COAL—SYNTHETIC CRUDE OIL FROM COAL. See Fuels, synthetic.

COAL CONVERSION PROCESSES

CARBONIZATION

Carbonization of coal is an old established industry by which coal is subjected to destructive distillation in a heated retort in the absence of air. Coke is the main product, though tar, light oil, ammonia liquor, and coke-oven gas are also produced. Coke is used principally as a fuel and reductant in the blast furnace for ironmaking. The other products are further refined into commodity chemicals such as ammonium sulfate, benzene, toluene, xylene, naphthalene, pyridine, phenanthrene, anthracene, creosote, road tars, roofing pitches, and pipeline enamels. The coke-oven gas is a valuable heating fuel used mainly within steel plants. Hence the carbonization industry is closely tied to the steel industry (see Iron; Steel).

In the United States, most carbonization is conducted in the high temperature range of 900–1200°C; medium (750–900°C) and low temperature carbonization are rare. The latter processes are more common in Europe and Asia where they are used principally in the manufacture of special products such as smokeless fuel, or reactive reductants for certain metallurgical processes. Carbonization operations are conducted almost entirely in slot-type ovens; only slightly more than 1% of the total coke in the United States is produced in beehive ovens. Horizontal retorts and continuous vertical retorts, used chiefly to produce gas in Europe, have virtually disappeared.

The carbonization industry has passed through several phases since its primitive development in England near the end of the 16th century. The history of carbonization has been fully described (1).

In 1619 Dudley discovered that certain coals after "charking" produced a "coak," which substituted for wood charcoal, improved blast-furnace performance. Darby, a Quaker ironmaster, developed the first commercially successful carbonization process in 1709. Carbonization of coal in iron retorts was applied for the production of an illuminating gas for homes and street lighting about 1800.

Coke is produced by the partial combustion of bituminous coal. The original conical piles of coal covered with earth or coke dust evolved into a hemispherical brick retort, or beehive oven, about 1840. In the mid-1850s German developers began work on refractory retorts designed to recover the chemical by-products of carbonization. The first successful by-product oven, built in Germany in 1881, led to the spectacular development of the German iron, steel, and coal-tar chemical industries. At Syracuse, New York, in 1893, Semet-Solvay constructed the first coke ovens in the United States, principally to recover ammonia for use in the Solvay soda-ash process (see Ammonia). The first successful ovens built in the United States to provide coke for ironmaking were constructed by the Koppers Company at Joliet, Illinois, for the United States Steel Corporation.

World War I prompted the development of an aromatic chemical industry in the United States since these chemicals were no longer available from Germany. It was also apparent at this time that by-product coke ovens produced a more uniform coke that enhanced blast-furnace performance. Thus, recovery ovens made important strides after World War I. Also after the war, several coke ovens were built to supply gas for public utilities in many large cities as well as some smaller ones. Some coke was used as household fuel and some converted by utilities into gas in water-gas sets. The demand for manufactured gas virtually disappeared when natural gas from the fields in the Southwest became available by pipeline transport in the 1940s and 1950s. Some coke plants were shut down and most remaining plants were converted to the manufacture of foundry coke. The carbonization industry also was no longer the dominant source of aromatic chemicals, as the petroleum industry became a very large producer of products such as toluene, benzene, xylene, and naphthalene.

The history of the carbonization industry is reflected in the statistics given in Table 1. Growth of the carbonization industry in the United States is reflected in the following statistics. In 1910, ca 100,000 beehive ovens produced about 30 million metric tons of coke. Beehive oven production declined after that as the by-product coke oven took over. In 1955, 26,143 coke ovens in 85 plants produced about 68 million metric tons of coke. Vertical slot-type ovens with about 67 million metric tons capacity accounted for 16,039 of these ovens. Variations in the activity of the steel industry have caused considerable fluctuations in coke production. The maximum production of coke, ca 72 million tons, occurred in 1951 and over 90% was produced in slot-type ovens. The total value of U.S. coke and chemical products has increased steadily to >6 billion dollars in 1976.

The total world production of coke in 1976 amounted to more than 367 million metric tons. As shown in Table 2, the United States produced 14.4% of the world total, second to the USSR at 23.0%. Table 3 shows production trends in the United States over the last 20 years including amounts of coal carbonized, hot metal produced and coke consumed. A decline in the coke rate, because of improved blast-furnace practice, as well as a decline in hot metal production, caused a decline in the amount of coal carbonized during the first few years of this period. Later, as hot metal production gradually increased, the amount of coal carbonized also increased in spite of continued improvement in blast-furnace performance. The reduction in coke rate during this period can be attributed to more efficient, larger furnaces, agglomerated burdens, higher iron contents of the ores, improved coke, higher top pressures, higher blast temperatures, oxygen enrichment, supplementary fuels, and other factors. As blast-furnace practice improves, a continued rise in hot metal production and decline in the amount of coke required to produce a metric ton of hot metal is expected.

Table 1. Historical Statistics of the Carbonization Industry in the United States[a]

Year	No. ovens		Coke production, million t			Coke yield, %[b]	Average value of coke at plant, $/t[b]	Total value at plant, million $[c]		
	Slot Type	Beehive	Oven Coke	Beehive	Total			Total oven coke	Total beehive coke	All coal-chemical products
1880		12,372		3.0	3.0	63.7	2.19		7	7
1890		37,158		10.4	10.4	63.9	2.23		23	23
1900	1,085	57,399	1.0	17.6	18.6	63.9	2.55	2	45	c
1910	4,078	100,362	6.4	31.4	37.8	66.1	2.63	25	75	8
1920	10,881	75,298	27.9	18.6	46.5	67.4	10.22	317	182	101
1930	12,831	23,907	41.0	2.5	43.5	68.7	4.81	209	9	157
1940	12,734	15,150	49.0	2.8	51.8	70.1	5.29	269	13	156
1950	14,982	17,708	60.7	5.3	66.0	69.9	14.80	918	78	293
1960	15,323	7,583	51.0	0.9	51.9	70.3	20.18	1,060	15	303
1965	14,357	3,433	59.1	1.5	60.6	70.2	18.56	1,129	25	311
1970	d	d	59.6	0.8	60.4	69.0	32.89	1,882	17	293
1976	13,374	1,100	52.4	0.5	52.9	68.9	104.00	4,976	44	920

[a] Ref. 2.
[b] Weighted average of oven coke and beehive coke.
[c] Value of coke-oven products revised beginning 1916.
[d] Not available.

Table 2.   World Production of Coke in 1976, Millions of Metric Tons[a]

Country	Production	% of total
USSR	84.37	23.0
United States	52.92	14.4
Japan	47.41	12.9
FRG	32.92	9.0
People's Republic of China	28.03[b]	7.6
Poland	18.93	5.2
United Kingdom	15.75	4.3
France	11.31	3.1
South America	3.69	1.0
other	71.82	
Total World Production	*367.153*	

[a] Ref. 2.
[b] Estimated.

Table 3.   Hot Metal Production and Coke Consumption in the United States, Millions of Metric Tons[a]

Year	Hot metal produced	Coke rate, t/t hot metal	Coke consumed[b] Blast furnace	Total	Total coal carbonized[b]
1955	70.6	0.880	62.1	69.0	97.7
1960	61.1	0.770	46.3	51.6	73.8
1965	80.6	0.665	53.6	59.3	86.5
1970	83.3	0.634	52.8	57.3	87.5
1973	91.8	0.600	55.1	59.7	85.4
1976	78.8	0.594	46.8	51.6	76.8

[a] Ref. 3.
[b] Ref. 2.

Coals for Coking

It is general practice in the United States to blend two or more types of coking coals. Usually a high-volatile coal is blended with a low-volatile coal at a 90:10 to a 60:40 blend, respectively. Medium-volatile coals have also been used, either singly or in a three-component blend with high- and low-volatile coals. As previously explained, to produce a suitably strong coke, it is necessary to charge a mixture with a reactives-to-inert ratio as near as possible to the optimum for the coals blended (4). Geographic location, coal costs, and freight rates are important factors. Less manipulation of the coal blend is also possible if the coals are captive. About 55% of the coals used for blast-furnace coke production are captive. The coals must also be blended in order to have a satisfactory chemical analysis in the coke. Impurities of ash and sulfur should be as low as possible and in a range tolerable to the blast furnace operator. The composition of the coke ash is also needed in the calculation of the burden of the furnace.

A larger amount of the low-volatile coal, reduced coking rate, and incorporation of 5–15% of an inert material such as anthracite fines or coke breeze are customary in the production of foundry coke. The resultant coke is large and has a high shatter strength and reduced reactivity.

Carbonization pressure is an important consideration in the selection of coals for use in coke ovens. Coals develop pressures that are exerted on the walls as a result of volatile products attempting to escape through the plastic layers, particularly if the layers are viscous. High-volatile coals that attain high fluidity in the plastic state develop little pressure, low-volatile coals with a very viscous plastic state generate very high pressures, and blends of the two develop intermediate pressures. The development of excessive pressures by some coals or blends of coals can distort or damage the brickwork of coke ovens during coking. In view of the large initial investment of coke ovens, as well as their costly repair, it is necessary to select and use coal blends that develop safe pressures. Tests on dangerous coals have determined that ca 10 kPa (1.5 psi) is the maximum safe pressure (5–6). Movable-wall test ovens that assess the pressure-developing properties of coals are used extensively to monitor coal blends intended for use in coke ovens and have resulted in a drastic reduction in the occurrence of oven damage by excessive carbonization pressure (7). It is therefore important that a coal blend that meets the requirement of producing a satisfactory coke, with acceptable analysis, also meets the safe pressure criterion.

In 1976, West Virginia produced the largest amount of coking coal in the United States (32%), followed by Pennsylvania (30%), Kentucky (14%), Illinois, Alabama, Colorado, Utah, Virginia, Oklahoma, and New Mexico. Table 4 lists typical coals from these states that are used to produce metallurgical coke. It also gives average proximate and ultimate analyses of these coals, which indicate the wide variety of coking coals employed.

The *Minerals Yearbook* for 1976 (2) shows that the average coal mixture charged into coke ovens that year in the United States contained 66.0% high-volatile, 13.1% medium-volatile, and 20.9% low-volatile coals, with volatile matter contents on the average of 34.9, 26.4, and 17.4%, respectively. These are averages of coals charged predominantly for blast-furnace coke manufacture and the composition of the blend as well as of the individual coals in the blend varies widely. Typical proximate and ultimate analyses of coal blends for both blast-furnace and foundry coke manufacture in the United States at present are given in Table 5. The foundry coal mixture, containing more low-volatile coal component and an inert carbon additive, is leaner than the blast-furnace mixture and contains less ash and sulfur.

Coal Preparation

Coals arrive at the coke plant by rail, ship, or other means, and are unloaded into separate stockpiles according to high-, medium- or low-volatile classification or are conveyed directly to the coal-blending plant, mixed, crushed, and stored in the storage bunkers near the ovens. Each coal is processed separately up to the mixing step in a Bradford breaker where it is reduced to a top size of ca 25 mm, and rock, wood, metal, and other extraneous materials are removed. The coals are placed in two or more mixer bins from which they are withdrawn onto a horizontal conveyor belt. An adjustable gate meters a constant volume of each coal per unit time onto the belt in the proportion needed for the required blend. Weightometer or vibrating feeders are used in some plants to meter the coals by weight. The mixer belts convey each coal to a common hopper which feeds to a hammermill for final pulverization and mixing. The mixed coal is pulverized to 75–85% <3 mm. Many plants spray some oil (0.1–1.0%) or water, or both, onto the coal just before entering the hammermill in order to control the bulk

Table 4. Properties of Typical U.S. Coals Used to Produce Metallurgical Coke, Analyses on Wet Basis[a]

State, county	Bed	Moisture as received	Volatile matter	Fixed carbon	% Ash	Sulfur	Hydrogen	Carbon	Nitrogen	Oxygen	Free-swelling index
Alabama											
Jefferson	Pratt	2.7	24.7	65.9	6.7	1.5	5.0	79.7	1.5	5.6	7.5
	Mary Lee	4.2	27.6	59.9	8.3	0.8	5.1	76.2	1.6	8.0	7.5
Colorado											
Las Animas	Frederick	3.9	29.7	57.8	8.6	0.6	5.4	75.8	1.6	8.0	9.0
Illinois											
Franklin	No. 6	8.0	35.4	50.0	6.6	0.9	5.5	70.6	1.7	14.7	4.5
Saline	No. 5	6.4	34.4	52.3	6.9	1.7	5.4	71.2	1.6	13.2	
Kentucky											
Floyd	Elkhorn No. 1	3.3	38.6	53.7	4.4	0.8	5.6	77.3	1.6	10.3	4.5
	Elkhorn No. 3	3.3	38.1	56.4	2.2	0.5	5.6	80.3	1.5	9.9	
Harlan	High Splint	4.1	35.4	52.0	8.5	0.5	5.3	74.0	1.6	10.1	5.0
Letcher	Hazard No. 4	3.0	35.6	52.1	9.3	0.9	5.2	73.2	1.6	9.8	5.5
New Mexico											
Colfax	Raton	2.0	37.2	49.6	11.2	0.8	5.3	72.1	1.3	9.3	
Oklahoma											
Haskell	Stigler	2.1	25.8	67.3	4.8	0.5	5.0	82.1	1.8	5.8	8.0
Le Flore	Upper Hartshorne	2.7	18.3	69.4	9.6	0.8	4.3	77.9	1.7	5.7	8.0
Pennsylvania											
Cambria	Lower Kittanning	1.1	16.3	75.1	7.5	1.4	4.3	82.6	1.3	2.9	9.0
Greene	Pittsburgh	2.1	35.4	57.0	5.5	1.1	5.3	78.2	1.6	8.3	9.0
Washington	Pittsburgh	1.6	37.7	55.6	5.1	1.3	5.4	78.5	1.5	8.2	7.5

Table 4 (*continued*)

State, county	Bed	Moisture as received	Volatile matter	Fixed carbon	Ash	Sulfur	Hydrogen	Carbon	Nitrogen	Oxygen	Free-swelling index
						%					
Utah											
Carbon	Sunnyside	3.5	38.9	52.3	5.3	1.1	5.6	74.5	1.6	11.9	5.0
Virginia											
Tazewell	Jewell	2.8	25.0	69.0	3.2	0.7	5.1	84.1	1.4	5.5	9.0
Wise	Clintwood	2.0	31.8	62.2	4.0	0.8	5.3	81.0	1.8	7.1	7.5
West Virginia											
Boone	Hernshaw	1.6	34.7	58.5	5.2	0.8	5.3	80.5	1.6	6.6	5.5
Fayette	No. 2 Gas	2.3	33.5	58.6	5.6	0.8	5.3	79.2	1.5	7.6	8.0
Kanawha	Dorothy	2.0	34.5	59.4	4.1	0.6	5.4	81.1	1.4	7.4	
Logan	Eagle	1.6	31.9	61.0	5.5	0.6	5.1	80.6	1.6	6.6	
	Powellton	2.5	33.7	61.1	2.7	0.6	5.4	82.5	1.5	7.3	
McDowell	Pocahontas No. 3	2.4	16.6	75.8	5.2	0.7	4.5	84.2	1.2	4.2	8.5
Mingo	L. Cedar Grove	1.8	35.3	58.6	4.3	0.6	5.4	80.7	1.5	7.5	
Nicholas	Sewell	2.4	27.3	66.9	3.4	0.8	5.1	82.3	1.7	6.7	8.0
Raleigh	Pocahontas No. 3	1.7	15.9	75.7	6.7	0.8	4.3	82.9	1.1	4.2	8.5
Wyoming	Pocahontas No. 3	1.6	17.4	74.5	6.5	0.6	4.4	83.7	1.2	3.6	8.5

[a] Ref. 8.

290

Table 5.   Typical Characteristics of Coal Charged into Coke Ovens, Dry Basis

	Blast-furnace coke, wt %	Foundry coke, wt %
moisture	6.0	6.0
volatile matter	30.5	23.0
fixed carbon	63.5	72.0
ash	6.0	5.0
carbon	78.8	81.7
hydrogen	5.3	5.1
nitrogen	1.7	1.6
oxygen	7.2	5.9
sulfur	1.0	0.7

density of the coal in the ovens. Thus uniform charge weights, heating, and coke quality are obtained. In some plants the coals are passed individually through the Bradford breaker, hammermill, and mixer bins. The pulverized coals are then blended in the desired proportions and mixed by passage over riffle splitters or paddle, twin-screw, or other mixers.

The prepared coal blend containing about 6% moisture is conveyed to the larry bin with ca 24-h operating capacity, located at one end of the coke oven battery. The bin is usually sectioned to minimize coal segregation, with rows of hoppers at the bottom provided with gates for feeding the 3 to 5 hoppers of the larry car. Though most plants in the United States use some variation of the systems described above, other countries may have very different methods of preparation. In Europe, coke plants are commonly located at the coal mines. European plants generally pulverize the coal blend to a greater degree (85–95% <3 mm) and charge coal of lower volatile matter (23–25%), but with higher moisture content (9–10%) (9). Japanese plants also pulverize the coal more extensively and use high moisture coal, but the volatile matter is ca 28–30%.

Coking Mechanism

When coal is charged into a coke oven, the coal particles adjacent to the heated walls melt and fuse, lose much of their volatile materials, and finally solidify into semi-coke. Heat conducted from the walls into the coal mass generates plastic layers of molten coal which progress from each wall toward the center plane of the oven. In each plastic layer there is evolution of volatile matter which, during its passage out of the plastic layer, leaves a residue with a porous semi-coke structure; this structure is further devolatilized to form a hard cellular-structured coke. During most of the coking cycle, there is newly-formed coke adjacent to the walls, along with two plastic layers of coal, and unreacted coal between the plastic layers. The coke layers at the walls grow wider and the width of the unconverted coal between the layers is correspondingly reduced. The plastic layers meet and then disappear near the end of the coking cycle; the coal is thus completely converted into coke. The coke in the oven is usually soaked to reduce the volatile matter content of the entire mass to ≤1%. Soaking also allows the coke cake to shrink sufficiently so that it can be pushed readily out of the oven.

The preceding description of the transient coking mechanism is very general. The temperature attained and the rate of travel of the plastic layers depend upon the coal, particularly upon the rate of heat transfer to the coal. There is little travel or migration of the coal in solid or liquid form but there is a progression of the liquid phase from each wall toward the center. Figure 1 presents a typical progression of temperatures in a coal charge during carbonization. This chart depicts the rapid temperature increase at the wall and slow temperature rise in the center. Figure 2 (10) is a plot of the rate of travel of the plastic layers as a function of their distance from the wall: the rate is rapid at the walls, reaches a minimum during most of the coking time, and accelerates when the plastic layers approach the center of the oven.

Coking coal is a very heterogeneous material, even from one bed of coal. The temperatures of softening and resolidification vary considerably from coal to coal as well as with a coal from a single source. Every coal is composed of petrographic entities each of which has different thermal characteristics. These can be generally classified as reactives, which soften and resolidify during carbonization, and inerts, which do not melt. The reactives are further classified as vitrinites, exinites, and resinites; the inerts as fusinites and micrinites. Some semi-fusinites possess both properties. The

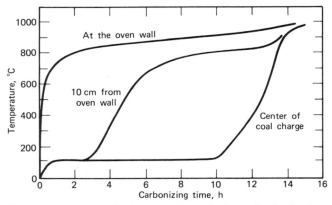

Figure 1. Temperature progression in a coal charge during carbonization in a coke oven.

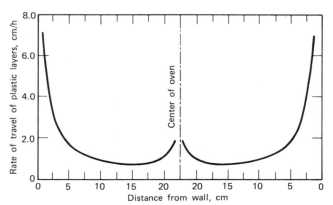

Figure 2. Rate of travel of fusion zone across oven.

vitrinites occur in large amounts in coking coals and soften and become fluid, the degree of fluidity varying with the rank and source of the coal. They release considerable volatile matter during the coking process and solidify into a residue with a moderately high yield. Exinites exhibit a higher degree of fluidity, release a large amount of volatile matter, and leave a highly swollen residue of low yield. The inert fusinites undergo little or no decomposition, and micrinite decomposes but remains an inert material as far as coking is concerned. These residues are obtained in high yield.

The melting and fusion phase of the coal particles during carbonization is the most important part of the coking of coal. The degree of melting and degree of assimilation of the coal particles into the molten mass, especially the inert particles, determine the characteristics of the coke produced. A coal rich in vitrinite produces a weak spongy coke; a coal deficient in vitrinite has insufficient plastic properties necessary for good coke formation. To produce the strongest coke from a particular coal or coal blend, there is an optimum ratio of reactive to inert entities (4). The reactives melt, flow around and coat nonmelting particles or fuse to less fluid particles, then solidify and form the bond. The reactives should possess sufficient fluidity to flow between and around other particles. The inerts are thus assimilated into the plastic matrix and act as aggregates. Two or more coals must be blended in such a manner that the optimum reactive–inert ratio is attained or approached as closely as possible within economical limits.

Coke Ovens

Industrial coal is commonly carbonized in a narrow slot-type chamber constructed of silica brick. Many such chambers are arranged side by side; 85 or more ovens have been included in a single battery. The walls between the ovens have heating flues in which gas, usually part of the gas production of the plant, is burned to provide heat for carbonization. The coal in the ovens is thus heated indirectly through the silica brick walls. Until recently, a typical coke oven chamber was ca 12 m long, 3.7 m high, and 0.5 m wide, and held ca 15 metric tons of coal. Many newer ovens, however, are about 14 m long, 6 m high, and 0.5 m wide, and hold ca 31 t of coal. Both ends of the oven are closed with removable self-sealing doors. Each oven has 3 to 5 charging holes on top for filling the oven with coal. Larry cars, filled with coal from a storage bunker, travel on top of the battery and discharge the coal into the selected oven through charging holes. A long leveller bar is inserted through a small opening in the door at one end of the oven to level the coal charge. In a recent development, some oven batteries are outfitted with pipeline charging systems by which dried or preheated coal is conveyed through a pipe as a dense-phase fluid in steam and charged into the appropriate oven; preheated coal has also been charged by mechanical conveyors, as well as by the so-called hot larry car.

At the top of one or both ends of the oven, refractory-lined standpipes conduct the volatile carbonization products to a horizontal collecting main which conveys these products to the chemical recovery plant.

The oven walls are constructed of special tongue-and-groove silica brick shapes so as to make the wall structure strong and gas-tight. Silica brick is dimensionally stable and abrasion resistant at the temperatures employed. Once heated, an oven battery is usually maintained at or near operating temperature for the life of the ovens. Although the life of an oven battery is considered to be approximately 25 years, over one-half of the ovens in use in the United States in 1976 were more than 25 years old and about 15% were 45 years or older.

About 40% of coke-oven gas production is generally used for underfiring the oven walls. However blast-furnace gas can be used when the more valuable coke-oven gas is needed elsewhere for other applications, eg, soaking pits and reheating furnaces. Depending on the oven length and other design factors, an oven wall may contain 25 to 33 heating flues. These flues are fired with fuel from the gas main that runs the full length of the battery. A reversing machine controls the gas flow to the heating flues by turning on the gas in half the flue system of an oven, and shutting off the gas to the other half. After burning in one bank of flues for ca 30 min, the gas flow is reversed and the gas burns in the other bank of flues. The spent combustion gases pass through the off-flues and enter the regenerators under the ovens where most of the residual heat is given up. The flue gases then pass into the stack flue, and to the chimney at the end of the battery. Air for combustion of the fuel gas is preheated by passing through the regenerators. Some regenerators are designed so that lean gases, such as blast-furnace gas or producer gas, can be preheated also. The fuel gas and the hot combustion air enter the bottom of the vertical flues through separate ducts. After burning upwardly in the on-flues and imparting heat to the walls, the combustion gases are conveyed to the downflow set of off-flues.

It would be impractical to describe each of several coke-oven designs in detail beyond the preceding general description. More details of such designs are given in reference 11. As an example, Figures 3 and 4 present longitudinal and transverse sections of the heating system of a 6-m high Koppers-Becker combination underjet oven, which represents a large percentage of the coke ovens in the United States. It is characterized by several horizontal or bus flues at the top of the vertical flues connected to crossover flues which transport the combustion gases from one wall, over the top, to the flue system in the other wall of the oven. Gas burns in all the flues in one wall then passes into the horizontal flues, through the crossover flues, and down the off-flue in the other wall. Another feature, as shown in Figure 4, below the regenerators is a duct between adjacent flues by which the heating gas aspirates a controlled amount of spent combustion gas from the off-flue. The mixture of rich heating gas (coke-oven gas) and flue gas produces a long flame in the on-flue that extends to the top of the flue and thus provides uniform heating from bottom to top. This waste-gas recirculation also minimizes the deposition of carbon in the flues and eliminates the need for decarbonizing by air.

The flue temperature and the resultant coking time are different for blast-furnace coke and for foundry coke manufacture. Generally, the coking time for blast-furnace coke is 16–22 h, and for foundry coke the coking time is 24–36 h. The flue temperature, usually measured optically at the base of the flue, may be as high as 1400°C, and the wall temperature in the oven chamber, 1200°C for blast-furnace coke. For foundry coke, these temperatures are correspondingly less. When preheated coal is charged, the coking time may be reduced as much as 20–35%.

Energy Considerations

With the recent awareness of an energy crisis and increased cost of all forms of energy, more attention is being given to the amount of energy required for coking. In the past, coke-oven design was based on a requirement of ca 2.4–2.7 MJ/kg (1050–1150

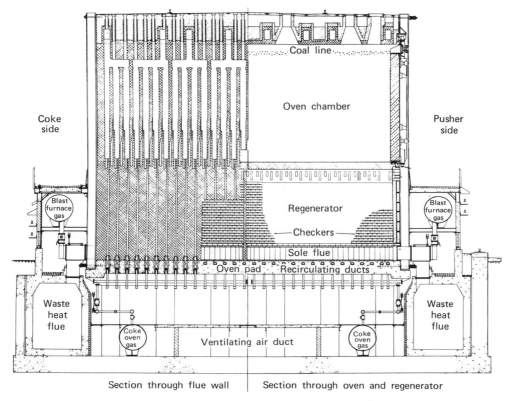

Figure 3. The heating system of a 6-m high Koppers-Becker combination underjet oven.

Btu/lb) of wet coal when underfiring with coke-oven gas and carbonizing at a rate of ca 25 mm/h. The change to higher capacity ovens, faster coking rates, and higher temperature coke to minimize emissions as required by air pollution regulations, has resulted in higher underfiring requirements. An increase in coking rate to 28.6 mm/h and in final coke temperature from 1010 to 1038°C increases underfiring ca 114 kJ/kg (49 Btu/lb) of wet coal (12). A further increase in coke temperature to 1093°C results in a further increase of 128 kJ/kg (55 Btu/lb). A change in coal moisture from 6 to 9% increases the underfiring by almost the same amount. A compensating effect for the heat required for coking is the net exothermic heat of reaction released during car-bonization, which can amount to as much as 465 kJ/kg (200 Btu/lb) for good coking coals. Lower reaction exotherms may occur with the use of poorer coking coals in many plants. Table 6 shows a typical heat balance in a coke oven per kilogram of coal car-bonized, and the distribution of the sensible heat in the products as actually measured in a coke-oven battery.

Oven Operation

The coal blend is charged into hot ovens according to a planned schedule. The covers on the standpipes are closed and the valves in the collecting mains are opened so that the volatile products from the coal are discharged directly into the collecting

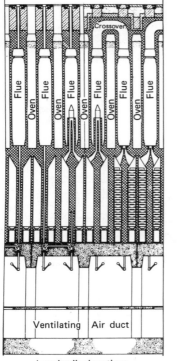

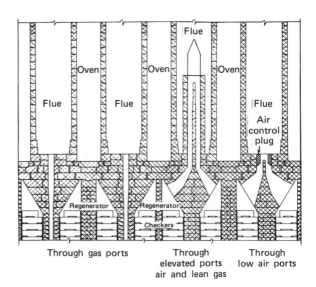

Enlarged sections at base of flue

Longitudinal sections

Figure 4. Longitudinal cross section of a battery of 6-m high Koppers-Becker combination underjet ovens.

mains. Steam jet aspirating nozzles are generally used during charging to maintain a slight suction in the oven, pull the gas evolved into the collecting mains, and prevent emission of smoke or dust. Many older batteries have only one collecting main, but current design practices usually provide one on each side of the battery to minimize the chance of charging emissions escaping into the atmosphere. Upon completion of charging, the coal is leveled by passing the leveling bar through a leveler door near the top of the pusher-side door, and moving the bar back and forth across the top of the coal charge several times. Any excess coal is withdrawn by the bar, and falls down a chute into a small bin on the pusher machine. The lids are placed on the charging holes, either by hand or by a lid-lifter mechanism, and are sealed.

The coal is carbonized for a length of time proportional to the flue temperature, oven design, and characteristics of the coal. Heating is adjusted so that every oven charge is coked out in approximately the same amount of time. Coal is charged into each oven and coke is pushed out according to a planned schedule. A pusher ram on rails along the pusher side of the battery removes the coke from the oven. The ovens are not pushed in order from one end of the battery to the other but are pushed according to one of several series. One of the most popular, for example, is the minus 10 series in which the ovens are numbered consecutively except for 10, 20, 30, etc, and the ovens are pushed in the order: 1s (1, 11, 21, 31, etc), 3s, 5s, 7s, 9s, 2s, 4s, 6s, 8s, etc. One of the main considerations for a pushing schedule is the requirement that ovens

Table 6.   Typical Heat Balance in a Coke Oven [a,b]

	kJ/kg [c]	%
Heat output		
coke, 1063°C	1127	39.3
gas	369	12.8
water vapor, 660°C	300	10.4
tar and light oil, 660°C	95	3.3
stack gas, 269°C	528	18.4
radiation and convection	455	15.8
Total	*2874*	*100.0*
Heat input		
coal	2.5	0.1
air	0.0	0.0
fuel gas	2.5	0.1
combustion of fuel gas (gross)	2415	99.8
Total	*2420*	*100.0*

[a] Ref. 11.
[b] Exothermic heat of carbonization 454 kJ/kg (195 Btu/lb).
[c] To convert kJ/kg to Btu/lb, divide by 2.324.

adjacent to that being pushed are nearly one-half complete in carbonization. When an oven is pushed in the above series, carbonization in the oven on one side is ca 55% complete and in the oven on the other side, ca 45% complete.

The coke is pushed into a quenching car for transport to a quenching station where the coke is sprayed with thousands of liters of water to cool it below its ignition temperature. After draining, the quenched coke is deposited on an inclined brick wharf where it is further cooled by ambient air.

Coke Preparation and Properties

The quenched or cooled coke is conveyed from the wharf to the screening station for separation into desired sizes. Generally, the desired blast-furnace size is 75 × 20 mm. Coke larger than 75 mm may be crushed to the desired size. Coke smaller than ca 20 mm, called breeze, is used for iron ore sintering, boiler firing, and as an "inert" in foundry coke manufacture, electric smelting, chemicals manufacture, and other purposes. Foundry coke is separated into various double-screened sizes larger than 75 mm. Smaller coke is used for other applications.

The typical chemical and physical properties of blast furnace and foundry cokes are given in Table 7. Foundry coke generally has lower ash and sulfur values to meet the more stringent requirements of the foundryman. The mean size of furnace coke is ca 50 mm; foundry coke is considerably larger. Most foundry coke producers market several coke sizes with narrow top-to-bottom sieve ranges. The smallest size is usually ca 75–100 mm and is used in small cupolas. Large cupolas may use coke as large as 200 × 300 mm. For furnace coke, the average tumbler test stability factor (ASTM D-3402) is about 55, but may range from 45 to 63. Whereas the tumbler test evaluates furnace coke, the shatter test (ASTM D-3038) assesses the strength of foundry coke. Typical foundry coke has shatter values of ca 90 and 98 on 75 mm and 50 mm sieves, respectively.

The chemical and physical properties (moisture, ash, and sulfur contents, size

Table 7.   Typical Properties of Cokes, Dry Basis

	Blast-furnace coke	Foundry coke
moisture, %	10.0	2.0
volatile matter, %	1.0	0.7
fixed carbon, %	91.0	92.3
ash, %	8.0	7.0
sulfur, %	0.8	0.6
mean size, mm[a]	50	140
tumbler stability factor[b]	55	
tumbler hardness factor	68	
shatter on 75 mm[c]		90
shatter on 50 mm[c]		98
porosity, %	50	45

[a] Sizes vary, especially for foundry coke.
[b] ASTM D-3402.
[c] Full length pieces tested (ASTM D-3038).

distribution, and strength) of coke used in a blast furnace should be as uniform as possible. The noncarbon impurities should be as low as economically feasible and tumbler-test strength as high as possible. The huge, modern furnaces that are now being constructed are very sensitive to changes in coke quality and require extra strength in the coke. Weaker coke degrades into smaller coke and fines during handling on the way to the furnace, as well as in its descent through the furnace; this leads to poor gas permeability, and hence poor furnace operation. There has been a gradual trend over the past thirty years to improve coke quality. The average ash content for the industry has decreased from ca 11.5% in the mid-1940s to ca 8% at present, and the sulfur content from ca 0.85 to 0.73%; the tumbler test stability factor has increased from ca 40 to 55; and the hardness factor has increased from ca 65 to 68. Improved coke quality along with the adoption of new techniques of preparing burdens, higher top pressures, higher blast temperatures, and oxygen enrichment, have resulted in a significant increase in blast-furnace productivity. However, any further improvements in coke properties will be difficult because of the depletion of the better quality coals and will require significant technological advances in the beneficiation of coals, as well as in the manufacture of coke.

Recovery of Chemical Products.   The gases and vapors that evolve during the carbonization process are treated in the chemical recovery plant; a typical flow sheet is shown in Figure 5. Volatile products leaving the coke oven pass through the standpipes and goosenecks into the collecting main. The collecting main contains a water-sealed valve at each oven for sealing it from the main when coking is completed and the oven is ready for the push. In the gooseneck, flushing liquor containing dissolved salts of ammonia from the aqueous condensates in the system is sprayed into the gas stream to cool the gas from ca 700 to <100°C. In some cases, two collecting mains, one on each side, are provided to draw off the gases and to minimize the atmospheric pollution while coal is charged into the oven. The initial cooling of the hot gases by the flushing liquor spray condenses much of the tar from the gas. The mixture of tar and liquor flows with the gas through the collecting mains into the suction main and then into a downcomer where the tar and liquor are separated from the gas. A butterfly

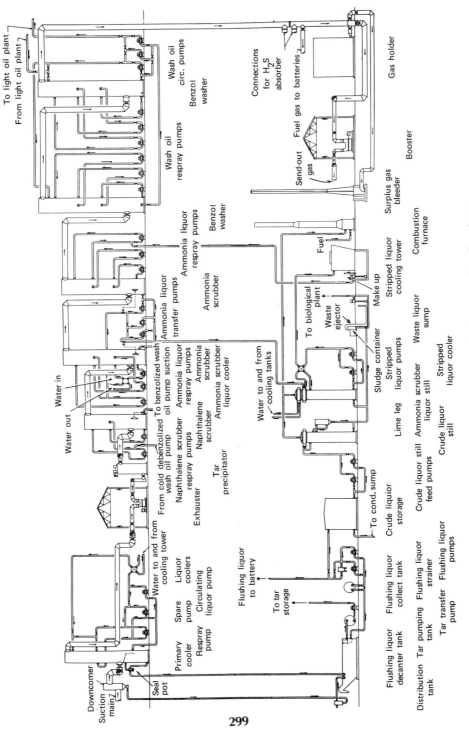

Figure 5. A coal-chemical recovery plant flow sheet.

To light oil plant

From light oil plant

Wash oil circ. pumps

Benzol washer

Connections for H_2S absorber

Send-out Fuel gas to batteries gas

Gas holder

Wash oil respray pumps

Surplus gas bleeder

Booster

Ammonia liquor respray pumps

Benzol washer

Fuel

Stripped liquor cooling tower

Combustion furnace

Ammonia scrubber

Ammonia liquor transfer pumps

To biological plant

Make up

Water in

Ammonia liquor respray pumps

Ammonia scrubber

Waste ejector

Sludge container

Waste liquor sump

From cold debenzolized wash oil pump suction

To benzolized wash oil pump

Water out

Naphthalene scrubber respray pumps

Ammonia scrubber liquor cooler

Water to and from cooling tanks

Lime leg

Stripped liquor pumps

Ammonia scrubber liquor still

Stripped liquor cooler

Exhauster

Naphthalene scrubber

Crude liquor still

Tar precipitator

Crude liquor storage

Crude liquor still feed pumps

To cond. sump

Water to and from cooling tower

Liquor coolers

Flushing liquor to battery

Flushing liquor collect. tank

Flushing liquor strainer

Spare pump

Circulating liquor pump

To tar storage

Flushing liquor pumps

Primary cooler

Respray pump

Flushing liquor decanter tank

Distribution Tar pumping Flushing liquor tank tank tank

Tar transfer pump

Flushing liquor pumps

Downcomer

Suction main

Seal pot

299

valve between the two mains controls the gas pressure in the collecting mains so that slight positive pressure prevents air leakage into the ovens. The tar and liquor flow to a decanter for separation. Part of the liquor is recycled to the goosenecks and collecting mains and the remainder is sent to the ammonia recovery plant. The tar from the decanters is pumped to tar storage tanks.

The uncondensed gases from the top of the downcomer flow to the primary coolers where they are further cooled to ca 25–35°C. The primary cooler is either an indirect-type tubular cooler, where cooling water flows inside the tubes, or a direct type where the gas is cooled by sprays of recirculated cooled ammonia liquor; additional tar is condensed and collected. The gas passes through exhausters, usually the steam-driven centrifugal type, which push the gas through the rest of the chemical recovery plant. From the exhauster, the gas proceeds to electrostatic precipitators where the residual tar is removed (see Air pollution control methods).

Economic conditions and environmental restrictions have caused changes in the gas treatment facilities of coke plants. Unprofitable chemicals are no longer recovered. In most older plants, gas from the primary coolers and vapors from the ammonia still are passed through a saturator or ammonia absorber where ammonia is absorbed by contact with a saturated solution of ammonium sulfate containing ca 5% of free sulfuric acid. The ammonium sulfate crystals formed are separated from the liquor and dried in a high-speed centrifuge. An acid separator then removes any entrained droplets of solution containing sulfuric acid, and the gas flows through the final cooler for cooling to 20–30°C. The gas is usually cooled by direct contact with water, resulting in condensation of naphthalene from the gas; the naphthalene may be skimmed off the water in a sump. In some cases the naphthalene is dissolved by passing the water through a bath of tar in the base of the final cooler. Alternatively, a petroleum oil fraction may be used instead of water to cool the gas and to absorb the naphthalene. The naphthalene is recovered from the oil by steam stripping.

One notable change in some modern plants in the method of recovery of ammonia is depicted in Figure 5. After removal of the naphthalene, the ammonia is washed from the coke-oven gas with a recirculated aqueous solution. The absorbing solution is pumped to the ammonia scrubber liquor still where it is distilled to produce ammonia vapors which are incinerated in the combustion furnace.

The gas is passed into the light oil scrubbers where petroleum wash oil in packed towers or spray chambers countercurrently removes light oil from the gas. Steam distillation removes the light oil from the enriched wash oil. The crude light oil is a mixture of approximately 70% benzene, 15% toluene, 8% xylene, and 7% higher homologues. The recovered crude light oil is sold to petroleum processors or refined and the fractions sold. In older refining plants, the light oil is first treated with concentrated sulfuric acid to remove thiophene and unsaturated compounds, and is then distilled to separate the benzene, toluene, and xylene fractions. In some larger plants, the crude light oil is refined by hydrogenation and solvent extraction or by hydrogenation and dealkylation, producing a highly refined product equal in quality to that made by the petrochemical industry.

In new plants, as well as in many older plants, the gas must be further treated to remove the hydrogen sulfide, which may amount to 7–14 g/m³ (3–6 g/SCF). Since the gas is burned as fuel, there are increasing environmental regulations to reduce H_2S levels in order to reduce the atmospheric pollution by SO_2. The most common process for H_2S removal is the vacuum carbonate process, in which the H_2S is absorbed in an aqueous solution of sodium carbonate and sodium bicarbonate by passage of the gas up through a packed tower with countercurrent flow of the absorbing solution. By boiling the foul solution under vacuum the H_2S is released and can be converted to elemental sulfur in a Claus kiln, or to sulfuric acid in a contact acid plant. Other processes used for H_2S removal include the Stretford and the Carl Still processes. The former recovers the H_2S by absorption with a proprietary aqueous solution containing anthraquinone-2,7-disulfonic acid, sodium vanadate ($NaVO_3$), citric acid, and sodium carbonate, after HCN removal by a polysulfide wash. In the Still process the ammonia recovered from the coke-oven gas is used to desulfurize the coke-oven gas. As in the case of the vacuum carbonate process, the recovered H_2S can be converted to elemental sulfur or to sulfuric acid (see Sulfur recovery).

The crude ammonia liquor, resulting from water distilled from the coal, contains free ammonia in the form of salts of weak acids and fixed ammonia in the form of chloride and thiocyanate. The free ammonia can be liberated by simple boiling, but lime or caustic soda addition is needed to liberate the fixed ammonia. The ammonia liquor is fed to an ammonia still where it is mixed with lime or caustic, stripped with steam, and concentrated by dephlegmation into a vapor containing up to about 25% ammonia. These vapors are then added to the gas in the saturators and the ammonia is converted to ammonium sulfate or is incinerated along with the ammonia vapor from the ammonia scrubber still.

In the relatively new Phosam process the coke-oven gas passes into an absorber where weak ammonium phosphate solution is contacted with the gas. Rich ammonium phosphate solution leaving a second absorber in series with the first, after detarring, is steam-stripped. The vapors are fractionated to produce anhydrous ammonia.

The waste liquor after ammonia removal generally cannot be released to streams because it contains phenols and other contaminants. The phenols can be removed by extraction with aromatic solvents and recovered as sodium phenolates by contact with sodium hydroxide (see Phenol). In most new plants the waste liquor is treated by bacterial oxidation using the activated sludge process to remove the phenols and produce an effluent of low oxygen demand.

At one plant where pure hydrogen is required for synthesis of anhydrous ammonia, an alternative method of recovering coal chemicals was installed. A cryogenic process cools the raw carbonization gases to very low temperatures stepwise, removing various condensates at each step, until only hydrogen remains as a gas (see Cryogenics; Hydrogen).

Yields and Properties of Products. Typical yields of the products of carbonization for 1976 are given in Table 8. These are average yields of the products of the entire United States for a wide variety of coals and under different conditions of carbonization. Though the average yields have not changed appreciably over many years, the value of the products has changed rather drastically. In 1976, the value of the coke represented ca 82.5% of the total value of the products; in 1961 it represented ca 75%. This demonstrates how the value of the chemical products has decreased in comparison. Also, coke has increased in value almost five times during this inflationary period, about the same as the value of coal, but the value of the chemical products has only tripled in the same period. This relationship could change substantially if coke-oven gas values are related to replacement fuel cost based on oil or natural gas at free market prices. Some changes have already occurred in recent months. With a total value for coal products of $79.23 per metric ton and a coal cost of $48.68/t, the difference has

Table 8.   Products Obtained per Metric Ton of Coal Carbonized; Average Yield and Value, 1976[a]

Product	Yield	Value, $/t 1976[a]	Value, $/t 1961[b]
coke, %	68.36	65.35	13.72
breeze, %	5.73	1.79	0.37
gas, m³	340.00	5.29[c]	1.73[c]
tar, L	31.79	5.02	1.42
light oil, L	11.01	1.40	0.73
ammonium sulfate, kg	8.33	0.38	0.35
Total		79.23	18.32
Average value of coal per metric ton		48.68	10.79

[a] *U.S. Bureau of Mines Minerals Yearbook,* 1976.
[b] *U.S. Bureau of Mines Minerals Yearbook,* 1961.
[c] Surplus gas.

to cover labor, maintenance supplies, interest on investment, amortization, and profit.

Coke plants produce crude tar in quantities of 25–38 L/t (6–9 gal/net ton) of coal depending on the type of coal and the carbonizing conditions. Some coke plants "top" the crude tar, producing a distillate containing low-boiling tar acids and bases together with naphthalene and a residue called soft pitch. Some steel plants sell the distillate and use the pitch as fuel. Some larger plants and tar processors process their own, or purchased tar into tar-based products, including creosote oil, cresylic acid, cresols, naphthalene, phenol, xylols, pyridine, quinoline, and medium and hard pitch. The pitches are processed into electrode binder pitch, roofing pitches, and road tars (see Carbon; Tar and pitch).

Crude light oil is recovered in coke plants in quantities of 10–14 L/t of coal, also depending on the type of coal and the carbonizing conditions. Some coke plants refine light oil into benzene, toluene, and xylene, but most sell the light oil to processors, usually petroleum refiners, who refine it along with their own stock (see Petroleum). Benzene and toluene fractions must be pure and free of thiophene. Benzene (qv) is refined now to a 5.4°C or higher freezing point.

Coke-oven gas is a mixture of several gases, but methane and hydrogen are ca 85% of the total volume. A typical analysis is given in Table 9. The gross heating value of the gas is 18.6–21.6 MJ/m³ (500–580 Btu/ft³), about one-half that of natural gas (see Gas, natural). As mentioned before, this gas is used for heating coke ovens, various furnaces in the steel mill, and under boilers for steam raising. A small amount of coke-oven gas is used in city gas mains and other industrial applications. The yield of gas also varies with the type of coal carbonized and the carbonizing conditions, and can range from 300 to 375 m³/t of coal carbonized (see Fuels, synthetic).

Pollution Aspects

In recent years control of atmospheric pollution in the vicinity of the coke plant has been considered essential and efforts have been made to find practical solutions to that very difficult problem. Consequently, several systems for pollution control have been under development. These attempt to control continuous emissions from doors and lids, and intermittent emissions that result from charging the coal into the ovens,

Table 9. Typical Analysis of Coke-Oven Gas, Vol %

hydrogen sulfide	0.6
carbon dioxide	1.6
illuminants[a]	2.9
oxygen	0.2
hydrogen	56.3
carbon monoxide	5.8
methane	29.1
ethane	1.4
propane	0.1
nitrogen	2.0
sp gr	0.36
calorific value, MJ/m^3 [b]	20.9

[a] Ethylene, propylene, butylene, acetylene.

[b] To convert MJ/m^3 to Btu/ft^3, multiply by 26.87.

and pushing and quenching the coke. The charging emissions are considered the most severe, amounting to ca 60% of the total emissions. Pushing the coke provides ca 30%, and quenching ca 10% of emissions.

Conventional charging practice has been improved in some plants by the adoption of a sequential charging practice: coal from the larry-hoppers is dropped into the oven in a predetermined sequence, each lid is replaced when a hopper is emptied; as the last hopper is emptied the levelling bar is put into operation. Other plants have adopted modified systems using either double collecting mains or single mains and jumper pipes.

Under the sponsorship of the American Iron and Steel Institute a larry car system was developed by which a single operator in an environmentally-controlled cab is able to operate all the necessary battery top procedures by remote and automatic means. The openings through which emissions can occur are minimized and suction is maintained at all times in the free space along the top of the oven. Each hopper has drop sleeves that form seals with the charging hole rings and oscillating butterfly feeders that control the coal flow and shut off the coal flow at the appropriate time, leaving a residue of coal in the hopper to serve as a seal.

Scrubber cars have also been developed. Coal is discharged into the oven through a telescopic spout and emissions are collected by suction in an annular duct around the spout. These gases are ignited, cleaned, and then discharged to the atmosphere.

Coal preheating technology has been developed in recent years and three methods of charging the hot coal have been adopted. In one method, the preheated coal is conveyed into the oven through a sealed system as a dense-phase fluid with steam as the carrier. In another system, the coal is conveyed by a Redler-type conveyor; the third system is a modified hot larry car.

At least three types of emission control systems for pushing emissions are under development. In one system the coke guide is enclosed, and a hood equipped with scrubbers and fans covers the hot car; this equipment traps and collects the emissions. In another system, the hood and guide enclosure is connected to a duct by which the emissions are scrubbed in a static washer. In a third system the coke is discharged through an enclosed guide into a closed hot car and the emissions are scrubbed out by a washer mounted on the hot car.

Attempts have also been made to control coke quenching emissions by equipping the conventional quenching station with mist suppressors, and by the development of mobile quenchers, continuous quenching systems, and dry coke cooling plants.

Recent Technology

Since the mid-1800s coke ovens have progressed from beehives, with low-unit capacity, poor quality control, no chemicals recovery, and serious air pollution disadvantages, to the by-product ovens of 3-m height, and 12-t capacity, operating at about 25 mm/h coking rate, and then to the present ovens of 6-m height and 34-t capacity, operating at 30 mm/h coking rate. The modern coke oven plant with its ancillary equipment is a sophisticated facility improved in recent years to increase efficiency and productivity, maintain unit cost of product at low levels, and reduce coke-oven emissions to air and streams. Many innovations have been incorporated within the coke plant to attain these objectives. Some of these developments are as follows: (1) use of denser silica refractories in the walls so that reduced flue temperature is required for fast coking rates, or faster coking rates can be obtained at the same flue temperatures (13); (2) higher and longer oven chambers so that larger coal charges can be made (14); (3) thinner refractory walls for more rapid heat transfer, especially toward the coke side where additional heat is required because of increased oven chamber width (13); (4) improved heating systems to heat the enlarged flue area more uniformly (14–15); (5) preheating of coal charges to attain more rapid coking rates and improve the coking properties of poorly-coking, marginal coals (16); (6) charging of preheated coal via dense-phase fluid transport in a pipeline, conveyor charging, or by hot larry car; these are intended to reduce emissions during charging (17); (7) charging wet coal by conventional larry car, practicing sequential charging or stage charging to minimize emissions to the atmosphere, or using cars equipped with washers which trap emissions during charging (18); (8) blending coals by weight rather than by volume so as to reduce the effect of moisture and size variations on the proportioning of the blend (9); (9) control of bulk density of the coal blend in the oven by moisture and/or oil addition and the aid of a nuclear gage (9); (10) removal of deposits from door jambs, door sealing edges, and goosenecks by mechanical cleaning apparatus (18); (11) interlocking devices on the pusher machine and quenching car so that they are operationally coordinated; (12) use of air-conditioned cabs on oven machinery to reduce exposure of workers to emissions (18); (13) use of coke-side hoods, travelling quench car hoods, and hooded coke guides to collect and reduce emissions to the atmosphere (18); (14) mobile coke quenching machines as well as continuous coke quenchers that contain and remove the emissions from the coke quenching operation (18) (dry coke quenching systems are used in the USSR and Japan); (15) advanced understanding of the coking process and the requirements for making good coke (19); (16) partial briquetting of coal charges to improve coke strength and permit use of marginal coking coals in the blend to the ovens (20); (17) recycling of pulverized coke breeze into the oven via the coal blend (21); and (18) advancement of three or more formcoke processes to the demonstration plant stage (22).

Some recent advances listed above are in initial stages of development. The most promising technology will continue to be improved during the next few years. Oven design considerations to promote more uniform heating for ovens of higher capacity will continue. Preheating coal charges and charging them into ovens presents a number of problems, particularly that of coal carry-over. A solution to this problem is imperative for more universal adoption of the technology. In addition, partial briquetting of coal charges will probably be developed further (see Pelleting and briquetting). Both preheating and partial briquetting have the potential of using poorly coking coals which may become necessary when premium coals are exhausted. The advancement of formcoke technology should continue and the adoption of one or more processes should be made, particularly in regions where suitable coking coals are scarce.

BIBLIOGRAPHY

"Carbonization" in *ECT* 1st ed., Vol. 3, pp. 156–178, by W. O. Keeling, Koppers Co., Inc., and F. W. Jung, Research Consultant; "Carbonization" in *ECT* 2nd ed., Vol. 4, pp. 400–423, by C. C. Russell, Koppers Co., Inc.

1. C. S. Finney and J. Mitchell, *Trans. AIME* **13,** 285, 373, 425, 501, 559 (1961).
2. *Coke and Coal Chemicals*, Bureau of Mines Minerals Yearbooks, U.S. Dept. of Interior, Superintendent of Documents, Washington, D.C. 20402.
3. "Total Blast Furnace Production of Pig Iron and Ferroalloys" and "Coke Consumption by Use," American Iron and Steel Institute, Annual Statistical Reports; *Disposal of Coke,* U.S. Bureau of Mines.
4. N. Schapiro and co-workers, *Proc. Blast Furnace, Coke Oven, and Raw Materials Committee, AIME* **20,** 89 (1961); N. Schapiro and R. J. Gray, *J. Inst. Fuel* **37,** 234 (1964); J. A. Harrison, *Proc. Ill. Min. Inst.* 17 (1961).
5. C. C. Russell, M. Perch, and J. F. Farnsworth, *Ironmaking Proc. Metall. Soc. AIME* **8,** 32 (1949).
6. C. C. Russell, M. Perch, and H. B. Smith, *Proc. Blast Furnace, Coke Oven, Raw Materials Committee, AIME* **12,** 197 (1953).
7. H. E. Harris, W. L. Glowacki, and J. Mitchell, *Proc. Blast Furnace, Coke Oven, Raw Materials Committee Proceedings AIME* **19,** 336 (1960).
8. J. G. Walters, W. H. Ode, and L. Spinetti, *U.S. Bur. Mines Bull.,* 610 (1963).
9. J. D. Doherty and J. A. DeCarlo, *Blast Furn. Steel Plant* **55,** 141 (1967).
10. W. P. Ryan, *Am. Gas Assoc. Proc.* **7,** 861 (1925).
11. F. Denig in H. H. Lowry, ed., *Chemistry of Coal Utilization*, John Wiley & Sons, Inc., New York, 1945, Chapt. 21.
12. H. A. Grosick, E. J. Helm, and J. M. Airgood, *Iron Steel Eng.* **53,** 27 (1976).
13. E. J. Helm, *Ironmaking Proc. Metall. Soc. AIME* **25,** 53 (1966).
14. D. Wagener, *The Coke Oven Managers' Year-Book*, 243 (1974).
15. H. Schürhoff, *Ironmaking Proc. Metall. Soc. AIME* **30,** 102 (1971).
16. M. Perch and C. C. Russell, *Blast Furn. Steel Plant* **47,** 591 (1959); H. W. Jackman and R. J. Helfinstine, *Ill. State Geol. Surv. Circ.,* 423 (1968), 434 (1968), 449 (1970), and 453 (1970).
17. D. G. Marting and R. F. Davis, Jr., *Ironmaking Proc. Metall. Soc. AIME* **31,** 174 (1972); C. E. McMorris, *Ironmaking Proc. Metall. Soc. AIME* **34,** 330 (1975); D. A. Cooper and co-workers, *Ironmaking Proc. Metall. Soc. AIME* **35,** 450 (1976); C. Flockenhaus and co-workers, *Ironmaking Proc. Metall. Soc. AIME* **33,** 382 (1974).
18. W. D. Edgar, *Iron Steel Eng.* **49,** 86 (1972); F. C. Voelker, Jr., *Iron Steel Eng.* **52,** 57 (1975).
19. J. D. Clendenin in J. H. Strassburger, ed., *Blast Furnace—Theory and Practice,* Gordon and Breach Science Publishers, New York, 1969, pp. 325–436; Coke Oven Managers Association, Third Carbonization Science Lecture, London, Nov. 3, 1971.
20. T. Matsuo, "Recent Developments in Coking Processes Utilizing Low Grade Coals" *Southeast Asia Iron and Steel Institute Manila, Philippines, Seminar* (1976); T. Ikeshima and co-workers, "Developments in Metallurgical Coke Production and Application of Those Developments to Latin American Coals," *Latin American Iron and Steel Congress, Lima, Peru* Sept., 1975.
21. T. M. Larimer and co-workers, *Ironmaking Proc. Metall. Soc. AIME* **34,** 160 (1975).
22. E. Ahland and co-workers, *Ironmaking Proc. Metall. Soc. AIME* **31,** 285 (1972); E. Ahland and co-workers, *International Iron and Steel Congress* Paper 1.2.2.7, Dusseldorf, 1974.

General References

H. H. Lowry, ed., *Chemistry of Coal Utilization,* Suppl. Vol., Chapt. 10, 11, John Wiley & Sons, Inc., New York, 1963.

H. H. Lowry, ed., *The Chemistry of Coal Utilization*, John Wiley & Sons, Inc., New York, 1945, Chapt. 25 and 31.

P. J. Wilson and J. H. Wells, *Coal, Coke, and Coal Chemicals,* McGraw-Hill Book Co., Inc., New York, 1950.

F. M. Gentry, *The Technology of Low-Temperature Carbonization,* The Williams & Wilkins Co., Baltimore, Md., 1938.

C. C. Russell, *Am. Gas Assoc. Annu. Rep.* **29,** 733 (1947).

J. H. Strassburger, ed., *Blast Furnace—Theory and Practice,* Gordon and Breach Science Publishers, New York, 1969.

M. Perch and R. E. Muder in J. A. Kent, ed., *Riegel's Handbook of Industrial Chemistry,* Van Nostrand Reinhold Co., New York, 1974, Chapt. 8.

MICHAEL PERCH
Koppers Co., Inc.

DESULFURIZATION

Coal is a major source of energy for the United States and will continue to be so for many years. However coal is "dirty," ie, it contains substantial amounts of sulfur, nitrogen, and mineral matter, including significant quantities of toxic impurities such as mercury, beryllium, and arsenic. During combustion these materials enter the environment as sulfur oxides, nitrogen oxides, and compounds of toxic metals, and thus constitute a health hazard through atmospheric and food-chain consumption. Therefore, coal must be cleaned either before, during, or after combustion to prevent deterioration of the environment as the U.S. economy shifts its major energy source from gas and oil to coal. This article presents brief descriptions of various schemes concerned with desulfurization of coal prior to combustion (for desulfurization after combustion, see Sulfur recovery; Air pollution control methods).

Physical and chemical coal cleaning are being studied extensively in an effort to provide a clean source of solid energy from coal and at the same time reduce pollution of the environment by sulfur oxides, nitrogen oxides, and toxic metals. There is considerable literature on physical coal cleaning since it is an older proven technology for cleaning coal.

U.S. coals may contain any or all of four petrographic subclasses or lithotypes. *Vitrain* is a bright coal characterized by a relatively low specific gravity and carbon content. It is moderately friable. *Clarain* is similar to vitrain in most respects but has differences in friability and breaking characteristics. *Durain* is a dull coal with a higher specific gravity and carbon content, and lower friability than vitrain or clarain. *Fusain* is a black coal with a high specific gravity and carbon content. It is highly friable and difficult to wet. See the preceding article Coal for complete discussion of coal structure and composition.

Impurities in Coal

The three categories of potential air pollutants found in coal are sulfur, ash, and nitrogen (as a source of NO). Sulfur and ash are associated with the mineral and organic portions, and nitrogen most likely with the organic matter.

Sulfur is an undesirable constituent of all coals. It is present in amounts ranging from traces to >10 wt%. Most commercial coals of the eastern United States contain 0.5–4.0 wt % S (1) whereas the western coals contain <1 wt % S.

Sulfur is present in coal as sulfate, pyritic, and organic sulfur. Sulfate sulfur is of minor concern as its concentration in coal is only ca 1%. It is present as indiscrete particles of gypsum and copperas ($FeSO_4.7H_2O$) and in the mineral jarosite [(Na, K), $Fe_3(SO_4)_2(OH)_6$]. The iron sulfates are important only in weathered coals and, like all sulfate sulfur, are a minor problem with respect to atmospheric pollution.

Inorganic sulfur is found primarily as iron disulfide (FeS_2) in pyrite and/or marcasite. The two minerals have the same chemical composition but different crystalline forms. Pyrite, the most commonly reported form of the two, is ubiquitous in coal.

Pyritic sulfur occurs in coal as veins, lenses, fiber-bundles, nodules, or balls, and pyritized plant tissue. The pyritic particles may be macroscopic or microscopic. Because much pyritic sulfur is present as microscopic particles, physical desulfurization is difficult.

Concentrations of pyritic sulfur vary widely within the same deposit and along the same seam, sometimes even within the same mine. Concentrations normally range from 0.2 to >3% (sulfur basis), depending on the location. For example, coal taken from Seam 4A in Ohio, but from different mines, Meigs 1 and Meigs 2 in Declarion County, contained 4.90 and 1.77% pyritic sulfur, respectively.

Organic sulfur occurs chemically bound to the organic part of coal structure (2). Concentrations range from ca 0.3–2.4 wt %, constituting ca 20–85% of the total sulfur content of coal. Groups present include: (1) mercaptan or thiol, RSH′; (2) sulfide or thio ether, RSR′; (3) disulfide, RSSR′; and (4) aromatic systems containing the thiophene ring (3).

Concentrations of organic sulfur also vary widely within the same seam, deposit, or mine, eg, the organic sulfur content of three coal samples taken from Ohio Seam 6 in Coshocton and Muskingum Counties was 1.55, 1.04, and 1.94% each, respectively. This variation is not only characteristic of Pennsylvania and Ohio coals, but is observed in coals throughout most of the United States.

Theoretically, organic sulfur cannot be removed from coal unless the chemical bonds holding the sulfur are broken or the organic sulfur compound is extracted. Thus, the amount of organic sulfur present defines the lowest limit to which a coal can be cleaned by physical methods (see Sulfur compounds).

No definite relationship between the organic and pyritic sulfur contents of coal has been established. In the United States, the organic sulfur content of coal may be 20–80% of the total sulfur. The variation of organic sulfur content of a bed from top to bottom is usually small. Pyritic sulfur may vary greatly.

Nitrogen, like sulfur, is most likely linked to the coal structure and is therefore not removable by physical cleaning. Eastern coals average ca 1.4% N, with a range of 0.7–2.5 wt %. The nitrogen content of several Ohio coals, taken from locations throughout the state, varies from 0.79–1.4 wt %.

Coal ash is derived from the mineral content of coal upon combustion or utilization. The minerals are present in coal as discrete particles, cavity fillings, and aggregates as the sulfides, sulfates, chlorides, carbonates, hydrates and oxides or both. Overall concentration may range from a few to more than 40%.

The degree of de-ashing possible by physical cleaning depends on the distribution of the mineral matter in the coal. In some cases, a considerable amount of mineral matter can be removed; in other cases, where the mineral matter is distributed throughout the coal as microscopic particles, de-ashing by physical cleaning is not practical.

Physical cleaning and chemical cleaning are employed in the desulfurization of coal. Physical cleaning, a proven technology for removal of the mineral matter, including some sulfur, is currently practiced throughout the coal industry. Chemical cleaning, which removes a major portion of the sulfur and a significant portion of the mineral matter, is in its early stages of development and is not yet practiced commercially. If successful, however, chemical cleaning alone or in combination with physical cleaning could provide a clean source of energy from coal.

Physical Coal Cleaning

Physical coal cleaning is a proven technology for upgrading raw coal by physical removal of associated impurities. Coal has traditionally been cleaned to meet certain market requirements as to size and ash and moisture contents. However, governmental regulations of atmospheric emission of sulfur oxides from coal combustion have been focused on sulfur reduction. Physical coal cleaning processes grind coal to release impurities; the fineness of the coal governs the degree to which the impurities are released.

Current industrial and laboratory physical coal cleaning methods can be divided into four broad categories based on the physical properties that effect the separation: gravity, flotation, magnetic, and electrical methods. Gravity separation depends upon the specific gravity difference between coal and its impurities. The processes are relatively simple and the category includes most conventional coal cleaning methods such as jigging, tabling, dense-medium processing, hydrocloning, and air classifications.

Flotation (qv) effects separation by the difference in surface characteristics between coal and its impurities. Froth flotation depends upon the selective adhesion of air bubbles. Oil agglomeration depends on the wetability differences.

Electrical methods employ electrical charges and magnetic forces to effect separation. Electrophoretic separation depends on the difference in electrical charge of the particles produced by various mechanisms in an electric field in air.

Magnetic separation relies on the difference in magnetic susceptibility of coal and mineral matter, either in air or aqueous slurries.

Commercial coal cleaning is currently limited to gravity separation with minor application of froth flotation methods. Table 1 summarizes the types of processes and equipment used in coal cleaning since 1942. Jigging represents the largest portion of coal cleaning but dense-medium processes and concentrating tables are becoming more popular and froth flotation is beginning to play an important role.

Table 1. Preparation of Coal by Type of Equipment, % of Clean Coal Produced, Year[a]

Washer type	1942	1952	1962	1972
jigs	47.0	42.8	50.2	43.6
dense-medium processes	8.8	13.8	25.3	31.4
concentrating tables	2.2	1.6	11.7	13.7
flotation			1.5	4.4
pneumatic	14.2	8.2	6.9	4.0
classifiers	7.4	8.5	2.1	1.0
launders	13.1	5.2	2.2	1.9

[a] Ref. 4.

The potential of a given coal for desulfurization by physical cleaning is usually expressed in terms of washability data. Recently, the U.S. Bureau of Mines compiled the washability data of over 455 U.S. coal samples (5). Some important findings from this study are:

(1) Coal from all regions of the United States (455 samples) contained on the average of 1.91% total sulfur, and 29.2 kJ/g (12,574 Btu/lb) heat value. Without preparation, 14% of the raw coal could meet the current Federal sulfur emission standard of 516 g SO_2/GJ (1.2 lb SO_2/million Btu).

(2) If all the coals were upgraded at a specific gravity of 1.60, the clean coal products would contain on the average 7.5% ash, 0.85% pyritic sulfur, 2.0% total sulfur, and 31.4 kJ/g (13,530 Btu/lb) at an average energy recovery of 93.8%. If a 50% energy recovery were acceptable, then 32% of the samples could be upgraded to meet the SO_2 emission standard when crushed to 1.4 mm (14 mesh) top size.

(3) In general, cleaner coal could be separated as the sample is crushed to the smaller top sizes. However, the effect of crushing on sulfur reduction varies widely depending on the coal seams.

(4) The Northern Appalachian Region coal (227 samples) was especially amenable to pyritic sulfur reduction upon crushing to finer size. Only 4% of the raw coal samples would meet the emission standard. For a 50% energy recovery, 31% of the samples would meet the standard when crushed to 1.4 mm (14 mesh) top size. In particular, coals from the Upper and Lower Freeport and Upper Kittanning coal beds averaged an 80% reduction in pyritic sulfur at a 70% Btu recovery when crushed to 1.4 mm (14 mesh) top size, and separated gravimetrically.

(5) Sixty-three percent of the Southern Appalachian Region coal (35 samples) would comply with the federal sulfur emission standard at an energy recovery of 50% when crushed to 1.4 mm (14 mesh) top size.

(6) Pyrite is not readily liberated from Alabama Region coal (10 samples). Although 30% of the raw coal would meet the emission standard, only 40% would meet the standard at a 50% energy recovery when crushed to 1.4 mm (14 mesh) top size.

(7) The Eastern Midwest Region coal (45 samples) contained an average of 1.65% organic sulfur and thus generally could not be upgraded to meet the Federal sulfur emission standard.

(8) The Western Region coal (44 samples) contained 0.68% total sulfur and would thus meet the federal sulfur emission standard as mined. However, these are generally subbituminous coals of high inherent moisture and low energy content.

Conventional Coal Washing. In a modern physical cleaning plant, coal is typically subjected to: (1) size reduction and screening; (2) separation of impurities; and (3) dewatering and drying.

Size reduction is accomplished in rotary or roll crushers. The extent of size reduction depends on the type of coal processed and the desired product characteristics. It is well-known that more impurities are liberated as the size of coal is reduced. However, because the costs of preparation rise exponentially with the amount of fines to be treated, there is an economic optimum in size reduction (qv).

Coal is screened either wet or dry to separate various size fractions resulting from size reduction. The raw coal is commonly divided into coarse, intermediate, and fine sizes: course sizes are those larger than a specified size in the range of 1.3–0.6 cm; intermediate sizes have an upper limit in the range of 1.3–0.6 cm and a lower limit in the range of ca 625–325 μm (28–48 mesh); and fine sizes are those smaller than a specified size of ca 625–325 μm.

A variety of devices for the three individual size groupings are used to separate ash and pyrite from coal. In coarse coal circuits, coarse coal is cleaned with one or more pieces of gravity separation equipment such as jigs, launders, or heavy-medium vessels; intermediate size coals are usually cleaned with concentration tables or heavy-medium cyclones; and fine size coals, with froth flotation or hydrocyclones.

The product coal of any wet separation process must be dewatered or completely dried depending on mode of transportation and end use. Coarse coal can be dewatered by natural drainage with dewatering screens or bucket elevators. Intermediate size coal is dewatered with sieve bends and centrifuges; some thermal drying may also be required. With fine coal, dewatering requires not only more complicated mechanical techniques, such as centrifugation or vacuum filtration, but also thermal drying, to obtain an acceptable moisture content.

There are currently more than 450 physical coal cleaning plants throughout the United States with a total capacity of more than 360 million metric tons of raw coal per year. Table 2 summarizes the status of coal cleaning plants operated in 1975. Some plants use only one unit process, others use a series of cleaning processes. The capacity of individual plants varies widely from <180 t/d to >22,750 t/d.

The conventional coal cleaning process can remove ca 50% of pyritic sulfur and 30% of total sulfur. However, sulfur reduction by physical cleaning varies considerably depending on the washability of coal, unit processes employed, separating density, and energy recovery.

The following classes of coal preparation are defined in order to compare sulfur reduction potential at different levels of preparation. *Level 1:* breaking for top size control only, with limited removal of coarse refuse and trash. *Level 2:* coarse beneficiation through washing to plus 0.95 cm material only after crushing to a top size of 3.8–12.7 cm. *Level 3:* deliberate beneficiation through washing of all >625 μm (+28 mesh) material after crushing to a top size of 3.8–12.7 cm. *Level 4:* full beneficiation through washing of all size fractions including <625 μm (−28 mesh) material in addition to Level 3.

Table 3 shows the beneficiation of the sulfur reductions possible by physical cleaning of coal from various regions of the United States. The values are estimated from the data of float-sink analysis by the U.S. Bureau of Mines. These data indicate hypothetical enhancement of coal quality which can be achieved by beneficiation. Actual values vary with each installation, reflecting coal seam characteristics, mining procedures, and specific beneficiation processes selected.

Table 2. Physical Coal Cleaning Plants Operated by States for 1975[a]

State	Est. total production coal technology, 1000 t	No. of coal cleaning plants	No. of coal cleaning plants for which capacity data reported	Reporting plants Total daily capacity, t	Reporting plants Est. annual capacity[b], 1000 t	Heavy media washers	Size	Flotation units	Air table	Washing tables
Alaska	19,437	22	10	36,832	9,208	8	10	6	1	12
Arkansas	608	1	0			1				1
Colorado	7,410	2	0			2		1		
Illinois	53,752	33	20	124,081	31,021	17	20	4	1	1
Indiana	22,609	7	6	38,102	9,526	2	5	1		1
Kansas	515	2	2	3,447	862		2			
Kentucky	132,267	70	48	222,898	55,724	43	27	16	4	20
Maryland	2,533	1	0					1		
Missouri	4,568	2	1	3,175	794		2			
New Mexico	8,384	1	1	5,443	1,361	1		1		
Ohio	42,214	15	13	93,215	23,306	6	11		1	
Oklahoma	2,513	2	1	499	127	1	1			
Pennsylvania (anthracite)	4,536	24	14	11,794	2,948	21	4	4		3
Pennsylvania (bituminous)	73,345	66	50	258,559	64,642	30	19	16	20	15
Tennessee	8,432	5	4	7,729	1,932	1	1	1	2	
Utah	5,987	6	4	20,956	5,239	2	4	2	2	
Virginia	33,113	42	29	130,228	32,559	26	15	9	8	25
Washington	3,357	2	1	18,144	4,536	1	1			
West Virginia	99,791	152	113	523,791	130,949	104	55	59	12	55
Wyoming	21,405	1	1	544	136				1	
Total	*547,776*	*456*	*318*	*1,499,437*	*374,870*	*266*	*177*	*121*	*52*	*133*

No. of plants using various cleaning methods (spanning Heavy media washers, Size, Flotation units, Air table, Washing tables)

[a] Ref. 6.
[b] The estimated annual capacity values for the reporting plants were calculated from the daily capacity values by assuming an average plant operation of 250 days per year (5 days per week for 50 weeks per year).

311

Table 3.   Sulfur Reduction by Physical Coal Cleaning

Energy	Level 1 beneficiation	Level 2 beneficiation	Level 3 beneficiation	Level 4 beneficiation
recovery, %	99	95	90	85
Pyritic sulfur reduction, %				
Northern Appalachian	10	33	47	54
Southern Appalachian	15	35	44	48
Alabama	10	32	38	39
Eastern Midwest	20	45	54	59
Western Midwest	15	33	41	45
Western	8	30	33	33
U.S. average	*13*	*35*	*46*	*52*
Total sulfur reduction, %				
Northern Appalachian	5	20	28	33
Southern Appalachian	2	6	9	9
Alabama	3	9	12	12
Eastern Midwest	6	22	29	32
Western Midwest	5	21	25	29
Western	2	8	11	11
U.S. average	*6*	*20*	*25*	*29*
Emission, kg SO_2/GJ (lb SO_2/10⁶ Btu)				
Northern Appalachian	1.9 (4.5)	1.5 (3.5)	1.3 (3.1)	1.2 (2.9)
Southern Appalachian	0.6 (1.5)	0.6 (1.4)	0.6 (1.3)	0.6 (1.3)
Alabama	0.9 (2.0)	0.9 (2.0)	0.9 (2.0)	0.9 (2.0)
Eastern Midwest	2.5 (5.8)	2.0 (4.7)	1.8 (4.2)	1.7 (4.0)
Western Midwest	3.4 (8.0)	2.9 (6.8)	2.7 (6.2)	2.5 (5.9)
Western	0.5 (1.1)	0.4 (0.9)	0.4 (0.9)	0.4 (0.9)
U.S. average	*2.0 (4.7)*	*1.6 (3.8)*	*1.2 (2.9)*	*1.2 (2.7)*

Two-Stage Froth Flotation.   Froth flotation has long been used as a beneficiation method for fine coals, usually <625 μm (−28 mesh). The process consists of agitating the finely divided coal and mineral suspension with small amounts of reagents in the presence of water and air. The reagents assist the formation of small air bubbles that collect the hydrophobic coal particles and carry them to the surface. The hydrophilic mineral matter is wetted by water and drawn off as tailings (see Flotation).

Recently, a novel two-stage froth flotation process was developed by the U.S. Bureau of Mines to remove pyrite from fine size coals (7). In the first stage, coal is floated with a minimum amount of frother (methylisobutyl carbinol) while coarse, free pyrite and other refuse are removed as tailings. In the second stage, coal is suppressed with a coal depressing agent (Aero Depressant 633), while fine size pyrite is floated with a pyrite collector (potassium amylxanthate).

The two-stage froth flotation process has been demonstrated in a 0.45 t/d-capacity pilot plant. Negotiations to install a full-scale prototype of 11 t/h-capacity in an existing coal cleaning plant are reportedly underway.

The pilot-plant data showed that up to 75% of pyritic sulfur could be removed from the Lower Freeport coal (<470 μm or −35 mesh) at about 60 wt % recovery.

Oil Agglomeration.   The use of a water-immiscible liquid, usually hydrocarbons, to separate coal from the impurities is an extension of the principles employed in froth flotation. The coal surface is preferentially wetted by the hydrocarbons while the water-wetting minerals remain suspended in water. Hence, separation of two phases produces a clean coal containing some oil and an aqueous suspension of the refuse containing some coal.

The National Research Council of Canada (8) recently developed a spherical oil agglomeration process for cleaning coal fines in two steps: flocculation and balling. In the flocculation stage, a small amount of light oil, <5% is added to a 20–30% coal slurry in a high-speed agitator to form microagglomerates. In the balling stage, a heavy, less expensive oil is added to a rotating pelletize-disc to form strong spherical balls. The spherical oil agglomeration process has been incorporated into the coal fine recovery circuit of a western Canadian preparation plant (8).

Laboratory batch experiments have shown that about 50% of the pyritic sulfur was removed from the Canadian coal ground to less than 50 μm at >90% energy recovery (9). The oil requirements were given as: light fuel oil for flocculation, 3–5 wt % of solid feed; and heavy oil for pelletizing, 5–10 wt % of solid feed.

High-Gradient Magnetic Separation. A high-gradient magnetic separator uses electromagnets to generate a magnetic field and remove mineral components, especially pyrite, from either an aqueous suspension of finely ground coal or dry powder. The separator consists of a column packed with magnetic stainless steel wool or screens inserted in the base of a solenoid magnet.

The General Electric Company (10), in conjunction with the Massachusetts Institute of Technology and Eastern Associated Coal, is attempting to establish the technical feasibility of removing inorganic sulfur from dry coal powders at commercially significant rates. In addition, Indiana University (11) is investigating the use of high-extraction magnetic filters for the beneficiation of coal slurry containing fines <74 μm (−200 mesh). A magnetic filter of 2.13 m can process up to 91 metric tons of raw coal per hour.

Test data from the high-gradient magnetic separation of dry coal powders showed that as much as 57% of total sulfur can be removed from eastern coals <325 μm (−48 mesh) with a magnetic field intensity of 5.09 MA/m (64,000 Oe) and flow velocity of 2.8 cm/s. The laboratory tests of the Indiana University project indicated that up to 93% of the inorganic sulfur can be removed from the coal slurry containing 90% of <510 μm (−325) mesh sizes with a magnetic field intensity of 1.6 MA/m (20,000 Oe) using 3 passes at a 30 s retention.

Heavy Liquid Separation. Heavy liquid separation is a practical extension of the laboratory float-sink test. Crushed raw coal is immersed in a static bath of a heavy liquid with a density between that of clean and reject coal. The float material is recovered as clean coal product and the sink material is rejected as refuse.

The heavy liquid for coal separation does not react with the coal product or reject material. The specific gravity of the liquid can be regulated by certain additives. The used liquid is completely recovered by draining and evaporating from the product coal and reject material.

The heavy liquid method of coal cleaning is not new. In 1936, a 45 t/h pilot plant using chlorinated hydrocarbons was built by the DuPont Company. However, the high costs of heavy liquids and toxic effects of the vapors have prohibited the commercialization of heavy liquid processes.

Otisca Industries recently reported the development of an anhydrous heavy liquid for gravity separation of coal (12–13). The heavy liquid is presumably a fluorocarbon with a boiling point of 24°C, a heat of vaporization of 180 J/g (43.1 cal/g), and a specific gravity of 1.50 at 16°C.

The Otisca process is reportedly capable of a near theoretical separation (which can be obtained in the laboratory in a float-sink test). The data showed that about 44% of total sulfur was removed from 4 mm × 0 size coal at 74 wt % recovery. Misplaced material was 0.5 ± 0.25% under normal operating conditions. Typical process conditions are: 15–20°C at 101 kPa (1 atm minus one inch of water) and a slurry concentration of 15–25 wt % solids.

Chemical Comminution.   The chemical comminution process involves permeation of certain low-molecular-weight compounds throughout existing faults, pores, and other discontinuities in coal to weaken and disrupt the interlayer forces. The chemical selectively affects the coal but not the mineral matter associated with it. The net result is the selective fracture of coal along the bedding planes, mineral constituents, and mineral boundaries. The fragmented coal and unaffected mineral matter can then be separated by some conventional cleaning process. Therefore, chemical comminution is a physical cleaning process.

The chemicals that appear to have the greatest comminution ability are ammonia and methanol. By varying the chemical or reaction conditions, eg, pressure, temperature, exposure time, moisture content, etc, the fragmentation of several ranks of coal can be effectively controlled. Chemical comminution of coal has been studied by the Syracuse University Research Corporation since 1971 (14). A 36 t/d capacity pilot plant using chemical comminution has been designed by Catalytic, Inc., and the application of chemical comminution technique for underground mining has been proposed.

Bench-scale studies indicate that chemical comminution is capable of liberating a comparable amount of ash and more pyrite than mechanical crushing to the same size. The float-sink test for the Upper Freeport sample showed that 96.4 wt % pyrite was removed from the chemically comminuted coal at 1.3 specific gravity, and 90 wt % pyrite was removed from the coal mechanically crushed to <1.4 mm (−14 mesh) at the same specific gravity.

Fine Grinding and Classification.   Raw coal is ground to extremely fine sizes and an air classifier separates the heavy impurity particles from the light coal particles.

Fine grinding of coal followed by an air classification has long been studied. The U.S. Bureau of Mines has investigated the use of centrifugal separators for removing pyrite from coal pulverized to power-plant fineness [70% <74 μm (−200 mesh)] on a laboratory scale (15). Although the study showed moderate success in sulfur reduction, it indicated that the sulfur reduction obtained did not justify the requisite large and expensive equipment (see Size classification; Size reduction).

Chemical Coal Cleaning

Chemical coal cleaning is not practiced commercially. Theoretically, however, chemical cleaning of coal to produce an environmentally acceptable solid fuel as a clean source of energy is technically feasible. For example, pyritic sulfur may be removed from coal as water-soluble sulfates, sulfides, and/or elemental sulfur. Since pyritic sulfur is present in coal as a distinct phase, its extraction may be achieved without disturbing the coal structure.

Organic sulfur, reportedly a part of the coal molecule, is not readily susceptible to attack by chemical reagents. Certain organic sulfur compounds found in coal, eg, thiophenes, are very stable and resistant to attack; others, eg, mercaptans, react readily. Thus, removal of organic sulfur from coal could involve extraction of the portion of the coal combined with the sulfur, or rupture of the carbon–sulfur bond to convert the organic sulfur to a more readily extractable form.

In chemical cleaning, raw coal is treated with a reagent or reagents that react with and liberate pollutant-forming constituents. Chemical cleaning can be used alone or in combination with physical coal cleaning and has the potential to significantly increase the amount of coal reserves that could be used directly as a solid fuel with little or no pollution control. Because a solid fuel containing a sulfur content equal to or less than 516 g SO_2/GJ (1.2 lb SO_2/10^6 Btu) can be produced by chemical cleaning, many processes are under development. Some of these have been developed through the miniplant stage; others are still in the laboratory stage. Several processes are: (1) Air Oxidative Leaching by Pittsburgh Energy Research Center, U.S. Bureau of Mines, Pittsburgh, Pennsylvania; (2) Nitrogen Dioxide Oxidative Cleaning by KVB Engineering, Inc., Tustin, California; (3) Ferric Salt Leaching by Systems Group of TRW, Redondo Beach, California; (4) Carbonyl-Activated Magnetic Cleaning by Hazen Research Institute, Golden, Colorado; (5) Hydrothermal Leaching by Battelle's Columbus Laboratories, Columbus, Ohio; (6) Promoted Oxidative Leaching by Atlantic Richfield Co., Harvey, Illinois; (7) Microwave Desulfurization of Coal by General Electric, Valley Forge, Pennsylvania; and (8) Desulfurization of Coal by Chlorinalysis by Jet Propulsion Laboratory, California Institute of Technology, Pasadena, California.

Physical coal cleaning has been practiced commercially for years. In contrast, chemical coal cleaning is in an embryonic stage of development. A good reserve of cheap gas and petroleum, and the absence of pollution standards undermined any serious pursuit of a practical chemical cleaning process in the past. However, the situation has changed. The reserves of gas and petroleum are dwindling and expected to be depleted within the next 50 to 75 years. For the immediate future, the United States must turn to coal as its major source of energy, and the coal must be clean. Physical separation of the sulfur is inadequate. At best, only a portion of the pyritic sulfur and none of the organic sulfur can be removed without high coal losses. Chemical cleaning can achieve essentially complete removal of the pyritic sulfur from most coals and up to 40–50 wt % of the organic sulfur from some coals. Several processes that can achieve this degree of cleaning are discussed in the following paragraphs.

Oxidative Desulfurization of Coal. Basically, desulfurization of the coal is achieved by converting the sulfur to a water-soluble form by air oxidation. An aqueous slurry of coal ground to <74 μm (−200 mesh) and water is heated in a closed vessel under air as shown in Figure 1. Typical conditions are: 150–220°C at 1.5–10.3 MPa (220–1500 psi) pressure (air plus steam), a residence time of 60 min and a slurry concentration of 26% coal. Under these conditions, more than 95% of the pyritic sulfur and up to 40% of the organic sulfur is extracted from certain coals to produce clean solid fuels containing 215–775 kg SO_2/MJ (0.50–1.8 lb SO_2/10^6 Btu) (17).

This process, originally developed by Chemical Construction Company, is being studied extensively by the Pittsburgh Energy Research Center at Pittsburgh, Pennsylvania, under the support of the DOE. The process has been developed through the laboratory stage into the pre-pilot stage. A small continuous unit capable of processing ca 25 kg/h of coal is in operation.

Nitrogen Dioxide Oxidative Cleaning. This process employs both dry and wet operations as shown in Figure 2. The dry, pulverized coal with a top size of ca 0.65 cm is treated with an oxidant, gaseous NO_2, at ca 120°C and 342 kPa (35 psia) for 1 hour,

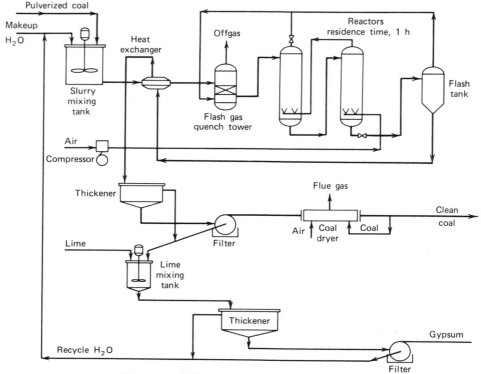

Figure 1.   Oxidative desulfurization of coal (16).

to convert the sulfur to an alkali soluble form. The coal is then washed with aqueous sodium hydroxide solution to dissolve the sulfur compounds. The waste effluent is treated with lime to regenerate the caustic values for recycle and the gypsum goes to disposal. Gaseous sulfur dioxide and sulfur trioxide generated during the NO_2 treatment are removed by a gas scrubbing system. The active oxidant, NO_2, is regenerated in the reactor by the oxidation of NO in the feed gas. Other gases in the feed gas are NO, N_2, and NO_2.

KVB has studied the applicability of the process for 8–10 coals on the laboratory scale and has just completed a pilot-plant program. The process is capable of removing essentially all of the pyritic sulfur and up to ca 40% of the organic sulfur from a Lower Kittanning coal (18).

Ferric Salt Leaching.   The ferric salt leaching process (Fig. 3) is based on the oxidation of pyritic sulfur with ferric sulfate. The leaching reaction is highly selective to pyritic sulfur with 60% of the pyritic sulfur converted to ferrous sulfate and 40% converted to elemental sulfur. The process removes up to about 95% of the pyritic sulfur contained in the coal by heating an aqueous slurry of coal ground to 1.96–0.14 mm (10–100 mesh) and ferric sulfate at 90–130°C. The leach solution is regenerated at similar conditions using air or oxygen. Elemental sulfur, a reaction product, is separated from the coal by solvent extraction or vaporization. The by-product elemental sulfur and iron sulfates may be marketed, stockpiled, or disposed of in a landfill (19).

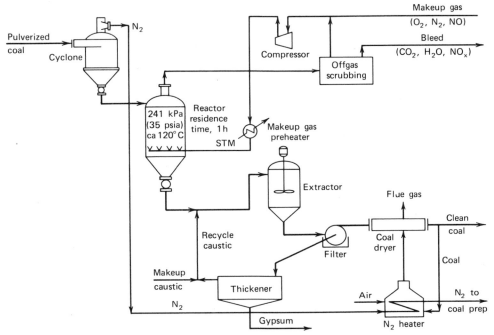

Figure 2.   Nitrogen dioxide oxidative cleaning (16).

This process, under development at the Systems Group of TRW, Inc., has received extensive study in the laboratory with the support of the EPA. This study revealed that the process is capable of extracting 80–95% of the pyritic sulfur, but none of the organic sulfur, from most coals. Therefore, the process should effectively reduce the sulfur content of most coals with an organic sulfur content of ≤ 516 kg SO_2/MJ (1.2 lb $SO_2/10^6$ Btu) to meet Federal sulfur emission standards for new sources (20–21). Further development of the ferric salt leaching process is being conducted at TRW's Capistrano Test Site under the sponsorship of the EPA. A reactor test unit has been constructed that is capable of processing 7.3 t coal/d, and evaluating the critical parameters: particle size effects, regeneration of spent leachant, and liquid–solid separation. The unit was placed on stream during the first half of 1978 (22).

Promoted Oxidative Leaching.   The promoted oxidative leaching process is reportedly similar to a process under development by the Pittsburgh Energy Research Center and another that has been studied by Kennecott Copper Company. The primary difference is that a proprietary oxidation promoter is added to the aqueous slurry to increase the rate of desulfurization. Desulfurization is achieved by heating an aqueous slurry of coal, water, and a promoter under an overpressure of air or oxygen. The desulfurized coal is separated from the product slurry and washed with water to remove the soluble iron sulfates. No other process conditions, eg, leachant regeneration, have been reported.

This process is capable of extracting more than 95% of the pyritic sulfur from all coals and up to 50% of the organic sulfur from certain coals (23). It was invented by and is being developed by Atlantic Richfield through a joint venture with Electric Power Research Institute (EPRI). A small continuous unit capable of processing ca 0.45 kg coal/h is in operation.

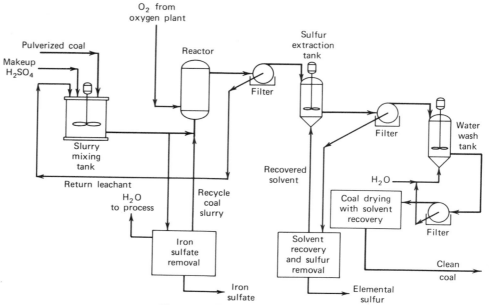

Figure 3. Ferric sulfate leaching (16).

Microwave Desulfurization. Microwave desulfurization of coal is one of two processes that use an alkali to desulfurize coal. The other process, hydrothermal leaching, uses water along with the alkali.

In microwave desulfurization, a mixture of coal and an alkali, eg, sodium hydroxide, is heated at ca 250°C (bulk temperature) for ca 30 s to promote reaction of the sodium hydroxide with the sulfur contained in the coal. The mixture is prepared by drying an aqueous slurry of coal (<0.15 mm or −100 mesh) and alkali. Desulfurized coal is separated from the mixture by washing with water. No other processing steps have been reported but, most likely, the aqueous effluent containing the solubilized sulfur and alkali values are regenerated for recycle.

This process is under development by General Electric's Reentry and Environmental Division with the support of the EPA. Reportedly, 95% of both pyritic and organic sulfur can be removed from certain coals by this scheme. Specific coals include Lower Kittanning Seam, Pittsburgh Seam, Kentucky 9 Seam, and Clarion County (24) (see Microwave technology).

Wet Oxygen Leaching. The wet oxygen leaching process (Fig. 4), which has been developed through the laboratory stage by Kennecott Copper Corporation, is similar to that being developed by Pittsburgh Energy Research Center and ARCO. In this case, oxygen is the oxidant used to convert the pyritic sulfur to water-soluble iron sulfates. A portion of the organic sulfur can be removed with coal losses. Desulfurization is achieved by heating an aqueous slurry containing 20% coal ground to <0.15 m (−100 mesh) at ca 130°C under an oxygen overpressure of ca 2 MPa (300 psi) for ca 2 h. The resulting slurry is separated, and the coal fraction washed with water. The water from both the washing and leaching operations is neutralized with lime or limestone. The solids are separated from the water and sent to a disposal area; the water is recycled.

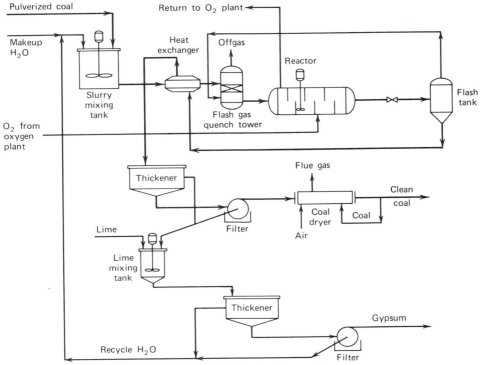

Figure 4. Wet oxygen leaching (16).

This process is applicable for pyritic sulfur removal from coals of different rank, lignite to semianthracite. More than 90% of the pyritic sulfur is removed from these coals under the conditions described above (25–26).

Hydrogen Peroxide–Sulfuric Acid Leaching. Although some processes use oxygen, air, or ferric sulfate as the oxidant, this process uses hydrogen peroxide. Desulfurization is achieved by heating a slurry consisting of coal ground to <0.55 mm (−32 mesh), 7–17% H_2O_2, and 0.01–0.5 N H_2SO_4 for ca 1 h at <100°C.

This process has been studied by the DOE at Laramie Energy Research Center, Laramie, Wyoming, in cooperation with the University of Wyoming. It is effective in extracting up to ca 90% of the pyritic sulfur, but none of the organic sulfur, from several types of coal (27).

Fluidized-Bed Air Oxidation. Treatment of coal with air at elevated temperatures is effective in removing a significant portion of the sulfur. Air, alone or in combination with steam, is passed through a fluidized bed of coal (Fig. 5) heated to 350–550°C (28).

This concept is several years old. Oxley used a fluid-bed process to study pyritic sulfur gasification rates in the presence of excess air and found preferential sulfur oxidation at temperatures of ca 550°C (28). Blum and Cinda successfully carried out desulfurization at 380°C using an air stream mixture in a 15:85 ratio (29) and Sinha and Walker considered promising their study of sulfur removal from seven U.S. coals at temperatures of 350–450°C (30). This approach is currently under study at Texas A&M University. The oxidation concept as discussed above, employing aqueous solutions under elevated pressure, is also being explored by several organizations.

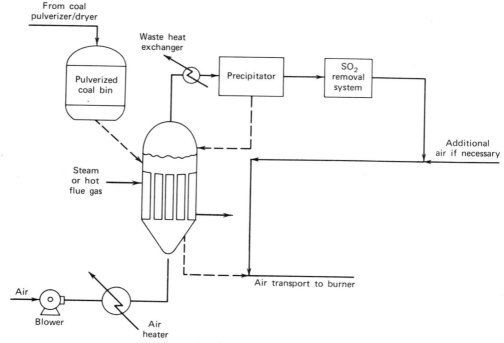

Figure 5. Preferential oxidation of coal (16).

This process is capable of removing essentially all of the pyritic sulfur using a mixture of steam and air. During the treatment, the organic sulfur might be oxidized to a more extractive form, and thus be removed by volatilization, or by posttreatment with, eg, an alkali. The final product may be a char, depending on the treatment conditions.

Hydrodesulfurization. This process is directed at the production of solid fossil fuel, most likely a char because of the desulfurization temperatures, that can be consumed directly in existing equipment in an environmentally acceptable manner. Desulfurization is achieved by heating ground, pretreated coal at 760–815°C for as much as 2 h. The coal is pretreated in an oxidizing atmosphere to destroy caking properties and improve sulfur removal.

This process can remove 97–100% of the pyritic sulfur and up to 88% of the organic sulfur to produce a char containing ca 0.6% sulfur (dry basis) at 815°C. Organic sulfur is the majority, ca 80% of the total sulfur contained in the product (31).

Development of this process has progressed through the laboratory to the project development unit (PDU) stage. A 25.4-cm fluidized-bed unit that can be fed continuously at rates of 11.3–90.7 kg/h is now in operation. IGT in Chicago, Illinois, under the sponsorship of the EPA, is conducting this research.

Chlorinalysis. A relatively new process uses chlorine to remove the sulfur from coal. The coal is contacted with chlorine gas in methylchloroform at ca 75°C. After separation of the coal from the slurry, the solvent is recovered by distillation. The chlorinated coal is washed with water and finally dechlorinated by heating at ca 300–350°C.

This process is still in the laboratory stage of development at the Jet Propulsion Laboratory of the California Institute of Technology, Pasadena, California. The process is supposedly capable of extracting ca 90% of the pyritic sulfur and up to 70% of the organic sulfur from some coals. It could thus extend reserves of low-sulfur solid fuels by making a greater portion of the high organic sulfur coals available as a clean source of energy. However, much research and development is needed to overcome the technical problems such as chlorine retention by the coal, and chlorine regeneration and recycle (32).

Hydrothermal Leaching. In hydrothermal leaching, extraction of pyritic sulfur from a wide variety of coals and a portion of the organic sulfur from some coals is achieved by heating an aqueous alkaline slurry of coal of <1.4 mm (−14 mesh), water, and the selected leachant at elevated temperatures and the corresponding steam pressure (Fig. 6). The resulting slurry is separated and the coal fraction washed with water. The aqueous solutions from both the washing and leaching operations are regenerated by, eg, the CO₂–CaO process. The evolved hydrogen sulfide is converted or to sulfuric acid. The regenerated leachant is recycled.

This process is applicable for pyritic sulfur removal from lignite, subbituminous and bituminous coals, and removal of a significant portion of the organic sulfur from some coals, depending on the processing conditions (33–34).

Further treatment of the desulfurized coal with dilute acid removes a significant portion of the ash (35). Hydrothermally treated coals are potential sources of clean solid fuels and improved feedstocks for gasification and liquefaction operations and the production of metallurgical coke.

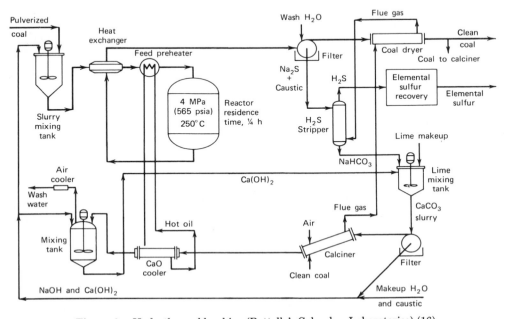

Figure 6.   Hydrothermal leaching (Battelle's Columbus Laboratories) (16).

Table 4.   Cost Comparison of Chemical Coal Cleaning Processes for 7250 t/d Plant [a]

	Air oxidative leaching	NO₂ oxidative cleaning	Ferric salt leaching	Wet oxygen leaching	Hydrothermal leaching
Plant capital cost (10^6 $)	130.5	67.9	130.9	155.2	103.4
Annual operating cost, 1976 (10^6 $) [b]	51.2	58	52.5	48.8	53.5
Fuel cost (dry basis, $/t of clean coal)	21	25	21	21	24
$/GJ [c]	0.80	0.93	0.78	0.77	0.98

[a] Ref. 16. Costs are fourth quarter 1976 U.S. dollars. There are no economic data available for other processes because they are early stages of development.

[b] Costs include coal preparation, desulfurization, and compaction.

[c] To convert $/GJ to $/$10^6$ Btu, divide by 1.054.

The process has been studied at Battelle Memorial Institute since the late 1960s and at the U.S. Bureau of Mines since the early 1970s. It has been scaled up to a miniplant scale at Battelle, and a conceptual process layout for a 365-t/h commercial plant has been conducted under the support of the Battelle Energy Program. Study of the combustion characteristics of the treated coals in laboratory (0.45 kg/h) and multi-fuel (13.6 kg/h) combustion units and assessment of the environmental impact of converting to coals cleaned by the hydrothermal coal process has been conducted under the support of the EPA.

The process is capable of converting a variety of coals, lignite, subbituminous and bituminous, to clean, solid fuels that meet Federal sulfur emission standards for new sources. More than 90% of the pyritic sulfur from most coals studied and up to ca 50% of the organic sulfur from certain coals has been extracted using a mixture of sodium hydroxide and lime as the leachant system. Desulfurization is achieved under the following set of processing conditions: leachant system, NaOH–CaO; temperature, 200–350°C; pressure (steam), 1.8–17.3 MPa (250–2500 psig); time, 5–30 min; coal grind, <100 μm (−150 mesh); NaOH concentration, 6–10%; and CaO concentration, 5% (coal basis).

During desulfurization, the coal is impregnated with an alkaline sulfur scavenger, such as calcium, which captures additional sulfur during combustion. This feature increases the number of coals that can be utilized as a source of environmentally acceptable solid fuels. Therefore, removal of a portion of the sulfur before combustion and additional sulfur during combustion could provide a scheme for controlling sulfur emissions at various levels (36). Hydrothermal treatment of coal also improves its combustion characteristics in terms of ignition and reactivity. For example, the ignition temperature of a Pittsburgh coal was reduced from 426 to 344°C, and the burnout temperature was reduced from 585–600 to 470°C (37).

Economic Aspects

Most of the chemical cleaning processes are in early stages of development. Some economic information is shown in Table 4 for five processes that were developed prior to 1977. Additional economic information is available from reference 16.

BIBLIOGRAPHY

1. H. H. Lowry, *The Chemistry of Coal Utilization*, John Wiley & Sons, Inc., London, 1945, p. 425.
2. J. W. Leonard and D. R. Mitchell, *Coal Preparation*, AIME, New York, 1968, pp. 1–45.
3. P. H. Given and W. F. Wyss, *Br. Coal Util. Res. Assoc. Mon. Bull.* **25**, 166 (1961).
4. A. W. Deurbrock and P. S. Jacobsen, "Coal Cleaning State of the Art," paper presented at *Coal Utilization Symposium—SO₂ Emission Control, National Coal Conference, Louisville, Kentucky*, Oct. 1974.
5. J. A. Cavallaro, M. T. Johnston, and A. W. Deurbrock, "Sulfur Reduction Potential of the Coals of the United States," *U.S. Bur. Mines Rep. Invest. RI 8118* (1976).
6. *1976 Keystone Coal Industry Manual, Directory of Mechanical Coal Cleaning Plants and Directory of Mines*, McGraw-Hill, New York, 1976.
7. K. J. Miller, *U.S. Bur. Mines Rep. Invest. RI 7822* (1973); *Trans. AIME* **258**, 30 (1975).
8. C. E. Capes, A. E. McIlhinney, and R. D. Coleman, *Trans. AIME* **247**, 233 (1970).
9. C. E. Capes and co-workers, *Can. Inst. Min. Metal.* **66**, 88 (1973).
10. P. S. Jacobsen and A. W. Deurbrock, Physical Coal Cleaning Contract Research by the U.S. Bureau of Mines," paper presented at *Energy and the Environment Proceedings of the 4th National Conference*, Cincinnati, Ohio, Oct. 1976, pp. 364–371.
11. M. M. Murray, "High Intensity Magnetic Cleaning of Bituminous Coal," paper presented at *2nd Symposium on Coal Preparation*, Louisville, Kentucky, Oct. 1976.
12. D. V. Keller, Jr., C. P. Smith, and E. F. Burch, "Demonstration Plant Test Results of the Otisca Process Heavy Liquid Beneficiation of Coal," paper presented at the *Annual SME-AIME Conference*, Atlanta, Georgia, March, 1977.
13. U.S. Pat. 3,941,679 (March, 1976), C. D. Smith and D. V. Keller, Jr. (to Otisca Ind.).
14. R. S. Datta and P. M. Howard, Characterization of the Chemical Comminution of Coal, Syracuse University Research Corporation, Final Report, prepared for DOE, March, 1978.
15. W. T. Akel and co-workers, *U.S. Bur. Mines Rep. Invest. RI 7732* (1973).
16. R. R. Oder and co-workers, Bechtel Corporation, San Francisco, Calif., "Technical and Cost Comparison For Chemical Coal Cleaning Processes," *Coal Conference, American Mining Congress, Pittsburgh, Penn.*, May, 1977.
17. S. Friedman, R. B. LaCount, and R. P. Warzenski, "Oxidative Desulfurization of Coal," *173rd National Meeting of ACS, New Orleans, La.*, March 21–25, 1977, pp. 100–105.
18. U.S. Pat. 3,909,211 (Sept. 30, 1975), A. F. Diaz and E. D. Guth (to KVB Engineering Inc.).
19. J. W. Hamersmo, M. L. Kraft, and R. A. Meyers, *Am. Chem. Soc. Div. Fuel Chem. Prepr.* **22**(2), 73 (1977).
20. E. P. Hensley and W. F. Nekervis, "Conceptual Design of a Commercial Scale Plant for Chemical Desulfurization of Coal," Dow Chemical Co., Midland, Mich., prepared for the EPA, Parts 1 and 2, Nov., 1974.
21. E. P. Koutsoukas and co-workers, "Final Report—Program for Bench Scale Development of the Process for the Chemical Extraction of Sulfur from Coal," *EPA 600/2-76-143a and 143b*, Environmental Protection Agency, May, 1976.
22. R. A. Meyers and co-workers, *Proceedings of the Fourth National Conference on Energy and the Environment*, Cincinnati, Ohio, Oct. 3–7, 1976.
23. L. Beckberger, Atlantic Richfield Company, private communication, 1977.
24. Discussion with Peter D. Zavitsanos, General Electric Company, 1977.
25. J. C. Agarwal and co-workers, "Coal Desulfurization: Costs/Processes and Recommendation," *Division of Fuel Chemistry, 167th National Meeting of ACS*, Los Angeles, Calif., April, 1974.
26. U.S. Pat. 3,960,513 (June 1, 1976), J. C. Agarwall and co-workers (to Kennecott Copper Corp.).
27. E. B. Smith, *Am. Chem. Soc. Div. Fuel Chem. Prepr.* **20**(2), 140 (1975).
28. J. H. Oxley, Ph.D. Thesis, Carnegie Institute of Technology, 1956.
29. I. Blum and V. Cinda, *Pop. Romine Inst. Energ. Studii* **11**, 325 (1961).
30. R. K. Sinha, and P. L. Walker, Jr., *Fuel* **51**, 125 (1972).
31. D. K. Fleming, R. D. Smith, and M. R. Y. Aquino, *Am. Chem. Soc. Div. Fuel Chem. Prepr.*, **22**(2), 45 (1977).
32. P. S. Ganguli and co-workers, *Am. Chem. Soc. Div. Fuel Chem. Prepr.*, **21**(7), 118 (1976).
33. E. P. Stambaugh and co-workers, " Environmentally Acceptable Solid Fuels by the Battelle Hydrothermal Coal Process," *Proc. 2nd Symposium on Coal Utilization, National Coal Association and Bituminous Coal Res., Inc., Coal Conference and Exposition*, Oct., 1975, p. 250.

34. U.S. Pat. 4,055,400 (Oct. 25, 1977), E. P. Stambaugh and co-workers (to Battelle Development Corp.).
35. E. P. Stambaugh and co-workers, " Clean Fuels by Battelle Hydrothermal Coal Process," *American Power Conf., 38th Annual Meeting, Chicago, Ill., April 20–22, 1976.*
36. E. P. Stambaugh and co-workers, *Proceedings of Fourth National Conference on Energy and the Environment, Cincinnati, Ohio, October 3–7, 1976.*
37. E. P. Stambaugh, "Review of Hydrothermal Coal Process," *Proceedings of 173rd National Meeting of ACS, New Orleans, La.,* March 21–25, 1977, EPA-600 17-78-068, April, 1978.

EDGEL STAMBAUGH
Battelle Memorial Institute

MAGNETOHYDRODYNAMICS

In the magnetohydrodynamic (MHD) method of electrical power generation, an electrically conducting fluid is caused to flow in a duct through a transverse magnetic field. When electrodes are placed along the sides of the duct at appropriate locations, electrical power can be extracted from the system. This results from the induced electric field in the moving fluid.

The basic idea of MHD generation was reported over one hundred and forty years ago by Faraday (1). He performed laboratory experiments with moving mercury streams and also investigated electromotive forces associated with the flow of the River Thames beneath the Waterloo Bridge in the magnetic field of the earth. The first serious attempt at engineering utilization of MHD was made by Karlovitz (2) in the period 1936–1945.

The electrically conducting fluid for MHD generation may be a liquid metal or a hot ionized gas. The latter method is the more direct and efficient for the conversion of thermal to electrical energy in commercial application. This article therefore is concerned with generation of power by the MHD process from a stream of hot conducting gas. Such a stream may consist of the products of combustion of a fossil fuel, in particular the products of combustion of coal or a coal-derived fuel such as char. The hot gas stream might also be a clean gas flowing in a closed loop, with heating provided by an externally fired coal combustion chamber.

The MHD process in its simplest form is illustrated in Figure 1. The hot ionized gas flows from the chamber at the left through the MHD generator duct, which is provided with a transverse magnetic field of flux density B (tesla, T). If the flow speed is u (m/s), the transverse induced electric field is of magnitude uB (V/m). With electrodes placed as shown, there is a transverse current density of magnitude j_y (A/m^2) which contributes to the total current I flowing to the load circuit. Actually the generator process is more complex than shown here, and a modification of the electrode arrangement is desirable.

In a power plant the MHD generator may use either a closed system or an open system. In either system the gases exhausting from the generator are very hot, and it is desirable to recover their heat. A large part of the exhaust heat can be used to

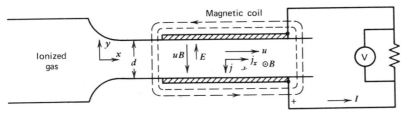

Figure 1.   Simple MHD generator, showing E-field due to electrodes; induced field uB; terminal voltage $V = Ed$ and current density components; and the total current, I.

preheat the combustion air; then a further utilization can be realized in a bottom-plant, which may be either a gas turbine system or a conventional steam power plant. Figures 2 and 3 show the basic arrangements for the MHD closed and open cycles. In principle, the closed cycle could draw heat from a nuclear reactor rather than a coal-fired combustor (see also Plasma technology).

Generator Fundamentals

When electrons move in a conducting medium in the y-direction through a magnetic field in the z-direction, they experience an electric field (Hall field) in the x-direction. Because of this Hall effect there will be an axial current component, j_x, as well as the transverse component, j_y, in the simple generator of Figure 1. The effect of j_x on the generator performance is not great if the magnetic field is low and the operating pressure in high. For a negligible Hall effect the useful current density component, j_y, is as follows:

$$j_y = -\sigma(uB - E) \tag{1}$$

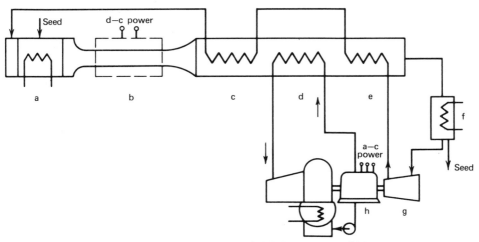

Figure 2.   Closed-cycle MHD plant: a, gas heater; b, MHD generator; c, high temperature recuperator; d, steam generator; e, low temperature recuperator; f, cooler and seed trap; g, helium compressor; and h, steam turbo generator (feed heaters, water cooling lines, and coal combustor not shown).

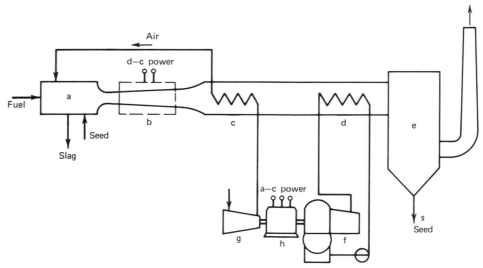

Figure 3.   Open-cycle MHD plant: a, coal combustor; b, MHD generator; c, air preheater; d, steam generator; e, gas cleaner; f, steam turbine; g, air compressor; h, a-c generator (feed heaters and water cooling lines not shown).

where σ is the electrical conductivity of the gas in $(\Omega \cdot m)^{-1}$ and E is the transverse electric field (in y-direction) due to the electrodes. The field $E - uB$ is the magnitude of the field in the y-direction that would be seen by an observer moving with the gas at speed u. If $K = E/uB$ (the generator coefficient), the current density can now be expressed simply in terms of its magnitude j, is as follows:

$$j = \sigma uB(1 - K) \tag{2}$$

When the current, j, flows at right angles to the magnetic field, an upstream Lorentz force of magnitude jB, N/m^3 acts on the gas. The rate of doing work in pushing the gas against this force, per unit volume, is

$$juB = jE + j^2/\sigma \tag{3}$$

The first term on the right side member of equation 3 is the electrical work done per unit volume; the second term is the rate of dissipation of energy per unit volume by joule heating. All terms are in W/m^3. The electrical efficiency of the process can be expressed:

$$\eta_e = \frac{jE}{juB} = K \tag{4}$$

The power generated per unit volume, P', is (from the foregoing relations):

$$P' = jE = \sigma u^2B^2K(1 - K) \tag{5}$$

P' is maximum when $K = 0.5$, ie, when the internal resistance of the generator, $d/\sigma Lb$, is equal to the load resistance, Ed/jLb, where L is active duct length and b is breadth of the channel.

The simple equation 5 indicates the level of conductivity necessary in a practical

MHD generator. If $K = 0.75$ (ca 75% efficiency) and $B = 5$ T (50,000 G), with $u = 800$ m/s, then:

$$P' = 3 \times 10^6 \, \sigma \qquad (6)$$

A gas electrical conductivity σ of 5 $(\Omega \cdot m)^{-1}$ gives a power density of 15 MW/m³, which is quite acceptable. Conductivities of 5 $(\Omega \cdot m)^{-1}$ are realizable with hot combustion products containing a small percentage of potassium or cesium seeding. The seeding provides an easily ionizable species.

The effects of the Hall current j_x have so far been neglected. In a more complete discussion of the situation expressions are derived for current density and electric field components at a local region in the channel where velocity is $\bar{u}$, magnetic induction is $\bar{B}$ and conductivity is σ. The bars denote vector quantities. See also references 3–4.

Expressions for the electrical conductivity and electron current density are written in terms of electron number density and electronic charge:

$$\sigma = n_e \mu_e e \qquad (7)$$

$$\bar{j}_e = -n_e \bar{v}_e e \qquad (8)$$

where n_e = electron density, m⁻³; μ_e = electron mobility, m²/v·s; e = electron charge, 1.602×10^{-19} C; and $\bar{v}_e$ = electron mean drift velocity, m/s. Equation 8 carries a minus sign because current is by convention the flux of positive charges.

The electric field relative to the moving gas is as follows:

$$\bar{E}' = \bar{E} + \bar{u} \times \bar{B} \qquad (9)$$

However, since the electrons move at mean drift velocity, $\bar{v}_e$, the field $\bar{E}''$ in a reference frame moving at the velocity $\bar{v}_e$ relative to the gas is as follows:

$$\bar{E}'' = \bar{E} + \bar{u} \times \bar{B} + \bar{v}_e \times \bar{B} \qquad (10)$$

The vector diagram Figure 4 illustrates these vector quantities. The field $\bar{E}''$ is the field which, when multiplied by the electrical conductivity, σ, gives the current density in the gas. (Strictly, a slight correction must be made to the scalar conductivity but

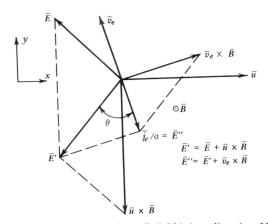

Figure 4.   Vector diagram in x,y plane. B-field is in z-direction. No ion-current.

the correction can be neglected in this case). The electron current density then becomes:

$$\bar{j}_e = \sigma(\bar{E} + \bar{u} \times \bar{B} + \bar{v}_e \times \bar{B}) \tag{11}$$

Equations 7 and 8 are now used to obtain the current density equation in the form:

$$\bar{j}_e = \sigma(\bar{E} + \bar{u} \times \bar{B}) - \mu_e \bar{j}_e \times \bar{B} \tag{12}$$

It is possible to solve equation 12 for the three components of $\bar{j}_e$ for a case that is simplified by choice of the coordinate axes. When the x-axis is the flow direction, the z-axis is in the $\bar{B}$-direction, and $E_z = 0$: $u_x = u$, $u_y = u_z = 0$, and $B_z = B$, $B_x = B_y = 0$, and for j_{ex}, j_{ey}, j_{ez}:

$$\left. \begin{array}{l} j_{ex} = \dfrac{\sigma}{1 + \beta^2} [E_x - \beta(E_y - uB)] \\[2mm] j_{ey} = \dfrac{\sigma}{1 + \beta^2} [E_y - uB + \beta E_x] \\[2mm] j_{ez} = 0 \end{array} \right\} \tag{13}$$

The parameter β, the Hall coefficient, is equal to $\mu_e B$. It is a measure of the relation of the Hall field $v_e B$ (magnitude of $\bar{v}_e \times \bar{B}$ since $\bar{v}_e$ is normal to $\bar{B}$) to the field E''. Thus if θ is the angle between vectors $\bar{E}'$ and $\bar{E}''$ (or $\bar{j}_e/\sigma$), as shown in Figure 4, $\tan \theta$ is $v_e B \sigma / j_e$. By equations 7 and 8:

$$\tan \theta = \mu_e B = \beta \tag{14}$$

A further refinement is made in the current equations when currents resulting from ion drift are included. The mobility of the ions is designated μ_i, and the ionic conductivity and current density are expressed:

$$\sigma_i = n_i \mu_i e \tag{15}$$

$$\bar{j}_i = n_i \bar{v}_i e \tag{16}$$

where $\bar{v}_i$ is the mean drift velocity of the ions and we assume singly charged ion species. The ratio of ion mobility to electron mobility, μ_i/μ_e, is approximately the square root of the ratio of electron mass to ion mass. For cesium seeding μ_i/μ_e is about 0.0022, and for potassium, 0.004. The resulting small drift velocity of the ions means that the ion Hall field $v_i B$ is negligibly small in cases of interest in MHD generators, so that ionic Hall effects can be neglected. The ion current density, on the other hand, is sometimes of importance. Since the ion current Hall effects are neglected, the Ohm's law formulation for the ion current becomes:

$$\bar{j}_i = \sigma_i (\bar{E} + \bar{u} \times \bar{B}) \tag{17}$$

When the components of $\bar{j}_i$ are combined with the components of $\bar{j}_e$ as given by equation 13, the expressions for components of the total current $\bar{j}$ are:

$$\left. \begin{array}{l} j_x = \dfrac{\sigma}{1 + \beta^2} [E_x - \beta(E_y - uB)] + \sigma \dfrac{\mu_i}{\mu_e} E_x \\[2mm] j_y = \dfrac{\sigma}{1 + \beta^2} [E_y - uB + \beta E_x] + \sigma \dfrac{\mu_i}{\mu_e} (E_y - uB) \\[2mm] j_z = 0 \end{array} \right\} \tag{18}$$

The conductivity σ used in the foregoing equations is the scalar electronic conductivity as given by equation 7. A vector diagram for the system with ionic currents is shown in Figure 5.

There are four important types of MHD generator channels (Fig. 6) designated as follows: type a, continuous electrodes, $E_x = 0$; type b, segmented Faraday type, $j_x = 0$; type c, Hall type, $E_y = 0$; and type d, diagonal Faraday type, $j_x = 0$. For the continuous electrode case, $E_x = 0$, $X = 1 + \beta\beta_i$, $Y = 1 + \beta^2$ and $\beta_i = \mu_i B$ are substituted in equation 18. The expressions for E_x, E_y, K, j_x, j_y, η_e and P' are then as in Table 1; β_i/β, compared to unity, can be neglected.

For the segmented Faraday case, column (b) of Table 1, the power density P', which is derived from $j_y E_y$, is smaller for type a than for type b by the ratio X^2/Y (again neglecting β_i/β compared to unity). The improved power density obtained by using finely segmented electrodes rather than continuous electrodes is a strong factor in favor of the segmented construction. The penalty of continuous electrodes could be a reduction of power density by a factor of 0.1–0.2.

For the Hall generator, type c, the essential characteristic is that $E_y = 0$ since the opposite electrodes are short circuited. The generator coefficient is defined differently in this case, in keeping with the fact that the power current is axial rather than transverse; K' is the ratio of the E_x field at some load condition to the E_x field at open circuit. Thus $K' = 1$ at open circuit. The power density is generally lower in the Hall generator than in the segmented Faraday type. The electrical efficiency, η_e, is greatest near short circuit (small K'), and large β values are favorable (3–4).

The diagonally connected segmented electrode generator, type d, can under ideal conditions exhibit as high efficiencies and power densities as a segmented Faraday generator. However, there is almost invariably some mismatch which either causes lack of current continuity or alters the ideal diagonal connection angle. As a result, diagonally connected generators rarely attain optimum performance throughout the whole channel. In spite of this, the diagonally connected generator enjoys considerable popularity because of the great simplicity in the load circuits. This is especially true of the diagonally conducting wall type (DCW), in which the diagonal connections are extended electrode bars running diagonally across the side walls between their respective electrode pairs. The DCW generator has been investigated by Dicks, Shanklin

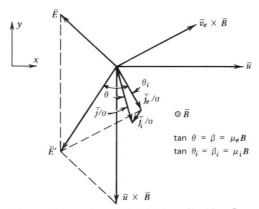

Figure 5.   Vector diagram in x,y plane. B-field is in z-direction. Ion current j_i included.

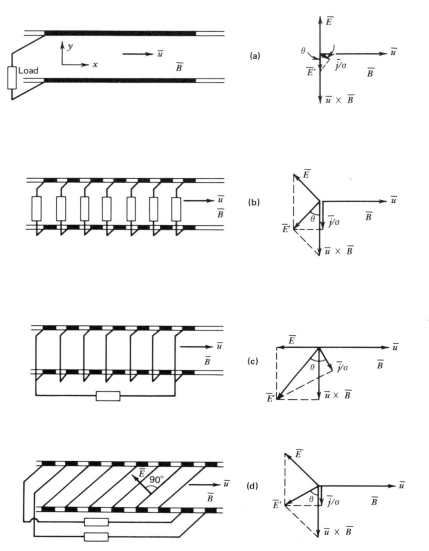

Figure 6.   Types of MHD generator. (**a**) Continuous electrodes, (**b**) segmented Faraday, (**c**) Hall generator, (**d**) series, or diagonally, connected generator.

and Brogan and co-workers (5–7). Ideally, the diagonally connected electrode pairs should be sets that normally have the same potential, so that the slant angle α becomes $\tan E_y/E_x$ or $\tan K/\beta(1 - K)$. Originally, studies were made of diagonal connections by Montardy (8).

The electrodynamic equations given apply to any small local region within the channel. The MHD duct flow relations are necessary to describe the behavior of the channel as a whole. The flow through the duct is accompanied, in general, by changes in pressure, temperature, velocity, density, enthalpy, entropy, conductivity, and gas composition. Many of these variables are interrelated. The situation is further com-

Table 1. Formulas for Field and Current Components, Power Density, and Efficiency

	Continuous electrodes (type a)	Segmented Faraday (type b)	Hall generator (type c)
E_x	0	$-\beta uB(1 - K)/X$	$-\beta uBK'/X$
E_y	KuB	KuB	0
K (def)	E_y/uB	E_y/uB	$-E_x X/uB\beta$
j_x	$\sigma\beta uB(1 - K)/Y$	0	$\sigma\beta uB(1 - K')/Y$
j_y	$-\sigma XuB(1 - K)/Y$	$-\sigma uB(1 - K)/X$	$-\sigma uB(X^2 + K'\beta^2)/XY$
η_e	K	K	$\beta^2 K'(1 - K')/(X^2 + K'\beta^2)$
P'	$\sigma u^2 B^2 XK(1 - K)/Y$	$\sigma u^2 B^2 K(1 - K)/X$	$\sigma u^2 B^2 \beta^2 K'(1 - K')/XY$

plicated by internal friction and thermal transfer. Real MHD channels are also characterized by such phenomena as leakage currents and nonuniformities of plasma properties.

A preferred way of dealing with this very complex situation is to carry out an analysis based on assumptions that are reasonably valid, but which, at the same time, lead to simplifications. Special investigations can then be made of aspects ignored in the simplified treatment.

A one-dimensional, hydraulic type of analysis or an analysis based on a boundary layer treatment can be used. Since the former is much simpler, and since it appears to err, for the most part, on the safe side while being sufficiently accurate for many practical applications, this method is the one that is presented here. More exact treatments can be found in the technical literature (9).

A few of the flow parameters may be established by design choice. One is the type of velocity variation, another is the heat loss (or, alternatively, the wall temperature). Certain parameters are specified at the outset, such as combustor pressure and temperature, gas type, duct inlet velocity and outlet total pressure, allowed Hall field E_x max, and magnetic induction B. The generator length then becomes a dependent variable. If a certain generator length is desired, one must be prepared either to alter B to achieve it or to revise the initial pressure assumption and carry out a revised duct calculation. These procedures can, of course, be computerized but direct human participation is a valuable factor.

For analysis of the flow in open-cycle MHD systems, a detailed study must first be made of the equilibrium gas compositions at various temperatures and pressures, of the state properties like enthalpy, density, entropy and sonic velocity, and of transport properties such as electron mobility, electrical conductivity, viscosity, and thermal conductivity. The conservation equations for momentum, energy, and mass, along with the gas law and the current equation, can then be used to analyze the flow in the duct. An example of this treatment is given in reference 10.

The equations below assume a segmented Faraday generator, and neglect ion currents. Flow is in the x-direction. The pressure p is assumed uniform over any cross section normal to x. Quantities T, ρ, σ, u, s, and h symbolize average values over the cross section:

Current:
$$j = \sigma uB(1 - K) \tag{19}$$

Momentum:
$$\rho u \frac{du}{dx} + \frac{dp}{dx} = -jB\left(1 + \frac{f}{\xi(1 - K)}\right) \tag{20}$$

Energy:
$$\rho \frac{d}{dx}\left(h + \frac{u^2}{2}\right) = -jBK(1 + \lambda) \qquad (21)$$

Mass:
$$\rho u A = m \qquad (22)$$

Gas Law:
$$\rho = \frac{pw}{RT} \qquad (23)$$

Hall field:
$$E_x = \mu u B^2 (1 - K) \qquad (24)$$

where ρ = density, kg/m^3; p = pressure, Pa; h = specific enthalpy, J/kg; w = molecular weight; R = universal gas constant = 8.314 J/(mol·K); f = friction factor; λ = heat loss ratio (assumed); r_h = hydraulic radius, m; m = mass flow rate, kg/s; A = flow area, m^2; j = magnitude of current density, A/m^2, in y-direction; E_x = Hall field, V/m; μ = electron mobility, m^2/s·V; σ = electrical conductivity, $(\Omega \cdot m)^{-1}$; B = magnetic induction, T; K = generator coefficient E_y/uB; u = flow velocity, m/s; ξ = the interaction parameter $2r_h \sigma B^2/\rho u$; and the heat loss ratio, λ, is the ratio of heat loss per unit volume to power generated per unit volume.

   To carry out the analysis of the flow, it is convenient to assume a pattern of speed variation in which the ratio of kinetic energy drop to isentropic enthalpy drop is constant per unit length. This ratio is designated by C (note isentropic enthalpy drop is $-dp/\rho$):

$$C = \rho u\, du/dp \qquad (25)$$

A constant velocity design is obtained by taking $C = 0$. For $C \neq 0$, from equations 20 and 21:

$$dx = \frac{-(1 + C)dp}{\sigma u B^2 (1 - K + f/\xi)} \qquad (26)$$

$$dh = \frac{dp}{\rho}\left\{\frac{K(1 + \lambda)(1 + C)}{1 + f/\xi[(1 - K)]} - C\right\} \qquad (27)$$

Equation 26 permits calculation of generator length by integration. Equation 27 allows one to carry out a numerical analysis of the generator process. With the aid of a Mollier diagram (or equivalent computer stored information), one can assume Δp, determine $-\Delta p/\rho$ (isentropic enthalpy drop) and calculate Δh from 27. The state point at the end of the incremental change is determined by the final values of h and p. Average values of σ, u, and B^2 for the interval would be used in calculating Δx by equation 26. In the calculations K, λ, and C are assumed. Friction factor, f, may be assumed at a constant (typical) value, or it may be evaluated from point to point in terms of the Reynolds number. Selection of λ as a design choice implies that subsequent calculation should be made to determine the wall temperature corresponding to the assumed heat loss. The interaction parameter ξ can also be calculated point-by-point as the analysis proceeds down the duct. The interaction parameter is a dimensionless quantity related to the ratio of the work done against the electromagnetic body force per unit volume over distance r_h (N·m/m^3 or J/m^3), to the kinetic energy per unit volume (J/m^3) of the plasma.

   In some instances it is desirable to analyze the generator starting at the outlet and working back toward the inlet. This may be true especially in analyses of part load conditions (11). Another variant of the calculation procedure arises when area, A, is assigned as a function of x. It is then necessary to dispense with the assumption that C is constant, and to calculate u point-to-point by successive approximations.

The local generator efficiency is the ratio of increment of output power to isentropic heat drop. The increment of output power is as follows:

$$\Delta P = -\frac{(\Delta h + \Delta(u^2/2))}{1 + \lambda} \tag{28}$$

The incremental isentropic heat drop is $-\Delta p/\rho$. Reference to equations 25 and 27 indicates that:

$$\eta_e = \frac{K(1 + C)}{1 + f/\xi(1 - K)} \tag{29}$$

For a constant velocity design, $C = 0$ and η_e is simply:

$$\eta_e = \frac{K}{1 + f/\xi(1 - K)} \tag{30}$$

A plot of this efficiency vs K and f/ξ (Fig. 7) is instructive (3). To achieve good efficiency f/ξ should be smaller than 0.05. For very large machines $f/\xi \approx 0.01$. For a given f/ξ the efficiency goes through a maximum at some optimum K value. With K above the optimum, the generator becomes long and frictional losses are excessive. With K too small, the ohmic heating losses are excessive.

The total power generated is given by:

$$P = \frac{m(h_3 - h_6)}{1 + \lambda} \tag{31}$$

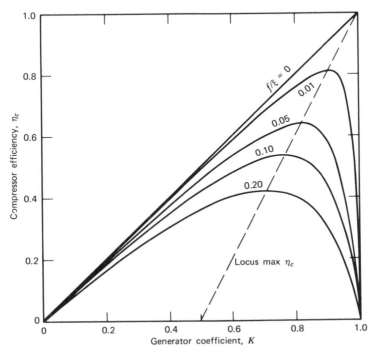

Figure 7. Conversion efficiency for constant velocity MHD generator, in terms of generator coefficient K, friction factor f, and interaction parameter ξ.

where h_3 is inlet total enthalpy (combustion chamber; see Fig. 8) and h_6 is total exit enthalpy, both per unit mass. In this notation h_4 inlet gas specific enthalpy at the duct inlet and h_5 the outlet gas specific enthalpy.

Electrical Conductivity

The electronic conductivity in the gas is given by equation 7. The theoretical treatment of the problem involves calculation of mobility, μ_e, and electron density, n_e.

The fairly satisfactory theory (12) is outlined here (see also ref. 13). The electron mobility is calculated taking account of dependence of collision frequency, ν, on electron speed in the integral expression:

$$\mu_e = -\frac{4\pi e}{3m_e} \int_0^\infty \frac{v^3}{\nu} \cdot \frac{d}{dv} F(v)\,dv \tag{32}$$

$$F(v) = \left(\frac{m_e}{2\pi kT}\right)^{3/2} \exp\left(\frac{-m_e v^2}{2kT}\right) \tag{33}$$

The collision frequency, ν (s^{-1}) of the electrons with various species is a function of electron speed, v (m/s), the distribution of which is given by equation 33, where e = electronic charge, 1.602×10^{-19} C; m_e = electron mass, 9.11×10^{-28} g; $F(v)$ = speed distribution function; k = Boltzmann constant, 1.38×10^{-23} J/K; and T = gas temperature, K.

The collision frequency involves collisions with many species, including neutral particles and ions; ν_k is the collision frequency of an electron with the kth neutral species, and ν_{ei} is the collision frequency of electrons and ions. If there are s species of neutral particles, the total collision frequency is:

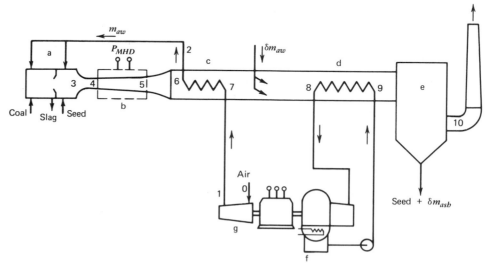

Figure 8. Simplified MHD–steam plant system diagram showing numbering of stations and secondary air admission δm_{aw}. Components are as follows: a, two-stage coal combustor; b, MHD generator; c, air preheater; d, steam generator; e, bag house; f, steam turbogenerator; g, air compressor. δm_{aw} is admitted to complete the combustion of unburned fuel.

$$\nu = \sum_1^s \nu_k + \nu_{ei} \tag{34}$$

The frequency, ν_k, depends on the electron energy:

$$\nu_k = (10^{11}\, p_k/T)[f_k(U)] \tag{35}$$

where p_k is the partial pressure in kPa of the kth species and U is the electron energy $m_e v^2/2e$. The functions f_k depend on experimental data. Bibliographical information is summarized by Pepper (14). Much of the experimental work on electron collisions was done by Frost and Phelps (15) and by Hake and Phelps (16). The electron-ion collision frequency can be approximated by the Spitzer-Härm (17) formulation for a completely ionized gas:

$$\nu_{ei} = \frac{3.64 \times 10^{-6}}{T^{1.5}}\, n_e \ln[(1.27 \times 10^7)T^{1.5}/n_e^{0.5}] \tag{36}$$

The electron density is calculated by taking account not only of the electrons released by ionization of the alkali atoms, but also of the electrons lost from the plasma by attachment to certain species to form negative ions. Although there is, in general, just one type of positive ion, namely K^+ or Cs^+, there may be a number of different negative ion species, among them OH^-, O^-, NO_2^-, C_3^-, CN^-, NO^-, and C_2^-. Usually OH^- is the most important. Electron affinities, for some of the negative ion-forming species are: 2.93×10^{-19} J (1.83 eV) for OH, 3.68×10^{-19} J (2.30 eV) for NO_2, 0.144×10^{-19} J (0.09 eV) for NO, and 2.34×10^{-19} J (1.46 eV) for O. The ionization potentials V_i are 6.95×10^{-19} J (4.34 eV) for K and 6.23×10^{-19} J (3.89 eV) for Cs.

The steps involved in calculation of n_e are indicated. α_1, α_2, α_3, are the negative ion-forming species, and $n_{\alpha 1}$, $n_{\alpha 2}$, $n_{\alpha 3}$, ..., their number densities per cubic meter. Let n_A be the number density of free alkali atoms, where A may be potassium or cesium. The terms $n_{\alpha 1}^-$, $n_{\alpha 2}^-$, etc, and n_A^+ represent the individual ion number densities.

The gas is, in general, electrically neutral in domains of dimensions larger than the Debye length. Therefore an equation expressing charge neutrality can be written as follows:

$$n_e + n_{\alpha 1}^- + n_{\alpha 2}^- + n_{\alpha 3}^- + \text{etc} = n_A^+ \tag{37}$$

In terms of partial pressures:

$$[e^-] + [\alpha_1^-] + [\alpha_2^-] + \text{etc} = [A^+] \tag{38}$$

The ionization equilibrium equations are:

$$\left.\begin{array}{ll} A \rightleftharpoons A^+ + e^- & K_e = [A^+][e^-]/[A] \\ \alpha_1^- \rightleftharpoons \alpha_1 + e^- & K_1 = [\alpha_1][e^-]/[\alpha_1^-] \\ \alpha_2^- \rightleftharpoons \alpha_2 + e^- & K_2 = [\alpha_2][e^-]/[\alpha_2^-] \end{array}\right\} \tag{39}$$

and by rearrangement of equation 38, an expression for $[e^-]^2$ is:

$$[e^-]^2 = \frac{[A]K_e}{1 + [\alpha_1]/K_1 + [\alpha_2]/K_2 + \text{etc}} \tag{40}$$

Thus if the equilibrium constants and the equilibrium values of the partial pressures of the various species are known, $[e^-]$ and n_e can be expressed:

$$n_e = \frac{n}{p}\,[e^-] \tag{41}$$

Where p is the total pressure and n is the total number of particles per unit volume:

$$n = 0.7339 \times 10^{28}\, p/T \tag{42}$$

The equilibrium constants are given by Saha type equations:

$$\log K_e = 2.5 \log T - 6.48 - 5040\, V_i/T \tag{43}$$

$$\log K_1 = 2.5 \log T + \log 2\gamma_1 - 6.48 - 5040\, E_1/T \tag{44}$$

$$\log K_2 = 2.5 \log T + \log 2\gamma_2 - 6.48 - 5040\, E_2/T \tag{45}$$

where γ_1, γ_2, etc, designate the ratio of statistical weight for the ion ground state to that of the neutral particle ground state (18).

The partial pressures $[A]$, $[\alpha_1]$, $[\alpha_2]$, etc, are actually those that remain after partial ionization. The question may therefore arise as to whether to consider the ionization equilibria in determining these quantities. Only a very trivial error results from calculating the partial pressures on the basis of chemical equilibrium of the neutral species alone.

From $[e^-]$ and n_e one can calculate $[A^+]$, $[\alpha_1^-]$, $[\alpha_2^-]$, etc, by equation 39, and the number densities can be obtained by equations similar to 41.

Although more exact theoretical treatments that take account of small corrections to the tensor components of the electron mobility can be formulated, the theory given here is generally adequate for MHD generator calculations. Improved accuracy might be obtained with better values of collision frequency and electron affinities.

Using the theory given above and generally accepted values for the physical parameters, the electrical conductivity can be calculated for various gas mixtures. Results are shown in Figure 9.

Experimental values of electrical conductivity of seeded combustion product plasmas are scarce; refs. 19–21 include such data, but often shortcomings in the experimental determination lead to uncertainty in the results. Recent work in this field in the U.S.S.R. is of interest (22). Results of tests on MHD channels are generally consistent with calculated values of electrical conductivity, as indicated, eg, in ref. 23.

Applications in Power Plants

Fuel. The fuels of primary interest for MHD power generation are coal and coal derivatives, including char and gasified coal. Attention centers on coal because of its plentiful supply and moderate cost. Coal and char are fuels admirably suited for MHD power generation. Data on several coals are given in Table 2.

The factors that determine the suitability of a fuel for MHD application are the capability of giving both good electrical conductivity and high flame temperature. The factor most conducive to high conductivity is a low ratio of hydrogen to carbon. Abundance of hydrogen is disadvantageous for two reasons. First, in the presence of oxygen and alkali atoms, it promotes a higher concentration of potassium or cesium hydroxide, thereby diminishing the concentration of alkali atoms and reducing the numbers of positive ions (K^+ or Cs^+) and free electrons. This reduction of n_e lowers electrical conductivity. The second unfavorable effect of hydrogen is also related to an adverse effect on n_e. The hydrogen tends to promote a greater concentration of OH, ie, OH^-.

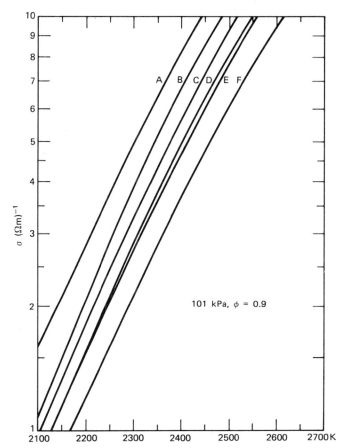

Figure 9. Electrical conductivity of products, various fuels at 101 kPa (1 atm) and ϕ = 0.90.

Curve	Fuel	Seed, wt
A	char	0.5% Cs
B	coal	0.5% Cs
C	coal	0.7% K
D	coal	0.5% K
E	distillate oil	0.5% Cs
F	natural gas	0.5% Cs

The second favorable fuel characteristic of thermodynamic nature is the high flame temperature. Fuels that give a higher flame temperature for a given air preheat temperature are more desirable on account of both beneficial effects on conductivity and increase in enthalpy drop attainable in the generator. Char, anthracite, and coke are superior MHD fuels. Next are bituminous and sub-bituminous coals, followed in turn by lignite, fuel oil, and natural gas (24). Electrical conductivity comparisons are given in Figure 9.

Preheat Temperature Requirements. The air preheat temperature required to give a certain flame temperature can be calculated from the following equation:

$$\frac{Q_{add}}{N_c w_c} = \frac{m_{aw}}{m_c} (h_{2\ aw} - h_{0\ aw}) - \lambda_1 \theta_c \qquad (46)$$

Table 2. Composition of Selected Coals

		West Pennsylvania bituminous	Low S char	Montana Rosebud[a]	Illinois no. 6[a]
proximate, %	moisture	4.3		22.7	8.9
	volatile	26.5	3.6	38.0	41.7
	FC[b]	65.6	87.1	50.7	45.8
	ash	7.9	9.3	11.2	12.5
ultimate	ash	7.9	9.3	11.2	12.5
(dry), %	C	80.6	85.2	67.5	68.5
	H	4.9	1.3	4.5	4.8
	S	2.6	0.8	1.1	3.6
	N	1.4	1.7	1.0	1.3
	O	2.6	1.7	14.7	9.2
atom composition	C	1.0	1.0	1.0	1.0
(atom ratio,	H	0.724	0.182	0.790	0.841
C = 1) daf[c]	O	0.024	0.015	0.163	0.101
	N	0.015	0.017	0.013	0.016
	S	0.012	0.004	0.006	0.020
HHV[d], daf[c] J/kg[e] θ_c		36.3×10^6	33.5×10^6	30.2×10^6	32.9×10^6

[a] These coals are similar to those assumed in design studies of the Montana Engineering Test Facility, U.S. Department of Energy.
[b] FC = fixed carbon.
[c] daf = dry, ash-free state.
[d] HHV = higher heating value.
[e] To convert J to cal, divide by 4.184.

where Q_{add} is the heat added to system above the reference state per mole of fuel mixture to bring the products to the state (p_3, T_3); N_c is the number of moles of combustible (eg, dry, ash-free coal) per mole of fuel mixture; λ_1 is the fractional heat loss ratio; w_c = equivalent molecular weight of the combustible; m_{aw}/m_c = ratio of moist air to combustible, weight basis; $h_{2\,aw}$ = specific enthalpy of moist air at T_2; and $h_{0\,aw}$ = specific enthalpy of moist air at reference temperature T_0.

The quantity Q_{add} is also based on the thermodynamic relation:

$$Q_{add} = H_3 N - H°_{fF} - X H°_{fox} \qquad (47)$$

where N = moles of products at state p_3, T_3 per mole fuel mix; $H°_{fF}$ = heat of formation of one mole of fuel mixture, J/mol; $H°_{fox}$ = heat of formation of one mole of oxidant mixture, J/mol; X = moles of oxidant mixture per mole of fuel mixture; and H_3 = molar enthalpy of products at state 3, sensible heat plus heat of formation, J/mol.

The calculation cannot be completed without detailed knowledge of the gas composition. Usually kinetic effects can be neglected and the flame temperature calculation based on the gas composition at chemical equilibrium. Such equilibrium calculations, as well as calculations of molar enthalpies, should be made at the outset of MHD thermodynamic investigations (see, eg, refs. 13 and 24). Use of JANAF tables for basic data is recommended (25). One completes the calculation of the preheat temperature by first assuming a set of flame temperatures, T_3, at some chosen pressure p_3, and then applying equation 46 to find $h_{2\,aw} - h_{0\,aw}$. The preheat temperature,

T_2, can be found from the known enthalpy properties of the moist air. The quantity, $Q_{add,}$ would have previously been determined for the given fuel and equivalence ratio θ by computerized calculation for many pressures and temperatures.

The ratio m_{aw}/m_c is determined by stoichiometric considerations. Let ϕ signify the ratio of the number of moles of oxidant X to the stoichiometric number of moles of oxidant corresponding to one mole of fuel X_s:

$$\phi = X/X_s \tag{48}$$

The ratio m_{aw}/m_c is expressed as

$$\frac{m_{aw}}{m_c} = \frac{N_a w_a \phi X_s}{N_c w_c} (1 + \delta) \tag{49}$$

where N_a = moles of dry air per mole of oxidant mixture; w_a = molecular weight of dry air, 28.97; and δ = kg moisture/kg dry air.

In calculating X_s, one assumes complete combustion to H_2O, CO_2, K_2O, Cs_2O, and SO_2, and takes account of the oxygen introduced with the fuel, seed, and moisture as well as the oxygen in the air and in the CO_2 in the air. X_s is determined when the fuel, oxidant, and seed material (eg, K_2CO_3) are specified. (It is advantageous in formulating the fuel and oxidant mixtures to place equal mass fractions of seed in each mixture. A change in ϕ will then maintain the seeding ratio unchanged.)

The fuel mixture would, in general, include the ash-free combustible (eg, dry, ash-free coal), the water in the coal, the appropriate seed fraction, and the water (if any) introduced with the seed. The oxidant mixture would, in general, include the dry air constituents, the moisture in the air, the seed fraction, and the moisture in the seed. Excessive moisture in the fuel is disadvantageous because of the harmful role of the hydrogen and its adverse effect on the flame temperature. Very wet coals should be dried to a moisture content of 5 or 6% to give good MHD performance; however, further drying than this is hardly worthwhile.

More refined calculations may also include the appropriate amount of ash constituents, corresponding to the fraction carried over into the MHD channel from the combustor. The exclusion of the ash only gives an imperceptible effect on the thermodynamic properties. On the other hand, for more precise knowledge of the electrical conductivity, the ash constituents and their physical states should be included in the gas calculations. One may observe, however, that the errors in theoretical calculation of conductivity associated with uncertain values of collision frequencies are undoubtedly much larger than those resulting from neglect of a small concentration of ash constituents. When there is massive ash carry-over, as from a nonslagging combustor, it is more likely that ash constituents will have a significant influence on conductivity.

The relations given above for calculation of preheat temperature are slightly altered when oxygen enrichment is applied (see below). Also, if the fuel or seed is preheated, additional terms must be introduced, with positive sign, in the right hand side of equation 46.

Preheat and flame temperatures are given for several fuels in Figure 10. Char is a superior MHD fuel from the standpoint of flame temperature.

Oxygen Enrichment. Oxygen enrichment may be used to avoid a costly investment in air preheaters. For a given flame temperature the air preheat temperature can then be reduced, even to the level of no preheat at all if that condition is desired. The penalty

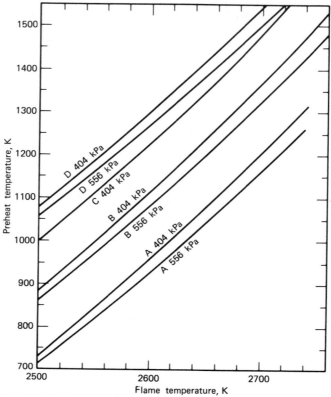

Figure 10. Air preheat temperature required to give a certain flame temperature. Various fuels, at $\theta = 0.90$. Fuel A, char (seed/wt, 0.5% Cs); B, coal (seed/wt, 0.5% Cs); C, coal (seed/wt, 0.7% K); and D, coal (seed/wt, 0.5% K). Heat loss 3.5% for A, B, and D; 2.5% for C. To convert kPa to atm, divide by 101.

is a reduction of generation efficiency because the ratio of fuel burned to mass flow of gas in the MHD generator is increased. For short-burst generators, or generators applied for peaking or emergency service, oxygen enrichment may be attractive to reduce capital cost.

The other aspect of oxygen enrichment is the possibility of increasing flame temperature at a given air preheat temperature. In this case the enthalpy drop of the gases in the MHD duct can be increased, and the efficiency of the plant improved. There are essentially two reasons why one does not invariably resort to oxygen enrichment. First, it is undesirable to operate with the duct materials at a higher temperature than that necessary to give acceptably good performance. Secondly, the economic penalty of the cost of the oxygen (or of the equipment and energy to produce it) discourages the use of extensive enrichment. Some optimum level of oxygen enrichment may well be found, perhaps at a very low value such as 1–2% of the air, which will be helpful and advantageous.

The modified form of the preheat equation with oxygen enrichment is

$$\frac{Q_{add}}{N_c w_c} = \frac{m_{aw}}{m_c}\,(h_{2\ aw} - h_{0\ aw}) + \frac{m_{ox}}{m_c}\,(h'_{2\ ox} - h_{0\ ox}) - \lambda_1 \theta_c \qquad (50)$$

where m_{ox}/m_c is mass of raw oxygen per unit of combustible mass; $h'_{2\,ox}$ is the specific enthalpy of oxygen at oxygen preheat temperature T'_2; θ_c is the heating value.

Calculations for a particular fuel were made showing the effects on preheat and flame temperatures of various levels of oxygen enrichment. Results are shown in Figure 11.

Selection of Parameters.   Deciding upon the maximum temperature to use in the MHD power plant involves various trade-offs. It is possible to increase plant efficiency by increasing the combustion chamber temperature, but this necessarily implies higher air preheat temperature and more costly air preheaters. Problems in the combustor, ducting, and MHD generator walls are also magnified by temperature increases.

Metal strength considerations set a practical upper limit of 1350–1400 K on the air preheat temperature. Use of silicon carbide tubes in the high temperature section of the recuperator may make possible attainment of preheat temperatures to 1500 K, but even to reach 1400 K requires special design features. This preheat temperature yields a flame temperature of about 2700 K, with minor variations depending on coal type. Using direct-fired regenerative (storage type) heaters, it may be possible, with some further development, to preheat the air to 1700 K, which could give a flame temperature over 2800 K. This would potentially give an efficiency improvement of about 2%, but the gain is reduced by heat losses to the flow control valves of the re-

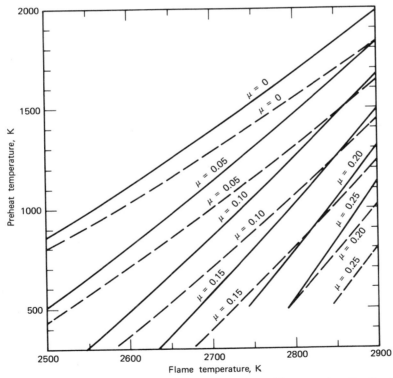

Figure 11.   Required air preheat temperature with oxygen enrichment. μ is ratio of moles of raw oxygen to moles of dry air. Fuel: toluene with char additive; seeding by K_2CO_3 (0.5% K in gas); $\theta = 0.95$. —, 303 kPa; – –, 606 kPa. To convert kPa to atm, divide by 101.

generators and blow-down losses during cyclic operation of the regenerative beds. Also the added cost of the heaters offsets the potential fuel savings. A further possible variation is to use indirect firing of the regenerative air heaters in order to keep slag and seed out of the matrix. The preheat temperature can then be pushed upward toward 1900 K, and flame temperatures are 2850–2900 K. The thermodynamic penalties are rather severe in this case since the energy of fuel fired in the external heater is not utilized as effectively as that fired in the main combustor.

These considerations lead some investigators to favor, for the first generation of MHD power plants, the use of a direct-fired recuperator with air preheat to ca 1350 K and a maximum temperature for coal burning of 2700 K. This heating level can yield power plant efficiencies greater than 50% for large installations (see MHD Power System Analysis below).

Selection of the maximum temperature leads directly to assignment of a maximum pressure. Selection of the combustion chamber pressure, p_3, depends on the power level of the plant, the desired generator duct length, and the magnetic field strength. For a generator of about 500 MW electrical rating, a length of 16–18 m might be acceptable, for a smaller unit of 60 MW, ca 7–8 m would be more appropriate. Some iteration of the design calculations is often necessary before the appropriate p_3 level is identified. A further guide is established by the criterion that the L/d ratio for the channel should be between 6 and 10. Experience indicates that for the 500 MW example and for $T_3 = 2700$ K, the pressure, p_3, becomes 500 or 600 kPa (5 or 6 atm) if $B = 6$ T; and for the 60 MW case, a value for p_3 of ca 400 kPa (4 atm) is more appropriate.

The assigned exit pressure, after the diffuser at the outlet of the MHD duct, also influences the selection of p_3. The heat exchanger and steam generator system may be designed to run at pressures above or below atmospheric. Thus diffuser exit pressure p_6 might be either about 1.16 kPa (1.15 atm) or 0.96 kPa (0.95 atm). Still another variation is to use a low pressure system throughout, with a powerful induced-draft blower ahead of the stack. The lower pressure can reduce wall heat fluxes and lower the required magnetic field. A disadvantage is the increase of flow area.

The specification of channel inlet velocity must be made with due consideration of the pressure recovery in the diffuser as well as overall enthalpy drop per unit of channel length. One generally finds that for base load plants, with maximum temperature around 2700 K, an inlet velocity of 750–800 m/s is appropriate.

The decision on seed concentration is based partly on economic factors and partly on technical considerations. It is advantageous to use enough alkali seed material to combine with any sulfur introduced with the fuel. There is a strong tendency for formation of K_2SO_4 (or Cs_2SO_4). This has the fortunate consequence of greatly reducing or eliminating SO_2 discharges in the stack gas (26). It is necessary to recover the seed material in any case, and so doing removes sulfur from the system. It is only necessary to process the K_2SO_4 to separate the sulfur and form K_2CO_3 for reintroduction into the system (27). The unit processes involved are fairly well known. The amount of seed needed to accomplish sulfur removal is indicated in Figure 12 (see Sulfur recovery).

For coals having less than 3% sulfur (in the daf coal), a seeding level of 0.7 wt % potassium in the gas is sufficient, assuming a weight ratio of gaseous products to daf coal of 10 (Fig. 12). Use of 0.5 wt % potassium might be acceptable in cases of lower sulfur coals. The conductivity varies approximately as the square root of the seeding level. From the standpoint of securing adequate conductivity in the gas working medium, a seed level of 0.5–1% is sufficient.

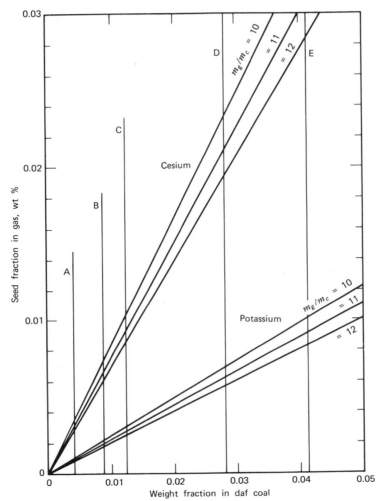

Figure 12. Seeding level needed to combine with sulfur in the fuel. A, Utah coal; B, OCR char; C, Montana S.B. coal; D, West Pennsylvania bituminous coal; and E, Illinois no. 6 coal. m_g/m_c = mass of gas/mass of daf coal.

For low sulfur coals or low sulfur chars, it may be possible to use cesium seeding. In that case 0.5 wt % cesium may be sufficient because, as indicated above, the lower ionization potential of cesium leads to significantly better conductivity in the seeded gases. It might be thought that the economics of cesium seeding would be very unfavorable; however, when good gas clean-up is applied with about 99.8% particulate capture, and when pollucite ore is used as make-up material, the operating cost increment due to cesium seeding is no higher than with potassium.

Air Preheaters. Air preheaters are classified as follows: (*1*) regenerative, or storage-type, heaters (direct or indirect-fired); (*2*) recuperators (direct and indirect-fired radiant type, and direct and indirect-fired convective type).

Direct firing uses the heat of the MHD exhaust stream directly. Indirect firing uses a separate, generally clean, fuel and combustion system to supply heat.

Indirect-fired heaters are considered useful chiefly for test facilities and interim installations until the direct-fired heaters can be fully developed. Use of indirect firing in practical MHD power plants tends to render the plant noncompetitive.

Direct-fired regenerative, or storage-type, heaters are receiving a great deal of attention at the present time (28–30). The well-known cowper stoves used in the steel industry are a familiar example. Cycling times of 10–15 min are usually considered appropriate. Use of at least three units helps to reduce transitional temperature excursions during change-over. For large plants with very high air flows, use of several units in parallel may be desirable in order to reduce pressure drop. Alumina or magnesia are usually considered as bed materials. Preheats to 1644 K with direct fired regenerators are anticipated. An example of an MHD plant utilizing regenerative air heaters is given in ref. 30.

If somewhat lower air preheat temperatures (1200–1400 K) are sufficient, radiant recuperators may be used. The air heating pipes are generally arranged in a cylindrical array surrounding a large open space (Fig. 13). The hot combustion products flow through the central region and radiate heat to the air tubes. The tubes must be made of a superior heat resistant steel of good creep rupture strength at high temperature and superior oxidation and corrosion resistance. Tube life can be improved by local atmospheric control and protection. In the radiant recuperator, flow of air and hot

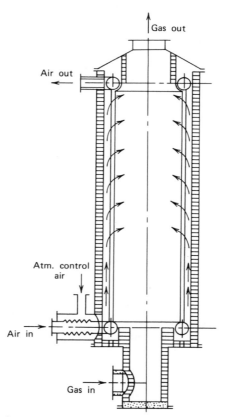

Figure 13.   Schematic diagram of high temperature radiant recuperator.

gas is usually parallel; ie, the hot air leaves adjacent to the hot gas discharge end. This helps prevent excessive metal temperatures at the hot air outlet ends of the heating tubes. Examples of recuperative radiant heaters are given in refs. 31–33.

Direct-fired convective recuperators can be used for the low temperature section of the air heater. All particulate materials are by this time in solid form and are not expected to cause serious fouling problems. Deposits on tubes can be periodically removed by soot-blowing or shot-blasting techniques. Design of convective air heaters follows established engineering procedures. Direct-fired convective heaters might be used in the MHD system for air temperatures below 1000 K. The air could go from there to a radiant type heater. It is often found advantageous to dispense with convective air heating and do all preheating of the air in a radiant recuperator. Air preheaters proposed by some investigators are combinations of several types of heater (30).

The Combustion Process. Although a clean fuel could be made from coal by gasification or solvation refining, coal or char is burned directly. If the combustion process can be carried out with maximum rejection of the ash, the problems in the MHD duct and heat exchange system will be considerably simplified. The goal might be at least 90% ash-rejection, leading to 10% carry-over. A discussion of combustion techniques for MHD plants is given in ref. 34.

A process well-suited to coal combustion with ash rejection as slag is that used in the cyclone furnace (35–36). The wall is protected primarily by a layer of frozen slag, which also reduces the heat flux. The vortex flow in the chamber is also conducive to efficient combustion, since continuous ignition is maintained. Strong radial accelerations promote separation of particles and slag droplets and tend to bring the unreacted oxidant (cool air) into contact with coal particles on the chamber wall. There would normally be considerable ash vaporization in a high-temperature vortex coal combustor but this can be minimized by use of a two-stage configuration.

In the two-stage cyclone combustor, one may either run the first stage as a gasifier with about 0.6 stoichiometric air ($\phi = 0.6$) and admit the balance of the air in the second stage (37), or one might operate with excess air in the first stage and inject additional fuel in the second stage (38). Since the slag is tapped off in the first stage, the fuel supplied to the second stage should be largely ash-free. The first approach, that of the gasifier–combustor (Fig. 14), has advantages of directness and simplicity. The second approach can be accomplished by use of a coal carbonizer, or devolatilizer (Fig. 15). The fuel gas from this unit would be burned in the second stage and the char residue in the first stage (34).

The gasifier (first-stage) component bears some resemblance to the Ruhrgas vortex slagging gasifier (39). A principal difference is operation at elevated pressure (400–700 kPa or 4–7 atm). Another difference is that the air is preheated to 1300–1400 K. Both factors are expected to have a favorable effect on the gasification process. The second-stage component may receive the secondary air near the throat section that leads from the first stage. The walls of the second stage are essentially nonslagging, and might be built with magnesia or zirconia arch-brick elements designed to run close to the flame temperature.

Sizing of the first stage of the two-stage combustor can be based on a thermal input of 3 MW/m^2 of internal surface per 101 kPa (1 atm) pressure. This value is consistent with loadings used in cyclone furnaces for large steam generators (36) (see Steam). The linear relation of loading to pressure is based on the concept that the burning rate

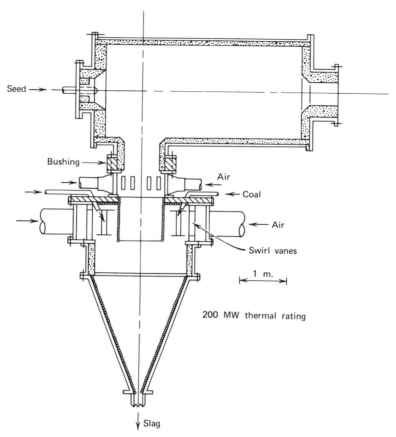

Figure 14.   General arrangement of two-stage vortex gasifier–combustor.

of an individual coal particle is independent of pressure, whereas an increase in pressure permits burning more particles per unit time in a given space. Use of an area rather than a volume basis arises from the fact that in a vortex chamber the coal particles are gasified largely in a layer near the walls. Moreover, area loading is consistent with cyclone furnace experience.

The idea of the two-stage cyclone coal combustor for MHD systems was first introduced in Westinghouse engineering studies in 1963–1965 (32,40–41). Both fuel-rich and excess-air first-stage concepts were envisaged.

Experimental work is currently underway at the Pittsburgh Energy Research Center (DOE) on a two-stage coal combustor with a vortex type first stage. This is a small-scale unit of 5 MW thermal rating (42). Another proposed installation (33), currently in the design stage, is a 190 MW (thermal) plant which would have a 20 t/h two-stage cyclone combustor of the general type illustrated in Figure 14. A single-stage vertical coal burning vortex chamber was built and tested by Tager in the Union of Soviet Socialist Republics. It was rated at 26 MW/m² at 101 kPa (1 atm) (43).

Other approaches might be considered for coal burning in the MHD power system, such as a rocket-type chamber with fuel and hot air injected by powerful jets into the burning space. Chambers of this type using oxygen rather than air have been built and

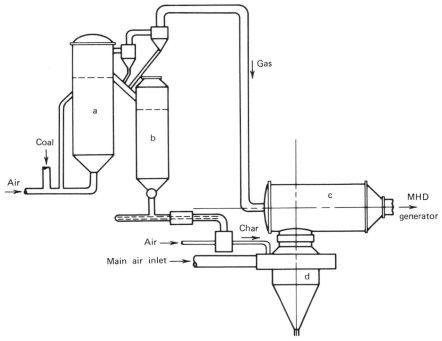

Figure 15. General arrangement of two-stage vortex combustor with excess air in first stage. a, carbonizer; b, char hopper; c, combustor second stage ($\phi \approx 0.9$); d, combustor first stage ($\phi \approx 1.5$).

operated with 100% of the slag passing downstream to the MHD generator (44). Some technical groups also plan to investigate such turbulent reaction chambers with air firing and slag separation.

MHD Power System Analysis. The MHD generator must not only perform in a satisfactory manner but also must be integrated into the complete power plant system. The bottom plant must be matched to the MHD topping unit. Adequate provision must be made for exhaust-heat utilization, pollution control, and seed recovery. For example, a coal-fired MHD generator could have potassium seeding, a direct-fired air preheater, and no oxygen enrichment. The coal chosen for the example is a Western Pennsylvania bituminous, the characteristics of which are given in Table 2. Combustion is carried out at equivalence ratio $\phi = 0.90$ to minimize NO formation, and secondary combustion takes place further downstream to bring ϕ to 1.025 (see Burner technology).

The general arrangement of the plant is shown in Figure 16. The exhaust gases go into the steam generator after leaving the air preheater. Note that a secondary, small bottom plant, Plant R, has been introduced after the steam boiler. The purpose of this is to absorb heat from the exhaust stream to the maximum extent, while permitting the steam power plant to use a full complement of regenerative feed water heaters.

The choice of the plant parameters often requires several trials. In Figure 16 the design is based on an assumed MHD generator output of 500 MW (dc power). The generator length is <18 m, and the maximum Hall voltage <3800 V/m. Assumed data are as follows: fuel higher heating value (daf coal), $\theta_c = 36.3$ MJ/kg (15,620 Btu/lb);

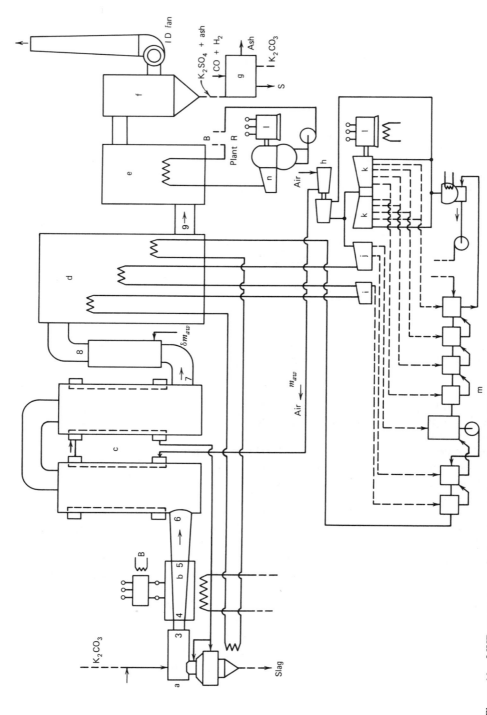

Figure 16. MHD–steam power plant arrangement. a, combustor; b, MHD generator; c, air preheaters; d, steam generator; e, NH_3 boiler; f, bag house; g, seed process plant; h, air compressor; i, high pressure turbine; j, intermediate pressure turbine; k, low pressure turbine; l, a-c generator; m, feed-water heaters; and n, NH_3 turbine [24.1 MPa (3500 psi); turbine steam inlet, 538°C; reheat temperature, 538°C].

equivalence ratio in combustor, $\phi = 0.900$; equivalence ratio after secondary air, $\phi = 1.025$; combustor pressure, $p_3 = 535$ kPa (5.3 atm); compressor outlet pressure, $p_1 = 586$ kPa (5.8 atm); combustor temperature, $T_3 = 2700$ K; combustor specific enthalpy, $h_3 = 646$ kJ/kg (278 Btu/lb); combustor heat loss (fraction of θ_c), $\lambda_1 = 0.035$; generator heat loss (fraction of P_{MHD}), $\lambda_2 = 0.10$; potassium seeding ratio, by weight, $r = 0.007$; generator inlet velocity, $u_4 = 750$ m/s; kinetic energy decrement factor, $C = 0.1$; average channel friction factor, $f = 0.006$; compressor efficiency, $\eta_c = 0.90$; diffuser pressure recovery ratio, $\eta_D = 0.75$ (in this analysis, the definition of η_D is based on an ideal inlet dynamic pressure related to the average velocity at the diffuser inlet, rather than the center line value); generator coefficient, $K = 0.81$; inverter efficiency, $\eta_i = 0.985$; Turbogenerator efficiency, $\eta_g = 0.985$; magnetic field at inlet, $B = 6.25$ T; max allowed Hall field, $E_{x\ max} = 3800$ V/m; and assumed moisture ratio in air (wt), $\delta = 0.01$.

The calculated air preheat temperature in this case is 1316 K. The air preheating can be carried out in a direct-fired radiant type heater that embodies design features ensuring adequate protection of tube surfaces from seed and ash constituents. A small transpiration flow of air around and between the rows of tubes is used for this purpose (31,33).

Ratios of several flow quantities can be calculated: moist air to daf coal, $m_{aw}/m_c = 10.80$; dry K_2CO_3 seed to daf coal, $m_s/m_c = 0.15$; water in coal to daf coal, $m_{wc}/m_c = 0.03$; ash in coal to daf coal, $m_{ash}/m_c = 0.086$; coal as fired to daf coal, $m_{cf}/m_c = 1.12$; ash carry-over to daf coal, $\delta m_{ash}/m_c = 0.0086$; secondary air to moist air supplied, $\delta m_{aw}/m_{aw} = 0.139$; gas flow in generator to daf coal, $m_g/m_c = 11.97$; gas flow after sec. air to daf coal, $m_g^*/m_c = 13.47$; and ratio of potassium to gas flow m_g, $m_K/m_g = 0.007$.

After preparation of tabulated values for the composition and properties of the combustion products at various temperatures and pressures, one can calculate the gas states and electrical parameters through the MHD generator. The generator process ends at $p_5 = 81$ kPa (0.80 atm), this value being chosen to assure a diffuser exit pressure, p_6, close to 101 kPa (1 atm). The channel length is less than 18 m.

The exit kinetic energy (181 kJ/kg or 77.9 Btu/lb) combined with the enthalpy h_5 yields the outlet total specific enthalpy (h_6) of -278 kJ/kg (-120 Btu/lb). The generated power is given by:

$$P_{MHD} = m_g \frac{(h_3 - h_6)}{1 + \lambda_2} \tag{51}$$

The enthalpy drop, $h_3 - h_6$, is 924 kJ/kg (398 Btu/lb) and for $P_{MHD} = 500 \times 10^6$ W, $m_g = 595.08$ kg/s. All the flow quantities of interest are: gas flow in generator, $m_g = 595$ kg/s; daf coal, $m_c = 49.7$ kg/s; moist air sent to combustor, $m_{aw} = 536.5$ kg/s; secondary air to complete combustion, $\delta m_{aw} = 74.5$ kg/s; gas flow after second air injection, $m_g^* = 669.6$ kg/s; water introduced with coal, $m_{wc} = 1.49$ kg/s; K_2CO_3 seed, $m_s = 7.36$ kg/s; ash in coal fired, $m_{ash} = 4.26$ kg/s; coal as fired, $m_{cf} = 55.4$ kg/s; carried over slag, $\delta m_{ash} = 0.426$ kg/s; and total potassium in gas stream, $m_K = 4.17$ kg/s.

At the exit of the diffuser the pressure becomes 101.7 kPa (1.004 atm) instead of 108.6 kPa (1.072 atm); the temperature is 2238 K and the entropy is 9.2 kJ/(kg·K) [4.0 Btu/(lb · C)]. Enthalpy is -278 kJ/kg (-120 Btu/lb).

From the mass flow, density, and velocity, the inlet and outlet areas can be calculated:

$$A_4 = 1.560 \text{ m}^2 \quad A_5 = 7.253 \text{ m}^2 \brace$$
$$d_4 = 1.249 \text{ m} \quad d_5 = 2.693 \text{ m} \brace \tag{52}$$

The transverse dimensions, d, given above are for square cross sections.

For the compressor, moist air at 298 K and 101 kPa (1 atm) is compressed to 590 kPa (5.8 atm) with a compressor efficiency of 0.90. This efficiency is commensurate with values reported for axial flow compressors for large gas turbines (45). The compressor outlet state becomes $T_1 = 511.5$ K, at $p_1 = 588$ kPa (5.8 atm). Compressor work is 217.6 kJ/kg (93.8 Btu/lb), or $P_c = 116.7$ MW (compressor power).

The air preheater must heat 536.53 kg/s of moist air from 511.5 K to 1316 K; this requires 913 kJ/kg (393 Btu/lb) air, or $Q_a = 490$ MW total heat input.

On the hot gas side of the air heater, the temperature drops from the diffuser outlet value $T_6 = 2238$ K to the value T_7. The gas stream enthalpy at the heater exit is given by:

$$h_7 = h_6 - \frac{Q_a}{m_g} = -1.101 \text{ MJ/kg} \tag{53}$$

The corresponding temperature is $T_7 = 1670$ K.

At this point it is appropriate to introduce the secondary air. (A portion of it would previously have been introduced in the heater for atmospheric protection of the tubes.) Gas m_g at $\phi = 0.90$ and state 7, mixes with air δm_{aw} at assumed temperature 400 K and enthalpy 105 kJ/kg (45.2 Btu/lb) to give the gas flow after the second injection of air m_g^* at $\phi = 1.025$ and enthalpy h_8; $h_8 = -967$ kJ/kg (-416 Btu/lb) and $T_8 = 1820$ K. The temperature increases by combustion of the CO.

The exhaust gas stream, m_g^*, now gives up heat, Q_{s1}, to the steam boiler. This heat is given by:

$$Q_{s1} = m_g^*(h_8 - h_9) \, 0.995 \tag{54}$$

where enthalpy h_9 corresponds to an assumed temperature leaving the steam generator of 600 K. Thus $h_9 = -2588$ kJ/kg (-1114 Btu/lb). The factor 0.995 assumes a 0.5% loss from the boiler to the surroundings:

$$h_8 - h_9 = 1621 \text{ kJ/kg} \tag{55}$$

$$Q_{s1} = 1080 \text{ MW} \tag{56}$$

Other sources of heat for the steam are Q_{s2} from the combustor wall cooling and Q_{s3} from the generator duct walls. (The diffuser (ceramic walls) is assumed to be adiabatic.) The heat Q_{s2} involves the usable part of the 0.035 $\theta_c m_c$ combustor heat loss. This is assumed to be 0.03 $\theta_c m_c$ or 54 MW. (The portion 0.005 $\theta_c m_c$ is wasted as slag heat loss, after recovering a portion of the heat from the slag.) Note that the fuel heating value, θ_c, is 36.3 MJ/kg (15,620 Btu/lb) of daf coal. The heat Q_{s3} is 0.1 P_{MHD} or 50 MW. The total heat given to the steam is:

$$Q_s = Q_{s1} + Q_{s2} + Q_{s3} = 1184 \text{ MW} \tag{57}$$

The secondary bottom plant, using ammonia or an organic working fluid, has been mentioned. Its heat input Q_R consists of several parts Q_{R1} recovered from inverter cooling, Q_{R2} recovered from the hydrogen coolers of the turbo-generator, and Q_{R3} from the exhaust gas stream, m_g^*, as its temperature falls from 600 K to the stack temperature of 400 K. This latter point is called state 10. Then $h_{10} = -2805$ kJ/kg (-1209 Btu/lb) and $h_9 - h_{10} = 217$ kJ/kg (93.4 Btu/lb), and $Q_{R3} = m_g^*(h_9 - h_{10}) = 145.3$ MW. The lost heat (inverter) is $Q_{R1} = 7.5$ MW, and the hydrogen coolers yield $Q_{R2} = 6.3$ MW. Thus $Q_R = Q_{R1} + Q_{R2} + Q_{R3} = 159.1$ MW.

With an efficiency of conversion of Q_R to electric power of 28% in the plant R the power from the secondary bottom plant is:

$$P_R = \eta_R Q_R = 0.28 \times 159.1 = 44.55 \text{ MW} \tag{58}$$

The power produced in the steam bottom plant depends on the turbine efficiency η_t of that plant. This is the ratio of turbine shaft power to heat input to the steam, before deduction of auxiliary power requirements. When $\eta_t = 0.46$, a realistic value, the turbine shaft power is:

$$P_T = \eta_t Q_s = 544.6 \text{ MW} \tag{59}$$

From this the compressor power (P_c) is subtracted and the difference multiplied by the generator efficiency (η_g) to yield the turbo-generator electrical power P_g:

$$P_g = \eta_g(P_T - P_c) = 421.5 \text{ MW} \tag{60}$$

The power output from the MHD generator is $\eta_i P_{MHD}$, and the total plant electrical power output becomes:

$$P_E = P_g + \eta_i P_{MHD} + P_R = 958.5 \text{ MW} \tag{61}$$

Auxiliary power requirements are estimated at 30.5 MW. After this value is subtracted from P_E:

$$P_{E \text{ net}} = P_E - P_{aux} = 928.1 \text{ MW} \tag{62}$$

The energy input with the coal Q_{th} is:

$$Q_{th} = \theta_c m_c = 1804 \text{ MW} \tag{63}$$

On this basis the plant efficiency becomes:

$$\eta = \frac{P_{E \text{ net}}}{Q} = 0.5144 \tag{64}$$

There is, however, one further correction. The seed is recovered, in part, as K_2SO_4. This must be converted to K_2CO_3. The steps and processes involved for this conversion involve the application of known techniques. This matter has been discussed by Bergman (27) and others. A reasonable estimate is that the seed recovery process leads to a degradation of the plant efficiency of ca 1.1% based on 2.8% S in coal. Subtracting the 1.1%, the plant efficiency is:

$$\eta_{net} = 0.503 \tag{65}$$

It is possible to realize efficiencies over 50% with a maximum temperature of 2700 K. Analyses made with Montana Rosebud Coal give results similar to these.

Inversion to AC Power. The inversion from d-c power to a-c power for the multielectrode Faraday generator requires, in general, as many inverters as there are electrode pairs. Thus for an electrode pitch of 4 cm and a channel length of 12 m there would be 300 inverter circuits, each of which would need to float at its particular level of Hall voltage.

The inverter circuits present no unusual complications. The inverters could be built integrally with the transformers, each inverter supplying power to a portion of the primary winding. Proposed designs are discussed in references 32 and 46. Special measures may have to be taken to ensure an approximately sinusoidal wave form. The experience on the Russian U-25 generator has been instructive; in that installation great importance is being attached to the electrical interface with the Moscow power grids. Engineering work on inverter circuitry at the Component Development and Integration Facility (CDIF) in Montana is reported in reference 47.

One of the major advances in the area of dc to ac inversion for MHD has been the concept, recently proposed by Rosa, for load consolidation (48–49). Essentially, his idea is to introduce d-c emf between adjacent electrodes in such a way as to bring leads from two or more adjacent electrodes to the same potential. The power source for the emf would come from a power take-off from the main a-c output circuit, the power in the feedback lines being inductively coupled to rectifiers that produce the d-c incremental voltages. Thus power is circulated in the electrical output circuitry but without dissipation. The circulated power per electrode pair is $I \cdot \Delta V_H$ where I is the electrode current and ΔV_H is the difference of Hall voltage between adjacent electrodes. The load consolidation concept can be applied to many adjacent electrodes. The number of inverter circuits could be reduced by a factor of ten by this means.

Load consolidation can effect great simplification of the MHD plant electrical system and reduce costs. The relative attractiveness of segmented Faraday generators compared to diagonally-conducting wall (DCW) generators is increased. It should be pointed out, however, that the consolidation technique can also be applied profitably to DCW generators (49).

Chemical Regeneration. Chemical regeneration, as applied in MHD systems, is the process of recovering energy from the hot exhaust system by storage of chemical enthalpy in fuel components. This was first proposed by Carrasse (50) and was also discussed by Sheindlin, Shumyatsky, and co-workers (51). Char-burning systems involving chemical regeneration are discussed in ref. 52.

The types of reactions that might be considered are the following:

$$CH_4 + Q \rightarrow C + 2\,H_2 \tag{66}$$

$$CH_4 + H_2O + Q \rightarrow 3\,H_2 + CO \tag{67}$$

$$C + CO_2 + Q \rightarrow 2\,CO \tag{68}$$

$$C + H_2O + Q \rightarrow CO + H_2 \tag{69}$$

The third and fourth reactions represent a simplified formulation of the reaction of coal and the exhaust stream gases. The coal can be rapidly devolatilized with evolution of vaporized fuel constituents, and the resulting char can react with CO_2 or H_2O to yield CO and H_2.

The rapid devolatilization of coal by hot exhaust gases from the MHD generator is discussed in ref. 53. The gaseous fuel so produced could be cooled, cleaned, compressed, and reheated for firing in the main MHD combustion chamber. Alternatively, it could be fired at atmospheric pressure in a separately fired air preheater, which would thus be free of slag and seed.

The use of chemical regeneration makes possible a gain of three or more efficiency points in the MHD system without increase of air preheat temperature or oxygen enrichment. It is therefore viewed as one of the principal avenues for further improvement of the MHD combined cycle.

Control of Effluents and Environmental Effects.   The MHD combined cycle plant, which may reach efficiencies over 50%, is attractive in that it has low thermal discharge. This reduces heat evolution at the power plant site and also reduces cost of the heat dissipation equipment. Table 3 indicates the amount of heat evolved from several types of power plants of 1000 MW electrical output.

The discharge of particulate materials is a second important environmental effect. In the coal-fired MHD plant there are arrangements in which not more than 10% of the ash is carried downstream. This is typically ca 0.1 wt % of the gas flow. The seed material, on the other hand, when converted to K_2SO_4 amounts to about 1.4% of the gas flow. Total particulates are thus ca 1.5% of the flow. By use of a good bag-house or electric precipitation cleaning system, 99.7% of this particulate material can be captured, leaving discharged particulates of only 0.004% or 40 ppm in the discharged gas flow. The particulate discharge might be even less with superior filtration. Good particulate recovery is essential to minimize cost of seed make-up material. The stack discharge in this example is less than 0.013 kg/GJ (0.03 lb/10^6 Btu) (see Air pollution control methods).

By carrying out the fuel-rich combustion at ϕ values less than 0.95 and allowing a sufficiently slow cooling rate at ca 1800 K, the nitric oxide concentration can also be brought to a satisfactorily low value. When the secondary air δm_{aw} is introduced to complete the combustion, no further increase of NO concentration takes place, because the temperature is so low (see Burner technology).

Reactions of principal concern during the cooling process are the following:

$$O + N_2 \rightleftharpoons NO + N \qquad k_1 = 2 \times 10^{-11} \text{ cm}^3/\text{s} \tag{70}$$

$$N + O_2 \rightleftharpoons NO + O \qquad k_2 = 1.1 \times 10^{-14} \, T \exp{(-6.25/RT)} \text{ cm}^3/\text{s} \tag{71}$$

$$N + OH \rightleftharpoons NO + H \qquad k_3 = 6.8 \times 10^{-11} \text{ cm}^3/\text{s} \tag{72}$$

Equations 70 and 71 are the classical Zeldovich reactions (54), and equation 72 is of particular importance in fuel-rich mixtures. A good discussion and review of the entire NO problem in MHD generation has been given by Pepper (14). Attention is directed also to an earlier study by Pepper and Kruger (55).

The rate of formation of NO, $\dot{n}_{NO}$, can be expressed in terms of the rate constants for the above reactions and the several equilibrium concentrations:

$$\dot{n}_{NO} \approx 2 \, (1 - \alpha^2) \, \frac{R_1}{1 + \alpha K_1} \tag{73}$$

Table 3.   Heat Evolution from Selected Power Plants of 1000 MW Rating

Plant type	Efficiency	Thermal rating, MW	Heat rejected, MW
conventional steam	0.395	2667	1615
PWR nuclear	0.32	3125	2125
near term MHD	0.50	2000	1000
advanced MHD	0.56	1770	770

where $\alpha = [NO]/[NO]_e$; $K_1 = R_1/(R_2 + R_3)$; $R_1 = k_1[N]_e[NO]_e$; $R_2 = k_2[N]_e[O_2]_e$; $R_3 = k_3[N]_e[OH]_e$; $[\quad]_e$ = equilibrium partial pressure. When approximately two seconds are allowed for cooling to 1800 K, as might be the case with high temperature air pre-heaters of large volume, and with a fuel-rich combustion at about $\phi = 0.90$, the final NO concentration can be held to a value below 300 ppm. Some estimates as low as 200–240 ppm have been given (56). Nitric oxide emission is customarily expressed in terms of thermal input to the power plant. For the MHD–steam plant, a stack concentration of 300 ppm leads to 0.099 kg NO/GJ (0.23 lb NO/10^6 Btu), less than one third the EPA emission standard of 1977.

Removal of sulfur dioxide from the power plant exhaust is accomplished in the MHD system by the action of the alkali seed material (see Selection of Parameters). Since the potassium (or cesium) has a strong tendency to form K_2SO_4 when SO_2 and adequate oxygen are present, the sulfur can to a large extent be bound up as sulfate. The gas clean-up process, which involves use of a bag-house or electrostatic precipitation, removes the sulfate from the system.

The seeding level needed to effect a stoichiometric match with the sulfur present in the coal is indicated in Figure 12. The possibility of moderate sulfur oxide emissions can also be considered. Let x be the number of kilograms of sulfur per kilogram of daf coal, and y be the permitted sulfur discharge in kg/kg daf coal. Then $x - y$ is the amount of sulfur that must be removed. For example, if $x = 0.028$ and $y = 0.008$, then $x - y = 0.020$ kg sulfur per kg daf coal, and this must be removed from the system. Figure 12 shows that 0.5% potassium seeding should be sufficient. However, considerations of electrical conductivity would lead to use of a seeding level closer to 0.7%.

The K_2SO_4 formed, as the sulfur is captured by the alkali seed, is processed to remove sulfur as potassium returns to the system. The various steps have been discussed in detail by Bienstock and Bergman (26–27). The seed–ash mixture caught in the gas cleaner is first washed to remove the K_2SO_4 and other alkali salts in aqueous solution. The ash residue can be reintroduced in the first stage of the combustion chamber to salvage the small amount of potassium it may contain. Alternatively, this ash residue may be discarded, or marketed as a soil conditioner. The K_2SO_4 solution next goes to an evaporator or crystallizer where solid K_2SO_4 is formed. At this point the stream may be split, the portion $x - y$ being retained for further processing and the portion y being returned to the combustion chamber. Thus usually only a fraction of the K_2SO_4 needs to be processed. The processing required is reaction at elevated temperature with synthesis gas, $CO + H_2$, or some alternative mixture of CO nd H_2 to form H_2S, CO_2 and H_2O. The H_2S can then be isolated and conveyed to a Claus reactor for production of elemental sulfur (see Sulfur recovery). Bergman discusses the processes and assesses their thermodynamic and economic impact (27). Consideration of the several small but significant thermodynamic penalties leads to the conclusion that a bituminous coal of 3% sulfur content entails an efficiency penalty of 1.2 percentage points; a low sulfur western coal of 1.1% S would have a penalty of only 0.2 percentage points. These values are based on permitted SO_2 discharge of 0.516 kg/GJ (1.2 lb/10^6 Btu) thermal input. Figure 17 from Bergman summarizes the efficiency penalty. The quantity y referred to above has a value 0.0093 kg S/kg daf coal when the permitted discharge is 516 MJ/kg, and the heating value is 36.31 MJ/kg. The latter figure corresponds to 27.5 kg daf coal/GJ (64 lb/10^6 Btu).

In the example of the MHD/steam power plant discussed above, the fuel supply

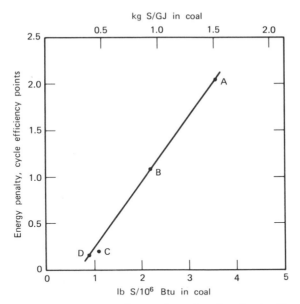

Figure 17.   Efficiency penalty due to seed processing as function of coal sulfur content. Coal A, Illinois no. 6 sub-bituminous; B, Pittsburgh high volatile bituminous; C, Montana sub-bituminous; D, North Dakota lignite (27).

rate was 49.7 kg/s of daf coal. The value of x for the coal was 0.028. For the assumed permissible sulfur oxide emission and $y = 0.0093$, $x - y = 0.0187$ kg S/kg daf coal. Thus 0.929 kg/s sulfur must be removed, and 5.07 kg/s K_2SO_4 must be processed with the synthesis gas.

Generator Construction Features.   The duct or channel of an MHD generator for long-term use has four main functions: to provide a safe and secure means of containing and carrying the hot plasma stream to provide for current collection at the electrodes and transmission of these currents to the external load circuits; to provide for adequate internal and external electrical insulation; and to ensure good durability of all internal parts in contact with the hot plasma stream. Generators for short burst operation may be designed on the heat-sink principle, which leads to a configuration quite different from that of long duration channels with cooled walls. These requirements should be met with minimal electrical and thermodynamic losses. This article describes some of the general approaches used to attain the desired objectives, and indicate sources of more detailed information.

The internal construction of a generator includes electrodes, insulators, insulating walls and cooling means; and the external casing construction provides the main structural support and gas-tight envelope. A good discussion of the internal generator construction problems has been given by Rosa (57).

Three general approaches to internal generator construction have been used or proposed: (1) electrode bar segments and insulating walls, with connection as in Figure 6**b**; (2) electrode bar segments and diagonal bar wall elements to form a diagonally conducting wall with electrical connections as in Figure 6**d**; and (3) the DCW generator with window frame construction. Each type of construction requires an array of water cooled bars or blocks that are electrically insulated from each other and carry either the facing material selected for contact with the hot plasma or a layer of liquid slag formed from the ash in the coal.

Consider first arrangement (1) with segmented electrodes and insulating side walls. The electrodes would usually be installed with one water cooled bar per electrode segment. The bars would be insulated from each other and from ground, and water connections must also be insulated. Since condensed water containing dissolved seed is a good conductor, it is important that any regions where electrical leakage may occur be made to operate at a temperature high enough to prevent moisture condensation.

Electrode surfaces may be designed to run cool (600–900 K) or hot (over 1900 K). Electrodes at intermediate or semi-hot temperature (1500–1700 K) have also been investigated. The electrode face temperature is controlled by using either an extended cross section to increase the thermal conductance path, or ceramic caps, or facings on the electrode bars. With hot electrodes the ceramic cap which is 5–8 mm thick becomes in effect a ceramic electrode. Hot electrodes of a noble metal (Pt or Pt/Rh) have also been proposed.

Cool or semi-hot electrodes, when used with coal combustion products, would be expected to have a coating of slag—the slag in contact with a cold electrode, frozen, and the outer slag layer molten and continually flowing downstream. An equilibrium thickness would be established for which the rate of run-off is balanced by the rate of deposition of new slag on the surface. A number of studies have been made of these electrode slag layers (58–59). Insulation must be provided between the adjacent electrodes. Mica, alumina or boron nitride strips have been used for this purpose.

Two electrode constructions are shown in Figure 18. These constructions were investigated at the Avco-Everett Research Laboratories, Inc. The platinum-sheathed copper proved to be superior to the T-head Inconel electrode.

Although cool electrodes offer some advantages in material durability, they are not without problems. In general, due to local Hall effects, current tends to concentrate at the upstream edges of the anodes and the downstream edges of the cathodes. This causes likely sites for arc initiation. Cool electrodes, generally, are more susceptible to arcing problems because of the cool boundary layer. Current passage near the electrode faces is only possible by way of local small arcs, or arc spots, which may cause damage or may trigger interelectrode breakdown. Fortunately the slag layer helps to protect the electrode from arc damage.

If the interelectrode insulation is also designed to operate at a low temperature, the alumina or BN strips need be only 1.5–2.0 mm thick. With such thin insulating spacers it has been found that the interelectrode voltage must be held to a low value of 40–50 V. Since the Hall field in the generator may be 30 to 40 V/cm this requires relatively narrow electrodes with a cool electrode construction. With an electrode pitch of 12 mm, eg, there may be one thousand electrodes in a channel 12 m long. The problems of construction, insulation and water cooling are therefore enormous.

With cool electrodes, current from or to the plasma must pass through a cool gas boundary layer, across the gas–slag interface, through the slag layer (liquid and frozen) and from the slag into the electrode metal. In the cool gas layer and the slag layer there is danger of current instabilities, ie, current concentration leads to excessive local joule heating, which reduces the local resistivity and further increases the current concentration. Breakdown may result.

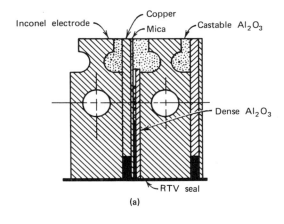

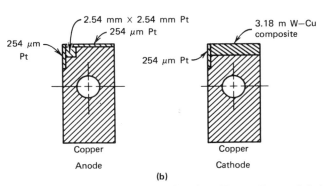

Figure 18. Cool electrode construction as used at Avco-Everett Research Laboratories in (**a**) 100 hour test, and (**b**) electrode design based on 250-h power generation test. Courtesy of Avco-Everett Research Laboratories (60).

Cool electrodes have two other disadvantages. One is the excessively high potential drop, which may be 250–300 V for the combined anode and cathode drop. Another is the high thermal loss. The thermal flux may run over 2 MW/m², and this can comprise a sizeable fraction of the generated power. Although this heat loss can be recaptured for the bottom steam plant (at lower efficiency of utilization), it diminishes the electrical conductivity of regions of the flowing plasma.

Hot electrodes avoid some of the disadvantages of cool electrodes. Much work on hot electrodes has been done in the Union of Soviet Socialist Republics (61) and in the U.S.–U.S.S.R. cooperative MHD program (62). For hot electrodes either an insulating ceramic holder may be used, with conducting ceramic insets, or ceramic electrodes can be mounted on an array of water cooled bars of the kind described above. The latter construction is favored at this time. A possible configuration is shown in

Figure 19. The electrodes would be 5–8 mm thick, to provide a surface temperature over 1800°C. The interposed insulators of MgO or possibly MgAl$_2$O$_4$ would also run with high surface temperature. They could be of 12–14 mm thickness. Note, from the figure, that the electrode elements help to retain the insulator elements. The second arrangement shown in Figure 19 uses platinum electrode face caps with a ceramic barrier that can be designed to cause the Pt or Pt–Rh to run at ca 1700°C.

The insulator elements may be of a width approximately equal to the width of the electrode faces in the flow direction. Such a 1:1 ratio is close to the optimum, as has been pointed out by Dzung (63). Dzung also calculates the penalty on effective plasma resistivity of insufficiently fine segmentation. This arises from the current concentrations mentioned above, which are associated with local Hall effects. Good design would dictate a pitch to channel width ratio of not over 0.02, or if some penalties are acceptable, not over 0.04.

The problems with hot electrodes involve possible loss of material by vaporization, interelectrode current leakage, thermal stress cracking, transmission of the sizeable heat flux, and collection of the electric current. Materials such as ZrO$_2$ stabilized with about 12% Y$_2$O$_3$ are suitable for operation around 1800–1900°C. When used in subdivided short segments (as opposed to a relatively long electrode bar) the problems

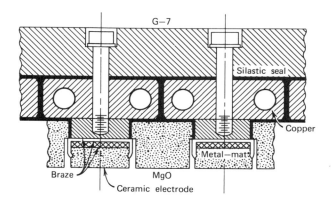

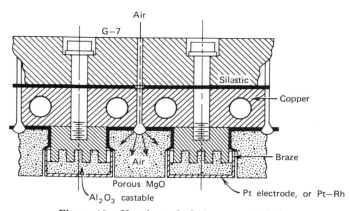

Figure 19.   Hot electrode design concepts (33).

of thermal distortion and stresses due to the heat flux can be circumvented. Platinum leads or other metallic collector means can be brazed to the backs of the electrode blocks. There are other promising hot electrode materials based on various forms of lanthanum chromite. The first of these to achieve prominence was $La_{0.84}Sr_{0.16}CrO_3$. A more recent, and superior material is an aluminum-substituted form of lanthanum magnesium chromite: $La_{0.95}Mg_{0.05}Cr_{0.5}Al_{0.5}O_3$ (64).

Zirconia–yttria electrodes conduct current largely by ion (O^{2-}) transfer rather than by electrons. This tends to deplete the oxygen in the material and cause structural weakening. It may be necessary to supply oxygen to the rear faces of the cathodes, which are the most vulnerable. Zirconia–ceria electrodes with more than 50% ceria are better electronic conductors but have poor durability at high temperatures. Electrodes based on lanthanium chromite, on the other hand, have good electronic conductivity, and the conductivity at low temperatures (where the connection to the electric lead wire must be made) is much better than that of zirconia (65).

Figure 20 shows a qualitative picture of the potential distribution across the channel as it might be observed by appropriately designed voltage probes. The potential drop, δV_a or δV_c, is the difference between the ideal electrode potential V_a^* or V_c^* and the actual value at the surface. The potential drop is due in part to the fact that in the region of low velocity near the electrode wall there is no induced emf; hence there is a gas layer that presents a simple resistive barrier to current flow. This barrier is augmented by the concentrations of current lines near the electrode edges. Another factor is that the relatively low temperature of the boundary layer increases its resistivity. Further, there will be polarization drops very close to the electrode surfaces. These drops are associated with the transport of charges to or from the surfaces. At the cathodes, the drop is larger because of the necessity for electron emission. Use of hot electrodes reduces the combined voltage drop, $\delta V_a + \delta V_c$, to less than 50 V.

Several investigators have looked for ways to minimize the current concentrations near the electrode edges. One approach is to use triangular electrodes that can be de-

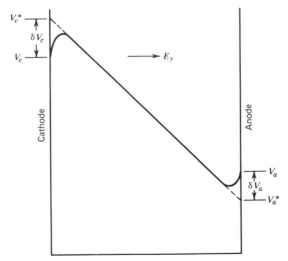

Figure 20.   Schematic of potential distribution across MHD channel, indicating potential drops δV_a, δV_c.

signed to present a local potential distribution at the electrode face just matching that in the plasma; ie, the potential gradient along the face matches the Hall field, E_x. This design is discussed by Rosa (66). A similar approach, using semiconductor electrodes, proposed by Croitoru (67). These modifications to the segmented electrode design could help avoid some of the erosion problems near the electrode edges; the adverse effect on plasma effective conductance is not improved because of additional resistive losses within the electrodes themselves. The current concentration effects can be reduced to some degree by designing the electrodes so that the anodes collect more current near the downstream edges and the cathodes collect more current near the upstream edges. Designed temperature patterns can be used for this purpose.

There are other important differences in phenomena at anodes and cathodes. At the cathodes the alkali ions are driven toward the surface. Electrochemical problems are likely to produce corrosion of the surface material. The combination of high temperature, chemical reaction, and transmission of electric current forms a particularly severe environment. The anodes are less aggressively attacked, but arcing is potentially more damaging. Owing to the fact that any arc between adjacent electrodes will have current flow in the upstream direction. Therefore on the anode wall the $\bar{j} \times \bar{B}$ force on the arc current will drive it directly into the wall, melting the wall material.

The insulating side wall often has a peg wall design, using water cooled metal blocks, all insulated from each other and with insulated water connections. Because of the many thousands of water cooled elements that must be used, the construction is very complex and chances of water leaks and electrical breakdown are considerable. One variation is to apply ceramic capping to the faces of the pegs or blocks by flame spraying or vapor deposition. Slag also deposits over the faces of the side wall blocks if they are designed to run at cool temperature.

The large heat losses from such water cooled side wall blocks can be reduced by use of hot side walls. The water cooled individual metal blocks can be capped by ceramic blocks 12–15 mm thick, held in place by brazing or other means. In this retention means one must assure that there is a satisfactory transmission of the heat flux to the water cooled metal. With the hot ceramic side wall construction, the supporting water cooled blocks would be individually covered with silicone rubber elastomer to enhance their electrical isolation. Figure 21 shows a possible hot side wall construction. Electrical leakage through the magnesia blocks is trivial. Although absorbed seed compounds reduce electrical resistivity (68), this reduction can be accommodated in the

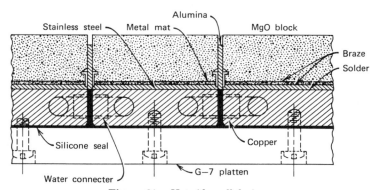

Figure 21.   Hot side wall design.

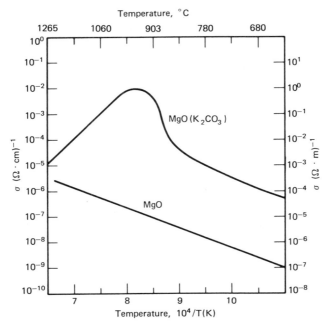

Figure 22. Electrical conductivity of MgO wall material in presence of K_2CO_3 seed (68). MgO > 97%; 12–14% porosity.

design in such a way as to achieve satisfactory performance. Figure 22 indicates the resistivity reduction. The design and use of hot side walls and hot electrodes is still actively under development in MHD programs, and proposed solutions to the problems they involve still remain to be verified in long-duration generator tests on a meaningful scale.

It may be advantageous to introduce gas infiltration on the insulating side walls for atmospheric protection. A very slight flow of warm air may be admitted between adjacent ceramic wall blocks. This prevents entry of seed and ash constituents in the slots and provides a low-conductivity gas layer which discourages both j_x and j_y leakage currents along the face of the wall.

With a hot wall construction, similar to that shown in Figure 19, Hall fields up to 4000 V/m should be achievable. The potential difference between adjacent electrodes may be made as high as 80–100 V, if insulator widths are 15–20 mm (69). Electrode pitch may be 3–4 cm, and the total number of electrodes is far less than in a cool electrode system.

Use of diagonal conducting bars on the side walls has been mentioned. These would be set at an angle, α, to the horizontal so as to conform to the equipotential lines:

$$\alpha = \tan (E_y/E_x) = \tan K/[\beta(1 - K] \tag{74}$$

Another requirement, for ideal operation of the DCW system is that current continuity be preserved. This means that j_ybs for any electrode be the same as the j_ybs for the diagonally opposite electrode to which it is connected, where j_y is local transverse current density, b is electrode length, and s is electrode pitch. Usually j_y,

b, and s depend on x, but with some measure of design choice included. In any case, attainment of the ideal geometrical and physical conditions in the DCW generator is extremely difficult, and as a result the ideal condition $j_x = 0$ is rarely achieved. At best, it could be achieved only at one design point condition. Even with such shortcomings there is considerable interest in DCW generators because of the simplification of the load circuits, as indicated schematically in Figure 6d.

In the DCW generator with diagonal connecting bars in contact with the plasma, construction of the bars and their interposed insulation follows lines very similar to that used for the electrode walls.

In the DCW generator with window frame construction, the electrodes and diagonal bars are made up in a single frame structure, with internal water passages in all four sides of the frame, and with means of screwing adjacent frames together with insulating screws. Such MHD passages have been used by Dicks (5) and have been discussed also by Shanklin (6). Brogan (7) has recently completed several designs of window frame DCW generators for clean fuel use (no slag). Most DCW duct designs have been of the cool electrode type. The frames are quite thin, on the order of 10 mm. Moreover, they are usually of nonuniform thickness (tapered) and may be of trapezoidal shape. The construction is thus very complicated.

One advantage of the DCW window frame generator, other than simplification of load circuits, is that the assembly of the frames comprises the structural body of the MHD duct, and no other casing structure is necessary unless one wants a secondary external envelope for extra safety and security.

The casing structure may be used for MHD generators other than the window frame DCW type. These generators require a strong, gas-tight casing that serves to support the internal structure with electrical isolation of the parts. The casing must be constructed to prevent any current leakage in the axial direction. The upstream end of the casing will run at full Hall potential of 20–30 KV. The casing can be made of a glass-reinforced elastomer, though the transverse flange design is difficult. Alternatively, the casing can be fabricated of segmental metal sections, each about 1 m long, mutually insulated by silicone rubber gaskets. Each segment would consist of four panels (two electrode wall panels and two sidewall panels). The inner surfaces of the casing segments are lined with silicone rubber sheet, and the internal electrode and wall parts are mounted on water cooled elements attached in turn to glass–mica or glass–elastomer platens. The casing segments are bolted together with insulated bolts. The importance of insulating these transverse flange joints cannot be overemphasized. For minimal space requirements the ducts should be of nearly square cross section.

The platens mentioned above can contribute considerably to the simplicity of the design. A number of fiber glass plates about 30×30 cm, on which would be mounted 36 water-cooled metal blocks each about 5×5 cm, can be used for the side wall. These blocks would all be insulated from each other by silicone rubber sealant and would have interconnected water cooling passages. Each water-cooled metal block would carry its own ceramic (MgO) facing block to comprise the hot wall. In this way, the designer has to cope with only one pair of external water connections for each 30×30 cm platen, rather than with one for each 5×5 cm block. Arrays of 25 or 30 electrode bars can also be assembled in platen construction with great simplification of the water cooling circuits. When maintenance work is done on the generator, the technician should replace an entire platen. The platen assemblies are fabricated in advance by quantity production methods and are available for fast overhaul and maintenance operations.

The Magnetic Field.   The magnetic field for large MHD generators is normally provided by a superconducting (SC) magnet. To maintain the superconducting state in the winding, the temperature should be held close to that of liquid helium, ca 4 K. Slightly higher temperatures could be used but the gas of next highest boiling point is hydrogen and the temperature of boiling hydrogen, 21 K, is too high (see Superconducting materials).

The winding consists of a composite material of superconductor such as Nb–Ti, Nb_3Sn, and copper. The copper serves to stabilize the conductor, providing adequate heat absorption for the joule heating that results from a local excursion into the normal conducting state. The winding coils are then installed in rigidly restrained layers on a cylindrical mandrel. Magnets of various forms are considered for MHD generators as indicated in Figure 23. Simple solenoid coils can be used; these are often considered feasible for small MHD generators because of their simplicity of construction. A second type is the race track configuration, in which two flat, oval coils are placed on either side of the duct to provide the transverse field. The third type of SC magnet for MHD application is the saddle coil type, in which two saddle-shaped coils are placed along the channel, with the longitudinal portion of the conductors lying generally just externally to the electrode walls. The cylindrical mandrel that supports the coils also supports the clamping loads that must be applied by a system of yokes and tie rods to restrain the windings and prevent the conductors from flying outwardly under the action of the $\bar{j} \times \bar{B}$ forces. These holding forces become very large, and may be of the order of 1400 t/m.

The entire magnet structural assembly must be bathed in liquid helium and contained in a Dewar vessel with vacuum layer thermal barriers. With suitable design and precautions against inward heat leakage, the refrigeration power requirements for the SC magnet system can be kept to a very low level of a few hundred kilowatts even for a generator of hundreds of megawatts rating.

Superconducting magnet systems for MHD generators have been investigated and applied on a moderate scale in the United States and abroad. A Japanese instal-

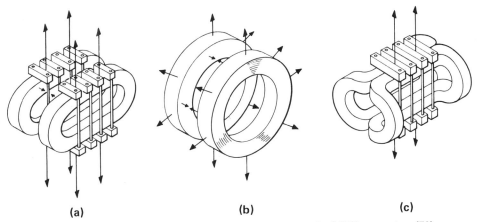

          (a)   (b)   (c)

Figure 23.   Various forms of superconducting magnets for MHD generators (70).

lation is described in ref. 71. The Argonne National Laboratories have designed and built a number of magnets, one of which was sent to the Union of Soviet Socialist Republics in 1976 (70, 72). Other noteworthy magnet studies are discussed in references 87 and 88.

It is important in the generator design to provide for a casing of minimal external dimensions so that the inner warm bore of the magnet can be made as small as possible. Channels with square outlet are preferable to channels of rectangular outlet section for this reason. The warm bore is then very slightly larger than the external diagonal dimension of the outlet of the MHD channel. Since the channel is generally tapered, the warm bore tube may also be of conical form. This construction is in keeping with the desirability, which frequently arises, of making the B-field at the generator inlet higher than at the outlet on account of Hall field limitations. In other cases, a nearly uniform field is appropriate, and in that case, a cylindrical warm bore tube may be used, which is a structural simplification. The additional clearance space between casing and warm bore tube thereby provided can be used to good advantage to accommodate coolant lines, electrical power leads, instrument lines, etc.

Magnetic field levels generally considered are of the order of 6 to 7 T (60,000 to 70,000 gauss) at the inlet for large central station generators, 5 to 6 T for smaller pilot plants, and ca 4 T for smaller machines designed to run at partially subatmospheric pressure.

Economic Aspects

A power plant can be evaluated economically in terms of capital cost in \$/kW or in terms of energy cost, mills (10^{-3} \$) per kilowatt hour. Such evaluations do not necessarily form the sole basis for evaluation of a plant. For example, an unusually high efficiency is of value in terms of national energy conservation and low thermal discharge, entirely apart from monetary power cost (see Power generation).

It must also be kept in mind that estimates of absolute equipment or energy costs are of little significance unless they can be related to costs of alternative methods of electrical energy production. Uniform cost analyses and criteria of acceptability are employed.

One problem that arises in such analyses is that escalation of costs goes on steadily, and it becomes especially difficult to isolate costs in time. This is further complicated by differences in completion time for different plants.

These questions and many others were dealt with in the Energy Conversion Alternatives Study undertaken under guidance of NASA, NSF, and ERDA in 1974–1975 (73). Although this study has limitations, it generated some important comparative results. The MDH–steam combined cycle plant compared favorably with other systems such as advanced coal-fired steam plants, coal-fired gas turbine combined cycles, fuel-cell systems, metal-vapor cycles, and others. The MHD position is further enhanced if one gives extra credit for the unusually high efficiency it might deliver.

Other economic evaluations and comparisons include that made at Westinghouse, under sponsorship of the Office of Coal Research, in 1965. An 800 MW MHD–steam combined plant was compared with a similarly analyzed conventional steam plant. At the much lower 1965 costs, the conventional plant was estimated at \$98.75/kW; the coal-fired MHD–steam plant was estimated to lie in the range \$94.94–112.15/kW; and a char-fired MHD–steam plant was estimated as \$88.40–102.26/kW.

In a cost comparison by Pepper (14) in 1974, particular emphasis was put on the economic aspects of $(NO)_x$ clean-up. Pepper's comparison was \$245/kW for the MHD–steam plant (2000 MW thermal) and \$192/kW for the conventional plant. The MHD plant had efficiency 51.8% and the conventional plant 39.15%.

It is possible to make a current cost comparison of an MHD–steam combination plant and a conventional steam power plant, both 928 MW, by starting with common estimates for the equipment common to both plants and estimating the costs of additions, deletions and modifications arising when the MHD topping plant is added. Figures supplied by the firm Burns and Roe, Inc. have been adjusted for a 1977 version 600 MW steam power plant to match the 928 MW plant. In accord with data in ref. 74 there is about a 4% decrease in the \$/kW for each 100 MW increase in plant capacity, the rule applying for plants over 400 MW.

The MHD–steam plant considered is essentially like that reviewed above, with a calculated station heat rate of 7151 kJ/kW·h (6785 Btu/kW·h). The conventional plant is assumed to have a station heat rate of 9224 kJ/kW·h (8751 Btu/kW·h). Results of the cost estimates are given in Tables 4 and 5. The steam-bottom plant in the MHD system has about 545 MW turbine shaft power, 117 MW of which is required to drive the compressor. In the contingency figure, 10% was used for MHD items and 5% for conventional plant items.

The close agreement of capital cost figures in the comparison is striking. This is somewhat accidental because of uncertainties in cost estimates of large components like the magnet or the air heater. Note, however, that even a 50% error in the magnet cost estimate would affect the overall cost by less than 3%.

Note that the values given in Tables 4 and 5 do not include escalation allowance. However, this would presumably be about the same for the two plants, since estimated construction times are similar.

The estimation of plant capital cost is only one part of total energy cost. One is interested in the cost of the electrical energy in \$/kW·h. This may be expressed as:

$$C = (C_c + C_f + C_{om}) \times 1000 \tag{75}$$

Table 4. MHD System Cost Estimates

500 MW MHD generator	Millions of dollars
compressor	6.00
MHD duct	3.75
combustor	4.00
magnet	20.00
air heater	16.00
inverters	15.00
bag house	10.00
seed processing	16.00
diffuser	1.00
secondary air system	1.00
MHD auxiliaries	2.00
foundations	8.00
instruments	8.00
piping	4.00
Total	*114.75*

Table 5.   Comparative Costs 928 MW Plants (Millions of Dollars)

Item	MHD–steam	Conventional steam
land	2.24	2.96
structure and improvements	25.15	35.10
boiler house and steam gen.	118.75	207.90
turbo-generator systems	38.84	60.66
switch gear and structures	18.83	18.83
services, furnishings, fixtures	1.81	2.38
station equipment	3.67	3.67
Total equipment cost, nonMHD	*209.29*	*331.50*
44 MW Rankine bottom plant	6.60	
Total of MHD items (Table 6)	*114.75*	
Total direct costs	*330.64*	*331.50*
contingency	22.26	16.55
fee	16.53	16.55
engineering and overhead	33.06	33.10
Total	*402.49*	*397.70*
interest (10%)	40.25	39.72
Total capital cost	*442.74*	*437.42*
$/kW	*477.09*	*470.83*
station heat rate, kJ/kW·h (Btu/kW·h)	7151 (6785)	9224 (8751)

where C_c is the capital component of energy cost in mills/kW·h, C_f is the fuel component of energy cost, and C_{om} is the operation and maintenance component. The annual interest and fixed charge rate is r. F is the utilization factor (kW·h generated as a fraction of kW·h generated in 8760 h at rated station capacity), and C_c is expressed as:

$$C_c = \frac{1000\, r\ (\$/\text{kW})}{8760\, F} \tag{76}$$

If $r = 0.14$ and $F = 0.8$ this expression gives:

$$C_c = 0.02 \times (\$/\text{kW}) \tag{77}$$

The factor C_f is formulated as:

$$C_f = c_f\ (\text{HR}) \times 10^{-5} \tag{78}$$

where c_f is fuel cost in ¢/GJ. For example for 7151 kJ/kW·h (6785 Btu/kW·h) heat rate (HR) and c_f 57¢/GJ (60¢/10^6 Btu), $C_f = 4.07$ mills/kW·h, and for 9224 kJ/kW·h (8751 Btu/kW·h), the heat rate $C_f = 5.25$ mills/kW·h.

The term C_{om} is taken as 0.9 mills/kW·h for the conventional plant and 1.0 mills/kW·h for the MHD combination, because of an extra 0.1 mill allowance for seed make-up (a conservative estimate). The energy costs are summarized in Table 6.

Note that there is a saving of slightly under 1 mill/kW·h in energy cost for the MHD combination. For more expensive fuel the cost benefit for the MHD system becomes more pronounced.

The small cost benefit for the MHD system is really of minor significance compared to two other major advantages: (*1*) the very appreciable reduction in station

Table 6. Energy Cost Comparison for 57¢/GJ (60¢/10⁶ Btu) Fuel in 928 MW Plants, mills/kW·h (1 mill = 10^{-3} U.S. Dollar)

	MHD–steam plant	Conventional steam
capital component, C_c	9.54	9.42
fuel component, C_f	4.07	5.25
operation and maintenance, C_{om}	1.00	0.90
overall energy cost, C	14.61	15.57

heat rate which greatly aids fuel conservation, and (2) the possibility afforded to burn coal at very low pollution levels.

Alternative Arrangements of the MHD Power Plant

Disk Generator. The disk generator has been investigated experimentally and theoretically by Louis (75), Klepeis (76), and others. The basic idea involves a radially symmetrical configuration, in which flow takes place radially between nearly flat disk-shaped walls. The magnetic field is in the axial direction (Fig. 24). It is generally also considered advisable to impart a swirl component to the velocity. Laboratory experiments with shock-tube driven disk generators confirm in principle the soundness of the concept (76).

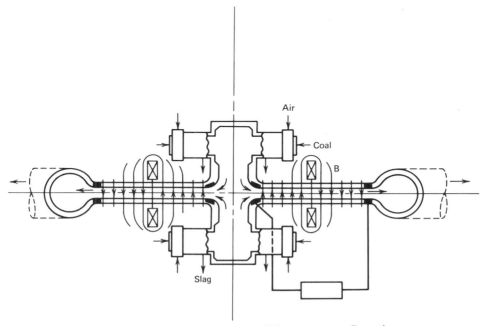

Figure 24. General arrangement of disk generator configuration.

Theoretical studies indicate the conditions of operation for disk generators applied for practical power generation. Generally, the flow is supersonic and the generator functions as an impulse device, ie, kinetic energy is converted to electrical energy.

The chief merit of the disk generator is the simplicity of the electrode system. The disk walls are primarily insulating walls and ring electrodes are used only at the inlet and outlet. When there is no azimuthal flow (no swirl) the generator becomes essentially a Hall machine, with azimuthal currents in closed loops in the plasma. When swirl is present, the azimuthal currents are still present, but the generator functions more as a diagonally connected unit. The absence of an internal electrode structure permits the use of higher electric fields in the radial direction (corresponding to Hall field in a linear generator); this in turn allows use of higher magnetic fields in the lower pressure areas. As a result, the power density may be moderately higher in a disk generator.

Another cited advantage of the disk generator is the possibility of using a Helmholz coil-, or solenoid-type magnet. Such a magnet coil provides its own restraining forces by hoop tension, and the superconducting magnet design is considerably simplified.

Offsetting the above advantages is the disadvantage that the disk generator presents a much larger ratio of surface to volume than does the linear generator. The area-to-volume ratio for a 500 MW linear generator is about 2 m^{-1}; for a corresponding disk generator it might be 10 m^{-1}. This leads to a large relative increase in frictional dissipation and heat loss. It follows that generally competitive performance with a disk generator can only be achieved in units of very large capacity. Whereas a satisfactory and competitive linear generator of 100 MW could be designed, this would hardly be possible with a disk generator; in fact, for the latter, competitive performance requires sizes of about 800 MW or larger (77). In spite of the above mentioned limitations, interest continues in the disk generator and several projects are being supported.

Coal-Plex Processes. The Coal-plex concept has been suggested by Squires (78) as a means of more effectively using our coal resources. The coal is processed in a carbonizer, where volatile materials are driven off and the coal is partly pyrolyzed to form a relatively clean gaseous fuel. This off-gas can be used either as a clean fuel or as feedstock for various chemical processes. The nonvolatile char residue can be used as fuel for power generation (Fig. 25) (see Fuels, synthetic—gaseous).

This procedure is particularly attractive for MHD power generation because the char is a superior fuel for MHD because of its low H–C ratio. It may also turn out that much of the sulfur will be removed from the char in the pyrolysis step. In that case cesium seeding may be used with a further marked increase in electrical conductivity and related benefits.

Use of char as fuel makes possible considerable simplification of the MHD plant, permitting preheat temperatures as low as 1100–1150 K and attainment of efficiencies several percentage points higher than in coal-fired MHD plants. The subject is discussed in more detail in refs. 34 and 79. The pyrolysis gas provides a valuable and economical source of hydrocarbon fuel components.

Gas Turbine Bottom Plants. The MHD generator can be used effectively in conjunction with a gas turbine or air turbine plant. Several versions may be considered (Fig. 26).

The first arrangement, Figure 26a, involves use of a hot air turbine. Air is com-

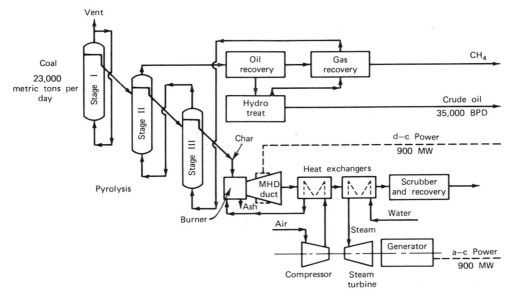

Figure 25. Coal-plex combination plant for production of synthetic fuel and generation of power by MHD (38). BPD = barrels per day.

pressed (with intercooler), then heated by an exhaust gas heat exchanger to an appropriate turbine inlet temperature of 1350 K. Pressure at this point is ca 3.6 MPa (36 atm). The air is allowed to expand through the air turbine, exhausting at a pressure of 600 kPa (6 atm). Further heating takes place in a low pressure heater, until 1350 K is again reached. At this point the air enters the combustion chamber where coal is burned, and the process from this point on is that of a typical coal-fired MHD generator. At the exit of the diffuser (downstream from the generator) the hot gas enters the high pressure and low pressure air heaters referred to above. Finally, the residue of thermal energy may be used for steam generation in a steam power plant. In this arrangement there is a three-plant combination, involving air turbine, MHD generator, and steam plant. An efficiency of 53–54% is attainable even though preheat temperature does not exceed 1350 K.

The second arrangement (Fig. 26**b**) uses the air turbine as part of the bottom plant system. The air turbine is simply an open Brayton cycle, with the air heated by the exhaust stream and expanding in the turbine. The low pressure air stream is directed into the main exhaust stream. A steam-bottom plant is also used to further recover heat from the exhaust stream. In a recent paper (80) this type of system was combined with a bottoming plant using Freon R-11.

A third gas turbine–MHD combination, shown in Figure 26**c**, involves the usual coal-fired MHD system with the exhaust gas going first to a high temperature air preheater, and then to the heater of a closed cycle gas turbine. The closed cycle gas turbine can be operated at super atmospheric pressure, perhaps 3 MPa (30 atm). The components thus become very small and compact.

Figure 26**c** indicates a dual bottom plant using waste heat from the gas turbine loop. This dual bottoming plant involves a small steam plant plus a small Rankine

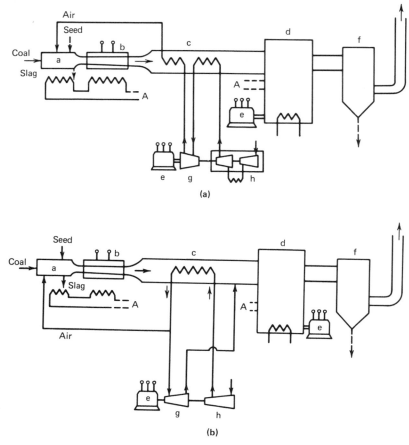

Figure 26.   Application of gas turbines with MHD systems. (**a**) Hot air turbine discharging to MHD combustor. (**b**) Air turbine–steam turbine combination as bottom plant. a, combustor; b, MHD generator; c, air heaters; d, steam plant; e, a-c generator; f, gas cleaner; g, air turbine; h, air compressor. (**c**) Closed-cycle (helium) turbine with steam plant combination as bottom plant. Diagrams simplified to show principal features. a, combustor; b, MHD generator; c, air heaters; d, He heaters; e, He recuperator; f, NH_3 boiler; g, He low temperature heater; h, gas cleaner; i, air compressor; j, He turbines; k, He compressors; l, a-c generators; m, steam boiler; n, NH_3 power plant; p, steam power plant; q, He cooler.

cycle ammonia plant. Though there are many components in this installation, all except the MHD generator employ established technologies. The efficiency of the combination was calculated to be 56% (81). Without the ammonia plant the efficiency drops to 52.5%.

Status and Assessment of MHD Power Generation

Development programs underway in the United States, Union of Soviet Socialist Republics, Poland, Canada, and Japan aim toward commercial power generation by MHD. These investigations are concerned with resolution of various problems and with implementation of plans for pilot plant construction. In fact, in the Moscow In-

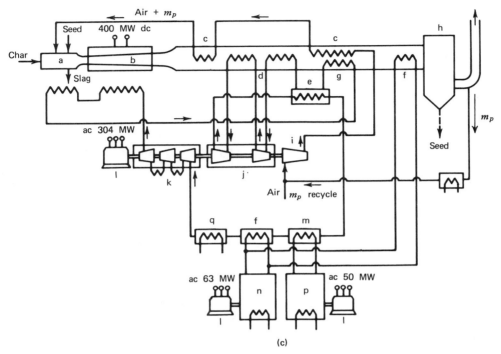

Figure 26.   (*continued*).

stitute of High Temperatures a plant of pilot-scale proportions, the U-25, has been operated in a number of different versions over the past five years. The generator has attained a level of 20 MW electrical output, and was operated continuously for 10 d at a lower load (82).

The main problem of MHD has to do with the design of electrode and wall structures for long duration of operation with coal combustion products. This problem is partly one of identifying the best materials, and partly one of establishing a durable mechanical design. A good discussion is given in ref. 57. Considerable attention has been given to electrode designs for metal temperatures below 1000 K. More difficult problems occur when one wishes to operate with higher surface temperatures in order to reduce electrical and thermal losses. Material studies are underway at Westinghouse (83), the NBS, and at MIT (84). Hot electrode investigations have been made on the U-02 machine in Moscow (61), and also on that same installation under the guidance of the joint U.S.–U.S.S.R. cooperative MHD effort (62). Hot electrode studies were underway at Westinghouse till 1977. It is expected that the several studies in the United States and Union of Soviet Socialist Republics on hot electrodes will yield acceptable designs within a period of about 2–3 years.

A major generator and components test facility is being constructed in Butte, Montana. The CDIF generator will be of about 50 MW rating. A second installation called the Engineering Development Facility (EDF) is essentially a pilot plant with a thermal rating of 250 MW. The Montana installations and the objectives of the program are discussed in a paper by Jackson (85).

Progress in MHD channel development is being made at the Avco Everett Laboratories. A generator of 200–300 kW output was operated for durations of approximately 100 h (86). Some electrode damage was experienced but steps were indicated to remedy the problems. Slag has been introduced to simulate coal firing. Recent runs have extended over 250 hours.

Cold wall generators under actual coal firing conditions have been run at the University of Tennessee Space Institute. This is the first installation where coal products have actually been sent through the generator.

Extensive small-scale studies have been made at Stanford University, focusing on electrode phenomena, voltage drops, interelectrode breakdown and slag layers.

Other components for the MHD system are also being investigated in the United States. Regenerative heaters are under study at FluiDyne and coal combustion chambers at the Pittsburgh Energy Research Center of the U.S. Department of Energy. Using innovative construction means, heat exchangers of direct-fired recuperator type can be made of metals for air temperatures to 1400 K. Direct-fired regenerative heaters to 1644 K are being constructed under a current U.S. Department of Energy program.

A large 5-T (50,000 gauss), 25-t superconducting magnet was recently completed at the Argonne National Laboratories under the U.S.–U.S.S.R. cooperative MHD program. It was shipped to Moscow for test in the U25-B where it has been installed and preliminary tests have been made. There appears to be no reason why a magnet of 5–6 T, suitable for a generator of pilot plant proportions (6–7 m long, 40–60 MW generator output), could not be designed and built at the present time. Experience with such magnets will serve to guide design and construction of the larger magnets that will be needed for commercial base-load power plants.

The matter of pollution control and seed recovery can be handled in the MHD system by application of known technologies. However, operating experience, and experience in integrating the seed system with the rest of the MHD–steam plant is badly needed.

The tasks of power inversion, arrangement of load circuits, integration with the a-c station output, and handling part load and transient conditions call for application of the best engineering technology, plus actual operating experience. Here again, experience with the Montana EDF and other pilot plant facilities will be invaluable; it may finally resolve the question of use of the diagonally connected generator vs the segmented Faraday multielectrode system. Recent developments in load consolidation systems are of interest in this connection (48–49).

Nomenclature

A	flow area
B	flux density, magnetic field, magnetic induction
C	capital component
c	cost
d	transverse dimension
daf	dry, ash-free
DCW	diagonally conducting wall
e	electron charge
E	transverse electric field
E_x	Hall field
f	friction factor

F	utilization factor
FC	fixed carbon
$F(v)$	speed distribution function
h	specific enthalpy
HHV	higher heating value
j	current density
k	Boltzmann constant, 1.38×10^{-23} J/K
K	generator coefficient
L	active duct length
m	mass flow rate
MHD	magnetohydrodynamics
n	density, number per volume
$\dot{n}$	rate of formation
p	pressure
P'	power per unit volume

Subscripts

a	dry air
add	added
aw	moist air
c	combustible; compressor
cf	coal as fired
f	fuel
fF	fuel mixture, heat of formation
fox	oxidant mixture, heat of formation
g	gas; generator
h	hydraulic
om	operation and maintenance
ox	oxygen
R	plant R
s	seed
t	turbine
T	turbine shaft
wc	water in coal
0	reference

Superscript

$*$	after second injection of air
Q	heat
R	universal gas constant, 8.314 J/(mol K)
r_h	hydraulic radius
u	flow speed
$\bar{u}$	velocity
$\bar{v}$	mean drift velocity
w	molecular weight
α	mole ratio at equilibrium
δ	kg moisture/kg dry air
ρ	density
η	efficiency
θ	angle between vectors
θ_c	heating value
λ	heat loss value
μ	mobility
ν	collision frequency
ξ	interaction parameter
ϕ	equivalence ratio
σ	electrical conductivity

BIBLIOGRAPHY

"Power from Coal by Gasification and Magnetohydrodynamics" in *ECT* 2nd ed., Suppl. Vol., pp. 217–249, by Stewart Way, Westinghouse Electric Corp.

1. M. Faraday, *Experimental Researches in Electricity*, Series I, Quaritch, London, Eng., 1839; *Phil. Trans. Roy. Soc.* 15 (Jan. 12, 1832).
2. B. Karlovitz, *Third Symp. on Engineering Aspects of MHD, Rochester, N.Y., 1962*; U.S. Pat. 2,210,910 (Aug. 13, 1940), B. Karlovitz and D. Halasz.
3. S. Way, *Proc. Eleventh International Congress on Applied Mechanics, Munich, FRG, Aug., 1964*.
4. C. P. Harris and J. D. Cobine, *J. Eng. Power. Ser. A.* **83**, 392 (1961).
5. J. B. Dicks, Jr., and co-workers, *Diagonal Conducting Wall Generators, AFAPL-TR-67-25*, University of Tennessee Space Institute, Tullahoma, Tenn., 1967; Y. C. L. Wu and co-workers, *Proc. 13th Symp. on Engineering Aspects of MHD, Stanford University, Mar., 1973*.
6. R. V. Shanklin, III, *J. Am. Inst. Aeronaut. Astronaut.* **7**, 975 (1969).
7. T. R. Brogan, J. A. Hill, and A. M. Aframe, *Proc. Sixth International Conference on MHD Electrical Power Generations, Washington, D.C., 1975; Open Cycle Generators and Systems, CONF-750601-P1*, Vol. 1, U.S. Dept. of Energy, Washington, D.C., 1975.
8. A. de Montardy, *First International Symp. on MHD Electric Power Generation, Newcastle on Tyne, Eng., 1962*, paper 19.
9. S. T. Demetriades and co-workers, *Twelfth Symp. on Engineering Aspects of MHD, Argonne, Ill., 1972*; E. D. Doss and co-workers, *Fourteenth Symp. on Engineering Aspects of MHD, Tullahoma, Tenn., 1974*.
10. S. Way, *Fourth International Symp. on MHD Electrical Power Generation, Warsaw, Pol., 1968*.
11. S. Way, *Third International Conference on MHD Electrical Power Generation, Salzburg, Austria, 1966*.
12. L. S. Frost, *J. Appl. Phys.* **32**, 2029 (1961).
13. S. Way, *AIChE–I Chem E Symp. Ser. No. 5*, Institute Chemical Engineering, London, Eng., 1965.
14. J. W. Pepper, *Effect of Nitric Oxide Control on MHD-Steam Power Plant Economics and Performance, SU-IPR Report No. 614*, Institute for Plasma Research, Stanford University, Dec. 1974.
15. L. S. Frost and A. V. Phelps, *Phys. Rev.* **127**, 1621 (1962).
16. R. D. Hake and A. V. Phelps, *Phys. Rev.* **158**, 158 (1967).
17. L. Spitzer and R. Härm, *Phys. Rev.* **89**, 977 (1953); L. Spitzer, *Physics of Fully Ionized Gases*, Interscience Publishers, Inc., New York, 1956.
18. R. Rosa, *Magnetohydrodynamic Energy Conversion*, McGraw Hill Book Co., New York, 1968, pp. 18–22.
19. G. Hoover, Masters Thesis, University of Pittsburgh, Pittsburgh, Pa., 1964.
20. P. E. English and T. D. Rantell, *Br. J. Appl. Phys. Ser. 2* **2**, 1215 (1969).
21. K. Wissel, "Beitrag zur Bestimmung der Elektrischen Leitfähigkeit von Verbrennungsplasmen aus Ruhrkohlen in Hinblick auf ihren Einsatz in MHD-Generatoren," Thesis, Aachen, FRG, 1972.
22. I. M. Gaponov and L. P. Poverezhskiy, *Study of the Electrical Conductivity of a Plasma in Combustion Products, ERDA-tr-186*, Academy of Science of the U.S.S.R., Institute of High Temperatures, 1975 (U.S. Dept. of Energy).
23. S. Way, *2nd Symp. on Engineering Aspects of MHD, Philadelphia, Pa., 1961; Engineering Aspects of Magnetohydrodynamics*, Columbia University Press, New York, 1962.
24. S. Way and co-workers, *J. Eng. Power* **87**, 125 (Apr. 1965).
25. *NSRDS-NBS-37, JANAF Thermochemical Tables*, 2nd ed., NBS, Washington, D.C.
26. D. Bienstock and co-workers, *Environmental Aspects of MHD Power Generation*, Intersociety Energy Conversion Engineering Conference, Boston, Mass., Aug. 1971.
27. P. Bergman and co-workers, *16th Symp. on Engineering Aspects of MHD, Pittsburgh, Pa., 1977*.
28. FluiDyne Engineering Corp., *Report on the Air Preheater System for the MHD Component Development Integration Facility*, Contract 31-109-38-3453, U.S. Dept. of Energy, Washington, D.C., July 1976.
29. F. A. Hals and co-workers, *Proc. 15th Symp. on Engineering Aspects of MHD, Philadelphia, Pa., 1975*; F. A. Hals and co-workers, *Proc. 14th Symp. on Engineering Aspects of MHD, Tullahoma, Tenn., 1974*.

30. *Evaluation of Phase 2 Conceptual Designs and Implementation Assessment Resulting From the Energy Conversion Alternatives Study (ECAS), NASA TMX73515,* NASA Lewis Research Center, Washington, D.C., Apr. 1977.

31. C. Keller, *Trans. Am. Soc. Mech. Eng.,* (1949).

32. D. Q. Hoover and co-workers, *Feasibility Study of Coal Burning MHD Generation, Contract 14-01-001-476,* Westinghouse Research Laboratories, Pittsburgh, Pa., 1966.

33. S. Way, "A Proposed 40 MWe MHD Pilot Plant," paper submitted to Intersociety Energy Conversion Engineering Conference, San Diego, Calif., Aug. 1978.

34. S. Way, "Combustion Aspects of MHD Power Generation," *Combustion Technology, Some Modern Developments,* Academic Press, Inc., New York, 1974.

35. H. Seidl, *Proc. Joint Symp. on Combustion, Institution of Mechanical Engineers,* London, Eng., 1955.

36. H. Seidl, *Eleventh Coal Science Lecture, Inst. of Civil Engineers, Publ. Gazette No. 46,* British Coal Utilization Research Assoc., Leatherhead, Eng., Oct. 1962.

37. S. Way, W. E. Young, and T. C. Tsu, *Combustion Systems, Progress Report No. 3, Contract 14-01-001-476,* Office of Coal Research, Westinghouse Research Laboratories, Pittsburgh, Pa., 1965.

38. W. E. Young and co-workers, *Intersociety Energy Conversion Engineering Conference, Am. Soc. of Mech. Eng., Miami Beach, Florida, 1967.*

39. F. Nister, *Coke Gas,* 54 (Feb. 1957).

40. S. Way, W. E. Young, and B. W. Swanson, Westinghouse Research Labs, internal company report, Nov. 5, 1963.

41. U.S. Pat. 3,358,624 (Dec. 19, 1967), S. Way (to Westinghouse).

42. R. Wright, *4th Annual International Conf. on Coal Gasification, Liquifaction and Conversion to Electricity,* Univ. of Pittsburgh, Aug. 1977.

43. S. A. Tager and co-workers, *5th International Conf. on MHD Electrical Power Generation, Munich, FRG, Apr. 1971.*

44. J. B. Dicks and co-workers, *14th Symp. on Engineering Aspects of MHD, Tullahoma, Tenn., Apr. 1974.*

45. V. de Biasi, *Gas Turbine World,* 9 (Nov. 1975).

46. D. T. Beecher, *Contract NAS3-19407 (NASA), Energy Conversion Alternatives Study (ECAS), Westinghouse Phase I Final Report,* Volume VIII, Sec. 9, Westinghouse Research Laboratories, Pittsburgh, Pa., Feb. 12, 1976.

47. *CDIF-MHD Generator System Conceptual Design, Contract 31-109-38-3635,* Systems Research Laboratories, Inc., Dayton Ohio, Feb. 1977, S-e Section II, pp. II-207 to II-214.

48. R. Rosa, *Proc. 15th Symp. on Engineering Aspects of MHD, Philadelphia, Pa., 1976.*

49. A. Lowenstein, *17th Symp. on Engineering Aspects of MHD, Stanford University, Calif., Mar., 1978.*

50. J. Carrasse, *Third International Symp. on MHD Electrical Power Generation, Salzburg, Austria 1966.*

51. A. E. Sheindlin, B. Y. Shumyatsky, and A. G. Sokolsky, *Tenth Symp. on Engineering Aspects of MHD, Cambridge, Mass., 1969.*

52. S. Way, *J. Eng. Power* **93,** 345 (July 1971).

53. S. K. Ubhayakar, R. Gannon, and D. Stickler, *Development Program for MHD Power Generation-Coal Devolatilization, Contract EX-76-C-01-2015 (Energy Research and Development Administration),* Avco-Everett Research Laboratory, Everett, Mass., Apr. 1977.

54. J. Zeldovich, *Acta. Physicochim. URSS* **21,** 577 (1946).

55. J. W. Pepper and C. H. Kruger, *13th Symp. on Engineering Aspects of MHD, Stanford University, Mar. 1973.*

56. Y. Mori and co-workers, *Proc. Sixth International Conference on MHD Electrical Power Generation, Washington, 1975,* (CONF-750601-P1, U.S. Dept. of Energy).

57. R. J. Rosa, *Conf. on High Temperature Sciences Related to Open-Cycle Coal Fired, MHD Systems,* Argonne National Laboratory, Argonne, Ill., Apr. 4–6, 1977.

58. R. J. Rosa, *Fifth International Conference on MHD Electrical Power Generation, Munich, FRG 1971.*

59. M. E. Rodgers, P. C. Ariessohn, and C. H. Kruger, *Proc. 16th Symp. on Engineering Aspects of MHD, Pittsburgh, Pa., 1977.*

60. A. Demirjian and co-workers, *Development Program for MHD Power Generation, Contract EX-76-*

C-01-2015 (Energy Research and Development Administration), quarterly reports Apr. 1976–June 1976; *MHD Generation Component Development, Contract EF-77-01-2519* (U.S. Dept. of Energy), quarterly reports Jan. 1978–Mar. 1978, Avco-Everett Research Laboratory, Inc., Everett, Mass; *Chem Week,* 35 (Feb. 14, 1979).

61. V. A. Kirillin and co-workers, *12th Symp. on Engineering Aspects of Magnetohydrodynamics, Argonne, Ill., 1972;* E. K. Keler, *Third U.S.–U.S.S.R. Colloquium on MHD Electrical Power Generation, Moscow, U.S.S.R., Oct. 1976* (MHD-Division, U.S. Dept. of Energy).

62. W. R. Hosler, ed., *Final Report on Joint U.S.–U.S.S.R. Test of U.S. MHD Electrode Systems in U.S.S.R. U-02 MHD Facility (Phase I),* Energy Research and Development Administration, Washington, D.C., 1975; *Phase II Final Report, Contract E(49-18)2248,* Westinghouse Research Labs., Pittsburgh, Pa., Dec. 1977.

63. L. S. Dzung, *Brown Boveri Rev.* **53,** 238 (1965).

64. H. U. Anderson and co-workers, *Conference on High Temp. Sciences Related to Open Cycle Coal Fired MHD Systems, Argonne National Laboratories, Argonne, Ill., Apr. 4–6, 1977.*

65. D. B. Meadowcroft, *Br. J. Appl. Phys.* **2,** 1225 (1969).

66. R. Rosa, *Magnetohydrodynamic Energy Conversion,* McGraw-Hill Book Co., New York, 1968.

67. Z. Croitoru, *Rev. Gen. Ectr.* **74,** 873 (1965).

68. G. P. Telegin and co-workers, *High Temp. High Pressures* **8,** 199 (1976).

69. J. K. Koester and W. Unkel, *14th Symp. on Engineering Aspects of MHD, Univ. of Tenn. Space Inst. Tullahoma, Tenn, Apr. 1974.*

70. D. B. Montgomery and co-workers, *Proc. Sixth International Conf. on MHD Electrical Power Generation, Washington, D.C., June 1975,* Vol. 4, pp. 115–130.

71. K. Fushimi, *Proc. 13th Symp. on Engineering Aspects of MHD, Stanford Univ, Univ., Mar. 1973.*

72. J. R. Purcell and co-workers, *Proc. Sixth International Conf. on MHD Electrical Power Generation, Washington, D.C., 1975,* pp. 143–154.

73. G. R. Seikel and co-workers, *15th Symp. on Engineering Aspects of MHD, Philadelphia, Pa., May 1976.*

74. H. H. Petersen, *Power Eng.* (4), 48 (1963).

75. J. F. Louis, *J. Am. Inst. Aeronautics Astronautics* **6,** 1674 (1968).

76. J. E. Klepeis and J. F. Louis, *Proc. 14th Symp. on Engineering Aspects of MHD, Tullahoma, Tenn., 1974.*

77. J. E. Klepeis and J. F. Louis, *Proc. First Colloquium according to U.S.–Soviet cooperation in the field of MHD Power, Feb., 1974* (U.S. Dept. of Energy); J. E. Klepeis and co-workers, *Proc. Sixth International Conference on MHD Electrical Power Generation, Washington, D.C., 1975* (U.S. Dept. of Energy, CONF-750601-P1).

78. A. M. Squires, *Science* **169,** 821 (Aug. 28, 1970).

79. S. Way, *North American Fuel Technology Conference, Ottawa, June 1970.*

80. K. D. Annen and R. H. Eustis, *Proc. 16th Symp. on Engineering Aspects of MHD, Pittsburgh, Pa., 1977.*

81. S. Way, *U.S. German Workshop on MHD Systems and Economic Aspects, Essen, FRG, June 1972.*

82. *ERDA-tr-182, Studies at the U-25 Facility (IV), Experimental Studies of the MHD Generator of the U-25 Facility with the ID Channel,* Moscow Institute of High Temperatures, 1976 (U.S. Dept. of Energy).

83. W. E. Young and co-workers, *Development, Testing, and Evaluation of MHD Materials and Component Designs, Annual Report 1976, Contract E (49-18)-2248,* Westinghouse Research Laboratories, Philadelphia, Pa., 1976.

84. J. F. Louis, *Critical Contributions in MHD Power Generators, Work Performed Under Contract EX-76-C-01-2215,* Mass. Inst. of Technology, 1976, quarterly reports, F.E. 2215-1, 2, 3 . . . .

85. W. D. Jackson, *Ninth Fluid and Plasma Dynamics Conference, San Diego, Calif., July 1976,* paper no. 76-309; W. D. Jackson and co-workers, *12th Intersociety Energy Conversion Engineering Conference, Washington, D.C., Aug. 1977.*

86. S. Petty and co-workers, *Proc. 15th Symp. on Engineering Aspects of MHD, Philadelphia, Pa., 1976.*

General References

R. Rosa, *Magnetohydrodynamic Energy Conversion,* McGraw-Hill Book Co., New York, 1968.

Y. W. Sutton and A. Sherman, *Engineering Magnetohydrodynamics,* McGraw-Hill Book Co., New York, 1965.

J. B. Heywood and G. J. Womack, *Open Cycle MHD Power Generation,* Pergamon Press, London, Eng., 1969.
R. Bünde and co-workers, *MHD Power Generation,* Springer, Berlin, Ger., 1975.

STEWART WAY
MHD Consultant

COAL GAS. See Coal conversion Processes; Tar and pitch.

COAL TAR. See Coal conversion processes; Tar and pitch.

COATED FABRICS

A coated fabric is a construction that combines the beneficial properties of a textile and a polymer. The textile (fabric) provides tensile strength, tear strength, and elongation control. The coating is chosen to provide protection against the environment in the intended use. A polyurethane might be chosen to protect against abrasion or a polychloroprene (Neoprene) to protect against oil (see Urethane polymers; Elastomers, synthetic).

Textile Component

The vast majority of textiles that are used for coating are purchased by the coating company. Several large textile manufacturers have divisions that specialize in industrial fabrics. These companies perform extensive development on substrates and can provide advice on choosing the correct substrate, hand samples, pilot yardage, and ultimately production requirements. A listing of suppliers of industrial fabrics can be found in *Davison's Textile Blue Book* (1).

Fiber. For many years cotton (qv) and wool (qv) were used as primary textile components, contributing the properties of strength, elongation control, and esthetics. Although the modern coated fabrics industry began by coating wool to make boots, cotton has been used more extensively. Cotton constructions, including sheetings, drills, sateens and knits, command a major share of the market. Cotton is easily dyed, absorbs moisture, withstands high temperature without damage, and is stronger wet than dry. This latter property renders cotton washable; it can also be drycleaned because of its resistance to solvents (see Drycleaning).

Polyester, by itself and in combination with cotton, is used extensively in coated fabrics. Polyester produces fibers that are smooth, crisp, and resilient. Since moisture does not penetrate polyester, it does not affect the size or shape of the fiber. Polyester resists chemical and biological attack. Because of its thermoplastic nature, the heat required for adhesion to this smooth fiber can also create shrinkage during coating (see Polyester fibers; Fibers, man-made and synthetic).

Nylon is the strongest of the commonly-used fibers. Since it is both elastic and resilient, articles made with nylon will return to their original shape. There is a degree of thermoplasticity so that articles can be shaped and then heated to retain that shape. Nylon fibers are smooth, very nonabsorbent, and will not soil easily. Nylon resists chemical and biological action. Nylon substrates are used in places where very high strength is required. Lightweight knits and taffetas are thinly coated with polyurethane or poly(vinyl chloride) and used extensively in apparel. In coating, PVC does not adhere well to nylon (see Polyamides).

Rayon and glass fibers are the least used because of their poor qualities. Rayon's strength approaches that of cotton but its smoother fibers make adhesion more difficult. Rayon has a tendency to shrink more than cotton, which makes processing more difficult. Glass fibers offer very low elongation, very high strength, and have a tendency to break under compression. Therefore, glass fabric is only used where support with low stretch is required and where the object is not likely to be flexed. For instance, glass might be used to support a lead-filled vinyl compound for sound dampening (see Glass; Insulation, acoustic).

Textile Construction.   There are many choices in textile construction. The original, and still the most commonly used, is the woven fabric. Woven fabrics have three basic constructions: the plain weave, the satin weave, and the twill weave. The plain weave is by far the strongest because it has the tightest interlacing of fibers; it is used most often (2). Twill weaving produces distinct surface appearances and is used for styling effects. Because it is the weakest of the wovens, satin weave is used principally for styling. Woven nylon or heavy cotton are used for tarpaulin substrates. For shoe uppers, and other applications where strength is important, woven cotton fabrics are used.

Knitted fabrics are used where moderate strength and considerable elongation are required. Where cotton yarn formerly dominated the knit market, it has recently been replaced by polyester–cotton yarn, and polyester yarn and filament. Where high elongation is required, nylon is used. Knits are predominantly circular jersey; however, patterned knits are becoming more and more prevalent. When a polymeric coating is put on a knit fabric, the stretch properties are somewhat less than that of the fabric. Stretch and set properties are important for upholstering and forming. The main use of knit fabrics is in apparel, automotive and furniture upholstery, shoe liners, boot shanks—any place elongation is required.

Many types of nonwoven fabrics are used as substrates (3) (see Nonwoven textiles). The wet web process gives a nonwoven fabric with paperlike properties; low elongation, low strength, and poor drape. When these substrates are coated, the papery characteristics show through the coating, and fabric esthetics are not satisfactory. The nonwovens prepared by laying dry webs, compressing by needle punching, and then impregnating from 50–100% with a rubbery material, resemble in many respects split leather. These materials are used for shoe liners (see Leather-like materials). It is difficult to achieve uniformity of stretch and strength in two directions as well as a smooth surface; therefore, a high quality nonwoven of this type is very expensive. Spunbonded nonwovens are available in both polyester and nylon in a range of weights. The strength qualities are very high and elongation is low. Since they are quite stiff, these materials are used where strength and price are the major considerations. A lightly needled, low density nonwoven was marketed in 1970 and in the last three years has gained prominence in the coated fabrics industry. It is used in weights from 60–180 g and can be prepared from either polypropylene or polyester fibers. The light needling combined with careful orientation of the fibers and selection of the fiber length gives very good strength and more balanced stretch. Optionally, a thin layer of polyester-based polyurethane foam can be needled into the nonwoven to improve the surface coating properties. The furniture upholstery market was the first to accept this product. A 0.4 mm poly(vinyl chloride) skin on this nonwoven replaced expanded PVC on knit fabric at approximately half the cost. The finished product is softer, plumper, and may have better wear characteristics. A major automobile company has introduced this type of seating in one line of cars, and apparel manufacturers have exhibited interest in these types of construction.

Post Finishes of the Textile Component. The construction that results from either weaving or knitting is called a greige good. Other steps are required before the fabric can be coated: scouring to remove surface impurities; and heat setting to correct width and minimize shrinkage during coating.

Optional treatments include: dyeing if a colored substrate is required; napping of cotton and polyester–cotton blends for polyurethane coated fabrics; flame resistance treatments; bacteriostatic finishes for hygienic applications; and mildew treatments for applications in high humidity (see Textiles).

Polymer Component

Rubber and Synthetic Elastomers. For many years coated fabrics consisted of natural rubber (qv) on cotton cloth. Natural rubber is possibly the best all-purpose rubber but some characteristics such as poor resistance to oxygen and ozone attack, reversion and poor weathering, and low oil and heat resistance, limit its use in special application areas (see also Elastomers, synthetic).

Polychloroprene (Neoprene) introduced in 1933 rapidly gained prominence as a general purpose synthetic elastomer having oil, weather and flame resistance. The introduction of new elastomers in solid or latex form was accelerated by World War II. Currently, in addition to natural rubber and polychloroprene, other polymers in use include: styrene–butadiene (SBR), polyisoprene, polyisobutylene (Vistanex), isobutylene–isoprene copolymer (Butyl), polysulfides (Thiokol), polyacrylonitrile (Paracril), silicones, chlorosulfonated polyethylene (Hypalon), poly(vinyl butyral), acrylic polymers, polyurethanes, ethylene–propylene copolymer (Royalene), fluorocarbons (Viton), polybutadiene, polyolefins, and many more. Copolymerizations and physical blends make the number available staggering (see Copolymers; Olefin polymers; Polymers containing sulfur; Acrylic ester polymers; Vinyl polymers; Fluorine compounds; Acrylonitrile polymers; Silicon compounds; Urethane polymers).

In fact, the number of commercially available polymers in use is well over 1000. In each class there are several variations manifesting a wide range of properties. DuPont supplies about 24 types of polychloroprene. B. F. Goodrich supplies about 140 types of acrylonitrile elastomers (Hycar) and the same holds true for all the other types of coating polymers.

Most elastomers are vulcanizable; they are processed in the plastic state and cross-linked to provide elasticity after being put into final form. With the number of elastomer coatings available today almost any use requirement can be met. If there are limitations, they lie in the areas of processability and cost.

Elastomers are applied to the textile by either calendering or solution coating. Thin coatings are applied from solution and thicker coating by direct calendering.

A natural rubber-based formulation is shown in Table 1.

Table 1.   A Typical Natural Rubber Compound

Component	Parts
smoked sheet	100.00
stearic acid	1.00
ZnO	3.00
agerite white antioxidant	0.50
P-33 black	10.00
$CaCO_3$	75.00
clay	50.00
sulfur	0.75
methyl zimate ⎱ accelerators	0.25
telloy ⎰	0.50
Total	*200.00*

SBR (styrene–butadiene rubber) has replaced natural rubber in many applications because of price and availability. It has good aging properties, abrasion resistance and flexibility at low temperatures. A typical SBR-based formula is shown in Table 2.

Neoprene offers resistance to oil, weathering, is inherently nonburning and is processable on either calenders or coaters. The cost of Neoprene and its reduced availability in recent years have led to the development of substitutes. Nitrile rubber–PVC blends and nitrile rubber–EPDM (ethylene–propylene–diene monomer) perform on an equivalent basis. The blends do not discolor like Neoprene, and light-colored decorative fabrics can be made.

A typical Neoprene-based formulation is shown in Table 3.

This mixture can be calendered or dissolved in toluene to 25–60% solids for coating.

Isobutylene–isoprene elastomer (Butyl) has high resistance to oxidation, resists chemical attack and is the elastomer most impervious to air. These properties suggest its use for protective garments, inflatables, and roofing.

Chlorosulfonated polyethylene (Hypalon) resists ozone, oxygen, and oxidizing agents. In addition it has nonchalking weathering properties and does not discolor, permitting pigmentation for decorative effects.

Table 2.   A Typical SBR Compound

Component	Parts
SBR	100.0
processing aid	5.0
stearic acid	2.5
ZnO	3.0
agerite white antioxidant	0.5
tackifier	20.0
$CaCO_3$	75.0
P-33 black	10.0
clay	75.0
sulfur	2.5
methyl zimate ⎱ accelerators	0.5
tuex ⎰	
Total	*294.0*

Table 3.   A Typical Neoprene Formulation

Component	Parts
polychloroprene	100.0
stearic acid	1.5
antioxidant	2.0
MgO	4.0
clay	66.0
SRF black	22.0
circo oil	10.0
petrolatum	1.0
ZnO	5.0
ethylenethiourea	0.5
Total	*212.0*

Nitrile elastomers (acrylonitrile–butadiene copolymers) have high resistance to oils at up to 120°C. If higher temperature protection is required, a polyacrylate elastomer can be employed up to 200°C.

Polyurethane.   Polyurethanes have a number of important applications in coated fabrics. The most striking is footwear uppers because polyurethanes are lighter weight than vinyl polymers and have better abrasion resistance and strength. Polyurethane-coated fabrics can be decorated to look like leather (qv). Earlier attempts to produce poromerics (coatings that transmit moisture much like leather) were not commercially successful because, although they approach leather in cost, they did not match it in comfort. However, poromerics are still available. Most of the urethane-coated fabrics are used in women's footwear where styling is important and lightweight is desirable. These products usually consist of 0.05 mm of polyurethane on a napped woven cotton fabric. The result is a lightweight product 0.88 mm thick that has good abrasion and scuff resistance (4). Urethane-coated fabrics have not been successful in either men's or children's shoes because greater toughness is required.

Low-weight coatings of polyurethane on very low-weight nylon fabric produce products suitable for apparel. This lightweight product, used for windbreakers and industrial clothing, resists water, provides thermal insulation, and has good drape. Coatings of urethane on heavier nylon structures are used for industrial tarpaulins to provide protection from the elements and extreme toughness. Polyurethane coatings have had limited application to furniture upholstery and practically none on automobile seating.

Poly(vinyl Chloride).   By far the most important polymer used in coated fabrics is poly(vinyl chloride). This relatively inexpensive polymer resists aging processes readily, resists burning, and is very durable. It can be compounded readily to improve processing, aging, burning properties, softness, etc. In addition, it can be decorated to fit the required use. PVC-coated fabrics are used for window shades, book covers, furniture upholstery, automotive upholstery and trim, wall covering, apparel, conveyor belts, shoe liners, and shoe uppers. These few uses require millions of meters of coated fabrics each year and demonstrate the diverse properties of PVC coatings. Tables 4 and 5 show typical PVC formulations.

Table 4.   A Typical Compound for Calendering PVC

Component	Parts
poly(vinyl chloride) resin (calender grade)	100.00
epoxy plasticizer	5.00
dioctyl phthalate	35.00
polymeric plasticizer	35.00
BaCdZn stabilizer	3.00
TiO_2 (pigment)	15.00
calcium carbonate (filler)	20.00
stearic acid (lubricant)	0.25
Total	*213.25*

Table 5.   A Typical Plastisol PVC Formulation

Component	Parts
poly(vinyl chloride) resin (dispersion grade)	100.00
epoxy plasticizer	4.00
dioctyl phthalate	70.00
BaCdZn stabilizer	2.50
lampblack (pigment)	2.00
calcium carbonate (filler)	25.00
lecithin (wetting agent)	1.00
Total	*204.50*

Processing

Coated fabrics can be prepared by lamination, direct calendering, direct coating or transfer coating (see Coating processes). The basic problem in coating is to bring the polymer and the textile together without altering undesirably the properties of the textile. Almost any technique in applying polymers to a textile requires having the polymer in a fluid condition, which requires heat. Therefore, damage to the synthetic or thermoplastic fabric may occur.

Calendering.   The polymer is combined in a Banbury mill with a filler, stabilizing agents, pigments, and plasticizers and brought to 150–170°C. The mixture ("compound") temperature is adjusted on warming mills and calendered directly onto a preheated fabric. The object is to get the required amount of adhesion without driving the compound into the fabric excessively, which would cause a clothy appearance and lower the stretch and tear properties of the coated fabric.

Coating.   Coating operations require a much more fluid compound. Rubbers are dissolved in solvents. In the case of PVC, fluidity is achieved by adding plasticizers and making a plastisol. If lower viscosity is required, an organosol is made by adding solvent to the plasticized PVC. After the ingredients are mixed and brought to a coating head, the mixture is applied by either knife, knife-over-roll, or reverse roll coaters. Unless the fabric is very dense, or a high degree of penetration is desired, the coating cannot be placed directly on the textile. Transfer coating limits the penetration into the fabric. The mixture is coated directly on the release paper and penetration is limited either by the viscosity of the coating or partial solidification (gelling) of the coating prior to application of the textile. Most polyurethane coated fabrics are transfer coated. Expanded vinyl-coated fabrics consist of a wear layer, an expanded layer, and the textile substrate. The wear layer is coated on release paper (see Abherents) and gelled. A layer of vinyl-based compound containing a chemical blowing agent such as azodicarbonamide is applied. The fabric is placed on top of the second layer and sufficient heat is applied to decompose the blowing agent causing the expansion (5).

Lamination. In lamination a film is prepared by calendering or extrusion. It is adhered to the textile at a laminator either with an adhesive or by sufficient heat to melt the film (see Laminated and reinforced plastics).

Post Treatment. Coated fabrics can be decorated by printing with an ink. Usually the appearance of a textile or leather is the goal. The inks are applied as low-solid solutions by metal rotogravure rolls. Warm air drying is carried out in an oven. Because the ink dries rapidly, multiple print heads can be used (see Ink).

If a textured surface is desired, the coated fabric is heated to soften it and pressure is applied by an engraved embossing roll. Printing usually precedes embossing so that a flat surface is presented for printing. Special effects are obtained by embossing first and then printing or wiping the high points (see Printing processes).

The final layer is called the slip. Most coatings are tacky enough to stick to themselves (block) during stacking or rolling. The main purpose of the slip is to prevent blocking (see Abherents). Slips can be formulated as shown in Table 6 to improve abrasion resistance, seal the surface, adjust color and adjust gloss. Slips are low-solid solutions that are applied by metal rotogravure rolls. Air drying leaves about 200 g of solids per 100 m^2 of coated fabric.

Economic Aspects

Poly(vinyl chloride) is the principal polymer employed (see Vinyl polymers). In the United States alone the consumption of PVC for calendered-coated fabrics was 28,000 metric tons in 1975 and 45,000 t in 1976. Coating of paper and textiles consumed 75,000 t in 1976.

The United States consumption of polyurethane for fabric coatings was 4500 t in 1975 and 5000 t in 1976 (6).

Table 6. A Typical Slip Formulation for PVC-Coated Fabric

Component	Parts
vinyl chloride–vinyl acetate copolymer	100.00
polymethacrylate resin(s)	96.00
vinyl stabilizer	2.00
silica gel	18.00
methyl ethyl ketone	620.00
xylol	350.00
Total	*1186.00*

Table 7. Uses of Coated Fabrics [a]

Substrate	Coating	Use
nylon tricot nylon sheeting cotton sheeting	PVC, PU, SBR, Neoprene	clothing
polyester–cotton sheeting	PVC	
napped cotton drill	PU	shoe uppers
nonwovens (high density)	PU	
nonwovens (medium high density)	PVC	shoe liner-insoles
cotton knits	PVC	
polyester nonwovens (light density)	PVC	
polypropylene nonwovens (light density)	PVC	
cotton knits	PVC	furniture upholstery
polyester knits	PVC	
polyester–cotton knits	PVC	
napped cotton drills	PU	
nylon Helanca knits	PVC	
cotton single knits	PVC	
polyester–cotton pattern knits	PVC	auto upholstery
polyester nonwoven (light density on PU foam)	PVC	
polyester nonwoven (light density)	PVC	
polyester stitched nonwoven	PVC	landau tops
polyester knit	PVC	
asbestos nonwoven	PVC	floor covering
polyester spunbonded	PVC PE	wallcovering
cotton sheeting glass scrim rayon scrim	PVC, SBR, Neoprene, silicone rubber, etc	tapes
polyester drill	PVC, SBR, Neoprene, natural rubber	hospital sheeting
absorbent cotton	PVC	Band-Aids
nylon scrim	PVC	window shades
cotton scrim	PVC	wallpaper
cotton sheeting polyester sheeting polyester–cotton sheeting	acrylic	lined drapes
glass scrim polyester scrim	lead-filled PVC, barytes-filled PVC, barytes-filled SBR	acoustical barriers
paperlike nonwovens	PVC	air and oil filters
dyed rayon drill	expanded PVC	soft-side luggage
rayon drill cotton drill polyester–cotton drill polyester drill nonwovens	PVC	luggage
nylon woven polyester woven	PVC, Neoprene, Hypalon, PU	tarpaulins
nylon woven cotton woven	PVC	awning
nylon scrim polyester scrim	PVC, EPDM, Hypalon, Butyl	pond and ditch liner
glass woven polyester woven	Neoprene, PVC, Hypalon	air supported structures

[a] PVC = poly(vinyl chloride); PU = polyurethane; EPDM = ethylene–propylene–diene-modified rubber; SBR = styrene–butadiene–rubber; PE = polyethylene.

Health and Safety Factors

Some materials used in coating operations have been identified by the United States government as being hazardous to the workers' health.

Even when coatings are applied by extrusion and calendering, consideration should be given to handling the materials, evolution of gases during heating and post-finishes. For instance, there are strict regulations on exposure to vinyl chloride monomer. Emptying bags or bulk transfer must be monitored. The regulations do not apply to the handling or use of fabricated products made from poly(vinyl chloride).

When a coating machine is employed, attention must be given to exposure of the operator to the solvents.

In addition, particulate irritants such as asbestos (qv), pigments, and reactive chemicals are often involved.

Coating operation should not be initiated without consulting the *Federal Regulations on Occupational Safety and Health Standards, Subpart Z, Toxic and Hazardous Substances* (7).

Uses

Table 7 lists uses of coated fabrics and demonstrates how combinations of textiles and polymers can give significantly different products.

BIBLIOGRAPHY

"Coated Fabrics" in *ECT* 1st ed., Vol. 4, pp. 134–144, H. B. Gausebeck, Armour Research Foundation of Illinois Institute of Technology; "Coated Fabrics" in *ECT* 2nd ed., pp. 679–690, by D. G. Higgins, Waldron-Hartig Division of Midland-Ross Corporation.

1. *Davison's Textile Blue Book,* Davison Publishing Co., Ridgewood, N.J., 1977.
2. N. J. Abbott, T. E. Lannefeld, and R. J. Brysson, *J. Coated Fibrous Mater.* **1,** 4 (July 1971).
3. S. P. Suskind, *J. Coated Fibrous Mater.* **2,** 187 (Apr. 1973).
4. H. L. Gee, *J. Coated Fabr.* **4,** 205 (Apr. 1975).
5. W. G. Joslyn, *Rubber Age* **106,** 49 (Feb. 1974).
6. *Mod. Plast.* **54,** 49 (Jan. 1977).
7. *Code of Federal Regulations, Title 29,* Chapter XVII, Section 1910.93 of Subpart G redesignated as 1910.1000 at 40 FR23072, U.S. Government Printing Office, Washington, D.C., May 28, 1975.

General References

F. J. Beaulieu and M. D. Troxler, "Substrates for Coated Apparel Applications," *J. Coated Fibrous Mater.* **2,** 214 (Apr. 1973).
"1977 Manmade Fiber Deskbook," *Mod. Text.* **2,** 16 (Mar. 1977).
"Generic Description of Major U.S. Manmade Fibers," *Mod. Text.* **58,** 17 (Mar. 1977).
"Names and Addresses of U.S. Manmade Fiber Producers," *Mod. Text.* **58,** 30 (Mar. 1977).
"77–78 Buyers Guide," *Text. World* **127,** (July 1977).
R. M. Murray and D. C. Thompson, *The Neoprenes,* E. I. du Pont de Nemours & Co., Inc., Wilmington, Del., 1963.
M. Morten, *Rubber Technology,* Van Nostrand Reinhold Co., New York, 1973.
J. Bunten, "Performance Requirements of Urethane Coated Fabrics," *J. Coated Fabr.* **5,** 35 (July 1975).

D. Popplewell and L. G. Hole, "Urethane Coated Fabrics," *J. of Coated Fabrics* **3**, 55 (July 1973).
H. A. Sarvetnick, *Polyvinyl Chloride,* Van Nostrand Reinhold Co., New York, 1969.
"Manufacturing Handbook and Buyer's Guide 1977/78," *Plast. Technol.* **23**, (Mid-May 1977).
Davison's Textile Blue Book, Davison Publishing Co., Ridgewood, N.J., 1977.
Rubber Red Book, Palmerton Publishing Co., New York, 1977.

FRED N. TEUMAC
Uniroyal, Inc.

COATING PROCESSES

Plastic film-forming materials may be applied as coatings in many different processes. Coatings may be applied for either functional purposes (eg, waterproofing, flameproofing, mildewproofing, abrasion resistance, rust resistance, reflection, insulation, adhesion, or coating impermeable to gases and liquids) or decorative purposes (eg, gold coloration on aluminum foil or wrapping paper).

During the nineteenth century vegetable oils and rubber latexes were applied to textiles to produce waterproof garments. In addition, starch and protein (casein) coatings for paper were developed, followed by the utilization of cellulose esters and phenolic resins. In the past few decades many new coating techniques were introduced to deal with the myriad of new plastic materials.

In addition to coated, shaped rigid articles such as automobile bodies, furniture frames, machinery, etc, the range of coated web or sheet materials includes a wide spectrum of products such as artificial leather, garment interlinings, impermeable products such as shower curtains, book bindings, paper and paperboard, a variety of tapes, coated metal strips for building and furniture applications, hard surface phenolic and melamine laminations for furniture and insulating uses, floor and wall coverings, food packaging and other coated films, and coated plywood and fiber panels.

Classification and Definitions

Although coating processes may be categorized as premetered and postmetered, or as brush, roller, knife, and spray, such distinctions cannot be clearly made in many of today's applications of plastics in suspension, melt, solution, or powder form. Application of polymerized coatings from gaseous or liquid monomers further complicates any classification system. Table 1 gives a list of data to be considered in the selection of a coating process (see also Film deposition techniques; Paint; Powder coating; Coatings, industrial; Coatings, marine; Coated fabrics; Coatings, resistant; Radiation curing).

Definitions. Across-machine direction: 90° web direction.

Air knife or air doctor coater: applies a surplus of coating to a substrate and removes the excess by an adjustable precision air impingement.

Applicator: device that applies coating to a substrate.

Back-up roll: substrate support roll for application of coating.

Blade coater: applies a predetermined amount of excess coating to a substrate and removes the excess by adjustable unit pressure on a flexible steel doctor blade.

Coating: liquid mixture, solution, or suspension that is applied by the applicator. Sometimes referred to as color.

Coat weight: dry weight of coating retained by the substrate, in g/m^2. Some companies may still identify coat weights as pounds per ream ($1\ lb/3000\ ft^2 = 1.62\ g/m^2$).

Deckle: device or result of varying the width of coating retained on a substrate.

Doctor blade: thin flexible blade, 0.25–1.50 mm, used to trowel, remove, or meter coating.

Dwell: distance between the applicator and metering device.

Film-split pattern: fracture of the miniscus effect caused by two rotating rolls and determined by the properties of the coating.

Gravure coater: applies an excessive amount of coating to a gravure roll, removes the excess by a doctor blade; surface tension transfers most of the coating to the substrate.

Gravure roll: a mechanically or chemically etched roll whose etching cell size determines the amount of coating to be applied to a substrate.

Meniscus bead coater: transfer of coating from the applicator to the substrate is caused by surface tension between the applicator and the substrate.

Metering: method to remove excess coating from a substrate during or after application of coating.

Metering rod coater: applies an excessive amount of coating to a substrate and removes the excess by a Mayer rod.

Mayer rod: small driven rod, 6–12 mm in diameter, used to meter coating. Rod can be smooth, wire wound, or grooved to control final coat weight.

Nip: area where two rolls are close together or touching.

Offset gravure coater: same as gravure coater, except coating is applied to a resilient transfer roll before final application to the sheet.

Premetering: mechanical rejection of surplus coating before the metering device. On die applicators this can be an internal orifice restriction.

Profiling: incremental method of adjusting deviation of final coating weight required in across-machine direction.

Reverse roll coater: all coating rolls rotate the same direction; therefore, their relative metering nip areas rotate in reverse to each other. The actual application of coating is in the reverse direction of the web.

Speed: velocity of substrate as it passes through the process, measured in m/min.

Substrate or web: material that supports the applied coating, such as paper, film, foil, wovens, nonwovens, or endless belt.

TIR: total indicator runout, measured in millimeters or micrometers, is the deviation of a roll as read from the total movement of a dial indicator.

Processes

A coating process is the application of a liquid to a traveling web or substrate. The primary substrates include paper and paperboard grades, films, foils, nonwovens, and wovens. The coating principles are better suited to continuous webs than to short individual sheets. In general, there is an ideal coater arrangement for any given end

Table 1. Guide to Selection of Coating Processes

Coating method	Base material[a]	Coating composition[b]	Usual coating speed, m/min	Viscosity range, mPa·s (= cP)	Wet-coating thickness range, μm	Advantages	Disadvantages	Normal temp operating range, °C
air knife	b,d	R,T,X	15–600	1–500	25–60	minimizes scratches, profile contour coater, easy to change coat weight	noisy, dusty, tension sensitive high viscosity	20–25
brush	b,c,e,f,g	R,S,X,Z	30–120	100–2,000	50–200	contour or smooth application not tension sensitive	noisy, high maintenance slow speed high viscosity	20–25
calender	a,b,d,e	U,V,W	5–90		100–500	laminates fabric to plastic coatings wet or smooth, glossy finish	slow speed, high cost	above 180 100
cast-coating	a,b,d	Q,R,S,T,V,Y					slow speed required	
curtain	a,b,c,d,e,f	R,S,V,X,Z	20–400	100–2,000	25–250	sheet feeding, low viscosity, contour, low coating waste	slow speed, high viscosity speed sensitive	20–180
die fountain	a,b,d,e	R,S,T,V,X,Y,Z	3–600	300–100,000	60–2,500	wide viscosity range wide coat weight range, dead end supply	low coat weight, backing roll wear	20–200
dip	a,b,d,e,f,g	R,S,V,X,Y,Z	15–200	100–1,000	25–250	saturates web, wets both sides	low viscosity, slow speed low coat weight	20–60
extrusion	a,b,d,e	T,U,V,W	20–900	30,000–50,000	12–50	high speed, uniform, low coating loss, smooth finish	sensitive to viscosity changes	20–320
fibrous belt	b,c,e,f	R,T,X,Z	30–150	100–2,000	25–200	contour, sheet fed	slow speed, messy scratches, high volume coating supply	20–25
flexible blade	a,b	R,S,T,V,X,Y,Z	120–1,200	50–20,000	60–150	high speed, leveling, smooth finish, wide range of viscosity coating weight		20–65
floating knife	a,b,d	R,S,T,V,X,Y,Z	3–30	500–5,000	50–250	simple, precision, high coating, weight smooth finish minimum strike through	tension sensitive, low speed streaks	20–150
gravure	a,b,d,e	R,S,T,V,X,Y,Z	2–450	100–1,000	12–50	simple, coat wide range of webs, low coat weight	difficult to change coating weight film split, high viscosity	20–120
kiss roll	a,b,c,d,e,f	R,S,V,X,Z	30–300	50–1,000	25–120	simple, easy to vary coat weight and web weight	high viscosity, film split, tension sensitive	20–120

Method	Substrate[a]	Composition[b]						Temp, °C
knife over blanket	a,b,d	R,S,T,V,X,Y,Z	3–30	500–5,000	50–250	low coat weight wide range web thickness	tension sensitive, strike through maintenance and clean up, low speed	20–25
knife over roll	a,b,c,d,e	R,S,T,U,V,X,Y,Z	3–120	1,000–10,000	50–2,500	precision, high coating weight, high viscosity	splices, low speed, streaks	20–180
meniscus	d,e		3–60	1–50	60–25	smooth, precision, simple, low coating weight, photographic quality	high viscosity, slow speed, speed sensitive	10–25
offset gravure	a,b,d,c,e	R,S,V,Y,Z	3–600	1–1,000	5–25	low coating weight, high speed, precision	high coating weight hot melt	20–120
pressure roll	a,b,c,d,e,f	R,S,T,U,V,X,Y	30–700	100–5,000	25–120	coat both sides web, saturate, simple	pattern, high coating weight or viscosity	20–60
reverse roll	a,b,c,d,e,f	R,S,T,V,X,Y,Z	3–600	50–20,000	50–800	wide range of coating weights, viscosity and speeds independent of web	high cost, high maintenance, low coat	20–150
reverse smoothing roll	a,b	R,T,X	3–600	1,000–5,000	25–80	simple, minimizes film split	tension sensitive high viscosity	20–150
sprays								
airless spray	a,b,c,d,e,f,g	S,T,V,X,Y,Z	3–90	8–50 s in no. 4 Ford cup	2–250	less dusting, higher solids	low coat weight	room
air spray	a,b,c,d,e,f,g	S,T,V,X,Y,Z	3–90	8–50 s in no. 4 Ford cup	2–250	low coat weight	plugging	room
electrostatic	a,b,c,d,e,f,g	S,T,V,X,Y,Z	3–90	8–50 s in no. 4 Ford cup	2–250	minimum coating loss, coating uneven shapes, low coat weight	high coating weight, high voltage	room
in situ polymerization	a,b,c,d,e,f,g	Y,Z	undetermined	liquid or vapor	6–2.5	absence of solvents, surface coating low coat weight		above room
powdered resin	a,b,c,e,f,g	Q	3–60		25–250	irregular shapes such as auto frames	high voltage	above room

[a] a, woven and nonwoven textiles; b, paper and paperboard; c, plywood and pressed fiber boards; d, plastic films and cellophane; e, metal sheet, strip, or foil; f, irregular flat items; g, irregularly shaped.

[b] Q, powdered resin compositions; R, aqueous latexes, emulsions, dispersions; S, organic lacquer solutions and dispersions; T, plastisol and organosol formulations; U, natural and synthetic rubber compositions; V, hot-melt compositions; W, thermoplastic masses; X, oleoresinous compositions; Y, reacting formulations; (eg, epoxy and polyester); Z, plastic monomers.

product; however, most machines are required to produce many different products and coating thicknesses and the machine is therefore usually a compromise for several applications.

All coating machines contain application and metering devices. These devices are usually located adjacent to each other, but in knife or in die coaters they can be one and the same. In order to ensure that a given quantity of coating is retained on a substrate, it is necessary to incorporate the metering principle.

Coaters can be tension sensitive and tension insensitive. Tension-sensitive coaters (eg, the metering rod) simply means the ability of the coating station to maintain coating weight is dependent on ability of the process line to maintain a constant substrate tension. As the value of tension decreases, the retained coating on the web usually increases. It is not recommended that web tension be altered to control coating weight and, needless to say, all tension-sensitive coating heads should be equipped with automatic tension-control devices. Tension-insensitive coating stations have the inherent ability to maintain coat weight even when web tensions vary. The supported knife blade and gravure coaters are good examples.

Table 2 illustrates factors that must be considered when attempting to maintain coating weight with various basic methods.

Brush Coating. Present-day bristle brushes evolved from vegetable fibers such as rice roots. Today's synthetic nylon fiber brushes succeed a long line of animal fibers such as Chinese boar, horsehair, Siberian mink, skunk, camel, squirrel, and badger. Synthetic fiber bristles may be selected to obtain the desired flexibility, taper, density, surface tension, diameter, and uniformity. The hand brush technique has been extended to permit continuous webs or sheets to be uniformly coated. Figure 1 illustrates the design of a simple brush-coating machine in which the coating is applied by a rotating cylindrical or roller brush and then leveled and spread by transverse oscillating brushes as the substrate is supported against them. These machines operate at speeds up to 100 m/min and can apply up to 45 g/m^2. Brush-coating methods are best suited to slow-drying coatings such as oleoresinous paints, starch, latexes, or emulsions. They permit the application of a flawless coating, such as required on playing cards, or the uniform coverage and penetration on uneven surfaces, such as required on rough fiber panels. Brush-coating machines have been replaced largely by air-knife coaters.

Advantages: smooth application; contour coater; and not tension sensitive.

Disadvantages: noisy; high maintenance; slow speed; high viscosity; and difficult to clean.

Knife Coating. The simplest, least expensive, and yet one of the best coating methods is rigid knife coating, either unsupported or supported, with the latter being the more accurate.

Basically the coating is pumped or even-hand ladled into a trough of variable width supported at the bottom by the web, with end dams at each side, and a precision metering knife at right angles to web direction. Coat-weight control is maintained by adjusting the clearance and pressure between the knife and the web. For optimum uniformity across the machine, incremental profiling adjustment screws are provided to deflect the knife at right angles to the web. The more precise the relationship of knife to web, the more precise the results.

The unsupported knife coater, shown in Figure 2, has the advantage when coating open fabric type webs, where coating strike-through is desired or cannot be prevented, such as plastisols on a jute substrate for automotive floor mats. Its main deficiency

Table 2. Factors Affecting Final Coating Weight

Coater	Web absorptivity	Viscosity, tension	Solids, %	Dwell	Web speed	Applicator speed	Metering roll speed	Roll gap	Blade angle	Blade pressure	Blade gap	Tension
air knife	X	X	X	X	X	X	X	X	X	X[a]	X	X
brush	X	X	X	X	X	X[a]	X	X[a]			X	
calender		X			X			X[a]				
cast coating	X	X	X	X	X[a]	X[a]	X	X	X	X	X	X
curtain		X	X		X[a]	X						scrape type
die fountain	X	X	X	X[a]	X	X		X[a]	X	X	X	
dip	X	X	X	X[a]	X			X	X	X	X	
extrusion		X	X	X[a]	X[a]	X						
fibrous roll	X	X	X	X	X	X[a]	X	X	X	X[a]		
flexible blade	X	X	X	X	X	X	X	X	X	X[a]	X	
floating knife	X	X	X	X	X				X		X[a]	X
gravure	X	X	X		X		X	X	X	X, etch[a]		
kiss roll	X	X	X		X	X[a]	X	X				X
knife over blanket	X	X	X		X				X	X[a]	X	
knife over roll	X	X	X		X				X	X	X[a]	
meniscus	X	X	X		X	X	X	X[a]				
offset gravure		X	X	X	slight	X	X[a]	X[a]	X	X	X	
pressure roll	X	X	X	X	X	X[a]	X[a]	X				
reverse roll		X	X		X	X[a]	X[a]	X				
reverse roll smoothing	X	X	X	X	X	X[a]						X[a]

[a] Primary coat with adjustments.

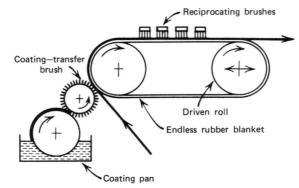

Figure 1.   Brush-coating machine.

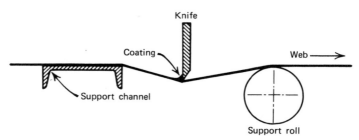

Figure 2.   Unsupported knife.

is that it is tension sensitive and requires that the web have substantial tensional strength on high viscosity coatings.

A full-width endless driven belt to support and pull the substrate through the knife area can overcome the side effect of the knife drag; such a coater can still be considered as tension sensitive.

In the knife-over-bed coater a ground cast-iron support bed in conjunction with a profiling knife blade can be adjusted to practically form a perfect gap, thereby eliminating the TIR effect of an eccentric backing roll. Again, this coater requires that the web withstand the dragging effect on low-coat weight applications. The suction apron or driven vacuum belt is required to maintain a steady process speed on the substrate with contacting the coated side.

The knife-over-roll coater (see Fig. 3) is probably the most common of the rigid knife coaters owing to its compactness and simplicity. The driven back-up roll is usually made of precision metal for heavy coat weights and covered with resilient material for low coat weight as rubber type coverings are more adaptable to gage variations in the substrate or TIR tolerance. Special stripping effects can be obtained by adding more end dams and coating supplies across the knife holder assembly.

Advantages: low cost; simple, compact; precision $\pm 1\%$ variation; heavy coat weights, 0.012–2.5 mm wet; high viscosities up to 100,000 imPa·s (= cP); levels rough surfaces; low solvent loss; and produces uniform total web cross section.

Disadvantages: streaking or scratches; passing slices; messy on web breaks; contacting-type edge dams; coating weight changes with web caliper or profile; and speeds above 300 m/min.

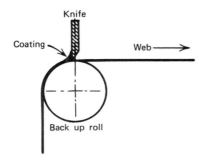

Figure 3.   Knife over roll.

Inverted Knife.   This version is essentially a knife over an unsupported web coater except that the coating is applied to the web by a variable speed applicator roll. The knife metering device is 0.25–1.5 mm thick and is used primarily for low-coat-weight asphalt applications or starch–latex fillers for open fabrics that are used for lumber wrap or wallpaper materials, respectively. Resilient edge wipes at each end of the applicator roll allows deckling in for dry edges when desired.

Advantages: simple, compact; low cost; low viscosity on open webs; low coat weights; and simple deckle control.

Disadvantages: tension sensitive; scratches on smooth webs; high viscosity; and speeds above 300 m/min.

Inverted Unsupported Die.   The unsupported inverted die coater, see Figure 4, is a spin-off of the inverted-knife coater and is better suited for high viscosity coatings and hot melts.

Unlike the knife-over-roll coater, this applicator and metering device is pressurized and its end dams are completely internal and do not contact the web. Optional hot smoothing knives can also be incorporated in the web path after the die to minimize patterns or to reduce coat weight. The primary advantage of this coater is that on hot melts a minimum reservoir of coating is required at the head where the coating is uniformly extruded against the web by a constant displacement pump of variable speed. The die applicator can be heated by steam, oil, or zoned resistance electric heaters.

Advantages: high viscosities; low coat weights; internal deckle; uniform appli-

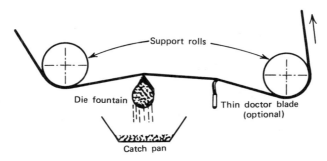

Figure 4.   Inverted unsupported die coater.

cation; uniform temperature; high speed, over 300 m/min; no recirculation of coating required; and easy to change web widths.

Disadvantages: tension sensitive; high tension required; and high coat weight difficult to achieve.

Supported Die Fountain. Addition of a driven back-up or support roll eliminates the disadvantages of the unsupported die fountain. However, this backing roll can become a maintenance and efficiency problem, especially when running a variety of web widths on the same roll.

Advantages: viscosity, 500–200,000 mPa·s (= cP); coat weight, 7–70 g/m²; speeds, not limited; internal deckles; uniform application; uniform temperature; and no recirculation required.

Disadvantages: backing roll maintenance due to wear; and backing roll heat buildup.

Air-Knife Coater. The air-knife coater, also known as the air-doctor coater, was patented in 1930 by the S. D. Warren Company (see Fig. 5). Unlike other coaters, the air-knife coater follows the contour of the substrate resulting in an end product that retains most of its original texture after the coating has dried.

Air knives are the metering devices that are so positioned that they simply blow off surplus coating that has been applied by any one of several methods. However, on some applications, such as gelatins, the air knife is used only as a leveling device and does not remove coating.

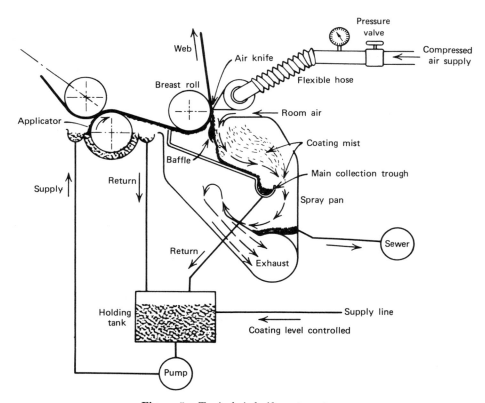

Figure 5.   Typical air-knife coater setup.

The air knife consists of a main body, a pair of adjustable stainless lips that form a precision orifice through which compressed air impinges against the coated sheet, supported by a breast roll that may or may not be driven. The air supply to the air knife is provided by a nonpulsating impeller-type compressor whose output pressure is controlled by adjustable valves. Weight is controlled mainly by air pressure, decreasing with increasing pressure. Increasing applied coating or process speed also requires increasing air pressure to maintain coat weight.

Air knives usually do not require more than 41 kPa (6 psi) pressure at 0.11 m^3/(min·cm) [10 ft^3/(min·in.)] However, this volume, passing through a 0.75 mm orifice, results in a high degree of atomized coating at the impact point some 0.5 mm away. This coating, which must be prevented from fouling the orifice, dusting into the room, or escaping to the sewer, is collected in a device called the spray pan.

Air that exits the knife builds up in temperature and tends to dry the coating prematurely, thus when requirements exceed 20 kPa (3 psi), heat exchangers are necessary to reduce this heat of compression.

Good examples of air-knife applications are coating of embossed papers and fabrics and applications of poly(vinylidene chloride) or other shear-sensitive coatings that are prone to streaking.

Operating speeds exceed 650 m/min and coat weights are applied up to 35 g/m^2 in aqueous systems.

Advantages: contour coater; easy to change coat weight; excellent coating weight profile; gentle leveling device; minimizes streaking or scratches; easy-to-change web width; and handles many substrates and coatings.

Disadvantages: noisy; dirty; low coating weight; and tension sensitive, depending on application system.

Fibrous Belt Coating. The familiar example of sheep's wool or synthetic fleece-covered hand-painting rolls is an excellent illustration of the coating method preferred for the application of latex and emulsion coatings. Figure 6 illustrates a coating arrangement to apply latex and emulsion coatings to pressed acoustical fiber ceiling panels. The endless felt blanket is wetted with coating from the pan and then surface scraped to remove excess coating prior to contact with the board surface. Such equipment frequently carries oscillating distributing brushes against the wet coated surface to ensure coating coverage on high and low surface areas.

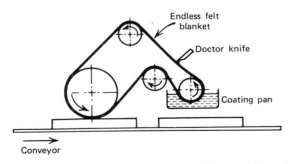

Figure 6.   Coating blanket arrangement for pressed board.

Metering Rod Coater. This coater (see Fig. 7), also known as a Mayer rod coater utilizes a driven single-roll applicator to coat in either direction and is followed by a driven metering rod assembly to control weight. Normal rotation of the rod is in the reverse direction to web travel and rod smoothness selection controls coat weight. The rod is rotated in order to clean it, increase its life, and in some cases vary the speed with varying coat weight. The actual metering rod can be smooth, wire wound, or machined with annular grooves.

This tension-sensitive coater is commonly used for low-solids, low-viscosity coatings such as used in poly(vinylidene chloride), carbon paper, silicone release papers, and various precoaters. Coat weights range from 1.5–10 g/m^2, and speeds are as high as 300 m/min.

Additional Mayer rods may be used to progressively meter the coating by going from coarse to smooth finishes. This is common on applicators that require absolute coverage on low coat weights.

Advantages: low cost; and simple, compact.

Disadvantages: Mayer rod wear; rod holder wear; speed above 300 m/min; high viscosity; and high cost weight.

Blade Coating. *Flexible-Blade Coaters.* The first flexible-blade coater was introduced in 1945 and is known today as the puddle coater (Fig. 8). A puddle-type pigment coater was developed at Fraser Paper approximately 25 years ago for waxed

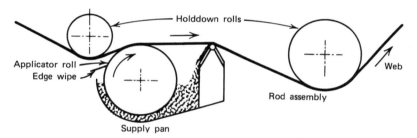

Figure 7. Metering rod coater.

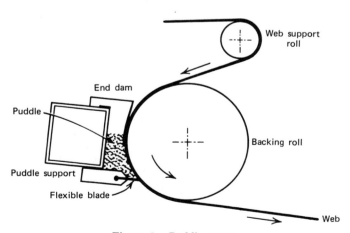

Figure 8. Puddle coater.

breadwraps. Subsequently it was adapted for the application of pigmented coatings on publication-grade papers.

These coaters, also known as trailing-blade coaters, apply aqueous coatings and occasionally volatile solvents. Blade coaters have the unique feature of troweling in the low areas in the fiber formulation of a web of paper or board, thus producing a coated surface that has excellent smoothness and printing qualities. However, if the web is poorly formed, the blade coater can provide a mottled appearance owing to the additional coating in low and porous areas. Often these areas do not appear until the substrate has been calendered. If a base substrate has a high tendency for mottle, such as the felt side on a board sheet, the final coat should be applied by an air knife coater which follows the contour of the sheet and disguises the mottle.

The backing roll is usually covered with resilient material and is driven at web speed to stabilize the web and draw it past the blade. A replaceable blade is rigidly clamped at one end, whereas the unsupported end is forced against the substrate to support the coating pond and meter the amount applied to one side of the substrate. Blade pressures usually do not exceed 1400 N/m (8 pounds per linear inch) at the tip. The blades are usually of Swedish hardened spring steel, 0.30–45 mm and operate at 45° tangent to the roll. Weight is usually adjusted by varying blade thickness, angle, and blade-to-substrate pressure. Coat weights obtained by the puddle-type coater are 7–12 g/m^2; operating speeds are as high as 1300 m/min.

Advantages: high speed; low coat weight; and short dwell.

Disadvantages: high coating loss on web breaks; backing roll wear from dikes; poor coat weight running adjustment; poor operator vantage point; difficult to change doctor blade; difficult to filter coating effectively; and difficult to go off coat.

Roll Applicator Blade Coater. A second-generation blade coater, known as the flooded-nip, inverta-blade, or roll applicator (Fig. 9). This coater isolates the application from the blade assembly which enables the coating rejected by the blade to be refiltered before being reapplied. The roll applicator also eliminates wear of the backing

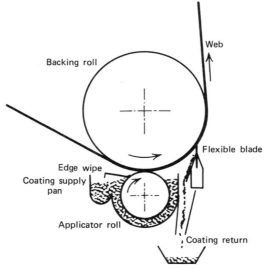

Figure 9.   Roll applicator blade coater.

roll owing to edge deckles because the coating deckle width is controlled by edge wipes on the applicator roll.

Unlike the puddle coater, the blade clamp and blade pressure are usually pneumatically controlled. The coating weight profile across the machine is accomplished by individual profile screws located approximately every 10 cm. This coater is capable of running at 1300 m/min and has a coating weight range of 5–18 g/m^2.

Advantages: application easily interrupted; quick blade change; adjustable coating weight; good operator accessibility; rejected coating refiltered; profile adjustment across the machine; and wide range of blade angles from 0–50°.

Disadvantages: applicator roll maintenance; high coating loss on web breaks; high viscosity; prone to skip coating at high speeds; high rate of coating supply required; and low coat weights.

Some applications require a more precise applicator system and this can be accomplished by using the two-roll system. However, the top speed of the coater may be reduced to 700 m/min.

Flexiblade. The Flexiblade coater was developed by The Black Clawson Company in 1958. It is similar to the puddle type coater, except that the coating pond is located at the bottom and is under pressure. It incorporates a pressure transducer in the coating pond to automatically control supply pressure. Multicoating ports allow for recirculating coating to ensure homogeneous coating. Although this coater suffers faults similar to the puddle coater, namely wear of the backing roll, it introduces many new concepts that are more or less standard on today's coaters. These include a pneumatic blade clamp, pneumatic tube–blade loading, and across-machine profilers for quality control. Coat weights range from 7–14 g/m^2 and the operating speed is as high as 1100 m/min.

Advantages: application easily interrupted; minimum coating loss; wide range of viscosity; good coating weight control; and short dwell.

Disadvantages: wear of backing rolls at end dams; leaking end dams; difficult to change blade angle; difficult to filter; difficult to control solids; low coating weight; and prone to skip coating.

Fountain Blade Coater. Another version was introduced by the Black Clawson Co. in 1963. The fountain blade coater provides noncontacting deckles, prefiltration of all coating applied, adjustable dwell, adjustable blade angle, and an oscillating blade coater. The early designed fountains featured a slot or coating chamber that had to run within a few thousandths of the backing roll. Figure 10 shows a typical fountain blade coater installed on a machine with a 3-m-wide board.

Coating is pumped into multiple-feed ports directly from a filtered adjustable supply pump. The coating rejected at the blade is refiltered before reapplication so that blade scratches and solids build-up are minimized.

The latest improvement is the jet applicator which actually extrudes the coating through a substrate to applicator gap of approximately 25–50 mm. This offers the advantages that the die orifice can be unplugged while running without going off coat and less coating is forced into the sheet because of low contact pressure. In operation these jet fountain coats require half of the coating supply that the puddle, roll, or wide-slot fountains require. Fountain blade coaters run as high as 1500 m/min and have a coat weight range of 2 to 90 g/m^2. Speed and viscosity are only limited by the requirement that the rejected coating from the blade flow down a pipe.

Advantages: wide coating weight range; no speed limitation; wide range of vis-

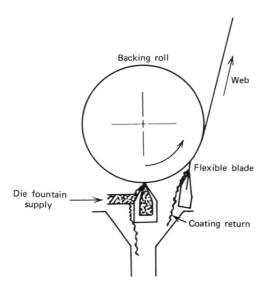

Figure 10.   Fountain blade coater.

cosity; less concern for equipment deflection; less coating penetration; less skip coating; adjustable dwell; applies and premeters; can be unplugged without going off coat; web splice passes applicator; maintains uniform viscosity and temperature; and minimum coating loss on web breaks.

Disadvantages: plugging.

Billblade Coater.   In 1968, AB Inventing of Sweden introduced the Billblade coater to upgrade the quality of coating at the size press area of a paper machine. This coater is very similar to the puddle coater, except that two similar or different coatings can be applied to a sheet simultaneously. However, only one side of the sheet has the smoother blade coating characteristic; the other side has basically the same smoothness as a size press and is used more or less for curl control.

Figure 11 shows a more versatile version including metering and transfer rolls. To redirect the flow of the coated web, a driven cooling roll is required below the coater on the coated side of the roll applicator before entering a dryer section. The water that condenses on the surface of the roll minimizes coating build-up or adverse effects on the wet coating. The coater can also be operated as a single-side coater. Coating weight range is approximately 10–12 g/m^2 and the operating speed is at 1300 m/min.

Advantages: curl control; handling of two formulas simultaneously; compact; high speed; and short dwell.

Disadvantages: viscosity limited; same as puddle; and one wet-coated side touches a support roll before drying.

Twin Blade Coater.   The latest blade coater arrangement is the twin-blade concept which consists of two fountain-blade coaters back to back without a backing roll. Each of the components of the applicator and blade assemblies are essentially the same as conventional flexible blade coaters.

The web is drawn through the coater in a vertical path with one blade on either side and it is essential that the web tensile withstand the drag created by two opposing blades. The blade angles are usually less than 35° and may be unbeveled or bent.

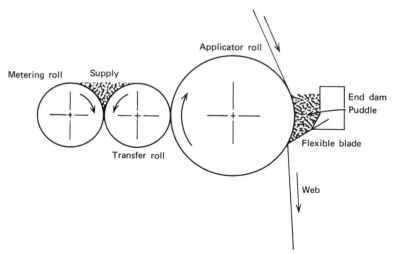

Figure 11.   Billblade with transfer rolls.

Coating weights range from 4–15 g/m^2 per side at an operating speed of up to 300 m/min.

Advantages: simultaneous two-blade surfaces; compact; and wide range of viscosity.

Disadvantages: keeping blade tips together; high web drag; and difficult to remove coating scratches.

Supported Rod Coater.   In another version of the blade coater the flexible blade metering device is replaced by a rotating metering rod supported in a flexible holder. However, this attachment is not common on puddle-type coaters because of edge dam requirements. The coater performs much the same as the metering rod coater except that the addition of the backing roll makes the coater not sensitive to tension and that the coat weight is adjusted by blade pressure. Operating speeds are as high as 300 m/min; coating weights range from 6–12 g/m^2.

Advantages: minimizes coating scratches on rough substrates; coating weight easily adjustable; and not tension sensitive.

Disadvantages: machine-direction film-split pattern; maintenance of rod and holder; rod speed above 300 m/min rod; high viscosity; and high coat weight.

Meniscus Coater.   One of the simplest coating machines used today on films that have stringent optical requirement utilizes the meniscus principle. Figure 12 shows one of the early basic designs known as a bead coater.

The web is supported by an idler-type back-up roll so that the web is a few micrometers above the level of coating in a constant-level pan. To start coating, the pan is usually raised to make contact with the web and then retracted to form the meniscus. The pan design is very critical and some employ a vacuum to minimize air currents. The coater has a very slow operating speed of only about 10 m/min on watery, thin liquids.

Advantages: excellent precision application; produces good optical qualities with minimum pattern; does not scratch substrate or undercoat; does not distort or elongate substrate; low coating weight; low viscosity; and clean room qualities.

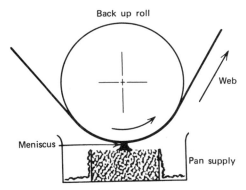

Figure 12. Pan-type meniscus coater.

Disadvantages: slow speed; high viscosity; coat weight changes with speed, viscosity; sensitive to room air currents; sensitive to air bubbles and ripples; high coating weight; low speed; and different width back-up roll for each web width.

The basic meniscus coater can be refined by adding an additional variable-speed applicator roll or rolls. This feature improves flexibility in adjusting coat weight or varying web speed and increases the operating speed up to 40 m/min.

A modified meniscus roll coater incorporates a die–fountain-type applicator with a variable-speed applicator roll to create an adjustable meniscus. The fountain can be deckled to control the desired coating width and minimize the effects of air currents and viscosities.

Kiss-Roll Coating.   There are many types of kiss-roll coaters. They are usually tension sensitive and require a postmetering or smoothing device such as a metering rod or air knife. Generally, the applicator (kiss) roll operates in the web direction; however, it can also operate in the reverse direction if additional coating is required. At slow speed, low viscosity, or high coating-weight requirements, the single-roll system is used (see Fig. 13). Addition of a second and third roll permits operation at higher speed and high viscosity.

The two-roll system is very common as an adhesive applicator for laminating absorbent-type kraft papers; however, a machine-direction film-split pattern (see Fig. 14) is usually evident.

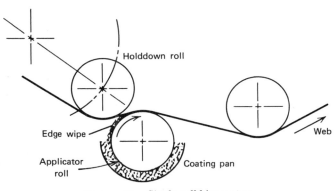

Figure 13. Single-roll kiss coater.

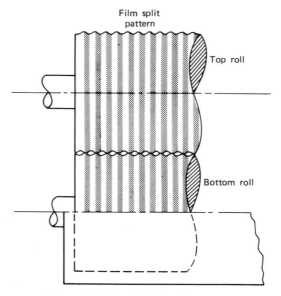

Figure 14.   Typical film split pattern.

Size Press Coater.   Size-press coaters are used to apply aqueous sizing materials to fabrics, but more specifically on machines making paper and paperboard. Since most sizing aids such as starch are intended to penetrate the sheet, the coater is somewhat crude in the sense that it applies a surplus and like a wringer-type washer squeezes the excessive amount out of the sheet.

The two-roll size press can be vertical or horizontal (see Fig. 15), or inclined. In each version, one roll is covered with resilient material, the other one is metal and is being driven. Nip pressures of up to 2.9 MPa (425 psi) allows the first roll to be driven by the second. Sometimes both rolls are driven to aid in threading the coater.

Advantages: simple; compact; easy to maintain; and can add size to either or both sides.

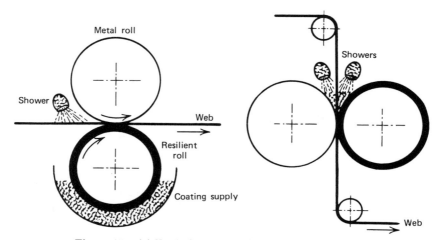

Figure 15.   (**a**) Vertical-size press; (**b**) Horizontal-size press.

Disadvantages: film split pattern; poor coating weight control; reduces caliper; and low solids, low coating weight.

Calender Water Boxes. Paperboard surface characteristics are improved by adding water boxes to the steel roll calender stack on the board-making machine (Fig. 16). Drying cans may be lined up after the stack to remove the excess moisture. The typical water box allows a surplus of liquid to contact the calender roll surface and reject the surplus at each nip following each water box. The solutions used are 3–5% solids content and can impart release holdout, pick resistance, and curl control to sheets of paperboard or folding carton. The water boxes are usually portable and can be installed or removed while the machine is running.

Advantages: simple; inexpensive; improves end product; reduces curl; and can treat either or both sides.

Disadvantages: corrosion of rolls covered by solutions; usually slows machine process down because of additional drying requirements; reduces caliper; and increases bulk density.

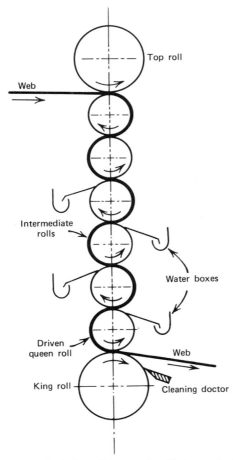

Figure 16. Typical 8-roll iron stack with four water boxes.

Transfer-Roll Coater. The transfer-roll coater, also known as the Kimberly-Clark-Mead (KCM), upgrades the performance of a conventional two-roll size press so that size solutions and coatings can be applied in a single station at higher solids and/or coat weights. It can be designed to coat the wire side, felt side, or both sides simultaneously on a paper or board machine.

In general, the rolls are each driven by separate drives at web speed; however, coating-weight control can be adjusted by operating the metering roll at only 50% of web speed. Coating weight is also altered by varying nip pressures.

The metering rolls are usually elastomer covered and nip lightly against the stainless steel transfer roll, which in turn is nipped against the resilient applicator rolls that are as hard as bowling balls. When coating only one side, the back-up roll is stainless; however, when coating both sides at once, both applicator rolls act as back-up rolls and are elastomer covered.

In operation, coating is pumped into the nip of each metering and transfer roll so that the surplus simply runs out the ends of the nip and is collected with funnel-type troughs that return the coating to the supply. Transfer roll coaters are capable of operating at 1000 m/min applying coat weights of 0.4–20 g/m^2.

Advantages: high speed; adjust coating weight while running; high solids; high viscosity; can apply two different formulas simultaneously; and easy to thread.

Disadvantages: roll split pattern on coatings; very cumbersome and expensive; and roll deflection compensation is required to ensure uniform coating weight across the machine.

Gravure Coating. *Direct Gravure.* Direct gravure coaters are the most accurate of the roll-type coaters. The coating weight is usually controlled by proper selection of the gravure roll etch. This coater is well suited to solvent, aqueous, and 100% solids-type coatings of viscosity below 500 mPa·s (= cP) (see Printing processes).

As shown in Figure 17, the substrate passes through a nip formed by the driven gravure roll and an elastomer-covered backing roll. The former usually runs in a pan of coating or is equipped with a fountain-fed applicator, followed by a flexible steel

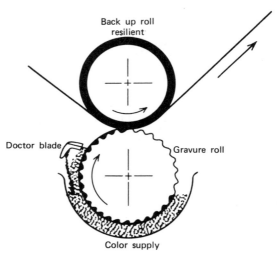

Figure 17. Direct gravure coater.

doctor blade before the nip. The film-split pattern formed at the nip or from the gravure cell structure can be minimized with a post-smoothing roll attachment. Gravure coaters operate at speeds up to 300 m/min and apply coating weights of 1–40 g/m² on almost any smooth substrate.

The most common cell patterns are the trihelical, quadrangular, and pyramid (see Fig. 18). They are applied to a steel roll by mechanical knurling or chemical etching, and can be obtained from the supplier or roll engraving according to the type, size, and depth of etch required for a specified coating weight with a given percent solids.

The direct reverse-roll gravure was developed in the 1970s. The coating is supplied to the gravure roll in a puddle-type trailing-blade fashion, and the gravure roll applies the metered coating in a reverse-roll action against the web and driven back-up roll. This arrangement is best suited for low viscosity coatings and offers coating weight adjustment by varying the speed of the gravure roll.

Advantages: simple to operate; compact; easy to maintain coating weight; coats a wide range of substrate materials; low solvent or heat loss; not tension sensitive; and gives low coating weights.

Disadvantages: difficult to change coating weight except on reverse application; backing roll is undercut for substrate width; gravure roll maintenance and life; doctor blade maintenance; high viscosity; film-split pattern; and coating weights change with speed.

Offset Gravure. In offset or indirect gravure (see Fig. 19) a steel back-up roll is added above the direct-gravure roll arrangement allowing an additional film split before the application to the substrate. This minimizes coating patterns and reduces coat weight.

In earlier versions, such as flexographic presses, the three rolls were geared together to maintain a desired speed relationship and coat weights were controlled with

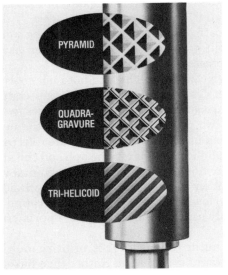

Figure 18. Common cell patterns. Courtesy of Pamarco Inc.

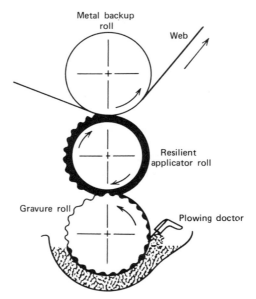

Figure 19.   Offset gravure.

direct gravure coaters. Today machines are usually equipped with individual variable-speed drives to each roll (offset differential gravure coaters). This feature allows using a deeper etch than generally required. Furthermore, running the gravure roll at a speed slower than normal provides the coater with adjustable coat weight control. Varying the speed of the elastomer-covered applicator roll practically eliminates coating patterns on light coat weights; surprisingly this change does not affect coating weight to any appreciable degree. A gravure roll operating at 10% and an applicator roll at 75% of line speed are not unusual.

The offset gravure coater can operate at speeds of 1000 m/min and provides a coat weight range of 0.02–7 g/m².

Advantages: low coating weight; can adjust coating weight while running; high speed; minimal film-split gravure-cell patterns; higher viscosity; and good coating weight stability with process speed changes.

Disadvantages: high coating weight; and hot melt cools on resilient roll.

Simultaneous Offset Gravure.   Both sides of a substrate can be simultaneously coated by a symmetrical coater arrangement. It features two gravure rolls in the outermost positions and two resilient covered rolls in the upper position; the latter function as applicator and back-up rolls. The gravure rolls can be driven at a variable speed; whereas the two applicator rolls nip the web and run at web speed. The vertical web lead is the most common arrangement, but horizontal web leads are also used.

This arrangement offers the advantage of coating both sides at one coating station. However, it is limited to low viscosities and web speed applicator rolls that can form a pattern. It is commonly used for the application of primers or precoats to a plastic web.

Transfer Printing.   During the early 1970s, printing multi- or single-color designs on fabrics was an important development. Rather than flexographic printing directly onto the fabric with its shrinkage and register problems, the patterns are heat transferred from a release sheet. The release sheet consists of a base paper with a release coating or a web of foil.

The web is then printed on the release side by the direct gravure method, dried, and rewound. The roll of preprinted release is then sent on to the fabric convertors for unwinding and pressure heat lamination of the preprint to the fabric. The fabric is then cooled and rewound and the release sheet is rewound and discarded. The precision of rotogravure printing with its half tone and over printing of endless color possibilities has brought a new dimension to the fabric industry.

Reverse-Roll Coating. *Three Rolls Vertical.* The vertical or in-line reverse-roll coater utilizes the principles of the reverse-roll kiss coater in addition to a back-up roll above the applicator roll to eliminate the tension-sensitive application.

Figure 20 shows the simplest version of reverse-roll coating where roll TIRs are not critical. This principle is widely used in the metal coating industry since the rubber covered applicator roll is gentle on the prime coat. The bottom roll is equipped with a coating supply, gravure pattern, doctor blade, and variable-speed reversible drive. The middle applicator roll is elastomer covered, drives at a variable speed in reverse direction to the web, and is lightly nipped to the gravure roll and the web which is supported by the driven back-up roll. The coating weight is controlled by adjusting the speed of the gravure roll and the coating pattern is controlled by the applicator roll speed.

Alternatively, the bottom and the top rolls may be rubber covered, and the middle roll may be steel with a hydrophilic covering.

Coating speeds of up to 300 m/min are common with coat weights up to 35 g/m^2.

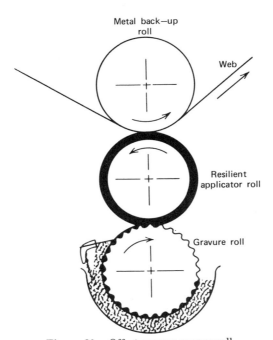

Figure 20. Offset gravure reverse roll.

Advantages: open design; simple; wide range of coating weights, process speed, and viscosity; applies a uniform coating even with substrate caliper variations; and not tension sensitive.

Disadvantages: solvent evaporation; deterioration of the resilient roll caused by solvents; solids buildup; contamination buildup; scratches on sensitive substrates; new top roll for each web width; promotes air bubbles; and strike-through on open webs.

Nip-Fed Coaters.   The three-roll nip-fed coater (Fig. 21) consists of a rubber-covered back-up roll and precision metal-applicator rolls. The web enters the coater through the nip of the low-position backing roll and leaves at the metering roll side of the coater. Coating is contained in a nip formed by the metering and applicator rolls and dams with teardrop-shaped edges. Coating weight is controlled by the combination of metering roll gap and applicator roll speed-up. Coating film-split patterns are minimized by varying the speed of the metering roll.

Metering and applicator rolls and bearings must be of utmost precision to ensure coating uniformity. It is common to find coasters with a 1-μm TIR for each metal roll. These rolls are usually manufactured from stable chilled cast iron and are supported by preload super-precision antifriction bearings. They are usually chrome plated to increase the wear characteristics.

A die-fountain applicator can be added to a conventional nip-fed coater arrangement. The die fountain can supply the coating nip, thereby eliminating the teardrop-edge dams or it can apply directly to the web at the applicator roll and bypass the metering roll. Typical applications for three-roll nip-fed coaters on most types of webs are adhesives, silicones, plastisols, organisols, and lacquers. The nip-fed reverse-roll coater is capable of operating at 250 m/min with coat weights up to 50 g/m^2.

Advantages: low solvent loss; wide range of coating weights, viscosity, and speed; improved resilient roll life; maintains coat weight even with caliper variations; and not tension sensitive.

Disadvantages: end-dam maintenance and wear on coating rolls; coating streaks by particles in the nip; danger of clashing and damaging metering and applicator rolls; poor operator vantage point when adjusting coater; TIR maintenance; promotes air bubbles; backing-roll strike-through on open webs; and backing-roll maintenance.

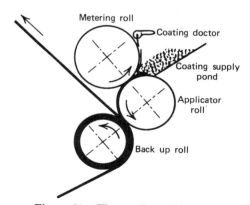

Figure 21.   Three-roll nip-fed coater.

Pan-Fed Coaters. The pan-fed reverse-roll coater, similar in principles and end uses to the nip-fed coater, was developed in the 1950s by Sorg Paper to apply a wear-resistant coating to paper webs. The coating supply is located under the metering and applicator roll system with the back-up roll on top. The pan-fed system with a three-roll applicator is limited to speeds of 200 m/min, and a four-roll coater to 300 m/min because of splash and feed problems.

Pan-fed reverse-roll coaters were employed for zinc oxide coatings, magnetic tapes, plastisols, silicones, thermosensitive copy or chart papers, and typewriter ribbons.

Advantages: good operator vantage point; and building block for many types of applications.

Disadvantages: splash; solvent loss; induces air bubbles; difficult to filter; and TIR maintenance.

Fountain-Fed Coater. The fountain-fed reverse-roll coater was developed in the late 1960s to improve the speed, splash, and filtering of the nip-fed and pan-fed coaters.

A die fountain has adjustable internal deckles that prevent the coating from contacting the roll ends where centrifugal force can produce splash or flinging at speeds over 200 m/min. Furthermore, for applications of stripe coating, the fountain can be restricted in areas where coating is not required. Coating speeds can be increased to 300 m/min.

Advantages: good operator vantage point; minimum splash; minimum air bubbles; widest range of viscosity; and good filter capability.

Disadvantages: TIR maintenance; fountain plugging or buildup; solvent loss; and backing roll maintenance.

The latest version of reverse-roll coatings, known as direct-fountain feed with reverse-roll metering is shown in Figure 22. The fountain applies the coating directly to the web supported by the applicator roll. Then the substrate and coating both pass through the nip of the metering and applicator rolls eliminating the need for the resilient back-up roll. The coating weight control is the metering roll gap with the pattern being controlled by metering the roll speed. On extremely wet films, it is not uncommon to operate without the metering roll feature. The fountain-fed reverse-roll arrangement gives excellent results on zinc oxide and solvent silicone coatings at speeds up to 600 m/min.

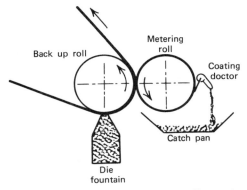

Figure 22.   Direct-fountain reverse-roll metering.

Advantages: high speed; no back-up roll; easy to change web widths; low coating weight; and low roll maintenance.

Disadvantages: open nip to pass splice; and fountain plugging.

Saturator-Web Impregnation. Saturators are webs designed to control the amount of residual coating solution evenly distributed throughout the cross-sectional area of a substrate. Final coat weights are often referred to in percent pick-up; eg, if the raw sheet weighs 100 g/m², 100% pick-up indicates that the end product weighs 200 g/m².

Saturating-grade substrates include porous or unsized papers, creped paper, woven fiberglass, canvas, and nonwoven and cotton linter. The substrate becomes only the carrier for a functional compound applied in a solution or suspension to exhibit wet strength, abrasion resistance, electrical properties, rigidity, and transparency. End products of saturators include masking tape, counter tops, electrical circuit board, fish poles, and air and oil filters, to name a few.

The three basic methods of saturating include dip and scrape, dip and squeeze, and reverse roll. Dip-and-squeeze saturators closely resemble dip-and-scrape saturators; however, the squeeze method is preferred because it is not tension sensitive, is less harsh on substrates, and provides for a more precise coating control. Maximum coating weight is obtained by increasing the dwell time in the pan, decreasing the squeeze-roll pressure, and increasing percent solids concentration. For substrates that can produce a blotter effect, the reverse-roll method is considered the most accurate, most suitable to adjust coat weight, and the fastest means of applying a saturate. Process speeds for the dip-and-squeeze and reverse-roll systems can exceed 300 m/min.

Figure 23 shows a modern saturator used for countertop laminate products (see Laminated and reinforced plastics).

Calender Coating. Many plastic compositions can be softened by heat and formed into self-supporting sheets by squeezing between heated iron calender rolls. Although this process was used during the nineteenth century for rubber compounds and for more than fifty years in the linoleum industry, it has been greatly refined for handling newer plastic materials to produce very precise film thicknesses. Figure 24 illustrates a typical calender-coating line in which a support web is pressed against the formed plastic sheet to obtain a coated product. The calender-roll arrangement may be stacked in line or may be offset, as shown, in a Z formation. The offset arrangement is generally preferable to a stacked design because the roll deflection or bending effect due to the separating force between the rolls is not cumulative and a change of roll nip pressure may be effected at any nip without significant reflection on the nip pressure or roll

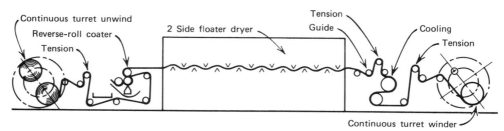

Figure 23.   Typical high speed impregnator.

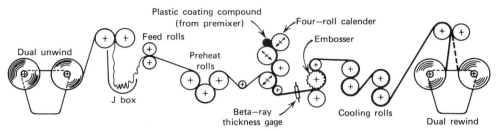

Figure 24. Z calender-coating arrangement.

deflection in any other nip. Changes in roll nip pressures and consequent change in roll deflections give uneven roll clearances across the roll nip face which, in turn, result in uneven film thickness across the coating width.

At lower operating speeds the rolls are heated to maintain the proper temperature. At higher speeds (ca 100 m/min), the rolls must be cooled to remove the heat created by shear friction in the rolling bank of plastic which is formed at the metering-roll nips. Since the roll pressures required for heavy plastic masses are very high, the roll construction must be very heavy. Attempts to heat and cool the roll surface of a heavy chilled-iron calender roll from a center-core opening are inefficient and not very effective. The rolls should therefore be designed with laterally drilled holes under the outer roll periphery to provide channels for heat-transfer liquids.

In the operation of a calender-coating line, a beta-ray scanning gage should measure coating thickness continuously. The data obtained may be used to automate the roll-compensating devices. Accuracy of control of coating thickness may be maintained within 5%.

To improve adhesion of calender coatings to the substrate, a priming coat is usually applied to the substrate before the calender coating. Such priming coats, improve coating-bond strength by a spread-out process and minimize coating penetration at the coating nip. Excessive coating penetration into porous textile substrates may cause stiffness and poor tear resistance.

Immediately following the coating operation the coated web passes to an embossing nip where a pattern is pressed into the hot plastic surface. Plastic coating calenders may range in width up to 3.7 m for floor-covering products and in operating speed up to 100 m/min.

Cast Coating. Many film-forming materials can be applied to continuous substrates by cast-coating methods, ie, the process which includes contact of a wet coated surface with a smooth polished drum or metal belt during the drying or fusing operation (see Fig. 25). This process gives flat-coated surfaces for decorative effects or for subsequent printing operations (see Printing processes).

Precast Coating. In the precast-coating process, a metered coating is deposited on a metal drum or metal belt followed by contact with the substrate and transfer of the coated material from the substrate, resulting in a smooth coated finish.

The precast-coating process includes many variations to meet special requirements. Multiple coats may be applied before contact with the substrate. Support surfaces such as release-coated paper and coated textile webs may be used to replace the metal belt. The casting support surface may be printed or embossed to transfer these effects to the cast coating. The principal advantages of the precast coating process

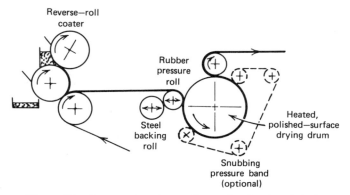

Figure 25.   Drum cast coater; coating metered onto substrate.

are accurate control of coating thickness and production of a smooth-bridged coated surface regardless of the roughness or irregularity in substrate.

Powdered-Resin Coating.   Powdered thermoplastic resins (particle size 25–200 μm) may be applied by flame-spray coating methods, principally on metallic substrates. For proper bonding the substrate should be cleaned and preheated to the required temperature. The powdered resins (such as polyethylene, epoxy resins, polysulfide polymers, nylon, etc) can be delivered axially into the flame, or molten to the substrate by an oxygen–propane flame blast or may be entrained in one of the gas components. Flame temperature may range to 540°C. Residence time in the flame is too short for serious thermal degradation. A typical hand spray unit may apply powdered polyethylene over a 4.5 m^2 area in 0.125 mm thickness in 1 h. Approximately 6.8 m^3 of oxygen and 4.5 kg of propane are consumed in 1 h of operation. The requirement that the substrate be at elevated temperature (ca 200°C) limits the scope of application in a commercial process. Nonporous films can be applied as thin as 0.125 mm by flame spraying.

Powdered resins can also be applied by electrostatic attraction by spray gun delivery. The powder particles are delivered from the spray gun under pressure with a high-voltage, low-amperage charge. No flame is required for the electrostatic spray gun to coat any electrically grounded metal object with a covering of thermoplastic powder. The process is completed by fusion. Owing to the electrical insulating characteristics of the deposited powder, the total coating thickness that is electrostatically attracted is limited to approximately 0.250 mm. Application of electrostatic powder coating to interior surfaces difficult to reach is not recommended.

Powdered thermoplastic resins such as polyolefins, poly(vinyl chlorides), polystyrenes, epoxy resins, etc, by the process for textiles is illustrated in Figure 26. In carpet backing a heavy coating is required in order to improve stiffness and permit thermoplastic moldability to conform to shaped spaces such as automobile floors. Applications to textiles may improve stiffness or web strength in nonwoven textiles. Deposition of the powdered resin may be from a hopper or reciprocating screen or by flour sifter and other techniques. Powdered polyolefin resins of very small particle size (8–30 μm) can be dispersed in water or organic liquids to permit usual coating techniques such as air-knife, roller, or knife coating followed by a fusing operation.

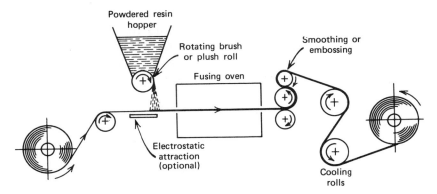

Figure 26.   Powdered-resin coating on textile substrate.

Dip Coating.   Irregularly shaped objects are frequently coated by immersion in a coating composition, and then removed, drained, and dried or baked. Coating thickness is determined by the end use. Tears or drops of coating at the bottom of dip-coated articles may be removed by electrostatic attraction as the article is moved along the conveyor.

The plastisol dip-coating process may be used when the article to be coated can be safely heated to the fusion temperature 176–205°C. As the heated article is immersed in a tank of plastisol, the coating adjacent to the hot surface is first thinned as it is heated and then gelled as it fuses. This viscosity minimum is passed very quickly. At some distance from the heated surface a layer of plastisol is reduced in viscosity but does not arrive at the gelation temperature. This layer and material at a distance from the heated surface, drains away leaving a gelled coating when the article is withdrawn from the tank. Additional subsequent fusion is usually required in a heated oven or under infrared radiant heaters. This process is widely used on irregularly shaped articles such as tool handles, toys, and automotive parts. The plastisol dip tank is built with water-cooled double walls to prevent overheating and arranged for hydraulic raising and lowering at automatically controlled rates. Figure 27 shows a dip-coating line. Application of water-based coating materials by a dip process frequently employs the electrostatic coating method using a positive electrode connection to the

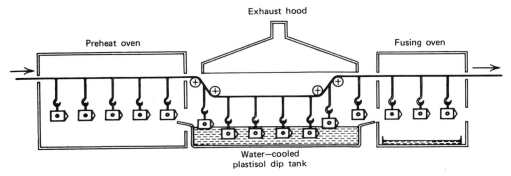

Figure 27.   Dip-coating line.

immersed conducting-metal article which then attracts the negatively charged paint particles from the coating solution in a manner similar to that in electroplating. This process gives good coverage to inaccessible areas and sharp edges.

A modified form of dip coating is obtained when the coating composition is poured or allowed to flow over the substrate. This operation depends on gravity flow as in dip coating. Production-scale flow-coating equipment may be conveyorized with the work pieces passing multiple-coating delivery nozzles at the top of the article to be coated. Excess coating drains to recovery tanks for recirculation. As in the case of dip-coated articles, tears or drops remaining on the bottom edges may be removed electrostatically before drying or baking.

Special precoated steels are used for the corrosion protection of cars including the standard galvanized variety and a new type consisting of two coatings, one containing chromium, the other zinc. They are applied and bonded to the steel before it is formed into parts. Plastic is used for such parts as grill opening panels, wheel splash shields, and bumper sight shields. The precoated steels are used on selected body panels.

Spray Coating. Plastic coatings may be applied by spraying on irregularly shaped and compound curved or sharp-edged surfaces. The principal distinction in spray-gun design depends on the method used to obtain atomization of the coating liquid.

Airless Spraying. In pressure atomization or airless spraying the coating is delivered to the nozzle under very high pressure and atomization results from dispersion in air as it leaves the nozzle. Figure 28 illustrates a typical pressure coating nozzle. Fluid pressure may range from 10–20 MPa (1500–3000 psi). The nozzle aperture ranges from 0.18–1.2 mm in diameter for various viscosities of coating. The nozzle is frequently constructed with a tungsten carbide insert to minimize abrasive wear. Nozzle contours are available to deliver round hollow, round solid, or fan-shaped spray patterns. The trigger arrangement for a pressure atomizer may require only a slight touch to move the needle valve from its seat. The fluid-coating pressure completes the retraction motion hydraulically. This rapid-opening feature minimizes sputtering. The valve is reseated under the same hydraulic pressure to prevent drip.

Air Atomization. Air atomization spray-nozzle design depends on the impingement of an air stream against the coating to break it into small droplets and carry it to the surface to be coated. Figures 29 and 30 illustrate two arrangements for external and internal air mix, respectively. Most spray-nozzle operations require the fluid coating to be delivered through the nozzle under pressure rather than by suction from its supply source. Figure 31 illustrates suction delivery of the coating directly from the fluid nozzle by vacuum created by the annular air delivery around it. As the coating

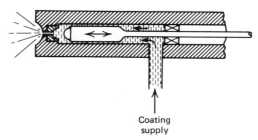

Coating
supply

Figure 28.   Pressure spray-coating nozzle; hydraulic airless atomization.

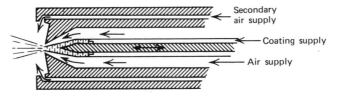

Figure 29. External air-mix spray.

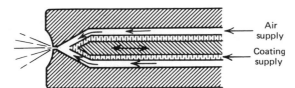

Figure 30. Internal air-mix spray.

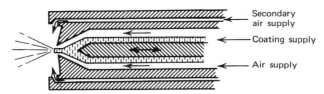

Figure 31. Coating delivery by nozzle vacuum.

travels in the annular air-ring path it is struck by secondary air nozzles for the second stage of atomization. Additional air from two horns or wings diametrically opposed to impinge at 90° from the secondary air jets shapes the spray into the fan shape which is usually desired. Suction delivery of fluid is not recommended for coating reservoirs with capacities larger than 1 L.

Spray-coating fluids are usually delivered by constant-displacement pumps or by other means to assure a smooth, pulseless flow. Pressure of delivery of a pressure feed should not excees 124 kPa (18 psi). Higher pressures result in streams of higher velocity which, in turn, require higher atomizing air pressures and are inefficient in operation. In practice, the fluid-flow valve is operated wide open and control of coating flow is obtained by changing valve size or by adjusting fluid pressure. For nozzles using suction feed the supply rate of coating is usually controlled by partial opening of the coating valve.

To reduce the coating viscosity, the coating fluid may be delivered to the nozzle through heated, jacketed pipes. With hot coatings this operation minimizes the amount of solvent required and speeds drying time. Coating temperature may be maintained within ±0.5°C.

Nozzle openings for air-atomized coatings range from about 1 mm in diameter for thin coating compounds to 3.5 mm in diameter for higher viscosities. Air volume required depends on many factors such as coating viscosity, nozzle size, rate of application required, air pressure, etc. Air volumes usually range from about 0.08–0.8 m^3 at 344 kPa (50 psi).

Coating materials are packaged in pressurized containers using liquefied fluorocarbon propellants. These are gases at room temperature and pressure; consequently they tend to push the coating materials from the can and to expand and to atomize and propel them toward the object to be coated. The average aerosol paint container carries approximately 50% propellant and 50% paint material (see also Aerosols).

Multicomponent Spraying. Multiple-component spray nozzles may be arranged for internal mixing or external mixing of components as shown in Figures 32 and 33, respectively. Materials of short pot life such as epoxy resins, polyurethanes, polyesters, etc, may be mixed and applied as coatings in this manner.

Multiple-component spray nozzles are usually arranged for rapid solvent flushing for clean-up. Coatings with viscosities as great as 5000 mPa·s (= cP) may be handled in multicomponent nozzle applications. The total fluid flow per nozzle may be as great as 48 L/min (13 gal/min). For intimate mixing of multiple-coating components inside the spray nozzle, some designs are available using a motorized high-shear self-cleaning rotor in the delivery passage.

To deliver spray coating uniformly on continuous flat sheets or continuous webs, the spray nozzles may be mounted as multiple units reciprocating laterally across the web as it travels under them. The nozzles are equipped with automatic spray shut-off triggers which stop the coating when the guns are decelerating beyond the edge of the sheet and when they are accelerating again to meet the edge of the sheet. Alternatively, the spray nozzles may be mounted in a horizontal rotating fixture above the web to obtain a uniform coating pattern (see Fig. 34).

When coating irregularly shaped objects in spray booths, it is common practice to revolve the object on a turntable so that the spray may strike from many angles. Spray booths are usually constructed with exhaust systems designed to trap or wash out substantially all of the overspray vapors that would otherwise be vented to the

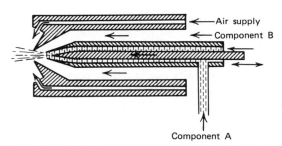

Figure 32.   Multiple-component spray nozzle, internal mix.

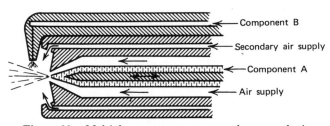

Figure 33.   Multiple-component spray nozzle, external mix.

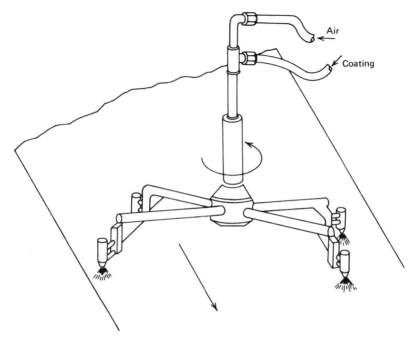

Figure 34.   Rotary multiple-spray nozzle coating head.

air. Spray booths may be small with a single opening, or may be totally enclosed for continuous conveyorized operation, or may even be tunnel-shaped mounted on floor tracks to permit operation on large objects such as railway cars, trucks, etc.

Electrostatic Coating.   Many coating materials of suitable dielectric constant may be electrically polarized so that they are attracted to a grounded or oppositely charged surface. This principle (electrophoresis) has been used for many years in electrostatic spraying equipment. The coating particles may be generated by airless or air-atomized spray nozzles, centrifugal force, or electric attraction. Practical applications extend from paint or plastic solutions and suspensions to hot-melt compositions and powdered resins.

Spray nozzles or guns are specially constructed for electrostatic spraying with an electrode extension in the center of the fluid nozzle (Fig. 35) to create a high voltage which delivers a charge to the atomized coating. Alternatively, coating may be sprayed centrifugally from a charged spinning cone or nozzle (Fig. 36) to present charged coating particles to the substrate. A suitable electric field may also be generated within

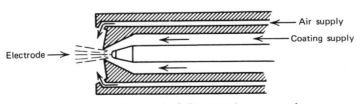

Figure 35.   Air-atomized electrostatic spray nozzle.

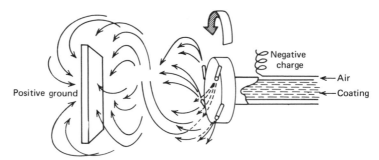

Figure 36.   Electrostatic spray from spinning nozzles.

a high-voltage wire frame or grid through which sprayed particles may travel to acquire a charge.

The article to be coated must be electrically conducting and supplied with a ground or opposite electrical charge to attract the charged spray particles. This arrangement can be made conveniently when the articles to be sprayed are traveling on a conveyor chain and hangers. Automobile bodies and frames are electrostatically coated with good economy of paint.

Substantially all of the electrostatically charged paint particles may be delivered to the substrate in a suitably designed installation. Charged particles travel around corners and are attracted to areas impossible to reach with conventional spray-coating techniques.

Continuous webs of nonconductive material such as textiles and paper may be electrostatically coated by using a grounded roller or plate supporting the web to attract coating to that area where it is intercepted by the web. To deliver an electrostatically charged coating for wide and uniform application, such as on paper and textiles, the coating may be presented from a flat charged plate or die. Electrostatic attraction is sufficient to pull the coating from the sharp die edge and to atomize it, because all the coating particles tend to repulse each other as they carry the same polarity. Further work on electrostatic coating of flat sheet is directed toward use of spray-generating equipment of higher capacity for delivery of coating in order to permit higher processing speeds.

Powdered plastics may be electrostatically sprayed by entraining them in an air stream and delivering them through a charged field to a grounded or oppositely charged area. After delivery, the coated article or surface must be fused unless it can be maintained at the plastic fusion temperature during coating. To obtain maximum adhesion of coating, the substrate is usually prime-coated with a suitable adhesive surface. As an example of electrostatic powder coatings, phenolic resin powder is coated on kraft paper and laminated for electrical-insulation purposes (see Powder coating).

Polymerization Coatings Formed *in Situ*.   Many monomers can be polymerized directly on a substrate to form a uniform surface coating. Several different methods are used: (*1*) magnetically contained glow discharge, (*2*) exposure to an electron beam, and (*3*) exposure to ultraviolet radiation (see Radiation curing).

Glow discharge is illustrated in Figure 37. Polymerizable gases or vapors are introduced into a chamber evacuated to 100 Pa (0.75 mm Hg). A minimum vapor pressure of 33.3 Pa (250 μm Hg) at room temperature is required to sustain the discharge. An electrical charge of 300–1000 V potential is applied to a substrate or to a metallic

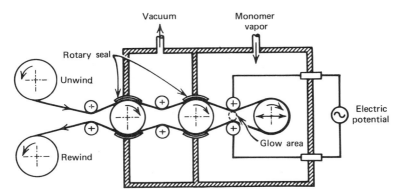

Figure 37. Coating by *in situ* polymerization induced by glow discharge.

support under it. Materials such as chloroform, isobutylene, toluene, or hexachlorobenzene are drawn to the electrode, where they coalesce and polymerize on the substrate surface.

Coating speeds on lineal webs up to 300 m/min have been reported. Glow discharge is superior to air–corona discharge since less energy is required. Coating thicknesses from 20–25 nm or greater are reported obtainable by this process. Further advantages of the glow-discharge coating system include absence of solvents and consequent solvent absorption by the substrate, coating film thicknesses above 0.5 μm substantially free of pinholes, and coating formation is confined to the surface of the substrate and does not penetrate as with use of ultraviolet irradiation.

Magnetically contained glow-discharge employs an external electroless radiofrequency power generator operating at 0.4 MHz. Electron dispersion is inhibited by magnetic guiding solenoids as it passes through the coils of the generator toward the substrate. Growth rates of 160 nm/min for polystyrene coating have been reported. Among other advantages, this is a cooler process than glow discharge. By accuracy of the magnetic focus the field can be controlled more closely, and by using lower energy input a polymer of greater uniformity can be obtained.

Electron-beam-induced polymerization uses a low-energy electron beam on preselected areas of a substrate. This equipment requires a higher vacuum than does glow discharge which results in higher equipment costs and slower rate of polymer formation. Using Van de Graaff electron accelerators, coatings have been cured on plyboard and wallboard at satisfactory production rates on conveyorized equipment.

In the third process, the substrate is subjected to ultraviolet irradiation in the presence of vapors of a monomer such as butadiene. This permits application to preselected areas instead of the overall application obtained in glow-discharge apparatus. A tendency for ultraviolet radiation to penetrate the substrate may cause deterioration in some materials.

It has been reported that polytetrafluoroethylene can be coated onto a substrate by vaporization at 480–815°C in a vacuum chamber of ca 0.1 Pa (1 μm Hg) with condensation on the adjacent substrate. The polytetrafluoroethylene is reported to be depolymerized into constituent monomers and dimers at the operating temperature which is attained in an electrically heated tungsten container. Coatings 10–12 μm thick obtained by polymerization on the surface of aluminum are resistant to attack by salt spray, acid, and caustic (see Film deposition techniques).

Curtain Coating. In the curtain-coating process the coating composition is delivered in a falling sheet or curtain to the substrate, which is moved through the curtain at a controlled rate. The fluid coating composition is delivered to an enclosed slotted flat die where the coating emerges as a falling film or sheet (see Fig. 38). Its thickness is controlled by careful regulation of fluid pressure at the nozzle and precise lateral adjustments of the discharge-slot opening. The vertical distance of the coating head above the substrate can be adjusted. The falling curtain of coating is protected from stray air movements by transparent enclosure sheets. Coating thicknesses as low as 12 μm are possible when coating with lacquers or with low-viscosity wax melts, and as 20 μm with hot-melt compositions of higher viscosity. Use of multiple or tandem coating heads offers the possibility of application of an extended range of heavier coating thicknesses as well as of multiple layers of different coating compositions in a single pass. Air-bubble entrapment may occur in the case of a gravity-applied continuous coating over an impermeable substrate. Bubbles may also be caused by moisture vaporization from the substrate. Remelting of the coating may minimize the bubble defects. Curtain-coating equipment of the slotted-die (orifice) design is capable of operation at lineal substrate speeds up to 500 m/min.

Curtain-coating equipment is also available in which the falling curtain is generated by overflow from an open weir. The coating is delivered to the open weir uniformly across its width by a pipe with diffuser jet openings. As the coating overflows the low side of the weir it travels down a short flat skirt before dropping. By precise control of the rate of delivery of coating to the weir, the thickness of the falling curtain is adjusted. As with the slotted die, hot-melt coatings can also be applied by the open

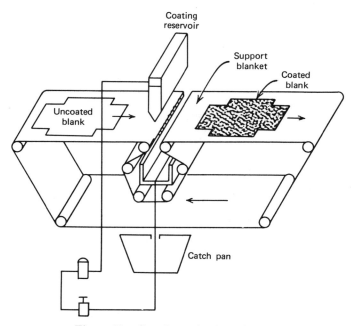

Figure 38.   Curtain coating from die slit.

weir. Since there is no close restriction to flow as in the slotted die, the open weir does not tend to form scratches or coating streaks due to crusting or coating hang-up in the slot opening. When applying hot-melt coating formulations the coating supply is held in a reservoir at a temperature that does not thermally degrade the material during its residence. As the coating is pumped to the weir it passes through heat exchangers to bring it to the optimum coating viscosity. Weir-type equipment is recommended for operation at substrate speeds up to 400 m/min. All coating composition not carried away on the coated surface falls into a collection trough for recirculation.

Curtain coating is adaptable for coating on irregularly sized sheets such as slotted cut-out corrugated carton blanks as well as on continuous substrates. Coatings may also be applied to uneven geometric shapes such as square or triangular blocks.

The principal limitation of this process is that the coating composition must be sufficiently fluid to fall freely and sufficiently cohesive to present a continuous film to the substrate.

Extrusion Coating. Extrusion coating is the process of melting, metering, combining, and cooling a molten homogeneous sheet of thermoplastic material onto a continuous substrate such as paper, paperboard, film, foil, or fabric. Today's expanding food packaging industry is the direct result of packaging improvements that can be attained by altering the surface and physical characteristics of a flexible web by the extrusion coating process (see Film and sheeting materials).

Some foods such as milk, cheese, and cold cuts require that the packaging material exhibit specific heat seal, vapor, and oxygen barriers to obtain an extended shelf life (see Barrier polymers). A good example is offered by Tetra Pak of Sweden with a report that liquid milk can be preserved for up to six months without refrigeration. The main characteristic of their plastic-coated milk carton is that it can be heat sealed through the milk area thus eliminating air in the package.

A typical extrusion coater consists of an extruder to melt and pump the thermoplastic fluid, a heated die to uniformly distribute the plastic into a vertical curtain, and a laminating station to combine the plastic with the substrate and then cool the laminate. Figure 39 shows a basic extrusion coating station that applies polymers such as polyethylene, polypropylene, surlyn, nylon, etc. As the plastic leaves the die at approximately 175°C, depending on the resin being extruded, it is approximately 0.5 mm thick. It is then elongated owing to the pulling effect caused by the pressure nip and the substrate which is moving at the higher velocity. The elongation effect also reduces the width of the extrudate approximately 2–6 cm and the film thickness to approximately 12–25 μm before it makes contact with the substrate. Uniform temperature control of the plastic and nonsurging fluid pressure before the coating exits the die is very important to the success of the coating. Variations lead to irregularities in the curtain thickness both in the machine direction and across. Stable irregularities in the across machine direction can be minimized by mechanical profiles on the die metering lip. The extruded film width is adjustable by external deckles to block off the exit of the die.

The extruded film usually first makes contact with the substrate. The laminating nip then helps to promote bonding before chilling the molten plastic. The driven chill roll is chromium or nickel plated and can have a mirror, matte, or an embossed surface. Once the extruded film passes through the laminating nip, it takes on the finish of the chill roll. The chill roll is 60–90 cm in diameter, double shell, spiral fluted, and utilizes refrigerated water to reduce the film temperature to ca 65°C in 120° wrap of the chill

Figure 39.   A basic extrusion coating station.

roll before the film is stripped. Extrudate bond to the substrate usually is insufficient by the pressure nip alone and adhesion-promotion primers are usually applied to the web in line before extrusion coating. Priming can be electrostatic, chemical, or in the form of ozone treatment.

Coating weights are controlled by varying line and extruder speed by keeping the extruder constant. However, in many cases the chill roll capacity limits the maximum thickness to be obtained. Extrusion coating lines operate at speeds up to 1000 m/min, and can consume 1400 kg of plastic per hour per extruder and apply 10–30 g/m^2 of coating.

Coextrusion or multiple extruders feeding one die provides the unique capability of producing multiple layers of different resins to produce superior functional properties with many economical advantages such as using inexpensive resin as the core of a three-ply extrudate with the outer plies being more expensive but also much thinner than they can normally be extruded alone. A typical coextruded film could be 5-μm of surlyn, a 15-μm core of polyethylene, and a 5-μm surlyn outer layer.

DRYING SYSTEMS

Drying systems are a very important part of the coating process since once a coating has been properly applied to a substrate, it must be conveyed, dried, and cured before the coated side can be contacted again. The drying system is based on the available fuel for the heat source, strength or stability of the coated substrate and the maximum rate at which evaporation and curing can be accomplished without detrimental effect on the end product (see Drying).

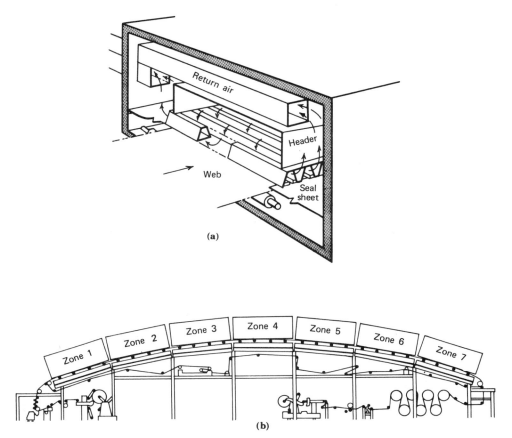

Figure 40. (a) Cross section of air impingement dryer. (b) High speed air-knife coating line with single side felt-supported impingement drying with can after dryers.

The most common heat sources for drying include steam, oil, dual gas (natural and propane) and/or oil, electrical radiation, ultraviolet light, and high frequency generation. Drying can be contacting or noncontacting.

In general, contacting dryers are in the form of steam-heated rotating cast-iron drums that lead the substrate through the drying or curing process in a serpentine fashion, drying each side alternately. The moisture evaporated by the drums is removed by the exhaust hood. In the coating process, drum drying is used in the final stages for curing or reducing sheet cockle. Drying rates can be increased by incorporating felt or a fabric belt to press the sheet against the drum. Normal coating drying rates for drums are 7.5 kg per square meter of dryer per hour. Heated air impinged against the coated side of the substrate is by far the most common means for drying coated papers or films. Thus the coated side is not contacted during the drying process but a drying rate of up to 50 kg per square meter of dryer per hour is obtained.

As shown in Figure 40, hot air driven by high-speed fans is forced through nozzles that are 10–20 cm apart and a few millimeters from the sheet, directly on the coated surface. This air eventually heats the coating and causes evaporation which then can be exhausted from the dryer area by slots located between each dryer nozzle. Air ve-

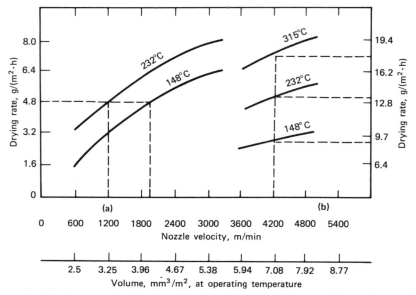

Figure 41. Drying rates for water-based coatings; (**a**) air-knife rates, solids under 50%; (**b**) blade and rod rates, solids over 50%. All rates are maximum. Reduced rates shown when (*1*) coating weights above 10 g/m² (dry) are applied; (*2*) slow release binders such as casein, glucose, or protein are used; (*3*) coating which have mud-cracking or skimming-over tendencies are applied. These rates are based on slot-type nozzles which run continuously across web width and are located no less than 2.5 cm from the web surface; the distance between nozzles does not exceed 10 cm.

locities, 900–4500 m/min at the exit from the nozzle, are determined by the ability of the coating to remain completely undisturbed during the drying process.

Convection or impingement-type dryers are available in the supported or the flotation type. The supported type, as shown in Figure 40, has air impingement on one side with web support in the form of drums, rolls, belts, or conveyors on the un-coated side. In flotation or two-side drying, the sheet actually floats through the drying process between staggered top and bottom nozzles that are 3–5 mm away from the sheet. Thus drying action is provided from both sides which in fact improves the drying rate by 25% over single-side impingement. This two-side drying effect reduces web distortion, binder migration, and maintenance.

Radiant-type drying in the form of gas or electric infrared emitter systems can be utilized as a drying system or integrated at the entrance to convection systems to raise the coating temperature to aid in evaporation or at the exit to promote cur-ing.

High frequency or dielectric drying was first applied to the paper-coating field in the mid 1960s but is little used because of high energy cost. It is mostly used today for final moisture profiling at the exit of the convection dryer on paper or heavy cross-section webs such as plywood.

Figure 41 illustrates typical drying rates for aqueous coating applications for air knife (low viscosity) and blade (high viscosity) coatings with regard to temperature of air, velocity of air at the nozzle, and volume of air on a square meter basis.

Figure 42 indicates that in application of silicone at 100% solids at a coating weight of 0.9 g/m², the dryer has only a time–temperature cure requirement. This curve would

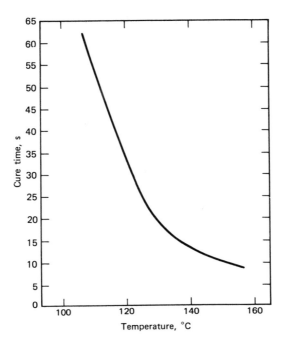

Figure 42. Cure time vs temperature of 100% solids (no evaporation required). Conditions: 250 g/280 m² surface silicone; the air circulating oven is calibrated to ±1.5°C. Courtesy of The Dow Chemical Company.

be additive to any solvent evaporation that may have been required to other than 100% solids.

BIBLIOGRAPHY

General References

G. L. Booth, *Coating Equipment and Processes,* Lockwood Publishing, New York, 1970.
Modern Plastics Encyclopedia, McGraw-Hill, New York, 1977.
TAPPI Monogr. Ser. (38), "Pigments;" (37), "Adhesives;" "Extrusion Coating;" "Blade Coating," Technical Association of The Pulp & Paper Industry, Atlanta, Georgia; a source of current coating technology for the paper and plastics industry. This technology is available in the form of a monthly magazine, conferences, seminars, films, recordings, and periodically upgraded monographs. The data is from actual field experience provided by suppliers and users of coating process lines.
A. N. Leadbetter, *Backsizing of Tufted Carpet,* Dow Chemical U.S.A., Midland, Mich., Nov. 4, 1976.
PVDC Emulsions—Applications and Handling Techniques, W. R. Grace & Co., New York, pp. 35–38.
Z. Tadmor and I. Klein, *Engineering Principles of Plasticating Extrusion,* Van Nostrand Reinhold Co., New York, 81–84.
Lockwood's Directory of Sources of Supply for the Paper and Allied Trades, Vance Publishing Corporation, New York, published yearly.
Walden's ABC Guide—The Complete Paper Directory of Manufacturers & Convertors, Walden Mott Corporation, Oradell, N.J., published yearly.
E. Miller and D. Taft, *Fundamentals of Powder Coating,* Society of Manufacturing Engineers, Dearborn, Mich., 1974.
Expanding Market for Electrostatic Coating, Gorham International, Gorham, Maine, 1975.

"Continuous Vacuum Metalizing," *Mod. Plast.* (Dec. 1977).

A. S. Mujumdar and W. J. M. Douglas, *Impingement Heat Transfer—A Literature Survey,* Pulp & Paper Research Institute of Canada, Pointe Claire, Quebec, Can., 1972, pp. 86–91.

L. G. Janett, *Drying of Coatings and a Summary of Methods,* Midland Ross Corporation, New Brunswick, N.J.

"Gas Phase Polymerizations," *Plast. Technol.* **10,** 9 (Feb. 1964).

"Graft Polymerization on Cellulose," *U.S. Dept. Comm. Office Tech. Serv. PB Rept. 181,580.*

A Guide to Literature and Patents Concerning Polyvinyl Chloride Technology, 2nd ed., Society of Plastics Engineers, Stamford, Conn., 1963, pp. 179–201, 241–300.

"Industrial Paints," *Chem. Eng. News* **42**(41), 100 (Oct. 12, 1964).

Kaiser Aluminum Foil, 1st ed., Kaiser Aluminum and Chemical Sales, Inc., Chicago, Ill., 1958, pp. 80–161.

Machinery and Equipment for Rubber and Plastics, 2nd ed., Rubber World, New York, 1963, pp. 73–78, 245–258.

"The Paint Makers," *Chem. Eng. News* **42**(6), 80 (Feb. 8, 1964).

Pigmented Coating Processes, Monograph No. 28, Technical Association of The Pulp and Paper Industry, New York, 1964, pp. 1–89.

Plastics Engineering Handbook, 3rd ed., Reinhold Publishing Corp., New York, pp. 250–285.

Processing of Thermoplastic Materials, Reinhold Publishing Corp., New York, 1959, pp. 380–404.

R. M. Brick and J. R. Knox, "Coating by Glow Discharge," *Mod. Packaging* **38,** 123 (Jan. 1965).

D. E. J. Cunningham, "Plastics Coatings from Powders," *Rubber Plast. Age* **45**(9), 1054–1055 (Sept. 1964).

A. S. Dawe and J. A. Kinn, "The Finishing Story," *Can. Paint Varnish* **31**(11), 24 (1957); **31**(12), 32 (1957); **32**(2), 31 (1958); **32**(3), 36 (1958); **32**(4), 26 (1958); **32**(5), 39 (1958).

J. H. Gorrell, "New Way to Coat Plastics and Metals," *Plast. Technol.* **10,** 45 (Oct. 1964).

R. J. Jacobs, "Selecting Coating Methods," *Pap. Film Foil Convertor* **37**(2), 47 (1963); **37**(3), 50 (1963); **37**(4), 62 (1963); **37**(5), 60 (1963); **37**(6), 60 (1963); **37**(7), 53 (1963).

R. Mosher, *Technology of Coated and Processed Papers,* Chemical Publishing Co., Inc., New York, 1952, pp. 294–460.

H. F. Payne, *Organic Coating Technology,* Vol. II, John Wiley & Sons, Inc., New York, 1961.

J. Pomeraniec, U. G. Shapiro, and H. Mark, "Electrostatic Web Coatings," *Mod. Plast.* **41,** 133 (Sept. 1963).

H. R. Simonds, *Encyclopedia of Plastics Equipment,* Reinhold Publishing Corp., New York, 1964, pp. 66–80, 94–100.

Stanley C. Zink
The Black Clawson Co.

COATINGS, INDUSTRIAL

Industrial coatings production in 1978 for the U.S. market was about 1.4×10^9 L (3.7×10^8 gal) valued at $\$2.2 \times 10^9$ and was about equal to the production of trade sales coatings (see Paint). The industry is a mature one, growing little in volume yearly.

Coatings are available as liquid or fusible compositions. Table 1 gives a classification of coatings formulations. The liquids are usually aqueous or organic solutions. The coatings are applied by the user to the substrates, allowed to flow out smoothly chiefly by forces of surface tension, and then cured to the final solid form.

An industrial organic coating usually consists of an organic binder, pigments, a carrier liquid (sometimes omitted), and various additives. The binder is a polymer of high molecular weight which may or may not be cross-linked. The pigments, which may be organic or inorganic, contribute primarily to opacity and color, in addition to durability, hardness, adhesion, and particular rheological properties of the coating in fluid form. If the pigment content of a coating exceeds a critical value, the coating will become porous and brittle (an advantage when sanding is required). The relative volume concentration of pigment to binder at which porosity begins is called the critical pigment volume (see Pigments).

Lacquers contain high molecular weight binders that may be linear or branched but not cross-linked and do not change upon application and further treatment.

Lacquers that have the binder in solution also have high viscosity and are usually applied by spraying with a volume of solvent five to eight times that of the coating. Lacquers in which the molecules are dispersed as multimolecular particles are lower in viscosity and can be sprayed at higher concentrations.

When lacquers are dissolved in a volatile solvent their molecules are loosely intertwined and in constant motion. As the solvent evaporates, the polymer molecules draw closer to each other and, after sufficient evaporation, the film will be dry and hard.

When lacquers are dispersed (as opposed to dissolved) in a volatile solvent, the polymer particles (each particle containing many molecules closely packed) float freely in the fluid medium. As the solvent evaporates the particles move closer together; when sufficient solvent is evaporated, the particles touch and may coalesce if they are soft enough. At this stage they must flow sufficiently to cover the pigment particles and to form smooth films if high gloss is desired.

For lacquers applied from solution, the lowest molecular weight material is used that will yield an acceptable coating. When the lacquers are applied from dispersion in organic liquids or water, the effect of high molecular weight upon viscosity may be negligible. However, it reduces flow after the dispersion has coalesced. Therefore, high molecular weight material in a dispersed lacquer reduces gloss.

The binder can be incorporated into the coating composition in the form of low molecular weight materials that react after application to form a high molecular weight barrier. Such coating compositions are known as enamels. Enamel binders can be dissolved or dispersed in water or in an organic solvent.

After application of a smooth coating, the solvents are removed by heating. In the case of enamels, polymerization is then initiated by heat, atmospheric oxygen, water vapor, or exposure to an electron beam or ultraviolet radiation depending on the binder.

Table 1.   Formulation Possibilities[a] **of Synthetic Resins in Coatings Formulation**[b]

Vehicle system	Uses
Class 1: Vehicles containing oil-modified alkyds or other polymers containing drying oil	
(1) oxidizing alkyd resins (sometimes mixed with oleoresinous varnishes)	(1) architectural enamels, house paints, interior paints, flat wall paints, baking and air-drying undercoats and enamels for machinery, prefab housing structural units, and other factory products
(2) alkyd and phenoplast alkyd and nitrocellulose alkyd and chlorinated rubber alkyd and polystyrene alkyd and diisocyanate alkyd and vinyl and epoxy	(2) air-drying or low-temperature baking undercoats and enamels (for metal products) that have more plasticlike film properties than is possible to attain with alkyds alone
(3) alkyd and aminoplast alkyd and aminoplast and epoxy alkyd and silicone	(3) similar uses as above, but where a high premium is placed on color retention, and superior chemical and heat resistance
(4) oil-modified epoxy resins and aminoplast	(4) air-drying or baking-type undercoats or enamels; improved baking enamels and undercoats
Class 2: Vehicle systems containing no alkyd or drying oil	

(5) vinyl acetals and/or
    phenolic
    allylaminoplast } and 2,4,6-trimethylolphenyl
    alkyd   ether
    epoxy
	(5) chemically resistant baking undercoats and enamels

(6) phenoplasts (with or without epoxy, or vinylacetal, or aminoplast)	(6) thermosetting undercoats and/or enamels for high corrosion protection, especially in thin films (not resistant to discoloration); room-temperature setting mastics for corrosion and abrasion protection
(7) polyester and triazine resin allyl polyester silicone thermosetting acrylics complex amino resins some other polyesters	(7) chemical and discoloration resistant, glossy, clear films and pigmented baking enamels for metallic and nonmetallic production goods (thermosetting film formers)
(8) vinyl acetate–chloride, copolymers vinylidene or vinyl chloride– acrylonitrile copolymers butadiene copolymers acrylic copolymers poly(vinyl acetate)	(8) thermoplastic lacquers, baking or air-drying emulsion paints for production finishes on nonmetallic goods such as acoustical board or molded plastics, fire-retardant and corrosion-protective mastics; exterior house paints and interior decorators' paints of emulsion type
(9) nylons some cellulosic esters and ethers polyurethanes polytetrafluoroethylene poly(vinyl acetals) saturated polyesters unsaturated polyesters and styrene epoxy and polyamide copolymers of ethylene or propylene	(9) special type of coatings, potting compounds, mastics, etc, for electrical insulation and corrosion protection

[a] Commercial formulations are generally proprietary.
[b] Courtesy *Chemistry in Canada.*

428

It is difficult to form lacquers with molecules that are not cross-linked and are of sufficiently high molecular weight from low molecular weight materials. In the formulation and cure of enamels, extensive cross-linking occurs.

Cross-links limit swelling in solvents. Lacquers without cross-links are dissolved or infinitely swollen in good solvents; enamels with tight cross-links swell little; enamels with loose cross-links swell more.

Lacquers have certain advantages over enamels. They remain soluble and combine well with an additional coat containing the same binder, whereas enamels do not always adhere to a previous coat of the same composition. Although lacquers dissolve on application of a good solvent, they do not wrinkle or lose adhesion as enamels sometimes do. On the other hand, lacquers applied from solution require larger volumes of solvent than enamels. Up-to-date lacquers applied from nonaqueous dispersion are highly swollen and are partially dissolved and dispersed materials.

Enamels have certain advantages over lacquers, ie: when highly cross-linked they are not thermoplastic; they do not flow (or soften excessively) on exposure to heat; they are not softened or permeated as readily by solvents; and they can generally be applied at higher concentration.

Although most industrial coatings are applied from solution or dispersion in liquids, coatings can also be applied from powders containing pigment, binder, and additives. The powder particles are large compared to those in aqueous or organic dispersion. The deposit on the substrate is initially a porous mass that is melted and flowed together at an elevated temperature. Air must be expelled during the molten stage. The molecular weight must be particularly low to permit sufficient flow for satisfactory appearance without the aid of solvent. Powders are usually of the enamel type. Additives are used as catalysts for cure reactions and to control the smoothness of the coating. The flow of the coating slows upon melting as the enamel components react. The formulation must be adjusted to flow out sufficiently before the film is cross-linked.

Coatings are classified as primers that are applied directly to a substrate or as topcoats. The latter are applied over a primer and are usually the last coat. Intermediate coats are called sealers. In some uses, one coat satisfies all requirements.

Primers are usually pigmented slightly below the critical pigment volume. In some cases pigments give opacity and color. Extenders are added for economy. Topcoats are usually pigmented well below the critical pigment volume, particularly where high gloss is required.

Pigments are ground as concentrated dispersions (or mill bases) of high pigment-to-binder ratio in a grinding vehicle, usually a polymer solution. The dispersions of pigment are diluted with additional polymers and other ingredients to form the final lacquer or enamel. Ball, pebble, attrition, or sand mills are used for grinding and dispersing opaque pigments and extenders; two-roll (or rubber) mills and dough mixers are used for the dispersion of the finest transparent pigments, ie, those that absorb light without scattering it. Aluminum pigment is stirred into a mill base. For industrial coatings, few pigments can be dispersed finely enough at low viscosity with high shear equipment.

Titanium dioxide is the preferred white opaque pigment. It is generally used in the rutile form and is surface-treated with silica and alumina for easy dispersion and to shield the binder from the catalytic degradation by titanium dioxide in uv light.

Selection of colored pigments is based on light absorption and scattering characteristics, contribution to durability, and the effects on rheological properties.

Binders are grouped into certain overlapping classes such as acrylic, vinyl, alkyd, polyester, etc. The structure of the binder molecules and the forces operating between the molecules determine the properties. For example, polyester molecules containing esters of linear aliphatic acids are easily hydrolyzed on exposure to water whereas those containing esters of acids branched on an α carbon are not; polymers containing hydrogen atoms α to a benzene ring are easily oxidized as compared to those having no α hydrogens; polyethers containing hydrogen on the α carbon are apt to be oxidized on exposure to air and light whereas aromatic polyethers are stable.

Application

Industrial coatings are generally applied with specialized machinery representing a large investment. Coating properties must be controlled carefully to ensure continuous and satisfactory application (see Coating processes).

Rheological Problems. Sagging occurs when (1) the film applied is too thick, (2) the yield value is too low, and (3) the viscosity and yield value drop in the early stages of solvent evaporation or cure. Drops in viscosity may occur because of a temperature rise or change of solvent composition.

Popping, evident as small intact or broken bubbles, occurs when gases are liberated in a film by a chemical reaction or boiling. Ordinarily, the gases are dissolved in the film and diffuse to the surface. If the film becomes too rigid or the rate of gas formation is too great, the gas cannot diffuse fast enough and the concentration of gas exceeds its solubility. Bubbles occur in nucleated supersaturated solutions of gas.

Craters or shallow dimples appear in coatings because of local reduction in surface tension of the liquid film. The surface is pulled to regions of higher surface tension and the liquid adjacent to the surface is pulled with it.

Benard cells are hexagonal patterns that are produced by surface-tension forces. The patterns indicate flow as a result of surface-tension variation. Pigments segregate or orient themselves along the flow lines and may produce effects on gloss, color, and durability.

During application by roller, ridging occurs. The film does not form a flat surface parallel to the sheet but develops elevated ridges parallel to the machine direction. Ridging has been attributed to instabilities caused by surface-tension forces and may also be related to tensions produced by extensional forces.

Spattering is the formation of fine droplets carried in the air during roller coating. This is the result of formation and destruction of coating threads as the roller leaves the wet surface.

Vibration of rollers produces lines parallel to the axis of the rollers.

Spraying usually produces a more or less grainy surface, that is, minute depressions and elevations result in imperfect specular reflection. Some of these deviations are caused by large spray droplets that splash where they hit. Some are caused by Benard cells, and some by particularly fine spray droplets that arrive at the surface too dry (high in viscosity) to flow out smoothly within the available time at the existing temperature. Smooth flow-out is opposed by the rheological properties that are introduced to prevent sagging and pigment settling.

Although it is difficult to produce perfect smoothness in commercial coatings, it is possible to reduce the forces of surface tension that cause roughness, or to operate in gloss ranges where changes in roughness are not readily discerned by the eye.

Properties

Table 2 gives comparison ratings of properties and relative costs of coating for-mulations.

Stability. *Settling.* Pigments settle at rates depending on their particle size and the viscosity of the medium. Coarse pigments such as extenders and metallic flakes settle rapidly compared to the finer white and colored pigments. Settling results in a hard cake difficult to disperse that has no yield value if the dispersion is deflocculated. Flocculation and settling to a loose structure yields compositions easy to redisperse before use.

Freeze–Thaw. Aqueous compositions may freeze and crystals of ice may sepa-rate. If the dispersion of the pigment or vehicle is lyophobic (eg, a latex), it will not be easily redispersed upon thawing. Addition of glycols lowers the freezing point. Use of a sufficient amount of dispersion agent, properly selected, and a latex that does not coalesce at the freezing point improves stability.

Chemical Resistance. In the case of enamels, reactions take place upon storage, perhaps the very ones that are designed to cross-link the films. These reactions may lead to changes of viscosity and application properties.

Dispersions or solutions of polymers in water may hydrolyze and consequently decrease in molecular weight and viscosity. Thus, drying rates may be reduced and final film properties affected. Hydrolysis is to be expected in the case of polyesters or alkyds.

Aluminum pigment reacts with water in aqueous finishes unless treated with inhibitors.

Color. Mixtures of pigments are often used to provide a desired color. Occa-sionally the color changes on storage because of flocculation. This can be prevented by adequate formulation. Certain binders discolor on exposure to light, eg, poly(vi-nylidene chloride) and its copolymers; others yellow on standing in the dark, eg, lin-seed-oil alkyd resins (qv); and still others yellow on heating and become bleached in sunlight, eg, acrylonitrile copolymers (qv). Vinyl polymers (qv) of acrylic esters, sty-rene, acrylamide, acrylic acid, and condensation polymers employing melamine–formaldehyde (see Amino resins) and oil-free alkyd resins are relatively color sta-ble.

In certain industrial plants the coatings are circulated continuously through pipes at low viscosity using centrifugal pumps. Separation of pigments in circulatory lines by centrifugation can occur resulting in color change.

Good color matching is important in automobile finishes. Metallic appearance varies strongly with angle of illumination and observation and any flow that might occur in the fluid coating before solidification occurs. Metallic colors are highly sen-sitive to the technique of application. Matching must be satisfactory over a range of angles of illumination and observation, and measured spectrophotometrically (see Color).

Smoothness or Gloss. Liquids applied to a plane surface tend to adopt a smooth contour. As solvent evaporates, the coating shrinks as a liquid with a smooth surface to form a solvent-laden solid. As the solid loses more solvent it shrinks toward the substrate. This shrinkage in the solid state causes roughness in the surface to match roughness in the substrate.

Table 2. Comparison of the Relative Highs and Lows of Properties and Cost for Eighteen Commercial Coating Formulations[a]

Dominant resin type in coating	Probable range of baking temp, °C	Relative ratings of important film properties							Relative cost of a 0.05 mm thickness of coating[b]
		Quality of whiteness after production baking to fix film	Resistance to additional discoloration when aged at max baking temp in column 2	Acid and alkali resistance	Blister resistance in hot deionized water	Resistance to damage by impact	Resistance to abrasion	Resistance to solvents and food stains	
oxidizing alkyd	100–120	5	2	2	4	7	4	3	1.00
oxidizing alkyd and melamine and/or urea	100–150	6	4	4	4	6	5	5	1.30
nonoxidizing alkyd and melamine	140–170	10	7	5	6	4	5	4	1.50
nonoxidizing alkyd and urea	140–170	8	5	3	4	4	4	3	1.35
vinyl chloride–acetate copolymer	150–200	8	4	6	3	9	10	6	1.50
acrylic-type copolymers	120–190	10	9	5	6	7	6	4	4.00
styrenated alkyds (oxidizing)	80–120	7	4	3	3	6	3	2	1.10
phenolic	160–200	3	5	5	9	4	8	8	1.70
epoxy	190–200	5	7	8	8	4	9	8	2.00
epoxy and melamine	120–170	6	8	7	6	5	7	8	1.50
melamine and ethyl cellulose	135–170	8	6	6	5	3	6	6	1.50
polyurethane and alkyd	65–135	4	2	5	6	8	8	5	1.60
silicone	190–225	10	10	9	8	5	5	4	10.00
silicone and alkyd	175–200	10	8	7	7	6	6	4	7.00
allyl ester copolymers	160–190	10	9	8	6	2	8	9	6.00
polyamide (nylon) 0.25 mm flame spray	>290	2	6	4	10	6	10	4	5.00
polytetrafluoroethylene (Teflon), flame spray	>340	4	10	10	8	10	10	10	13.00
poly(chlorofluoroethylene) (Kel-F)	>315	4	9	9	9	10	10	10	11.00

[a] On a rating scale of 0–10, 10 represents the best performance and 0 represents failure. A rating of 4–5 represents fair to good.
[b] Alkyd equals 1.00. Courtesy of Finish.

432

Films can be smooth when applied over a rough surface only when there is no shrinkage after the film becomes immobile. This condition can be approached with high-solids enamels and with powder coatings (qv), but not with the more common formulations.

When extender particles are dispersed in a coating, the particles may protrude through the surface on drying, resulting in a rough surface. Such extender particles confer more roughness as the coating continues to shrink after solidification. Therefore, high-solids coatings tend to be glossy and do not yield readily to flatting with pigments.

Metallic Luster. The typical appearance of metallics is caused by the orientation of a metal such as aluminum flake in planes parallel to the surface. This occurs because the film shrinks after it has become solid, ie, after the flakes are kept from randomizing their orientation. Accordingly, lacquers show the highest metallic luster and enamels or powder coatings the poorest.

Durability. Ultraviolet radiation, water, and elevated temperatures accelerate the degradation of coatings. Resistance to commonly encountered chemicals (eg, fuel and oils in the case of automobiles, food stains in refrigerators, and hydrogen sulfide in certain can coatings) should be high.

Organic coatings may fail by cracking, usually accompanied by gradual erosion and the loss of adhesion and gloss. Cracking is attributed to shrinkage because of polymer loss, accompanied by reduction of molecular weight and brittleness.

Pigments affect durability in a number of ways: (1) Certain pigments, notably untreated titanium dioxide, accelerate the degradation of vehicles. (2) Others appear to prolong coating life because the intensity of uv light is reduced in the bulk of the film. Carbon black and iron oxides may function in this way. (3) Film degradation exposes pigments, and some are washed away. The effect of the change of refractive index upon exposure alters the scattering and absorption quality resulting in a change of color even if pigment is not lost. Gloss or smoothness also changes.

Adhesion. For certain uses coatings must be easily detached from the substrate; for example, strippable coatings of plasticized poly(vinyl chloride) for temporary protection. For most of the industrial coatings, a high level of adhesion to the substrate is desirable.

Many organic coatings consist of two or more layers applied directly to a substrate or to an inorganic conversion firmly adhering to the substrate. A single layer of organic coating is frequently insufficient. For example, a finish for steel used outdoors might employ a topcoat formulated for appearance and durability, and a primer layer or a layer next to the substrate formulated for low permeability and protection against corrosion. In some cases, the topcoat binder does not adhere to the undercoat binder but chips off or peels under stress. In such cases, increasing the pigment content of the primer to near or above the critical pigment volume improves the intercoat adhesion. Adhesion depends not only on the force applied to the interface but also on the rheological and mechanical properties of the surfaces. Stresses, film thicknesses, temperature, and humidity affect adhesion (see also Adhesives).

Permeability. For protection against water, the solubility of water in the vehicle and the diffusion coefficient should be low. The solubility is increased by the presence of hydroxyl, carboxyl, amino, and other polar groups within the polymer. The diffusion coefficient is reduced by increased packing of binder molecules; eg, by crystallization. Water passage through pigmented films is often increased by the presence of hydrophilic pigments that provide aqueous pathways along pigment surfaces, and by poor pigment dispersion which provides air-filled spaces between close-packed pigment particles not wetted by vehicle. Although permeability to water is generally reduced by increasing pigmentation, it increases dramatically when the critical pigment volume is exceeded.

Hardness, Flexibility.   In general, an industrial coating should be resistant to mechanical damage of coated articles. However, coatings often fail by cracking and chipping, the former often occurring between substrate and coating. A coating should be hard and flexible over the temperature range to which the article will be subjected.

Hardness is tested by the magnitude of a depression left by a weighted diamond-point. Flexibility is measured by bending a coated sheet over a conical mandrel and observing the curvature at which cracking appears. Tests of tensile strength and elongation of free film yield some information on flexibility.

Polymers harden as the temperature is reduced. At the glass transition temperature any single polymer or copolymer passes from a brittle to a plastic state. The plastic state has a rather short temperature range within which it is tough. The glass transition temperature may be changed by choice of monomers in the binder.

Enamels are best formulated for flexibility when they are in a rubbery range and loosely cross-linked, ie, long chains between cross-links. Tightening of cross-linking reduces staining and swelling by solvents, but makes the films more brittle and harder (see Hardness).

Corrosion Protection.   Iron and aluminum are corroded by electrolytic processes that involve reactions and diffusion of ions and oxygen between anodic and cathodic areas of the metal. For corrosion to occur rapidly an adequate supply of water must be available at the surface to carry these ions. The diffusion of oxygen is also important.

In order to protect the sheet metal against corrosion before painting, the metal is first cleaned. Then an inorganic coating is deposited, largely zinc or iron phosphate on iron or aluminum oxide on aluminum; a chromate rinse may also be applied. These coatings provide an inorganic hard surface to which paint or other coatings adhere strongly. The inorganic coatings alone give only brief protection.

Organic coatings also afford excellent corrosion resistance. This type of coating is usually hard, adherent, and hydrophobic and may contain corrosion-inhibiting pigments, notably zinc chromate. These coatings reduce the corrosion rate yet the mechanisms by which they do so are not known in detail. It seems that (1) reduction of the number of ions and restriction of water reduces electrical conductivity, (2) adherent and impervious oxide films are produced by the chemical reaction of inhibiting pigments, and (3) strong adhesion prevents the formation of a water layer that might bridge anodic and cathodic areas. Suitable topcoats provide good appearance, in addition to protection against corrosion and mechanical damage.

Organic coatings containing close-packed zinc particles in contact with steel protect the steel; the zinc is corroded instead (see Corrosion).

GOVERNMENT REGULATIONS

Air

In 1974, the EPA set national air quality standards for particulates, sulfur dioxide, carbon monoxide, photochemical oxidants, and hydrocarbons (see Air pollution). The standards for photochemical oxidants and hydrocarbons are most significant for the coatings industry. The concentration of photochemical oxidants is limited to 160 $\mu g/m^3$ as the maximum average concentration for one hour; this 1-h concentration is not to be exceeded more than once per year. Hydrocarbon standards are similarly set at 100 $\mu g/m^3$. The intent of antipollution regulations is to eliminate the emission of volatile organic compounds, all of which are regarded as ultimately photoreactive. Most guidelines or plans follow the Los Angeles Rule 66 and its amendments of August and November 1972. Rule 66 limits the solvent emissions from plants in the area. Such solvents are routinely emitted into the atmosphere from plants that use organic coatings dissolved in organic solvents. Rule 66 distinguishes between solvents that are photochemically reactive (defined below) and those that are not. It also distinguishes between emissions that are liberated after contact with a flame or without contact with a flame. Table 3 gives the limitations and exemptions provided in Rule 66.

It is recognized that all solvents will, in fact, react ultimately on exposure to uv light and oxides of nitrogen. Current rules limit "photochemically active" to certain types of solvents that had yielded appreciable amounts of oxidants in short-term experiments. These are: solvents containing more than 20% in total by volume of either group 1 (olefinic or cycloolefinic hydrocarbons, alcohols, ketones, esters, ethers, or aldehydes), group 2 (aromatic hydrocarbons with 8 or more carbon atoms per molecule, except ethylbenzene), or group 3 (a combination of ethylbenzene, ketones having branched-chain hydrocarbon structures, toluene, or trichloroethylene). They also include any solvent that contains more than 5% of group 1, more than 8% of group 2, and more than 20% of group 3.

The limitations do not apply if pollution abatement equipment is used (incinerators, absorbers) that removes at least 85% of the vapors which would otherwise be emitted (see Air pollution control methods; Incinerators).

The effect of the air-pollution regulations is to encourage the development of high-solids and water-based finishes to replace solvent-based finishes in industrial coatings although almost all contain substances that yield volatile organic effluents to some degree. New binder systems and new compounding and application procedures have been developed.

Table 3. Limitations and Exemptions Provided in Rule 66

Process stage	Exemptions of nonphotochemically active solvents	Limitations of photochemically active solvents
application and flash-dry areas, baking area	1361 kg/d and 204 kg/h	18 kg/d and 3.63 kg/h
no flame contact	1361 kg/d and 204 kg/h	18 kg/d and 3.63 kg/h
with flame contact	6.8 kg/d and 1.36 kg/h	6.8 kg/d and 1.36 kg/h

Some users of industrial coatings have converted from direct-fired to indirect-fired ovens to take advantage of the more liberal emission standards for indirect-fired ovens (see Furnaces; Solvents, industrial).

High-Solids Finishes. Unlimited use of paints is allowed when solvent volume does not exceed 20%, provided that (1) the volatiles are not photochemically reactive; (2) more than 50% by volume of the volatiles are evaporated before entering an oven, and (3) the organic solvent vapors do not contact flame.

Unlimited use of paints is allowed when organic solvent content does not exceed 5% by volume, with a volatile content that is not photochemically reactive, and when the organic solvent does not contact flame.

To be exempt from most current regulations, coating materials must contain, in their volatile portions, no more than 20% by volume of a nonphotochemically active mixture in an organic solvent, as defined in the regulations. The binder must be used in solution because there is an insufficient volume of solvent to separate the particles of binder (except under very restrictive conditions of particle size). The polymer must be of low molecular weight to meet viscosity requirements and must be applied in enamel form.

Alkyd resins (qv) and oil-free polyesters (qv) with appropriate use of cross-linking nitrogen resins are the best choices. Hot application is optional.

Water-Based Finishes. Unlimited use is allowed provided that the volatile content is only water and organic solvents; the organic solvents do not exceed 20% of the volatiles and the volatile content is not photochemically reactive; and provided that the organic solvent does not contact flame.

Water-based coatings have been used in trade sales for a long time (see Paint). In these finishes the binder, notably poly(vinyl acetate) and acrylic copolymers stabilized with surfactants, is dispersed in water. Under the impact of air pollution regulations, water-based coatings are being used again in a variety of industrial operations.

Water-based coatings using a polymer latex are difficult to redisperse or dissolve after drying (see Latex technology). Solutions of polymers in water in the form of amine or ammonium salts of polymeric acids can be redispersed easily. Water-dispersible coreactants such as certain urea–formaldehyde or melamine–formaldehyde compositions may be used (see Amino resins).

Frequently, coatings meet air pollution standards by using mixtures of water and organic solvents.

Ultraviolet and Electron-Beam Cured Coatings. Coatings containing monomeric materials and that are cured by polymerization under uv or electron-beam irradiation qualify as high solids finishes under air pollution regulations because only small amounts of volatile monomers are liberated. Electron-beam curing has had only a poor reception in industrial coating operations because of high investment cost; uv curing is used in coating of papers and wood (see Radiation curing).

The Environmental Protection Agency has issued guidelines for the States setting limits on the emission of organic vapors from painting operations in terms of grams of organic vapors per liter of paint. The limits vary with the industry but do not distinguish between organic chemical species. The States are expected to conform by Jan. 1, 1982.

Water

The Water Pollution Control Act amendment of 1972 extends the Federal authority to all United States waters. Industrial plants within a given industry were instructed to eliminate pollutants from their effluents by July 1, 1977. Higher standards will be required by EPA if they judge that such higher standards can be achieved at low cost. By July 1, 1983 industrial plants must reduce pollutants according to the best feasible measures so that progress is made towards elimination of effluent pollutants by 1985 (see Water pollution).

The new law increases the cost of compounded coatings because cleaning of equipment between production batches yields materials that must be eliminated in addition to unusable by-products and leftover materials.

Health and Safety

NIOSH has developed stringent standards for maximum permissible concentrations of many materials employed in the organic coatings industry. Asbestos, chromates, lead, cadmium, and mercury are some of the hazardous materials used, and an intensive search is going on for safe substitutes. Residual vinyl monomers of polymers, particularly vinyl chloride and acrylonitrile (qv), are being reduced greatly.

The FDA issues regulations governing materials that may contact food. The regulations are given in The Code of Federal Regulations, Title 21, Subchapter B, Parts 175, 176, and 177. The composition, mode of treatment, and permitted use are given for a wide variety of coatings and adhesives. Generally, a new coating must pass a series of extraction tests before obtaining FDA approval.

The Toxic Substances Control Act of 1977 is meant to protect the public from toxic substances that might be introduced into the environment. By May 1, 1978 an inventory list of all manufactured materials had to be submitted by producers and was published in rough draft (Candidate list of Chemical Substances, EPA Office of Toxic Substances). The complete list will be published in 1979. New manufactured materials require adequate testing and have to meet toxicity standards (see Industrial hygiene and toxicology; Regulatory agencies).

Uses

Automotive Finishes. Coatings for the automobile industry amount to about $300 million per year. Coatings for automobiles must achieve satisfactory levels of appearance, durability, and corrosion protection. In addition, topcoats are formulated to give a high gloss. Close control of color is important since different parts may be coated in different plants and assembled after painting. The shading of topcoats is, therefore, often controlled by spectrophotometers or electronic colorimeters (see Color). Repairs involve additional applications. Heating and coating must be resistant to baking and stable in storage.

Coating Process. The partly assembled car is bathed or showered with hot aqueous phosphoric acid in an automotive coating containing dissolved zinc and other metals, rinsed with dilute chromic acid and water, and then dried leaving a conversion coating of zinc or iron phosphate. Surfaces within box-like structures are likely to be coated incompletely (see Metal surface treatment).

Subsequent coating operations are limited to temperatures below 177°C to avoid buckling of the metal and movement of the solder. Primers are applied and baked, sealers may be applied, and topcoats are applied and partially dried. Drying removes most of the solvent. During this drying operation, metallic particles become oriented parallel to the surface. The surface may then be sanded very lightly to remove rough spots and then the automobile is heated to melt the topcoat (ie, evaporate more solvent) causing the coating to flow and remove the fine scratches introduced by sanding.

Repairs required after the final assembly operation involve sanding, usually a small area, recoating that area, and baking with an ir lamp.

The interiors of some cars are coated with lacquers containing an acrylic copolymer made with small amounts of certain monomers to obtain adhesion to the undercoat.

Primers.   Steel corrodes rapidly upon exposure to water, particularly in the presence of chlorides. The increased use of salt on snow and ice on the roads increases the corrosion of automobiles. Special corrosion problems are also introduced by connections between dissimilar metals. For example, the contact between sections made of galvanized steel and adjacent nongalvanized steel. The electrical nature of the contact causes paint films to lose adhesion (cathodic disbonding) from the adjacent nongalvanized steel, and rusting is accelerated on the nongalvanized area (see Corrosion). Coatings must be resistant to this type of failure.

Automobiles have areas of steel that are not exposed and are inside channels or recesses where they can, nevertheless, get wet and corrode. These areas are now receiving special attention. For example, new primer systems (especially by electrodeposition) are being introduced, and galvanized and precoated steel are primed with an additional coating of zinc, chromic acid, and hydroxyl-bearing polymers.

Corrosion protection is the main function of the primer, in addition to smoothing out small imperfections. A primer-surfacer is used when the metal is rough enough to require a thicker coating that can be sanded to eliminate irregularities.

Some of the metal parts (eg, front fenders, hoods) are dipped or flow coated with a primer bath using a binder of alkyd and epoxy resins (qv) dissolved in an organic solvent. The primer is usually black, and contains rust-inhibitive pigment below the critical pigment volume. The dipping tanks may contain large amounts of primer and are constantly being aerated. Antioxidants prevent drying or bodying of the paint in the tanks. The primers must be of proper thickness, usually 7.5–25 μm thick. The coating is then baked to harden. Dipping processes work poorly in recessed areas because the solvent vapors remain too long, and the coating film is dissolved again.

Some of the rougher metal, notably that used on bodies, should be coated with a solvent-borne primer-surfacer pigmented close to the critical pigment volume to allow smooth sanding. However, such films are porous and are, therefore, preceded by a thin film of flash primer, ie, a thin coating of low pigment concentration that does not harden before the application of the primer-surfacer.

Some of the metal in certain plants is coated after conversion coating by anodic electrodeposition (eg, with polybutadiene- or epoxy-based binder) and baked. Electrodeposition provides means of meeting air-pollution regulations while reaching recessed areas with an adequate coating, whereas dipping processes do not. Currently, cathodic processes are being developed that, unlike anodic processes, do not attack the conversion coatings and can result in better surface protection.

Topcoats. The binder used for solution lacquer topcoats consists mainly of poly(methyl methacrylate), and also contains cellulose acetate butyrate, a small amount of alkyd plasticizer, and a copolymer of methyl methacrylate with an acrylate. Pigment dispersions used with lacquers may contain copolymers with varying amounts of polar monomers to aid in dispersion of pigments. The mixture of polymers used in the binder confers a longer plastic range than pure poly(methyl methacrylate). The amount of each pigment used depends on its ability to hide the substrate and give the desired color. Pigment concentrations close to the minimum required improve the gloss and hold down costs. A concentration of aluminum pigment of about 2% achieves the particular familiar metallic appearance.

The binders used for dispersion lacquers are also composed largely of poly(methyl methacrylate) and are stabilized against flocculation by graft polymers whose segments are more nearly compatible with the hydrocarbon mixtures that comprise the bulk of the organic solvent used (see Methacrylic polymers).

The primer surfacer is sanded lightly and coated with an acrylic lacquer or enamel. For use with acrylic lacquers, surfaces that are coated by anodic electrodeposition or with dip-primer are also coated with a "sealer" or intermediate coat of a special copolymer that will adhere to the undercoat, and to which a lacquer will adhere. The sealer is not baked separately and the topcoat is applied directly to the wet sealer. The acrylic enamels often require no sealer.

A water-based acrylic enamel topcoat can be used to meet air-pollution standards. It is composed of a mixture of an acrylic resin (see Acrylic ester polymers) and a melamine–formaldehyde resin, both of which have good resistance to ultraviolet light.

Some plants apply acrylic enamels, dissolved or dispersed in organic solvents, for their topcoats. Although enamels cannot be applied as smoothly as lacquers, they are resistant to the aromatic solvents contained in gasoline.

Lacquer topcoats can be repaired by sanding a small section to a feather edge and applying a primer coat followed by a lacquer topcoat that overlaps the primed area. The patch will then be barely visible. Enamels, particularly if overbaked, tend to require sanding in order for an extra coat to adhere. They require repair of whole panels rather than spot repairs not only because of adhesion problems, but also because the metallic particles cannot orient over the feathered edge as they do over the bulk of the patch.

Metallic lacquers that solidify while still containing a large amount of organic solvent dry with higher luster than the lower molecular weight enamels.

Enamel coats ordinarily need no polishing. However, metallic finishes cannot flow out to yield perfect gloss, and clear unpigmented coats of a colorless vehicle are sometimes applied over the topcoats to improve smoothness. Such clear coats also protect aluminum in topcoats from attack by acid and alkali. Metallic finishes are almost always topcoated with clear finishes on European cars.

Today, automobiles are constructed with plastic parts coated with the same color as the outside of the car. Thin plastic parts may be made of thermoplastic polyurethanes and bulkier parts of thermoset microcellular foam polyurethanes. These plastics are relatively soft and resilient and cannot be heated above ca 120°C. They are coated with a lacquer-type primer consisting of a polyether urethane and are topcoated with an enamel finish to match the color of the rest of the car. Most of the topcoats are made with binders of hydroxy-functional polyester urethanes that are cross-linked with a melamine–formaldehyde resin. Others are acrylic resins cross-linked with melamine–formaldehyde resin and contain a plasticizer (see Urethane polymers; Amino resins).

The coatings for flexible substrates cannot easily match the metallic colors. The durability requirements for the metal parts are difficult to meet with the soft, flexible resins that must be used over relatively soft, tough substrates. Ordinary coatings for the metals would not tolerate the flexing expected for the soft parts, particularly at low temperatures.

The two-package enamel systems that react at room temperature (limited pot-life) are particularly useful for this purpose.

The use of powder coatings (qv) in automobile manufacture is limited to wheels in a few plants.

Coil Coating.   Consumption of coil coatings is estimated at ca 80 ML (ca 21 million gal) or $120 million annually. For articles fabricated from sheet metal by bending or forming, precoated metal is often used. Strips of metal 1000 m long and about 1.8 m wide are uncoiled and fed through machines that apply conversion and organic coatings in continuous motion at speeds of up to 100 m/s, followed by baking. Steel may be coated with zinc, either electrogalvanized or hot-dipped, for corrosion protection. Usually a conversion coating is applied, dried off rapidly, and followed by a primer coat of about 5 μm and baking. The metal is cooled, a topcoat of 20–25 μm is applied, and the metal is then baked. Then it is cooled rapidly and rolled up again. Coatings are sometimes applied to both surfaces of the sheet. Application of organic coatings is usually by reverse roller coating (see Coating processes).

Typical uses for strip-coated metal are aluminum siding, ceiling panels, radiator enclosures, and building panels. The strip-coated metal is formed into final shapes by bending and stretching. The coatings must be capable of withstanding the resulting stresses at various temperatures.

For those end uses requiring long exposure to uv radiation (eg, residential siding), special attention must be paid to durability characteristics of the vehicles and pigments used.

Many coil coatings are applied from organic solvents. Frequently the solvents are incinerated to meet air-pollution standards.

Primers.   Primers are used on galvanized and cold-rolled steel for exterior exposure or when water-borne finishes are applied to steel or aluminum.

Primers for less critical applications are often based on alkyd binders. Epoxy esters are mainly used for more demanding applications. Acrylic latex primers have been used for fluoropolymer-based topcoats and for thermoset acrylics.

Topcoats.   Organic solvent-based topcoats may be thermoset acrylic resins containing alkoxymethylacrylamide and durable acrylic monomers, or melamine–formaldehyde resins.

A latex-based composition may comprise a binder mixture of an aqueous solution of a low molecular weight acrylic resin or a high molecular weight latex based on alkoxymethyl acrylamide and durable acrylic monomers (see Acrylamide; Acrylamide polymers).

Poly(vinyl chloride)-based organosols or plastisols formulated to withstand uv degradation and applied in heavy coats of 75–125 μm over an epoxy ester-based primer can be durable and cheap, and also meet the forming and embossing requirements (see Vinyl polymers).

Silicone-modified oil-free polyesters and silicone alkyds cross-linked with melamine–formaldehyde are often used on steel building panels (see Silicon compounds).

Fluoropolymers based on poly(vinylidene fluoride) and poly(vinyl fluoride) are the most durable binders, but are not widely used (see Fluorine compounds, organic).

A fast-growing product of the coil-coating industry is the weldable zinc-rich coating based on zinc, chromic acid, and hydroxylated polymers named Zincromet (Diamond Shamrock Co.). The expanded use of this material in automobile manufacture is responsible for the production of 640,000 metric tons of coated sheet steel in 1976, accounting for 15% of total prefinished coil production.

Can Coatings. The can industry uses about 140 ML (37 million gal) of coatings per year worth about $100,000,000.

Some food cans are coated on the outside with printed paper wrappings and others with organic coatings. The insides are coated to prevent corrosion.

To be useful in food cans, the inside coatings must be capable of being sterilized with steam in the presence of the food without damage. This is a difficult challenge for most coating materials.

To be useful in beer containers, the coating must contribute very little to taste and must not absorb excessive amounts of the traces of flavoring materials present in beer.

Cans may be made of 2 or 3 pieces. The 3-piece type is mostly used for food containers, the 2-piece type for beer and beverage containers.

Table 4 gives some of the coating materials used in the canning industry. Production lines are highly automated. A typical baking schedule for a catalyzed epoxy–urea–formaldehyde coating and 2-piece cans is 20 s at 149°C; for a polybutadiene, 8 min at 207°C; and for an epoxy end-enamel, 8 min at 215°C.

New methods and materials are constantly investigated, eg, anodic electrodeposition to produce uniform coatings with minimum air pollution.

Both electrodeposition and wash coating with aqueous resin dispersions offer opportunities for recovery of solids and disposal of wastes in dilute form to meet pollution regulations. Meanwhile many plants have incineration or similar methods to dispose of polluting organic vapors.

Table 4. Materials for Can Coating

| Polymer system | 3-Piece cans | | 2-Piece cans | | Can-ends | | |
	Tin free	Tin plate	Al	Tin plate	Tin free	Tin plate	Al
poly(vinyl chloride)	inside	topcoat	inside	spray coat			coil or sheet coat
epoxy–phenolic resins	complete coat						
alkyd white base coat	outside	decoration	outside	decoration			
epoxy–urea–formaldehyde			inside	spray coat	sheet	coat	coil or sheet coat
oleoresins[a]		pattern coat			sheet	coat	
polybutadiene	pattern	coat[b]					

[a] With or without zinc oxide.

[b] Pattern coat avoids areas to be welded or soldered.

Appliance Finishes. *Refrigerators and Freezers.* About 9 million refrigerators and freezers are made in the United States per year consuming about 6.8 ML (1.8 million gal) of coating. Generally, a one-coat organic solvent-based finish is applied electrostatically over a conversion coating of iron phosphate; some two-coat systems are also used for the outer surface. The most popular topcoats are thermoset acrylics based on copolymers containing some styrene, simple acrylate and methacrylate esters, hydroxyl-functional acrylates, acrylic or methacrylic acid, and alkyl ethers of methylacrylamide or melamine–formaldehyde resins. White is the predominant color, usually a composition of about 90 parts of titanium dioxide pigment to 100 parts of binder. Primers are based on polyester–melamine–formaldehyde.

When refrigerators and freezers are assembled, a polyurethane foam is injected into the insulation space. Excess foam that comes in contact with the topcoat should not adhere strongly or mar the coating.

Washing Machines. About 5 million washing machines are built in the United States annually using ca 1.4 ML (370,000 gal) of coating. Resistance to water and detergent solutions is a requirement that makes a primer especially important. Epoxy-based materials are generally used for primers with a zinc phosphate or iron phosphate conversion coating. Electrodeposited epoxy primer coatings are becoming more popular. The topcoats are sprayed thermoset acrylics or oil-free polyesters cross-linked with melamine–formaldehyde and baked typically for 20 min at 176°C.

Dishwashers. About four million dishwashers are produced annually using ca 1 ML (ca 260,000 gal) of coating. Many of these have no side panels and slide under a counter.

The inside linings are often a thick (300 μm) layer of poly(vinyl chloride) plastisol (for alkali resistance) applied over epoxy–phenolic primer. Powder coatings are also used for these thick layers.

Outer surfaces are coated like washing machines with a thermosetting acrylic system applied over an epoxy primer.

Miscellaneous Industrial Coatings for Metal. A considerable quantity of sheet metal is first bent into shape and then painted with coatings that do not require high levels of resistance to water and long-term exterior durability; eg, shelving, closets, metal office furniture, and lighting reflectors. Usually a single coat of an alkyd–urea–formaldehyde based finish is applied. The oil–acid part of the alkyd may be derived from soya oil, dehydrated castor oil, coconut oil, or tall oil, depending on heat discoloration characteristics, cost, and other qualities (see Alkyd resins). Application may be by hand or automatic spray and may be electrostatic for reduced losses.

Wood Coatings. *Flatstock.* The market for industrial wood coatings can be divided into prefinished flat stock, ie, wood fashioned into large, flat sheets by a variety of processes, and wood furniture usually coated after assembly (see Wood; Laminated and reinforced wood). Prefinished flat stock consists largely of plywood made of hardwood accounting for about 50% of the coatings used, hardboard (eg, Masonite) accounting for 20%, and particle-board, ie, compacted and cemented wood chips that may or may not be overlaid with a wood veneer accounting for only 10%.

The annual United States market for flat stock coatings is estimated at about $100 million for a consumption of ca 120 ML (ca 32 million gal) of which about half is used for wall panelling. Coatings for hardboard and particle-board account for about $30 million.

The coating of flat stock is characterized by high line speeds of 60–90 m/min, extensive automation, and the predominance of curtain and roller coating. Spray methods are used less.

Alkyd–urea–formaldehyde finishes are the most commonly used (40%) topcoats for prefinished flat stock. Nitrocellulose accounts for only about 20% (see Cellulose derivatives). Both of these are being replaced by water-based latexes of acrylic copolymers, amounting to 10% of the market in 1973 and 50% in 1978.

Alkyd–urea finishes typically contain 50–60% oil-modified alkyd resin and 40–50% urea–formaldehyde resin dissolved in hydrocarbon solvents at a concentration of 480 g/L (4 lb/gal).

The aqueous finishes are largely latex types, dispersions of copolymer of acrylates and methacrylate esters and their acids in water, and are made by emulsion polymerization. Such dispersions are usually stabilized by ammonia and surface-active agents. The latex particles coalesce as water and amine are evaporated, a slow process when relative humidity is high. Heat is often applied to hasten evaporation (see Emulsions; Latex technology).

Interior plywood panels may be subjected to a long series of automated steps, including sanding and shaping, adding highly pigmented filler, coating with a tinted sealer (to fill pores and protect the filler), printing with a suitable pattern (often a wood-grain pattern), and finally coating with an alkyd–urea clear enamel.

Fillers for particle board include ultraviolet-cured styrene–polyesters, acrylic, and polyurethane resins. Most particle board is supplied uncoated and only a small percentage is filled (see Fillers).

About 40% of hardboard is supplied prefinished either by a prime coat or with a suitable topcoat. Grains resembling wood are printed on before the final clear topcoat.

Wood Furniture. The market for wood furniture finishes amounts to about $200 million annually and consumes ca 20 ML (ca 53 million gal) of finishes, about 75% of which is based primarily on nitrocellulose. In contrast to flat stock coating, furniture coating is characterized by many manual operations. Furniture is coated in many small factories and a great variety of procedures and formulas are used.

Finishes often rely for their decorative effect on the appearance of the fibrous structure of the wood itself. This appearance is enhanced by the addition of transparent coatings that penetrate pores, and pigments that fill and color pores and also improve the color uniformity of the wood. Nitrocellulose has advantages over other vehicles in its ease of application and drying, and in the degree to which it emphasizes the natural wood patterns.

In general, the wood is stained to a uniform desired color (bleaching may be a necessary first step); stains may be applied from water or solvent. A typical stain would contain about 1% of a dye mixture in methanol and might contain a small amount of less volatile solvent.

A sealer coat is then applied and sanded. The sealer might be a 15% nitrocellulose-based vehicle (eg, 0.25 s nitrocellulose, maleinized acid rosin, coconut oil alkyd, dioctyl phthalate); 1% colloidal silica (for filling, flatting, and transparency); 1% zinc stearate (for easy sanding); and mixtures of alcohol, ester and hydrocarbon solvents appropriate for the spray system and ambient temperature in the factory.

A wiping stain may be applied instead of the separate stain and sealer. The wiping stain may contain mixture of iron oxide pigments and others to produce the appropriate color, 1% aluminum silicate, 2% pigment, and 0.5% linseed oil, and hydrocarbon solvents. This stain is applied and the excess wiped off.

Shellac-type sealers may contain 10% binder (50% shellac and 50% other film formers such as nitrocellulose or poly(vinyl butyral) and 90% solvent. Transparent pigments such as silica or zinc stearate can be added to improve sanding properties.

The finish coat is usually nitrocellulose based. The binder may contain about 35% nitrocellulose and 65% of a mixture of nitrocellulose (0.5 s) and plasticizers, which may include simple esters, polyesters, and esters of rosin.

Urea–formaldehyde resins, acid catalyzed to permit low-temperature drying, are second to nitrocellulose in the market. A typical formulation would consist of 45% butylated urea–formaldehyde resin, and 55% plasticizers (nonvolatile basis) dissolved in aromatic hydrocarbon, alcohol, and ketone solvents. Butyl dihydrogen phosphate is a favored catalyst permitting the resin to become cross-linked at 60°C.

An important requirement of the finishing system is that repairs can be made to the coating without revealing damage and repair through a change of color or gloss where the repaired area abuts on the undamaged area.

BIBLIOGRAPHY

"Coatings (Industrial)" in *ECT* 1st ed., Vol. 4, pp. 145–189, by H. C. Payne, American Cyanamid Co.; "Coatings, Industrial" in *ECT* 2nd ed., Vol. 5, pp. 690–716, by William von Fischer, Consultant, and Edward G. Bobalek, University of Maine.

General References

H. F. Payne, *Organic Coating Technology,* Vols. I and II, John Wiley & Sons, Inc., New York, 1954.
A. G. Roberts, *Organic Coatings, Properties, Selection and Use, Building Science Series,* Vol. 7, National Bureau of Standards, U.S. Govt Printing Office, Washington, D.C., 1968.
E. Singer, *Fundamentals of Paint, Varnish and Lacquer Technology,* The American Paint Journal Co., St. Louis, Mo., 1957.
W. von Fischer and E. G. Bobalek, *Organic Protective Coatings,* Reinhold Publishing Co., New York, 1953.
R. R. Myers and J. S. Long, *Treatise on Coatings,* Marcel Dekker Inc., New York, 1967–1975.
T. C. Patton, *Paint Flow and Pigment Dispersion,* Interscience Publishers Inc., a division of John Wiley & Sons, Inc., New York, 1964.
R. D. Deanin, *Polymer Structure, Properties and Applications,* Cahner's Books, Boston, Mass. 1972.
F. W. Billmeyer, Jr., and M. Saltzman, *Principles of Color Technology,* Interscience Publishers Inc., a division of John Wiley & Sons, Inc., New York, 1966.
F. W. Billmeyer, Jr., and J. G. Davidson, "Color and Appearance of Metallized Paint Films-Characterization" in *J. Paint Tech.* **46**(593), 31 (1974).
Federation Series on Coatings Technology, Federation of Societies for Coatings Technology, Philadelphia, Pa.
Basic Coatings Technology Program, Federation of Societies for Coatings Technology, Philadelphia, Pa., 1973.
A. S. Gardon and J. W. Prane, eds., *Non-Polluting Coatings and Coating Processes,* Plenum Press, New York, 1973.
Guide to U.S. Gov't Specifications, National Paint and Coatings Association, Washington, D.C.
Index of Federal Specifications and Standards, General Services Administration, U.S. Government Printing Office, Washington, D.C.

Journals

Am. Paint Coatings J., American Paint Journal Co., St. Louis, Mo.
Mod. Paint Coatings, Palmerton Publishing Co., New York, N.Y.
Color Eng., Technology Publishing Corp., Los Angeles, Calif.
Chem. Mark. Rep., Schnell Publishing Co., New York, N.Y.

Chem. Week, McGraw-Hill Publications, New York, N.Y.

Chem. Eng. News, American Chemical Society, Washington, D.C.

J. Coatings Tech., Federation of Societies for Coating Technology, Philadelphia, Pa.

J. Am. Oil Chem. Soc., American Oil Chemists' Society, Champaign, Ill.

Ind. Finish., Hitchcock Publishing Co., Wheaton, Ill.

Mater. Perform., National Assoc. of Corrosion Engineers, Houston, Texas.

Met. Finish., Metals & Plastics Publications, Inc., Hackensack, N.J.

Finish. Highlights, Special Technical Publications, Oxnard, Calif.

Abst. Rev., National Paint and Coatings Assoc., Washington, D.C.

World Surface Coat. Abstr., Paint Research Assoc., Teddington, Middlesex, England.

Prod. Finish., Gardner Publications, Cincinnati, Ohio.

Appl. Manuf., Cahner's Publishing Co., Denver, Colo.

Automot. Ind., Chilton Co., Radnor, Pa.

Current Industrial Reports M 28 F—Paint Varnish & Lacquer, U.S. Bureau of Census, Washington, D.C., Code 28516.

Census of Manufactures, U.S. Bureau of Census, Washington, D.C., Code 28517, 1947, 1954, 1958, 1963, 1967, and 1972.

Annual Survey of Manufactures, U.S. Bureau of Census, Washington, D.C., Code 28519.

Annual Sales Survey, National Paint and Coatings Assoc., Washington, D.C.

Coatings II, Skeist Labs., Inc., Livington, N.J., Nov. 1974.

Water-Borne Coatings and Non-Aqueous Dispersion Coatings, DeBell and Richardson, Inc., in collaboration with H. S. Holappa and Associates, Linfield, Mass, 1974.

Kline's Guide to the Paint Industry, 4th ed., C. H. Kline & Co., Fairfield, N.J., 1975.

SEYMORE HOCHBERG
E. I. du Pont de Nemours & Co., Inc.

COATINGS, MARINE

Ships, offshore working platforms, and onshore waterfront structures are damaged by contact with the harsh marine environment. This damage results in shutdown of operations, dry-docking of vessels, and costly repairs (see Corrosion). Control of this destructive action is best achieved through a program of (*1*) selection of the materials most resistant to deterioration, (*2*) design to minimize conditions favorable to corrosion, and (*3*) effective utilization of protective coatings and/or cathodic protection (an electrical method of preventing metal corrosion in a conductive medium by placing a charge on the item to be protected) to deter corrosion. Protective (anticorrosive) coatings impart protection to the substrate by forming a barrier to the water, salt, and oxygen which accelerate corrosion. Thus, the thickness, impermeability, and integrity of a film of coating are of prime importance in its ability to provide corrosion control. Although protection of steel is of top priority, appearance of the coating may also be important. Therefore, naval vessels are given a color to provide camouflage and coatings on fixed offshore structures should provide optimum visual detection.

The attachment and growth of marine fouling organisms (mostly barnacles, tunicates, hydroids, marine plants, and bryozoa) to ship bottoms are also economically important (1). Larval forms attach themselves to nonmoving surfaces and may grow quite rapidly, especially in warm waters. Ship-drag caused by fouling can result in reduced speed, limited maneuverability, and increased fuel consumption. Fixed structures may undergo increased drag, reduced freeboard, clogging of seawater intake lines, or coating damage as a result of marine fouling. Currently, the one proven method employed in the prevention of marine fouling is the use of biocidal chemicals in special antifouling paints; these chemicals are gradually released into the seawater at the paint surface to provide continuous fouling control. Thus, two types of coating are described: protective (anticorrosive) coatings for corrosion control, and antifouling paints for controlling the attachment and growth of marine fouling organisms.

Coating Composition

Some of the first steel ships were coated with a product made by blending red lead, turpentine, and vegetable or fish oils. The modern synthetic coatings differ greatly from these, but still have the same three common ingredients:

Solvent: An organic liquid that dissolves the binder and thins the product to brushing, rolling, or spraying consistency. The solvent, which is lost by evaporation, does not remain in the cured film.

Binder: An organic film-forming solid or liquid that converts to a continuous solid film upon curing. The chemical nature of the binder determines the generic type of the coating.

Pigment: A solid material that imparts color, opacity (thus protecting the organic binder from deterioration by sunlight), and in some cases (eg, red lead and zinc chromate) corrosion inhibition (see Pigments; Corrosion).

The first marine coatings were cured by air oxidation of drying oils to a solid film after the solvent had evaporated (see Driers). This process was slow and the protection was of limited duration. Modification of such oleoresinous coatings into alkyds made them more durable and faster drying. Alkyd coatings are not suited for damp or immersed environments since they are not stable in aqueous environments, however, they are still used frequently on shipboard compartments, decks, and vertical surfaces above the water and splash zones. Silicone-modified alkyds have improved durability and gloss retention and are, therefore, used on exterior marine atmospheric areas (see Alkyd resins; Silicon compounds).

Coatings used later were lacquers that, upon solvent evaporation, deposited a continuous film of pigmented solid binder. They were easy to touch up and repair since the cured film could be softened by the solvent used in the touch-up topcoat, thereby resulting in an excellent intercoat bonding. Lacquers provided much greater durability than the alkyd paints and could be used in areas that received continuous immersion in seawater. Vinyls and chlorinated rubbers are two lacquers that find considerable marine use today. Relatively inexpensive coal tar coatings find occasional marine use on mechanically cleaned surfaces although their relatively soft films can be penetrated by barnacles unless fouling-resistant biocides are added.

Latex coatings are somewhat similar to lacquers in that the dispersed binder particles coalesce during the evaporation of the water solvent to form a film (see Latex technology). Acrylic and vinyl–acrylic latexes find limited use on offshore platforms, but may find more use if environmental and safety regulations greatly restrict the amounts of organic solvents used in coatings. All organic solvents in paints may be greatly reduced in the near future since it is currently thought that they contribute to the production of photochemical smog.

Coatings that cure by chemical reaction of two component parts are the most widely used in submerged marine applications. For example, epoxies, coal tar-epoxies, urethanes, and polyesters are durable and resistant to water, solvent, and chemicals. When properly formulated they can provide excellent protection to steel in severe marine environments.

Surface Preparation Requirements

To be successfully applied, the first marine coatings required only minimum preparation of steel surfaces because the vegetable or fish oils in the coatings wetted the incompletely cleaned surfaces well enough to provide adequate bonding. However, in a severe marine environment their service lives were limited. Modern synthetic marine coatings provide much longer protection, but require both complete cleanliness and surface profile (tooth) in order to obtain adequate bonding of prime coats. The time-honored tradition of chipping and wire brushing deteriorated coatings from ships is no longer used. This method has been replaced by high-speed abrasive (sand, grit, or shot) blasting by conventional air-pressured equipment or by newer equipment that requires centrifugal force to propel the abrasive (see Abrasives). Several standards (2–4) have been established to determine if a steel surface is properly prepared for coating; that of the Steel Structures Painting Council (SSPC) (2) is most frequently used in the United States. The necessary level of surface preparation for tight bonding of coatings varies with the generic type, the severity of the environment, and the desired length of protection. Manufacturers of similar coatings do not agree on a single level of surface preparation for their products, some being much more cautious than others. The relatively new and widely used inorganic zinc coatings are probably the most demanding for optimum preparation of steel surfaces. Thus, the suppliers of these coatings commonly recommend white-metal blast cleaning (SSPC No. 5) or near-white blast cleaning (SSPC No. 10) when they are used in a marine environment. Manufacturers of vinyls, chlorinated rubbers, epoxies, coal tar-epoxies, and urethanes indicate that commercial blast cleaning (SSPC No. 6) is usually satisfactory. The surface preparation requirements stated by the coating manufacturer or specification should be met in full.

Recently, several shipyards and other steel construction and repair facilities have been cited by local air pollution control agencies for emitting particulates (dust and other fine solids) into the air during abrasive blasting of steel. Actions taken by coating applicators to eliminate such plumes from the atmosphere include: (1) the use of hard, sharp, and properly sized abrasives that produce an adequate surface with minimum emission of particulates, (2) blasting inside a building or under a temporary shroud, and (3) using equipment that automatically moves across a regular surface (the area being blasted can be completely enclosed) and picks up and recycles the spent abrasive. Several different systems that use water to keep dust at a minimum are currently being investigated. Also, powdered dry ice is being studied as an abrasive (see Air pollution).

Application of Marine Coatings. Modern synthetic coatings have strict application requirements (5–6). Multipackage systems must be mixed and thinned in exactly the proportions specified by the manufacturer. Specified times between mixing and application (induction times), times between multiple coats, and pot-life limitations must all be met. Spraying, particularly airless spraying, is the fastest and most commonly used method for applying the coating (see Coating processes). This usually results in a very pleasing appearance. Before spraying it is good practice to round or smooth all welds, sharp edges, and corners, and to fill crevices and other structural features that are difficult to coat and that are susceptible to accelerated corrosion. After these areas are cleaned, a primer coat should be brushed into them before a full prime coat is sprayed on the overall area.

On offshore platforms or waterfront structures, periodic wire brushing and touch-up painting of areas with localized coating damage may be a more effective way to achieve continuous protection from corrosion. For wire-brushed surfaces it is best to use cheaper oleoresinous, alkyd, or coal tar coatings that are more tolerant of this method of surface preparation. For repair of localized coating damage all loose paint, rust, and other contaminants must first be removed. The same types of prime, intermediate, and topcoat are then normally applied to the cleaned steel and overlapped onto the surrounding coating.

For optimum performance in immersion service, all marine coatings should have a minimum dry film thickness of 250 μm (200 μm may be effective for epoxies) which is best achieved in at least three coating applications. Different pigmentation in each coat will avoid overlapping holidays (discontinuities) in the total system. Dry film thicknesses are usually determined with magnetic gages, and holidays are detected with low-voltage holiday detectors.

Manufacturers of marine coatings always have printed information available on use of their products. This information includes recommendations on the equipment used, mixing of components, time and temperature requirements, coverage rate at a recommended dry film thickness, and good application practices. New application equipment used in mixing the two components at the gun head and/or utilizing heat may be necessary if limitations on the amounts of organic solvents that can be used in marine coatings result in viscous products that cannot be sprayed with presently used equipment.

Protective (Anticorrosive) Coatings

The more durable and commonly used types of marine protective coatings are discussed separately here. There are specific recommendations for coatings applied to ships (5–7), offshore fixed platforms (8), fleet moorings (9), and waterfront structures (10). Table 1 indicates the zones (atmospheric, splash, or immersed) for which suppliers recommend their marine coatings.

Vinyls. The first marine vinyl coatings had a very low build rate (25–50 μm per coat) and required many coats to achieve a 250-μm dry film thickness. Later formulations have a much higher rate of film-build (100 μm or more per coat). Since vinyls are lacquers they are readily touched up or recoated after weathering. The organic solvents in vinyls may become a deterrent to their use if air quality standards become more restrictive (see Vinyl polymers).

Table 1. Sources of Marine Protective Coatings

Name of supplier	Alkyd	Silicone alkyd	Vinyl	Chlorinated rubber	Coal tar	Acrylic or vinyl acrylic	Epoxy	Coal tar epoxy	Urethane	Polyester	Zinc inorganic	Zinc rich epoxy	Zinc rich chlorinated rubber
Advanced							abc[a]		abc	abc	abc	abc	
Ameron	a	a	abc	abc			abc	abc	ab		abc	ab	
Carboline	ab	a	abc	abc	bc		bc	bc	a	bc	abc	abc	
Cook	a	a	abc	ab		a	abc	abc	a	abc	abc	abc	ab
Devoe	a	a	abc	abc		ab	abc	abc	abc		ab	ab	
Farboil	a	a	abc	abc			abc	c	a	a	abc	abc	abc
Koppers	a	a			c		abc	abc	abc		abc		
Porter	a	a	abc	ab	bc		abc	abc	ab	ab	abc	abc	
Pro-Line	a	a	abc	ab	abc	ab	abc	abc	ab	abc	ab	ab	ab
Reliance	a	a	ab	abc		a	abc	bc	ab		abc	ab	
Rustoleum	a	a	ab				ab	ab				ab	
Wisconsin							abc	abc	a	bc	a	a	

[a] Recommended for use in: a = atmospheric zone; b = splash zone; or c = immersed zone.

Chlorinated Rubbers. The use of chlorinated rubber coatings on marine structures has increased markedly in the last few years. They have a fairly high build rate (75 μm or more) and, like vinyls, are easily repaired. They also have good low temperature application and curing properties. However, they utilize strong organic solvents which may prove to be a deterrent to their use (see Elastomers, synthetic—neoprene; Butadiene; Rubber).

Epoxies. Two-component epoxies, notably the polyamide-cured epoxies, find much use on steel marine structures (see Epoxy resins). One such formulation (MIL-P-24441) is unique in that it is somewhat tolerant of incompletely cleaned steel and can be used on almost all shipboard steel surfaces (7), including the interiors of potable water tanks. Polyamide-cured epoxies require less stringent steel surface preparation than the amine-cured epoxies which are less commonly used for marine work. Amine-cured epoxies may have bloom or sweat of the amine catalyst to the surface of applied films before curing. This may hinder adequate bonding of topcoats unless removed by solvent after curing. Epoxies cure to a hard finish that presents a surface with no tooth and is, therefore, difficult to topcoat. As a result, in multiple coat systems topcoats are applied before the undercoat completely cures (as specified by the manufacturer). Similarly, antifouling paints must be applied to the incompletely cured final epoxy coat. To topcoat completely cured epoxies, a fog coat (thinned coat of the topcoat) is first applied before a full topcoat is sprayed. Epoxies chalk freely in sunlight. Although this does not affect the protection of the steel, it does present an additional topcoating problem for weathered epoxy coatings. The chalk must be removed by sweep abrasive blasting or light sanding before a fog coat and a full topcoat are applied. The low temperature application limit of epoxies varies greatly with the formulation.

Coal Tar-Epoxies. Polyamide-cured coal tar epoxies, such as SSPC No. 16, are also widely used on marine structures because they are especially impermeable to water. A coal tar-epoxy catalyzed with a low molecular weight amine is especially resistant to an alkaline environment, such as occurs on cathodically protected surfaces. Some coal tar-epoxies become brittle in direct sunlight, but others with aluminum pigmentation or other special variations are less susceptible to this deterioration. Coal tar-epoxies have topcoating problems similar to those of epoxies, and they are generally more difficult to topcoat than epoxies. Also, they are available only in black or a slight variation of black (see also Coal, carbonization).

Urethanes. Two-component (polyester, polyether, or acrylic-curing) urethanes give a tough, durable, smooth finish that presents topcoating problems similar to those of epoxies. The first marine urethane coatings chalked and yellowed rapidly in sunlight. This is not true of the newer aliphatic urethanes (11) which have excellent weathering properties, abrasion resistance, and good low temperature application and curing characteristics. Because of their outstanding weathering characteristics, they are frequently used over epoxy primers, particularly in areas exposed to direct sunlight (see Urethane polymers).

Polyesters. Polyester coatings for marine use are most frequently used with fiberglass reinforcement to impart strength and rigidity. Fiberglass-reinforced polyester construction can be used to produce strong, but lightweight boats, buoys, and surfboards. Epoxy and urethane coatings can also be applied to fiberglass. On steel ships and fixed marine structures, glass fibers or flakes in a high-build polyester coating produce a tough, durable, abrasion-resistant surface (see Laminated and reinforced plastics; Polyesters).

Inorganic Zincs. Marine inorganic zinc coatings are available in different packaged forms and in formulations that cure by different mechanisms (12). The two-package, self-curing, solvent-thinned products (alkyl silicates) are probably the most widely used (see Silicon compounds). In all cases, a well-formulated zinc inorganic coating of 75–125 μm dry film thickness can provide excellent long-lasting abrasion and corrosion resistance to steel in a mild atmospheric environment even when not topcoated. Therefore, they are commonly used as preconstruction primers for steel plate in an automated shop system (centrifugal abrasive blasting and spray priming of steel). After the steel has been fabricated into a finished product in the field, localized areas of damaged coating are cleaned (preferably by abrasive blasting) and spot primed with an inorganic zinc coating before one or more topcoats are applied. Inorganic zincs deter corrosion through a form of sacrificial cathodic protection in which the zinc in the coating is corroded rather than the steel substrate. In seawater, the loss of zinc is too rapid for long-lasting protection of the steel. Therefore, zinc coatings are almost always topcoated for marine use even when not immersed. Vinyls, epoxies, and urethanes (polyether- or acrylic-curing, not polyester-curing), are commonly used as topcoats. Alkyds and other coatings that are not stable to an alkaline environment should not be used as topcoats.

Zinc-Rich Organics. Many zinc-rich organic coatings, notably zinc-rich polyamide-cured epoxies, are available for marine use. They do not have the same abrasion resistance as some of the inorganic zincs, but they can tolerate a lower level of surface preparation and are more easily topcoated. Again, the high zinc content provides a form of cathodic protection to the steel.

Specialized Coatings. Several protective coatings, mostly solvent-free epoxies, are marketed for application underwater to steel surfaces (13). Some are so viscous (like putty) that they must be hand-applied by divers; other thinner products can be applied underwater by brush or rollers. These coatings are most frequently used to repair localized areas of coating damage.

Powder coatings (qv) (14) are strong, abrasion-resistant, insulating, durable, protective coatings that are applied to metal components of different sizes and configurations at shops where optimum surface preparation and application conditions are possible. They can give outstanding protection to pumps, valves, and other critical components of operational systems where abrasion, wear, chemical or solvent attack, or rapid corrosion presents a serious problem.

Plastic-coated steel electrical conduits and fittings that perform well in a marine atmospheric environment are available under military specification MIL-C-29169.

Petrolatum-coated tapes (15) find many uses on ships, waterfront structures, and offshore platforms. They are readily applied over wirebrushed steel and can, therefore, be used on difficult to clean and coat structures such as piping. The petrolatum paste can provide protection to wire ropes and cables in marine atmospheric environments. Protection of the steel is provided by a thick insulation that separates it from the hostile environment.

Antifouling Paints

Many methods for preventing marine fouling or for removing it from ships have been proposed (1). Currently, diver-operated or remotely controlled equipment that utilizes rotating brushes or water jets is being used to remove fouling growth from ships (16). Their economic and practical effectiveness in a scheduled fouling removal program has not been proven to date.

Many different biocides have been tested in paints as deterrents to marine fouling (see Industrial antimicrobial agents). Most have not imparted long-lasting properties; others are not being used because of health, safety, or environmental hazards. The latter include organic mercury, lead, and arsenic compounds. Currently, the only two biocidal materials that are used extensively in antifouling paints are cuprous oxide and organotin compounds (usually tributyltin oxide or tributyltin fluoride) (17). On steel surfaces it is always necessary to apply an insulating protective coating between the steel and a cuprous oxide-containing antifouling paint to avoid rapid galvanic (dissimilar metal) corrosion of the steel. Aluminum is even more chemically active than steel so that cuprous oxide-containing antifouling paints are never used on it; however, the organotin-containing antifouling paints can be used safely on aluminum. (Similarly, a red lead-containing primer is never used on aluminum surfaces.) Although organotin-containing antifouling paints present no galvanic corrosion problems, special precautions must be taken while spraying and handling them because of health and environmental hazards. Many manufacturers are marketing antifouling paints containing both cuprous oxide and organotin to obtain a broader spectrum of fouling control (see Tin compounds).

Table 2. Sources of Antifouling Paints and Generic Type of Paint Binders

Name of supplier	Vinyl	Epoxy	Hydrocarbon	Chlorinated rubber	Urethane	Coal tar-epoxy	Alkyd	Oil	Resin tar
Advanced	c,oᵃ	c,o			c,o				
Ameron	o		c	o					
Carboline	c,o								c
Devoe	c,o	m	c,m						
Farboil		o	c	o			c		
Koppers	c,o,m							c	
Porter				o		o			
Pro-Line	c,o,*	o				o			m
Reliance	c								
Wisconsin		o							

ᵃ Note that c = cuprous oxide biocide; o = organotin biocide; m = mixed biocides; and * = markets water-based as well as solvent vinyl product.

Most commercial antifouling paints use a vinyl binder, although products with other binders are also marketed (see Table 2). Rosin or some other leaching agent must be added to cuprous oxide formulations to permit its controlled release into seawater where it is lethal to fouling larva forms. Organotins usually do not require leaching agents to dissolve slowly in seawater.

Recent research has lead to the development of organometallic polymers for use in antifouling paints (18). Controlled release of organometallic biocide by dissociation from the polymer in seawater may extend the normal 2–3-yr period of fouling resistance to as long as 5 years. Another recent development (19–20) is the formulation of antifouling paints that can be applied underwater. Research continues to be conducted into safer biocides that are effective in fouling control, nontoxic fouling repellents, and low surface energy (slippery) coatings that will not permit the attachment of fouling organisms.

A sheet material (ca 2 mm thick) of black neoprene rubber (see Elastomers, synthetic, neoprene) impregnated with tributyltin is currently marketed (21–22) for bonding with adhesive to structures placed in seawater. Because the sheet is so much thicker than an antifouling coating system (usually about 100 μm) it has a larger reservoir of biocide that can result in longer lasting fouling control.

Table 3. Sources of Underwater-Applicable Coatings

Name of firm	Method of application	
	Hand or glove	Brush or roller
Advanced	yes	yes
Koppers	yes	no
Pro-Line	yes	no
Sta-Crete	yes	no
Sika	yes	no

Table 4. Addresses of Suppliers

Supplier	Address
Advanced Coatings and Chemicals	2213 North Tyler Ave. South El Monte, Ca. 91733
Ameron Corrosion Control Division	201 North Berry Street Brea, Ca. 92621
Carboline Company	350 Hanley Industrial Ct. St. Louis, Mo. 63144
Cook Paint and Varnish Company	P. O. Box 389 Kansas City, Mo. 64141
Devoe and Raynolds Company, Inc.	5850 Hollis Street Emeryville, Ca. 94608
Farboil Paint Company	8200 Fischer Road Baltimore, Md. 21222
Hempel's Marine Paints, Inc.	25 Broadway New York, N.Y. 10004
International Paint Company	17 Battery Place, North Room 1150 New York, N.Y. 10004
Koppers Chemicals and Coatings	480 Frelinhuysen Ave. Newark, N.J. 07114
Mobil Chemical Company	150 East 42nd Street New York, N.Y. 10017
Napko Company	5300 Sunrise Street Houston, Tx. 77021
Porter Coatings	P. O. Box 1439 Louisville, Ky. 40201
Pro-Line Paint Manufacturing Company	2646 Main Street San Diego, Ca. 92113
Reliance Universal, Inc.	P. O. Box 1113 Houston, Tx. 77001
Rustoleum Corporation	2301 Oakton Street Evanston, Il. 60204
Sika Chemical Corporation	Lyndhurst, N.J. 07071
Sta-Crete, Inc.	893 Folsom Street San Francisco, Ca. 94107
Wisconsin Protective Coating Corp.	P. O. Box 3396 Green Bay, Wi. 54303

Sources of Materials

Sources of the protective coatings, antifouling paints, and underwater-applicable coatings are listed in Tables 1, 2, 3, and 4, respectively.

BIBLIOGRAPHY

1. *Marine Fouling and Its Prevention,* Woods Hole Oceanographic Institution and United States Naval Academy, Annapolis, Md., 1952.
2. *Systems and Specifications, Steel Structures Painting Manual,* Vol. 2, Steel Structures Painting Council, Pittsburgh, Pa., 1969.

3. *Visual Standard for Surfaces of New Steel Airblast Cleaned with Sand Abrasive,* NACE Standard TM-01-70, National Association of Corrosion Engineers, Houston, Tx., 1970.

4. *Abrasive Blasting Guide for Aged or Coated Steel Structures,* Society of Naval Architects and Marine Engineers, Report No. TR-4-9, New York, 1969.

5. "The Painting of Steel Vessels for Salt Water Service," *Good Painting Practice, Steel Structures Painting Manual,* Vol. 1, Steel Structures Painting Council, Pittsburgh, Pa., Chapt. 11, 1966.

6. *Coating Systems Guide for Hull, Deck and Superstructure,* Society of Naval Architects and Marine Engineers, Report No. TR-4-10, New York, 1973.

7. "Preservation of Ships in Service (Paints and Cathodic Protection)," *Naval Ships Technical Manual,* Naval Ships Command System, Washington, D.C., 1970, 9190.

8. M. Cashman, *Ocean Ind.* **12**(6), 106 (May 1976).

9. *Mooring Maintenance,* NAVFAC NO-124, U. S. Naval Publications and Forms Center, Philadelphia, Pa., 1973.

10. *Maintenance of Waterfront Facilities,* NAVFAC NO-104, U. S. Naval Publications and Forms Center, Philadelphia, Pa., 1977.

11. A. H. Roebuck and G. C. Cheap, "Corrosion/77," in *Recent Advancements in Marine Coatings,* Paper No. 99, National Association of Corrosion Engineers, Houston, Tx., 1977.

12. D. H. Gelfer, *J. Paint Tech.* **47,** 43 (1975).

13. R. W. Drisko, in ref. 12, pp. 40–42.

14. S. B. Levinson, *J. Paint Tech.* **44,** 37 (1975).

15. W. D. Parker and W. H. Yeigh, *Mater. Perform.* **11,** 31 (1972).

16. H. S. Preiser, *Energy (Fuel) Conservation Through Underwater Removal and Control of Fouling on Hulls of Navy Ships,* David Taylor Naval Ships Research and Development Center, Report 4543, Annapolis, Md., 1975.

17. U.S. Pat. 3,167,473, (Jan. 3, 1963), J. R. Leebrick (to M&T Chemicals Inc.).

18. J. A. Montemarano and E. J. Dyckman in ref. 12, pp. 59–61.

19. R. W. Drisko, *Protective Coatings and Antifouling Paints That Can Be Applied Underwater,* 9th Offshore Technology Conference, 1977 Proceedings, Dallas, Tx., 1977, pp. 419–421.

20. R. W. Drisko, *Antifouling Paints That are Applied Underwater,* Proceedings of 1977 International Controlled Release Pesticide Symposium, Oregon State University, Corvallis, Or., 1977.

21. J. M. D. Woodford, *Underwater Marine Coatings—Marine Biocidal Rubbers Containing Organotin Toxics,* Defense Standards Laboratories Report 496, Australian Defense Scientific Service, Defense Standards Laboratories, Maribyrnong, Victoria, Australia, 1972.

22. G. A. Janes, *Polymeric Formulations for the Control of Fouling on Pleasure Craft,* Proceedings of 1975 International Controlled Release Pesticides Symposium, Wright State Univ., Dayton, Oh., Sept. 9–10, 1975.

General References

Antifouling Marine Coatings, Noyes Data Corporation, 1973.

Coating Systems Guide for Hull, Deck, and Superstructure, Society of Naval Arch. & Marine Engineering Report No. TR-4-10, 1973.

Abrasive Blasting Guide for Aged or Coated Steel Structures, Society of Naval Arch. & Marine Engineering Report No. TR-4-9, 1969.

J. Miles Sharpley and A. M. Kaplan, eds., *Proceedings of the 3rd International Biodegradation Symposium,* Applied Science Publishers, Ltd., London, 1976; 4th Symposium Berlin, 1978, to be published.

Proceedings of the 4th International Congress on Marine Corrosion and Fouling, Centre du Recherche et D'Etudes Oceanographiques, Boulogne, 1976.

R. W. Drisko
U.S. Naval Civil Engineering Laboratory

COATINGS, RESISTANT

Resistant or high-performance coatings or linings are specialty products used to give long-term protection under difficult corrosive conditions to industrial structures such as chemical plants, paper plants, atomic power plants, food plants, tanks, tank cars, barges, the interior of ships transporting chemicals or strongly corrosive products, etc. This contrasts with paint, which is used for general appearance and shorter term protection against milder atmospheric conditions, and industrial coatings, ie, coatings that are applied to manufactured products, such as refrigerators, washing machines, bicycles, automobiles, etc, during the manufacturing process (see Paint; Coatings, industrial; also Coating processes; Coatings, marine).

A high-performance coating or lining is one that goes beyond paint in adhesion, toughness, resistance to continuing exposure to industrial chemicals or food products, resistance to water or sea water, and to weather and high humidity. It is designed for difficult exposures and to prevent serious breakdown of an industrial structure even though there may be abrasion damage or holidays (gaps) and imperfections in the coating. It must be inert and noncontaminating to materials with which it is in contact; must be dense and have a minimum of absorption of contacting materials; have a high resistance to the transfer of chemicals through the coating such as various anions and cations; must be able to expand and contract with the surface over which it is applied; must maintain generally good appearance even though subject to severe weather or chemical conditions. It must be and do these things for a sufficient period of time so as to be economically feasible and justify its price and applications costs.

The following definitions will be used in this article.

A *resistant coating* is a film of material applied to the exterior of structural steel, tank surfaces, conveyor lines, piping, process equipment or other surfaces which is subject to weathering, condensation, fumes, dusts, splash or spray, but is not necessarily subject to immersion in any liquid or chemical. The coating must prevent corrosion or disintegration of the structure by the environment.

A *resistant lining* is a film of material applied to the interior of pipe, tanks, containers or process equipment and is subject to direct contact and immersion in liquids, chemicals, or food products. As such, it must not only prevent disintegration of the structure by the contained product, but must also prevent contamination of it. In the case of a lining, preventing product contamination may be its most important function.

The function of a high-performance coating or lining is to separate two highly reactive materials, ie, to prevent strongly corrosive industrial fumes or actual liquids, solids, or gases from contact with the reactive structure or underlying surface. The concept that a coating is a very thin film separating two highly reactive materials brings out the vital importance of the coating and its need to be completely continuous in order to fulfill its function. Any imperfection in the coating becomes a focal point for corrosion and breakdown of the structure or a focal point for the contamination of the contained liquid. The relatively thin, continuous film concept takes on even greater importance when it is understood that these protective coatings are applied to very large areas of structural steel, tank surfaces and similar areas. Many thousands of square meters may be involved in a single coating use.

There are two basic methods by which coatings or linings protect the surface (1).

The first is based on the principle of impermeability. The coating must have excellent adhesion and be inert to chemicals and impervious not only to air, oxygen, water, and carbon dioxide, but also to the passage of ions and to the passage of electrons or electricity. Such a coating prevents corrosion of steel by interrupting or providing a block to the normal processes necessary for the corrosion (Fig. 1) (1).

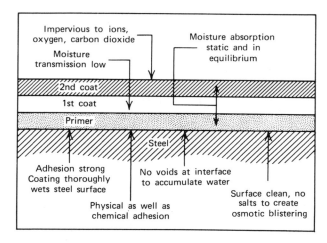

Figure 1.   Impervious coating.

The second method uses anodically active or inhibitive pigments in the primer or in the coating to regulate corrosion (see Pigments). Corrosion is prevented not necessarily by the nature of the binder film, but by the use of pigments which, when subject to moisture or humidity, ionize sufficiently to react with (Fig. 2) (1) or cathodically protect the steel or metal surface (Fig. 3) (1) to maintain it in a passive state (see Corrosion and corrosion inhibitors). This takes advantage of the water absorbed by ionizing the active pigment and forming a passive layer on the steel surface. The active pigments are metal salts of various chromates such as zinc chromate, lead chromate, and strontium chromate. The action of anodically active pigments is to cathodically protect the underlying steel surface. The pigment is powdered metallic

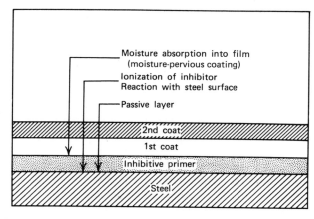

Figure 2.   Inhibitive coating.

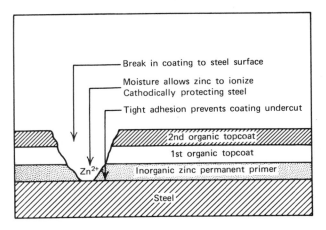

Figure 3.   Inhibition by zinc primer.

zinc and it may be incorporated into either organic or inorganic vehicles. Zinc metal may be used also in the form of galvanizing or metallic zinc spray.

These methods in practice have both been proven to be effective. In many areas they overlap in usefulness. However, there is a distinction between them.

The inert impervious system performs best as a lining where it is subject to continual moist, wet or immersion conditions and where it is subject to little or no physical abrasion. Such areas are underwater sewage structures, water tanks, petroleum tanks, chemical tanks or tank cars, food tanks, wine or beer storage tanks, etc.

Inhibitive coatings perform best in areas where the coating is subject to weathering, atmospheric conditions, high humidities or chemical fumes. Such uses are generally for the exterior maintenance of structures or tanks where the coating may be subject to physical abrasion and coating damage. Under these conditions the inhibitive pigments and particularly the cathodic protection provided by zinc aid materially in the prevention of corrosion.

Components of Resistant Coatings and Linings

The similarity to paint extends to the essential ingredients of the coating or lining. The general composition of a coating or lining is shown in Figure 4.

The principal functions of these ingredients are as follows:

Binder: binds or glues the pigments together in a homogeneous film; provides adhesion by preferentially wetting the substrate; adheres to itself to prevent delamination of subsequent coats; prevents penetration of aggressive atmosphere and chemicals; and may impart flexibility and abrasion resistance to the final film.

Color-carrying pigments: impart opacity and hide substrates; impart decorative color; and provide protection for the binder by preventing penetration of actinic rays below the coating surface.

Inert and reinforcing pigments: increase hardness and tensile strength of the film; increase chemical and atmospheric exposure resistance; aid in the adhesion of the primer; improve the bonding surfaces for the following coats; may be used to build film thickness; and contribute to viscosity control and thixotropy.

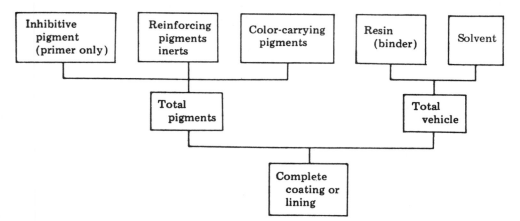

Figure 4.   The general composition of a coating or lining.

Inhibitive pigments (primer only):   reduce corrosion of the substrate; react with the substrate to provide a passive surface; cathodically protect the substrate and may aid in the improvement of the primer adhesion.

Solvents: dissolve the binder; control viscosity and workability of the pigment–binder combination; transport the pigment-binder combination to the substrate; and aid in wetting the substrate.

Binders.   Four general types of binders are used to formulate resistant coatings and linings for the chemical processing, food, petroleum, marine industries, etc. These include lacquers, coreacting coatings, condensation coatings and inorganic binders (Table 1).

In order to produce a film that will perform satisfactorily in a given environment after application, the coating or lining must convert to a dense, solid membrane. Some materials can accomplish this simply by releasing their solvents but others must go through a series of complicated chemical reactions, sometimes requiring the application of heat after evaporation of the solvent.

The ability of a resin to form a dense, tight film is directly related to its molecular size and complexity. The polymers that are capable of forming such films simply by the evaporating of solvent are initially of very high molecular weight and are not capable of further chemical reaction. Because of their large molecular size, these polymers must be kept in dilute solution, and coatings based on them have a low volume of solids. Resins of low molecular weight, although requiring chemical conversion to attain polymer structures of suitable size, have the advantage of being able to produce higher-solids combinations.

Color-Carrying Pigments.   Colorants are used in resistant coatings and linings for appearance. In the category of high-performance resistant materials every care must be taken to ensure proper resistance by the colorant. Many have poor resistance despite excellent color. Colorants are therefore limited in resistant coatings and linings (Table 2).

Table 1. Binders Commonly Found in Resistant Coatings and Linings[a]

Binder type	Generic type	Resistant properties				Temperature, °C	Primary use
		Alkali	Acid	Water	Weather		
lacquer (air dry thermoplastic)	vinyl chloride–vinyl acetate copolymer	E	E	E	E	to 65	resistant intermediate and topcoats
	polyacrylates	F	F	F	E	to 65	resistant topcoats
	chlorinated rubber	E	E	E	G	to 60	resistant intermediate and topcoats
coreacting (air dry thermoset)	epoxy–amine cure	E	G	G	F	to 93	resistant coatings and linings
	epoxy–polyamide	E	F	G	G	to 93	resistant coatings and linings
	urethane (2 package)	G	F	G	F	to 120	abrasion resistant coatings
	urethane (moisture cure)	G	F	G	G	to 120	abrasion resistant coatings
	urethane–aliphatic isocyanate	G	F	G	E	to 120	weather and abrasion resistant topcoats
condensation (requires added heat to cure) (baked)	phenolic	P	E	E	F	to 120	chemical and food resistant lining
	epoxy–phenolic	G	E	E	F	to 120	chemical and food resistant lining
	epoxy–powder coating (requires high heat to fuse and cure)	G	G	G	F	to 93	pipe coating and lining
inorganic	zinc silicate (air cure)	P	P	G	E	to 315	permanent primer or single-coat weather resistant coating
	glass (fused to metallic substrate)	F	E	E	E	to 260	chemical and food resistant lining

[a] Rating: E, Excellent; G, Good; F, Fair; P, Poor.

459

Table 2. Color-Carrying Pigments Found in Resistant Coatings and Linings[a]

Color	Type of colorant		Resistant properties			
	Organic	Inorganic	Alkali	Acid	Exterior Durability	Heat resistance
red	toluidine red		P	P	G	P
	monastral red		G	G	G	G
		iron oxide red	G	G	G	G
		dmium red	F	F	G	E
orange		iron oxide	G	G	G	G
		molybdenum orange	G	F	G	P
yellow		yellow iron oxide	G	G	G	G
		nickel titanate	G	G	G	G
		chrome yellow	F	F	G	P
green	phthalocyanine green		E	E	E	E
		chrome oxide	G	G	G	G
blue	phthalocyanine blue		E	E	E	E
black	lamp black		G	G	G	G
	carbon black		G	G	G	G
white		iron oxide	G	G	G	G
		titanium dioxide	G	G	G	G
metallic		aluminum flake	P	P	E	E
		zinc dust	P	P	E	E
		stainless-steel flake	E	G	E	E
		lead flake	E	E	E	G

[a] E, Excellent; G, Good; F, Fair; P, Poor.

Reinforcing Pigments. Reinforcing pigments are common to almost all resistant coatings and linings. Most are inert and generally improve the adhesion, impermeability and physical characteristics of the coatings or linings (Table 3).

Table 3. Reinforcing Pigments Found in Resistant Coatings and Linings[a]

| Generic type | Common name | Resistant characteristics | | | | Physical characteristics |
		Alkali	Acid	Water	Weather	
magnesium silicate	talc;	F	G	E	E	fibrous-platelike
	asbestine; asbestos					fibrous
barium sulfate	barytes	G	G	G	G	cubical, heavy
silica	diatomite;	P	E	E	E	porous,
	silica flour					hard, sharp
						crystals
aluminum silicate	clay	F	G	F	G	platelike
potassium–aluminum silicate	mica	G	G	G	G	platelike, used to reduce moisture vapor transfer

[a] E, Excellent; G, Good; F, Fair; P, Poor.

Inhibitive Pigments. Inhibitive pigments are not common to all resistant coatings or linings. They are primarily used in coatings for atmospheric conditions rather than for immersion and only in primers (Table 4).

Table 4. Inhibitive Pigments Found in Some Primers for Resistant Coatings and Linings

Pigment	Solubility in 1 L H_2O at equilibrium, g of CrO_3
zinc chromate	1.1
strontium chromate	0.6
basic zinc chromate	0.02
barium chromate	0.001
lead chromate	0.00005
red lead	essentially insoluble
zinc powder	provides cathodic protection to substrate

Solvents. The solvents are divided into five groups as shown in Table 5. In practice, blends of solvents are used in order to achieve proper evaporation rate, viscosity, conformance with California's Rule 66 (see Air pollution) and atomization of the coating. In many linings heat must be used to completely remove the solvent from the lining for maximum resistance.

The actual thickness of the coating or lining deposited from the solvent is extremely important since the moisture–vapor transfer rate through a coating is directly proportional to the thickness. Thickness is also important from the physical standpoint in that many industrial structures are subject to abrasion. Each resistant coating or lining has its own optimum thickness.

The coating or lining thickness may vary from a few micrometers (ca 0.1 mil) for a lining in a food can to chemical resistant coatings of 0.25–0.38 mm (10–15 mils) thickness, to chemical resistant linings in a copper leaching plant of 1.5–6.4 mm (60–250 mils) thickness.

Table 5. Common Solvents Used in Resistant Coatings and Linings

Class	Examples	Rate of evaporation			Los Angeles County APCD, Rule 66 restriction
		Slow	Medium	Fast	
aliphatic hydrocarbon	dipentene	X			none (any aromatic content lowers toluene or xylol use by equivalent amount)
	mineral spirits		X		
	VM&P naphtha			X	
alcohols	octyl, amyl, n-butyl	X			none
	isopropyl, ethyl		X		
	methyl			X	
aromatic hydrocarbon	Solvesso—150	X			same as xylol
	Solvesso—100		X		same as xylol
	xylene			X	8 vol % max
	toluene			X	20 vol % max
esters	cellosolve acetate	X			none
	butyl acetate		X		
	ethyl acetate			X	
ketones	methyl amyl ketone	X			none
	methyl isobutyl ketone		X		20 vol % max
	methyl ethyl ketone			X	none

Manufacture

The development of the modern resistant coating and lining is an outgrowth of the paint industry. The differences among paint, industrial coatings, and the resistant coatings are in the formulation and use of the ingredients. Each supplier of resistant coatings has developed a specific formulation through research into a specific corrosion or contamination problem and the selection of each ingredient which provides the coating with its optimum properties. Except for a very few government specifications that outline a specific formula, each manufacturer's resistant coatings are proprietary and may not be similar to any other coating manufacturer's product. As the ingredients in high-performance coatings and linings are an outgrowth of paint technology, the same is true of manufacturing techniques. In general, the same equipment and manufacturing procedures are used for paint (qv) or industrial coatings.

Application of Resistant Coating Systems

There is no perfect coating formulation to satisfy the demands of every environment or set of application conditions. In order to obtain the resistant properties required from a coating or lining, manufacturers often rely on a system approach to obtain the optimum thickness and properties from the coating. The coating or lining system may involve a single material applied in several coats as in the case of a baked phenolic lining in a food processing vessel. It may consist of a primer and one or more topcoats as a coating for structural steel in a chemical plant. Or, the system may be a primer, then an intermediate coat for building thickness, and finally, a finish or seal coat to provide impermeability and suitable appearance as in linings for many critical chemical industry structures.

The characteristics provided by each part of a coating system are as follows:

Primer. Although all parts of the coating or lining are vital to its success in any given environment, the primer is the base on which the remainder of the film is built and as such is critical to the success of the system. A primer provides: (*1*) good adhesion to the surface to be protected when the latter has been cleaned or prepared according to specification to prevent osmotic blistering of the coating or lining; (*2*) a satisfactory bonding surface for the next coat; (*3*) the ability to retard the spread of corrosion from discontinuities such as pinholes, holidays, or breaks in the coating film; (*4*) enough chemical and weather resistance by itself to protect the surface for a time in excess of that anticipated before application of the next coat in the system; and (*5*) where used as a tank lining or in areas of strong chemical fumes, chemical resistance equivalent to the remainder of the system.

Intermediate Coats. Intermediate coats may be required in a system to provide one or more of the following properties: (*1*) adequate film thickness for the coating or lining system; (*2*) a uniform bond between the primer and the topcoats; and (*3*) a superior barrier to moisture vapor and aggressive chemicals.

Finish Coats. The finish coat is the first barrier to the environment such as weather, fumes or chemical solutions. The visible finish coat: (*1*) must be pleasing in appearance as well as resistant; (*2*) may serve to provide a nonskid surface, a matrix for antifouling agents or other specialized purposes where the barrier to the environment is primarily a function of the body or primer coat; and (*3*) the weather, chemical and abrasion resistance of the finish coat must be sufficient in any environment to ensure its remaining intact and providing protection to the substrate.

Each coating or lining system has its own particular properties as developed by a specific manufacturer and although it is sometimes expedient to mix coating system components from several manufacturers, this practice cannot be recommended for any critical lining or coating use.

There are two essentials that must be included in any coating or lining process. These are surface preparations and coating or lining application. This becomes more evident when it is realized that both the surface preparation and application processes are manual. This is one of the major differences between resistant and industrial coatings. Industrial coatings may have both surface preparation and application fully automated.

Surface Preparation. The preparation of the surfaces to receive high performance coatings is extremely important, particularly for immersion service since any contamination of the surface can be a focal point for loss of adhesion or for osmotic blistering. The second purpose of surface preparation is to abrade the surface and increase the surface area so that maximum adhesion is obtained. This is particularly important as most high performance coatings are also of high molecular weight which means that wetting of the surface or penetration of the surface is more difficult. Increasing the surface area by blasting or other means vastly increases the surface area for proper adhesion. The best possible surface preparation is a key to the proper performance of resistant coatings.

The Steel Structures Painting Council (SSPC) through its specifications (2) recognizes ten methods of preparing steel for coating (Table 6). Of these, four are abrasive blasting procedures which are also recognized by the National Association of Corrosion Engineers (NACE) (3).

Where any resistant coating is to be used for any critical purpose, the level of surface preparation for steel should not be less than SSPC-10-63 or NACE no. 2 near-white blast cleaning.

Table 6. Surface Preparation Standards

SSPC	NACE	Description
SP-1-63		solvent cleaning
SP-2-63		hand tool cleaning
SP-3-63		power tool cleaning
SP-4-63		flame cleaning of new steel
SP-5-63	no. 1	white-metal blast cleaning
SP-6-63	no. 3	commercial blast cleaning
SP-7-63	no. 4	brush-off blast cleaning
SP-8-63		acid pickling
SP-9-63T		weathering followed by blast cleaning
SP-10-63	no. 2	near-white blast cleaning

Surface preparation for the application of resistant coatings over concrete is entirely different from steel (4). Concrete provides a very uneven and variable surface; it is porous and contains air and moisture, and cast concrete surfaces are full of pinholes and air and water pockets. Unless these are eliminated, any resistant coating applied over the surface will fail rapidly. The primary method of surface preparation is to trowel a cement plaster over the surface of the concrete to be coated; another and more modern method is to apply a resinous surfacer by spray and trowel sufficiently thick to eliminate the imperfections over the concrete.

The resinous surfacer also acts as a primer for the resistant coatings that follow.

At present there is no industry standard available for the surface preparation of concrete. Each resistant coating has its own application characteristics. Application instructions for each material are available from the manufacturer and should be followed carefully to obtain a coating or lining with long-lasting properties. It is because of the care required, particularly in the lining of tanks, tank cars, or ships for the transportation of chemicals or food, such as caustic soda, wine, or liquid sugar, that many companies have turned to application specialists in order to obtain the quality of lining required.

Economic Aspects

Resistant coatings or linings represented about 9% of the overall paint or coatings market of approximately 2 billion dollars in 1977. In 1975 there were approximately 45.5 ML (12×10^6 gal) of resistant coatings (chlorinated rubber, epoxy, vinyl, urethane, acrylic, inorganic zinc, bituminous) used for maintenance by the manufacturing, utilities, water and sewage, marine, offshore highway, and railroad industries (5) (Table 7). All maintenance paint sold in the United States in 1975 amounted to 348 ML (92×10^6 gal). For the total resistant coatings market, new construction in the same industries using the same coatings must be included. It is estimated by the resistant coatings industry that addition of the new construction volume, which is ca 50% of the maintenance coating market, gives a total resistant coating market of nearly 75 ML (20×10^6 gal). The estimated total value of the resistant coatings market is $200,000,000. Although this is a relatively small volume, the economic importance of resistance coatings and linings is quite large and their importance cannot be over-

Table 7.   Resistant Coatings Used for Maintenance, ML [a]

Industry	Chlorinated rubber	Epoxy	Vinyl	Urethane	Acrylic	Inorganic zinc	Bituminous
manufacturing	2.27	5.3	4.92	2.27	2.27		
gas and electric utilities	0.64	0.38					
water and sewage	0.075	0.265	0.227				
railroads	0.227	0.568		0.151	0.34		
marine	3.59	9.42	5.0			2.5	2.2
offshore	0.189	0.605	0.53	0.416			
Total	*6.99*	*16.54*	*10.68*	*2.84*	*2.61*	*2.5*	*2.2*

[a] To convert ML to millions of gallons, divide by 3.785.

emphasized in the protection of costly industrial structures and the prevention of contamination of products that come in contact with the coating.

The economics of high-performance coating systems are very closely tied to the cost of application and surface preparation. The three items, materials, surface preparation, and application cannot be separated since the materials are of no value until they are properly applied and form an effective coating or lining. The actual economics in most protective coating or lining systems must be related to the replacement cost of the structure or the value of the product which would be contaminated if the lining of the container was improper. For example, an unlined tank in a tanker carrying gasoline or clean refined oil products deteriorates at a rate such that the bulkheads in the tank must be replaced after approximately seven years of service. The replacement cost today is over $1,000/m^2 (ca $100/ft^2). On the other hand, a resistant lining can be applied at a cost of approximately $50/m^2 (ca $5/ft^2) and will provide a corrosion-free life to the tank from 15 to 20 years. At the end of the 15-yr period there would be no major metal loss in the tank. Under these conditions $50/m^2 is a relatively low cost to ensure an essentially new tank at the end of a 15-yr period. In another example, the cost of lining for a 37,500-L (10,000-gal) wine tank car with approximately 100 m^2 (ca 1000 ft^2) of internal surface with a resistant coating which is noncontaminating to the wine would be approximately $40/m^2 (ca $4.00/ft^2). Wine is extremely sensitive to iron contamination so that even a small unprotected area would cause discoloration and bad taste. At a value of $1.32/L ($5/gal) there is little question that the resistant coating provides an economic service. However, this again depends on the application of the product. Without thorough application and 100% coverage by the coating it is of little value.

There are many factors that enter into the economics of a resistant coating. The actual material in many cases is a relatively small part of the actual cost since each application is different. The structural steel in a chemical plant several stories high has a very high cost of application. Tank exteriors in the same plant are a relatively low cost application because of the plain surfaces involved. In this case, the material costs would be a higher proportion of the total applied cost. There is a useful discussion of chemical plant coatings in reference 6. The actual cost of the coating of steel tank exteriors was a small proportion of the total applied cost. In the case of the high performance coating in ref. 6, the cost of the material alone for the initial application was 25% of the total initial cost. In the case of the low-cost system, the material amounted to only 14% of the initial cost. However, if these same materials had been applied to the interior surfaces of tanks or on elevated structures, the material costs would have dropped to 10% or less of the total applied cost. This indicates that a compromise on material quality is not economical because of the relatively small cost of the material as compared to the overall installation.

The material cost of some typical high-performance coatings is shown in Table 8.

Table 8.   Typical High-Performance Coating Material Costs, 1977

	Cost/m² per 25 µm thickness, $	Cost/ft² per mil of thickness, $	Cost/m² per recommended thickness, $	Cost/ft² per recommended thickness, $
4-coat water tank, vinyl	0.376	0.035	2.26	0.21
6-coat vinyl food lining	0.495	0.046	4.95	0.46
2-coat high-build vinyl (ext)	0.258	0.024	2.58	0.24
2-coat polyamine–epoxy tank lining	0.204	0.019	2.04	0.19
3-coat polyamide–epoxy (exterior)	0.247	0.023	3.44	0.32
3-coat epoxy-phenolic tank lining	0.430	0.04	5.16	0.48
2-coat coal tar–epoxy	0.118	0.011	1.93	0.18
inorganic zinc	0.398	0.037	1.18	0.11

A partial list of U.S. manufacturers of resistant coatings is: Ameron, Protective Coating Division; Carboline Company; Devoe and Reynolds Company; DuPont Company; Matcote Company; Mobil Chemical Co.; Napko Corporation; Plas-Chem Coating Company; Porter Paint Company; Prufcoat Division; and Wisconsin Protective Coating Company.

Resistant Coating Testing and Evaluation

Inspection both during and after the application is essential. Evaluation starts with surface preparation for steel; the surface should be prepared either to a white or near-white condition by abrasive blasting. Photographic standards showing the various grades of surface preparation are available. These visual standards are an aid in making certain that the surface preparation is proper. No such visual standards are available for concrete surfaces; here the written description in the job specification must be followed and used as a guide by the inspector.

Test Equipment.   During the application of a protective coating, there are three pieces of equipment to aid in the inspection process (7). The first of these is a wet-film thickness gage; such a gage is put out by the Nordsen Corporation, Amherst, Ohio. This instrument provides a quick check of the wet film thickness as the coating is being applied. Once the coating is dry, a dry-film thickness gage may be used. At this stage it is preferable to use a nondestructive instrument and there are several available (see Non-destructive testing). These are based on a magnetic principle and are therefore only operable on steel or iron surfaces. One of the most practical and easiest to use is the Mikrotest Gage distributed by the Nordsen Corporation. These instruments measure the film thickness in micrometers or mils and can be used on complex surfaces. The third type of instrument used to determine quality of application is the holiday detector. There are two types. First, the relatively high-voltage tester which, when passed over the coated surface, arcs when there is a pinhole or imperfection in the coating. These are used primarily on thick film coatings. They indicate an imperfection either by the visual arc or by a bell which rings as the instrument is passed over the imperfection. The second type, used for thinner film coatings, is based on a relatively low voltage and uses a damp sponge to provide contact with the coating surface. As the sponge passes over an imperfection, electrical contact is made with the underlying surface and a bell rings. The two instruments provide a positive indication where coating imperfections exist and both can be used on either metal or concrete. Both types are manufactured by the Tinker and Rasor Company, San Gabriel, Calif.

On nonmagnetic surfaces the Tooke gage can be used to determine coating thicknesses. It is a destructive test. The coating is cut with an accurate tip providing a uniform groove and the coating depth is measured by the width of the cut using a small microscope incorporated into the equipment. This equipment is distributed by Ken Tator Associates, Coraopolis, Pennsylvania. Visual inspection aided by the above instruments will provide a good measure of the integrity of the coating.

Coating and Lining Evaluation.   The real measure of a coating's effectiveness is a test under actual operating conditions. There are a number of laboratory tests that can be used for screening purposes which indicate the effectiveness of a coating or lining. Where a single factor of corrosion or contamination is involved, laboratory tests can be definitive, such as where a coating or lining is in contact with a single product, such as sugar syrup, or where a coating would be subject to splash and spillage of materials such as sodium hydroxide. Where there are a large number of variables, however, laboratory tests can only be comparative between the materials and only indicative of the coating or lining resistance. There are several published test methods for coatings and linings; these are listed in Table 9.

From a purely practical standpoint, it should be emphasized that if a protective coating is going to be used, a firm specification and good inspection procedures to assure compliance with the specification are essential. Coating manufacturers have such detailed engineering specifications available covering all important points of application and control (inspection) for their products.

Resistant Coating Types

Resistant coatings are classified by binder types in Table 1. The important materials in the resistant category and a brief description of the principal coatings and linings follows.

Vinyl Coatings.   Vinyl resin coatings are solidly entrenched as one of the standard specialty coatings that will perform where other materials have failed. They have one of the broadest and most useful ranges of properties of any of the coating types yet developed. Resins have been made by copolymerizing various vinyl molecules to form high molecular weight, soluble resins which may then be incorporated into a resistant coating (see Vinyl polymers). Two principal types of vinyl coatings that are used in protective coatings are vinyl chloride–vinyl acetate copolymers composed of 87% vinyl chloride and 13% vinyl acetate, and vinyl–acrylic copolymers. Coatings made from these materials form a very tight, dense film. The coatings are always soluble in their own solvents, allowing easy repair. Since they are thermoplastic, they are heat sensitive and can only be used at temperatures below 65°C. Vinyl resin coatings have their principal use in industrial applications where their inertness is of special value. They are used for coatings and linings of tanks, pipelines, petrochemical plants, well heads,

Table 9. Methods of Testing Coatings and Linings[a]

NACE Standard TM-01-74, Laboratory Methods for the Evaluation of Protective Coatings Used as Lining Materials in Immersion Service

ASTM Standard B-117-73, Salt Spray (Fog) Testing

ASTM Standard B-287-74, Acetic Acid–Salt Spray (Fog) Testing

ASTM Standard B-368-68 (1973), Copper Accelerated Acetic Acid–Salt Spray (Fog) Testing (CASS Test)

ASTM Standard D-610-68 (1974), Preparation of Steel Panel for Testing Paint, Varnish, Lacquer, and Related Products

ASTM Standard D-611-64 (1973), Evaluating Degree of Rusting on Painted Steel Surfaces

ASTM Standard D-659-74, Evaluating Degree of Chalking of Exterior Paints

ASTM Standard D-660-44 (1970), Evaluating Degree of Checking of Exterior Paints

ASTM Standard D-661-44 (1970), Evaluating Degree of Cracking of Exterior Paints

ASTM Standard D-662-44 (1970), Evaluating Degree of Erosion of Exterior Paints

ASTM Standard 714-56 (1974), Evaluating Degree of Blistering of Exterior Paints

ASTM Standard 772-47 (1970), Evaluating Degree of Flaking (Scaling) of Exterior Paints

ASTM Standard D-822-60 (1973), Recommended Practice for Operating A Light-Water-Exposure Apparatus (Carbon Arc Type) for Testing Paint, Varnish, Lacquer and Related Products

ASTM Standard D-823-53 (1970), Producing Films of Uniform Thickness of Paint, Varnish, Lacquer, and Related Products on Test Panels

ASTM Standard D-870-54 (1973), Water Immersion Test of Organic Coatings on Steel

ASTM Standard D-968-51 (1972), Test for Abrasion Resistance of Coatings of Paint, Varnish, Lacquer, and Related Products by the Falling Sand Method

ASTM Standard D-1005-51 (1972), Measuring the Dry Film Thickness of Organic Coatings

ASTM Standard D-1014-66 (1973), Conducting Exterior Exposure Tests of Paints on Steel

ASTM Standard D-1186-53 (1973), Measuring of Dry Film Thickness of Nonmagnetic Organic Coatings on a Magnetic Base

ASTM Standard D-1212-70, Measurement of Wet Film Thickness of Organic Coatings

ASTM Standard D-1400-67, Measurement of Dry Film Thickness of Nonmetallic Coatings of Paint, Varnish, Lacquer, and Related Products Applied on a Nonmagnetic Metal Base

ASTM Standard D-1653-72, Test for Moisture Vapor Permeability of Organic Coating Films

ASTM Standard D-1654-61 (1974), Evaluation of Painted or Coated Specimens Subject to Corrosive Environments

ASTM Standard D-1735-62 (1973), Water Fog Testing of Organic Coatings

ASTM Standard D-2197-68 (1973), Test for Adhesion of Organic Coatings

ASTM Standard D-2247-68 (1973), Testing Coated Metal Specimens at 100% Relative Humidity

ASTM Standard D-3258-73, Test for Porosity of Paint Films

ASTM Standard D-3276-73, Recommended Guide for Paint Inspection (this is an outline for inspection and testing procedures)

ASTM Standard D-3359-74, Measuring Adhesion by Tape Test

ASTM Standard D-3361-74, Recommended Practice for Operating Light and Water Exposure Apparatus (unfiltered carbon arc type) for Testing Paint, Varnish, Lacquer, and Related Products Using the Dew Cycle

ASTM Standard D 3363-74, Test for Film Hardness by the Pencil Test

ASTM Standard E-376-69, Recommended Practice for Measuring Coating Thickness by Magnetic Field or Eddy Current (electromagnetic test methods)

ASTM Standard G-8-72, Test for Cathodic Disbonding of Pipe Line Coatings

ASTM Standard G-12-72, Nondestructive Measurement of Film Thickness of Pipe Line Coatings for Steel

---

[a] Ref. 8.

offshore drilling rigs, paper plants, food processing equipment, water and sewage plants and in many similar areas.

Chlorinated Rubber Coatings. Chlorinated rubber is a resin made from unvulcanized natural rubber that has been allowed to react with chlorine. The chlorine adds to the double bonds of the rubber molecule giving a product with a very high wt % of chlorine (see Elastomers, synthetic, neoprene; Butadiene). The final product is a very hard, tough resin which is soluble in aromatic solvents. Chlorinated rubber cannot be used alone as a coating resin but must be modified with other resinous materials to plasticize it. To accomplish this the chlorinated rubber is modified with other low molecular weight but pliable chloride-containing resins. These plasticize the chlorinated rubber molecule without materially reducing its highly resistant characteristics. Chlorinated rubber-base coatings, eg, the vinyl coatings, dry by evaporation only so that the properties of the basic resin do not change during the drying process. Chlorinated rubber coatings are very satisfactory for the interior or exterior surfaces of concrete, either submerged or in the atmosphere. They are primarily used as maintenance coatings, although one of the larger uses is a coating for concrete swimming pools. Their greatest use is in industrial plants and marine areas where their hard, tough, chemically resistant properties combined with the ability to be easily repaired and recoated make them particularly useful.

Epoxy Coatings. There are several types of epoxy coatings (see Epoxy resins). Each has different properties; however, each can also be included in the resistant coating category. They are all of the coreacting resin type. The basic epoxy resin, which is essentially a bisphenol–epichlorohydrin condensation product (bisphenol diglycidyl ether), is combined with other resinous materials or chemicals that react with the epoxy groups. These materials cure by internal linkage only, and do not require oxygen from the air to dry or set. Thus heavy coating thicknesses can be achieved even in a single application (150–3000 μm). This is highly advantageous and one of the reasons for their rather wide popularity. The principal types of epoxy coatings are as follows.

Amine-Cured Epoxies. Amines or polyamines react very readily with the epoxy resin. Some of them react extremely rapidly, almost to the point of being explosive, whereas others are very slow and require heating. From a coating standpoint, amines are used that will provide a workable pot life and cure without the addition of heat. To form a coating, the epoxy resin and the amine (supplied in separate containers) are mixed just prior to application. The amine-cured epoxy coatings are the most chemically resistant of the ambient-temperature cure variety. They are somewhat brittle. They tend to chalk readily when exposed to the weather. They are used primarily for the more difficult chemical problems in industry where an air-dry coating is required.

Polyamide-Cured Epoxies. Here the epoxy resin reacts with a separate polyamide resin, both of which can be of approximately the same semiliquid consistency. This combination is somewhat less chemically resistant than the amine-cured epoxy. On the other hand, it has much better weather resistance. The resulting coating is nonbrittle and somewhat flexible. They have excellent resistance to alkali and to water. They should not be used where exposed to strong bacterial growth, such as under sewage conditions. The use of this type of material is broad, including maintenance coatings for most industries such as chemical industry, paper industry, marine, atomic power, food, etc.

Phenolic–Epoxies. These materials are a combination of phenolic and epoxy resins, usually cured by additional heat. As such, they are primarily used for tank linings or for direct exposure to various chemicals. When heat cured, they are the most resistant of the epoxy coatings and are widely used for exposure to solvents, vegetable and animal oils, fatty acids, foods, and alkalis. They are not resistant to oxidizing agents such as sodium hypochlorite.

Coal Tar–Epoxies. Coal tar–epoxy coatings are a fortunate combination of two materials that provide the best features of both components. The coal tar improves the water resistance of the epoxy and the epoxy improves the thermal plasticity of the coal tar as well as the weather resistance (see Tar and pitch). One of the secrets of a good coal tar–epoxy coating is the choice of the coal tar pitch to be used. Inasmuch as pitches are natural products, there is a wide variability in the coal tar pitch so that some of the derived resins react more readily than others with the amine or polyamide curing agents. As these materials are chemically reactive, the amine and resin are supplied in separate packages and mixed just prior to use. The resulting coatings have good chemical resistance; reasonable weather resistance (they do chalk to a brown color); excellent resistance to fresh and salt water, brine, hydrogen sulfide, and generally to both acidic and alkaline conditions. They are extremely durable coatings for the protection of concrete and metal, either submerged or nonsubmerged, against rather severe corrosive action. They cannot be used as a decorative coating inasmuch as the coal tar is black and, except for a reddish-brown, they can only be used in areas where black is satisfactory. They have a general use throughout the chemical industry. They are also very widely used in the marine industry both on ships and on offshore structures (see Coatings, marine).

Phenolic Coatings. Phenolic coatings as described here are the high-baked phenolic products primarily used as linings for various containers or chemical equipment (see Phenolic resins). The coatings used are in the pure phenolic category. These materials require very careful application and curing in order to be effective but, properly applied and cured, the resistance is excellent to almost all acidic conditions as well as water up to close to boiling temperature. Solvent resistance is excellent including both chlorinated and oxygenated solvents, fatty acids and aromatic hydrocarbons. They are widely used throughout the food industry as a lining for containers such as beer storage tanks, wine storage tanks and sugar syrup tanks since they are very inert and nontoxic when properly cured. They are not satisfactory under alkaline conditions.

Polyurethane Coatings. There are two principal types that can be considered under the resistant coatings category. These are the coreactable polyurethane, and the moisture-reacted polyurethane. The urethane prepolymers contain isocyanate (NCO) groups that react without catalyst with any active hydrogen group. The hydrogen group may be a hydroxyl, an amine or even water (see Urethane polymers).

The coreactable, or so-called two-package, polyurethane coatings are two component systems; one component containing the polyisocyanate resin (usually a pigmented urethane prepolymer–isocyanate adduct) and the second consisting of a hydroxyl-bearing material such as a polyester, polyether, or other active hydrogen-bearing polymer. These materials can vary from an extremely hard and glasslike coating to soft, rubbery elastomers. Chemical resistance is good; however, it is not comparable to epoxy amines, vinyls, or chlorinated rubber. Solvent resistance is good and the polyurethane coatings with aliphatic isocyanates as cure coreactants display excellent weather resistance, gloss, and color retention. They retain their original appearance without chalking or other change for a much longer period than other resistant coating types. Their abrasion resistance is outstanding and with these properties one of their uses is as an aircraft coating where they must withstand extreme weather conditions as well as extreme abrasion by sand, rain, and hail.

The moisture-cured polyurethane coatings are polyurethane prepolymers that contain free isocyanate groups which cure by reacting with atmospheric moisture. No separate catalyst is needed. This one-component system is extremely popular because of its convenience. It has most of the good properties of the two-component coreactable system; on the other hand, curing difficulties may arise where these materials are used under conditions of low humidity. The greatest use of these polyurethane coatings is for their exceptional abrasion resistance where they are used as coatings for floors, drive areas, and other areas exposed to excessive traffic. They have been used as clear coatings for plywood–concrete forms and exterior wood finishes.

Inorganic Zinc Coatings.   These materials are included in the resistant coatings category inasmuch as they are the most weather resistant coatings available today that can be applied by conventional spray paint equipment to extremely large steel structures. These materials are produced by reaction of either an alkali metal silicate or a hydrolyzed ethyl silicate with metallic zinc. The product of either reaction is quite similar in that the zinc reacts with the silicic acid forming a zinc silicate matrix around the zinc particles. The reaction product can be considered a cold galvanizing; however, under many marine conditions the resistance of the inorganic zinc coating is much superior to galvanizing and will last several times as long under the same conditions. The inorganic zinc coatings have revolutionized the maintenance of structures in the marine industry (see Coatings, marine). This applies both to the shipping industry as well as offshore installations. In many cases, the inorganic zinc coating is used alone as a single coating only 75–100 μm in thickness for decks, superstructures, coating of equipment, tanks, and in many cases all of the above-water area of offshore platforms. These materials are also used as the lining for refined-oil tankers and in such service have lasted for 20 or more years without serious corrosion to the tank surface. In one application in Australia, a 400 km, 760 mm pipeline was coated and installed above ground over grassland, salt marshes, and along the sea coast (see Pipelines). After 30 years of exposure the pipe was corrosion free and had not lost any coating thickness. The inorganic zinc coatings have excellent solvent resistance and have been used as a lining for many solvent tanks, including gasoline and petroleum products of all types, aromatic solvents, oxygenated solvents, chlorinated solvents, and many others. They are exceptionally resistant to solvents but should not be exposed to strong acid or alkali which could react with the zinc metal.

Miscellaneous.   There are many other materials that have been used to make coatings that can be considered in the resistant category. These materials are polyester, furan, and silicone coatings, and rubber, vinyl sheet, and glass linings. The uses of these materials are very specialized compared to the ones described above and therefore cannot be described in detail here.

Uses of Resistant Coatings

In order to better make a comparison of resistant coatings, the following is a specific series of uses where one or more of the above products are outstanding.

Abrasion Resistance. This is a rather narrow field of use; however, when it is considered in conjunction with chemical exposure, food exposure, or personal safety, it is very important. The outstanding coatings in this category are the polyurethanes. The abrasion resistance of these materials is much superior to any of the other primary resistant coatings. Their use on aircraft where the coating is subject to abrasion by sand on take-offs and landings, and to rain, ice and hail in the air demonstrates this property well.

Bacterial and Fungal Resistance. Of all the materials listed, undoubtedly the vinyl coatings have the broadest resistance to biological activity since they are completely reacted prior to their application, have very high molecular weights, are very inert, and do not act as a nutrient for either bacteria or fungi. The outstanding example of the use of this material for severe biological conditions is to protect concrete surfaces under sewage conditions. Here, bacterial slimes and fungi form on all surfaces, and even after exposure of 20 or more years, vinyl linings have shown no evidence of disintegration, nor have they allowed any concrete corrosion.

Chemical Resistance. The coating type with the broadest overall chemical resistance is the vinyl coating. Vinyl coatings will withstand exposure to strong acids, salts, strong alkali, and oxidizing materials. This property is demonstrated by their successful use in chemical, paper, sewage, food processing, and atomic power plants, as well as aboard ship and in offshore structures. Vinyl sheet linings have been used for electrolytic zinc and copper tanks where both acid concentrations and temperatures are high. They have also been used for strong oxidizing conditions such as linings for sodium hypochlorite manufacturing and storage tanks (see Film and sheeting materials).

Acid Resistance. This is a very broad category and there are several specific conditions to be considered. For acid fumes, under chemical plant or similar conditions, vinyl or epoxy–amine coatings are most suitable and have been used successfully for many years. Where such exposure must be combined with weather resistance, the vinyl coatings and vinyl–acrylic coatings have given good results.

To withstand acid splash or spillage a heavy duty coating of either the vinyl or epoxy–amine type may be used. For these conditions the coating should be of a substantial thickness of >0.4 mm. There are extensive exposures of both of these coatings to just such conditions in the chemical industry.

For immersion under strong acid conditions, heavy rubber (qv) or vinyl sheet are required in order to assure a completely pinhole-free lining. Thin coatings are extremely difficult to apply, even in multiple coats, to obtain a pinhole-free condition. As a lining for acidic food products, the high-baked phenolic coatings have been most successful. Elevated temperatures may also be involved under such conditions and only a highly cross-linked coating such as a baked phenolic is suitable. There are many areas in between the two extremes just described and specialized versions of vinyls and epoxies have been successful, depending on the conditions.

Alkali Resistance. One use for resistance to strong alkali is the lining for storage containers and transportation tanks containing 50 or 73% liquid sodium hydroxide. Here the containers must be maintained at above 72°C in order to prevent crystallization of the sodium hydroxide. There are hundreds of tank cars dedicated to this service. The most successful coatings have been medium-temperature-cure epoxy–phenolics and a special coating developed for this purpose composed of multiple coats of neoprene latex (see Elastomers, synthetic). For lesser exposures of alkali, the epoxy–polyamides have provided excellent resistance. For protection of spillage areas of strong caustic, coal tar–epoxies have performed well. Alkali resistance is one of the strong characteristics of epoxy coatings.

Water Resistance.   Water resistance must be considered under the overall heading of chemical resistance, inasmuch as water is one of the most damaging of all chemicals to various coatings and linings. The reason is that all resinous materials have specific moisture absorption and moisture vapor transfer rates. Coatings that are not firmly adherent to a substrate act as semipermeable membranes. Moisture also has a very marked effect on the adhesion of various materials and the retention of that adhesion over a period of time. When water vapor passes through a coating or lining to the interface between the substrate and the lining itself, the moisture may release the adhesion of the resinous material to the substrate. All coatings and linings, therefore, are not fully resistant to continuous water exposure. One of the oldest and best known of the water-resistant linings, particularly for steel water pipe, are the hot-applied coal tar–enamels. These are solid coal tar–pitches that are applied to the interior surfaces of pipe as a hot melt. Coal tar–epoxy coatings are also very water resistant and have been used as linings for water tanks as well as pipe. They have also been used extensively in the marine field for protection against sea water. Vinyl coatings have been used for many years as a completely nontoxic and noncontaminating lining for potable water tanks. They were also one of the first of the coatings to be used as linings for storage areas for fissionable materials. In this case water is used as a radiation shield. Polyamide–epoxy coatings have been used in more recent years as coatings for water areas in and around atomic reactors and for atomic fuel storage areas (see below).

Solvent Resistance.   The need for a resistant material in this catogory is primarily as a lining. The material with the broadest solvent resistance is the inorganic zinc coating. This coating has been used extensively as a lining for gasoline, diesel oil, lubricating oil, jet fuel, toluene, xylol, styrene, acetone, methyl isobutyl ketone, and similar storage or transportation tanks. Baked phenolic linings have been extensively used as tank lining for solvent conditions involving fatty acids and the more active lower molecular weight organic acids such as acetic, propionic and butyric acids. Medium-temperature-cure epoxy–phenolics have been used extensively as tank linings aboard ships for transporting large quantities of materials such as methyl alcohol and other high volume solvents.

Contamination-Resistant Linings.   One of the primary uses for linings for tanks is to protect the product contained, rather than to protect the container from the product. In many cases, of course, it is necessary to protect each from the other.

Chemical Linings.   One large use of a lining for the protection of the product is storage and transportation of sodium hydroxide. One of the uses of this product is for the manufacture of rayon and any iron contamination in the caustic causes red discoloration of the rayon fiber (see Rayon). This being the case, all containers from the time that the sodium hydroxide is manufactured to its ultimate use, must be either lined to protect the sodium hydroxide or be of a nonreactive material, such as a stainless alloy. Medium-temperature-cure epoxy–phenolic linings have been used in hundreds of tank cars as well as ships for the transportation of liquid caustic soda. One of the most difficult lining applications is in ship tanks that are used to transport many different chemicals. There are many ships used specially for chemical transport and the materials protected by the lining may range from ethyl alcohol, to Brazil nut oil, engine oil, styrene monomer, tallow, and even whiskey. There are several hundred liquid commodities transported and any one may follow the other in the tank. Epoxy–phenolic coatings are most successful in this service.

Food Linings. A very large area where contamination is a factor is in the food industry. In this case, linings are required for protection of the food from bacteria and fungus contamination as well as from chemical contamination. One of the materials that is shipped in large quantity and which is very sensitive to iron contamination is wine (qv). There are upwards of 10,000 tank cars dedicated to wine service, all of which must be thoroughly lined to prevent any iron coming in contact with the wine. Should this occur, the wine picks up a metallic taste as well as a blue coloration. Bulk transportation of wine was originally made possible by a specialized, five-coat, vinyl coating. Epoxy–phenolic as well as high-baked phenolic coatings are also now used for this purpose. Beer (qv) is another large volume food product that must be protected during storage. High-baked phenolic and glass linings are commonly used for this purpose. Milk and other food products that require processing or storage at elevated temperatures are stored in either stainless steel containers or glass-lined tanks. Fats and oils, as well as their fatty acid derivatives, many times require lined containers (see Fats and fatty oils). Here epoxy–phenolic or glass-lined containers are used, depending upon the temperatures involved.

Coatings Requiring Decontamination. Large volumes of resistant coatings are used in atomic energy plants to prevent contact of radioactive materials with concrete or metal tanks, structures, or equipment (see Nuclear reactors). The coating must be extremely inert so as not to absorb the radioactive material, react with it or be disintegrated by it. One of the primary properties of the coating must be that it can easily be decontaminated. Epoxy–polyamide coatings are used for this purpose at the present time because of their good radiation resistance and ease of decontamination. Decontamination testing measures the ratio of the original beta–gamma activity vs the activity after decontamination (Decontamination Factor or DF number). The four major coating types that have been rated (9) are: epoxy–polyamide; modified phenolic; vinyl; and chlorinated rubber (Table 10).

Friction-Resistant Coatings. Friction resistance is a rather unique property and it is extremely important where large structures, buildings or bridges are bolted or riveted together. There are only a very few coatings that can be used between two sections of steel and which make a satisfactory joint from an engineering standpoint. Most organic coatings do not have a sufficiently high coefficient of friction to be used on the mating steel surfaces (faying surfaces) of a riveted or bolted structure. Following is a list showing the friction resistance of some coatings and other typical surfaces used on faying surfaces in the construction of bolted or riveted structures (10):

Surface conditions	Coefficient of friction
solvent-based inorganic zinc coating	0.52
water-based inorganic zinc coating	0.47
rusted surface	0.48
sandblasted surface	0.47
mill scale surface	0.30
rust-preventing paint	0.11
red lead paint	0.06

Any surface that has a coefficient of friction less than a sandblasted surface is not considered satisfactory for a bolted or riveted joint. As shown, the only resistant

Table 10. Radiation Tolerance of Resistance Coatings[a,b]

Coating	Maximum allowable radiation dose in air, MGy[c]	
	On steel	On concrete
chlorinated rubber	1	1
epoxy–amine	10	10
epoxy–coal tar	5	5
epoxy–polyamide	100	NA[a]
inorganic silicate finish	100	100
inorganic zinc	200	NA[b]
epoxy phenolic	100	100
silicone (baked)	100	NA[d]
urethane	5	60
vinyl	1	

[a] Ref. 11.

[b] Severe exposure, greater than 45 MGy; moderate exposure, 5–45 MGy; light exposure, less than 5 MGy.

[c] Gray: the SI unit of absorbed radiation; for most organic material one Roentgen = 0.01 Gy (= 1 Rad). To convert MGy to rad, multiply by 10^8.

[d] NA, not available.

coatings that have these properties are the inorganic zinc coatings. They have been used for such surfaces in many areas and provide not only a satisfactory friction-resistant surface, but also provide a corrosion-resistant joint between the two steel surfaces.

Heat-Resistant Coatings. Heat resistance can be divided into two different categories: dry heat and wet heat.

Dry Heat. Coatings used for dry heat conditions are those applied to the exterior of furnace breachings, stacks, the exterior of chemical reactors, high temperature piping and similar structures. The most difficult condition exists on those structures where the temperature varies from ambient up to several hundred degrees. The stacks on diesel motors are a good example. As soon as the motors are stopped, the temperature of the stacks drop to ambient, moisture condenses and rusting starts. One coating used for many such areas is an aluminum-pigmented silicone. The limitations of the silicone–aluminum coating is 315–425°C. Above 370°C there is little truly protective film left. Thus if the stack or reactor varied from 370°C to ambient temperature, corrosion would soon take place creating an unsightly condition. A combination of coatings, however, has proven to be quite satisfactory to 480°C. This is the application of a base coat of inorganic zinc, top coated with silicone–aluminum. Under these conditions, the inorganic zinc coating provides the corrosion resistance and the aluminum–silicone, the desired appearance. Inorganic zinc coatings without topcoats have been used extensively on stacks and furnace breachings and have been effective in the 315–370°C range for long periods. Combinations of silicone resins (see Silicon compounds) and alkyd resins (qv), primarily pigmented with aluminum, have been effective up to approximately 230°C.

Wet Heat. The boiling temperature of water represents the critical temperature for coatings. Above this temperature there are very few coatings effective for any extended period. For salt solutions, sugar syrups, or other nonacidic chemicals, the temperature at which a coating can be effective is increased by the concentration of the dissolved material. The alkali-resistant coatings (see above) have been used in concentrated sodium hydroxide up to 110°C for long periods. The processing of foods is one area where temperatures above boiling may be encountered. Here glass-lined vessels have been most effective, although baked phenolic coatings have proven satisfactory, depending on the food and the process. Baked phenolics and thermally formed epoxy–phenolics have been used for water in the 90–100°C range. Vinyl or epoxy coatings are seldom recommended above the 70°C range.

Soil-Resistant Coatings. There are many underground structures that require protection. Pipelines and tanks have the greatest surface area requiring protection. Here the coatings must withstand continuous moisture exposure, working of the soil owing to the absorption and desorption of moisture, earth movement, and heavy traffic. Coatings must also be resistant to soils of different electrical resistivity since corrosion cells are easily established in soil. In the chemical industry new installations are often made where previous structures have been removed and soils may contain high concentrations of chemicals such as acids or alkali. Chemical resistance then becomes important. The most widely used coating for buried pipe is a hot-applied coal tar–enamel, reinforced with a layer of glass fiber followed by a final layer of asbestos felt impregnated with the hot-applied coal tar. Such a coating is approximately 3 mm in thickness. Thickness, under these conditions, is important to withstand the abrasion from earth movement and the penetration of rock points through the coating owing to the weight of the pipe and the back fill. Thinner coatings, such as coal tar–epoxies, have been used under certain circumstances with effectiveness. More recently, powdered epoxy coatings have been applied to pipelines by fusing the epoxy on the surface (see Powder coatings). Both of these latter materials, however, lack useful thickness.

Weather-Resistant Coatings. The greatest single use of coatings is for the protection of structures from the atmosphere. Resistant coatings are primarily used for severe atmospheric conditions combined with chemical fumes, salt deposits or corrosive dusts. Under most circumstances, weathering is a limiting factor to the life of the coating. The most light-resistant, color stable and chalk-resistant coatings for exterior surfaces are the aliphatic isocyanate-cured polyurethane coatings and the vinyl–acrylics. These materials have extended periods of life—several years under extreme weather and chemical fume conditions without appreciable change in color, gloss, or coating integrity. Although not decorative, the coatings with the longest life in severe weather are the inorganic zinc coatings. These have been exposed to severe weather conditions for periods of up to 30 years without any loss of coating integrity, coating thickness, or corrosion. The combination of the inorganic zinc coating and one of the above organic coatings as a topcoat has proven to be the best answer to atmospheric corrosion in chemical plants, refineries, etc.

Health and Safety Factors

There are two areas where there may be health and safety hazards involved with protective coatings. The first is in the manufacture of the coating; the second is in the application of the coating. The hazards are similar inasmuch as they are primarily involved with the solvent used in the coating. There are three health and safety factors to be considered both during manufacture and application. These are: air pollution; health hazards to the individual; and fire and explosion hazards. In the manufacturing plant, all three of these factors are relatively easily controlled through proper factory operation under regulations established by OSHA and the EPA.

On the other hand, the application of the coatings on site, which is where the majority of protective coatings are used, presents a very different control problem which, if overlooked, can lead to catastrophe. Air pollution (qv) is probably the easiest factor to control in the field inasmuch as many areas have specific requirements for the type and the amount of solvent that can be used in the manufacture of protective coatings. The Los Angeles Air Pollution Control District Rule 66 has the most stringent requirements and regulations patterned after Rule 66 are being adopted by many cities, states, and nations. Most protective coating manufacturers supply coatings that conform to Rule 66.

The greatest hazards are created when protective coatings are applied to the interior of closed or confined spaces such as the interior of atomic plants, enclosed manufacturing areas, underground sumps, storage tanks, tank trucks, ship tanks, tank cars, or water tanks. There are three hazards in the application of coatings: fire or explosion, physiological damage to the applicator from inhalation of solvents, and dermatitis caused by contact with some of the solvents or resins. Dermatitis is the least of these hazards as it can be effectively controlled by the proper use of protective clothing. A compressed-air headmask plus gloves and overalls will normally prevent individual contact with any irritating chemicals. The hazard due to toxic vapors from the solvents in the coatings can also be reduced to a negligible point through the use of proper ventilation. First by ventilation of the enclosed area through the use of suction fans to remove the solvents; and second by the use of compressed-air masks for the individual whereby a constant supply of clean air under light pressure is received.

Fire and explosion is the greatest hazard from the standpoint of loss of property and personnel. Many resins used in resistant coatings have a very limited flammability or are self-extinguishing as in the case of chlorinated rubbers and vinyls. The key to the elimination of fire hazard is the proper ventilation of the enclosed space so that the solvent vapor concentration never reaches the lower explosive limit. Most of the solvents used in protective coatings have a lower explosive limit above 1 vol % of solvent in the air. Full ventilation must be maintained during the entire drying period. Another important factor is that most of the solvents used in protective coatings are heavier than air. Therefore, they tend to concentrate in the lower areas of any enclosed space. Suction blowers should be used to draw the solvent-laden air from the lowest point of the enclosed space to prevent solvent build up in the lower area to above the explosive limit.

General safety rules that must be observed in the application of coatings to enclosed areas are discussed in ref. 12.

New Coating Developments

As a consequence of the new and increasing governmental environmental standards as well as health and safety requirements, there are three avenues of development that will have a strong influence on the future of protective coatings. The first is the use of water-based coatings. These will be based on water-soluble, water-dispersed or water-emulsion coatings. Currently there are many water-dispersed or latex paints on the market (see Latex technology). However, these materials do not have the physical or chemical properties for resistant coatings. A few manufacturers have water-based epoxy coatings on the market. However, much more development work is required to make them generally as satisfactory as the solvent-based epoxies. The second approach is that of solvent-free coatings or 100% solids coatings. There are some coatings available today that are close to being 100% solid or solvent free. Most of these materials still contain 10–15% solvent. Future work is required on the development

Table 11. Properties of Resistant Coatings[a]

Uses of resistant coatings	Lacquer coatings				Coreactable coatings		
	Vinyl chloride–acetate copolymer	Vinyl–acrylate copolymer	Chlorinated rubber	Coal tar–pitch (hot melt)	epoxy–amine	epoxy–polyamide	Coal tar–epoxy
abrasion resistance	G	G	G	F	G	G	G
bacterial and fungal resistance	E	G	G	G	G	NR	G
chemical resistance	BSR	G	BSR	G	G	G	G
acid-oxidizing	S or F	F	S or F	NR	NR	NR	NR
-nonoxidizing	S or F	F	S or F	F	F	F	F
-organic	I fatty acid	F	dissolves in fatty acids	NR	Spillage, F	F	NR
alkali	G	D	G	G	G	G	G
salts-oxidizing	Splash, S	F or D	F or D	NR	F	F	F
-nonoxidizing	I, G	F or D	I, G	sea water, G	I, G	I, G	I, G
solvent-aliphatic	E	F	G	NR	E	G	G
-aromatic	swells	dissolves	dissolves	NR	G	NR	NR
-oxygenated	dissolves	dissolves	dissolves	NR	NR	NR	NR
water	VGI	G	VGI	E	VGI	VGI	VGI
moisture	low		low	low	low	low	low
permeability							
contamination of contacting materials							
food	[c]		water, G	water, G	G	G	NR
chemical	VG		G		G	G	F
decontamination	VG				VG	VG	NR
friction resistance (faying surfaces)							
heat resistance, °C							
wet	48	38	38	48	48	48	48
dry	65	65	60	65	95	95	95
radiation resistance, Gy[d]	10^6		10^6		10^7	10^8	5×10^6
soil resistance				E			G
weather and light resistant	G, properly pigmented	E	G, properly pigmented	NR	heavy chalking	G	chalking
principal hazard application	solvent F	solvent F	solvent F	coal tar F	dermatitis solvent F	solvent F	dermatitis solvent F

[a] G, Good; VG, Very Good; E, Excellent; NR, Not Recommended; BSR, Broad-Spectrum Resistance; I, Immersion; S, Spray; F, Fumes; D, Dusts.
[b] G, primer required; critical for immersion.
[c] G, odorless; tasteless; nontoxic.
[d] To convert gray to rad, multiply by 100.

of low-viscosity reactive resins which, when combined into a coating, can be applied without the use of any solvent. Much work is being done in this area and effective 100% solid coatings should be available in the near future. The third approach is the use of inorganic materials as coatings. Inorganic zinc coatings have been the start of this development. These materials have proven to be effective; however, they are all based on the reactivity of metallic zinc which limits their effectiveness where chemical resistance and appearance are factors. Research and development is presently being done on cold-applied inorganic coatings that will have properties similar to ceramic enamel (see Enamels).

Urethane (2 package)	Urethane Moisture cure	Urethane–aliphatic isocyanate cure	Condensing coatings			Inorganic coatings		
			Baked phenolic	Epoxy–phenolic	Epoxy powder	Water-base	Solvent-base	Fused ceramic
E	E	E	G, brittle	G	G	VG, metallic	VG, metallic	VG, brittle
			E	E	G			E
G	G	G	VG	VG	G	NR	NR	VG
S or F	F	S or F	NR	splash, F	NR	NR	NR	I, G
S or F	F	S or F	G, ambient temp.	splash, F	splash, F	NR	NR	I, G
splash or F	F	splash or F	G, ambient temp.	splash, F	splash, F	NR	NR	I, G
S or D	D	S or D	NR	E	G	NR	NR	NR
NR	NR	NR	NR	F	F	NR	NR	I, G
I, G[b]	splash	splash	I, G	I, G	splash, F	G, marine splash, S	G, marine splash, S	I, G
E	VG	E	E	E	E	E	E	E
E	VG	E	E	E	E	E	E	E
NR	NR	NR	E	alcohols, G, ketones, NR	NR	E	E	E
[b]		[b]	E	E	I, G	G	G	E
medium, low	medium, low	medium, low	low	low	low			low
			[c]	G, non-toxic				[c]
			VG	G				G
VG	G	VG						G
						E coef. of friction 0.47	E coef. of friction 0.52	
38		38	82	82	48			95
120	120	120	120	120	95	370	315	260
5×10^6	5×10^6	5×10^6		10^8		10^8	10^8	10^8
					thin film-care required for backfill	NR	NR	NR
G, yellows	G	E, color and gloss retention	NR		G, chalks	E >20 yr	E >20 yr	E >20 yr
solvent F	solvent F	solvent F	phenol, solvent F	solvent F	none	none, water base	solvent F	none

Table 11 allows a quick comparison of the properties of the resistant coatings described above.

BIBLIOGRAPHY

1. F. L. LeQue, *Marine Corrosion, Causes, and Prevention*, John Wiley & Sons, Inc., New York, 1975, pp. 291–297.
2. J. D. Keane, *Steel Structures Painting Manual*, Vol. 2, Steel Structures Painting Council, Pittsburgh, Pa., 1973.

3. *NACE Standard T.M.-01-70, Visual Standard for Surfaces of New Steel Air Blast Cleaned with Sand Abrasive,* National Association of Corrosion Engineers, Houston, Tex., 1970.
4. C. G. Munger, *Coating Concrete Surfaces in Nuclear Plants, Material Performance,* Houston, Tex., May 1976, pp. 31–35.
5. Private study, *Maintenance Paints, III—A Market and Economic Study of the United States Markets by Smith, Stanley & Co., Inc. and American Paint Journal Co.,* Ameron Corporation, Brea, Calif., 1976.
6. F. P. Helms, *Paper presented at 2nd Annual Corrosion Short Course, NACE, Philadelphia, Pa., Sept. 1964.*
7. K. B. Tator and K. A. Trimber, *Plant Eng.* **28,** 155 (Sept. 19, 1974); *ibid.,* 73 (Oct. 3, 1974).
8. *Annual Book of ASTM Standards, Paint Tests for Formulated Products and Applied Coatings,* ASTM, Philadelphia, Pa., Part #27.
9. C. D. Watson and G. A. West, *Mater. Prot.* **6,** 44 (Feb. 1967).
10. W. H. Munse, *Static & Fatigue Tests of Bolted Connections,* private report, Ameron Corporation, Brea, Calif., Mar. 1961.
11. *ANSI Standard N5.12-1973, Protective Coatings (Paints) for the Nuclear Industry,* American National Standards Institute, Washington, D.C.
12. C. G. Munger, *Coatings and Their Safe Application,* Vol. 14, Marine Section National Safety Council, Washington, D.C., 1964.

General References

NACE TPC Publication #2, Coatings and Linings for Immersion Services, National Association of Corrosion Engineers, Houston, Tex., 1973; an excellent treatise providing properties and comparisons of all major generic types of coatings and linings.
J. E. Rench and co-workers, *Fundamentals of Exterior Maintenance Coatings,* Internal Publication Napko Corporation, Houston, Tex.; good basic information.
Corrosion Control Principles and Methods, internal publication, Ameron, Protective Coatings Division, Brea, Calif.; a text for a basic course in protective coatings.
A short course in practical paint technology to assist consulting and maintenance engineers, Koppers Chemicals and Coatings, Pittsburgh, Pa.; a very good resume of coatings and application including resistance charts.
Federation Series on Coatings Technology, Federation of Societies for Coatings Technology, Philadelphia, Pa.; *Unit 1—Introduction to Coatings Technology,* 1973; *Unit 14—Silicone Resins for Organic Coatings,* 1970; *Unit 15—Urethane Coatings,* 1970; *Unit 18—Phenolic Resins,* 1971; *Unit 19—Vinyl Resins,* 1972; *Unit 20—Epoxy Resins in Coatings,* 1972.
Properties and Uses of Parlon Chlorinated Rubber, Hercules, Inc., Wilmington, Del., 1971; good reference for formulations and properties of chlorinated rubber.
Dow epoxy resins in coating applications, Technical Data Report No. 35-0 plus No. 35-0-1, etc, The Dow Chemical Company, Midland, Mich. 1970, epoxy coating formulation information.
Chemistry for Coatings, Mobay Chemical Corporation, Plastics & Coatings Division, Mar. 1976; this is a good reference for urethane coatings.
Johnson, *Polyurethane Coatings,* Noyes Data Corporation, Park Ridge, N.J., 1972.
Potter, *Epoxide Resins,* Iliffe Books, London, Eng., 1971.
Army TM 5-618, Navfac MO-110, Air Force AFM 85-3, Paints and Protective Coatings, Superintendent of Documents, U.S. Government Printing Office, Washington, D.C., 1969.
R. C. Snogren, *Handbook of Surface Preparations,* Palmerton Publishing Co., New York, 1974.
J. Bigos, *Steel Structures Painting Manual,* Vol. 1, Steel Structures Painting Council, Pittsburgh, Pa., 1966.
P. Weaver, *Industrial Maintenance Painting,* 3rd ed., NACE, Houston, Tex., 1970.
P. E. Baedenmann, "The Systems Approach to Application of Protective Coatings: How to Achieve Longer Life at Less Cost," *Pulp Pap.* **43,** 116 (1969).
ANSI Standard N 101.2-72, Protective Coatings for Light Water Nuclear Reactor Containment Facilities, American National Standards Institute, Washington, D.C., 1972.
ANSI Standard N 101.4 1972, Quality Assurance for Protective Coatings Applied to Nuclear Facilities, American National Standards Institute, Washington, D.C., 1972.
NACE Standard RP-02-72, Recommended Practice-Direct Calculation of Economic Appraisals of Corrosion Control Measures, National Association of Corrosion Engineers, Houston, Tex., 1972.

E. N. Hiestand and R. E. Hicks, *A study of conditions for the safe application of Amercoat Coatings,* private report for Ameron Corporation, Battelle Memorial Institute, Columbus, Ohio, Sept. 1953.

"Rule 66—After 8 Years, Los Angeles Society for Coatings Technology," *J. Coat. Technol.* **48**(613) (Feb. 1976).

W. D. Fruzel, "Correlation of the Efforts of the Dew Cycle Weatherometer and Florida Exposure on Highly Durable Coatings," *J. Paint Technol.* **44**, (Nov. 1972).

P. Whiteley, G. W. Rothwell, and J. Kennedy, "Deterioriation of Paint Films by Water," *J. Oil Color Chem. Assoc.* **58**, (June 1975).

C. J. Hicks, "Chlorinated Rubber Fights Corrosion," *Mod. Paints Coat.* **66**(12), (Dec. 1976).

I. Metil, "The Importance of Galvanic Protection in Zinc Rich Coatings," *Paint Varn. Prod.* **63**, (Jan. 1973).

"Coal Tar Epoxy Protects Jersey Sewage Plant," *Paint Varn. Prod.* **62**, (Mar. 1972).

CHARLES G. MUNGER
Consultant

COBALT AND COBALT ALLOYS

COBALT

Cobalt, a transition series metal with atomic number 27, is a metallic element that is similar to silver in appearance.

Cobalt was used as a coloring agent by Egyptian artisans as early as 2000 BC. Cobalt-colored lapis or lapis lazuli was used as an item of trade between the Assyrians and Egyptians. In the Greco–Roman period cobalt compounds were used as ground coat frit and coloring agents for glasses. The common use of cobalt compounds in coloring glass and pottery led to their import to China during the Ming Dynasty under the name of Mohammedan blue.

The ancient techniques for mining cobalt and the use of cobalt compounds were forgotten during the dark ages. However, in the 16th century mining techniques became widely known through the works of Georgius Agricola, the German mineralogist. At that time cobalt was supplied as smalt or zaffre. The latter was a cobalt arsenide of sulfide ore that was roasted to yield a cobalt oxide. When fused with potassium carbonate to form a type of glass, zaffre became smalt. At the same time the ability of cobalt to color glass blue was rediscovered. Metallic cobalt was isolated in 1735 by a Swedish scientist named Brandt, and was established as an element in 1780 by Bergman.

Cobalt and cobalt compounds have expanded from strict use as coloring agents in glasses and ground coat frits for pottery to drying agents in paints and lacquers, animal and human nutrients, electroplating materials, high-temperature alloys, high-speed tools, magnetic alloys, alloys used for prosthetics, and uses in radiology. Cobalt is also used as a catalyst for hydrocarbon refining. The most timely use of this last application is the synthesis of heating fuels (1–2).

Occurrence

Cobalt is the 30th most abundant element on earth and comprises approximately 0.0025% of the earth's crust (3). It occurs in mineral form as arsenides, sulfides, and oxides; trace amounts are also found in other minerals of nickel and iron as substitute ions (4). Cobalt minerals are commonly associated with ores of nickel, iron, silver, bismuth, copper, manganese, antimony, and zinc. Table 1 lists the major cobalt minerals and some of their properties. A complete listing of cobalt minerals is given in reference 5.

The world's largest cobalt reserves are in Zaire, Zambia, Morocco, Canada, and Australia. Together the ores of these countries contain well over one-half of the world cobalt supply. The richest deposits are in Zaire and Zambia. The reserves of Canada and Australia comprise approximately one-fourth of the world supply. Smaller but commercially practical ore bodies also exist in the U.S.S.R., Finland, Uganda, and the Philippines.

The largest cobalt producing mine in the world is a national company in Zaire, La Générale des Carriers et des Mines du Congolaise des Mines (Gecomines). In addition to cobalt, this mining complex also produces copper and zinc (6). Zaire has been producing cobalt since 1914 and has been the world leader of cobalt ore production since 1940.

Cobalt deposits of Zaire are either cobaltiferrous or cobalt–copper ores. A small amount of these deposits occur as high-grade oxidized ore. As much as 0.4% cobalt is contained in malachite ($CuCo_3.Cu(OH)_2$) deposits as $CuO.2Co_2O_3.6H_2O$. The oxidized cobalt minerals that have been found in these deposits are heterogenite, stainierite, asbolite, sphaerocobaltite, carrollite, cobaltite, cattiertite, cobalto–vaesite ($NiCoS_2$), siegenite, and selenio–seigenite ($(Co,Ni)_3(Si,Se)_4$).

In Zambia, with a production second only to Zaire, cobalt is mined by Nchanga Consolidated Copper Mines Ltd. (NCOM) and Roan Consolidated Ltd. (RCM). The cobalt minerals found there include minnaeite, carrollite, and cobaltiferrous pyrite.

In Moroccan deposits, cobalt occurs with nickel in the forms of smaltite, skutterudite, and safflorite.

In Canadian deposits, cobalt occurs with silver and bismuth. Smaltite, cobaltite, erythrite, safflorite, linnaeite, and skutterudite have been identified as occurring in these deposits.

Australian deposits are associated with nickel, copper, manganese, silver, bismuth, chromium, and tungsten. In these reserves, cobalt occurs as sulfides, arsenides, and oxides.

The Colombian Government is now studying apparently large deposits of cobalt ore in Cerro Matoso. There are also large deposits in Cuba.

In the United States, cobalt is found at Blackbird, Idaho, and the Grace and Cornwall mines in Pennsylvania. Although the reserves at the Blackbird mine are considered large, operations were suspended in 1971 because of a decline in demand for cobalt.

At the Blackbird mine, cobalt occurs with chalcopyrite, gold, silver, and nickel. The Pennsylvania deposits occur with magnetite as cobaltiferrous pyrite. In 1976 the United States produced only small quantities of cobalt domestically, however, imported cobalt was refined at an AMAX plant at Port Nickel, La., which has a production capacity of ca 22 metric tons/yr. This plant is expected to increase its nickel

Table 1. Important Cobalt Minerals and Some of Their Properties

Mineral	CAS Registry No.	Chemical formula	Crystalline form	Approximate hardness, Mohs	Density, kg/m³	Cobalt, %	Location	References
arsenides								
smaltite	[12044-42-1]	$CoAs_2$	cubic	6.0	6.5	23.2	United States Canada Morocco	5
safflorite	[12044-43-8]	$CoAs_2$	orthogonal	5.0	7.2	28.2	Morocco Canada	5
skutterudite	[12196-91-7]	$CoAs_3$ $(Co, Ni)As_3$ $(Co, Ni, Fe)As_3$	cubic	6.0	6.5	20.8	Ontario Morocco	5-6
sulfides								
carrollite	[12285-42-6]	$CuCo_2S_4$ $CuS.Co_2S_3$ Co_3S_4	cubic	5.5	4.85	38.7	Zaire Zambia	5-6
linnaeite	[1308-08-3]	So_3S_4	cubic	5	4.5	48.7	Zaire	5-6
siegenite	[12174-56-0]	$(Co, Ni)_3S_4$				26.0	United States	5
cattierite	[12017-06-0]	CoS_2 $(Co, Ni)S_2$ $(Ni, Co)S_2$	cubic				Zaire	6
arsenide–sulfide								
cobaltite	[1303-15-7]	$CoAsS$	cubic	6	6.5	35.5	United States Canada Australia	5-6
oxide								
asbolite	[12413-71-7]	$CoO.MnO_2.4H_2O$	ore	1–2	1.1		Zaire Zambia	5-6
erythrite	[149-32-6]	$3CoO.As_2O_5.8H_2O$	ore	2	3	29.5		5-6
heterogenite	[12323-83-0]	$CuO.2Co_2O_3.6H_2O$ $Co_2O_3.H_2O$; also pres. $CuO.2Co_2O_3.3H_2O$ $CoO.2Co_2O_3.6H_2O$ $CoO.3Co_2O_3.CuO.7H_2O$	ore	4	3.5	57	Zaire	5-6
sphaerocobaltite	[14476-13-2]	$CoCO_3$	ore			49.6	Zaire Zambia	5-6

and cobalt production by 3% by 1980. Furthermore, Roland F. Beers Inc. is investigating large deposits in Maine.

Future Sources.   Lateritic ores (7) are becoming increasingly important as a source of nickel, with cobalt as a by-product. In the United States, laterites are found in Minnesota, California, Oregon, and Washington. Deposits also occur in Cuba, Indonesia, New Caledonia, the Philippines, Venezuela, Guatamala, Australia, Canada, and the U.S.S.R.

The laterites can be divided into three general classifications: (*1*) Iron nickeliferrous limonite which contains approximately 0.8–1.5 wt % nickel. The nickel to cobalt ratios for these ores is typically 10:1. (*2*) High-silicon serpentinous ores that contain more than 1.5 wt % nickel. (*3*) A transition ore between type 1 and type 2 containing about 0.7–0.2 wt % nickel and a nickel to cobalt ratio of approximately 50:1.

Laterites found in the United States (8) contain 0.5–1.2 wt % nickel with the nickel occurring as the mineral goethite. Cobalt occurs in the lateritic ore with manganese oxide at an estimated wt % of 0.06 to 0.25 (9).

A potentially significant source of cobalt is the ocean floor. Deep Sea Adventures of Virginia and several Japanese firms are evaluating locations off the coast of Baja, California, as possible sites for mining cobalt-bearing nodules from approximately 4.0 km under the ocean surface. A similar off-shore mining venture in the Atlantic Ocean is being planned by Kennecott Copper. The nodules will be removed from the ocean bottom either by a dredging pipe or by dredging buckets. If a dredging pipe is used, cleaning and crushing of the nodules will occur as they are picked up (6) (see Ocean raw materials; also Manganese).

The nodules occurring as ferromanganese deposits can be grouped into four classes: (*1*) hydrogeneous deposits formed by slow precipitation of Mn and Fe from seawater; (*2*) hydrothermal deposits occurring as a result of volcanic activity; (*3*) halmyrolytic deposits resulting from submarine weathering; and (*4*) diagenetic deposits arising from the movement of manganese from a reducing to oxidizing environment. It has been estimated that sea nodules are being formed at a rate of 10 million tons/yr in the Pacific. However, there is no proof that the newly formed nodules occur in areas that can be exploited as mine sites. Elemental composition of sea nodules varies with location. Nodules found in the western and southern Pacific are rich in cobalt but lean in nickel and copper. Compared with the cobalt production from Zaire and Zambia, the quantity derived from sea nodules is small.

Properties

The electronic structure of cobalt is $3d^7 4s^2$. At room temperature the crystalline structure is close-packed-hexagonal (cph), the α (or ϵ) form, with lattice parameters of $a = 0.2501$ nm and $c = 0.4066$ nm. Above approximately 417°C, a face-centered cubic (fcc) allotrope, the γ (or β) form, becomes the stable crystalline form with a lattice parameter of $a = 0.3544$ nm. The mechanism of the allotropic transformation has been well described (5,10–12). Cobalt is magnetic up to 1123°C (see Magnetic materials). At room temperature the magnetic moment is parallel to the c-direction. The physical properties are listed in Table 2.

Many different values for room-temperature mechanical properties can be found in the literature. The lack of agreement depends, no doubt, on the different mixtures of α and γ phases of cobalt present in the material. This, on the other hand, depends

Table 2. Properties of Cobalt

Property	Value		
Thermal			
atomic weight	58.93		
transformation temperature, °C	417		
heat of transformation, J/g[a]	251		
mp, °C	1493		
latent heat of fusion, J/g[a]	259.4		
bp, °C	3100		
latent heat of vaporization, J/g[a]	6276		
specific heat, J/(g·°C)[a]			
15–100°C	0.442		
molten	0.560		
coefficient of thermal expansion, per °C			
cph room temperature	12.5		
fcc at transformation temperature	14.2		
thermal conductivity at room temp, W/(m·K)	69.16		
thermal neutron absorption, Bohr atom	34.8		
Electrical and magnetic			
resistivity, at 20°C[b]	6.24		
Curie temperature, °C	1121		
saturation induction, $4\pi I_S$, T[c]	1.870		
permeability			
initial, μ_0	68		
max, μ_m	245		
residual induction, T[c]	0.490		
coercive force, A/m	708		
Selected mechanical			
Young's modulus, GPa[d]	211		
Poisson's ratio	0.32		
hardness, diamond pyramid (Vickers)	*99.9% Co*		*99,98% Co[f]*
at 20°C	225		253
at 300°C	141		145
at 600°C	62		43
at 900°C	22		17
strength of 99.9% cobalt, MPa[e]	*as cast*	*annealed*	*sintered*
tensile	237	255	679
tensile yield	138	193	302
compressive	841	808	
compressive yield	291	387	

[a] To convert J to cal, divide by 4.184.
[b] Conductivity = 27.6% of International Annealed Copper Standard.
[c] To convert T to gauss, multiply by 10^4.
[d] To convert GPa to psi, multiply by 145,000.
[e] To convert MPa to psi, multiply by 145.
[f] Zone refined.

on the impurities present, the method of production of the cobalt, and the treatment. These discrepancies appear not only with regard to the metal by itself, but also with regard to the alloys such as Stellites, cemented carbides, and hard-facing alloys.

The hardness on the basal plane of cobalt depends on the orientation and extends between 70 and 250 HK. Among the interesting facts about cobalt is its resistance to loss of properties when heated to fairly high temperatures. This explains the use of cobalt in high-temperature alloys of the superalloy type. Another valuable property of cobalt is its good work-hardening characteristics.

Whereas finely divided cobalt is pyrophoric, the metal in the massive form is not attacked by air or water at temperatures below approximately 300°C. Above this temperature, it is oxidized by air. Cobalt combines readily with the halogens to form halides. It combines with most of the nonmetals when heated or in the molten state. Although it does not combine directly with nitrogen, it decomposes ammonia at elevated temperatures to form a nitride. It reacts with carbon monoxide above 225°C to form the carbide Co_2C. Cobalt forms intermetallic compounds with many metals, such as Al, Cr, Mo, Sn, V, W, and Zn.

Metallic cobalt dissolves readily in dilute H_2SO_4, HCl, or HNO_3 to form cobaltous salts. Like iron, cobalt is passivated by strong oxidizing agents, such as dichromates and HNO_3. It is slowly attacked by NH_4OH and NaOH.

Cobalt cannot be classified as an oxidation-resistant metal. Scaling and oxidation rates of unalloyed cobalt in air are twenty-five times those of nickel. The oxidation resistance of Co has been compared with that of Zr, Ti, Fe, and Be. Cobalt in the hexagonal form (cold-worked specimens) oxidizes more rapidly than in the cubic form (annealed specimens) (3).

The scale formed on unalloyed cobalt during exposure to air or oxygen at high temperatures is double-layered. In the range of 300 to 900°C, the scale consists of a thin layer of Co_3O_4 on the outside and a CoO layer next to the metal. Co_2O_3 may be formed at temperatures below 300°C. Above 900°C, Co_3O_4 decomposes and both layers, although of different appearance, are composed of CoO only. Scales formed below 600°C and above 750°C appear to be stable to cracking on cooling, whereas those produced at 600–750°C crack and flake off the surface.

Processing

Sulfide Ores. In the Zairian ores, cobalt sulfide as carrolite, is mixed with chalcopyrite and chalcocite. For processing, the ore is finely ground and the sulfides are separated by flotation (qv) with frothers. The resulting products are leached with dilute sulfuric acid to give a copper–cobalt concentrate. The concentrate is then used as a charge in an electrolytic cell and the copper is removed. Because the electrolyte becomes enriched with cobalt, solution from the copper circuit is added to maintain a desirable copper concentration level. After several more steps to remove copper, iron, and aluminum, the solution is treated with milk of lime to precipitate the cobalt as the hydroxide.

Zambian copper sulfide ores are leaner in cobalt and are, therefore, concentrated twice. In the first stage, the bulk of the copper ore is floated off in a high-lime circuit. The carrolite is not carried by the lime, but is recovered in a second pass using different flotation methods. The second concentration, which contains 25% Cu, 17% Fe, and 3.5–4% Co, undergoes a sulfatizing roasting to convert the cobalt to water soluble cobalt sulfate. The iron and copper in the concentrate form insoluble oxides and sulfates. After roasting, the matte is leached with water and filtered. After removing the last of the copper, milk of lime is added to precipitate the cobalt hydroxide which is filtered and dissolved in sulfuric acid. The resulting solution is then used as an electrolyte from which cobalt is electrodeposited. The cobalt thus produced is marketed either in a granulated form or still attached to the cathode (see Extractive metallurgy).

Arsenic-Free Cobalt–Copper Ores. The arsenic-free cobalt–copper ores of Zaire are treated by smelting. Ores such as heterogenite with high cobalt content can be sent directly to an electric furnace in lump form. The fines must be sintered before being charged. Smelting the cobalt–copper feed along with lime and coke produces a slag and two alloys.

Recoverable amounts of cobalt and copper (10–15 wt %) are contained in the slag. A white alloy and a red alloy are formed in smelting (see Table 3). The slag is poured into a water-cooled slag hole for further processing. The red alloy is tapped from the bottom whereas the white alloy is poured from the ladle. The white alloy is cast into ingots, and the red alloy is further refined with a smelter process. The slag is then returned to the electric furnace as feed. Cobalt oxide can be obtained by further processing the white alloy to remove copper and iron as sulfates and calcining cobalt as the carbonate.

Table 3.   Composition of Cobalt Alloys Obtained from Cobalt–Copper Ores, wt %

Constituent	White alloy	Red alloy
cobalt	42	4.5
copper	15	89
iron	39	4
silicon	1.6–2	

The red alloy is sent to a copper refinery. Slag obtained from this operation contains approximately 15 wt % Co and is returned to the electric furnace to be mixed with the cobalt–copper feed.

Arsenic Sulfide Ores. The high-grade ores of Morocco are magnetically separated to give arsenides and oxides (see Magnetic separation). Ores that are most concentrated in the arsenides are subjected to an oxidizing roast to remove the arsenic. The concentrates are used as feed in a blast furnace, resulting in speiss, matte, and perhaps boullion. The cobalt-containing speiss is then crushed and roasted. The roasted speiss is treated with sulfuric acid and the solids are removed and roasted. After the iron compounds are precipitated with sodium chlorate and lime, the cobalt–nickel solution is treated with sodium hypochlorate to precipitate a cobalt hydrate.

Pressure-acid leaching was used to extract cobalt from Blackbird mine ores before its closing in 1974. The cobalt-rich mineral cobaltite was separated from chalcopyrite and pyrite by controlling the pH of the solution containing the minerals. The slurry concentrate was fed into an autoclave at 3.55 MPa (500 psig) and 190°C. The final slurry contained cobalt, nickel, and iron sulfates. The solution was purified to remove the iron and was autoclaved at 190°C under a hydrogen atmosphere of 5.6 MPa (800 psig) in the presence of a catalyst. The result was a very fine cobalt powder which was subjected to a seeding process to produce cobalt granules. Leaching methods are also used in the refinement of lateritic ores.

Lateritic Ores. The process used at the Nicaro plant in Cuba requires that the dried ore is roasted in a reducing atmosphere of carbon monoxide at 760°C for 90 min. The reduced ore is cooled and discharged into an ammoniacal leaching solution. Nickel and cobalt are held in solution until the solids are precipitated. The solution is then thickened, filtered, and steam heated to eliminate the ammonia. Nickel and cobalt are precipitated from solution as carbonates and sulfates. This method (8) has several disadvantages: (*1*) a relatively high reduction temperature and a long reaction time; (*2*) formation of nickel oxides; (*3*) a low recovery of nickel and the contamination of nickel with cobalt; and (*4*) low cobalt recovery. Modifications to this process have been proposed but all include the undesirably high 760°C reduction temperature (9).

A similar process has been devised by the U.S. Bureau of Mines (8) for extraction of nickel and cobalt from United States laterites. The reduction temperature is lowered to 525°C and the holding time for the reaction is 15 min. An ammoniacal leach is also employed, but oxidation is controlled, resulting in a high extraction of nickel and cobalt into solution. Mixers and settlers are added to separate and concentrate the metals in solution. Organic strippers are used to selectively remove the metals from the solution. The metals are then removed from the strippers. In the case of cobalt, spent cobalt electrolyte is used to separate the metal-containing solution and the stripper. Metallic cobalt is then recovered by electrolysis from the solution. Using this method 92.7 wt % nickel and 91.4 wt % cobalt have been economically extracted from domestic laterites containing 0.73 wt % nickel and 0.2 wt % cobalt (8).

Deep Sea Nodules.   Metal prices will influence the type of extraction process used for sea nodules which are typically rich in manganese. However, most of the mining and refining expenses must be met by the price of commodity metals such as copper and cobalt rather than by an abundant metal like manganese. It has been suggested that a method be used that would selectively remove cobalt, nickel, and copper, leaving manganese stored in the tailings for future use (13–14). This involves smelting of the reduced new nodules. The result is a manganiferrous slag and an alloy containing the metals. The alloy is converted into a matte by oxidation and sulfidation. An oxidative pressure leach is used to produce a purified solution from which the metal is obtained (13).

As rich deposits are depleted, it will become necessary to find more efficient ways to extract metals. Research is continuing in the areas of pyrometallurgy, hydrometallurgy, and electrometallurgy to find more efficient extraction techniques (15–18).

Analysis

The detection and determination of cobalt traces is of concern in such diverse areas as soils, plants, fertilizers, stainless and other steels (for atomic energy equipment), high-purity fissile materials (U, Th), refractory metals (Ta, Nb, Mo, and W), and semiconductors. Useful techniques are spectrophotometry, polarography, emission spectrography, flame photometry, x-ray fluorescence, activation analysis, tracers, and mass spectrography, chromatography, and ion exchange (19) (see Analytical methods).

For colorimetric or gravimetric determination 1-nitroso-2-naphthol can be used. For chromatographic ion exchange, cobalt is isolated as the nitroso-R-salt complex. The cyanate complex is used for photometric determination and the thiocyanate for colorimetry. A rapid chemical analysis of alloys, powders, and liquids, employing x-ray spectrography has been developed. With this method cobalt, in a concentration of 10–60%, may be analyzed in a few minutes with accuracies of the order of ±1%.

Economic Aspects

Table 4 gives the U.S. consumption of cobalt between 1967 and 1978.

During recent years the market price of commercial-grade metallic cobalt has risen slowly and unevenly to about $15/kg. In May of 1978, as a consequence of a political and military crisis in Zaire, the production of cobalt in that country came to a standstill. The world price of the metal immediately increased at least four-fold. It is expected that the Zairian mines will remain inoperative into 1979. This situation, in addition to heavy recent stockpiling by the U.S.S.R., assures a continuing price rise for the foreseeable future.

Uses

As Metal. The largest consumption of cobalt is in metallic form in magnetic alloys, cutting and wear-resistant alloys, and superalloys. Alloys in the last group are used for machine components requiring high strength performance as well as corrosion and oxidation resistance, usually at high temperatures. It has been proposed that cobalt be substituted for nickel in certain types of stainless steel (21).

During World War II, German scientists developed a method of hydrogenating solid fuels to remove the sulfur by using a cobalt catalyst (see Coal). Subsequently, various American oil refining companies used the process in the hydrocracking of crude fuels (see Catalysis). Cobalt catalysts are also used in the Fisher-Tropsch method of synthesizing liquid fuels (22–23) (see Fuels, synthetic).

Cobalt–molybdenum alloys are used for the desulfurization of high-sulfur bituminous coal, and cobalt–iron alloys in the hydrocracking of crude oil shale (qv) and in coal liquefaction (6).

As Salts. The second largest use of cobalt is in the form of salts. These have their largest application as charges for electroplating baths (see Electroplating) and as highly effective driers for lacquers, enamels, and varnishes (see Driers). Addition of cobalt salts to paint greatly increases the rate at which the paint hardens (see Paint). Cobalt oxide colors glass pink or blue depending upon how the CoO_x molecule complexes within the glass. The pink colors are formed in boric oxide or alkali–borate glasses. The cobalt concentration determines the intensity of the color obtained, eg, glass used in making foundryman's goggles requires 4.5 kg of cobalt per ton of glass. By contrast, the glass used in making decorative bottles requires only about 280 g/t. Cobalt is also

Table 4. U.S. Consumption of Cobalt, Metric Tons[a]

Form	1967	1970	1973	1975
metal	5266	4607	6405	4174
oxide	297	284	303	169
scrap	54	31	120	155
salts and driers	722	1186	1619	1302
total	*6339*	*6108*	*8501*	*5800*

[a] *U.S. Bureau of Mines Minerals Yearbook.* Note that the total metric tons for 1976 were 5143; for 1977, 6544; and for 1978 (estimated January–April (20)), 2177. Prices as of March 1978 (20), $/kg: electric cathodes, granules, 15.10; powder, 37–149 μm (100–400 mesh), 19.64; extrafine powder, 24.87; and metallurgical oxides, 14.81.

used to decolorize soda-lime-silica glass. Pottery enamels react very similarly since the fusible enamels are forms of fusible glasses. Colors varying from blue to black can be obtained, depending upon the oxides added to the frit to improve the adherence of porcelain enamel to sheet metal. Combinations of cobalt compounds are also used as ceramic pigments ranging in colors of violet, blue, green, and pink (24–26).

Radioactive cobalt (^{60}Co), produced by bombarding stable ^{59}Co with low energy neutrons, has application in radiochemistry, radiography, and food sterilization (27–29) (see Radioisotopes).

Cobalt is an essential ingredient in animal and plant nutrition. It has been shown that animals deprived of cobalt show signs of retarded growth, anemia, loss of appetite, and decreased lactation. Dressing the top soil of pastures with cobalt increases the cobalt content of the vegetation. Cobalt is now known to be necessary in the synthesis of vitamin B_{12} in domestic animals other than ruminants (see Vitamins). A lack of vitamin B_{12} has also been linked to anemia in man (see Mineral nutrients). Cobalt is also used as the target material in electrical x-ray generators (see X-ray technology).

COBALT ALLOYS

Pure metallic cobalt has a solid-state transition at approximately 417°C. When certain elements such as Ni, Mn, or Ti are added, the fcc phase is stabilized. On the other hand, adding Cr, Mo, Si, or W stabilizes the cph-phase. If the fcc phase is stabilized, the energy of crystallographic stacking faults (single unit cph inclusions that impede mechanical slip within the fcc matrix) is high, whereas stabilizing the cph-phase produces a low-stacking fault energy. With high stacking fault energy, only a few stacking faults occur; thus, the ductility of the alloy is high. When foreign elements are dissolved throughout the matrix lattice, the mechanical slip is generally impeded. This results in increased hardness and strength, a metallurgical phenomenon known as solid solution hardening.

Mechanical properties may be influenced by the addition of alloying elements, of which carbon is the most effective, to the cobalt base metal. The carbon forms various carbide phases with the cobalt and the other alloying elements. The presence of carbide particles is controlled in part by certain alloying elements such as chromium, nickel, titanium, and manganese that are added during melting. The distribution of the carbide particles is controlled by heat treatment of the solidified alloy.

Thus, cobalt alloys are strengthened by solid-solution hardening and by the solid-state precipitation of various carbides and other intermetallic compounds. Minor-phase compounds, when precipitated at grain boundaries, tend to prevent slippage at those boundaries thereby increasing creep strength at high temperatures. Aging and service under stress at elevated temperature induces some of the carbides to precipitate at slip planes and at stacking faults thereby providing barriers to slip. If carbides are allowed to precipitate to the point of becoming continuous along the grain boundaries, they will often initiate fracture. A thorough discussion of the mechanical properties of cobalt alloys is given in references 30–31 (see Carbides; Refractories).

Binary Alloys With Cobalt as Solute or Solvent.   Table 5 lists typical binary alloys with cobalt as the solvent, and Table 6 lists binary alloys with cobalt as the solute.

Many of the investigated binary systems are high-temperature alloys (qv), such

Table 5. Binary Cobalt Alloys, Cobalt as Solvent

System	References	System	References
Co–Cr	32[a], 33, 34[a], 35	Co–Pd	32[a], 34[a], 39, 59–60
Co–Cu	32[a], 34[a], 36–38	Co–Pr	33, 39
Co–Dy	39, 33[a], 40	Co–Pt	32[a], 33, 61
Co–Er	39[a], 33, 41–42	Co–Pu	33[a], 39[a], 62
Co–Fe	32–34[a], 39[a], 43	Co–Re	32[a], 39, 34[a], 62
Co–Ga	39[a], 44–45	Co–Rh	32
Co–Gd	32–33[a], 46–47	Co–Ru	32
Co–Ge	32[a], 33, 39, 48–49	Co–S	32[a], 33, 39, 63
Co–H	32–33, 39	Co–Sb	32[a], 39[a], 64–65
Co–Hf	39[a], 50–51	Co–Sc	33, 39
Co–Hg	32, 39	Co–Se	32[a], 39[a], 66–67
Co–Ho	33, 39	Co–Si	32[a], 68
Co–In	39[a], 52	Co–Sm	33, 39
Co–Ir	32[a]	Co–Sn	32[a], 69–70
Co–K	32	Co–Ta	32[a], 39[a], 71
Co–La	39	Co–Tb	33, 39
Co–Li	32[a]	Co–Tc	33
Co–Lu	39	Co–Te	32, 39[a]
Co–Mg	32–33[a], 39[a], 53	Co–Th	32, 39
Co–Mn	32[a], 33, 34[a], 39	Co–Ti	32–34[a], 39[a], 72
Co–Mo	32–34[a], 39[a], 54–55	Co–Tl	32
Co–N	33–34	Co–Tm	32, 59
Co–Na	32, 39	Co–U	32[a], 39[a], 73
Co–Nd	32, 39	Co–V	32[a], 39[a], 74
Co–Ni	32[a], 34[a], 56	Co–W	32[a], 34[a], 39
Co–O	32–33	Co–Y	32[a], 75
Co–Os	32[a], 57	Co–Yb	33
Co–P	32[a], 58	Co–Zn	32[a], 34[a]
Co–Pb	32[a], 34[a], 39[a], 58	Co–Zr	32, 33–34[a], 39

[a] References represent phase diagram.

Table 6. Binary Cobalt Alloys, Cobalt as Solute

System	References	System	References
Ag–Co	32	C–Co	32[a], 33, 39
Al–Co	32[a], 33, 39, 77–79	Ca–Co	32[a]
As–Co	32[a], 81–82	Cb–Co	33[a], 83
Au–Co	32[a], 39, 82	Cd–Co	32, 39
B–Co	32[a], 39, 83	Ce–Co	32[a], 33, 39, 85
Bi–Co	32[a], 84		

[a] References represent phase diagram.

as Co–Cr, Co–Fe, Co–Mo, Co–Ta, Co–Ti, Co–V, and Co–W. More recently, rare earth alloys such as SmCo$_5$ [*12017-68-4*] have become prominent because of their outstanding permanent magnetic properties.

Economically, multicomponent systems are desirable and often tailored to serve a specific purpose. For instance the cobalt-based cemented carbide is Widia, and the multicomponent magnet alloys Remalloy, Cunico, Hiperco, Alnico I [*12605-54-8*], and Perminvar [*12605-04-8*]. The Stellite series of alloys, based on the Co–Cr system, are used as hardfacing materials where extreme wear resistance is required.

Superalloys. Most superalloys designed for high-stress/high-temperature service consist of a nickel-rich matrix with various alloying elements added. However, there is a group of superalloys based upon cobalt instead of nickel. They were developed to improve turbo-superchargers on aircraft engines but later found application in steam turbines and gas turbines as well. Nickel-base superalloys deteriorate rapidly in use with fuel containing appreciable quantities of sulfur. Cobalt-base alloys such as X-40, X-45, and Haynes-Stellite alloy 31 have been widely used in jet aircraft engines and gas turbines because of their resistance to sulfidation. Furthermore, their high micro-structural stability makes them resistant to failure at the elevated service temperatures. Most cobalt superalloys have an fcc matrix. The deliberately added minor phases usually consist of one of several carbides that serve to strengthen the alloy. Topologically closed-packed phases such as σ, π, μ, or the Laves phases and occasionally a cph phase are, for the most part, deleterious, and tend to make the alloy brittle at the high-service temperatures. Therefore, these phases are avoided by careful composition control and by suitable heat treatment of the alloy.

It has been found that elevated temperature aging usually affects carbide precipitation in a way that improves the rupture strength of the alloy. Most heat treatments that are given to wrought cobalt-base superalloys consist of recrystallization or stress-relief treatments. Cast cobalt-based superalloys are usually homogenized or solution treated at 870–980°C, and then aged. Homogenization treatments can be at temperatures as high as 1425°C depending on the alloy composition. Aging takes place over longer periods of time between 730 and 800°C. Proper heat treatments, some at several temperatures, can also improve weldability of superalloys.

Permanent-Magnetic Alloys. Cobalt is a common constituent of many types of magnetic alloys including precipitation-hardened alloys, quench-hardened steels, alloys with ordered structures, cold-worked alloys, and single-domain powder magnets. Magnetic materials (qv) with cobalt additions characteristically retain their magnetism at higher temperatures and have higher coercive force at room temperature. Typical cobalt-containing magnetic alloys include Fe–Co–Mo, Fe–Ni–Al–Co, Fe–Ni–Cu–Co, Co–Pt, and Fe–Ni–Co–Mn, in addition to Sm–Co alloys. An extensively used composition is $SmCo_5$ (86).

Soft Magnetic Alloys. In applications such as generators, motors, and static transformers, materials are needed that will not retain much of their magnetism when the applied magnetic field is removed. Fe–Ni–Co or Fe–Co–V combinations are typical of this kind of magnet. Commercially important alloys in this group are Permendur, Supermendur, Hiperco, and Perminvar (1,14,87).

Special Applications. Cobalt is added to both molybdenum and tungsten to make high-speed cutting tools that remain hard at elevated temperatures. However, cobalt additions often make the tool alloys more susceptible to thermal shock and vibration while in service. Cobalt is also added to W–Cr hot-work tool steels to increase the toughness and strength of the alloy at high temperatures. Cobalt is added to sintered tungsten carbides where it serves as a binder. In this application, cobalt acts as a wetting agent so that the carbide particles are effectively bound together. Under service conditions where impact and thermal cycling occur, Co–Cr–W alloys are used as hardfacing and wear resistant alloys. The corrosion resistance of certain Co–Cr alloys makes them useful to the aircraft industry. One of the earliest commercial alloys, known as Vitallium [12629-02-6], was used to make dental castings (see Dental materials). Its composition is as follows. 56–68% Co, 25–29% Cr, 5–6% Mo, 1.8–3.8% Ni, 0–1% Mn, 0–1% Si, and 0.2–0.3% C, (ASTM A567-1).

BIBLIOGRAPHY

"Cobalt and Cobalt Alloys" in *ECT* 1st ed., Vol. 4, pp. 189–199, by G. A. Roush, Mineral Industry; "Cobalt and Cobalt Alloys" in *ECT* 2nd ed., Vol. 5, pp. 716–736, by F. R. Morral, Cobalt Information Center, Battelle Memorial Institute.

1. R. S. Young *Cobalt, American Chemical Monograph Series,* No. 149, American Chemical Society, Rhinehold, New York, 1960.
2. *Cobalt Monograph,* Cobalt Information Center, Brussels, 1960.
3. A. H. Hurlich, *Met. Progr.* **112**(5), 67 (Oct. 1977).
4. C. S. Hurlbut Jr., *Dana's Manual of Mineralogy,* 17th ed., John Wiley & Sons, Inc., New York, 1966.
5. J. W.Christian, *Proc. R. Soc.* **206A,** 51 (1951).
6. *U.S. Bureau of Mines Minerals Yearbook,* Vol. III, 1971, 1974.
7. C. Chandra, *Characterization of Lateritic Nickel Ores by Electron-Optical and X-Ray Techniques,* Ph.D. Thesis, University of Denver, Denver, Colorado, 1976.
8. R. E. Siemens, *Process for Recovery of Nickel from Domestic Laterites,* paper presented at the 1976 Mining Convention, U.S. Bureau of Mines, 1976.
9. L. F. Power and G. H. Geiger, *Miner. Sci. Eng.* **9**(1), 32 (1977).
10. J. B. Hess and C. S. Barrett, *Trans. Am. Inst. Min. Met. Eng.* **194,** 645 (1952).
11. H. Bibring and F. Sebilleall, *Rev. Met.* **52,** 569 (1955).
12. A. Seeger, *Z. Metallkunde* **47,** 653 (1956).
13. R. Sridhar, W. E. Jones, and J. S. Warner, *J. Met.* **28**(4), 32 (1976).
14. J. C. Agarwal and co-workers in ref. 13, p. 24.
15. M. Wadsworth, *J. Met.* **28**(3), 4 (1976).
16. P. Duby in ref. 15, p. 8.
17. P. Tarassoff in ref. 15, p. 11.
18. M. G. Manzone in ref. 15, p. 16.
19. C. Tombu, *Cobalt* **20,** 103; **21,** 185 (1963).
20. H. R. Millie, ed., *Minerals and Materials,* U.S. Bureau of Mines, March, 1978, p. 28.
21. A. Granville, *Miner. Sci. Eng.* **1**(3), 170 (July 1975).
22. Ger. Pat. 1,012,124 (July 11, 1957), B. Lopmann.
23. U.S. Pat. 3,576,734 (Apr. 27, 1971), H. L. Bennett (to Bennett Engineering Co.).
24. M. A. Aglan and H. Moore, *J. Soc. Glass Tech.* **39,** 351 (1955).
25. J. Berk and J. de Jong, *J. Am. Ceram. Soc.* **41,** 287 (1958).
26. J. C. Richmond and co-workers *J. Am. Ceram. Soc.* **36,** 410 (1953).
27. A. Charlesby, *Nucleonics* **14,** 82 (Sept. 1956).
28. E. B. Darden, E. Maeyens, and R. C. Bushland, *Nucleonics* **12**(10), 60 (1954).
29. A. E. Berkowitz, F. E. Joumot, and F. C. Nix, *Phys. Rev.* **98,** 1185 (1954).
30. C. T. Sims in C. T. Sims and W. C. Hagel, eds., *The Superalloys,* John Wiley & Sons, Inc., New York, 1972, p. 145.
31. N. J. Grant and J. R. Lane, *Trans. ASM* **41,** 95 (1949).
32. M. Hansen, *Constitution of Binary Alloys,* K. AnDerko, ed., McGraw-Hill Book Co., New York, 1958.
33. F. A. Shunk, *Constitution of Binary Alloys—Second Supplement,* McGraw-Hill Book Co., New York, 1969.
34. *Metals Handbook,* Vol. 8.
35. A. R. Elsa, A. B. Westerman and G. K. Manning, *Trans. AIME* **180,** 579 (1949).
36. W. Koster and E. Wagner, *Z. Metallk.* **29,** 230 (1937).
37. U. Haschimoto, *Nippon Kinzoku Gakkai-Shi* **1,** 19 (1937).
38. G. Tammann and W. Oelsen, *Z. Anorg. Chem.* **186,** 260 (1930).
39. R. P. Elliott, *Constitution of Binary Alloys—First Supplement,* McGraw-Hill Book Co., New York, 1965.
40. J. D. Wood and G. P. Conrad II in K. S. Varres, ed., in *Rare Earth Research II,* Proceedings of the Third Conference 1963, Gorden and Breach, New York, 1964, p. 209.
41. B. Love, *WADD Technical Report 61-123,* 1961, p. 179.
42. *Ibid.,* 60–74, 1960, p. 226.
43. W. C. Ellis and E. S. Greiner, *Trans. ASM* **29,** 415 (1941).

44. K. Schubert and co-workers, *Z. Metallk.* **50,** 534 (1959).
45. W. Koster and E. Horn, *Z. Metallk.* **43,** 333 (1952).
46. V. F. Novy, R. C. Vickery, and E. V. Kleber, *Trans. AIME* **221,** 588 (1961).
47. E. M. Savitskii, V. F. Tereknova, and I. V. Burov, *Russ. J. Inorg. Chem.* **7,** 1332 (1962).
48. A. Sieverts, *Z. Physik. Chem.* **60,** 169 (1907).
49. A. Sieverts and H. Hagen, *Z. Physik. Chem.* **A169,** 237 (1934).
50. W. L. Larsen, W. H. Pechin, and D. E. Williams, *U. S. Atomic Energy Commission IS-700,* 1963, M34–M36.
51. *Ibid., IS-900,* 1964, M37–M38.
52. A. N. Khlapova, *Khim. Redkikh Elementov Akad. Nauk SSSR* 1 115 (1954).
53. J. F. Smith and M. J. Smith, *Trans. ASM* **57,** 337 (1964).
54. T. J. Quinn and W. Hume-Rothery, *J. Less-Common Met.* **5,** 314 (1963).
55. U. Raydt and G. Tammann, *Z. Anorg. Chem.* **83,** 246 (1913).
56. W. Broniewski and W. Pietrik, *Compt. Rend.* **201,** 206 (1935).
57. W. Koster and E. Horn, *Z. Metallk.* **43,** 444 (1952).
58. S. Zemczuzny and J. Schepelew in ref. 46, p. 245.
59. G. Grube and H. Kastner, *Z. Electrochem.* **42,** 156 (1936).
60. G. Grube and O. Winkler, *Z. Electrochem.* **41,** 52 (1935).
61. E. Gebhardt and W. Koster, *Z. Metallk.* **32,** 253 (1940).
62. F. W. Schonfeld in A. S. Coffinberry, W. N. Miner, eds., *The Metal Plutonium,* Univ. of Chicago Press, 1961, p. 240.
63. K. Friedrich, *Metallurgie* **5,** 212 (1908).
64. K. Lewkonja, *Z. Anorg. Chem.* **59,** 305 (1908).
65. K. Lossew, *Zhur. Russ. Fiz-Khim. Obshchestva* **43,** 375 (1911).
66. F. Bohm and co-workers, *Acta Chem. Scand.* **9,** 1510 (1955).
67. L. D. Dudkin and V. I. Vaidanich, *Sov. Phys. Solid State* **2,** 1384 (1961).
68. K. Lewkonja, *Z. Anorg. Chem.* **59,** 327 (1908).
69. *Ibid.,* p. 294.
70. S. F. Zemczuzny and S. W. Belynski in ref. 68, p. 364.
71. M. Korchynsky and R. W. Fountain, *Trans. AIME* **215** 1033 (1959).
72. H. S. Wallbaum, *Arch. Eisenhuttenw.* **14,** 521 (1940–1941).
73. W. K. Noyce and A. H. Daane, *Ann. Rev. Nucl. Sci.* **1,** 448 (1952).
74. W. Koster and H. Schmid, *Z. Metallk.* **46,** 195 (1955).
75. Ref. 41, *60-74,* 1960, p. 226.
76. (a) J. Schramm, *Z. Metallk.* **30,** 10, 122, 131, 327 (1938); (b) **33,** 46 (1941).
77. W. L. Fink and H. R. Freche, *Trans. AIME* **99,** 141 (1932).
78. Ref. 76 (b), p. 381.
79. A. J. Bradley and G. C. Seager, *J. Inst. Met.* **64,** 81 (1939).
80. K. Freidrich, *Metallographie* **5,** 150 (1980).
81. E. Raub and P. Walter, *Z. Metallk.* **41,** 234 (1950).
82. W. Wahl, *Z. Anorg. Chem.* **66,** 60 (1910).
83. W. Koster and W. Mulfinger in ref. 76(a), p. 348.
84. A. Lewkonga in ref. 82, p. 60.
85. R. Vogel, *Z. Metallurgie* **38,** 97 (1947).
86. H. E. Chandler and D. F. Baxter, Jr., *Met. Progr.* **113**(1), 41 (1978).
87. W. E. Wallace and L. V. Cherry in E. V. Kleber, ed., *Rare Earth Research,* The Macmillan Co., New York, 1961, p. 211.

F. PLANINSEK
JOHN B. NEWKIRK
University of Denver

COBALT COMPOUNDS

Cobalt is similar to its neighbors, iron and nickel, in the periodic table (1–3). In nearly all its compounds it exhibits a valence of +2 or +3. The stable divalent form is not subject to appreciable hydrolysis in aqueous solutions whereas, trivalent cobalt compounds are powerful oxidizing agents that are mostly unstable. In the complexed state cobalt(II) is relatively unstable and is readily oxidized to cobalt(III) by ordinary oxidants. An extremely large number of complex ions have been identified, most of which are quite stable in aqueous media. Cobalt has a formal valence of +1 only in a few complex nitrosyls and carbonyls. Tetravalent cobalt exists solely in fluoride complexes and in one unusual series of binuclear peroxo compounds. However, many oxidation states have been reported (see Table 1).

The amines of cobalt were discovered by Werner (4) in 1894, and served as the basis for the formulation of the coordination theory in inorganic chemistry (see Coordination compounds).

Cobalt salts are usually made from the hydroxide or carbonate or by dissolving fine cobalt powder in acid. Table 2 gives typical analyses of a group of commercial cobalt salts (see also ref. 5).

Properties and uses of many cobalt compounds are listed in Table 3. Other compounds that may be available in semicommercial quantities, or that have been made for specific purposes by the manufacturers, are listed in Table 4.

Complex Cobalt Compounds

The innumerable coordination compounds of cobalt(III) exhibit substantial diversity in their coordination number, geometric structure, and stability, and in many aspects of their chemistry (1,7). The most common coordination number is 6.

Historically, the ammines of cobalt(III) have dominated the chemistry of cobalt complexes and their influence on the development of chemistry has been substantial. The important donor atoms (in order of decreasing tendency to complex) are nitrogen, carbon in cyanides, oxygen, sulfur, and the halogens. Divalent cobalt exhibits a coordination number of either four or six, whereas that of the trivalent cobalt ion is invariably six.

Table 1.   Cobalt Ion Valencies and CAS Registry Nos.

Ion valency	CAS Registry No.	Ion valency	CAS Registry No.	Ion valency	CAS Registry No.
1−	[16727-18-7]	9+	[22374-31-8]	18+	[12663-75-1]
1+	[16610-75-6]	10+	[25879-24-7]	19+	[12663-76-2]
2+	[22541-53-3]	11+	[18973-68-7]	20+	[12663-77-3]
3+	[22541-63-5]	12+	[20573-04-0]	21+	[54603-39-3]
4+	[20499-79-0]	13+	[26445-27-2]	22+	[54603-31-5]
5+	[20499-80-3]	14+	[20573-05-1]	23+	[54603-22-4]
6+	[20499-81-4]	15+	[20573-06-2]	24+	[52488-14-9]
7+	[20508-39-8]	16+	[12595-93-6]	25+	[12663-78-4]
8+	[14841-23-7]	17+	[12663-74-0]	26+	[58246-85-6]

Table 2. Typical Analyses of Commercial Cobalt Compounds

Assay, %	Acetate	Carbonate	Hydrate	Oxide Ceramic grade	Oxide Chemical grade	Hydrated sulfate
Analysis						
cobalt	23.50	46.00	61.00	71.70	56.00	21.00
nickel, max	0.10	0.15	0.20	0.40	0.20	0.10
iron, max	0.04	0.10	0.10	0.13	0.20	0.04
manganese, max	0.02	0.05	0.05	0.01	0.05	0.02
silica, max				0.10	0.30	
copper, max	0.02	0.05	0.05	trace	0.04	0.02
sulfur, max					0.50	
calcium oxide, max				0.22	0.10	
moisture at 105°C				trace	31.00	
alkali and alkaline earth sulfates,	0.40	0.80	0.80			0.30
hydrochloric acid-insoluble matter,		0.05	0.05			
water-insoluble matter, max	0.05					0.10
apparent sp gr	0.960	0.835	0.350			1.114

Ammines. By adding excess ammonia to a cobalt salt and exposing it to air, oxidation occurs and brown solutions form that become pink on boiling. The solutions contain complex cobaltammines, eg, $[Co(NH_3)_6]Cl_3$ [*10534-89-1*], $[Co(NH_3)_5Cl]Cl_2$ [*13859-51-3*], and $[Co(NH_3)_5H_2O]Cl_3$ [*13820-80-9*]. These solutions show none of the reactions of cobalt.

The chelates with bis(salicylaldehyde)ethylenediamine derivatives and the types of diamines have unusual oxygen-carrying properties; for example, some of these compounds absorb and release oxygen so readily that they have been used in the purification of oxygen and have been proposed for its production (see Chelating agents).

Cyanides. If KCN is added to a solution of cobalt salt, reddish-brown cobaltous cyanide, $Co(CN)_2 \cdot 3H_2O$ [*26292-31-9*] precipitates. However, with excess cyanide this compound redissolves forming a red solution of potassium cobalt(II) cyanide (potassium hexacyanocobaltate, $K_4(Co(CN)_6)$ [*14564-70-6*]. If a little HCl or acetic acid is added to this solution and the solution is boiled in the presence of oxygen, oxidation occurs and potassium cobalt(III) cyanide [*13963-58-1*], $K_3(Co(CN)_6)$ forms. Single crystals of this compound have been made in fairly large amounts for use in lasers (qv).

Other Complexes. The hexaaquo complex, $(Co(H_2O)_6)^{2+}$, is pink, but introduction of a halide into the coordination sphere to give the complex anion, $(CoX_4)^{2-}$, produces a blue color. The substitution occurs stepwise, and many mixed aquo–halo complexes are known. The blue color is much more intense than the pink, and a relatively small concentration of the blue complex can completely mask the color of a much larger concentration of pink complex. Thus, cobalt(II) halides in dilute aqueous solutions are normally pink, but on heating or in concentrated solutions they may turn blue. This behavior is the basis for "sympathetic (invisible) inks" and desiccant indicators.

Table 3. Properties and Uses of Cobalt Compounds

Cobalt compound	CAS Registry No.	Formula and synonyms	Crystalline form, color, refractive index	Density	mp, °C	Solubility in grams per 100 mL Water Cold	Hot	Other solvents	Uses
(III) acetate	[917-69-1]	$Co(C_2H_3O_2)_3$	octahedral, green, hyg		dec 100	hyd readily		sol glacial acetic acid	catalyst for cumene hydroperoxide decomposition
(II) acetate tetrahydrate[a]	[71-48-7]	$Co(C_2H_3O_2)_2 \cdot 4H_2O$	red–violet, monoclinic	1.705	−4 H_2O, 140	sol	sol	sol alcohol	driers for lacquers and varnishes, sympathetic inks, catalysts, mineral supplement, anodizing stabilizer malt beverages
acetylacetonate[a]	[14024-48-7]	$Co(C_5H_7O_2)_3$	black, monoclinic	1.43	241				vapor plating of cobalt nickel and cobalt base alloys
aluminate[a]	[13820-62-7]	(approx) Thenard's blue $CoAl_2O_4$	blue, cubic			insol	insol		
(II) ammonium sulfate hexahydrate	[13586-38-4]	$CoSO_4(NH_4)_2SO_4 \cdot 6H_2O$	ruby red, monoclinic, 1.490, 1.495, 1.503	1.90		20.5 (20°C)	45.4 (80°C)	insol alcohol	catalyst, plating
(II) orthoarsenate octahydrate	[24719-19-5]	$Co_3(AsO_4)_2 \cdot 8H_2O$	violet–red, monoclinic, 1.626, 1.661, 1.669	3.178	dec	insol	insol	sol dilute NH_4OH	light blue color for painting on glass and porcelain, coloring glass catalyst
arsenic sulfide	[12254-82-9]	nat cobaltite CoAsS	gray–reddish	6.2–6.3	dec				
(II) benzoate tetrahydrate	[17875-31-4]	$Co(C_7H_5O_2)_2 \cdot 4H_2O$	gray–red leaf		−4 H_2O, 115	v sol			catalyst
boride, mono-	[12006-77-8]	CoB	prisms	7.25		dec	dec	sol HNO_3 aqua regia	ceramals
(II) bromate hexahydrate	[13476-01-2]	$Co(BrO_3)_2 \cdot 6H_2O$	red, octahedral			45.5		sol NH_4OH	

497

Table 3 (*continued*)

Cobalt compound	CAS Registry No.	Formula and synonyms	Crystalline form, color, refractive index	Density	mp, °C	Solubility in grams per 100 mL Water Cold	Hot	Other solvents	Uses
(II) bromide	[7789-47-7]	$CoBr_2$	green, hexahydrate, deliquescent	4.909	678 (in N_2)	66.7 (50°C)	68.1 (97°C)	77.1 alcohol 58.6 CH_3OH; sol ether acetone	hydrometers, catalyst
(III) bromide	[15605-72-8]	$CoBr_3$							
carbide	[12011-59-5]	Co_3C							
carbonate	[7542-09-8]	$Co_2(CO_3)_3$	green						hydrometers, catalyst stable in glycerol or dry form
(II) carbonate[a]	[513-79-1]	nat spherocobaltite $CoCO_3$	red, trigonal, 1.855, 1.60	4.13	dec	insol	insol	insol NH_3, sol acid	pigments, ceramics, trace mineral supplement feed, temperature indicator, catalyst preparation of cobalt compounds
(III) carbonate, basic[a]	[7542-09-8]	$2CoCO_3 \cdot Co(OH)_2 \cdot H_2O$	violet–red prisms			insol	dec	sol $(NH_4)_2$-CO_3	preparation of cobalt compounds
carbonyl, tetra	[15226-74-1]	dicobalt octacarbonyl $[Co(CO)_4]_2$ or $Co_2(CO)_8$	orange crystalline dark brown micro-crystalline	1.73	51	insol	insol	sl sol alcohol, CS_2 ether	catalyst
carbonyl, tri	[19212-11-4]	tetracobalt dodecacarbonyl, $[Co(CO)_3]_4$ or $Co_4(CO)_{12}$	black crystalline				sl sol	sl sol acid benzene, dec Br	catalyst
(II) chloride, anhydrous	[7646-79-9]	$CoCl_2$							

498

Name	CAS Registry Number	Formula	Color, crystalline form	Density	Mp, °C	Solubility in water, cold	Solubility in water, hot	Solubility in other solvents	Uses
(II) chloride, hexahydrate[a]	[7791-13-1]	$CoCl_2 \cdot 6H_2O$	red, monoclinic	1.924	86	76.7	190.7	v sol (blue color) alcohol; sol acetone; 0.29 ether	sympathetic inks, barometers, absorb poison gas and NH_3, electroplating, flux for magnesium refining, solid lubricant, dye mordant, catalyst, foam stabilizer in beer
(II) chromate	[24613-38-5]	$CoCrO_4$	gray–black crystalline		dec	insol	dec	sol acid, NH_4OH	green tints in ceramics
(II) citrate dihydrate	[18727-04-3]	$Co_3(C_6H_5O_7)_2 \cdot 2H_2O$	rose red		−2 H_2O, 150	0.8	sol		vitamin preparations, therapeutics agents
(II) cyanide dihydrate	[542-84-7]	$Co(CN)_2 \cdot 2H_2O$	buff anhydrous blue–violet pwd	anhydrous 1.872	−2 H_2O, 280	0.00418	sol	sol KCN, HCl, NH_4OH	catalyst
potassium hexacyanocobalt(III)	[13963-58-1]	$K_3Co(CN)_6$	yellow, monoclinic	1.906	dec	sol	sol	insol alcohol	suggested microwave studies (pure and electronic grades available)
(II) ferricyanide	[15415-49-3]	$Co_3[Fe(CN)_6]_2$	red needles			insol	insol	sol NH_4OH; insol HCl	
(II) hexacyanoferrate(II)	[4049-81-1]	$Co_2Fe(CN)_6 \cdot xH_2O$	gray–green			insol	insol	sol KCN; insol HCl	
disodium ethylenediaminetetraacetate	[68867-22-1]	$CoNa_2(C_{10}H_{12}N_2O_8) \cdot H_2O$	amorphous pink pwd			sol	sol	insol alcohol	support of other chelating compounds in medicinal and tree-spray preparation
(II) fluoride	[10026-17-2]	CoF_2	pink, monoclinic	4.46	ca 1200	1.5	sol	sl sol acid; insol alcohol, ether, benzene	
(III) fluoride	[10026-18-3]	CoF_3	brown, hexagonal	3.88	dec to Co(OH)$_3$			insol alcohol, ether, benzene	fluorinating agent

Table 3 (continued)

Cobalt compound	CAS Registry No.	Formula and synonyms	Crystalline form, color, refractive index	Density	mp, °C	Solubility in grams per 100 mL Water		Other solvents	Uses
						Cold	Hot		
(III) fluoride, dihydrate	[54496-71-8]	$CoF_2 \cdot 2H_2O$	α: red, rhombic octahedral β: rose or pwd	2.192	dec 200	sol	sol	insol alcohol	catalyst in organic reactions
fluosilicide(II)-hexahydrate	[15415-49-3]	$CoSiF_6 6H_2O$	pink, trigonal, 1.382, 1.387	2.113		118.1			ceramics, possible source of fluorine
(II) formate dihydrate	[6424-20-0]	$Co(CHO_2)_2 \cdot 2H_2O$	red crystalline	2.129	−2 H_2O, 140	5.03			catalyst
(III) hydroxide[a]	[1307-86-4]	$Co(OH)_2$	rose red rhombic	3.597	dec	0.00032		sol acid, NH_4 salts; insol alkali	paints, preparation of cobalt compounds, catalysts, storage batteries, lithographic printing inks
(III) hydroxide trihydrate	[21041-93-0]	$Co_2O_3 \cdot 3H_2O$	black–brown pwd	4.46	dec	0.00032		sol acid; insol alcohol	
(II) iodate	[13455-28-2]	$Co(IO_3)_2$	black–violet needles	5.008	dec 200	0.45	1.33	sol HCl, HNO_3; hot H_2SO_4	
(II) iodide (α) stable	[15238-00-3]	CoI_2	black hexagonal, hygrosopic	5.68	515 (vacuum)	159	420	v sol alcohol, acetone	moisture indicator
linoleate	[14666-96-7]	$Co(C_{18}H_{31}O_2)_2$	brown, amorphous			insol		sol alcohol, ether, acetone	driers for paints, varnishes (particularly enamels and white paints)
lithium cobaltite	[12190-79-3]	$LiCoO_2$	dark blue pwd						ceramics
(II) maleic hydrazide	[63307-78-8]	$\left[\underset{\|}{\overset{O}{C}} - CH = CH - CHC - \underset{\|}{\overset{O}{\ }} NHN \right]_2$ brown–pink pwd				insol	insol		stable in air

Name	CAS Registry No.	Formula	Appearance	Sp gr	mp, °C	bp, °C	Solubility in water	Solubility in other solvents	Uses
naphthenate			purple liquid, (6% Co)	0.966					driers for paints and varnishes
(II) nitrate[a] hexahydrate	[10026-22-9]	$Co(NO_3)_2 \cdot 6H_2O$	red, monoclinic, 1.52	1.87	55–56	133.8	0.217	100.0 alcohol; sol acetone; sl sol NH_3	pigments, sympathetic inks, decoration for stoneware and porcelain, hair dyes, feed supplement, Vitamin B_{12} preparation, catalysts
(III) nitrate	[15520-84-0]	$Co(NO_3)_3$	green, hygroscopic crystal					reacts vigorously with organic solvents	
octoate (ethylhexanoate)	[136-52-7]	$Co(C_8H_{15}O_2)_2$	blue liquid (12% Co)	1.013					driers, whiteners
nitrosyldicarbonyl	[12021-68-0]	$Co(NO)(CO)_2$	cherry red liquid	1.05			insol	sol alcohol, ether, acetone, benzene	
(II) oleate[a]	[14666-94-5]	$Co(C_{18}H_{33}O_2)_2$	brown, amorphous					sol alcohol, ether, oils, benzene	drier for paints and varnishes
(II) oxalate	[814-89-1]	CoC_2O_4	white or reddish	3.021	dec 250		insol	sol acid, NH_4OH	temperature indicator, hydrous form for preparation of catalysts
(II) oxide[a]	[1307-96-6]	CoO	green–brown cubic	6.45	1935		insol	sol acid; insol alcohol, NH_4OH	glass decorating, coloring and whitener, drier for paints and varnishes, semiconductor, powder for sintered Co_6W_6C
(II, II, III) oxide[a]	[1308-06-1]	Co_3O_4	black cubic	6.07	transition to CoO, 900–950		insol	v sl sol acid; insol aqua regia	enamels, semiconductors, grinding wheels
(II) perchlorate	[13455-31-7]	$Co(ClO_4)_2$	red needles	3.327	100	115		sol alcohol, acetone	chemical reagent

Table 3 (continued)

Cobalt compound	CAS Registry No.	Formula and synonyms	Crystalline form, color, refractive index	Density	mp, °C	Solubility in water per 100 mL		Other solvents	Uses
						Cold	Hot		
(II) orthophosphate octahydrate[a]	[10294-50-5]	$Co_3(PO_4)_2 \cdot 8H_2O$	reddish pwd	2.769	−8 H_2O, 200	sl sol		sol mineral acid, H_3PO_4; insol alcohol	glazes, enamels, pigments, plastic resins
phosphide	[12643-12-8]	Co_2P	gray needles	6.4	1386	insol	insol	sol HNO_3, aqua regia	
potassium nitrite hydrate	[17120-39-7]	$K_3Co(NO_2)_6 \cdot 1.5H_2O$	yellow, tetragonal		dec 200	0.089	sl sol	v insol alcohol, methane	oil and water color pigment, paint for glass and porcelain, rubber colorant. Fishers yellow; in Co analysis
(II) propionate	[1560-69-6]	$Co(CH_3CH_2COO)_2$	pink pwd						
resinate		$Co(C_{44}H_{52}O_4)_2$	brown–red, amorphous pwd			insol	insol		drier for paints, enamels, varnishes; lustrous coating for chinaware, pottery, textiles, catalysts
(II) selenate, pentahydrate	[14590-19-3]	$CoSeO_4 \cdot 5H_2O$	ruby red, triclinic	2.512	dec	v sol		sol HNO_3, aqua regia; insol alkali	
selenide, mono	[1307-99-9]	$CoSe$	yellow, hexagonal	7.65	red heat				
selenite dihydrate	[19034-13-0]	$CoSe_2O_3 \cdot 2H_2O$	blue–red amorphous rhombic			insol	insol	sol H_2SeO_3	

502

(II) orthosilicate silicide, di-	[12017-08-2] [12017-12-8]	Co$_2$SiO$_4$ CoSi$_2$	violet crystalline rhombic	4.63 5.3	1345 1277	insol	insol	sol dil HCl insol H$_2$SO$_4$; sol hot HCl	driers for paints and varnishes
(II) orthostannate	[12139-93-9]	Co$_2$SnO$_4$	cubic greenish–blue	6.30					driers
(II) stearate	[13586-84-0]	Co(C$_{17}$H$_{35}$CO$_2$)$_2$					sol alkali		vitamin preparations, therapeutic agents
(II) succinate tetrahydrate	[23788-77-6]	Co(C$_4$H$_4$O$_4$)·4H$_2$O	violet, crystalline			sl sol		sol alkali	
(II) sulfamate (amino sulfonate) trihydrate	[16107-41-3]	Co(NH$_2$SO$_3$)$_2$·3H$_2$O				sol	sol	insol alcohol	electroplating
(II) sulfate	[10124-43-3]	CoSO$_4$	dark bluish, cubic	3.71	dec 735	36.2	83	1.04 CH$_3$OH insol NH$_3$ alcohol	ceramics
(II) sulfate, heptahydrate[a]	[10026-24-1]	nat bieberite CoSO$_4$·7H$_2$O	red–pink, monoclinic 1.477, 1.483, 1.489	1.948	96.8	60.4	67	2.5 alcohol; 54.5 CH$_3$OH	pigments for porcelain; glazes, plating, feed supplement, catalysts storage batteries, drier for inks
(II) sulfate, monohydrate[a]	[10124-43-3]	CoSO$_4$·H$_2$O	red crystalline, 1.603, 1.639, 1.683	3.075	dec	sol	sol		same as for CoSO$_4$·7H$_2$O
(III) sulfate hydrate	[13478-09-6]	Co$_2$(SO$_4$)$_3$·18H$_2$O	blue crystal						stable in diluted sulfuric acid
sulfide, mono-	[1317-42-6]	nat sycoporite, CoS	reddish, silver–white, octahedral	5.45	>1116	0.00038		sl sol acid	catalyst for hydrogenation or hydrodesulfurization
(III) sulfide, sesqui sulfide	[1332-71-4]	Co$_2$S$_3$	black, crystalline	4.8				dec acid, aqua regia	
tallate		(varying composition)	purple-violet liquid (6% Co)	0.975					driers for paint and varnishes

Table 3 (continued)

Cobalt compound	CAS Registry No.	Formula and synonyms	Crystalline form, color, refractive index	Density	mp, °C	Solubility in grams per 100 mL Water		Other solvents	Uses
						Cold	Hot		
thiocyanate trihydrate	[3017-60-5]	$Co(SCN)_2 \cdot 3H_2O$	rhomb		-3 H_2O, 105	sol		sol alcohol, CH_3OH ether	humidity indicator
orthotitanate	[12017-38-8]	Co_2TiO_4	greenish–black, cubic	5.07– 5.12				sol conc HCl; sl sol dil HCl	
(II) tungstate	[12640-47-0]	$CoWO_4$	blue–green, monoclinic	8.42		insol		sol hot conc acid; sl sol dilute acid	light-sensitive varnishes, drier for enamels, inks, paints, and varnishes, antiknock agents
Cobalt complexes									
hexamminecobalt(III) chloride)	[10534-89-1]	$Co(NH_3)_6Cl_3$	wine–red, monoclinic	1.710	$-1NH_3$, 215	5.9	12.74	sol conc HCl; insol alcohol NH_4OH	
triethylenediaminecobalt(III) chloride trihydrate	[10241-04-1]	$Co[C_2H_4(NH_2)_2]_3$ $Cl_3 \cdot 3H_2O$	brown, prisms	1.542	256; $-3H_2O$, 100	v sol			
ammonium tetranitrodiammine(III) cobaltate	[13600-89-0]	Erdmann's salt $NH_4[Co(NH_3)_2(NO_2)_4]$	reddish, pale brown, rhomb, 1.78, 1.78, 1.74	1.876					
aquapentammine cobalt(III) chloride (roseo)	[18194-88-2]	$[Co(NH_3)_5 \cdot H_2O]Cl_3$	brick red crystalline	1.7	dec 100	24.87		sl sol HCl, insol alcohol	

ᵃ Commercially available (6).

Table 4. Other Cobalt Compounds

Compound	CAS Registry No.	Formula	Uses and remarks
acetylene dicobalt nonacarbonyl	[18177-59-8]	$C_4H_2O_2Co_2(CO)_7$	
ammonium cobalt(II) phosphate	[14590-13-7]	$NH_4CoPO_4 \cdot H_2O$	red to violet powder or monoclinic lamella, ceramic pigment, plant fertilizer in Co-analysis
barium cobalt(III) cyanide heptahydrate	[60970-90-3]	$Ba_3(Co(CN)_6)_2 \cdot 7H_2O$	
cobalamine(cyanocobalamin, Vit. B-12)	[68-19-9]	$C_{63}H_{88}CoN_{14}O_{14}P$	hygroscopic dark red crystals; darkens at 210–220°C; hematopoietic vitamin; nutritional (growth and antianemic factor)
cobalt(II) cobalt(III) cyanide tetrahydrate	[12548-03-7]	$Co_2(CN)_5 \cdot 4H_2O$	

Economic Aspects

The 1977 prices of selected cobalt compounds are given in Table 5, and U.S. consumption in various applications in Table 6. Rising energy costs, as well as unstable conditions in Zaire, the world's main supplier of cobalt, have resulted in a sharp price increase. However, U.S. consumption of cobalt compounds has more than doubled since 1964.

Table 5.   Selected Cobalt Compounds, 1979 Prices[a]

Cobalt compound	$/kg	Cobalt compound	$/kg
acetate	5.52 (2.29)[b]	oxide	
carbonate, pwd	28.82 (3.19)[b]	metal grade, 75–76% Co	38.26
chloride	5.88	72–73% Co	37.05
hydrate	12.82 (4.33)[b]	70–71% Co	36.08
naphthenate, liq, 6% Co	4.58	sulfate	
nitrate	12.40	cryst	4.73
phosphate, pwd, 31% Co	2.97	monohydrate	7.66
resinate, fused	0.85	tallate, 6% Co	2.45

[a] Ref. 6.
[b] 1963 prices in parentheses.

Table 6.   U.S. Consumption of Cobalt Compounds, Metric Tons

Form	1964	1974	1976
pigments	104.5	96	103.5
catalysts		689	724
ground coat frits	299.5	66.5	45
glass decolorizers		25.5	15
salts and driers, lacquers, varnishes, paints, inks, pigment, enamels, glazes, feed, electroplating, etc	639.5	1817.5	1992.5
other	274	75.5	3.5
Totals	*1317.5*	*2770.0*	*2883.5*

Health and Safety

Cobalt salts in sufficiently large doses can irritate the gastrointestinal tract and cause nausea, vomiting, and diarrhea. In man single oral doses as low as 500 mg of the chloride have provided nausea and vomiting; however, dosages as high as 1200 mg/d over a period of 6 wk have been given to individuals, where tolerated, without evidence of toxic symptoms.

Cobalt salts are used in the treatment of anemia, in some cases together with iron or manganese salts. The oral dosage range has been estimated at approximately 0.25–1.0 mg elemental cobalt per kg body weight. In infants and children, much larger doses (up to 12.5 mg/kg) have not produced recognized toxic effects.

Externally, cobalt may produce dermatitis (8). It is also suspected as a carcinogen of the connective tissue and lungs.

Cobalt Deficiency. Cobalt is one of the important trace elements necessary for nutrition of cattle and sheep (9). Although cobalt deficiency is known by many names, eg, bush sickness in New Zealand, salt sickness in Florida, and pine sickness in Scotland, the symptoms are strikingly similar. Lack of appetite for feed appears to be the initial symptom followed by a long, rough, hair coat, scaliness of skin, muscular incoordination, gauntness, loss of flesh, pale mucous membranes, decreased milk flow, retarded growth, and in some cases, constipation or diarrhea. It is mostly eliminated through the digestive tract and only very little is distributed throughout the tissues. However, on passing through the stomach and colon, cobalt, possibly by modifying enzymic reactions through catalysis, substantially affects the appetite of the animal and the amount of hemoglobin present in the blood (see Mineral nutrients; Pet and other livestock feeds).

Cobalt affects the digestive and colonic system, but only in ruminants, and nonruminants apparently require no cobalt in their diet. A dose of 1–5 mg of cobalt per day is adequate for sheep and cattle, and doses as high as 50 mg/d appear to exhibit no toxicity.

Uses

Cobalt oxides are a source of metallic cobalt powder which is used to prepare alloys and cemented carbides by powder-metallurgy techniques.

Catalysts. Cobalt, like the other transition elements, is an effective catalyst for many organic and inorganic reactions (2). In general, its catalytic behavior is similar to that of iron and nickel; for some reactions, however, it is superior. Its advantages may be in the nature of increased yields, better selectivity, slower poisoning, or more desirable physical and chemical properties of the products. The catalytic reactions include hydrodesulfurization of petroleum; reforming gasoline; hydroformylation; Fischer-Tropsch synthesis; hydrogenation and dehydrogenation; fluorination of hydrocarbons; polymerization (such as butadiene, etc); oxidation (xylenes to toluic acid, oxidation of hydrogen cyanide in gas masks or of carbon monoxide in automobile exhaust); dehydration; CS_2 production; H_2S production; nitrile synthesis; amination of olefins; and reductions with borohydrides.

Cobalt molybdates or mixed oxides (Co–Mo, Ni–Co–Mo) are used in petroleum hydrogenation, desulfurization, denitrification, and hydrocracking. Cobalt carbonyls are used as promoters for organic peroxides in reinforced plastics.

Objectionable odors are eliminated by the use of cobalt oxide catalysts. An autoemission catalyst has been patented (10) containing Co (see Exhaust control, automotive).

Ceramic Colors and Pigments. According to their use ceramic pigments are designated as body stain, underglaze stain, glaze stain, overglaze color, and ceramic color, and there are cobalt compounds that generally fit into all of these categories. Their instability when in contact with molten silicates such as enamels or glazes may, however, restrict their utility in some cases. The typical colors that may be obtained are as follows: *Violet* is produced by $Co_3(PO_4)_2$ [18475-47-3]. However, when heated to about 800°C or above in the presence of silica, it turns into cobalt silicate blue. *Blue.* Ultramarine is produced by cobalt silicate [20731-36-6]. The cobalt aluminate, $CoO \cdot Al_2O_3$ [1333-88-6], which is more neutral, is also known as Thenard's blue [1345-16-0]. Cobalt stannate is Cerulean blue [6546-12-5]. *Green.* Cobalt chromite

($CoO \cdot Cr_2O_3$) [*12016-69-2*] is a very strong green. A series of blue–greens, including turquoise green, can be obtained with mixtures of $CoO \cdot Al_2O_3$ and $CoO \cdot Cr_2O_3$. Another green, Rinmann's green, is obtained by heating mixtures of cobalt oxide and ZnO. *Pink* is produced by cobalt oxide and MgO, but it must be stabilized. *Black.* For an intense black, a substantial proportion of cobalt is mandatory. For instance, a good underglaze black might contain 35% iron oxide, 32% cobalt oxide, 13% nickel oxide, 12% manganese oxide, and 7% chromium oxide (see Colorants for ceramics; Pigments).

The paint and varnish industry makes use of these common pigments: Aureolin or cobalt yellow, potassium cobalt(III) nitrite [*13782-01-9*], $K_3Co(NO_2)_6$, which turns blue if baked. Cobalt-bearing ultramarine blue is one of the pigments and extenders that is most frequently used in the manufacture of printing inks. Cobalt oxide is used in black silicon enamels where service temperatures exceed 315°C.

Cobalt is a constitutent in several patented metal-containing dyes.

Cobalt Soaps.   These salts have enjoyed a very wide use as catalysts to accelerate the drying or oxidation of linseed, soybean, and similar unsaturated oils. (The word drying has come to be used to describe the process by which the liquid unsaturated oils change to elastic films and the word drier to describe the catalyst used to hasten this process.) These soaps are marketed either as liquids or as solids (11) (see Driers and metallic soaps).

Enamels.   Cobalt oxides and salts are employed in the vitreous enameling industry to provide color and promote adherence of enamel to steel. A cobalt-containing enamel has also been used as a solid electrolyte for a heat-activated reserve cell (see Batteries; Enamels).

Porcelain enamels are fusible alkaline–borosilicate glasses, and the colors obtained are therefore, identical to those expected in glasses of this type.

Many theories have been advanced on the tendency of cobalt to promote the adherence of enamel to steel, but a generally accepted explanation is still lacking. From 0.2 to 3% cobalt oxide may be incorporated with the ground coat or *frit,* the usual concentration being 0.5–0.6%. The coating is usually 0.0025–0.0075 mm thick and is fired for 10 min at 825–870°C. The compound mixture known as frit is the typical blue enamel on bath tubs and kitchen ware, later coated with a white cover coat. The adherence increases as the cobalt oxide content of the enamel is raised to 1% (12).

Most of the cobalt consumed in the ceramic industry is not for producing blue, but rather white. In white domestic ware, small quantities of Fe_2O_3 and TiO_2 give a yellow tint if it is not neutralized by the blue imparted by the presence of small amounts of the cobalt oxide. This has become a large outlet for cobalt in the ceramic field (see Ceramics).

Hygrometric Indicators.   Cobalt salts, especially the chlorides, have proved useful as visual indicators of humidity; the blue anhydrous form becomes pink when the humidity is sufficiently high. Cobalt thiocyanate and other organic salts have also been recommended for the purpose.

Trace Element in Animal Feed.   Cobalt may be administered as the sulfate, carbonate, chloride, acetate, and nitrate, or as any compound that is soluble or can be rendered soluble in the animal's stomach (see under Health and Safety). The cobalt compound may be included in salt blocks or mixed in with prepared feeds, or if water-soluble salts are used, may be put in fertilizer so that the animal secures it in the forage. Although the plant does not seem affected, the cobalt is readily assimilated into the plant and is then available for the animal (9).

The rumen pellet or cobalt bullet developed in Australia offers an effective and novel method of supplying cobalt. The pellet, which looks like an oversized vitamin pill, consists mainly of bonded CoO (95%). It is inserted into the throat of the animal by means of a suitable tool. The cobalt in a 5-g pellet for sheep is considered sufficient to release cobalt to supply the animal's requirements for six months to one year.

Miscellaneous Uses.   Cobalt in very small amounts is often used in the manufacture of electronic devices.

Cobalt chelates have been investigated for tonnage production of oxygen (13) and for special-purpose oxygen. Some cobalt chelates have been tested for a dry photographic process (14).

Cobalt acetate is used for sealing in aluminum anodizing solutions. The oxide gives a bronze seal in dichromate solutions and promotes accelerated etching in aluminum chemical milling. Cobalt salts may be used in mordants for textiles, as intensifiers in light-sensitive solutions to form images on porcelain, ceramic, and metal printing plates, and for heat-resistant decalcomania transfers (see Printing processes). They are used in glass dosimeters to help measure gamma rays and electron fields. Cobalt inorganic salts and complex salts, such as cobalt pyridine thiocyanate and other related compounds, are used as chemical temperature-indicating salts (15–16).

Cobalt salts may be used as antifoggants for photographic emulsions. They are also used as a hardening agent for silicone resins in the hardening bath in nylon manufacture, and to improve the properties of lubricating oils. Cobalt(II)–cobalt(III) cyanide [12548-03-7] is introduced in electric flashbulbs for indicating water vapor penetration into the bulb. Invisible ink, cosmetics (eg, hair dyes, etc), and high antiknock gasolines are other possible uses of cobalt salts.

Calcium fluoride held by a binder (a mix containing CoO, B_2O_3, and BaO), stirred in a water slurry and sprayed on the surface of some superalloys including René 41, was found a satisfactory lubricant up to 1000°C (8).

Cobalt amine azides have been found to ignite at a low drop height (see Explosives).

Cobalt may be useful in recovering at least 95% of the cesium in concentrated waste from nuclear reactors (see Nuclear reactors).

Cobalt salts are being used as accelerators and activators for catalysts, fuel cells, and batteries. A Co–Mo alloy electroplated on dies improves their performance. A sandwich plate consisting of layers of copper, cobalt, nickel, and with a thin layer of microcracked chromium has been suggested for automobile bumper sections made of aluminum.

Cobalt oxide has been suggested for air-deodorizing filters, for exhaust treatment (17), and for the production of sulfuric acid from combustion of waste gases (17), as well as for the production of synthetic gems for watches (18), electric contacts, thermisters, varistors (19), and piezo electric ceramics (20) (see Gems, synthetic).

BIBLIOGRAPHY

"Cobalt Compounds" in *ECT* 1st ed., Vol. 4, pp. 199–214, by S. B. Elliott, Ferro Chemical Corporation and Carl Mueller, General Aniline & Film Corporation; "Cobalt Compounds" in *ECT* 2nd ed., Vol. 5, pp. 737–748, by F. R. Morral, Cobalt Information Center, Battelle Memorial Institute.

1. L. Gmelin, *Handbuch der Anorganischen Chemie,* 8th ed., No. 58, Verlag Chemie, GmbH, Weinheim, Germany, Pt. A, Suppl. 1, 1961, pp. 471–886; Pt. B, Suppl. 1, 1963, p. 314; Pt. B, Suppl. 2, 1964, p. 821.

2. R. S. Young, ed., *Cobalt: Its Chemistry, Metallurgy and Uses,* Reinhold Publishing Co., New York, 1960; *Cobalt Monograph,* Centre d'Information du Cobalt, Brussels, Belgium, 1960, pp. 140–160, 433–506.

3. D. Nicholis, *The Chemistry of Iron, Cobalt and Nickel, Pergamon Text in Inorganic Chemistry,* Vol. 24, Pergamon Press, N.Y., 1975.

4. *Alfred Werner and Cobalt Complexes, Werner Centennial ACS Monograph Series, Advances in Chemistry Series,* Vol. 62, 1966.

5. *Kolthoff's Encyclopedia of Industrial Chemical Analysis,* Vol. 10, 1970, pp. 327–347; R. S. Young, *The Analytical Chemistry of Cobalt,* Pergamon Press, 1966; I. V. Pyatnitskii, *Analytical Chemistry of Cobalt,* Israel Program for Scientific Translations, Jerusalem, 1966.

6. *Chem. Mark. Rep.,* (Jan 15, 1979).

7. A. E. Martell and M. Calvin, *Chemistry of the Metal Chelate Compounds,* Prentice Hall, Inc., Englewood Cliffs, N.J., 1952.

8. N. Irving Sax, *Dangerous Properties of Industrial Materials,* 4th ed., Van Nostrand Reinhold Co., 1975.

9. E. J. Underwood, *Trace Elements in Humans and Animal Nutrition,* Academic Press, Inc., New York, 1971.

10. U.S. Pa. 3,897,367 (July 29, 1975), A. Lander (to E. I. du Pont de Nemours & Co., Inc.).

11. R. P. Ware, *Dryers, A Prior Art Survey,* Circular 755, National Paint, Varnish, & Lacquer Assoc., Inc., Scientific Section, Washington, D.C., May 1952; H. Hurlston Morgan, *The Use of Cobalt and Other Metals as Driers in the Paint and Allied Industries,* The Mond Nickel Co., London, England.

12. A. Petzold and H. Betzer, *Glas-Email-Keramo Tech.* **9,** 287 (1958); K. Bates, *Enameling, Principles & Practice,* Funk & Wagnall, 1974.

13. R. F. Stewart and co-workers, *U.S. Bureau of Mines Information Circular* 7906, 1959.

14. Belg. Pat. 614,064 (Aug. 20, 1962) (to Gevaert Photoproducts, N.V.).

15. S. P. Gvozdov and A. A. Erunova, *Izv. Vyssh. Uchebn. Zaved. Khim. Khim Tekhnol.* **5,** 154 (1958).

16. K. T. Wilke and W. Opfermann, *Monatsber. Deut. Akad. Wiss. Berlin,* 5(8–9), 587 (1963).

17. Jpn. Kokai 76-08,164 (Jan. 22, 1976), S. Nagai, H. Mizune, A. Kashiwaya, and M. Miura (to Ube Industries).

18. Jpn. Kokai 75-145,405 (Nov. 21, 1975), K. Sugiyama and S. Kurihara.

19. Ger. Offen. 2,453,065 (May 15, 1975), J.E. May (to General Electric Co.).

20. Jpn. Kokai 74-45,119 (Nov. 8, 1969), N. Ohuchi and M. Nishida (to Matsushito Electric Industries); Jpn. Kokai 75-09,231 (April 17, 1969); 75-09,232 (April 10, 1975); 75-09,233 (April 10, 1975), T. Ohno, T. Tsubonshi, and M. Takahasi (to Nippon Electric Co.).

F. R. MORRAL
Consultant

COCHINEAL.   See Colorants for foods, drugs, and cosmetics.

COCOA.   See Chocolate and cocoa.

COCONUT OIL.   See Fats and fatty oils.

COFFEE

Coffee was originally consumed as a food in ancient Abyssinia and was presumably first cultivated by the Arabians in about 575 AD (1). By the sixteenth century, it had become a popular drink in Egypt, Syria, and Turkey. The name coffee is derived from the Turkish pronounciation, kahveh, of the Arabian word gahweh, signifying an infusion of the bean. Coffee was introduced as a beverage in Europe early in the seventeenth century and its use spread quickly. In 1725, the first coffee plant in the western hemisphere was planted on Martinique in the West Indies. Its cultivation expanded rapidly and its consumption soon gained the wide acceptance it enjoys today.

Modern Coffee Production

Commercial coffees are grown in tropical and subtropical climates at altitudes up to ca 1800 m; the best grades are grown at high elevations. Most individual coffees from different producing areas possess characteristic flavors. Commercial roasters obtain preferred flavors by blending—mixing the varieties before or after roasting. Colombian and washed Central American coffees are generally characterized as mild, winey-acid and aromatic; Brazilian coffees as heavy body, moderately acid, and aromatic; and African robusta coffees as heavy body, neutral, slightly acid, and slightly aromatic. The premium coffee blends contain higher percentages of Colombian and Central American coffees.

Economic Importance of Coffee

Coffee has been commercially significant for about 165 years. It is second only to petroleum in importance as an article of international trade. The total world exportable production of green coffee in the 1975–1976 growing season was 52.6 million bags, for which growing countries received $4.2 billion. Distribution is shown in Table 1.

A disastrous frost in Brazil in mid July 1975 that destroyed many coffee trees, followed by a combination of drought, floods, and civil unrest in other producing countries, caused the price of green coffee to escalate from less than $2.20/kg prior to the news of the frost, to a peak of about $6.60/kg in April 1977.

United States imports from producing countries in 1965 totaled 21.7 million bags of green coffee equivalent: 20.3 million bags of green coffee, 0.3 million of roasted coffee, and 1.1 million of soluble coffee (2), and were valued at $1.7 billion. The United States imports more than 65% of its coffee from countries in the Western Hemisphere. The major coffee importing countries are listed in Table 2. Coffee retained in producing countries for domestic consumption or stockpiling accounts for the differing figures of total coffee production and coffee imports.

In 1975, producer and consumer members of the International Coffee Organization approved the 1976 International Coffee Agreement, combining elements of prior agreements and attempts to achieve stable prices through export quotas adjusted by indicator price change. The agreement was intended to achieve even prices without sacrificing producer incentives (2). However, such economic sanctions are at present unnecessary owing to the high price of green coffee.

Table 1. World Production of Green Coffee, in 1975–1976[a]

Rank	Country	Exportable production[b]	%
1	Brazil	15,000	28.5
2	Colombia	6,450	12.3
3	Ivory Coast	4,580	8.7
4	Uganda	2,778	5.3
5	Indonesia	2,105	4.0
6	Mexico	2,056	3.9
7	El Salvador	1,914	3.6
8	Guatemala	1,710	3.3
9	Cameroon	1,472	2.8
10	Ethiopia	1,375	2.6
11	Costa Rica	1,283	2.4
12	Kenya	1,221	2.3
13	Angola	1,140	2.2
14	Zaire	900	1.7
15	Madagascas	853	1.6
16	India	785	1.5
17	Tanzania	778	1.5
18	Honduras	698	1.3
19	Dominican Republic	689	1.3
20	Ecuador	658	1.3
	others	4,127	7.9
	Total	*52,572*	*100.0*

[a] Ref. 2.
[b] Thousands of 60-kg bags.

Table 2. World Imports of Green Coffee in 1975[a]

Country	Imports[b]	%
United States	20,289	34.9
Canada	1,386	2.4
other Western Hemisphere	860	1.5
FRG	5,702	9.8
France	5,295	9.1
Italy	3,398	5.9
Netherlands	2,502	4.3
United Kingdom	1,535	2.6
other western Europe	10,019	17.3
eastern Europe	2,989	5.1
Africa	1,074	1.9
Asia and Oceania	3,033	5.2
Total	*58,082*	*100.0*

[a] Ref. 2.
[b] Thousands of 60-kg bags.

Processing and Packaging

Green Coffee Processing.   The coffee plant is a relatively small tree or shrub, often controlled to a height of 3 to 5 meters, belonging to the family *Rubiaceae*. *Coffea arabica* accounts for 69%; *Coffea robusta*, 30%; and *Coffea liberica* and others, 1% of world production (3). Each of these species includes several varieties. After the spring rains, the plant produces white flowers. About six months later, the flowers are replaced by fruit, approximately the size of a small cherry. The ripe fruit is red or purple.

The outer portion of the fruit is removed by curing: yellowish or light green seeds, the coffee beans, remain. They are covered with a tough parchment and a silvery skin known as the spermoderm. Each cherry normally contains two coffee beans.

Curing is effected by either the dry or wet method. The dry method produces so-called natural coffees; the wet method, washed coffees. The latter coffees are usually more uniform and of higher quality.

Dry curing is used in most of Brazil and in other countries where water is scarce in the harvesting season. The ripe cherries are spread on open drying ground and turned frequently to permit thorough drying by the sun and wind. Sun-drying usually takes two to three weeks depending on weather conditions. Some producing areas use hot air, indirect steam, and other machine-drying devices. When the coffee cherries are thoroughly dry, they are transferred to hulling machines which remove the skin, pulp, parchment shell, and silver skin in a single operation.

In wet curing, freshly picked coffee cherries are fed into a tank for initial washing. Stones and other foreign material are removed. The cherries are then transferred to depulping machines which remove the outer skin and most of the pulp. However, some pulp mucilage clings to the parchment shells that encase the coffee beans. Fermentation tanks, usually containing water, remove the last portions of this pulp. Fermentation may last from twelve hours to several days. Because prolonged fermentation may cause development of undesirable flavors and odors in the beans, some operators use enzymes to accelerate the process.

The beans are subsequently dried either in the sun or in mechanical dryers. Machine drying continues to gain popularity in spite of higher costs because it is faster and independent of weather conditions. When the coffee is thoroughly dried, the parchment is broken by rollers and removed by winnowing. Further rubbing and winnowing removes the silver skin to produce ordinary green unroasted coffee, containing about 12–14% moisture.

Coffee prepared by either the wet or dry method is machine-graded into large, medium, and small beans by sieves, oscillating tables, and airveyors. Damaged beans and foreign matter are removed by handpicking, machine separators, electronic sorters, or a combination of these techniques. Commercial coffee is graded according to the number of imperfections present, black beans, damaged beans, stones, pieces of hull, or other foreign matter. Processors also grade coffee by color, roasting characteristics, and cup quality of the beverage.

Chemical Composition of Green Coffee.   Coffee varies in composition according to the type of plant region from which it comes, altitude, soil, and method of handling the beans. As shown in Table 3, differences are greater between species, eg, arabica vs robusta (African) than within the same species grown in different regions, eg, Colombian vs Brazilian arabicas.

The lower oil, trigonelline, and sucrose contents are typical of robusta beans as

Table 3.   Typical Analyses of Green Coffee Types, % [a]

Variety	H_2O	Oil	Total nitro-gen	Ash	Caffeine	Chloro-genic acid[b]	Trigo-nelline	Protein	Reducing sugar	Sucrose	Total carbo-hydrate
robusta African	11.5	7.0	2.5	3.8	2.06	4.7	0.76	11.4	0.40	4.2	35.0
arabica Colombian	13.0	13.7	2.1	3.4	1.10	4.1	0.94	10.5	0.17	7.2	34.1
arabica Brazilian	11.0	14.3	2.2	4.1	1.01	4.1	1.24	11.1	0.27	7.1	32.0

[a] Ref. 4.

[b] Chlorogenic acid values vary somewhat with the method of analysis used.

is the higher caffeine content (see Alkaloids). Green coffee contains little reducing sugar but a considerable quantity of carbohydrate polymers. The polymers are mainly mannose with varying percentages of glucose, arabinose, and galactose (see Biopolymers; Carbohydrates).

Effects of Roasting on Major Components.   Green coffee has no desirable taste or aroma; these are developed by roasting. Many complex physical and chemical changes occur during roasting including the obvious change in color from green to brown, and a large increase in bean volume. As the roast nears completion, strong exothermic reactions produce a rapid rise in temperature, usually accompanied by a sudden expansion, or puffing of the beans, with a volume increase of 50–100%. However, this behavior varies widely among coffee varieties because of differences in composition and physical structure.

Table 4 shows the most significant and well-established chemical changes that occur in green coffee as a result of roasting. The major water-soluble constituents of green coffee are protein, sucrose, chlorogenic acid, and ash, which together account for 70–80% of the water-soluble solids. Most sucrose disappears early in the roast. Reducing sugars are apparently formed first and then react rapidly so that the total

Table 4.   Average Composition of Green and Roasted Coffee

Constituents	Green, % db[a]	Roasted, % db[a,b]
hemicelluloses	23.0	24.0
cellulose	12.7	13.2
lignin	5.6	5.8
fat	11.4	11.9
ash	3.8	4.0
caffeine	1.2	1.3
sucrose	7.3	0.3
chlorogenic acid	7.6	3.5
protein (based on nonalkaloid N)	11.6	3.1
trigonelline	1.1	0.7
reducing sugars	0.7	0.5
unknown	14.0	31.7
Total	*100.0*	*100.0*

[a] Dry basis.

[b] Not corrected for dry-weight roasting loss, which varies from 2 to 5%.

amount of sugar decreases as the roast nears completion. The sugar reactions, dehydration and polymerization, form high molecular weight water-soluble and water-insoluble materials. The formation of carbon dioxide and other volatile substances as well as the loss of water, account for most of the 2–5% dry-weight roasting loss.

Roasting essentially insolubilizes the proteins, which constitute 10–12% of green coffee, and 20–25% of the fraction soluble in cold water. The flavor and aroma of roasted coffee are probably due in large part to breakdown and interaction of the amino acids derived from these proteins. Analyses of the amino acids present after acid hydrolysis in both green and the corresponding roasted coffee show marked decreases in arginine, cysteine, lysine, serine, and threonine in Colombian and Angola robusta types after roasting. The amounts of glutamic acid and leucine for both coffee types, and in the robusta, phenylalanine, proline, and valine increase with roasting (5). Cysteine is the probable source of the many sulfur compounds found in the coffee aroma (see Amino acids).

About 15–40% of the trigonelline is decomposed during roasting. Trigonelline is a probable source of niacin, which reportedly increases during roasting, and of the potent aromatic nitrogen ring compounds, such as pyridine, found in roasted coffee aroma. However, the pyrazines, oxazoles, and thiazoles, also components of the coffee aroma, are probably products of protein breakdown. Caffeine is relatively stable, and only small amounts are lost by sublimation during roasting.

The chlorogenic acids 3-caffoylquinic acid (chlorogenic acid), 4-caffoylquinic acid, cryptochlorogenic acid , and 5-caffoylquinic acid (neochlorogenic acid) occur at least in part as the potassium caffeine chlorogenate complex. They decompose in direct relationship to the degree of roast. Table 5 shows the changes that occur in these acids during roasting.

Apparently, chlorogenic acids modify and control reactions that occur during the roast and are particularly important to the decomposition of sucrose.

Table 5. Changes in Chlorogenic Acids During Roasting, %[a]

	Santos				Colombian			
	Green	Light roast	Medium roast	Dark roast	Green	Light roast	Medium roast	Dark roast
chlorogenic acid	5.56	2.90	1.96	1.11	3.77	2.74	2.16	0.93
neochlorogenic acid	0.88	1.59	1.02	0.63	0.60	1.53	1.16	0.49
isochlorogenic acid	0.41		0.24					

[a] Ref. 6.

Glycerides of linoleic and palmitic acids along with some glycerides of stearic and oleic acids, make up the 7–16% fat content of coffee (see Fats; Carboxylic acids). Some cleavage of glycerides and some loss of unsaponifiables occur during roasting. Table 6 details these losses (7).

Table 6. Characteristics of Oil from Green and Roasted Coffee, %

	Green			Roasted		
	Oil	FFA[a]	Unsaponifiables	Oil	FFA[a]	Unsaponifiables
santos	12.77	0.78	6.56	16.05	2.70	6.10
robusta (Indonesia)	9.07	1.00	6.40	11.27	1.99	5.65

[a] As oleic; FFA, free fatty acid.

Aroma. The volatile flavor components of roasted coffee, though present in minute quantities, are extremely significant. Knowledge of the composition of natural flavors and aromas, eg, coffee, has been advanced by recent improvements in methods of isolation and fractionation and in the sensitivity of instruments that determine and measure chemical compounds. As reported in a 1928 patent, Staudinger (8) used classical chemical methods to isolate and identify twenty-six compounds in coffee aroma. In a summary of the literature up to 1957, Lockhart (9) listed forty-two compounds identified in coffee aroma. In 1960, Rhoades (10) reported additional compounds identified by gas chromatography and mass spectrometry, and in 1963, Merritt (11) brought the total to almost 100. Since that time, primarily through the work of Stoll (12), Stoffelsman (13), and Vitzthum (14), the list has been expanded to almost 500 compounds.

Freshly roasted ground coffee rapidly loses its fresh character when exposed to air, and within a few weeks develops a noticeable stale flavor. The mechanism for the loss of freshness is not known (15) but is presumably caused by volatilization of the aroma and polymerization of some of the aromatic compounds (see also Flavors and spices).

Roasting Technology. The main processing steps in the manufacture of roasted coffee are blending, roasting, grinding, and packaging. Green coffee is shipped in bags weighing from 60 to 70 kg. Prior to processing, the green coffee is dumped and cleaned of string, lint, dust, hulls, and other foreign matter. Coffees from different varieties or sources are usually blended before or after roasting.

Roasting by hot combustion gases in rotating cylinders requires 8–15 minutes. The bean charge absorbs heat at a fairly uniform rate and most moisture is removed during the first two-thirds of this period. As the temperature of the coffee increases rapidly during the last few minutes, the beans swell and unfold with a noticeable cracking sound, like that of popping corn, indicating a reaction change from endothermic to exothermic. This stage is known as development of the roast. The final bean temperature, 200–220°C, is determined by the blend, variety, or flavor development desired. A water or air quench terminates the roasting reaction. Most, but not all, of any added water is then evaporated.

Most batch roasters operate at air temperatures of about 425–490°C, with forced recirculation of air. Continuous roasters, at air temperatures of 375°C or lower, obtain good heat transfer with very high air velocity (a large proportion of the air is recirculated) and require only 5–8 minutes to complete a roast. However, if the green coffee used in continuous roasters varies greatly in physical properties such as bean size, density, and moisture content, the beans will not roast uniformly.

Theoretically, about 315 kJ (300 Btu) is needed to roast 0.45 kg of coffee beans. However, the air recirculation rate determines the thermal efficiency achieved. Older roasters that do not use recirculation may have an efficiency rate as low as 25%, requiring as much as 1260 kJ (1195 Btu). Roasters with complete air recirculation have an efficiency rate of 75% or more, requiring about 420 kJ (400 Btu).

Most roasters are equipped with temperature controls and automatic quenches that control the roasting operation. The bean temperature, correlated to the color of ground coffee measured by a photometric reflectance instrument, determines the quench end point of a roast. At the final bean temperature, the firing shuts down automatically, followed by water spraying for a timed period, and finally, discharge of the coffee.

Air must be circulated through the beans to remove excess heat before the finished and quenched roasted coffee is conveyed to storage bins. Residual foreign matter, such as stones and tramp iron, which may have passed through the initial green coffee cleaning operation, must be removed before grinding. This is accomplished by an air lift adjusted to such a high velocity that the roasted coffee beans are carried over into bins above the grinders, and heavier impurities left behind. The coffee beans flow by gravity to mills where they are ground to the desired particle size.

Grinding. Roasted coffee beans are ground to improve extraction efficiency in the preparation of the beverage. Particle size distributions ranging from about 1100 μm average (very coarse) to about 500 μm average (very fine) are tailored by the manufacturer to the various kinds of coffee makers used in households, hotels, restaurants, and institutions.

Most coffee is ground in mills that use multiple steel cutting rolls to produce the most desirable uniform particle size distribution. After passing through cracking rolls, the broken beans are fed between two more rolls, one of which is cut or scored longitudinally, the other, circumferentially. The paired rolls operate at differential speeds to cut, rather than crush, the coffee particles. A second pair of more finely scored rolls, installed below the main grinding rolls and running at higher speeds, is used for finer grinds.

Packaging. Most roasted and ground coffee sold directly to consumers in the United States is vacuum-packed in metal cans; 0.45, 0.9, or 1.35 kg are optimal, although other sizes have been used to a limited extent. After roasting and grinding, the coffee is conveyed, usually by gravity, to weighing-and-filling machines that achieve the proper fill by tapping or vibrating. A loosely set cover is partially crimped. The can then passes into the vacuum chamber, maintained at about 3.3 kPa (25 mm Hg) absolute pressure, or less. The cover is clinched to the can cylinder wall and the can passes through an exit valve or chamber. This process removes 95% or more of the oxygen from the can. Polyethylene snap caps for reclosure are placed on the cans before they are stacked in cardboard cartons for shipping. A case usually contains 10.9 kg of coffee, and a production packing line usually operates at a rate of 250–350 0.45-kg cans per minute.

Vacuum-packed coffee retains a high quality rating for two years or more. The slight loss in fresh roasted character that occurs is due mostly to chemical reactions with the residual oxygen in the can.

Coffee vacuum-packed in flexible, bag-in-box packages has gained wide acceptance in Europe. The inner liner, usually a preformed pouch of plastic-laminated foil is placed in a paperboard carton that helps shape the bag into a hard brick form during the vacuum process (16). The carton also protects the package from physical damage during handling and shipping. This type of package provides a barrier to moisture and oxygen almost as good as that of a metal can.

Inert gas flush packing in plastic-laminated pouches, although less effective than vacuum-packing, can remove or displace 80–90% of the oxygen in the package. These packages offer satisfactory shelf life and are sold primarily to institutions.

Some coffee in the United States, and an appreciable amount in Europe, is distributed as whole beans which are ground in the stores or by consumers in their homes. Whole-bean roasted coffee remains fresh longer than unprotected ground coffee and retains its fresh roasted flavor for several days.

Modified Coffees. Coffee substitutes, which include roasted chicory, chick peas, cereal, fruit, and vegetable products, have been used in all coffee-consuming countries.

Though consumers in some locations prefer the noncoffee beverages, they are generally used as lower-cost beverage sources, rather than as coffee. The coffee shortage created by a severe Brazilian frost in July 1975 has resulted in increased use of these materials as extenders with coffee in the United States, Canada, and Europe.

Chicory is harvested as fleshy roots which are dried, cut to a uniform size, and roasted. Chicory contains no caffeine and on roasting, develops an aroma compatible with that of coffee. It gives a high yield, about 70%, of water-soluble solids with boiling water and can also be extracted and dried in an instant form. Chicory extract has a darker color than does normal coffee brew.

The growing technology for the processing and use of roasted cereals and chicory is evidenced by the introduction on the market of coffees extended with these materials.

Instant Coffee

Instant coffee is the dried water-extract of roasted, ground coffee. Though used in army rations during the Civil War, instant coffee did not become a popular consumer item until after World War II. Improvements in manufacturing methods and product quality as well as a trend toward the use of convenience foods, account for this rise in popularity.

The patent literature on instant coffee products and processes dating back to 1865 (17), is very extensive (18).

Beans for instant coffee are blended, roasted, and ground as they are for regular coffee. Roasted, ground coffee is then charged into columns called percolators through which hot water is pumped to produce a concentrated coffee extract. The extracted solubles are dried, usually by spray or freeze drying, and the final powder is packaged in glass jars at rates of up to 200 jars per minute.

Blends.   Coffee is blended to achieve desired flavor characteristics. The concepts used to blend green coffees for regular roasted coffee may also be applied to blends for instant coffee. Most soluble coffee blends contain Brazilian, Central American, Colombian, and African robusta coffees. Some soluble coffees are manufactured in producing countries for export.

Roasting.   The batch or continuous-type roasters used for roasted and ground coffee are also used for instant coffee. The degree of roast can be varied somewhat, depending upon the varieties of coffees and the blend composition, to develop the desired characteristics of flavor and aroma.

Grind.   Grind is adjusted to suit the type of percolation used and is generally coarser than the regular grind for vacuum or bag-packed coffee (19). Coarser products avoid the development of excessive pressures in the percolator hydraulic system. The amount and distribution of very fine particles must also be controlled as they may interfere with uniform extraction. However, a grind that is too coarse necessitates longer time periods and higher temperatures for adequate extract concentrations and solubles yields.

Extraction.   Commercial extraction equipment and conditions have been designed to obtain the maximum soluble yield and acceptable flavor. In most processes, the water-soluble components in roasted coffee are first extracted with boiling water at atmospheric pressure. Additional solubles are then removed by pressure extraction at higher temperatures, thus hydrolyzing hemicelluloses and other components of the roasted coffee to water-soluble materials (20).

The factors influencing extraction efficiency and product quality are: (*1*) grind of coffee; (*2*) temperature of water fed to the extractors and temperature profile through the system; (*3*) percolation time; (*4*) ratio of coffee to water; (*5*) premoistening or wetting of the ground coffee; (*6*) design of extraction equipment; and (*7*) flow rate of extract through the percolation columns (19) (see Extraction).

Cylindrical percolators with height-to-diameter ratios ranging from 7:1 to 4:1 are common. They are usually operated in series as semicontinuous units of five to ten percolators, with the water flowing countercurrent to the coffee. The ground coffee may be steamed or wetted with water or coffee extract; this supposedly improves extraction (21). Feed water temperatures range from 154 to 182°C, and unless the columns are heated, the temperature drops so that the extract effluent will have cooled to 60–82°C. The effluent extract temperature may be reduced by water cooling in plate heat exchangers to minimize flavor and aroma loss prior to drying.

The extract is removed from the percolators and stored in insulated tanks until dried. The extract solubles yield is calculated from the extract weight and soluble solids concentration as measured by specific gravity or refractive index. Yield is controlled directly by adjusting the weight of soluble solids removed and depends primarily upon the properties of the coffee, operating temperatures, and percolation time. Soluble yields of 24–48% on a roasted coffee basis are possible. Robusta coffees give yields about 10% higher than arabica coffees (19).

High solubles concentrations are desirable to reduce evaporative load in drying and provide good flavor retention. Percolate concentrations are usually maintained in the range of 20–30% soluble solids for good flavor quality. Some processors concentrate solubles by vacuum evaporation prior to drying. The concentrated percolate may also be clarified by centrifuging prior to drying so that the dry product will be completely free of insoluble fine particles.

The flavor of instant coffee can be enhanced by recovering and returning some of the natural aroma lost during processing. The aroma constituents may be collected from coffee grinders or percolator vents, or may be obtained by concentrating the percolate. Many patents have been issued in the past ten years on the separation, collection, and transfer of aroma from roasted coffee to instant coffee (17).

Drying. The following factors are important criteria for good instant coffee drying processes: (*1*) minimum loss or degradation of flavor and aroma; (*2*) free-flowing particles of desired uniform size and shape; (*3*) suitable bulk density for packaging requirements; (*4*) desirable product color; and (*5*) moisture content below 4.5%. Operating costs, product losses, capital investment and other economic aspects must be considered in selecting the drying process (see Drying).

Spray Drying. This process is most often used for drying instant coffee. Atomization is usually obtained with pressure nozzles. Selection of nozzles and nozzle combinations is based upon properties of the extract pressures used, desired particle size, and bulk density and capacity requirements. The flow of hot air is usually concurrent with the atomized extract spray. Most processors prefer to use low inlet air temperatures (200–260°C) for best flavor quality. Outlet air is 107–121°C. Spray dryers are usually constructed of stainless steel and must be provided with adequate dust collection systems, such as cyclones or bag filters (19).

The particles of dried instant coffee are collected from the conical bottom of the spray dryer through a rotary valve and conveyed to bulk storage or packaging bins. Processors may screen the dry product to obtain a uniform particle size distribution.

Agglomeration. Most instant coffees have been marketed in a granular-appearing form since the mid-1960s. Previously, most soluble coffees were marketed as small spherically-shaped particles.

The granular appearance is usually achieved by fusing small spray-dried particles with steam and a low level of heat in a tower similar to a spray-drying tower. Other methods using a continuous belt are described.

Freeze Drying. Freeze drying became commercially important in the United States in 1964, although it was introduced a few years earlier in Europe. In 1976, freeze dried instant coffee represented about 12% of all coffee consumed and about 40% of all instant coffee consumed (22).

Freeze drying occurs at much lower temperatures than spray drying. Sublimation of ice crystals to water vapor under a very high vacuum [about 67 Pa (500 μm Hg) or lower] removes most of the water. Heat input is controlled to give maximum end-point temperatures between 38 and 49°C (23). Drying times are significantly longer than for spray drying (24).

Freeze dried coffees have lower extraction yields and better retention of volatile aromatics than do spray dried coffees and are thus considered quality instant coffees.

Packaging. In the United States, instant coffee for the consumer market is usually packaged in glass jars containing from 56 to 340 g of coffee. Larger units for institutional, hotel, restaurant, and vending-machine use are packaged in bags and pouches of plastic or paper. In Europe, instant coffee is packaged in paper, plastic, glass jars, and, frequently, plug-closure metal containers with foil liners.

Protective packaging is primarily required to prevent moisture pick-up. The flavor quality of regular instant coffee changes very little during storage. However, the powder is hygroscopic and moisture pick-up can cause caking and flavor impairment. Moisture content should be kept below 5%; packaging rooms are usually conditioned to a relative humidity of 50% or less.

Recently, many instant coffee producers in the United States have incorporated natural coffee aroma in coffee oil in the powder. These highly volatile and chemically unstable flavor components necessitate vacuum or inert-gas packing to prevent aroma deterioration and staling from exposure to oxygen (see also Packaging materials).

Decaffeinated Coffee

Decaffeinated coffee was first developed on a commercial basis in Europe about 1900. The basic process for decaffeinating coffee is described in a 1908 patent (25). Green coffee beans are steamed to increase the moisture content to at least 20%. The additional water and heat separate the caffeine from its natural complexes and aid its transport to the surface of the beans. An organic solvent extracts the caffeine from the wet beans. The beans are again steamed to remove the solvent, and then dried and roasted (24).

Decaffeination processes cause some changes in the beans that subsequently affect roasting and development of flavor. However, a water extraction process patented in 1943 is supposedly more rapid and efficient and causes less damage to the quality of the coffee flavor than the original process (26). In this process, caffeine is dissolved from the green beans by an equilibrium water extract in a countercurrent series of percolators. The caffeine is then removed from the water extract by liquid–liquid extraction with organic solvent and the water extract is recycled.

The decaffeinated beans are rinsed with water to remove soluble solids from the surface, then dried and roasted. According to the patent, trichloroethylene is used to extract caffeine from the water extract, but since December 1977, most processors have used methylene chloride (see Chlorocarbons). The caffeine is purified by water extraction, recrystallization, and absorbents to meet USP specifications for its use in pharmaceuticals and soft drinks (19) (see Alkaloids).

Roasted and ground decaffeinated coffee is vacuum-packed for consumer use or made into soluble coffee powder by the methods previously described. Consumption of decaffeinated coffees, regular and instant, has increased during recent years and currently accounts for about 30% of all instant coffees, including freeze dried, sold in the United States and Europe.

BIBLIOGRAPHY

"Coffee" in *ECT* 1st ed., Vol. 4, pp. 215–223, by L. W. Elder, General Foods Corp.; "Coffee, Instant" in *ECT* 1st ed., Suppl. 2, pp. 230–234, by H. S. Levenson, Maxwell House Division of General Foods Corp.; "Coffee" in *ECT* 2nd ed., Vol. 5, pp. 748–763, by R. G. Moores and A. Stefanucci, General Foods Corp.

1. W. A. Ukers, *All About Coffee*, 2nd ed., Tea and Coffee Trade Journal, New York, 1935, pp. 1–3.
2. *Annual Coffee Statistics 1975*, Publication No. 39, Pan American Coffee Bureau, New York, pp. 1–3, 5, 21, 29, 57, A-39.
3. W. C. Struning, private communication, Pan American Coffee Bureau, New York, Aug. 1977.
4. A. Stefanucci and K. Sloman, private communication, Technical Center, General Foods Corp.
5. H. Thaler and R. Gaigl, *Z. Lebensm. Unters. Forsch.* **120**, 449 (1963). ·
6. J. R. Feldman, W. S. Ryder, and J. T. Kung, *J. Agric. Food Chem.* **17**(4), 733 (1969).
7. L. Gariboldi and A. Carisano, *J. Sci. Food Agric.* **15**, 619 (1964).
8. U.S. Pat. 1,696,449 (Dec. 25, 1928), H. Staudinger (to Internationale Nahrungs-und Genussmittel A.G.).
9. E. E. Lockhart, *Chemistry of Coffee*, Publication No. 25, The Coffee Brewing Institute, Inc., New York, 1957, p. 10.
10. J. W. Rhoades, *J. Agric. Food Chem.* **8**, 136 (1960).
11. C. Merritt, Jr., and co-workers, *J. Agric. Food Chem.* **11**, 152 (1963).
12. M. Stoll and co-workers, *Helv. Chim. Acta* **50**, 628 (1969).
13. J. Stoffelsma and co-workers, *J. Agric. Food Chem.* **16**, 1000 (1968).
14. O. G. Vitzthum and P. Werkoff, *J. Food Sci.* **39**, 1210 (1974); *J. Agric. Food Chem.* **23**, 510 (1975).
15. R. Radtke, W. Mohr, and R. Springer, *Z. Lebensm. Unters. Forsch.* **119**, 293 (1963); **128**, 321 (1965); **129**, 344 (1966).
16. A. L. Brody, Arthur D. Little, Inc., Food and Flavor Section, Cambridge, Mass., *Flexible Packaging of Foods*, CRC Press, Division of the Chemical Rubber Co., Cleveland, Ohio, 1970, pp. 41–42.
17. U.S. Pat. 48,268 (June 20, 1865), Gale.
18. N. D. Pintauro, *Coffee Solubilization. Commercial Processes and Techniques*, Noyes Data Corp., Park Ridge, N.J. 1975.
19. H. Foote and M. Sivetz, *Coffee Processing Technology*, Vol. 1, AVI Publishing Co., Westport Conn., 1963, pp. 19, 332–334, 320–340, Chapt. 11; M. Sivetz, *Coffee Processing Technology*, Vol. 2, AVI Publishing Co., Westport, Conn., 1963, p. 213.
20. U.S. Pat. 2,324,526 (July 20, 1943), M. R. Morgenthaler (to the Nestle Co., Inc.).
21. U.S. Pat. 3,549,380 (Dec. 22, 1970), J. M. Patel and D. A. Stang (to Procter & Gamble).
22. *Coffee Drinking in the United States*, survey by National Coffee Association and Pan American Coffee Bureau, 1976.

23. U.S. Pat. 3,438,784 (Apr. 15, 1969), W. P. Clinton, G. B. Ponzoni, and J. Mahlmann (to General Foods Corp.).
24. M. Sivetz, *Coffee,* Coffee Publication, 1977, Chapt. 8, pp. 45–46; Chapt. 5, p. 29.
25. U.S. Pat. 897,763 (Sept. 1, 1908), J. F. Meyer, L. Roselius, and K. H. Wimmer.
26. U.S. Pat. 2,309,092 (Jan. 26, 1943), N. E. Berry and R. H. Walters (to General Foods Corp.).

General References

J. R. Feldman, W. S. Ryder, and J. T. Kung, *J. Agric. Food Chem.* **17**(4), 733 (1969).
W. A. Ukers, ed., *All About Coffee,* Tea and Coffee Trade Journal, New York, 1935.
O. G. Vitzthum in O. Eichles, ed., *Kaffee and Coffein,* Springer-Verlag, Berlin, 1976.
F. L. Wellman, *Coffee (Botany, Cultivation and Utilization),* Interscience Publishers, a division of John Wiley & Sons, Inc., New York, 1961.
H. E. Foote and M. Sivetz, *Coffee Processing Technology,* Vol. 1, and M. Sivetz, *Coffee Processing Technology,* Vol. 2, AVI Publishing Co., Westport, Conn., 1963.

A. Stefanucci
W. P. Clinton
M. Hamell
General Foods Corporation

COGENERATION.   See Energy management.

COKE.   See Coal, coal conversion processes.

COKE OVEN GAS.   See Coal, coal conversion processes.

COLLIDINES (TRIMETHYLPYRIDINES, ETC).   See Pyridine and pyridine bases.

COLOR

Color can be defined as that part of the visual experience that deals with aspects of objects perceived other than their size, shape, and surface texture. This definition, purposely vague, emphasizes the perceptual, and therefore highly personal, nature of color: color is what we see. Light (electromagnetic radiation in the visible region, roughly 400–700 nm) (Fig. 1) either falls directly on the eye, or is seen after it is modified by an object. Neural impulses are created in the eye and sent to the brain. There, and not before, they are interpreted as color. One cannot know exactly what another sees as color, but through common experience and agreed terminology the basis for developing a science of color is produced (1).

Interactions among the source of light, the object, and the eye and brain must always be considered in understanding the visual experience. However, color is not the totality of visual experience. Many other perceptions contribute to the appearance of objects (2), including such properties as size and shape, translucency, gloss, surface uniformity, and metallic character (see General References). This article describes the applications of color in business, science, and industry (3).

The Variables of Perceived Color

Most of the literature on color assumes that only three variables are required for a complete description of color; a common set (defined below) is hue, lightness, and saturation. It has long been recognized, but rarely mentioned, that this is an oversimplification, correct only for unrelated colors seen as self-luminous stimuli bearing no relation to the background or surround. For related colors, whose appearance depends upon their surround, four or five variables are required (4–5).

Listed here are the perceptual terms that describe the four variables of related colors and the three of unrelated colors. They are the terms that denote important attributes of sensations of light and color (4). Described later are the corresponding psychophysical terms denoting objective measures of physical variables related to the magnitudes of these attributes, and psychometric terms denoting these objective measures scaled so that equal scale intervals represent approximately equal perceived differences in the attribute considered.

Considered first are the achromatic attributes present in all colors, including the whites, grays, and blacks that do not possess a hue. For both unrelated and related colors, the perceptual term for one such attribute is brightness, the attribute of a visual

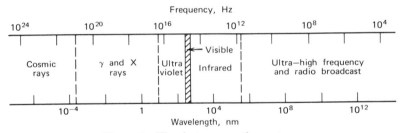

Figure 1. The electromagnetic spectrum.

sensation according to which an area appears to emit more or less light. For related colors, a second attribute (which, together with two chromatic attributes to be defined, provide the four variables of perceived color in the general case) is lightness, according to which an area appears to reflect a greater or smaller fraction of incident light. (The word transmit could replace reflect throughout this discussion.)

Of the chromatic attributes, the most easily defined perceptual term is hue, which denotes whether the color is red, yellow, green, blue, etc. More difficult and controversial are the terms denoting the strength of the chromatic response by which the hue is recognized. Hunt proposed a new perceptual term, colorfulness, for the attribute of a visual sensation according to which an area appears to exhibit more or less chromatic color (4). When this amount is judged relative to the brightness, a relative colorfulness results for which Hunt recommends the term saturation in its usual meaning with respect to color. Saturation tends to remain constant as the level of illumination is changed, since changes in colorfulness are accompanied by corresponding changes in brightness. Saturation is therefore an important attribute for the recognition of objects. Finally, for a related color, colorfulness can be judged relative to the average brightness of its surroundings; for the resulting relative colorfulness Hunt recommends the term perceived chroma.

These three measures of the strength of the chromatic response, colorfulness, saturation, and perceived chroma, are not independent but can be interrelated if the lightness and brightness are known. The two achromatic terms are independent, however. Because lightness, as well as saturation (and perceived chroma), tends to remain constant as the level of illumination changes, the visual task of recognizing objects usually involves hue, lightness, and saturation, and these are the most important variables of perceived color.

Physical Aspects of Color

Sources of Light. Light is one type of electromagnetic radiation, limited to the visible region of the spectrum (Fig. 1). Sources of light emit power, and terms related to this power (radiometric terms) and to the response it creates in the eye (photometric terms) are described by the adjectives radiant and luminous. Radiometric quantities are often considered as a function of the wavelength of the radiation, and are designated by the adjective spectral. Photometric quantities are obtained by multiplying the corresponding spectral radiant quantity by the eye's response to power, the spectral luminous efficiency function $V(\lambda)$, and integrating the product over the wavelengths of the visible spectrum. The total radiant power emitted by a light source is the radiant flux, measured in watts (W). The corresponding luminous flux is measured in lumens (lm) (683 lm = 1 W). Other radiometric and photometric quantities of importance are the flux in watts per steradian, sr (unit solid angle, shown in Fig. 2) (radiant intensity, W/sr), or luminous intensity (candela, cd, or lm/sr), flux per unit area (irradiance, W/m^2, or illuminance, in lux, lx, or lm/m^2), and flux per unit area and solid angle (radiance, $W/(sr \cdot m^2)$ or luminance, cd/m^2 or $lm/(sr \cdot m^2)$) (6).

The effect of a light source on color is best described by its spectral power distribution. Typical examples are shown in Figure 3 for two common sources, an incandescent lamp (labeled A) and natural daylight (labeled D_{65}). These are examples of continuous distributions, whereas for many arc lamps used for illumination, such as the sodium and mercury arc lamps, the power is concentrated in a few narrow lines

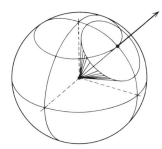

Figure 2. Unit of solid angle, one steradian.

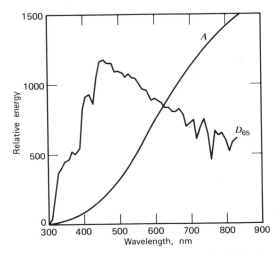

Figure 3. Relative spectral power distributions of CIE standard illuminants A and D_{65} (7).

at specific wavelengths. In fluorescent lamps, mercury vapor excites fluorescence of a phosphor by the absorption of the mercury lines in the ultraviolet; the spectral power distributions of these lamps consist of a combination of the continuous distribution of the phosphor and the line distribution of the mercury vapor.

The incandescent source A in Figure 3 is typical of the class of materials known as perfect radiators, or black bodies, whose spectral power distributions can be calculated from their temperature, called their color temperature, by Planck's law. The color temperature of source A is 2854 K. Other sources that have colors similar to those of perfect radiators but different spectral power distributions may be described by a correlated color temperature corresponding to that of the perfect radiator nearest in color. The correlated color temperature of D_{65} is 6500 K. The term illuminant is used to describe D_{65}, whose spectral power distribution is known by definition but which is not readily available (since actual daylight is quite variable) as a real source.

For critical visual judgments of color, standard sources are defined and made available; for the equivalent measurements to be described, the corresponding spectral power distributions are tabulated and referred to as standard illuminants.

Interaction of Light with Objects.   Of the many forms of the interaction of light with objects, only three are of major interest for their influence on the colors of objects: absorption, scattering, and reflection.

Absorption.   Absorption is the process in which radiant power is utilized to raise molecules in the object to higher energy states. In their return to the ground state, the molecules ultimately dissipate the power as heat. Spectrally selective absorption is responsible for color effects in transparent objects; if only absorption is present, ie, there is no scattering to cause reflection, opacity or translucency cannot exist (except for opacity produced by complete absorption of all flux incident on the object, in which case it is black).

Fluorescence.   Fluorescence is a process in which absorption is followed by re-radiation of the flux at only moderately longer wavelengths, still (as far as color effects are concerned) in the visible spectral region. The absorption, however, can occur in the ultraviolet region, leading to visible fluorescence in amounts dependent on the spectral power of the source in the ultraviolet region. The phenomenon is widespread through the use of fluorescent brightening agents with textiles, paper, detergents, and the like (see Brighteners, fluorescent).

Scattering.   Scattering is the interaction in which radiation is redirected without change in wavelength. Atoms, molecules, or colloidal particles when irradiated generate small amounts of radiation moving in directions determined by the size of the particle and the polarization of the radiation. The scattered radiation may be observed directly and may be spectrally selective, leading, eg, to the blue color of the sky. In denser systems, the scattered radiation may be rescattered many times; multiple scattering provides translucency and (especially if some absorption is present) opacity to objects.

Reflection.   On the macroscopic scale, radiation is reflected at the boundaries of objects where a change in refractive index occurs. The laws of reflection at a smooth surface, exemplified by a mirror, are well known and lead to the phenomenon of specular reflection or gloss. Reflection from a rough surface, in which reflected radiation propagates in a variety of directions, is termed diffuse reflection. Diffuse reflection can also result from scattering within the body of an object, the scattered radiation passing through the surface of the object. This diffuse internal scattering is the predominant source of color effects in most translucent and opaque objects.

Terms Describing Reflection and Transmission.   In describing quantitatively the interactions of light with matter leading to color effects, the following terms are useful. Reflectance is the ratio of reflected flux to incident flux. More commonly measured is the reflectance factor, the ratio of flux reflected from the test object to that reflected from a similarly irradiated agreed standard, normally the perfect reflecting diffuser which reflects all the flux incident on it such that its reflected radiance is equal in all directions. For fluorescing reflectors, it is customary to speak of total radiance factor, the sum of the reflected and fluoresced radiance relative to the radiance reflected from the perfect reflecting diffuser. The modifiers *diffuse* or *specular* apply to the reflectance terms, and the modifier *spectral* applies to all the above terms, as required. Terms describing transmission are similarly defined, the most important being transmittance, the ratio of transmitted to incident flux. Transmittance factor, in which the reference is the perfect transmitting diffuser, may be used to describe diffuse transmittance. Transmittance without diffusion is preferably described as regular rather than specular transmittance. The term internal transmittance is used to describe the quantity of interest to the chemist, the flux regularly transmitted through a test object relative to that transmitted through a similar nonabsorbing, nonscattering object taken as the reference. These and many other terms relating to the radiometric and photometric properties of materials are defined in ref. 8.

Color Vision. Although it is of paramount interest to physiologists and theorists, a detailed knowledge of color vision is not required for an understanding of color science; the following few facts suffice. Light falling on the retina of the eye (Fig. 4) is detected, for normal levels of illumination, by three types of cone receptors, leading to photopic or color vision. In dim light, a fourth type of receptor, the rods, is activated (scotopic vision) but at the higher light levels considered in this article the rods normally play no role in color vision (10). The details of the photosensitive pigments in the cones, their mode of action, and their spectral responsivity curves have not been unequivocally resolved. It is known that the cones are most numerous in the central (foveal) region of the retina which subtends an angle of about 2° and is used when the eye fixates an object. Within the central foveal region the absence of blue cones and macular pigment absorption lead to color-vision responses somewhat different from those outside the fovea.

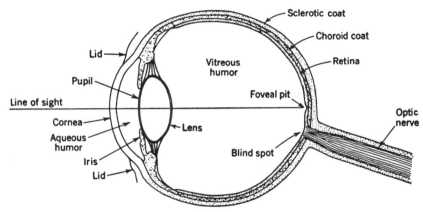

Figure 4. Cross section of the human eye (9).

Beyond the retina, the neural impulses transmitting color-vision information to the brain appear to traverse complex networks. For many purposes it is convenient to think in terms of derived opponent signals (yellow–blue, red–green, light–dark) instead of the three signals derived from the respective cones. The details of these stages, and the subsequent interpretation of the signals in the brain, are largely speculative. What is well established and of major importance, however, is the trichromacy of color vision, ie, the need for only three signals to explain all known facts of photopic vision (as noted above, at sufficiently high levels of illumination rod vision does not contribute).

Color-Vision Deficiencies. About 8% of the population, almost all males, has some degree of color-vision deficiency, characterized by difficulty in recognizing colors or the confusion of dissimilar colors. Most of these people suffer from one of two forms in which red and green are confused (protanomaly, deuteranomaly); a very small minority confuses yellow and blue (tritanomaly) or lacks color vision entirely. There is no cure for the deficiencies. A number of tests for them are available, the simplest of which are series of plates in which numbers or patterns can be distinguished depending on the presence and severity of the deficiency. Attempts to relate all the known types of color-vision deficiency to anomalies in the retina (such as the lack of one type of cone) or the neural pathways to the brain have not been fully successful.

Metamerism. A major corollary of the trichromacy of color vision is that many different stimuli can lead to the same perceived color (Fig. 5). The stimulus is described in the general case by the product of the spectral power distribution curve of a source and the spectral reflectance curve of an object. The set of three signals, here represented by one only, results from the product of the spectral stimulus and the spectral responsivity curves of the particular observer involved. It is obvious that a vast number of different stimuli could lead to the same color. Such stimuli are called metamers, and the phenomenon metamerism.

If two objects have different spectral reflectance curves which, when combined with the spectral power distribution of a specific source and the spectral responsivities of a specific observer, give rise to the same color, ie, to two colors which match, then the objects form a metameric pair. In general, a change in either the source or the observer characteristics will destroy the color match. In the former case, illuminant metamerism is said to result: two objects which match to a given observer under specified illumination no longer match when the illumination is changed. In the latter case, observer metamerism results. The recognition and avoidance of metamerism plays a major role in the industrial coloring of materials.

Other phenomena are related to metamerism. Color constancy refers to the change in the appearance of a color as the illumination or other conditions change. Color rendering describes the ability of a given test illuminant to produce the same color effects as a reference illuminant. Chromatic adaptation results when the responsivities of the eye are altered so as to preserve the colors of objects as seen under specified conditions. To a large extent these represent aspects of color science that are not yet fully understood.

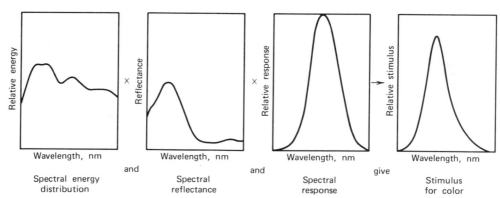

Figure 5. The stimulus perceived as color is made up of the spectral power (or, as here, energy) curve of a source times the spectral reflectance (or transmittance) curve of an object times the appropriate spectral response curves (one shown here) of the eye (11).

Colorants. Colorants (qv) are chemical substances added to materials to produce color effects. They are commonly classified as dyes (qv) or pigments (qv). The distinction between dyes and pigments is muddied by conflicts among tradition, methods of use, and chemical and physical properties of colorants. For the purposes of this article, dyes are substances that are molecularly dispersed, often in aqueous solution, transferred to a material, and bound to it by intermolecular forces to provide color effects by spectrally selective absorption processes. Pigments are larger than molecular-particle size, held in place by their corresponding low mobility, and usually scatter as well as absorb light. Traditionally, dyes were organic in nature and pigments inorganic, but many exceptions exist, notably the widespread modern use of organic pigments.

The dependence of absorption and scattering on the properties of pigments should be mentioned. In such products as paints and plastics, the use of highly scattering pigments is customary to obtain opacity. (In contrast, opacity in papers and textiles results from scattering by the dyed fibers.) Scattering power depends upon particle size of the pigment (Fig. 6) and is a maximum when the pigment particle diameter is about half the wavelength of light. Scattering power also increases with the refractive index of the pigment relative to that of the surrounding medium. Thus maximum scattering occurs with inorganic pigments of high refractive index and appropriate average particle size.

Because of their low refractive index, organic pigments scatter relatively little light, and a second consideration becomes important. This is absorptive power, known in the trade as strength. As shown in Figure 6, this increases continuously with decreasing particle size down to very small sizes compared to the optimum for scattering. Few if any organic pigments are used for their scattering power, and most are made to have small particle size for maximum strength. Consequently, many organic pigments scatter so little as to be almost transparent. Compromises are sometimes necessary, since the lightfastness of pigments generally becomes poorer with decreasing particle size.

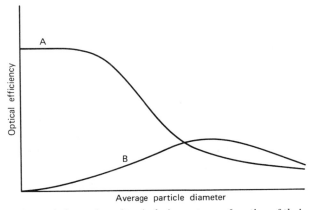

Figure 6. Scattering and absorption of typical pigments as a function of their particle size (12). A, absorption (color strength); B, scattering power.

Colorant Mixing. The mixing of dyes and pigments to produce desired colors is a major objective of color technology, applying to art and design as well as to the coloring of materials in industry.

Simple-Subtractive Mixing. This term is reserved for the mixing of colorants in transparent systems in which only absorption is important. Transmittance T may be related to absorbance A which is linearly related to the concentration c of the colorant and the thickness b of the specimen by a mixing law, the Beer-Lambert law:

$$\log (1/T) = A = \sum_i a_i b c_i$$

where a is the specific absorbance or absorptivity. The quantities T, A, and a_i are functions of the wavelength, and the Beer-Lambert law applies only when these quantities are constant—usually to data at one wavelength at a time. Because of the trichromacy of color, use of no more than three properly selected colorants at a time is required to achieve a desired color effect (see Color photography).

Complex-Subtractive Mixing. The Beer-Lambert law does not suffice for the more complex situation in which both scattering and absorption are important. The most widely used law for this case, commonly called turbid-medium theory, is that formulated by Kubelka and Munk (3,13). It applies to both translucent and opaque specimens, but it is written here only for the simpler case of complete opacity. The reflectance factor R is related to Kubelka-Munk scattering and absorption coefficients K and S, and these in turn to corresponding specific coefficients and the concentrations of the colorants:

$$\frac{(1 - R)^2}{2R} = \frac{K}{S} = \frac{\sum_i c_i k_i}{\sum_i c_i s_i}$$

Since the specimen is opaque, its thickness is not considered. Two cases may be distinguished. (1) For textiles and papers, in which the fiber structure provides the scattering, nonscattering dyes are used. The equation can be simplified with only the term for the fiber substrate remaining in the denominator on the right. Three dyes at a time are sufficient to produce the desired color effect; thus the numerator on the right contains three terms for the dyes and one to account for the absorption of the substrate. (2) The other case is for paints and plastics in which a highly scattering pigment is used to obtain opacity. All terms in the equation are retained except for pigments where scattering is negligibly small. Four pigments are required, three to satisfy the trichromacy of color and one to ensure that opacity is obtained. Typically, two chromatic pigments plus white and black may be used. For both cases, the quantities R, K, S, k_i, and s_i are dependent upon the wavelength, and the Kubelka-Munk equations apply only to data at one wavelength at a time.

Spectral Reflectance Curves and Color. Although the importance of the source and observer cannot be neglected, it is possible to make useful qualitative correlations between spectral reflectance curves and the colors they represent, seen in white light by normal observers. Some examples are shown in Figure 7.

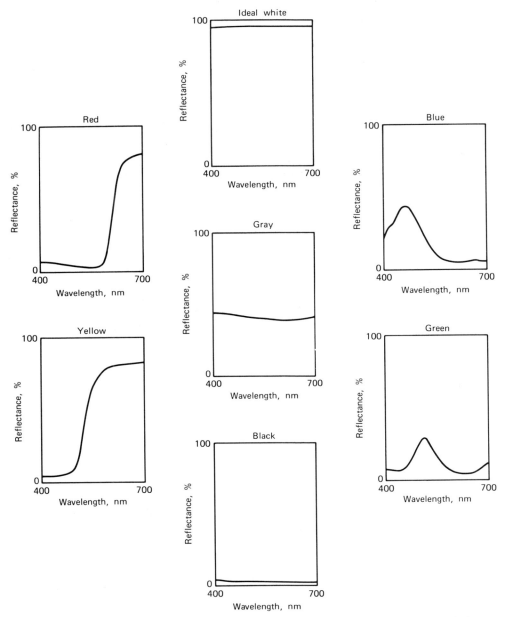

Figure 7. Typical spectral reflectance curves of colored objects with their color names (11).

531

Additive Mixing and Primary Colors. In addition to the laws of subtractive mixing, it is possible to mix color stimuli, such as colored lights, additively. Color television is an important application of additive color mixing. The laws of additive mixing (Grassmann's laws) (14) are quite different from those of subtractive mixing; the spectral character of the stimuli need not be known to predict results, which are quite different in the two cases (see Color photography, instant).

If primary colors are defined as sets of three colors that combine to produce the largest gamut of mixture colors, the primaries for additive mixing are red, blue, and green. Additive mixtures include: red and blue, which give purples including magenta; blue and green, giving blue–greens including cyan; and green and red, which give yellows. Mixtures of the three primaries, in the proper proportions, give white. The primary colors for subtractive mixing are magenta, cyan, and yellow. Subtractive mixtures include: magenta and cyan, which give blues; cyan and yellow, giving greens; and yellow and magenta, which give reds. Mixtures of the three primaries, in the proper proportions, give black.

Color-Order Systems

The major color-order systems provide a substantial body of data against which the results of color measurement and color technology may be tested.

Munsell System. Developed by the artist A. H. Munsell in the early 1900s, the Munsell system (15) utilizes the three equally-spaced perceptual variables Munsell hue, Munsell value (lightness), and Munsell chroma, arranged in the cylindrical coordinate system illustrated in Figure 8. Scales of the three Munsell variables are reproduced approximately in Plate I. Munsell hue is designated by combinations of letters representing five principal hues, Red, Yellow, Green, Blue, and Purple, preceded by a digit between 1 and 10 to give a 100-step hue scale. Munsell value, the correlate of lightness, is scaled with 0 at black and 10 at white. Munsell chroma starts at zero for the achromatic (neutral) gray series and extends out to the limits set by perception,

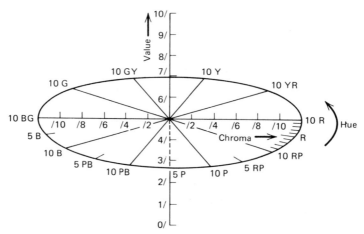

Figure 8. The coordinate system of the Munsell color-order system (11). R, red; Y, yellow; G, green; B, blue; and P, purple.

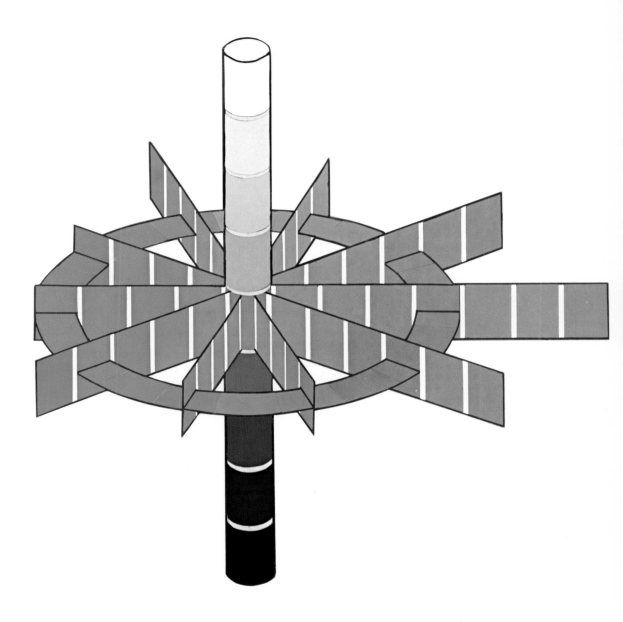

Plate I. Scales of the Munsell color-order system: the circle of hues, a neutral value scale, and a chroma scale in the red hue region. The reproduction is approximate only. Courtesy of Munsell Color and the Philatelic Foundation.

or by the colorant system used for real samples, in equal steps of size such that 2 chroma steps equal 1 value step (and one hue step at chroma 5). The nature of color vision and colorants is such that for yellow hues, higher chromas can be achieved for lighter than for darker colors, whereas the reverse is true for blue hues.

The Munsell system is illustrated by a collection of approximately 1500 painted samples, the *Munsell Book of Color* (16). Maintained to very close tolerances and anchored to the results of color measurement, the *Munsell Book of Color* is used worldwide as a basis for color specifications and as a prime example of equal visual spacing of colors.

Universal Color Language. The Munsell system is also an integral part of a universal color language (16) designed to provide standard verbal or numerical descriptions of color at various levels of complexity. Such descriptions, in readily understood terms, provide color specifications for a wide variety of situations. The simplest level (Level 1) of the universal color language describes color in terms of ten simple hue names plus black, gray, and white (Table 1). Level 2 adds intermediate hue names, also shown in Table 1, to a total of 29.

Level 3 of the language consists of the ISCC-NBS color system, a scheme developed by a Problems Subcommittee of the Inter-Society Color Council in collaboration with the NBS. Here the color solid is divided into 267 regions, a few of which are shown in Figure 9, each designated by the combination of a hue name with adjectives denoting lightness and saturation. The centers of the 267 regions are illustrated by the ISCC-NBS centroid colors (17).

Level 4 of the universal color language consists of the approximately 1500 samples of the *Munsell Book of Color;* level 5 of the language, of the ca 100,000 colors to which interpolated Munsell designations can be given on the basis of visual comparison to the book samples; and the highest level, 6, of the several million samples that can be specified by careful color measurement.

Table 1. Generic Hue Names (Levels 1 and 2) and Intermediate Hue Names (Level 2) of the Universal Color Language, Including Centroid Numbers and ISCC-NBS Abbreviations [a]

Generic hue names	Intermediate hue names
2. pink, Pk	26. yellowish pink, yPk
11. red, R	34. reddish orange, rO
48. orange, O	40. reddish brown, rBr
55. brown, Br	66. orange–yellow, OY
82. yellow, Y	74. yellowish brown, yBr
107. olive, Ol	95. olive-brown, OlBr
115. yellow–green, YG	97. greenish yellow, gY
139. green, G	125. olive-green, OlG
178. blue, B	129. yellowish green, yG
218. purple, P	160. bluish green, bG
263. white, W	169. greenish blue, gB
265. gray, Gy	194. purplish blue, pB
267. black, Bl	207. violet, V
	237. reddish purple, rP
	247. purplish pink, pPk
	254. purplish red, pR

[a] Ref. 17.

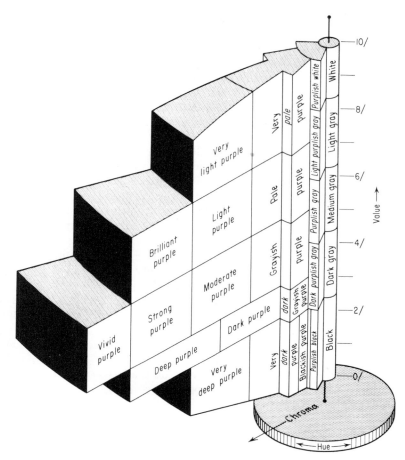

Figure 9. The purple section of the color solid at level 3 of the universal color language, showing ISCC-NBS color names (18).

Other Systems. *OSA Uniform Color Scales System.* The Optical Society of America's Committee on Uniform Color Scales has completed work extending over three decades leading to a color-order system (19) considered by many to be the ultimate in equal visual spacing. The system is illustrated by 500 painted colors (20). The coordinates of the system are based on the opponent theories of color vision, describing redness–greenness, yellowness–blueness, and lightness. The sampling of color space is unique in that each color is surrounded by twelve others equally visually spaced from it: six at the same lightness, three at higher lightness, and three at lower lightness, in a close-packed array.

Natural Color System. Developed in Sweden, this system is based on the ability of observers to select four unitary hues: a red seen to contain no yellow or blue, a yellow seen to contain no red or green, and similarly a green and a blue (21). The hues of other colors can be scaled in terms of the amounts of two neighboring unitary hues they contain; this scaling does not lead to equal visual spacing around the hue circle. Coordinates of saturation and darkness (the inverse of lightness) complete the system. No collection of samples based on the Natural Color System is currently available.

Ostwald System. The German chemist Ostwald developed a color-order system (22) early in this century based on mixing full colors (the highest chroma pigments he could find) systematically with white, black, and both white and black. This system, unlike others described above, is closed in that there is no way to include a new, higher chroma pigment should it be found. Since the colors obtained depend intimately on colorant properties, there can be no definitive collection illustrating the Ostwald system. Popular among artists and designers, the system has been used to produce several editions of the Container Corporation's *Color Harmony Manual,* no longer published. The Ostwald system is not equally visually spaced.

Commercial Colorant-Based Systems. Many paint companies, producers of printing inks, and others, offer aids to sales in the form of color-order systems based on the systematic mixing of a limited number of colorants. These closed, proprietary systems generally do not exemplify basic principles of color science, but are devised to meet merchandising needs.

Single-Number Color Scales. In restricted regions of color space, it is possible to devise scales in which systematic color effects can be described by a single number. A typical example is the degree of yellowness associated with chemical purity in a wide variety of products ranging from petroleum to beer. Many such single-number color scales were studied and correlated by a Problems Subcommittee of the Inter-Society Color Council, and a summary was published (23).

Basic Colorimetry: The CIE System

Colorimetry, the quantitative science of color, exists in two parts: basic colorimetry, which sets in numerical terms the criteria for the identity of two color stimuli and is concerned with color matching; and advanced colorimetry, which is concerned with color differences, color appearance, and other difficult and largely unsolved problems in color science (24). Basic colorimetry does not depict color sensations, only whether two stimuli do or do not match. In the extreme case, advanced colorimetry includes methods of assessing the appearance of color stimuli presented to the observer in complicated surroundings as they might occur in everyday life.

Basic colorimetry rests on methodology and standardization introduced by the International Commission on Illumination (Commission Internationale de l'Éclairage—CIE) beginning in 1931. CIE recommendations over the intervening years have been conveniently summarized (25). The CIE methodology is based on the trichromacy of vision and states that two stimuli produce the same color if each of three tristimulus values X, Y, and Z are equal for the two:

$$X = k \int S(\lambda)R(\lambda)\bar{x}(\lambda)d\lambda$$

$$Y = k \int S(\lambda)R(\lambda)\bar{y}(\lambda)d\lambda$$

$$Z = k \int S(\lambda)R(\lambda)\bar{z}(\lambda)d\lambda$$

$$k = 100/\int S(\lambda)\bar{y}(\lambda)d\lambda$$

where $S(\lambda)$ is the spectral power distribution of the illuminant, $R(\lambda)$ the spectral reflectance factor of the object, and $\bar{x}(\lambda)$, $\bar{y}(\lambda)$, and $\bar{z}(\lambda)$, defined further below, are color-matching functions of the observer. The quantity k is defined so that tristimulus value Y for the reference for $R(\lambda)$, usually the perfect reflecting diffuser, is 100. (Similar equations could, of course, be written for transmittance, but are omitted for simplicity.) The equations express the schematic diagram of Figure 5 in quantitative terms.

Standard Observer.   The terms $\bar{x}(\lambda)$, $\bar{y}(\lambda)$, and $\bar{z}(\lambda)$ were derived by the CIE from data obtained in visual experiments using the foveal region of the retina, in which observers matched the colors of the spectrum by mixtures of three selected lights termed primaries, usually red, green, and blue. The data were transformed to correspond to unreal primaries in order to achieve mathematical simplification. The average results for human observers were defined as the CIE 1931 2° standard observer. In 1964 a slightly different set of data, obtained in experiments using the parafoveal region, was adopted as the CIE 1964 10° supplementary standard observer. The use of data for one of the standard observers in calculating tristimulus values is customary, and the tristimulus values then refer to the color as seen by that standard observer.

The transformation of data was arbitrary to the extent that many different but consistent sets of standard-observer data could be specified. The particular one recommended by the CIE in 1931 has the advantage that $\bar{y}(\lambda)$ was made identical to the spectral photopic luminous efficiency function $V(\lambda)$. Thus to a first approximation tristimulus value Y, often called the luminous reflectance, is a measure of the lightness of objects.

The CIE standard observer functions are shown in Figure 10. Values for the 1931

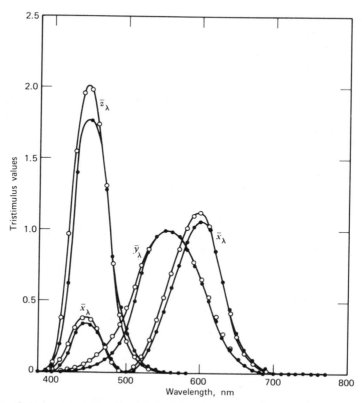

Figure 10.   Color matching functions of the CIE 1931 2° standard colorimetric observer, –●–; and the CIE 1964 10° supplementary standard colorimetric observer, –O– (11).

standard observer, at 20-nm intervals, are given in Table 2. More complete tables are given in ref. 25 and the General References, but the values given should suffice for most computations.

Table 2.   Color-Matching Functions of the CIE 1931 2° Standard Colorimetric Observer, Tabulated at 20-nm Intervals of Wavelength[a]

Wavelength, nm	$\bar{x}$	$\bar{y}$	$\bar{z}$
400	0.0143	0.0004	0.0679
420	0.1344	0.0040	0.6456
440	0.3483	0.0230	1.7471
460	0.2908	0.0600	1.6992
480	0.0956	0.1390	0.8130
500	0.0049	0.3230	0.2720
520	0.0633	0.7100	0.0782
540	0.2940	0.9540	0.0203
560	0.5945	0.9950	0.0039
580	0.9163	0.8700	0.0017
600	1.0622	0.6310	0.0008
620	0.8544	0.3810	0.0002
640	0.4479	0.1750	0
660	0.1649	0.0610	0
680	0.0468	0.0170	0
700	0.0114	0.0041	0

[a] Ref. 25, General References.

Standard Sources and Illuminants.   The CIE recommends the use of standard illuminants for the calculation of tristimulus values, which therefore (with the use of standard observer data) represent the colors of objects represented by reflectance factors $R(\lambda)$ when illuminated by a standard illuminant and seen by a standard observer. The 1931 CIE recommendations were in the form of three standard sources, whose spectral power distributions became the corresponding standard illuminants: source A (Fig. 3) is an incandescent lamp operating at a color temperature of 2856 K; sources B and C are obtained by combining source A with liquid filters providing light of the approximate quality of noon sunlight and north-sky daylight, respectively. However, sources B and C lack the ultraviolet content of natural daylight, and the increasing use of fluorescent brightening agents made it advisable for the CIE to add to its recommendation in 1964 a series of illuminants based on natural daylights of various correlated color temperatures, the most important of which is D_{65} (Fig. 3). Unfortunately there are not yet sources recommended by the CIE corresponding to the daylight D illuminants.

Throughout most of the history of basic colorimetry, illuminants A and C have been most widely used. It is the expectation of the CIE, however, that A and D_{65} should suffice. Values of relative spectral power for standard illuminants A, C, and D_{65} at 20-nm intervals are given in Table 3. More complete tables are given in ref. 25 and the General References, but the values given should suffice for most computations.

Chromaticity Diagram.   The most widely used graphical representation of the results of basic colorimetry is the CIE 1931 chromaticity diagram (Fig. 11) obtained by plotting the chromaticity coordinates:

$$x = X/(X + Y + Z)$$
$$y = Y/(X + Y + Z)$$

Table 3.   Relative Spectral Power Distributions of CIE Standard Illuminants A, C, and D_{65}, Tabulated at 20-nm Intervals of Wavelength[a]

Wavelength, nm	S_A	S_C	$S_{D_{65}}$
400	14.7	60.1	82.8
420	21.0	93.2	93.4
440	28.7	115.4	104.9
460	37.8	116.9	117.8
480	48.3	117.7	115.9
500	59.9	106.5	109.4
520	72.5	92.0	104.8
540	86.0	97.0	104.4
560	100.0	100.0	100.0
580	114.4	92.9	95.8
600	129.0	85.2	90.0
620	143.6	83.7	87.7
640	158.0	83.4	83.7
660	172.0	76.8	80.2
680	185.4	79.8	78.3
700	198.3	72.5	71.6

[a] Ref. 25, General References.

(The similarly defined chromaticity coordinate z is not independent since $z = 1 - x - y$.) The figure shows the location of the spectrum colors on the horseshoe-shaped spectrum locus, and the chromaticities of illuminants A, C, and D_{65} near the locus of blackbody chromaticities (Planckian locus). Purples, made by mixing red and blue light, do not occur in the spectrum. Their chromaticities lie on the line joining the ends of the spectrum locus, called the purple line.

The function of basic colorimetry is to define the conditions under which two colors match. In terms of the chromaticity diagram, it can be said that two colors with the same chromaticities x, y and the same luminous reflectance Y match. The chromaticity diagram thus provides no information about the appearance of such colors, or the difference in appearance of colors that do not match. Nevertheless, it is common practice to assume that locations on the diagram represent specific colors when illuminated by a standard illuminant (usually C) and observed by the 1931 standard observer whose vision is adapted to that illuminant. With these restrictions in mind, useful results based on the assumption can be obtained.

Relation to Metamerism.   In terms of the CIE system, the definition of metamerism is straightforward and revealing: two colors seen against a single background match if their tristimulus values are identical for a given illuminant and observer. If they no longer match (their tristimulus values are no longer identical) after a change to a different illuminant or observer, it follows that they must have different spectral reflectance-factor curves $R(\lambda)$. Conversely, metamerism may be avoided by requiring that the two objects have the same spectral character (see Color Measurement).

Psychophysical Color Terms.   Psychophysical color terms are variables of perceived color that denote objective measures of physical variables, ie, they identify stimuli which, when viewed on the same background, produce equal values of the perceptual variables (4). The psychophysical correlate of brightness is the luminance, the photometric term previously defined. Since lightness is relative brightness, it follows that its psychophysical correlate is the luminance factor $\beta = Y/Y_n$, where Y is the luminous

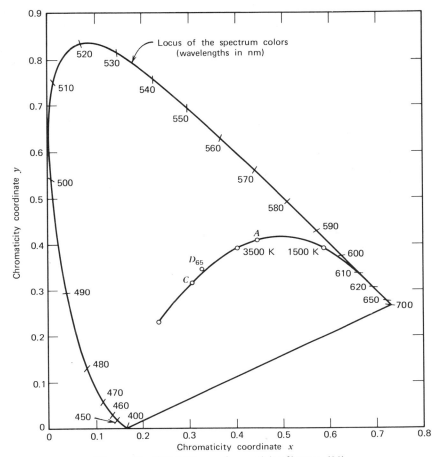

Figure 11. CIE 1931 x, y chromaticity diagram (26).

reflectance of the object and Y_n, normally 100, is the luminous reflectance of a suitably chosen reference white.

The psychophysical correlate of hue is dominant wavelength or complementary wavelength, whichever is appropriate, defined by the following operation on a chromaticity diagram. A line is drawn from the chromaticity of the illuminant through the sample chromaticity to the spectrum locus. The spectrum wavelength at that point is the sample's dominant wavelength, λ_d. If the line intercepts the purple line instead of the spectrum locus, it is continued back from the chromaticity of the illuminant to the spectrum locus, and the spectrum wavelength at the point of interception is the sample's complementary wavelength, λ_c.

There are no psychophysical terms correlating directly with the perceptual terms saturation, perceived chroma, or colorfulness. A related psychophysical quantity, excitation purity, is defined in terms of the above operation but has no corresponding perceptual concept; ie, excitation purity p_e is defined as the length of the line segment between the illuminant and sample points divided by the length of the segment between the illuminant point and the spectrum locus (or purple line).

Together, hue and one of the perceptual terms for the strength of chromatic response define the chromaticness of the sample, whose psychophysical correlate is the chromaticity, x, y.

Advanced Colorimetry

Whereas basic colorimetry is concerned with whether or not colors match, advanced colorimetry addresses such essentially unresolved problems of great complexity as the description of the differences between colors that do not match, and the appearance of colors for a variety of simple and complex viewing conditions.

Uniform Color Spaces. It has long been recognized that the 1931 CIE system is by no means equally spaced in terms of visual perception. Attempts to modify the CIE 1931 X, Y, Z or x, y, Y spaces to achieve improvement in this respect began in the 1930s and have continued to the present through a complex history (2–3). In 1976 the CIE recommended use, for uniformity of practice until more nearly uniform color spaces are derived, of one of the two spaces, CIELUV and CIELAB space (27).

CIE 1976 L*u*v* (CIELUV) Space. This transformation retains a feature of CIE 1931 space which is of importance to those concerned with color reproduction, such as the television, photographic, and graphic arts industries: In the CIE 1931 x, y diagram the chromaticities of additive mixtures of color stimuli lie on the straight line connecting the chromaticities of the stimuli themselves. The CIELUV chromaticity diagram is a linear transformation of the x, y diagram that retains this feature.

Chromaticity coordinates u', v' of CIELUV space are defined as follows:

$$u' = 4 X/(X + 15 Y + 3 Z)$$

$$v' = 9 Y/(X + 15 Y + 3 Z)$$

These coordinates are, by definition, the psychometric correlates of perceptual chromaticness and psychophysical chromaticity coordinates x, y. Lightness in the CIELUV system is scaled so as to correspond closely to Munsell value, and is designated psychometric lightness L^*:

$$L^* = 116 (Y/Y_n)^{1/3} - 16, \quad Y/Y_n > 0.01$$

Ref. 28 provides an alternative formula for $Y/Y_n < 0.01$. A three-dimensional space is generated as follows. First, the origin in the u', v' diagram is shifted to the chromaticity of the reference white, u'_n, v'_n; the resulting quantities $u' - u'_n$ and $v' - v'_n$, which can be either positive or negative, indicate difference from the neutral chromaticity axis which now has coordinates 0,0. Next, these shifted chromaticity coordinates are multiplied by $13 L^*$ to provide coordinates which, unlike u' and v', decrease to zero at the black point, at which X, Y, and Z are all zero. The three coordinates of the CIELUV space are thus L^* and

$$u^* = 13 L^* (u' - u'_n)$$

$$v^* = 13 L^* (v' - v'_n)$$

These are opponent coordinates in that positive values of u^* indicate redness (negative values, greenness); and positive values of v^* indicate yellowness (negative values, blueness).

CIE 1976 L* a* b* (CIELAB) Space. This space incorporates a nonlinear transformation of CIE 1931 space to yield opponent coordinates $a*$ and $b*$ that, just as $L*$, correlate reasonably well with Munsell coordinates:

$$a* = 500\,[(X/X_n)^{1/3} - (Y/Y_n)^{1/3}]$$

$$b* = 200\,[(Y/Y_n)^{1/3} - (Z/Z_n)^{1/3}]$$

The restrictions X/X_n, Y/Y_n, $Z/Z_n > 0.01$ apply, but see ref. 28. Positive values of $a*$ denote redness (negative values, greenness); positive values of $b*$ denote yellowness (negative values, blueness). Although the CIE makes no statement, CIELAB space is expected to be favored by most users concerned with the industrial coloring of such materials as paint, paper, plastics, and textiles (29).

Psychometric Color Terms. A complete list of psychometric color terms provides objective measures of physical variables evaluated so that equal scale intervals represent approximately equal perceived differences in the attribute considered (4). In addition to CIE 1976 psychometric lightness $L*$ and psychometric chromaticity u', v', there are the following:

CIE 1976 psychometric hue angle is the correlate of the perceptual quantity hue:

$$h^*_{uv} = \tan^{-1}(v*/u*); \quad h^*_{ab} = \tan^{-1}(b*/a*)$$

CIE 1976 psychometric chroma is the correlate of perceived chroma:

$$C^*_{uv} = (u*^2 + v*^2)^{1/2}; \quad C^*_{ab} = (a*^2 + b*^2)^{1/2}$$

Psychometric saturation is the correlate of the perceptual quantity saturation:

$$s_{uv} = 13\,[(u' - u'_n)^2 + (v' - v'_n)^2]^{1/2}$$

In addition, psychometric purity is defined just as (psychophysical) excitation purity except that the distances involved are measured on the u', v' chromaticity diagram.

An alternative set of psychometric quantities exists in principle in the coordinates Munsell hue, Munsell value, and Munsell chroma. However, with the exception of Munsell value, which is well represented by CIE 1976 psychometric lightness $L*$, the Munsell coordinates are not related to CIE coordinates in closed form. Graphical interpolation (30), which can be accomplished by computer program (31), is required to transform one to the other (Fig. 11).

Color Differences. Judgment of the direction and size of the difference in color between two objects is of considerable industrial importance. For more than forty years there have been many efforts to derive equations by means of which color differences may be calculated from CIE tristimulus values, in good agreement with visually perceived color differences. Despite a long and complex history (2,32), the search for a single equation providing good results for all colors has been totally unsuccessful, and it seems inevitable that this will remain an unsolved problem for a long time.

Visual Color-Difference Data. It is clear that a major part of the color-difference problem is the difficulty in generating precise, consistent, and pertinent visual data on color differences. In the past, such data arose from three types of experiments (2); studies of relatively large color differences among samples illustrating color-order systems based on equal visual perception, such as the Munsell system; studies of the spacing of colors adjunct to the development of uniform color spaces and chromaticity diagrams; and studies of very small color differences by measurement of the standard deviation of color-matching (33–34) or of color-difference perceptibility (35). Because of the nature of color vision and the difficulty of the psychometric scaling experiments required (36), the data are not precise. The studies of large and small color differences do not appear to be consistent, and it is not certain that the studies to date are pertinent to the industrial problem involving moderate color differences. Clearly, much research remains to be done in directions that are only becoming defined (37).

Color-Difference Equations. Sets of visual color-difference data are used to test equations by which color differences may be calculated, usually from CIE coordinates or derived psychometric quantities, but occasionally from other coordinates such as those of the Munsell system. Most of the equations are in Euclidean form, expressing color difference as the square root of the sum of squares of differences in three (assumed) orthogonal color coordinates; it is not certain, however, whether this is an adequate representation. Many equations are based on current or past theories of color vision, and it is hoped that their performance will improve as such theories become better. In recent years, approximately twenty different equations have been used to various extents, and no single one is favored by a majority of users. The linear correlation coefficients between calculated and perceived color differences are usually no higher than about 0.7 for any equation or data set, ie, they account for roughly half of the variability in the visual data.

CIE 1976 Recommendations. In 1976 (27,29) the CIE recommended use of one of two equations based on the uniform color spaces described above for uniformity of practice until improved equations are adopted. Using ΔE to represent color differences, the equations are:

$$\Delta E_{uv}^{*} = [\Delta L^{*2} + \Delta u^{*2} + \Delta v^{*2}]^{1/2}$$

$$\Delta E_{ab}^{*} = [\Delta L^{*2} + \Delta a^{*2} + \Delta b^{*2}]^{1/2}$$

Although the CIE has made no statement, it is expected that most industrial users of color-difference equations will adopt ΔE_{ab}^{*}, based on the CIELAB space, and only those who require use of a chromaticity diagram such as that associated with CIELUV space will prefer the use of ΔE_{uv}^{*}.

Other Unsolved Problems. *Chromatic Adaptation.* Color stimuli presented to the eye vary drastically with the conditions under which they are perceived, such as the spectral nature and luminosity of the illuminant. The perception of a given stimulus varies markedly with the nature of all the other stimuli in the visual field, yet the eye adapts to all these variations in such a way that the appearances of colors remain remarkably constant. A full understanding of color appearance requires solution of the complex problems of chromatic adaptation. Recent progress (38–39) is encouraging, but much work remains.

Indexes of Metamerism. A 1971 CIE recommendation (40) has established a special index of metamerism (change of illuminant), based on calculation of the color difference between two metameric samples, matching under a reference illuminant, when they are illuminated by a test illuminant. No recommendation has yet been made for an index of metamerism on change of observer. Although metamerism results from differences in the spectral nature of the samples, no attempts to formulate general indexes of metamerism based on these differences have been successful.

Indexes of Whiteness. Despite the successful derivation of many single-number color scales, including indexes of yellowness (41), none of many indexes of whiteness developed to date (42) has been successful enough to warrant a CIE recommendation. The problem is complicated by the necessity of including fluorescent white textiles and papers, and by strong and conflicting national and industry feelings on the nature of a preferred white which could be used as the basis of a whiteness scale, and of preferred hues in samples deviating from such a base white.

Color Measurement

In view of the earlier discussions of basic colorimetry and color mixing, the purpose of color measurement becomes clear: it is to provide a permanent, objective measure of the spectral character and color coordinates of samples. Some geometric considerations are appropriate before discussion of the instrumentation and methodology involved. An advantage of visual judgment of color and color difference not always appreciated is the ability of the observer to examine samples in a wide variety of geometric conditions of illumination and view. Instruments are, of course, limited in this respect. Only rarely does an instrument allow use of more than one geometric condition. In 1968, following industrial practice, the CIE recommended use of one or more of the four illuminating and viewing geometries for reflectance-factor measurement illustrated in Figure 12. To date, the use of the integrating sphere has been favored in spectrophotometers, and of 45°/0° (angle of illumination/angle of view, measured from the normal to the surface of the sample) geometry in tristimulus colorimeters, these being the two major classes of color-measuring instruments.

Spectrophotometry. Spectrophotometers for color measurement differ from visible-range analytical spectrophotometers in several respects. An obvious difference is the use of sample geometry suitable for the measurement of reflectance (usually diffuse reflectance, with an integrating sphere) and diffuse transmittance. A second requirement is that CIE tristimulus values should be computed simultaneously with the measurement; modern instruments invariably are coupled with a computer or equipped with a microprocessor to carry out these computations as well as control the operation of the instrument. Finally, the gently sloping spectral curves of most colored objects, combined with the requirement of high precision in reflectance (0.1–0.01% where possible) and tristimulus values (0.01–0.02) dictate substantially different operating parameters for the two types of spectrophotometers: for color measurement the spectral bandpass is typically broad, 5–10 nm or more, and more than usual attention is paid to photometric resolution and accuracy. With broad spectral bandpass, abridged spectrophotometry in which the spectrum is sampled only at intervals of, say, 20 nm has been favored (as in Tables 2 and 3) (see Analytical methods).

For the measurement of fluorescent samples, one should ideally use the equivalent of a spectrofluorimeter, with dual monochromators in the illuminating and viewing beams. No such instruments allowing reflectance measurement are commercially available. Alternatively it is necessary to illuminate the sample with polychromatic light, with appropriate spectral power distribution, and with the monochromator in the viewing beam. This measurement gives the spectral radiance factor. A full analysis of fluorescence requires additional measurements and calibrations (43) (see also Luminescent materials).

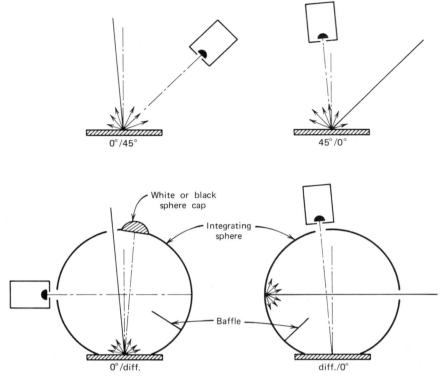

Figure 12. CIE recommended geometries for illumination and viewing for reflectance-factor measurement (9).

Typical Color-Measurement Spectrophotometers. Among the spectrophotometers designed for color measurement available in the United States in 1977 were the following: (*1*) The Diano-Hardy spectrophotometer, based on Hardy's design of the 1930s, was long the referee instrument for color measurement. It is a double-beam integrating sphere spectrophotometer with a Martens polarization photometer and a double-prism monochromator in the illuminating beam. A modification, the Diano-Hardy II, has direct sample illumination with a simulated daylight source and the monochromator in the viewing beam; all the following instruments have these features. (2) The Diano Match-Scan is a dual-beam integrating-sphere grating spectrophotometer using reflecting optics. (*3*) The Hunter Associates Laboratory D54 is a single-beam integrating-sphere spectrophotometer utilizing a circular interference wedge as a monochromator. (*4*) The Kollmorgen MS-2000 and MS-2045 are, respectively, an integrating-sphere and a 45°/0° spectrophotometer utilizing illumination with a pulsed-xenon arc simulating daylight, and a grating monochromator which, however, is utilized as an abridged spectrophotometer with 18 detectors spaced at 20 nm intervals across the spectrum. (*5*) The Zeiss DMC-26 is a versatile modular reflectance grating spectrophotometer offering a variety of geometries and measuring conditions.

Tristimulus Colorimetry. In tristimulus colorimeters, an optical analogue technique is used to produce instruments that allow direct reading in approximate CIE tristimulus values and derived coordinates. Instead of a monochromator, these instruments utilize four broadband filters which, when combined with the spectral power distribution of the source used and the spectral responsivity of the detector used, combine to give the equivalent of a CIE standard source, almost invariably C, and the 1931 CIE standard observer. Four filters are required, simulating, respectively, the $\bar{y}, \bar{z}$, and the two peaks of the $\bar{x}$ standard-observer functions. Because of limitations in filter glasses and other components, the absolute accuracy of these instruments is limited. The inaccuracies cancel, however, when results for nonmetameric samples of similar color are compared. They are thus properly used as, and often called, color-difference meters.

Typical Tristimulus Colorimeters. Among the tristimulus colorimeters available in the United States in 1977 were the following: (*1*) The Gardner Laboratory series, including the XL-20 (direct reading), the XL-23 (with microprocessor for simple transformations of color scales and calculations), and the XL-31 (coupled to a programmable minicomputer), utilizes 45°/0° geometry, although other versions with 0°/diffuse and special geometries are available. (*2*) The Hunter Associates Laboratory D25 series includes several measuring units with a variety of 45°/0° and 0°/diffuse geometries, and several processor units with a variety of computational capabilities. (*3*) The Kollmorgen MC-1010 is termed a colorimeter but is in fact identical in hardware to their MS-2045 but more limited in software. Unlike true tristimulus colorimeters, it is sensitive to the spectral character of the sample and can detect metamerism.

Other Instruments. In addition, there are spectroradiometers (Gamma Scientific, International Light, United Detector Technology), whose function is to measure the spectral power distributions of sources, and goniophotometers (Hunter Associates Laboratories, Zeiss), whose function is to measure luminous (rarely, spectral) reflectance as a function of illuminating and viewing angles. The latter class includes simple abridged instruments measuring specular reflectance, ie, glossmeters (Gardner Laboratory, Hunter Associates Laboratories).

Standardization and Appropriate Use of Instruments. No instrument can be expected to give satisfactory results without careful standardization and careful attention to the preparation of the samples used. For color measurement by spectrophotometry, particular care is required in calibration of wavelength (44) and photometric (45) scales and in the selection of a white reflectance standard calibrated in terms of the perfect reflecting diffuser (46). National standardizing laboratories offer sets of filters (47–48) and tiles (49) for checking performance and operating parameters. Special diagnostic tiles are available for tristimulus colorimeters (50). A collaborative reference program is available (51) to ensure continuing satisfactory performance of instruments.

Industrial Application of Color Science

Color Matching. A major concern of industries producing colored materials, eg, paint, paper, plastics, and textile industries, is to reproduce in their material the color effects desired by customers, stylists and designers. Avoidance of metamerism is a prime objective of this work, and although this can be accomplished by purely visual techniques (52), the fundamental approach is to reproduce the spectral reflectance curve of the sample to be matched. This requires the use of the same colorants, applied in the same manner in the same material as the original. Spectrophotometry is an indispensible tool for this work.

By utilization of the appropriate color-mixing laws within the framework of basic colorimetry, it is possible to formulate, ie, select colorants and determine the required concentrations by computer techniques based on the results of color measurement (53–54). A decade of experience has adequately demonstrated the feasibility of such techniques as aids to visual color matching, and the decision as to whether they should be utilized rests more often on economic than on technological considerations.

Color Control.   Once a matching formulation for a desired color effect has been achieved, questions of color control in production become important. Many of these involve no more than sound techniques of quality control. Often, color science enters only in the final assessment of the acceptability of the product. Although visual judgment must never be bypassed, instrumental color measurement provides a valuable objective record and, through the use of tolerance charts, a history of performance which can be a major aid in maintaining a high record of satisfactory production.

Color-tolerance charts may be used in the form of diagrams in which differences in color coordinates are plotted for successive production batches of a given color. If, eg, CIELAB space is selected as an approach to visual uniformity, two charts should be used, Δa^* vs Δb^* (where Δ indicates difference between standard and production batch) and ΔL^* vs either Δa^* or Δb^*, whichever spreads out the data more. As experience is gained with plotted data for both acceptable and unacceptable batches, it should be possible to draw tolerance figures or set limits within which a product acceptable for color may be expected. Because no color space is completely perceptually uniform, and because acceptability is a producer–consumer relationship which may not directly reflect color-difference perceptibility, the tolerance figures cannot be expected to be entirely regular, or the same from color to color. However, the experience gained with the use of a set of color-tolerance charts should make each future application easier.

Color Reproduction.   Color reproduction is the satisfactory reproduction of the colors in a complex scene by such means as photography, television, or the graphic arts (55). Although the nature of these processes allows the production of wide gamuts of color with the use of a very few colorants, it sets severe limits on what can be accomplished. Thus the objectives of color reproduction must be carefully defined and assessed to obtain optimum results (56).

Other Applications.   *Color in the Lighting Industry.*   The application of color science by lighting engineers and architects requires more attention. New light sources are being adopted, often for reasons of energy conservation, which produce novel or unusual color effects in merchandising (57). Means of assessing these effects require further study (58).

Colors for Signalling.   The use of colors for signal lights and signs is well established (59), but challenging problems remain, eg, in the colorimetry of retroreflective sign materials, in fluorescent signals, the avoidance of confusion by color deficients, and in international agreement.

Nomenclature

absorbance, A
Beer-Lambert Law, $\log (1/T) = A = \Sigma_i a_i b c_i$
color difference, ΔE
concentration, c
illuminance, lx (or lm/m^2)

irradiance, W/m^2
length, b
luminance, ed/m^2 (or lx/sr)
luminance factor, $\beta = Y/Y_n$
luminous flux, lm

luminous intensity, ed (or lm/sr)

luminous reflectance, Y

psychometric chroma, C^*

psychometric hue angle, h^*

psychometric lightness, L^*

psychometric saturation, s

radiance, $W/(sr \cdot m^2)$

radiant flux, W

radiant intensity, W/sr

reflectance factor, R

specific absorbance (absorptivity), a

transmittance, I

BIBLIOGRAPHY

"Color Measurement" in *ECT* 1st ed., Vol. 4, pp. 242–251 by G. W. Ingle, Plastics Division, Monsanto Chemical Company; "Color Measurement" in *ECT* 2nd ed., Vol. 5, pp. 801–812, by George W. Ingle, Monsanto Company.

1. D. B. Judd, *Color in Our Daily Lives,* NBS Consumer Information Series No. 6 (Superintendent of Documents Cat. No. CB. 53.6), U.S. Government Printing Office, Washington, D.C., 1975.
2. R. S. Hunter, *The Measurement of Appearance,* John Wiley & Sons, Inc., New York, 1975.
3. D. B. Judd and G. Wyszecki, *Color in Business, Science, and Industry,* 3rd ed., John Wiley & Sons, Inc., New York, 1975.
4. R. W. G. Hunt, *Color Res. Appl.* **2,** 55, 109 (1977).
5. R. M. Evans, *The Perception of Color,* John Wiley & Sons, Inc., New York, 1974.
6. F. E. Nicodermus, ed., *Self-Study Manual on Optical Radiation Measurements: Part I—Concepts, NBS Technical Note 910-1,* U.S. Department of Commerce, Washington, D.C., 1976, Chapters 1–3, esp. p. 58.
7. F. W. Billmeyer, Jr. in T. C. Laurin, ed. *The Optical Industry & Systems Directory,* 21st ed., Optical Publishing Company, Pittsfield, Mass., 1974, pp. E-10 to E-14.
8. *CIE Publication 38, Radiometric and Photometric Characteristics of Materials and their Measurement,* Bureau Central de la CIE, Paris, Fr., 1977 (U.S. National Committee CIE, c/o NBS, Washington, D.C.).
9. Ref. 3, p. 6.
10. P. W. Trezona, *Color Res. Appl.* **1,** 95 (1976).
11. F. W. Billmeyer, Jr., and M. Saltzman, *Principles of Color Technology,* Interscience Publishers, a division of John Wiley & Sons, Inc., New York, 1966.
12. L. Gall, *The Colour Science of Pigments,* BASF, Ludwigshafen, FRG, 1971.
13. P. Kubelka and F. Munk, *Z. Tech. Phys.* **12,** 593 (1931); P. Kubelka, *J. Opt. Soc. Am.* **38,** 448, 1067 (1948).
14. H. Grassmann, *Ann. Phys. Chem.* **89,** 69 (1853).
15. A. H. Munsell, *A Color Notation,* Munsell Color Company, Baltimore, Md., 1946–1961.
16. *Munsell Book of Color,* Munsell Color Company, Macbeth Division, Kollmorgen Corporation, Newburgh, N.Y., continuously available.
17. K. L. Kelly and D. B. Judd, *Color: Universal Language and Dictionary of Names, NBS Special Publication 440,* NBS Washington, D.C., 1977; A. F. Styne, *Color Res. Appl.* **1,** 79 (1977).
18. *ISCC-NBS Centroid Color Charts, Standard Reference Material No. 2106,* NBS, Washington, D.C.
19. D. L. MacAdam, *J. Opt. Soc. Am.* **64,** 1699 (1974).
20. *Uniform Color Scales Committee Samples,* Optical Society of America, Washington, D.C.
21. A. Hård, *Proc. 1st AIC Congress Color 69,* Musterschmidt, Göttingen, FRG, 1969, pp. 351–368.
22. E. Jacobson, *Basic Color: An Interpretation of the Ostwald Color System,* Paul Theobald, Chicago, Ill., 1948.
23. R. M. Johnston, *J. Paint Technol.* **43**(553), 42 (1971).
24. G. Wyszecki, *Proc. 2nd AIC Congress Colour 73,* Adam Hilger, London, Eng., 1973, pp. 21–51.
25. *CIE Publication 15, Colorimetry,* Bureau Central de la CIE, Paris, Fr., 1971 (U.S. National Committee CIE, c/o NBS, Washington, D.C.).
26. Ref. 3, p. 263.
27. *CIE Publication 15, Supp. 2, CIE Recommendations on Uniform Color Spaces, Color-Difference*

Equations, and Metric Color Terms, Bureau Central de la CIE, Paris, Fr., 1978 (U.S. National Committee CIE, c/o NBS, Washington, D.C.).

28. H. Pauli, *J. Opt. Soc. Am.* **66,** 866 (1976).
29. A. R. Robertson, *Color Res. Appl.* **2,** 7 (1977).
30. S. M. Newhall, D. Nickerson, and D. B. Judd, *J. Opt. Soc. Am.* **33,** 385 (1943).
31. W. C. Rheinboldt and J. P. Menard, *J. Opt. Soc. Am.,* **50,** 802 (1960).
32. G. Wyszecki in J. J. Vos, L. F. C. Friele, and P. L. Walraven, eds., *Color Metrics,* AIC/Holland, Soesterberg, The Netherlands, 1972, pp. 339–379.
33. D. L. MacAdam, *J. Opt. Soc. Am.* **33,** 18 (1943); W. R. J. Brown and D. L. MacAdam, *J. Opt. Soc. Am.* **39,** 808 (1949).
34. G. Wyszecki and G. H. Fielder, *J. Opt. Soc. Am.* **61,** 1135 (1971).
35. R. M. Rich, F. W. Billmeyer, Jr., and W. G. Howe, *J. Opt. Soc. Am.* **65,** 956, 1389 (1975).
36. W. S. Torgerson, *Theory and Methods of Scaling,* John Wiley & Sons, Inc., New York, 1958.
37. R. G. Kuehni, *Color Res. Appl.* **2,** 187 (1977).
38. C. J. Bartleson, *Proc. 3rd AIC Congress Color 77,* Adam Hilger, Bristol, Eng., 1978, pp. 63–96.
39. L. M. Hurvich in ref. 38, pp. 33–61.
40. *CIE Publication 15, Supp. 1, Special Metamerism Index: Change of Illuminant,* Bureau Central de la CIE, Paris, Fr., 1972 (U.S. National Committee CIE, c/o NBS, Washington, D.C.).
41. F. W. Billmeyer, Jr., *Mater. Res. Stand.* **6,** 295 (1966).
42. E. Ganz, *Appl. Opt.* **15,** 2039 (1976).
43. D. H. Alman and F. W. Billmeyer, Jr., *Color Res. Appl.* **2,** 19 (1977).
44. D. H. Alman and F. W. Billmeyer, Jr., *J. Chem. Educ.* **52,** A281, 284, 315, 318 (1975).
45. C. L. Sanders, *J. Res. Natl. Bur. Stand.* **76A,** 437 (1972).
46. W. Budde, *J. Res. Natl. Bur. Stand.* **80A,** 585 (1976).
47. H. J. Keegan, J. C. Schleter, and D. B. Judd, *J. Res. Natl. Bur. Stand.* **66A,** 203 (1962).
48. K. L. Eckerle and W. H. Venable, Jr., *Color Res. Appl.* **2,** 137 (1977).
49. F. J. J. Clarke in ref. 24, pp. 346–350; F. J. J. Clarke, *Print. Technol.* **13,** 101 (1969).
50. R. S. Hunter, *J. Opt. Soc. Am.* **53,** 390 (1963).
51. *MCCA-NBS Collaborative Reference Program on Color and Color Difference,* NBS, Washington, D.C.
52. W. V. Longley, *Color Res. Appl.* **1,** 43 (1976).
53. R. G. Kuehni, *Computer Colorant Formulation,* Heath, Lexington, Mass., 1975.
54. E. Allen in ref. 38, pp. 181–197.
55. P. Kowaliski in ref. 38, pp. 129–152.
56. R. W. G. Hunt, *The Reproduction of Colour,* 2nd ed., Fountain Press, London, Eng., 1975.
57. W. A. Thornton, *Light. Des. Appl.* **6**(8), 46 (1976).
58. M. B. Halstead in ref. 38, pp. 97–127.
59. *CIE Publication 2.2, Colours of Signal Lights,* 1974; *CIE Publication 39, Surface Colours for Visual Signalling,* Bureau Central de la CIE, Paris, Fr., 1977 (U.S. National Committee CIE, c/o NBS, Washington, D.C.).

General References

F. W. Billmeyer, Jr., and Max Saltzman, *Principles of Color Technology,* Interscience Publishers, a division of John Wiley & Sons, Inc., New York, 1966; valuable for those with no previous background in color science, particularly useful in an annotated bibliography covering the period up to 1965; a supplement to date is available from the author of this article.

R. S. Hunter, *The Measurement of Appearance,* John Wiley & Sons, Inc., New York, 1975; the only book available dealing with the broad topic of appearance, including color.

D. B. Judd and G. Wyszecki, *Color in Business, Science and Industry,* 3rd ed., John Wiley & Sons, Inc., New York, 1975; a classic now in its third edition, considerably expanded and better than ever.

W. D. Wright, *The Measurement of Colour,* 4th ed., Van Nostrand, New York, 1969; concerned with the principles, methods, and applications of colorimetry, but not (as the title might suggest) with reflectance measurement.

G. Wyszecki and W. S. Stiles, *Color Science—Concepts and Methods, Quantitative Data, and Formulas,* John Wiley & Sons, Inc., New York, 1967; perhaps the most important book on color science to appear in recent years from the standpoint of completeness.

FRED W. BILLMEYER, JR.
Rensselaer Polytechnic Institute

COLOR AND CONSTITUTION OF ORGANIC DYES.   See Cyanine dyes;
   Color.

COLORANTS FOR CERAMICS

Although ceramic colorants have been in use for many centuries, it was not until about 1800 that some order and understanding of the chemistry of ceramic colorants began to be developed (1). Even though today there exists a great body of information about color in ceramics, many aspects remain to be studied. An alphabetical list of each ceramic colorant, and its CAS Registry Number, referred to in the text is given at the end of the article.

Colorant Technology

Ceramic colorants can be divided into two broad groups: colorants used in conjunction with clay-based products such as dinnerware, tile, sanitaryware, etc, or colorants used in the manufacture of glass (usually containers) (see Ceramics; Glass). Because many clay-based products are coated with a glaze which is most frequently glassy in nature, there is obviously an overlapping involved. In addition one must also consider the coloring of porcelain enamels which are essentially vitreous coatings firmly bonded to a metal substrate such as steel, cast iron, copper, or aluminum (see Enamels). One can identify several distinct applications for colorants in the manufacture of ceramic products. These applications include: ceramic bodies; engobes; underglaze decoration; colored glazes; overglaze decoration; porcelain enamels; colored glasses (other than glazes or porcelain enamels); forehearth colors (intensely colored glass added to molten uncolored glass); and glass enamels (inorganic coatings, usually pigmented, applied on and fused to a glass substrate).

Although the same element or elements may be used to develop a given color in different types of products, many distinctly different colorant systems are used in the ceramic industry. In addition to the manufacture of ceramic products in which manufactured ceramic colorants are added in controlled quantity, many clay-base products are made from naturally occurring materials that develop a characteristic color during firing (2). Color effects can be controlled by adjusting the firing temperature, by proper selection of kiln atmosphere, by the addition of diluents, and by finer grinding. These comments apply most particularly to products such as bricks used in architectural applications.

Color in Glass.   In the manufacture of glass, wherein silica is the major ingredient, the presence of iron as an impurity imparts a decided colored tint under normal conditions (3). Since this discoloration is generally considered to be undesirable, decolorizers are used as components of the glass batch (3). If the iron can be oxidized through the use of additives such as As_2O_3, CeO_2, or MnO_2, the coloring effect is considerably diminished through chemical means. Alternatively, the effect of the iron can be negated through the addition of ingredients that themselves produce complementary colors in the glass. Materials such as selenium, cobalt oxide, neodymium oxide, and manganese dioxide are used for this physical decolorization; the addition of one of these materials produces a slightly darker but more neutrally colored glass which is more acceptable visually.

In clear glasses color is produced primarily by transition elements, alone or in combination (4). In effect, the glass, most often a silicate composition, acts as a solvent for the solute coloring ions which produce a color characteristic of the element(s) involved. For example, the addition of Fe_2O_3 to a glass (or its presence as an impurity) leads to the presence of Fe^{2+} and Fe^{3+} ions; a preponderance of Fe^{2+} ions in fourfold coordination with oxygen results in a pure blue color, and Fe^{3+} in sixfold coordination with oxygen causes a yellow-brown tint. Mixtures of the ions, of course, produce intermediate colors.

Color can also be produced by developing a colloidal suspension in a glass matrix (3). Metals such as copper (or possibly Cu_2O), gold, and silver, and pigmenting materials such as cadmium sulfoselenides are known to produce strong colors and are used commercially to this end. In systems of this nature a thermal process is employed (4). The glass is melted, formed, and cooled under controlled conditions (annealing) as is the case in all glass products. This is followed by a reheating to a temperature where nuclei of the elements or compound are formed; then the temperature is adjusted to permit these nuclei to grow to an optimum size for the development of a particular color. The nucleation and growth treatment is known as striking.

Translucent or opaque glasses, most often white, can be produced by the precipitation of a crystalline phase during the cooling of a glass, by separation of an immiscible vitreous phase, or by the precipitation of a crystalline phase from the glass during a secondary heating similar to the striking process mentioned above (3). In such instances the separated phase produced, having a different index of refraction than the vitreous matrix, creates the desired opacity. The greater the difference in indexes of refraction, the greater will be the degree of opacity. Opacified glass systems often contain fluorine or phosphates and develop a characteristic milky-white color in the resultant product. Titanium dioxide is an ingredient commonly used to develop a white opaque condition as a result of partial crystallization. In vitreous systems used as coatings, opacity is produced by similar means or by the inclusion of inert materials such as TiO_2, SnO_2, or $ZrSiO_4$ in the glassy matrix.

Physical Properties

Temperature Effects.   Both temperature and firing atmosphere can have marked effects on color (1). The latter is usually controllable. Temperature can be a limiting factor in the production of colors. It must be taken into account not only in the manufacture of the ceramic colorants, but also in uses for which there may be a maximum-use temperature. This condition implies, therefore, that a given colorant system, which might be designed for a glass enamel system fired at, say 570°C, would not necessarily be suitable for use as a body colorant. This temperature constraint can tend, therefore, to limit a widespread applicability of a given colorant system, especially one such as a combination of Cd, S, and Se. This limitation is particularly evident in glasses where combinations of these elements are used to produce red, orange, and yellow hues, depending on the amounts of each element present. As originally processed at ca 1500°C with decomposition of the cadmium sulfoselenides, the glass shows no definite color. If the same material is reheated at a lower temperature, however, the sulfoselenides can recombine to produce readily recognizable, intense colors (4). The temperatures involved in the manufacture of ceramic products preclude the use of organic colorants.

Matrix Effects. In addition to temperature, the chemical composition of the matrix in which the color is developed can have a major effect on the color produced (5). This is particularly true if the matrix is essentially vitreous as opposed to being predominantly crystalline. The presence of certain oxides may have extremely deleterious effects on some colorant systems. For example, combinations of Cr and Sn to produce colors ranging from pink through maroon work best in SrO-free or ZnO-free glazes having a high CaO content (5). On the other hand, combinations of Cr, Fe, and Zn, used to produce certain brown colors, work best with zinc-bearing glazes (5). Thus when the use of ceramic colorants is contemplated it is always advisable to think not only of the colorant system to be employed, but also of the matrix in which it will be accommodated.

Particle Size. The average particle diameter and particle size distribution of ceramic colorants can have a considerable effect on the color of ceramic products (1). Generally, if a colorant is reduced to too small a size, the resultant color will be weakened; the colorant will be more readily attacked by the matrix with a probable total loss of color. Commercial colorants are sold in a ready-to-use condition, but with the recognition that some additional comminution may result from the preparation of the total ceramic system. Although it is difficult to generalize owing to the wide range of compositions utilized as ceramic colorants, an average particle diameter of $1-7$ μm can be considered typical even though some red and gray colorants have average diameters <1 μm. When completely dissolved in a glass, the size of the colorant particles becomes less important.

Manufacture and Processing

Historically, ceramic colorant manufacturers have produced a wide variety of compositions in order to meet specific customer demands. On the other hand, consumer preferences for pastel colors, coupled with the existence of colorants that permit easy blending to produce differences in color tints, has enabled manufacturers to reduce the number of individual pigmenting systems somewhat. By combining ZrO_2 and SiO_2 with sources of vanadium, praseodymium, and iron, one can produce blue, yellow, and pink hues, respectively. These end-components can be mixed to yield a continuing gradation of color from one end-member to another (6). Blending of some other colorant systems is also possible.

A number of different designations is used for colorants for ceramics in the different branches of the ceramic industry. Colorants may be called color or enameling oxides even though the compounds formed may be zirconates, antimonates, or spinels. They may also be called glaze stains or body stains as well as underglaze or overglaze colors or glass colors. Although there is a difference in nomenclature, there can be a similarity in elemental make-up, perhaps with different proportions of the components.

Because of the inherent relatedness of ceramic colorants, it is possible to utilize similar manufacturing methods in their preparation. Even though individual unit operations may be varied, the basic process involved is reasonably uniform for all colorants (1). Close control is exercised in the selection of raw materials, which may be metallic oxides or some salts of appropriate metals, plus normally inert substances (from the point of view of not developing color by themselves) such as silica, alumina, zirconia, calcium carbonate, clays, borax, etc. Certain mineralizers, such as some of the alkali or alkaline earth halides, may be added to promote the formation of desired crystalline species in a subsequent thermal treatment. Each ceramic colorant is a separate entity, and the raw materials and their appropriate proportions are selected with this thought in mind. Depending on the colorant being prepared, both very narrow and relatively broad impurity tolerances are encountered.

Following a weighing operation, the raw materials are thoroughly blended. Although this blending may be accomplished in the dry state, a wet ball-milling may be used in order to achieve not only a more thorough mixing but also some size reduction. Wet milling also more readily permits the use of aqueous solutions of some metallic salts. These solutions also serve to impregnate carriers such as the inert substances mentioned earlier, so that when the water is subsequently eliminated, the maximum possible homogeneity has been attained (1). Besides the mixing that occurs during wet milling, some size reduction also takes place. This reduction is most desirable because smaller diameter particles with correspondingly larger surface areas react more readily during the calcination operation (see Size reduction).

The end-use coloring capabilities of the raw material mixtures are developed through a high temperature calcining operation (6). Depending on the composition involved, the temperatures utilized are 500–1400°C with kiln atmospheres oxidizing, neutral or reducing. During calcination volatile materials are driven off, and the colorant is developed through a sintering reaction. High temperatures are necessary not only to develop the colorants initially but also to stabilize them for later use. With some colorant systems, care must be taken not to overfire mixtures, or the desired end-effects will not be achieved. In addition to temperature, time is also a factor that must be taken into account. For economic reasons, including a concern for energy conservation, as short a time as possible to promote a complete chemical reaction is desirable. Calcination may be carried out on a continuous basis when large quantities of colorants are involved, or colorant compositions may be heated individually in periodic furnaces. Continuous firing is usually done in tunnel kilns, although lower temperature calcinations can be carried out in metal rotary-tube furnaces (see Furnaces). The containers used for holding the reactants may be refractory crucibles or saggers; these may be used open and exposed to the kiln atmosphere or they may be covered. Materials used for containers must be heat resistant and nonreactive with the materials being calcined. Often a set of containers is used only with a given family of colorants in order to prevent contamination from one mix to another. Considerable care must be taken in loading the refractory containers because too large a quantity in a given holder can affect packing density, hence thermal conductivity and degree of calcination. Further, attention must be given to the volatilization of some ingredients, especially strong colorants such as Cr_2O_3, which can collect on the kiln walls and contaminate other mixtures fired later. Some color concentrates (ie, colorants for forehearth colors) may be prepared in a glassy form by completely melting the raw material mixture and quenching it in air or water to produce a frit. Subsequent handling of such materials is usually confined to size reduction.

Following calcination the colorant is ground or milled to the optimum size desired. A variety of equipment, including ball mills, hammer mills, pan grinders, etc, is used. Size reduction may be carried out wet or dry (1). Wet milling is often used in instances where soluble materials may be present. If this operation is not used, or if soluble materials are not washed out, color variations may develop (1). In either case, grinding is continued until a desired average particle diameter or particle size distribution is attained. If wet-grinding is utilized, the material must be dried, usually in hot-air tray dryers or rotary drum dryers (see Drying). Dried materials are crushed (or powdered) and screened to break up aggregates. Finally, colorants may be blended either with other colorants to produce an intermediate hue (eg, the ZrO_2–SiO_2 family) or with inert materials, such as silica, alumina, kaolin, etc, to produce different tints or tones of the same hue.

The ultimate control utilized in the manufacture of ceramic colorants is the final appearance in an application. As noted earlier, colorants must be compatible with the matrices in which they will be used; therefore, a given colorant will, eg, be evaluated against a color reference standard in a specific glaze composition. Although a visual color match may be sufficient for some purposes, it is preferable to use some type of colorimeter or spectrophotometer for a quantitative determination of the parameters used for describing the color produced (see Color).

Economic Aspects

Ceramic colorants are seldom manufactured by the ultimate user. Owing to the limited market and the variety and complexity of the systems involved, only a small number of manufacturers (eg, Ferro Corp., Fusion Ceramics Inc., General Color and Chemical Co., Harshaw Chemical Co., Hercules Inc. (Drakenfeld), Mason Color and Chemical, Pemco Products Group (SCM Corp.), and The O. Hommel Co.) produce colorants for the ceramic industry; in addition, some colorants are imported. The proprietary nature of the industry precludes any definitive information, but an estimate of about 3000 metric tons annually seems reasonable. This figure does not include some of the same colorant systems that may be used in nonceramic applications (eg, plastics, rubber, vinyl tile, etc), premixed glass enamel systems, or oxides such as Cr_2O_3 and TiO_2 which are consumed in large quantities for ferrites (qv) or pigmentation for paper and paint, respectively (among their other applications). The average cost of ceramic colorants is ca $7.70/kg, but costs range from ca $4–40/kg or higher, depending on the elemental composition and the processing required during manufacture (7). The most expensive colorants are those containing gold or combinations of cadmium and selenium for the development of red hues. Darker reds are more expensive than lighter colors. Colorants for use in the glass industry are usually obtained in the same form as the materials used by the manufacturers of ceramic colorants. Some colorant manufacturers do, however, produce color concentrates for sale to the glass industry for use as forehearth colors (7). These materials are intensely colored glasses in themselves, and are blended with colorless molten glass in discharge channels of glass tanks. This technique permits the production of one or more colored glasses from a single source without the need for individual furnaces for each colored product made.

Health and Safety Factors

There are no serious problems with health or safety in the use and manufacture of ceramic colorants. Preventive measures to avoid inhalation of fine particulate matter should, of course, be observed. Some materials containing lead may be used in conjunction with ceramic colorants, but in such instances the lead is normally prereacted with silica to form a lead–silicate glass which, if used according to the manufacturer's recommendations, should not pose any great difficulties (5) (see Lead compounds, industrial toxicology). Materials containing toxic elements, such as mercury or tellurium, that can produce colors are not used commercially. Cadmium compounds, although hazardous, are used widely with proper precautions observed (5). Other good colorants containing uranium, which were used prior to World War II, are no longer widely available. Depleted uranium oxide is available but colorants produced from it are somewhat radioactive and unsafe for use (5).

Glazes used with colorants should be lead-safe. This is particularly important when using colorants such as those made from lead–antimony or cadmium–sulfur–selenium combinations.

Colorant Systems

A wide variety of elements is available from which to develop inorganic pigments for use in the ceramic industry. Cataloging these alone or in combination poses some difficulties, for the same element or combination of elements can—depending on ratios of elements, time, and temperature of calcination, atmosphere, matrix effects, and many other factors—yield considerably different visual effects. Because of this complexity, it seems best to consider colorants on the basis of end-hue produced rather than on an element-by-element basis. Most specific colorant compositions are considered proprietary, even though many formulations are published in the technical literature. Some of these, however, represent laboratory curiosities or patented ranges and do not always reflect commercial reality. Presented below are the component elements most widely used for a given hue by manufacturers of ceramic colorants; most of them are shown with their CAS Registry Numbers in the list at the end of the article.

Black. Most black ceramic colorants are formed by a calcination of oxides (usually) of selected elements that develop a spinel structure when heated. If spinel is considered to be of the form $A^{(2+)}B_2^{(3+)}O_4$, the divalent ion may be cobalt, manganese, nickel, iron, or copper, and the trivalent element either iron or chromium, or both (1). Because Cr_2O_3 and Fe_2O_3 form a solid solution, complex spinels, such as $Co(Cr,Fe)_2O_4$, may be developed during calcination. A copper–chrome spinel produces an excellent low temperature black (1) but is limited to applications <800°C. Generally a balance of the elements noted (excluding copper) is preferable to using any given pair. The selection of one or another can control the shade of black, ie, blue–black from a preponderance of cobalt, green–black from a preponderance of chromium, etc. Some use–matrix ingredients such as ZnO can also affect the shade, so zinc-bearing glazes for use in conjunction with the colorant are not advised (1). Most black pigments are stable to temperatures up to ca 1300°C.

In combination as spinels, the elements noted are used in most ceramic compositions except colored glasses. In this instance the loading of the glass with any of these oxides individually will produce a black glass, but manganese oxide alone is the colorant most often employed (3). A deep purple can be observed in slivers of the colored glass, but in normal thicknesses a satisfactory black color is apparent. It is usually preferable to use mixtures of chromium, iron, and manganese in approximately equal proportions; these elements are taken into solution by the vitreous matrix and produce a dense black color (4).

Blue. Cobalt is the major element used in producing blue colors for the ceramic industry. It has been used for centuries, and because of its intense coloring effect and the appealing colors produced will undoubtedly be used for a long time to come (1). In the past 25–30 yr another major colorant system for producing blue colors (zirconium–silica–vanadium) has been developed; but since the color tones produced are considerably different, both major systems enjoy considerable popularity.

In the preparation of a calcined pigment, cobalt is added as cobalt oxide, Co_3O_4, or in the nitrate, carbonate, or sulfate form. It is commonly added to aluminum oxide, aluminum hydroxide, and/or silica to form a cobalt aluminate spinel or cobalt silicate (6). Other additives, such as feldspar, calcium carbonate, lead oxide (1:1 or 3:4), and alkali nitrates are used in conjunction to adjust color tones or enhance brilliance. Both cobalt aluminate and cobalt silicate are used in a number of different applications because of their inherent stability (normal in-use temperature maximum is ca 1300°C) and intense coloring power. Generally a lesser amount of cobalt oxide is used with alumina than is used with silica.

Less intense blue colors tending towards turquoise can be obtained by a combination of zirconium, silicon, and vanadium oxides to which is often added an alkali halide such as NaF to serve as a mineralizer; this halide promotes the formation of zircon ($ZrSiO_4$) during calcination in the temperature range of 800–950°C. The vanadium required is often added as ammonium vanadate. The composition used to produce blue colorants is ZrO_2, 35–80%; SiO_2, 55–10%; V_2O_5, 3–17% (1). Alkali oxides are also added as a part of the colorant composition. This system is used extensively because a variety of tints can be obtained by varying: suspension matrices, atmospheric conditions during calcining, the type of zirconia used, etc.

Cobalt oxide can be used alone in many coloring applications, but preference is indicated for either the aluminate or silicate combination. This choice does not hold true, however, in the glass industry where cobalt oxide is preferred because of its ready solubility in siliceous glasses. Amounts of cobalt oxide as low as 0.01% produce decided colors (4). Appearances are modified by the alkali content of the glass, with potassium-bearing glasses being bluer than sodium-containing glasses. Boron in the glass shifts the color to a pinkish blue.

Blue colors in glass, including some glazes, can also be developed by the addition of copper oxide (3). Characteristically, the resultant color is bluish-green, somewhat between those colors produced by cobalt and chromium oxides when they are used separately.

Brown. A wide range of brown colors can be produced from combinations of chromium, iron, and zinc, most usually from equal parts of the oxide forms, although metal salts of these elements can be used as well (6). In order to produce modified tones or shades, additives such as alumina, manganese oxide, nickel oxide, and silica can be added. Calcination temperatures of ca 1250–1300°C promote the formation of spinels which are brown. Some chrome–iron–tin oxide combinations are also produced, but these are of far lesser importance than the chrome–iron–zinc combinations which produce a wide palette of colors and can be easily controlled in use.

In the manufacture of brown glasses, nickel oxide, or combinations of chromium and manganese have been used, but iron–sulfur combinations are much more common. It is presumed that the familiar brown–amber color is caused by iron sulfide (the iron present either as an impurity or an intentional additive) and sodium polysulfides which are stabilized by the addition of carbonaceous materials such as sugar, starch, and carbon (4). Various shades of color from deep brown to black can be produced as a function of the iron content.

Gray. Gray colors in ceramics can be obtained by use of one of the black colorants mentioned above with a suitable inert diluent (6). This practice can, however, lead to inhomogeneities so that it is preferable to use a specially compounded mixture in which the diluent is a part of the calcine, with the coloring elements similar to those used for preparing blacks. Beyond this approach, however, other combinations of elements may be preferred, eg, a blend of SnO_2 and Sb_2O_3 (1). This combination may also contain some vanadium or manganese for slight hue or value variations. Such mixtures are highly stable when calcined at 1250°C and are widely used. Cobalt, nickel, and zirconium; cobalt, nickel, and titanium; tin, cobalt, and nickel; or cobalt, chromium, and iron combinations also produce satisfactory gray colors. Some excellent gray colors can be obtained from simple mixtures of chromium and iron oxides (5).

Gray colors can be obtained in glass by introducing NiO. This oxide, which produces a smoky gray color, is used in the production of art glass. Silver halides are used in the production of some gray optical glass (3). Color centers are formed by striking the glass after its initial annealing; the number and size of the colloid particles present are critical in developing the desired color.

Green. Chromium oxide is the main ingredient in most green ceramic colorants (1) and is used to prepare most green-colored glasses (3). Although Cr_2O_3 alone can be used in a number of applications in ceramics, its tendency to volatilize (1) can lead to many problems; therefore, it is usually added during calcination with other ingredients to develop a greater stability. It is often used with cobalt oxide alone to produce a blue–green color or in combination with cobalt and aluminum oxides; these mixtures not only have good stability, but in addition the presence of small amounts of cobalt tends to produce modified shades of bright green as well. Combinations of calcium oxide with chromic oxide are also used for color modification. Silica may also be used as an ingredient in this colorant series. As is true with some other colorants, chrome-bearing combinations work best in a zinc-free environment (5).

A green colorant can be obtained directly by calcining a mixture of zirconium oxide, silica, and a source of vanadium, but it is preferable to blend blue and yellow colorants containing vanadium which has reacted with zirconia and silica (blue) or with zirconia (yellow). These colors are cleaner than those produced with Cr_2O_3 or cobalt chromite. Often a zirconia–praseodymium yellow is used; however, a tin–vanadium yellow can be substituted.

Green glasses can be obtained using only chromium oxide in solution in a siliceous matrix, although the Cr_2O_3 is not as soluble as some other colorants (4). The presence of Sb_2O_3 in the glass helps to preserve the true green color; without it the chromic oxide could convert to the anhydride and develop a yellow–orange color (3). Since this tendency exists, small amounts of copper can be added to a chrome-bearing glass to stop transmission in the red wavelength range. This composition modification is often utilized in the preparation of green signal lenses.

Copper, as noted earlier, develops a blue–green color in soda–lime–silica glass systems. The best copper color is obtained in glasses with a high percentage of magnesium oxide (4).

Pink. Several basic colorant families are available for use in the manufacture of products requiring a pink coloration. These can be enriched, darkened, or lightened through the use of additives. Probably the oldest composition series is a combination of chromium, tin, and calcium to form the so-called chrome–tin pinks. Depending on the ratio of chromium and tin oxides, plus the presence of calcium and other ingredients, eg, borax, silica, clay, etc, hues ranging from pink through crimson to lilac can be obtained from the basic coloring elements (5). Reaction products formed are tin sphene doped with Cr_2O_3 (pinks) or chromium-doped tin oxide (orchids and lilacs). The majority of pigments produced from this series is stable up to 1350°C.

Combinations of chromium and aluminum oxides will also yield a pink to ruby color (1). High calcination temperatures and the presence of a mineralizer such as boric acid are required to produce colors in this series. ZnO may also be associated with this combination. Another series, producing colors similar to the chrome–alumina pigments, is obtained from manganese–alumina combinations (6). These pigments are widely used in applications where higher processing temperatures are encountered, eg, in ceramic bodies or in underglaze decorations.

Another series, which yields colors ranging from peach to coral as well as shades approaching true pink, involves the use of iron in combination with zirconium and silicon. It has the composition range: ZrO_2, 35–80%; SiO_2, 55–10%; Fe_2O_3, 0.5–20% (7). Calcination temperatures of 700–1000°C depend on the composition. These pigments are most often used as glaze colorants with a zircon opacifier which helps to stabilize the color. They are also used with other zircon-base colors to produce the blended hues mentioned previously.

Very delicate pinks can be produced in completely (or nearly so) vitreous systems by colloidal suspensions of gold or selenium (3). Silver is also sometimes used in conjunction with gold. Gold is usually added as a batch component in the form of gold chloride, and through subsequent melting, cooling, and striking a colloidal suspension of gold is formed which gives a characteristic pink color to the glass. Excessive growth of the colloid destroys the ruby tint desired. Selenium, mentioned earlier as a decolorizer in glass, can also be used to produce a pink tint. It has been reported that elemental selenium must exist in company with selenides and/or selinite groups in order to produce its characteristic color. The presence of potassium oxide in the base glass composition, as opposed to the more often used sodium oxide, enhances the pink color.

Red. Iron oxide is one of the oldest red colorants and also one of the most powerful (1). In addition to its continuing use today as a naturally associated component in many clay bodies (brick, sewer-pipe, drain tile, etc), it is used as an additive to color otherwise noncolored ceramic bodies when a brick red hue is desired. It is also used as a main component for some reds in underglaze decorations. With appropriate adjustments in composition, some of the colorant systems described to produce pinks can also be used for reds, especially the chrome–tin and chrome–aluminum series (occasionally these colors are considered to be deep pinks or dirty maroons, rather than red).

By far the most prevalent red colorant series for lower temperature applications, ie, overglaze colors, glass enamels, porcelain enamels, and colored glazes, is a family of pigments based on cadmium, selenium, and sulfur (1). This series can also develop excellent maroon, yellow, and orange hues. Although inorganic pigments in this series are among the most expensive, they provide the means for obtaining at low temperatures bright, strong colors that are impossible to obtain by any other means. Unfortunately, they have the limitation of an upper-use temperature range of 800–900°C, and the systems are quite sensitive to temperature variations during manufacture and the firing of the desired products; therefore, close control is essential at all stages of their handling.

The colorants consist of solid solutions of CdS and CdSe. The bright yellow of CdS is shifted through orange to red and maroon by replacing sulfur with selenium (1). Dark reds are obtained at a combination of $CdS_{0.65-0.7}Se_{0.35-0.3}$ and the deepest maroon at about $CdS_{0.55}Se_{0.45}$. The cadmium sulfoselenide colorant systems are made from ingredients such as sulfur, elemental selenium, and cadmium carbonate or oxalate, precipitated from an alkaline solution and calcined at 600–700°C in the absence of air. Lower temperatures are used for light, bright colors. Because the fumes of cadmium and its compounds are poisonous, safety precautions during preparation and use are advisable.

Cadmium sulfoselenide colors are also used in a completely vitreous matrix to produce bright red and orange colors (4). A complete solution of the elements takes place during melting with the result that on cooling a relatively colorless glass results. When reheated, however, the cadmium sulfoselenide compounds are formed and give a decided coloration to the glass. A high selenium content promotes the most intense red, and with lesser quantities the color shifts through orange towards yellow. Such glasses are often designated selenium ruby glasses and are used for signal lenses, among other purposes. The presence of ZnO in the glass composition is necessary as a means of sulfur retention.

Red glasses can also be obtained through the use of gold or copper (4). Both of these systems develop as a result of a colloidal suspension of the metals in the vitreous matrix. For color development, striking of the glass is necessary; however, reheating at a high temperature causes an increase in size of the colloids with a resulting loss of color.

White.   In ceramic bodies white achromatic colors are developed as a result of the use of materials relatively free from impurities. However, most so-called white bodies have a slight cream tint resulting often from minor quantities of iron and titanium dioxide that are often associated with kaolins, ball clays, feldspar, and silica, the usual ingredients in most whiteware products. Iron is also a contaminant in many other materials as well; therefore, unless highly purified raw materials are used, some color will be apparent. Where maximum whiteness is desired, bodies are generally covered with a white glaze opacified with ingredients such as tin oxide, titanium dioxide, or zirconium silicate, all of which can be obtained in sufficient purity to eliminate color effects in the body (1). In glass enamels, whiteness is attained by adding TiO_2 to the low temperature glass coating. In porcelain enamels, a titania-bearing frit is used which, on remelting to bond the coating to the substrate metal, results in the crystallization of TiO_2, producing a bright, white surface (7).

Depending on the crystalline form of TiO_2 that develops, anatase or rutile, a yellowish-white or bluish-white cast may be observed. Porcelain enamels can also be opacified with SnO_2 and ZrO_2. In the past, antimony oxide was also used for this purpose, but it has been supplanted almost completely by the other oxides mentioned.

Opaque, white glasses were described earlier. Fluorine-bearing compounds used to provide opacity include fluorspar, cryolite, and sodium hexafluorosilicate (3).

Yellow.   A number of different compounds are in commercial use to produce yellow colors in ceramics; one of the most widely used is cadmium sulfide (5). The previous comments concerning cadmium sulfo-selenide reds are equally applicable here, except that the yellow pigments ordinarily do not contain selenium.

Another combination that gives somewhat deeper tones approaching orange is

Table 1. Alphabetical List of Ceramic Colorants Referred to in the Text

Colorant	CAS Registry No.	Molecular formula
alumina	[1344-28-1]	Al_2O_3
aluminum cobalt oxide	[1333-88-6]	Al_2CO_4
aluminum hydroxide	[21645-51-2]	$Al(OH)_2$
ammonium vanadate	[7803-55-6]	NH_4VO_3
anatase	[1317-70-0]	TiO_2
antimony oxide	[1309-64-4]	Sb_2O_3
arsenic oxide	[1327-53-3]	As_2O_3
borax	[1303-96-4]	$Na_2B_4O_7.10H_2O$
cadmium carbonate	[513-78-0]	$CdCO_3$
cadmium oxalate	[814-88-0]	CdC_2O_4
cadmium selenide	[1306-24-7]	$CdSe$
cadmium selenide sulfide	[11112-63-3]	$CdSSe$
cadmium selenide sulfide	[12214-12-9]	Cd_2SSe
cadmium sulfide	[1306-23-6]	CdS
calcium oxide	[1305-78-8]	CaO
cerium oxide	[1306-38-3]	CeO_2
chromium cobalt oxide	[12016-69-2]	Cr_2CoO_4
chromium copper oxide	[12018-10-9]	Cr_2CuO_4
chromium iron oxide	[12068-77-8]	Cr_2FeO_4
chromium manganese oxide	[12018-15-4]	Cr_2MnO_4
chromium nickel oxide	[12018-18-7]	Cr_2NiO_4
chromium oxide	[1308-38-9]	Cr_2O_3
cobalt carbonate	[513-79-1]	$CoCO_3$
cobalt iron oxide	[12052-28-7]	$CeFe_2O_4$
cobalt nitrate	[10141-05-6]	$Co(NO_3)_2$
cobalt oxide	[1308-06-1]	Co_3O_4
cobalt silicate	[13455-33-9]	$Co(SiO_4)_2$
cobalt sulfate	[10124-43-3]	$CoSO_4$
copper iron oxide	[12018-79-0]	$CuFe_2O_4$
copper oxide	[1317-39-1]	Cu_2O
iron manganese oxide	[12063-10-4]	Fe_2MnO_4
iron nickel oxide	[12168-54-6]	Fe_2NiO_4
iron oxide	[1309-37-1]	Fe_2O_3
iron oxide	[1317-61-9]	Fe_2O_4
iron sulfide	[12068-85-8]	FeS_2
manganese oxide	[1313-13-9]	MnO_2
magnesium oxide	[1309-48-4]	MgO
neodymium oxide	[1313-97-9]	Nd_2O_3
nickel oxide	[1313-99-1]	NiO
potassium oxide	[12136-45-7]	K_2O
rutile	[1317-80-2]	TiO_2
silica	[7631-86-9, 11126-22-0]	SiO_2
silver nitrate	[19582-44-6]	$AgNO_3$
silicon oxide	[7631-86-9]	SiO_2
sodium oxide	[1313-59-3]	Na_2O
strontium oxide	[1314-11-0]	SrO
tin oxide	[18282-10-5]	SrO_2
titanium oxide	[13463-67-7]	TiO_2
uranium oxide	[1344-59-8]	U_3O_8
vanadium oxide	[1314-62-1]	V_2O_5
zinc oxide	[1314-13-2]	ZnO
zircon	[14940-68-2]	$ZrSiO_4$
zirconium oxide	[1314-23-4]	ZrO_2
zirconium silicate	[10101-52-7]	$ZrSiO_4$

made from titanium, antimony, and chromium. This mixture, which may also include some nickel for minor modification, is used primarily for body colors and underglaze decoration because it is stable to ca 1300°C; however, it is also used as a coloring agent for porcelain enamels and glass enamels (7).

A combination of lead and antimony oxides (6) or metallic salts that is stable to ca 1100°C is useful in most ceramic applications in that very strong yellow colors can be achieved. Because some of the antimony may be volatilized during calcination, the pigment may be stabilized with small quantities of cerium and aluminum oxides (tin oxide is also useful for this purpose). Diluents may be added as toners in use, as well as small quantities of Fe_2O_3 to shift the resultant color towards orange.

Tin–vanadium yellows are usually prepared from tin oxide and a compound, such as ammonium vanadate, by calcining at 1100–1200°C (TiO_2 is often added to modify the resultant color). Sn–V systems are used extensively in glazes for ceramic bodies where a lemon-yellow appearance is desired. They are also used extensively in underglaze colors even though they are sensitive to firing conditions.

A companion system, also widely used as a glaze colorant, is made from zirconium oxide and ammonium vanadate (6). It may ultimately supplant the Sn–V yellow, for it can be made at a somewhat lower cost and has a wide firing range. A ratio of 95% ZrO_2:5% NH_4VO_3, calcined at a temperature of 1250°C, produces a pigment that leads to a pronounced yellow color. Because the pigment is somewhat sensitive to titanium, lead and boron, these elements should not be included when the Zr–V system is used. This system is also sensitive to particle diameter but relatively insensitive to firing conditions. Unlike the Zr–V–Si system for producing blue colors, where zircon formation is the key to the development of color, Zr–V yellows are developed with zirconium dioxide alone. If small amounts (±1%) of indium, gallium, or yttrium oxides are added, orange colors can be developed in the Zr–V system.

The last system comprises a combination of zirconium, silicon, and praseodymium (6). These pigments are similar to the Zr–Si–V blues described above, except that a substitution of Pr for V produces a strong, bright yellow color. Essentially the same quantities of the reactants are used in both series along with a mineralizer. Pigments of this family are most frequently used as underglaze colors and in colored glazes.

The main colorant for producing a very pure yellow in glass is CdS which is characterized by a sharp absorption edge (4). Low concentrations of silver, added as $AgNO_3$, also produce an unstable yellow color (4) if the glass is reheated in a manner similar to that applied to the Cd–SSe compositions mentioned earlier. When appropriate control of the oxidizing character of the melt is maintained, iron will produce a yellow-tinted glass if maintained in the Fe^{2+} ionic state (3). Finally, a light amber color can be produced from iron and sulfur in the presence of carbon (7).

An alphabetical list of ceramic colorants referred to in the text is given in Table 1.

BIBLIOGRAPHY

"Colors for Ceramics and Glass" in *ECT* 1st ed., Vol. 4, pp. 276–287, by W. A. Weyl and R. R. Shively, Jr., The Pennsylvania State College; "Colors for Ceramics" in *ECT* 2nd ed., Vol. 5, pp. 845–856, by W. A. Weyl, The Pennsylvania State University.

1. K. Shaw, *Ceramic Colours and Pottery Decoration*, 2nd ed., Maclanen and Sons, London, Eng., 1968.

2. R. W. Grimshaw, *The Chemistry and Physics of Clays,* 4th ed., Wiley-Interscience, New York, 1971, pp. 905–928.

3. S. R. Scholes and C. H. Greene, *Modern Glass Practice,* 7th ed., Cahners Publishing Co., Boston, Mass., 1975, pp. 302–329.

4. N. J. Kreidl in F. V. Tooley, ed., *The Handbook of Glass Manufacture,* Vol. II, Books for Industry, Inc., New York, 1974, pp. 976–987.

5. C. W. Parmalee and C. G. Harman, *Ceramic Glazes,* 3rd ed., Cahners Books, Boston, Mass., 1973, pp. 460–512.

6. L. E. Valdés-Gámez, *Color in Ceramic Glazes,* M.S. Thesis, Alfred University, Alfred, N.Y., 1971.

7. A. Burgyan and H. Lowery in T. C. Patton, *Pigment Handbook,* Vol. 2, John Wiley & Sons, Inc., New York, 1973, pp. 371–421.

EDWARD E. MUELLER
Alfred University

COLORANTS FOR FOODS, DRUGS, AND COSMETICS

The use of synthetic organic colorants (color additives) in food products in the United States was first regulated by an act of Congress on August 2, 1886. This act authorized the addition of coloring matter to butter. A second act on June 6, 1896, recognized coloring matter as a legitimate constituent of cheese (1).

In the Appropriations Act of May, 1900 for the Department of Agriculture (USDA), Congress recognized the use of coloring matter as a problem that might affect the health of the nation. The allocation of funds for the general expense of the Bureau of Chemistry included the following item: "To enable the Secretary of Agriculture to investigate the character of proposed food preservatives and coloring matters to determine their relation to digestion and health and to establish the principles which should guide their use." Under this authority, the Secretary of Agriculture issued several food inspection decisions relating to the coloring of foods.

The Federal Food and Drugs Act of June, 1906, brought the use of coloring matters in foods under government supervision. Previously, "harmless colors" were available and generally used, but the publicity given to the indiscriminate use of questionable products resulted in government legislation. For example, toxic colorants such as chrome yellow [1344-37-2] (lead chromate) and red lead [1314-41-6] (Pb_3O_4) had been used in foods and many cases of poisoning had been diagnosed as overindulgence in sweets (2).

Certified Colorants

Food Inspection Decision No. 76 of July 13, 1907 created certified colorants in an attempt to end the indiscriminate use of impure and unpermissible coloring matters in foods (3).

Seven colorants were accepted as harmless based upon a study of the colorants then in use for the coloring of foods. This investigation included a detailed and exhaustive search of the literature concerning the chemistry and toxicity of these coal-tar colorants, a study of the law of various countries and states regarding their use, and many chemical investigations in the Bureau of Chemistry laboratories (3). The following colorants were permitted. *Red shades:* Amaranth [915-67-3] (formerly FD&C Red No. 2), Ponceau 3R [3564-09-8] (formerly FD&C Red No. 1 and Ext. D&C Red

No. 15), and Erythrosine [16423-68-0] (now FD&C Red No. 3); *Orange shade:* Orange I [523-44-4] (formerly FD&C Orange No. 1 and Ext. D&C Orange No. 3); *Yellow shade:* Naphthol Yellow S [846-70-8] (formerly FD&C Yellow No. 1, now External D&C Yellow No. 7); *Green shade:* Light Green SF Yellowish [5141-20-8] (formerly FD&C Green No. 2); and *Blue shade:* Indigo Disulfo Acid [860-22-0] (now FD&C Blue No. 2).

Provision was made for the addition of new colorants under the Bureau of Chemistry's rules of selection (3). Colorants were subsequently added only after "appropriate pharmacological and toxicological tests had proven them harmless" (4). These color additives and their year of admission are:

Tartrazine [1934-21-0] (now FD&C Yellow No. 5), 1916; Yellow AB [85-84-7] (formerly FD&C Yellow No. 3), 1918; Yellow OB [131-79-3] (formerly FD&C Yellow No. 4), 1918; Sudan 1 [842-07-9], and Butter Yellow [60-11-7], (both removed from the list in 1919), 1918; Guinea Green B (formerly FD&C Green No. 1) [4680-78-8] 1922; Fast Green FCF [2353-45-9] (now FD&C Green No. 3) 1927; Ponceau SX [4548-53-2] (now FD&C Red No. 4) 1929; Sunset Yellow [2783-94-0] (now FD&C Yellow No. 6) 1929; and Brilliant Blue [2650-18-2] FCF (now FD&C Blue No. 1) 1929.

A system of voluntary certification was set up under the coal-tar colorant regulations of the 1906 act (3–4). Manufacturers submitted representative samples of batches of colorants to the FDA for analysis and approval. Batches which met the specification for identity and purity were assigned lot numbers by the FDA.

The certified color business was firmly established in the United States by 1907–1914. However, because it depended upon other countries for its supply of basic raw materials and intermediates such as 2-naphthol, phthalic anhydride, and dimethylaniline, a large proportion of the coal-tar colorants used were imported.

World War I abruptly halted the supply of raw materials and finished dyes, and forced the United States to produce coal-tar intermediates and colorants of sufficient purity to meet the standards set up by the USDA (3). Eventually, the supply and purity of the necessary intermediates increased and by 1937 ca 230 metric tons of certified food colorants were produced in the United States (2).

Federal Food, Drug, and Cosmetic Act of 1938. By 1938 the inadequacies in the law and regulations of 1906 were apparent. Changes were essential to increase protection of public health and welfare. These changes were effected in colorants used by the drug and cosmetic industries as well as in food colorant applications.

On June 25, 1938, Congress passed the Federal Food, Drug, and Cosmetic Act of 1938 which became effective on January 1, 1940 (5). This act dictated that only certified coal-tar colorants could be used in foods, drugs, and cosmetics. The use of any uncertified coal-tar colorant was considered an adulteration and, as such, a misdemeanor punishable by law.

Three groups of colorants were listed: (1) FD&C colorants were certifiable for use in coloring foods, drugs, and cosmetics. (2) D&C colorants were certifiable for use in ingested and externally applied drugs and cosmetics but not in foods. They constituted a group of dyes and pigments considered suitable for use in drugs and cosmetics for internal use if in contact with mucous membranes or ingested only occasionally. (3) External D&C colorants constituted a group not certifiable for use in products intended for ingestion, but considered suitable in externally applied products. They were not permitted for use on any mucous membrane.

Color Additive Amendments of 1960. This law amended the Food, Drug, and Cosmetic Act of 1938 (6). Under this new law the Secretary of Health, Education, and Welfare (HEW) is required to list separately color additives for use in foods, drugs, and cosmetics to the extent that these listed colorants are suitable and safe when used in accordance with published regulations.

Under the 1938 law coal-tar colors could not be used in foods, drugs, and cosmetics unless they were listed by the FDA as "harmless and suitable for use." The law also called for certification of batches of these listed colorants with or without harmless diluents.

A Supreme Court decision (7) defined harmless as harmless regardless of the quantity of coal-tar color being used. Thus the FDA had to decertify a colorant if any amount or concentration of the colorant caused harm, even if a lesser amount related to its actual use level was safe. Because of this "harmless *per se* principle" the FDA removed seven FD&C colorants from the list, and began to remove several D&C color additives, many of which were used in the production of drug and cosmetic products.

The need for a new law to permit the continued use of colorants was clearly stated by the Secretary of HEW in 1959 (8).

Provisions of the Color Additive Amendments of 1960. The 1960 act brought about some important changes: (*1*) Uniform criteria of admissibility. The law did away with the differences in legal requirements and treatment between the so-called coal-tar colorants and other colorants. (*2*) Safety of use principle. The requirement that colorants under the condition of use specified in the regulations are safe replaced the harmless *per se* interpretation formerly used. (*3*) Certification and exemptions from certification. The new law provided for listing and certification of batches of colorants as required under the old law, but also permitted the Secretary of HEW to grant exemptions from the requirement of certification where certification was not necessary to protect the public's health. (*4*) Effective date and transitional provisions. The bill became effective upon enactment and provided for provisional listings, pending completion of the scientific investigations needed as a basis for making determinations as to listing of such color additives under the new permanent provisions of the bill.

The Color Additive Amendments of 1960 also provided that the Secretary of HEW, in determining whether a colorant should be listed for use in foods, drugs, and cosmetics, consider the scientific data that establish safety under condition of use. Other considerations are probable consumption, cumulative effect, safety factors, and availability of any needed practicable methods of analysis for determining the identity and quantity of the pure dye, intermediates, and impurities contained in the colorants as well as the amount of colorant in or on any such food, drug, cosmetic, any substance formed in or on such food, drug, or cosmetic because of the use of the colorant.

The producers and users of the colorants were required to supply the necessary scientific data for permanent listing. The great expense of the pharmacological and chemical experiments necessary for inclusion in a petition for permanent listing limited the work to colorants economically important to the food, drug, and cosmetic industries. Many previously certifiable colorants were delisted by default (9–12).

The Color Additive Amendments of 1960 provided a $2\frac{1}{2}$ year grace period to colorants that were commercially available before it was enacted. These were provisionally listed so that testing could be completed to meet the new requirements of proof of safety. The amendments further provided that the Secretary of HEW had the power to postpone the original closing date "for such period or periods as he finds necessary to carry out the purposes" of the law.

The Cosmetic Toiletry and Fragrance Association (CTFA) (formerly the Toilet Goods Association), the Certified Color Manufacturers' Association (CCMA) (formerly the Certified Color Industry Committee), and the Pharmaceutical Manufacturers' Association, representing the manufacturers and users of color additives, immediately started to provide the required data. As joint petitioners they began submitting completed colorant petitions in 1965. They filed the last petition in 1968 and additional data requested in 1969.

At present, many years after the passage of the act, final determination has not been made on many of the provisionally listed colors. Action to permanently list the colorants has been plagued by postponements, court challenges, and new toxicity testing requirements. For example, at the onset of the toxicity testing program, in addition to the chronic rat studies required of all the colorants, 7-yr dog-feeding studies were initiated on four of the FD&C colorants to see if this longer term chronic feeding study would give more information on chronic toxicity than the 2–2.5-yr rat-feeding study then and still in use (13).

In March 1966, the FDA informed the Toilet Goods Association that it would not grant permanent listing to any of the provisionally listed certified colorants used for cosmetics without information about the formulation in which the colorants were used. The FDA held further consideration of permanent listing in abeyance until 1969 when the courts ruled that the agency could not require premarket clearance of cosmetics through its colorant requirements. In effect, "Color additives were held hostage" (13) until this ruling.

Status of Approved Colorants

All approved colorants are classified by the FDA as either permanently or provisionally listed. They are further classified as those that require certification and those exempt from certification.

Table 1 (14) shows certification figures for each product listed as an acceptable color additive during a five year period.

Permanently Listed Colorants Exempt from Certification

Tables 2–4 show all presently approved colorants that are permanently listed and exempt from certification. The list is divided into three categories of use: foods, drugs, and cosmetics, and includes the Code of Federal Regulations (CFR) section number and the use restrictions for each product.

Physical Properties of Certified Colorants

Table 5 gives the solubilities and fastness properties of the FD&C, D&C, and External D&C colorants.

Chemical Classification of Certified Colorants

Survey of Certified Food, Drug, and Cosmetic Colorants. Certified colorants may be classified into groups according to their chemical structures.

Nitro Colorants. There is the only one nitro dye certified. It is External D&C Yellow No. 7.

Table 1. Metric Tons of Colorants Certified for Sale in U.S. Annually, Fiscal Years Ending June 30

Colorants	1973	1974	1975	1976	1977
FD&C (Primary)					
FD&C Blue No. 1 [2650-18-2]	54.9	72.2	71.9	47.6	79.3
FD&C Blue No. 2 [860-22-0]	28.6	40.2	38.5	30.0	40.0
FD&C Green No. 3 [2353-45-9]	1.84	2.35	9.15	2.92	1.35
FD&C Yellow No. 5 [1934-21-0]	468	585	631	554	529
FD&C Yellow No. 6 [2783-94-0]	459	452	469	349	486
Orange B	14.2	8.64	14.1	9.28	18.1
FD&C Red No. 2[a] [915-67-3]	446	410	625	109	
FD&C Red No. 3 [16423-68-0]	104	130	153	113	246
FD&C Red No. 4[b] [4548-53-2]	7.22	12.8	15.9	1.89	2.53
FD&C Red No. 40 [25956-17-6]	256	331	358	522	646
Citrus Red No. 2 [6358-53-8]	0.743	0.225	5.77	0.795	
Subtotal	*1840*	*2040*	*2390*	*1740*	*2050*
FD&C (Lakes)					
FD&C Blue No. 1 [15792-67-3]	21.4	30.5	26.1	13.3	44.5
FD&C Blue No. 2 [16521-38-3]	8.57	15.5	16.2	14.8	33.7
FD&C Yellow No. 5 [12225-21-7]	197	224	228	170	340
FD&C Yellow No. 6 [15790-07-5]	78.3	85.4	107	53.6	187
FD&C Red No. 2[a] [12227-62-2]	15.8	22.7	18.5	1.76	
FD&C Red No. 3 [16423-68-0]	83.0	77.0	82.4	92.5	130
FD&C Red No. 40 [25956-17-6]	12.4	15.2	15.0	20.9	39.0
Subtotal	*416*	*470*	*493*	*367*	*774*
D&C (Primary)					
D&C Violet No. 2 [81-48-1]	0.921	0.227		2.19	1.23
D&C Blue No. 4 [6371-85-3]					
D&C Blue No. 6[c] [482-89-3]	0.114	1.38	0.106	0.443	0.492
D&C Blue No. 9 [130-20-1]					
D&C Green No. 5 [4403-90-1]	1.96	6.09	8.47	2.54	1.31
D&C Green No. 6 [128-80-3]	0.902	1.12		1.34	0.299
D&C Green No. 8 [6358-69-6]	3.68	5.35	3.42	3.64	4.50
D&C Yellow No. 7 [2321-07-5]	0.181	0.216	0.902	0.419	
D&C Yellow No. 8 [518-47-8]	1.07	2.90	2.77	1.72	1.27
D&C Yellow No. 10 [8004-92-0]	6.41	7.16	7.94	5.60	8.16
D&C Yellow No. 11 [8003-22-3]	0.937	1.56	0.232	2.42	2.23
D&C Orange No. 4 [633-96-5]	1.74	1.94	0.464	0.215	1.22
D&C Orange No. 5 [596-03-2]	3.21	1.78	2.28		
D&C Orange No. 10 [576-24-7]		0.694	0.238		
D&C Orange No. 11 [33239-19-9]					
D&C Orange No. 17 [3468-63-1]		1.13	2.79		0.977
D&C Brown No. 1 [1320-07-6]					
D&C Red No. 6 [5858-81-1]					0.246
D&C Red No. 8 [2092-56-0]		0.940	0.444	0.434	0.206
D&C Red No. 10[d] [1248-18-6]		0.234		0.726	
D&C Red No. 17 [85-86-9]	0.038	0.187	0.321		0.597
D&C Red No. 19 [81-88-9]	1.25	2.85	1.94	1.60	0.907
D&C Red No. 21 [15086-94-9]	1.54	3.88	1.99	3.55	0.898
D&C Red No. 22 [17372-87-1]	1.08	1.44	1.78	0.979	2.29
D&C Red No. 27 [13473-26-2]	0.166	0.877		0.279	0.405
D&C Red No. 28 [18472-87-2]	0.321	0.653	0.418		
D&C Red No. 30 [2379-74-0]		0.289			
D&C Red No. 31 [6371-76-2]		1.21			

Table 1 (*continued*)

Colorants	1973	1974	1975	1976	1977
D&C Red No. 33 [3567-66-6]	0.434		0.786	2.84	1.75
D&C Red No. 36 [2814-77-9]	2.75	3.59	3.11	0.478	2.22
D&C Red No. 37 [6373-07-5]	0.968		0.555		0.266
Subtotal	*29.7*	*47.7*	*40.9*	*35.2*	*33.7*
External D&C (Primary)					
Ext. D&C Violet No. 2 [4430-18-6]			0.355	0.401	
Ext. D&C Green No. 1 [10401-67-9]	0.254	0.806	0.899		
Ext. D&C Yellow No. 1 [587-98-4]	4.48	4.85	7.08	4.38	2.94
Ext. D&C Yellow No. 7 [846-70-8]	0.841			0.953	
Subtotal	*5.57*	*5.66*	*8.33*	*5.73*	*2.94*
D&C (Lakes)					
D&C Blue No. 1 [2650-18-2]		0.885	1.10	0.401	
D&C Yellow No. 5 [1934-21-0]	6.74	13.7	10.2	5.30	9.62
D&C Yellow No. 6 [2783-94-0]	1.93	3.18	1.23	2.66	1.12
D&C Yellow No. 10 [8004-92-0]	0.501	2.98	0.615	3.31	1.13
D&C Orange No. 4 [633-96-5]	1.88		1.02	0.454	0.454
D&C Orange No. 5 [4372-02-5]	2.57	1.44	5.65	2.49	2.13
D&C Orange No. 17 [3468-63-1]	6.76	9.71	5.70	7.57	15.7
D&C Red No. 2 [915-67-3]	0.755		0.237		
D&C Red No. 3 [12227-78-0]	4.13	2.19	4.94	3.72	2.24
D&C Red No. 6 [5858-81-1]	14.9	15.8	7.71	16.2	18.1
D&C Red No. 7 [5858-81-1]	15.0	22.3	17.1	13.7	14.8
D&C Red No. 8 [2092-56-0]	0.846	1.45	0.292	0.765	0.762
D&C Red No. 9 [2092-56-0]	29.7	23.8	37.8	16.1	41.8
D&C Red No. 10 [1248-18-6]	2.05	2.18	3.35		0.465
D&C Red No. 11 [1103-39-5]	0.855	2.57	1.82	0.516	0.732
D&C Red No. 12 [1103-38-4]	1.48	3.65	2.67	1.22	2.03
D&C Red No. 13 [6371-67-1]	1.94	1.98	2.39	3.46	0.570
D&C Red No. 19 [1326-03-0]	7.55	6.60	4.71	5.29	3.80
D&C Red No. 21 [16508-80-8]	8.47	6.30	8.11	6.33	2.69
D&C Red No. 27 [13473-26-2]	4.24	3.00	4.66	3.30	0.349
D&C Red No. 30 [2379-74-0]	5.03	6.12	14.5	7.30	9.54
D&C Red No. 33 [6222-44-2]				0.215	
D&C Red No. 34 [6417-83-0]	1.25	5.92	3.57	2.85	0.335
D&C Red No. 36 [2814-77-9]	1.46	0.435	1.86		
Subtotal	*120*	*136*	*141*	*103*	*128*
External D&C (Lakes)					
Ext. D&C Yellow No. 7 [518-47-8]	0.688	0.633	0.331	1.10	
Subtotal	*0.688*	*0.633*	*0.331*	*1.10*	
Total	*2410*	*2700*	*3070*	*2250*	*2990*

[a] On February 12, 1976, the FDA terminated the provisional listing of FD&C Red No. 2 for use in foods, drugs, or cosmetics.

[b] On September 23, 1976, the FDA terminated the provisional listing of FD&C Red No. 4 for use in maraschino cherries and ingested drugs. FD&C Red No. 4 is now permanently listed for use in externally applied drugs and cosmetics.

[c] On Dec. 13, 1977, the FDA terminated the provisional listing of D&C Blue No. 6. D&C Blue No. 6 is now permanently listed for suture use only (see Sect. 74.1106 CFR).

[d] On December 13, 1977, the FDA terminated the provisional listing of D&C Red No. 10, 11, 12, and 13. External D&C Yellow No. 1, and External D&C Green No. 1.

Table 2. Permanently Listed Food Colorants Exempt from Certification

Colorant	CFR sect. no.	Use restrictions[a]
annatto extract	73.30	none
dehydrated beets (beet powder)	73.40	none
ultramarine blue [1317-97-1]	73.50	may only be used for coloring salt intended for animal feed; not to exceed 0.5 wt % of the salt
canthaxanthin [514-78-3]	73.75	may not exceed 66 mg/kg of solid or 66 mg/L of liquid food
caramel	73.85	none
β-apo-8'-carotenal [1107-26-2]	73.90	may not exceed 33 mg/kg of solid or 33 mg/L of liquid food
β-carotene [7235-40-7]	73.95	none
cochineal [1260-17-9]		
cochineal extract (carmine) [1390-65-4]	73.100	none
toasted, partially defatted, cooked cottonseed flour	73.140	none
ferrous gluconate [299-29-6]	73.160	coloring ripe olives only
grape skin extract (enocianina)[b]	73.170	may be used only for coloring still and carbonated drinks, ades, beverage bases, and alcoholic beverages
synthetic iron oxide [1309-37-1], [1309-38-2] (2:3)	73.200	coloring of dog and cat foods only; not to exceed 0.25 wt % of the finished food
fruit juice	73.250	none
vegetable juice	73.260	none
dried algae meal	73.275	may only be used for coloring chicken feed[c]
tagetes (aztec marigold) meal and extract	73.295	may only be used for coloring chicken feed[c]
carrot oil	73.300	none
corn endosperm oil	73.315	may only be used for coloring chicken feed[c]
paprika	73.340	none
paprika oleoresin	73.345	none
riboflavin [83-88-5]	73.450	none
saffron	73.500	none
titanium dioxide [13463-67-7]	73.575	may not exceed 1 wt % of the food
turmeric [458-37-7]	73.600	none
turmeric oleoresin	73.615	none

[a] No colorant may be used in foods for which standards of identity have been promulgated under Section 401 of the Federal Food, Drug, and Cosmetic Act, unless the use of added color is authorized by such standards.

[b] See "heavenly blue" morning glories (15).

[c] This colorant may be used in chicken feed only if it meets the tolerance limitation for ethyoxyquin in animal feed prescribed in Section 573.380 of the Code of Federal Regulations (see Pet and other livestock feeds).

Azo Colorant. This group includes the greatest number of colorants on the certified list. They are characterized by the presence of the azo functional group $+N{=}N+$ and can be separated into four types: (*1*) the unsulfonated compounds that are insoluble in water but soluble in aromatic solvents and oils, eg, D&C Red No. 17; (*2*) the insoluble pigments that contain a sulfonic acid group in the ortho position that is converted by a permissible precipitant into an insoluble metal salt, eg, D&C Red No. 7, 9, and 34; (*3*) the soluble azo dyes that contain one or more sulfonic acid or

Table 3. Permanently Listed Drug Colorants Exempt from Certification

Colorant	CFR sect. no.	Use restrictions
alumina (dried aluminum hydroxide) [1344-28-1]	, 73.1010	none
chromium–cobalt–aluminum oxide	73.1015	may only be used for coloring linear polyethylene surgical sutures, USP; may not exceed 2 wt % of suture material
ferric ammonium citrate [1185-57-5]	73.1025	may only be used in combination with pyrogallol (as listed in Sect. 73.1375) for coloring plain and chromic catgut sutures for use in general and ophthalmic surgery; the complex may not exceed 3 wt % of the suture material
annatto extract	73.1030	none
calcium carbonate [471-34-1]	73.1070	none
canthaxanthin [514-78-3]	73.1075	may be used in ingested drugs only
caramel	73.1085	none
β-carotene [7235-40-7]	73.1095	none
cochineal extract (carmine)	73.1100	none
potassium sodium copper chlorophyllin (chlorophyllin–copper complex) [11006-34-1]	73.1125	may only be used for coloring dentifrices that are drugs; may not exceed 0.1 wt %
dihydroxyacetone [96-26-4]	73.1150	may only be used in externally applied drugs intended solely or in part to impart a color to the human body
synthetic iron oxide [1309-37-1], [1309-38-2] (2:3)	73.1200	may not exceed prescribed or recommended daily dosage of 5 mg, calculated as elemental iron
ferric ammonium ferrocyanide (iron blue) [14038-43-8]	73.1298	external use only
chromium hydroxide green [12182-82-0]	73.1326	external use only
bismuth oxychloride [7787-59-9]	73.1162	external use only
logwood extract	73.1410	coloring nylon 66, nylon 6, and silk, nonabsorbent sutures in general or ophthalmic surgery; may not exceed 1 wt % of suture
chromium oxide greens [1308-38-9]	73.1327	external use only
guanine [73-40-5] (pearl essence)	73.1329	external use only
pyrogallol [87-66-1]	73.1375	may only be used in combination with ferric ammonium citrate (as listed in Sect. 73.1025), for coloring plain and chromic catgut sutures for use in general and ophthalmic surgery; the complex may not exceed 3 wt % of suture material
pyrophyllite [12269-78-2]	73.1400	external use only; not in area of the eye
mica [12001-26-2]	73.1496	external use only
talc [14807-96-6]	73.1550	none
titanium dioxide [13463-67-7]	73.1575	none
aluminum powder [7429-90-5]	73.1645	external use only
bronze powder [7440-50-8]	73.1646	external use only
copper powder [7440-50-8]	73.1647	external use only
zinc oxide [1314-13-2]	73.1991	external use only

Table 4. Permanently Listed Cosmetic Colorants Exempt from Certification

Colorant	CFR sect. no.	Use restrictions
annatto [8015-67-6]	73.2030	none
carmine	73.2087	none
β-carotene	73.2095	none
disodium EDTA-copper [14025-15-1]	73.2120	shampoo use only
potassium sodium copper chlorophyllin (chlorophyllin–copper complex)	73.2125	may be used for coloring dentifrices only; may not exceed 0.1 wt %[a]
dihydroxyacetone [96-26-4]	73.2150	may only be used in externally applied cosmetics intended solely or in part to impart a color to the human body
guaiazulene [489-84-9]	73.2180	external use only; not in area of the eye
henna [83-72-7]	73.2190	hair coloring only; not in area of the eye
iron oxides	73.2250	none
ferric ammonium ferrocyanide (iron blue)	73.2298	none
chromium hydroxide green	73.2326	external use only; including eye area
bismuth oxychloride	73.2162	coloring cosmetics; generally including eye area
chromium oxide green	73.2327	external use only; including eye area
guanine (pearl essence)	73.2329	none
pyrophyllite	73.2400	external use only; not in area of the eye
mica	73.2496	none
titanium dioxide	73.2575	none
aluminum powder	73.2645	external use only
bronze powder	73.2646	none
copper powder	73.2647	none
ultramarines	73.2725	external use only; including eye area
manganese violet [10101-66-3]	73.2775	none
zinc oxide	73.2991	none

[a] This colorant may be used only in combination with the following substances: water, glycerol, sodium carboxymethyl cellulose, tetrasodium pyrophosphate, sorbitol, magnesium phosphate (tribasic), calcium carbonate, calcium phosphate (dibasic), sodium N-lauroyl sarcosinate, artificial sweeteners that are generally recognized as safe or that are authorized under Sub-Chapter B drugs, flavors that are generally recognized as safe or that are authorized under Sub-Chapter B drugs, preservatives that are generally recognized as safe or that are authorized under Sub-Chapter B drugs.

carboxy groups to produce their water solubility. The sulfonic acid group is generally in the meta or para position to the azo group, eg, FD&C Red No. 40 and D&C Orange No. 4 (see Azo dyes); and (4) the unsulfonated pigments are precipitated directly on coupling and contain no groups capable of salt formation, eg, D&C Red No. 36, and D&C Orange No. 17. These pigments have some solubility in organic solvents.

Triphenylmethane Colorants. This is a group of sulfonated dyes (derivative of the corresponding basic dye) containing two or more sulfonic acid groups, eg, FD&C Blue No. 1 and FD&C Green No. 3.

Xanthene (Fluoran) Colorants. The xanthene group is characterized by the structure:

The xanthene colorants can be divided into two subgroups, acidic and basic types, eg, an acidic type is D&C Red No. 27 and a basic type is D&C Red No. 19.

Quinoline Colorants.   There are only two quinoline dyes certified by the FDA. They are D&C Yellow No. 11 which is oil soluble and D&C Yellow No. 10 which is water soluble and a mixture of the mono and disulfonated derivatives of D&C Yellow No. 11.

Anthraquinone Colorants.   The certified anthraquinone colors fall into three subgroups: (1) the sulfonated acid types are the water soluble dyes, eg, D&C Green No. 5; (2) the unsulfonated type are the oil soluble dyes which can be converted to the acid type by sulfonation, eg, D&C Green No. 6; and (3) the hydroxyanthraquinone types that have two members on the certified list, the oil soluble D&C Violet No. 2 and the water soluble External D&C Violet No. 2 (Ext. D&C Violet No. 2 is made by the sulfonation of D&C Violet No. 2) (see Dyes, Anthraquinone).

Indigoid Color Additive.   There are three certified colorants in this group: D&C Blue No. 6, D&C Red No. 30, and FD&C Blue No. 2. D&C Blue No. 6 and D&C Red No. 30 are pigments. FD&C Blue No. 2 is a water soluble colorant made by sulfonation of D&C Blue No. 6.

Pyrene Color Additive.   D&C Green No. 8 (water soluble sulfonated pyrene) is the only pyrene colorant in the certified list.

Table 5.   Solubilities and Fastness Properties of Certified Colorants[a]

Official FDA name	H_2O	Glycerol	Methanol	Ethanol	Petroleum jelly	Toluene	Stearic acid	Oleic acid	Mineral oil
FD&C colorants									
Blue No. 1	S	S	S	S	C	I	C	C	C
Blue No. 2	S	S	SS	SS	I	I	I	I	I
Green No. 3	S	S	S	M	I	I	I	I	I
Yellow No. 5	S	S	SS	SS	IE	IE	IE	IE	IE
Yellow No. 6	S	S	S	SS	I	I	I	I	I
Orange B	S	S	SS	SS	IE	IE	IE	IE	IE
Red No. 3	S	S	S	S	IE	I	IE	IE	IE
Red No. 4	S	S	SS	SS	IE	I	IE	IE	IE
Red No. 40	S	S	S	SS	I	I	I	I	I
Citrus Red No. 2	I	SS	SS	SS	S	S	S	S	S
D&C colorants									
Violet No. 2	I	Ia	SS	SS	S	S	S	S	S
Blue No. 4	S	S	S	S	C	I	C	C	C
Blue No. 6	IU	D	I	I	D	Ia	D	D	D
Blue No. 9	IU	ID	Ia	I	D	Ia	D	D	D
Green No. 5	S	S	S	SS	IE	I	IE	IE	IE
Green No. 6	I	Ia	SS	SS	M	S	M	M	M
Green No. 8	SF	SSF	SSF	SSF	Ia	I	Ia	Ia	I
Yellow No. 7	IBF	SSF	SF	SS	D	I	D	D	D
Yellow No. 8	SF	SF	SF	M	IE	I	IE	IE	IE
Yellow No. 10	S	S	M	SS	I	I	I	I	I

Certified Colorants. Table 6 gives all the currently listed certified colorants. The names shown in parentheses following the official FDA name are the best-known trade names. The classifications listed are those designated by the FDA, and include the chemical class and the color index number. The chemical names are those given in the Code of Federal Regulations (16).

Analyses of Food, Drug, and Cosmetic Colorants

Comprehensive reviews of the methods of analysis used for identification of food, drug, and cosmetic colorants are presented in references 17–19.

Permanently Listed Colorants Subject to Certification

Tables 7–9 show all presently approved colorants that are permanently listed and subject to certification. The list includes the CFR section number and use restrictions for each product.

| | | | | | | | Fastness properties | | | | | |
Mineral wax	Ethyl ether	Acetone	CH$_3$COOBu	Light	10% Acetic acid	10% HCl	10% NaOH	0.9% Physiol salt solution	5% FeSO$_4$	5% Alum	Oxidizing agents	Reducing agents
C	I	Ia	I	3	5	4g	4	6	4r	4	2	1
I	I	I	I	1	6	5	4	6	4	J	2	4
I	I	I	I	3	5	5	2b	6	3y	4	2	1
IE	I	I	I	5	5	5	5	6	d	4	3	1
I	I	I	I	3	5	5	5	6	4	4	3	1
IE	I	Ia	I	5	5	5	5	6	d	4	3	1
IE	Ia	SS	I	3	2pya	1py	6	6	p	p	3	1
IE	I	I	I	6	6	5	5	6	z-p	p	3	1
I	I	I	I	3	5	5	5	6	4	4	3	1
S	SS	SS	M	3	16	12k	15	I	14	14	3	1
SW	SS	SS	S	4	5I	5I	5I	6I	4I	4I	2	1
C	I	Ia	I	3	5	5	4	6	4	4	2	1
D	I	I	I	6	71	51	L6U	I	I	I	6	U
D	I	I	I	7	71	51	6IU	I	I	I	6	U
IEW	I	SS	I	5	5	5	5	5	4	4	3	2
M	SS	SS	S	4	5L	5I	6I	I	I	I	3	2
I	Ia	Ia	Ia	2	I	I	5	6	4d	4d	3	3
D	SS*	S	I	2	I	I	S6	I	I	I	3	3
IE	Ia	kIa	I	3	3p	3p	6	6	z-p	p	3	3
I	Ia	SS	I	3	5	5	4r	6	z	4	2	5

Table 5 (*continued*)

Official FDA name	H_2O	Glycerol	Methanol	Ethanol	Petroleum jelly	Toluene	Stearic acid	Oleic acid	Mineral oil
									Solubilities
Yellow No. 11	I	SS	S	S	S	S	S	S	S
Orange No. 4	S	S	S	M	IE	I	IE	IE	IE
Orange No. 5	IB	SS	S	M	D	I	D	D	D
Orange No. 17	I	D	Ia	Ia	D	I	D	D	D
Brown No. 1	S	S	S	SS	IE	I	IE	IE	IE
Red No. 6	S	S	SS	Ia	I	I	I	I	I
Red No. 7	I	D	Ia	Ia	D	I	D	D	D
Red No. 8	I	D	Ia	Ia	D	I	D	D	D
Red No. 9	I	D	Ia	Ia	D	I	D	D	D
Red No. 17	I	SS	SS-M	SS	S	S	S	S	S
Red No. 19	SF	SF	SF	SF	I	I	IC	SS (hot)	IC
Red No. 21	IBF	Da	SS	SS	D	I	D	D	D
Red No. 22	SF	SF	SF	SF	IE	I	IE	IE	IE
Red No. 27	IB	Da	SS	SS	D	I	D	D	D
Red No. 28	S	S	S	S	IE	I	IE	IE	IE
Red No. 30	IU	D	I	I	I	Ia	D	D	D
Red No. 31	M	SS	SS	SS	I	I	I	I	I
Red No. 33	S	S	SS	SS	I	I	I	I	I
Red No. 34	I	I	Ia	I	D	I	D	D	D
Red No. 36	I	D	Ia	Ia	D	I	D	D	D
Red No. 37	Ia	SS	SF	SF	IE	S	S	S	IEG
Red No. 39	Ia	M	M-S	S†	I	Ia	I	SS	I
External D&C colorants									
Violet No. 2	S	S	SS	SS	I	I	I	I	I
Yellow No. 7	S	S	M	SS	I	I	I	I	I

a Abbreviations

a = may bleed or stain, very sparingly soluble.

B = insoluble in water, soluble in aqueous alkaline solution.

b = turns much bluer in hue.

C = practically insoluble, but is useful in nearly neutral or slightly acid emulsions.

c = at 25°C.

D = practically insoluble, but may be dispersed by grinding and homogenizing. Solid mediums (waxes) should be softened or melted before or during the grinding.

d = hue becomes duller or darker.

E = practically insoluble in the fatty acid, oil, or wax, but is useful in coloring slightly alkaline aqueous emulsions.

F = solution is usually fluorescent.

G = soluble or dispersible in oils and waxes, when 10–25% of a fatty acid is present.

g = turns much greener in hue.

I = insoluble.

J = tends to thicken or gel the solution.

k = turns brownish in hue.

L = turns orange in hue.

M = moderately soluble (less than 1%).

m = turns scarlet in hue.

n = slowly or on standing for some time.

p = dye precipitated as heavy-metal salt or color acid.

| | | | | | Fastness properties | | | | | | | |
Mineral wax	Ethyl ether	Acetone	CH$_3$COOBu	Light	10% Acetic acid	10% HCl	10% NaOH	0.9% Physiol salt solution	5% FeSO$_4$	5% Alum	Oxidizing agents	Reducing agents
S	S	S	S	2	1	5I	Iw	I	I	I	2	5
IE	I	Ia	I	5	5	5	2m	6	J-p	J-p	3	3
D	M	S	I	2	4aI	4I	Sr	I	I	I	3	3
D	Ia	Ia	I	5	5I	Id	Idr	I	I	I	3	2
IE	SS	SS	I	3	5	5	6sly	6	p	p	3	1
I	I	Ia	I	5	5	4	4d	6	p	p	3	1
D	I	Ia	I	6	5I	4I	5I	I	4Id	4I	3	1
D	I	Ia	I	6	6I	4Id	4Id	I	4Id	4I	3	1
D	I	I	I	6	6I	4I	4I	I	4Id	4I	3	1
S	SS	SS	M	3	5L	4Id	5I	I	4Id	4I	3	1
IC	S*	SF	I	3	5	5	2p	6	6	6	3	5
D	M*	S	I	2	3I	3I	5Sr	I	Id	4I	4	4
IE	Ia	SS	I	2	2py	Ipy	5	6	3d	2y	4	4
D	Ia	SS	I	2	3I	3	5Sr	I	I	I	4	4
IE	Ia	SS	I	3	2p	4p	6	6	z	p	4	4
D	Ia	Ia	Ia	6	7I	I	6IU	I	I	I	5	u
I	I	Ia	Ia	5	5	4	5	6	p	p	3	1
I	I	I	I	5	6	3z	5	6	4	4	3	1
D	I	D	D	4	5I	4	4I	I	I	I	3	1
D	I	Ia	D	6	6I	4d	4d	I	4d	4	3	1
IEG	S*	SF	SS	3	6I	5I	Ia	Ia	I	Ia	3	5
Ia	S	S	SS	2	Sy	Sx	6Sx	I	4ald	I	3	3
I	I	SS	I	5	5	5	5	6	4z	4	3	2
I	I	M	I	4	5	5	5	6	zd	4	3	3

r = turns redder in hue.
S = dissolves (soluble 1% or more).
SS = sparingly soluble (less than 0.25%).
sl = slightly.
t = dye destroyed, or solution goes practically colorless.
U = in alkaline reducing vats a soluble leuco compound forms.
v = turns violet in hue.
W = not fast to prolonged storage in some waxes.
w = becomes tinctorially weaker.
x = turns yellow in hue.
y = turns yellower in hue.
z = hazy or cloudy.
* = practically colorless.
† = not suitable for acid solutions.
1 = very poor fastness.
2 = poor fastness.
3 = fair fastness.
4 = moderate fastness.
5 = good fastness.
6 = very good fastness.
7 = excellent fastness.

Table 6. Certified Colorants[a]

Official FDA name and trade name	Classification CI name and number	Struc-ture no.	Chemical name and CAS Registry No.	Manufacturing process	Shade and typical applications
FD&C Blue No. 1 (Brilliant Blue FCF)	triphenylmethane; CI Food Blue 2, CI No. 42090	(1)	disodium salt of ethyl[4-[p[ethyl(m-sulfobenzyl) amino]-α-(o-sulfophenyl)ben-zylidene]-2,5-cyclohexadien-1-ylidene] (m-sulfobenzyl)-ammonium hydroxide inner salt [2650-18-2], with smaller amounts of the isomeric disodium salts of ethyl[4-[p-[ethyl(p-sulfophenyl) amino]-α-(o-sulfophenyl) benzylidene]-2,5-cyclohexadien-1-ylidene] (p-sulfobenzyl)ammonium hydroxide inner salt and ethyl-[4-[p-[ethyl(o-sulfobenzyl)-amino]-α-(o-sulfophenyl)ben-zylidene]-2,5-cyclohexadien-1-ylidene](o-sulfobenzyl) ammo-nium hydroxide inner salt	condensation of benzaldehyde-o-sulfonic acid with α-(N-ethylanilino)toluenesulfonic acid (benzylethylaniline-sulfonic acid)	*Greenish blue:* *Food:* A, C, E, F, G, H, J, L *Drugs:* C1, E1, F1, J1, I1 *Cosmetics:* E2, F2, J2, K2, (G2)
FD&C Blue No. 2 (Indigotine, Indigotin IA)	indigoid; CI Food Blue 1, CI No. 73015	(4)	disodium salt of 5,5′-disulfo-3,3′-dioxo-Δ²,²′-biindoline [860-22-0] with smaller amounts of the isomeric disodium salt of 5,7′-disulfo-3,3′-dioxo-Δ²,²′-biindo-line	sulfonation of indigo	*Deep Blue:* *Foods:* G, H, J, L *Drugs:* A1, C1, E1
FD&C Green No. 3 (Fast Green FCF)	triphenylmethane; CI Food Green 3, CI No. 42053	(2)	disodium salt of ethyl[4-[p-[ethyl(m-sulfobenzyl)am-ino]-α-(o-sulfo-p-hydroxy-phenyl)benzylidene]-2,5-cyclo-hexadien-1-ylidene](m-sulfobe-nzyl)ammonium hydroxide inner salt [2353-45-9]	condensation of p-hydroxybenzaldehyde-o-sul-fonic acid with α-N-ethyl-anilino]toluenesulfonic acid	*Bluish Green:* *Foods:* A, E, F, G, H, L, C *Drugs:* C1, E1, F1, I1, J1 *Cosmetics:* F2, K2, J2

574

Name	Structure	Preparation	Uses
FD&C Red No. 3 (Erythrosine, Erythrosine Bluish) xanthene; CI Food Red 14, CI No. 45430	(5) disodium salt of 9(o-carboxyphenyl)-6-hydroxy-2,4,5,7-tetraiodo-3H-xanthen-3-one [568-63-8] with smaller amounts of lower iodinated fluoresceins	iodination of fluorescein (D&C Yellow No. 7)	*Bluish Pink:* *Foods:* C, E, H, J, L *Drugs:* C1, E1, I1, J1 *Cosmetics:* K2, (G2)
FD&C Red No. 40 (Allura Red AC) monoazo; CI Food Red 17, CI No. 16035	(15) disodium salt of 6-hydroxy-5-[2-methoxy-5-methyl-4-sulfophenyl)azo]-2-naphthalenesulfonic acid [25956-17-6]	coupling diazotized 5-amino-4-methoxy-2-toluenesulfonic acid with 6-hydroxy-2-naphthalenesulfonic acid	*Yellowish Red:* *Foods:* A, C, E, F, G, H, J, K, L *Drugs:* C1, E1, F1, H1, I1, J1 *Cosmetics:* (G2), K2
FD&C Red No. 4 (Ponceau SX) monoazo CI Food Red 1, CI No. 14700	(27) disodium salt of 3-[2,4-dimethyl-5-sulfophenyl)azo]-4-hydroxy-1-naphthalenesulfonic acid [4548-53-2]	coupling of diazotized 1-amino-2,4-dimethylbenzene-5-sulfonic acid with 1-naphthol-4-sulfonic acid	*Yellowish Red:* *Drugs:*[b] C1, J1 *Cosmetics:*[b] K2, A2
Citrus Red No. 2 monoazo; CI No. 12156	(16) 1-(2,5-dimethoxyphenylazo)-2-naphthol [6358-53-8]	coupling of diazotized 2,5-dimethoxyaniline with 2-naphthol	*Scarlet Foods:* B
FD&C Yellow No. 5 (Tartrazine) pyrazolone; CI Food Yellow 4, CI No. 19140	(29) trisodium salt of 4,5-dihydro-5-oxo-1-(4-sulfophenyl)-4-[(4-sulfophenyl)azo]-1H-pyrazole-3-carboxylic acid [1934-21-0]	condensation of phenylhydrazine-p-sulfonic acid with oxalacetic ester, coupling of the product with diazotized sulfanilic acid, then hydrolysis of the ester with NaOH; or condensation of phenylhydrazine-p-sulfonic acid with dihydroxytartaric acid	*Greenish Yellow:* *Foods:* A, E, F, G, H, J, K, L *Drugs:* C1, E1, H1, F1, I1, J1 *Cosmetics:* (H2) C2, E2, F2, (G2) J2, K2, (I2)

Table 6 (*continued*)

Official FDA name and trade name	Classification CI name and number	Structure no.	Chemical name and CAS Registry No.	Manufacturing process	Shade and typical applications
FD&C Yellow No. 6 (Sunset Yellow FCF)	monoazo; CI Food Yellow 5 CI No. 15985	(17)	disodium salt of 6-hydroxy-5-[4-sulfophenyl)azo]-2-naphthalene sulfonic acid [2783-94-0]	coupling of diazotized sulfanilic acid with 2-naphthol-6-sulfonic acid	*Reddish Yellow* *Foods:* A, E, F, G, H, L, J *Drugs:* C1, E1, F1, H1, I1, J1 *Cosmetics:* E2, F2, (G2) J2, K2
Orange B	pyrazolone	(30)	disodium salt of ethyl 4,5-dihydro-5-oxo-1-(4-sulfophenyl)-4-[(4-sulfonaphthyl)azo]-1*H*-pyrazole-3-carboxylate [15139-76-1, 53060-70-1]	condensation of phenylhydrazine-*p*-sulfonic acid with oxalacetic ester, coupling of the product with diazotized naphthionic acid	*Reddish Yellow:* *Foods:* D
FD&C Lakes	same as straight colorant		same as straight colorant	Al or Ca salt extended on a substratum of alumina	
D&C Blue No. 4 (Alphazurine FG, Erioglaucine)	triphenylmethane; CI No. 42090	(3)	diammonium salt of ethyl[4-[p[ethyl(*m*-sulfobenzyl)-amino]-α-(*o*-sulfophenyl)benz-ylidene]-2,5-cyclohexadien-1-ylidene](*m*-sulfobenzyl)ammo-nium hydroxide inner salt [37307-56-5]	same as FD&C Blue No. 1	*Bright Greenish Blue:* *Drugs:* C1, J1 *Cosmetics:* E2, J2, K2
D&C Blue No. 9 (Carb-anthrene Blue, Indanthrene Blue GCD)	anthraquinone vat; CI Vat Blue 6, CI No. 69825	(31)	7,16-dichloro-6,15-dihydro-5,9,14,18-anthrazinetetrone [130-20-1]	chlorination of indanthrene	*Dull Greenish Blue:* *Drugs:* A1
D&C Green No. 5 (Alizarin Cyanine Green F, Alizarin Cyanine Green)	anthraquinone; CI Acid Green 25, CI No. 61570	(32)	disodium salt of 2,2′-[9,10-dihydro-9,10-dioxo-1,4-anthra-cenediyl)diimino]bis[5-methylbenzenesulfonic acid] [4403-90-1]	condensation of leucoquinizarin with *p*-toluidine and sulfonation	*Dull Bluish Green:* *Drugs:* C1, E1, J1, *Cosmetics:* A2, E2, F2, J2, K2, (G2)

576

Name	CI	No.	Chemical	Method	Uses
D&C Green No. 6 (Quinizarin Green SS)	anthraquinone; CI Solvent Green 3, CI No. 61565	(33)	1,4-di-*p*-toluidinoanthraquinone [128-80-3]	condensation of leucoquinizarin with *p*-toluidine	*Dull Bluish Green: Drugs:* D1, G1, K1 *Cosmetics:* A2, B2, D2, L2
D&C Green No. 8 (Pyranine Concentrated)	pyrene; CI No. 59040	(34)	trisodium salt of 8-hydroxy-1,3,6-pyrenetrisulfonic acid [6358-69-6]	sulfonation of pyrene to tetrasulfonic acid, salting out with NaCl, hydrolysis in NaOH solution, addition of formic acid, and salting out with NaCl	*Yellowish Green: Cosmetics:;* E2
D&C Yellow No. 7 (Fluorescein)	fluoran; CI Acid Yellow 73, CI No. 45350	(13)	fluorescein [518-45-6]	condensation of resorcinol with phthalic anhydride in the presence of ZnCl₂ or H₂SO₄	*Greenish Yellow: Cosmetics:* A2
D&C Yellow No. 8 (Uranine)	xanthene; CI Acid Yellow 73, CI No. 45350	(14)	disodium salt of fluorescein [518-47-8]	conversion of D&C Yellow No. 7 to the Na salt	*Greenish Yellow: Cosmetics:* F2, E2, A2 *Drugs:* C1
D&C Yellow No. 10 (Quinoline Yellow WS, Quinoline Yellow)	quinoline; CI Acid Yellow 3, CI No. 47005	(35)	mono- and disodium salts of the 6-mono- and 6,5′-disulfonic acids of 2-(2-quinolinyl)-1,3-indandione [8004-92-0] and [38615-46-2], respectively	sulfonation of D&C Yellow No. 11	*Greenish Yellow: Drugs:* C1, E1, F1, I1 *Cosmetics:* A2 (G2) (H2) K2
D&C Yellow No. 11 (Quinoline Yellow SS, Quinoline Yellow Spirit Soluble)	quinoline; CI Solvent Yellow 33, CI No. 47000	(36)	2-(2-quinolinyl)-1,3-indandione [8003-22-3]	condensation of quinaldine with phthalic anhydride in the presence of ZnCl₂	*Greenish Yellow: Cosmetics:* B2, A2, L2 *Drugs:* D1
Ext D&C Yellow No. 7 (Naphthol Yellow S)	nitro; CI Acid Yellow 1, CI No. 10315	(37)	disodium salt of 8-hydroxy-5,7-dinitro-2-naphthalenesulfonic acid [846-70-8]	nitration of the di- or trisulfonic acids of 1-naphthol or the nitroso compound of the 2,7-disulfonic acid	*Greenish Yellow; Cosmetics:* (H2) J2

Table 6 (*continued*)

Official FDA name and trade name	Classification CI name and number	Structure no.	Chemical name and CAS Registry No.	Manufacturing process	Shade and typical applications
D&C Brown No. 1 (Resorcin Brown)	disazo; CI Acid Orange 24, CI No. 20170	(38)	sodium salts of 4-[[3-[(dialkylphenyl)azo]-2,4-dihydroxyphenyl]azo]-benzenesulfonic acid; the alkyl group is principally the methyl group [1320-07-6]	coupling of diazotized sulfanilic acid with resorcinol and coupling of the product with diazotized 2,4-xylidine in dil alk soln	*Light Orange–Brown:* *Drugs:* C1, J1 *Cosmetics:* K2, J2
D&C Orange No. 4 (Orange II)	monoazo; CI Acid Orange 7, CI No. 15510	(18)	monosodium salt of 1-p-sulfophenylazo-2-naphthol [633-96-5]	coupling of diazotized sulfanilic acid with 2-naphthol	*Bright Orange:* *Drugs:* J1 *Cosmetics:* A2, K2, (H2) (12)
D&C Orange No. 5 (Dibromofluorescein)	fluoran; CI No. 45370:1	(6)	4,5-dibromo-3,6-fluorandiol [596-03-2]	bromination of fluorescein (D&C Yellow No. 7)	*Reddish Orange:* *Cosmetics:* G2
D&C Orange No. 10 (Diiodofluorescein)	fluoran; CI No. 45425:1	(7)	4,5-diiodo-3,6-fluorandiol [518-40-1]	iodination of fluorescein D&C Yellow No. 7	*Reddish Orange:* *Cosmetics:* G2
D&C Orange No. 11 (Erythrosine Yellowish Na, Erythrosine Yellowish)	xanthene; CI No. 45425	(8)	disodium salt of 9-o-carboxyphenyl-6-hydroxy-4,5-diiodo-3-isoxanthone [6371-82-0]	conversion of D&C Orange No. 10 to the Na salt	*Red* *Drugs:* K1 *Cosmetics:* L2
D&C Orange 17 (Permatone Orange or Permanent Orange)	monoazo; CI No. 12075	(19)	1-(2,4-dinitrophenylazo)-2-naphthol [3468-63-1]	coupling of diazotized 2,4-dinitroaniline with 2-naphthol	*Bright Orange:* *Cosmetics:* G2
D&C Red No. 6 (Lithol Rubin B)	monoazo; CI Pigment Red 57 CI No. 15850	(20)	disodium salt of 4-(o-sulfo-p-tolylazo)-3-hydroxy-2-naphthoic acid [5858-81-1]	coupling of diazotized 6-amino-m-toluenesulfonic acid with 3-hydroxy-2-naphthoic acid	*Medium Red:* *Cosmetics:* (G2) (12) *Drugs:* C1, E1

D&C Red No. 7 (Lithol Rubin B Cal) monoazo CI Pigment Red 57, CI No. 15850	(21)	calcium salt of 4-(o-sulfo-p-tolylazo)-3-hydroxy-2-naphthoic acid [5281-04-9]	heating of D&C Red No. 6 with CaCl₂	*Bluish Red:* *Cosmetics:* G2, I2 *Drugs:* E1
D&C Red No. 8 (Lake Red C) monoazo; CI Pigment Red 53, CI No. 15585	(22)	monosodium salt of 1-(4-chloro-o-sulfo-5-tolylazo)-2-naphthol [2092-56-0]	coupling of diazotized 2-amino-5-chloro-p-toluene-sulfonic acid with 2-naphthol	*Orange:* *Cosmetics:* B2, G2 *Drugs:* E1, G1
D&C Red No. 9 (Lake Red C Ba) monoazo; CI Pigment Red 53, CI No. 15585	(23)	barium salt of 1-(4-chloro-o-sulfo-5-tolylazo)-2-naphthol [5160-02-1]	boiling of D&C Red No. 8 with BaCl₂	*Red–Orange:* *Cosmetics:* G2, I2
D&C Red No. 17 (Toney Red, Sudan III) disazo CI solvent Red 23, CI No. 26100	(24)	the colorant D&C Red No. 17 is principally 1-[[4-(phenylazo)-phenyl]azo]-2-naphthalenol [85-86-9]	coupling of diazotized aminoazobenzene with 2-naphthol	*Dull Red:* *Drugs:* D1, G1, K1 *Cosmetics:* A2, B2, D2, I2, L2
D&C Red No. 19 (Rhodamine B) xanthene; CI Basic Violet 10, CI No. 45170	(9)	3-ethochloride of 9-o-carboxyphenyl-6-diethylamino-3-ethylimino-3-isoxanthene [81-88-9]	fusion of *m*-diethylaminophenol with phthalic anhydride and treatment of the base with dil HCl; or treatment of fluorescein chloride under pressure with diethylamine	*Magenta:* *Cosmetics:* A2, F2, (G2) (H2) *Drugs:* (I2) K2
D&C Red No. 21 (Tetrabromofluorescein) fluoran; CI Solvent Red 43, CI No. 45380A	(39)	2,4,5,7-tetrabromo-3,6-fluor-andiol [15086-94-9]	bromination of fluorescein (D&C Yellow No. 7)	*Bluish Pink:* *Cosmetics:* G2
D&C Red No. 22 (Eosin YS, Eosine G) xanthene; CI Acid Red 87, CI No. 45380	(10)	disodium salt of 2,4,5,7-tetrabromo-9-o-carboxyphenyl-6-hydroxy-3-isoxanthone [548-26-5]	conversion of D&C Red No. 21 to the Na salt	*Yellowish Pink:* *Drugs:* C1, F1 *Cosmetics:* A2

Table 6 (*continued*)

Official FDA name and trade name	Classification CI name and number	Structure no.	Chemical name and CAS Registry No.	Manufacturing process	Shade and typical applications
D&C Red No. 27 (Tetrachlorotetrabromofluorescein)	fluoran; CI Solvent Red 48, CI No. 45410A	(40)	2,4,5,7-tetrabromo-12,13,14,15-tetrachloro-3,6-fluorandiol [2134-15-8]	condensation of resorcinol with tetrachlorophthalic anhydride and bromination	*Bluish Pink:* *Cosmetics:* G2
D&C Red No. 28 (Phloxine B)	xanthene; CI Acid Red 92, CI No. 45410	(11)	disodium salt of 2,4,5,7-tetrabromo-9-(3,4,5,6-tetrachloro-o-carboxylphenyl)-6-hydroxy-3-isoxanthone [4618-23-9]	conversion of D&C Red No. 27 to the Na salt	*Bluish Pink:* *Drugs:* C1, F1, J1 *Cosmetics:* K2, F2
D&C Red No. 30 (Helindone Pink CN)	indigoid; CI Vat Red 1, CI No. 73360	(41)	6,6'-dichloro-4,4'-dimethyl-thioindigo [2379-74-0]	oxidation of 6-chloro-4-methyl-thioindoxyl; or chlorination of 4,4'-dimethylthioindigo	*Bluish Pink:* *Drugs:* E1 *Cosmetics:* A2, G2, H2, I2
D&C Red No. 31 (Brilliant Lake Red R)	monoazo; CI Pigment Red 64, CI No. 15800	(25)	calcium salt of 3-hydroxy-4-(phenylazo)-2-naphthalenecarboxylic acid [6371-76-2]	coupling of diazotized aniline with 3-hydroxy-2-naphthoic acid and conversion to the Ca salt	*Bluish Pink:* *Cosmetics:* H2, G2
D&C Red No. 33 (Acid Fuchsin D or Naphthalene Red B; Fast Acid Fuchsine B)	monoazo; CI Acid Red 33, CI No. 17200	(28)	disodium salt of 8-amino-2-phenylazo-1-naphthol-3,6-disulfonic acid [3567-66-6]	coupling of diazotized aniline with 8-amino-1-naphthol-3,6-disulfonic acid in alk soln	*Dull Bluish Red:* *Drugs:* C1, F1, J1 *Cosmetics:* (G2), (H2), (I2), A2, K2, E2, F2, J2
D&C Red No. 34 (Deep Maroon or Fanchon Maroon; Lake Bordeaux B)	monoazo; CI Pigment Red 63, CI No. 15880	(42)	calcium salt of 3-hydroxy-4-[(1-sulfo-2-naphthalenyl)azol-2-naphthalenecarboxylic acid [6417-83-0]	coupling of diazotized 2-naphthyl-amine-1-sulfonic acid with 3-hydroxy-2-naphthoic acid and conversion to the Ca salt	*Maroon:* *Cosmetics:* I2

Colorant	Class / CI	No. / Chemical name	Method of manufacture	Shade / Use
D&C Red No. 36 (Flaming Red)	monoazo; CI Pigment Red 4, CI No. 12085	(26) 1-(o-chloro-p-nitrophenylazo-2-naphthol [2814-77-9]	coupling of diazotized 2-chloro-4-nitroaniline with 2-naphthol	*Blazing Red:* *Cosmetics:* I2, G2
D&C Red No. 37 (Rhodamine B-Stearate)	xanthene; CI Solvent Red 49, CI No. 45170B	(12) 3-ethostearate of 9-o-carboxyphenyl-6-diethyl-amino-3-ethylimino-3-isoxanthene [6373-07-5]	same as for D&C Red No. 19, except treatment of the base with stearic acid	*Bluish Pink:* *Cosmetics:* A2, B2 *Drugs:* B1, G1, K1
D&C Red No. 39 (Alba Red)	monoazo; CI Pigment Red 100, CI No. 13058	(43) 2-[[4-[bis(2-hydroxyethyl)-amino]phenyl]azo]benzoic acid [6371-55-7]	coupling of diazotized anthranilic acid with N,N-(β,β'-dihydroxydiethyl) aniline	*Dark Bluish Red* *Drugs:* B1
D&C Violet No. 2 (D&C Blue No. 3 or Alizurol Purple SS)	anthraquinone; CI Solvent Violet 13, CI No. 60725	(44) 1-hydroxy-4-[(4-methylphenyl)-amino]-9,10-anthracene-dione [81-48-1]	condensation of quinizarin with p-toluidine; or condensation of 1-hydroxy-4-halogenoanthraquinone with p-toluidine	*Dark Bluish Violet:* *Drugs:* A1, D1, G1 *Cosmetics:* L2, B2, A2
Ext D&C Violet No 2 Alizurol Purple, Alizarin Irisol R	anthraquinone; CI Acid Violet 43, CI No. 60730	(45) sodium salt of 2-[(9,10-dihydro-4-hydroxy-9,10-di-oxo-1-anthracenyl)amino]-5-methylbenzenesulfonic acid [4430-18-6]	sulfonation of D&C Violet No. 2	*Bluish Violet:* *Cosmetics:* A2, K2, E2, J2
D&C Lakes	same as straight colorant	same as straight colorant	straight colorant extended on substratum of alumina, blanc fixe, gloss, white, clay, titanium dioxide, zinc oxide, talc, rosin, aluminum benzoate, calcium carbonate, or any combination of two or more of these	

Table 6 (*continued*)

Official FDA name and trade name	Classification CI name and number	Structure no.	Chemical name and CAS Registry No.	Manufacturing process	Shade and typical applications
Ext. D&C Lakes	same as straight colorant		same as straight colorant	straight colorant extended on substratum of alumina, blanc fixe, gloss white, clay, titanium dioxide, zinc oxide, talc, rosin, aluminum benzoate, calcium carbonate, or on any combination of two or more of these	

a Abbreviations

Foods

A = gelatin desserts
B = orange skins
C = maraschino cherries
D = sausage castings
E = ice cream and frozen desserts
F = carbonated beverages
G = dry powdered drinks
H = candy and confectionery products that are oil and fat free
I = candy and confectionery products that contain fats and oils
J = bakery products and cereals
L = puddings

Drugs

A1 = sutures
B1 = germicidal solutions
C1 = aqueous solutions
D1 = oil solutions
E1 = tablets
F1 = capsules
G1 = ointments
H1 = mouthwash
I1 = toothpaste
J1 = oil in water emulsions
K1 = water in oil emulsions

Cosmetics

A2 = soap
B2 = suntan oils
C2 = hair-waving lotions
D2 = hair oils and pomades
E2 = shampoos
F2 = bath salts
G2 = lipsticks and rouges
H2 = face powders and talcums
I2 = nail lacquers
J2 = hair rinses
K2 = oil in water emulsions
L2 = water in oil emulsions

b External only

582

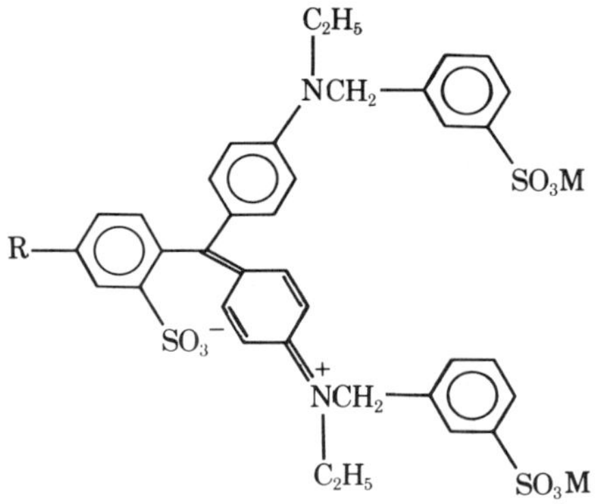

(**1**) R = H, M = Na
(**2**) R = OH, M = Na
(**3**) R = H, M = NH$_4$

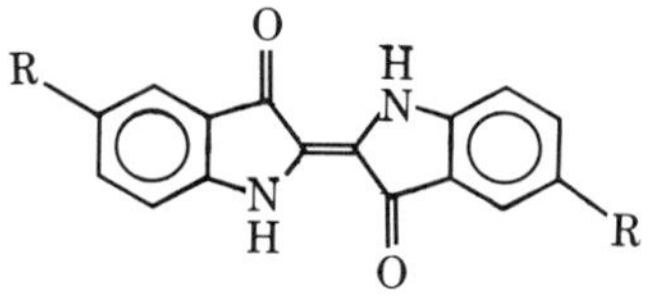

(**4**) R = SO$_3$Na

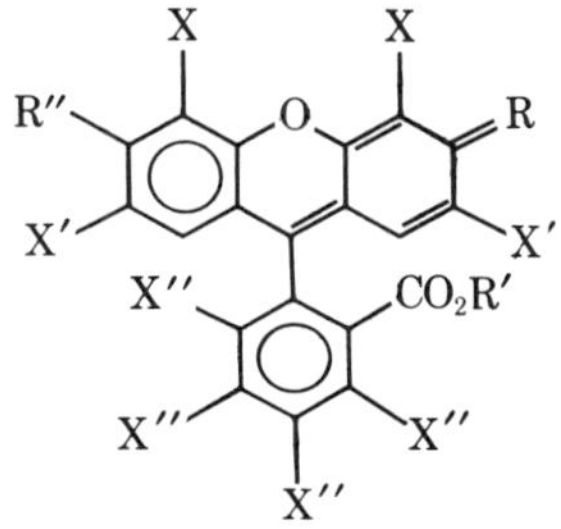

(**5**) R = O, R′ = Na, R″ = NaO
X = X′ = I, X″ = H
(**6**) R = O, R′ = H, R″ = OH
X = Br, X′ = X″ = H
(**7**) R = O, R′ = H, R″ = OH
X = I, X′ = X″ = H
(**8**) R = O, R′ = Na, R″ = NaO
X = I, X′ = X″ = H
(**9**) R = N$^+$(C$_2$H$_5$)$_2$Cl$^-$, R′ = H,
R″ = N(C$_2$H$_5$)$_2$
X = X′ = X″ = H

(**10**) R = O, R′ = Na, R″ = NaO
X = X′ = Br, X″ = H
(**11**) R = O, R′ = Na, R″ = NaO
X = X′ = Br, X″ = Cl
(**12**) R = N$^+$(C$_2$H$_5$)$_2$ C$_{17}$H$_{35}$CO$_2^-$,
R′ = H, R″ = N(C$_2$H$_5$)$_2$
X = X′ = X″ = H
(**13**) R = O, R′ = H, R″ = OH
X = X′ = X″ = H
(**14**) R = O, R′ = Na, R″ = NaO
X = X′ = X″ = H

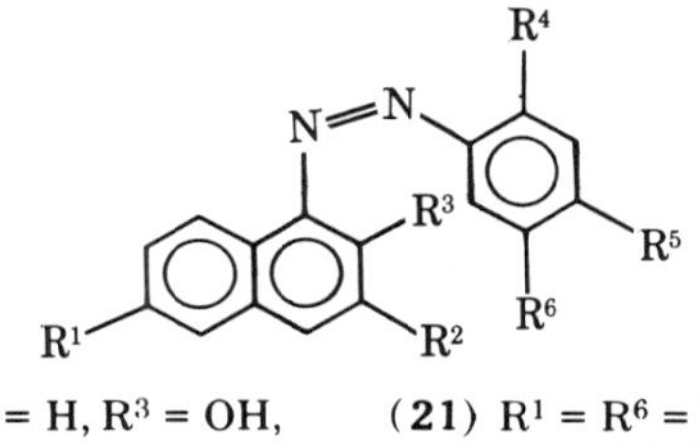

(**15**) R^1 = R^5 = SO$_3$Na, R^2 = H, R^3 = OH,
R^4 = OCH$_3$, R^6 = CH$_3$
(**16**) R^1 = R^2 = R^5 = H, R^3 = OH,
R^4 = R^6 = OCH$_3$
(**17**) R^1 = R^5 = SO$_3$Na, R^2 = R^4 = R^6 = H,
R^3 = OH
(**18**) R^1 = R^2 = R^4 = R^6 = H, R^3 = OH,
R^5 = SO$_3$Na
(**19**) R^1 = R^2 = R^6 = H, R^3 = OH,
R^4 = R^5 = NO$_2$
(**20**) R^1 = R^6 = H, R^2 = CO$_2$Na, R^3 = OH,
R^4 = SO$_3$Na, R^5 = CH$_3$

(**21**) R^1 = R^6 = H, R = CO$_2$Ca/2, R^3 = OH
R^4 = SO$_3$Ca/2, R^5 = CH$_3$
(**22**) R^1 = R^2 = H, R^3 = OH, R^4 = SO$_3$Na,
R^5 = Cl, R^6 = CH$_3$
(**23**) R^1 = R^2 = H, R^3 = OH, R^4 = SO$_3$Ba/2,
R^5 = Cl, R^6 = CH$_3$
(**24**) R^1 = R^2 = R^4 = R^6 = H, R^3 = OH,
R^5 = —N = N—C$_6$H$_5$
(**25**) R^1 = R^4 = R^5 = R^6 = H, R^2 = CO$_2$Ca/2,
R^3 = OH
(**26**) R^1 = R^2 = R^6 = H, R^3 = OH,
R^4 = Cl, R^5 = NO$_2$

583

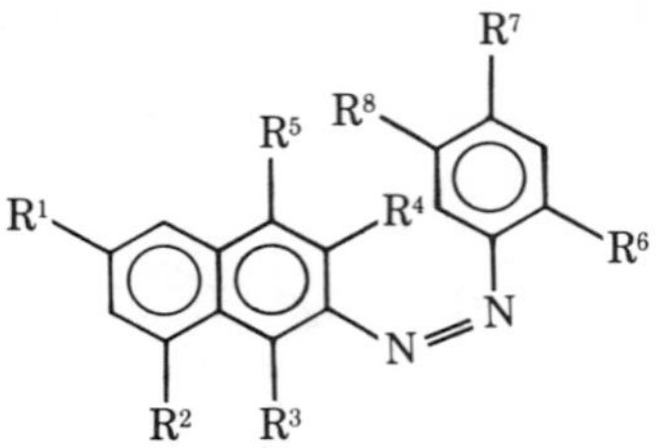

(27) $R^1 = R^2 = R^4 = H$, $R^3 = OH$,
$R^5 = R^8 = SO_3Na$, $R^6 = R^7 = CH_3$

(28) $R^1 = R^4 = SO_3Na$, $R^2 = NH_2$,
$R^3 = OH$, $R^5 = R^6 = R^7 = R^8 = H$

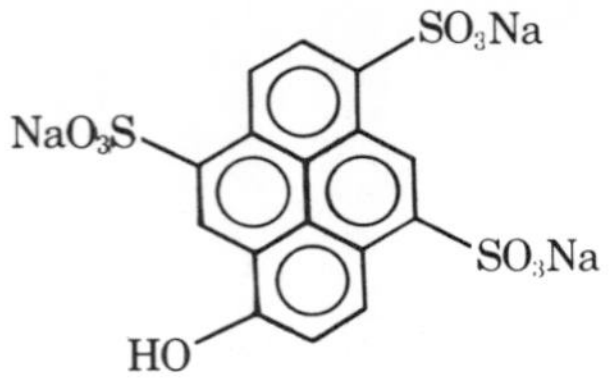

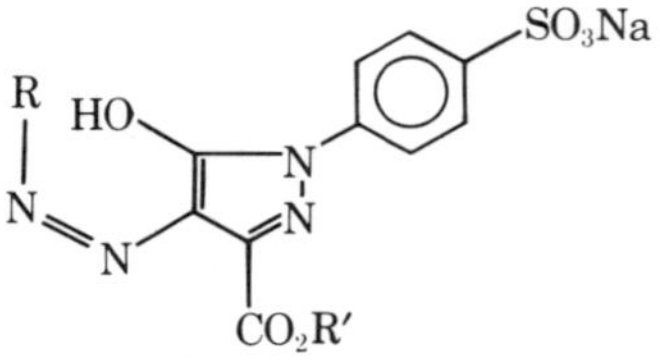

(29) $R = $  —⟨⟩— SO_3Na, $R' = Na$

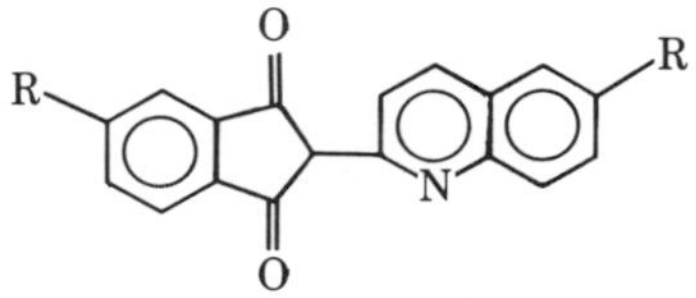

(30) $R = $ —⟨⟩ SO_3Na, $R' = CH_2CH_3$

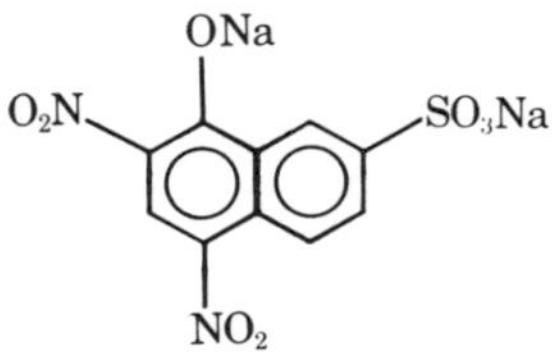

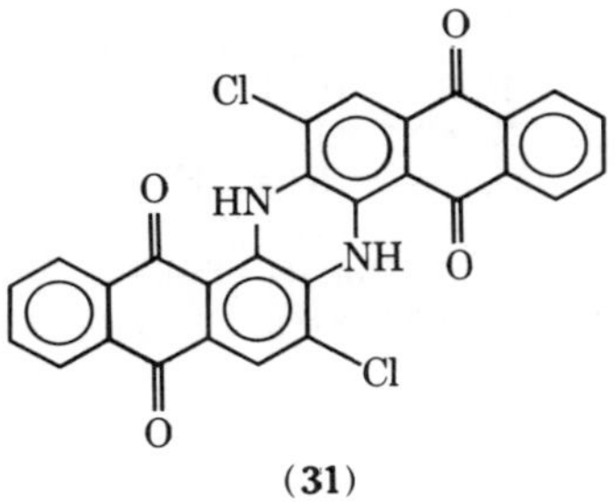

(31)

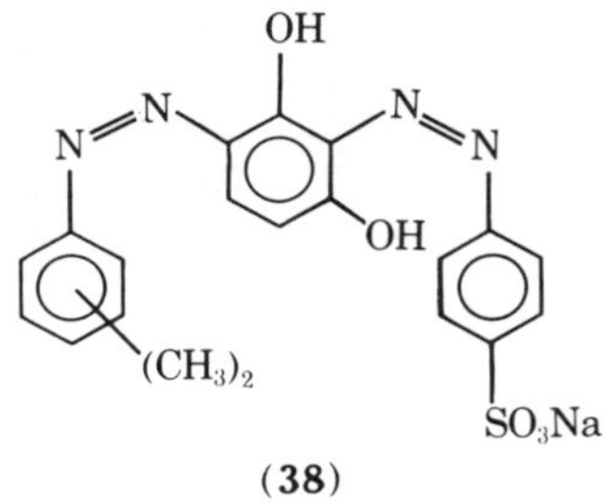

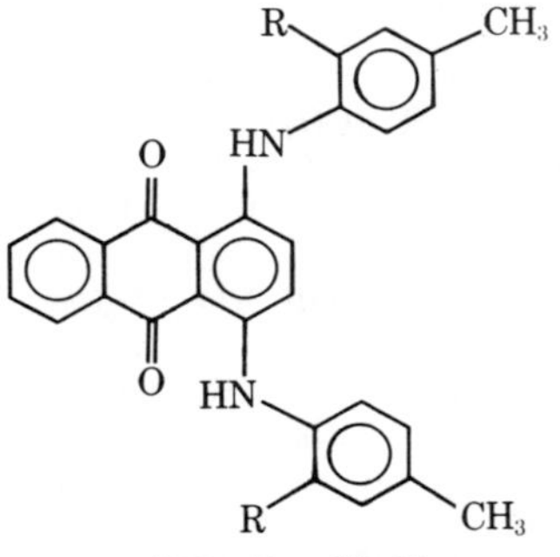

(32) $R = SO_3Na$
(33) $R = H$

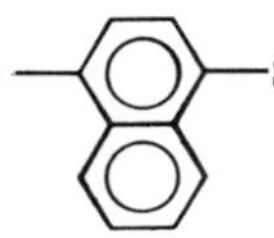

(34)

(35) $R = H$ or SO_3Na
(36) $R = H$

(37)

(38)

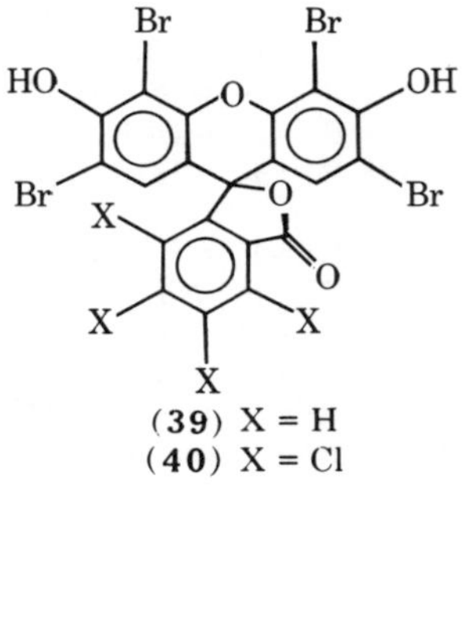

(39) $X = H$
(40) $X = Cl$

584

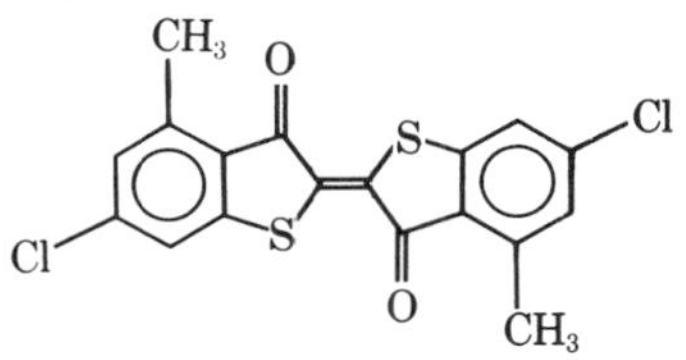

(41)

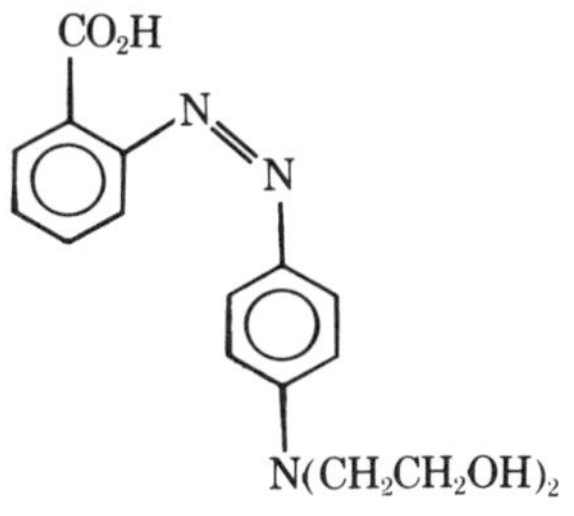

(43)

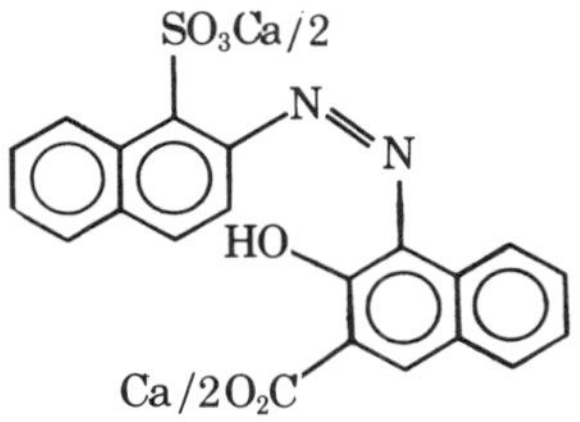

(42)

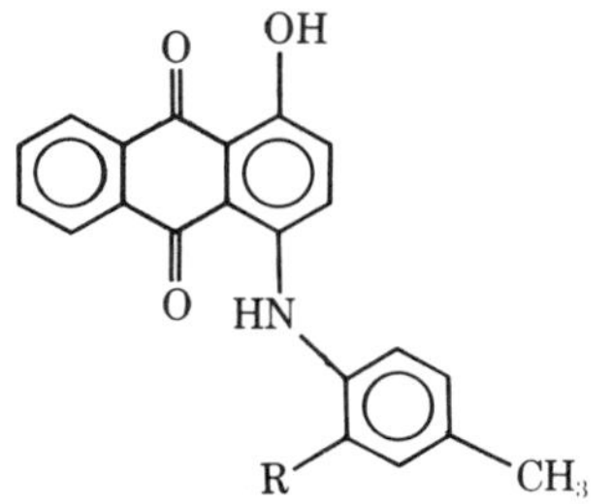

(44) R = H
(45) R = SO$_3$Na

Table 7. Permanently Listed Food Colorants Subject to Certification

Colorant	SFR sect. no.	Use restrictions[a]
FD&C Blue No. 1[b]	74.101	none
Orange B	74.250	may only be used for coloring casings or surfaces of frankfurters and sausages; may not exceed 150 ppm by weight of finished food
Citrus Red No. 2	74.302	may only be used for coloring orange skins; may not exceed 2 ppm by weight of whole fruit
FD&C Red No. 3[b]	74.303	none
FD&C Red No. 40	74.340	none
FD&C Red No. 40 Lake	74.340	none
FD&C Yellow No. 5[b]	74.705	none[c]

[a] No colorant may be used in foods for which standards of identity have been promulgated under Section 401 of the Federal Food, Drug, and Cosmetic Act, unless the use of added color is authorized by such standards.

[b] Lake provisionally listed under 81.1a and 82.51.

[c] The *Federal Register* of February 4, 1977 (20) proposed that all food containing FD&C Yellow No. 5 be required to declare its presence on the label.

Provisionally Listed Colorants Subject to Certification

Tables 10 and 11 list all presently approved colorants that are provisionally listed and subject to certification. The list is divided into two categories: foods, and drugs and cosmetics. The tables also include the Code of Federal Regulations section number, use restrictions for each product, and the closing date for each product.

Two cosmetic colorants are provisionally listed and exempt from certification. They are provisionally listed for cosmetic use only and are shown in Table 12.

Table 8. Permanently Listed Drug Colorants Subject to Certification

Colorant	CFR sect. no.	Use restrictions[a]
phthalocyaninato (2−) copper	74.1045	may only be used to color polypropylene sutures for general and ophthalmic surgery; may not exceed 0.5 wt % of the suture
FD&C Blue No. 1 (also see provisional listings 82.101 and 81.1a)	74.1101	ingested use only
FD&C Blue No. 2 (also see provisional listings 82.102) and 81.1a: ingested use only	74.1102	may only be used to color nylon 66 (the copolymer of adipic acid and hexamethylenediamine); surgical sutures for general surgery; may not exceed 1 wt% of the suture
D&C Blue No. 4	74.1104	external use only
D&C Blue No. 6 (Lakes; also see provisional listing 82.1106)	74.1106	not to exceed 0.2 wt % of poly(ethylene terephthalate) surgical suture for general surgical use; not to exceed 0.25 wt % of the suture material for coloring plain or chromic collagen absorbable sutures for general surgical use; not to exceed 0.5 wt % for coloring plain or chromic collagen absorbable sutures for ophthalmic surgical use; not to exceed 0.5 wt % for coloring polypropylene surgical sutures for general surgical use
D&C Blue No. 9	74.1109	may only be used to color cotton and silk surgical sutures for general and ophthalmic surgery; may not exceed 2.5 wt % of the suture
D&C Green No. 5 (also see provisional listings 82.1205 and 81.1b: no restrictions)	74.1205	may only be used for coloring nylon 66 and nylon 6 (the homopolymer of ϵ-caprolactam); nonabsorbable surgical sutures for use in general surgery; may not exceed 0.6 wt % of the suture
D&C Green No. 6 (also see provisional listings 82.1206 and 81.1b: no restrictions)	74.1206	may only be used for coloring poly(ethylene terephthalate) or polyglycolic acid surgical sutures for general and ophthalmic surgery; when used with poly(ethylene terephthalate) may not exceed 0.75 wt % of suture; when used with polyglycolic acid may not exceed 0.1 wt % of the suture
D&C Green No. 8	74.1208	external use only; may not exceed 0.01 wt% of finished product
D&C Orange No. 4	74.1254	external use only
FD&C Red No. 3	74.1303	ingested use only
(also see provisional listings 82.303 and 81.1a: no restrictions)		
FD&C Red No. 4	74.1304	external use only
D&C Red No. 17	74.1317	external use only
D&C Red No. 31	74.1331	external use only
D&C Red No. 34	74.1334	external use only
(also see provisional listings 82.1334 and 81.1b: external use only)		
D&C Red No. 39	74.1339	may only be used to color quaternary ammonium-type germicidal solutions intended for external application only; may not exceed 0.1 wt % of the finished product
FD&C Red No. 40	74.1340	none
D&C Violet No. 2	74.1602	external use only; may also be used for coloring glycolic–lactic acid polyester (USAN polyglactin 910) synthetic absorbable sutures for use in general and ophthalmic surgery; in sutures the colorant may not exceed 0.2 wt % of the suture
FD&C Yellow No. 5	74.1705	ingested use only[b]
(also see provisional listings 82.705 and 81.1a: no restrictions)		
D&C Yellow No. 7	74.1707	external use only
External D&C Yellow No. 7	74.1707a	external use only

Table 8. (*continued*)

Colorant	CFR sect. no.	Use restrictions[a]
D&C Yellow No. 8	74.1708	external use only
D&C Yellow No. 11	74.1711	external use only

[a] No colorant is certified for use in the area of the eye. In addition, no color additive in this table is certified for use in injectable drugs or surgical sutures unless specifically stated for such use.

[b] The *Federal Register* of February 4, 1977 (21) proposed that the use of FD&C Yellow No. 5 in drugs be declared in the form of a precautionary label statement, ie "this product contains FD&C Yellow No. 5 which may cause allergic-type reactions in certain susceptible individuals." Also proposed was that FD&C Yellow No. 5 not be permitted in analgesic, anti-histaminic, cough and cold, oral nasal decongestant, and anti-asthmatic drugs.

Table 9.   Permanently Listed Cosmetic Colorants Subject to Certification

Colorant	CFR sect. no.	Use restrictions[a]
D&C Blue No. 4	74.2104	external use only
D&C Brown No. 1	74.2151	external use only[b]
D&C Green No. 8	74.2208	external use only
		may not exceed 0.01 wt % of the finished product
FD&C Red No. 4	74.2304	external use only
D&C Red No. 17	74.2317	external use only
D&C Red No. 31	74.2331	external use only
D&C Red No. 34	74.2334	external use only
(also see provisional listings 82.1334 and 81.1b: external use only)		
D&C Orange No. 4	74.2254	external use only
FD&C Red No. 40	74.2340	none
D&C Violet No. 2	74.2602	external use only
External D&C Violet No. 2	74.2602a	external use only
D&C Yellow No. 7	74.2707	external use only
External D&C Yellow No. 7	74.2707a	external use only
D&C Yellow No. 8	74.2708	external use only
D&C Yellow No. 11	74.2711	external use only

[a] No colorant in this table is certified for use in the area of the eye.

[b] D&C Brown No. 1 is an approved colorant, but is not presently manufactured because FDA specifications make the processing of this color too complicated.

Colorant Regulations

The Code of Federal Regulations of April 1978 contains the most recent version of the colorant regulations. Additional changes were made in later *Federal Registers,* the most recent on December 13, 1977.

Some of the regulations important to users of food, drug, and cosmetic color additives, with CFR numbering, are as follows:

General Provisions. *CFR Section 70.3—Definitions.* (f) A colorant (color additive) is any material not exempted under section 201 (t) of the Act, that is a dye, pigment, or other substance made by a process of synthesis or similar artifice, or extracted, isolated, or otherwise derived, with or without intermediate or final change of identity from a vegetable, animal, mineral, or other source. When added or applied to a food,

Table 10.   Colorants Subject to Certification and Provisionally Listed for Use in Foods

Colorant	CFR sect. no.	Closing date	Use restrictions
FD&C Blue No. 2	81.1a	Jan. 31, 1981	ingested use only
FD&C Green No. 3	81.1a	Jan. 31, 1981	none
FD&C Yellow No. 6	81.1a	Jan. 31, 1981	none
Lakes (FD&C) (except FD&C Red No. 40 Lake)	82.51	(same as straight color)	

drug or cosmetic or to the human body, the colorant is capable (alone or through reaction with another substance) of imparting a color thereto. Substances capable of imparting a color to a container for foods, drugs, or cosmetics are not color additives unless the customary or reasonably foreseeable handling or use of the container may reasonably be expected to result in the transmittal of the color to the contents of the package or any part thereof. Food ingredients such as cherries, green or red peppers, chocolate, and orange juice that contribute their own natural color when mixed with other foods are not regarded as color additives; but where a food substance such as beet juice is deliberately used as a color, as in pink lemonade, it is a color additive.

(g) For a material otherwise meeting the definition of colorant to be exempt from section 706 of the Act, on the basis that is used (or intended to be used) solely for a purpose or purposes other than coloring, the material must be used in a way that any color imparted is clearly unimportant insofar as the appearance, value, marketability, or consumer acceptability is concerned. (It is not enough to warrant exemption if conditions are such that the primary purpose of the material is other than to impart color).

(i) The term safe means that there is convincing evidence that establishes with reasonable certainty that no harm will result from the intended use of the colorant.

(j) The term straight color means a colorant listed in Parts 71 and 81 of this chapter, and includes lakes and such substances as are permitted by the specifications for such colorants. (Part 71—Colorant petitions; Part 81—General specifications and general restrictions for provisional colorant for use in foods, drugs, and cosmetics).

(k) The term mixture means a colorant made by mixing two or more straight colors, or one or more straight colors and one or more diluents.

(l) The term lake means a straight color extended on a substratum by absorption, coprecipitation, or chemical combination that does not include any combination of ingredients made by simple mixing process.

(m) The term diluent means any component of a colorant mixture that is not of itself a colorant and has been intentionally mixed therein to facilitate the use of the mixture in coloring foods, drugs, or cosmetics or in coloring the human body. The diluent may serve another functional purpose in the food, drug, or cosmetic, as for example sweetening, flavoring, emulsifying, or stabilizing, or may be a functional component of an article intended for coloring the human body.

(n) The term substratum means the substance on which the pure color in a lake is extended.

(o) The term pure color means the color contained in a colorant, exclusive of any intermediate or other component, or of any diluent or substratum contained therein.

(s) The term area of the eye means the area enclosed within the circumference of the supra-orbital ridge and the infra-orbital ridge, including the eyebrow, the skin below the eyebrow, the eyelids and eyelashes, and conjunctival sac of the eye, the eyeball and the soft aerolar tissue that lies within the perimeter of the infra-orbital ridge (see Color).

(u) The hair dye exemption in section 601 (2) of the Act applies to coal-tar hair dyes intended for use in altering the color of the hair and which are, or which bear or contain, color additives derived from coal-tar with the sensitization potential of causing skin irritation in certain individuals and possible blindness when used for dyeing the eyelashes or eyebrows. The exemption is permitted with the condition that the label of any such article bear conspicuously the statutory caution and adequate directions for preliminary patch-testing. The exemption does not apply to coloring ingredients in hair dyes not derived from coal tar, and it does not extend to poisonous or deleterious diluents that may be introduced as wetting agents, hair conditioners, emulsifiers, or other components in a color shampoo, rinse, tint, or similar dual-purpose cosmetic that alter the color of the hair.

(v) The terms externally applied drugs and externally applied cosmetics mean drugs or cosmetics applied only to external parts of the body and not to the lips or any body surface covered by mucous membrane.

70.11 (a) Different colorants may cause similar or related pharmacological or biological effects, and in the absence of evidence to the contrary, those that do so are considered to have additive toxic effects.

General Specifications. All colorants must be free from impurities other than those listed below to the extent that such impurities can be avoided by good manufacturing practices.

General Specifications for Straight Colors—CFR Section 82.5, FD&C Colorants (with CFR numbering). (I) Lead (as Pb), not more than 0.001%. (II) Arsenic (as As_2O_3), not more than 0.00014%. (III) Heavy Metals (except Pb and As, not more than a trace (by precipitation as sulfide).

D&C and External D&C Colorants. (I) Lead (as Pb), not more than 0.002%. (II) Arsenic (as As_2O_3), not more than 0.0002%. (III) Heavy Metals (except Pb and As), not more than 0.003% (by precipitation as sulfides). (3) Soluble barium (in dilute HCl, not more than 0.05% (as $BaCl_2$).

Section 82.51—FD&C Lakes. (2) Must be prepared from previously certified color. Soluble chlorides and sulfates (as sodium salts), not more than 2.0%. Inorganic matter, insoluble HCl, not more than 0.5%. Only alumina substratum. Only aluminum or calcium salts.

Section 82.1051—D&C Lakes. (2) Ether extracts, not more than 0.5%. Soluble chlorides and sulfates (as sodium salts), not more than 3.0%. Intermediates, not more than 0.2%. (a) Substratum may be alumina, blanc fixe, gloss white, clay, titanium dioxide, zinc oxide, talc, rosin, aluminum benzoate, calcium carbonate, or any combination of these. (I) Salts may be prepared using sodium, potassium, aluminum, barium, calcium, strontium, or zirconium radicals.

Section 82.2051—External D&C Lakes. (2) Ether extract, not more than 0.5%. Soluble chlorides and sulfates (as sodium salts), not more than 3.0%. Intermediates, not more than 0.2%. (2) Substratum and salts same as Section 82.1051—D&C Lakes.

Table 11. Colorants Subject to Certification and Provisionally Listed for Use in Drugs and Cosmetics

Colorant	CFR sect. no.	Closing date	Use restrictions[a]
FD&C Blue No. 1[b]	81.1a	Jan. 31, 1981	none
FD&C Blue No. 2	81.1a	Jan. 31, 1981	ingested use only
FD&C Green No. 3	81.1a	Jan. 31, 1981	none
FD&C Yellow No. 5[b]	81.1a	Jan. 31, 1981	none
FD&C Yellow No. 6	81.1a	Jan. 31, 1981	none
FD&C Red No. 3[b]	81.1a	Jan. 31, 1981	none
Lakes (FD&C)	82.51	(same as straight color)	
D&C Green No. 5	82.1205	Jan. 31, 1981	none
D&C Green No. 6	82.1206	Jan. 31, 1981	none
D&C Yellow No. 10	82.1710	Jan. 31, 1981	none
D&C Red No. 6	82.1306	Jan. 31, 1981	none
D&C Red No. 7	82.1307	Jan. 31, 1981	none
D&C Red No. 8	82.1308	Jan. 31, 1981	may not exceed 6% pure dye when used in lipstick; may not exceed 0.75 mg per daily dosage of pure dye when used in internally used drugs, mouthwashes, dentifrices and proprietary drug products[c]
D&C Red No. 9	82.1309	Jan. 31, 1981	may not exceed 6% pure dye when used in lipsticks and lip products
D&C Red No. 19	82.1319	Jan. 31, 1981	same restrictions as D&C Red No. 8
D&C Red No. 21	82.1321	Jan. 31, 1981	none
D&C Red No. 22	82.1322	Jan. 31, 1981	none
D&C Red No. 27	82.1327	Jan. 31, 1981	none
D&C Red No. 28	82.1328	Jan. 31, 1981	none
D&C Red No. 30	82.1330	Jan. 31, 1981	none
D&C Red No. 33	82.1333	Jan. 31, 1981	same restrictions as D&C Red No. 8
D&C Red No. 34	82.1334	none established	external use only
D&C Red No. 36	82.1336	Jan. 31, 1981	may not exceed 3% pure dye when used in lipstick; may not exceed 1.7 mg per daily dosage of pure dye when used in internally used drugs, mouthwashes, dentifrices and proprietary drug products
D&C Red No. 37	82.1337	Jan. 31, 1981	external and ingested drug use; ingested drugs may not exceed 0.75 mg per daily dosage
D&C Orange No. 5	82.1255	Jan. 31, 1981	same restrictions as D&C Red No. 8
D&C Orange No. 10	82.1260	Jan. 31, 1981	none
D&C Orange No. 11	82.1261	Jan. 31, 1981	none

Table 11. (*continued*)

Colorant	CFR sect. no.	Closing date	Use restrictions[a]
D&C Orange No. 17	82.1267	Jan. 31, 1981	same restrictions as D&C Red No. 9
D&C Lakes	82.1051	same as straight color	

[a] These color additives may not be used for surgical sutures unless so specified and may not be used in the area of the eye.
[b] Permanently listed for ingested drug use. See Table 8.
[c] No combination of the colors D&C Orange No. 5 and 17 and D&C Red No. 8, 9, 19, 33, and 36 may exceed 6% in lipstick.

Table 12.　Cosmetic Colorants Provisionally Listed and Exempt for Certification

Colorant	CFR sect. no.	Closing date	Use restrictions
caramel	81.1 (g)	Jan. 31, 1981	none
lead acetate	81.1 (g)	April 30, 1979	for use only as a color component in hair dyes
ferric ferrocyanide	81.1 (g)	Nov. 30, 1978	external use including eye area

Health and Safety Aspects

The following are several case histories of delisted colorants. They are examples of the current regulatory climate and are useful indicators of the problems encountered by manufacturers of color additives.

FD&C Red No. 2. In 1968, 1969, and 1970, reports of studies in the USSR on amaranth (22–24) raised concerns about possible carcinogenic and reproductive effects of FD&C Red No. 2 and other food colors. Evaluation of the USSR work by scientists in the United States, Canada, United Kingdom, the WHO and the FAO (25–30) showed this work to be faulty and conclusions incorrect because FD&C Red No. 2 was never tested in the USSR, no tumors were reported in the control rats, a highly unusual result in view of the normal spontaneous incidence of cancer in rats, and the cancer types found in the experiment were those types usually found in old rats and were neither organ- nor tumor-specific.

For years studies have been conducted with FD&C Red No. 2 to establish its safety. Acute and chronic studies with laboratory animals from the early 1900s to the 1960s revealed no adverse effects as a result of feeding the color additive (27).

In an attempt to help clarify the situation, the FDA called upon the Committee on Food Protection of the NAS/NRC to assess the data on the safety of FD&C Red No. 2. In the opinion of the committee (27–28), "restrictions on its usage was not warranted."

However, the FDA in view of the doubts raised by the USSR studies requested the petition sponsors to provide reproduction and teratology studies for all provisionally listed color additives intended for ingestion. The FDA decided to do the chronic rat feeding study on FD&C Red No. 2 in its own labs.

A task force of the Inter-Industry Color Committee completed the requested teratogen and multi-generation reproduction studies and submitted the results to the FDA in March, 1974.

A Toxicological Advisory Committee met in November, 1975, to assess the safety of FD&C Red No. 2 (13). The committee did not resolve the issue because the FDA's chronic rat feeding study mentioned above was "botched." There had been a mix-up in the diets fed to some of the rats and many of the tissues were badly decomposed, making microscopic examination difficult. The pathologist who conducted the study, however, concluded that "there were no apparent adverse effects produced in rats when continuously fed FD&C Red No. 2 in their diets at dose levels of 0.003, 0.03, and 3% for 131 weeks."

Whether FD&C Red No. 2 is a cancer-causing agent still remains unresolved. However, on Feb. 12, 1976, the FDA terminated the provisional listing of FD&C Red No. 2. The color additive was banned not because it was demonstrated beyond a doubt that it is a carcinogenic agent, but because of existing uncertainties about its safety. The Commissioner concluded that the adverse implications of the FDA study cannot be ignored. In light of continuing public concern and the serious new questions about carcinogenesis raised by Gaylor's analysis, the Commissioner concluded that a study adequate to dispel all such questions must be performed before the color additive can be demonstrated to be safe as required by the act.

FD&C Red No. 2 (amaranth) is still permitted to be used in Canada and many other countries. The Canadian Health Protection Branch reviewed the results of the FDA study and concluded (30), "The recent decision by the FDA of the United States to propose a ban on the use of Amaranth in that country raised legitimate concern among Canadian consumers over its use in this country. This concern however is not substantiated by the available scientific evidence."

The CCMA requested a hearing to reinstate FD&C Red No. 2 on the basis of experimental evidence. The hearing was granted and began in April, 1977. Much data and expert testimony were presented to show that FD&C Red No. 2 is not a carcinogenic agent. Nevertheless, the administrative law judge denied the CCMA's request. The CCMA has appealed the decision to the FDA Commissioner. As of February 1979, no decision has been made.

FD&C Red No. 4.   On December 11, 1964, the FDA terminated the listing of FD&C Red No. 4 for use in food and ingested drugs and cosmetics (31). This action was based on a 7-yr dog-feeding study that showed adverse effects in the urinary bladder and adrenal glands; chronic tests on rats and mice showed no effects.

In 1965 the FDA returned FD&C Red No. 4 to the provisional list for use in maraschino cherries at a maximum level of 150 ppm (32) and in ingested drugs on a restricted short term basis (33). The Commissioner concluded that this limited use presented "no potential for harm to the public."

The question of whether the adverse effects noted were caused by catheterization was raised following a study completed in 1968–1969. The FDA decided that an additional study was necessary before a final determination could be made concerning the safety of this color additive.

By mid-1976 no new study on FD&C Red No. 4 was available or underway. Hence, on September 23, 1976 (34), the FDA delisted the color for use as a food or ingested drug or cosmetic additive. FD&C Red No. 4 is permanently listed for external use in drugs and cosmetics.

The trade association of the maraschino cherry industry requested an administrative hearing on FD&C Red No. 4. The administrative law judge ruled against the industry.

Again, the Canadian Health Protection Branch did not conclude that FD&C Red

No. 4 constituted a danger to the public and did not restrict the colorant's use in Canada. However, based on the U.S. action, many other countries, including the EEC (European Economic Community or Common Market), prohibit its use.

Carbon Black. The FDA voiced concern about the safety of channel or impingement carbon black (qv). Their concern was generated by the possibility that extractable polynuclear aromatic hydrocarbons (PNAs) may be present in carbon blacks as a by-product of the manufacturing process. Some PNAs are known carcinogens. Toxicity studies on carbon black have disclosed no adverse effects. However, on September 23, 1976, the FDA banned the use of carbon black in foods, drugs, and cosmetics (35) because the absence of PNAs in the colorant could not be demonstrated. There is no analytical method sufficiently sensitive to detect extractable PNAs in carbon black at extremely low levels (ppb) because of the nature of the carbon black particle.

Carbon black is still an approved color additive in many countries, including Canada.

FD&C Yellow No. 5. On February 4, 1977, the FDA proposed that FD&C Yellow No. 5 (tartrazine) be identified specifically in food products by ingredient labeling. It further proposed that the colorant be banned from the following categories of over-the-counter and prescription drugs that are allergy related: analgesic drugs, anti-histaminic drugs, cough and cold preparations, anti-asthmatics, nasal decongestants, nonsteroidal anti-inflammatory drugs and glucocorticoid drugs. Drugs not affected by the ban would be required to carry on their label the warning statement: "This product contains FD&C Yellow No. 5 which may cause allergic-type reactions in certain susceptible people" (36).

The proposed order allowed 60 d for interested parties to respond to the FDA. The FDA will issue new regulations concerning the use of FD&C Yellow No. 5 after they complete the review of comments from the public and industry. It was proposed that industry have one year to comply with any new regulations.

For the external use of FD&C Yellow No. 5, the FDA has stated: "There are no reports of reactions to FD&C Yellow No. 5 from external applications and accordingly the use of the color additive in externally applied drugs and cosmetics is not considered to present a likelihood of allergic-type response" (37). A final order has not yet been issued on FD&C Yellow No. 5 as of February 1, 1979.

Terminated Colorants. The Code of Federal Regulations of April, 1978 summarized past terminations of provisionally listed colorants. The FDA wrote a statement explaining the termination of Ext. D&C Yellow No. 9 (formerly FD&C Yellow No. 3) and Ext. D&C Yellow No. 10 (formerly FD&C Yellow No. 4).

"These colors cannot be produced with any assurance that they do not contain β-naphthylamine as an impurity. Although it has been asserted that the two colors can be produced without the impurity named, no method of analysis has been suggested to establish the fact. β-Naphthylamine is a known carcinogen; therefore, there is no scientific evidence that will support a safe tolerance for these colors in products to be used in contact with the skin. The Commissioner of Food and Drugs, having concluded that such action is necessary to protect the public health, hereby terminates the provisional listing of Ext. D&C Yellow No. 9 and Ext. D&C Yellow No. 10" (14).

The FDA has established a policy with the preceding statement: just because a questionable chemical substance cannot be detected in a color additive does not mean that the substance is not present. As of October 3, 1978, this principle is being applied to Orange B.

The colorant industry is in a dynamic state. The FDA is questioning the chemical analysis and/or toxicity of several presently listed colorants. Concerning the colorants with questions of chemistry, the FDA stated: "Of the 72 provisionally listed color additives, 15 cannot be permanently listed at this time because complete chemistry data are lacking to establish specifications for them. The chemistry data lacking on these 15 colors consist of sufficiently precise analytical methods and other information to enable the FDA to identify and define the color additives more accurately than currently available data permit."

The *Federal Register* of February 4, 1977 (38), established October 31, 1977, as a closing date for nine of these colors. Six of the original 15 had received FDA approval on chemistry based on the data supplied by industry. As of December 13, 1977, only 3 of the color additives had unresolved chemistry questions. They are: D&C Red No. 6; D&C Red No. 7; and D&C Red No. 30.

The FDA postponed the closing date on these color additives until October 31, 1978, and required the chemistry question be resolved by July 31, 1978. The initial chemical questions on D&C Red No. 6 and D&C Red No. 7 have been resolved; however the FDA has presented a new type of "elementary" question, ie, do these colors have a possible content of *o*-toluidine. The CCMA has submitted the requested chemical data on D&C Red No. 30. As of March 1, 1979 there has been no response by the FDA.

The FDA has expressed concern about toxicity studies conducted on color additives during the 1950s and 1960s. After careful study of the data available, the FDA concluded (39): "Of the 72 provisionally listed color additives, 31 cannot be permanently listed because the available toxicity data—though suggestive of no diverse effects—are derived from studies that do not meet contemporary standards generally accepted within the scientific community as minimum requirements for the design and conduct of toxicological studies to establish safety."

The *Federal Register* of December 13, 1977 revised this list to 25 color additives and postponed the closing date of provisional listing to January 31, 1981 (40).

These 25 color additives are: FD&C Yellow No. 5; FD&C Yellow No. 6; D&C Yellow No. 10; FD&C Red No. 3; D&C Red No. 6; D&C Red No. 7; D&C Red No. 8; D&C Red No. 9; D&C Red No. 19; D&C Red No. 21; D&C Red No. 22; D&C Red No. 27; D&C Red No. 28; D&C Red No. 30; D&C Red No. 33; D&C Red No. 36; D&C Red No. 37; FD&C Green No. 3; D&C Green No. 5; D&C Green No. 6; FD&C Blue No. 1; FD&C Blue No. 2; D&C Orange No. 5; D&C Orange No. 17; and caramel. By the closing date of January 31, 1981, chronic rat-feeding studies must be conducted and evaluated on these colorants. An evaluation of caramel by a mouse skin painting study was ordered to be completed by March 17, 1978. The Cosmetic Trade Association decided not to make the latter test because of their confusion in determining the kind of caramel to be used.

To further complicate the color additive situation, the Health Research group, a Ralph Nader-affiliated organization asked the U.S. District Court for the District of Columbia to order the FDA to immediately withdraw the extended closing date for the colorants requiring toxicity studies. The suit contended that any extension of the closing date for provisionally listed color additives violated the Food, Drug, and Cosmetic Act and represented a hazard to the public health.

In a counter action, the Department of Justice, the CTFA, and the CCMA asked the Federal District Court for the District of Columbia for a summary judgement dismissing the suit on the grounds that the action taken by the FDA was both "reasonable and lawful," and that the new studies are required not to eliminate existing doubts about the safety of these color additives, but to assure that any final actions taken by the FDA on these colorants are based on "research meeting the most exacting standards of modern toxicology."

In late September, 1977, the Federal District Court issued a summary judgement dismissing the suit.

The FDA has indicated that it is contemplating a cyclic review program for all permanently listed colorants (41). This program would be designed to continually review the permanently listed colorants in light of contemporary scientific standards.

The future of natural colorants also remains uncertain. The FDA has indicated that natural does not necessarily assure the color is safe (see Food toxicants, naturally-occurring). The FDA will require toxicology studies and more complete information about the chemistry of these natural colorants (see also Dyes, natural; Food additives).

BIBLIOGRAPHY

"Colors for Foods, Drugs, and Cosmetics" in *ECT* 1st ed., Vol. 4, pp. 287–313, by Samuel Zuckerman, H. Kohnstamm & Co., Inc.; "Colors for Foods, Drugs, and Cosmetics" in *ECT* 2nd ed., Vol. 5, pp. 857–884, by Samuel Zuckerman, H. Kohnstamm & Co., Inc.

1. H. O. Calvery, *Am. J. Pharm.* **114,** 324 (1942).
2. W. C. Bainbridge, *Ind. Eng. Chem.* **18,** 1329 (1926).
3. U.S. Department of Agriculture, Board of Food & Drug Inspection, *Food Inspection Decision 76,* July 13, 1907.
4. U.S. Department of Agriculture, *FDA Service and Reg. Announcement, Food & Drugs No. 3,* Sept. 1931.
5. *Public Law No. 717,* 75th U.S. Congress, 1938.
6. *Public Law No. 618,* 86th U.S. Congress, 1960.
7. U.S. Supreme Court, 358 U.S. 153, Dec. 15, 1958.
8. *Rept. 795,* 86th U.S. Congress, 1st Session, Senate Calendar No. 807, Aug. 21, 1959, p. 3.
9. Colorants Regulations issued under Title 21 (Code of Federal Regulations) Part 8 (Color Additives) of the Federal Food, Drug & Cosmetic Act as ammended (see ref. 14 for full description of this Act). Issued as per *Fed. Regist.* **28,** 317 (1963).
10. *Ibid.,* 2674 (1963).
11. *Ibid.,* 7424 (1973).
12. Color Additives Transitional Regulations, reissue of Color Additives Regulations, Jan. 31, 1964. As per *Fed. Regist.* **28,** 14311 (1963).
13. *Color Additive FD&C Red No. 2,* Toxicological Advisory Committee Minutes, Nov. 1975.
14. U.S. Department of HEW, FDA, *Report on Certification of Color Additives,* Fiscal years 1973, 1974, 1975, 1976, 1977.
15. *Chem Week,* **86** (Oct. 25, 1978).
16. *Code of Federal Regulations,* Parts 1–99, April 1, 1977, p. 218.
17. *Encyclopedia of Chemical Analysis,* Vol. 10, John Wiley & Sons, Inc., New York, 1970, pp. 447–547.
18. W. Horwitz, ed., *Official Methods of Analysis of the Association of Official Analytical Chemists,* 12th ed., AOAC, Washington, D.C., 1975.
19. K. Vankateraman, ed., *The Analytical Chemistry of Synthetic Dyes,* Vol. 8, Academic Press, New York, 1975.
20. *Fed. Regist.* **42,** 6835 (1977).
21. *Ibid.,* 6837 (1977).
22. Baigusheva, *Vopr. Pitan.* **27,** 46 (1968).

23. M. M. Andrianova, *Vopr. Pitan.* **29,** 61 (1970).
24. A. I. Shtenberg and Y. V. Gavsilenko, *Vopr. Pitan.* **29,** 66 (1970).
25. British Industrial Biological Research Association, *Significance of Recent Studies on Amaranth,* BIBRA Reports, PXXIX, 1972.
26. C. J. Kokoski, *FDA Memorandum of Conference,* Nov. 18, 1971, p. 2.
27. *Report of Ad Hoc Subcommittee on the Evaluation of Red No. 2,* Committee on Food Protection, Food and Nutrition Board, NAS/NRC, 1972.
28. *Report of the Ad Hoc Advisory Group on FD&C Red No. 2 (Amaranth),* NAS/NRC, 1974.
29. H. G. Wilms, *FDA Memorandum,* March 20, 1975, pp. 1–2.
30. Canadian Government Press Release, 1976–12, Feb. 2, 1976.
31. *Fed. Regist.* **29,** 16983 (1964).
32. *Ibid.,* **30,** 10289 (1965).
33. *Ibid.,* 13056 (1965).
34. *Ibid.,* **41,** 41852 (1976).
35. *Ibid.,* 41857 (1976).
36. *Ibid.,* **42,** 6835 (1977).
37. *Ibid.,* 6836 (1977).
38. *Ibid.,* 6997 (1977).
39. *Ibid.,* **41,** 41862 (1976).
40. *Ibid.,* **42,** 62503 (1977).
41. *Ibid.,* **41,** 41863 (1976).

General Reference

T. E. Furia, ed., *Current Aspects of Food Colorants,* CRC Press, Inc., Cleveland, 1977.

SAMUEL ZUCKERMAN
JOSEPH SENACKERIB
H. Kohnstamm & Co., Inc.

COLORANTS FOR PLASTICS

There are three general types of colorants for plastics: dyes, and inorganic and organic pigments. Dyes (qv) are soluble under the conditions of use but must be completely dissolved, leaving no color streaks and little or no haze. Pigments are insoluble and consist of particles that must be dispersed by physical means (see Pigments).

Resins (plastics) fall into two categories: thermoplastics and thermosets. For thermoplastics, the polymerization reaction has been completed, the materials are processed at or close to their melting points, and scrap may be reground and remolded, eg, polyethylene, polypropylene, poly(vinyl chloride), acetal, acrylics, ABS, nylons, cellulosics, and polystyrene. In the case of thermoset resins, the chemical reaction is only partially complete when the colorants are added, and is concluded when the resin is molded. The result is a nonmeltable cross-linked resin which cannot be reworked, eg, epoxy resins, urea–formaldehyde, melamine–formaldehyde, phenolics, and thermoset polyesters. The method of coloring differs depending upon whether the resin is a transparent acrylic or an opaque ABS and also upon the desired transparency or opacity of the final product.

There is possibility of a chemical reaction between a plastic and a colorant at processing temperatures. More often, allowance must be made for additives in the plastic, such as antioxidants (qv), flame retardants (qv), ultraviolet light absorbers (see Uv absorbers), and fillers (qv). Thermal stability is a factor. The final use of the colored resin often dictates the selection of colorant. Obviously, colorants must be evaluated for the user's processing conditions in view of the subsequent fate of fabricated parts. A list of colorants and their CAS Registry Numbers is provided at the end of the article.

Dispersion and Stability

Colorants are chosen for their hue, among other reasons. In matching a shade submitted by a customer, it is desirable to use the same colorant combination so that the match will be valid under all types of illumination (see Color). The process of dispersion of a dry pigment into a liquid vehicle or molten plastic may be visualized as taking place in two steps. These are the breaking up of the pigment agglomerates into the much smaller ultimate particles, and then the displacement of air from the particles to obtain a complete pigment-to-vehicle interface (1). The usual method is to coat the colorant particles onto the surface of the resin granules by simple mixing in a drum tumbler, double cone blender, or ribbon blender. The colorants are intimately mixed by passing over a heated roller mill, a Banbury sigma-bladed mixer, or an extruder. The important point is to have sufficient molten resin at hand to coat fresh exposed surfaces. The colored resin is then cut or ground into small cubes or granules for extrusion or molding. To test for completion of dispersion and colorant stability, the colored resin is put through several cycles. If the color becomes stronger, there is an indication that dispersion was not complete the first time. If the color weakens or becomes duller, it is certain that reaction or decomposition is taking place. The task of pigment dispersion may often be simplified by use of a predispersed colorant concentrate in a resin or by a liquid color (a similar dispersion in a compatible fluid). These two materials may often be used directly in a screw-injection molding machine.

In some cases, it is possible to displace the water from the surface of a pigment in an aqueous presscake simply by blending with an organic liquid. The liquid preferentially wets the pigment, particularly if it is organic, displacing the water which may be poured off. The process is called flushing (2). In this way, the pigment is dispersed in an organic medium without ever having been dried and without the need to redisperse it. The process is widely used in the paint and and printing ink industries. It is useful for plastics where the flushing vehicle is a plasticizer which is intended for incorporation into a resin such as poly(vinyl chloride).

Once the colorant has been incorporated into the plastic, more tests are needed. Lightfastness may be determined by exposure to a carbon arc or xenon light. Outdoor weathering includes exposure to the moist climate of Florida and the hot dry conditions of Arizona. It is often found that the masstone (dark) colors are more weatherable than light shades. Performance requirements may include examination of loss of tensile strength or change in brittleness. The addition of a colorant to a resin may change melt flow, mold shrinkage, electrical conductivity, or impact resistance. In evaluating colorants, it is essential that uncolored controls be run under the same conditions.

Dyes

Dyes should be checked for migration, sublimation, and heat stability before use. These precautions are particularly important for plasticized resins.

Azo Dyes. The Colour Index classifications of dyes depend more upon their historical early use than upon their structures, eg, Oil Orange is named Solvent Yellow 14, and a yellow for synthetic fibers is Disperse Yellow 23.

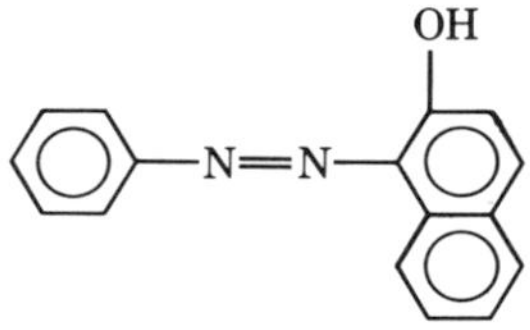

Solvent Yellow 14

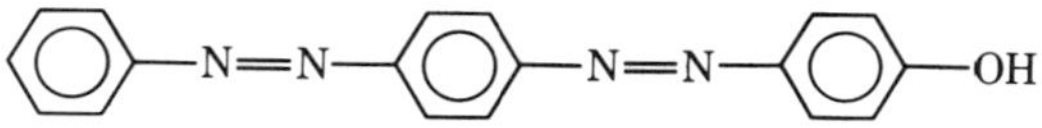

Disperse Yellow 23

In general, the azo colors are useful for coloring polystyrene, phenolics, and rigid poly(vinyl chloride). Many are compatible with poly(methyl methacrylate) but in this case the weatherability of the resin far exceeds the life of the dyes. Among the more widely used azo dyes are Solvent Yellows 14 and 72, Orange 7, and Reds 1, 24, and 26 (see Azo dyes).

Azo acid dyes, of which Metanil Yellow is an example, are stabilized by sulfonic acid groups and also have affinity for phenolic resins.

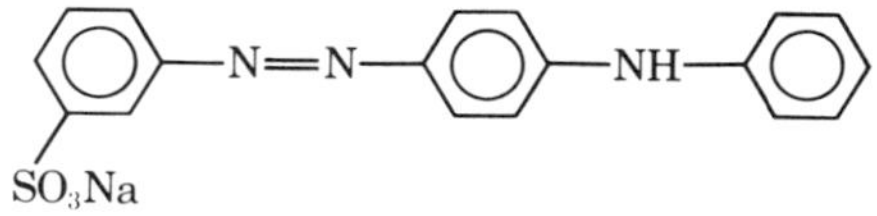

Metanil Yellow (Acid Yellow 36)

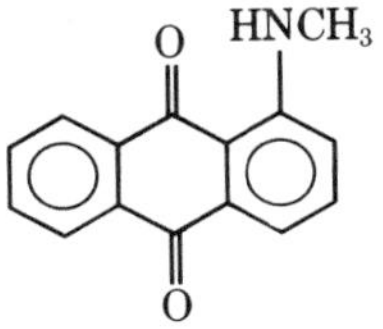

Solvent Red 111

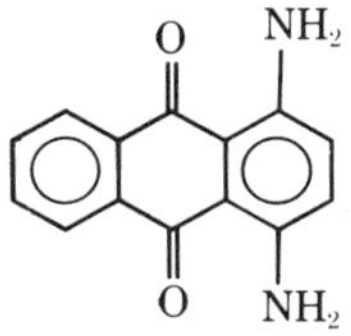

Disperse Violet 1

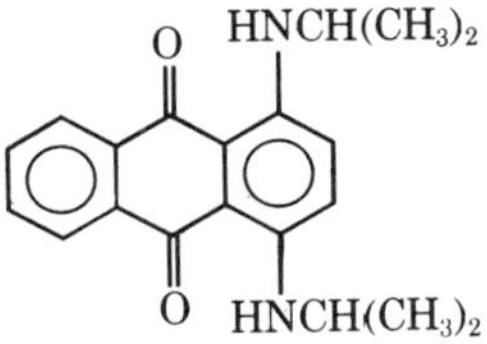

Solvent Blue 56

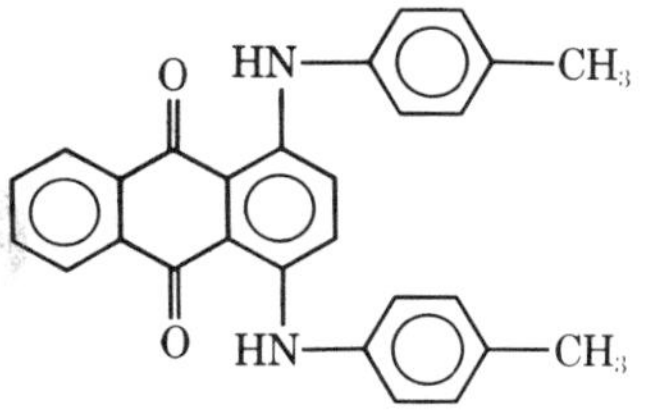

Solvent Green 3

Anthraquinone Dyes. Anthraquinone dyes have much superior weatherability and heat stability than the azos, but at higher cost. Typical examples are Solvent Red 111, Disperse Violet 1, Solvent Blue 56, and Solvent Green 3 (see Dyes, anthraquinone).

The anthraquinones are very useful in acrylics and are compatible with polystyrene and cellulosics. Solvent Red 111 has a special affinity for poly(methyl methacrylate) as the red in automobile taillights; exposure for a year in Florida or Arizona produces only a very slight darkening. Again, acid types are useful for phenolics.

Xanthene Dyes. This class is best represented by Rhodamine B. It has high fluorescent brilliance but poor light and heat stability; it may be used in phenolics. Sulfo Rhodamine is stable and is useful in nylon 66. Other xanthenes used in acrylics, polystyrene, and rigid poly(vinyl chloride) are Solvent Green 4, Acid Red 52, Basic Red 1, and Solvent Orange 63.

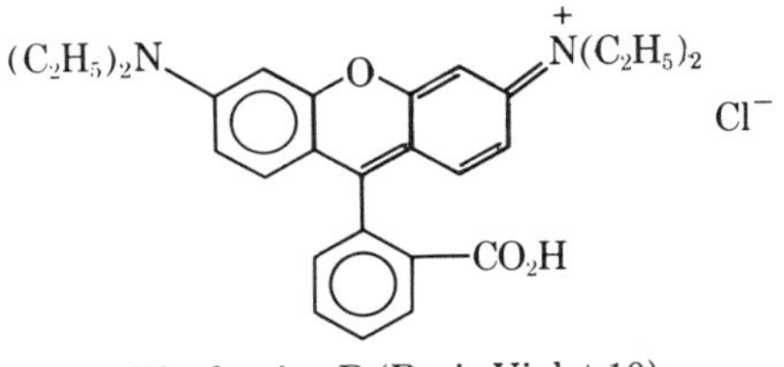

Rhodamine B (Basic Violet 10)

Azine Dyes. Azine dyes (qv) include induline and nigrosines. They produce jet blacks unobtainable with carbon black. This was particularly true of Induline Base in nylon before its manufacture was discontinued because of a carcinogenic impurity (4-aminobiphenyl) (3). The nigrosines are used in ABS, polypropylene, and phenolics.

Other Dyes. Brilliant Sulfoflavine, Acid Yellow 7, is an amino ketone having bright greenish fluorescence used in acrylics and nylon. Solvent Orange 60 is a perinone dye for acrylics, the turn-signal amber. It has good light and heat stability for ABS, cellulosics, polystyrene, and rigid poly(vinyl chloride). Two basic triphenylmethane dyes, Methyl Violet and Victoria Blue B, are suggested for phenolics although their heat and light stability is minimal.

The quinoline yellows are quite stable and may be used in ABS, polycarbonate, acrylics, polystyrene, and nylon.

Inorganic Pigments

White. Titanium dioxide is by far the most widely used white pigment for plastics. The rutile crystalline modification has the high refractive index, 2.76, and causes strong scattering of light and high opacity. The untreated pigment apparently contains reactive centers; the weatherability is greatly increased by coating with alumina and silica (4). The anatase variety is slightly whiter, has a lower refractive index, and blocks out less ultraviolet light. It is not recommended for outdoor use. The refractive index of titanium dioxide varies with wavelength, being higher in the blue region of the spectrum. For this reason, more blue light is scattered and the resulting white is slightly yellow. To match Illuminant C or to produce a dead white, a slight amount of shading with blue or violet pigments is necessary. Typical particle size of TiO_2 is 0.25–0.3 μm.

Light fixtures are fabricated from transparent plastics such as the acrylics. It is desirable to transmit as much light as possible but to conceal the source, either a filament or fluorescent tube. For this purpose, zinc oxide, zinc sulfide, and barium sulfate are used as diffusing pigments. All have relatively high refractive indexes, and the particle sizes are 1–4 μm. The resulting lenses may transmit as much as 85% of the light and are known as lighting whites. Aluminum silicate is used as an extender pigment in thermoset polyester resins; natural calcium silicate, wollastonite, is used in polyethylene, vinyls, and thermoset resins.

Black. Carbon black (qv) is another outstanding pigment for plastics. It has the unique property of blocking uv, visible, and ir radiation. A common use is to improve weatherability of polyethylene sheeting for outdoor use. At one time, the best pigmentary carbon black was prepared by burning natural gas in a deficiency of air and impinging the luminous flame upon water-cooled iron channels. The resulting channel black contained about 5% volatiles and was acidic. Today, furnace black is produced by partial combustion of aromatic residual oils. Volatiles are at a minimum and the product may be used in all plastics. If the ultimate degree of dispersion is required, it may be necessary to prepare a concentrate by first dispersing the carbon black into the resin on a roll mill, Banbury mixer, or extruder, followed by final dilution to the desired level with further mechanical work. When observed under an electron microscope, some carbon blacks are seen as discrete particles, others form chains of platelets. The latter are said to have high structure; they are slightly less jet and glossy but have higher electrical conductivity. In acrylics and other highly transparent plastics used in architecture, small amounts of carbon black are added to reduce light and heat transmittance. At such levels, slightly more of the longer wavelengths of light are transmitted and black imparts a bronze or golden hue. Carbon black is available as powder, pellets, and in the form of aqueous and non-aqueous dispersions, as well as concentrates.

When it is desirable to use a weak black, bone black may be substituted for carbon. It is manufactured by calcining animal bones and contains approximately 85% calcium phosphate and calcium carbonate. Black iron oxide, Fe_3O_4, is stable up to 150°C.

Copper chromite black, $Cu(CrO_2)_2$, is inert to all but rubberlike compositions, and has been calcined to 600°C.

Iron Oxides. In addition to the black iron oxide mentioned above, there are several yellow, brown, and red oxides, both natural and synthetic. As a class, they provide inexpensive but dull, lightfast, chemically resistant, nontoxic colors. The natural products are known as ocher, sienna, umber, hematite, and limonite. These include varying amounts of several impurities; in particular, the umbers contain manganese.

The synthetic iron oxides are much purer and have less variation in composition. Red oxide, Fe_2O_3, has excellent bleed, chemical, heat, and light resistance and is nontoxic. The yellow hydrate, $Fe_2O_3 \cdot x\,H_2O$, is useful up to 175°C, where it loses water and becomes red. Both of these pigments protect resins by screening ultraviolet light. Brown oxide, a mixture of ferrous and ferric oxides, $(FeO)_x (Fe_2O_3)_y$, is useful for producing woodgrain effects in plastics. There are also two mixed oxides, $ZnO.Fe_2O_3$, and $MgO.Fe_2O_3$, that are stable, nontoxic tans.

The natural iron oxides are recommended for use in cellulosics and phenolics, the synthetics for cellulosics, polyethylene, flexible vinyls, and all thermosets.

Chromium Oxide Greens. Chromium oxide, Cr_2O_3, is dull but is the most weatherable green pigment. It is stable up to 1000°C and is chemically resistant. The infrared reflectance resembles that of chlorophyll, making it a good camouflage color. Where durability is more important than brightness, it is used in cellulosics, polyethylene, and vinyls. The hydrated form, $Cr_2O_3.2H_2O$, has a much more brilliant shade but is dehydrated at about 500°C. Its use is largely confined to the cellulosics; phthalocyanine green is much to be preferred in other resins.

Iron Blue and Chrome Green. Iron Blue or Prussian Blue is ferric ammonium ferrocyanide, $FeNH_4Fe(CN)_6$. It is coarse in texture, difficult to grind, and withstands only a few minutes at 175°C. It has some use in low density polyethylene.

Not to be confused with chromium oxide green, above, is chrome green, a blend in varying proportions of chrome yellow (see below) and iron blue of the general formula $PbCrO_4.PbSO_4.FeNH_4Fe(CN)_6$. Both iron blue and chrome green are stable to mild acids but not to alkalis, even calcium carbonate. Chrome green cannot be used in children's toys and is blackened by sulfur. It is used in thermoset polyesters and in polyethylene leaf bags.

Violet. There are several inorganic violet pigments. Manganese violet or mineral violet, $NH_4MnP_2O_7$, is weak and has poor alkali and heat resistance. It may be used in vinyls. Among cobalt compounds, the phosphate, $Co_3(PO_4)_2$, is little used; $CoLiPO_4$, which is stronger and more stable than manganese violet, as well as other cobalt and manganese complexes are available. Ultramarine violet is discussed below. As a class, the inorganic violets are weak and difficult to disperse. Suppliers of inorganic violets, as well as many other metal oxide combination pigments, include Ferro, Harshaw, Drakenfeld, and Shepherd.

Ultramarine Pigments. Ultramarine is a reddish blue with a clean, brilliant shade having good durability except to acids. It is manufactured by calcining a mixture of china clay, soda ash, sodium sulfate, carbon, silica, and sulfur at 800°C. The crude product is ground and washed. It is a complex aluminum sulfosilicate. Extraction with hydrogen chloride or chlorine produces a violet containing less sodium. Even though ultramarine has only 7% of the tinting strength of phthalocyanine blue, it is considerably redder in shade and has wide use in plastics. The violet, which is much weaker, is useful for tinting whites.

Blue, Green, Yellow, and Brown Metal Combinations.   There are a number of inorganic oxide pigments that are indispensable for coloring transparent or translucent high temperature plastics. Their properties have been summarized (5). The pigments were originally developed for ceramics but have been improved in recent years for use in plastics. The ingredients are mixtures of metal salts that are calcined at 800–1300°C. The resulting oxides are ground, washed, dried, and pulverized to fine powders. The products are insoluble in solvents and all resins, have excellent resistance to heat, light, and chemical attack as well as low oil absorption, good dispersibility, and little or no tendency to migrate. Compared to organic pigments and dyes, the oxide pigments are at a serious disadvantage in terms of brightness and, usually, cost.

The inorganic blues start with cobalt aluminate, which has the nominal formula $CoO.Al_2O_3$. More commonly, the chemical compositions vary. As chromium is added, the blue becomes greener and moves into the turquoise region. Other blues may contain silicon, zinc, titanium, tin, or aluminum, in addition to cobalt. Barium manganate, $BaMnO_4$, is a weak greenish blue with a high specific gravity of 4.85. It has mild oxidizing properties and should not be used without rigorous preliminary tests. However, it does have a spectral curve similar to that of green-shade phthalocyanine blue and is occasionally useful in making nonmetameric matches (see Color).

A wide variety of greens ranging from bluish to yellow in shade is based on cobalt in combination with chromium, aluminum, titanium, nickel, magnesium, antimony, or zinc. These are brighter than the chromium oxides.

Inorganic yellow oxide combinations may contain lead, antimony, tin, nickel, or chromium. They are classed as yellows rather than browns, but they are dull compared to the cadmium yellows.

Brown combinations usually contain iron with chromium zinc, titanium, or aluminum. A few without iron contain chromium, antimony, tin, zinc, manganese, or aluminum. They range from light tans to dark chocolate. The shades are not as red as ferric oxide, but the browns are far superior to hydrated iron oxide in brightness and thermal stability.

Lead Chromates and Molybdates.   The lead chromates appear in several shades of yellow. The primrose and lemon are solid solutions of lead sulfate in the chromate and have the stable monoclinic structure. The medium shade contains no sulfate. Chrome orange is a compound with lead oxide, $PbCrO_4.PbO$. Molybdate orange is a combination of lead chromate and sulfate with the molybdate, $PbMoO_4$. These pigments have the advantages of opacity, brightness, and low cost. They are sensitive to acids, alkalies, and hydrogen sulfide. Pigments containing lead are banned from many uses and hexavalent chromium is a suspected carcinogen. Use is confined to vinyls and low temperature polyolefin and polystyrene resins except that chrome orange is too sensitive to acids for use in vinyls.

Conventional chromates and molybdates are stable only up to 200°C. Silica encapsulation increases the stability to 300°C (6). These pigments are readily shaded. Chrome yellow mixed with molybdate orange is brighter than chrome orange. The molybdates may be shaded with organic reds such as lithol rubines and quinacridones. With the current trend away from colorants containing heavy metals, the lead pigments are being replaced by organics.

Cadmium Pigments.   The cadmiums constitute a continuous series from green shade yellow through orange and red to maroon (7). They may be used in most plastics, and are stable at 325°C and up to 550°C for short periods. Resistance to chemicals is good. Lightfastness is satisfactory indoors, improving from yellow to red. The full tones are strong and bright but light tints, especially reds, become dull.

Chemically, the pigments are cadmium sulfides and selenides which have been precipitated, dried, and calcined at 650°C to convert to the hexagonal crystalline form. The greenest yellows contain 25 mol % of zinc sulfide as a solid solution in cadmium sulfide. Pure cadmium sulfide, CdS, is orange. Selenium is added to produce reds until at 50 mol % a maroon results. The cadmium selenide, CdSe, is also in solid solution. As the proportions of the ingredients are varied, a large number of shade gradations are possible. One supplier markets 14 principal shades: primrose, lemon, golden, and dark yellow; light, medium, and dark orange; orange–red; extra light, light, medium light, medium, and dark red; and maroon. One reason for this seeming proliferation of types is that single pigments are brighter than mixtures. In the list cited, the lemon yellow is brighter than a mixture of primrose and golden yellows.

The above types are known as CP (chemically pure) cadmiums. With the development of other uses for cadmium and selenium, costs have risen substantially in recent years. Some cost reduction may be obtained by use of the cadmium lithopones. These have the same relative shades but have been coprecipitated onto about 60% of barium sulfate. The resulting extensions give better money value, if the higher pigment loading can be tolerated, with no loss in properties.

A third form of cadmium pigments is the mercury–cadmiums, sold under the Hercules trademark Mercadium. Mercuric sulfide, HgS, forms solid solutions up to about 20 mol % with the oranges, reds, and maroon. The heat stability is improved up to 370°C and the costs are somewhat lower than the CP grades. The mercury–cadmiums are slightly more reactive, but have excellent bleed resistance.

The improved organic pigments and dyes developed in recent years have displaced the cadmiums from the lower temperature plastics. For the high temperature engineering plastics (qv), eg, nylon and acetals, there are no substitutes for the CP cadmium pigments.

Titanate Pigments. When a nickel salt and antimony oxide are calcined with rutile titanium dioxide at just below 1000°C, some of the added metals diffuse into the titanium dioxide crystal lattice and a yellow color results. In a similar manner, a buff may be produced containing chromium and antimony, a green with cobalt and nickel, and a blue with cobalt and aluminum. These pigments are relatively weak but have extreme heat resistance and outdoor weatherability, eg, the yellow is used where a light cadmium could not be considered. They are compatible with most resins.

Pearlescent Pigments. The technology of producing pearlescent pigments has passed through three stages. The original natural pearl essence was obtained from fish scales and was expensive. Later, basic lead carbonate, bismuth oxychloride, and lead hydrogen arsenate were used. The preferred process is to coat mica with layers of titanium dioxide.

The pearlescent or nacreous pigments owe their effects to the partial transmittance and partial reflection of light from the multiple coating layers. Since the layers have about the thickness of the wavelengths of visible light, some wavelengths interfere and colors are produced. Basic lead carbonate is supplied as a 70% suspension in dioctyl phthalate. Use is chiefly in vinyls. Coated mica may be used in any transparent or translucent plastic, particularly vinyls and polyethylene. The shade is sometimes modified by the incorporation of red iron oxide. Other pleasing effects are produced by carbon black or colored pigments in trace amounts.

Metallic Pigments. Aluminum is a reactive metal that is protected from chemical attack by a natural thin oxide film. It is attacked by alkalis and acids under oxidizing conditions. Commercial flakes are 0.1–2.0 μm thick and are generally sold in paste form in a hydrocarbon or plasticizer medium. Dry aluminum dust forms explosive mixtures with air. For paint use, the flakes are coated with stearic acid to produce a leafing grade which tends to line up with the platelets parallel to the surface. At the higher temperatures of plastics, the stearic acid has little effect; however, there is usually some orientation during molding. The metal protects plastics by reflecting ultraviolet light.

Aluminum imparts a grayish cast to plastics. In transparent systems, dyes or pigments may be added to impart various colors. Such systems often exhibit flop, a drastic change in shade or lightness between 90° and low angle viewing. In the former case, more relatively uncolored light is reflected from the oriented metal flakes. At a low viewing angle, more colored resin is visible and less light is reflected from the edges of the aluminum.

A few plastics can tolerate aluminum much more readily than bronze flakes. It is possible to formulate a bronze color using only aluminum and suitably transparent yellow, orange, and red dyes or pigments.

Bronze powders are available for use in plastics in various compositions and particle sizes. Copper is seldom used alone; all bronzes contain zinc. The names and zinc contents of the common types are: pale gold, 8%; rich pale gold, 15%; rich gold, 30%. The shades become less red as zinc is increased. The finer particle size grades are generally preferred.

Some plastics tend to react with zinc to give a very undesirable bubbling effect when molded. There is also a tendency to darken due to tarnishing by the action of sulfides. Tarnish-resistant bronzes are available in which the particles have been coated with a transparent resin (8).

Organic Pigments

It is more difficult to discuss the uses of organic pigments than inorganics in plastics. The wide range of chemical types of pigments compounded with the multiplicity of resins makes any general statement open to many questions. One trend in the industry is certain: with faster molding speeds and higher temperature resins, the demand is for increasing thermal stability. As plastics replace metals in automobiles, eg, service temperatures and weatherability requirements are sharply increased. Pigment suppliers are responding with improved products.

There is usually no doubt about the identity in composition of inorganic pigments; with organics, compositions are often uncertain. For an almost complete list of organic pigments see ref. 9. References to pigment Green B and a few other types which find only trivial application in plastics are omitted. Most suppliers identify their pigments with both the CI designations and common names, eg, PY 1 indicates Pigment Yellow 1 (Hansa Yellow G).

Monazo Pigments. The Hansa Yellows are prepared from aromatic amines that have been diazotized and coupled to acetoacetanilide or 1-phenyl-3-methylpyrazolone. Hansa Yellows G and 4R are typical; in general, they lack the thermal and light stability required of plastics. They are also not used in plastics because of their tendency to bleed and crock. However, there is one exception, Permanent Yellow FGL (PY 97)

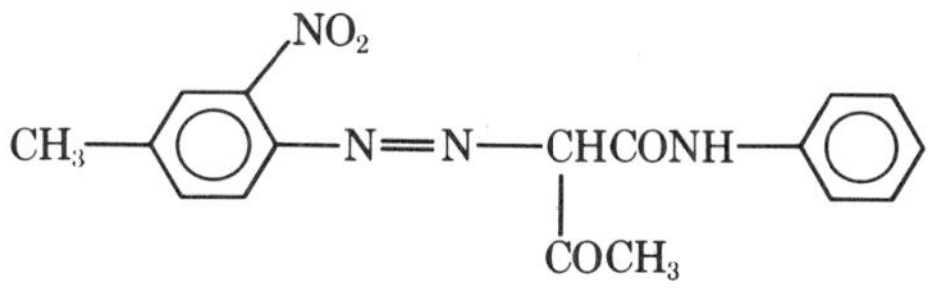

Hansa Yellow G (PY 1)

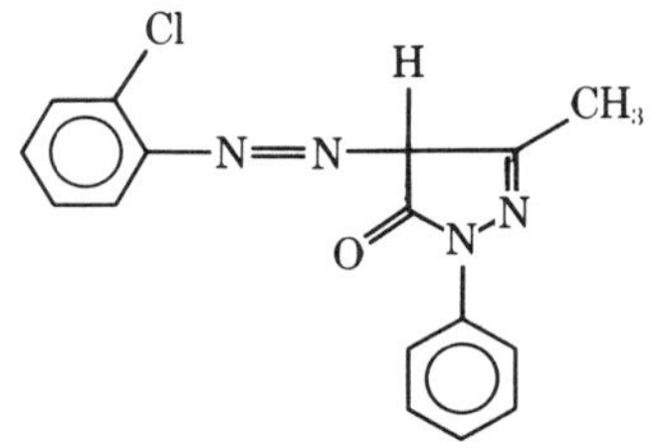

Hansa Yellow 4R (PY 60)

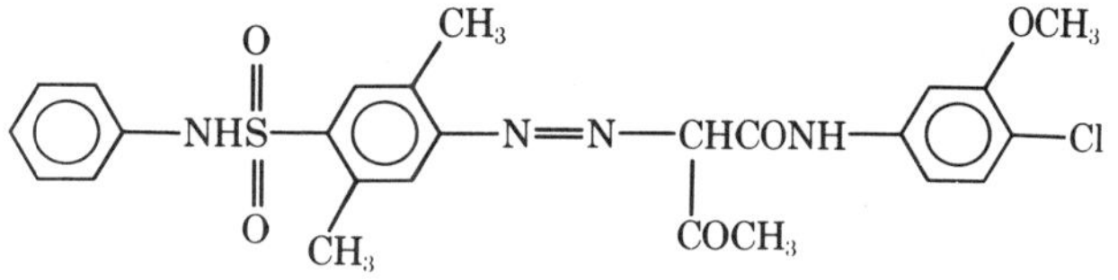

Permanent Yellow FGL (PY 97)

which is recommended for polystyrene, polypropylene, and rigid poly(vinyl chloride).

The toluidine reds are substituted amines coupled to β-naphthol. The rubines are sulfonated, coupled to BON (3-hydroxy-2-naphthoic acid) and precipitated as calcium or barium salts. Neither class has the required heat, light, and sublimation fastness for use in plastics.

The naphthol reds are phenyl amides of the rubines, usually with substituents on both the phenyls. PR 23 is an example. They are not used in plastics with the possible exception of three recent additions, PR 150, 170, and 210 listed in *Modern Plastics Encyclopedia* (10) as naphthols of undisclosed structure and suggested for use in acrylics, polyethylene, polypropylene, and polystyrene.

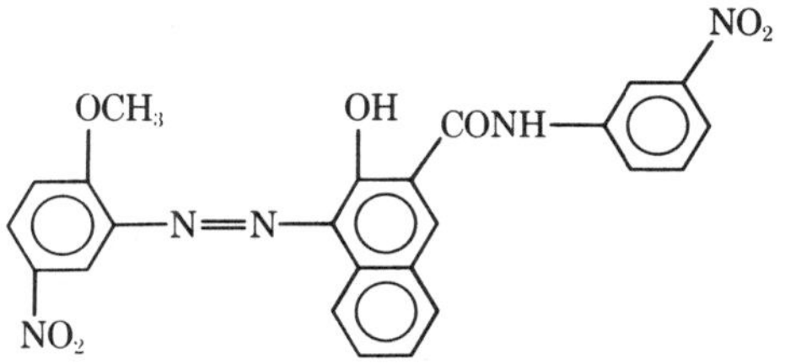

Pigment Red 23

Nickel Azo Yellow, PG 10, is a departure from the above pigments in that it owes its fastness to a metal atom. In this case the metal is chelated; in many other azo pigments the metal forms a salt with a carboxylic or sulfonic acid and precipitates the pigment, eg, Permanent Red 2B.

In PG 10, p-chloroaniline is coupled to 2,4-dihydroxyquinoline, then converted to a metal complex. The resulting pigment is green in heavy shades, yellow in lighter tints, and has very good fastness. It is suitable for cellulosics, polyethylene, and poly(vinyl chloride). It is demetallized by acids with loss of its fastness properties.

Lake Red C, PR 53, is an example of a pigment that has been made insoluble by a heavy metal. In this case the metal is barium and the symbol $Ba^{2+}/2$ indicates that one atom precipitates two molecules. Other metals used are calcium, strontium, manganese, and aluminum. This pigment is used in polystyrene.

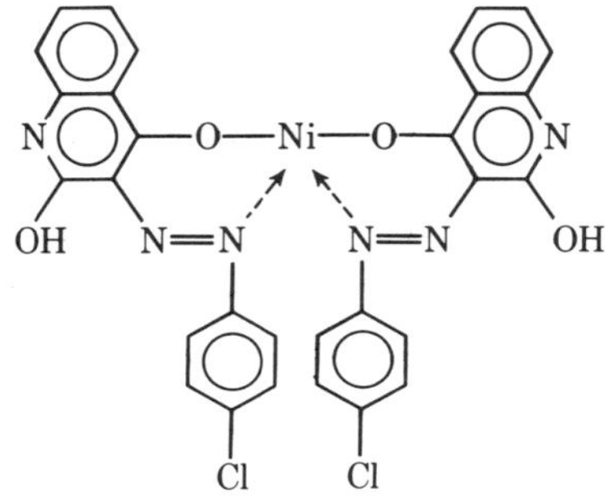

Nickel Azo Yellow (PG 10)

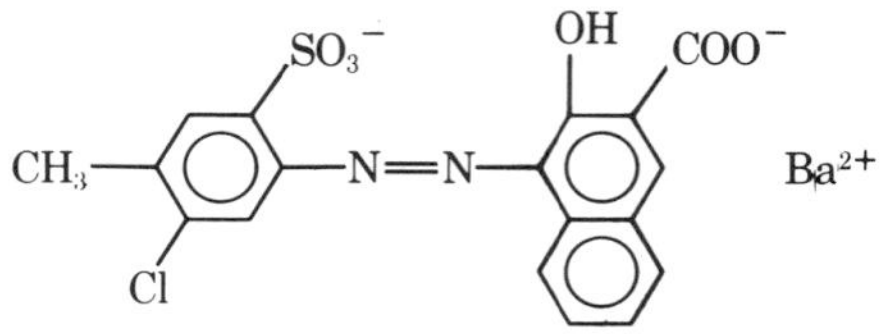

Permanent Red 2B (PR 48)

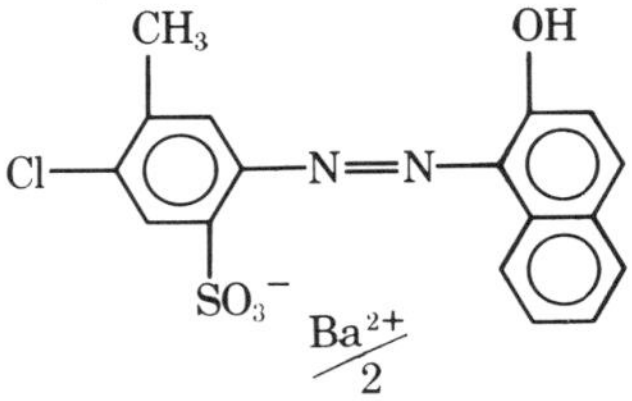

Lake Red C (PR 53)

Permanent Red 2B, PR 48, is a similar but much more useful pigment in which the amine has been coupled to BON. The pigment is sold as a precipitate of calcium, barium, or manganese. Although lightfastness is only fair, Permanent Red 2B is used extensively in vinyls, polyethylene, polypropylene and cellulosics. The barium salt has good resistance to migration. The manganese salt may not be used in rubber-containing compositions such as ABS and high impact polystyrene.

Pigment Scarlet 3B Lake is formed from the dye Mordant Red 9 which is mixed with freshly precipitated aluminum hydroxide, then precipitated on this base with barium chloride and zinc oxide. The resulting lake or extension has a bright, clear shade, free of bronzing. It is finding increasing use in cellulosics, polyethylene, polystyrene, and rigid vinyls.

The benzimidazolone pigments have been developed recently. Conventional coupling agents, such as acetoacetic acid used in Hansa Yellows or β-naphthol of the naphthol reds are joined to the benzimidazolone grouping. These reagents are coupled with amines, often containing methoxyl groups. The resulting pigments have good fastness to light, solvents, migration, and heat to 330°C. They are used in vinyls, polyethylene, thermoset polyesters, and have been suggested for ABS. The shades extend from greenish yellow to brown to bluish red. In the formulas, the arrows indicate the points of azo coupling.

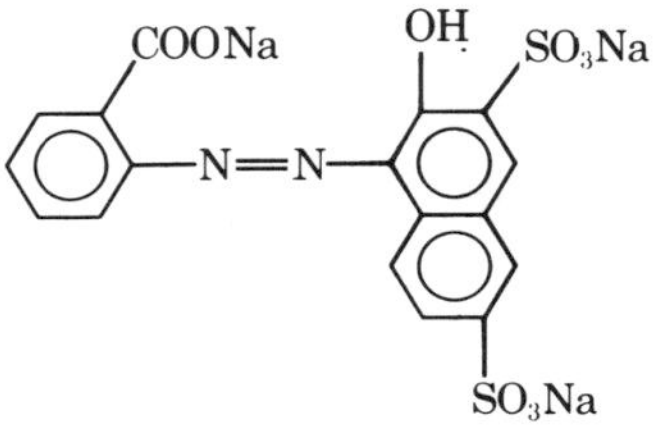

Mordant Red 9

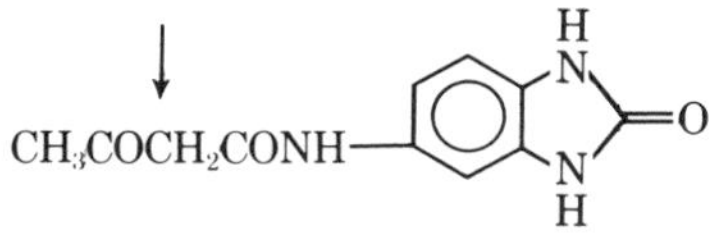

5-Acetonacetylaminobenzimidazolone

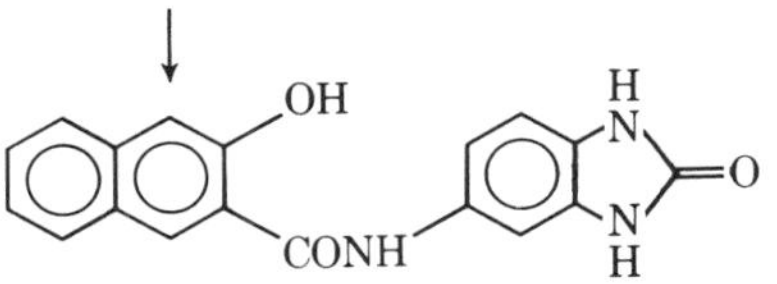

5-(3'-Hydroxy-2'-naphthoylamino)benzimidazolone

Disazo Pigments. The diarylide yellows and oranges are derivatives of benzidine (qv) coupled to two moles of substituted acetoacetanilide. The pigments are better known as benzidines; they are properly called disazo because of the twinned azo linkages, eg, Benzidine Yellows AAMX, AAOT, AAOA, and HR Yellow (PY 13, 14, 17, and 43). Yellows AAMX and AAOT are used in flexible vinyls. AAOA also colors polyethylene and polypropylene. These three differ only slightly in shade. Benzidine HR Yellow is redder than any of these and is compatible with all of the above thermoplastics.

All of these examples start with dichlorbenzidine. If the chlorines are replaced by methoxyls, the pigments are called dianisidines, eg, Dianisidine Orange GG (PO 14).

The Pyrazolone pigments may be considered as combinations of the hansas and the diarylides. Dichlorbenzidine is coupled to 1-phenyl-3-methylpyrazolone or one of its derivatives. The unsubstituted compound is Pyrazolone Orange (PO 13), and the 4'-methyl derivative is PO 34. Both are useful in cellulosics, polyethylene, polystyrene, vinyls, phenolics, and polyesters.

Replacement of the chlorines of the benzidine component of PO 13 with methoxyls gives Dianisidine Red, PR 41, used in cellulosics and vinyls.

Of some historic interest in Dianisidine Blue, PB 25, the coupling product of dianisidine with Naphthol AS. It is resistant to bleeding in oils, fats, and waxes and is compatible with vinyls. It has largely been replaced by phthalocyanine blue.

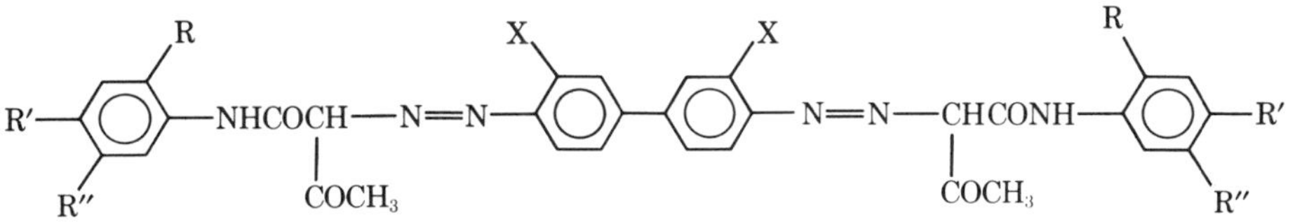

Benzidine Yellow AAMX, R = R' = CH$_3$, R'' = H, X = Cl (PY 13)

Benzidine Yellow AAOT, R = CH, R' = R'' = H, X = Cl (PY 14)

Benzidine Yellow AAOA, R = OCH$_3$, R' = R'' = H, X = Cl (PY 17)

Benzidine HR Yellow, R = R'' = OCH$_3$, R' = H, X = Cl (PY 43)

Dianisidine Orange GG, R = R' = CH$_3$, R'' = H, X = OCH$_3$ (PO 14)

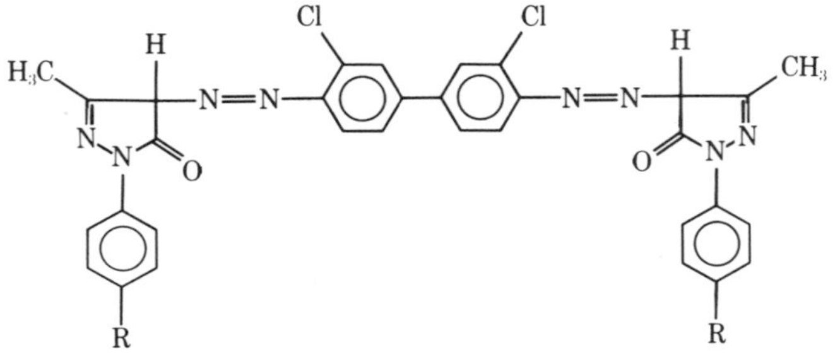

Pyrazolone Orange (PO 13), R = H
Pigment Orange 34, R = CH$_3$

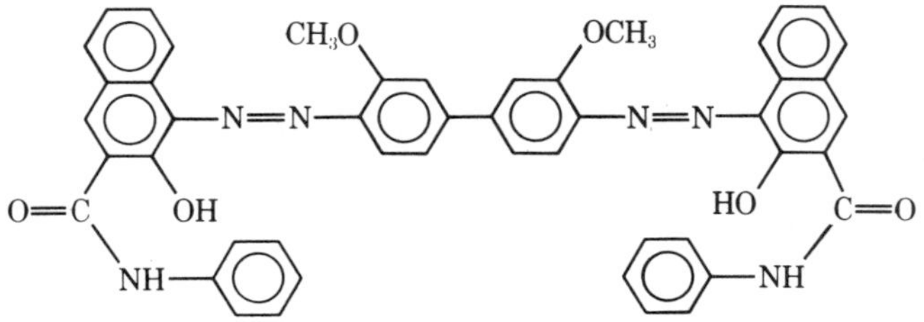

Dianisidine Blue (PB 25)

Disazo Condensation Pigments.   Another advance in azo pigment technology was made by Ciba-Geigy with the development of a new class of disazo pigments trademarked Cromophtal. In the manufacture of the benzidine pigments, both azo groups are diazotized and the entire molecule is synthesized at once. The condensation series accomplishes approximately the same thing in two steps, a diazo reaction followed by the combination of two azo dyes into one high molecular weight pigment. To illustrate the reactions, aniline is diazotized and coupled to BON. The product is converted to the acid chloride and condensed with benzidine.

The reaction is much smoother and more complete than the alternative of first allowing benzidine to react with BON and then coupling. With the large number of amines available for the first condensation, and the multitude of diamines, the possible

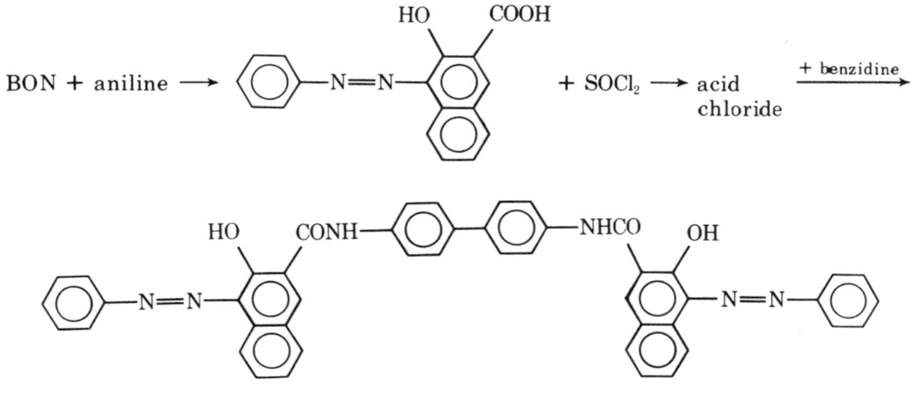

Disazo Pigment

combinations run into the thousands. Of these, seven were on the market by 1973, ranging in shade from yellow to orange, red, and brown. These are PY 93, 94, 95, PO 31, PR 144, 166, and PBr 23.

The disazo condensation pigments have double the molecular weights of the ordinary types. This factor greatly increases their stability and intensifies the shades. They have displaced the cadmium pigments from some uses on the basis of better money value and the lead chromates and molybdates because of better potential toxicity. Specifically, they are recommended for vinyls, polyolefins, polystyrene, ABS, cellulosics, and SAN.

Quinacridone Pigments. The quinacridones have the formula:

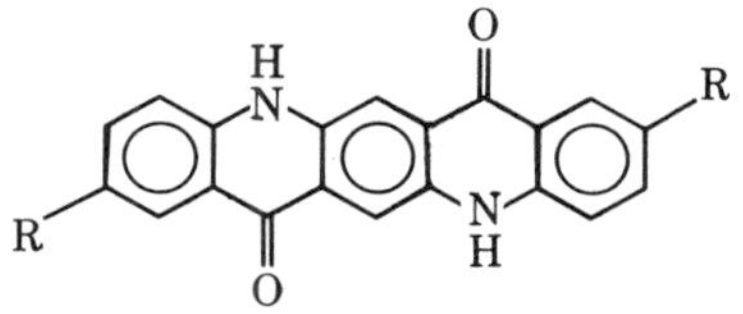

Pigment Violet 19, R = H
Pigment Red 122, R = CH$_3$

Although the chemical structures were known, it remained for Struve to develop practical methods for synthesis and to elucidate the optimum physical forms (11). The violets fill the void in the color gamut where the inorganics are inadequate. The quinacridones may be used in all resins except nylon-66 and polystyrene. They are stable up to 400°C and show excellent weatherability. One use is to shade phthalocyanines to match Indanthrone Blue. In carpeting, the quinacridones are recommended for polypropylene, acrylonitrile, polyester, and nylon-6 filaments. Predispersions in plasticizers (qv) are used in thermoset polyesters, urethanes, and epoxy resins (12).

Dioxazine Violet. Carbazole Dioxazine Violet is prepared by the reaction of two moles of 2-amino-*N*-ethylcarbazole with chloranil. This violet may be used in most plastics for shading phthalocyanine blues, since it has comparable light fastness. At relatively high temperatures, it may be subject to slow decomposition.

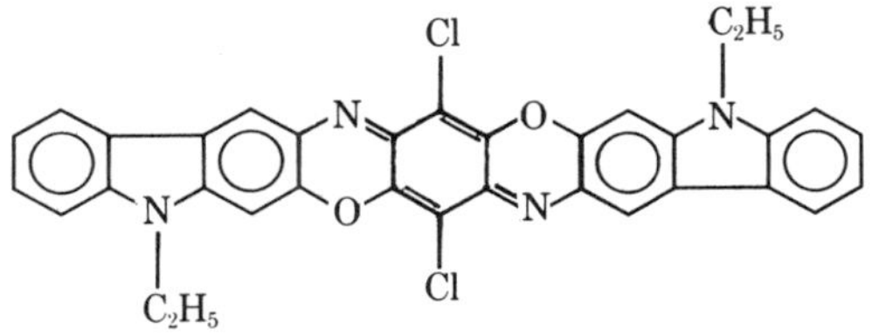

Carbazole Dioxazine Violet (PV 23)

Vat Pigments. Of the many anthraquinone vat dyes developed for textiles, only a few are used in plastics. The trend was started in the 1950s when Vesce of Harmon Colors adapted them to automotive finishes (13). As a class, the vat pigments have good light and heat fastness and virtually no tendencies to migrate or bleed. They are expensive and are not stable under reducing conditions. The following are the more commonly used vat pigments. Flavanthrone Yellow (PY 112), not over 200°C; Anthrapyrimidine Yellow (PR 108), not for nylon; Pyranthrone Orange (PO 40); Perinone Orange (PO 43); Brominated Anthranthrone Orange (PR 168), vinyl only; Brominated Pyranthrone (PR 197); Anthramide Orange (VO 15); Indanthrone Blue, Red Shade (PB 64); and Isoviolanthrone Violet (PV 31 or PV 33).

The formula for Isoviolanthrone Violet illustrates the complicated structures of the vat pigments.

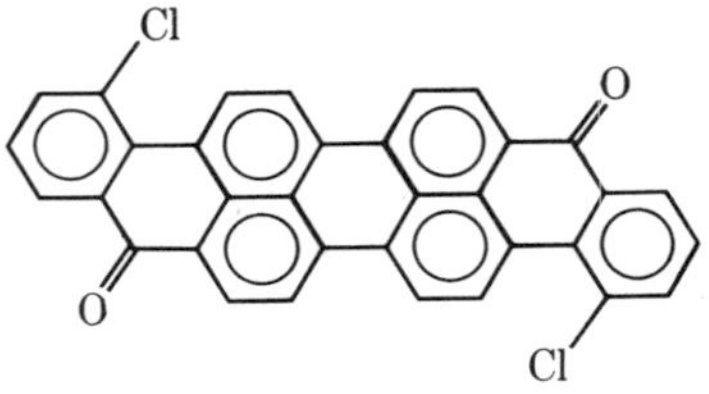

Isoviolanthrone Violet (PV 31)

Perylene Pigments. The Perylenes are a class of red and maroon pigments. In the general formula, R may represent a simple alkyl, methyl, or a substituted phenyl, eg, PR 123, R = *p*-ethoxyphenyl.

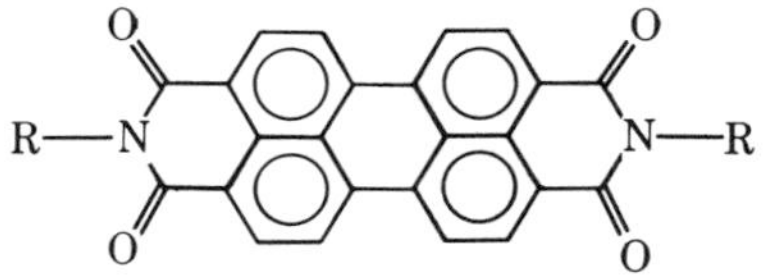

Pigment Red 123 R = ⟨◯⟩—OC_2H_5

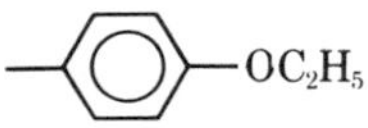

These pigments are recommended for most plastic systems because of their excellent stability to chemicals, bleeding, and light. They are widely used in vinyls, polyethylene, polypropylene, and cellulosic plastics. The Colour Index classes are listed as PR 123, 149, 179, and 190.

Thioindigo Pigments. The thioindigos are red and violet pigments developed for textiles. Two red-violets (PR 88 and PR 198) are recommended for plastics because of their excellent fastness properties.

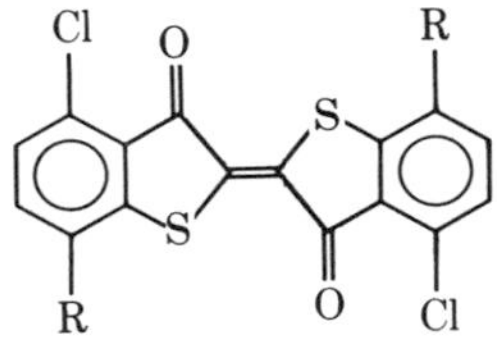

Pigment Red 88 R = Cl
Pigment Red 198 R = CH_3

Phthalocyanine Pigments. Phthalocyanine blue was discovered accidently; Linstead elucidated the structure (14). The common form is the copper derivative, prepared from four moles of phthalonitrile and a copper compound or, more commonly, from phthalic anhydride, urea, a copper salt, and catalyst.

Many other metal salts have been prepared, but none equals the copper compound in purity of hue. The copper is so tightly bound that the pigment may be sublimed in a vacuum at 500°C. It is compatible with rubber, provided any excess unbound copper has been removed. The crude pigment from the reactor must be reduced to its optimum particle size by grinding with salt or in a solvent, or by dissolving in sulfuric acid and precipitation.

There are two crystalline modifications of copper phthalocyanine: alpha, and the

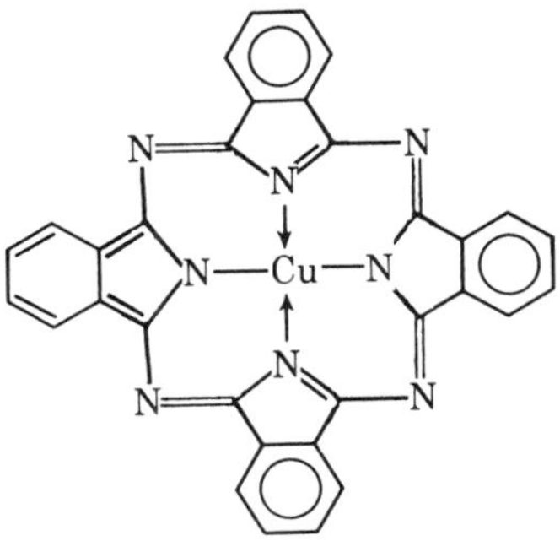

Phthalocyanine Blue, PB 15

slightly greener beta. The latter is the more stable form. In the presence of aromatic solvents or heat, the red shade may become greener, although much of it is stabilized to prevent this tendency. A third blue product, metal-free phthalocyanine, is greener and duller and is rarely used in plastics.

Because of their clear bright shades, fastness, and chemical inertness, the phthalocyanines may be used in virtually any resin. Their very high strength often makes predispersion into a concentrate advisable. A well dispersed pigment is completely transparent in acrylics. Very light tints with titanium dioxide in nylon are not reliable as to shade. There is also a tendency for a shift toward the green shade as temperatures are increased. Phthalocyanines may retard the curing of thermoset resins.

In phthalocyanine green, the aromatic hydrogens have been replaced by chlorine. It is in all respects an excellent pigment for plastics and many other uses. The chlorinated product is a blue green. Replacement of the chlorine by bromine gives much yellower pigments designated 2Y, 3Y, 6Y, up to 8Y, depending upon the degree. The chlorinated product contains 48% halogen, the brominated may have as much as 53%. As with the blues, dispersion must be complete to obtain maximum tinctorial strength and to avoid streaking and specking.

Tetrachloroisoindolinones. The recently developed tetrachloroisoindolinones are yellow, orange, and red pigments of Ciba-Geigy, marketed under the trade name Irgazin. They are difunctional amines stabilized by two tetrachloroisoindolinene units (15).

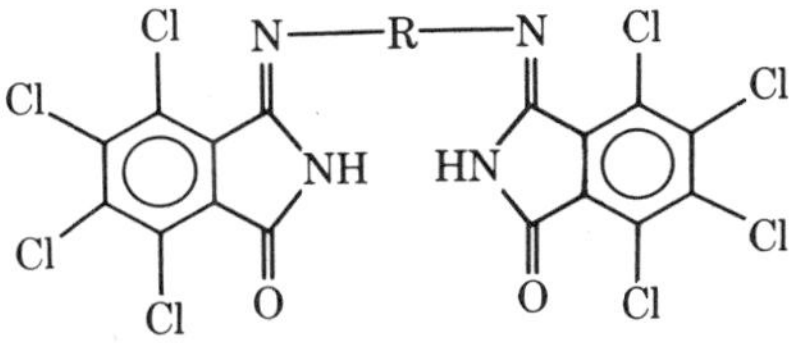

Tetrachloroisoindolinones,

eg, R =

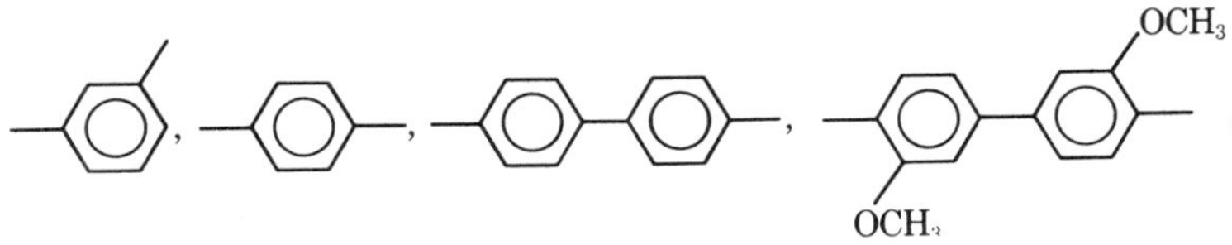

The tetrachloroisoindolinones have excellent resistance to bleeding and light and are stable up to 290°C. They are suggested for ABS and polypropylene and may be used in thermosets.

Fluorescent Pigments. Fluorescent pigments or dyes depend upon their ability to absorb light at one wavelength and to re-emit it in a narrow intense band at a longer wavelength (see Luminescent materials). The dyes used include the rhodamines, which emit pink, and aminonaphthalimides which are bright greenish yellow. To obtain maximum effect, the dyes are dissolved in brittle resins at low concentrations. The colored resins are then ground to powders and used as pigments. The brightness of such a combination far exceeds that of any pigment alone.

Fluorescent dyes do not have lightfastness. Their use in plastics is confined to the lower temperature resins, vinyls, polyethylene, polystyrene, and acrylics, at maximum temperatures of 200°C.

Economic Aspects

The cost of the coloring process should be calculated as soon as a formula is projected, and compared against that of any alternative colorant systems that may be available.

Estimates of the annual consumption of colorants for plastics over a 5-year period are shown in Table 1 (16). Information on prices is given in the Dyes and Pigments articles.

Table 1.   United States Consumption of Colorants for Plastics[a], Metric Tons

Colorant	Year 1972	Year 1977
Dyes		
nigrosines	1,140	1,500
oil solubles	610	650
anthraquinones	218	210
metal complex types	155	155
others	120	115
	2,243	*2,630*
Inorganic pigments		
titanium dioxide	86,000	92,000
carbon black	23,500	26,900
iron oxides	3,334	3,530
cadmiums	2,300	2,680
chrome yellows	2,594	2,750
molybdate oranges	1,780	1,750
others	1,540	1,510
	97,548	*104,220*
Organic pigments		
phthalocyanine blues	1,450	1,660
phthalocyanine greens	950	930
organic reds	1,110	1,140
organic yellows	190	200
others	490	520
	27,690	*31,350*
Total	*127,481*	*138,200*

[a] Ref. 16.

Table 2. Alphabetical List of Colorants for Plastics Referred to in the Text

Colorant	CAS Registry No.	Formula
5-acetoacetylaminobenzimidazolone	[26576-46-5]	
Acid Red 52	[3520-42-1]	
Acid Yellow 7	[2391-30-2]	
Acid Yellow 36	[587-98-4]	
aluminum silicate	[14504-95-1]	$(Al_2(SiO_3)_3)$
Anthramide Orange	[2379-78-4]	
Anthrapyrimidine Yellow	[4216-01-7]	
barium manganate	[7787-35-1]	$(BaMnO_4)$
barium sulfate	[7727-43-7]	$(BaSO_4)$
Basic Red 1	[989-38-8]	
Basic Violet 10	[81-88-9]	
Benzidine Yellow HR	[20139-68-8]	
Benzidine Yellow AAMX	[5102-83-0]	
Benzidine Yellow AAOA	[4531-49-1]	
Benzidine Yellow AAOT	[5468-75-7]	
black iron oxide	[1317-61-9]	(Fe_3O_4)
Brominated Anthanthrone Orange	[4378-61-4]	
Brominated Pyranthrone	[1324-33-0]	
brown iron oxide	[1309-38-2]	$(FeO)_x.(Fe_2O_3)_4$
cadmium selenide	[1306-24-7]	$(CdSe)$
cadmium sulfide	[1306-23-6]	(CdS)
calcium silicate	[10101-39-0]	$(CaSiO_3)$
Carbazole Dioxazine Violet	[6353-30-1]	
carbon black	[1333-86-4]	
channel black	[1333-86-4]	
chrome green	[1308-38-9]	
chrome orange	[1344-38-3]	
chrome yellow	[1344-37-2]	
chromium oxide	[1308-38-9]	(Cr_2O_3)
chromium oxide dihydrate	[12182-82-0]	$(Cr_2O_3.2H_2O)$
cobalt aluminate	[1333-88-6]	$(CoO.Al_2O_3)$
cobalt lithium phosphate	[13824-63-0]	$(CoLiPO_4)$
cobalt phosphate	[13455-36-2]	$(Co_3(PO_4)_2)$
copper chromite black	[12018-10-9]	$(Cu(CrO_2)_2)$
Dianisidine Blue	[5437-88-7]	
Dianisidine Orange GG	[6837-37-2]	
Dianisidine Red	[6505-29-9]	
Disperse Violet 1	[128-95-0]	
Disperse Yellow 23	[6250-23-3]	
Flavanthrone Yellow	[475-71-2]	
furnace black	[1333-86-4]	
hematite	[1317-60-8]	(Fe_2O_3)
5-(3′-hydroxy-2′-naphthoylamino)-benzimidazolone	[26848-40-8]	
Indanthrone Blue, red shade	[81-77-6]	
Induline Base	[8004-98-6]	
iron blue	[14038-43-8]	
iron oxide hydrate	[11100-07-5]	$(Fe_2O_3.x\,H_2O)$
Isoviolanthrone Violet	[81-28-7]	
Lake Red C	[5160-02-1]	
lead chromate	[7758-97-6]	$(PbCrO_4)$
limonite	[1317-63-1]	(Fe_2O_3)
magnesium oxide-iron oxide	[12068-86-9]	$(MgO.Fe_2O_3)$
manganese violet	[10101-66-3]	$(MnHP_2O_7.NH_3)$

Table 2. (*continued*)

Colorant	CAS Registry No.	Formula
mercuric sulfide	[1344-48-5]	(HgS)
methyl violet	[8004-87-3]	
Metanil Yellow	[587-98-4]	
Mineral Violet	[10101-66-3]	$(MnHP_2O_7 \cdot NH_3)$
Molybdate Orange	[12656-85-8]	
Mordant Red 9	[1836-22-2]	
Nickel Azo Yellow	[51931-46-5]	
ocher	[1309-37-1]	(Fe_2O_3)
Perinone Orange	[4424-06-0]	
Permanent Red 2B	[3564-21-4]	
Permanent Yellow FGL	[12225-18-2]	
phthalocyanine	[574-93-6]	
Phthalocyanine Blue	[147-14-8]	
Phthalocyanine Green	[1328-53-6]	
Pigment Blue 15	[147-14-8]	
Pigment Blue 25	[10127-03-4]	
Pigment Blue 64	[81-77-6]	
Pigment Brown 23	[57972-00-6]	
Pigment Green 10	[61725-51-7]	
Pigment Orange 13	[3520-72-7]	
Pigment Orange 14	[6837-37-2]	
Pigment Orange 31	[12286-58-7]	
Pigment Orange 34	[15793-73-4]	
Pigment Orange 40	[128-70-1]	
Pigment Orange 43	[42612-21-5]	
Pigment Red 41	[6505-29-9]	
Pigment Red 48	[7585-41-3]	
Pigment Red 53	[5160-02-1]	
Pigment Red 108	[4216-01-7]	
Pigment Red 122	[980-26-7]	
Pigment Red 123	[23108-89-2]	
Pigment Red 144	[5280-78-4]	
Pigment Red 149	[12225-02-4]	
Pigment Red 150	[56396-10-2]	
Pigment Red 166	[12225-04-6]	
Pigment Red 168	[4378-61-4]	
Pigment Red 170	[12236-67-8]	
Pigment Red 179	[5521-31-3]	
Pigment Red 190	[6424-77-7]	
Pigment Red 197	[1324-33-0]	
Pigment Red 210	[61932-63-6]	
Pigment Scarlet 3B Lake	[1836-22-2]	
Pigment Violet 19	[1047-16-1]	
Pigment Violet 23	[6358-30-1]	
Pigment Violet 31	[1324-55-6]	
Pigment Violet 33	[1324-17-0]	
Pigment Yellow 1	[2512-29-0]	
Pigment Yellow 13	[5102-83-0]	
Pigment Yellow 14	[5468-75-7]	
Pigment Yellow 17	[4531-49-1]	
Pigment Yellow 43	[20139-68-8]	
Pigment Yellow 60	[6407-74-5]	
Pigment Yellow 93	[5580-57-4]	
Pigment Yellow 94	[5580-58-5]	

Table 2 (*continued*)

Colorant	CAS Registry No.	Formula
Pigment Yellow 95	[5280-80-8]	
Pigment Yellow 97	[12225-18-2]	
Pigment Yellow 112	[475-71-2]	
prussian blue	[14038-43-8]	$(Fe_2(Fe(CN)_6)_3)$
Pyranthrone Orange	[128-70-1]	
Pyrazalone Orange	[15793-73-4]	
red iron oxide	[1309-37-1]	(Fe_2O_3)
Rhodamine B	[81-88-9]	
sienna	[1309-37-1]	(Fe_2O_3)
Solvent Blue 56	[14233-37-5]	
Solvent Green 3	[128-80-3]	
Solvent Green 4	[81-37-8]	
Solvent Orange 7	[3118-97-6]	
Solvent Orange 60	[61969-47-9]	
Solvent Orange 63	[54578-43-7]	
Solvent Red 1	[1229-55-6]	
Solvent Red 24	[85-83-6]	
Solvent Red 26	[4477-79-6]	
Solvent Red 111	[82-38-2]	
Solvent Yellow 14	[824-07-9]	
Solvent Yellow 72	[61813-98-7]	
Sulfo Rhodamine	[2609-88-3]	
Sulfoflavine	[2391-30-2]	
tetrachloroisoindolinone, R = 4,4′ = biphenylene	[4988-75-4]	
tetrachloroisoindolinone, R = 3,3′-dimethoxy-4,4′-biphenylene	[5507-73-3]	
tetrachloroisoindolinone, R = 1,3-phenylene	[5507-72-2]	
tetrachloroisoindolinone, R = 1,4-phenylene	[5590-18-1]	
titanium dioxide	[1317-80-2]	(TiO_2)
Ultramarine Dark Blue	[57455-37-5]	
Ultramarine Violet	[12769-96-9]	
umber	[12713-03-0]	
Vat Orange 15	[2379-78-4]	
Victoria Blue B	[6786-83-0]	$(CaSiO_3)$
wollastonite	[14567-51-2]	
zinc oxide	[1314-13-2]	(ZnO)
zinc oxide–iron oxide	[12063-19-3]	$(ZnO.Fe_2O_3)$
zinc sulfide	[1314-98-3]	(ZnS)

Health and Safety Factors

Dyes may sublime under operating conditions and show up in dust collectors. Migration is tested for by making a sandwich of alternately colored and uncolored pieces under a weight at an elevated temperature.

Toxicity of colored plastics is a complex and rapidly changing subject. For food packaging, an elaborate series of extraction tests must be run to demonstrate that neither colorants nor other additives can migrate into the food. Only titanium dioxide, iron oxides, and ultramarine blue are exempt from this requirement at present. A few pigments are certified for use in drugs and cosmetics and a very few dyes and alumina hydrate lakes are certified as food colors. Both classes are among the least thermally stable azo types. Each batch must be tested and certified by the FDA (17). In the past decade, colorant manufacturers have eliminated not only colorants of known toxicity, but also some made from toxic intermediates (see Colorants for food, drugs, and cosmetics). At the same time, it has been shown that some resin monomers are toxic on long time exposure. Handlers of colorants, additives, and resins should use maximum practical ventilation to protect their workers.

Table 2 is an alphabetical list of the colorants for plastics referred to in this article.

BIBLIOGRAPHY

1. T. B. Reeve and W. L. Dills, *Principles of Pigment Dispersion in Plastics, 28th ANTEC Preprint,* Society of Plastics Engineers, Inc., Greenwich, Conn., 1970, p. 574.
2. R. J. Kennedy and J. F. Murray, *Internal Pigmentation of Low Shrink Polyester Molding Compositions with Flushed Pigments, 33rd ANTEC Preprint,* Society of Plastics Engineers, Inc., Greenwich, Conn., 1975, p. 148.
3. C. A. Frankhauser, private communication, American Cyanamid Co., Sept. 1977.
4. D. A. Holtzen, *Plast. Eng.* **33**(4), 43 (1977).
5. N. J. Napier, *High-Temperature Synthetic Inorganic Pigments, 31st ANTEC Preprint,* Society of Plastics Engineers, Inc., Greenwich, Conn., 1973, p. 397.
6. T. B. Reeve, *Plast. Eng.* **33**(8), 31 (1977).
7. W. G. Huckle, G. F. Swigert, and S. E. Wiberley, *I&EC Prod. Res. Dev.* **5**, 362 (1966).
8. H. C. Felsher and W. J. Hanau, *J. Paint Technol.* **41**(534), 354 (1969).
9. T. C. Patton, ed., *Pigment Handbook,* Wiley-Interscience, New York, 1973.
10. J. Agranoff, ed., *Modern Plastics Encyclopedia,* Vol. 53, McGraw-Hill Book Co., New York, 1976, p. 678.
11. U.S. Pat. 2,844,484 (July 22, 1958), W. S. Struve (to E. I. du Pont de Nemours & Co., Inc.).
12. H. F. Bartolo, *Quinacridone Pigments in Plastics, 27th ANTEC Preprint,* Society of Plastics Engineers, Inc., Greenwich, Conn., 1969, p. 518.
13. V. C. Vesce, *Off. Dig. Fed. Soc. Paint Technol.* **28**(II), 1 (1 56); **31**(II), 1 (1959).
14. R. P. Linstead, *J. Chem. Soc.,* 1016, 1022, 1031 (1934).
15. Ref. 9, p. 699.
16. *Mod. Plast.* (9), 49 (1975); (9), 54 (1977).
17. M. J. Dunn, *Colors for Food Packaging—Conforming with FDA Regulations, SPE Color and Appearance Div. Preprint,* Society of Plastics Engineers, Inc., Greenwich, Conn., 1974, p. 1.

General References

Colour Index, and its Additions and Amendments, 3rd ed., Society of Dyers and Colourists, London, Eng., and American Association of Textile Chemists and Colorists, Durham, N. C. It now consists of seven volumes. The *Colour Index* was originally written by and for the textile industry but pigments are receiving increasing attention.

Textile Chemist and Colorist—Buyer's Guide, American Association of Textile Chemists and Colorists, Durham, N. C. annual. It lists American-made dyes and pigments by trade name, Colour Index generic number and formula number, if available (no formulas are given).

K. Venkataraman, ed., *The Chemistry of Synthetic Dyes,* Academic Press, Inc., New York, Volumes I and II were published in 1952, Volume III in 1970, and there are now eight volumes. The most comprehensive treatise on dyes, emphasis is on chemistry, not applications.

H. A. Lubs, ed., *The Chemistry of Synthetic Dyes and Pigments,* Reinhold, N.Y., 1955; written by DuPont research chemists and is highly authoritative as of that date.

T. C. Patton, ed., *Pigment Handbook,* Wiley-Interscience, New York, 1973; written by a number of specialists in the pigment field. The three volumes describe pigment chemistry and applications.

D. B. Judd and G. Wyszecki, *Color in Business, Science, and Industry,* 3rd ed., John Wiley & Sons, Inc., New York, 1975; probably the best single book on color.

F. W. Billmeyer, Jr., *Textbook of Polymer Science,* 2nd ed., Wiley-Interscience, New York, 1971; an excellent book on polymer chemistry and physics.

J. Fradas, ed., *Plastics Engineering Handbook of the Society of the Plastics Industry, Inc.,* 4th ed., Van Nostrand Reinhold, New York, 1976; equipment to handle plastics is well described.

Periodicals which occasionally include articles on coloring plastics include: *Modern Plastics,* monthly, with encyclopedia edition, McGraw-Hill Book Co., New York; *Plastics Engineering,* monthly; *Color Research and Application,* quarterly.

THOMAS G. WEBBER
Consultant

COLOR MEASUREMENT.   See Evaluation of dyes—Instrumental methods; Color.

COLORIMETRY AND FLUOROMETRY.   See Color; Dyes—Evaluation of dyes, instrumental methods; Analytical methods.

COLOR PHOTOGRAPHY

Color photography is a process by which light of varying wavelength, intensity, and location, collected and focused by a lens system, can initiate chemical reactions leading to a relatively permanent record in two-dimensional space of these variations. This article describes the development of this process and the physical, chemical, and physiological principles that are involved. For a more detailed discussion of the photographic process see refs. 1–12 (see also Photography; Color photography, instant).

A complete survey of the early history of color photography is given by Wall (13) and the more recent history by Friedman (14). The preliminary work leading to color photography was done by physicists and physiologists rather than by chemists, who were particularly active in the beginnings of monochrome photography. Although the development of commercial color photography followed black-and-white photography by almost a century, very early workers considered a full color record to be their goal.

In 1611 a physicist, deDominis, demonstrated that red, green, and violet are fundamental colors from which all colors can be compounded (see in ref. 13). Fifty-five years later Newton performed his well-known experiments relating these fundamental color sensations to the solar spectrum. In addition, Newton attributed the color of reflecting objects to the differential absorption and reflection of different parts of the incident light. He showed that sunlight, eg, is a composite of all of the components of the visible spectrum. Objects appear white or colored or black as they reflect all, or portions, of these components, or none (see Color).

The Young-Helmholtz theory of color vision (1790), which treated the visual system as having effectively three types of receptors sensitive to different portions of the visible spectrum, clarified still further the relationship between physical phenomena and physiological responses (15).

Between 1855 and 1861 Maxwell (16) applied the ideas of Newton, Young, and Helmholtz to produce the first color photograph. He recorded the red, green, and blue light reflected from a still life by exposing, separately and successively, three negatives, each through the appropriately colored filter. From the developed silver-image negatives he made positives. Each positive was thus an inverse image in silver of the blue, green, or red content of the still life. Therefore, projection of the images in register on a white screen, each through the appropriate blue, green, or red filter, produced the sensation of natural color. This classic experiment was long questioned, since there were no panchromatic-sensitive photographic emulsions available in Maxwell's time. Evans showed that the experiment worked only because of accidental factors of which Maxwell was probably unaware (17). Figure 1 illustrates the principle responsible for Maxwell's success. Blue, green, and red light beams can be adjusted to form white and, by implication, all of the colors to which the human eye responds. The figure also shows the formation of yellow, magenta, and cyan colors, whose significance was recognized later by others.

Maxwell's system, and all practical systems of color photography since, have fulfilled at least the following two requirements: (1) the sensitivity of the photographic material must be such that objects producing different colors and different lightnesses form latent images that can be differentiated; (2) the chemical or physical process that makes the latent image visible must maintain these differences and translate them into material images that will modulate some form of general illumination to cause the observer or other sensor to perceive the desired result.

Much of color photography may be called pictorial in that the end result is a picture that appears natural. In such photography, the sensitive materials must in some way approximate the spectral sensitivity of the eye, and the viewing system must provide a gamut of colors similar to those of the observer's experience.

In technical uses, the image frequently need not be natural to convey information. Energy outside the visible spectrum may be involved: eg, ultraviolet or infrared. Colors need not be reproduced in a natural or pleasing fashion. The reproduction may be designed for presentation to the human observer or for presentation to some other sensor such as a photoelectric cell.

After Maxwell's demonstration, active minds of a hundred years ago began inventing other systems by which practical color photography might be achieved. One of the most active of these investigators was DuHauron (18) who, in 1862 and subsequent years, described several important systems. His three-color camera, the chromoscope, provided a means for obtaining the three separation negatives that are the basis for most additive color systems. He also first described the screen plate, inspired by the then-current school of French painters known as pointillistes, who laid down juxtaposed, minute dots of color. In its photographic form this method has appeared as Dufay (Dufay Ltd.) and Lumière screen processes.

Collen (13) in 1865 presented the now generally used idea of three very thin layers,

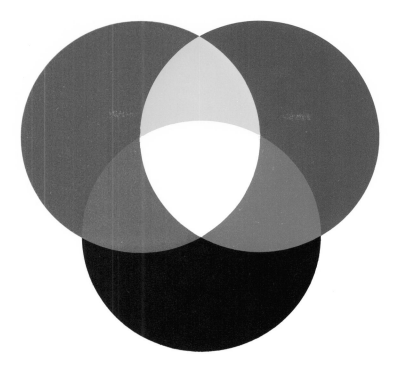

Fig. 1. Maxwell's arrangement (additive).

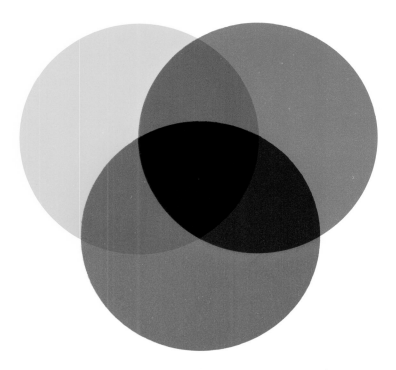

Fig. 2. Superimposed dyes (subtractive).

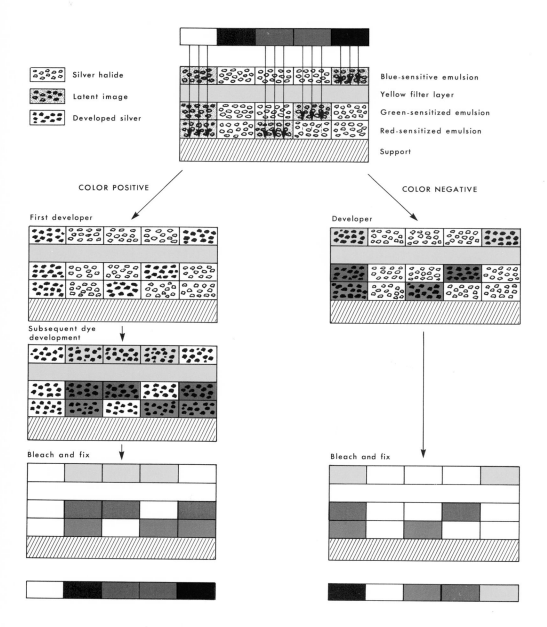

Fig. 3. Subtractive processes.

each separately sensitive to a different region of the visible spectrum. These layers could be exposed to make the negative separations and converted to positives in a system that anticipated current dye bleach processes. Collen recognized the practical problems and stated that all would be solved when chemists discovered the necessary materials.

Chemistry has been the senior partner among the cooperating sciences behind color photography. Color photographic materials, themselves products of chemical technology, produce color photographs by a series of intricate chemical processes. During the time of exposure, a photochemical reaction occurs that makes possible further complicated sequences of chemical reactions. The end result is a color photograph whose color and lightness are produced by the presence or absence of yellow, magenta, and cyan dyes.

The Light-Sensitive System.   The first requirement for a color-photographic material is that its sensitivity be such that objects producing different color sensations and lightnesses will form latent images that can be differentiated. The fundamental light-sensitive element in most color-photographic materials, just as in most black-and-white materials, is a silver halide or mixture of silver halides, silver chloride, silver bromide, or silver iodide, dispersed as crystals or grains in a medium such as gelatin. For the chemistry of the formation of the latent image see Photography.

For color photography a silver halide must be sensitive to all regions of the visible spectrum and sometimes beyond. Pure silver halides are sensitive only to ultraviolet and blue light, according to the fundamental photochemical principle of Grotthus and Draper that only light rays that can be absorbed can produce chemical changes. This color-blind property of silver salts was one severe limitation to color processes. Vogel (13) in 1873 opened the door to a range of spectral sensitivities when he discovered that certain dyes were capable of imparting a sensitivity to silver halide grains in the regions where the dyes absorb light. The manner of his discovery is interesting. Observing that certain color-blind plates had unexpected sensitivity in the yellow–green region, he determined that a yellow dye, coralline, that was put into the emulsion to reduce halation was responsible. He, as well as others, immediately investigated other dyes. DuHauron found that chlorophyll from crushed ivy leaves gave red sensitivity. The imagination of these early chemists is exemplified by C. Cros, who in 1879 tried chlorophyll, tincture of black currants preserved in brandy, aqueous extract of marshmallow leaves, and even bullock's blood!

Additive Color Processes.   Experiments with light and color and human vision have established that for full color reproduction the spectrum must be separated into three components, generally in the red, green, and blue light regions. The identity of the separate components must be maintained long enough so that they can be translated into the physical or chemical systems that serve to modulate three components of the viewing illuminant to give the sensation of full color to the observer.

In the generally used system of three silver halide emulsions color-sensitized with dyes, there are many different ways in which this three-component analysis can be achieved. The simplest in concept and the first historically is that involving three separate films or plates, each of which carries one of the three color records of all of the elements in the picture. These physically separated color records are handled separately in the subsequent operations and are combined only in the final, viewing stage to produce the color image as in Maxwell's classic experiment. The separation negatives for such a procedure can be made from still subjects by three successive exposures, each through a separate color filter onto a separate emulsion-coated material. Alternatively (and necessarily if the subject is moving), the exposures can be made simultaneously if an optical method is employed of splitting the light into three paths by means of prisms and mirrors. The negatives for the Technicolor process (Technicolor Corporation) were formerly made this way. A third variant, also useful in cinematography, is that of producing in rapid succession the three color separations, either on three successive frames of a motion picture film or in three different areas on a single frame. The system using three successive frames may employ during exposure a rotating color filter that is successively red, green, and blue and is synchronized with the moving photographic material. The positive color images may be treated as three separate still pictures or may be combined by presenting them rapidly enough, taking advantage of persistence of vision. Such a system is not very satisfactory, however, since it leads to visual fatigue.

The mechanical problems associated with handling three separate materials, one for each record, put severe limitations on their use. Systems designed for more convenient film handling placed the three records in one material. One approach is to have the individual records randomly or regularly arrayed as fine units in a single layer. For example, in the screen-type processes a random or regular array of red, green, and blue filters coated over or under a panchromatic-sensitive silver halide emulsion is used to provide a side-by-side spatial distribution of the three color elements. If an exact relationship between the developed image in the light-sensitive layer and the filter through which the exposure was made is maintained for viewing or projecting, a true color picture can result. In this situation the size of the color filter units is small enough so that they are not resolved by the eye, and color mixture results. The Dufay and Lumière color screen plate processes are examples. If the sensitive layer and the filter layer are not an integral unit but are two separate materials, extreme care and precision are necessary in subsequent presentation to ensure that the filter units and the corresponding developed images can be accurately aligned. Current color television systems, which are examples of these additive color principles, use juxtaposed multicolor phosphor elements in the display tubes to generate the gamut of colors (see Special Applications).

A different approach is the lenticular process. A system of lenticules, small separate cylinders of minute dimensions (25 per mm), is placed between the sensitive material and the exposing light. The lenticules are usually obtained by embossing the support of the sensitized material, since registration of the developed image relative to the lenticules must be maintained between exposure and projection.

A banded tricolor (red, green, and blue) filter placed in the nodal plane of the lens is oriented so that the bands are parallel to the lenticule bands. Since the lenticules will focus red, green, and blue at different sites, the three-color records will consist of side-by-side latent image bands corresponding to the amount of light transmitted by the banded filter. The bands are developed to silver images by a reversal process. Display of the colored images results from projecting the film back through a system that has a banded multicolor filter oriented relative to the lenticules exactly as it was during the original photography. Although the system is relatively simple as a photographic process, it uses light very inefficiently both in exposure and projection, its resolution is limited by lenticule size, and it requires a fairly precise optical system for good color reproduction.

These systems depend more heavily on mechanical and physical processes than on chemistry. For example, the lenticular system involves no more chemistry than does black-and-white photography but it requires considerable mechanical ingenuity and precision in the preparation and location of the elements. However, the more recent Polavision (Polaroid Corp.) system of movies combines considerable chemical ingenuity as well (see Color photography, instant).

Subtractive Color Processes.   Three separation negatives or positives have been the basis of many practical photographic processes, especially those involving the transfer of dyes. In such processes separation negatives are used to prepare three separate gelatin positive-relief matrices. A subtractive dye corresponding to each record is imbibed into each of the three matrices and transferred in register to a mordanted receiving material. This technique is called imbibition printing. The Technicolor motion picture process uses this technique; it depends on very precise registration systems to obtain the required degree of superposition. Similarly, the Kodak dye-transfer system, in which the dyes are transferred to a white reflecting support, produces color prints. It is important to note that the dyes used in these transfer processes are not blue, green, and red but their subtractive complements—yellow, magenta, and cyan (Fig. 2).

The most common way of handling three color-recording emulsions is to place them in a multilayer arrangement. This introduces several chemical problems but avoids many mechanical ones and adds much to the convenience for the user. The usual layer order (see Fig. 3) places the blue-sensitive emulsion, the first to be exposed, on top. This is primarily because the natural sensitivity of silver halide is to blue and violet light; therefore, all the records will have some blue sensitivity. By putting the layer intended to carry the blue record on top, a yellow filter may be placed beneath it to prevent the blue light from affecting the other, also blue-sensitive, layers. The layers below the yellow filter can then be spectrally sensitized in the green and red regions, respectively, to provide those records.

In the multilayer system, the sensitizing dyes must be attached relatively firmly to the silver halide grain so that they do not wander from layer to layer, causing unwanted sensitivities. In some processes the sensitizing dye must stay on the grain through at least part of the processing if, for example, any selective reexposing has to be done as part of the processing cycle. Another requirement that color processes put on sensitizing dyes is that they must be compatible with the many other chemicals, eg, the color-forming couplers, which might be put in the film to form image dyes (see under Sensitizing Dyes below).

Most current color processes, then, depend on silver halide grains that can be selectively sensitized to red, green, or blue and placed in separate layers so that the three records can maintain their identity. The rest of the color-image-forming process consists of using some reactions associated with the developing silver grain or the silver grain itself to form, destroy, or modify dyes, generally yellow, magenta, and cyan. The latter dyes are termed *subtractive primaries* in that each absorbs or subtracts one of the colors that are the basis of color vision. Thus yellow absorbs blue but transmits green and red; magenta absorbs green but transmits blue and red; and cyan absorbs red but transmits blue and green. The combination of yellow in superposition with magenta will, therefore, appear red; yellow with cyan will appear green; and magenta with cyan will appear blue. The three subtractive colors in equivalent amounts will reproduce a range of grays through black, depending upon the densities of the three dye images (see also Color).

Current Basis for Color Photography

Sensitizing Dyes. The natural sensitivity of silver halide crystals to ultraviolet and to blue light can be extended to green and red by the adsorption of sensitizing dyes to the crystal surfaces. The sensitizing dyes absorb green and red light and transfer the energy to the silver halide substrates. In this way the silver halide can form a latent image of light to which it is otherwise insensitive.

The most widely used sensitizing dyes are of the cyanine class (see Cyanine dyes). These consist of heterocyclic moieties linked by conjugated systems of atoms. An example of this class is:

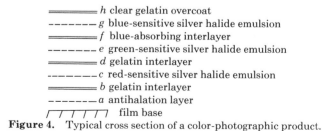

When $n = 0$ the dye is yellow and sensitizes silver halide to blue; this is not really advantageous because silver halide is naturally sensitive to blue. When $n = 1$ the dye, a carbocyanine, is magenta (absorbing green) and sensitizes silver halide to green; the natural blue sensitivity of silver halide remains. When $n = 2$ the dye, a dicarbocyanine, is cyan (absorbing red) and sensitizes silver halide to red. If the chain is lengthened further, sensitization is extended into the infrared. Cyanine dyes are symmetrical and exist as resonance hybrids of many forms. A large number of such dyes and combinations involving two different heterocyclic moieties in the same dye have been exploited for their sensitization effects (see also Dyes, sensitizing).

Photographic Emulsions. As in black-and-white photography, specially sensitized silver salts are precipitated in gelatin media with which they form emulsions (see Emulsions; Photography). In order to make use of the ability of such silver salts to differentiate a scene into three-color records, they must be prepared in separate emulsions, which are most commonly coated as separate layers, in superposition on a supporting film. A variety of other layers are frequently used; a typical color-photographic material is illustrated in Figure 4. The layers are coated in the order $a–h$. Layer a contains a black or colored material to prevent any spurious images that might result if light were to be totally reflected from the film base–air interface (halation). Layer f contains a yellow colorant that prevents blue light from reaching layers e and $c;$ the yellow color is destroyed during processing. The total thickness of the combined layers is of the order of 15–40 μm; the separate dyes are formed within this sandwich of permeable gelatin layers.

========	h clear gelatin overcoat
--------	g blue-sensitive silver halide emulsion
========	f blue-absorbing interlayer
--------	e green-sensitive silver halide emulsion
========	d gelatin interlayer
--------	c red-sensitive silver halide emulsion
========	b gelatin interlayer
--------	a antihalation layer
⌐ ⌐ ⌐ ⌐ ⌐ ⌐	film base

Figure 4.   Typical cross section of a color-photographic product.

The color records obtained by exposure of these emulsions may be processed to a color negative, from which a color positive can be obtained by printing onto a similar emulsion package, followed by color processing. Alternatively, the color positive may be obtained by reversal processing, for example by black-and-white negative development of exposed silver halide to silver, followed by color development using the residual silver halide to influence dye formation (Fig. 3). Or the color positive may be obtained directly by other chemical reactions.

In most subtractive color-photographic processes the dye image is produced by oxidative coupling as described by Fischer in 1912 (19). In this process the exposed silver salts are developed to metallic silver as in black-and-white photography. The developing agent, generally an *N,N*-dialkyl-*p*-phenylenediamine, becomes oxidized and capable of coupling to form dyes with substances that possess active methine (—CH=) or methylene (—CH$_2$—) groups (see Polymethine dyes).

A number of color-photographic products are based on such chemistry. Classifications can be made with respect to the manner in which the dyes are formed. In one class of product the dye formation involves couplers incorporated in the emulsions during manufacture; a single color developer serves to produce the three dye images. In the other class the couplers and the developers are contained in three separate processing solutions, one for each dye. Both classes impose rigid requirements on the couplers. Couplers that are incorporated must be compatible with the emulsion and must not diffuse or form dyes that diffuse. Couplers that are contained in the developer formulations must be completely removable by washing but must form nondiffusible dyes during processing (see below).

Color Development. *Oxidative Coupling.* The chemistry of color photography involves the conversion of the latent images into dye images. Fischer (19–20) reported that *N,N*-dialkyl-*p*-phenylenediamines are useful developing agents whose oxidation products are capable of forming dyes in proportion to the amount of exposure to a given region of the spectrum. The second component of the dye is the coupler, the choice of which determines the color obtained with a given developing agent. Fischer's proposal is given schematically as follows:

exposed silver salt + developer → oxidized developer

oxidized developer + coupler → dye

In the case of a specific developer, *N,N*-diethyl-1,4-phenylenediamine, and a specific coupler, 1-naphthol, the overall reaction can be written as:

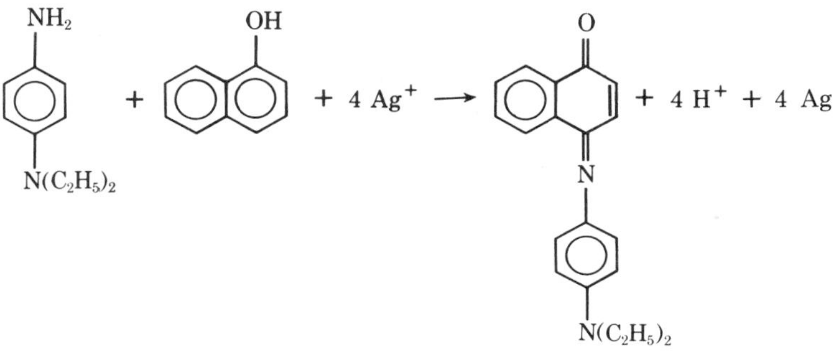

a cyan dye [*2363-99-7*]

The color of this indoaniline dye is cyan. From 2-benzoylacetanilide as the coupler, a yellow azomethine dye results; and from cyanoacetylcoumarone, a magenta dye, also of the azomethine type, is formed.

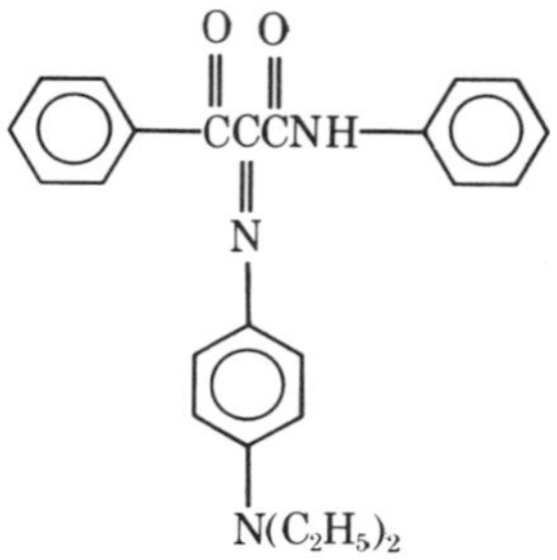

a yellow dye [4754-85-2] 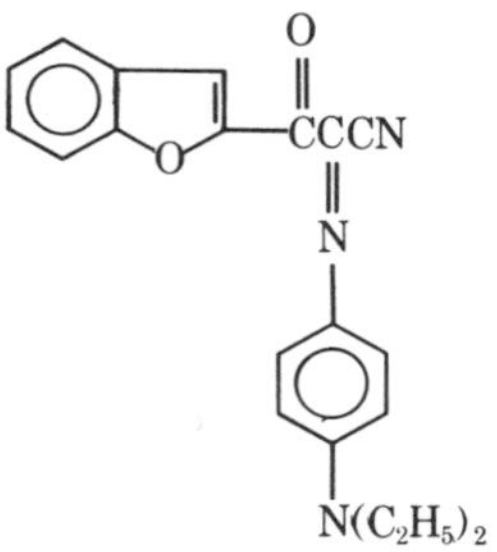   a magenta dye [1680-62-2]

These are dyes of the types described by Fischer. The variety of couplers and developers capable of these reactions will be discussed later in this article.

Mechanism.   The overall color development reaction proposed by Fischer involves a four-electron change and the couplers are termed four-equivalent because four silver ions are reduced for each molecule of dye formed. For a number of years the exact course of the reaction was unknown, although it was well established that a one-electron change resulting in the formation of the semiquinone of the developing agent was involved. Ultimately it was shown that the semiquinone produces quinone diimine by disproportionation (21–22). These reactions are illustrated as follows:

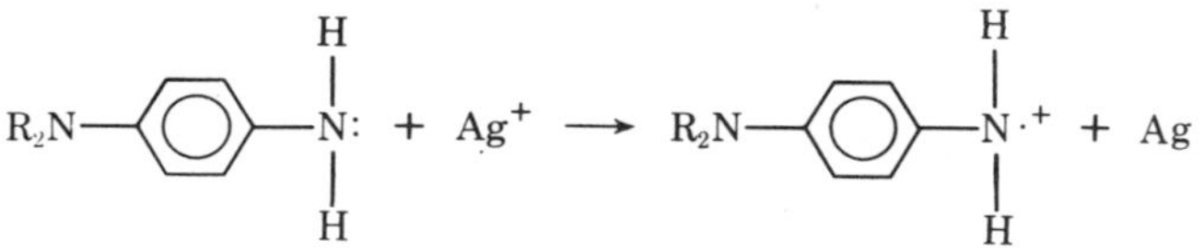

semiquinone

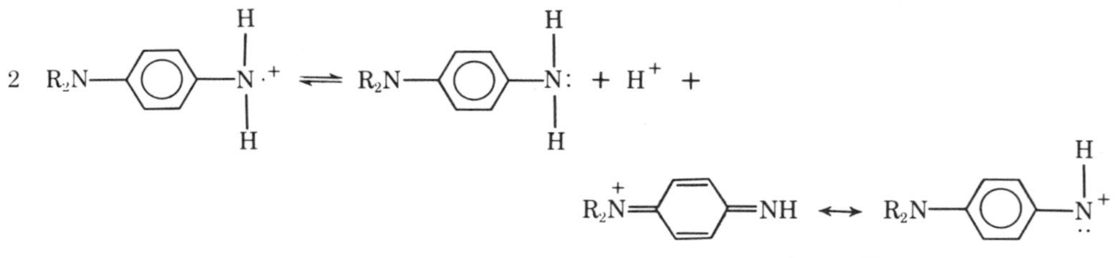

quinone diimine

The semiquinone and the quinone diimine are resonance-stabilized hybrids for which many contributing forms can be written.

The coupler in the alkaline processing system forms the anion, which couples with the quinone diimine to form a leuco dye.

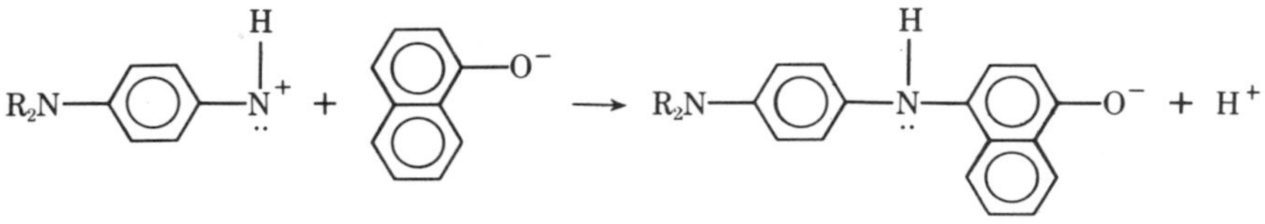

leuco dye anion

The leuco dye requires further oxidation, which is probably accomplished by quinone diimine. A molecule of developer is regenerated in the process.

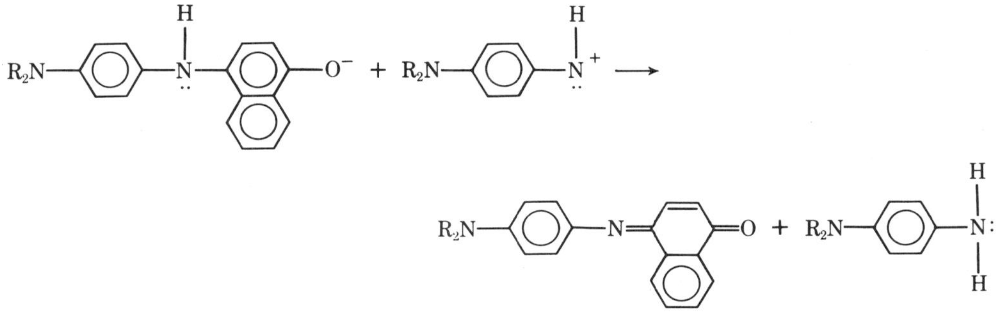

A modification of the Fischer reaction involves couplers that contain a substituent in the coupling position. When the substituent is halogen, sulfo, alkoxy, etc, it is eliminated during the dye-forming reaction and, since only one molecule of quinone diimine is involved, only two ions of silver are required per molecule of dye formed; such couplers are termed two-equivalent.

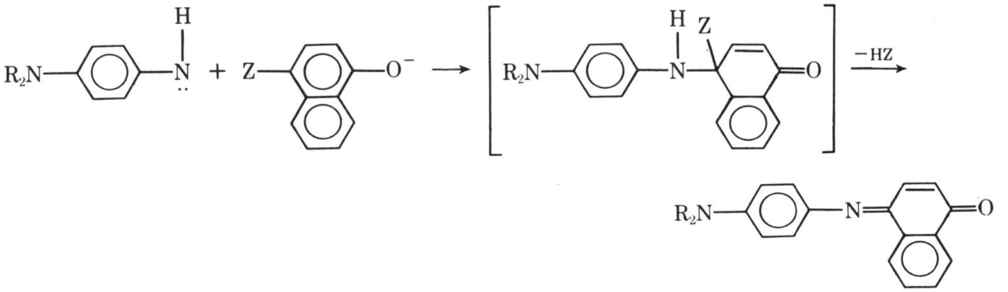

When the substituent in the coupling position is arylazo, the reaction with oxidized developing agent produces the desired dyes with simultaneous release of nitrogen and an aryl compound from the azo moiety as by-products (1).

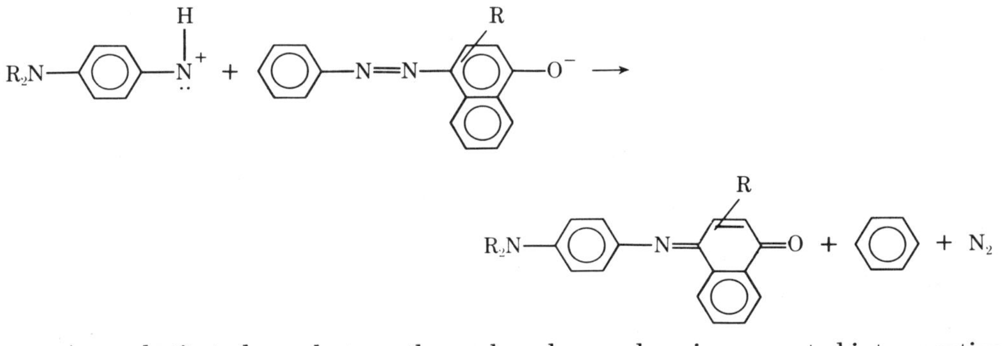

Azo-substituted couplers are themselves dyes and are incorporated into negative materials for color correction. In exposed regions of the negative the arylazo group is eliminated during coupling and the normal dye results. In unexposed regions, which do not develop, the azo dye persists in the negative and provides a color-correcting mask which can compensate for undesirable absorption characteristics of the image dye during printing (see below under Color Fidelity—Masking).

Side Reactions.   The dye-forming reaction is extremely rapid, but side reactions can occur. A prominent side reaction is that with alkali. Quinonelike organic molecules are susceptible to addition reactions (see Quinones) and quinone diimine is no exception. Generally the adduct loses the dialkylamino group, forming a quinone monoimine, 4-imino-2,5-cyclohexadien-1-one, from which, eg, a red indophenol dye forms instead of the expected cyan indoaniline dye.

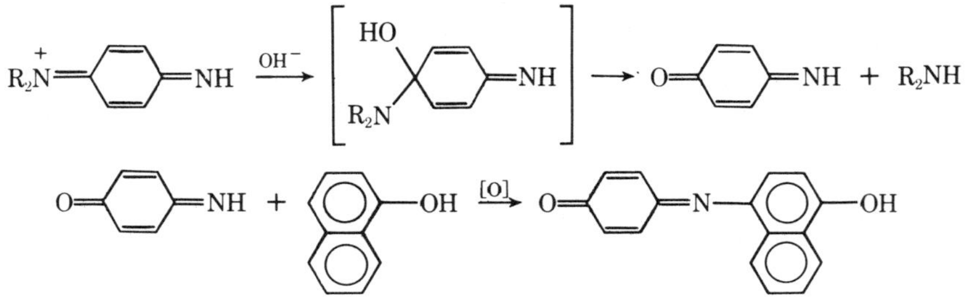

red indophenol dye [52200-93-8]

The quinone monoimine can have another complicating influence; since it is more diffusible than quinone diimine, it can move away from the development site and cross-oxidize unoxidized developer to yield quinone diimine and dye in remote areas.

Quinone diimine can also react with sulfite that is present in the developer solution and give the sulfonate of the developing agent.

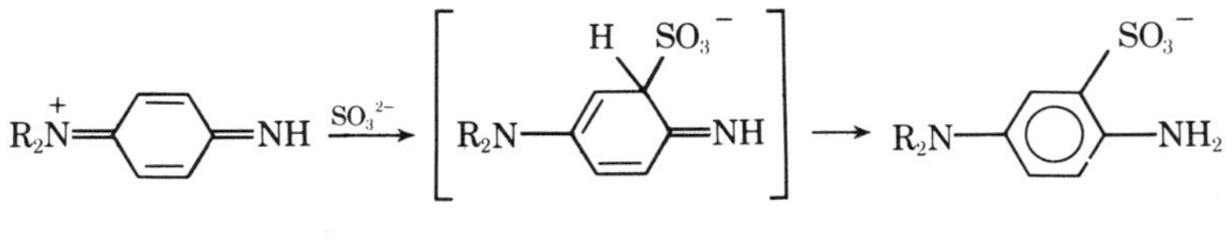

These reactions must be controlled so that they do not seriously affect the quality of the developed dye image.

Bleaching and Fixing.   After the development is completed the silver image is removed. This is done by reoxidation of silver to silver ions (bleaching) and subsequent dissolution and removal of the silver ions (fixing).

Agents suitable for the oxidation include ferric ion [as in ferric chloride, in ferric complexes with ethylenediaminetetraacetic acid (EDTA), and in potassium ferricyanide], benzoquinone, etc. The silver image is particularly susceptible to this heterogeneous reaction because of its finely divided structure, and bleaching is accomplished rapidly. The dyes that have been formed must, of course, be stable to the bleaching agents. Frequently this oxidation helps to complete the oxidation of the leuco dye. Finally, all silver salts are removed, as in black-and-white photography, by complex formation with sodium thiosulfate—the familiar hypo—followed by thorough washing.

Color-Forming Agents

Yellow Couplers. Yellow couplers give dyes with maximum absorption (λ_{max}) between 400 and 500 nm. The compounds suitable for this purpose contain an active methylene group that is generally not part of a ring. The methylene group is usually activated by at least one carbonyl group. The other activating group may also be carbonyl or any of a number of other groups that involve $\rangle C{=}O$, $\rangle C{=}N{-}$, etc.

The simplest yellow couplers are the β-diketones and β-ketoacylamides, eg:

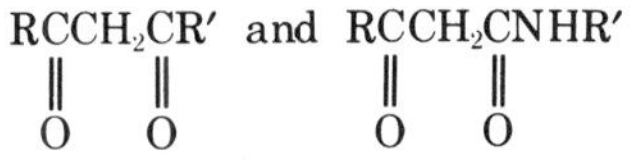

$$\underset{\substack{\| \quad \| \\ O \quad O}}{RCCH_2CR'} \quad \text{and} \quad \underset{\substack{\| \quad \| \\ O \quad O}}{RCCH_2CNHR'}$$

By far the most important class of yellow couplers consists of benzoylacetanilides,

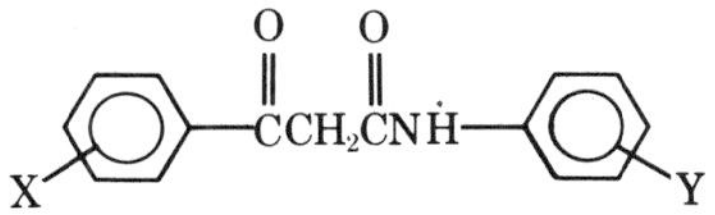

Ease of preparation and versatility of substituent effects are features of this system. The benzoylacetanilides give azomethine dyes having their maximum absorptions at about 450 nm. By contrast, the β-diketone dibenzoylmethane gives a dye with λ_{max} at about 480 nm. The latter is an orange–yellow.

The substituents X and Y may be chosen to modify the hue of the dye or to make the coupler and dye suitable for a selected system or process. It is reported that for hue purposes alkoxy or aryloxy in any position of the benzoyl ring (X) causes hypsochromic shifts in the dye absorption and steepens the long-wavelength side of the absorption curve (23). A further improvement results if the ortho position of the benzoyl nucleus is so substituted and a chlorine atom occupies the ortho position of the anilide ring (Y); in this case the extinction coefficient of the dye is markedly increased (24).

The presence of an o-alkoxy group in the anilide ring is also said to produce a hypsochromic shift and to steepen the absorption curve. The net effect is a more nearly lemon-yellow hue (25).

For processes in which the couplers are dissolved in the developer, either X or Y or both are small so that the couplers are readily diffusible through the emulsion. For products involving couplers incorporated in the emulsions, one substituent can be used for hue purposes and another for ballast and solubilization. Typical examples are:

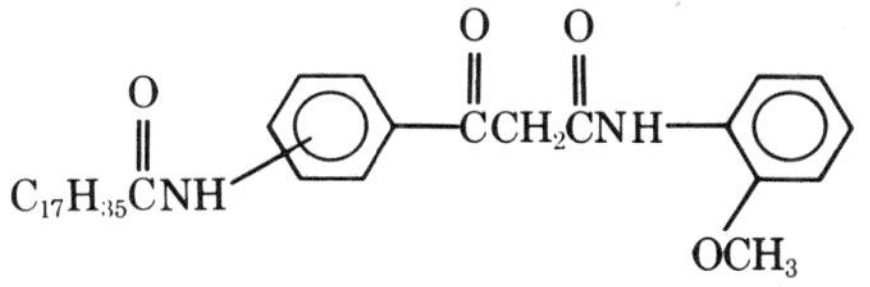

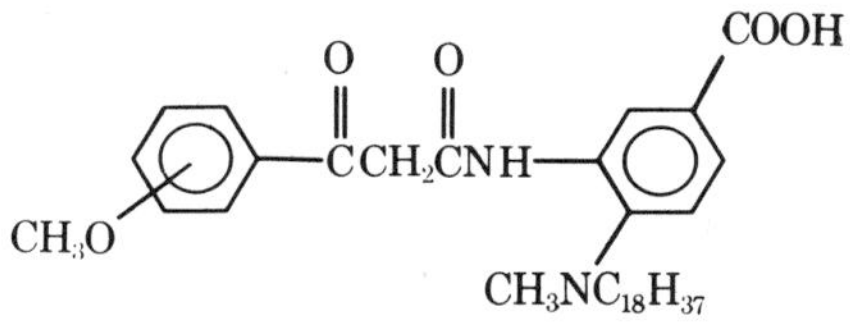

A great variety of other amines have been used for the preparation of benzoylacetamides; in general, when heterocyclic amines are used, bathochromic hue shifts are observed.

A number of patents have been obtained on derivatives of active methylene groups that form dyes by elimination of the substituent group. Examples are:

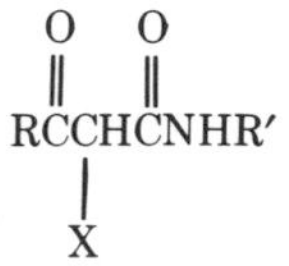

$$\underset{\underset{X}{|}}{RC\overset{\overset{O}{\|}}{C}H\overset{\overset{O}{\|}}{C}NHR'}$$

where X is halogen, SO$_3$H, COOH, COCOOH, COR″, —CH$_2$—, —CHR″—, —S—, and —S—S—.

Another class of yellow to orange dye-formers are the cyanoacetyl compounds, eg, CNCH$_2$CONHR. Closely analogous types are:

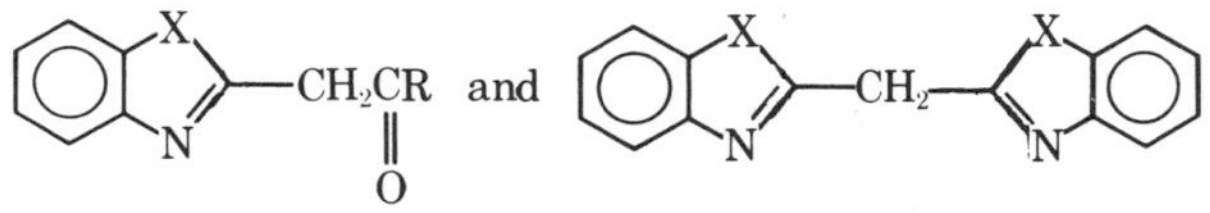

where X is O or S.

A class that apparently does not possess an active methylene group consists of 2,1-benzisoxazolin-3-ones, which couple with ring-opening to form yellow dyes as follows (26):

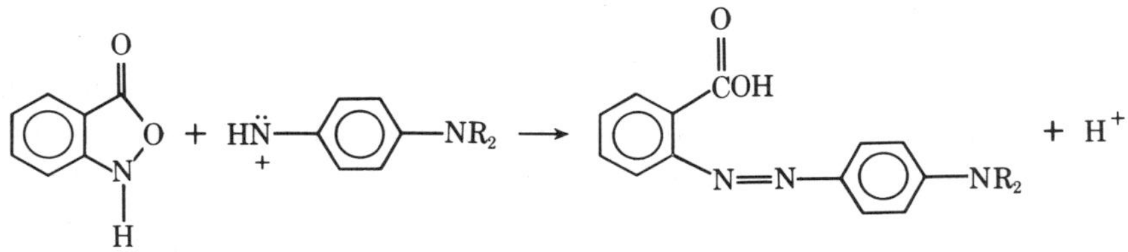

Benzisoxazolinones bear a structural relationship to 3-indazolinones, which give mesionic magenta dyes without ring opening (27).

Magenta Couplers. Magenta couplers are selected to give dyes whose maximum absorption occurs between 500 and 600 nm. Yellow dyes generally have considerable absorption in the ultraviolet region short of 400 nm; also, magenta dyes generally have some absorption in the blue, between 400 and 500 nm. Such secondary absorption of the magenta dyes is kept to a minimum by selection of the coupler type and its substituents or, in the case of negative materials, is compensated for by color masking (see under Color Fidelity—Masking).

Selection of a suitable magenta coupler is facilitated by the wide variety of types available. In general, the magenta dye formers are similar to the yellows in structure except that they give dyes that are shifted in hue as a result of substitution. Such substitution may involve cyclization, which makes the active methylene group part of a ring structure, or it may involve substituents that extend or intensify the conjugation of the dye structure.

Among the latter class are the cyanoacetyl types. The following examples illustrate the effects of conjugation and substitution on the absorptions of the dyes in emulsions:

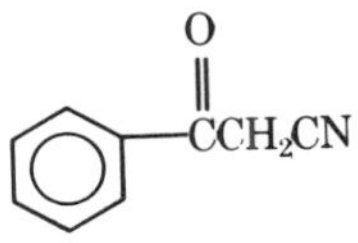

dye λ_{max} 516 nm

[614-16-4]

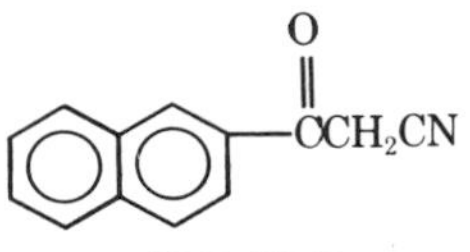

dye λ_{max} 520 nm

[613-57-0]

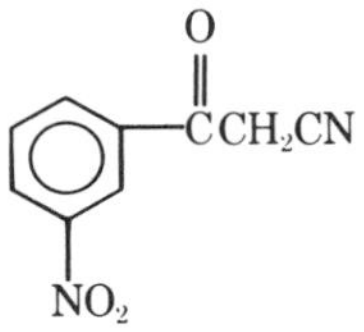

dye λ_{max} 526 nm

[21667-64-1]

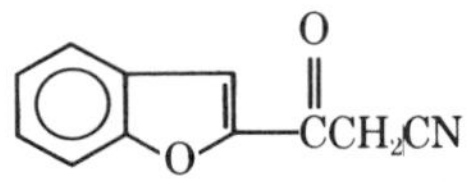

dye λ_{max} 528 nm

[5149-69-9]

The most important class of couplers for magenta dye formation contains the derivatives of 2-pyrazolin-5-ones, especially the 1-aryl-3-alkyl-(or 3-acylamido-)-2-pyrazolin-5-ones. The pyrazolinones are characterized by rapid coupling to give brilliant dyes with fairly sharp absorption curves of λ_{max} at 530–550 nm. Research on pyrazolinones has been extensive with many publications and patents. The range of structural variations can be illustrated by the following examples:

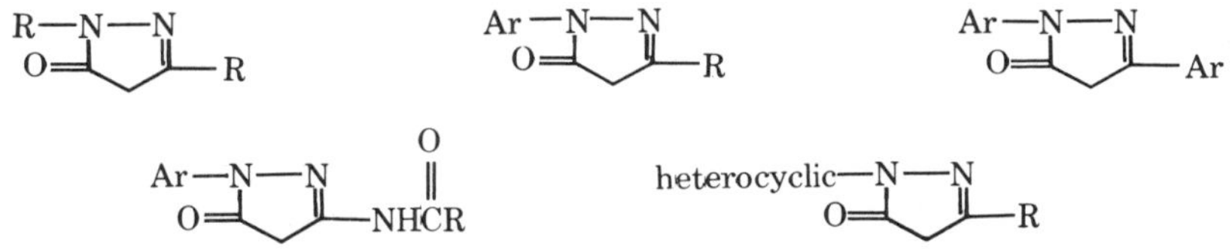

The first examples of 1-aryl-3-amino-2-pyrazolin-5-ones, which are important intermediates for practical magenta couplers, were at first incorrectly described as derivatives of 1-aryl-3-hydroxy-2-pyrazolin-5-imine (28). Later the structure was shown to be 1-aryl-3-amino-2-pyrazolin-5-one (29).

1-aryl-3-hydroxy-2-pyrazolin-5-imine

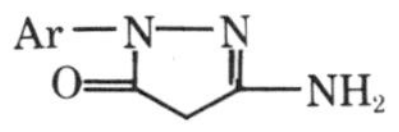

1-aryl-3-amino-2-pyrazolin-5-one

Substitution in the 1- and 3-positions of the pyrazolinone by alkyl, aryl, acylamino, and heterocyclic groups provides a great variety of structural variations for magenta couplers.

Hydrophilic character can be imparted similarly by either or both substituents.

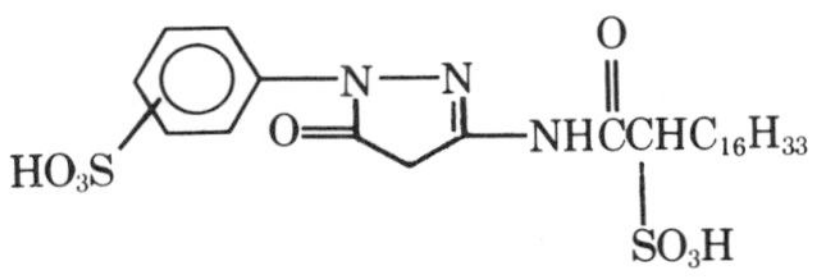

An interesting class of magenta couplers, the 3-indazolinones, apparently form dyes that are mesoionic (27).

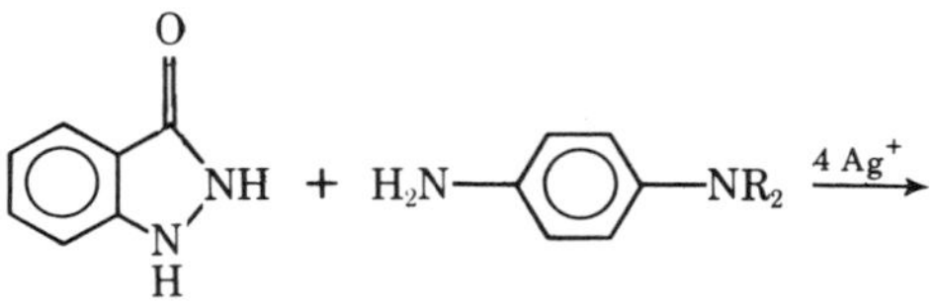

[7364-25-2]

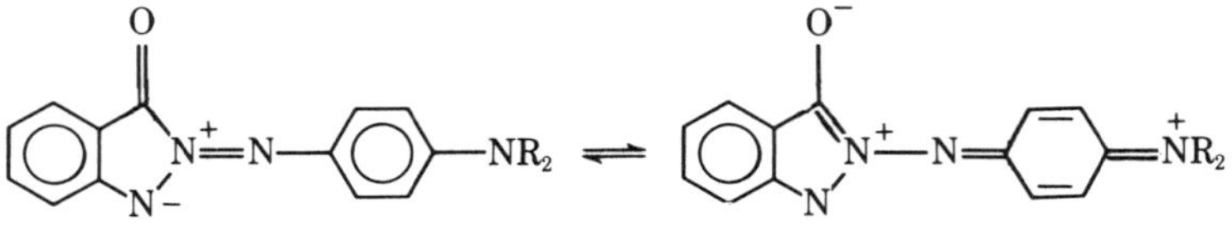

These dyes have desirably low blue and low red absorption.

Cyan Couplers. Cyan couplers give indoaniline dyes (see Dyes and dye intermediates) with λ_{max} generally between 600 and 700 nm; in almost all cases considerable absorption is noted beyond 700 nm, outside the visual range. Cyan dyes generally also absorb a considerable amount of green light and usually some blue light. It is frequently desirable to apply color-corrective schemes to such dyes in the negative so that unwanted green and blue absorptions do not give seriously distorted colors in the print materials (see also under Color Fidelity—Masking).

Cyan dyes are usually obtained by coupling development involving phenols and α-naphthols. The dyes of simple phenols are generally not sufficiently stable to heat and light to be practical. However, the introduction of halogen and alkyl groups yields couplers that give dyes of enhanced stability (30).

Another type of cyan-dye-forming phenol contains an acylamido group in the 5-position and yields heat-stable dyes. The short λ_{max} of such dyes is adjusted to longer wavelengths by a 2-perfluoroacylamido group, as in:

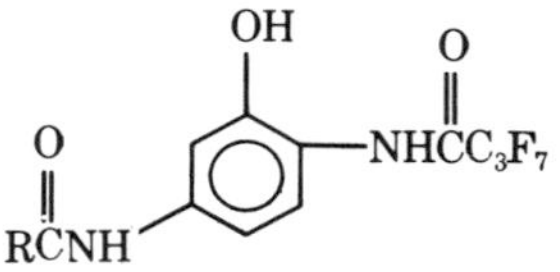

which gives a dye with λ_{max} at about 670 nm, a shift of about 25 nm over the nonfluorinated analogue (31).

The naphthols of greatest importance as couplers are derivatives of 1-hydroxy-2-naphthoic acid (see Naphthalene derivatives). For oil-phase incorporation, hydrophobic derivatives such as the following appear frequently in the patent literature.

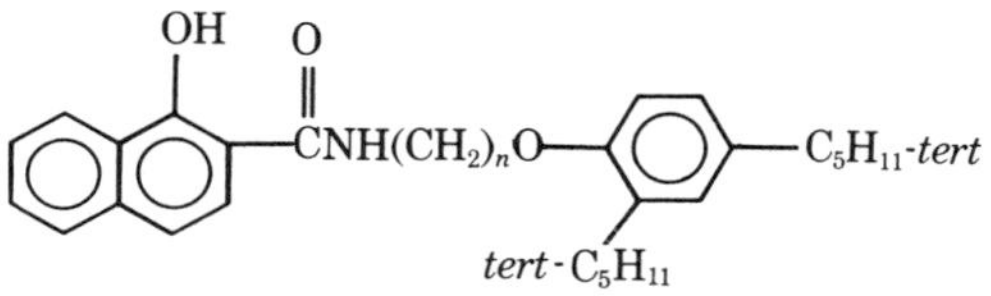

$n = 3$ [41434-22-4]

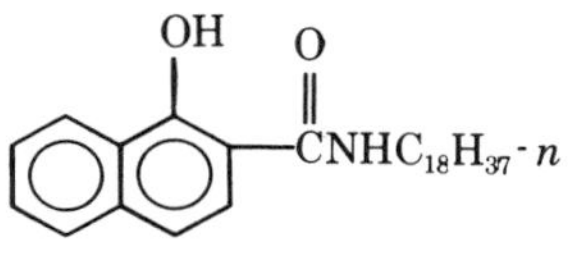

[20043-92-9]

For water-phase incorporation, hydrophilic derivatives such as the following have been patented.

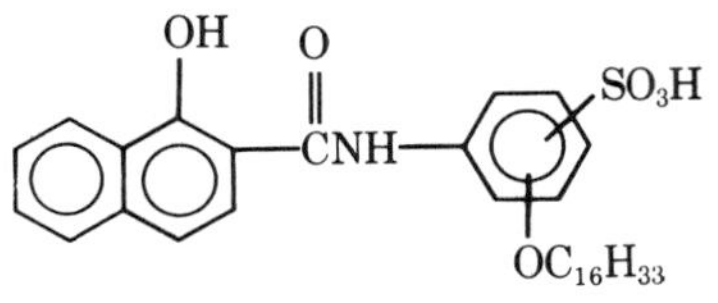

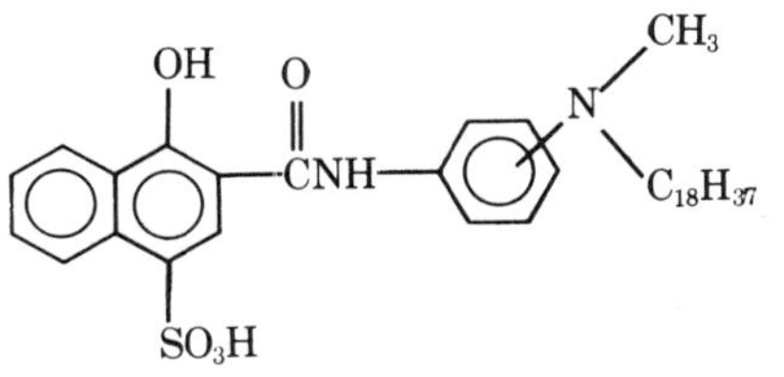

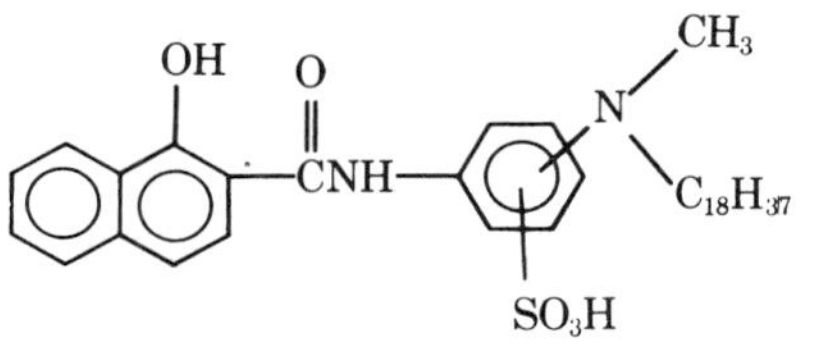

H_5C_2\N/CH_2CH_2OH phenyl NH_2 [92-65-9]

H_5C_2\N/CH_2CH_2NHSO_2CH_3 phenyl CH_3 NH_2 [92-09-1]

Bu\N/(CH_2)_nSO_3H phenyl NH_2 $n = 2,3,4$ or higher

Color Developers. Important properties of the color-developing agents are their solubilities, their reactivities, and their effect on the hue of the dyes (see below). Most of the simple N,N-dialkyl-1,4-phenylenediamines are sufficiently soluble in alkaline processing solutions to perform adequately. Where additional solubilization is desirable, hydrophilic groups are introduced as in the following:

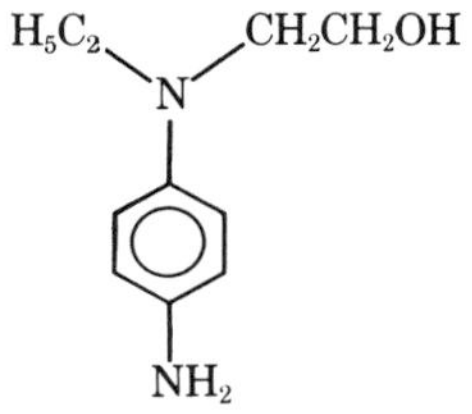

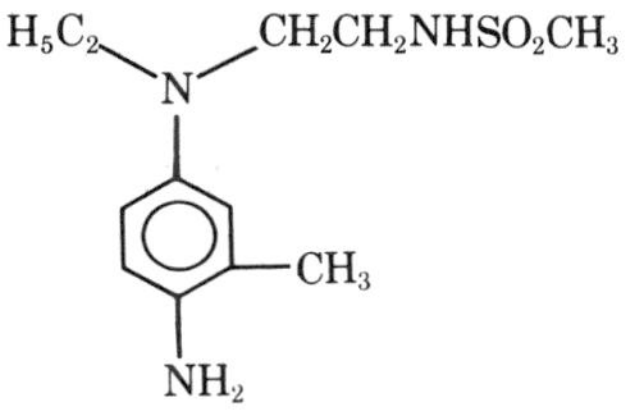

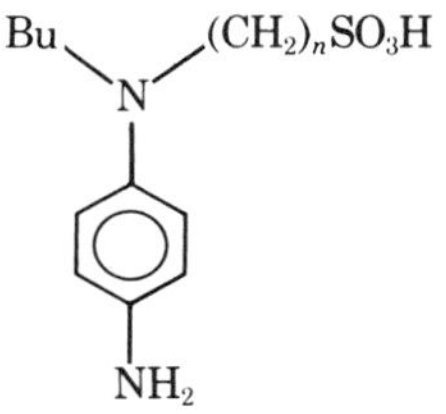

Solubilizing groups on the benzene ring greatly reduce the reactivity of the developing agent. On the other hand, electron-donating groups, such as alkyl, alkoxy, or alkylamino, in the position ortho to the primary amino groups greatly enhance developer activity, as illustrated by the following series (32).

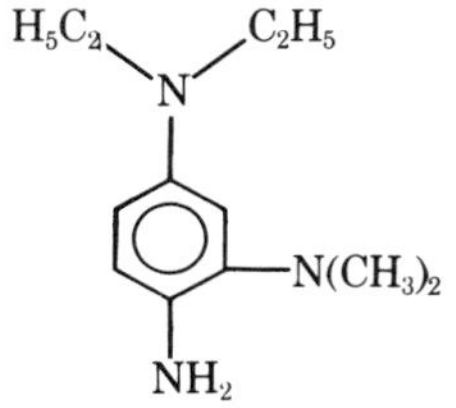

[68854-60-4]	[148-71-0]	[2359-46-8]	[93-05-0]
dev. rate: min^{-1}	$\gg$0.80	>0.71	>0.44
2.2			

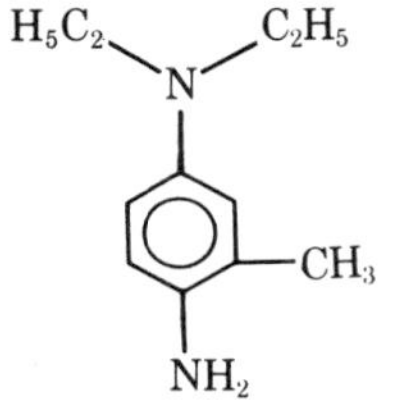

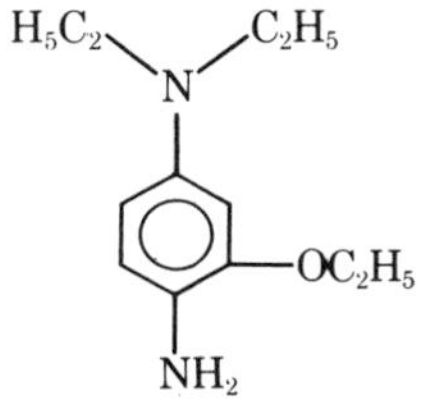

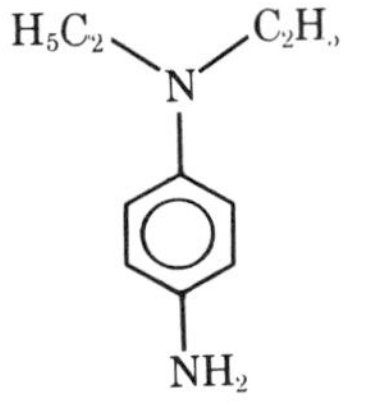

In general, increased development rate is directly related to the ease of electrochemical oxidation of the developer.

A similar enhancement of development rate is noted as alkyl substituents on the tertiary nitrogen atom become more electron-donating.

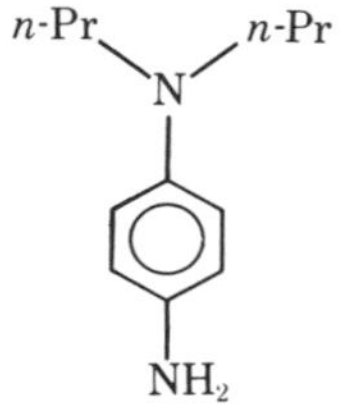

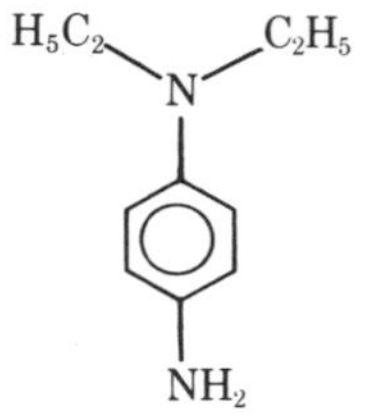

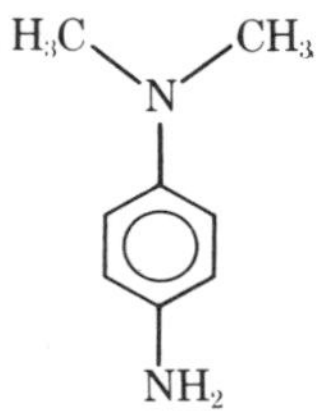

[49645-22-9]
dev. rate: min^{-1}
0.53

[93-05-0]

>0.44

[99-98-9]

>0.30

The rate of reaction of oxidized developers with couplers (coupling efficiency) varies only slightly for compounds differing by large amounts in developing activity.

Color developers are notoriously allergenic and must be handled with care (see Amines, aromatic, phenylenediamines). However, the incidence of allergenic reactions, resulting in skin irritation similar to that caused by poison ivy, can be reduced by certain substituents. Most prominent among these substituents are the β-hydroxyethyl group (33) and the β-methylsulfonamidoethyl group (34), as illustrated above.

Other methods for reducing allergenic toxicity involve protective groups on the primary amino function that are eliminated during or just prior to processing. Examples of these are a phthalimide derivative (35), which releases developer by reaction with hydrazine or hydroxylamine; a dialkylaminophenylglycine (36), which probably releases a developer oxidation product during processing; a sulfohomologue (37) of the glycine; and the so-called sulfur dioxide complexes (38) of the developer base. These latter protective devices promote safe handling of the dry materials and stock solutions. Alkaline solutions of such agents still require careful handling.

Modification of the Dye Hue

Dye hues can be modified by the introduction of substituents into either of the dye-forming components, viz, coupler or developer.

Within each of the classes of yellow, magenta, and cyan dyes considerable variation of hue is possible and, indeed, necessary in order to produce hues suitable for a given use. This can be achieved by introduction of suitable substituents into the coupler molecule. The data that follow are given for dyes in methanol; dyes in photographic emulsions are similarly affected by substituent changes. The λ_{max}, in nanometers, indicates the position of maximum absorption of the dyes.

Table 1.   Effect of Substituents on λ_{max} of Some Yellow Azomethine Dyes in Methanol[a]

	A	B	CAS Registry Number	λ_{max}, nm
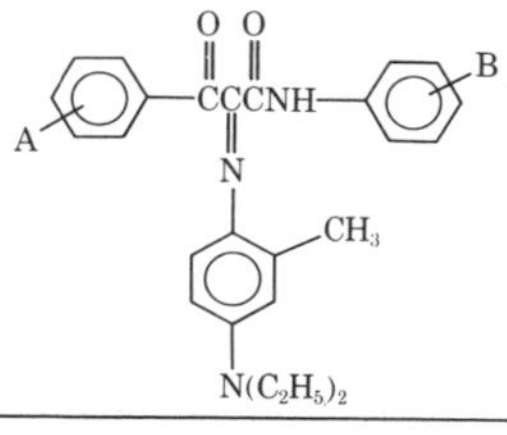	o-OCH$_3$	H	[68854-62-6]	434
	H	H	[4754-82-9]	448
	H	m-NO$_2$	[68854-64-8]	455
	m-NO$_2$	H	[68854-63-7]	472

[a] Ref. 39.

In a series of yellow dyes derived from benzoylacetanilide, the λ_{max} in methanol has been varied from 434 to 472 nm depending on the nature of the substituents A and B and their positions (39) (see Table 1).

Similarly, the effects of substituents on λ_{max} of dyes derived from pyrazolinones (40) (magenta) and phenols (41–42) (cyan) have been reported.

From studies of cyan dyes derived from 1-naphthols substituted in the 2-position (43) and in the 5-, 6-, 7-, or 8-positions (44), it is evident that steric as well as electronic factors influence λ_{max}.

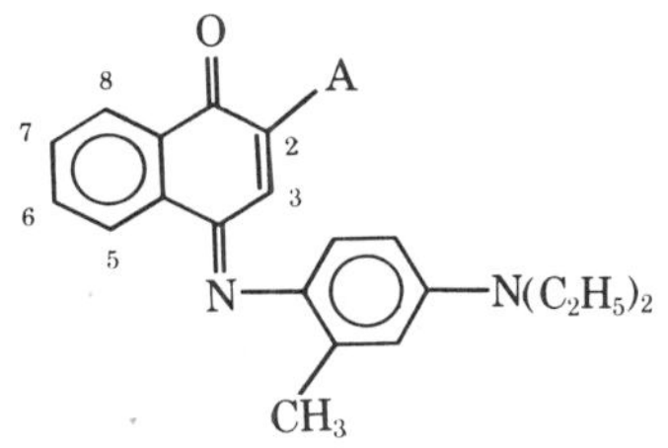

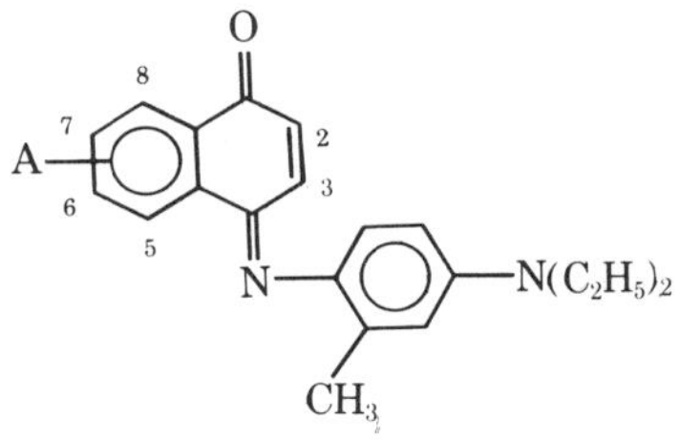

The hues obtainable from a given coupler can be varied greatly by the choice of the developing agent. For example, the indoaniline dyes from o-cresol are reported to vary in absorption maxima in acetonitrile–phosphate buffer from 505 to 700 nm depending upon the developer selected (45).

Electronic as well as steric factors are responsible for these variations in λ_{max} to the extent that they increase or decrease the contribution by form **B,** or other charge-separated forms, to the resonance hybrids of the dyes.

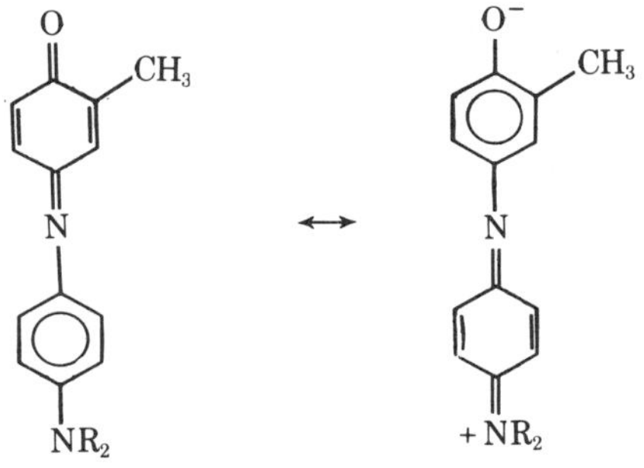

Increased contribution by **B** in the excited state of the dyes lengthens the wavelength of maximum absorption and vice versa.

Other Systems Proposed for Chromogenic Photography

Azine Dyes. The azine dye (qv) system for color photography was proposed by Agfa in 1947 and described at length in 1953 (46). The system appears not to have been used commercially. Azine dyes form by cyclization and further oxidation of indoaniline and azomethine dyes (see Polymethine dyes) which contain amino or anilino functions either on the developer moiety or on the coupler in the position adjacent to the azomethine linkage. The color-forming reactions are as follows (see also Chromogenic materials):

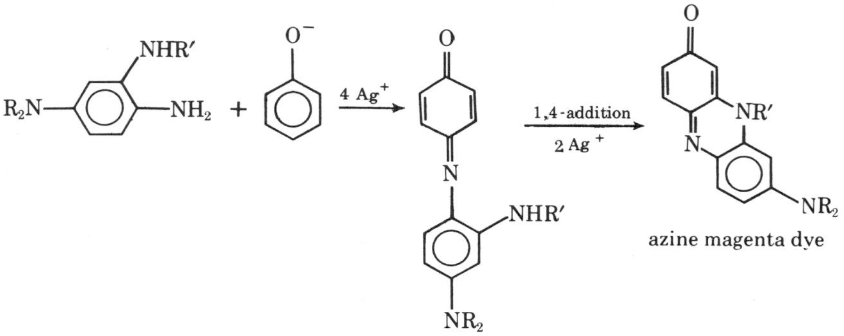

indoaniline cyan dye

azine magenta dye

This sequence of reactions requires development of six silver ions per molecule of dye formed. It is likely that the leuco dyes do not react directly with the latent image sites but are cross-oxidized by an oxidation carrier.

Yellow dyes were obtained by cyclization of the azomethine dyes, possibly as follows (the constitution of the yellow dye was not proved conclusively):

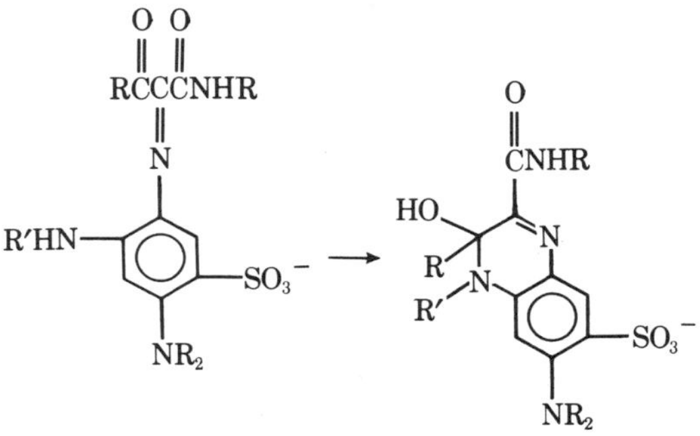

azine yellow dye

Careful adjustment of the pH of the photographic element was necessary to produce a permanent yellow image.

The cyan dye is also difficultly accessible in this system. Apparently the maximum absorption is about 630 nm and the absorption is also relatively high in the green and the blue regions of the spectrum.

Amidrazone Dyes.   Hünig and co-workers (47) proposed a scheme for forming azo dyes by oxidative condensation of heterocyclic hydrazones with phenols or aromatic amines.

Azo Dyes.   Azo dyes (qv) are said to result by the developing and coupling behavior of β-arylsulfonhydrazides (48).

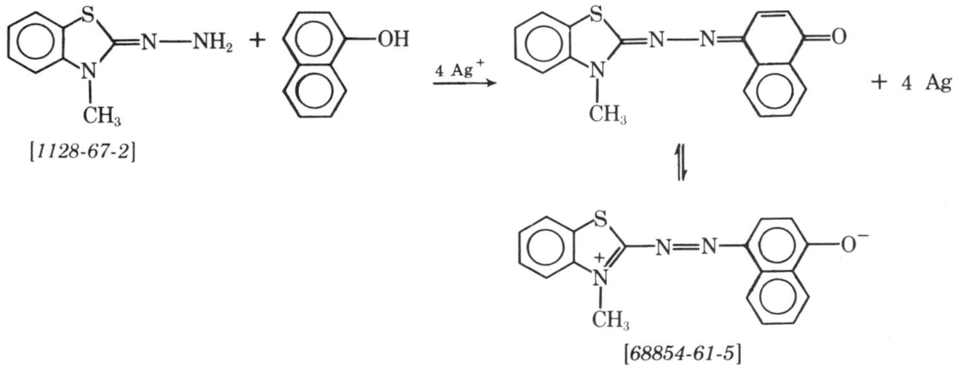

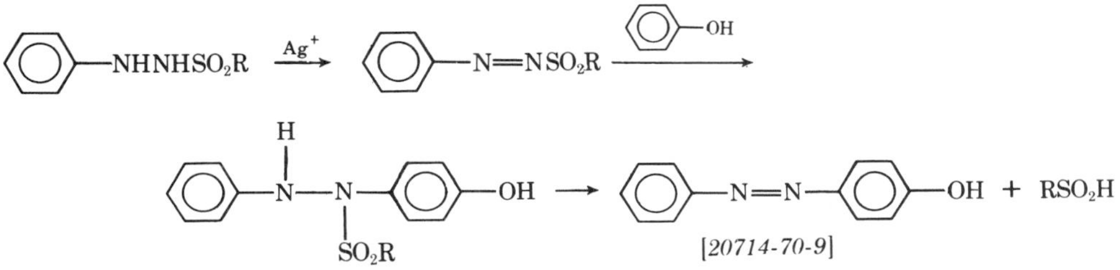

[1128-67-2]

[68854-61-5]

[20714-70-9]

Practical Chromogenic Color Processes

Developer-Soluble Couplers. In a black-and-white reversal process, the exposed image is first developed to a negative silver image; after that image is removed, the positive image in residual silver halide is developed to a positive silver image. The positive silver halide can also be developed chromogenically to form a positive dye image, without removal of the negative silver image. Two musicians, L. Mannes and L. Godowsky (49), used this principle to devise a color process. Their ideas were developed into the Kodachrome film and process marketed by Eastman Kodak Company in 1935. The original Kodachrome film was not very different from a multilayer coating of black-and-white emulsions; the color formers (couplers and developers) were constituents of a lengthy and complicated process.

Because the film had excellent image structure (fine grain and good sharpness) it could be used in the small cameras popular with untrained photographers, and because it was a reversal process (lightness and color are reproduced in their correct values) the camera film could be directly projected after processing. These factors made Kodachrome film popular for home movies and slides, and it, along with the Agfacolor film introduced by Agfa, A.G. (now Agfa-Gevaert, A.G.) also in 1935, sparked the explosive growth of mass-market color photography.

The Agfacolor film, like most of the many color films that have succeeded it, used color couplers that were incorporated into the emulsion layers. Kodachrome film, on the other hand, depends on soluble couplers in three different color-developing solutions to obtain the dye images.

The Kodachrome process, as well as most current chromogenic reversal color processes, used a first developer to reduce the negatively exposed silver halide to silver, without dye formation (Fig. 3). A carefully designed variant of conventional black-and-white developers serves this purpose. If this first development is complete, ie, if all of the exposed silver halide is disposed of in this manner, the residual silver halide is a complete, positive record of the original exposure. Chromogenic development of this residual halide produces positive dye images. The developed silver from both negative and positive development is removed by bleaching and fixing.

The problem in reversal color development is to maintain the identity of the three color records as they are developed to separate dye images. In the early version of the Kodachrome process this was effected by a differential dye-bleaching technique that depended on a slow rate of diffusion of the bleach into the superimposed layers. In the first chromogenic development, the residual (positive) silver halide was developed in all of the layers (which had been reexposed to make them developable) to form a cyan dye. This produced the desired image for the red-sensitive layer, but produced an unwanted cyan dye image in the other layers. The film was washed and dried; then a slowly penetrating bleach diffused into the blue and green record layers to oxidize the silver back to silver bromide and decolorize the unwanted dye in those layers. Depth of penetration was controlled by regulating the viscosity and the time the film remained in this bath. Bleaching was stopped and the residual halide (now only in the blue- and green-sensitive layers) was developed in a magenta-coupling developer. Again the film was washed and dried, and the bleach was used to rehalogenize and remove dye from the blue-sensitive layer only, leaving the desired magenta image in the middle layer. The final set of operations developed yellow dye in the only layer remaining with developable silver halide. In this ingenious but tedious manner a correct color image was obtained.

Later (1938) the process was changed to use the technique of selective reexposure. With the discovery of sensitizing dyes that remained on the silver halide grains during processing, it became possible to reexpose the residual positive silver halide selectively, first with red light, through the support. A color developer containing cyan-dye-forming coupler produces the desired cyan image only in the red-sensitive layer. A subsequent exposure from the emulsion side, with blue light, makes only the blue-sensitive residual halide developable. This is developed to the desired yellow positive image in a color developer containing a yellow-dye-forming coupler. The magenta image is then formed from the residual silver halide in the other sensitive layer by a magenta-dye-forming chromogenic developer. The halide in that layer is made developable by chemical fogging. Bleaching and fixing of the developed silver completes the process. Although this later process requires careful control of reexposure and of the several development stages, it is much more practical than the selective bleach process first used. The type of processing used for Kodachrome film was used for other still and motion picture film, almost all of which now use incorporated couplers (11).

The complexity of the Kodachrome process precluded its operation by the casual photographer. A substantial capital investment in precise machinery and accurate chemical control are required to obtain the desired results. An attractive feature of this system is that the emulsions, since they consist of very little more than conventional black-and-white emulsions, may be coated in quite thin layers, leading to a high level of sharpness. This sharpness is necessary for home movies, eg, which are magnified many times on projection. The greatest use of this process at the present time is for 8-mm and super-8 amateur motion pictures or for 35-mm still transparencies. In the past some professional motion picture uses were also based on this process, but most of these have been replaced by simpler processes.

Incorporated Couplers. A simplification in the process is effected when the different dye-forming couplers are incorporated into the appropriate sensitized layers. Incorporation is effected by several methods. In one method the coupler is water-soluble by virtue of a hydrophilic group, eg, SO_3H or $COOH$, yet is nondiffusible by virtue of a long aliphatic chain. Such couplers were used in the 1935 Agfacolor film and are used in current Agfacolor and Agfachrome materials. After World War II the Agfacolor process became free for general use and was applied by many manufacturers for their first commercial color-reversal and color-negative materials, eg, Ferraniacolor, Fujicolor, Gevacolor, Oriental Color, Pakolor, Sakuracolor, Telcolor, and Valcolor films. Most of these manufacturers have since changed (11) to the type of incorporation exemplified by Kodak Ektachrome film, which involves couplers that are nondiffusible by virtue of long aliphatic chains or combinations of short aliphatic chains and aromatic systems without hydrophilic groups. Such couplers are dissolved in high-boiling nonpolar organic solvents and dispersed in the silver halide emulsions. The couplers and the resulting dyes are thus present in tiny hydrophobic globules distributed evenly within the emulsion layers. Materials based on this method are Kodacolor, Kodak Vericolor, and Eastman color negative films, as well as Kodak Ektacolor film papers. Materials produced by other manufacturers using this method include Fujicolor, GAF Color Print, new Gevacolor, Sakuracolor II N100, and 3M Color Print films and the associated color papers. Reversal films using this method are Eastman Ektachrome motion picture films and Kodak Ektachrome still films as well as Fujichrome, 3M Color Slide, and Sakuracolor R100 films (9,11).

Color materials containing incorporated couplers can be processed in a reversal mode or in a negative–positive mode (Fig. 3). In the reversal mode, negative silver images are formed from the exposed silver halides, leaving positive images in unreacted silver halide. The positive color images are obtained in one chromogenic bath containing a developing agent whose oxidation product couples with each of the three couplers to form the desired dye in the appropriate amount in the appropriate layer. A bleaching and fixing operation to remove the developed silver completes the process. In the negative–positive mode the initial development of the latent images is accomplished by a color developing agent whose oxidation product reacts with coupler to form dyes complementary to the colors reflected by the subject. Bleaching of the silver and fixing complete the preparation of the negative. Exposure of the print material is modulated by the negative. Following color development, bleaching, and fixing, the print reproduces the colors of the subject.

The Eastman color motion-picture film system is a good example of such a film system. The original scene is recorded on the color negative. A print for record or editing purposes, or in some limited cases for actual sale, is made by photographic printing directly onto a positive material that produces a color image in proper lightnesses and hues. Where many prints are desired and special photographic operations must be introduced, it is quite common to produce a duplicate negative. In the Eastman color system this is done by printing the camera negative onto an intermediate positive material, having a contrast of one, and then printing that positive record onto the same material, to produce a negative that is a close match to the original negative. Various multiple-printing and other specialized photographic techniques may be introduced in the preparation of the positive and duplicate-negative stage so that the duplicate negative can contain all of the desired picture information. Since there is a degradation in image quality in duplication, the original negative is frequently used for as much as possible of the final picture image and the duplicate negative is interspliced with it only where necessary. This composite negative containing all of the desired picture in proper sequence is then continuously printed onto the positive color print film to produce the prints that are shown in theaters. A reversal-color intermediate film is also frequently used for these processes. It is used where complex special effects that demand two film stages are not present, and when the economy and quality advantages of having only one duplicating step allow it to replace the two-stage duplicating system just described. It is particularly useful when only film-size changes are desired to produce printing masters for contact printing. Duplicate negatives of a size different from that of the original may also be produced, eg, if the original is in 35-mm positive format.

Color motion-picture prints can also be produced from a reversal original. These originals are usually in 16-mm format and are used in the nontheatrical field for educational, advertising, and documentary purposes.

Reflection Prints.   By coating the proper amounts of emulsion and color formers on white reflecting supports instead of on transparent film, it is possible to make reflection-print materials. If the first operation is chromogenic development of a transparent color film, a negative color image is formed. Exposure of reflection materials to this negative image, after chromogenic development, provides prints of positive aspect. If a black-and-white developer is used first, and then is followed by a color-developer process, it is possible to obtain a direct reversal.

Nonchromogenic Color Photography

A number of color-photographic processes depend upon the black-and-white development of exposed silver salts followed by chemical reactions that result in the destruction of a preformed dye (dye bleach, see below) or in the transfer of a dye out of the emulsion to a mordanted receiver (see Color photography, instant).

Cibachrome Process.   A dye-bleach process, proposed by Schinzel (50) in 1905, became commercially important in 1963 when Ciba Photochemical Company introduced a product, Cilchrome Print (now called Cibachrome print), for producing color prints (51). In dye–bleach materials, yellow (52), magenta (53), and cyan dyes (54) are incorporated in separate layers that contain sensitized silver salts as in other color-photographic products.

The following are some of the dye types patented by Ciba:

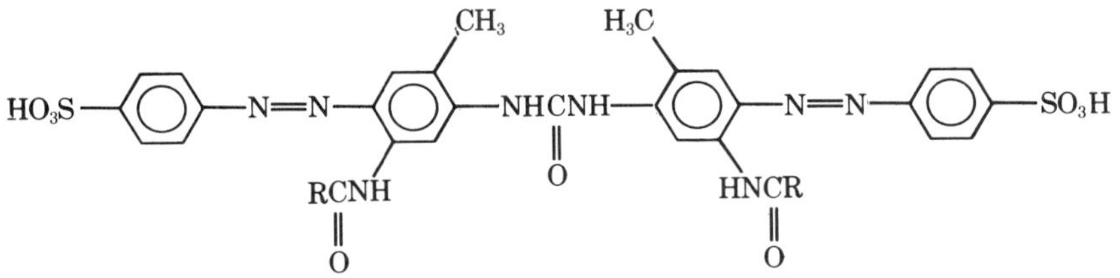

yellow dye

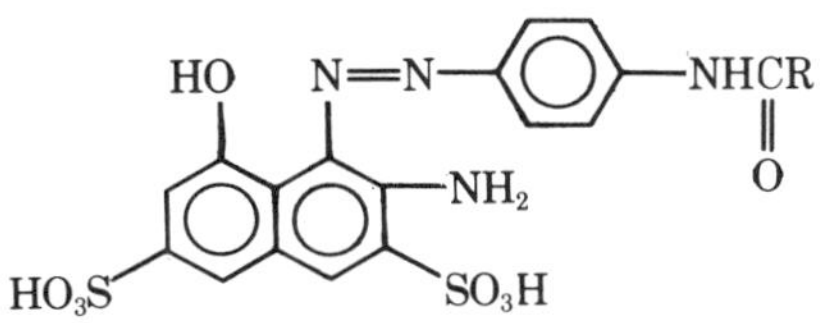

a magenta dye

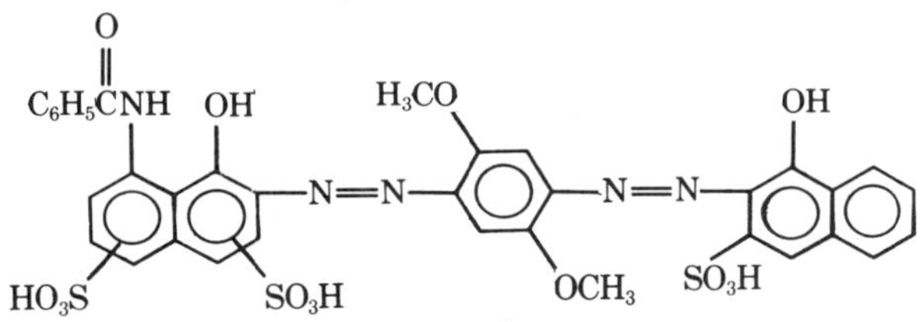

a cyan dye

Black-and-white development results in a silver image, which, in a subsequent step, initiates the destruction of dye in its vicinity. Thus exposure to red light results in a silver image that can be used to destroy cyan dye, leaving magenta and yellow for red reproduction. The dye destruction is aided by addenda such as phenazine and thiourea. Phenazine probably serves as a hydrogen carrier; thiourea, by forming a stable complex with silver ions, removes the latter and helps to drive the reaction to completion. The sequence of reactions may be represented as follows:

$$Ag^+ + developer \rightarrow Ag$$

$$R\text{---}N{=}N\text{---}R' + 2\,H^+ + 2\,Ag \rightleftharpoons RHN\text{---}NHR' + 2\,Ag^+$$

$$RHN\text{---}NHR' + 2\,H^+ + 2\,Ag \rightarrow RNH_2 + R'NH_2 + 2\,Ag^+$$

Cibachrome print film is designed for the preparation of prints from existing transparencies. Because of the light-absorbing nature of the dyes, which are present at the time of exposure, speeds are too low for camera use.

Quality Characteristics of Color Materials

Color Fidelity—Masking. The fidelity of color reproduction has been another important area occupying the attention of color-photographic chemists. Practical color-photographic systems fall short of perfect color reproduction for a number of reasons, one of which is that there are no perfect coloring materials with which to work. Even the very best dyes that are available for use in color photography have absorption curves showing so-called unwanted absorptions. Thus the yellow dyes, which should absorb only in the blue region, have fairly strong absorption in the green region from 500 to 600 nm. The magenta dyes all absorb some of the blue and red as well as the desired green light and, similarly, the red-absorbing cyan dyes have unwanted absorptions in the green and blue regions.

Over the years, dye chemists have been able to evolve improved dyes for color photography, ie, dyes with lower unwanted absorptions, and these improvements have been an important factor leading to better color reproduction. But even though further improvements in dyes will undoubtedly be made, color-photographic image dyes will always have some degree of unwanted absorption. Accordingly, methods have had to be found to compensate for these defects.

The method used in Kodacolor film, Eastman color motion-picture negative film, and many other negative films involves the use of colored couplers (55–56). In this method, the couplers incorporated in the green- and red-sensitive layers are colored azo compounds, each having a color matching the unwanted absorption of the image dye which it produces on coupling with oxidized color developer. Thus the yellow-colored coupler in the green-sensitive layer absorbs as much light in the blue region as does the magenta dye produced from it. After color development the combination of the negative magenta dye image and the positive image of unreacted yellow-colored coupler affords an equal absorption of blue light in all areas. This is optically equivalent to a magenta dye image having no blue-light absorption overlaid with a uniform density to blue light, which requires only that the blue exposure be increased when a print is made. The same principles apply to colored couplers used to correct for the absorption deficiencies of cyan dyes. In this case, a reddish-colored coupler is employed which has the same absorption of blue and green light as has the cyan dye formed from it. There are also examples of color negative films wherein masks are produced in the bleach step (Agfa, Gevaert, and Agfa-Gevaert) but the most common masking method is that based on azo derivatives of couplers (9).

Another method for color correction in negatives involves couplers known as development-inhibitor-releasing (DIR) couplers (57). Such couplers release chemical species that inhibit development; ie, species that stop or slow down the development process. Thus the imagewise release of such agents from a cyan dye-forming coupler can, among other reactions, inhibit development in the green-recording layer sufficiently to reduce the formation of magenta dye to compensate for the green absorption of the cyan dye. This is called an interlayer–interimage effect.

Since colored couplers impart a strong hue to highlight areas of the image, they cannot be used in photographic materials to produce images that are viewed directly. Hence in these cases other means have to be used to compensate for the unwanted absorptions of image dyes. With many color-reversal films, eg, compensation is achieved by the introduction of development interimage effects (58) at the black-and-white development step. These are complicated chemical interactions by which the degree of development in one layer of a multilayer film is made dependent upon the development taking place in other layers. By proper control of these effects which influence the relative amounts of the dyes formed, compensation for unwanted absorption is often achieved.

Dye Stability.   Most of the dyes formed by color development belong to the indoaniline and azomethine classes of dyes. Dyes of these types were known long before the discovery of color development, but none of them gained any practical importance in the dye industry because of their relative instability. Despite this deficiency, these dyes are increasingly successful in color photography, largely for these reasons: (*1*) the conditions of storage of photographic materials are usually not severe; (*2*) photographic materials need not withstand the cleaning and ironing required by textiles; (*3*) structural modifications of the dyes, introduced by changes in developers and couplers, have led to higher stability; (*4*) the dyes in the color images are embedded within a layer of gelatin or other carrier which affords some protection from outside influences; and (*5*) the physical structure of the photographic materials has been adjusted to confer marked improvements in the stability of the dye images.

Usually, the manufacturer of any given color product has established the conditions of processing that will lead to the maximum dye stability. Any deviation from the recommended procedures can have disastrous results with regard to dye stability (59).

Color Balance. Absolute sensitivity to light, or photographic speed, is of major concern in both monochrome and color photography. In color photography, in addition, the relative speeds of the red, green, and blue records are important, since the color-sensitive emulsions must be balanced for the spectral quality of the exposing illumination, as well as for its total energy. Although human vision is able to adapt to and perceive as a white color such extremes of illumination as daylight, with an effective color temperature of the order of 6000 K, and incandescent light, with an effective color temperature below 3000 K, color film records these illuminants as different colors. Thus a color film balanced to produce acceptable pictures under daylight illumination produces rather orange pictures under tungsten illumination. When other than the specified illuminant prevails, corrective measures may involve use of specific color filters during exposure or control of color balance in making a print. It is possible to design the spectral sensitivity of a color film to minimize the color balance shift with illuminant change. This has been done in high-speed color films such as Kodacolor 400 film and Kodak Ektachrome movie film, type G, which are used under various illuminants.

Speed. The earliest color photographic processes were suitable only for photographic still pictures or for reproduction purposes. By the 1930s enough speed had been realized to make color motion pictures practical. This speed level is described by ASA (American Standards Association, now ANSI) exposure indexes (60) from 5 to 16 (8/10–13/10 DIN, Deutsche Industrie Normen). By the 1960s reversal and negative color film having ASA indexes of the order of 100 ASA (21/10 DIN) were common.

By 1977 ASA 400 was available, and higher indexes have been achieved for specialized purposes by special processing techniques. The effective speeds of color-photographic materials have generally been lower than those of monochromatic photographic materials of the same image structure.

Grain. When color-photographic materials based on color development were relatively new, the view was widely held that color-photographic images were essentially grainless in structure because they were composed of transparent dyes as opposed to the opaque silver clumps arising in black-and-white images. It is now generally known that this is not true, that the morphology of the dye image is determined to a large part by the location of the silver halide and, therefore, that most dye images have a random particulate structure.

The graininess problem can actually be more troublesome in complex color materials than it is in black-and-white films (61). Color development produces oxidized developer at the developing silver bromide grain; diffusion of this material away from the point of formation can change the size of the grain units. Thus dye clouds that form can be considerably larger than the developed silver grains.

One method of reducing graininess involves restricting the length of the path that the oxidized developer can traverse by the addition of a soluble competing coupler, which forms a soluble and diffusible dye that escapes from the emulsion layer either during development or in subsequent processing steps (62). Another method for reducing graininess involves DIR couplers. By use of excess silver halide and repression of some of the development, centers of smaller size result and, consequently, more dye clouds of smaller size (57,63).

Sharpness.   The ability to reproduce detail and crisp edges in a picture is another problem more pronounced in color photography than in black-and-white photography. Light entering a multilayer coating is scattered by the turbid gelatin–silver halide medium. The light is scattered progressively further as it penetrates deeper into a multilayer structure. As a consequence, a cyan dye image tends to be less sharp than the overlying magenta image, and the latter less sharp than the outermost yellow dye image. One important means for decreasing this optical scattering is to reduce the thickness of the coating structure; it is largely for this reason that efforts continue toward finding ways to make emulsion layers as thin as possible. The average multilayer color film of the 1970s is 40% as thick as that of the 1940s.

In a color picture the magenta dye image carries the greatest proportion of sharpness and graininess information, because the eye has its greatest sensitivity to detail in the green region of the spectrum. Thus, in general, the image structure of the magenta layer determines the visual image structure of the neutral image. For the same reason, the yellow image is least important in this respect. Some color films take advantage of these facts by having the magenta layer on top to avoid the scattering of light by overlying layers during exposure (64).

In addition to the optical turbidity effects, there are also chemical factors that reduce the sharpness of color images. One mechanism that can cause the deposition of dyes some distance from an exposed grain involves the solution of some of the silver halide during development. This mobile silver halide can migrate to a site distant from that of the original exposure. A second process that may produce dye at some distance from the developing grain results from long-lived but noncoupling intermediates that may diffuse and cross-oxidize developer at some distance from the original development site. The secondary oxidized developer then couples to form dye. A third, and common, phenomenon is the diffusion of oxidized developer away from the developing grain center before coupling.

The causes just cited, in addition to degrading sharpness, can also degrade color reproduction and give unwanted color images if the movement of the oxidation product or the silver ion is from one color record to another. This color contamination is frequently minimized by the use of spacer or barrier interlayers between the image-forming layers (65).

Two-Color Systems

Because of the complexity and cost of arriving at a natural three-color photographic image, several processes were derived that depend on just two color records.

Since it is impossible to obtain a full gamut of colors with only two records, a compromise in color quality was made in two-color systems for pictorial photography. Two-color systems were also developed for data-recording uses, where a full range of colors was not needed.

The first color process sold under the name Technicolor used only two color separation negatives from which a two-dye print was made by imbibition printing (see p. 621). Cinecolor Corporation's Cinecolor process yielded another two-color motion picture print derived from two color-separation negatives. The two spectral regions recorded in these negatives were usually the blue–green and the orange–red. The blue–green separation negative modulated the generation of an orange dye in the print, and exposure through the reddish separation negative controlled cyan dye formation in the print. A proper combination of the cyan and orange could produce a neutral tone, although obtaining an achromatic, dark neutral was frequently impossible with these processes.

As technology improved, both the Technicolor and Cinecolor processes evolved into three-color systems. This has generally been the case; with time the customer demands the level of color quality obtainable from three-color systems. When one full-color system is available, it is difficult for two-color systems to compete, even at lower cost.

Oscillograph recording papers have been marketed that contain two-color records, usually a cyan and a magenta or red dye, produced from a yellow and a green record, respectively. These products are intended to aid in identification of multiple traces made from multichannel recorders. Traces can be produced in the two colors separately, or in combination.

Special Applications of Color Photography

Color Television. A major use of color motion-picture film is in television broadcasting. Besides the broadcasting of films originally made for theater showing, many entertainment programs are produced primarily for television markets. Thousands of short film commercials are produced each year. The negative–positive film systems are usually used. However, reversal films are used in recording and televising news events, owing to the simplicity of a single-film system. More sensitive films with shorter, more convenient processes have evolved for this use.

Another use of color films for color television is that of recording the transient television signals for more permanent records and optical display. This operation is called color kinescope recording. Although special recorders have been constructed to write on the film with three modulated laser beams (see Lasers), the more general technique is to photograph a special television display on one (shadowmask) (66) or three (trinoscope) (67) cathode ray tubes.

Infrared-Recording Color Film. Another interesting use of color photography is illustrated by Kodak Aerochrome infrared film. Here the specialized problems of high-altitude aerial photography call for a material different from those designed for conventional pictorial uses. The scattering of blue light by the atmosphere is such that at high altitudes there is really no blue record that is worth photographing. This film dispenses with the blue record, uses a green record that is converted to a yellow dye and a red record that is converted to a magenta dye, and adds an infrared record that is converted to a cyan dye. Since infrared radiation is scattered even less than the visible and since natural foliage has characteristic reflectivity in the infrared region, this extra record is useful for the identification of objects on the ground. Although the colors presented from such a photograph are not natural, a photo-interpreter learns by experience to read this new language (see Infrared technology).

Uses and Economic Factors in Color Photography

Amateur photography, which is about one third of the total photographic business, is now largely color photography. The major form, reflection prints, reached over five and a half billion prints per year in the United States by the mid 1970s according to Wolfman (68), and 85% of these were in color. (This compares with a total of one billion prints per year ten years earlier, with about 60% in color.) This increase in volume and trend toward color is expected to continue with the introduction of new, convenient materials for both negative–positive and instant print photography. Transparencies for the amateur are all in color, and more than a billion of these are made each year. Home movies, now largely in the super-8 format, are practically all in color.

In professional or commercial use of color photography one finds a wide variety of applications. Entertainment and television motion pictures have been described earlier. Movies for educational sales, motivational, and technical purposes are a large business; most of this is now in color. Color negatives and transparencies are the starting point for graphic reproduction in display advertising and in magazines. Color portraits are both a big business and a commonplace decoration. Identification photos are increasingly in color. Mapping, surveying, record taking, and plan drawings now use color photography to take advantage of the extra information that the spectral dimension gives. A most dramatic use of color photography in the past decade has been the color photographs taken in space, either directly on photographic materials or reconstituted from signals beamed back to earth.

The rate of growth of photography, as measured by the gross national photo product (68), has been about double that of the gross national product; much of this growth can be attributed to the expanding use of color.

BIBLIOGRAPHY

"Color Photography" under "Photography" in *ECT* 1st ed., Vol. 10, pp. 577–584, by T. H. James, Eastman Kodak Company; "Color Photography" in *ECT* 2nd ed., Vol. 5, pp. 812–845, by J. R. Thirtle and D. M. Zwick, Eastman Kodak Company.

1. T. H. James, ed., *The Theory of the Photographic Process,* 4th ed., Macmillan Publishing Co., Inc., New York, 1977.
2. J. M. Sturge, ed., *Neblette's Handbook of Photography and Reprography,* 7th ed., Van Nostrand Reinhold Company, New York, 1977.
3. P. Kowaliski, *Applied Photographic Theory,* John Wiley & Sons, Inc., New York, 1972.
4. W. C. Guida and D. J. Raber, *J. Chem. Ed.* **52,** 622 (1975).
5. J. Eggers, "The Chemistry of Photographic Development," in W. F. Berg, ed., *Photographic Science,* Focal Press, London, Eng., and New York, 1963.
6. R. W. G. Hunt, *The Reproduction of Colour,* 3rd ed., Fountain Press, London, Eng., 1975.
7. J. Bailey and L. A. Williams, "The Photographic Development Process," in K. Venkataraman, ed., *The Chemistry of Synthetic Dyes,* Vol. IV, Academic Press, Inc., New York, 1971, pp. 341–387.
8. G. Koshofer, *Br. J. Photogr.* **113,** 562, 606, 644, 738, 824, 920 (1966); *ibid.* **114,** 128 (1967).
9. G. R. Koshofer, *Br. J. Photogr.* **123,** 409, 460, 524, 568, 610, 658, 696, 736, 840, 918, 962, 1006, 1050, 1096, 1142, 1165 (1976).

10. R. T. Ryan, *A Study of the Technology of Color Motion Picture Processes Developed in the United States,* University Microfilms, Ann Arbor, Mich., 1967.
11. *Br. J. Photogr. Ann.* (1977).
12. A. Weissberger, *Am. Sci.* **58,** 648 (1970).
13. E. J. Wall, *History of Three-Color Photography,* American Photographic Publishing Co., Boston, Mass., 1925.
14. J. S. Friedman, *History of Color Photography,* 2nd ed., Focal Press, New York, 1968.
15. R. W. Burnham, R. M. Hanes, and C. J. Bartleson, *Color: A Guide to Basic Facts and Concepts,* John Wiley & Sons, Inc., New York, 1963, p. 180.
16. J. C. Maxwell, *Trans Roy. Soc. Edinburgh* **21,** 275 (1857); *Proc. Roy. Soc.* **60,** 404 (1859).
17. R. M. Evans, *Sci. Am.* **205,** 118 (1961).
18. A. DuHauron, *La Triplice Photographique des Couleurs et l'Imprimerie,* Gauthier-Villars, Paris, Fr., 1897, p. 448.
19. Ger. Pat. 253,335 (Nov. 12, 1912), R. Fischer.
20. R. Fischer and H. Siegrist, *Photogr. Korresp.* **51,** 18 (1914).
21. L. K. J. Tong and M. C. Glesmann, *J. Am. Chem. Soc.* **79,** 583 (1957).
22. J. Eggers and H. Frieser, *Z. Elektrochem.* **60,** 372 (1956).
23. Ger. Pat. 744,265 (Jan. 17, 1944), W. Schneider, A. Frohlich, and W. Tappe (to Agfa, Inc.).
24. U.S. Pat. 2,407,210 (Sept. 3, 1946), A. Weissberger, C. J. Kibler, and P. W. Vittum (to Eastman Kodak Co.).
25. Brit. Pat. 800,108 (Aug. 20, 1958), F. C. McCrossen, P. W. Vittum, and A. Weissberger (to Eastman Kodak Co.).
26. Brit. Pat. 778,089 (July 3, 1957), J. M. Woolley (to Imperial Chemical Industries, Ltd.).
27. J. Jennen, *Ind. Chim. Belge* **16,** 472 (1951); *Chim. Ind.* **67**(2), 356 (1952).
28. M. Conrad and A. Zart, *Ber.* **39,** 2282 (1906).
29. H. D. Porter and A. Weissberger, *J. Am. Chem. Soc.* **64,** 2133 (1942).
30. U.S. Pat. 2,367,531 (Jan. 16, 1945), I. F. Salminen, P. W. Vittum, and A. Weissberger (to Eastman Kodak Co.).
31. U.S. Pat. 2,895,826 (July 21, 1959), I. F. Salminen, C. R. Barr, and A. Loria (to Eastman Kodak Co.).
32. R. L. Bent and co-workers, *J. Am. Chem. Soc.* **73,** 3100 (1951).
33. U.S. Pat. 2,108,243 (Feb. 15, 1938), B. Wendt (to Agfa, Inc.).
34. U.S. Pat. 2,193,015 (Mar. 12, 1940), A. Weissberger (to Eastman Kodak Co.).
35. Brit. Pat. 775,692 (May 29, 1957), D. W. C. Ramsay (to Imperial Chemical Industries, Ltd.).
36. Brit. Pat. 783,887 (Oct. 2, 1957), J. F. Willems (to Gevaert, Ltd.).
37. Fr. Pat. 1,049,864 (Jan. 4, 1954) (to Gevaert, Ltd.).
38. Brit. Pat. 626,958 (July 25, 1949), D. H. O. John and G. T. J. Field (to May and Baker Ltd.).
39. G. H. Brown and co-workers, *J. Am. Chem. Soc.* **79,** 2919 (1957).
40. *Ibid.,* **73,** 919 (1951).
41. P. W. Vittum and G. Brown, *J. Am. Chem. Soc.* **68,** 2235 (1946).
42. *Ibid.* **69,** 152 (1947).
43. C. R. Barr and co-workers, *Photogr. Sci. Eng.* **5,** 195 (1961).
44. A. P. Lurie and co-workers, *J. Am. Chem. Soc.* **83,** 5015 (1961).
45. S. Hunig and P. Richters, *Ann.* **612,** 282 (1958).
46. W. A. Schmidt and co-workers, *Ind. Eng. Chem.* **45,** 1726 (1953).
47. S. Hunig and co-workers, *Angew. Chem.* **70,** 215 (1958); Ger. Pat. 963,297 (May 2, 1957), S. Hunig (to C. Schleussner Fotowerke G.m.b.H.).
48. W. A. Schmidt in W. Eichler, H. Frieser, and O. Helwich, eds., *Wissenschaftliche Photographie (Ergebnisse der Intern. Konf. fur Wissenschaftliche Photographie, Koln, 1956),* Verlag Dr. O. Helwich, Darmstadt, Ger., 1958, p. 456.
49. L. D. Mannes and L. Godowsky, *J. SMPE* **25,** 65 (1935).
50. K. Schinzel, *Br. J. Photogr.* **52,** 608 (1905).
51. R. Wartburg, *Camera Appl. Photogr.* (3), 31 (1964).
52. Brit. Pat. 922,568 (Nov. 30, 1961), (to Ciba Corp.).
53. Ger. Pat. 1,139,378 (Nov. 8, 1962), P. Dreyfuss (to Ciba Corp.).
54. Ger. Pat. 1,161,135 (Jan. 8, 1963) (to Ciba Corp.).
55. W. T. Hanson, Jr., and P. W. Vittum, *J. Photogr. Sci. Am.* **13,** 94 (1947).
56. W. T. Hanson, Jr., *J. Opt. Soc. Am.* **40,** 166 (1950).

57. C. R. Barr, J. R. Thirtle, and P. W. Vittum, *Photogr. Sci. Eng.* **13,** 74 (1969).
58. W. T. Hanson, Jr., and C. A. Horton, *J. Opt. Soc. Am.* **42,** 663 (1952).
59. R. O. Gale and A. L. Williams, *J. SMPTE* **72,** 804 (1963); R. K. Schafer, *BKSTS J.* **54,** 286 (1972).
60. *American Standard PH 2.21-1972* (for color reversal films) American Standards Association Inc., New York, 1972; *American Standard PH 2.27-1965 (R 1971)* (for color negative films), 1971.
61. D. Zwick, *J. Photogr. Sci.* **11,** 269 (1963).
62. U.S. Pat. 2,689,793 (Sept. 21, 1954), W. R. Weller and N. H. Groet (to Eastman Kodak Co.).
63. L. E. Beavers and D. H. Nelander, *J. Opt. Soc. Am.* **62,** 1403 (1972).
64. W. T. Hanson, Jr., *J. SMPTE* **58,** 223 (1952).
65. V. A. Bogolyubskii and co-workers, *Usp. Nauchn. Fotogr. Akad. Nauk SSSR Otd. Khim. Nauk* **8,** 61 (1962).
66. K. G. Lisk and C. H. Evans, *J. SMPTE* **80,** 801 (1971).
67. *Ibid.* **83,** 719 (1974).
68. A. Wolfman, *1976–1977 Wolfman Report on the Photographic Industry in the United States,* ABC Leisure Magazines, Inc., New York, 1977.

General Reference

J. R. Thirtle, *Chemtech,* 25 (Jan. 1979).

JOHN R. THIRTLE
DAAN M. ZWICK
Eastman Kodak Company

COLOR PHOTOGRAPHY, INSTANT

Instant color photography comprises the group of color processes that result in finished color photographs within moments after exposure of the film. Processing is accomplished in a single step, which takes place rapidly under ambient conditions. The processes are essentially dry and the reagent is provided as a part of the film unit. In most cases the reagent is applied by mechanical action of the camera, film holder, or other processing device immediately after the film is exposed.

Whereas lengthy, multistep darkroom processes used for non-instant color films require great precision in both time and temperature control, the instant processes do not involve darkroom handling and operate over a wide range of temperatures. Timing, when required, applies to only a single step, and the integral instant processes are completely self-timing.

For a further account of Polaroid instant photography, covering both black-and-white and color processes, see the article "One-Step Photography" (1).

Principles of Instant Photography

The broad requirements of instant photography were described by Land in 1947 when he introduced the first one-step print process (2) and discussed the problems of devising a camera and process that together would provide a finished print shortly after exposure. The principles set forth are fundamental to all instant systems, from the first sepia and black-and-white processes to the color processes considered here.

Film and Process Design. To be practical for handheld cameras, a film requires sufficient photographic speed to permit short exposures at small apertures. For this reason Land worked with a silver halide emulsion, taking advantage of its high sensitivity to light and the enormous amplification provided by the development process. He described a one-step process that entailed the spread of a viscous reagent between two sheets, one bearing an exposed silver halide emulsion and the other an image-receiving layer, as both sheets were drawn out of a camera through a pair of pressure rollers, as illustrated in the schematic drawing of Figure 1. A sealed pod attached to one of the sheets ruptured to release the viscous reagent, which spread to form a thin layer between the two sheets, temporarily bonding them together. The action of the reagent produced concomitantly a negative image in the emulsion layer and a positive image in the image-receiving layer. After about a minute, the two sheets were stripped apart to reveal the positive image.

Land outlined several classes of silver halide processes that could provide useful one-step positive images according to this scheme. In addition to the soluble-silver-complex process, which has been the basis for Polaroid black-and-white films as well

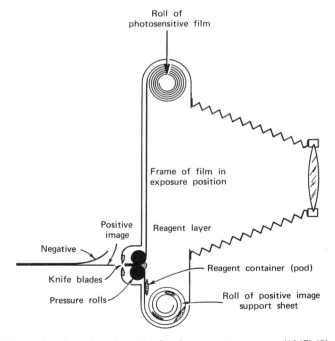

Figure 1.   An early schematic plan for a one-step camera (1947) (2).

as the Polavision additive color film, he described processes in which images were derived from unoxidized developer or from oxidized developer, and color-developer processes analogous to each (2). The instant color processes illustrate the formation of images of each of these types.

One-Step Cameras and Processors.   The earliest one-step cameras used roll film and completed the processing inside a dark chamber within the camera (3). The first instant color film, Polacolor, was provided in roll film format to fit these cameras. Later, roll film cameras (4) and flat pack film cameras (5) were equipped for drawing film and sheet between the processing rollers and out of the camera before processing was completed. In some pack cameras non-rotating spreaders have been used in place of rollers (6).

Fully automatic processing was introduced in the Polaroid SX-70 camera, which ejects each picture unit automatically, passing it between motorized processing rollers immediately after exposure (Fig. 2) (7–8). Kodak instant cameras include both motorized and hand-cranked roller models.

Processing rollers that are not integral with a camera are used in specialized Polaroid film holders (9). The 4×5 holder is equipped with a dark slide and built-in processing rollers. Both the 10×12 x-ray film holder and the 8×10 Polacolor film holder are inserted after exposure into automatically operated processors that contain motorized processing rollers.

Polaroid's instant color motion picture film, Polavision Phototape, is contained in sealed cassettes, and processing takes place within the film cassette after its insertion into the Polavision player (10) (see The Polavision Process below).

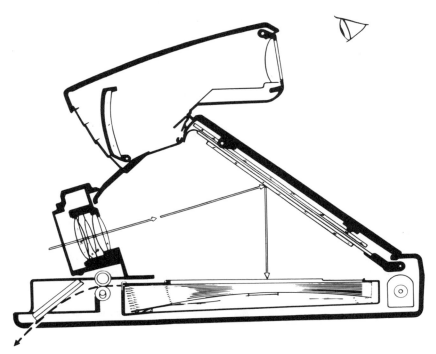

Figure 2.   Schematic section of Polaroid SX-70 camera (1972) showing light path during exposure and subsequent path of exposed film unit between processing rollers and out of camera (8).

Film Configuration. Processing of Polaroid black-and-white and Polacolor films is terminated after a specified time by stripping apart the pair of sheets that have passed together through processing rollers. Films handled in this way are described as peel-apart materials. The positive image carried by the separated image-receiving sheet has the correct geometric orientation for direct viewing.

An integral print film comprises two sheets permanently secured as a single unit. The image-forming layers are located on the inner surfaces of the two sheets, one of which is transparent. The image is viewed through the transparent sheet against a reflective white pigment layer within the film unit.

Integral processes may be designed to provide exposure of the negative layers and viewing of the print through the same surface (11) or through opposite surfaces (12) of the film unit. Polaroid SX-70 film units are exposed and viewed through the same surface, and correct image orientation is obtained with a single mirror reversal in the camera, an arrangement chosen for compactness in optical design. Kodak PR-10 film units are exposed and viewed through opposite surfaces, so that image orientation is correct either without optical reversal in the camera or with two mirror reversals.

Reagents for Instant Photography. *Viscous Reagents.* In each instant process, an essential component is the viscous reagent. The term reagent is used here in the sense of an entire processing composition rather than a single reactive component. In designing the first reagent system, Land determined that the volume of strong developer required to develop a black-and-white negative corresponded to a layer less than $1/10$ of a mm (13), and he recognized that such a quantity of reagent could be contained in a small sealed pod (14), thereby permitting the use of fresh reagent for each picture unit. The concept of essentially dry processing works by using a highly viscous fluid reagent and restricting the amount to just that needed to complete the image-forming reactions for one picture.

The viscosity of the reagent is increased by the addition of water-soluble polymeric thickeners (15). Suitable polymers include hydroxyethyl cellulose and the alkali-soluble salts of carboxymethyl cellulose and carboxymethyl hydroxyethyl cellulose. The use of high molecular weight grades of such polymers gives suitably high viscosity at low concentrations of polymer and the low solids content often facilitates rapid image transfer. The high viscosity of the reagent makes possible accurate metering to form a layer of uniform thickness which is also suitable as a temporary adhesive for the two sheets during processing. The viscous reagent layer may further serve as a protective colloid before and during processing. The layer may either remain on the surface of the print or be removed from the print surface by stripping away with the other sheet.

Fresh reagent in a sealed pod for each picture permits the use of more highly reactive reducing agents and more strongly alkaline conditions than would be feasible in a tank or tray process, as the reagent may be kept oxygen-free from the time it is sealed in the pod until after image formation is complete.

In addition to high molecular weight polymer, reducing agents, and alkali, the viscous reagent may contain reactive species that participate in image formation, deposition, and stabilization. Coatings on either of the two sheets may also contain reactive components in order to isolate materials from one another prior to processing and locate each where it is compatible and stable.

Pods. The reagent-filled pod attached to one of the two sheets must be carefully designed both for containing and discharging its contents (14). The pod lining should be inert to strong alkali and other components of the reagent, and the pod should be impervious to oxygen and water vapor over long periods. When the pod is passed through processing rollers, the seal must rupture with a peeling separation to avoid explosive bursting, so that reagent will be released along one edge and spread uniformly, starting as a fine bead and forming a smooth, thin layer. The amount of reagent contained in each pod is particularly critical with integral film units having limited capacity to conceal surplus reagent.

Formation of Transfer Images. *Images Initially Present.* In investigating mechanisms for one-step photography, Land identified within the conventional photographic process an entire family of images resulting from reactions that vary from point to point as a consequence of exposure and development. These images, shown in Table 1, represent possible starting points for transfer processes leading to useful positive images.

In the black-and-white silver processes, as well as in additive color processes utilizing silver image transfer, images *4, 6,* and *8* of Table 1 (the undeveloped silver, the unoxidized developer, and the unused alkali) transfer and react within the image-receiving layer to form the positive silver image. In the instant color processes using dye developers, molecules which are both image dyes and photographic developers, image *6* (the unoxidized dye developer) transfers to form a positive dye image in the image-receiving layer; in color development and redox dye-release systems image *5* (an oxidized developer) forms or releases the final positive dye image.

Methods of Color Reproduction. The formation of a three-color photograph starts with a set of three latent images. Each image results from exposure to light from about a third of the visible spectrum, so that one image represents the red components, one the green components, and one the blue components of the original scene. The three images may be formed in a single layer or in separate layers of a multilayer film (see also Color).

In the non-instant systems of color photography that produce positive images, the final image viewed is located in the layer or layers where the image has been recorded. In the instant-color positive systems, the three color records, whether initially formed in a single layer or in three separate layers, produce the final picture in a single, separate image-receiving layer.

Table 1.   Images Present in Silver Halide Emulsion Upon Exposure and Treatment with a Developer[a]

1	the exposed grains of the latent image
2	the unexposed grains of the latent image
3	the developed silver
4	the undeveloped silver halide
5	the oxidized developer
6	the developer that is not oxidized
7	the neutralized alkali
8	the alkali that is not neutralized
9	the hardened gelatin
10	the unhardened gelatin

[a] Ref. 13.

Given the variety of initial images in Table 1 and the option of producing color records in single or multiple layers, there are numerous possibilities for design of a complete color process and its color-forming components. Some of the principal alternatives are described below. For a more complete discussion of classical principles of color photography and for information on their application to non-instant systems, see Color photography.

In *subtractive color* photography the three color records are formed in three separate silver halide emulsions sensitive to red, green, and blue light, respectively. Processing results in positive images in complementary dyes: the red record produces an image in cyan dye, ie, minus red; the green record, in magenta dye, ie, minus green; and the blue record, in yellow dye, ie, minus blue. By analogy, an ordinary black-and-white image uses a black "dye," ie, minus white. When white light passes through the set of cyan, magenta, and yellow dye images, the dyes subtract red, green, and blue light, respectively, from white light, so that the combined image reproduces the original colors.

In *additive color* photography the three color records are produced by the exposure of a silver halide emulsion sensitive to red, green, and blue light through a layer containing an array of red, green, and blue filter elements. The exposed emulsion is processed to form a black-and-white positive silver transparency comprising black-and-white image records of the three color exposures. Projected together with the colored filter elements, the images add to reproduce the colors of the original scene.

Both additive and subtractive systems have been utilized in instant color films. The Polavision instant transparency system is based on an additive color process, and the Polacolor, SX-70, and PR-10 instant print systems use subtractive color processes.

Dye Imaging Systems.   There are several ways among the dye image systems to utilize images formed initially in a set of silver halide emulsions sensitive to red, green, and blue light, respectively. The dyes may be formed *in situ* by a chromogenic process as, for example, by color development, or preformed dyes may be transferred from layers of the negative to the image-receiving layer. The dyes or dye precursors may be initially mobile in alkali, in which case the process will immobilize them as images, or they may be initially immobile in alkali and mobilize to form images during processing.

Positive-working transfer dye processes produce dye densities in the image-receiving layer in inverse relation to development of silver in the emulsion layers; conversely, negative-working transfer dye processes produce dye density in the receiving layer in direct proportion to silver development in the emulsion.

With positive-working dye processes, the use of a negative-working emulsion, ie, an emulsion in which the exposed grains develop to form a negative silver image, results in the formation of a positive dye transfer image. When negative-working dye processes are used, positive dye images may be obtained by using direct reversal, or direct positive, emulsions which develop unexposed rather than exposed grains to form positive silver images. Alternatively, the development of exposed silver halide in a negative-working emulsion may control the development of unexposed or prefogged silver halide, which in turn controls the release of dyes to form positive dye images (16–17).

Each of the subtractive instant color films is based on preformed dyes. Polaroid's

Polacolor and SX-70 dyes are initially mobile in alkali and are used in a positive-working sense in conjunction with an emulsion that develops negative silver images. Kodak's PR-10 film uses initially immobile dyes in a negative-working sense in conjunction with an emulsion that forms direct positive silver images.

Commercial Instant Color Processes

The Polacolor Process.   Polacolor, the first instant color film, was introduced by Polaroid Corporation in 1963. An improved Polacolor film, introduced in 1975, was designated Polacolor 2. For convenience, the earlier Polacolor product will be referred to as Polacolor 1. The film was the first based on the use of the dye developer, a single molecule comprising both a preformed image dye and a silver halide developer (18).

The Polacolor process produces subtractive multicolor prints comprising positive images in terms of cyan, magenta, and yellow dye developers. The dye developers that form the positive image transfer from a multilayer color negative to an image-receiving layer concomitantly with the formation of negative silver images and negative dye developer images within the layers of the negative. Image formation is based on the immobilization of dye developers by oxidation in areas where exposed silver halide grains are developed, and on diffusion of dye developers from areas that are unexposed and hence free from oxidation.

Each dye developer is initially located in a layer just behind the layer of silver halide emulsion that will control it during processing. As shown in Figure 3, the negative comprises the yellow dye developer behind the blue-sensitive emulsion, the magenta dye developer behind the green-sensitive emulsion, and the cyan dye developer behind the red-sensitive emulsion, with interlayers separating the three units. During exposure, each of the dye developers functions both as an antihalation dye for the emulsion layer in front of it and as a filter dye protecting the emulsion layer or layers behind it from unwanted exposure. Thus the yellow dye developer protects the green- and red-sensitive emulsions from exposure to blue light, and the magenta dye developer protects the red-sensitive emulsion from exposure to green light.

The Polacolor receiving sheet comprises three active layers (19), represented schematically in Figure 3. The outermost layer is a polymeric image-receiving layer containing a mordant that immobilizes image dyes. Next is a polymeric timing layer which controls by its permeability and thickness the length of time that the process will continue at high pH. Beyond the timing layer is an immobile polymeric acid layer. The timing layer and the acid polymer layer operate in conjunction to provide a controlled drop in alkalinity of the system.

At the start of processing, the viscous reagent is spread to form a thin layer, temporarily laminating the exposed negative to the image-receiving sheet. As layers of the two sheets absorb liquid, they swell and the thickness of the reagent layer diminishes as the two sheet surfaces move closer together. Within the negative, development of silver halide grains in exposed areas in each of the three emulsion layers is accompanied by oxidation and immobilization of the contiguous dye developer. In unexposed areas, the dye developers, which remain unoxidized and thus still mobile in alkali, diffuse through layers of the negative to the image-receiving sheet. Figure 3 represents the structures during exposure and after processing, and the micrographs of Figure 4 are cross section views of the Polacolor 2 negative before and after processing.

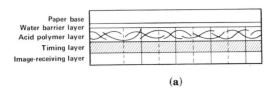

(a)

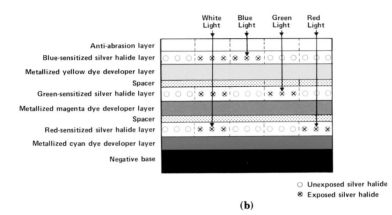

○ Unexposed silver halide
⊗ Exposed silver halide

(b)

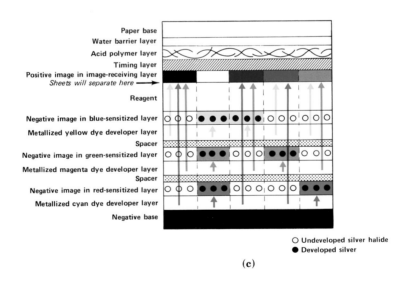

○ Undeveloped silver halide
● Developed silver

(c)

Figure 3. Schematic sections of Polacolor 2 components. (**a**) image receiving sheet before processing; (**b**) negative during exposure; and (**c**) negative and positive sheets during image formation. The two sheets are laminated together by the viscous reagent and stripped apart 60 s later. During processing silver halide develops in the exposed areas of each emulsion layer and the associated dye developer is oxidized and immobilized; in unexposed areas the dye developers diffuse from layers of the negative to form a positive color image in the image-receiving layer. When the two sheets are stripped apart, the reagent layer adheres to the negative, which is discarded, and the positive print is ready for immediate viewing.

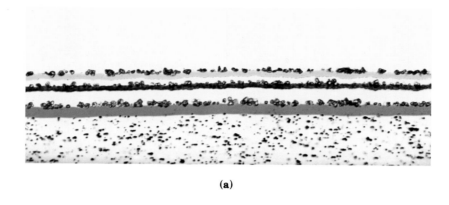

(a)

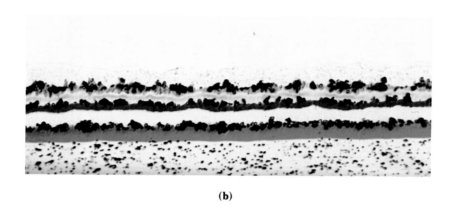

(b)

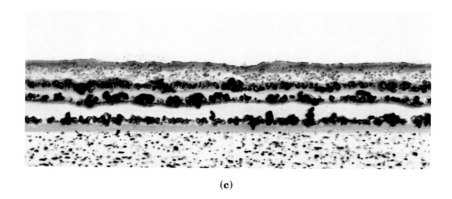

(c)

Figure 4. Cross sections of Polacolor negative (**a**) before processing, (**b**) and (**c**) after processing. (**b**) is from a fully exposed area, corresponding to a minimum density area of the print, and (**c**) is from an unexposed area, where dye has transferred to produce maximum density in the print (1000×).

Polacolor processing takes approximately 60 s. Within this period many image-forming and image-stabilizing reactions proceed almost simultaneously. When image formation is terminated by stripping apart the negative and positive sheets, the print is ready for viewing; its surface is almost dry, and the pH of the image layer is rapidly approaching neutrality. The surface dries quickly to a hard gloss, and the finished print is durable and stable.

Chemistry of Image Formation and Stabilization.   The sequence of reactions responsible for image formation and stabilization begins as alkali in the reagent permeates the layers of the negative, ionizing each of the three dye developers (eq. 1) and

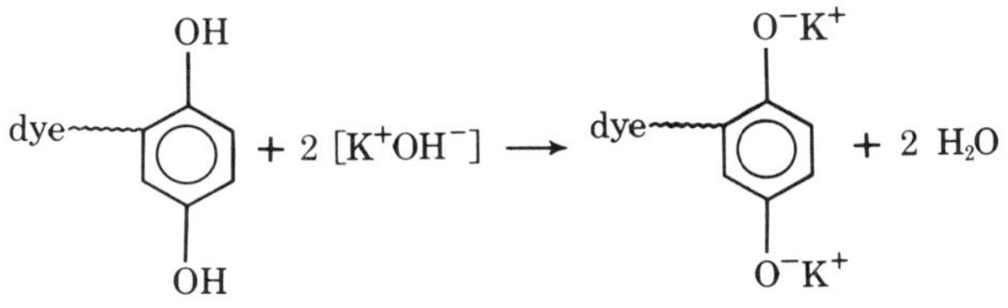

$$\text{dye} - \overset{\text{OH}}{\underset{\text{OH}}{\bigcirc}} + 2\ [K^+OH^-] \longrightarrow \text{dye} - \overset{O^-K^+}{\underset{O^-K^+}{\bigcirc}} + 2\ H_2O \qquad (1)$$

an auxiliary developer, 4′-methylphenylhydroquinone, which may be present in one or more layers of the negative (eq. 2). The ionized developers are capable of diffusing

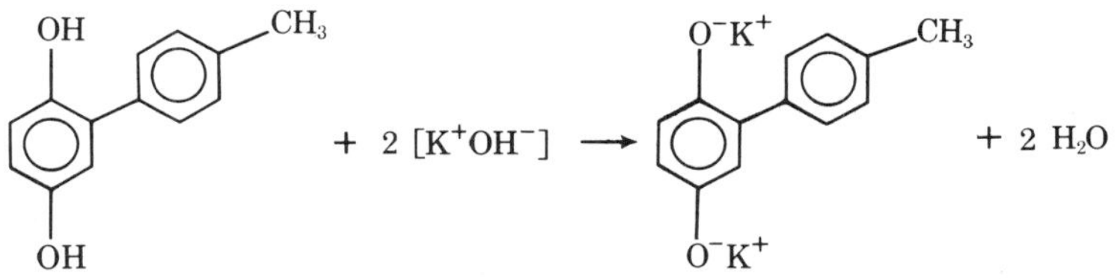

$$\overset{\text{OH}}{\underset{\text{OH}}{\bigcirc}}\!\!-\!\!\bigcirc\!-CH_3 + 2\ [K^+OH^-] \longrightarrow \overset{O^-K^+}{\underset{O^-K^+}{\bigcirc}}\!\!-\!\!\bigcirc\!-CH_3 + 2\ H_2O \qquad (2)$$

from their original locations. The auxiliary developer is smaller and more mobile than the dye developers and can more rapidly reach the emulsion layers of the negative, initiating development of silver halide grains in exposed areas of each emulsion layer and acting as electron transfer agent between the slower moving dye developers and the immobile silver halide. Transfer of an electron to a silver ion reduces the silver and generates the semiquinone ion radical of the auxiliary developer (eq. 3). In turn,

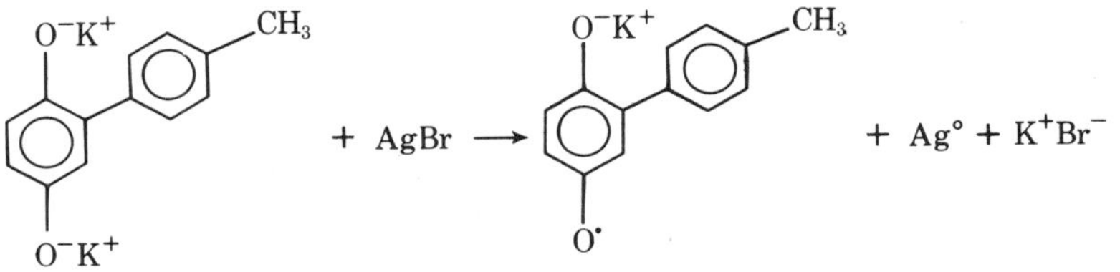

$$\overset{O^-K^+}{\underset{O^-K^+}{\bigcirc}}\!\!-\!\!\bigcirc\!-CH_3 + AgBr \longrightarrow \overset{O^-K^+}{\underset{O^\cdot}{\bigcirc}}\!\!-\!\!\bigcirc\!-CH_3 + Ag^\circ + K^+Br^- \qquad (3)$$

a dye developer molecule of the adjacent layer transfers an electron to the semiquinone, returning the auxiliary developer to its original state and leaving the dye developer in the semiquinone state (eq. 4). Further oxidation of the semiquinone leads to the quinone state of the dye developer, and reaction with quaternary salts included in the reagent can aid in the immobilization of oxidized dye developer (20).

Within the receiving sheet the acid polymer layer, acting as an ion exchanger, forms an immobile polymeric salt with the alkali cation and returns water in place of alkali. Capture of alkali by the polymer molecules prevents salt deposition on the print surface, and the dye developers become immobile and inactive as the pH of the system is reduced.

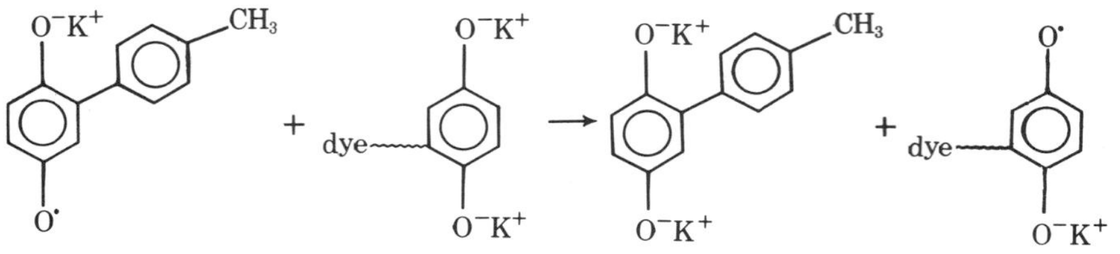

$$ (4) $$

Dye Developers. Image formation with dye developers depends on their mobility in alkali in reduced form and their immobilization as a result of oxidation. In addition to being able to diffuse as image components from layers of the negative to the positive image-receiving layer, the dye developers must be stable and inert in the negative before processing. After deposition in the image-receiving layer and after final equilibration of the pH of the system, the dye must have suitable spectral absorption characteristics and stability to light.

Although other developer moieties can be used, the requirements of a developer for incorporation into dye developers are particularly well fulfilled by hydroquinones. Under neutral or acidic conditions hydroquinones are very weak reducing agents and their weakly acidic phenolic groups are not very solubilizing. In alkali, however, hydroquinones are readily soluble, powerful developing agents. Dye developers containing hydroquinone moieties have many solubility and redox characteristics in alkali similar to those of the parent hydroquinones.

Among the first dye developers synthesized and studied by Rogers and his associates were simple azohydroquinones, with hydroquinone and a coupler attached directly through an azo group to form the chromophore, and hydroquinone-substituted anthraquinones (Fig. 5). (In describing dye developers the term chromophore signifies the whole color-providing moiety or molecule, rather than a specific functional group.) These dye developers were very active developing agents and diffused readily to form excellent positive images. However, because the hydroquinone was an integral part of the chromophore, the colors of these dye developers shifted with pH change and with changes in the oxidation state of the hydroquinone.

To solve the problems of unwanted color shifting, "insulated" dye developers with unconjugated groups as insulation between the chromophore and developer moieties were designed and synthesized (21). Dye developers of this type formed excellent images; the chromophore spectra were not rendered pH-sensitive by the hydroquinone moiety and were relatively independent of its state of oxidation.

In the course of developing the Polacolor and SX-70 processes, many insulated dye developers were synthesized and studied to achieve optimum combinations of solubility, diffusion properties, spectral characteristics, and light stability. The insulating linkage, chromophore, and developer moiety can each be varied. Substituents on the developer modify development and solubility characteristics, and substituents on the chromophore modify the spectral properties of the dye. Among the azo dye developers, the selection and positioning of substituents in the chromophore provides great flexibility in color and light stability. Pyrazolone dyes are particularly versatile yellow chromophores; the yellow dye developer used in Polacolor 1 was based on a chromophore of this class. The study of azo dyes derived from 4-substituted 1-naphthols led to the magenta chromophore used in the Polacolor 1 magenta dye developer (see Azo dyes).

Yellow:

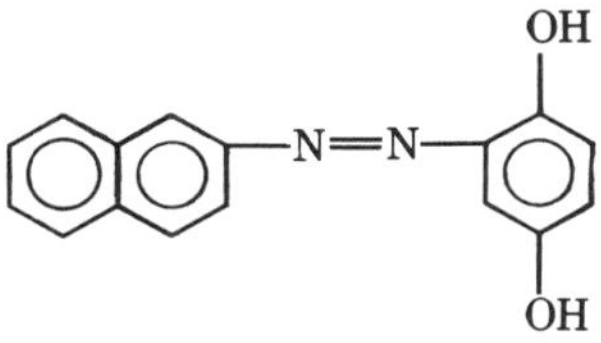

OH

$-N=N-$

OH

[68975-52-0]

Magenta:

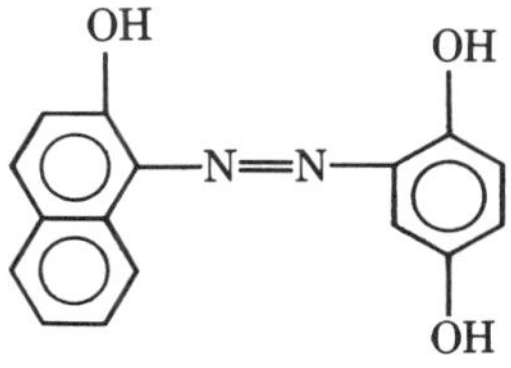

OH   OH

$-N=N-$

OH

[52603-19-7]

Cyan:

OH

O   NH—

OH

OH

O   NH—

OH

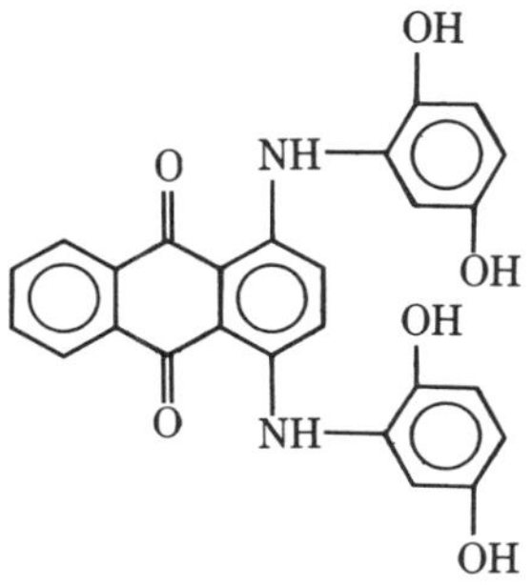

[68975-53-1]

Figure 5.   Early dye developers (1).

Reference 22 describes several cyan and magenta chromophores based on substituted anthraquinones. The synthesis and study of the properties of these dyes resulted in the design of the Polacolor 1 cyan dye developer, which contains a 1,4,5,8-tetra-substituted anthraquinone as chromophore (see Dyes, anthraquinone). The structures of the three Polacolor 1 dyes are illustrated in Figure 6.

The metallized dye developers used in the SX-70 and Polacolor 2 films (8) are insulated dye developers that contain metallized dyes (23) (Fig. 7). The images formed by these dye developers are characterized by very high light stability (24).

The metallized cyan dye developer is based on a copper phthalocyanine pigment (23,25) (see Phthalocyanine compounds). Incorporation of developer groups converts the pigment into an alkali-soluble dye developer. A study of the properties of chromium complexes of azo and azomethine dyes was basic to the design of the magenta and yellow metallized dye developers (26). Chromium can form complexes with either one or two azo dye molecules per chromium atom, and comparison of such dyes showed the 1:1 complexes to have greater spectral purity. A colorless, modified acetylacetone ligand was introduced to occupy two of the remaining coordination sites in order to

Yellow:

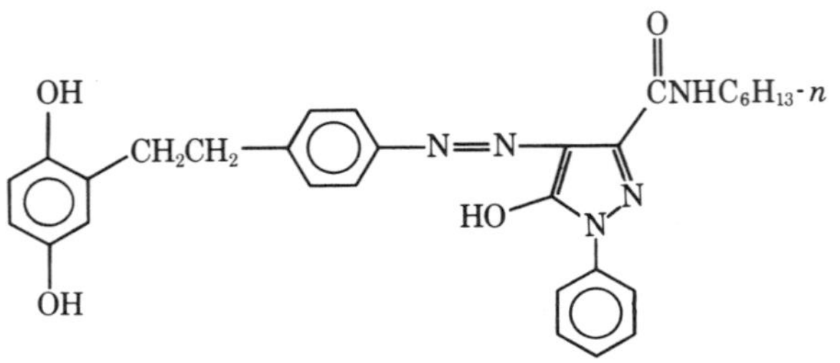

[14848-08-9]

Magenta:

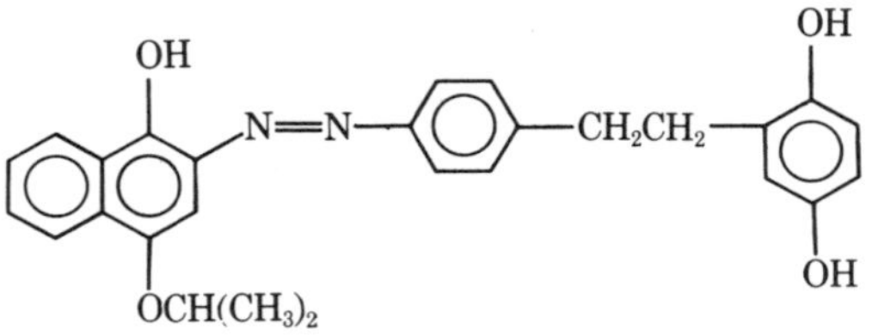

[1180-52-5]

Cyan:

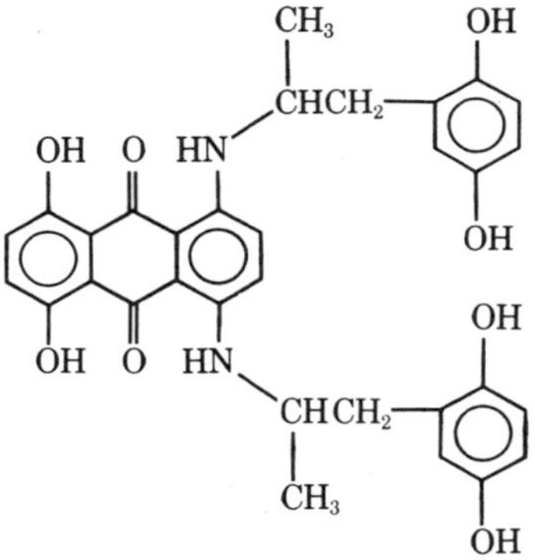

[2498-16-0]

Figure 6. Dye developer with insulating linkages between developing groups and chromophore, as used in Polacolor 1 (1963).

minimize reactions that might otherwise take place between the metallized chromophore and gelatin as the dye diffuses through gelatin-containing layers of the negative. The colorless ligand may also serve as a convenient site for the developer moiety.

The flexibility in design of hydroxynaphthylazopyrazolone dyes, from which the magenta dye developer in Figure 7 was derived, is shown by the general structure (1) in which X, Y, and Z are substituents that may be varied to control the spectral absorption of the chromophore. Its absorption is also influenced to a lesser extent by the choice of colorless ligand (23). The yellow dye developer for SX-70 and Polacolor 2 was similarly derived from the 1:1 chromium complex of an o,o'-dihydroxyazomethine dye.

Yellow:

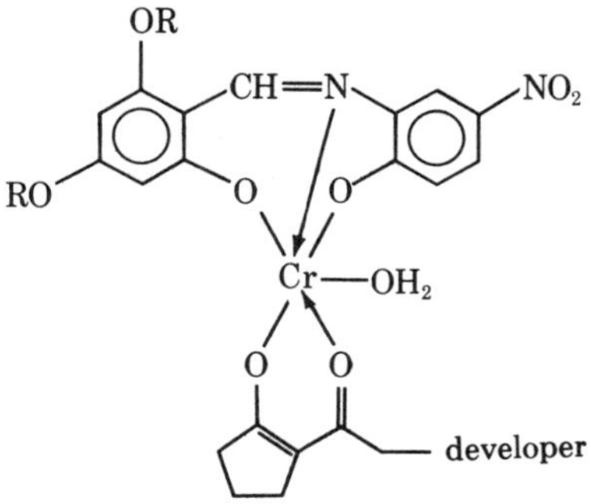

Magenta:

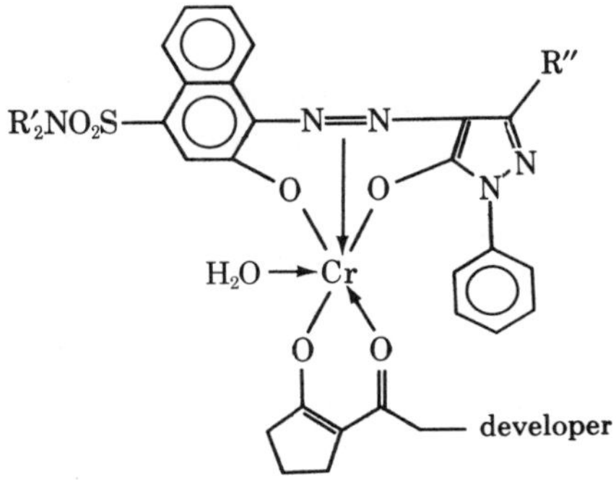

Cyan:

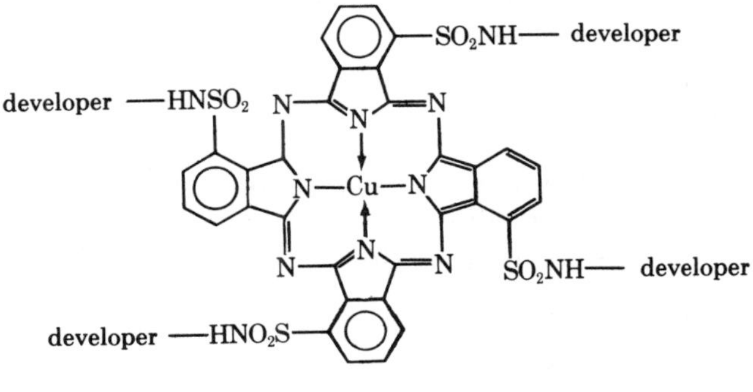

Figure 7.   Metallized dye developers used in SX-70 film (1972) and Polacolor 2 film (1975).

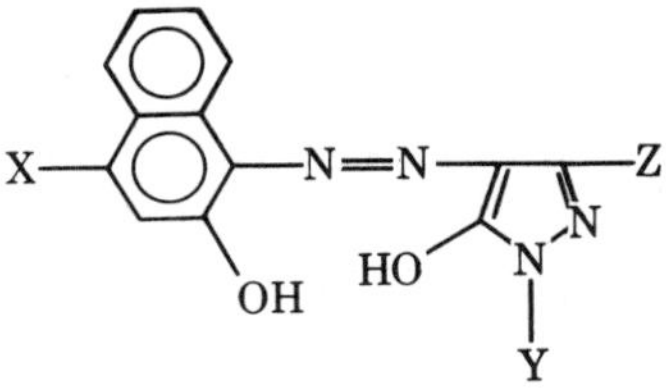

(1)

Additional types of dye developers are described below. For a detailed discussion of dye developer chemistry, the comprehensive review by Bloom, Green, Idelson and Simon (22), is a valuable reference.

Multilayer Negatives. In the first three-color tests with dye developers, the film was constructed in a screen-like configuration: the three color image-forming elements were arrayed side by side in an arrangement that had been used with moderate success in earlier work with color development processes (27).

To use negatives with multiple continuous layers, Land and Rogers studied and devised hold-release mechanisms to keep each dye developer in close association with the emulsion designated to control it until development had progressed substantially. In addition to the design and selection of dye developers characterized by a high development rate relative to the diffusion rate, useful measures include the introduction of small quantities of highly mobile auxiliary developers and the intercalation of temporary barrier layers between the three monochrome systems comprising the negative (28). Furthermore, since alkali and auxiliary developers diffuse more rapidly than dye developers, the dye developer may undergo oxidation and immobilization in regions adjacent to the exposed silver halide before the major migration of the dye developer from the regions adjacent to the unexposed silver halide.

Depending on their composition, barrier layers can function simply as spatial separators or can provide specified time delays by swelling at controlled rates or undergoing reactions such as hydrolysis or dissolution. Suitable barrier materials include cellulose esters and water-permeable polymers such as gelatin and poly(vinyl alcohol). Preproduction dye developer negatives used a combination of cellulose acetate and cellulose acetate hydrogen phthalate as barrier layers. The images produced from these negatives were outstanding in color isolation, color saturation, and overall color balance (see Color). The Polacolor 1 negative had water-coated barrier layers of gelatin (29). The SX-70 and Polacolor 2 negatives use water-coated barrier layers comprising combinations of polymeric latexes and water-soluble polymers (30) (see Latex technology; Resins, water-soluble).

Auxiliary Developers. The use of auxiliary developers as electron transfer agents, although not necessary with alkali-mobile dye developers, is usually advantageous. A developer with mobility greater than that of a dye developer can quickly reach the emulsion layers of a multilayer structure to initiate development, and the oxidation product of the auxiliary developing agent can in turn oxidize and immobilize the dye developer associated with that emulsion (18). Some auxiliary developers that have been used in this way are Phenidone, Metol, and substituted hydroquinones, such as 4'-methylphenylhydroquinone (31), either alone or with Phenidone (32).

Utilization of Polacolor Films. Polacolor 2 films are manufactured in several formats outlined in Table 2. The film is balanced for daylight exposure and has an

Table 2. Polacolor 2 Films

Designation[a]	Format	Print size, cm	Image area, cm	Balance
type 58 (4 × 5)	sheet	10.5 × 13	9 × 11.5	daylight or electronic flash
type 808 (8½ × 10¾)	sheet	22 × 27	19 × 23	daylight or electronic flash
type 88 (3¼ × 3⅜)	pack	8.3 × 8.6	7 × 7.3	daylight
type 108 (3¼ × 4¼)	pack	8.3 × 10.8	7.3 × 9.5	daylight
type 668 (3¼ × 4¼)	pack	8.3 × 10.8	7.3 × 9.5	daylight or electronic flash

[a] Values in parentheses correspond to print size in inches.

equivalent ASA speed of 75. Processing can be carried out satisfactorily at temperatures from 16–38°C, with slight exposure compensation recommended at the extremes.

Handheld cameras using Polacolor pack films range from inexpensive amateur models to professional models with great flexibility. Pack films are also used in a variety of cameras and camera backs for professional and industrial applications. Polacolor 4×5 films are used in holders that fit many specialized instruments as well as 4×5 view cameras; the 8×10 film holder fits standard 8×10 equipment. Applications of these Polacolor films, as well as photography with Polacolor film in larger formats, are discussed under Uses below.

The SX-70 System. From the beginning, a principal goal in instant photography, as stated by Land, has been "to remove the manipulative barriers between the photographer and the photograph . . . . so that the photographer by definition need think of the art in the *taking* and not in *making* photographs" (2). This goal was realized in 1972 with Land's introduction of the SX-70 automatic camera and integral film system (7–8). Used in cameras that control exposure automatically and eject each film unit through processing rollers immediately after exposure, the SX-70 film provides images that require no timing and no peeling apart. Once the exposure has been made, the photographer is free to proceed with his next photograph or to watch the image materialize from its initially invisible state to become a full color print. The entire process takes place within the film unit under ambient conditions.

The integral film format introduced new concepts in processing chemistry and new film components. The need for the processed film unit to contain all of the reaction products along with the final color image imposed new and unusual requirements in film and pod structure and in stabilization.

Film Structure. The SX-70 picture unit is a fully integral multilayer structure with no air spaces before or after processing (Fig. 8). The two outer layers are sheets of polyester whose structural symmetry ensures that the pictures stay flat under all conditions. The upper polyester sheet is transparent and the lower one is opaque black. Light passes through the upper sheet to expose the layers of the negative during the taking of the photograph and the picture is viewed through the same sheet. A durable, low index of refraction, quarter-wave antireflection coating on the outer surface of the transparent polyester minimizes flare during exposure, increases the efficiency of light transmission in both the exposure of the negative and the viewing of the final image, and permits seeing the light reflected from the image proper with a minimum of surface luster (33).

The positive image-forming components of the SX-70 film units are metallized dye developers. The sequence of dye developers, silver halide emulsions, and other layers is shown in Figure 9. The inner surface clear polyester sheet through which the negative layers are exposed bears an acid polymer layer, a timing layer, and an image-receiving layer. There is a weak bond between the image-receiving layer and the top layer of the negative.

SX-70 Image Formation. The viscous SX-70 reagent is contained in a small pod concealed within the wide border of the film unit. As an exposed film unit is ejected through the processing rollers and out of the camera, the pod bursts and the viscous reagent forces apart the top negative layer and the image-receiving layer, rupturing the weak bond between them and forming in its place a new layer comprising water, alkali, white pigment, polymer, and other processing addenda (Fig. 9). Alkali quickly

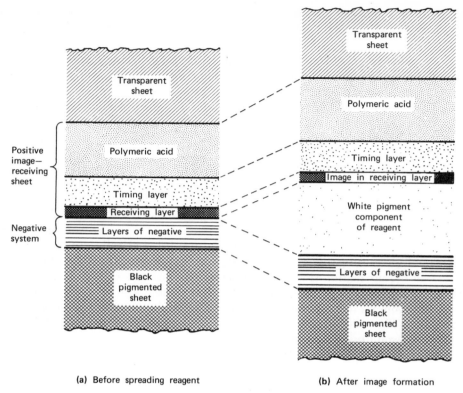

(a) Before spreading reagent   (b) After image formation

Figure 8.   Schematic cross sections of SX-70 film (**a**) before and (**b**) after development (8).

permeates the layers of the negative. In areas that have been exposed, silver halide is reduced and the associated dye developer is oxidized and immobilized. The auxiliary developing agent 4'-methylphenylhydroquinone acts as an electron transfer agent or "messenger" developer. In areas that have not been exposed, the dye developers, ionized and solubilized by the action of the alkali, migrate through overlying layers of the negative to reach the image-receiving layer above the new stratum of pigment-containing reagent. The transferred image is seen through the transparent polyester sheet by light reflected from the white pigment. The micrograph of Figure 10(**a**) shows an unprocessed SX-70 negative in cross section; 10(**b**) and (**c**) are cross sections comparing the distribution of dyes in a fully exposed and an unexposed region of a processed SX-70 negative.

Opacification.   An important aspect of the SX-70 system is that the developing film unit is ejected while still light sensitive. Hence both surfaces must be opaque to ambient light. However, the top surface, through which the camera exposure is made, cannot become opaque until after that exposure has taken place. Protection against further exposure through this surface is provided by the combined effects of opacifying dyes and pigment included in the viscous reagent layer (34). The opposite surface is protected by the support layer of black polyester negative.

The opacifying dyes are phthalein indicators that have high extinction coefficients at very high pH levels. Incorporated in the viscous reagent, they render it opaque when first spread, then lose their color as the pH of the system is reduced. The opacifying

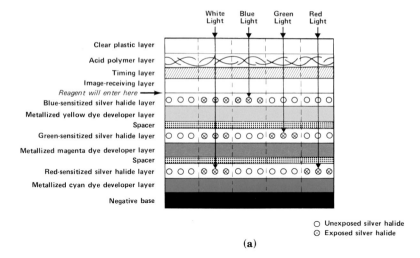

○ Unexposed silver halide
⊗ Exposed silver halide

(a)

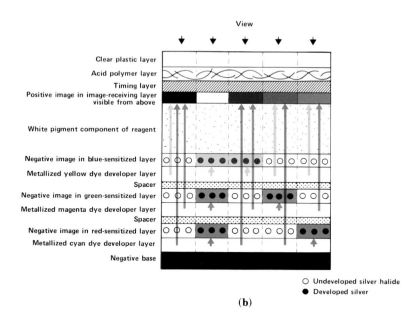

○ Undeveloped silver halide
● Developed silver

(b)

Figure 9. Schematic sections of SX-70 integral film unit (**a**) during exposure and (**b**) after reagent has been spread to form a new stratum within the integral structure. Light passes through the transparent upper sheet to expose the photo sensitive layers, and the color image is viewed through the same surface by light reflected from the white pigment component of the reagent. The color image is formed by dye developers transferred from unexposed areas of the negative through the pigment-containing reagent to the image-receiving layer. Cf Figure 13.

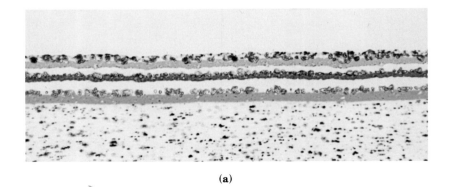

(a)

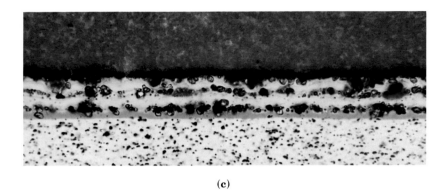

(b)

(c)

Figure 10. Cross sections of SX-70 negative (**a**) before processing, (**b**) after processing, in an area that had received full exposure, corresponding to a minimum density area of the print, and (**c**), in a processed unexposed area, corresponding to maximum print density. The dark layer at the top of (**b**) and (**c**) is the white pigment, which appears dark by transmitted light (1000×).

dyes and the pigment together have a synergistic optical effect as the highly reflective titanium dioxide pigment greatly lengthens the optical path through the layer and thus increases the total absorption by the dye contained in the layer. The result is that the negative is fully protected, even in bright sunlight where the total exposure to ambient light may be more than a million times greater than the original camera exposure. Even though the transmission density of the reagent layer is initially very high, over 6.5, the reflection density is low because of the high reflectivity of the pigment. Thus the SX-70 image may be seen against the reagent layer well before the opacifying dyes have lost their color.

The opacifying dyes used in the SX-70 film represent a new class of phthalein indicator dyes (35) designed to have unusually high pK_a values (see Hydrogen-ion concentration). The solution of such an indicator is highly colored when the dye is at or above its pK_a value, and becomes progressively less colored as the pH is reduced below the indicator's pK_a value. The very high pK_a values of these opacifying indicators are induced by hydrogen-bonding substituents located in juxtaposition with the hydrogen that is removed by ionization (36). Figure 11 illustrates the range of pK_a values of a series of phthalein opacifying indicators with such substituents.

Utilization of SX-70 Film. Cameras using SX-70 film include both folding single-lens reflex models, designated as SX-70 cameras, and non-folding models with separate camera and viewing optics. All of the cameras are motorized. Immediately after exposure the integral picture unit is automatically ejected from the camera. Pictures may be taken as frequently as every 1.5 seconds.

The SX-70 picture unit is 8.9 × 10.8 cm with a square image area approximately 8 × 8 cm. Ten of these picture units and a flat battery about the size of a picture unit are contained in the SX-70 film pack, along with an automatically ejected opaque cover sheet (see also under Uses).

SX-70 film is balanced for daylight exposure, and its equivalent ASA speed is 150. The temperature range for development is approximately 7–35°C. With slight exposure compensation, the range may be extended to 38°C.

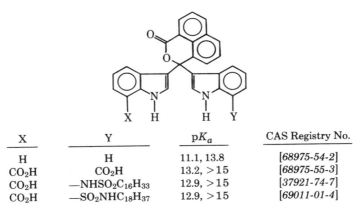

X	Y	pK_a	CAS Registry No.
H	H	11.1, 13.8	[68975-54-2]
CO_2H	CO_2H	13.2, >15	[68975-55-3]
CO_2H	$-NHSO_2C_{16}H_{33}$	12.9, >15	[37921-74-7]
CO_2H	$-SO_2NHC_{18}H_{37}$	12.9, >15	[69011-01-4]

Figure 11. A series of naphthalein indicator dyes, illustrating the effects of hydrogen-bonding substituents on pK_a (1).

The PR-10 Process.   Hanson described the technical aspects of the PR-10 integral color print film and its components in 1976 (37–38). The film system is based on a negative-working dye release process using preformed dyes. Transfer of dyes is initiated by the oxidized developing agent formed in areas where silver halide grains are undergoing development. The oxidized developing agent reacts with alkali-immobile dye-releasing compounds, which in turn release mobile image dyes that diffuse to the image-receiving layer. A positive print is formed with emulsions that yield positive silver images directly by developing only those grains that have *not* received exposure.

Film Structure.   The principal components of the PR-10 integral film unit are shown schematically in Figure 12. A clear polyester support, through which the positive image will be viewed, has on its inner surface an integral imaging receiver, which includes both an image-forming section and an image-receiving section. The image-forming section includes direct positive emulsion layers, dye-releasing layers, and scavenger layers; the image-receiving section includes the image-receiving layer and opaque layers. The pod contains a viscous reagent and an opacifier. The transparent polyester cover sheet, through which the sensitive layers are exposed, has on its inner surface polymeric timing layers, one of which contains the precursor of a development inhibitor, and a polymeric acid layer.

Process Description.   After the film unit has been exposed through the cover sheet, it is passed between processing rollers. Viscous reagent is released from the pod and spread between the cover sheet and the sheet bearing the integral imaging receiver, thus initiating the sequence of reactions that develop the unexposed silver halide grains, release the dyes, and form the positive image in the image-receiving layer (37–38). Processing takes place outside the camera under ambient conditions. As the positive image forms, it is seen against a layer of titanium dioxide covering an opaque carbon layer behind the image-receiving layer. The developing film is protected from

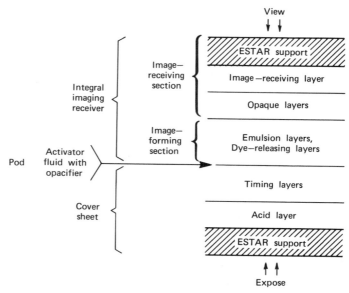

Figure 12.   Schematic section showing principal components of Kodak PR-10 film (37).

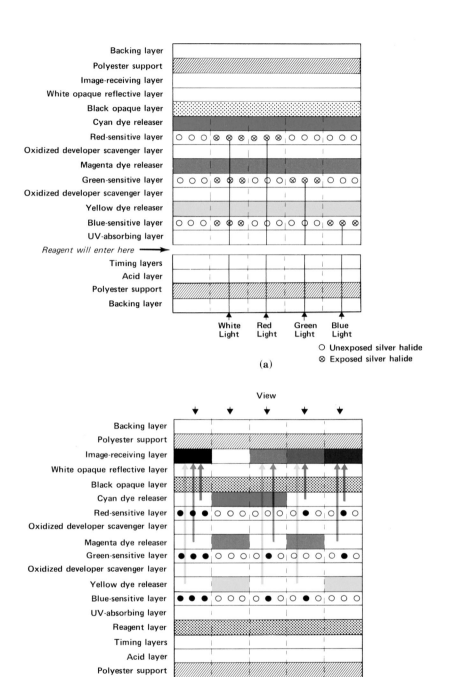

Figure 13. Schematic sections showing structure of the PR-10 film unit during exposure (**a**) and after spreading of the reagent (**b**). The photosensitive layers are exposed by light passing through a transparent cover sheet, and the color image, formed of dyes that transfer through the opaque pigment layers to the image-receiving layer, is viewed through the opposite surface of the film unit. The development of unexposed grains of the direct positive emulsions is accompanied by release of the positive image-forming dyes. Cf Figure 9.

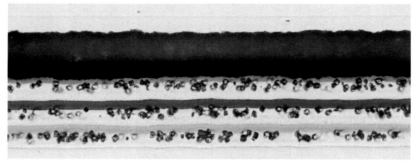

(a)

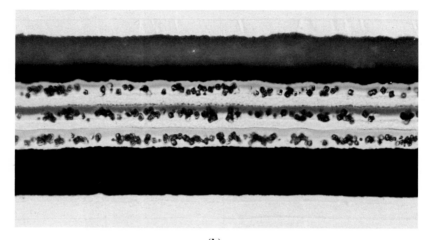

(b)

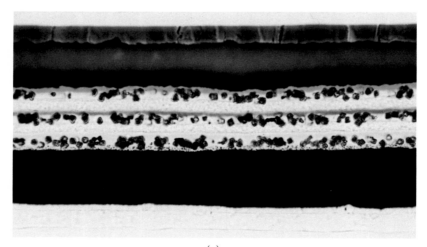

(c)

Figure 14. Cross sections of Kodak PR-10 film (**a**) before processing, (**b**) and (**c**) after processing. (**b**) is from an exposed area, showing minimum print density; (**c**) is from an unexposed area, showing maximum print density (1000×).

ambient light exposure through the viewing surface by the titanium dioxide and carbon pigment layers of the image-receiving section. Exposure through the opposite surface is prevented by a carbon black layer formed by spreading the viscous reagent. Neutralization to terminate processing is effected by the polymeric acid layer of the cover sheet; the initiation of this reaction is controlled by the rate of permeation of the overlying polymeric timing layers. Mobility of the transferred dyes is also reduced by their interaction with a mordant contained in the image-receiving layer. A development inhibitor released from one of the timing layers by the alkaline hydrolysis of its precursor assists in restraining further development and consequent additional dye release.

The schematic cross section drawings of Figure 13 show in further detail how the subtractive dye image is produced. Light entering the image-forming section of the film unit through the cover sheet encounters in turn the blue-sensitive emulsion over the yellow dye, the green-sensitive emulsion over the magenta dye, and the red-sensitive emulsion over the cyan dye. Silver halide development in unexposed areas of each emulsion results in dye release. Thus the amount of each dye that migrates to form the final image is related inversely to exposure. The micrograph of Figure 14(**a**) shows an unprocessed PR-10 film unit in cross section, and 14(**b**) and (**c**) compare dye distributions in minimum and maximum density areas of a processed film.

Image-Forming Reactions.   The dye-releasing reactions are initiated by the oxidized developing agent formed in the course of silver halide reduction. The developing agent, a derivative of 1-phenyl-3-pyrazolidone (37–38), acting as an electron transfer agent, gives up electrons to the developing silver halide grains and is regenerated to its reduced form by taking electrons from the dye releaser. The oxidized dye releaser then undergoes hydrolysis to yield a mobile dye (39).

The dye releasers used in PR-10 film are alkali-immobile p-sulfonamidophenols, with dye attachment through the sulfonamido group. Equation 5 illustrates the oxidation of such a dye releaser by oxidized developing agent and the alkaline hydrolysis of the resulting quinonimide to form a mobile sulfamoyl-solubilized dye and an immobile quinone. The developing agent regenerated by electron transfer may recycle to reduce more silver ions.

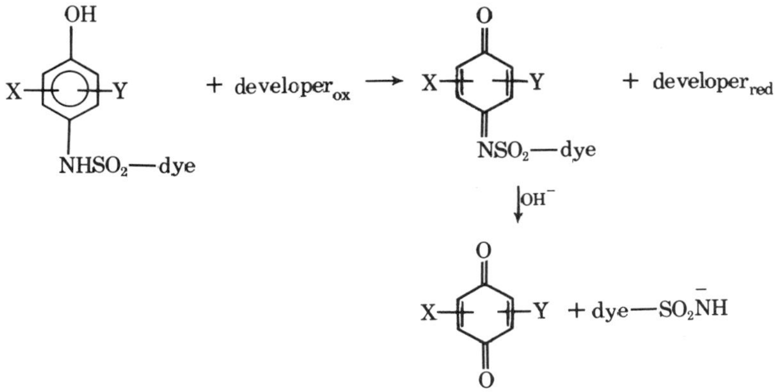

X,Y = immobilizing groups

Image Dyes. In describing the design of the PR-10 image-forming components, Hanson discussed the effects of substituents in each part of the p-sulfonamidophenol dye releaser molecule (37–38). Substituents on the p-sulfonamidophenol moiety influence dispersion, coating, and diffusion properties, as well as the ease of oxidation and the rate of release of dye by the resulting quinonimide. Fused ring substitution, for example, leads to p-sulfonamidonaphthol dye releasers that are more readily oxidized than are the analogous p-sulfonamidophenols.

The PR-10 image dyes are anionic forms of monoazo dyes derived from naphthol or pyrazolone couplers. Hanson described studies of the specific effects of electron-withdrawing and electron-donating groups as substituents, with the resulting selection of substituents to achieve low pK_a values and desired hues. Low pK_a values enable the dyes to be used in anionic form, thus providing for their immobilization after transfer by interaction with the cationic mordant present in the image-receiving layer. The hue of the dye is also influenced by its pK_a value.

General structural formulas for the PR-10 image dyes are shown in Figure 15. The effects of substituents on the pK_a and hue of 4-aryl-1-naphthol magenta dyes are linear functions of the Hammett sigma values of substituent groups (37–38). Variation of substituents in the 2-position of the naphthol ring affects both hue and pK_a, whereas substituents in the arylazo ring affect primarily the hue. Hue and pK_a may thus be manipulated somewhat independently to provide optimum characteristics. The cyan dyes, which are nitro-substituted 4-aryl-1-naphthols, may be similarly

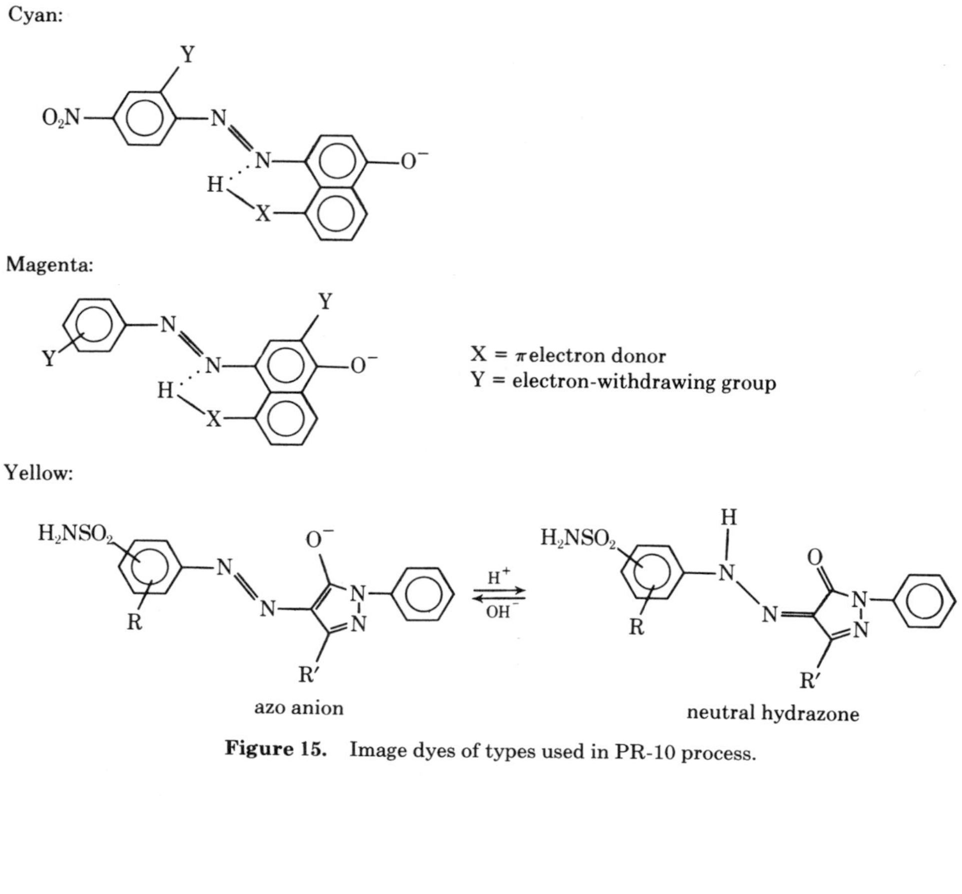

Cyan:

Magenta:

X = π electron donor
Y = electron-withdrawing group

Yellow:

azo anion   neutral hydrazone

Figure 15. Image dyes of types used in PR-10 process.

adjusted by selection of substituents. In the case of the yellow dyes, which are aryl azopyrazolone dyes, pKa values low enough to maintain the anionic species were not readily obtained. The dye was therefore designed, again by selection of substituents of appropriate Hammett sigma values, so that its anionic and neutral forms would be of similar hues.

Direct Positive Emulsions. The direct positive, or direct reversal, emulsions used in the PR-10 film to obtain positive images with the negative-working dye release process comprise silver halide grains that form latent images internally, rather than at the surface. The reversal process is effected during processing by the action of a nucleating or fogging agent present in the emulsion layer and a surface developer.

The emulsion reversal mechanism may be explained in terms of the fate of conduction band electrons (37–38). Photoelectrons generated during exposure are trapped preferentially inside the grain, forming an internal latent image which then acts as a monitor for conduction electrons provided by the nucleating agent. In an exposed grain the internal latent image is an efficient trap for electrons provided by the nucleating agent, causing the internal latent image centers to grow larger. Since these grains have no surface latent image, they do not develop in a surface developer. On the other hand, grains that have not been exposed trap conduction electrons from the nucleating agent, at least temporarily, on the grain surface, thus forming fog nuclei that initiate development in the surface developer. Since the unexposed grains are developed and the exposed grains are not developed, a positive image is formed.

Utilization of PR-10 Film. PR-10 instant print film is used in Kodak instant cameras, which include battery-operated motorized models and lower cost hand-cranked models. The pictures begin to develop as they are ejected from the camera and development proceeds over several minutes under ambient conditions. The recommended temperature range is 16–38°C, with slight exposure compensation at the higher temperatures.

The PR-10 film is balanced for daylight exposure, with an equivalent ASA speed of 150. The picture units are 9.7 × 10.2 cm and the image measures 6.7 × 9.0 cm.

The Polavision Process. Land demonstrated and described Polavision, the first instant motion picture system, in 1977 (10). The Polavision system is based on a very fine additive color screen and an integral silver image transfer film, with automatic processing that is essentially dry. The integral 8-mm Polavision film is provided in a sealed cassette, and the film is exposed, processed, viewed, and rewound for further viewing without leaving the cassette (Fig. 16).

In the Polavision camera the film is exposed through the film gate of the cassette as it moves from the supply reel to the take-up reel. After the film has been exposed, the cassette is removed from the camera and inserted into the Polavision player (Fig. 17). Insertion of the cassette activates the player, which provides power to turn the reels. The film, which carries all of the signals, immediately instructs the player to rewind. As the exposed film starts to rewind onto the supply reel it automatically peels a foil cover from a small cavity to release a miniscule amount of a viscous reagent, which is applied to the moving film surface through a fine slot. The film pauses for a few seconds at the end of rewinding to allow for completion of image formation and then winds forward in viewing mode, with light directed by the cassette's prism through the film onto the screen of the player.

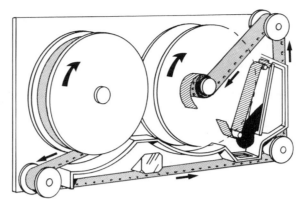

Figure 16.   The Polavision cassette. As the first rewind begins, the film peels a foil cover from the small cavity at right, releasing a very small amount of viscous reagent. The film is coated with a reagent layer approximately 10 μm thick as it passes the precision slot and rewinds at about 200 cm/s (10).

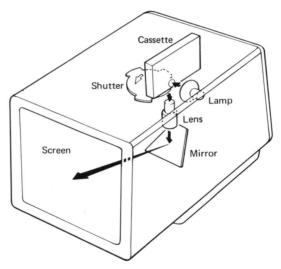

Figure 17.   The Polavision cassette in place in the player. Black arrows indicate the light path during projection (10).

Polavision Film. Additive Color Screen; Temporary Lenticules.   The Polavision additive screen comprises a microscopically fine pattern of red, green, and blue filter stripes. To produce lines fine enough to be essentially invisible when projected, Land devised a new process using temporary lenticules (Fig. 18) (40). The film base is embossed to form fine lenticules on one surface, and a layer of dichromate-treated gelatin on the opposite surface is exposed through the lenticules to form hardened line images. After washing away the unhardened gelatin, the lines that remain are dyed. The process is repeated to complete the array of alternating red, green, and blue stripes. As indicated in Figure 18, the lenticules are removed after they have been used to form lines.

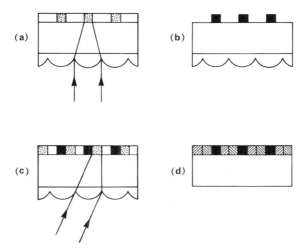

Figure 18. Schematic representation of use of temporary lenticules: (**a**) exposure of film to form first line image in dichromate-treated gelatin; (**b**) after removal of unexposed, unhardened gelatin and dyeing of first line; (**c**) exposure to form second line image; (**d**) completed screen, with dyed triplets of red, green, and blue lines. Lenticules have been removed (10).

The Polavision linemaking system permits continuous screen production, and the frequency of the lines is determined by the lenticule frequency. Polavision film base uses 590 lenticules per cm to provide 1770 color lines per cm, lines so fine that they are not visible on the player screen under normal viewing conditions.

Silver Image Formation and Stabilization. The integral additive transparency is based on the rapid, one-step formation of a stable, neutral, positive black-and-white image of high covering power behind the fine pattern of colored lines. The negative image formed at the same time is of much lower covering power and thus so inconsequential optically that it does not have to be removed to permit viewing the positive image (41–42).

To achieve good color it is necessary for the emulsion resolution to be significantly higher than the resolution of the color screen. This requirement is fulfilled by using a fine-grained, high resolution silver halide emulsion and coating the emulsion in a thin layer containing a minimal amount of silver. The high-covering-power positive image is a mirror-like deposit formed in a very thin image-receiving layer (42). The negative image formed in the emulsion layer has only about $1/10$ the covering power of the positive image.

The Polavision film structure and the formation of a Polavision color image are illustrated in Figure 19 (10). The color screen is separated from the image-receiving layer by an alkali-impermeable barrier layer. Over the image-receiving layer is the light-sensitive silver halide emulsion, and over the emulsion is a layer containing antihalation dyes and a stabilizer precursor (43).

During exposure, red, green, and blue records form behind the respective color stripes. As shown in Figure 19, for example, exposure to red light affects only the silver halide grains behind the red lines.

Processing is initiated by the application of a very thin layer of reagent to the film surface as the film rewinds in the cassette. As the exposed silver halide grains develop *in situ*, the unexposed grains dissolve and the resulting soluble silver complex migrates to the positive image-receiving layer, developing there to form the thin, compact positive image layer. Figure 20 shows an unprocessed Polavision film with 1770 color lines per cm and Figure 21 shows a portion of a red image area, with high density, high covering power positive silver over the blue and green lines, and low density, low covering power negative silver over the red lines.

The difference in covering power of negative and positive silver is further illustrated by the electron micrographs of Figure 22, which compare the structures of the two types of image silver at much higher magnification. The optical neutrality of the positive image throughout its density range is also determined by its structure. At each density the positive image silver is in the form of a nearly continuous array of aggregates which act collectively as an efficient reflecting layer for light in the visible range. Figure 23 is a schematic representation of the summation of negative and positive image densities, showing the decreasing attenuation of light by the negative image as positive image density increases.

Both during and after the formation of the negative and positive images there are additional important reactions. In the uppermost layer, the antihalation dyes are decolorized and stabilizer is released to diffuse through the layers and stabilize the developed silver images. The oxidation product of the developing agent, tetramethyl reductic acid [2,3-dihydroxy-4,4,5,5-tetramethyl-2-cyclopenten-1-one] (44), meanwhile undergoes alkaline decomposition to form colorless, inert tetramethylsuccinic acid (eq. 6).

Utilization of Polavision Film. Polavision cassettes (Polavision Phototape Cassette, Type 608) are used in the Polavision instant movie camera and player. Exposure in the Polavision camera is controlled automatically by an electric eye circuit, which is activated just prior to exposure by a switch built into the camera grip. The film is exposed and viewed at 18 frames per second. Film processing is initiated by inserting the cassette into the Polavision player, and about 90 s later the full color motion picture appears automatically on the player screen. At the end of each viewing the film instructs the player to turn off the lamp and to rewind to be ready for the next viewing.

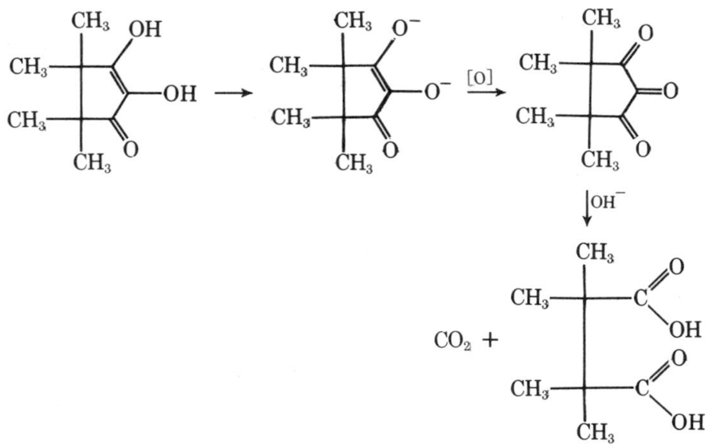

$$(6)$$

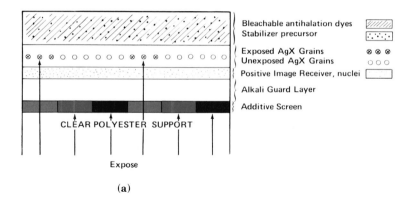

Bleachable antihalation dyes
Stabilizer precursor

Exposed AgX Grains
Unexposed AgX Grains
Positive Image Receiver, nuclei

Alkali Guard Layer

Additive Screen

CLEAR POLYESTER SUPPORT

Expose

(a)

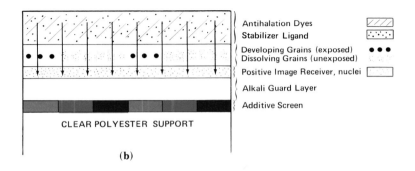

Antihalation Dyes
Stabilizer Ligand

Developing Grains (exposed)
Dissolving Grains (unexposed)
Positive Image Receiver, nuclei

Alkali Guard Layer

Additive Screen

CLEAR POLYESTER SUPPORT

(b)

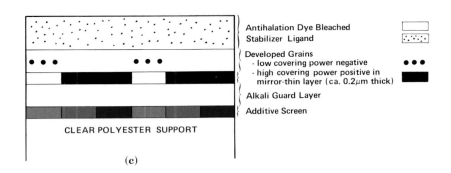

Antihalation Dye Bleached
Stabilizer Ligand

Developed Grains
- low covering power negative
- high covering power positive in mirror-thin layer (ca. 0.2 μm thick)

Alkali Guard Layer

Additive Screen

CLEAR POLYESTER SUPPORT

(c)

Figure 19. Schematic representation of Polavision process. (**a**) shows the film during exposure to red light, with only red lines transmitting light to the emulsion. In (**b**) processing has been initiated by the application of a very thin stratum of viscous reagent. Arrows indicate penetration by reagent. Exposed grains are reduced *in situ,* forming the negative image; concomitantly, unexposed grains dissolve, and the soluble silver complex migrates and undergoes reduction at nucleation sites in the image-receiving layer to form the positive image. (**c**) represents the final image, as viewed in the player at completion of the 90 s processing.

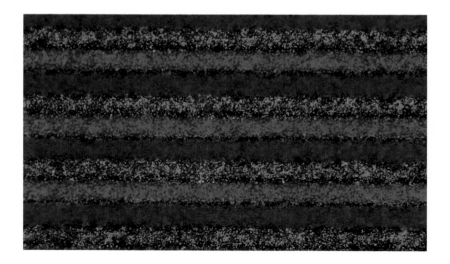

Figure 20.   Polavision film before processing (1000×).

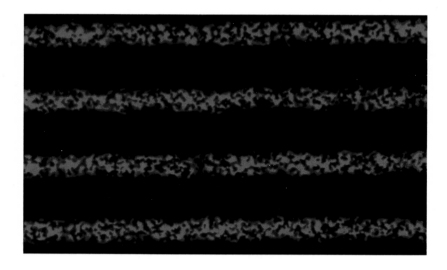

Figure 21.   A red image area. Low covering power negative image overlying red lines readily transmits light, and high covering power positive image blocks transmission through blue and green lines (1000×).

Figure 22.   Comparison of undeveloped grain (left); exposed and developed grain (center); and silver of a single unexposed grain that has transferred and developed to form a positive image deposit (right) (10).

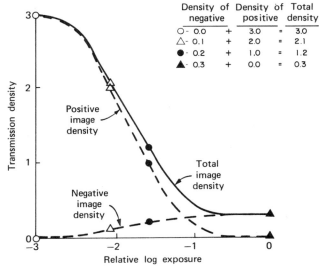

Density of negative		Density of positive		Total density
○- 0.0	+	3.0	=	3.0
△- 0.1	+	2.0	=	2.1
●- 0.2	+	1.0	=	1.2
▲- 0.3	+	0.0	=	0.3

Figure 23.   Throughout the characteristic curve the total image density is the summation of negative and positive image densities, as shown here schematically. Attenuation of light by the negative corresponds to the difference between the positive image density and the total image density (10).

A cassette part moved by the film at the end of the first rewinding indicates to the player when a processed film is re-inserted that it has already been processed and is simply to be viewed.

Additional Materials and Mechanisms

In addition to the systems described in the preceding sections, many other approaches to instant color image formation have been proposed and studied in conjunction with the design of instant color processes.

Color Development. In a color development process the image dye is formed by the interaction of two components: the oxidation product of a color developing agent, usually a substituted p-phenylenediamine; and a color coupler. Land proposed one-step color processes based on dye formation by color development as early as 1947 (2), and much of the early work of Land and Rogers involved such processes (1).

Land used a pair of mobile color components, which could be located initially in the emulsion, in a separate layer, or in the reagent. During development, as immobile images of dye and silver formed in exposed areas of the emulsion layers, unused coupler and unoxidized developing agent transferred as image components from unexposed areas to an image-receiving layer containing a colorless, immobile oxidizing agent. There the transferred unoxidized developing agent underwent oxidation, and reaction of its oxidation product with the coupler produced a positive dye image (45). In one version, a fogged silver halide emulsion served as both oxidizing agent and image-receiving layer.

Rogers later used preformed dyes, each bearing the same coupler moiety, so that all would react at the same rate with the oxidized developing agent to effect immobilization in areas of exposure (46). Transfer images with preformed dyes of this type thus eliminated some of the problems associated with disparate reaction rates of three unlike coupling reactions.

Dye Developers. The section on Polacolor above described the non-insulated dye developers used in early work toward one-step color systems and discussed the transition from non-insulated to insulated dye developers (21), with particular emphasis on the dye developers used in Polacolor and SX-70 films. This section surveys additional dye developers and dye developer components that have been investigated.

Color-Shifted Dye Developers. The image formed by a dye developer is designed to absorb light of the color that its associated emulsion records. It is sometimes desirable to coat the dye developer and the silver halide in the same layer, and under these circumstances the dye developer's absorption of light to which the silver halide is sensitive may interfere with efficient exposure. Also, in some cases the use of a dye developer having an extended absorption tail may result in light filtration that is detrimental to the exposure of an underlying silver halide emulsion. It is advantageous in such situations to shift the spectral absorption of a dye developer temporarily, usually to shorter wavelength, in order to avoid unwanted loss of light during exposure, and to regenerate the desired image color afterward (18).

Compound (2) is a magenta azo dye developer that has been color-shifted by acylation (47). In the form shown, its color is a weak yellow. Restoration of the chro-

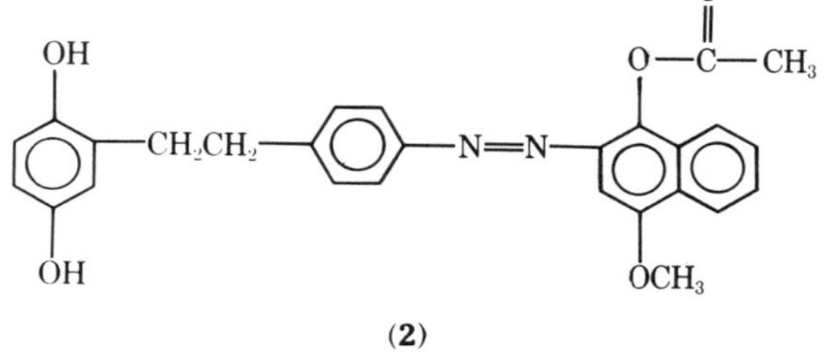

(2)

[16044-30-7]

mophore's green absorption is accomplished by alkaline hydrolysis during the development process. Similarly, acylation of α-amino groups in an anthraquinone dye developer, using readily hydrolyzable acyl groups such as trifluoroacetyl groups, produces temporary hypsochromic shifts that can be reversed by alkaline hydrolysis (48). Anthraquinone dye developers may also be used in the leuco form during exposure, the dye color being generated afterward by oxidation (49).

Indophenol dye developers color-shifted by protonation are described by Bush and Reardon (50). Compound (3) is a color-shifted cyan indophenol dye developer

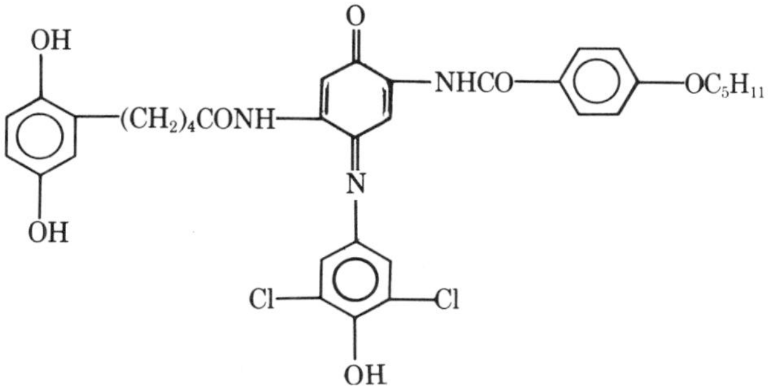

(3)

[50695-79-9]

that shows little color when the chromophore is in the protonated form. In alkali the dye ionizes to the indophenoxide, which is the colored form, and in this state it may be stabilized by association with a quaternary mordant in the receiving layer.

Dye developers incorporating the reduced forms of azomethines as dye precursors have been described by Lestina and Bush for use in processes that provide for later oxidation to the colored forms of the azomethine dyes (51).

Another approach involves the formation of dye images from colorless oxichromic developers, leuco azomethines stabilized against premature oxidation by acylation and linked to developer moieties (51), as in (4). The transferred images are oxidized to colored form either by aerial oxidation or by oxidants present in the receiving layer.

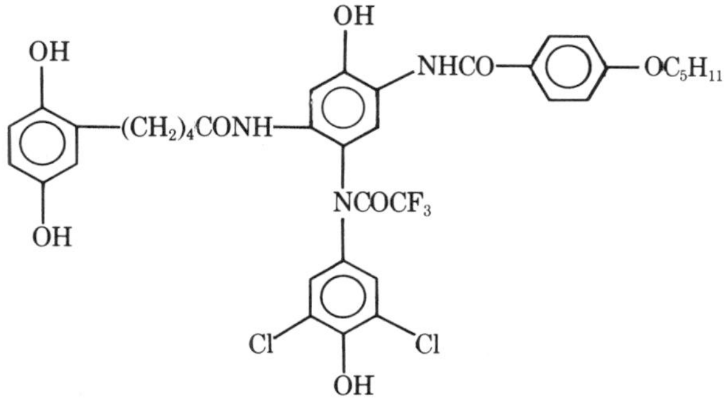

(4)

[50481-86-2]

Other Dye Developers. Other classes of dyes that have been studied as moieties in dye developers include rhodamine dyes, (see Triphenylmethane and related dyes), azamethine dyes indophenol dyes and naphthazarin dyes (22). Cyanine dyes (qv), although not generally suited for use as image dyes, have also been incorporated in dye developers (52) (see Dyes and dye intermediates).

Several types of functional groups have been found useful as insulating linkages between the dye and developer moieties of dye developers (22). For example, there are hydroquinone dye developers with sulfonyl, oxy and thio linkages (11). Redox dye-releaser compounds may contain the same hydroquinone-dye linkages but are rendered non-diffusible by ballast groups on the hydroquinone moiety (see Dye Release Processes below). Other examples of useful linkages include carboxamido groups, acyl groups, and amides. The attachment of two dyes to a single developer by amide linkage has also been described (53).

Dye developers may be provided with initial low mobility by the incorporation of a moiety such as an ester group that hydrolyzes during processing to yield a more mobile species (48).

Images from Colorless Dye Precursors. Color images may be produced from initially colorless image-forming compounds. The use of colorless compounds avoids loss of light by unwanted absorption during exposure, and the colored form is generated during processing.

One such system includes a coupler-developer, consisting of a color coupler and a color developer moiety linked together by an insulating group for use as a developing agent (54). Compound (5) is an example. In areas where silver halide undergoes de-

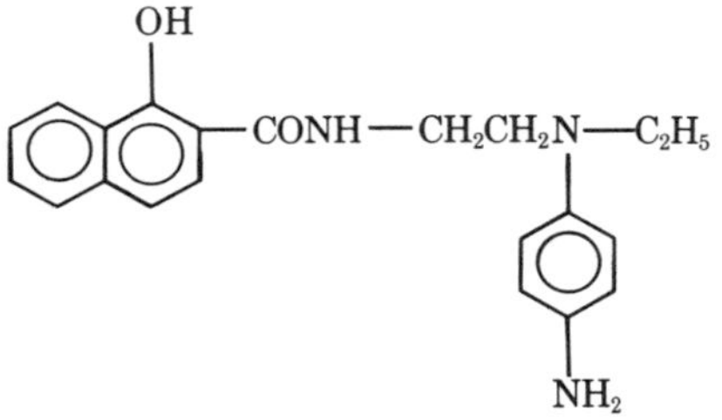

OH

$\bigcirc\bigcirc$ —CONH—CH$_2$CH$_2$N—C$_2$H$_5$

$\bigcirc$

NH$_2$

(5)

[41683-23-2]

velopment the coupler-developer is oxidized and immobilized by intermolecular coupling. In undeveloped areas the coupler-developer transfers to the receiving layer, where oxidation leads to intermolecular coupling to form a positive dye image.

In another transfer process a leuco indophenol is used both to develop a silver image in the negative and to form the inverse, positive dye image in the receiving layer (55).

Colorless triazolium and tetrazolium bases have also been proposed as image dye precursors (56). In unexposed regions, where silver halide is not undergoing development, unused developing agent reduces such a base to its colored form, rendering it immobile. In developing regions the base may migrate to a receiving layer and there undergo reduction to form a negative dye image.

Dye Imaging by Barrier Formation. One of the earliest image transfer mechanisms investigated by Rogers used bubbles formed during oxidation of a developing agent such as 1,2-bis-(phenylsulfonyl)hydrazine to prevent the transfer of a soluble dye through the exposed area of the developing negative (57). In unexposed areas the dye is transferred to form a positive image in the receiving layer.

Another barrier-forming mechanism is based on a polymeric coupler that reacts with oxidized color developer in exposed areas to produce a negative image in the form of an impermeable membrane. The polymeric membrane then acts as an image monitor to control the diffusion of positive image-forming dye to a receiving layer (28).

Dye Release Processes. A number of dye dropping or dye release processes provide image-related release of dyes or dye precursors that are initially immobile or of low mobility in alkali. One of the advantages of a process of this type is that the released species may themselves be unreactive with respect to other components of the negative. Dyes may thus diffuse through the layers of the negative to reach the image-receiving layer without undergoing unwanted reactions.

Dye release may relate either directly or inversely to the image-related reduction of silver halide. Release of the dye or dye precursor may be accomplished or initiated by the oxidized developing agent or the unoxidized developing agent or by alkali or silver salts having an image-derived distribution as the result of development. Examples of each of these routes to dye release are described in this section.

Release by Oxidation. Dyes or dye precursors may be released from alkali-immobile compounds as a result of interaction with the oxidized form of a mobile developing agent. Mobility of the developing agent is important because both the silver halide and the dye or dye precursor in the initial state are immobile in alkali and thus cannot interact directly. The mobile developing agent may participate directly in dye release or it may act as an electron transfer agent between silver halide and a dye releaser that is initially a reducing agent. With alkali-mobile image-forming species, such as the alkali-mobile dye developers, the use of electron transfer by auxiliary developers is optional, as noted earlier, whereas with alkali-immobile species the use of an electron transfer agent may be essential to the process.

The image-related release of a diffusible dye or dye precursor formed as a product of the oxidation of a dye containing a developing agent moiety was described by Rogers in 1966 (58). The process depends on the preferential transfer of an oxidation product having greater mobility than the unoxidized species. Compound (6), for example, a bis-sulfonylhydrazide, upon oxidation releases (7), a smaller, more mobile dye.

Several release mechanisms initiated by mobile, oxidized developing agents are

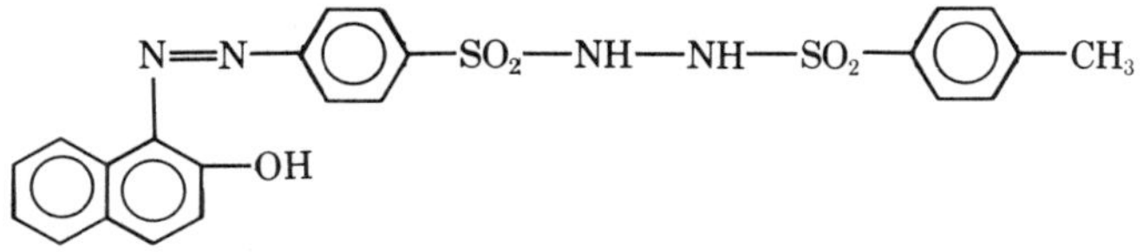

$$N=N-\bigcirc-SO_2-NH-NH-SO_2-\bigcirc-CH_3$$

(6)

[13251-03-1]

$$N=N-\bigcirc-SO_3H$$

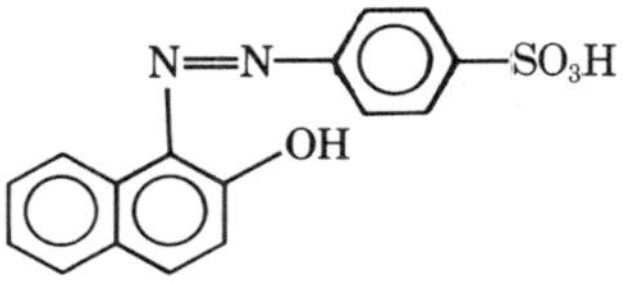

(7)

[573-89-7]

based on oxidative labilization of a linkage that would otherwise be alkali-stable, together with alkaline hydrolysis.

The use of images in terms of oxidized developer to mobilize dyes initially immobilized through a sulfonamide linkage was disclosed by Polaroid in 1969 (17,59). According to one concept based on coupling reactions described earlier (60), an *N,N'*-dialkyl-*p*-phenylenediamine was employed as color developing agent, together with an immobile coupler linked through a sulfonamide group to a dye (17). As shown in equation 7, development and color coupling lead to ring closure and concomitant

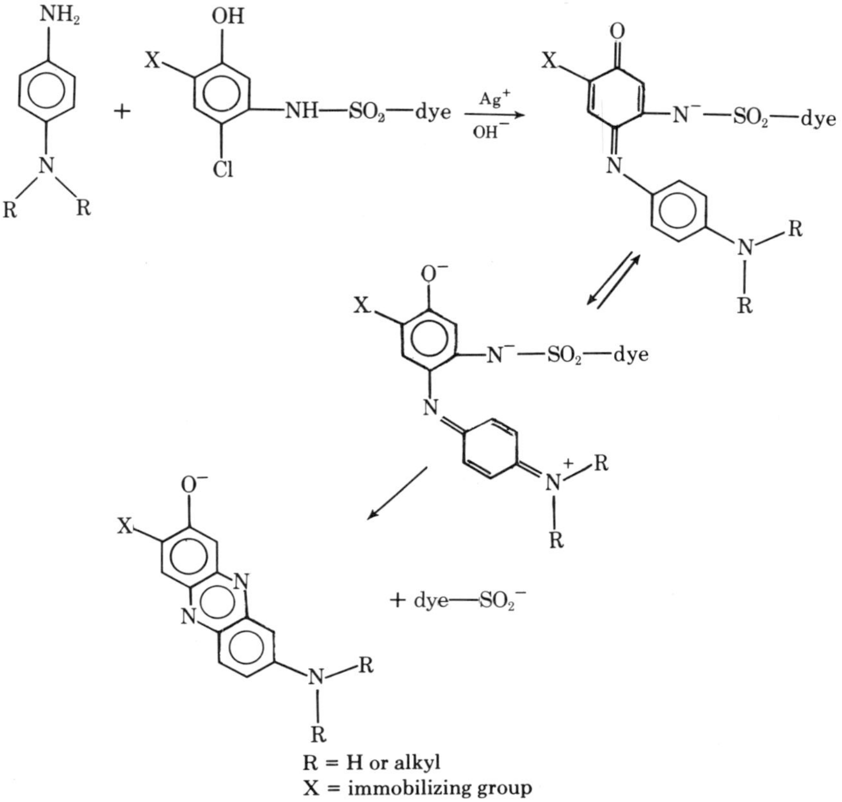

$$R = H \text{ or alkyl}$$
$$X = \text{immobilizing group}$$

release of an alkali-soluble dye. A second concept utilizes a more active primary developer, such as Metol or Phenidone, and the ready oxidation of a low mobility substituted 4-hydroxydiphenylamine, to which an image dye is linked through a sulfonamide group. Oxidation and hydrolysis result in ring closure and elimination of the alkali-soluble dye (eq. 8).

These processes are negative-working; dye is released where silver halide development takes place. Positive images may be obtained by using a direct reversal emulsion. Alternatively, both the immobile dye and silver-precipitating nuclei are included in a layer adjacent to the emulsion layer, and the reagent includes a silver halide solvent (17). In unexposed regions, soluble silver complex diffuses into the layer containing the nuclei and the immobile dye. Development of silver is catalyzed by the nuclei, and the resulting oxidized developing agent cross-oxidizes the dye releaser,

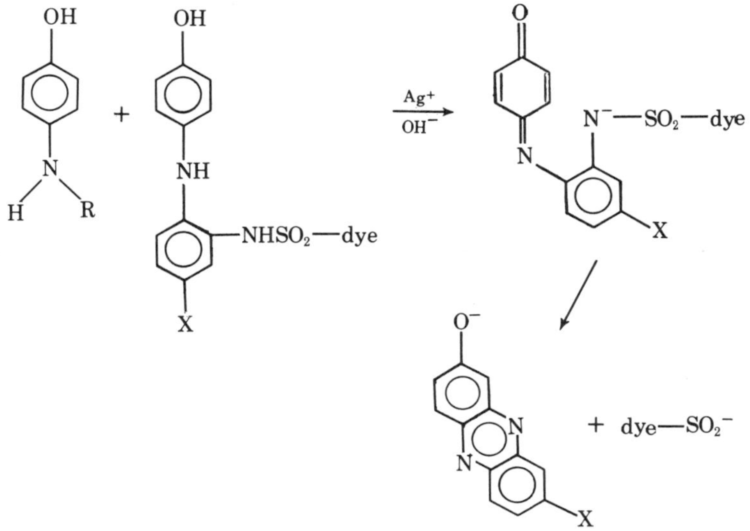

$$\tag{8}$$

R = H or alkyl
X = immobilizing group

as in equation 8. The oxidized dye releaser undergoes ring closure and the mobile dye is released for transfer to a receiving layer. The transferred dye image in this case is a positive.

Immobile p-sulfonamidophenol dye release compounds of the type used in the Kodak PR-10 film were disclosed by Fleckenstein and Figueras in 1972 (39). These compounds undergo image-derived oxidation to quinonimides through interaction with an oxidized developing agent, followed by alkaline hydrolysis to release soluble, diffusible dye images (eq. 5). Similarly, immobile p-sulfonamidoanilines, such as structure (8), may be used as redox dye releasers.

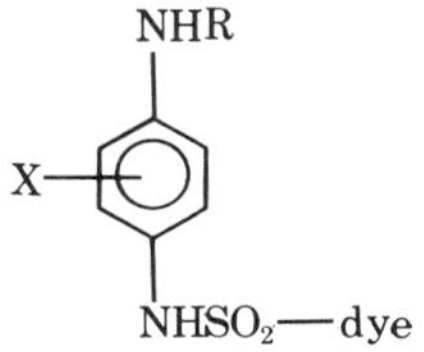

(8)

R = H or alkyl
X = immobilizing group

Further examples of image-derived labilization and release by hydrolysis include redox dye releasers containing "ballasted" hydroquinones with sulfonyl, oxy, and thio linkages, as in (9). Image-related oxidation to form quinones, followed by alkaline hydrolysis, releases mobile image dyes (61). Corresponding hydroquinone derivatives *without* ballast are dye developers that transfer in non-developing regions to form positive images (17). A dye release system has been described that yields positive images based on immobile benzisoxazolone dye releasers such as (10) (62). With these compounds, oxidation prevents cleavage. In alkali, the heterocyclic ring opens, forming

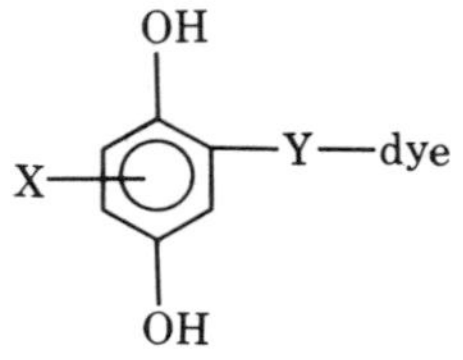

OH

X—⬡—Y—dye

OH

(9)

X = immobilizing group
Y = S, O or SO$_2$

a hydroxylamine, and in unexposed areas, where silver halide is not developing, the hydroxylamine cyclizes to form a second benzoxazolone (11), eliminating the mobile dye moiety. In exposed areas, silver halide is reduced by a mobile developing agent which in turn transfers electrons to oxidize the hydroxylamine to the non-releasing species (12) (eq. 9).

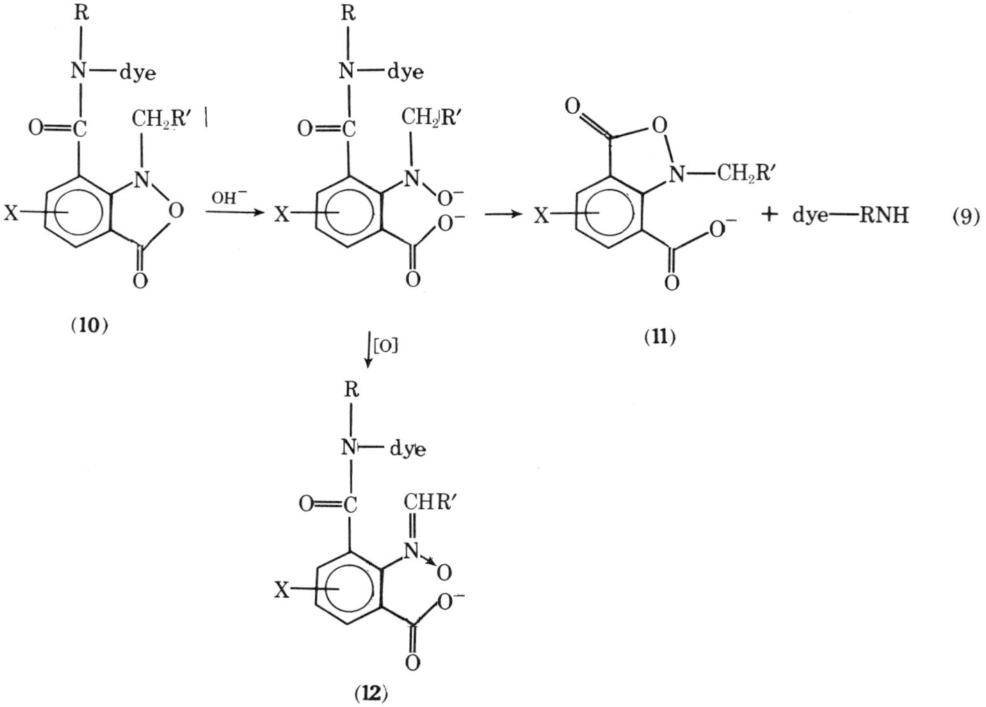

(10) $\xrightarrow{\text{OH}^-}$ → ... + dye—RNH   (9)

(11)

$\downarrow$ [O]

(12)

R,R' = alkyl
X = immobilizing group

Another positive-working release by cyclization, illustrated by equation 10, starts with an immobile hydroquinone dye releaser (13). Cyclization and dye release take place in alkali in areas where silver halide is not undergoing development, whereas in areas where silver halide is being developed the oxidized form of the mobile developing agent oxidizes the hydroquinone to its quinone (14), which does not release the dye (63).

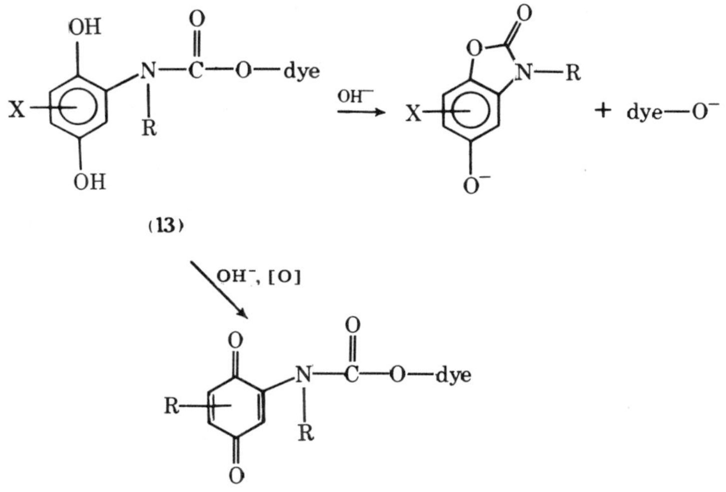

(13)

(10)

(14)

R = alkyl
X = immobilizing group

Displacement by Coupling. The coupling reactions of color developing agents have been described as release mechanisms for initially immobile image dyes or dye precursors (64). The releasing couplers may have substituents in the coupling position that are displaced in the course of the coupling reaction with an oxidized color developer. In one process, the eliminated substituent is the immobilizing group, so that the dye formed by coupling is rendered mobile. Compound (15) is an example of a coupler

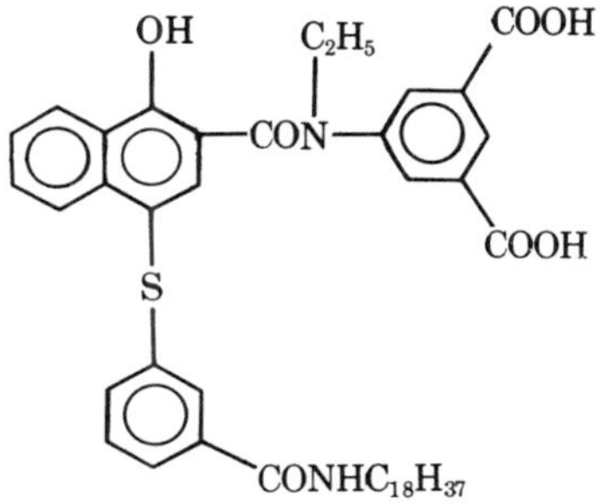

(15)

[5135-15-9]

that splits off an immobilizing group in this manner. In a related process, the coupler itself is the immobile moiety, as in (16), and the substituent that is split off is a mobile dye.

Redox dye release systems based on immobile sulfonylhydrazones, as in structure (17), were disclosed by Puschel and co-workers in 1971. These compounds react with oxidized *p*-phenylenediamine developing agents to release soluble dye moieties (65). To minimize stain the *p*-phenylenediamine developing agent may itself be rendered immobile and used in conjunction with a mobile electron transfer agent (66).

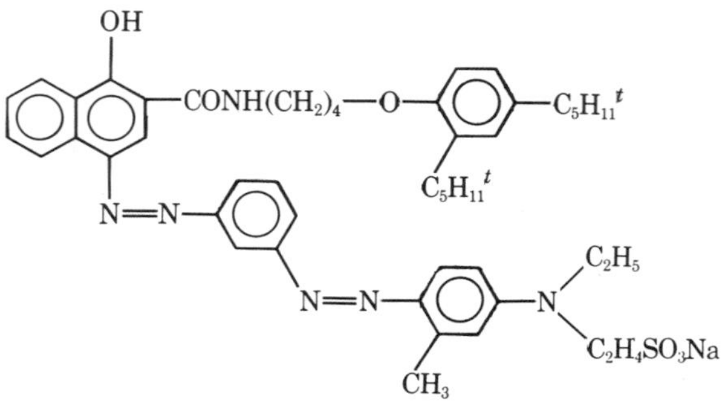

(16)

[4137-16-0]

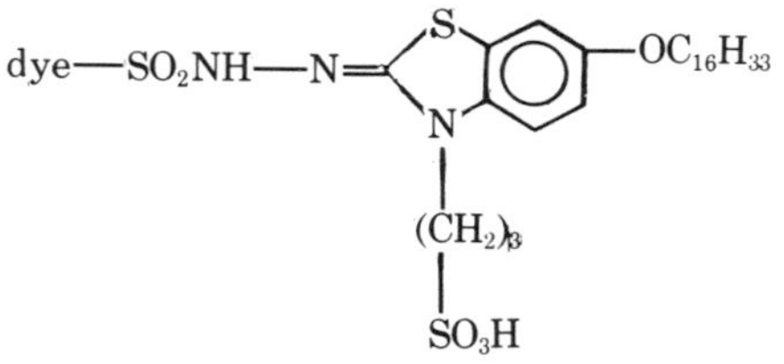

(17)

Release by Unexhausted Developing Agent. The image-forming dye release processes described so far have begun with images in terms of oxidized developing agents. Alternatively, dye release may be effected by unoxidized silver halide developing agent, taking advantage of its image-related distribution (56). An example is the image-derived reduction of triazolium and tetrazolium bases used as temporary mordants for acid dyes. Reduction of the bases in regions where silver halide is not developing releases the dyes for transfer to a receiving layer.

The image-related distribution of unexhausted developing agent may also be used to reduce and solubilize an indophenol dye initially present in a negative. The reduced and solubilized form of the dye may then transfer and undergo oxidation in a receiving layer to form a positive dye image.

Release by Silver-Assisted Cleavage. The use of a reagent containing both a silver halide developing agent and a silver halide solvent produces an image-related distribution of a soluble silver complex in the unexposed area. This silver may be used to effect a cleavage reaction that releases a dye or a dye-forming compound (67). An example is the silver-assisted cleavage of dye-substituted thiazolidine compounds, as shown in equation 11. Since this mechanism releases dye in areas where silver halide is not undergoing development, the process yields positive dye transfer images directly with negative-working emulsions.

Release by Unexhausted Alkali. Another of the mechanisms that may be used to form and transfer positive images is based on the image-related distribution of alkali. Alkali exhaustion in areas where exposed grains are undergoing development results in a positive image in terms of alkali, and this image is used to effect image-related dissolution and transfer of an alkali-mobile dye (68).

$$\underset{\underset{\text{dye}}{|}}{\overset{\frown}{S}}\overset{\frown}{N}-X \xrightarrow[\text{OH}^-]{\text{Ag}^+} \text{dye}-C\overset{O}{\underset{H}{\diagup}} + \overset{\frown}{S}\overset{\frown}{\underset{\text{Ag}}{N}}H-X \qquad (11)$$

X = immobilizing group

Economic Aspects

The use of instant cameras and film by both amateurs and professionals has been expanding significantly in recent years. As in the entire photographic field, the principal growth is in color photography. The 1977–1978 Wolfman Report on the photographic industry stated that instant photography was the fastest growing segment of the domestic photographic market and would shortly reach this status worldwide (69). This report further predicted that foreign photographic manufacturers would enter the field competitively in the near future. Current domestic sales in the field of instant photography were estimated at 25% of total spending in the photographic field and 40% of the mass market portion.

According to a survey reported by the Photo Marketing Association, instant cameras would account for 38.9% of new camera purchases in the United States in 1978 (70). An independent domestic retail sales survey cited by Polaroid reported a 68% increase in sales of instant cameras in 1977, along with a 4% decline of conventional camera sales.

Health and Safety Factors (Toxicology)

An important aspect of photography that has received increasing attention is the environmental effect of effluents discharged by laboratories that process large quantities of color films. Another concern is the limited supply of clean water available for processing in many areas. The completely self-contained instant films avoid both of these problems. Conversely, the environment has little effect on processing of the self-contained films, which are used at ambient temperature and require no darkroom and no water supply.

No toxicological hazards have been associated with the normal use of instant color films. However, the manufacturers caution that direct contact with the highly alkaline processing fluids can cause alkali burns. The fluids are provided inside sealed pods and are not usually handled by the user. In the integral systems, the fluids are retained within the film unit and neutralized in the course of processing. In the peel-apart films, the fluids are rapidly rendered harmless by contact with air.

Uses of Instant Color Photography

At the time of its introduction in 1963, Polacolor presented significant new opportunities for photography in color. For the first time, the photographer could create a finished color photograph under ambient conditions, view and judge the results immediately, observing the subject and the print simultaneously, and take advantage of the observations to improve creative skills. In addition to providing new enjoyment for the amateur photographer, the instant color film quickly became a powerful medium for professional work, and over the years commercial and technical uses have grown exponentially. Polacolor film formats and corresponding cameras and film holders have been developed to fit a variety of specialized needs.

Many laboratory instruments and diagnostic machines now include built-in instant-film camera backs. An important aspect in professional applications of instant color photography is the rapid completion of color photographs without carrying out lengthy darkroom procedures or awaiting the services of a commercial processor. For example, photomicrography with instant color films enables the microscopist to work without interruption, document results quickly, and make and compare immediately images of specimens that are undergoing rapid change (Figure 24).

Specialized pack cameras permit the use of Polacolor film for rapid fabrication of portrait identification cards, widely used for drivers' licenses, student registration, and credit cards. The use of Polacolor passport photographs has become widespread since their introduction in 1975. About half of the new passports issued in the United States in 1978 contained such instant portraits.

Instant portrait studios using SX-70 integral color film were introduced in 1977. The fully automatic motorized equipment uses industrial batteryless SX-70 film packs (Type 708) to produce photographs in a consumer-operated self-portrait booth.

Large format Polacolor photography using 20 × 25 cm (8 × 10 in.) and 51 × 61 cm (20 × 24 in.) films was first demonstrated in 1976. The 8 × 10 in. system, commercialized in 1977, is used by professional photographers both to prepare prints for direct exhibition and to provide final art ready for reproduction. The immediate control of the print and the direct relationship of its density range to the requirements of the engraving and printing processes are particularly advantageous.

One of the most remarkable recent uses of instant color in large format is the making of full-size Polacolor replicas of paintings and tapestries, a technique of great value in professional museum work. Because they do not involve photographic reduction and enlargement, these 1:1 replicas accurately maintain the fine detail and the dimensional relationships of the original subjects. Photographs as large as about 1 × 2 meters are produced in a room-size museum camera that uses 152-meter production rolls of negative and receiving sheet (71) (Fig. 25) (see Fine art authentication and preservation).

Medical applications include photomicrographs of stained tissue and scintillation camera records of the brain and other organs in the presence of radioactive isotopes (Fig. 26) (see Radioactive drugs and tracers). For more information on professional applications, see the periodical *Close-Up*.

BIBLIOGRAPHY

"Nonchromogenic Color Photography" under "Color Photography" in *ECT* 2nd ed., Vol. 5, pp. 837–838, by J. R. Thirtle and D. M. Zwick, Eastman Kodak Company.

1. E. H. Land, H. G. Rogers, and V. K. Walworth in J. M. Sturge, ed., *Neblette's Handbook of Photography and Reprography*, 7th ed., Van Nostrand Reinhold, New York, 1977, pp. 258–330.
2. E. H. Land, *J. Opt. Soc. Am.* **37,** 61 (1947).
3. U.S. Pat. 2,435,717 (Feb. 10, 1948), E. H. Land (to Polaroid Corporation); U.S. Pat. 2,455,111 (Nov. 30, 1948), J. F. Carbone and M. N. Fairbank (to Polaroid Corporation).
4. U.S. Pat. 3,382,788 (May 14, 1968), V. K. Eloranta (to Polaroid Corporation).
5. U.S. Pat. 2,495,111 (Jan. 17, 1950), E. H. Land (to Polaroid Corporation); U.S. Pat. 3,161,122 (Dec.

(a)

(b)

Figure 24. Reflected light microscopy: (**a**) macrophotograph of an integrated circuit (18×); (**b**) petrographic section of amphibole gneiss, photographed in polarized light with first order red compensator (50×).

Figure 25. Portions of large 1:1 replica photographs, comparing front (L) and back (R) sides of medieval tapestry, *The Martyrdom of Saint Paul* (ca 1460). Comparison of the two images indicates that the front has lost considerable color and detail through fading. Study of the back reveals much of the original detail and shows the complexity of the weaving, as well as some repair work. Reduced approximately 50% for reproduction here. Courtesy of Boston Museum of Fine Arts, Francis Bartlett Fund, No. 38.758.

Figure 25. Continued.

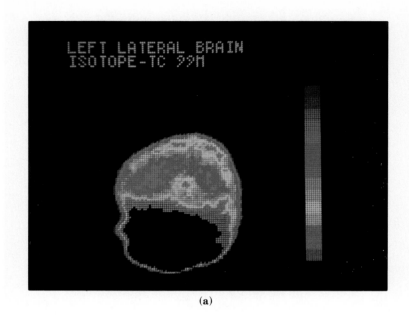

(a)

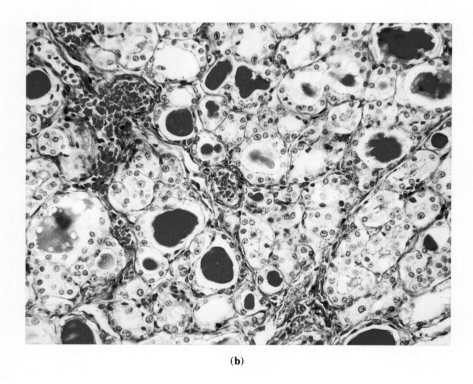

(b)

Figure 26. Medical applications of instant color photography. (a) scintillation camera record of human brain scan following uptake of a radioactive isotope. Digital count is color coded by computer, then displayed and photographed in color for diagnostic study. (b) micrograph, stained section of human thyroid (200×).

15, 1964), J. A. Hamilton (to Polaroid Corporation); U.S. Pat. 3,079,849 (Mar. 5, 1963), R. R. Wareham (to Polaroid Corporation).

6. U.S. Pat. 3,485,155 (Dec. 23, 1969), E. H. Land and V. K. Eloranta (to Polaroid Corporation).
7. E. H. Land, *Photogr. Sci. Eng.* **16,** 247 (1972).
8. E. H. Land, *Photogr. J.* **114,** 338 (1974).
9. U.S. Pat. 2,933,993 (Apr. 26, 1960), A. J. Bachelder and V. K. Eloranta (to Polaroid Corporation).
10. E. H. Land, *Photogr. Sci. Eng.* **21,** 225 (1977).
11. U.S. Pat. 3,415,644 (Dec. 10, 1968), E. H. Land (to Polaroid Corporation).
12. U.S. Pat. 3,594,165 (July 20, 1971), H. G. Rogers (to Polaroid Corporation); U.S. Pat. 3,689,262 (Sept. 5, 1972), H. G. Rogers (to Polaroid Corporation).
13. E. H. Land, *Photogr. J.* **90A,** 7 (1950).
14. U.S. Pat. 2,543,181 (Feb. 27, 1951), E. H. Land (to Polaroid Corporation).
15. U.S. Pat. 2,603,565 (July 15, 1952), E. H. Land (to Polaroid Corporation).
16. U.S. Pat. 3,148,062 (Sept. 8, 1964), K. E. Whitmore and co-workers (to Eastman Kodak Company); U.S. Pat. 3,227,551 (Jan. 4, 1966), C. R. Barr, J. Williams, and K. E. Whitmore (to Eastman Kodak Company); U.S. Pat. 3,227,554 (Jan. 4, 1966), C. R. Barr, J. Williams, and K. E. Whitmore (to Eastman Kodak Company); U.S. Pat. 3,243,294 (Mar. 29, 1966), C. R. Barr (to Eastman Kodak Company).
17. U.S. Pat. 3,443,940 (May 13, 1969), S. M. Bloom and H. G. Rogers (to Polaroid Corporation).
18. U.S. Pat. 2,983,606 (May 9, 1961), H. G. Rogers (to Polaroid Corporation).
19. U.S. Pat. 3,362,819 (Jan. 9, 1968), E. H. Land (to Polaroid Corporation).
20. U.S. Pat. 3,146,102 (Aug. 25, 1964), W. J. Weyerts and W. M. Salminen (to Eastman Kodak Company); U.S. Pat. 3,173,786 (Mar. 16, 1965), M. Green and H. G. Rogers (to Polaroid Corporation); U.S. Pat. 3,253,915 (May 31, 1966), W. J. Weyerts and W. M. Salminen (to Eastman Kodak Company).
21. U.S. Pat. 3,255,001 (June 7, 1966), E. R. Blout and H. G. Rogers (to Polaroid Corporation).
22. S. M. Bloom and co-workers in K. Venkataraman, ed., *The Chemistry of Synthetic Dyes,* Vol. 8, Academic Press, New York, 1978, pp. 195–213.
23. E. M. Idelson and co-workers, *Abstract L-3, Annual Conf. of Soc. Photogr. Sci. Eng.,* Boston, Mass., May, 1974.
24. H. G. Rogers and co-workers, *J. Photogr. Sci.* **22,** 138 (1974).
25. U.S. Pat. 3,857,855 (Dec. 31, 1974), E. M. Idelson (to Polaroid Corporation).
26. M. Idelson and co-workers, *Inorg. Chem.* **6,** 450 (1967).
27. U.S. Pat. 2,968,554 (Jan. 17, 1961), E. H. Land (to Polaroid Corporation).
28. U.S. Pat. 3,345,163 (Oct. 3, 1967), E. H. Land and H. G. Rogers (to Polaroid Corporation).
29. U.S. Pat. 3,411,904 (Nov. 19, 1968), R. W. Becker (to Eastman Kodak Company).
30. U.S. Pat. 3,625,685 (Dec. 7, 1971), J. A. Avtges and co-workers (to Polaroid Corporation).
31. U.S. Pat. 3,192,044 (June 29, 1965), H. G. Rogers and H. W. Lutes (to Polaroid Corporation).
32. U.S. Pat. 3,039,869 (June 19, 1962), H. G. Rogers and H. W. Lutes (to Polaroid Corporation).
33. U.S. Pat. 3,793,022 (Feb. 19, 1974), E. H. Land, S. M. Bloom, and H. G. Rogers (to Polaroid Corporation).
34. U.S. Pat. 3,647,437 (Mar. 7, 1972), E. H. Land (to Polaroid Corporation).
35. S. M. Bloom, *Abstract L-2, Annual Conf. of Soc. Photogr. Sci. Eng.,* Boston, Mass., 1974.
36. U.S. Pat. 3,702,244 (Nov. 7, 1972), S. M. Bloom and co-workers (to Polaroid Corporation); U.S. Pat. 3,702,245 (Nov. 7, 1972), M. S. Simon and D. P. Waller (to Polaroid Corporation).
37. W. T. Hanson, Jr., *Photogr. Sci. Eng.* **20,** 155 (1976).
38. W. T. Hanson, Jr., *J. Photogr. Sci.* **25,** 189 (1977).
39. Fr. Pat. 2,154,443 (Aug. 31, 1972), L. J. Fleckenstein and J. Figueras (to Eastman Kodak Company); Brit. Pat. 1,405,662 (Sept. 10, 1975), L. J. Fleckenstein and J. Figueras (to Eastman Kodak Company); U.S. Pat. 3,928,312 (Dec. 23, 1975), L. J. Fleckenstein (to Eastman Kodak Company); U.S. Publ. Pat. Appl. B351,673 (Jan. 28, 1975), L. J. Fleckenstein and J. Figueras (to Eastman Kodak Company).
40. U.S. Pat. 3,284,208 (Nov. 8, 1966), E. H. Land (to Polaroid Corporation).
41. U.S. Pat. 2,861,885 (Nov. 25, 1958) E. H. Land (to Polaroid Corporation).
42. U.S. Pat. 3,894,871 (July 15, 1975), E. H. Land (to Polaroid Corporation).
43. U.S. Pat. 3,704,126 (Nov. 28, 1972), E. H. Land, S. M. Bloom, and L. C. Farney (to Polaroid Corporation); U.S. Pat. 3,821,000 (June 28, 1974), E. H. Land, S. M. Bloom, and L. C. Farney (to Polaroid Corporation).
44. U.S. Pat. 3,615,440 (Oct. 26, 1971), S. M. Bloom and R. D. Cramer (to Polaroid Corporation).
45. U.S. Pat. 2,559,643 (July 10, 1951), E. H. Land (to Polaroid Corporation); U.S. Pat. 2,661,293 (Dec. 1, 1953), E. H. Land (to Polaroid Corporation).

46. U.S. Pat. 3,087,817 (Apr. 30, 1963), H. G. Rogers (to Polaroid Corporation).
47. U.S. Pat. 3,307,947 (Mar. 7, 1967), E. M. Idelson and H. G. Rogers (to Polaroid Corporation).
48. U.S. Pat. 3,230,082 (Jan. 18, 1966), E. H. Land and H. G. Rogers (to Polaroid Corporation).
49. U.S. Pat. 3,135,606 (June 2, 1964), E. R. Blout and co-workers (to Polaroid Corporation).
50. U.S. Pat. 3,854,945 (Dec. 17, 1974), W. M. Bush and D. F. Reardon (to Eastman Kodak Company).
51. U.S. Pat. 3,880,658 (Apr. 29, 1975), G. J. Lestina and W. M. Bush (to Eastman Kodak Company); U.S. Pat. 3,935,262 (Jan. 27, 1976), G. J. Lestina and W. M. Bush (to Eastman Kodak Company); U.S. Pat. 3,935,263 (Jan. 27, 1976), G. J. Lestina and W. M. Bush (to Eastman Kodak Company).
52. U.S. Pat. 3,649,266 (Mar. 14, 1972), D. D. Chapman and L. G. S. Brooker (to Eastman Kodak Company); U.S. Pat. 3,653,897 (Apr. 4, 1972), D. D. Chapman (to Eastman Kodak Company).
53. U.S. Pat. 3,201,384 (Aug. 17, 1965), M. Green (to Polaroid Corporation); U.S. Pat. 3,246,985 (Apr. 19, 1966), M. Green (to Polaroid Corporation).
54. U.S. Pat. 3,537,850 (Nov. 3, 1970), M. S. Simon (to Polaroid Corporation).
55. U.S. Pat. 2,909,430 (Oct. 20, 1959), H. G. Rogers (to Polaroid Corporation).
56. U.S. Pat. 3,185,567 (May 25, 1965), H. G. Rogers (to Polaroid Corporation).
57. U.S. Pat. 2,774,668 (Dec. 18, 1956), H. G. Rogers (to Polaroid Corporation).
58. U.S. Pat. 3,245,789 (Apr. 12, 1966), H. G. Rogers (to Polaroid Corporation).
59. U.S. Pat. 3,443,939 (May 13, 1969), S. M. Bloom and R. K. Stephens (to Polaroid Corporation); U.S. Pat. 3,751,406 (Aug. 7, 1973), S. M. Bloom (to Polaroid Corporation).
60. W. A. Schmidt and co-workers, *Ind. Eng. Chem.* **45**, 1726 (1953); U.S. Pat. 2,414,491 (Jan. 21, 1947), V. Tulagin (to General Aniline & Film Corporation).
61. U.S. Pat. 3,725,062 (Apr. 3, 1973), A. E. Anderson and K. K. Lum (to Eastman Kodak Company); U.S. Pat. 3,698,897 (Oct. 17, 1972), T. E. Gompf and K. K. Lum (to Eastman Kodak Company); U.S. Pat. 3,728,113 (Apr. 17, 1973), R. W. Becker and co-workers (to Eastman Kodak Company).
62. Ger. Offen. 2,402,900 (Aug. 8, 1974), J. C. Hinshaw and P. B. Condit (to Eastman Kodak Company); Ger. Offen. 2,448,811 (Feb. 20, 1975), J. C. Hinshaw and P. B. Condit (to Eastman Kodak Company).
63. Belg. Pat. 834,143 (Apr. 2, 1976), D. L. Fields and co-workers (to Eastman Kodak Company).
64. Brit. Pat. 840,731 (July 6, 1960), K. E. Whitmore and P. M. Mader (to Kodak Limited); U.S. Pat. 3,227,550 (Jan. 4, 1966), K. E. Whitmore and P. M. Mader (to Eastman Kodak Company); ref. 16.
65. U.S. Pat. 3,628,952 (Dec. 21, 1971), W. Puschel and co-workers (to Agfa-Gevaert Aktiengesellschaft); Brit. Pat. 1,407,362 (Sept. 24, 1975), J. Danhauser and K. Wingender (to Agfa-Gevaert Aktiengesellschaft).
66. Ger. Offen. 2,335,175 (Jan. 30, 1975), M. Peters and co-workers (to Agfa-Gevaert Aktiengesellschaft).
67. U.S. Pat. 3,719,489 (Mar. 6, 1973), R. F. W. Cieciuch and co-workers (to Polaroid Corporation).
68. Brit. Pat. 860,233 (Feb. 1, 1961), (to Intl. Polaroid Corporation); Brit. Pat. 860,234 (Feb. 1, 1961), (to Intl. Polaroid Corporation).
69. *1977–78 Wolfman Report on the Photographic Industry in the United States,* ABC Leisure Magazines, Inc., New York, 1978.
70. *Photo Marketing* **53** (5), 30 (1978); *Instant Impact Survey,* Photo Marketing Association International, Jackson, Michigan, 1978.
71. E. H. Land in *Polaroid Corporation Annual Report for 1975,* Cambridge, 1976, and *Polaroid Corporation Annual Report for 1976,* Cambridge, 1977; L. Salmon, *A Medieval Tapestry in Sharp Focus,* Museum of Fine Arts, Boston, 1977.

VIVIAN K. WALWORTH
Polaroid Corporation

COLUMBIUM.   See Niobium.

COMPLEXING AGENTS.   See Chelating agents.

COMPOSITE MATERIALS

Composites are combinations of two or more materials present as separate phases and combined to form desired structures so as to take advantage of certain desirable properties of each component. The constituents can be organic, inorganic, or metallic (synthetic or naturally occurring) in the form of particles, rods, fibers, plates, foams, etc. Compared with homogeneous materials these additional variables often provide greater latitude in optimizing, for a given application, such physically uncorrelated parameters as strength, density, electrical properties, and cost. Furthermore, a composite may be the only effective vehicle for exploiting the unique properties of certain special materials, eg, the high strength of graphite, boron, or aramid fibers (qv).

Some measure of coarseness of the homogeneous constituent structures is needed for a meaningful definition of composite material. The term as used here assumes that the average dimension of the largest single homogeneous geometric feature, in at least one direction, is small relative to the size of the total body in that direction; in addition, it assumes that the dimensions of the minor constituent phase are sufficiently large so that its characteristic properties are substantially the same as if it were present in bulk. Thus, laminated safety glass or copper-clad stainless steel, although composite *structures,* are not considered to be composite *materials* (see Laminated materials, glass; Laminated and reinforced metals). At the other extreme, gold ruby glass, which contains submicroscopic gold precipitate particles, is also not considered in this article. The above definition in terms of size serves as a guideline. This is occasionally violated as in the case of certain particulate-filled materials that contain submicroscopic silica gel, carbon black, etc (see Fillers).

Arbitrary control of the geometry, and often composition, of the constituent phases within wide limits is also usually implied in the term composite. Thus, a glass-epoxy composite could consist of glass fibers, glass beads, powdered glass, glass flake, or foamed glass impregnated with various epoxy formulations (see Epoxy resins). Wood, although consisting of cellulose fibers bonded together with lignin (qv) and other carbohydrate constituents, is not usually considered to be a composite, since it does not have a structure capable of arbitrary variation. On the other hand, wood particle-board or wood-flour filled resins would be classed as composite materials (see Laminated and reinforced wood). Certain other structures, such as the oriented eutectics in which rods and plate structures can be produced by controlled solidification with some latitude in size and composition, comprise an intermediate case. Certain such alloy and ceramic systems are showing promise for advanced applications. It is often useful to treat such naturally produced heterogeneous materials as composites. However, these constitute a rather special and somewhat restricted part of the disciplines and technologies common to composite materials.

Composite materials consist of a continuous matrix phase that surrounds the reinforcing-phase structures. Possible exceptions are (*a*) a laminated stacking of sheets in which the phases are kept separated, and (*b*) two continuous interpenetrating phases, such as an impregnated sponge structure, in which it is arbitrary as to which phase is designated as the matrix. The relative role of the matrix and reinforcement generally fall into the following categories:

(*1*) The reinforcement has high strength and stiffness, and the matrix serves to transfer stress from one fiber to the next and to produce a fully dense structure.

(2) The matrix has many desirable, intrinsic physical, chemical, or processing characteristics, and the reinforcement serves to improve certain other important engineering properties, such as tensile strength, creep resistance, or tear resistance.

(3) Emphasis is placed on enhancing the economic attractiveness of the matrix, eg, by mixing or diluting it with materials that will improve its appearance, processability, or cost advantage while maintaining adequate performance.

The first category constitutes the high performance composites. High strength fibers are used in high volume fractions, with orientations controlled and tailored for optimum performance. Considerations such as system performance benefits often determine the range of applications of this class of composites.

In the remaining two categories, cost is the more immediate consideration. Category (2) emphasizes improving engineering properties to extend the range of usefulness and marketability of a given matrix; moderate concentrations of fibers, often as discontinuous random fibers, and of flake and certain particulate reinforcements are used. The reinforced plastics fall in this class. In category (3) the emphasis is somewhat the inverse, ie, how to make an otherwise attractive material less costly *per se,* or how to process the material at lower cost without unacceptable degradation of properties through the use of particulate, flake, or fibrous fillers and colorants. This category largely consists of the filled polymers (see Colorants for plastics; Fillers; Laminated and reinforced plastics).

A composite material, as defined, although itself made up of other materials, can be considered to be a new material having characteristic properties which are derived from its constituents, from its processing, and from its microstructure.

Properties

Composites typically are made up of the continuous matrix phase in which are embedded: (1) a three-dimensional distribution of randomly oriented reinforcing elements, eg, a particulate-filled composite; (2) a two-dimensional distribution of randomly oriented elements, eg, a chopped fiber mat; (3) an ordered two-dimensional structure of high symmetry in the plane of the structure, eg, an impregnated cloth structure; or (4) a highly-aligned array of parallel fibers randomly distributed normal to the fiber directions, eg, a filament-wound structure, or a prepreg sheet consisting of parallel rows of fibers impregnated with a matrix. Except in case (1) the properties of the composite structure viewed as a homogeneous average material are more complex than are the more familiar isotropic materials which require two independent constants, such as the Young's modulus and the Poisson ratio (the ratio of the strain in the direction of the responsible applied principal stress to the strain produced in the transverse direction), to specify their elastic response. The other types of composites (2, 3, and 4) require at least four independent constants, such as two Young's moduli and two Poisson ratios, for a comparable specification. These are needed to describe the dependence of the elastic response as affected by the orientation of the applied stress relative to that of the reinforcing fibers. These properties can be measured. Since composites in turn are often built up by laminating layers of composite sheets, these properties are needed to predict the overall response of the laminated structure. The fibers in each layer can be oriented differently from adjacent layers. If proper attention is not given to fiber symmetry, or the basic information needed for design is not available, peculiar effects can occur, such as a composite part twisting when a simple tensile load is applied. With an isotropic material, this would merely stretch the body. However, for purposes of designing optimum materials, it would be desirable to compute the properties of a composite considered as a homogeneous orthotropic material from the properties of the constituent matrix and reinforcements. The present state of analytical skills allows such predictions to be made with reasonable confidence in specialized cases. However, for many other situations, only upper and lower property bounds can be stated.

Micromechanics is the detailed study of the stresses and strains within a composite considered as a true heterogeneous system. This approach allows the effective average properties of the composite to be computed when the reinforcement has a simple geometric shape and is located in regular arrays. Such idealized models can be used to provide a semiquantitative framework for the behavior of real composite materials. Modeling of the properties of composites as a function of temperature, pressure, or other environments requires a corresponding knowledge of the behavior of the separate constituents plus that of their interactions, such as result from differences in thermal expansion. Much recent attention has been given to hygrothermal effects on the viscoelastic response in polymer matrix composites, ie, on the combined effect of temperature and moisture content on mechanical properties. Increasing water content decreases the stiffness of epoxies and other resins much as does increasing the temperature.

Prediction of Composite Properties. Certain properties, such as the colligative thermodynamic ones, can be accurately calculated from knowledge of the volume fractions and chemical composition of the constituent phases (see Thermodynamics). Other properties, such as thermal and electrical conductance and the elastic properties, can be calculated from idealized models which closely approximate real composite behavior. Other important properties, such as failure strength and fracture toughness, can only be approximated roughly.

Composite properties are often assumed to be representable by the rule of mixtures:

$$P = P_1V_1 + P_2V_2 + P_3V_3 + \ldots \tag{1}$$

in which P is the property value for the composite, and P_i and V_i are the property values and volume fractions of the ith phase. For a fully dense, two-component composite, the heat capacity and density are accurately given by the rule, where i has the values 1 and 2, $V_1 + V_2 = 1$, and P signifies either heat capacity or density. However, for the Young's modulus, E_{11}, of a continuous parallel fiber-reinforced composite in the direction of the fibers, one can only state rigorously that

$$E_{11} \geq V_fE_f + V_mE_m \tag{2}$$

although to a very good approximation the two sides can be taken to be equal. The subscripts f and m are now used to indicate fiber and matrix. However, if the fibers are parallel but are initially discontinuous or develop breaks, then V_f, the volume fraction of the fibers, must be replaced by βV_f where β is a constant less than 1 and reflects the fraction of the fiber rendered ineffective because of the loss of its ability to carry tensile load near a fiber end. The value of β depends on the fiber geometry, the elastic-deformation characteristics of the fiber and the matrix, and the interface. Thus within these limitations, equation 2 estimates one of the elastic constants needed. The Young's modulus transverse to the axis of the fibers is given by:

$$E_{22} \approx E_m(1 + \zeta\eta V_f)/(1 - \eta V_f) \tag{3}$$

where

$$\eta = (E_f - E_m)/(E_f + \zeta E_m), \tag{4}$$

and ζ is a constant determined by the fiber geometry. The major Poisson ratio, ie, the strain transverse to the direction of the fiber relative to the strain in the fiber direction when also stressed in the fiber direction, is approximated by:

$$\nu_{12} \approx V_f \nu_f + V_m \nu_m \tag{5}$$

and the minor Poisson ratio, ie, the ratio of strain in the fiber direction to the strain transverse when also stressed in that transverse direction, is given by:

$$\nu_{21} = \nu_{12} E_{22}/E_{11} \tag{6}$$

These equations permit the material parameters to be estimated from the fiber and matrix properties when the reinforcement consists of parallel fibers. For other reinforcement geometries, such as packing of spheres, these equations can be generalized using semiquantitative approaches in which ζ is allowed to take on geometrically dependent values that result from comparison with specialized micromechanical modeling or experiment.

Strength of Composites.   The relatively high strength of composites on an equal weight or cost basis is often a major contributing factor to the importance accorded this class of materials. Strength, although relatively easy to measure, is even more difficult to predict in the case of composites than for homogeneous materials. As a first approximation, the rule of mixtures for strength, S, of a fiber-matrix composite in the direction of the fiber is often used

$$S_{\text{composite}} = V_f \overline{S}_{\text{fiber}} + V_m \sigma^* \tag{7}$$

where σ^* is the stress in the matrix at the failure strain, and $\overline{S}_{\text{fiber}}$ is the mean fiber strength. This is an intuitive equation that is often useful as a starting point to estimate strength in the absence of better information. However, it has no theoretical basis, even as an upper or lower bound.

The possibility of transferring load from one reinforcing element to another means that the effect of a fiber break can be localized. Thus the average strength of a composite can exceed that of its constituent reinforcement. As noted, composites are usually anisotropic. In aligned or continuous filament composites, the strength in the direction transverse to the fibers is much less than that parallel to the fibers. The high strength possible in fiber composites is, in most cases, attained at the expense of an absolute weakening in strength in other directions (relative to the strength that the bulk reinforcement material would have). Isotropic chopped-fiber composites are limited in strength because the geometric interferences between fibers limit their packing fraction. For these reasons composites must be carefully designed to be strong where needed and to ensure that the stresses remain low in the other directions.

Quantitative considerations of strength require a criterion for failure. In composites this can mean the stress level at which detectable loss of structural coherence occurs, the maximum apparent stress the structure can sustain, or in some cases the maximum apparent strain to cause separation into two parts. The failure process depends strongly on how the material is stressed. A stress applied nearly parallel to the reinforcement requires fiber breakage in order to propagate a crack. However, as the stress is applied increasingly off-axis, first shear failure and then transverse tensile failure of the matrix controls strength. Under compressive loading several failure modes are possible.

Much effort has been directed to finding failure criteria, such as a limit to the local energy of distortion, in analogy to the Von Mises criterion used to predict when metals will yield under complex stress conditions. Although such approaches provide a framework for presenting failure-stress information, there is as yet no generally valid, convenient model.

The discipline of fracture mechanics, developed to predict failure in homogeneous ductile materials, provides another approach. This assumes an ability to detect and measure the most dangerous strength-impairing flaws. Fundamentally, this method requires a crack to be a simple topological surface, a condition that often is violated in composites as a result of multiple splitting. This approach has been most successful when the stress and orientation variables are kept simple.

Experimentation and micromechanical modeling have provided valuable insight into the major factors that influence the various failure modes and other mechanical properties. Resistance to crack propagation transverse to fibers or sheets depends on the matrix-reinforcement bond strength. If the bond is too strong, the composite is brittle, and if it is too weak, the composite becomes excessively weak in the transverse direction. Voids distinctly reduce shear and compressive strengths. Fiber misalignments and matrix-rich regions can initiate buckling under compressive loading. Hence, for reasons of material characterization and quality control, tensile, shear, and compressive strengths are routinely determined but require special specimen configurations to avoid measurement artifacts because of the strong material anisotropy. Flexural testing avoids some of the complications. Relatively long composite bars are often used to measure tensile strength, and short, stubby bars are used to measure the shear strength, often termed short beam shear or interlaminar shear strength. The literature on strength and testing is extensive (see General bibliography).

Fabrication Methods

The methods used to make composite materials and structures depend, among other factors, on the type of reinforcement, the matrix, the required performance level, the shape of the article, the number to be made, and the rate of production. The orientation and positioning of continuous filaments are controllable, whereas short fiber, flakes, or particulates are apt to be more randomly distributed. However, varying degrees of preferred orientation can be achieved by appropriate shearing action, electrical fields, etc.

Large diameter, single-filament materials, such as boron, silicon carbide, or wires, are often fed in precisely controlled, parallel arrays to form tapes of sheet materials. Complex computer-controlled machines for laying such tapes in desired overlap angles over complex surfaces have been built for fabricating aircraft structures, rocket casings, pipes, etc, where the highest possible performance is required. The matrix in these cases is usually polymeric and is added with the filament. However metals can also be used as the matrix for making tapes and sheets. This requires that the filaments be positioned while the molten matrix is allowed to infiltrate and solidify around the fibers, or while consolidation by diffusion bonding is taking place.

In the case of finer filaments, such as fiberglass, carbon fiber, or boron nitride fiber, bundles (tows or roving) of hundreds to tens of thousands of loosely aggregated fibers are handled as an entity. When these fibers are to be incorporated into a polymer matrix composite, it is usually convenient to form a semiprocessed, shapable, intermediate ribbon or sheet product known as prepreg in which the fibers are infiltrated by the resin. This impregnated material is further processed so that it can be handled conveniently. Typically, the fiber bundles are laid down in arrays along with a desired resin system in a controlled way. The structure is then rolled, combed, or otherwise handled to spread out the fibers as evenly as possible and with a uniform thickness. The impregnated system is then partly cured (B-staged) to fix the geometry while allowing enough shape relaxation (drape) and adherence (tack) to permit complex shapes to be built up from sheets of this prepreg material. The fibers can also be woven into cloth, infiltrated with resin, and handled as a prepreg. A variety of special weaves is available for composite usage.

Another approach is to form dry structures first, such as wire armatures, which are then impregnated with the matrix material. When standard shapes of uniform cross section, such as bars, rods, I-beams, or channels, are required, a process known as pultrusion can be used. The resin (thermoplastics can be used) and the continuous fiber are formed to the desired shape while pulling on the product and fibers to produce a highly aligned fiber arrangement as the composite is formed in the orifice region.

All of these continuous fiber methods are capable of yielding high quality, nearly ideal composites. The final structures must be carefully consolidated using various combinations of pressure and vacuum to eliminate porosity, to ensure complete coalescence of the matrix structure, and to avoid matrix-rich pockets and fiber misalignments. Large, expensive equipment is often required in the case of big structures. Typical examples of composite structures that have been made using the continuous fiber methods include automotive springs and frames, tires, pressure vessels, helicopter blades, aircraft airfoils and fuselage structures, spacecraft, boats, chemical plant equipment, and such sporting goods as skis, golf shafts, tennis racquets, and vault jump poles.

For many classes of applications, the ultimate in mechanical performance is not required, but complex shape and appearance are important considerations. A broad range of methods is available for the discontinuous fiber, flake, and particulate composites which can be used in these cases. Prepreg sheets can be made using chopped or discontinuous fiber reinforcement. This has the advantage of being moldable into double-curved shapes without buckling. The matrix resins can be either thermoplastic or thermosetting. The reinforced thermoplastic sheets are particularly adapted for the rapid pressing into shapes needed for automobile body structures.

Another method useful where larger shapes or lower production volume is needed is the spray-up technique. Special spray guns are available into which are fed the continuous filament roving. The fibers are chopped and dispensed along with controlled amounts of a suitable matrix resin to build a composite material; personal protection equipment is required. The spray is directed against a carefully prepared mold surface that is treated to release the composite shape after curing (see Abherents). Large structures, such as boat hulls, furniture, bath fixtures, and tanks, can be made. It is possible to incorporate continuous fiber structures, such as stiffers or ribs, or other fittings. Good surface finishes are achieved by the use of gel coats sprayed on the mold surface.

Molding compounds useful for injection, transfer, compression, or other similar types of force-flow-forming can be compounded by use of chopped fiber or other types of fillers. These generally have a lesser reinforcement content because of rheological considerations (see Rheological measurements). This method is used where small to medium repetitive, often complex shapes are required. Both thermoplastic and thermoset resins can be used as the matrix in forming such products as power tool casings, gears, and washer agitators.

Regardless of the fabrication process, the inherent anisotropy and materials combinations require close attention to such factors as residual stresses and defects arising from volumetric changes on polymerization, differences in thermal expansion, and post-processing creep. Although the methods may be relatively straightforward, details such as pressure and temperature history, post-curing cycles, formulation of the matrix, and surface condition of the reinforcement can be crucial to the production of composites of high quality.

Reinforcements

If a reinforcement is to improve the strength of a given matrix, it must be both stronger and stiffer than the matrix, and it must significantly modify the failure mechanism in an advantageous way, or both. The requirement of high strength and high stiffness implies little or no ductility and, thus, relatively brittle behavior. Brittleness is, in fact, not unusual in reinforcement materials. Such materials are often used in the form of filaments because flaws markedly affect their strength, and a fiber geometry more effectively enhances the amount of material unaffected by flaws than do sheet or particulate shapes. Furthermore, a characteristic length-to-thickness (aspect) ratio must be maintained in order to transfer effectively from one reinforcement element to the next. Fibers are convenient for meeting this requirement, and thus, comprise the most important class of reinforcing materials. However, other classes of composites in which strength is not the most important requirement frequently use powders, flakes, short chopped fibers, or other forms of reinforcing materials.

Strong natural fibers such as plant fibers (see Fibers, vegetable) silk (qv), and asbestos (qv) have long been known to have been used to make some of the first commercial high performance composites: varnish and lacquer-impregnated fabrics were used to cover early aircraft; cotton reinforcement of rubber was first used in the making of pneumatic tires; and phenol–formaldehyde impregnated cloth was used for automotive timing gears during the same era. However, the events that led to the rapid development of high performance composites as a new class of engineering materials were: (1) the emergence of thermosetting- and thermoplastic-synthetic polymer technologies, (2) the commercial availability of high-strength glass fiber in the late 1930s, and (3) the accelerated development of composites for military uses during World War II. To a considerable extent, the success of composites is inextricably linked to the development and availability of cost-effective, strong, stiff fibers.

Glass fiber, the first of the synthetic fibers, is produced in nature from air-blown strands of molten volcanic glass (see Glass). Egyptian, Roman, and Venetian artisans used glass fibers for decorative effects. Continuous filament-drawing originated in the late seventeenth century. Its potential for high strength was noted by Griffith in 1920 (1). The need for temperature-resistant electrical insulation led to the development just prior to World War II of E-type glass for use as winding and cloth insulation. Reportedly, the discovery of glass fiber-reinforced plastics (GFRP) resulted from the accidental spillage of a polyester resin onto glass fiber cloth (see Laminated and reinforced plastics). The cured composite was recognized as having attractive mechanical properties. The E-glass fiber, although not developed as a structural material, has become the most widely used of the high performance reinforcements, because of its low cost and its reproducibly good properties. It has subsequently grown into transportation, furniture, construction, industrial, recreational and other important segments of the economy.

The tire industry has provided another major impetus for the development of strong, economical, high performance fiber reinforcements (see Tire cords). Competition and materials advances have resulted in the progression: cotton, rayon, tensioned rayon, nylon, polyester, glass, steel, and aramid fibers. A second major factor stimulating the development of high stiffness-to-weight and strength-to-weight composites has come from the aerospace requirements (see Ablative materials). Weight reductions have compounding beneficial effects, eg, decreasing the mass that needs to be carried aloft also decreases the fuel requirements, the size of the propulsion system, and the need for massive load supporting structures. Finally, substantial impetus resulted from the materials science activities in the 1950s and 1960s, which demonstrated that very high strength was achievable over a wider range of materials than had been previously supposed. Theory and experiment indicated that strengths approaching 0.1 and even 0.2 times Young's elastic modulus (the proportionality constant between stress and strain in elongated structures) could be achieved if the material is perfect, ie, free of mobile dislocations, stress-raising notches, steps, or inclusions. By contrast, the strength of ordinary (imperfect) materials is typically a hundred-fold smaller. Fibers offered the greatest probability of achieving structural perfection, and record strengths were reported in rapid succession in filamentary crystals (whiskers) and in glass fibers (see Refractory fibers). Fiber composites were identified as the way to take advantage of the enormous strengths of whiskers. However, the problems associated with their production, handling, and conversion into composites have not been economically solvable to date. Whiskers, even if available, continue to cost ca $1–30/g, although projections of tenfold cost reductions have been made.

As a result, in spite of the great theoretical potential of whisker composites, attention since the mid-1960s has focused increasingly on continuous or mass-fabricated, high strength, high modulus fibers. These include glass, carbon and graphite, boron oxides, silicon carbide, and aramid filament.

The various types of fiber reinforcements listed in Table 1 provide a basis for comparisons. The data presented are approximate, ie, derived from various sources that may not use the same basis for measurement. Some fibers are available only as laboratory produced materials. The strengths cited often represent the maximum of what is practically possible, not necessarily what is typical of commercial material. The prices are those prevailing in 1977 or when the product was last available. In the case of metals, note that drawn wire is considerably more expensive than the bulk material. The noncontinuous or long-stranded types of reinforcements are presented in Table 2 according to their function. The cost of this class of materials in most cases is substantially less than in the case of the fiber materials. However, materials used in inorganic or metal composites, such as carbides, or refractory metal powders, can also be costly.

Nonstrand Reinforcements. A variety of reinforcements or fillers of plate-like or particulate geometries are frequently used in nonhigh-performance composites. These materials also function to extend the polymers, especially the more expensive engi-

Table 1. Characteristics of Candidate Reinforcing Fibers

Category	Material	E, GPa (10^6 psi)	S, GPa (10^3 psi)	Sp gr	Typical diameter, μm	$/kg
glass	vitreous silica	72 (10.5)	5.9 (850)	2.19	10	12–100
	E glass	72 (10.5)	3.4 (500)	2.54	10	0.75–1.00
	S glass	85 (12.4)	4.5 (650)	2.49	10	4.50–11
carbonaceous (material refers to starting process)	PAN[a] high strength	241 (35)	2.8 (400)	1.7–1.8	7	100–150
	PAN[a] high modulus	413 (60)	1.7 (250)	1.9–2.0	7	100–150
	pitch	345 (50)	1.4 (200)	2.0	7	20–50
	rayon, very high modulus	689 (100)	3.5 (510)	1.8	7	b
	rayon, high modulus	517 (75)	2.6 (380)		7	1,000
polymer	aramid	103–152 (15–22)	2.8 (400)	1.44	12	20–50
	olefin	0.7 (0.1)	0.62 (90)	0.97	3–500	1.5–2.5
	nylon	3.4 (0.5)	0.86 (125)	1.14	3–500	1.5–3.0
	rayon	6.9 (1)	1.1 (155)	1.52	3–500	0.50–2.50
inorganic	alumina (monocrystal)	510 (74)	3.4 (500)	3.96	250	20,000–150,000[c]
	alumina (polycrystal)	379 (55)	1.0 (150)	3.96	3	20–60
	alumina (whisker)	510 (74)	21 (3000)	3.96	1–10	(35,000 in 1966)[b]
	alumina silicates	103–138 (15–20)	1.4 (200)	2.5–2.6	10–15	2–5
	asbestos	172 (25)	1.4 (200)	3.2	0.02	0.5–2
	boron (tungsten core)	3.79 (55)	2.8 (400)	2.63	140	300–500
	boron nitride	55–76 (8–11)	0.38 (55)	1.85	7	2500[c]
	silicon carbide (carbon core)	482 (70)	2.8 (400)	3.2	100	250–1,000[c]
	silicon carbide (polycrystal)	44 (64)	6.2 (900)	3.2	5–25	b
	silicon carbide (whisker)	482 (70)	21 (3000)	3.21	1–10	1,200[c]
	silicon nitride (whisker)	379 (55)	14 (2000)	3.18	1–10	b
	zirconia (polycrystal)	427 (62)	1.4 (200)	4.84	3	20–65
metal	beryllium	221 (32)	1.3 (185)	1.84	75	27,000
	molybdenum	358 (52)	2.2 (320)	10.2	25	4,500–5,000
	steel	200 (29)	4.1 (600)	7.2	75	40–50
	tungsten	407 (59)	4.0 (580)	19.4	25	2,000–4,000

[a] PAN is polyacrylonitrile.
[b] Current price unavailable.
[c] Price highly subject to change.

Table 2.   Other Types of Reinforcements by Function

Polymer matrix
   Extenders—clay, sand, ground glass, wood flour
   Rheology control—mica, asbestos, silica gel
   Color—titanium dioxide, carbon, pigments
   Flame/heat resistance—minerals
   Heat distortion resistance—fibers, mica
   Shrinkage resistance—particulate minerals, beads
   Toughness—fibers, carbon black, dispersed rubber phase
Metal/inorganic matrix
   Hardness, wear resistance—metal, interstitial powders
   Creep resistance—oxide dispersions

neering types (see Engineering plastics). Minerals, such as clay, talc (qv), sand, mica (qv), and asbestos (qv), are often used because they are inexpensive and impart desired characteristics to the matrix/reinforcement combination during processing and in its use properties or both. These properties include thixotropy, strength, ease of finishing, creep, and heat distortion temperature. Table 3 gives as an example the effect of various types of filler on nylon 66. Other fillers include short, chopped fibers, which are often milled into the resin to produce a molding compound, ie, a mixture having adequate strength in thin sections, ears, etc, and rheological characteristics that permit flow in a die cavity under pressure. For this purpose chopped textile, glass, and carbon fibers and wires are used. Flake materials are often used to provide strength in sheet structures. Other fillers include carbon black, silica gel, hydrated alumina, titanium dioxide, and inorganic pigments (see Fillers). On a weight percentage basis, fine particulates increase the effective viscosity more than coarse materials. This is one of the factors involved in the selection of fillers for specific applications. Thus, the different materials can affect properties in many different ways. For this reason specialized formulations are usually developed to match the application.

Glass Fibers.   A wide variety of glasses can be used to make fibers by the old Modigliani process in which a thread is pulled away from a heated glass rod, much like pulling taffy (see Glass). However an enormous gain in productivity has been achieved in the more recent process in which a large number of fibers can be spun simultaneously by means of a heated platinum bushing having many small holes through which the glass can flow. The diameter of the holes, the temperature of the molten glass, and the rate of pulling determine the fiber diameter, which is ca 5–25 μm in commercial production. As the fibers are drawn, they are usually treated with sizing and coupling agents, ie, materials used to reduce strength degradation due to fiber–fiber abrasion, to transform a collection of loose parallel fibers into a coherent strand, and often to coat the fibers with agents that promote wetting and adherence to matrix resins. The properties of the glass fibers depend on the composition and to a lesser extent on processing history. Most (est 99%) of the glass fiber reinforcement is made from E glass. This glass is characterized by a range of compositions but has a very low alkali content, which results in a relative insensitivity to moisture, contributing to good electrical insulation and good strength retention over a wide range of conditions. The composition by weight of this type of glass is SiO_2 54 ± 2%, Al_2O_3 14 ± 2%, CaO + MgO 22

Table 3. Effect of Various Fillers/Reinforcements on Physical/Mechanical Properties[a] (As Exemplified by Nylon 66)

Property	Unreinforced	Glass fiber	Carbon (graphite)	Mineral	Carbon/ glass fiber[b]	Mineral/ glass fiber[b]	Glass bead
reinforcement content, wt %	0	40	40	40	20C/20G	20M/16.5G	40
specific gravity	1.14	1.46	1.34	1.50	1.40	1.42	1.44
tensile strength, MPa	83	214	276	103	234	121	90
(psi $\times 10^3$)	(12)	(31)	(40)	(15)	(34)	(17.5)	(13)
flexural modulus, GPa	2.8	11	23	7.6	19	6.5	5.5
(psi $\times 10^5$)	(4.0)	(16)	(34)	(11)	(28)	(9.5)	(8)
impact strength, notched/unnotched, J/m	48/320	139/1014	85/694	37/427	96/854	53/694	53/294
(ft·lbf/in.)	(0.9/6)	(2.6/19)	(1.6/13)	(0.7/8)	(1.8/16)	(1/13)	(1/5.5)
heat deflection temperature, 1.82 MPa (264 psi), °C	66	260	260	227	260	243	88
thermal expansion, 10^{-5} m/(m·°C)	8.1	2.5	1.4	5.4	2.1	4.5	3.6
mold shrinkage, %	1.5	0.4	0.4	0.9	0.5	0.7	
water absorption, 24 h, %	1.6	0.6	0.4	0.45	0.5	0.5	0.65

[a] Courtesy of *Plastics World*.
[b] G = glass fiber; M = mineral.

± 2%, B_2O_3 10 ± 3%, Na_2O + K_2O less than 2%, plus other minor constituents. The general utility and low cost of this glass make it an important engineering material.

The strength of E glass, as can be seen in Table 1, ranks high among available fibers. However, for many applications greater stiffness is required. Accordingly, there has been a search for high modulus glasses. A number of experimental glass formulations have been found that are 50% or more stiffer than E glass. Many of these contain BeO as a constituent and are very difficult to process. The most successful of the (relatively) high modulus formulations is S glass having a nominal weight composition SiO_2 65%, Al_2O_3 25%, MgO 10%. This glass is about 20% stronger and stiffer than E glass and has good chemical stability. For certain applications these improved properties warrant the five- to tenfold increase in price over E glass.

Ranking lower in terms of volume usage, but outstanding in terms of its intrinsic strength, corrosion resistance, and temperature capability, is silica glass fiber. The strength of virgin fiber at 77 K (bp, N_2) has been measured to reach the theoretically estimated limit of 0.2 times the Young's modulus, ie, 14 GPa (2×10^6 psi). The high softening temperature of 1100°C necessitates more difficult drawing conditions than for the above glasses. Accordingly, the cost is even greater than for S glass. Very reproducible fibers are used in aluminum matrix composites and for high temperature applications.

Carbonaceous Fibers.   Estimates of ultimate strength based on atomic bonding considerations and the observed strength of graphite whiskers indicate that graphite, in the direction of the basal plane, is among the strongest of known materials. Carbon fibers can be made by the controlled pyrolysis of organic fibers (see Carbon, carbon and artificial graphite). In general this process results in randomly oriented structure. However, in the early 1960s several approaches were recognized as leading to oriented fibers in which the basal plane of the graphitic structure is aligned with varying degrees of perfection parallel to the fiber axis.

The various carbonaceous fibers can be classified according to their starting materials, crystallinity, modulus, strength, density, etc. In all cases pyrolysis of an organic precursor is required. The requirement of removing the volatile decomposition products by diffusion limits the practical diameter of such fibers. Although fiber diameters somewhat in excess of 25 μm have been achieved, most fibers have diameters in the range 6–8 μm. For such small sizes filaments must be handled as bundles (tows), rather than as individual monofilaments. Commercially available tows contain ca 1000–60,000 fibers.

The first mass-produced, high performance carbon fibers were based on a rayon (qv) precursor, which was first pyrolyzed at relatively low temperatures and subsequently stress-graphitized. Commercial processes were developed by Union Carbide and by Hitco in the mid-1960s. In the pyrolyzed state the fiber has low strength and stiffness and appears to have a noncrystalline random structure using x-ray or electron microscopic techniques. By heating and stretching these filaments at 2200–2800°C, an oriented graphitic crystal texture develops. The strength and modulus values increase markedly with increasing degree of stretch, remaining approximately proportional to each other, ie, greater strength is accompanied by greater stiffness. Elastic moduli as great as 690 GPa (100×10^6 psi), corresponding to 70% of the value for graphite in the basal plane, have been achieved in experimental lots. The failure strain is ca 0.5%. Although much of the pioneering data on carbon fiber reinforced composites were generated using rayon-based fibers, this material has largely been supplanted commercially by the two grades described next.

Polyacrylonitrile (PAN) is a commonly available textile fiber and has been found to give high performance carbon and graphite fibers (see Acrylonitrile polymers). Both the process and the characteristics of these fibers differ from the cellulose-derived fibers, ie, stress graphitization is not required and the values of the modulus and strength of the resultant fibers are not simply related as a function of processing. The starting PAN polymer fibers are slowly oxidized at ca 150–300°C under conditions which stretch or, at least, restrain the fibers from shrinking. This transforms the polymer to give it a cross-linked ladder polymer structure. This material is no longer thermoplastic and can be heated to 1000°C to carbonize it or up to 3000°C to graphitize it, without requiring further tension in order to develop high strength and modulus. The stiffness increases monotonically with increasing final processing temperature. However, the strength passes through a maximum at about 1500°C. The absolute values depend on the starting fiber, initial degree of orientation, etc. This behavior allows fibers to be tailored to optimize various properties, such as strength, stiffness, or storable elastic energy. The latter property is related to impact toughness. The process for making these fibers is controllable and is convenient except for the long times of the order of many hours required to achieve the initial stretch oxidation. The basis for this technology can be traced in part to work by Shindo in Japan in 1960 (2) and to work at the Royal Aircraft Establishment (3) and at Rolls Royce in England (4) in the mid-1960s. PAN-derived fibers are being commercially produced by several manufacturers in the U.S., as well as the U.K., the Federal Republic of Germany, France, Japan, and the U.S.S.R.

Another approach uses pitch to produce fibers. These can be derived from a number of starting materials by pyrolysis carried out to the stage of yielding a liquid having a mesophase (planar, large fused-ring) structure. When this high-softening-temperature material is spun and then oxidized, a cross-linked, nonfusable fiber results that retains a high degree of molecular orientation. Such pitches have a high carbon content, hence, there is less problem in subsequently eliminating volatile decomposition products. Upon heating to carbonization or graphitization temperatures, an oriented, high performance fiber results. Stretch-graphitization can be used to further upgrade properties. This material appears to offer the greatest potential for achieving an economically low-cost product. Because the starting material is not a high volume fiber, to date the fiber is more variable than the rayon or PAN-derived fibers. Pitch-derived fibers are manufactured in the United States and Japan (see Tar and pitch).

Polymer Fibers. Almost any organic textile fiber can be incorporated into a composite structure. The carbon–carbon bond is very stiff, leading to a theoretical modulus of the order of 1000 GPa (ca 150×10^6 psi), based on the interatomic force constants for stretching. Against this as an upper bound, the stiffness of either natural or synthetic fibers is very low. This has been their major deficiency with respect to usage as reinforcements for high performance applications, even though failure strengths can be quite high. The cellulosic fibers exhibit the greatest stiffness and strength among the high volume, mass produced fibers (see Cellulose acetate and triacetate fibers). Because polymers typically exhibit complex time-dependent non-Hookean response, it is somewhat simplistic to use the concept of an elastic modulus, except as a limiting value at small strains for rapid loading conditions. In this sense, the highly oriented cellulose fibers, such as rayon and linen, have moduli that approach those of inorganic glass. Because of their relatively low stiffness, most of the organic fibers are used as reinforcements with even lower stiffness matrix materials, such as rubber and thermoplastics.

Recently, aramid fibers (qv) have been spun into high performance filaments that rival fiberglass in strength and have nearly twice the stiffness at about one-half of the density. These materials consist of highly aligned sheets of heterocyclic molecules. They have been considered to be intractable as solids. Hence, spinning from liquid crystal solution systems has been required (see Liquid crystals). These fibers have the unusual property of being strong in tension but relatively weak in compression. This is believed to be due to a microfibril rope-like structure of the individual filaments. These fibers are produced by the DuPont company under the trade name Kevlar and by Akzo as Arenka.

Inorganic Fibers. Whisker crystals have been made from a wide range of materials including Fe, Cu, Cr, Sn, Zn, Al_2O_3, Si, SiC, Si_3N_4, NaCl, among others. Typically whiskers exhibit outstandingly high strength and often have interesting magnetic, electrical, or other properties. They can be produced by various methods out of vapor or from condensed phases. A novel process for making low cost, very fine SiC whiskers involves the pyrolytic conversion of the silicon and the carbon contained in rice hulls. To be used in composites, they need to be collected, sorted, distributed into the desired configuration, etc. These practical considerations, their high cost, plus their noninfrequent degradation when processed into composites led to their displacement by continuous filament materials.

The most important of the continuous filaments is boron, which is produced from the thermal decomposition of BCl_3. This is deposited on a heated tungsten core having a typical diameter of 13 μm. Because of the high density and cost of the core, it is advantageous to build up boron layers to standard diameters of 100, 140, and even 200 μm. These fibers are stronger than carbon fibers, although on an equal weight basis, the strengths are about equal. Boron filaments can be made with uniform properties. They can be incorporated into both polymer and metal matrices, eg, aluminum. Their diameter, stiffness, and hardness require different fabrication techniques from those used with glass, carbon, or organic fibers. Individual filament positioning, the avoidance of sharp bends, and special cutting methods are used. The large diameter and perfection of the fiber stacking results in materials with high compressive strength. The principal drawback of this material is its relatively high cost, which is projected to be ca $150/kg in large volume. However, because of its special properties it will remain an important engineering reinforcement material.

Silicon carbide can be produced in a manner very similar to that for boron. The resulting fiber has comparable properties to boron, but is more dense. The main advantage of this fiber is that it remains strong to a higher temperature than boron. It can be used to reinforce aluminum and titanium matrices. A new process for making SiC has been developed in Japan based on the pyrolytic decomposition of a spun filament of a polycarbosilane having a C:Si ratio of unity. This process is analogous to making carbon filaments. Excellent properties have been reported.

Other methods for producing mainly oxide filaments include spinning or extruding a mixture of very fine-grained oxide with an organic binder. The binder is subsequently burned off and the oxide sintered or otherwise consolidated. Crystallizable glass filaments can subsequently be heat-treated to produce micaceous or mullite-dispersed fibers. Boron nitride filaments, which have properties somewhat like carbon fibers but are much more oxidation resistant, are made by a somewhat analogous process.

Filaments of B_2O_3 glass are spun and subsequently converted to BN by the action of ammonia. One of the most ingenious methods has been the crystal growth of shaped continuous filaments of aluminum oxide, using a bushing that allows a great variety of sizes and cross-sectional shapes to be produced (5). A difficulty has been the inclusion of small voids in the structure.

To bring costs down to the level where the derivative composites can compete with alternative materials requires high volume production. Fibers such as polycrystalline, Al_2O_3 and ZrO_2, which can be used in quantity and, reasonably high performance for high temperature insulation (qv) best meet these requirements. Silicon carbide boron fibers, or both, may have enough demand for use as premium reinforcement to remain commercially useful. Other fibers will undoubtedly continue to be available as specialty materials (see Boron; Boron compounds; Carbides).

Metal Filaments. Metallic filaments are conventionally made by wire drawing, ie, the mechanical reduction of continuous strands of metal through successive dies, often with intermediate heat treatments. The process is quite expensive for small wire diameters. Methods have been developed for melt-casting filaments with glass capillaries as the envelopes. Processes also have been developed for producing discontinuous strands of fine filaments by a melt-spinning process. However, these processes apparently have not produced commercially competitive materials to date.

Steel wire used for tire reinforcement comprises the major, large-scale application for high-performance metal filaments. Typically such wire is drawn from special, high quality steel rods having a low inclusion content and a starting diameter of 5.5 mm to final standard sizes of 0.38 and 0.25 mm (see Steel). Three to five such wires, each having a tensile strength of ca 2.8 GPa (400×10^3 psi), are twisted into strands and woven or otherwise positioned into lamellar arrays for use in tire manufacture. The wires are brass plated to promote adherence to the rubber matrix. The cost of 0.5–1.5 kg of steel reinforcement used per tire is ca $2/kg. Such wire is also used in enhancing the tensile strength of concrete (see Cement). Lower performance, coarser wire has long been used to produce a strong, flexible inner rim (bead) in tires (see Tire cords).

Matrices

Any solid that can be processed so as to embed and adherently grip a reinforcing phase is a potential matrix material. The polymers and metals have been the most successful in this role although cements, glasses, and ceramics have also been used. Thermosetting resins are particularly convenient because they can be applied in a fluid state, which facilitates penetration and wetting in the unpolymerized state, followed by hardening of the system at times and conditions largely controlled by the operator. Exothermicity, shrinkage, and the evolution of volatiles, if the polymerization is of the condensation type, are among the difficulties encountered using such resins.

Because of their cost, wide range of formulations, and generally good mechanical and electrical properties, the polyesters (qv) are an extensively used example of this class of resins. The epoxy systems are particularly useful because of their excellent adherence, low shrinkage, and freedom from gas evolution (see Epoxy resins). However, the epoxies are relatively expensive and are generally limited to service temperatures below 150°C. For many aerospace applications, higher use temperatures are required. Thermosetting systems having higher temperature capability tend to be resins that in their polymerization state contain nitrogen-heterocyclic moieties. Systems such as the imides, amide–imides, quinoxalines, imidazoles, etc, have been used. These are condensation–polymerization systems that are difficult to process; although stable at high temperature they have not succeeded in achieving mechanical properties in the composites that are as good as with the epoxy systems (see Embedding).

The reinforced thermoplastic resin systems are the most rapidly growing class of composites. In this class the focus is on improving the base properties of the resin to allow these materials to perform functionally in new applications or those previously requiring metals, such as die casting (see Engineering plastics). Thermoplastics may be of the crystalline or amorphous types. In the former type, the crystalline morphology may be significantly influenced by the reinforcement which can act as a nucleation catalyst. In both types there is a range of temperature over which creep of the resins increases to the point where it limits usage. The reinforcement in these systems can increase failure load as well as their creep resistance. Some shrinkage also occurs during processing, plus a tendency of a shape to remember its original form. The reinforcement can modify this response as well. Thermoplastic systems have advantages over thermosets in that no chemical reactions are involved causing release of gas products or exothermal heat. Processing is limited only by the time needed to heat, shape, and cool the structure, and the material can be salvaged or otherwise reworked. Solvent resistance, heat resistance, and absolute performance are not likely to be as good as with the thermosets.

Metal matrices offer even higher temperature capability, strength, and stiffness than is possible using polymers. Composite properties transverse to reinforcing fibers are superior, as is the fracture toughness. However, there is a penalty in terms of weight. More significantly, the fabrication of metal matrix composites is more difficult than in the case of polymers. Most metals react with fibers at elevated temperatures, especially in the molten state. Wetting is often uncertain. As a result it has been found necessary to coat fibers if melt infiltration is to be used, such as in the case of silica and boron filaments used with aluminum. Failure to do so causes marked degradation of strength of the fibers. Such reactions during processing can often be substantially reduced if the metal is retained in the solid state. This requires techniques such as diffusion bonding, roll bonding, and creep forming. These approaches depend on high temperature, pressure, and somewhat extended processing times. Conditions must be selected so as not to unduly damage the fibers mechanically.

Electrodeposition provides another technique for embedding reinforcement in a metal. In the case of metal wire-metal matrix composites, hot extrusion of a preform consisting of wires in the matrix can result in well-bonded material. To date aluminum, titanium, nickel, certain high temperature alloys, copper, and silver have been the most widely used. In moist, low temperature environments, galvanic couples between the reinforcement and the matrix can promote corrosion. At elevated temperatures reactions between the fiber and the matrix are difficult to avoid. Composites formed *in situ,* such as the oriented eutectics in which the fiber and the matrix are essentially in thermodynamic equilibrium at the time the total structure is formed by controlled solidification, offer a way of overcoming some of these problems.

Inorganic materials, such as glass, plaster, portland cement, carbon, and silicon, have been used as matrix materials with varying success. These materials remain elastic up to their point of failure and characteristically exhibit low failure strains under tensile loading but are strong under compression. Prestressed steel-reinforced concrete takes advantage of the latter characteristic.

The combined actions of various of the following factors in most cases have prevented the synthesis of high performance composites: (*1*) Chemical attack or dissolution of the reinforcement by the matrix. (*2*) Large thermal expansion mismatches. (*3*) The need for the modulus of the reinforcement to exceed that of the matrix. (*4*) Limited methods available for introducing and consolidating the matrix without introducing large, strength impairing, shrinkage effects.

In situ formation (and equilibration) of a ceramic matrix composite can be achieved in certain favorable cases using the solidification of eutectic compositions to achieve desired structures (see Ceramics). Another approach has been the melt infiltration of silicon into carbon fibers, which converts the carbon fibers into silicon carbide to yield a chemically stable system. Composites based on an inorganic matrix have received less attention than the other two matrix classes owing to the difficulties of fabrication and the relatively restricted range of promising systems identified to date. Reinforced concrete and fiberglass-reinforced plaster board represent the widest spread use of inorganic matrix composites.

Nomenclature

E	= Young's modulus
P	= property value
S	= strength
$\overline{S}$	= mean strength
V	= volume fraction
β	= efficiency factor determined by fiber geometry (fraction of ineffective fiber)
ζ	= a constant determined by fiber geometry
η	= a constant reflecting relative elastic behavior of fiber and matrix
ν_{12}	= major Poisson ratio
ν_{21}	= minor Poisson ratio
σ^*	= stress in matrix at failure strain

Subscripts

f	= fiber
m	= matrix
$1, 2, 3, \ldots$	= fraction i
11	= parallel to fiber
12 or 21	= ratio of fiber direction to transverse
22	= transverse to axis of fiber

BIBLIOGRAPHY

1. A. A. Griffith, *Philos. Trans. Roy. Soc. A.* **221,** 163 (1920).
2. A. Shindo, *Rep. of Gov't Res. Inst. Osaka* 317 (Dec. 1961).
3. Brit. Pat. 1,110,791, W. Johnson, L. N. Phillips, and W. Watt; S. Allen, G. A. Cooper, and R. N. Mayer, *Nature* **224,** 684 (1969).
4. A. E. Standage and R. Prescott, *Nature* **211,** 169 (1966).
5. H. E. LaBelle and A. I. Mlavsky, *Nature* **216,** 574 (1974).

General References

J. E. Ashton, J. C. Halpin, and P. H. Petit, *Primer on Composite Materials: Analysis,* Vol. III, Technomic Publishing Co., Inc., Stamford, Conn., 1969.
L. J. Broutman and R. H. Krock, eds., *Modern Composite Materials,* Addison-Wesley Publishing Company, Reading, Mass., 1967.
L. J. Broutman and R. H. Krock, eds., *Composite Materials,* Vols. 1–6, Academic Press, New York, 1974:

A. G. Metcalfe, ed., *Interfaces in Metal Matrix Composites,* Vol. 1; G. P. Sendeckyj, ed., *Mechanics of Composite Materials,* Vol. 2; B. R. Noton, ed., *Engineering Applications of Composites,* Vol. 3; K. G. Kreider, ed., *Metallic Matrix Composites,* Vol. 4; L. J. Broutman, ed., *Fracture and Fatigue,* Vol. 5; E. P. Plueddemann, ed., *Polymer Matrix Composites,* Vol. 6.
S. W. Tsai, J. C. Halpin, and N. J. Pagano, eds., *Composite Materials Workshop,* Vol. I, Technomic Publishing Co., Inc., Stamford, Conn., 1968.
J. E. Gordon, *The New Science of Strong Materials,* Penguin Books Inc., Baltimore, Md., 1968; see especially Chapt. 8, "Composite Materials."
J. R. Vinson and T. W. Chou, *Composite Materials and Their Use in Structures,* Halsted Press, Division of John Wiley & Sons, Inc., New York, 1975.
W. J. Renton, ed., *Hybrid and Select Metal Matrix Composites: A State of the Art Review,* American Institute of Aeronautics and Astronautics, New York, 1977.
W. B. Hillig, *New Materials and Composites, Science* **191**(4228), 773 (1976).

Journals and Proceedings of Conferences

Fiber-Strengthened Metallic Composites, Symposium Proceedings, ASTM STP No. 427, American Society for Testing and Materials, Philadelphia, Pa., 1967.
Metal Matrix Composites, Symposium Proceedings, ASTM STP No. 438, American Society for Testing and Materials, Philadelphia, Pa., 1968.
Composite Materials: Testing and Design, Symposium Proceedings, ASTM, Philadelphia: *First Conference,* ASTM STP No. 438, 1968; *Second Conference,* ASTM STP No. 497, 1972; *Third Conference,* ASTM No. 546, 1974; and *Fourth Conference,* ASTM STP No. 617, 1977.
Fracture Mechanics of Composites, Symposium Proceedings, ASTM STP No. 593, American Society for Testing and Materials, Philadelphia, Pa., 1975.
E. Scala, E. Anderson, I. Toth, and B. R. Noton, eds., *Proceedings of the 1975 International Conference on Composite Materials,* Vol. 1 and 2, The Metallurgical Society of The American Institute of Mining, Metallurgical and Petroleum Engineers (AIME), New York, 1976.
R. T. Schwartz and H. S. Schwartz, eds., *Fundamental Aspects of Fiber Reinforced Plastic Composites,* Conference Proceedings, John Wiley & Sons, Inc., New York, 1968.
F. W. Wendt, H. Liebowitz, and N. Perrone, *Mechanics of Composite Materials* (*Proceedings of the Fifth Symposium on Naval Structural Mechanics*), Pergamon Press Inc., Elmsford, N.Y., 1970.
D. Johnson, ed., *Composites,* a journal published by IPC Science and Technology Press Limited, Surrey, England; Vol. 1, 1968, and continuing quarterly.
S. W. Tsai, ed., *Journal of Composite Materials,* Technomic Publishing Co., Inc., Westport, Conn.; Vol. 1 (1967) and continuing quarterly.

W. B. HILLIG
General Electric Co.

COMPRESSORS.   See Fluid mechanics (transportation); High Pressure technology.

COMPUTERS

This is the computer age. In three decades the computer has become a pervasive influence in our lives, from energy control, to income tax return auditing, and to education. Yet, comparatively few people understand how computers work or how to make computers work. This has been the domain of the computer engineer and programmer who create computer systems that can be used by people unfamiliar with the details of computer technology. Thus, eg, an airline clerk can use a computer to make travel reservations, an accountant can enter financial data and receive a profit and loss analysis in return, and a scientist can provide empirical data and receive the coefficients of the curve-fitting polynomial. Nevertheless, some basic understanding of computer construction and operation provides insight into the use of computers so that they can be applied usefully and effectively to information processing problems.

The influence of computers is felt in almost every aspect of the chemist's or chemical engineer's job. In research, eg, computers control the operation of an x-ray diffractometer or chromatograph and analyze the resulting data; in oil field exploration they are used to analyze and interpret seismographs to aid in finding oil fields; in manufacturing they control chemical processes such as the synthesis of vitamins or the cracking of crude oil.

The principles on which computer operations are based have been known for over a century (1). In about 1833 Babbage proposed constructing a steam-driven Analytical Engine possessing many of the characteristics of the modern digital computer. In it, he planned to use punched cards to control the sequence of operation, an idea apparently inspired by the digital punched card control of Jacquard looms. The Analytical Engine was never built completely, primarily owing to the lack of technology. Hollerith's tabulating machine for the 1890 United States census represents the next significant application of digital computer principles. The census tabulating machines formed the basis for various accounting machines which drastically changed business operations.

The modern electronic digital computer is largely an outgrowth of military requirements but it evolved from these accounting machines. With government funding aid universities and other research centers developed the first electronic digital computers to calculate ballistic tables for artillery. These early successes called attention to the potential of computers in science and business, and an evolution started that is still continuing today.

It is not possible here to detail the numerous inventions and advances in computers since their modern beginning in the 1940s. In general, the advances have resulted in a rapid monotonic decrease in computing costs, accompanied by a simultaneous increase in performance. This rapid progress and some of the technology and programming advances that made it possible are summarized in Table 1. Without adjusting costs for inflation, the cost of computing has decreased by a factor of almost 73 in two decades; performance, measured in terms of computation time, has simultaneously increased by about the same factor. Thus the performance–cost ratio, a common measure for computer systems, has improved by over five thousand times! The net result is that computers have become economically feasible in thousands of applications, ranging from consumer products to sophisticated fundamental research.

Table 1.   Improvements in Digital Computers [a]

	1955	1960	1965	1975
user cost[b], \$	14.54	2.48	0.47	0.20
processing time[b], s	375	47	37	5
technology	vacuum tubes; magnetic cores; magnetic tapes	transistors; channels; faster cores; faster tapes	solid logic technology; large, fast disk files; new channels; larger, faster core memory; faster tapes	monolithic memory; monolithic logic; virtual storage; larger, faster disk files; new channels; advanced tapes
programming	stored program	overlapped input–output; batch processing	operating system; faster batch processing	virtual storage; advanced operating systems; multiprogramming; batch–on line processing

[a] Courtesy of IBM Corporation (2).
[b] A mix of about 1700 computer operations, including payroll, discount computation, file maintenance, table lookup, and report preparation; figures show costs of the period, not adjusted for inflation.

Computer Hardware

In the jargon of the computer industry, the term hardware refers to the physical equipment in a computer system—the electronic circuits, power supplies, mounting racks and chassis, and electromechanical equipment required to handle cards, magnetic tape, and the other media used with the system. To understand how hardware can perform mathematical and other information processing operations requires some understanding of information processing and data manipulation by electronic circuits (3–4). Such circuits are combined to form the hardware that constitutes a computer system.

Binary Numbers and Coding.   For problems involving quantities, the common numerical system based on powers of 10, the decimal number system, has evolved. For example, a quantity of one hundred and twenty-three objects is represented by the numeral 123. It can also be written as $1 \times 10^2 + 2 \times 10^1 + 3 \times 10^0$ to show that the digits in the numeral 123 are the coefficients of powers of 10, the base or radix of the decimal system. The quantity described by decimal 123_{10} can also be expressed as the coefficients of a series of terms of some other base raised to integer powers. If the base is 5, eg, the quantity $123_{10} = 4 \times 5^2 + 4 \times 5^1 + 3 \times 5^0 = 443_5$ (the quinary system).

Of particular interest in digital computers is the binary number system that uses the base 2. In this system the quantity $123_{10} = 1 \times 2^6 + 1 \times 2^5 + 1 \times 2^4 + 1 \times 2^3 + 0 \times 2^2 + 1 \times 2^1 + 1 \times 2^0 = 1111011_2$. In comparing this representation with that for base 5 and the base 10, note that the number of different symbols (numerals) needed in each system is exactly equal to the value of the base. In the decimal system, the ten symbols are 0, 1, . . ., 9; in the quinary system, the symbols are 0, 1, 2, 3, and 4; the binary system requires only two symbols, 0 and 1. This fact makes the binary system of special interest to computer designers because it is particularly easy to build electronic circuits to represent two states, such as on or off, or the presence or the absence of a voltage or current.

The binary number system can be used for the arithmetic operations addition, subtraction, multiplication, and division, and to indicate greater than, less than, and equal in much the same way as the decimal system. Figure 1 provides the arithmetic rules for binary arithmetic and an example of a multiple digit addition that includes carries.

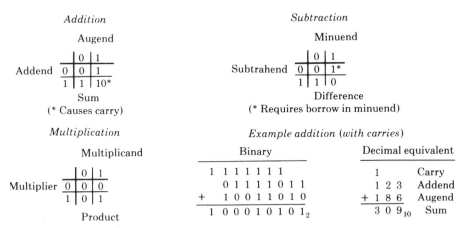

Figure 1. Binary arithmetic rules and example.

Computers must also process information in alphanumeric form, ie, information consisting of both numerals and alphabetic characters. This is done by assigning a binary code to each of the symbols to be represented in a manner similar to the childhood pastime of inventing secret messages by coding the alphabet into numerics. For example, the word DOG is coded as 041507 if A = 01, B = 02, C = 03, etc. A number of such codes are used in computers. One of the best known is the American Standard Code for Information Interchange (ASCII) (5). A sampling of ASCII codes for alphabetic and special symbols is shown in Figure 2. The computer has no way of determining if the datum on which it is operating is a code for an alphabetic symbol or the binary representation of a number; eg, the ASCII code for G = 1000111 is indistinguishable from the binary number $1000111_2 = 71_{10}$.

Logic. Logic is that branch of mathematics that deals with two states: true and false. A typical logic statement is "if A is true and B is true, then C is true." The logical operators involved in formal logic are *and, or, exclusive or* (*xor;* if A or B, but not both, is true, then C is true), and *inverse* or *not* (if A is true, then C is false). This type of mathematics involves only two states as inputs (variables) and outputs (results); namely, true and false. Similarly, the binary number system has only two states, 0 and 1. It is, therefore, possible to state all of the binary arithmetic operations in terms of formal logic statements. Further it can be shown that all of the operations required for information processing in a computer can be reduced to logical equations (6). The algebra associated with binary logical mathematics is called Boolean algebra after G. Boole.

The last step in hardware development is to show how these logical operators can be realized using electronic circuits. A simple logic circuit using two transistors is shown

Upper case letters

A	1	0	0	0	0	0	1
B	1	0	0	0	0	1	0
C	1	0	0	0	0	1	1
M	1	0	0	1	1	0	1
N	1	0	0	1	1	1	0
P	1	0	1	0	0	0	0

Lower case letters

a	1	1	0	0	0	0	1
b	1	1	0	0	0	1	0
c	1	1	0	0	0	1	1
m	1	1	0	1	1	0	1
n	1	1	0	1	1	1	0
p	1	1	1	0	0	0	0

Special symbols

*	0	1	0	1	0	1	0
(	0	1	0	1	0	0	0
/	0	1	0	1	1	1	1
#	0	1	0	0	0	1	1
=	0	1	0	1	1	0	1

Figure 2. Example alphanumeric coding using the American Standard Code for Information Interchange (ASCII) (7-bit version).

in Figure 3**a**. If the input is positive, transistor $T2$ is activated (turned on) and the output voltage, V_o, is approximately 0 V (logical state 0). If the input voltage, V_i, is approximately 0 V, $T2$ is off and the output voltage is $+V_{cc}$ volts (logical state 1). Thus the circuit provides the *not* logical function.

   In the circuit of Figure 3**b**, if one or both input voltages, V_1 and V_2, are approximately 0 V, the transistor $T2$ is not activated, and the output voltage, V_o, is $+V_{cc}$ volts. If both V_1 and V_2 are positive, $T2$ is turned on, and the output voltage, V_o, is approximately 0 V. If a positive voltage represents the logical state 1, and 0 V represents the logical state 0, this circuit corresponds to the logical equations:

$$V_o = \overline{V_1} + \overline{V_2} \text{ or } \overline{V_o} = V_1 \cdot V_2$$

where the overbar indicates the *not* operation, + represents *or*, and · represents *and.*

   Thus numbers and alphabetic information can be represented by binary numbers and binary numbers can be manipulated logically using electronic circuits. Digital computers are constructed of many such circuits, from about 2000 circuits in a small computer to 100,000 or more in a large computer. Present transistor technology makes it possible to construct (integrate) many of these circuits on a single chip of silicon approximately 3.8 × 3.8 mm square (7). Such a chip is called an integrated circuit. If the number of circuits on the chip is greater than 100, it is known as a large-scale integrated (LSI) circuit. Integrated and LSI circuits are used extensively in today's computers, and the rapidly increasing density of circuits per chip, mass consumption, and mass production techniques have resulted in a very low cost per circuit. Manufactured in quantity, an integrated circuit containing more than 2000 circuits cost less

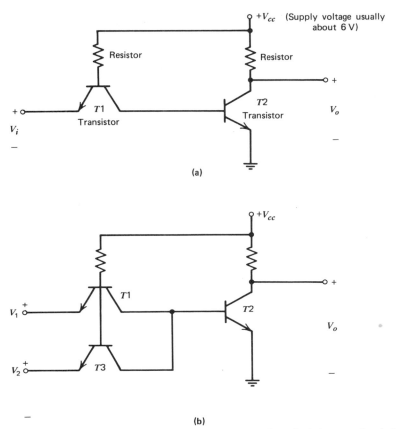

Figure 3. Electronic logic circuits; (**a**) a basic transistor–transistor-logic inverter (*not*) circuit; (**b**) a basic transistor–transistor-logic *and-invert* circuit. V_o, output voltage; V_i, input voltage.

than $10 in 1977. A storage chip containing 1000 bits (about 7000 transistors) costs about $2 (see Integrated circuits; Semiconductors).

Digital Computer Structure. To implement a digital computer, thousand of circuits must be combined into a structure for processing data (3–4). The typical structure for a small general-purpose computer shown in Figure 4 consists of a processing unit, a storage unit, and one or more peripheral units used primarily to transfer data to and from the processing unit and/or storage. Storage consists of electronic circuits that statically retain instructions specifying an operation to be performed and the operands on which the instruction is to operate. Both instructions and operands are coded in binary form and represented in storage as the presence or absence of a voltage in an electronic circuit corresponding to each bit in the coded representation.

Storage is organized into locations, each of which has a unique address. Conceptually, each storage location is like a mailbox corresponding to a street address in a city. Data or instructions can be stored in the location (mail delivered) and read from the location (mail dispatched). Each location stores one datum of information or one instruction. The number of locations available depends on the design of the system. A small general-purpose computer has about 2,000–130,000 storage locations. Large computers have as many as 6,000,000 or more storage locations.

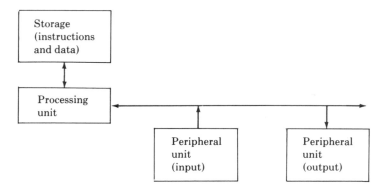

Figure 4. Basic digital computer organization.

The datum or instruction stored in each location is typically a fixed length "word," which may be subdivided into "bytes" (fractions of a word). In small general-purpose computers, the byte typically is 8 bits long and there are two bytes per word. In large computers, the byte is also 8 bits long but there are usually four or more bytes per word. Information in the computer is usually handled a byte or a word at a time, although multiple byte or word operations are provided in some machines.

The fact that the datum is of fixed length limits the numerical precision that can be expressed in a single word. For example, in a 16-bit word, the integers 0 to 65,535 ($= 2^{16} - 1$) can be directly expressed. Expression of more significant digits requires the use of more than one word. Quantities of larger magnitude can be expressed as a signed mantissa (m) and signed exponent (e) in scientific notation (ie, $m \times 2^e$). The mantissa can be stored as one word (or fraction of a word) and the exponent as another word (or fraction of a word). The mantissa and exponent are manipulated separately in mathematical operations. This form of computer arithmetic is known as floating-point arithmetic (8).

The instructions in storage indicate which operation is to be performed at a particular point in the processing. Typically, the instruction in a small general-purpose computer has the following format:

00000	mmm	aaaaaaaa
OP Code	Mod	Address/Operand

Here 00000 is the binary-coded representation of the operation (OP Code) such as *add, subtract,* or *transfer control.* The modifier (Mod) mmm indicates a modification of the operation (eg, *add to storage*), and the Address/Operand aaaaaaaa is either the storage address of the operand or the actual operand for the operation. A computer program, which consists of a sequence of such instructions, is kept in storage along with addresses and operands (data). During execution, each instruction is transferred from storage to the processing unit. The instruction is decoded and the processing unit retrieves the operand from storage (if necessary) and transfers it to appropriate registers in the processing unit. Particular circuits in the processing unit are then activated to perform the operation specified in the OP Code. After one instruction is executed, the next sequential instruction is retrieved from storage, and the cycle is repeated.

The processing unit typically can perform all the basic mathematical and relational operations, such as *add, subtract,* and *greater than.* It can also cause transfer of control to be effected; ie, it can cause the computer to execute a sequence of instructions starting at an arbitrary storage address, rather than continuing with the instruction stored in the next sequential location. This concept, called branching, is very important because it provides the capability to choose alternative actions conditionally based on the result of a computation or other operation. For example, it allows logical constructions such as "if the result is greater than zero, then do alternative 1, but if the result is less than or equal to zero, do alternative 2."

The peripheral units provide the system a means to input or output information or of storing large quantities (millions to hundreds of millions of words) of data or instructions (3,9). In the case of key-operated input devices, key depressions, such as on a typewriterlike keyboard, are coded by the input device into a binary-coded word or byte and are transferred to the processing unit or storage. Other media, such as punched cards or punched paper tape, are precoded by a preparation device (card or tape punch). In the case of output for use by a human user, the internal binary-coded data are converted into alphanumeric symbols by the output device for display on a cathode ray tube (CRT) or by hard-copy devices, such as a printer or typewriter.

Mass storage units, such as disk files or magnetic tape (qv), also serve as peripheral devices. Although several mass storage forms are available, magnetic media are most commonly used. Data are magnetically-recorded on the magnetic oxide coating of a metal or plastic disk or of a flexible plastic tape, similar to common audio magnetic tape. Data are recorded in binary format; conceptually, the presence of a magnetic spot indicates a binary 1 and the absence of a spot signifying a binary 0. Although access to data is slow (typically from 30 ms to many seconds) compared to data access from main storage (typically less than a microsecond), the peripheral mass storage devices provide storage at a small fraction of the cost per bit of main storage.

Computer Programming

The purpose of computer programming is to create a sequence of computer instructions that results in solution of a problem such as inverting a matrix or alphabetizing a list of names (10–12). The programming activity consists of a sequence of steps; define the problem; design a solution (an algorithm); convert the algorithm into a form for internal representation in storage by using a computer language (this step is known as coding and results in a program); test the program to ensure correct operation; document the program; and install it on the computer (ie, reads it into storage for subsequent use). Of these steps, all except coding are similar to those used in the solution of any engineering, scientific, or business data manipulation problem. Thus, even though coding typically involves less than 30% of the time required to prepare a program, it is the only step considered here.

Coding.   Coding involves converting the steps in the solution algorithm into a form acceptable to the computer (ie, binary coded instructions) by means of a computer language. Computer languages are similar to written human languages in that they consist of a set of symbols, a defined vocabulary of words (ie, allowable strings of symbols), syntax (rules for use), and semantics (the meaning of an expression). Computer languages can be categorized by level: low level languages resembling the internal representation of computer instructions, and higher level languages increasingly resembling written human languages.

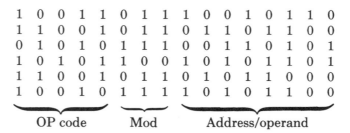

Figure 5.   Machine language program.

The machine-level language of the computer consists of the binary-coded instructions. A program is developed by writing a sequence of these instructions that realizes the desired solution. An example program is illustrated in Figure 5. This program can be recorded directly in the storage of the computer. When control is transferred to the first word in the program, the computer automatically executes the sequence of instructions represented by the binary words.

Clearly, this language is unnatural to a human. It is difficult to understand; memorizing 150 or 200 five-bit binary strings that represent the operation codes is difficult, and the probability of error in recording, copying, and manipulating the long strings of 1's and 0's is very high. For these reasons, machine language is rarely used directly by the programmer. It is, however, the form to which all programs must be converted before they can be executed by the computer.

The next higher language is called assembler language. In assembler language each of the OP Codes is represented by a mnemonic, an alphanumeric representation selected to aid human memory recall. For example, the machine representation of the operation for addition might be the binary string 01101. In assembler language, however, the programmer uses the mnemonic ADD to specify addition. Also, instead of having to specify the storage location of an operand as a binary number, the programmer selects an alphanumeric name for the operand. As a result, the programmer might write an instruction as:

ADD, 1, SUM

which causes the computer to add the contents of register 1 to the contents of a storage location which the programmer is calling SUM. A portion of an assembler language program is shown in Figure 6.

The assembler language still is unnatural as a human language, but it is easier to use since fewer symbols are required to express instructions, and mnemonics ease the memory recall problem. It also relieves the programmer of many bookkeeping chores, since mnemonics are used for functions and operand addresses.

In order for the assembler language program to be used by the computer, however, it must be translated into machine language form. Since each line of the assembler language program corresponds to a single machine instruction, this is largely a matter of substituting equivalent binary strings for mnemonic names and OP codes. This translation is done by another computer program (called an assembler) that processes the assembler language program (called the source program) to produce a series of machine instructions (called the object program) (13). In addition to this substitution operation, however, the assembler also assigns actual storage addresses for the pro-

Label	Op Code and modifier	Mnemonic address/operands and constants	Comments
	MVW	R6,R3	MOVE WORK AREA POINTER TO R3
	MVBI	8,R2	INIT. LOOP COUNTER TO 8
	MVBI	0,R5	INIT. RESULT TO 0
	MVBI	1,R4	INIT. R4 TO X**0 = 1
MAINLOOP	MVW	R4,(R3)	STORE POWER OF X IN WORK AREA
	MW	(R2,COEFF),R4	MULTIPLY BY CORRECT COEFFICIENT
	AW	R4,R5	ADD IN NEW TERM
	MVW	R0,R4	GET READY TO GENERATE NEXT POWER
	MW	(R3),R4	GET NEXT POWER OF X
	ABI	−2,R2	DECREMENT LOOP COUNTER
	BGE	R5,4,(R1)*	CONTINUE LOOP IF COUNTER = 0 OR MORE

Programmer assigned label

Combined Op Code and modifier

Mnemonic address/operands and constants

Programmer's comments (not part of instruction)

Figure 6. Assembly language program.

grammer's mnemonic addresses and performs tests to ensure, to the degree possible, that the language syntax has been followed.

Assembler languages are used widely on the small general-purpose computers often associated with industrial computer control applications (see Instrumentation and control). They are tedious to use but provide for detailed control over machine operation because of the one-to-one relationship with machine instructions. Since each computer design has a different instruction set, an assembler language program for one machine design cannot be transferred directly to a machine of a different design.

A third type of computer language is shown in Figure 7. This is a high-order procedural language using keywords, such as DO, IF, and WHILE, and conventional mathematical symbols to express logical or computational operations. As in assembler language, the programmer can assign arbitrary mnemonic names to operands and addresses. The language is much more natural than assembler language and, as a result, is more easily learned, easier to remember, and less susceptible to coding errors. Since the language is somewhat mathematical in appearance, the traditional training of engineers and scientists provides an additional advantage in learning it.

As in the case of assembler language, a high-order procedural language must be translated into a sequence of machine instructions by another computer program, known as a compiler or interpreter (14). In this case, however, each line or statement of high-level code typically yields many lines of machine instructions. For example, the language C = A + B might result in a machine instruction to fetch A from storage, another to fetch B, a third to add the two operands, and a fourth to store the result in the storage location reserved for C. A complex construction such as DO 10 UNTIL A.GT.B (ie, repeat the operations specified by the statements through the statement labeled 10 until the value of A is greater than that of B) might result in hundreds of words of machine code. High-level languages are relatively independent of machine design so that source programs can be transferred to machines of different design with minimum change.

High-order procedural languages are used extensively in commercial applications (specifically, the language COBOL) and, to a lesser extent, in scientific and control applications. Even higher order languages that closely resemble human written communication are available for specialized applications. For example, the question-and-answer dialogue used by airline reservation clerks at keyboard-display units is a high-order application language. Such languages are restricted to applications where their increased ease of use, reduced training requirements, and increased productivity for the user justify their high development cost.

GET LIST (XMIN)
XMAX = XMIN
LOOP: **GET LIST**(X)
IF X > XMAX
THEN XMAX = X
IF (X-XMIN) > 0 **THEN**
GO TO LABEL 4
ELSE GO TO LOOP

Figure 7.   High-level language program (words printed in bold face and mathematical symbols are defined in the language; all other names are programmer assigned).

The language translation programs used for converting source programs into machine form are usually provided by the computer vendor or by software development firms specializing in such tasks: they are rarely written by the user.

Operating Systems. Computer vendors usually also supply a control program, known as an operating system or executive for the system (11–12, 15). The operating system is a program that controls the execution of the user's programs and provides common facilities for use by a number of user programs. Its purpose is to assist in the efficient utilization of the machine and to relieve the user of tedious and routine details of machine operation.

Although operating systems differ in detail, most perform similar functions. Among these are scheduling program execution based on time-of-day, elapsed time, or the occurrence of an event external to the computer system. In the control of an industrial process or experimental apparatus, this is an important function since the process operation or experimental results are often dependent on timely response to the needs of the process or experiment.

It is also frequently necessary to schedule program execution at a particular time of day (eg, produce an activity log at 8 AM each morning) or after a specified time interval (eg, 30 s after actuating a valve, verify that it has changed position). Scheduling, either as a result of an external event or by time, is handled by the operating system and relieves the individual programmer of the detailed ordering of program execution.

The operating system also provides for easy and efficient use of computer resources. The resources consist of the various pieces of system hardware, the software, and the available time. For example, if a particular program requires an input device in order to complete its execution and that device is not available, the operating system suspends operation of the program and allows another program to utilize the processing unit until the input device becomes available. In this way, the processing unit is kept busy for the maximum amount of time, making more effective use of the system.

The operating system also provides services that are needed by a number of programs. For example, it keeps track of the time of day so that this information is available to any program. Similarly, it may provide access to a single program for use by all programmers which "dumps" the contents of storage onto a printing device for visual examination during testing. The availability of these common utility programs saves each individual programmer from having to write a program to perform the required operation.

Computer Applications

The number of computer applications has grown by orders of magnitude over the past several decades. Even considering only those associated with chemical technology, the list would easily exceed the entire length of this article. One valuable compilation of existing programs for chemical technology is given in ref. 16.

Although computers are often thought of as giant, super-fast calculating machines, a more accurate view is that they are information processors capable of manipulating numerical and non-numerical data. For the purposes of this article, applications are classified as information processing and industrial control.

Information Processing. Information processing, in its most general sense, involves the manipulation of data, both numeric and alphabetic, to provide for their organization according to a defined set of rules (eg, alphabetizing a set of names or ordering a list of quantities by magnitude), or to derive numerical results from a set of numeric data according to some mathematical equation (eg, calculating equilibrium conditions of a chemical mixture), or for empirically finding a solution by comparing many potential solutions to a set of rules or guidelines (eg, finding a solution to a transcendental mathematical equation or generating patterns until one satisfies the esthetic senses of an artist).

Not all problems are suited to solution by a digital computer. In general, problems that are well-suited to a digital computer involve one or more of the following:

(*1*) Relatively simple data manipulation but many repetitions: the expense of programming a computer rarely is justified in the case of a relatively simple data manipulation to be used on few occasions. For example, the occasional evaluation of the quadratic formula is done more economically using a hand calculator or pencil and paper. On the other hand, evaluating a tenth-order polynomial for 1000 values of the variable justifies the required programming effort.

(*2*) A complex calculation or data manipulation: it may be economical to use the computer to perform a complex task that would be extremely time-consuming by manual or other methods even if it is to be performed only once. Evaluation of the data from the ten-year United States population census is done best by computer because of the complexity of the analyses desired and the quantity of data. Similarly, the computer solution for 100 simultaneous difference equations is justified by the complexity and the number of operations that must be performed.

(*3*) A calculation or data manipulation that must be performed quickly: an outstanding attribute of the digital computer is the speed with which it can perform operations. In an airline reservation system, eg, the computer may search through millions of data records to determine the status of a flight or to find a particular passenger record. The speed of the computer makes it possible to conduct this search in a period short enough that the waiting customer does not become impatient. Similarly, a search of thousands of chromatograph patterns to identify a particular chemical compound can be performed quickly to satisfy the immediate needs of a researcher.

(*4*) A calculation or data manipulation involving vast amounts of data: large quantities of data become unmanageable by manual methods. For example, categorizing energy levels and time durations for the decay of a million atomic particles would be an almost impossible human task. Similarly, maintaining an up-to-date status of the inventory of 500 department stores or of the bank accounts of 50,000 people is, at best, difficult. The speed and accuracy of the computer makes such tasks practical.

Data Bases. Information processing applications increasingly involve manipulation and management of large quantities of information. This information is organized into a series of files which, taken collectively, are termed a data base (17). For example, a data base might include files of the production records, product production statistics, customer order status, and financial records.

The organization and control of such data bases currently is the subject of considerable research. Maryanski (18) provides an excellent review of this work and classifies data-base management systems into six categories ranging from a single data base on a single general purpose computer (least complexity) to a data base geographically distributed among a number of nonidentical computers utilizing nonidentical software (greatest complexity). Commercially-available systems generally fall into the two least complex categories (18). Thus we can expect continuing significant developments in data-base management systems (probably including specialized hardware) over the next five to ten years. Significant problems relating to such problems as optimum assignment of files and security remain to be solved before the more complex data base management systems become commercial realities (19) (see also Information retrieval).

Industrial Control Applications. Computers have been used since the late 1950s for controlling and gathering data from industrial processes (20). Applications range from the control of a single process unit, such as a reformer in an oil refinery, to integrated control of entire industrial complexes—a steel mill from ore carrier to finished steel products. Figure 8 shows a computer controlling an experimental material handler. Since processes involve quantities that are not inherently numerical (eg, flows, temperatures, pressures, etc), special peripheral equipment is needed to convert these non-numeric quantities to a form usable by the computer.

Figure 9 is a block diagram of a computer system suitable for controlling an industrial process. The storage unit, processor unit, and peripheral devices have the same function as those in Figure 4. The additional subsystems are those required to connect the computer to the external process so that the computer can have direct access to process parameters and can exercise direct control by actuating devices such as valves and switches. These four specialized subsystems are the analogue input subsystem, analogue output subsystem, digital input and interrupt subsystems, and digital output subsystem. Taken together, these subsystems are often called the process or sensor I/O (input–output) subsystems (9).

The analogue input subsystem provides the capability to measure voltages in the process or experiment under control (21). For example, the computer can measure

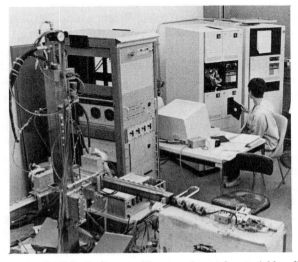

Figure 8.   IBM Series/1 controlling experimental material handler.

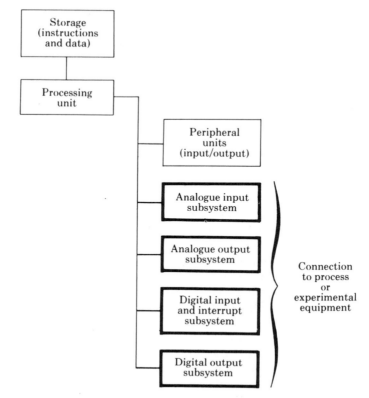

Figure 9. Basic industrial computer organization.

directly the output of a thermocouple or resistance temperature detector to determine temperatures; similarly, the computer can measure the output of a rotational or linear potentiometer to find a valve position or flue damper setting. The subsystem input is an analogue voltage, usually in the 0 to ±5 V range, and the output is a binary representation of the voltage value. Typically, the output is a 10 to 14 bit word.

The analogue output subsystem has the capability of providing a voltage between two limits, usually 0 to ±5 or ±10 V, to control process devices (22). It can be used, eg, to position a valve or to drive a strip chart recorder. The subsystem input is a binary number generated by the computer (usually 8–10 bits), and the output is an analogue voltage.

Process or experimental control also requires acquisition of discrete (ie, digital) binary data, such as whether a switch is on or off, or whether a valve is open or closed. The digital input and interrupt subsystem provides the capability of sensing these binary states. Fundamentally, the subsystem can determine whether an electrical switch is on or off. Other binary events are converted mechanically or by other means into the opening and closing of an electrical switch (23).

The interrupt portion of this subsystem appears electrically identical to the digital input portion (24). The difference is the manner in which the internal computer signals are handled. If an interrupt input is actuated, a signal is sent to the processor unit that causes the current program execution to be suspended. Another program, designed to respond to the interrupt, then is executed to service the interrupt; ie, the interrupt handling program takes the necessary action in response to the external event. When the interrupt has been serviced, the suspended program is reactivated and continues from the point of interruption. Servicing the interrupt may require only one or two milliseconds of processor unit time.

The use of interrupts allows quick response to external events in the process, even though these events may happen at unpredictable and (as far as the computer is concerned) unknown times. For example, suppose an open tank is being filled with a corrosive liquid. A level indicator near the top of the tank is actuated if there is danger that the tank will overflow. By using this level indicator as an interrupt input, the computer is notified of an impending overflow and can take corrective action.

There are required digital control actions in most processes. For example, a pump is to be turned on, a damper must be opened or closed, or an electrically-operated clutch is to be actuated. The digital output subsystem provides this capability (23). The subsystem provides a series of electrical switches (relays or transistors) that can be turned on and off by the computer system.

As an example of computer process control, consider a process that involves the flow of several liquid chemicals into a mixing tank, which is then heated according to a prescribed time–temperature profile. After a period of reaction time, the reaction mixture is transferred to storage tanks. Such a process could be controlled by a computer as follows: The computer first senses that the tank is empty by reading a digital input connected to a level indicator. It therefore opens valves controlling the flow of liquids into the tank. Based on analogue readings from flowmeters, the computer measures the quantity of each input flow and closes the appropriate valve when the correct amount of material has entered the tank. Concurrently, it monitors the level indicator input to ensure that the tank is not overfilled. After filling the tank, the computer actuates an agitator to mix the chemicals and turns on the heaters. The computer continually measures the temperature of the liquid in the tank and adjusts the heaters to obtain the prescribed time–temperature curve. Upon completion of the heating and reaction cycle, the computer opens valves to transfer the completed product to storage tanks, but only after it has verified that the storage tanks have enough unused volume to accept the product.

Process control computers increasingly are being interconnected with other computers to provide a form of distributed processing system (25). An example of a three-level hierarchical distributed system incorporating an industrial control computer is shown in Figure 10. The Level 1 computers provide the computer control at the loop level and must respond quickly to process events. (With the advent of microprocessor-controlled process equipment and instrumentation, one might consider these to be a Level 0 in the hierarchy providing even more detailed and specialized programmed function.)

The Level 2 computers perform functions that relate the units controlled by the Level 1 computers. These functions tend to have less stringent time response requirements and, often, greater computational requirements. For example, the Level 2 computers might provide production scheduling for sequential process steps (such as the chipper–digester–paper machine sequence in a paper mill) to ensure an orderly and uninterrupted flow of material between process units. This second level is sometimes called the supervisory level.

The third level can be considered a strategic or corporate management level

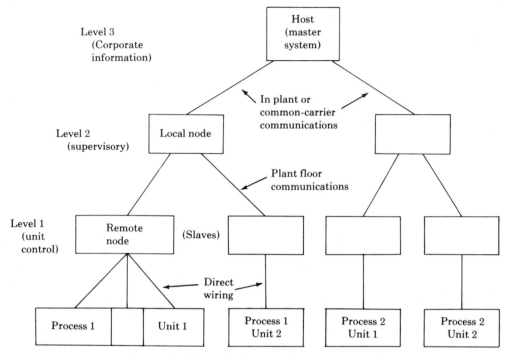

Level 3
   (Corporate
   information)

Host
(master
system)

In plant or
common-carrier
communications

Level 2
   (supervisory)

Local node

Plant floor
communications

Level 1
   (unit
control)

Remote
node

(Slaves)

Direct
wiring

Process 1   Unit 1

Process 1
Unit 2

Process 2
Unit 1

Process 2
Unit 2

Figure 10.   Hierarchical industrial computer network.

concerned with comprehensive, long-term control of a number of process units or an entire corporation. At this level, optimization of the entire business complex may be performed using summary data obtained from the lower-level systems and management information relating to business goals of the company. Optimization of a complex business is a difficult and time-consuming computation. However, the time constant of a business is long (weeks or months) so that the optimization calculation is required only at relatively long intervals.

The corporate management level generally is characterized as being a management information system (MIS) and involves many noncontrol functions (26). Typically, it includes the data base management system as one of its components. Depending on the design, the data base may be centralized at this level or may be distributed among the computers at lower levels.

Justification of Computer Use

Computers may represent a sizable investment and usually must be economically justified (27). The benefits of computer use can be tangible or intangible. Tangible benefits can be easily recognized and quantitatively estimated. For example, a tangible benefit might be increasing the throughput of a catalytic cracker in an oil refinery from 16,000 to 17,000 m^3/d (100,000–106,000 barrels per day). Intangible benefits are difficult or impractical to quantify. For example, increased process knowledge gained from better measurements by the computer typically yields improvements in operation that are difficult to attribute directly to computer control.

The source of benefits differs, depending on the particular process or experiment. The benefit may simply be increased production as in the above example, or it may be that a process that cannot be run manually—eg, one using a very expensive and potentially dangerous catalyst—becomes feasible. Table 2 lists some potential benefits that may be realized by using a computer.

Computers are used in industry and research to provide economic benefits. However, installing a computer is not an automatic guarantee of such benefits. In addition, the use of a computer entails some costs, such as the hardware, software development and maintenance, and operator training. For these reasons, an economic justification study should precede the decision to install a computer.

Recent Trends

The history of computers has been one of a rapid evolution in increased performance, increased functional capability, decreasing prices, and new technology. It is expected that this evolution will continue in the foreseeable future (19).

Microprocessors and Computer Networks. An important recent development has been introduction of the microprocessor or computer on a chip (28–29). An entire processor unit and some storage now can be fabricated on a single silicon chip approximately 3.8 × 3.8 mm. Typically, such a chip contains several thousand transistor circuits and, possibly, about 1000 8-bit bytes of read-only-storage (ie, storage in which information is fixed during manufacture). It sold in quantity for about $10 in 1978.

Table 2.   Example Potential Benefits of Industrial Computer Control

Benefit	Source of benefit
Tangible	
increased production	better or tighter control; fewer process disturbances; reduced downtime or rejected product; mathematical optimization techniques; advanced control techniques
more profitable product mix	increased yield of more valuable product; optimized product mix; quick response to requirements
improved quality control	reduced errors and human variance; less quality giveaway; less warranty cost
reduced maintenance	prevention of equipment overstress; prediction of impending failure; control and schedule preventive maintenance
Intangible	
increased process knowledge	better data with less variance; better model through computer techniques
greater safety	continuous, consistent, impersonal monitoring; immediate corrective computer action; assessment of interacting factors
smoother operation	fewer distrubances; immediate corrective action; predictive control techniques; a time-to-think environment

The chip represents the major electronic unit for an operable computer system. Such a microprocessor may have lower performance and function than a larger processor, but the device is nevertheless capable of satisfying many applications in industrial, research, and consumer products.

The advent of the microprocessor has spawned significant developments in instrumentation and control. It now is possible to incorporate a digital processor in an instrument to provide unit control (eg, controlling the valve sequence in a vapor-phase chromatograph) and to provide some data processing (eg, finding peak values, averages, and deviations) (30). As another example, microprocessors are built into some single-loop proportional-integral-derivative (PID) controllers used in the control of industrial processes (31). The primary implication of this trend is that some data processing can be performed in the instrument, thereby reducing the load on the central computer system.

There is increasing emphasis on digital communication methods between instruments and the central computer system. A related development is the current interest in computer networks and distributed processing (25,32). In distributed processing a given problem is solved by interconnected small processors, which may be located close together or distributed across several countries and interconnected by means of common carrier telephone facilities (33). Typical benefits claimed for distributed processing and networks are simplicity (since each node is dedicated to a single task), lower cost through the use of small, relatively simple computers at each node, availability of special computing resources (eg, a Fast Fourier Transform analyzer) within the network, and faster response time since some tasks are performed at the local node without communication delays.

During the current initial enthusiasm, distributed processing and networks sometimes appear to be the panacea for all computing problems. It is clear that they have advantages for problems that naturally lend themselves to functional partitioning for concurrent processing. It is equally clear that other problems are best solved using centralized systems (eg, systems that require a single data base). Although the trend toward distributed networks undoubtedly will continue, it also seems apparent that centralized systems will continue to be used. Thus distributed processing and networks provide alternatives for the system designer attempting to satisfy a computer application.

BIBLIOGRAPHY

"Computers" in *ECT* 1st ed., Suppl. 1, pp. 150–176, by Arthur Rose, Pennsylvania State University, T. J. Williams, Monsanto Chemical Co., and R. C. Johnson, Washington University; "Computers" in *ECT* 2nd ed., Vol. 6, pp. 25–34, by Leonard M. Naphtali, Polytechnic Institute of Brooklyn.

1. C. Eames and R. Eames, *A Computer Perspective,* Harvard University Press, Cambridge, Mass., 1973.
2. *The Computer Age,* IBM Corporation, Armonk, N.Y., 1976, form G-505-0029.
3. T. J. Harrison, *Handbook of Industrial Control Computers,* John Wiley & Sons, Inc., New York, 1972.
4. T. J. Harrison, *Minicomputers in Industrial Control—An Introduction,* Instrument Society of America, Pittsburgh, Pa., 1978.
5. *Code for Information Interchange, ANS X3.4-1968,* American National Standards Institute, New York, 1968.
6. H. T. Nagel, Jr., B. D. Carroll, and J. D. Irwin, *An Introduction to Computer Logic,* Prentice-Hall, Englewood Cliffs, N.J., 1975.

7. E. R. Hnatek, *A User's Handbook of Integrated Circuits,* John Wiley & Sons, Inc., New York, 1973.
8. F. J. Hill and G. R. Peterson, *Digital Systems: Hardware Organization and Design,* John Wiley & Sons, Inc., New York, 1973, Chapt. 13.
9. Ref. 4, Chapt. 3.
10. J. J. Donovan, *Systems Programming,* McGraw-Hill Book Co., New York, 1972.
11. A. J. Arthur in ref. 3, Chapt. 9.
12. Ref. 4, Chapt. 4.
13. Ref. 10, Chapters 2–5.
14. Ref. 10, Chapters 7–8.
15. A C. Shaw, *The Logical Design of Operating Systems,* Prentice Hall, Englewood Cliffs, N.J., 1974; ref. 10, Chapt. 9.
16. J. N. Peterson, C.-C. Chen, and L. B. Evans, *Chem. Eng.,* 69 (July 3, 1978); and earlier segments.
17. C. J. Date, *An Introduction to Data Base Systems,* Addison-Wesley, Reading, Mass., 1975.
18. F. J. Maryanski, *Computer (IEEE)* 28 (Feb. 1978).
19. T. A. Dolotta and co-workers, *Data Processing in 1980–1985,* John Wiley & Sons, Inc., New York, 1976.
20. C. L. Smith, *ACM Computing Surveys,* 211 (Sept. 1970).
21. Ref. 3, Chapters 3–4.
22. C. C. Liu in ref. 3, Chapt. 5.
23. J. P. Hammer in ref. 3, Chapt. 6.
24. Ref. 3, Sections 2.4, 9.3.2.1; ref. 4, pp. 117 ff., Chapt. 4.
25. A. van Dam and co-workers, *Computer (IEEE),* (Jan. 1978).
26. E. R. McLean and J. V. Soden, eds., *Strategic Planning for MIS,* John Wiley & Sons, Inc., New York, 1976.
27. L. S. Holmes, Jr., in ref. 3, Chapt. 11; T. M. Stout in ref. 4, Chapt. 5.
28. D. R. McGlynn, *Microprocessors—Technology, Architecture, and Applications,* John Wiley & Sons, Inc., New York, 1976.
29. B. Soucek, *Microprocessors and Microcomputers,* John Wiley & Sons, Inc., New York, 1976.
30. J. Tillotson and R. C. Chang, *Pittsburgh Conference on Analytical Chemistry and Applied Spectroscopy, Cleveland, Ohio, 1977,* paper 107.
31. G. F. Barnes, *Instrum. Technol.,* 47 (Dec. 1977).
32. R. P. Blanc and I. W. Cotton, *Computer Networking,* John Wiley & Sons, Inc., New York, 1976.
33. D. W. Davis and D. L. A. Barber, *Communication Networks for Computers,* John Wiley & Sons, Inc., New York, 1973.

THOMAS J. HARRISON
International Business Machines Corporation

CONCRETE. See Cement.

CONDIMENTS. See Flavors and spices.

CONGO COPAL. See Resins, natural.

CONSTANTAN. See Nickel and nickel alloys.

CONTACT LENSES

Contact lenses are optical devices that are placed over the cornea of the eye in such a manner that the lens remains on the eye's surface throughout blinking. The main purpose of contact lenses is correction of vision deficiencies; in this application they are called cosmetic lenses. Contact lenses can also be used medically for the treatment of certain corneal diseases; in such cases they are called therapeutic, or bandage lenses (1).

Glass contact lenses were first used at the end of the 18th century, but extensive development of the contact lens industry occurred only when a plastic that was lighter than glass and had excellent optical properties, poly(methyl methacrylate), became easily available after World War II. The industry in the United States now involves over $100 million annually. It is estimated that over 10 million people have purchased poly(methyl methacrylate) contact lenses. However, more than half of these purchasers do not continue to wear the lenses because of discomfort and adjustment difficulties. Some of the newer oxygen-permeable lenses and hydrogel soft lenses offer advantages over the hard poly(methyl methacrylate) lenses, particularly with regard to comfort, a shorter break-in period which accustoms the cornea to the presence of the lens, and an improved physiological environment for the cornea. Currently the contact lens industry is undergoing rapid change with new lenses made of a variety of materials appearing with increasing frequency (see Methacrylic acid and derivatives).

Contact lenses made of any material other than poly(methyl methacrylate) are regulated in the United States by the FDA. Before approving a new lens and its auxiliary systems (such as cleaning, disinfecting, and wetting solutions) for sale to the public, the FDA must be convinced of the safety and efficacy of the complete device. The FDA has stringent standards, and obtaining approval for a new contact lens requires several years of preclinical and clinical testing (see Regulatory agencies).

Contact lenses can generally be classified as hard lenses, flexible lenses, and soft hydrogel lenses. In all three types there are lenses that are oxygen permeable and in hard and flexible types there are lenses that are hydrophobic. An essential difference among the three types is the manner in which they fit the eye. Hard lenses and hydrophobic flexible ones require a relatively thick tear film between their posterior surface and the cornea of the eye which is a matter of considerable optical and physiological importance. Soft hydrogel lenses adhere closely to the cornea with a tear film of only capillary thickness between the lens and the cornea surface.

With any kind of contact lens the cornea's surface must be wet and oxygenated at all times to remain transparent and healthy. The cornea is an avascular tissue with an active aerobic metabolism. Oxygenation of the cornea surface is normally accomplished through the tear film that supplies the cornea surface epithelium with oxygen. When the eyelids are closed, oxygen is supplied to the cornea surface by the blood capillaries of the palpebral conjunctiva (the posterior part of the eyelids). When the cornea surface is deprived of oxygen, epithelium glycogen decreases and lactic acid production increases as a result of the anaerobic glycolysis (2). The cornea then swells and its surface becomes hazy. Thus, with any type of contact lens, disruption of oxygen supply to the cornea surface must be minimized, either by oxygen-rich tear exchange under the lens or by oxygen permeation through the lens, or both. Therefore, the main challenge in the art of fitting contact lenses is to provide oxygen to the cornea.

Nomenclature

Trade literature refers to contact lenses not only by the chemical name of the polymer and its acronym, such as PMMA for poly(methyl methacrylate) and HEMA or PHEMA for poly(hydroxyethyl methacrylate) hydrogel contact lenses; or by the trade name or registered name, such as Soflens, HydroCurve, PHP; or the manufacturer's or distributor's name, such as Bausch & Lomb, American Optical, but often by the name of the lens inventor or entrepreneur who developed it, such as Wichterle lens, deCarle lens. In the United States each kind of contact lens has an additional nonproprietary name assigned by the United States Adopted Name Council (USAN Council) (3). Except for the first FDA-approved hydrogel lens (Soflens of Bausch & Lomb) which was named polymacon, the USAN Council seems to use the stem -*filcon* in the names of all hydrogels for contact lens manufacture. Examples of USAN Council names are vifilcon A for Softcon lens (Warner-Lambert), and hefilcon A for the HydroCurve I lens (Continuous Curve) and Naturvue lens (Milton Roy), both of which are made of PHP polymer (trademark of Automated Optics, Inc.). A partial list of contact lens materials, trade names, USAN Council and other names, and manufacturers is given in Table 1.

Hard Contact Lenses

Poly(methyl methacrylate) Lenses. The excellent optical characteristics of poly-(methyl methacrylate), its resistance to discoloration, its light weight, and its good machining and polishing qualities, as well as the low incidence of irritation and allergic response in patients (5) have been important factors in the rapid expansion of the contact lens industry over the last three to four decades. Most poly(methyl methacrylate) contact lenses are actually cross-linked with a dimethacrylate comonomer and shaped by machining, but some lenses are produced from thermoplastic molding compounds.

Poly(methyl methacrylate) plastic lenses are practically impermeable to oxygen; the polymer's permeability is at least one to two orders of magnitude lower than those of the other hard contact lens material shown in Table 2 (see also Barrier polymers). For this reason, all oxygen that reaches the cornea under a poly(methyl methacrylate) lens comes from oxygen-rich tear exchange. These lenses must be specifically fitted to each patient so that normal blinking will permit continuous and adequate exchange of tears to take place under the lens.

Hard contact lenses made of thermosetting allyl resins, prepared by polymerization of diethylene glycol bis(allyl carbonate) and diallyl phthalate, were reported to have better thermal resistance and surface hardness than standard poly(methyl methacrylate) lenses (6) (see Allyl monomers and polymers).

Polymethylpentene Lenses. Other hard materials with improved oxygen permeability have been proposed for contact lenses. Poly(4-methyl-1-pentene), TPX, for example, is said to be about 650 times as permeable to oxygen as is poly(methyl methacrylate) (7); however, this may be an overstatement (see Table 2). The methylpentene polymer is crystal clear, the lightest of thermoplastic materials (sp gr 0.83), and resistant to heat (mp 240°C) and to many chemicals. However, the major drawback in its use for contact lenses is its rather poor wettability by tears. Thus, the contact angle of water on poly(methylpentene), $\theta_A = 103°$, is nearly twice as large as the water contact angle on poly(methyl methacrylate), $\theta_A = 65°$.

Table 1. Contact Lens Materials, Names, and Manufacturers (Excluding All Poly(methyl methacrylate) Hard Lenses and Some Others)

Material[a]	CAS Registry No.	Trademark	USAN Council and other names	Manufacturer
poly(hydroxyethyl methacrylate-co-ethylene dimethacrylate) (or, basically,	[25053-81-0]	Hydron		National Patent Development Corp. (U.S.)
poly(hydroxyethyl methacrylate))	[25249-16-5]	Geltakt	SpofaLens Ergon	Spofa, Ergon Co. (Czechoslovakia)
		Soflens	polymacon	Bausch & Lomb (U.S.)
		Weicon-38		Titmus-Eurocon (F.R. Germany)
		Hydrolens	Soft-Flex	Hydro Optics, Inc. (U.S.)
poly(hydroxyethyl methacrylate-co-2-ethoxyethyl methacrylate	[29403-23-4]	DuraSoft	Phemecol Phemfilcon	Wesley-Jessen Inc. (U.S.)
poly(vinylpyrrolidinone-g-hydroxyethyl methacrylate)	[29612-57-5]	Softcon	vifilcon A, Natura- lens,	American Optical Corp., Frigitronics (U.S.)
or (poly(hydroxyethyl methacrylate-co-methacrylic acid-co-vinylpyrrolidinone-co-ethylenedimethacrylate))	[35528-20-2]		Bionite, Isen, Griffin	
poly(hydroxyethyl methacrylate-co-vinylpyrrolidinone-co-ethylenedimethacrylate)	[36425-29-3]	HydroCurve I	hefilcon A	Soft Lenses, Inc. (U.S.)
		Naturvue		Milton Roy Soft Contact Lens, Inc. (U.S.)
			PHP	Automated Optics, Inc. (U.S.)
		Brucker Lens		Continuous Curve Contact Lens, Inc. (U.S.)
poly(hydroxyethyl methacrylate-co-methacrylic acid-co-)vinylpyrrolidinone)	[25068-64-8]	PermaLens	DeCarle	Global Vision Ltd. (U.K.)
		Permalens	perfilcon A	Cooper Laboratories, Inc. (U.S.)
poly(methyl methacrylate-co-vinylpyrrolidinone)-co-allyl methacrylate)	[25655-01-0]	Sauflon, Sauflex	lidofilcon A	Contact Lenses Manufacturing Ltd. (U.K.), Sauflon International (U.S.)
poly(hydroxyethyl methacrylate-co-methyl methacrylate-co-vinylpyrrolidinone-co-divinylbenzene)	[53626-53-2]	Aquaflex Aosoft	tetrafilcon A	UCO Optics, Inc. (U.S.) American Optical Corp. (U.S.)
poly(hydroxyethyl methacrylate-co-vinylpyrrolidinone-co-ethylenebis(oxyethylene)dimethacrylate)	[58503-81-4]	Hydralens	droxifilcon A	Ophthalmos, Inc. (U.S.)
poly(hydroxyethyl methacrylate-co-pentyl methacrylate-co-vinyl acetate-co-3-hydroxy-2-naphthyl methacrylate)	[68782-60-5]	N & N Lens	mafilcon A	N & N Menicon Inc. (U.S.)
poly(hydroxyethyl methacrylate-co-methyl methacrylate-co-ethylenebis(oxyethylene)dimethacrylate)	[54341-00-3]	Gelflex	dimefilcon A	Calcon Labs. (U.S.)
poly(hydroxyethyl methacrylate-co-methacrylic acid-co-	[33410-59-2]	Tresoft	ocufilcon A	Urocon International (U.S.)

722

Table 1 (*continued*)

Material[a]	CAS Registry No.	Trademark	USAN Council and other names	Manufacturer
ethylene dimethacrylate)		Ocu-Flex	ocufilcon B	Ocu-Ease Optical Products, Inc. (U.S.)
poly(hydroxyethyl methacrylate-*co*-*N*-(1,1-dimethyl-3-oxobu-tyl)acrylamide-*co*-1,3-pro-panediol trimethacrylate)	[68758-54-3]	HydroCurve II	bufilcon A	Burton, Parsons & Co., Inc. (U.S.) Soft Lenses, Inc. (U.S.)
poly(2,3-dihydroxypropyl methacrylate-*co*-methyl methacrylate)	[50450-03-8]	CS-Mem-branes	crofilcon A, CSI, CS-151	Corneal Sciences, Inc. (U.S.), Syntex Ophthalmics, Inc.
silicone rubber	[9016-00-6]	Silcon		Dow Corning (U.S.)
			Agfa Silicone	Danker & Wohlk, Inc. (U.S.)
cellulose acetate butyrate	[9004-36-8]	RX-56	CAB	Rynco Scientific (U.S.)
		Meso-lens	cobufocon A	Danker & Wohlk, Inc. (U.S.)
poly(methyl methacrylate)-*co*-3-[3,3,5,5,5-pentamethyl-1,1-bis[(pentamethyldisiloxanyl-oxy]trisiloxanyl]propyl methacrylate-*co*-methacrylic acid-*co*-tetraethylene glycol dimethacrylate)	[42994-11-6]	Polycon	Silafocon A	Guaranteed Contact Lens Co., Syntex, Ophthalmics, Inc. (U.S.)

[a] When the cross-linking agent is not given, it is ethylene dimethacrylate [25721-76-0] or a similar compound. Note that *co* = copolymer and *g* = graft polymer (4).

Hard Lens Wettability. In order for a contact lens to be comfortable, one of the most important requirements is that its surface must be uniformly wet by the tear film at all times. Good water wettability is also required for good optical performance of the lens and for its ability to remain clean of mucoid deposits in the eyes. To impart the desired surface characteristics to the normally relatively hydrophobic poly(methyl methacrylate) and other poorly wettable lenses, various specially prepared wetting solutions are commercially available. These solutions are reported to make contact lenses more comfortable because a hydrophilic lens is held better than a hydrophobic lens by the capillary action of tear film on the cornea (17). Temporarily soaking a lens in a wetting solution is reported to improve its wetting properties. For maximum comfort, however, this treatment must be repeated often. To increase the permanence of improved wettability, certain hydrophilic coatings have been developed (18–19). Chemical reactions on the surface layers of the lenses have also been used to render them more hydrophilic (20–22). However, none of these treatments appear to be practical since most hard lenses are still manufactured of essentially unmodified, pure poly(methyl methacrylate) (except for the possible presence of a small amount of comonomers).

Improved hard lens wettability has also been attempted by means of copolymerization of hydrophobic and hydrophilic monomers. Butyl or ethyl methacrylate copolymers with hydrophilic hydroxyethyl methacrylate and relatively large amounts of the cross-linking agent triethylene glycol dimethacrylate gave hard hydrophilic

Table 2.   Oxygen Permeability Coefficients (P_g) of Some Hard Contact Lens Materials at Room Temperature

Material[a]	CAS Registry No.	$P_g \times 10^4$, $\mu L(STP) \cdot cm/$ $(cm^2 \cdot h \cdot kPa)$[b]	References
poly(methyl methacrylate)	[9011-14-7]	0.27	7–8
poly(4-methyl-1-pentene)	[25068-26-2]	176	7
poly(4-methyl-1-pentene)		64.8	8
poly(4-methyl-1-pentene)		72.0	9
cellulose acetate butyrate	[9004-36-8]	15.1	8
cellulose acetate butyrate		10.4	10
cellulose acetate butyrate		15.4	11
cellulose acetate butyrate		14.0	12
cellulose acetate butyrate (RX-56)		31.5	13
poly(pentamethyldisiloxanylmethyl methacrylate-co-methyl methacrylate)	[52848-07-4]	7.95	14
poly(1,1,9-trihydrofluoro methacrylate-co-methyl methacrylate)	[68975-57-5]	1.58	15
poly(perfluoroalkyl alkyl methacrylate-co-methyl methacrylate)		ca 7.5–27	16

[a] Note that co = copolymer.
[b] To convert $\mu L(STP) \cdot cm/(cm^2 \cdot h \cdot kPa)$ to $\mu L(STP) \cdot cm/(cm^2 \cdot h \cdot mm\ Hg)$, divide by 7.5.

polymers that could be machined into contact lenses (23). Hard contact lenses with wettable surfaces were also prepared from copolymers of hydroxyethyl methacrylate (5–15 wt %) with methyl methacrylate (85–95 wt %) (24–25).

Fluorine-Containing Hard Lenses.   A series of fluorocarbon polymers and copolymers are said to be useful for the manufacture of contact lenses owing to their reportedly good wettability, and primarily owing to their low refractive index (n_D 1.37–1.39) which approximates the refractive index of tears and the cornea (26). Improved oxygen permeability is claimed as well for contact lenses made of copolymers of fluoroalkyl methacrylates with methyl methacrylate (15–16). Some preliminary results of the clinical trials of corneal lenses, probably made from one of these fluorine-containing copolymers, and coded CLP-2A by DuPont, have been reported (27) (see Fluorine compounds, organic).

Silicone-Containing Hard Lenses.   Another type of hard contact lens, about 15 times more oxygen permeable than poly(methyl methacrylate), is made from copolymers of a siloxane derivative, eg, (pentamethyldisiloxanyl)methyl methacrylate (35 parts) with methyl methacrylate (65 parts) (28–29). Polycon, made of a silicone derivative copolymerized with methyl methacrylate, is a contact lens currently under clinical testing by Guaranteed Contact Lens of Arizona, Inc., a subsidiary of Syntex (see Silicon compounds).

Cellulose Acetate Butyrate Lenses.   Cellulose acetate butyrate (CAB) has received wide attention in the last few years as an oxygen-permeable hard contact lens material (13,30). Cellulose acetate butyrate is not a single plastic material, but a family of plastics in which the ratio of acetyl and butyryl content may be varied considerably to produce a flexibility in the cellulose derivative ranging from hard to soft. One of the most commonly used cellulose acetate butyrate resins has about 13 wt % acetyl and 37 wt % butyryl, and 1–2 wt % unsubstituted hydroxyls; this is probably the material used most often to make CAB contact lenses. With regard to oxygen permeability, CAB offers substantial advantage over poly(methyl methacrylate) (Table 2). On the negative side, CAB contact lenses have a higher tendency than poly(methyl methacrylate) lenses to warp or deform, scratch and chip, and discolor. Dry CAB is about as wettable as poly(methyl methacrylate); however, after both polymers have been soaked in tears, CAB becomes somewhat more water-wettable than poly(methyl methacrylate).

Flexible Contact Lenses

Silicone Lenses. From the standpoint of corneal epithelial respiration, silicone rubber is an attractive contact lens material (31–33). It is highly permeable to oxygen owing to the flexibility of the Si–O–Si bonds which impart segmental mobility to the polysiloxane chains. A silicone rubber contact lens of usual thickness (up to about 0.4 mm) can easily maintain the normal respiratory process on the surface of the cornea without the tear exchange under the lens which is required with lenses of some other materials (34–35). Unfortunately, silicone rubber is strongly hydrophobic, and without a hydrophilic coating, contact lenses made of this material, or any hydrophobic material, are not only poor optical devices but also very uncomfortable.

Silicone lenses absorb lipophilic substances such as cholesterol and its esters which exist in the tear film (36–38). Not only can these absorbed substances alter the physical properties of the lens and render it uncomfortable, but it is also possible that a deleterious lipophilic substance such as an eye drug or cosmetic may accumulate in the lens and then be slowly released in the eye with potential harmful effects. This property of silicone lenses can be put to good use to dispense lipophilic drugs in a sustained manner onto the eye. Although the possible absorption of lipophilic substances by the silicone lenses from the environment must be kept in mind by fitters and wearers of these lenses, it certainly need not limit their use (see Pharmaceuticals, sustained-release).

Most of the oftquoted data on diffusion of gases through silicone rubber were obtained with clean, uncoated silicone rubber membranes. However, any surface modification that makes silicone rubber hydrophilic may affect its permeability characteristics as well. In the case of a wetted contact lens separated from the corneal surface by a finite tear layer, oxygen must dissolve in and diffuse through the tear film in front of the lens, and then through the lens itself into and through the tear film separating the lens from the cornea before it finally reaches the corneal surface. It has been reported that the transfer of water-dissolved oxygen through a polymer membrane can be considerably lower than the corresponding gas transfer (39) (Table 3). This difference is caused by boundary-layer resistance which is a more significant factor in highly permeable membranes such as silicone rubber than in less permeable materials.

There are widely divergent values reported for the permeability coefficients of penetrants through the same material (see Tables 2, 3, and 4). In the case of silicone rubber this divergence can be caused by the use of silicones with different amounts of fillers. In some cases it can simply be an experimental error.

The many attempts to improve the wettability of silicone contact lenses have been plagued with difficulties. However, it seems likely that a good wetting procedure involving an effective wetting solution or a lasting surface treatment will eventually be

Table 3. Oxygen Permeability Coefficient (P_g) of Hydrophobic Flexible Contact Lens Materials and Some Hydrophilic Silicones at Room Temperature

Material[e]	CAS Registry No.	$P_g \times 10^4$, μL(STP)·cm/(cm²·h·kPa)[a] G/G[b]	W/W[c]	References
poly(dimethylsiloxane)	[9016-00-6]	1750	308	40
poly(dimethylsiloxane)		1610		39
poly(dimethylsiloxane)		1350		41
poly(dimethylsiloxane)-RTV[d]		2330		42–43
silicone (RTV)[d]-g-vinyl-pyrrolidinone, dry		1450		43
silicone (RTV)[d]-g-vinyl-pyrrolidinone, hydrate		810		43
silicone–polycarbonate block copolymer		525		44
silicone–polycarbonate block copolymer (MEM-213)		428		41
polyethylene (sp gr 0.914)	[9002-88-4]	7.80		45
polyethylene (sp gr 0.964)		1.05		45

[a] To convert μL(STP)·cm/(cm²·h·kPa) to μL(STP)·cm/(cm²·h·mm Hg), divide by 7.5.
[b] G/G = oxygen gas on both sides of the membrane.
[c] W/W = oxygen dissolved in water on both sides of the membrane.
[d] RTV = room-temperature-vulcanizing.
[e] g = graft polymer.

developed (46–47). In any event, clinical testing of silicone lenses is already underway, and there is evidence of progress in improving the wettability and patient tolerance of these lenses (9,48–49).

Chemical modifications rendering the surface layers of silicone rubber contact lenses hydrophilic have been patented. One procedure uses ionized gases under glow discharge, or unsaturated hydrophilic compounds such as acrylic or crotonic acid followed by glow discharge (50).

Graft copolymers of poly(vinylpyrrolidinone) onto poly(dimethylsiloxane) have been prepared by the mutual irradiation technique utilizing high energy electrons generated by a Van de Graaff machine (51). The graft copolymers thus obtained are hydrophilic in proportion to the amount of poly(vinylpyrrolidinone) grafted onto silicone. In a somewhat similar procedure, a poly(diethylsiloxane) contact lens was exposed to irradiation from a Van de Graaff generator; the lens was then exposed to glycerol methacrylate, vinylpyrrolidinone, or hydroxyethyl methacrylate in solution (ie, in poly(ethylene glycol)) and irradiated again to give a contact lens with a water-wettable surface and good oxygen permeability (52). Another similar technique of grafting vinylpyrrolidinone onto silicone rubber to make hydrophilic contact lenses has been patented (42). This flexible and hydrophilic contact lens is made by irradiating an already cured room-temperature-vulcanizing (RTV) silicone rubber in the presence of oxygen. The irradiated rubber is then swollen with vinylpyrrolidinone monomer and heated to accomplish the chemical grafting of the hydrophilic vinylpyrrolidinone onto the silicone rubber. The swelling in water yields a ratio of absorbed water to grafted poly(vinylpyrrolidinone) that varies from 0.8–1.5. These hydrophilic

Table 4. Oxygen Permeability Coefficients (P_g) of Some Hydrogel Contact Lens Materials[a]

Hydrogel[b]	CAS Registry No.	H_2O, %	Temperature, °C	$P_g \times 10^4$, $\mu L(STP) \cdot cm/ (cm^2 \cdot h \cdot kPa)$[c]	References
poly(hydroxyethyl methacrylate)	[25053-81-0]	39	21	24.0	53
(polymacon)			23	6.83	54
			25	13.0	40
			35	29.7	53
			22	25.3	55
			37	27.0	55
(Phemecol)		30	25	8.10	56
poly(hydroxyethyl methacrylate-co-vinylpyrrolidinone) (hefilcon A)	[36425-29-3]	45	23	10.1	54
			22	25.7	55
			37	33.5	55
poly(vinylpyrrolidinone-g-hydroxyethyl methacrylate) (vifilcon A)	[35528-20-2]	55	23	23.0	54
			35	54.0	53
poly(2,3-dihydroxypropyl methacrylate-co-methyl methacrylate) (crofilcon A)	[50450-03-8]	40	22	19.3	55
			37	22.6	55
poly(hydroxyethyl methacrylate-co-N-(1,1-dimethyl-3-oxobutyl)-acrylamide-co-1,3-propanediol trimethacrylate) (bufilcon A)	[56030-52-5]	34	22	14.0	55
			37	13.8	55
poly(hydroxyethyl methacrylate-co-methacrylic acid-co-vinylpyrrolidinone) (Permalens)	[37017-46-2]	66	22	39.2	55

[a] For an extensive discussion of oxygen permeability of hydrogel contact lenses see reference 57.
[b] Note that co = copolymer, g = graft polymer (4).
[c] To convert $\mu L(STP) \cdot cm/(cm^2 \cdot h \cdot kPa)$ to $\mu L(STP) \cdot cm/(cm^2 \cdot h \cdot mm\ Hg)$, divide by 7.5.

silicone–vinylpyrrolidinone contact lenses have two of the most desirable properties for contact lenses, ie, improved wettability and high oxygen permeability (Table 3); however, it seems that so far none have achieved the high degree of comfort of the hydrogel lenses. The graft polymerization of poly(vinylpyrrolidinone) onto silicone rubber and its hydration result in a reduction of oxygen permeability to approximately one third of the level in ungrafted silicone rubber, but the hydrated silicone is still substantially oxygen-permeable.

Another reported type of hydrophilic and oxygen-permeable silicone contact lens has been obtained from interpenetrating network polymers obtained by cross-linking an RTV silicone mixed with hydrophilic monomers such as hydroxyethyl methacrylate, acrylamides, or vinylpyrrolidinone, with the addition of some cross-linking agents (58).

A soft, flexible, low-water-uptake contact lens material was obtained by esterification with alkyl alcohols, eg, 1-propanol, of a polymer of acrylic acid cross-linked with 1,4-butanediol dimethacrylate (59).

Fluorine-Containing Flexible Lenses. Fluorine-containing polymers with a refractive index approximating that of human tears were used to make soft, wettable contact lenses. The preferred polymers of the series described in a patent (60) are tetrafluoroethylene/perfluoro(methyl vinyl ether) [26425-79-6] and tetrafluoroethylene/perfluoro(2(2-fluorosulfonylethoxy)propyl vinyl ether) [26654-97-7].

Miscellaneous Flexible Lenses.   Also included among flexible hydrophobic contact lenses are potentially useful lenses made of 60–40% silicone–polycarbonate block copolymer that have good oxygen permeability (44). Molded lenses from a polyethylene [9002-88-4] with a low crystallinity degree which are, therefore, optically clear, have also been patented (61); in order to impart sufficient wettability, the lens surface was oxidized.

Hydrogel Contact Lenses

Some types of hydrogels such as gelatin and agar have been known for a long time, and in the 19th century, Herschel (62) had already proposed the use of jelly materials on the cornea for the correction of imperfect refractive powers of the eye. Wichterle and Lim first proposed the use of synthetic hydrogels for contact lenses in 1960 (63). In fact, these investigators had already applied for patents in the United States on the same concept around 1956; they were granted U.S. patents on cross-linked hydrophilic polymers and articles made from them, such as contact lenses, in 1961 (64) and 1965 (65).

The original work of Wichterle and Lim with synthetic hydrogels employed principally 2-hydroxyethyl methacrylate and some of the homologous esters of the glycol monomethacrylate series such as diethylene glycol monomethacrylate and tetraethylene glycol monomethacrylate. Slightly cross-linked (with a dimethacrylate of a glycol) copolymers of the higher glycol monomethacrylates and 2-hydroxyethyl methacrylate yielded transparent hydrogels that swell in water to a higher hydration than the hydrogels of 2-hydroxyethyl methacrylate. Slightly cross-linked 2-hydroxyethyl methacrylate transparent hydrogels swell in water to a maximum of ca 40% on a wet basis. Higher hydration hydrogels can be made by copolymerization of the 2-hydroxyethyl methacrylate with one or more hydrophilic comonomers. A review by Wichterle on the chemistry of hydrogels, with emphasis on poly(2-hydroxyethyl methacrylate) hydrogel, appeared in 1971 (66).

While Wichterle and Lim were pioneering the use of glycol methacrylate hydrogels as biomaterials and contact lenses, another group in the United States was developing the acrylamide hydrogels (67–68), which are chemically and physically related to the glycol methacrylate type of hydrogels. The acrylamide hydrogels were not originally intended for contact lens application, but have since been used for this purpose (see Acrylamide polymers).

Since the introduction of the glycol methacrylate hydrogels for contact lenses, there has been a continuous effort to develop other hydrogel contact lenses with improved properties. Some contain 2-hydroxyethyl methacrylate as the main component; in others it is only one of a number of major ingredients. A third type of hydrogel contact lens already in clinical use contains only monomers other than 2-hydroxyethyl methacrylate.

Hydrogels can be classified according to the chemical composition of the main ingredient in the polymer network regardless of the type or amount of minor components such as cross-linking agents and other by-products or impurities in the main monomer. Hydrogel contact lenses can be classified as: (1) 2-hydroxyethyl methacrylate lenses, (2) 2-hydroxyethyl methacrylate–vinylpyrrolidinone lenses, (3) hydrophilic–hydrophobic moiety copolymer lenses, and (4) miscellaneous hydrogel lenses. The chemical composition of most new contact lenses, hydrogels and others, is proprietary information. Some information on these lenses is available in technical, advertising, and professional literature. Most of the chemical descriptions of materials given in this article are based upon the patent, ophthalmic, and optometric literature.

In terms of physical and physiological properties, hydrogel lenses can be classified as low-hydration (ca 20–35% water), medium-hydration (ca 35–45% water), and high-hydration (up to ca 85% water) lenses. Medium-hydration lenses act physically and physiologically different when they are below about 0.1 mm in thickness; these can be regarded as a subclass of optical membranes or thin lenses.

In general, hydrogels for contact lenses are made either by the polymerization or copolymerization of hydrophilic monomers with a cross-linking agent, or by the copolymerization and cross-linking of a mixture of hydrophilic monomers or polymers with hydrophilic or hydrophobic monomers, to obtain a desired combination of physical properties and equilibrium swelling in water and ultimately in the aqueous phase of the tear film.

The most commonly used cross-linking agent—a minor ingredient in most hydrogel contact lenses—is ethylene dimethacrylate. This compound and small amounts of methacrylic acid are usually present in 2-hydroxyethyl methacrylate monomer as impurities or by-products that are difficult to remove completely from the main component.

Most hydrogels are obtained by bulk polymerization of the monomer mixture with the addition of a free radical initiator, or in some cases by solution polymerization with a free radical or redox initiator.

2-Hydroxyethyl Methacrylate Lenses. 2-Hydroxyethyl methacrylate lenses are the original and still most common hydrogel contact lenses. As described in the original patents (64–65,69), this hydrogel can be prepared from aqueous solutions of 2-hydroxyethyl methacrylate monomer containing small amounts of a cross-linking agent. Solution polymerization is currently being used in a unique contact lens fabrication technique developed by Wichterle (70) to obtain the "spun-cast" lenses (Soflens, Bausch & Lomb Co.). However, most 2-hydroxyethyl methacrylate lenses (Hydron, National Patent Development Corp.), as well as all other hydrogel contact lenses, are manufactured by standard machining and polishing techniques using dry polymer (xerogels) obtained by bulk polymerization.

Poly(2-hydroxyethyl methacrylate) hydrogels equilibrate in water at room temperature to ca 40% hydration, and to slightly below this hydration when equilibrated in physiological saline solution (0.9% sodium chloride, isotonic with tear film). Addition of other monomers to 2-hydroxyethyl methacrylate yields hydrogels of higher or lower hydration. Some workers have adjusted the two common by-products of 2-hydroxyethyl methacrylate, glycol dimethacrylate and methacrylic acid, to certain levels to obtain differently defined hydrogels (71). Modification of 2-hydroxyethyl methacrylate hydrogels was obtained by polymerizing this monomer, containing small amounts of methacrylic acid, in the presence of some polyethylene glycol (72). Irradiation using a ^{60}Co source of 2-hydroxyethyl methacrylate and a dimethacrylate cross-linking agent in vacuo yielded a gel directly machinable into a lens (73). 2-Hydroxyethyl methacrylate was polymerized with other hydroxyalkyl methacrylates, such as 1,4-butanediol monomethacrylate and 1,6-hexanediol monomethacrylate, by irradiating the mixture with a γ-ray source (74). The resulting polymers were soft and flexible in the nonswollen state but were cooled to a low temperature to be machined into a contact lens which was then hydrated.

Water-insoluble methacrylates have been copolymerized with 2-hydroxyethyl methacrylate in diverse proportions to obtain hydrogels useful for contact lenses. Thus, copolymers with glycidyl methacrylate (1) in the proportion of glycidyl:2-hydroxyethyl methacrylate, 10:90, yielded hydrogels of 47% water content, but in the proportion 50:50 of the water content dropped to 8.4% (75–77).

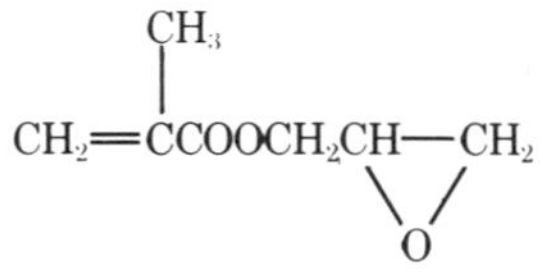

$$CH_2{=}\overset{\overset{\displaystyle CH_3}{|}}{C}COOCH_2CH\!\!-\!\!CH_2$$

(1)

Low-hydration hydrogel contact lenses were manufactured from copolymers of 2-hydroxyethyl methacrylate with methyl methacrylate (78). Another modification of basically 2-hydroxyethyl methacrylate hydrogels was obtained by copolymerizing isobutyl methacrylate with 2-hydroxyethyl methacrylate blended with their homopolymers (79).

To obtain hydrogels with at least 45% water at equilibirum swelling, copolymers of 2-hydroxyethyl methacrylate with small amounts of a dimethacrylate, some methyl or other alkyl methacrylate, and 1–10% methacrylic acid were obtained by bulk polymerization (80). The copolymer is then transformed to its alkali salt which is more hydrated, by at least 5 percentage points, than the free acid form at equilibrium swelling in water.

A contact lens material of about twice the tensile strength of pure 2-hydroxyethyl methacrylate hydrogel was obtained by copolymerizing this monomer with smaller amounts of n-pentyl methacrylate, vinyl propionate, and vinyl acetate (81). Other water-insoluble monomers copolymerized with substantially larger amounts of 2-hydroxyethyl methacrylate were isobutyl and cyclohexyl methacrylates, using trimethylolpropane trimethacrylate as the cross-linking agent (82); further modification was made by also adding a second hydrophilic monomer, diacetone acrylamide (N-(1,1-dimethyl-3-oxobutyl) acrylamide) (see (3) p. 734), to the above monomer mixture (83). Bufilcon A plastic, used in making HydroCurve II (Soft Lenses, Inc.) contact lenses, consists of a copolymer of 2-hydroxyethyl methacrylate (95.32%), diacetone acrylamide (4%), methacrylic acid (0.4%), and 1,1,1-trimethylpropane trimethacrylate (0.12%).

2-Hydroxyethyl Methacrylate-1-Vinyl-2-pyrrolidinone Lenses. Transparent hydrogels made of substantially pure 2-hydroxyethyl methacrylate can contain a maximum of about 40% water of hydration, but copolymers of 2-hydroxyethyl methacrylate with other, more hydrophilic monomers can produce hydrogels with higher levels of hydration. Vinylpyrrolidinone, an easily available hydrophilic monomer, was an early choice to copolymerize with 2-hydroxyethyl methacrylate. Currently, 1-vinyl-2-pyrrolidinone, like 2-hydroxyethyl methacrylate, has become one of the chemical compounds most often used as a principal ingredient in hydrogel contact lenses.

Hefilcon A, used in making HydroCurve I (Soft Lenses, Inc.) and Naturvue (Milton Roy Soft Contact Lens, Inc.) contact lenses, is also known as PHP (trademark of Automatic Optics, Inc.) (84) and is a random copolymer of 2-hydroxyethyl methacrylate (80%) and vinylpyrrolidinone (20%) cross-linked with ethyleneglycol dimethacrylate. This hydrogel at equilibrium swelling in water contains about 45% water

(85). Some of the patents issued to Seiderman, the inventor of PHP hydrogels, claim not only copolymers of vinylpyrrolidinone and 2-hydroxyethyl methacrylate, but also substances obtained by the polymerization of these monomers in the presence of poly(vinylpyrrolidinone) (86–89). It is probable that under polymerization conditions, the poly(vinylpyrrolidinone) is merely entrapped physically in the resulting cross-linked poly(2-hydroxyethyl methacrylate-*co*-vinylpyrrolidinone) gel; however, some of the homopolymer may also react by chain transfer with the copolymer gel. Polymerization of a substantial amount (70 mL) of 2-hydroxyethyl methacrylate and a small amount (1 mL) of tetrahydrofurfuryl methacrylate in the presence of poly-(vinylpyrrolidinone) was claimed to produce a graft copolymer that absorbed 47% water (90). Other materials of a chemical composition similar to PHP, but claimed to have a different structure, have also been patented. Thus, Ewell (91) has patented a hydrophilic contact lens material that he obtained by polymerizing 2-hydroxyethyl methacrylate in a mixture with poly(vinylpyrrolidinone) and a free radical initiator at monomer to polymer weight ratios ranging from 1:1 to 5:1.

Tetrafilcon A, which is used in making the Aquaflex (UCO Optics, Inc.) and Aosoft (American Optical Corp.) contact lenses has, in addition to its main ingredients 2-hydroxyethyl methacrylate (ca 82%) and vinylpyrrolidinone (ca 15%), small amounts of methyl methacrylate (ca 2%), and divinylbenzene (ca 0.5%) as the cross-linking agent. At equilibrium swelling, tetrafilcon A gel contains ca 40–43% water (92–93). A widely investigated contact lens material, vifilcon A, is used to make Softcon contact lenses (Softcon Products, Div. Warner-Lambert Co.). This material is a block or graft copolymer of 2-hydroxyethyl methacrylate onto poly(vinylpyrrolidinone) (94–97). The polymer is obtained by a procedure in which 2-hydroxyethyl methacrylate monomer is mixed with poly(vinylpyrrolidinone), and the mixture is then polymerized by free-radical initiators, irradiated to increase attachment between the two polymers, and finally treated with hydrogen peroxide so that the 2-hydroxyethyl methacrylate polymer will be further covalently bonded to the poly(vinylpyrrolidinone). However, it is evident that some poly(vinylpyrrolidinone) does not enter into chemical binding with the network but remains entangled, and will gradually leach from the hydrogel (98). Vifilcon A contact lenses hydrate to ca 50–55% water in distilled water, and ca 52–58% water in 0.9% sodium chloride solution (96). Because manufacturers upgrade the processes continually, it is possible that the latest vifilcon A contact lenses are made from an improved polymer. Thus, Hovey of American Optical Corp. (manufacturers of vifilcon A, Softcon lenses) has patented a process wherein monomeric vinylpyrrolidinone is substituted for poly(vinylpyrrolidinone) in the mixture with 2-hydroxyethyl methacrylate prior to polymerization (99). According to this polymerization procedure, which has much in common with the production method for PHP polymer, the use of the vinylpyrrolidinone monomer effects a considerably higher degree of chemical bonding than that achieved with poly(vinylpyrrolidinone) in 2-hydroxyethyl methacrylate, and reduces the possibility of the poly(vinylpyrrolidinone) leaching from the lens. Graft copolymerization of 2-hydroxyethyl methacrylate onto poly-(vinylpyrrolidinone) or onto vinylpyrrolidinone–vinyl acetate copolymer [25086-89-9], by means of a free radical initiator, di-*sec*-butyl peroxydicarbonate, has also been claimed by Leeds (100) as being useful for contact lenses.

The permeability of oxygen through hydrogel lenses is determined by their water content because the diffusion of oxygen occurs in the water-filled interstices of the gel. This permeability has a theoretical upper limit equal to the permeability to oxygen of a static layer of water of equal thickness. Thus, the oxygen permeability through hydrogels increases exponentially with increasing specific water content irrespective of the type of polymer forming the hydrogel network (57). Since the introduction of hydrogel contact lenses, efforts have been channelled principally toward developing lenses capable of transmitting greater amounts of oxygen to the cornea, thus enabling the lenses to be worn for longer periods of time. The two ways of accomplishing this are by increasing the water content of the lens or by decreasing its thickness. Most hydrogel lenses cannot be made very thin (below about 0.1 mm thickness) because the friability of the material increases chances of lens breakage, and also because most thin soft lenses tend to ripple with each blink causing unstable vision. So far most efforts to achieve higher oxygen permeability through hydrogel lenses have used hydrogels with the highest water content possible. One such lens is PermaLens (Global Vision Ltd., U.K.), which is apparently made by polymerizing a mixture of 60 parts of 2-hydroxyethyl methacrylate, containing about 1% glycol dimethacrylate and 3% methacrylic acid, and 40 parts of vinylpyrrolidinone (101–102). It is described as a cross-linked terpolymer of 2-hydroxyethyl methacrylate, methacrylic acid, and vinylpyrrolidinone. Thus, PermaLens is similar to PHP in chemical composition, but contains more water (>70%) at equilibrium swelling than does PHP (45%).

Lenses of Copolymers of Hydrophilic–Hydrophobic Moieties. Included in this group of hydrogel lenses are the Sauflons (Contact Lenses Manufacturing Ltd., U.K.), with water at equilibrium swelling varying from as low as ca 55% (Sauflex) to ca 85% and 70% (two types of Sauflon). These polymers are apparently made from a combination of hydrophilic polymer (poly(vinylpyrrolidinone)) or monomer (vinylpyrrolidinone) and a relatively hydrophobic monomer, methyl methacrylate, which are copolymerized by irradiation. By combining the proper proportions of hydrophilic and hydrophobic moieties in a cross-linked polymer, a variety of hydrogels can be obtained. One procedure for making Sauflons seems to start from a solution or slurry of poly(vinylpyrrolidinone) in methyl methacrylate monomer (103). The mixture is irradiated with a ^{60}Co γ-radiation source to yield a hard resin that can be machined into lenses, which are put into an aqueous solution to swell.

Cross-linked copolymers of alkyl acrylates or methacrylates and vinylpyrrolidinone produced by standard free radical initiation and heat have also been reported (104–105). Hydrogels of diverse hydration can be obtained by varying the proportions of the two types of monomers in the prepolymer mixture. A terpolymer of methyl methacrylate, vinylpyrrolidinone, and allyl methacrylate, with small amounts of glycol dimethacrylate, also yielded hydrogels useful for contact lenses (106). Likewise, transparent hydrogels with ca 74% water were obtained by copolymerizing methyl methacrylate (25 parts), vinylpyrrolidinone (70 parts), and N-vinylcaprolactam (5 parts), with small amounts of glycol dimethacrylate. The polymerization was initiated by azobisisobutyronitrile and heat (107).

Where the manufacture of Sauflon-type polymers seems to start with a polymer of vinylpyrrolidinone in methyl methacrylate monomer, other material has been described that starts from a mixture of poly(methyl methacrylate) in the monomer vinylpyrrolidinone (108). A small proportion of a cross-linking agent, diallyl succinate, plus a free radical initiator and benzoin ethyl ether as a photocatalyst is added to this mixture. The polymerization is carried out under ultraviolet light and then heated to yield a glassy transparent polymer which forms a gel having ca 79% water. Other variations of this type of polymer were obtained by dissolving 10 parts of poly(methyl methacrylate) in a mixture of 70 parts vinylpyrrolidinone and 20 parts methyl methacrylate with small amounts of glycol dimethacrylate and 5,5′-isopropylidenebis(N-vinylcaprolactam), and polymerizing the mixture with azobisisobutyronitrile and heat to obtain hydrogels of ca 72% hydration (109). High-water-content hydrogels were also obtained from reaction of 2.5 parts isotactic poly(methyl methacrylate), 12.5 parts syndiotactic poly(methyl methacrylate), 85 parts vinylpyrrolidinone, 2 parts dimethylsulfoxide, 0.5 parts diallyl succinate, and 0.03 parts azobiscyclohexanecarbonitrile under irradiation with a high-pressure mercury lamp, then curing under heat (110).

Copolymers of vinylpyrrolidinone (70 parts) with glycidyl methacrylate (1) (30 parts) yielded a hydrogel contact lens material with ca 53% water at equilibrium swelling (111). Higher-hydration (61–83% water) hydrogel materials of similar type were obtained when, in addition to glycidyl methacrylate (0.5–30%) and vinylpyrrolidinone (50–90%), 5–40% of methyl methacrylate, or another alkyl acrylate or hydrophobic vinyl monomer was added to the prepolymer mixture (112). The Union Optics Corp. holds patent rights to diverse hydrogels useful for contact lenses; these include hydrogels made from copolymers of glycidyl methacrylate and vinylpyrrolidinone (113–115), and from glycidyl methacrylate with 2-hydroxyethyl methacrylate (116–117).

Hydrogel contact lens materials with a high content of vinylpyrrolidinone and a small proportion of a hydrophobic monomer such as phenylethyl methacrylate, and cross-linked with allyl methacrylate, have a water absorption of 65–85% in physiological salt solution (118). Copolymers of vinylpyrrolidinone, methyl methacrylate, and glycerol monomethyl ether monomethacrylate, cross-linked with small amounts of allyl methacrylate, yielded hydrogels of 56% water of hydration (119).

High-water-content materials have been subject to various problems including fragility, reproducibility, and greater proneness to contamination (120). An alternative to high-water-content hydrogels is to make the lenses very thin in order to obtain high oxygen flux across the lens to the corneal surface. Thin lenses of hydrogels of medium hydration (up to 45%) can transmit oxygen to the cornea as well as can higher-hydration lenses, which must be thicker to conserve their integrity (57).

Oxygen flux, J [in μL(STP)/(cm. ꞎ)], across a lens is given by equation 1:

$$J = P_g \frac{\Delta p}{L} \tag{1}$$

where P_g is the oxygen permeability coefficient of the lens material [in μL(STP)·cm/(cm^2·h·kPa)]. Δp (in kPa) is the difference in partial pressure of oxygen between the front and back of the lens, and L (in cm) is the lens thickness. As the thickness of the lens decreases, the oxygen flux to the cornea increases.

According to Polse and Mandell (121), the minimum oxygen required by the epithelium of the cornea is about one-tenth of that available from the atmosphere at sea level (21.2 kPa or 159 mm Hg) when the eyes are open, and one-fourth of that available from the palpebral conjunctiva (7.33 kPa or 55 mm Hg) when the eyes are closed. Thus, at ca 2 kPa (15 mm Hg) oxygen partial pressure the cornea will have available for its consumption about 3.5 μL(STP)/(cm^2·h), which is the minimum required; the cornea will swell if less than this amount of oxygen is available. Direct comparison of the oxygen flux through a given contact lens, under both eye-open and eye-closed conditions, and the minimum oxygen required by the corneal surface, gives an indication of the probable physiological performance of such a lens under conditions of daytime wear and extended wear (sleeping with the lens in) (122).

The development of CS-151 Membrane lenses (made of crofilcon A by Corneal Sciences, Inc.) is based on the principle that thin lenses of not very high-water-content hydrogels (ca 40% hydration) can perform as well physiologically as more friable high-water-content hydrogel lenses. CS-151 Membranes are hydrogel thin lenses made of a copolymer of glyceryl methacrylate (2) and methyl methacrylate (123).

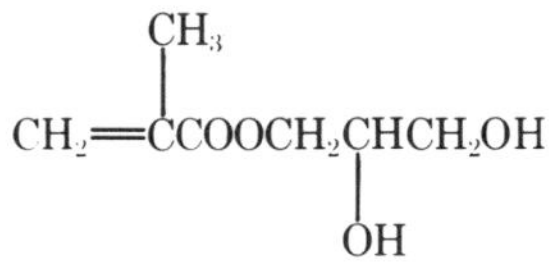

$$CH_2{=}\overset{\overset{\displaystyle CH_3}{|}}{C}COOCH_2\underset{\underset{\displaystyle OH}{|}}{C}HCH_2OH$$

(2)

The thickness of these lenses is ca 0.03–0.08 mm, depending on whether they are intended for prolonged (day and night) or day wear. Most other hydrogel contact lenses have thicknesses of ca 0.15–0.40 mm.

Miscellaneous Hydrogel Lenses. *Hydrogels Containing Acrylamide Derivatives.* In addition to the copolymers of diacetone acrylamide and isobutyl and cyclohexyl methacrylates with hydroxyethyl methacrylate that were noted previously (83), other acrylamido hydrogels potentially useful for contact lenses have been patented (see Acrylamide). These include a polymer of 2-acrylamido-2-methylpropanesulfonic acid cross-linked with ethylene dimethacrylate (124), copolymers of acrylamide and vinyl monomers having fluoroalkyl side groups (125), and copolymers of acrylamide, methyl methacrylate, and hydroxyethyl methacrylate (126). Strong, elastic hydrogels useful in contact lenses were made from acrylamide–acrylonitrile–acrylic acid copolymers (127).

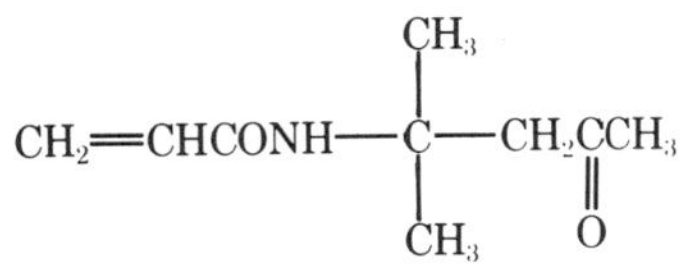

$$CH_2{=}CHCONH{-}\overset{\overset{\displaystyle CH_3}{|}}{\underset{\underset{\displaystyle CH_3}{|}}{C}}{-}CH_2\overset{\overset{\displaystyle}{\underset{\underset{\displaystyle O}{||}}{C}}}CH_3$$

(3)

Other hydrogel copolymers of methyl methacrylate with hydrophilic monomers of the acrylamide type have been patented. One of these is a terpolymer of acrylic acid, *N*-(1,1-dimethyl-3-oxobutyl)acrylamide (3), and methyl methacrylate (128). The acrylamide derivative (3) was also copolymerized with 2-hydroxyethyl methacrylate and vinyl acetate (129). Another hydrogel was obtained by copolymerization of *N,N*-dimethylacrylamide with methyl methacrylate or similar alkyl derivatives (130).

Hydrogels of Miscellaneous Composition. A hydrogel with 80% water absorption was obtained from reaction of 2-hydroxyethyl methacrylate, containing small amounts of ethylene glycol dimethacrylate, with the reaction product of poly(vinylpyrrolidinone) and methacryloyl chloride (131).

A soft polyurethane contact lens material with water absorption of about 30% was obtained when a polyethylene glycol was heated with isophorone diisocyanate (132).

Copolymerization of methyl methacrylate (80–90%) with acrylic acid (10–20%), and diverse cross-linking agents, followed by neutralization of the polymerized acrylic acid with a basic substance, such as ammonium hydroxide (133) or ethylenimine (134), yielded hydrogels useful for contact lenses. The depth of penetration of the neutralizing agent into the polymer determined the hydration of the lens (135).

Copolymers of olefins and hydrophilic monomers such as 2-hydroxyethyl methacrylate, hydroxypropyl acrylate, or vinylpyrrolidinone can be useful for the manufacture of hydrogel contact lenses. Thus, 2-hydroxyethyl methacrylate/1-octene copolymer [*39673-55-7*], 1:1 by weight, can be molded into a contact lens that swells in water to 40–60% hydration and has good oxygen permeability (136). Copolymers of ethylene glycol and diethylene glycol methacrylates, and possibly styrene, were allowed to react with trimellitic acid to yield contact lens materials with some water absorbency and good mechanical properties (137).

An allyloxy polymer derivative has been used to make Optamol (138) hydrogel contact lenses, and it appears to be a copolymer of allyl 2-hydroxyethyl ether and diallyl ether.

A different type of hydrogel that has also been proposed for the manufacture of contact lenses is the polyelectrolyte complex hydrogel, such as those from poly-(vinylbenzyltrimethylammonium chloride) [*26780-21-2*] and sodium poly(styrenesulfonate) [*9003-59-2*] (139–141). These polyelectrolyte complex hydrogels are transparent, flexible, and highly permeable to water. However, their dimensional instability, particularly on heating, makes them unlikely candidates for contact lens materials (see Polyelectrolytes).

Gel contact lenses were also made from the natural proteins of the crystalline lens of the eyes of warm-blooded animals (142).

Lenses With Hard Optical Centers and Soft Hydrophilic Peripheral Skirts. The biggest advantage of hydrogel contact lenses over other contact lenses is that they are more adaptable and comfortable when placed on the cornea. However, because these lenses conform to the cornea curvature and adjust to the particular configuration of the cornea, individuals with a significant degree of astigmatism often cannot be aided by soft hydrogel lenses. In order to obviate this drawback and yet retain the advantages of comfort and adaptability, some lenses contain a hard optical center and a soft hydrophilic peripheral skirt. So far these seem to be ideal conceptions that have not reached practical use, probably owing to the difficulty of maintaining the integrity of the lens at the interface between the hard and the soft materials. Another problem seems to be the maintenance of lens shape upon hydration of the portion attached to the rigid center.

A rigid, gas-permeable poly(4-methyl-1-pentene) contact lens to which a hydrophilic edge of poly(acrylic acid) was grafted has been described (143). A composite corneal contact lens with an optical portion of glass or hard plastic inserted in a hydrogel contact lens was also patented (144). Another patent provides a means of stabilizing the central optical portion of a soft contact lens with a rigid wire or other transparent ring device embedded in the lens body (145). Another invention uses a hydrogel marginal edge bonded to a hard contact lens material such as poly(methyl methacrylate) to reduce or eliminate the physical discomfort that wearers often experience with hard lenses (146).

The distortion on hydration of composite plastic contact lenses manufactured with hard centers and soft hydrophilic peripheral skirts was reduced by incorporation of a water-soluble component into the monomer mixture for making the skirt portion. Thus, a hard center consisted of a cross-linked 1,5-divinyloxy-3-oxapentane/2-hydroxyethyl methacrylate/2-methoxyethyl methacrylate copolymer [54688-52-7]. For the skirt portion, this mixture was allowed to react further with a water-soluble mixture of poly(oxyethylene glycol), 2-methoxyethyl and 2-hydroxyethyl methacrylates, and 1,5-divinyloxy-3-oxapentane, with a free radical initiator. The composite rod was used to machine contact lenses of center to outer-portion hardness ratio of 2:1 which was not altered on hydration (147–148).

Two-Layer Lenses. Another concept is realized in contact lenses made of two different layers. The layer facing the surface of the eye consists of a transparent plastic that is relatively soft and contains hydrophilic and reactive groups, eg, polyamines (149). The second layer which is applied immediately after the consolidation of the first layer is hydrophobic and hard. It contains groups that react with the plastic of the first layer to bond the layers together. Plastics for the second layer are of the epoxy type (see Epoxy resins). A similar invention of a corneal-type lens is composed of a hydrophilic liner that completely covers the posterior surface and extends around the edges of the anterior surface which itself is a relatively firm optical material (150).

Manufacture

Hard Contact Lenses. Starting from a polymer blank or "button" (generally made by cutting a rod of cross-linked acrylic resin) or from a molded semifinished lens, most hard contact lenses are made by standard methods of machining and polishing. Pressing–molding (or casting) and injection molding are also used to manufacture poly(methyl methacrylate) contact lenses, but these are usually not as good as the hardcrafted lenses (151–153). Copolymers of allyl diglycol carbonate (CR39) and maleic anhydride or glycidyl methacrylate were used to cast contact lenses or blanks from which lenses were made (154). These lenses were then surface-hydrolyzed by immersion in an alkaline solution to yield hard contact lenses with hydrophilic surfaces that were scratch resistant.

Silicone Contact Lenses. Silicone lenses are molded (31,155). Fabrication of these lenses has posed a major problem because silicone rubber is not amenable to good machining and polishing, and it has been difficult to obtain lenses with good edging (48). However, this difficulty seems to have been overcome and currently the principal problem with silicone lenses is surface coating rather than edge fabrication.

Hydrogel Contact Lenses. The most widely used system for the production of hydrogel contact lenses is a standard machining technique that is similar in many respects to the technique for manufacturing most hard contact lenses. The polymer is produced in the dry, hard state by bulk polymerization of a monomer or monomer mixture, a cross-linking agent, and a free radical initiator. It can be cast into a semifinished lens (156), or polymerized into rods which are cut into blanks and then lathe-cut into lenses. The hydrogel can also be obtained in the swollen state; it is then dried to a state where it can be ground and polished to the required optical shape. The finished lens is then reswollen by immersion in physiological saline solution (157). In general, the blanks or semifinished buttons are lathed, polished, and edged in the same manner as for a hard poly(methyl methacrylate) lens, except that in the manufacturing procedure for hydrogel lenses (at the xerogel state) an organic liquid or silicone oil, rather than water, must be used as the vehicle for the polishing powder.

A method for centrifugal casting of hydrogel contact lenses has been developed by Wichterle (70,158). A monomer solution containing a mixture such as 2-hydroxyethyl methacrylate with a small amount of a cross-linking agent and an initiator is polymerized in a mold rotating about its central axis. The procedure yields lenses already in a swollen state, which are then equilibrated in physiological saline solution to obtain the finished product. This procedure is used in the United States by Bausch & Lomb, Inc. to manufacture a contact lens under the trade name Soflens. For spun-cast lenses the curve that is used to determine the optical power of the lens is the concave surface, in contrast to the usual procedure which uses the front surface. The centrifugal casting technique was also employed in an invention to incorporate opaque materials in hydrogel lenses to simulate iris, pupil, and/or sclera in cosmetic lenses (159).

In general, lathe-cut lenses offer the advantage of a greater choice of parameters such as peripheral curves, optical and lenticular zone diameters, central and edge thicknesses, and peripheral finish of the lens. Spun-cast lenses, on the other hand, are cheaper to produce after an initially higher expenditure for equipment, but do not offer the variety of forms and shapes available in lathe-cut lenses.

Aspherical convex lenses that provide a paraboloidal front surface are required to correct deviations of the cornea from a spherical shape, as in the correction of an astigmatism not caused by the shape of the cornea. The standard methods for production of convex aspherical surfaces by machining are costly and not of practical use. Accordingly, Wichterle devised a procedure by which a cross-linked poly(2-hydroxyethyl methacrylate) blank (160) or a dry lens (161) was compressed above its glass transition temperature to set up internal stresses, the compressed blank was cooled below the transition point to fix the stress, the cold blank machined into a lens, and then the internal stresses were relaxed by reheating above the transition temperature or by swelling the lens, whereby the lens surfaces are deformed to the required shape.

A technique has been patented for producing polymer blanks molded with a convex surface to reduce the cutting and grinding steps of the conventional hard-lens lathe technique often used to make hydrogel lenses (162); a modification was then made to avoid the formation of strained polymer blanks (163).

Oxygenation of the Corneal Surface

The corneal surface can survive without insult with a lower concentration of oxygen in the tear film than is normally available to it from the atmosphere (pO_2 = 21.2 kPa or 159 mm Hg at sea level). The minimum concentration of oxygen tolerated by the cornea must be below the amount available to it during closed-eye condition, as during sleep, when oxygen is supplied to the corneal surface from the blood vessels of the palpebral conjunctiva (pO_2 = 7.3 kPa or 55 mm Hg). The corneal surface tolerates a minimum of ca 2 kPa (15 mm Hg) partial pressure of oxygen without swelling (164).

With all hard contact lenses and most hydrophobic flexible lenses, oxygenation of the corneal surface is obtained in part by a pumping mechanism that operates in the tear film during blinking. The tear film is normally oxygenated from the air; however, under a contact lens it must ideally be oxygenated across the lens. Thus, oxygen-permeable contact lenses perform better than do oxygen-impermeable lenses.

Hydrogel lenses are, in general, fitted closer to the cornea surface than other types of lenses, and they remain practically static on the cornea whereas other types move with the blink. Thus, the corneal surface depends more on the oxygen that passes across the lens with hydrogel lenses than with hard lenses. However, this point is still moot and a mechanism has been postulated for oxygenation of the cornea, at least in part, by means of a hypothetical capillary tear film that pumps tears between the lens and the corneal surface with the blink.

The oxygen flux across a contact lens of a given permeability coefficient and thickness is given in equation 1. The driving force for the oxygen to pass through the lens is the difference in the partial pressure of oxygen under and above the lens. Oxygen permeability coefficients of contact lens materials are given in Tables 2, 3, and 4. The units are selected to make the direct comparison of oxygen fluxes and oxygen consumed by the cornea possible; this is normally expressed in the ophthalmic literature in $\mu L(STP)/(cm^2 \cdot h)$. Thus, at its minimum requirement ($pO_2 = 2.0$ kPa or 15 mm Hg), the cornea consumes oxygen at a rate of 3.5 $\mu L(STP)/(cm^2 \cdot h)$ (57,122).

Economic Aspects

Sales of soft lenses for 1975–1977 and estimated sales for 1982 are shown in Table 5.

Table 5.   Manufacturers' U.S. Sales of Soft Lens Products, 1975–1977, and Market Share, 1975–1982[a]

	1975		1976		1977		1982
	Sales, 10^6 $	Market share, %	Sales, 10^6 $	Market share, %	Sales, 10^6 $	Market share, %	Market share, %
Bausch & Lomb		87.3	60.8	79.0	71.1	71.7	60
Soft Lenses	5.0	8.6	9.8	12.7	14.1	14.2	10–15
American Optical	2.0	3.4	3.0	3.9	6.0	6.0	10–15
other[a]	0.4	0.7	3.4	4.4	8.0	8.1	10–20
Total	*58.2*	*100.0*	*77.0*	*100.0*	*99.2*	*100.0*	*100*

[a] Ref. 165. Courtesy of Arthur D. Little, Inc.
[b] Includes Milton Roy, UCO, TRE-Optics, Coburn, and Wesley-Jessen, etc.

BIBLIOGRAPHY

1.   A. R. Gasset, *Contact Lenses and Corneal Disease*, Appleton-Century-Crofts, New York, 1976.
2.   C. D. Dohlman, ed., *Intern. Ophthal. Clin.* 8(3), (1968).
3.   M. C. Griffiths, ed., *USAN and the USP Dictionary of Drug Names*, United States Pharmacopeial Convention, Inc., Rockville, Md., 1976.
4.   R. B. Fox in N. M. Bikales, ed., *Encyclopedia of Polymer Science and Technology*, Vol. 9, Wiley-Interscience, New York, 1968, p. 336.
5.   P. Cochet and H. Amiard, *Intern. Ophthal. Clin.* 9(2), 292, 488 (1969).
6.   Y. Mizutani and K. Eguchi, *Nippon Ganka Gakkai Zasshi* **66,** 79, 379 (1962).
7.   U.S. Pat. 3,551,035 (Dec. 29, 1970), P. M. Kamath (to American Optical Corp., Southbridge, Mass.).
8.   M. Salame, *Polym. Prepr. Am. Chem. Soc. Div. Polym.* 8(1), 137 (1967).
9.   W. E. Long, *Contacto Intern. Cont. Lens J.* (5), 35 (1974).
10.   S. T. Hwang, C. K. Choi, and K. Kammermeyer, *Sep. Sci.* **9,** 461 (1974).

11. J. Agranoff, ed., *Modern Plastics Encyclopedia,* Vol. 52, No. 10A, McGraw-Hill Book Co., New York, 1975.
12. M. F. Refojo, F. J. Holly, and F. L. Leong, *Cont. Intraocular Lens Med. J.* **3**(4), 27 (1977).
13. N. O. Stahl, L. A. Reich, and E. Ivani, *J. Am. Optom. Assoc.* **45**, 302 (1974).
14. Braz. Pedido PI 74 03,534 (Jan. 6, 1976), N. G. Gaylord (to Polycon Laboratories, Inc.).
15. U.S. Pat. 3,808,179 (Apr. 30, 1974), N. G. Gaylord (to Polycon Laboratories, Inc.).
16. U.S. Pat. 3,950,315 (Apr. 13, 1976), C. S. Cleaver (to E. I. duPont de Nemours & Co., Inc.).
17. F. J. Holly and M. A. Lemp, eds., *Intern. Ophthal. Clin.* **13**(1), (1973).
18. A. L. Koven, *Eye, Ear, Nose Throat Mon.* **41,** 47 (1962).
19. H. Yasuda and co-workers, *J. Biomed. Mater. Res.* **9,** 629 (1975).
20. J. M. Deaton and J. E. Hodgkins, *Tex. J. Sci.* **17,** 125 (1965).
21. Ger. Offen. 1,255,950 (June 12, 1968), H. D. Gesser and R. E. Warriner.
22. E. Kanemitsu, *Ind. Chim. Belge* **32**(Spec. No.), 382 (1967).
23. U.S. Pat. 3,728,315 (Apr. 17, 1973), R. Gustafson.
24. Fr. Demande 2,181,065 (Jan. 4, 1974), G. H. Butterfield and G. H. Butterfield, Jr. (to Butterfield Laboratories, Inc.).
25. U.S. Pat. 3,948,871 (Apr. 6, 1976), G. H. Butterfield, Jr. and G. H. Butterfield (to Butterfield, G. H. and Son).
26. U.S. Pat. 3,542,462 (Nov. 24, 1970), L. J. Girard, W. G. Sampson, and J. W. Soper (to E. I. du Pont de Nemours & Co., Inc.).
27. D. Miller, P. White, and D. B. Hood, *Cont. Intraocular Lens Med. J.* **1,** 24 (1975).
28. Jpn. Kokai 75 87,184 (July 14, 1975), (to Polycon Laboratories, Inc.).
29. U.S. Pat. 3,808,178 (Apr. 30, 1974), N. G. Gaylord (to Polycon Laboratories, Inc.).
30. U.S. Pat. 3,900,250 (Aug. 19, 1975), E. J. Ivani (to Rynco Scientific Corp.).
31. U.S. Pat. 3,228,741 (Jan. 11, 1966), W. E. Becker (to Mueller Welt Contact Lenses, Inc.).
32. Fr. Pat. 1,532,820 (July 12, 1968), J. Mishler (to Dow Corning Corp.).
33. U.S. Pat. 3,341,490 (Sept. 12, 1967), D. F. Burdick, J. L. Mishler, and K. E. Polmanteer (to Dow Corning Corp.).
34. R. M. Hill and J. Shoessler, *J. Am. Optom. Assoc.* **38,** 480 (1967).
35. R. P. Burns, H. Roberts, and L. F. Rich, *Am. J. Ophthalmol.* **71,** 486 (1971).
36. R. M. Hill and J. E. Terry, *Arch. Ophthalmol. Paris* **36,** 155 (1976).
37. N. J. van Haeringen and E. Glasius, *Exp. Eye Res.* **20,** 271 (1975).
38. J. Moacanin and D. D. Lawson, *Biomater. Med. Devices Artif. Organs* **1,** 183 (1973).
39. H. Yasuda and A. Peterlin, *J. Appl. Polym. Sci.* **17,** 433 (1973).
40. H. Yasuda and W. Stone, Jr., *J. Polym. Sci. A-1* **5,** 2952 (1967).
41. *General Electric Co. Brochure GEA-8685A,* 2-70 (SM).
42. U.S. Pat. 3,700,573 (Oct. 24, 1972), J. Laizier and G. Wajs (to Commissariat à L'Energie Atomique, France).
43. Y. Pouliquen and co-workers, *Bull. Soc. Ophtalmol. Fr.,* Rapport Annuel, 58 (1974).
44. Fr. Demande 2,185,653 (Feb. 8, 1974), L. Stark and co-workers (to Biocontacts, Inc.).
45. J. Brandrup and E. H. Immergut, eds., *Polymer Handbook,* 2nd ed., Wiley-Interscience, New York, 1975, pp. III–234.
46. U.S. Pat. 3,350,216 (Oct. 31, 1967), D. E. McVannel, J. L. Mishler, and K. E. Polmanteer (to Dow Corning Corp.).
47. U.S. Pat. 3,954,644 (May 4, 1976), J. Z. Krezanoski and J. C. Petricciani (to Flow Pharmaceuticals, Inc.).
48. A. B. Rizzuti, *Ann. Ophthalmol.* **6,** 596 (1974).
49. T. N. Zekman and L. A. Sarnat, *Am. J. Ophthalmol.* **74,** 534 (1972).
50. Ger. Offen. 2,165,805 (July 5, 1973), P. Feneberg and U. Krekeler (to Agfa-Gevaert A.-G.); Fr. Demande 2,166,027 (Sept. 14, 1973).
51. H. Yasuda and M. F. Refojo, *J. Polym. Sci. A* **2,** 5093 (1964).
52. Fr. Demande 2,208,775 (June 28, 1974), E. W. Merrill.
53. I. Fatt and R. St. Helen, *Am. J. Optom.* **48,** 545 (1971).
54. F. J. Holly and M. F. Refojo, *J. Am. Optom. Assoc.* **43,** 1173 (1972).
55. *Dura Soft Contact Lens,* Wesley-Jessen Inc., Form No. DU250,5K076, 1976.
56. M. F. Refojo and F. L. Leong, *J. Membr. Sci.,* **4,** 415 (1979).
57. M. F. Refojo, in N. M. Bikales, ed., "Contact Lenses," *Encyclopedia of Polymer Science and Technology,* Suppl. 1, Wiley-Interscience, New York, 1976, p. 195.

58. Ger. Offen. 2,518,904 (Nov. 13, 1975), J. J. Falcetta, G. D. Friends, and G. C. C. Niu (to Bausch & Lomb, Inc.).
59. Ger. Offen. 2,262,866 (July 12, 1973), M. Shen, R. B. Mandell, and L. Stark (to Biocontacts, Inc.).
60. U.S. Pat. 3,940,207 (Feb. 24, 1976), A. E. Barkdoll (to E. I. du Pont de Nemours & Co., Inc.).
61. U.S. Pat. 3,431,046 (Mar. 4, 1969), T. J. Conrad, G. O. Dayton, Jr., and M. Arlin (to Studies, Inc.).
62. J. Fr. Herschel, cited by M. Rohr, *Die Brille als Instrument,* Vol. III, Aufl. 14, 1921; O. Wichterle, D. Lim, and M. Dreifus, *Cesk. Oftalmol.* **17,** 70 (1961).
63. O. Wichterle and D. Lim, *Nature* **185,** 117 (1960).
64. U.S. Pat. 2,976,576 (Mar. 28, 1961), O. Wichterle and D. Lim.
65. U.S. Pat. 3,220,960 (Nov. 30, 1965), O. Wichterle and D. Lim.
66. O. Wichterle, in N. M. Bikales, ed., "Hydrogels," *Encyclopedia of Polymer Science and Technology,* Vol. 15, Wiley-Interscience, New York, 1971, p. 273.
67. S. Raymond and L. Weintraub, *Science* **130,** 711 (1959).
68. W. M. Thomas, in N. M. Bikales, ed., "Acrylamide Polymers," *Encyclopedia of Polymer Science and Technology,* Vol. 1, Wiley-Interscience a division of John Wiley & Sons, Inc., New York, 1964, p. 177.
69. U.S. Pat. Re. 27,401 (reissued June 20, 1972), O. Wichterle and D. Lim (to Czechoslovak Academy of Sciences).
70. U.S. Pat. 3,408,429 (Oct. 29, 1968), O. Wichterle (to Czechoslovak Academy of Sciences).
71. U.S. Pat. 3,985,697 (Oct. 12, 1976), J. Urbach (to Uroptics International Inc.).
72. U.S. Pat. 3,951,528 (Apr. 20, 1976), H. R. Leeds (to Patent Structures, Inc.).
73. U.S. Pat. 3,854,982 (Dec. 17, 1974), R. Aelion and E. Ferezy (to Hydroplastics, Inc.).
74. U.S. Pat. 3,983,083 (Sept. 28, 1976), I. Kaetsu and co-workers (to Japan Atomic Energy Research Institute, and Tokyo Optical Company Ltd.).
75. Ger. Offen. 2,318,434 (Oct. 31, 1974), P. Stamberger (to Union Optics Corp.).
76. U.S. Pat. 3,758,448 (Sept. 11, 1973), P. Stamberger (to Union Optics Corp.).
77. U.S. Pat. 3,947,401 (Mar. 30, 1976), P. Stamberger (to Union Optics Corp.).
78. M. J. Popovich, *J. Am. Optom. Assoc.* **47,** 305 (1976).
79. Jpn. Kokai 75 129,648 (Oct. 14, 1975), H. Magatani, H. Atsuzawa, and Y. Kosaka (to Tokyo Contact Lens Research Institute).
80. U.S. Pat. 3,988,274 (Oct. 26, 1976), E. Masuhara, N. Tarumi, and M. Tsuchiya (to Hoya Lens Co., Ltd.).
81. Jpn. Kokai 76 56,893 (May 18, 1976), K. Tanaka (to Toyo Contact Lens Co., Ltd.).
82. U.S. Pat. 3,926,892 (Dec. 16, 1975), F. O. Holcombe, Jr. (to Burton, Parsons & Co., Inc.).
83. U.S. Pat. 3,965,063 (June 22, 1976), F. O. Holcombe, Jr. (to Burton, Parsons & Co., Inc.).
84. U.S. Pat. 3,721,657 (Mar. 20, 1973), M. Seiderman.
85. R. M. Hill and D. E. Linder, *Intern. Contact Lens Clinic,* 66 (Spring 1976).
86. U.S. Pat. 3,639,524 (Feb. 1, 1972), M. Seiderman.
87. U.S. Pat. 3,767,731 (Oct. 23, 1973), M. Seiderman.
88. U.S. Pat. 3,966,847 (June 29, 1976), M. Seiderman.
89. Brit. Pat. 1,339,726 (Dec. 5, 1973), M. Seiderman.
90. Brit. Pat. 1,330,727 (Dec. 5, 1973), M. Seiderman.
91. U.S. Pat. 3,647,736 (Mar. 7, 1972), D. G. Ewell (to Kontur Kontact Lens Co., Inc.).
92. USAN Council, *J. Am. Med. Assoc.* **236,** 189 (1976).
93. R. J. Morrison, *Intern. Contact Lens Clinic* **3,** 57 (1976).
94. U.S. Pat. 3,700,761 (Oct. 24, 1972), K. F. O'Driscoll and A. A. Isen (to Griffin Laboratories Inc.).
95. U.S. Pat. 3,822,196 (July 2, 1974), K. F. O'Driscoll and A. A. Isen (to Warner-Lambert Co.).
96. U.S. Pat. 3,841,985 (Oct. 15, 1974), K. F. O'Driscoll and A. A. Isen (to Warner-Lambert Co.).
97. U.S. Pat. 3,816,571 (June 11, 1974), K. F. O'Driscoll and A. A. Isen (to Warner-Lambert Co.).
98. M. F. Refojo, *Cont. Intraocular Lens Med. J.* 1(4), 36 (1975).
99. U.S. Pat. 3,839,304 (Oct. 1, 1974), R. J. Hovey (to American Optical Corp.).
100. Ger. Offen. 1,952,514 (Aug. 27, 1970), H. R. Leeds.
101. Ger. Offen. 2,205,391 (Aug. 24, 1972), J. T. DeCarle.
102. Ger. Offen. 2,426,147 (Dec. 19, 1974), J. T. DeCarle.
103. Ger. Offen. 2,312,470 (Sept. 27, 1973), P. W. Cordrey, J. D. Frankland, and D. J. Highgate (to Special Polymers Ltd.).
104. U.S. Pat. 3,532,679 (Oct. 6, 1970), R. Steckler.
105. A. L. Magnitskii and co-workers, *Vysokomol. Soedin. Ser. B* **17,** 298 (1975).

106. Brit. Pat. 1,391,438 (Apr. 23, 1975), P. W. Cordrey and W. Mikucki (to Contact Lens (Mfg) Ltd.).
107. Jpn. Kokai 75 140,594 (Nov. 11, 1975), T. Kunitomo and co-workers (to Toray Industries, Inc.).
108. Jpn. Kokai 74 102,790 (Sept. 27, 1974), S. Nagaoka, T. Kunitomo, and H. Tanzawa (to Toray Industries, Inc.).
109. Jpn. Kokai 75 133,292 (Oct. 22, 1975), T. Kunitomo and co-workers (to Toray Industries, Inc.).
110. U.S. Pat. 3,949,021 (Apr. 6, 1976), T. Kunitomo and co-workers (to Toray Industries, Inc.).
111. U.S. Pat. 3,772,235 (Nov. 13, 1973), P. Stamberger (to Union Optics Corp.).
112. Fr. Demande 2,239,486 (Feb. 28, 1975), (to Union Optics Corp.).
113. Jpn. Kokai 75 28,591 (Mar. 24, 1975), (to Union Optics Corp.).
114. Jpn. Kokai 75 28,590 (Mar. 24, 1975), (to Union Optics Corp.).
115. Brit. Pat. 1,430,300 (Mar. 31, 1976), (to Union Optics Corp.).
116. U.S. Pat. 3,947,401 (Mar. 30, 1976), P. Stamberger (to Union Optics Corp.).
117. U.S. Pat. 3,758,448 (Sept. 11, 1973), P. Stamberger (to Union Optics Corp.).
118. Ger. Offen. 2,529,639 (Jan. 22, 1976), J. G. B. Howes and co-workers (to Smith and Nephew Research Ltd.).
119. Ger. Offen. 2,503,755 (Oct. 9, 1975), J. G. B. Howes and co-workers (to Smith and Nephew Research Ltd.).
120. M. J. Barradell, *Contacto Intern. Cont. Lens. J.* (3), 33 (1975).
121. K. A. Polse and R. B. Mandell, *Arch. Ophthalmol.* **84,** 505 (1970).
122. M. F. Refojo, *Cont. Intraocular Lens Med. J.,* in press.
123. U.S. Pat. 3,957,362 (May 18, 1976), W. L. Mancini, D. R. Korb, and M. F. Refojo (to Corneal Sciences, Inc.).
124. U.S. Pat. 3,929,741 (Dec. 30, 1975), R. A. Laskey (to Datascope Corp.).
125. Ger. Offen. 2,541,527 (Mar. 25, 1976), R. Ensor, G. D. Pedley, and B. J. Tighe (to National Research Development Corp.).
126. Ger. Offen. 2,502,682 (July 24, 1975), J. R. Larke, D. G. Pedley, and B. J. Tighe (to National Research Development Corp.).
127. U.S. Pat. 3,812,071 (May 21, 1974), A. Stoy (to Ceskoslovenska Akademie Ved.).
128. U.S. Pat. 3,803,093 (Apr. 9, 1974), C. W. Neefe.
129. U.S. Pat. 3,813,447 (May 28, 1974), K. Tanaka, T. Mio, and T. Tanaka (to Toyo Contact Lens Co., Ltd.).
130. Ger. Offen. 2,416,353 (Oct. 24, 1974), H. S. Schultz (to Itek Corp.).
131. Jpn. Kokai 75 144,793 (Nov. 20, 1975), T. Nakashima and K. Takakura (to Kuraray Co., Ltd.).
132. Jpn. Kokai 75 83,468 (July 5, 1975), (to Frigitronics, Inc.).
133. Ger. Offen. 2,123,766 (Dec. 2, 1971), I. Blank (to Hydrophilics International, Inc.).
134. Ger. Offen. 2,239,206 (Mar. 15, 1973), I. Blank and J. Fertig (to Hydrophilics International, Inc.).
135. U.S. Pat. 3,728,317 (Apr. 17, 1973), I. Blank (to Hydrophilics International, Inc.).
136. Ger. Offen. 2,235,973 (Feb. 8, 1973), J. R. Larke and B. J. Tighe (to National Research Development Corp.).
137. Jpn. Kokai 72 44,033 (Nov. 7, 1972), H. Magatani, H. Atsuzawa, and S. Iwai (to Tokyo Contact Lens Research Institute).
138. Optalya S.A.R.L., Conoptica, S.A., and G. Nissel & Co., Ltd., U.K., *Optamol advertising literature.*
139. U.S. Pat. 3,271,496 (Sept. 6, 1966), A. S. Michaels (to Amicon Corp.).
140. U.S. Pat. 3,608,057 (Sept. 21, 1971), H. J. Bixler and M. A. Kendrick (to Amicon Corp.).
141. M. F. Refojo, *J. Appl. Polym. Sci.* **11,** 1991 (1967).
142. U.S. Pat. 3,553,299 (Jan. 5, 1971), H. Thiele and W. P. Soehnges.
143. U.S. Pat. 3,619,044 (Nov. 9, 1971), P. M. Kamath (to American Optical Corp.).
144. U.S. Pat. 3,488,111 (Jan. 6, 1970), A. A. Isen (to National Patent Development Corp.).
145. U.S. Pat. 3,933,411 (Jan. 20, 1976), A. E. Winner.
146. U.S. Pat. 3,489,491 (Jan. 13, 1970), C. P. Creighton.
147. Brit. Pat. 1,412,439 (Nov. 5, 1975), C. E. Erickson and A. N. Neogi (to Erickson Polymer Corp.).
148. Fr. Demande 2,213,964 (Aug. 9, 1974), C. E. Erickson and A. N. Neogi (to Erickson Polymer Corp.).
149. Belg. Pat. 616,333 (July 31, 1962), (to ESPE Fabrik Pharmazeutischer Praeparate G.m.b.H.).
150. U.S. Pat. 3,973,837 (Aug. 10, 1976), L. J. Page.
151. L. J. Girard and co-workers, eds., *Corneal Contact Lenses,* The C. V. Mosby Co., St. Louis, Mo., 1964.

152. J. Hartstein, *Questions and Answers on Contact Lens Practice,* 2nd ed., The C. V. Mosby Co., St. Louis, Mo., 1973.
153. P. Cordrey, *The Ophthalmic Optician,* Mar. 3, 1973, p. 230.
154. U.S. Pat. 3,221,083 (Nov. 30, 1965), H. D. Crandon (to American Optical Corp.).
155. J. L. Breger, *Opt.* **162,** 12 (1971).
156. U.S. Pat. 3,822,089 (July 2, 1974), O. Wichterle (to Ceskoslovenska Akademie Ved.).
157. Fr. Pat. 1,422,109 (Dec. 24, 1965), O. Wichterle (to Ceskoslovenska Akademie Ved.).
158. Brit. Pat. 990,207 (Apr. 28, 1965), (to Ceskoslovenska Akademie Ved.).
159. U.S. Pat. 3,557,261 (Jan. 19, 1971), O. Wichterle (to Ceskoslovenska Akademie Ved.).
160. U.S. Pat. 3,497,577 (Feb. 24, 1970), O. Wichterle (to Ceskoslovenska Akademie Ved.).
161. U.S. Pat. 3,542,907 (Nov. 24, 1970), O. Wichterle (to Ceskoslovenska Akademie Ved.).
162. U.S. Pat. 3,841,598 (Oct. 15, 1974), B. J. Grucza.
163. U.S. Pat. 3,894,129 (July 8, 1975), D. O. Hoffman and E. Z. Zdrok (to American Optical Corp.).
164. K. A. Polse and R. N. Mandell, *Arch. Ophthalmol.* **84,** 505 (1970).
165. *Update on the Market for Soft Contact Lenses in the United States, No. L771004,* Arthur D. Little, Inc., Cambridge, Mass., Oct. 25, 1977.

General References

R. B. Mandell, *Contact Lens Practice: Hard and Flexible Lenses,* 2nd ed., Charles C. Thomas, Springfield, Ill., 1974.
J. D. Andrade, ed., *Hydrogels for Medical and Related Applications,* ACS Symposium Series 31, Am. Chem. Soc., Washington, D.C., 1976.
M. Ruben, *Contact Lens Practice,* Williams & Wilkins Company, Baltimore, 1975.
M. Ruben, ed., *Soft Contact Lenses: Clinical and Applied Technology,* John Wiley & Sons, Inc., New York, 1978.

MIGUEL F. REFOJO*
Eye Research Institute of Retina Foundation
Department of Ophthalmology, Harvard Medical School

* Supported by USPHS grant EY-00327 from the National Eye Institute, National Institutes of Health.

CONTAINERS.   See Packaging materials.

CONTRACEPTIVE DRUGS

Control of fertility continues to be an important issue throughout the world even though the population growth rate has shown a steady decline in many countries, partly owing to the extensive use of oral contraceptives. The first of these products to be marketed was Enovid (1–2), a combination of norethynodrel (1) (a progestin) and mestranol (2a) (an estrogen), originally introduced in 1957 for the treatment of menstrual disorders.

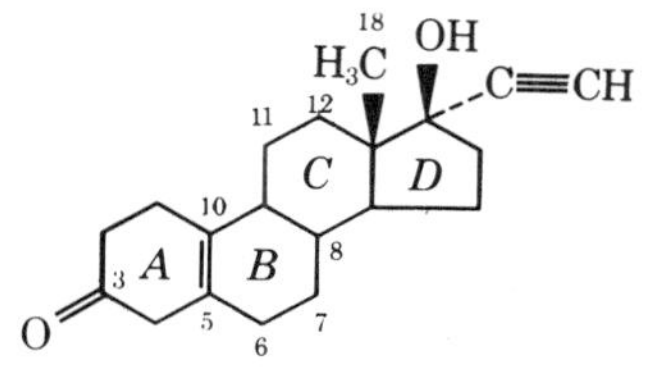

(1)

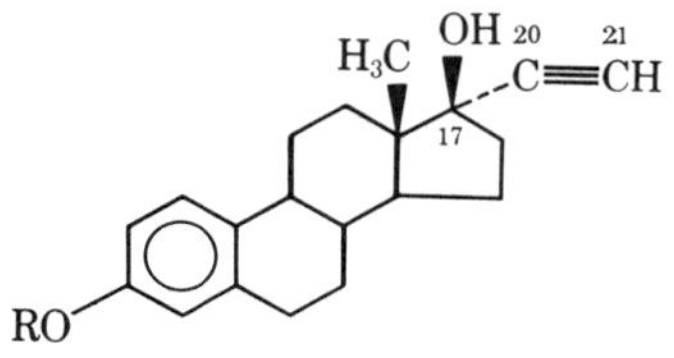

(2)

(a) R = CH_3

(b) R = H

(c) R =

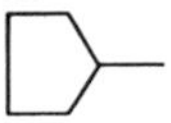

In 1960, the FDA approved its use for the cyclic control of ovulation. Because of its convenience, efficacy, and esthetic appeal, Enovid received wide acceptance as a means to regulate conception. Its success prompted the development of similar products. Today twelve major pharmaceutical companies based in eight countries produce and market approximately forty preparations of oral contraceptives. These products are used by about 55 million women throughout the world (3). The various methods of contraception used in the United States in 1977 are estimated to be:

Foams (female)	1 million
Condoms (male)	2.2 million
Diaphragms (female)	1–1.5 million
Intrauterine devices (female)	3 million
Oral contraceptives (female)	10 million
Sterilization (both male and female)	6–6.5 million

Worldwide (3), the estimates are:

Sterilization (both male and female)	80 million
Oral contraceptives (female)	55 million
Condoms (male)	35 million
Intrauterine devices (female)	15 million
Other	65 million
Abortion (annual incidence)	30–55 million

Oral contraceptives used by the majority of women today are each a combination of two steroidal substances, a progestin and an estrogen. The two substances are present in various ratios and act principally by inhibiting ovulation in normally cycling women (see Hormones; Steroids). A list of components of contraceptive drugs noted in this article with their CAS Registry Numbers is given at the end of the article.

It was known as far back as the beginning of this century that extracts of the corpus luteum could be used to inhibit ovulation. Intensive effects were made to isolate and identify the active principle. In 1934, the crystalline corpus luteum hormone, progesterone (3), was obtained (4). Shortly thereafter Makepeace (5) demonstrated that injection of progesterone into rabbits would inhibit ovulation. Later other investigators showed that a similar effect could be achieved in women who were given this hormone (2). Progesterone has other important physiological functions. For instance, it is essential for the maintenance of pregnancy as it prevents the spontaneous abortion of the implanted blastocyst. For this reason, it is often referred to as the pregnancy hormone. Thus, paradoxically progesterone can theoretically be employed for proceptive as well as contraceptive purposes.

In practice, however, the use of (3) is limited because the hormone is only weakly active when administered orally. In the mid-1930s an investigation was undertaken to modify the structure of the steroid molecule in the hope of obtaining oral activity. In 1937 Inhoffen and co-workers synthesized 17α-ethynyltestosterone [ethisterone (4a)] (6) and found it to possess greater oral activity than (3). Although the oral progestational activity of this substance was still of a low order of magnitude, in addition to its producing undesirable androgenic side effects, (4a) was introduced in Europe in 1941 as an oral progestational agent for the treatment of menstrual irregularities.

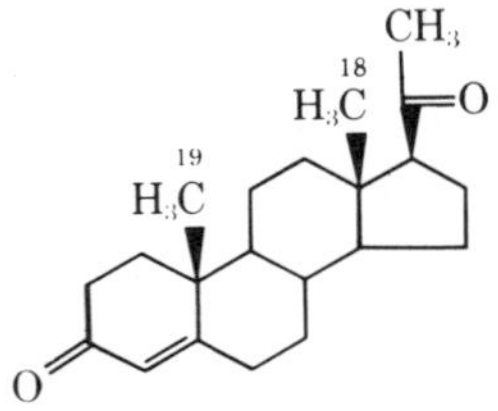

(3)

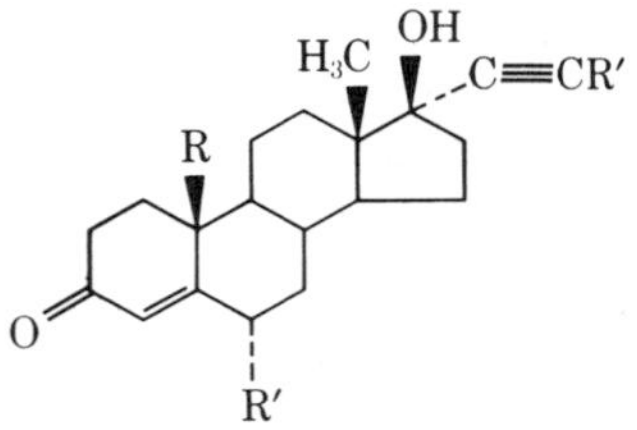

(4)

(a) R = CH$_3$, R′ = H
(b) R = R′ = H
(c) R = R′ = CH$_3$

In 1944, Ehrenstein prepared a resin that was thought to contain the 19-nor analogue (5) of (3) (7). The resin was obtained from the cardiac aglycone, strophanthidin (6), in a very low overall yield. Although it was obviously a mixture, the crude product was found to possess potent progestational activity in the rabbit when given by injection. It was as potent as the natural hormone, progesterone (4).

The unexpected findings that (4a) had oral activity, and Ehrenstein's 19-nor-progesterone (5) was a potent progestin when administered parenterally, led to in-

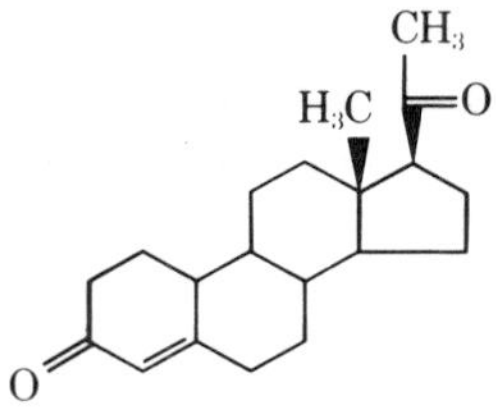

CH$_3$

H$_3$C $=$O

O

(5)

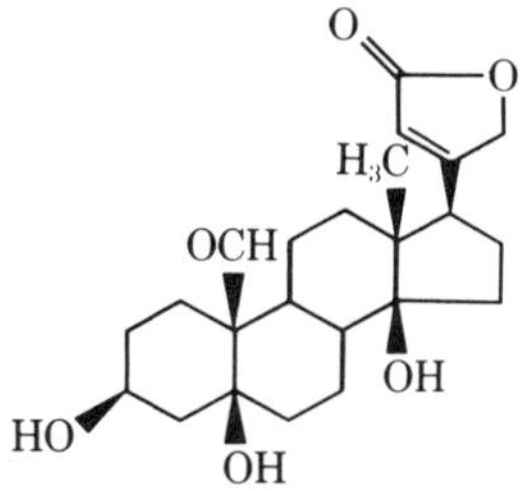

O

H$_3$C

O

OCH

HO

OH

OH

(6)

dependent efforts to combine the unique structural features of these two compounds. A series of 19-norsteroids was observed to possess potent oral progestational as well as other hormonal activity (7–8).

Although Ehrenstein had prepared his 19-nor compound from (6), a steroid that has a functional substituent at C-10, Colton and Djerassi found it more expeditious to begin their syntheses of the 19-norsteroids from the more abundant female sex hormone, estrone (7a) (8–9). They made use of the important process developed by Birch for the conversion of a phenolic ether to the corresponding dihydro derivative by reduction with metal and liquid ammonia (10).

Djerassi synthesized norethindrone [(4b), known also as norethisterone], the 19-nor analogue of ethisterone (4a), and Colton prepared norethynodrel (1), which is a double-bond isomer of norethindrone. In (4b) the nuclear double bond is in conjugation with the C-3 carbonyl group as in testosterone (8a); and in (1) the double bond is in the 5,10-position. Because of this, (1) produces less androgenic side effects than the 3-keto-Δ^4-compound (11). In practice, the foregoing progestins are administered with an estrogen on a 3 week on–1 week off regimen. This allows monthly withdrawal bleeding to occur. Their success as oral contraceptives led to their widespread adoption and to the synthesis of other progestins.

Although the combination of a progestin and estrogen is very effective in suppressing ovulation, certain undesirable side effects became apparent on widespread usage. Thromboembolic and related vascular disorders, as well as alterations in carbohydrate and lipid metabolism, have been reported in women taking contraceptive pills (12), particularly among those who are in the older age bracket. The estrogen component has been implicated in these disorders. For this reason, current practice suggests prescribing those products in which each pill contains 50 μg or less of estrogen (13). Estrogen is required for good cycle control. Reducing its quantity may result in a greater incidence of breakthrough bleeding.

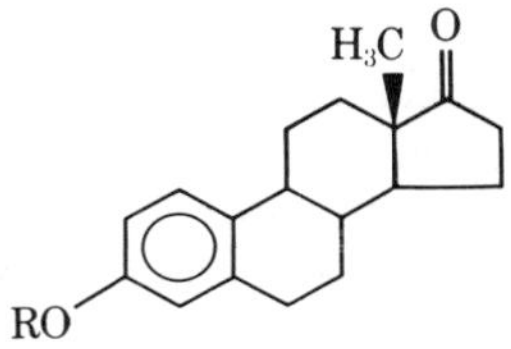

H$_3$C O

RO

(7)

(a) R = H
(b) R = CH$_3$

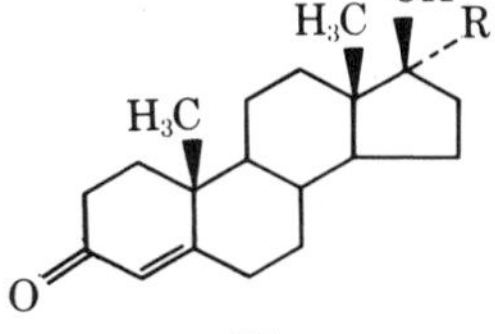

OH R

H$_3$C

H$_3$C

O

(8)

(a) R = H
(b) R = CH$_3$
(c) R = C≡CH

Because of the side effects reported with the progestin–estrogen combination pills, other approaches to contraception are being investigated. A recent report by the Ford Foundation (14) indicates that there are currently 230 promising scientific leads that could yield better contraceptives. However, a number of factors exert a restraining effect on the development of these leads. One of them is the need to spend large sums of money to develop the products, particularly to demonstrate in extensive animal studies and clinical trials that they meet the stringent safety requirements established for these products. Djerassi predicted in 1970 that unless the government participates in the development of new contraceptive agents and that certain novel steps are taken to improve government–industry interaction, the methods available for birth control in 1984 will not differ significantly from those used in 1970 (15).

STEROIDS

Ovulation and/or Implantation Suppression

Estrane Derivatives (19-Norsteroids). For the most part, the oral contraceptives that are currently in use contain either mestranol (2a) or ethinylestradiol (2b) as the estrogen component. The progestin component of the vast majority of oral contraceptives is also an estrane derivative. Norethindrone (4b) is the progestin present in most preparations, as for example, in Ortho–Novum, Norinyl, Modicon, Brevicon, Micronor, and Nor-Q.D. (see Steroids).

The initial syntheses of (4b) utilized estrone (7a) as the starting material (16–17). Successive methylation and hydride reduction transform it into estradiol 3-methyl ether (9). Reduction of (9) by a modified Birch reduction (10) affords the corresponding 1,4-dihydro derivative (10). When treated with ethylene glycol and p-toluenesulfonic acid, (10) is converted into the ketal (11). Chromic acid oxidation furnishes the 17-keto compound (12). Ethynylation at C-17 affords the 3-ketal (13) of norethynodrel. On treatment with acid, the ketal group is removed and the double bond migrates into conjugation with the carbonyl group to give norethindrone (4b) (17).

Acetylation of the hydroxyl group at C-17 of (4b) with either acetic anhydride or isopropenyl acetate is accompanied by enol acetylation at C-3 to afford the diacetate (14). Selective hydrolysis of the acetate group at C-3 can be accomplished under mild acid or alkaline conditions to give norethindrone 17-acetate (15) (18), which is the progestin component of Anovlar, Norlestrin, Loestrin, and Zorane.

Reduction of (4b) with lithium tri(t-butoxy)aluminum hydride affords 17α-ethynylestr-4-ene-3β,17β-diol (16) as the major product. The 3α-epimer is also produced, but only in small amounts. Acetylation of (16) furnishes ethynodiol diacetate (17) (19), the progestin present in Ovulen and Demulen.

Norethynodrel (1), the progestin component of Enovid, was also initially prepared from (9). Here, too, a Birch reduction was employed to produce the hydroaromatic system in ring A. More recently, however, a procedure that obviates the need to utilize a highly estrogenic substance as starting material has been adapted to the synthesis of both (1) and (4b) (20).

Addition of hypochlorous acid to dehydroisoandrosterone acetate (18a) affords the chlorohydrin (19). Oxidation with lead tetraacetate results in reaction of the nonactivated methyl group at C-10 to give the 6,19-epoxide (20). Successive hydrolysis,

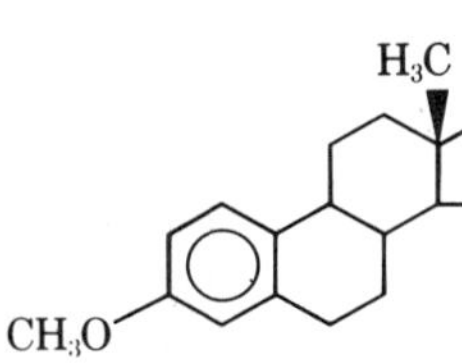

(9) [1035-77-4]

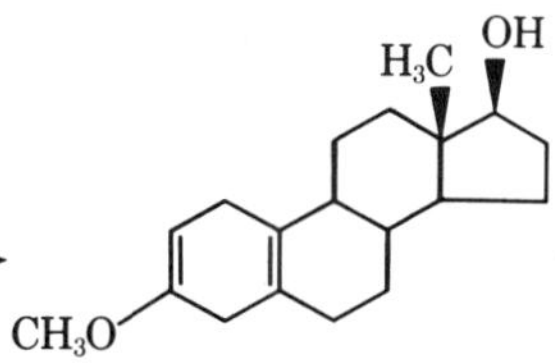

(10) [1091-93-6]

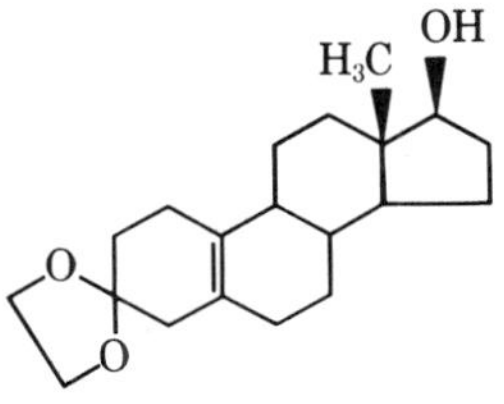

(11) [15342-09-3]

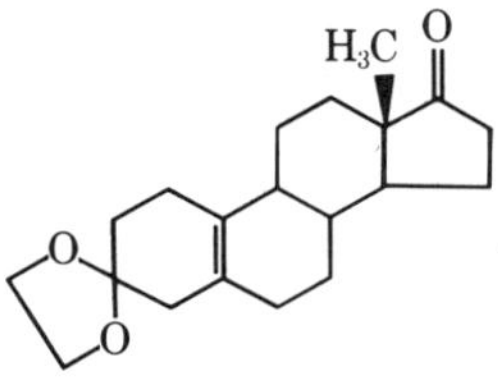

(12) [6193-99-3]

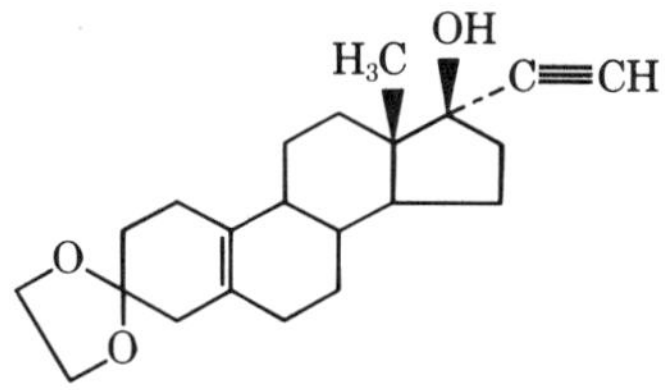

(13) [18314-02-8]

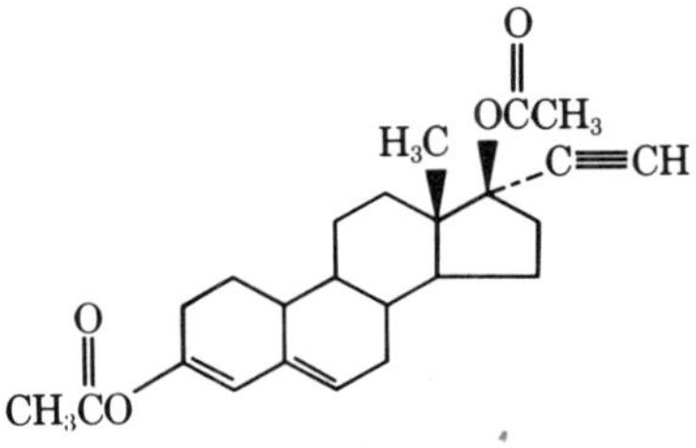

(14)

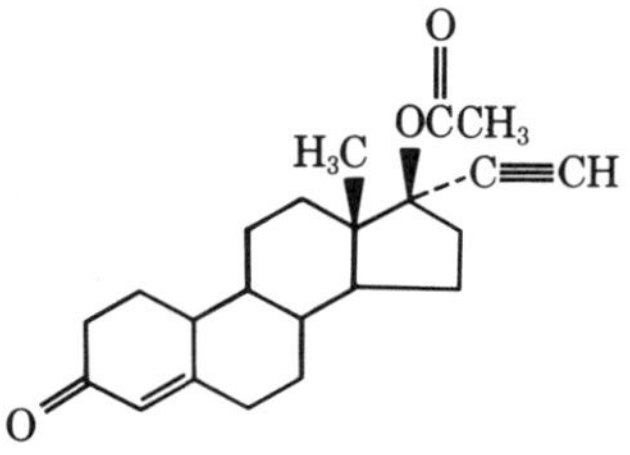

(15)

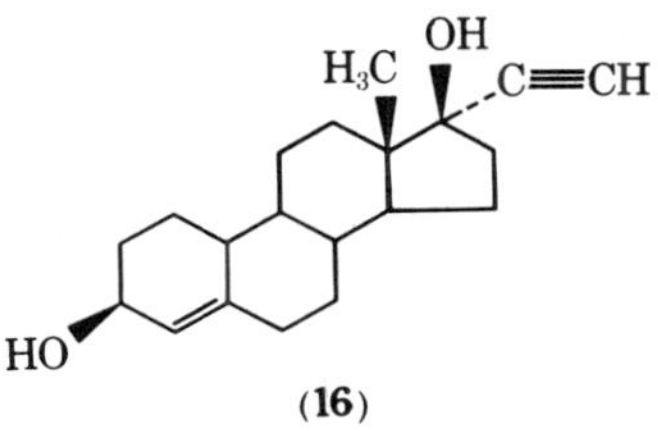

(16)

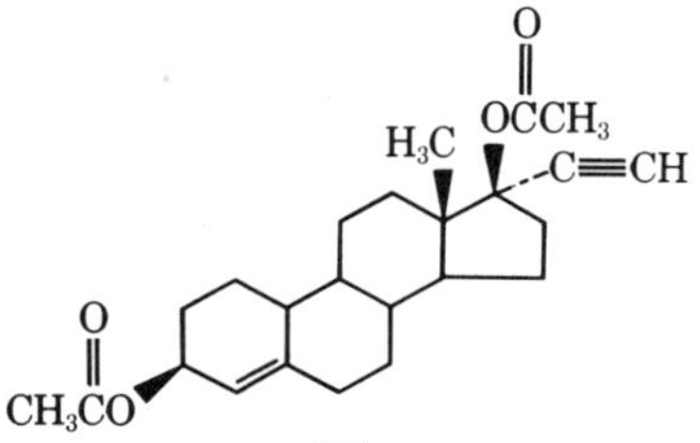

(17)

oxidation, and dehydrohalogenation furnish the 3-keto-Δ^4-compound (21). Cleavage of the epoxide ring between C-6 and the ether oxygen atom is readily achieved with zinc and acetic acid. The resultant compound, 19-hydroxyandrost-4-ene-3,17-dione (22), is oxidized and decarboxylated to give the $\Delta^{5(10)}$-derivative (23). The carbonyl group at C-3 can be selectively protected by ketalization with methanol in the presence of acid. Ethynylation at C-17 then yields the dimethyl ketal (24) of (1). Cleavage of the ketal group under gentle acidic conditions affords (1). When more vigorous acidic conditions are employed, (4b) is produced.

Norethynodrel (1) is transformed into (4b) in the presence of a mineral acid. However, unlike (4b), (1) readily undergoes oxygenation to afford the 10β-hydroperoxide derivative (25), which also possesses potent contraceptive properties (21).

Replacement of the C-13 methyl group of (4b) with an ethyl group affords norgestrel (26), the racemic term of which (26a) is the progestin component of Ovral. Because the C-13 ethyl group is not present in naturally occurring steroids, (26) cannot be economically prepared by partial synthesis from readily accessible steroids. Instead, total synthesis must be employed (22).

Many modifications of the norethindrone structure can be made without adversely affecting the biological activity. Acylation of the 17-hydroxyl group can enhance or prolong the effects of (4b). Thus, norethindrone enanthate (27) (Norigest) is utilized as an injectable contraceptive, given once every several months (23). Norethindrone

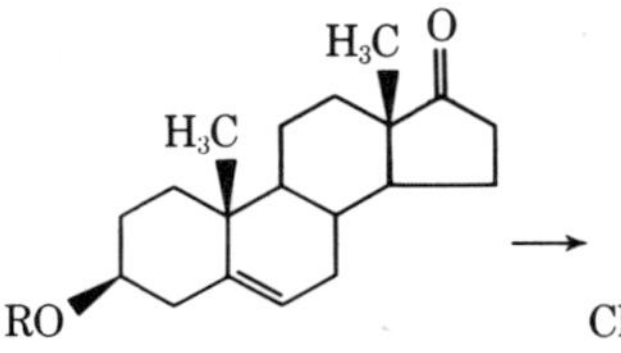

(18)
(a) R = CH₃CO
(b) R = H

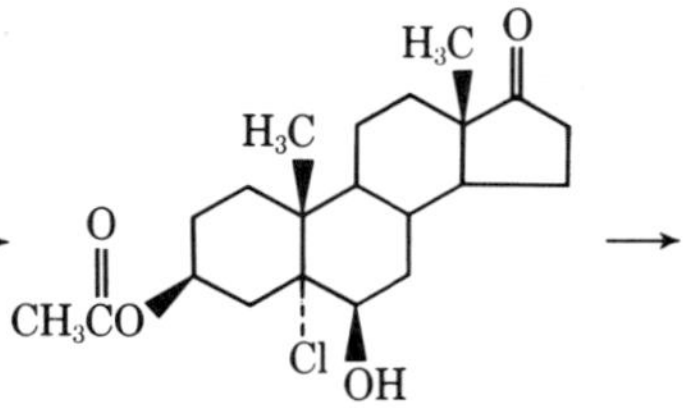

(19)

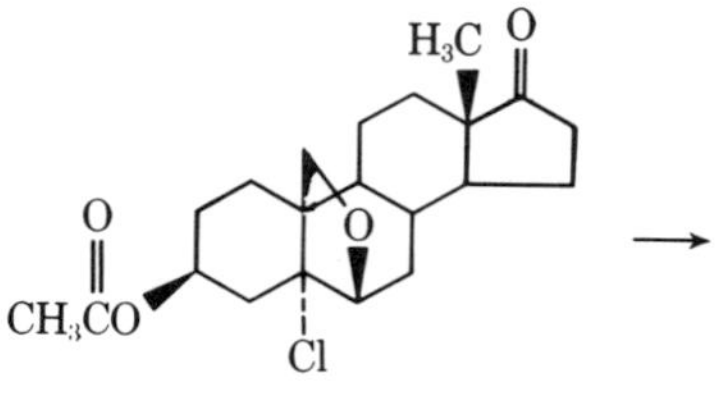

(20)

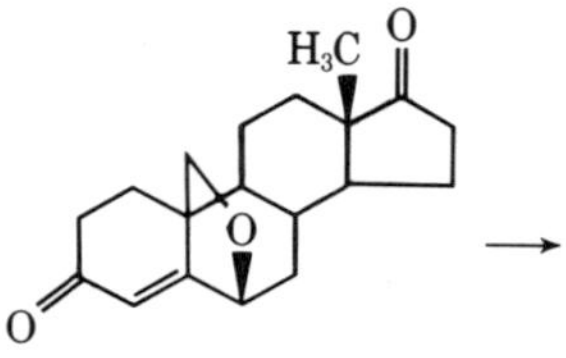

(21)

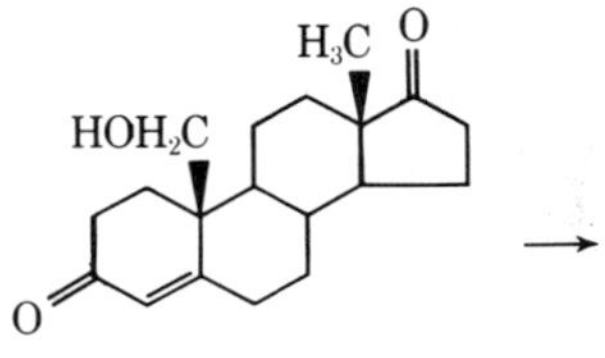

(22)

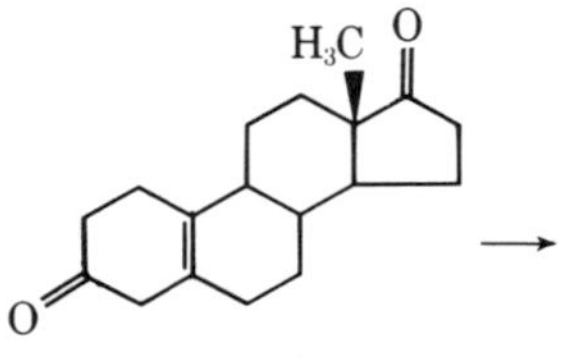

(23)

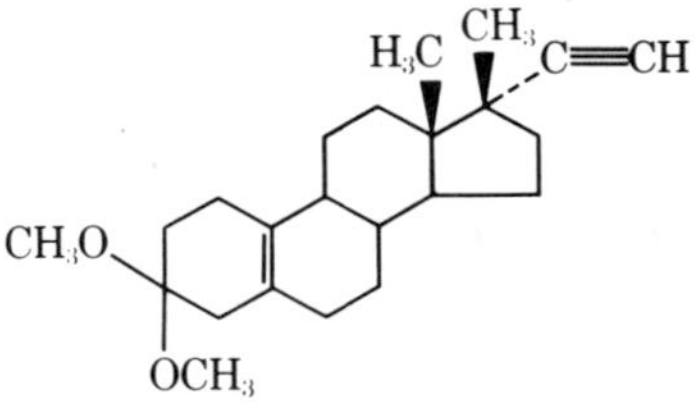

(24)

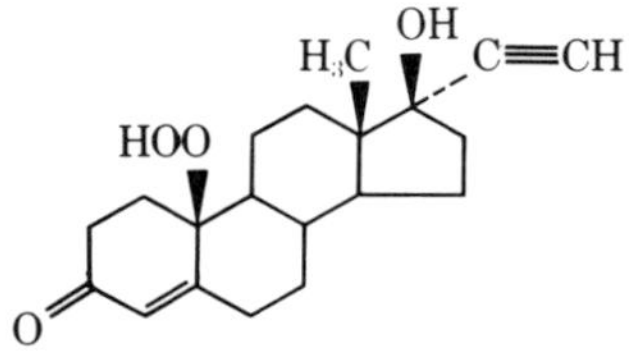

(25)

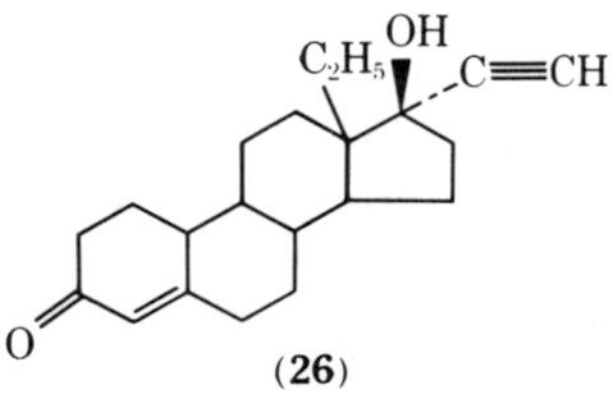

(26)
(a) dl-
(b) d-

acetate (15), the progestin component of a number of oral contraceptives (see above), has been further enhanced in activity by converting the 3-keto-Δ^4-system of (15) into a cyclopentyl enol ether to give quingestanol acetate (28). Combined with (2b), (28) is marketed in Mexico as Riglovis. By combining quingestanol acetate with quinestrol (2c), which is the 3-cyclopentyl ether of (2b), a long acting, orally effective contraceptive is obtained. This combination is taken orally once every four weeks and has been used in Latin America for the past nine years (24).

Replacement of the carbonyl function at C-3 of (15) with a β-acetoxy group furnishes ethynodiol diacetate (17), a potent progestin with estrogenic properties (25). As mentioned previously, (17) is the progestin component of Ovulen and Demulen. In the former product the estrogen component is (2a), and in the latter it is (2c).

Curiously, the removal of the carbonyl group at C-3 of norethindrone does not abolish the biological activity. The resultant product, lynestrenol (29) (26), when combined with (2a), is an effective contraceptive agent marketed as Lyndiol.

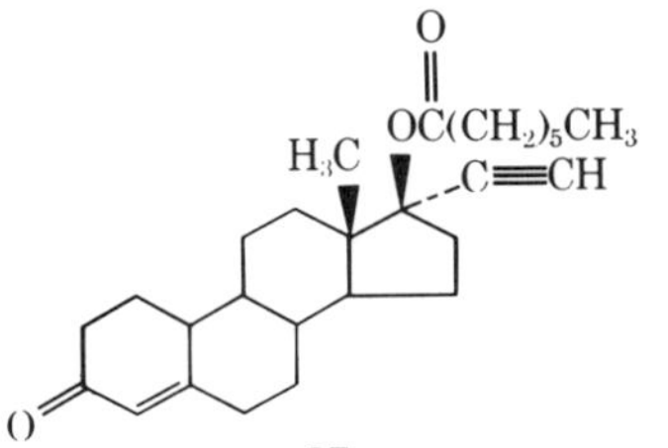

(27)

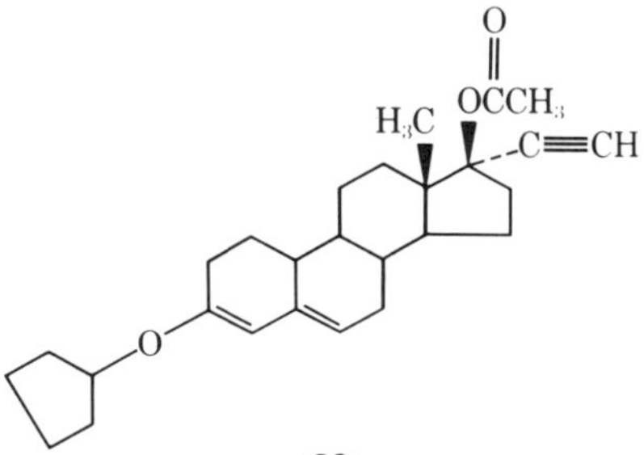

(28)

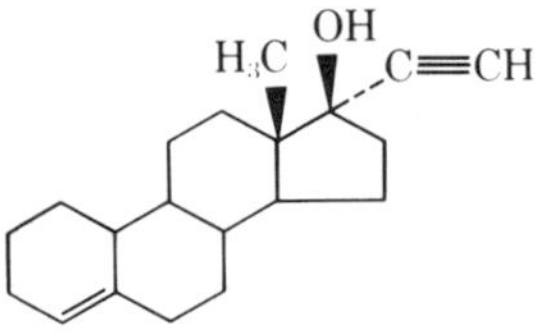

(29)

Despite differences in their structures, the cyclopentyl ether (28), the diacetate (17), and the deoxo derivative (29) are all effective progestins. There is evidence indicating that the three substances undergo metabolic transformations in the body to afford norethindrone (4b) which would account for the qualitatively similar biological effects that they display (27).

Extension of the conjugated system of norethindrone by insertion of additional double bonds enhances the antifertility activity of (4b). Both the 3-keto-$\Delta^{4,9}$-diene (30) and the 3-keto-$\Delta^{4,9,11}$-triene (31a) are more potent than the parent compound (4b) (28–29). The 3-keto-$\Delta^{4,9,11}$-triene is the progestin component of Planor, the contraceptive agent marketed in Europe whose estrogen component is mestranol (2a). The corresponding triene [R 2323 (31b)], in which the angular methyl group at C-13 is replaced by an ethyl group, has been in clinical trials as a once-a-week pill (30).

Insertion of a methyl group into certain sites of norethindrone (4b) will also result in increased progestational activity. Norgestrel (26), the potent progestin which is prepared by total synthesis, is the 18-methyl homologue of (4b). Addition of an axial

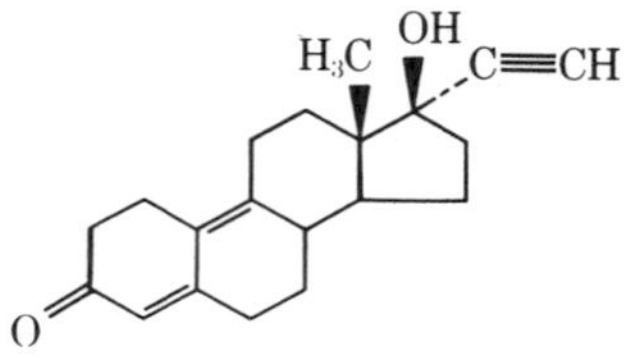

(30)

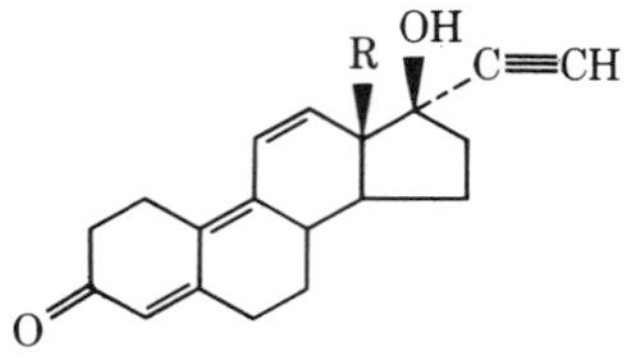

(31)

(a) R = CH$_3$

(b) R = C$_2$H$_5$

methyl group to either the 7α or the 11β-position of (4b) to afford (32) (31) and (33) (32), respectively, produces a substantial enhancement of the antifertility effect.

Although the addition of a single methyl group to either the 7α-, 11β-, or 18-position will increase progestational activity, attachment of a second methyl group to either one of the other two positions may have a detrimental effect by partially nullifying the effect of the first group (33).

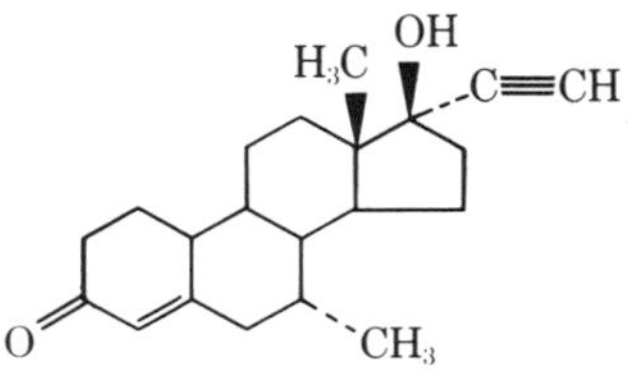

(32)

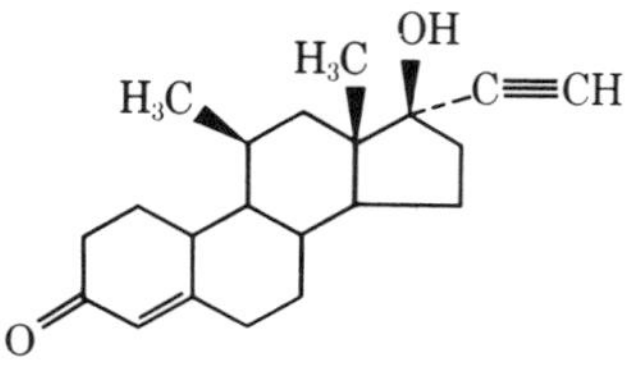

(33)

Attachment of a methylene group to C-11 and a methyl group to C-18 of lynestrenol (29) furnishes ORG 2969 (34). This compound is more potent than (29) and is currently being evaluated in the clinic (34). Another potent progestin that is presently undergoing clinical trials is norgestimate (35), the oxime of d-norgestrel acetate (35). The oxime (36) of norethindrone acetate (36) also possesses potent antifertility activity, as does the methoxime derivative (37) of norethindrone (37).

Replacement of the ethynyl hydrogen of (4b) with either a chloro, ethynyl, or a methyl group to afford (38a), (38b), or (38c), respectively, does not result in a diminution of the biological activity (28,38). If in addition to the methyl group at C-21, methyl groups are attached to the 6α- and 10β-positions, norethindrone is converted into the very potent progestin, dimethisterone (4c).

Allene analogues (allenologs) of (4a), norgestrel (26), and 17α-ethynylestr-4-ene-$3\beta,17\beta$,diol (16) have been reported to possess oral progestational potencies that are greater than that of norethindrone (39). The 17α-propadienyl steroids (39) can be prepared in a number of ways. In one method, 2-butyn-1-yl-magnesium bromide is added to a 17-keto steroid (40). Besides the main product (41), the allenolog (42) is also obtained. Another approach involves treating the 17-keto steroid (40) with 3-tetrahydropyran-2'-yloxy-1-propynylmagnesium bromide to afford (43). Reduction of (43) with lithium aluminum hydride yields the 17α-propadienyl steroid [(39), R = H].

The hydrogen atoms at the ring junctions in norethindrone (4b) are alternately β- and α-oriented (44). Thus, the backbone configuration of (4b) is $anti$–$trans$–$anti$–$trans$. Inversion of the configuration at C-8 gives rise to 8α-norethindrone (45)

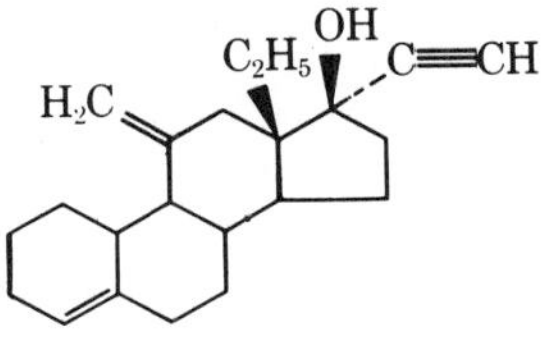

(34)

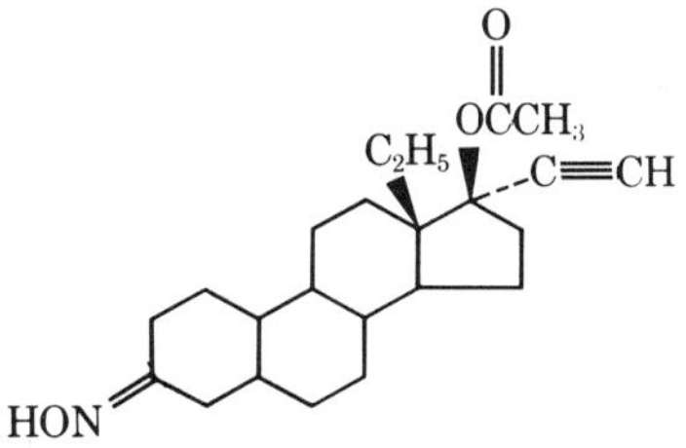

(35)

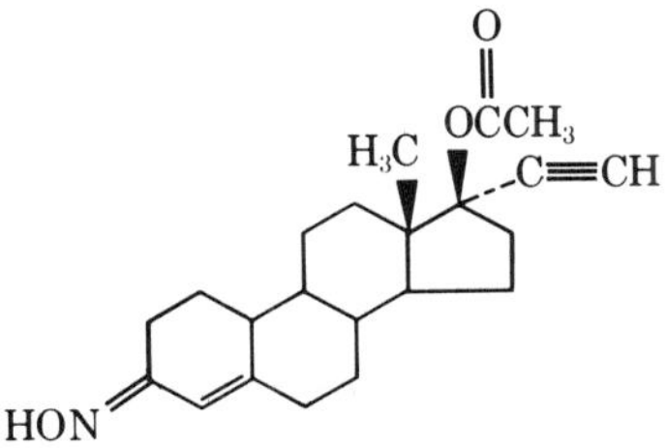

(36)

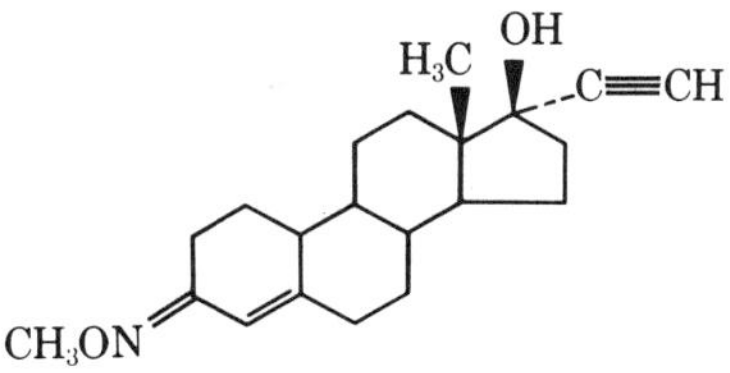

(37)

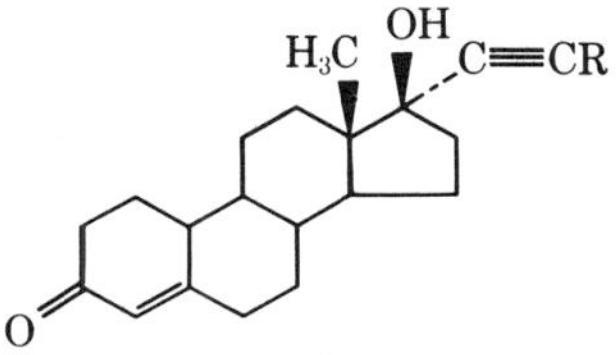

(38)

(a) R = Cl
(b) R = C≡CH
(c) R = CH₃

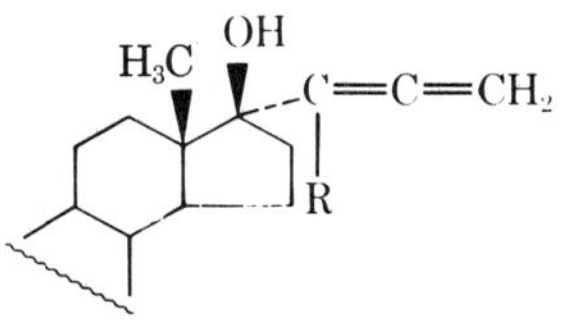

(39)

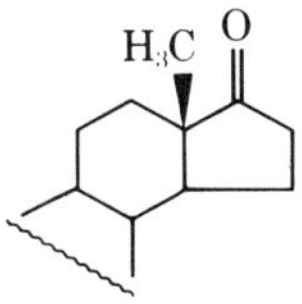

(40)

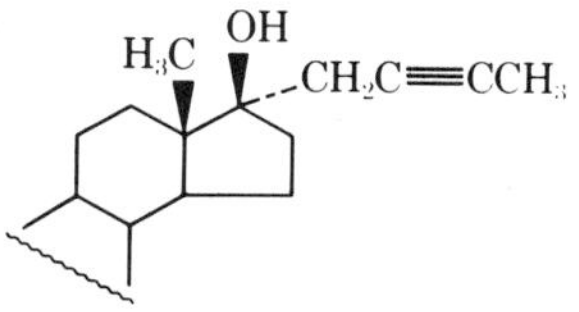

(41)

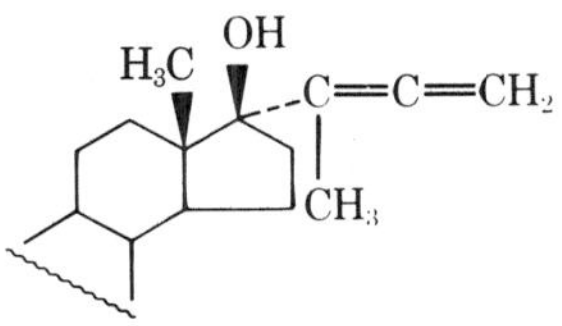

(42)

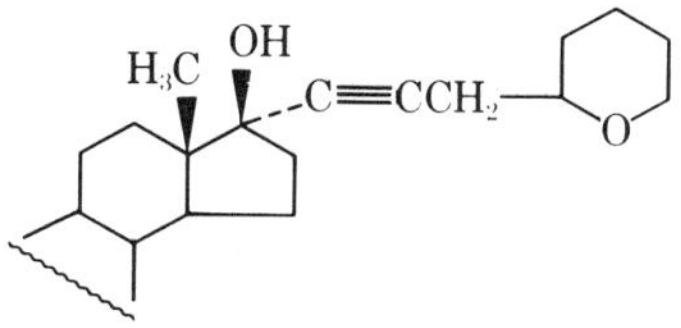

(43)

(40). This compound has the *anti–cis–syn–trans* backbone configuration. Its pro-gestational activity is considerably less than that of (4b). Inversion of the two chiral centers at C-9 and C-10 of (4b) to afford 9β,10α-norethindrone (46) having the backbone configuration of *anti–cis–anti–trans* abolishes the progestational effect altogether (41). In contrast, 17α-allyl-13β-ethyl-10α-methyl-3-oxo-$9\beta,10\alpha$-gon-4-en-17β-ol (**47**), which has the same configuration as (**46**), possesses considerable oral progestational activity (42).

Recent studies have shown that the 17β-hydroxyl group in norethindrone (**4b**) is not essential for progestational activity. A series of compounds was prepared in which the 17β-hydroxyl group was deleted from the molecule. These compounds, (**48**), (**49**), and (**50**), were found to be potent progestins. With the exception of (**48**), removal of the hydroxyl group at C-17 also eliminated the undesirable androgenic side effects present in the parent substances (43).

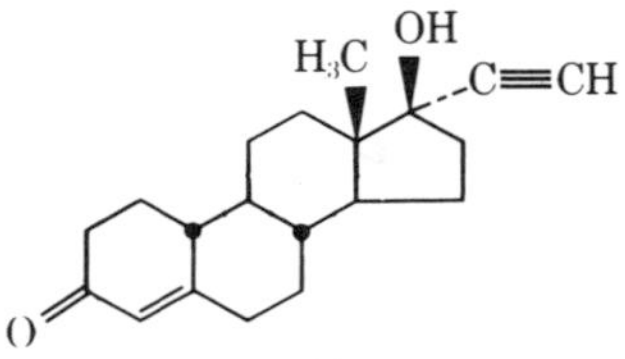

(44) 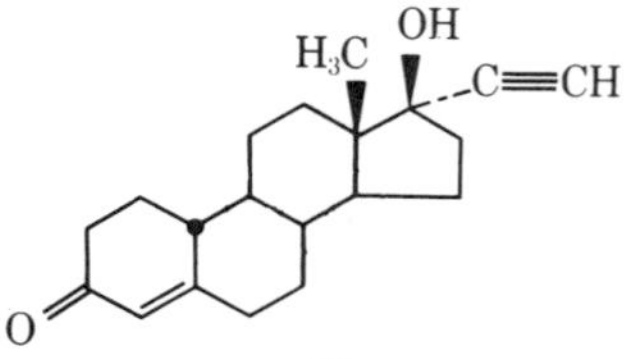   (45)

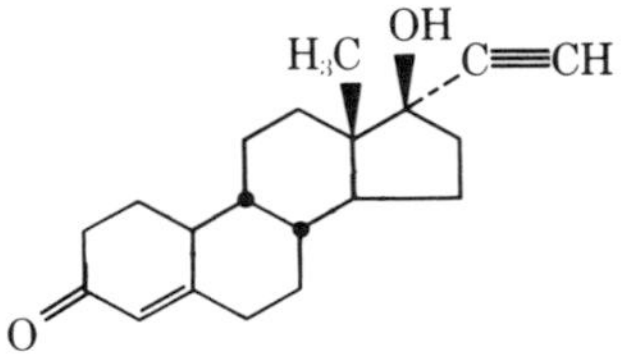

(46) 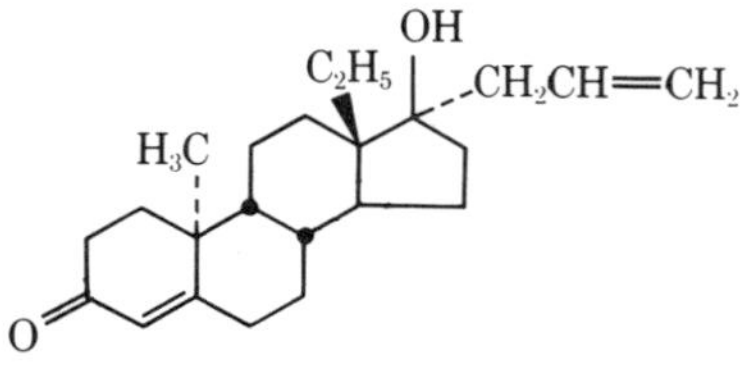   (47)

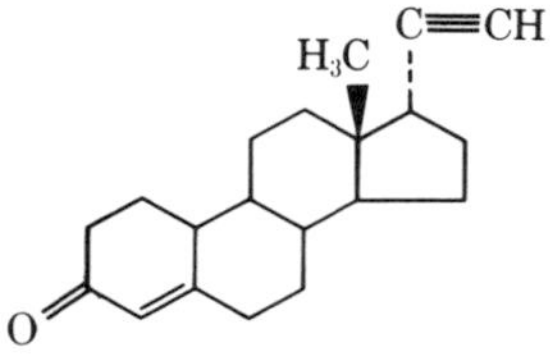

(48) 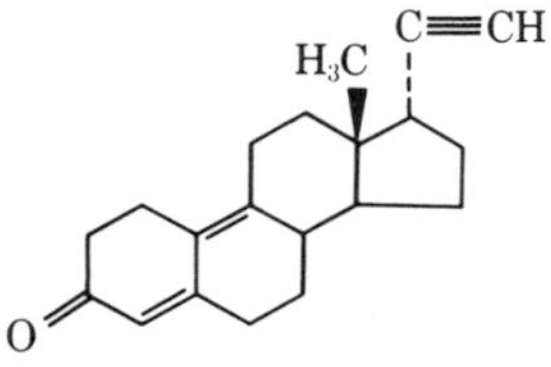   (49) 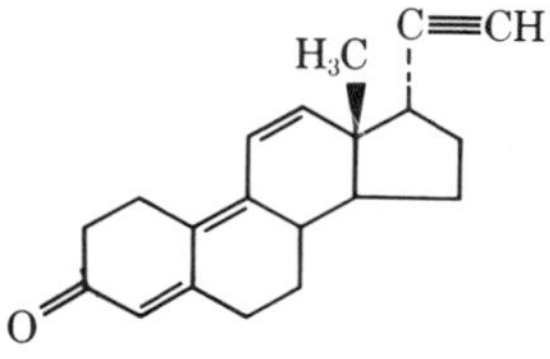   (50)

Progesterone-Like Structures.   Concurrently with the investigations in the 19-norsteroids, attempts were made to impart oral activity to progesterone (**3**) and to enhance its parenteral activity by introducing various groups into the molecule without alteration of the steroid skeleton. Among the earlier efforts was the insertion of an acyloxy group into the 17α-position, eg, 17α-acetoxyprogesterone (**51b**) is an orally-active progestin, and 17α-hydroxyprogesterone (**51a**) is inactive. The corresponding caproate (**51c**) has prolonged activity when administered parentally (44). It has been marketed as Delalutin.

As was the case with norethindrone (**4b**), further enhancement of activity could be achieved by inserting methyl and halogen groups into certain positions of the molecule and by extending the conjugated keto system with an additional double bond. 17α-Acetoxyprogesterone (**51b**) with a 6α-methyl group is a potent, orally active progestin known as medroxyprogesterone acetate (**52**) (45). It is the progestin component of Provest. Introduction of a double bond into the 6,7-position transforms (**52**) into megestrol acetate (**53**) (46). The latter progestin is present in a number of contraceptive products marketed in Europe. It is present in Pill #2, the oral contraceptive

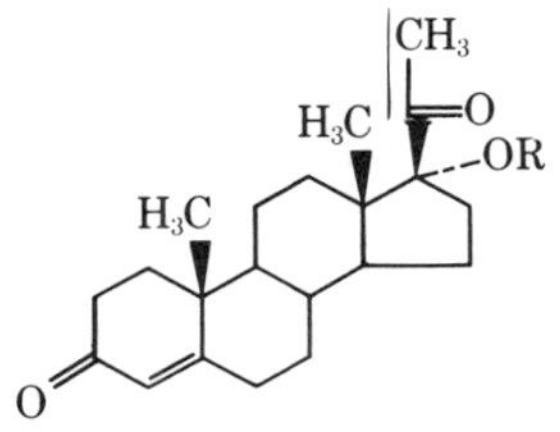

(51)

  (a) R = H
  (b) R = CH₃CO
  (c) R = CH₃(CH₂)₄CO

that is widely distributed in the People's Republic of China. Pill #1 is also used extensively in China. Its progestin component is (**4b**). In both pills, ethinylestradiol (**2b**) is the estrogen (47).

Other positions of the progesterone molecule (**3**) in which the presence of a methyl group enhances progestational activity are 16α and 17α. Thus, 17α-methylprogesterone (**54a**) is a very potent progestin (48–49), and (**54b**), in which an ethyl group is attached to C-17, has even greater activity. The presence of a moderate-size lipophilic group at the 17α-position contributes to progestational activity, conceivably because it promotes absorption or because it inhibits side chain degradation (50). Compounds in which both C-6 and C-17 are alkylated are probably more potent than the corresponding compound in which an alkyl group is present in just one of the two positions (48). Similarly, compounds having both C-6 and C-16 methylated are very potent progestins (51). Replacement of the 16α-methyl group with a methylene group does not adversely affect the biological activity. Attachment of a methylene group to C-16 of megestrol acetate affords melengestrol acetate (**55**), another potent progestin (52).

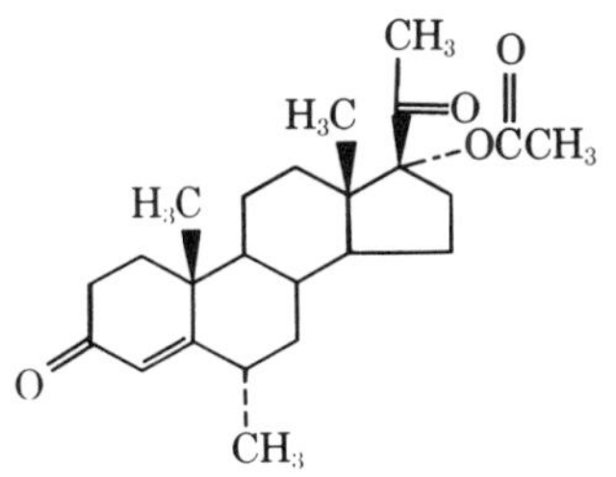

(52)

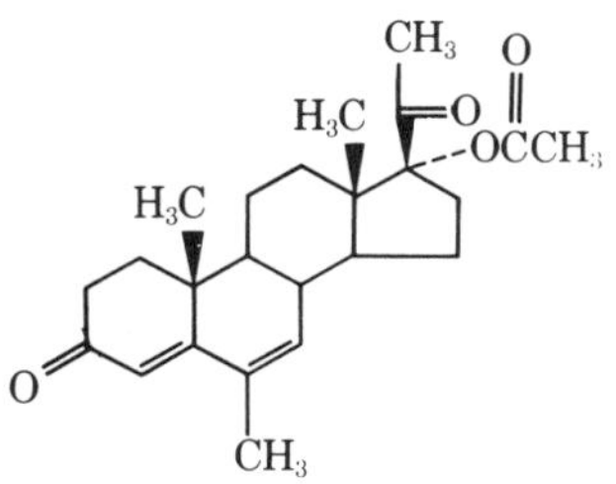

(53)

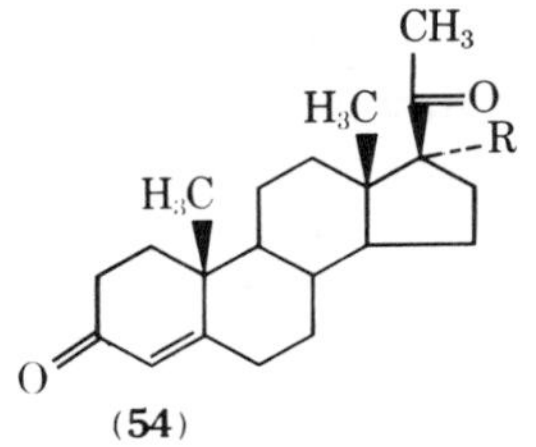

(54)

  (a) R = CH₃
  (b) R = C₂H₅

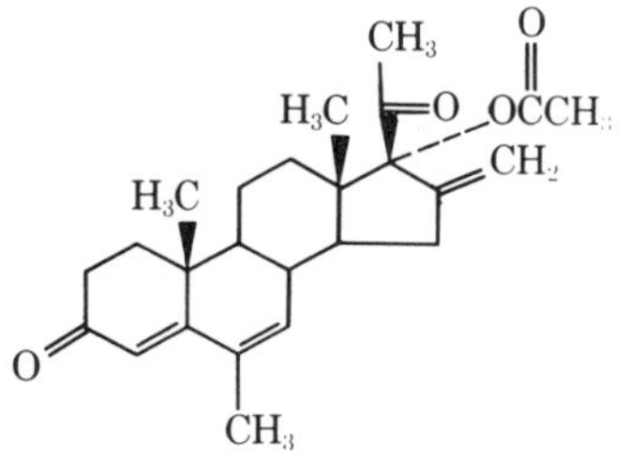

(55)

Enhancement of progestational activity can also be accomplished through the formation of halogenated derivatives. Attachment of a chlorine atom to either the 6- or 11-position, or a fluorine atom to either C-9 or C-21, generally results in greater potency. Indeed, among the most active progestational compounds reported to date are those in which chloro groups are attached to C-6 and C-11, but with the angular methyl group at C-10 missing, eg, (56) and (57). Compound (56) has been examined in the clinic. Unfortunately, the preliminary results do not confirm the contraceptive potential suggested by the animal studies (53).

Replacement of the methyl group at C-6 of megestrol acetate (53) with a chloro group furnishes chlormadinone acetate (58) (54), the progestin component of C-Quens, a sequential contraceptive product that was formerly marketed in the United States. This product is no longer sold in the United States because (58) produces benign mammary nodules in beagles (55). In similar long term studies, benign nodules have also been reportedly produced in beagles by medroxyprogesterone acetate (52), megestrol acetate (53), and even progesterone (3) (56). Since the significance of these findings in human usage is unclear, potential users should be warned that contraceptive drugs are not to be taken without proper medical guidance.

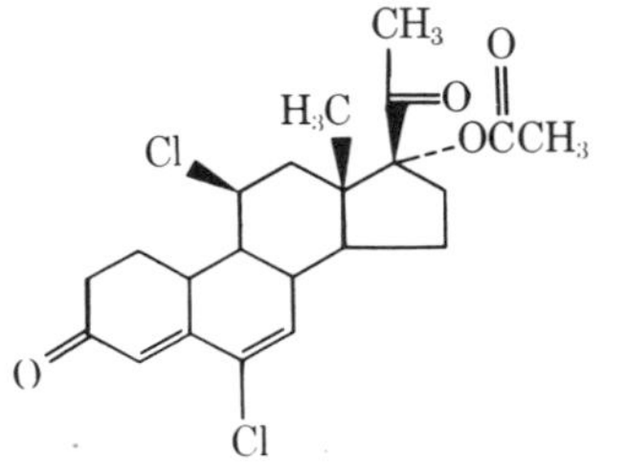

(56) 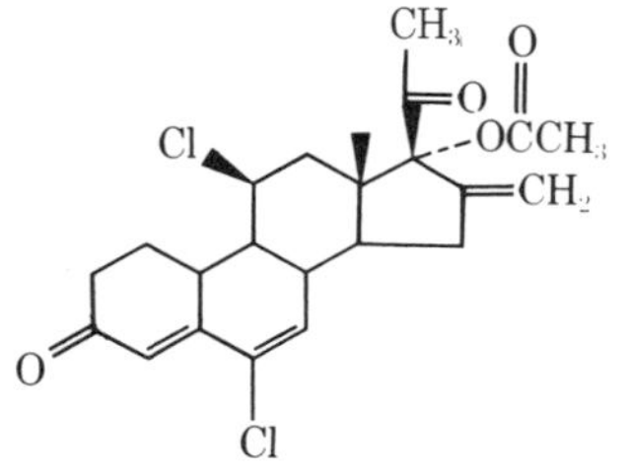   (57)

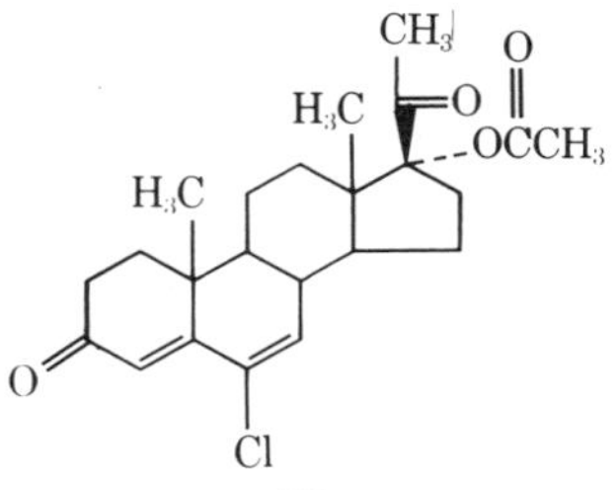

(58)

Although the presence of hydroxyl groups at both the 16α- and 17α-positions does not increase progestational activity, enhancement of potency can be accomplished by converting the 16,17-glycol to a dioxolane derivative by treatment with a ketone in the presence of a small amount of acid. 16α,17α-Dihydroxyprogesterone phenyl-acetonide (59), algestone acetophenide (57), is a cyclic ketal formed from 16α,17α-dihydroxyprogesterone and acetophenone. It is the progestin component of the injectable contraceptive, Deladroxate. The estrogen component of Deladroxate is estradiol 17-enanthate (60a).

The backbone configuration of progesterone (3) is anti–trans–anti–trans (61). Although inversion of one of the chiral centers will lead to a significant change in the shape of the molecule, it may not result in a complete loss of biological activity. Thus,

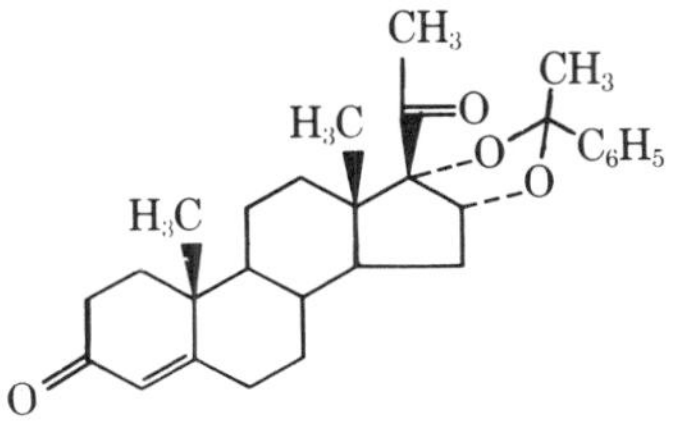

(59)

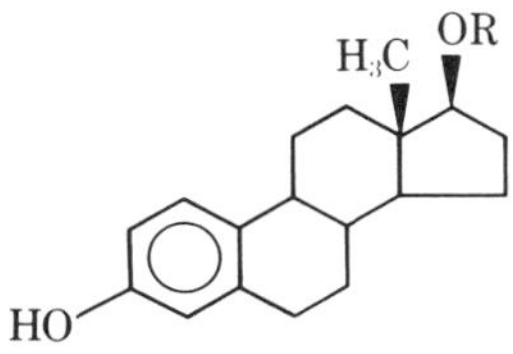

(60)

(a) R = CH₃(CH₂)₅CO

(b) R = H

8α-progesterone (62), which has the backbone configuration of *anti–cis–syn–trans*, has ca 30% the progestational activity of the natural hormone (3) (58). *dl*-$8\alpha,9\beta,10\alpha,14\beta$-Progesterone (63), which possesses the *anti–trans–anti–cis* backbone configuration, has 10% of the activity of progesterone in the Clauberg assay (59). The 19-norprogesterone originally prepared by Ehrenstein was later shown to have the isomeric configuration at C-14 and C-17, ie, the compound is $14\beta,17\alpha,19$-norprogesterone (64), and its backbone configuration is *anti–trans–syn–cis* (60).

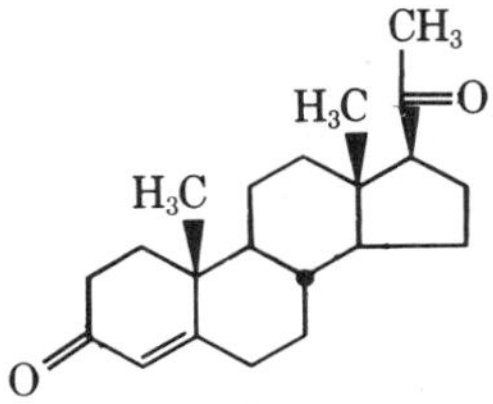

(61)

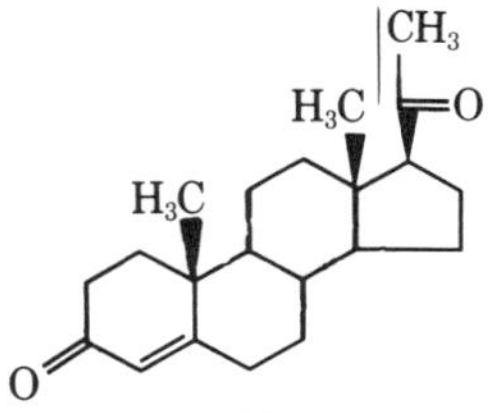

(62)

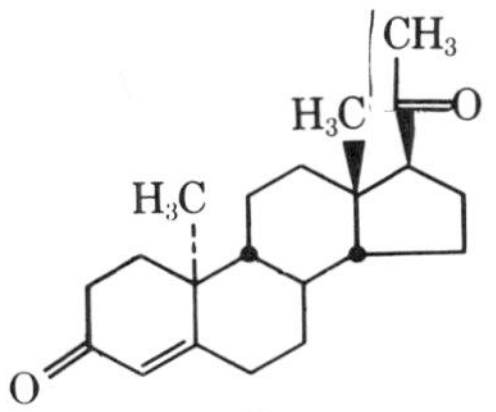

(63)

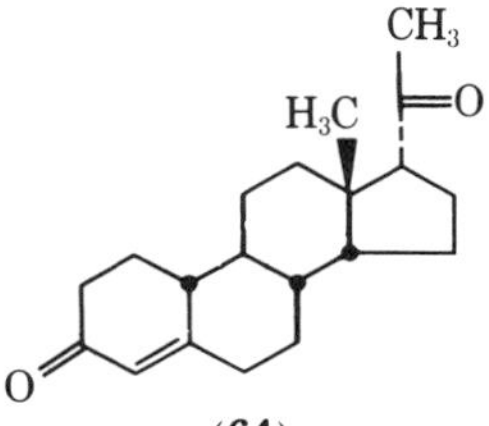

(64)

The potent, orally active progestin dydrogesterone (65) belongs to a class of compounds known as retrosteroids. In these steroids, the methyl group at C-10 and the hydrogen atom at C-9 are oriented α and β, respectively. The backbone configuration of the retrosteroids is *anti–cis–anti–trans*. Dydrogesterone (65) is marketed in Europe as Duphaston for the treatment of menstrual disorders. Given in combination with quinestrol (2c), the long acting estrogen that is stored in body fat, (65) has been observed to block pregnancy when given in a single dose on day 22 of a woman's menstrual cycle (61).

At one time it was thought that a high degree of structural and configurational specificity was required for hormonal activity. However, recent findings clearly indicate that the structural and configurational requirements for progestational activity are not as rigorous as they were once thought to be.

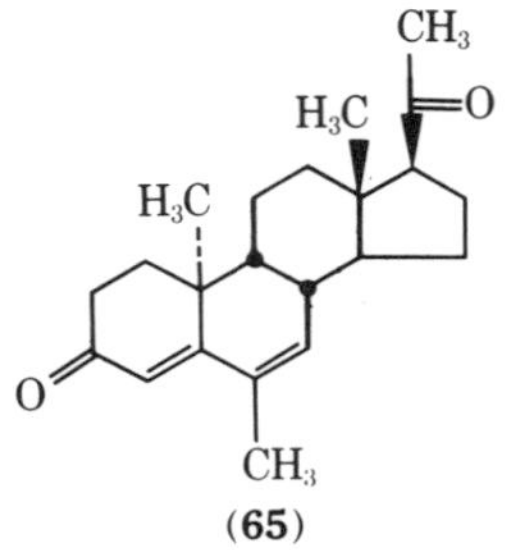

(65)

Postcoital Contraceptives. The progestin–estrogen combination pill produces its contraceptive effect mainly by inhibiting ovulation. In order to achieve 100% efficacy, strict adherence to the prescribed regimen is required. Women who do not engage in frequent sexual activity may not wish to take a pill every day. For those women who occasionally indulge in unprotected midcycle coitus or who are rape victims, a postovulatory, postcoital (morning-after) pill which prevents implantation would have considerable appeal.

Estrogens given in high doses will prevent implantation in women (62). They do not interfere with fertilization. Once implantation has occurred, estrogens will have no effect on gestation.

Agents that prevent implantation are known as interceptives. Generally they act by altering the rate of ovum or zygote transport. They may also act by luteolysis. Estrogens are luteolytic in certain species as evidenced by a reduction in plasma progesterone levels or a decrease in basal body temperature following postovulatory administration. Estrogen may also prevent implantation by altering the sensitive hormonal balance that is essential for synchronization of the uterine environment with endometrial implantation.

Both steroidal and nonsteroidal estrogens have been used postcoitally to prevent pregnancy in women. The orally active estrogens that have proved to be most effective are diethylstilbestrol (66) (however, see below), ethinylestradiol (2b) (5 mg/d), and conjugated equine estrogens (Premarin, 30 mg/d).

For maximum effectiveness, the estrogens should be taken within 24 h after coitus and no later than 72 h. Treatment should be continued for 5 consecutive days.

The extended period of treatment is necessary as the viability of the sperm in the female reproductive tract is 48–72 h, and ovulation may occur several days after coitus. If the time of ovulation is known precisely, then treatment can be limited to one or two days after release of the ovum.

To be effective, the estrogens must be given in relatively high doses. This may result in the production of side effects, such as edema, thrombophlebitis, menstrual irregularities, vomiting, and nausea. If ineffective doses are administered or if treatment is begun after implantation, fetal malformations may occur.

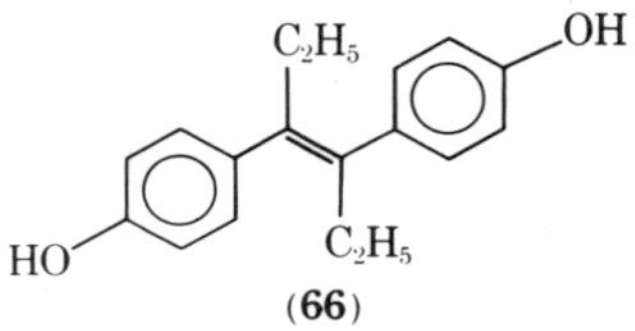

(66)

Herbst and co-workers observed that daughters of women treated with diethyl-stilbestrol (**66**) in the first trimester of pregnancy have a higher incidence of vaginal adenosis and adenocarcinoma (63). Consequently, it has been recommended that termination of pregnancy by surgical means be employed should the interceptive method fail.

Because of the reported side effects, efforts are being made to modify the structures of the estrogens in order to achieve a greater separation of antifertility and estrogenic activity (64). Analogues of ethinylestradiol (**2b**), in which the ethynyl hydrogen has been replaced by either an ethynyl (**67a**) or a trifluoromethyl group (**67b**), have an enhanced separation of antifertility and estrogenic activity (65).

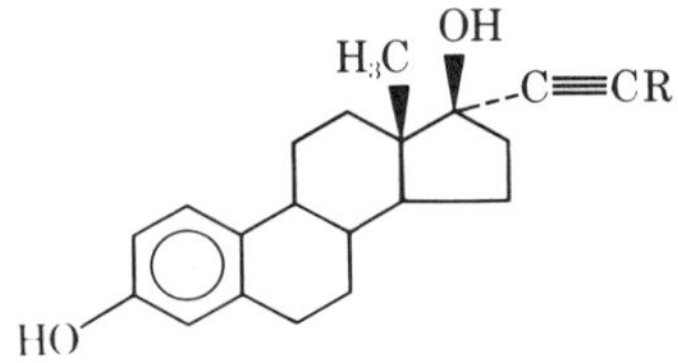

(**67**)

(a) R = HC ≡ C
(b) R = CF$_3$

Surprisingly, A-norandrostane-2α,17α-diethynyl-2β,17β-diol (**68a**), which lacks the aromatic ring system characteristic of the estrogens, is highly active in inhibiting implantation in the rat. In addition, it exhibits considerable potency in the immature mouse uterotrophic assay. The mechanism by which (**68a**) inhibits implantation involves alteration of ovum transport and uterine development (66).

In pursuing this lead, chemists in the People's Republic of China prepared the dipropionate of (**68a**) (67). Initially, pharmacological studies were performed on a mixture of epimers at C-2. Subsequently, the synthetic process was improved so that one epimer (**68b**) was obtained almost exclusively. Thereafter, studies were conducted on (**68b**). This compound has about 2.8% the estrogenic potency of ethinylestradiol (**2b**), as determined in the immature mouse uterotrophic assay. The A-nor compound exhibits antiprogestational activity in the rabbit. It inhibits ovulation or implantation or both in various species. In the clinic Anordrin (**68b**) is nearly 100% effective in preventing pregnancy. For human usage, Anordrin is administered orally. One tablet, containing 7.5 mg of Anordrin, is taken at mid-cycle, and an additional tablet is taken immediately following each intercourse thereafter during the cycle. In the unique socioeconomic system that has been established in China, husbands and wives are frequently separated for a considerable length of time because of their work. Thus,

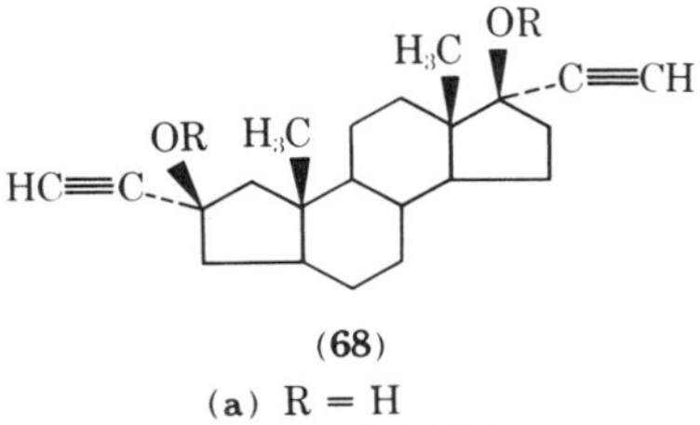

(**68**)

(a) R = H
(b) R = C$_2$H$_5$CO

the Chinese refer to Anordrin as the vacation pill, used when husbands and wives are able to be together.

One scheme that is utilized for the synthesis of Anordrin entails starting with 3β-hydroxy-5α-androstan-17-one (69) which is derived from the sapogenin, tigogenin (70a). Oxidative cleavage of ring A of the keto alcohol (69) can be achieved selectively with chromic acid at 60°C to afford the dicarboxylic acid (71). Cyclization of the latter compound with acetic anhydride and sodium acetate furnishes the diketone (72). Ethynylation with acetylene in tetrahydrofuran at 0°C, in the presence of powdered potassium hydroxide proceeds stereoselectively at both C-2 and C-17 to give (68a). Acylation of (68a) with propionic anhydride and p-toluenesulfonic acid at room temperature affords Anordrin (68b).

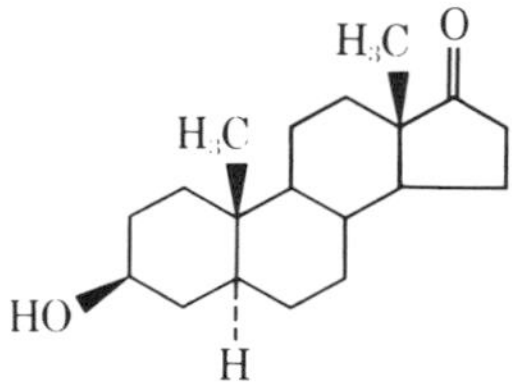

(69)

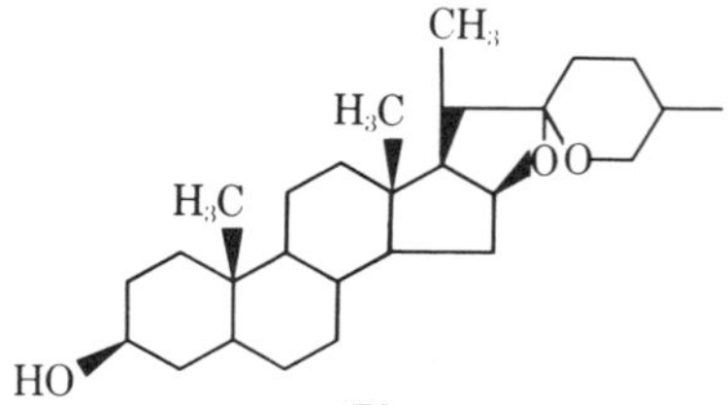

(70)
(a) 5α-H
(b) Δ^5-double bond

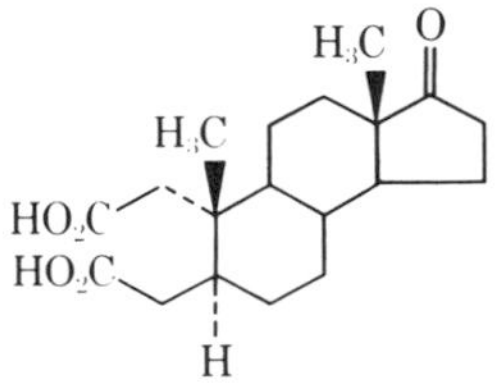

(71)

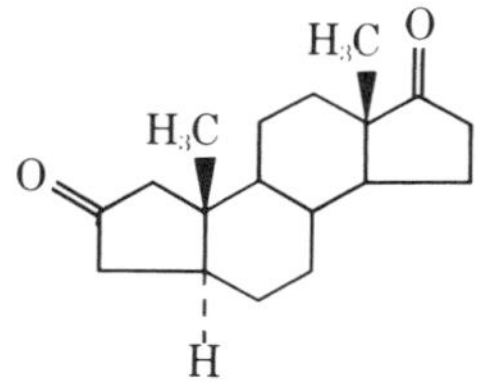

(72)

The synthetic estrogen diethylstilbestrol (66) is a derivative of stilbene. Many stilbene and dibenzyl derivatives, either occurring naturally (68) or produced synthetically, possess estrogenic activity of varying degree. The ovulation stimulant clomiphene citrate (73) is a weak estrogen. Its postcoital antifertility effect in several rodent species, as well as in the rabbit, has been determined. However, it has no effect on pregnancy when given postcoitally to the monkey (69).

Compounds in which the stilbene or dibenzyl moiety is part of a ring system have also been found to display postcoital antifertility effects, eg, U-11,100A (74) (70), U-11,555A (75) (70), and centchroman (76) (71).

Besides the estrogens, progestins such as norethindrone (4a), norgestrel (26), quingestanol acetate (28), and R 2323 (31b) are currently being studied in the clinic as potential postcoital contraceptives (72–77).

Progesterone Blockers. The binding of a hormone to its polypeptide receptor in the cytosol initiates the chain of events that results in the biological phenomenon observed. This has been demonstrated for progesterone (3), the pregnancy hormone, as well as for the estrogens, androgens, and corticoids (78).

The progesterone-receptor complex formed in the cytosol is translocated into the nucleus where it acts on the chromatin to promote the synthesis of deoxyribonucleic acid (DNA)-dependent mRNA (messenger ribonucleic acid). Through the process

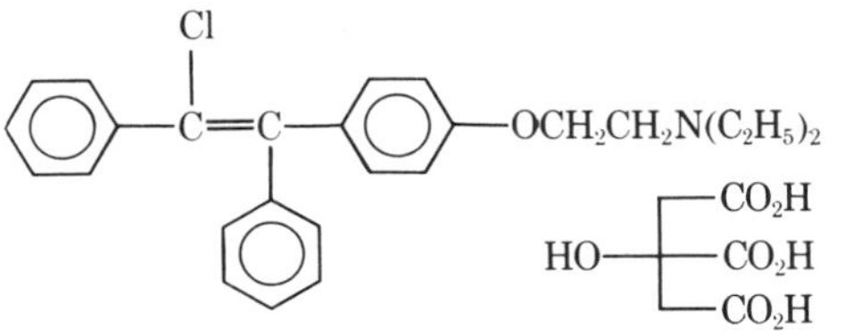

(73) 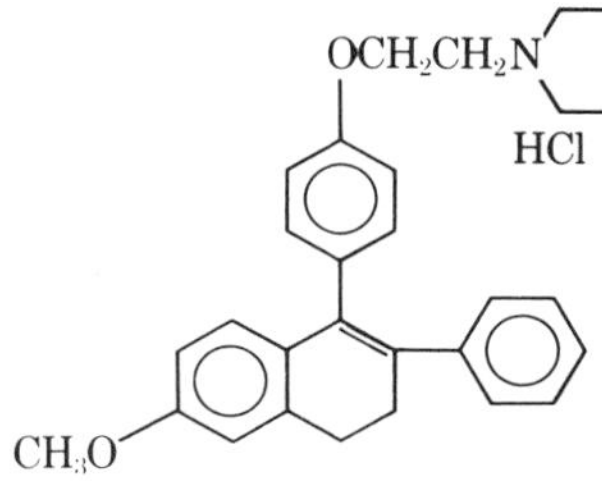   (74)

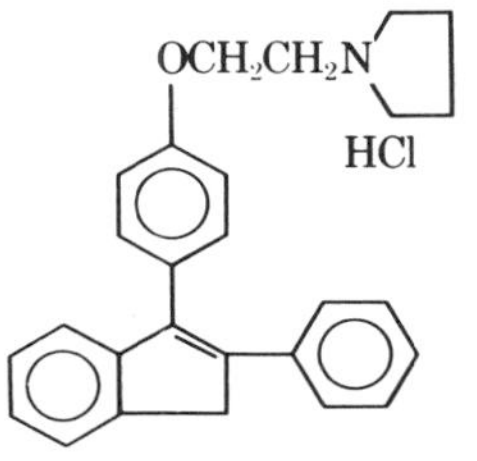

(75) 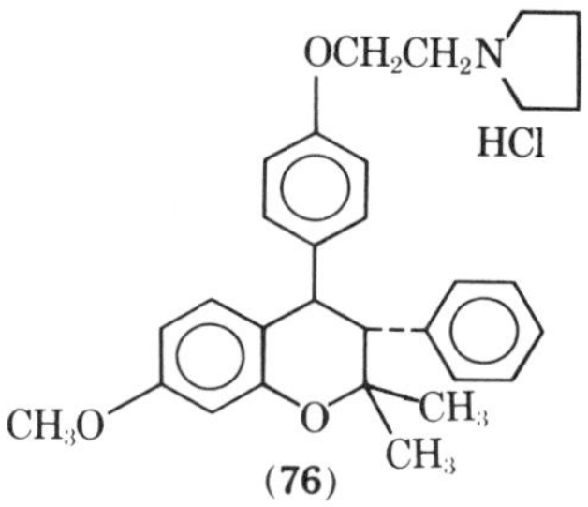   (76)

of translation, the latter induces the synthesis of proteins that are essential for implantation and maintenance of pregnancy.

Interference with the binding of (3) to its receptor may terminate pregnancy. The isolation of the progesterone receptor from human uteri, as well as from the uteri of several other species (79), has stimulated interest in finding compounds that would compete with (3) in binding to its receptor and that would not support pregnancy. Comparison of the relative binding affinity with *in vivo* results may also provide insight as to whether a compound undergoes metabolic transformation prior to eliciting the effects seen in the Clauberg and other *in vivo* assays (80).

The Clauberg assay is used to determine the extent to which a substance promotes the proliferation of the endometrium. Although compounds have been found that display high progestational activity in the Clauberg assay but do not support pregnancy in ovariectomized animals, the compounds that do maintain pregnancy are, without exception, active in this assay (80–81).

A good correlation was demonstrated between binding to the human progesterone receptor and activity in the Clauberg assay (80,82). For example, 17α-hydroxyprogesterone (51a) is essentially inactive in the Clauberg assay. It also binds poorly to the progesterone receptor. On the other hand, acetylation of the hydroxyl group of (51a) affords (51b) which exhibits substantial activity in the Clauberg assay and binds fairly well to the human progesterone receptor.

Besides the derivatives of (3), some 17α-ethynyltestosterone (8c) analogues also compete well with (3) in binding to its receptor. Marked enhancement in the binding affinity was observed when the angular methyl group at C-10 was removed. This lends support to a proposal that binding of the progestin to the receptor occurs preferentially on the β-face of rings A and B of the molecule (83–84).

Retroprogesterone (77) and dydrogesterone (78) are potent progestins. Both compounds have the 9β,10α-configuration. Because of the difference in configuration at C-9 and C-10, the shape of a 9β,10α-steroid is vastly different from that of progesterone (3) (85). Nevertheless, the binding affinities of (77) and (78) are comparable to that of (3) in the myometrium progesterone-binding proteins obtained from either the human or rabbit uterus. This suggests that either the progesterone receptor has a high degree of flexibility and, within limits, conformational changes do not lead to loss of activity, or that for progestational activity binding of the progestin to the receptor needs only to involve certain portions of the steroid molecule and not the entire structure.

Compounds are known that bind well to the progesterone receptor and yet are neither progestational or antiprogestational *in vivo* (50). These compounds may be inactive *in vivo* because of catabolism; 5α-pregnanedione (79a) is such a compound. Conceivably, the lack of *in vivo* activity is caused by metabolic degradation of the side chain or by poor absorption. The presence of an alkyl group at *C*-17 of progesterone has, on the one hand, been reported to enhance progestational activity by deterring the catabolism of the pregnane side chain and, on the other, by promoting absorption of the molecule (50,86).

17α-Ethyl-5α-pregnane-3,20 dione (79b) was prepared and tested. Although (79b) strongly binds to the progesterone receptor, it failed both to prevent implantation and to terminate pregnancy in the rat at doses of 10 mg/kg and 40 mg/kg, respectively (50).

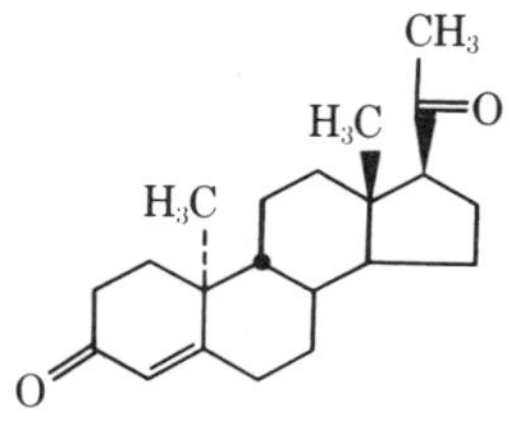

(77) 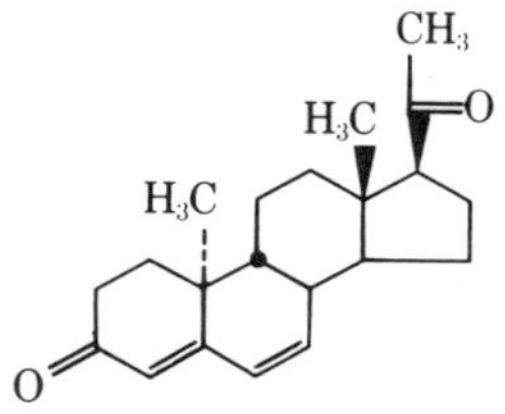   (78) 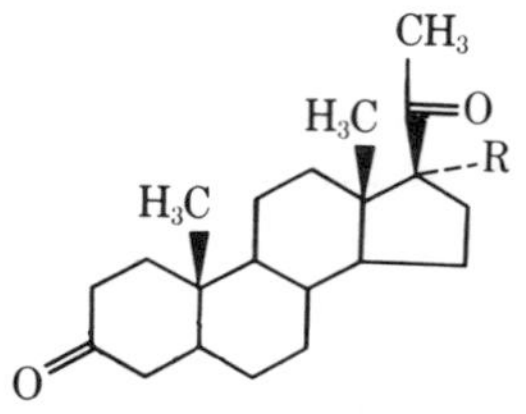   (79)
(a) R = H
(b) R = C$_2$H$_5$

13β-Ethyl-17α-ethynyl-17β-hydroxygona-4,9,11-trien-3-one [(31b), R 2323] has been in clinical trials as a once-a-week pill. In addition, its effect on pregnancy when administered on days 15, 16, and 17 of the menstrual cycle has also been studied in the clinic. In 2148 cycles of midcycle oral administration, the daily dose of 50 mg for three days produced a drug-failure pregnancy rate of 5%; (31b) exhibits marked antiprogesterone and moderate antiestrogenic activity. It does not maintain pregnancy in the castrated animal. R 2323 (31b) is believed to compete with (3) for the uterine cytosol progesterone-binding sites. As synthesis of the progesterone receptor is estradiol-induced, the antiestrogenic activity of (31b) would be expected to enhance the compound's contragestational effect (77).

Suppression of Spermatogenesis

Oral contraceptives for women inhibit ovulation by suppressing the mid-cycle surge of gonadotrophins. Similar suppression of gonadotrophins in males can be attained with the steroidal agents. Although spermatogenesis is inhibited as a result, it is often accompanied by loss of libido, potency, and secondary sexual characteristics. However, this can be avoided if an androgen is administered with the antigonadotrophic steroid. If the androgen is given at a high dose, a synergistic suppression of spermatogenesis is achieved as well (87).

Danazol (**80**) (88) is an agent that is employed in the treatment of endometriosis. It is devoid of estrogenic effects, and it has been evaluated as an oral contraceptive in women (89).

This compound is a 17α-ethynylandrostene derivative with an isoxazole ring fused to ring A of the steroid nucleus. When given alone in oral doses of 600 mg/d to men, (**80**) produces a slight and variable reduction in sperm concentration and a uniform reduction in serum luteinizing hormone (LH) and testosterone (**8a**) concentrations. However, when 10 mg of testosterone propionate (**81a**) is given intramuscularly three times a week in addition to (**80**), the sperm count is reduced more consistently, often to oligospermic levels. If the supplemental androgen is testosterone enanthate (**81b**), and if it is administered intramuscularly at a dose of 200 mg once monthly in addition to the daily administration of (**80**), the synergistic suppression of spermatogenesis is intensified, and the sperm count drops to less than 1 million/mL within two months (90).

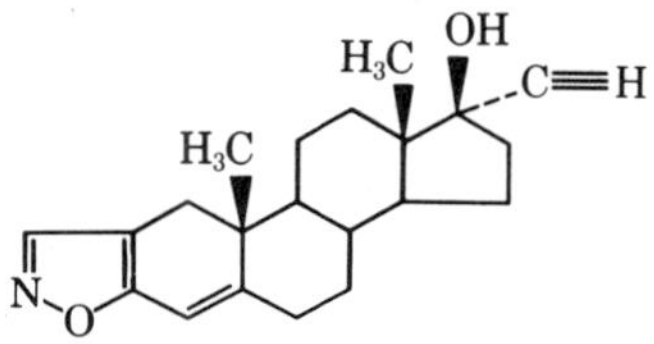

(**80**)

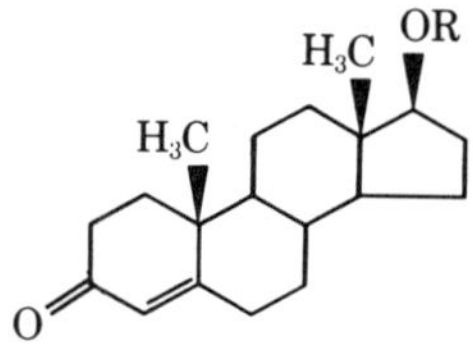

(**81**)

(a) R = C_2H_5CO
(b) R = $CH_3(CH_2)_5CO$
(c) R = $CH_3(CH_2)_9CO$

In another study, subdermal implants of long-acting testosterone Silastic capsules were used to prevent the major side effects produced by progestin-induced gonadotrophin inhibition in man (87). The progestins examined were norgestrienone (**31a**), R 2323 (**31b**), and norethindrone (**4b**). They were also administered in capsule form subcutaneously in the ventral aspect of the forearm. In addition to the implants, each subject received the progestin orally beginning either immediately or 8–12 weeks after implantation. The oral progestin was given at a dose of 50 mg one to three times a week. Each of the capsules inserted contained 40 or 50 mg of the progestin. Besides the 3 or 4 progestin capsules inserted, each subject also had 2 or 3 capsules of testosterone implanted. Each androgen capsule contained 23 mg of finely ground testosterone (**8a**). The results indicated that subjects who received (**8a**) and either (**31a**) or (**31b**) had a marked reduction in sperm count, without loss of libido or potency, several months following the initiation of the oral intake of the progestin. Subjects who received (**8a**) and (**4b**), on the other hand, did not exhibit a comparable drop in the sperm count.

The initial results with (**80**), (**31a**), and (**31b**) were encouraging. However, before large-scale testing of the progestin–androgen combinations can be undertaken, extensive toxicity studies in animals have to be conducted. This has led to a proposal that existing oral hormone products that have been approved for treatment of various disorders in males be studied for their effects on spermatogenesis (91).

A combination of 17α-methyltestosterone (**8b**) and ethinylestradiol (**2b**) is used for the treatment of osteoporosis and symptoms of the male climacteric. Tablets containing 10 mg of (**8b**) and 20 μg of (**2b**) were administered twice daily to healthy male volunteers. The androgen–estrogen combination inhibited spermatogenesis. Sperm count and motility decreased significantly at the 12th week of treatment. They returned to normal 35–40 weeks after cessation of treatment, thereby demonstrating the reversibility of the treatment.

Cyproterone acetate (**82**), the potent progestin with antiandrogen properties, has also been examined for its effects on reproduction in the male (92). In male rats, maximum reduction in fertility was observed after 5 weeks of daily treatment with 20 mg/kg of (**82**). Although there was a significant increase in testosterone (**8a**) levels, reduction in testes and epididymal weight, loss of libido, and atrophy of the seminal vesicles were observed. In normal men, daily doses of 10 and 20 mg of (**82**) were found to decrease sperm count. However, (**82**) also produced a concomitant drop in testosterone levels (93).

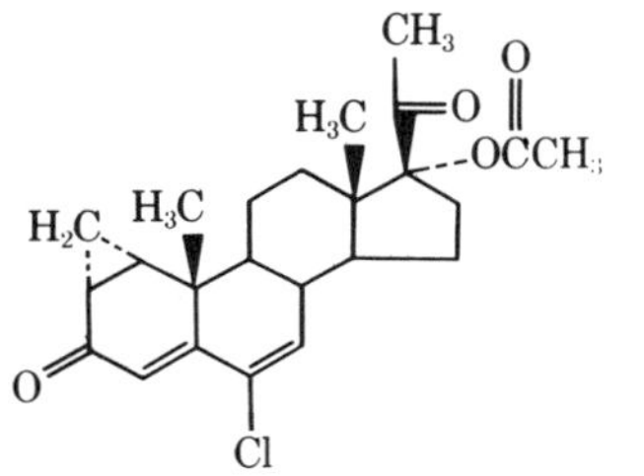

(**82**)

Because androgens suppress follicle-stimulating hormone (FSH) and luteinizing hormone (LH) levels, testosterone (**8a**) and its esters are being studied as potential male contraceptive agents. By maintaining a steady peripheral plasma level of (**8a**), inhibition of libido and potency can be avoided. The consistent concentration of (**8a**) in the peripheral plasma prevents the massive release of LH and FSH. Consequently, synthesis of (**8a**) in the Leydig cells of the testes, which is stimulated by the surge of LH, is suppressed and spermatogenesis is curtailed (94–95).

Testosterone (**8a**) and its esters, some of which are long-acting substances, are being tested in the clinic. In one study, 25 mg of (**8a**) was injected daily into normal men. After several months azoospermia was produced with no loss in libido or potency. Because (**8a**) is not active orally it is also administered by implanting Silastic capsules under the skin. Testosterone undecanoate (**81c**) is orally active, and it is currently undergoing clinical trials. 17α-Methyltestosterone (**8b**) and related 17α-substituted analogues are also orally active androgens. However, because they produce liver dysfunction or cholestasis, or both, at high doses (95), it appears unlikely that they would be given to men continuously or for long periods in order to suppress spermatogenesis. In addition to possible hepatic toxicity, concerns that long-term administration of androgens may stimulate development of the accessory sex glands and affect the cardiovascular system will have a tempering effect on the use of androgens as male contraceptives.

Manufacture

The commercial preparation of the progestin and estrogen components of the oral contraceptives can be accomplished by either partial or total synthesis. An ingenious procedure that makes use of carbon dating can be employed to determine whether one or the other form of synthesis had been utilized to prepare a specific product. The starting material used in partial synthesis is from the plant or animal kingdom and is of recent origin. Total synthesis, on the other hand, uses materials that are most likely derived from fossil fuels, either coal or petroleum (96).

In general, the starting material for partial synthesis is a sapogenin, as for example, diosgenin (**70b**), which is obtained mainly from the dioscorea plant. Degradation of the sapogenin side chain affords 16-dehydropregnenolone (**83**) (97). With appropriate modifications, (**83**) can be converted into compounds that resemble progesterone (**3**) in structure. Removal of the side chain of (**83**) furnishes dehydroisoandrosterone (**18b**). The latter compound is converted into ethynyltestosterone (**4a**) by successive ethynylation and oxidation, and it is used in a number of syntheses of norethindrone (**4b**) and structurally related compounds. In one synthesis, (**18b**) is transformed in several steps to estrone (**7a**) with expulsion of the 19-methyl group that is attached to C-10. Estrone (**7a**) is the estra-1,3,5(10)-triene derivative that was used in the initial syntheses of the 19-nonsteroids. In another approach, the hydroxyl group at C-3 of (**18b**) is first suitably protected. A 6β-hydroxyl group is then introduced into the molecule to give, for example, (**84**). The 19-methyl group and the 6β-hydroxyl group are *syn*-periplanar. Hence, the hydroxyl group at the 6-position can be used to convert the 19-methyl group to a functional derivative. Subsequent expulsion of the derivative of the methyl group, either as formaldehyde or carbon dioxide, provides the estr-4-ene system of norethindrone (**4b**) or the estr-5(10)-ene system of norethynodrel (**1**).

Oxidation of dehydroisoandrosterone (**18b**) yields androst-4-ene-3,17-dione (**85a**). The two carbonyl functions in (**85a**) differ in their reactivity. Consequently, reactions can proceed regioselectively, and (**85a**) can be utilized for elaboration of more complex steroids. For example, reaction with ethyl orthoformate in the presence of a catalytic

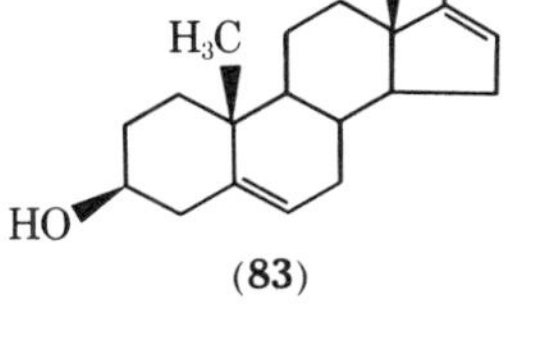

(**83**)

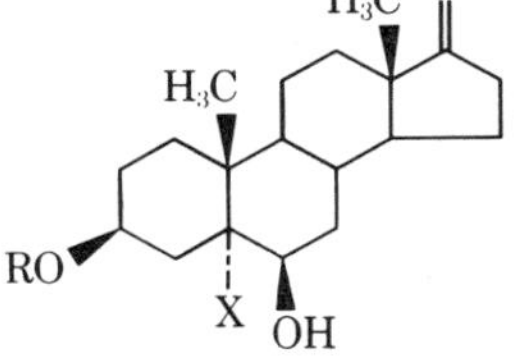

(**84**)

amount of p-toluenesulfonic acid affords an enol ether in which only the carbonyl group at the 3-position has entered into the reaction. Hydride reduction followed by acid cleavage of the enol ether gives testosterone (8a). In recent years, efficient fermentation processes have been developed for the preparation of (85a), starting from the abundant soya sterol, sitosterol (86) (98) (see Fermentation). These developments are of commercial importance as they allow processors to be less dependent on the discorea plant and permit the utilization of abundant and inexpensive sterols as starting materials.

(85)
(a) R = CH₃
(b) R = H

(86)

In contrast to partial synthesis, which affords a product free of its enantiomer, total synthesis as practiced in the past furnished products or intermediates that were racemic. At some stage in the synthesis a resolution with a chiral reagent was required to remove the undesired enantiomer which was generally discarded.

The Torgov-Smith process (99), when first applied to the synthesis of the 13-ethyl steroids, furnished dl-norgestrel (26a). To avoid the necessity of having to generate the undesired enantiomer, an approach was developed in which a chiral reagent could be employed to react selectively at a particular prochiral center present in an intermediate (or a substrate) to afford an optically active product having predominantly the desired configuration. A variety of microorganisms and chiral chemical reagents were found to be effective in the processes developed (100–102). As a result, asymmetric synthesis has become a useful means for the commercial production of optically active steroids (see Pharmaceuticals, optically active).

For the preparation of d(−)-norgestrel (26b), the seco diketone (87), which is an intermediate in the synthesis of dl-norgestrel (26a), can be reduced selectively by *Saccharomyces uvarium* in a fermentation process to afford the optically active hydroxy ketone (88) (100). Successive acetylation and cyclization provide the tetracyclic compound (89). Reduction of the double bond at the 14,15-position is readily achieved by catalytic hydrogenation to give (90a). Saponification yields the 17β-hydroxy compound (90b). Conversion of (90b) to (26b) can be accomplished in a manner identical to that employed for the preparation of the racemic compound. The lithium–ammonia reduction of (90b) reduces not only the 8,9-double bond but also the aromatic A-ring to afford (91). The hydroxyl function is converted into a carbonyl group by means of the Oppenauer oxidation to furnish (92). Ethynylation followed by acid cleavage of the enol ether system transform (92) into (26b).

Another application of asymmetric synthesis involves the generation of the optically active bicyclic system (93), which is characteristic of rings C and D of the steroids, from the achiral cyclopentanedione derivative (94). The transformation is an intramolecular aldol cyclization induced by an optically active amino acid. The amino acid can be either primary [eg, (S)-phenylalanine] or secondary [eg, (S)-proline] (101).

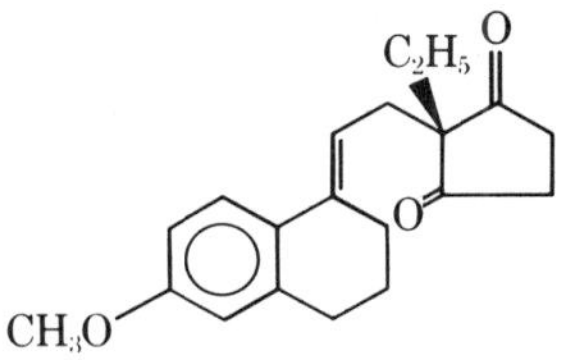

(87)
[830-92-0]

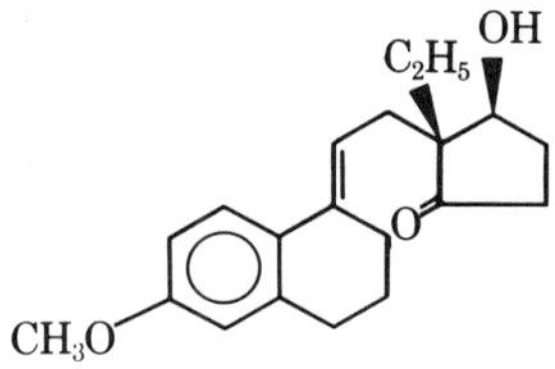

(88)
[14507-43-8]

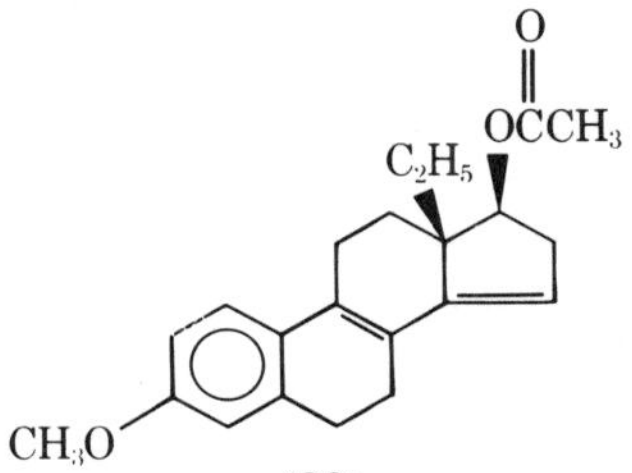

(89)
[2911-81-1]

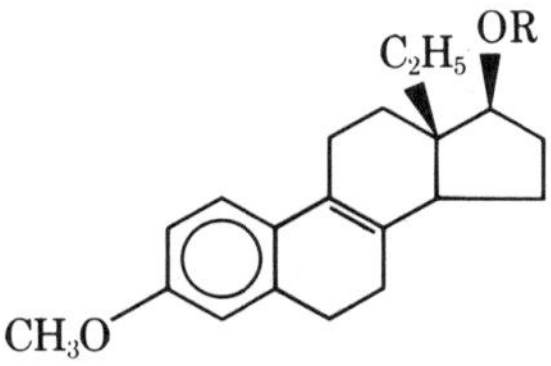

(90)
(a) R = CH₃CO
[19874-46-5]
(b) R = H
[7443-72-3]

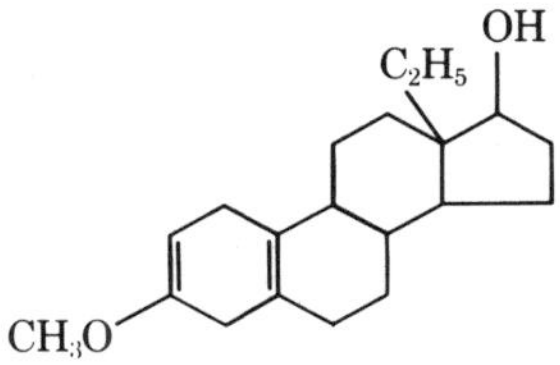

(91)
[14507-49-4]

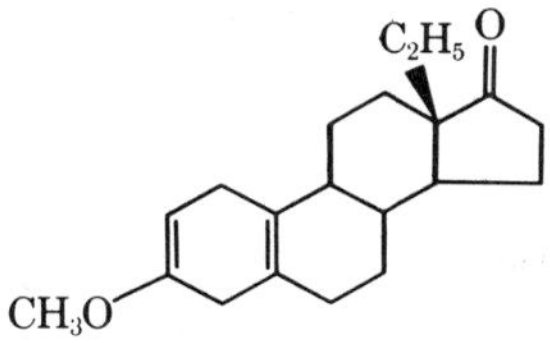

(92)
[2322-77-2]

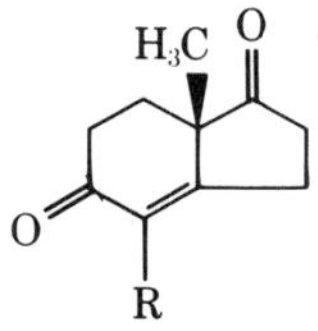

(93)

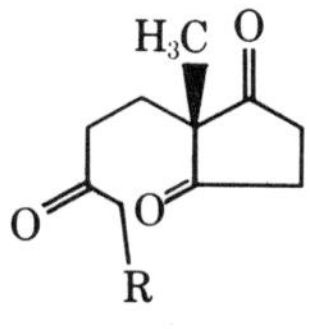

(94)

Further elaboration transforms (93) into 19-norandrost-4-ene-3,17-dione (85b), a key intermediate in several commercial syntheses of norethindrone (4b) and ethynyldiol diacetate (17).

Economic Aspects

Until recently diosgenin obtained from the Mexican species of the dioscorea plant was the chief starting material for the production of steroid drugs. With technical improvements made in microbiological transformations, however, processes involving the fermentation of abundant sterols, such as sitosterol and cholesterol, are now being widely used for the production of contraceptives. Efficiency achieved in total synthesis has not only made it economically possible to prepare steroids with an ethyl group attached to C-13 [eg, d-norgestrel (26b)], but also steroids with a methyl group at that position. As a result, production of steroids either involving fermentation or by total synthesis competes favorably with the production of steroids from dioscorea. Indeed, in some cases steroids prepared from the fermentation of sterols or by total synthesis are 30–60% less expensive than the same steroids derived from dioscorea (96).

Dioscorea is no longer in plentiful supply, and its collection in Mexico has become completely nationalized. The economic consequences of this action have been brilliantly analyzed by Djerassi (96). Although the immediate effect may be one of lower cost of the product to consumers in Mexico, in the long run the cost is expected to increase considerably because of smaller volume production.

In 1975, the U.S. Agency for International Development (AID) purchased 100 million cycle-equivalents of an oral contraceptive preparation for approximately $0.15/cycle. The cost of the same product for the private market was in excess of ten times the price paid by AID. Profits made in the private sector, together with large volume production, enabled the producer to make its product available to the government agency at a nominal cost.

Many steroidal contraceptive drugs are listed in Table 1, including virtually all those manufactured in the United States.

Table 1.   Composition of Some Steroidal Contraceptive Drugs

| Trade name | Manufacturer | Composition and structure no. | |
		Progestin	Estrogen
Anovlar	Schering AG	(15)	(2b)
Brevicon	Syntex	(4b)	(2b)
Deladroxate	Squibb	(59)	(60)
Demulen	Searle	(17)	(2c)
Enovid	Searle	(1)	(2a)
Loestrin	Parke-Davis	(15)	(2b)
Lo/Ovral	Wyeth	(26a)	(2b)
Lyndiol	Organon	(29)	(2a)
Micronor	Ortho	(4b)	none
Modicon	Ortho	(4b)	(2b)
Norigest	Schering AG	(27)	none
Norinyl	Syntex	(4b)	(2a)
Norlestrin	Parke-Davis	(15)	(2b)
Nor-Q.D.	Syntex	(4b)	none
Ortho-Novum	Parke-Davis	(4b)	(2a)
Ovcon	Mead Johnson	(4b)	(2b)
Ovral	Wyeth	(26a)	(2b)
Ovrette	Wyeth	(26a)	none
Ovulen	Searle	(11)	(2a)
Planor	Roussel-Uclaf	(31a)	(2a)
Provest	Upjohn	(52)	none
Riglovis	Warner-Lambert	(28)	(2b)
Zorane	Lederle	(15)	(2b)

POLYPEPTIDES

Luteinizing Hormone-Releasing Hormone (LH–RH) Analogues

Ovulation involves the interplay of the hypothalamus, pituitary, and ovaries. The pituitary controls the ovaries by means of luteinizing hormone (LH) and follicle-stimulating hormone (FSH). Secretion of LH and FSH by the pituitary gland is under the control of the hypothalamus. The hypothalamic chemotransmitter associated with this process is a decapeptide known as luteinizing hormone–releasing hormone (LH–RH) or as gonadotrophin-releasing hormone (GnRH). The amino acid (qv) sequence of LH–RH has been established as (Pyro)Glu–His–Trp–Ser–Tyr–Gly–Leu–Arg–Pro–Gly–NH_2 by Schally, Guillemin, and their associates (103–104). The isolation of the decapeptide hormone, 800/μg from 160,000 porcine hypothalami, and the subsequent elucidation of structure and synthesis provide another approach to fertility control (see Immunotherapeutic agents; Polypeptides).

Schally and his colleagues demonstrated that the progestin–estrogen combinations act on both the pituitary and the hypothalamus, but principally on the latter. They postulated that the contraceptive steroids cause a differential release of LH and FSH (103) (see Hormones).

Since 1971 immunologic studies have led to the development of antisera to LH–RH that inhibit ovulation in estrous-cycling rats and also produce abortion in pregnant rats (105). Numerous analogues of LH–RH have been synthesized. Some have proved to be agonists having potencies greater than that of the natural hormone, and others have been found to be antagonists. Both agonists and antagonists have been examined for their potential contraceptive usage. Agonists that are more potent than LH–RH either have a greater affinity for the pituitary receptor (106) or their half-life may be longer than that of LH–RH, which has been reported to be about four minutes (as computed from the first exponential portion of the disappearance curve of LH–RH in human plasma) (107).

LH–RH and stimulating analogues that cause the prolonged secretion of LH and FSH induce ovulation. They are also capable of interfering with pregnancy when administered postcoitally (108). Because of their hyperstimulating effect, the normal secretory pattern associated with pregnancy is altered. As a result, transport of the ovum through the oviduct may be either accelerated or retarded, the endometrium may not be in the correct state for implantation of the blastocyst, or the implanted blastocyst may be dislodged from the uterus.

The analogues of LH–RH were generally prepared by solid-phase methodology. One of the most potent agonists found was D-[Ala]^6-des-[Gly]^{10}-Pro^9-ethylamide-LH–RH. In this compound, the two glycine amino acids of LH–RH have been replaced, the one at the 6-position by D-alanine and that in the C-terminal position by the ethylamide function. This agonist is approximately thirty-six times more potent than LH–RH in inducing ovulation in proestrous rats treated with fluphenazine dihydrochloride (109). It is fifty to eighty times more potent in advancing ovulation in the diestrous rat. The antifertility effect of D-[Ala]^6-des-[Gly]^{10}-Pro^9-ethylamide-LH–RH, when administered in a single dose on the day before estrous, has been attributed to premature induction of ovulation (110).

In one study, the antigonadotrophin-releasing activity of a series of analogues was determined in rats. Among the analogues found to be potent inhibitors of LH–RH were [D-Phe^2-D-Leu^6]-LH–RH, [des-His^2-D-Leu^6]-LH–RH, and [D-Phe^2-D-Phe^6]-LH–RH (111). In another study, the antiovulatory activities of a series of LH–RH analogues substituted in positions 2 and 6 were examined. Several of these analogues were observed to block ovulation 100% in rats on the day of proestrus. The analogues were [D-Phe^2-D-Ala^6]-LH–RH, [D-Phe^2-D-(C$_6$H$_5$)Gly^6]LH–RH, [D-p-F-Phe^2-D-Ala^6]-LH–RH, and [D-Phe^2-2-CH$_3$-Ala^6]-LH–RH. They appear to suppress the proestrous serum LH surge and reduce the estrous morning FSH surge (112).

Another analogue that has been found to be a potent inhibitor of LH–RH was [D-Phe^2-Phe^3-D-Phe^6]-LH–RH. This compound inhibits the preovulatory surge of LH as well as the FSH surge when administered to female rats; in addition, it suppresses ovulation. Given to male rats, it suppresses LH and FSH release in response to LH–RH (113).

These results offer the promise that continued investigations will lead to the development of orally active analogues with prolonged half-lives which will be effective in the control of fertility. Perhaps they will not produce the side effects reported in association with the current contraceptive agents nor will they produce severe menstrual irregularities when used by women. If intended for men, it is hoped that inhibition of spermatogenesis will not be accompanied by suppression of libido. Otherwise steroid supplementation will be required.

Immunologic Approach

Active immunization is currently being explored as a means for controlling fertility (114). Identification of a polypeptide containing 30 amino acid residues, which is unique to the β-subunit of human chorionic gonadotrophin (HCG), has given rise to the hope that antibodies can be raised that would react specifically with HCG. This hormone is a glycoprotein that is normally produced by the placenta. It is excreted in the urine in large amount during the first trimester of pregnancy. Its main function in the early stages of pregnancy is to prolong the life of the corpus luteum and to stimulate the luteinized cells to produce progesterone which is necessary for the maintenance of pregnancy.

HCG is composed of an α- and a β-subunit. The amino acid sequence of the α-subunit of HCG is almost identical with that of human luteinizing hormone (LH). There is also considerable homology between the β-subunits of HCG and LH. However, the carboxyl terminus of the β-subunit of HCG contains a chain of approximately 30 amino acid residues that is not present in the β-subunit of LH (115).

Except for patients with hydatiform mole, choriocarcinoma, and certain other types of cancer, HCG is a hormone specifically released during pregnancy. Antibodies that are produced to neutralize the activity of HCG would be expected to terminate pregnancy. Experimentally, antibodies raised against HCG, which had been rendered antigenic, also cross-react with LH (116).

When the β-subunit of HCG was purified by an immuno-absorption technique, processed, and then conjugated to purified tetanus toxoid (TT), the resultant conjugate, "Pr–β–HCG–TT", was found to be antigenic in a variety of species. The conjugate elicited the formation of both anti-HCG and anti-TT antibodies. The anti-HCG antibodies reacted with both the β-subunit of HCG and intact HCG. The biological activity of HCG was neutralized by the antibodies as demonstrated in a number of assays.

Cross-reactivity with other human hormones, such as LH and FSH, was not significant. The antibody titers declined in the course of time, thus suggesting the response of the conjugate antigen was reversible. A repeated injection of the conjugate in the declining phase of antibody titers produced a booster response. Toxicity studies in mice, cats, rabbits, and monkeys indicated that Pr–β-HCG–TT was a safe substance (117). Currently, it is undergoing clinical trials in India, as well as in several other countries in Europe and Latin America.

Although the antibodies produced against the β-subunit of HCG cross-react to a very limited extent with LH, further reduction of cross-reactivity can conceivably be achieved by employing fragments of β-HCG to elicit the antibodies. These fragments can be obtained either synthetically or by enzymic cleavage of β-HCG. For the antibodies to be able to neutralize the activity of HCG, it appears likely that the peptide fragment employed must contain more than 23 amino acid residues. The C-terminal peptide residues 123–145 obtained from the trypsin digestion of reduced carboxymethylated asialo-HCG, when coupled to bovine serum albumin, produced antibodies that reacted immunologically with HCG, but the HCG–antibody complex formed retained the biological activity of the hormone (118).

PROSTAGLANDINS

The prostaglandins (qv) comprise another class of naturally occurring physiologically active substances that have been studied extensively for their effects on reproduction. Although it was known in 1930 that substances present in human semen, which were to be designated a few years later as prostaglandins, affected the motility of uterine strips *in vitro,* it wasn't until 1957 that they were shown to be long chain, oxygenated, unsaturated fatty acids (119).

The basic skeleton of the prostaglandins is prostanoic acid (95), a compound that contains 20 carbon atoms. A cyclopentane ring is present in the molecule, and two side chains are attached to this ring at contiguous carbon atoms. Prostanoic acid is a lipid-soluble substance. As with cholesterol (96), the main portion of the molecule is highly lipophilic. A hydrophilic group is attached to one end of the molecule. In cholesterol, this group is the hydroxyl function whereas in prostanoic acid it is the carboxyl group.

The six primary prostaglandins are PGE$_1$ (97), PGE$_2$ (98), PGE$_3$ (99), PGF$_{1\alpha}$ (100), PGF$_{2\alpha}$ (101), and PGF$_{3\alpha}$ (102). These substances are oxygenated derivatives of (95) in which one or more double bonds have also been introduced into the molecule.

In the PGE series, the oxygen function at C-9 is a keto group, and in the PGF series it is a hydroxyl group. Two additional hydroxyl groups are present in the primary prostaglandins, one at C-11 and the other at C-15.

(95)

(96)

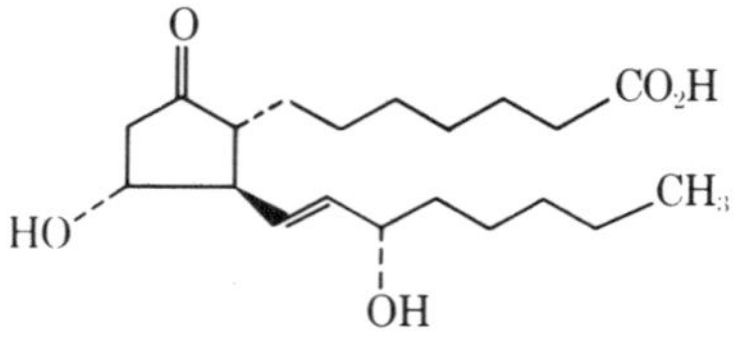

(97)

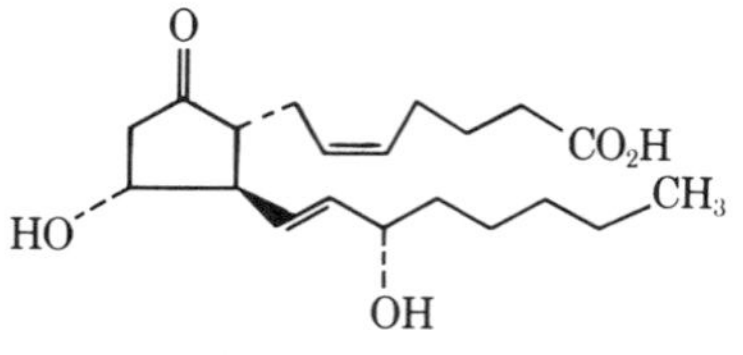

(98)

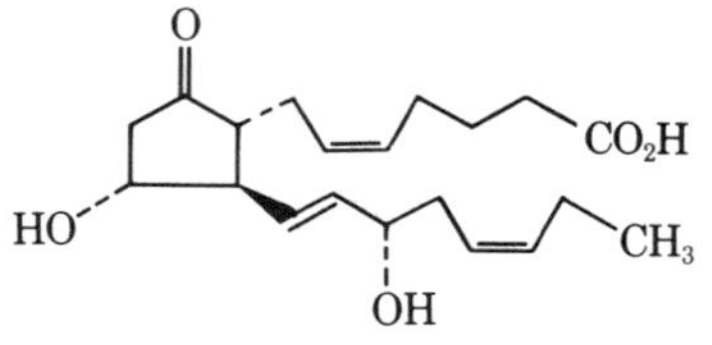

(99)

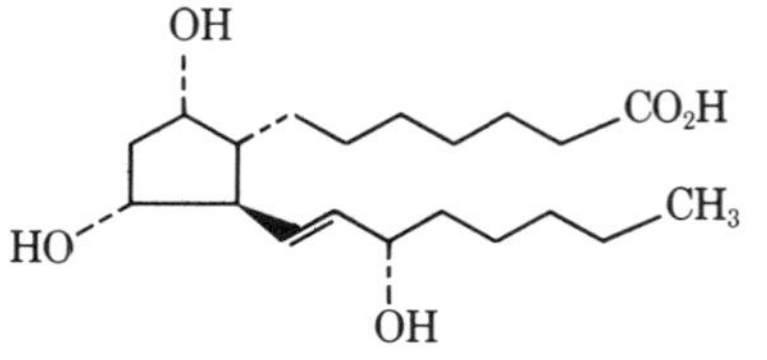

(100)

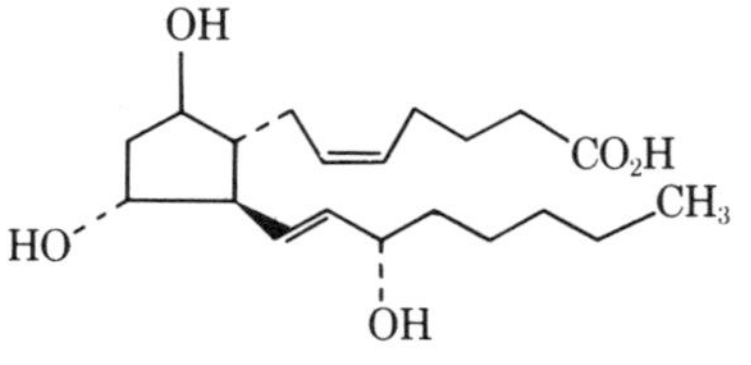

(101)

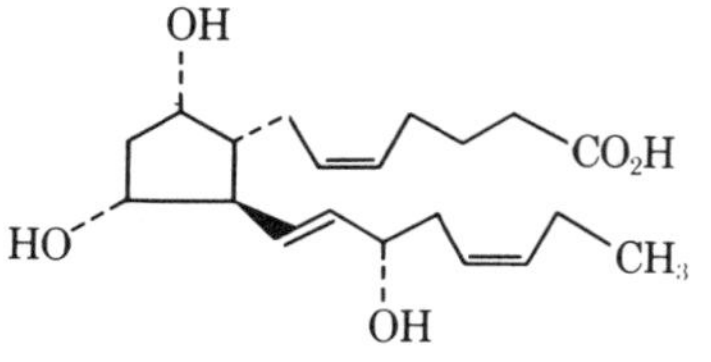

(102)

The junctions, C-8 and C-12, are chiral centers where the side chains are attached. In the PGF series, there are five chiral centers in the molecule. This gives rise to 2^5 or 32 stereoisomers. As with the steroids, the multiplicity of stereoisomers presents a challenge to the chemist to devise synthetic schemes that would furnish the desired isomer stereoselectively. In addition, it provides the biologist with an opportunity to determine the effect that stereochemistry has on a particular biological activity.

Diffraction and spectral studies suggest that the conformation of the active prostaglandins resembles a hairpin (120). Ramwell and Kury (121) have compared the biologically active, natural prostaglandins to a right-handed wedge in which activity is associated with right-handed chirality. Both ends of the wedge contain a hydrophilic group. The hydrophilic functional groups are on one side of the wedge and the hydrophobic groups are on the other.

The primary prostaglandins have a potent effect on smooth muscle contraction. In interacting with adenyl cyclase, they also modulate hormonal activity. At one time it was thought that $PGF_{2\alpha}$ (101) was the natural uterine luteolytic factor that was responsible for the demise of the corpus luteum (122). In several subprimate species (101) and a number of other prostaglandins have been found to cause the regression of the corpus luteum (123). There are conflicting reports on whether (101) produces luteolysis in humans and other primates (124). 15-Ketoprostaglandin $F_{2\alpha}$[15-keto $PGF_{2\alpha}$ (103)], a metabolite of $PGF_{2\alpha}$, has been reported to demonstrate luteolytic action in the rhesus monkey (125). 15-Methyl $PGF_{2\alpha}$ methyl ester (104b) has been observed to terminate pregnancy in humans when administered intravaginally (126). Although (104b) stimulates uterine contractions, termination of early pregnancy has been attributed in part to its luteolytic effect. The abortifacient activity of the prostaglandins can be ascribed to the effects that they exert on the uterus, either directly as a result of smooth muscle stimulation or indirectly via the ovary or pituitary.

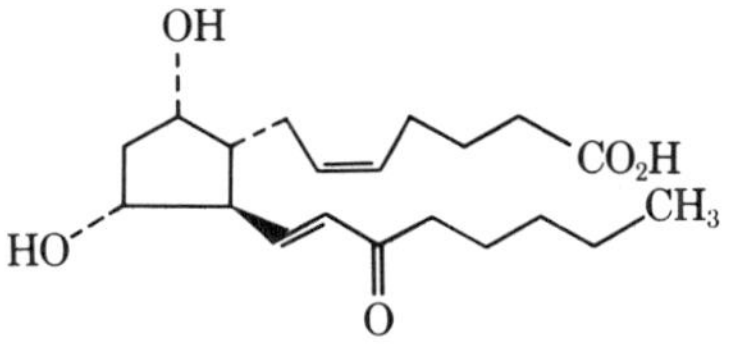

(103)

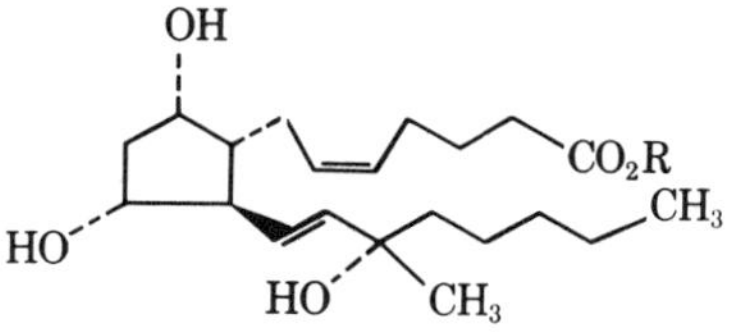

(104)

(a) R = H
(b) R = CH₃

Early studies of Karim indicated that both PGF$_{2\alpha}$ (101) and PGE$_2$ (98) can induce labor at term (127). Subsequent studies have confirmed Karim's original results. PGE$_2$ (dinoprostone) is marketed as an inducer of labor. PGF$_{2\alpha}$ (dinoprost tromethamine) is also on the market, but its use in clinical practice is currently limited to terminating second trimester pregnancy.

The natural prostaglandins display a broad spectrum of biological activities. Present efforts are directed toward the synthesis of analogues that would show a greater degree of separation in their biological action. Perhaps products can be developed that will affect reproduction and not have significant effects on the cardiovascular, pulmonary, gastrointestinal, and central nervous system.

The natural prostaglandins are readily inactivated as a result of metabolic transformations. One inactivating transformation involves the conversion of the secondary hydroxyl group at C-15 to a keto group. By attaching a methyl group to C-15 and thus converting the oxygen function at C-15 into a tertiary hydroxyl group, as in (104a), oxidation to a keto group can be avoided.

Another approach consists of alkylating C-16 so that the region about C-15 would be sterically hindered (128). This would interfere with the binding of the prostaglandin substrate to prostaglandin 15-hydroxydehydrogenase (PGDH), the enzyme system that catalyzes the oxidative process. As a result, the duration of action of the prostaglandin could be expected to be prolonged. Hence 16,16-dimethyl PGF$_{2\alpha}$ (105) and (104a) are more potent than (101) as luteolytic agents in the hamster (129).

Inhibition of PGDH as a means of prolonging the duration of action of endogenous prostaglandins is being examined in a number of laboratories. This manner of enhancing tissue function which is dependent on the local level of a specific prostaglandin, appears to have considerable appeal. It avoids the necessity of having to administer systemically the natural prostaglandins which are not orally active and which are likely to produce undesirable side effects (130).

2-(3-Methoxyphenyl)-5H-s-triazolo[5,1-a]isoindole (106) and 2-(3-methoxyphenyl)-5,6-dihydro-s-triazolo[5,1-a]isoquinoline (107) inhibit the PGDH enzyme system (131–132). The effects of the two heterocyclic compounds on pregnancy were studied in both the rat and hamster. The substances were administered after mating.

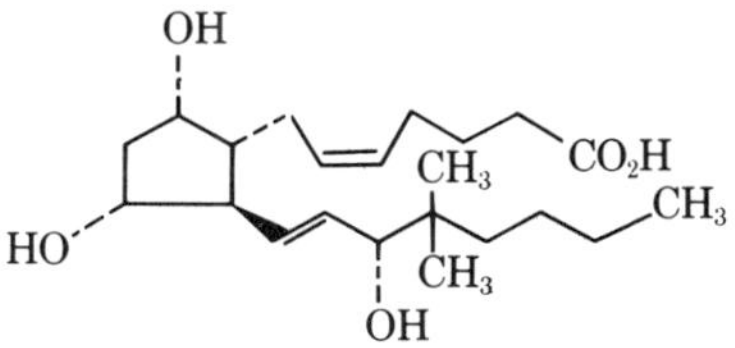

(105)

Each compound was found to be 100% effective in terminating pregnancy in the two species. Mechanism studies suggested that the primary site of action was the utero–placental complex. Besides the rodents, the pregnancy-terminating effects of the two compounds were also studied in the rhesus monkey and baboon. Given during the first two months of gestation, both compounds aborted the pregnancies. Normal menstrual cycles, mating behavior, and pregnancy rates returned after drug usage (132). In searching for other nonsteroidal luteolytic agents, Coombs and co-workers found that certain fused pyrazoles, such as (108), had the desired profile of activity in their animal testing systems (133). The pyrazole (108) bears a structural resemblance to the triazoloisoquinoline derivative (107).

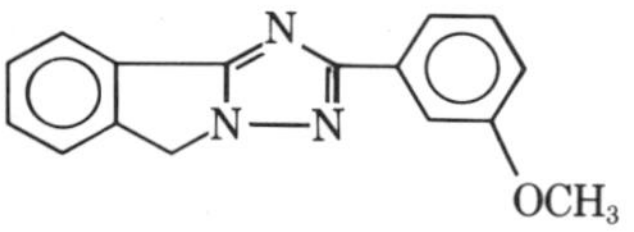

(106) 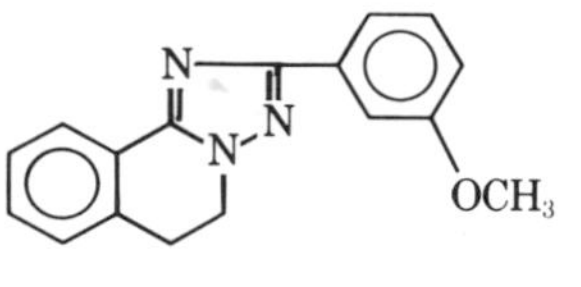   (107) 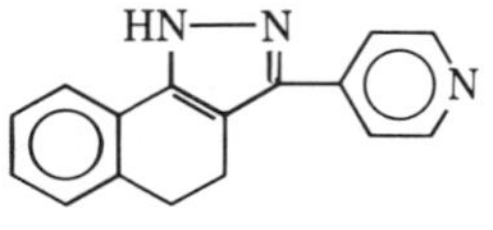   (108)

Systematic modification of the prostaglandin structure has led to the development of a novel series of analogues that possess very potent biological properties (134). In these analogues, an aryloxy group has replaced a portion of the lower side chain of the natural prostaglandins. Of particular interest are those analogues in which a p-fluoro, m-chloro, or m-trifluoromethyl group is attached to the phenyl ring. The compounds have been designated as ICI 79,939 (109), ICI 80,996 [cloprostenol (110)], and ICI 81,008 [fluprostenol (111)], respectively. The three compounds are from 100–200 times more potent than (101) in terminating pregnancy in the hamster when administered subcutaneously. When given orally, both (110) and (111) compare even more favorably with (101). ICI 79,939 (109) is a very potent smooth muscle stimulant, whereas (111) is only weakly effective. Of the three aryloxy derivatives (111) is the least toxic, and (109) is exceedingly toxic, producing lethality in rats at a relatively low dose. The side effects of (110) are between those of the other two compounds. Both (110) and (111) are available commercially for veterinary use, the former for estrous synchronization

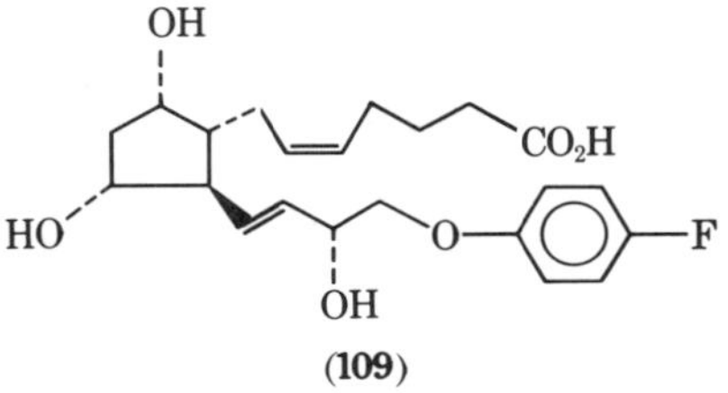

(109) 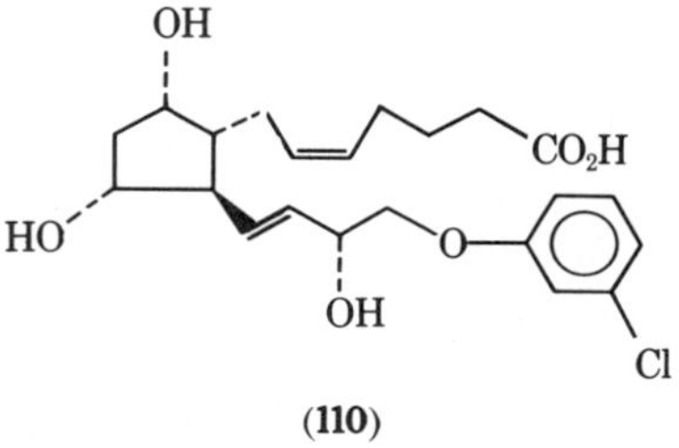   (110)

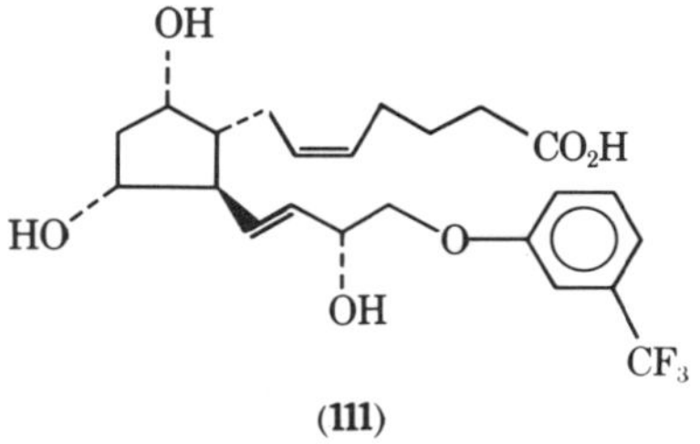

(111)

in cows, and the latter for treatment of persistent luteal function in horses. The extent to which (111) induces menstruation in women has been studied in the clinic. When applied as a vaginal gel or given transcervically in a single intrauterine dose of 400 μg, (111) is an effective menstrual inducer (135).

Synthetic Schemes

The synthetic scheme employed for the preparation of these compounds was an adaptation of one devised by Corey and co-workers (136). One of the key intermediates in the synthesis is a lactone aldehyde. It contains four chiral centers, all of which have the orientation of the natural prostaglandins.

Alkylation of the thallium salt of cyclopentadiene (112) with chloromethyl benzyl ether furnishes (113). When (113) is condensed with 2-chloroacrylonitrile, the bicyclo[2.2.1]heptene derivative (114) is obtained. In the presence of base, the chloro and nitrile groups are displaced to give the bicyclo ketone (115). Baeyer-Villiger oxidation converts (115) into the lactone (116). Cleavage of (116) furnishes the corresponding hydroxy acid (117), which can be resolved into its optical antipodes. Iodolactonization of (117) affords (118). After the hydroxyl group is acylated with p-phenylbenzoyl chloride, the iodo and benzyl groups are removed reductively with tributyltin hydride and by catalytic hydrogenation, respectively. The resultant alcohol (119) is oxidized to the aldehyde (120) with Collins reagent (chromium trioxide–pyridine complex).

In ICI's (Imperial Chemical Industries, Ltd.) variation of the Corey synthesis, cyclopentadiene (112) is successively formylated and acetylated to give the acetoxyfulvene (121) (137). A Diels-Alder reaction with 2-chloroacrylonitrile affords the bicycloheptene derivative (122). Hydrolysis of the enol–acetate group furnishes the aldehyde (123), which is protected as the dimethyl acetal (124). The conversion of (124) to the lactone (125) can be accomplished by employing steps analogous to those utilized by Corey and co-workers for the synthesis of (120). The lactones (120) and (125) are very versatile intermediates; both have been transformed into a variety of prostaglandins including the natural, optically active substances of both the E and F series.

Attachment of the two side chains to the cyclopentane ring is achieved with the appropriate phosphonium or phosphonate reagent on the aldehyde function. In order to attach the upper side chain to the ring by means of the Wittig reagent, the lactone, eg, (126), is first converted into the lactol (127), with the use of diisobutylaluminum hydride. The lactol (127) is a masked aldehyde, and it readily undergoes the Wittig reaction.

Other Methods of Contraception

Since the oral contraceptives are taken for long periods of time, much effort has gone into assessing the safety of these substances. As a result, women who are on the progestin–estrogen pill are medically the most scrutinized people in history. Although severe side reactions, including death, have been reported among users of the pill, the incidence is still relatively small, and the risk among nonsmoking women is less than that associated with pregnancy (3,138).

In seeking other means of controlling fertility, investigators are currently placing greater emphasis on safety than on efficacy. The argument has been advanced that

(112)
[542-92-7]

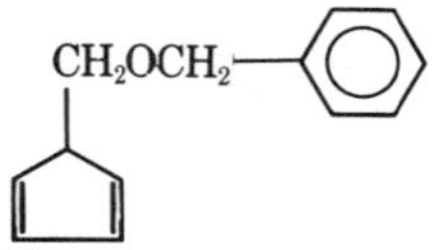

(113)
[39939-07-6]

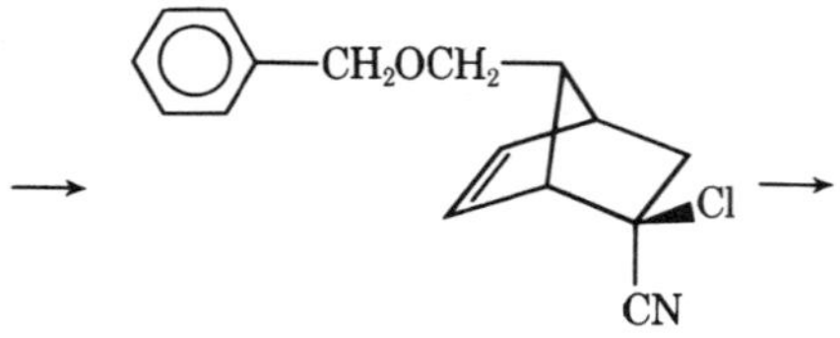

(114)
[50889-55-9]

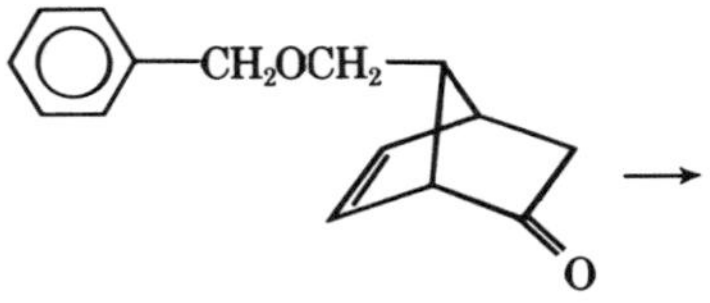

(115)
[56817-38-0]

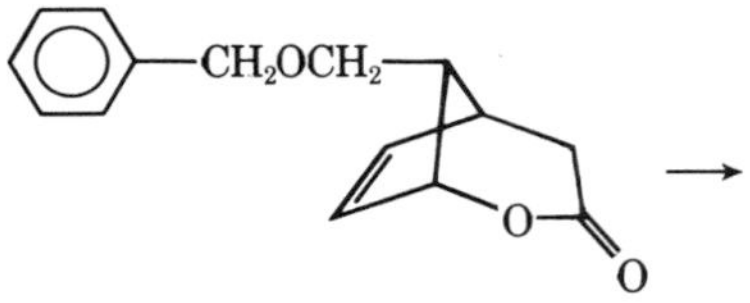

(116)
[50889-56-9]

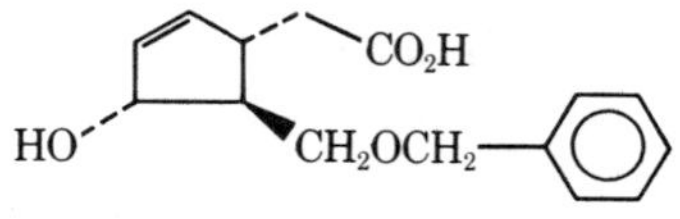

(117) [39507-49-8]

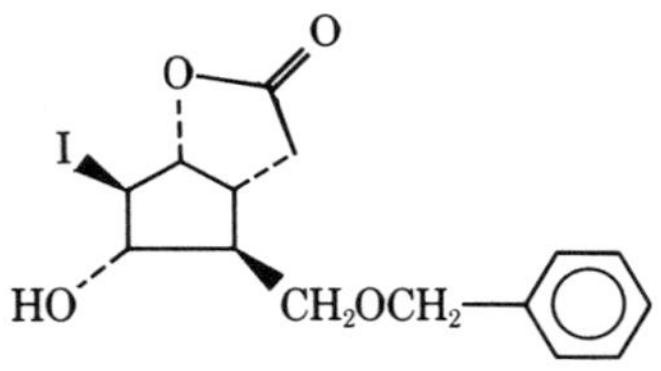

(118)
[31767-37-0]

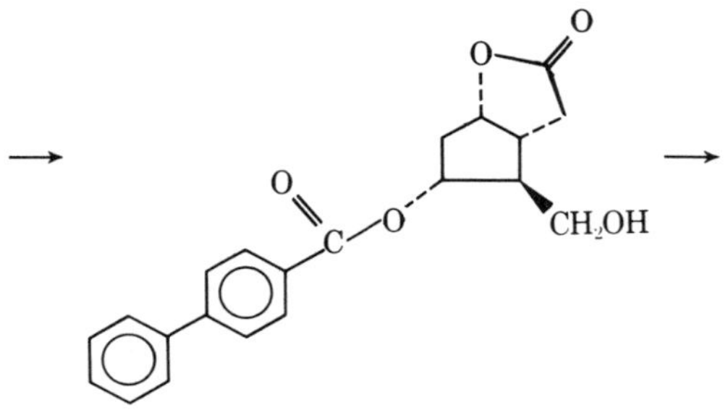

(119)
[58708-03-5]

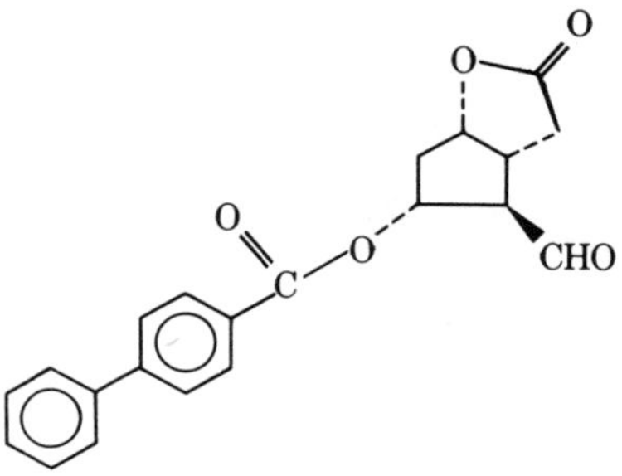

(120)

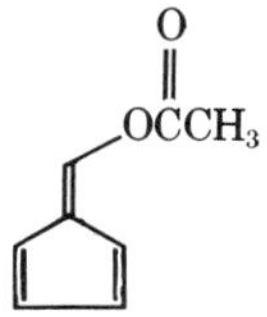

O
‖
OCCH₃

(121)
[699-15-0]

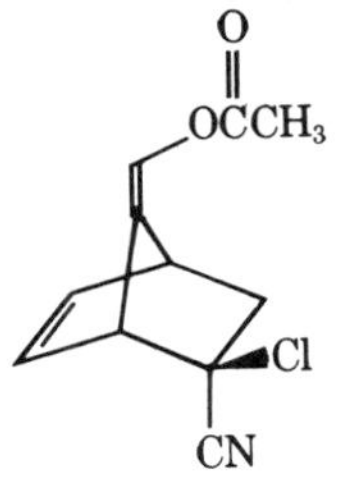

O
‖
OCCH₃

Cl
CN
(122)
[39746-59-3]

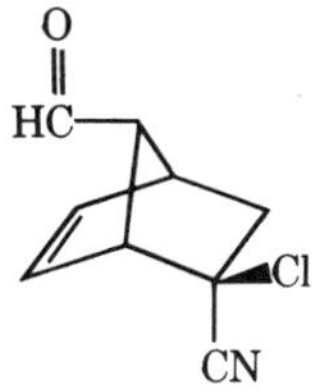

O
‖
HC

Cl
CN
(123)
[39746-60-6]

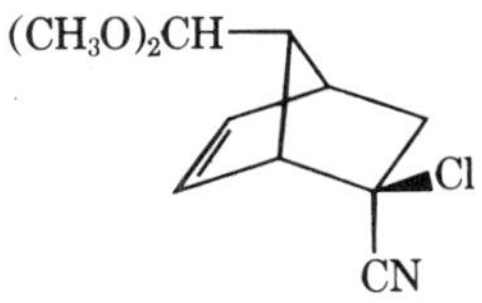

(CH₃O)₂CH

Cl
CN
(124)
[39746-70-8]

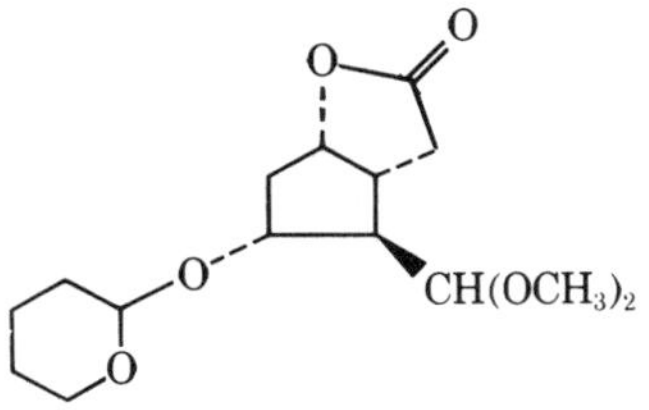

O
O

CH(OCH₃)₂
(125)

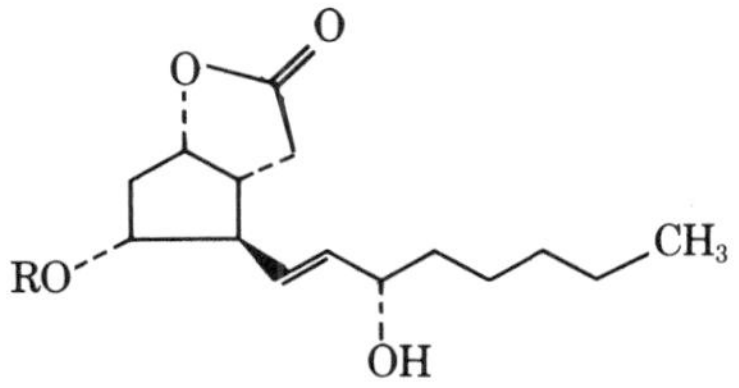

O
O

RO
CH₃
OH
(126)

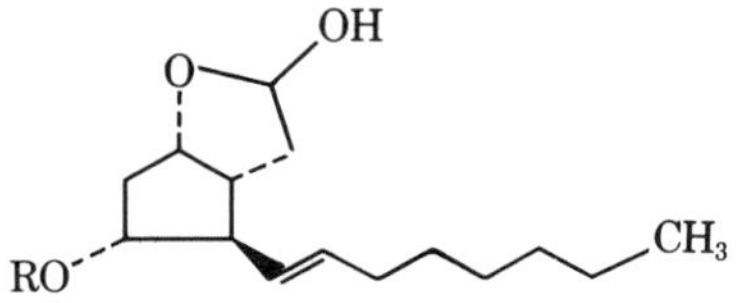

OH
O

RO
CH₃
(127)

with the legalization of abortion, a pregnancy failure rate of several percent in a contraceptive product that has been proven to be safe would be regarded as an acceptable risk by the great majority of users.

The contraceptive drugs considered so far are intended to disrupt the delicate hormonal balance that exists during the onset of ovulation or during pregnancy. These drugs are also intended to act systemically, as they are given either orally or by injection.

Other routes of administration are being explored. Medicated intrauterine devices (IUDs) have been developed that will release a small quantity of the drug daily in the uterus (see Pharmaceuticals, controlled-release). The copper IUD is highly effective in controlling fertility. One such device has been approved by the FDA to be left in the uterus for as long as 3 yr before replacement. The copper released from the IUD has been postulated, *inter alia,* to interfere with the biochemical processes that regulate implantation (139).

A progesterone-releasing IUD is also on the market, and it is also very effective in preventing pregnancy (140). The tiny amount of progesterone released daily is believed to act directly on the uterus and is quickly metabolized before it can get into general circulation. The amount of progesterone in the IUD is sufficient for the device to be left in the uterus for a little more than a year before it is replaced.

Although the intrauterine devices do not elicit the side effects attributed to the

Table 2. List of Components of Contraceptive Drugs and Intermediary Products

Structure No.	Name	CAS Registry No.
(1)	norethynodrel	[68-23-5]
(2) (a)	mestranol	[72-33-3]
(b)	ethinylestradiol	[57-63-6]
(c)	quinestrol	[152-43-2]
(3)	progesterone	[57-83-0]
(4) (a)	ethisterone	[434-03-7]
(b)	norethindrone or norethisterone	[68-22-4]
(c)	dimethisterone	[79-64-1]
(5)	19-norpregn-4-ene-3,20-dione	[472-54-8]
(6)	strophanthidin	[66-28-4]
(7) (a)	estrone	[53-16-7]
(b)	3-methoxyestra-1,3,5(10)-trien-17-one	[1624-62-0]
(8) (a)	testosterone	[58-22-0]
(b)	17α-methyltestosterone	[58-18-4]
(c)	17α-ethinyltestosterone	[434-03-7]
(14)	17α-19-norpregna-3,5-dien-20-yn-3,17-diol, diacetate	[2205-78-9]
(15)	norethindrone 17-acetate	[51-98-9]
(16)	17α-ethynylestr-4-ene-3β,17β-diol	[1231-93-2]
(17)	ethynodiol diacetate	[297-76-7]
(18) (a)	dehydroisoandrosterone acetate	[853-23-6]
(b)	dehydroisoandrosterone	[53-43-0]
(19)	5α-chloro-3β,6β-dihydroxyandrostan-17-one 3-acetate	[6557-16-0]
(20)	5α-chloro-6β,19-epoxy-3β-hydroxyandrostan-17-one	[2654-00-9]
(21)	6β,19-epoxyandrost-4-ene-3,17-dione	[6563-83-3]
(22)	19-hydroxyandrost-4-ene-3,17-dione	[510-64-4]
(23)	estr-5(10)-ene-3,17-dione	[3962-66-1]
(24)	17-hydroxy-19-nor-17α-pregn-5(10)-en-20-yn-3-one, dimethyl acetal	[19669-65-9]
(25)	17-hydroxy-10-hydroperoxy-19-nor-17α-pregn-4-en-20-yn-3-one	[1238-54-6]
(26) (a)	dl-norgestrel	[6533-00-2]
(b)	d-norgestrel	[797-63-7]
(27)	norethindrone enanthate	[3836-23-5]
(28)	quingestanol acetate	[3000-39-3]
(29)	lynestrenol	[52-76-6]
(30)	17-hydroxy-19-nor-17α-pregna-4,9-dien-20-yn-3-one	[14531-92-1]
(31) (a)	norgestrienone	[848-21-5]
(b)	R 2323	[16320-04-0]
(32)	17-hydroxy-7α-methyl-19-nor-17α-pregn-4-en-20-yn-3-one	[1162-60-3]
(33)	17-hydroxy-11β-methyl-19-nor-17α-pregn-4-en-20-yn-3-one	[18050-48-1]
(34)	13-ethyl-11-methylene-18,19-dinor-17α-pregn-4-en-20-yn-17-ol	[54024-22-5]
(35)	17-acetoxy-13-ethyl-18,19-dinor-17α-pregn-4-en-20-yn-3-one, oxime	[35189-28-7]
(36)	17-acetoxy-19-nor-17α-pregn-4-en-20-yn-3-one, 3-oxime	[20799-24-0]
(37)	O-methyloxime-17-hydroxy-19-nor-17α-pregn-4-en-20-yn-3-one	[58001-83-5]
(38) (a)	21-chloro-17-hydroxy-19-nor-17α-pregn-4-en-20-yn-3-one	[3124-70-7]
(b)	21-ethynyl-17-hydroxy-19-nor-17α-pregn-4-en-20-yn-3-one	[41983-59-9]
(c)	17β-hydroxy-17-(1-propynyl) estr-4-en-3-one	[7359-79-7]
(47)	17α-allyl-13β-ethyl-10α-methyl-3-oxo-8β,9β,10α,13β,14α-gon-4-en-17β-ol	[68906-31-0]
(48)	19-nor-17α-pregn-4-en-20-yn-3-one	[38673-42-6]
(49)	19-nor-17α-pregna-4,9-dien-20-yn-3-one	[68854-65-9]
(50)	19-nor-17α-pregna-4,9,11-trien-20-yn-3-one	[68854-66-0]
(51) (a)	17α-hydroxyprogesterone	[68-96-2]
(b)	17α-acetoxyprogesterone	[302-23-8]
(c)	17α-hydroxyprogesterone hexanoate	[630-56-8]

Table 2 (*continued*)

Structure No.	Name	CAS Registry No.
(52)	medroxyprogesterone acetate	[71-58-9]
(53)	megestrol acetate	[595-33-5]
(54) (a)	17α-methylprogesterone	[1046-28-2]
(b)	17α-ethylprogesterone	[1048-01-7]
(55)	melengestrol acetate	[2919-66-6]
(56)	17α-acetoxy-6,11β-dichloro-19-norpregna-4,6-diene-3,20-dione	[24432-00-6]
(57)	17α-acetoxy-6,11β-dichloro-16-methylene-19-norpregna-4,6-diene-3,20-dione	[24678-24-8]
(58)	chlormadinone acetate	[302-22-7]
(59)	algestone acetophenide	[1179-87-9]
(60) (a)	estradiol 17-enanthate	[4956-37-0]
(b)	estradiol	[50-28-2]
(64)	19-nor-14β,17α-pregn-4-ene-3,20-dione	[17554-45-9]
(65)	dydrogesterone	[152-62-5]
(66)	diethylstilbestrol	[56-53-1]
(67) (a)	17-(1,3-butadiynyl)estradiol	[2010-52-8]
(b)	trifluoromethylethynylestradiol	[2061-56-5]
(68) (a)	A-norandrostane-2α,17α-diethynyl-2β,17β-diol	[1045-29-0]
(b)	A-norandrostane-2α,17α-diethynyl-2β,17β-diol diproprionate	[56470-64-5]
(69)	3β-hydroxy-5α-androstan-17-one	[481-29-8]
(70) (a)	tigogenin	[77-60-1]
(b)	diosgenin	[512-04-9]
(71)	2,3-seco-androstan-17-one-2,3-dicarboxylic acid	[1165-38-4]
(72)	A-nor-5α-androstane-2,17-dione	[1032-12-8]
(73)	clomiphene citrate	[50-41-9]
(74)	U-11,100A	[1847-63-8]
(75)	U-11,555A	[68854-67-1]
(76)	centchroman	[31477-60-8]
(77)	retroprogesterone	[2755-10-4]
(78)	dydrogesterone	[152-62-5]
(79) (a)	5α-pregnanedione	[566-65-4]
(b)	17α-ethyl-5α-pregnanedione	[57154-66-2]
(80)	danazol	[17230-88-5]
(81) (a)	testosterone propionate	[57-85-2]
(b)	testosterone enanthate	[315-37-7]
(c)	testosterone undecanoate	[5949-44-0]
(82)	cyproterone acetate	[427-51-0]
(83)	16-dehydropregnenolone	[1162-53-4]
(85) (a)	androst-4-ene-3,17-dione	[63-05-8]
(b)	19-norandrost-4-ene-3,17-dione	[734-32-7]
(86)	sitosterol	[83-46-5]
(95)	prostanoic acid	[25151-81-9]
(96)	cholesterol	[57-88-5]
(97)	PGE$_1$	[745-65-3]
(98)	PGE$_2$	[363-24-6]
(99)	PGE$_3$	[802-31-3]
(100)	PGF$_{1α}$	[745-62-0]
(101)	PGF$_{2α}$	[551-11-1]
(102)	PGF$_{3α}$	[745-64-2]
(103)	15-keto PGF$_{2α}$	[35850-13-6]
(104) (a)	15-methyl PGF$_{2α}$	[35700-23-3]
(b)	15-methyl PGF$_{2α}$, methyl ester	[35700-21-1]
(105)	16,16-dimethyl PGF$_{2α}$	[39746-23-1]
(106)	2-(3-methoxyphenyl)-5H-s-triazolo[5,1-a]isoindole	[57170-08-8]
(107)	2-(3-methoxyphenyl)-5,6-dihydro-s-triazolo[5,1-a]isoquinoline	[55308-37-7]
(108)	4,5-dihydro-3-(4-pyridinyl)-1H-benz[g]indazole	[54569-87-8]

Table 2   (*continued*)

Structure No.	Name	CAS Registry No.
(109)	ICI 79,939	[40666-03-3]
(110)	ICI 80,996 (cloprostenol)	[40665-92-7]
(111)	ICI 81,008 (fluprostenol)	[55028-71-2]
(120)	[3aR-(3aα,4α,5β,6aα)]-4-formylhexahydro-2-oxo-2H-cyclopenta[b]-[1,1'-biphenyl]-4-carboxylic acid	[38754-71-1]
(125)	4-(dimethoxymethyl)hexahydro-5-[(tetrahydro-2H-pyran-2-yl)oxy]-2H-cyclopenta[b]furan-2-one	[51638-24-5]
(128)	5-thio-D-glucose	[20408-97-3]
(129)	gossypol	[303-45-7]
	acrosin	[9068-57-9]
	fluphenazine dihydrochloride	[146-56-5]
	FSH	[9002-68-0]
	HCG	[9002-61-3]
	LH	[9002-67-9]
	LH–RH	[9034-40-6]
	PGDH	[9030-87-9]

oral contraceptives, they may produce bleeding, be expelled, be associated with ectopic pregnancies, and be implicated in pelvic inflammatory disease (141).

Other approaches to terminating pregnancy include the use of antimetabolites and antimitotic agents. However, these substances are generally very toxic and are apt to produce genetic damage and fetal abnormalities. Thus, it appears unlikely that they will be approved by regulatory agencies for use in humans.

Release of LH–RH is under the control of neural transmitters as well as of ovarian steroids. Catecholamines have been reported to stimulate release of LH–RH and indoleamines are known to inhibit its release. Agents acting on the central nervous system (CNS), such as reserpine, meprobamate, chlorpromazine, and other phenothiazine tranquilizers, as well as α-methyldopa, haloperidol, pentobarbital, and related barbituates, are used to treat a variety of human disorders (see Psychopharmacological agents). Their effects on ovulation both in humans and in laboratory animals have also been examined (142). In some cases, there is evidence indicating that a particular CNS agent suppresses the activity of the hypothalamus whereas in others the compound appears to act directly on the ovary. Because of other effects, the CNS agents mentioned cannot be used for contraception. However, by modifying their structures, it may be possible to obtain compounds that would selectively inhibit the release of LH–RH and would not have undesired side effects.

The major effort in the control of fertility has been focused upon the female reproductive system. In part, this is because only one ovum is secreted by the woman whereas hundreds of millions of sperms are released by the man in a single episode. In addition, knowledge of the male reproductive system is not as extensive as that of the female.

Notwithstanding this, investigations are being conducted on male fertility with the view of developing male contraceptive agents. The approaches under study include disruption of the biochemical events associated with spermatogenesis.

Uptake of glucose by the testis is essential for development of the sperm. By blocking this process with 5-thio-D-glucose (**128**), a drastic reduction in the sperm count can be observed in mice and rats. As a result, the treated animals are rendered sterile. On cessation of treatment, the animals gradually regain their fertility (143).

Gossypol (**129**) is reported to be an effective male contraceptive (144) (see Terpenoids

(128)　　　　　　　　　　　　　　　(129)

Modification of the essential components of the epididymal fluid or other accessory secretion and disruption of epididymal sperm maturation offer additional methods for the control of male fertility. In another approach, specific agents are being sought that will block the protease, acrosin, present in the acrosome of the sperm. Inhibition of acrosin will cause the spermatozoan to lose its capacity to fertilize the ovum.

Concern for safety and the availability of abortion on demand in many regions have rekindled an interest in the spermicidal and barrier approach to fertility control. Spermicidal jellies and foams, as well as diaphragms and condoms, are acquiring an increasing number of users despite their double digit pregnancy failure rates (per 100 woman-years). In these techniques, there is a need for advanced preparation shortly before coitus, which could be bothersome. With the hormonal approach to contraception, such preparation is not necessary.

Table 2 lists the components of the contraceptive drugs and intermediary products referred to within this article.

BIBLIOGRAPHY

"Contraceptive Drugs" in *ECT* 2nd ed., Vol. 6, pp. 60–92, by F. B. Colton and P. D. Klimstra, G.D. Searle & Co.

1. J. Rock, C. R. Garcia, and C. Pincus, *Recent Progr. Horm. Res.* **13**, 323 (1957); G. Pincus and co-workers, *Am. J. Obstet. Gynecol.* **75**, 1333 (1958).
2. G. Pincus, *Vitam. Horm. N.Y.* **17**, 307 (1959); V. A. Drill, *Oral Contraceptives,* McGraw-Hill Book Co., New York, 1966.
3. P. T. Piotrow and C. M. Lee, *Popul. Rep. A* **1**, 1 (1974); M. P. Vessey and R. Doll, *Proc. R. Soc. London B* **195**, 69 (1976); B. Stokes, *Popul. Rep. J.* **20**, 639 (1978).
4. A. Butenandt, U. Westphal, and W. Hohlweg, *Z. Physiol. Chem.* **227**, 84 (1934); M. Hartmann and A. Wettstein, *Helv. Chim. Acta* **17**, 1365 (1934); W. M. Allen and O. Wintersteiner, *Science* **80**, 190 (1934).
5. A. W. Makepeace, G. L. Weinstein, and M. H. Friedman, *Am. J. Physiol.* **119**, 512 (1937).
6. H. H. Inhoffen and co-workers, *Ber.* **71**, 1024 (1938).
7. M. Ehrenstein, *J. Org. Chem.* **9**, 435 (1944); W. M. Allen and M. Ehrenstein, *Science* **100**, 251 (1944).
8. C. Djerassi, L. Miramontes, and G. Rosenkranz, *Abstract 18J*, 121st American Chemical Society Meeting, Milwaukee, Wis., 1952; C. Djerassi and co-workers, *J. Am. Chem. Soc.* **76**, 4092 (1954); U.S. Pat. 2,744,122 (May 1, 1956), C. Djerassi, L. Miramontes, and G. Rosenkranz (to Syntex, S. A. Mexico).
9. U.S. Pat. 2,655,518 (Oct. 13, 1953), F. B. Colton (to G. D. Searle & Co.); U.S. Pat. 2,691,028 (Oct. 5, 1954), (to G. D. Searle & Co.); U.S. Pat. 2,725,389 (Nov. 29, 1955), (to G. D. Searle & Co.).
10. A. J. Birch, *J. Chem. Soc.,* 367 (1950); A. L. Wilds and N. A. Nelson, *J. Am. Chem. Soc.* **75**, 5360 (1953); H. L. Dryden, Jr., G. M. Webber, and J. J. Wieczorek, *J. Am. Chem. Soc.* **86**, 742 (1964); and H. L. Dryden, Jr. in J. Fried and J. A. Edwards, eds., *Organic Reactions in Steroid Chemistry,* Vol. 1, Van Nostrand Reinhold Company, New York, 1972, p. 1.

11. L. J. Chinn and co-workers, *The Chemistry and Biochemistry of Steroids,* Vol. 3, Intra-Science Chemistry Reports, Intra-Science Foundation, Los Angeles, Calif., 1969, p. 15.

12. W. H. W. Inman and M. P. Vessey, *Brit. Med. J.* **2,** 193 (1968); M. P. Vessey and R. Doll, *Brit. Med. J.* **2,** 199 (1968); 651 (1969); R. K. Kalhoff, *Ann. Rev. Med.* **23,** 429 (1972); A. Fuertes-de la Haba and co-workers, *Obstet. Gynecol.* **38,** 259 (1971); N. Phillips and I. Duffy, *Am. J. Obstet. Gynecol.* **116,** 91 (1973); H. W. Ory, *J. Am. Med. Assoc.* **237,** 2619 (1977); W. D. Odell and M. E. Molitch, *Ann. Rev. Pharmacol.* **14,** 413 (1974); R. B. Wallace and co-workers, *Lancet* **2,** 11 (1977); E. Rice-Wray, *Contraception* **3,** 137 (1971).

13. L. Speroff, *Fertil. Steril.* **27,** 997 (1976); Committee on Safety of Drugs, *Brit. Med. J.* **2,** 231 (1970).

14. R. O. Creep, M. A. Koblinsky, and F. S. Jaffe, eds., *Reproduction and Human Welfare: A Challenge to Research,* The MIT Press, Cambridge, Mass., 1976.

15. C. Djerassi, *Science* **169,** 941 (1970).

16. C. Djerassi and co-workers, *J. Am. Chem. Soc.* **76,** 4092 (1954).

17. H. J. Ringold, C. Rosenkranz, and F. Sondheimer, *J. Am. Chem. Soc.* **78,** 2477 (1956).

18. J. Iriarte, C. Djerassi, and H. J. Ringold, *J. Am. Chem. Soc.* **81,** 436 (1959).

19. P. D. Klimstra and F. B. Colton, *Steroids* **10,** 411 (1967).

20. H. Ueberwasser and co-workers, *Helv. Chim. Acta* **46,** 344 (1963).

21. E. L. Shapiro, T. Legett, and E. P. Oliveto, *Tetrahedron Lett.* 663 (1964); A. S. Watnick and co-workers, *J. Endocrinol.* **33,** 241 (1965).

22. H. Smith and co-workers, *J. Chem. Soc.,* 4472 (1964).

23. J. Zañartu and C. Navarro, *Obstet. Gynecol.* **31,** 627 (1968); H. J. Gilfrich and co-workers, *Dtsch. Med. Wschr.* **94,** 2473 (1969).

24. B. Rubio, T. W. Mischler, and E. Berman, *Fertil. Steril.* **23,** 734 (1972); A. Larranaga and E. Berman, *Contraception* **1,** 137 (1970); M. Magueo-Topete and co-workers, *Fertil. Steril.* **20,** 884 (1969).

25. R. L. Elton, E. F. Nutting, and F. J. Saunders, *Acta Endocrinol.* **41,** 381 (1962); R. L. Elton, P. D. Klimstra, and F. B. Colton *Proc. Soc. Exptl. Biol. Med.* **121,** 1194 (1966).

26. M. S. de Winter, C. M. Siegmann, and S. A. Szpilfogel, *Chem. Ind. London,* 905 (1959).

27. C. E. Cook and co-workers, *J. Pharmacol. Exp. Ther.* **185,** 696 (1973); Y. Kishimoto and co-workers, *Xenobiotica* **2,** 237 (1972); A. Mazaheri, K. Fotherby, and J. R. Chapman, *J. Endocrinol.* **47,** 251 (1970).

28. J. H. Fried and co-workers, *J. Am. Chem. Soc.* **83,** 4663 (1961).

29. M. Perelman and co-workers, *J. Am. Chem. Soc.* **82,** 2402 (1960); L. Velluz and co-workers, *Compt. Rend.* **257,** 569 (1963).

30. E. Sakiz and co-workers, *Contraception* **14,** 275 (1976).

31. B. B. Phariss, *Contraception* **1,** 87 (1970).

32. J. S. Baran and co-workers, *Experientia* **26,** 762 (1970).

33. G. C. Buzby, Jr., C. R. Walk, and H. Smith, *J. Med. Chem.* **9,** 782 (1966).

34. L. Viinikka and K. Oulu, *Acta. Endocrinol.* **83,** 429 (1976).

35. J. E. Patrick and co-workers, *Pharmacologist* **18,** 153 (1976).

36. A. P. Shroff and co-workers, *J. Med. Chem.* **16,** 113 (1973).

37. J. Kärkkäinen, J. J. Ohisalo, and T. Luukkainen, *Contraception* **12,** 511 (1975).

38. J. N. Gardner and co-workers, *Steroids* **4,** 801 (1964); C. Burgess and co-workers, *J. Chem. Soc.,* 4995 (1962); U.S. Pat. 2,838,530 (June 10, 1958), F. B. Colton (to G. D. Searle & Co.); A. David and co-workers, *J. Pharm. Pharmacol.* **9,** 929 (1957).

39. E. E. Galanty and co-workers, *Experimentia* **28,** 771 (1972); M. Biollaz and co-workers, *J. Med. Chem.* **14,** 1190 (1971).

40. G. C. Buzby, Jr. and co-workers, *J. Med. Chem.* **9,** 338 (1966).

41. J. M. H. Graves and co-workers, *J. Chem. Soc.,* 5488 (1964).

42. S. J. Halkes and R. van Moorselaar, *Rec. Trav. Chim. Pays-Bas* **88,** 737, 752 (1969).

43. R. Bucourt and co-workers, *J. Steroid Biochem.* **5,** 298 (1974).

44. I. Segal, *Obstet. Gynecol.* **21,** 666 (1963).

45. J. C. Babcock and co-workers, *J. Am. Chem. Soc.* **80,** 2904 (1958).

46. B. Ellis and co-workers, *J. Chem. Soc.,* 2828 (1960).

47. C. Djerassi, *Stud. Fam. Plann.* **5,** 13 (1974).

48. R. Deghenghi, C. Revesz, and R. Gaudry, *J. Med. Chem.* **6,** 301 (1963).

49. M. J. Weiss and co-workers, *Chem. Ind. London,* 118 (1963).

50. R. M. Kanojia and co-workers, *J. Med. Chem.* **18,** 1143 (1975).

51. R. P. Graber and co-workers, *J. Med. Chem.* **7,** 540 (1964).
52. D. N. Kirk, V. Petrow, and D. M. Williamson, *J. Chem. Soc.,* 2821 (1961).
53. E. J. Bailey and co-workers, *J. Chem. Soc. D,* 106 (1970); M. E. Hill, G. H. Phillips, and L. Stephenson, *Third Int. Congress Hormonal Steroids Abstr.* 155, Excerpta Medica, #210, (1970); K. J. Childs and co-workers, *Steroids* **23,** 425 (1974).
54. H. J. Ringold and co-workers, *J. Am. Chem. Soc.* **81,** 3485 (1959).
55. G. R. Daniel, *Brit. Med. J.* **1,** 252, 303 (1970).
56. L. W. Nelson, W. W. Carlton, and J. H. Weikel, Jr., *J. Am. Med. Assoc.* **219,** 1601 (1972); K. Capel-Edwards and co-workers, *Toxicol. Appl. Pharmacol.* **24,** 474 (1973); M. E. Coleman, T. E. Murchison, and D. Frank, *Toxicol. Appl. Pharmacol.* **37,** 181 (1976); L. W. Nelson, J. H. Weikel, Jr., and F. E. Reno, *J. Nat. Cancer Inst.* **51,** 1303 (1973); M. Briggs, *Life Sciences* **1,** 275 (1977).
57. L. J. Lerner, D. M. Brennan, and A. Borman, *Proc. Soc. Exp. Biol. Med.* **106,** 231 (1961); W. S. Keifer, A. F. Lee, and J. C. Scott, *Am. J. Obstet. Gynecol.* **107,** 400 (1970).
58. C. Djerassi, A. J. Manson, and H. Bendas, *Tetrahedron* **1,** 22 (1957).
59. A. J. Solo, V. Kumar, and V. Alks, *Abstract Medi 100,* 172nd American Chemical Society Meeting, San Francisco, 1976.
60. G. W. Barber and M. Ehrenstein, *Ann. Chem.* **603,** 89 (1957); C. Djerassi, M. Ehrenstein, and G. W. Barber, *Ann. Chem.* **612,** 93 (1958).
61. A. D. Claman, *Am. J. Obstet. Gynecol.* **107,** 461 (1970).
62. J. M. Morris and G. van Wagenen, *Am. J. Obstet. Gynecol.* **115,** 101 (1973); Steroid Hormones, Pergamon Press, New York, 1961, p. 200.
63. A. L. Herbst and co-workers, *Am. J. Obstet. Gynecol.* **119,** 713 (1974); L. Herbst, H. Ullfelder, and D. C. Poskanzer, *N. Engl. J. Med.* **284,** 878 (1971).
64. C. W. Emmens and co-workers, *J. Pharmacol. Ecp. Therap.* **165,** 52 (1969); P. Queval and co-workers, *Chim. Therap.* **4,** (1969); L. J. Chinn and co-workers, *J. Med. Chem.* **17,** 351 (1974).
65. J. P. Bennett and co-workers, *Acta Endocrinol.* **53,** 443 (1966); Brit. Pat. 961,502 (June 24, 1964), P. Feather and V. Petrow (to British Drug Houses, Ltd.).
66. G. Pincus, U. K. Banik, and J. Jacques, *Steroids* **4,** 657 (1964).
67. C.-P. Ku and co-workers, *Sci. Sin.* **18,** 262 (1975).
68. N. R. Farnsworth and co-workers, *J. Pharm. Sci.* **64,** 717 (1975).
69. J. L. Thomson, *J. Reprod. Fertil.* **16,** 363 (1968); M. C. Chang, *Fertil. Steril.* **15,** 97 (1964); J. M. Morris and co-workers, *Fertil. Steril.* **18,** 18 (1967).
70. G. W. Duncan and co-workers, *Proc. Soc. Exp. Biol. Med.* **112,** 439 (1963); G. W. Duncan and co-workers, *Proc. Soc. Exp. Biol. Med.* **109,** 163 (1962); D. Lednicer, ed., *Contraception: The Chemical Control of Fertility,* Marcel Dekker, Inc., New York, 1969, p. 197.
71. S. K. Imam and co-workers, *Contraception* **11,** 309 (1975).
72. E. Kesserü and co-workers, *Contraception* **7,** 367 (1973); T. W. Mischler and co-workers, *Contraception* **9,** 221 (1974); K.-G. Nygren and co-workers, *Contraception* **9,** 249 (1974); A. Moggia and co-workers, *J. Reprod. Med.* **13,** 58 (1974).
73. K.-G. Nygren, E. D. B. Johansson, and L. Wide, *Contraception* **9,** 249 (1974).
74. E. Kesserü, A. Larrañaga, and J. Parada, *Contraception* **7,** 367 (1973).
75. T. W. Mischler and co-workers, *Contraception* **7,** 367 (1973).
76. A. Moggia and co-workers, *J. Reprod. Med.* **13,** 58 (1974).
77. G. Azadian-Boulanger and co-workers, *Am. J. Obst. Gynecol.* **125,** 1049 (1976); D. Philibert and J. P. Raynaud, *Contraception* **10,** 457 (1974).
78. L. Chan and B. W. O'Malley, *N. Engl. J. Med.* **294,** 1322, 1372, 1430 (1976).
79. W. W. Leavitt and co-workers, *Endocrinology* **94,** 1041 (1974); E. Milgrom and co-workers, *Endocrinology* **90,** 1071 (1972); P. D. Feil and co-workers, *Endocrinology* **91,** 738 (1972); B. W. O'Malley, D. O. Toft, and M. R. Sherman, *J. Biol. Chem.* **246,** 1117 (1971); R. G. Smith and co-workers, *Nature* **253,** 271 (1975); K. Kontula and co-workers, *J. Clin. Endocrinol. Metab.* **38,** 500 (1974); L. E. Faber, M. L. Sandmann, and H. E. Stavely, *J. Biol. Chem.* **247,** 5048 (1972); E. Milgrom and E.-E. Bauleiu, *Endocrinology* **87,** 276 (1970); L. Terenius, *Steroids* **23,** 909 (1974).
80. H. E. Smith and co-workers, *J. Biol. Chem.* **249,** 5924 (1974).
81. Z. S. Madjerrek in H. Tausk, ed., *International Encyclopedic Of Pharmacology and Therapeutics,* Pergamon Press, New York, 1971.
82. K. Kontula and co-workers, *Acta Endocrinol.* **78,** 574 (1975).
83. H. J. Ringold in C. A. Villee and L. L. Engel, eds., *Mechanism of Action of Steroid Hormones,* Pergamon Press, New York, 1961, p. 200.

84. J. L. McGuire, C. D. Bariso, and A. P. Shroff, *Biochemistry* **13,** 319 (1974).
85. W. L. Duax and D. A. Norton, *Atlas of Steroid Structure,* Plenum Publishing Corporation, New York, 1975, p. 510.
86. E. J. Plotz, *Brook Lodge Symposium on Progesterone,* Brook Lodge Press, Augusta, Mich., 1961, p. 91.
87. E. M. Coutinho and J. E. Melo, *Contraception* **8,** 207 (1973).
88. A. J. Manson and co-workers, *J. Med. Chem.* **6,** 1 (1963).
89. N. H. Lauersen and K. H. Wilson, *Obstet. Gynecol.* **50,** 91 (1977); N. H. Lauersen and K. H. Wilson, *Fertil. Steril.* **28,** 289 (1977).
90. R. D. Skoglund and C. A. Paulsen, *Contraception* **7,** 357 (1973).
91. M. Briggs and M. Briggs, *Nature* **252,** 585 (1974).
92. D. J. Back and co-workers, *J. Reprod. Fertil.* **49,** 237 (1977).
93. W. J. Bremmer and D. M. de Kretser, *N. Engl. J. Med.* **295,** 1111 (1976).
94. P. R. K. Reddy and J. M. Rao, *Contraception* **5,** 295 (1972).
95. J. Mauss and co-workers, *Acta Endocrinol.* **78,** 373 (1975).
96. C. Djerassi, *Proc. R. Soc. London B* **195,** 175 (1976).
97. R. E. Marker and co-workers, *J. Am. Chem. Soc.* **69,** 2167 (1947).
98. W. J. Marsheck, S. Kraychy, and R. D. Muir, *Appl. Microbiol.* **23,** 72 (1972); U.S. Pat. 3,684,657 (Aug. 15, 1972), S. Kraychy, W. J. Marsheck, and R. D. Muir (to G. D. Searle & Co.); U.S. Pat. 3,487,907 (Jan. 6, 1970), W. F. Van der Waard (to Koninklijke Nederlandsche Gist-En Spiritusfabriek, N.V., Netherlands); U.S. Pat. 3,684,656 (Aug. 15, 1972), W. F. Van der Waard (to Koninklijke Nederlandsche Gist-En Spiritusfabriek N.V., Netherlands).
99. S. N. Ananchenko and I. V. Torgov, *Tetrahedron Lett.,* 1553 (1963).
100. C. Rufer and co-workers, *Ann. Chem.* **702,** 141 (1967).
101. R. A. Micheli and co-workers, *J. Org. Chem.* **40,** 675 (1975).
102. N. Cohen, *Acct. Chem. Res.* **9,** 412 (1976); J. W. Scott and D. Valentine, Jr., *Science* **184,** 943 (1974).
103. A. V. Schally, *Am. J. Obstet. Gynecol.* **125,** 1142 (1976); A. V. Schally, *Science* **202,** 18 (1978); H. Matsuo and co-workers, *Biochem. Biophys. Res. Commun.* **43,** 1334 (1971).
104. R. Burgus and co-workers, *Compt. Rend. Acad. Sci. D* **273,** 1611 (1971); R. Guillemin, *Science* **202,** 390 (1978).
105. H. G. Madhwa Raj and N. R. Moudgal, *Endocrinology* **86,** 874 (1970).
106. A. J. Kastin and co-workers, *Int. J. Fertil.* **19,** 202 (1974).
107. W. R. Keye, Jr., J. R. Young, and R. B. Jaffe, *Obstet. Gynecol. Survey* **31,** 635 (1976); T. W. Redding and co-workers, *J. Clin. Endocrinol. Metab.* **37,** 626 (1973).
108. A. Corbin and C. W. Beattie, *Endocrinol. Res. Commun.* **2,** 445 (1975); A. Corbin and co-workers, *Fertil. Steril.* **28,** 471 (1977).
109. U. K. Banik and M. L. Givner, *J. Reprod. Fertil.* **44,** 87 (1975).
110. M. Fujino and co-workers, *Biochem. Biophys. Res. Commun.* **60,** 406 (1974); M. Fujino and co-workers, *Biochem. Biophys. Res. Commun.* **57,** 1248 (1974).
111. J. A. Vilchez-Martinez and co-workers, *Fertil. Steril.* **27,** 628 (1976).
112. C. W. Beattie and co-workers, *J. Med. Chem.* **18,** 1247 (1975).
113. A. de la Cruz and co-workers, *Science* **191,** 195 (1976).
114. V. C. Stevens, *Obstet. Gynecol.* **42,** 496 (1973); V. C. Stevens in E. Diczfalusy, ed., *Karolinska Symp.* **7,** 357 (1974).
115. R. B. Carlsen, O. P. Bahl, and N. Swaminathan, *J. Biol. Chem.* **7,** 6810 (1973); F. J. Morgan, J. Birken, and R. E. Canfield, *J. Biol. Chem.* **250,** 5247 (1975).
116. V. C. Stevens and C. D. Crystle, *Obstet. Gynecol.* **42,** 485 (1973).
117. G. P. Talwar and co-workers, *Proc. Nat. Acad. Sci. USA* **73,** 218 (1976); G. P. Talwar and co-workers, *Contraception* **13,** 131 (1976).
118. J. P. Louvet and co-workers, *J. Clin. Endocrinol. Metab.* **39,** 1155 (1974).
119. S. Bergström and J. Sjövall, *Acta Chem. Scand.* **11,** 1086 (1957); **14,** 1693 (1960); S. Bergström, *Science* **157,** 382 (1967).
120. G. T. Detitta, *Science* **191,** 1271 (1976); E. M. K. Leovey and N. H. Andersen, *J. Am. Chem. Soc.* **97,** 4148 (1975).
121. P. W. Ramwell and P. Kury, *Abstr. Seminar in Prostaglandins, INSERM, Fondation Royaumont (France),* 1973, p. 11, cited in P. Crabbé, *Chem. Br.* **11,** 132 (1975).

122. B. B. Pharris, *Rec. Prog. Horm. Res.* **28,** 51 (1972); B. B. Pharris, *Perspective Biol. Med.* **13,** 434 (1970); J. A. McCracken and co-workers, *Nature,* **238,** 129 (1973).

123. J. A. McCracken, M. E. Glew, and R. J. Scaramuzzi, *J. Clin. Endocrinol. Metab.* **30,** 544 (1970).

124. K. T. Kirton, B. B. Pharris, and A. D. Forbes, *Proc. Soc. Exp. Biol. Med.* **133,** 314 (1970); A. C. Wentz and G. S. Jones, *Obstet. Gynecol.* **42,** 172 (1973); R. N. Turksoy and H. S. Safaii, *Fertil. Steril.* **26,** 634 (1975).

125. J. W. Wilks, *Prostaglandins* **13,** 161 (1977).

126. C. P. Puri and co-workers, *Prostaglandins* **13,** 363 (1977); *Prostaglandins Suppl.* **12,** (1976).

127. S. M. M. Karim and G. M. Filshie, *Lancet* **1,** 157 (1970); S. M. M. Karim and G. M. Filshie, *Brit. Med. J.* **3,** 198 (1970); D. R. Tredway and D. R. Mishell, Jr., *Am. J. Obstet. Gynecol.* **116,** 795 (1973).

128. T. K. Schaaf, *Ann. Rep. Med. Chem.* **11,** 80 (1976).

129. B. J. Magerlein and co-workers, *Prostaglandins* **4,** 143 (1973); S. M. M. Karim and S. D. Sharma, *J. Obst. Gynecol. Brit. Commonwealth* **79,** 737 (1972).

130. S. E. Shaw and S. A. Tillson, *Steroid Biochem. Pharmacol.* **4,** 189 (1974).

131. L. J. Lerner, G. Galliani, P. Carminati, and M. C. Mosca, *Nature* **256,** 130 (1975).

132. L. J. Lerner, *Fertil. Steril.* **28,** 290 (1977).

133. R. Coombs and co-workers, *15th National Medicinal Chemistry Symp.,* Salt Lake City, Utah, 1976.

134. N. C. Crossley, *Prostaglandins* **10,** 5 (1975); D. Binder and co-workers, *Prostaglandins* **6,** 87 (1974).

135. A. I. Csapo and P. Mocsary, *Prostaglandins* **12,** 455 (1976).

136. E. J. Corey and co-workers, *J. Am. Chem. Soc.* **93,** 1489, 1490, 1491 (1971); E. J. Corey, *Ann. N. Y. Acad. Sci. Prostaglandins* **180,** 24 (1971).

137. E. D. Brown and co-workers, *J. Chem. Soc. Chem. Commun.,* 642 (1974).

138. C. Tietze, J. Bongaards and B. Schearer, *Fam. Plann. Perspect.* **8,** 6 (1976); C. Tietze, *Fam. Plann. Perspect.* **9,** 74 (1977); F. S. Jaffe, *N. Engl. J. Med.* **297,** 612 (1977).

139. G. Oster and M. P. Salgo, *N. Engl. J. Med.* **293,** 432 (1975); J. A. Zipper and co-workers, *Am. J. Obstet. Gynecol.* **105,** 1274 (1969); H. J. Tatum, *Clin. Obstet. Gynecol.* **17,** 93 (1974).

140. A. Scommegna and co-workers, *Obstet. Gynecol.* **43,** 769 (1974); V. A. Place and B. B. Pharris, *J. Reprod. Med.* **13,** 66 (1974).

141. S. D. Targun and N. H. Wright, *Am. J. Epidem.* **100,** 262 (1974).

142. D. de Wied and W. de Jong, *Ann. Rev. Pharmacol.* **14,** 389 (1974); L. J. Lerner in D. Lednicer, ed., *Contraception: The Chemical Control of Fertility,* Marcel Dekker, Inc., New York, 1969, p. 161.

143. J. R. Zysk and co-workers, *J. Reprod. Fertil.* **45,** 69 (1975).

144. *Chem. Eng. News,* 33 (Feb. 19, 1979).

General References

M. H. Briggs and M. Briggs, *Biochemical Contraception,* Academic Press, Inc., New York, 1976.

D. Lednicer, ed., *Contraceptions: The Chemical Control of Fertility,* Marcel Dekker, Inc., New York, 1969.

Contraceptives of the Future, The Royal Society of London, London, 1976.

J. P. Bennett, *Chemical Contraception,* Columbia University Press, New York, 1974.

R. Wiechert, "The Role of Birth Control in the Survival of the Human Race," *Angew. Chem. Int. Ed. Engl.* **16,** 506 (1977).

L. J. CHINN
F. B. COLTON
G. D. Searle & Co.

COOLANTS.

See Heat transfer media other than water; Refrigeration; Antifreezes and deicing fluids.

COORDINATION COMPOUNDS

Coordination compounds (or metal complexes) are important throughout chemistry and chemical technology. The chemistry of the metallic elements, which constitute 80% of the periodic table, is predominately coordination chemistry. Coordinating or complexing Lewis bases (electron-pair donors), called ligands, are often used commercially to modify the properties of the metals or metal ions; eg, metal deactivators in gasoline and complexes of copper and zinc in brass electroplating (qv). Ligands interact differently with different metal ions; therefore, analysis of metallic elements is possible through coordination followed by solvent extraction, spectroscopy, gravimetry, electrochemistry, etc (see Analytical methods). Some ligands are used to selectively extract metal ions from biological tissue (see Table 1). Most of these have been employed in analytical applications as well. Conversely, metal ions are sometimes

Table 1. Pharmaceutical Drugs Used for Metal Ion Removal[a]

Name	CAS Registry No.	Formula	Ion(s) removed
deferoxamine (desferal)[b]	[70-51-9]	$NH_2(CH_2)_5[-N-C(CH_2)_2CNH(CH_2)_5]_2-N-CCH_3$ (with HO, O groups)	Fe
cysteine	[52-90-4]	$HSCH_2-CH-COOH$, NH_2	Co
(ethylenedinitrilo)tetra-acetic acid (EDTA)	[60-00-4]	$(HOOCCCH_2)_2NCH_2CH_2N(CH_2COOH)_2$	Na, Al, K, Ca, Cu, Pb, U
D-penicillamine	[52-67-5]	$(CH_3)_2CCHCOOH$, HS NH_2	Cu
2,3-dimercapto-1-propanol (BAL)	[59-52-9]	$HSCH_2CHCH_2OH$, SH	Al, V, As, Cd, Sb, Au, Hg, Pb, Bi
diphenylthiocarbazone (dithizone)	[60-10-6]	C_6H_5-NH-N, CSH, $C_6H_5-N=N$	Zn
aurintricarboxylic acid	[4431-00-9]	(aromatic structure with HO, HO₂C, CO₂H, HO, CO₂H groups)	Be
8-hydroxyquinoline (8-quinolinol, oxine)	[148-24-3]	(quinoline structure with N, HO)	Fe

[a] Ref. 1.
[b] Ref. 2.

used to alter the properties of commercially important species through coordination; eg, the metallation of azo dyes (qv) to improve permanence or change color, or both, and the coordination of zinc to bactericides to modify their properties. Other examples are noted below.

The coordination compound often has properties entirely different from those of the ligand or the metal ion. For example, β-diketone complexes of the lanthanide elements have found applications as lasers, antiknock petroleum additives, and shift reagents for altering nuclear magnetic resonances. Coordinated metallic species serve as catalysts or intermediates in a wide range of organic and biochemical reactions. In fact, iron–heme-type chelates (1), such as the hemoglobins, myoglobins, and cyto-chromes, are essential to all higher forms of life. Biochemists have recently begun to study the large number of metalloenzymes that are important to life (see Enzymes). Conversely, chemists are now using metalloenzymes to produce chemicals that cannot be prepared easily (if at all) by other known routes.

Typically a coordination compound consists of a metal atom, ion, or other Lewis acid plus a number of electron-pair donors (ligands) coordinated to the metal. Ligands with two or more donor atoms coordinated to the same acceptor atom are *chelating ligands* such as those shown in Table 1 (see Chelating agents). The number of donor atoms coordinated to the same metal atom determine the dentate number of the ligand (bidentate, terdentate, quadridentate, etc, for 2, 3, 4, etc, donors to the same metal atom). The resultant ionic charge on the complex, if any, depends on the charges of the components. The coordination number is based on the total number of donor atoms that coordinate to the metal atom (see Tables 2 and 3). Common geometries for coordination numbers from two through nine are shown in Figure 1. *Bridging ligands,* designated with a μ, donate electron pairs to two or more metal atoms simultaneously. The precipitation of aqueous hydroxides is the result of polymeric hydroxide bridging (olation). The halides, carbon monoxide, EDTA, and many other ligands can also function as bridging ligands (see Carbonyls).

The *IUPAC Nomenclature of Inorganic Chemistry* (3) outlines the nomenclature. Coordination species are named by listing the ligands alphabetically with appropriate prefixes (di, tri, tetra, penta, hexa, etc, for 2, 3, 4, 5, 6, etc of the same ligand), followed by the name of the metal (again with prefixes if polynuclear) and the oxidation number in parentheses as a Roman numeral or Arabic zero (Stock nomenclature). The metallic ion is given an *ate* ending if the species is anionic. The Ewens-Bassett alternative of placing the overall charge, if any, as an Arabic number followed by a plus or minus sign is favored by some chemists to avoid oxidation-number ambiguities. If the ligand names are multicomponent or complicated, the prefixes bis, tris, tetrakis, etc, are used rather than di, tri, tetra, etc. In this case the ligand name is placed in parentheses (see Tables 2 and 3 for examples). Unsaturated organic ligands that donate π electrons

(1)

Table 2.   Some Common Coordination Species Possessing Unidentate Ligands

Name	CAS Registry No.	Formula	Coordination number	Typical use
dicyanoargentate(I)	[15391-88-5]	$[Ag(CN)_2]^-$	2	recovery of silver
tetraamminecopper(II)	[16828-95-8]	$[Cu(NH_3)_4]^{2+}$	4	dissolution of copper(II) in basic solution
tetracarbonylnickel(0)	[13463-39-3]	$[Ni(CO)_4]$	4	separation of nickel from assorted metals
hexafluorosilicate(IV)	[17084-08-1]	$[SiF_6]^{2-}$	6	dissolution of silicon(IV) in hydrolytic media

Table 3.   Examples of Less Common Coordination Numbers

Name	CAS Registry No.	Formula	Coordination number
tricyanocuprate(I)	[16593-63-8]	$[Cu(CN)_3]^{2-}$	3
pentacarbonylmanganate(−I)	[14971-26-7]	$[Mn(CO)_5]^-$	5
heptafluorozirconate(IV)	[27679-73-8]	$[ZrF_7]^{3-}$	7
octacyanotungstate(IV)	[18177-17-8]	$[W(CN)_8]^{4-}$	8
nonaaquaneodymium(III)	[54375-24-5]	$[Nd(H_2O)_9]^{3+}$	9

to the metal are given a *hapto* or η designation with a superscript to indicate the number of carbon or heteroatoms interacting at any one instant; eg, ferrocene [102-54-5] (2) is bis(η^5-cyclopentadieno)iron(II), indicating the sandwich structure of the two rings relative to the metal.

The quest for a theory that explains as many facts as possible and predicts the behavior of systems as yet unstudied—all in terms of a comprehensible model—has given rise to the use of valence bond, crystal field, and molecular orbital theories in the recent past. The symmetry separation of energy levels, which can in turn be observed in the chemical and physical properties of coordination compounds, is inherent in all three theories but is more apparent in the molecular orbital and crystal field theories. For details on symmetry and group theory, a number of excellent references are available (4). Molecular orbital treatments of complexes range from very simple semiempirical (Hückel, angular overlap, etc) through so-called *ab initio* methods (5). A simplified molecular-orbital energy-level diagram for an octahedral complex is shown in Figure 2. The twelve electrons designated by circles are the electron pair donors from the ligand donor atoms, whereas the electrons designated by x represent the *d* orbitals from a transition metal ion. For metal ions with more than three of the *d* electrons in the ion involved, choices of spin states exist which are important in kinetics, thermodynamics, spectroscopy, and magnetism of the octahedral species. In

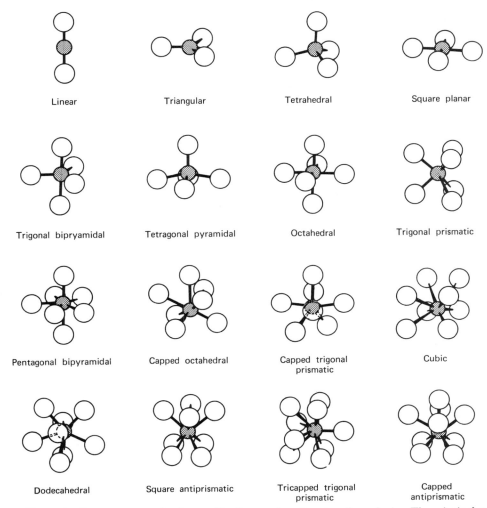

Figures labeled: Linear, Triangular, Tetrahedral, Square planar, Trigonal bipryamidal, Tetragonal pyramidal, Octahedral, Trigonal prismatic, Pentagonal bipyramidal, Capped octahedral, Capped trigonal prismatic, Cubic, Dodecahedral, Square antiprismatic, Tricapped trigonal prismatic, Capped antiprismatic.

Figure 1. Common geometries for coordination numbers from two through nine. The principal axis orientation is vertical for all as shown for the first diagram.

general, low-spin octahedral complexes have fewer antibonding electrons. Therefore, unless electron–electron repulsions predominate, the low-spin state is more thermodynamically stable. Furthermore, low-spin complexes are more kinetically inert.

Ligands with more than one electron pair can interact in a pi (π) sense as well as in the simple Lewis base electron-pair sigma (σ) sense. Examples of π donors include halide ions, oxygen donors, some nitrogen donors (NH_2^-, N_3^-, β-ketoimines, etc) and some sulfur donors (S^{2-}, RS^-, etc). Ligands such as the cyanide ion (cyano ligand) and carbon monoxide are considered to be π acceptors (or π acids) as a result of empty π^* levels that have appropriate symmetry for π-type overlap with filled or partially filled metal orbitals (eg, the t_{2g} level of octahedral complexes). This type of π interaction (also known as back donation) helps relieve the buildup of electron density on the central metal atom from the σ donation; therefore, such ligands stabilize low oxidation states.

Other types of bonding include σ donation by ligand π orbitals, as in the classical

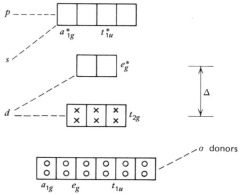

Figure 2. Simplified molecular orbital diagram for a typical octahedral d^6 complex. Δ = energy difference, a, e, and t = anti-(u, ungerade) or centrosymmetric (g, gerade) symmetry orbitals.

$Pt(\eta^2 CH_2$—$CH_2)Cl_3^-$ ion [*12275-00-2*] and sandwich compounds such as ferrocene. Another type is the delta (δ) bond, as in the $Re_2Cl_8^{2-}$ ion [*19584-34-8*], which consists of two $ReCl_4^-$ [*13569-71-6*] squares with Re—Re bonding and eclipsed chlorides. The Re—Re δ bond makes the system quadruply bonded and holds the chlorides in sterically crowded conditions. Numerous other coordination compounds contain two or more metal atoms with metal–metal bonds.

Properties

Stability.   The thermodynamic stability of coordination compounds in solution has been extensively studied (6). The equilibrium constants may be reported as stability or formation constants:

$$M + n\,L \rightleftharpoons ML_n$$

This has a cumulative stability constant β_n related to the activities a of the species by:

$$\beta_n = a_{ML_n}/(a_M a_L{}^n)$$

And the stepwise constant (K_n):

$$K_n = a_{ML_n}/(a_{ML_{n-1}} a_L)$$

Alternatively, instability or dissociation constants are sometimes used, and caution is necessary when comparing values from different sources (see Chelating agents).

Attempts have been made to arrange the interactions between metal ions and ligands. Whereas all metal ions interact in the gas phase more strongly with fluoride than with chloride, in aqueous solution the exceptions listed below occur, in addition to many organometallics.

copper(I)	palladium(II)	mercury(II)
silver(I)	platinum(II) and (IV)	thallium(I) and (III)
gold(I) and (III)	cadmium(II)	lead(II)

Ions that have the normal (class **a**) gaseous stability order of F $\gg$ Cl > Br > I also have N $\gg$ P > As > Sb and O $\gg$ S > Se > Te donor stability order (7). The inverse (class **b**) stability donor orders are F $\ll$ Cl < Br < I, N $\ll$ P > As > Sb, and O $\ll$ S $\approx$ Se $\approx$ Te. The inverse order has long been considered related to polarizability. Recently,

Table 4. Classification of Donors, Acceptors, and Solvents Based on Polarizability

Class **a**, nonpolarizable, hard	Borderline	Class **b**, polarizable, soft
Metallic ions		
H^+, Li^+, Na^+, K^+, Rb^+	Cs^+	
Be^{2+}, Mg^{2+}, Ca^{2+}, Sr^{2+}, Ba^{2+}		
Sc^{3+}, Y^{3+}, La^{3+}, (Lanthanides)$^{3+}$		
Ti^{4+}, Zr^{4+}, Hf^{4+}		
VO^{2+}, Cr^{3+}, UO_2^{2+}, Pu^{4+}		
Mn^{2+}, Fe^{3+}, Co^{3+}	Fe^{2+}, Co^{2+}	$Co(CN)_5^{2-}$, $Co(DMG)_2CH_3^+$
$Rh(NH_3)_5^{3+}$, $Ir(NH_3)_5^{3+}$		$Rh(CN)_5^{2-}$, $Ir(CN)_5^{2-}$, Rh^+, Ir^+
	Ni^{2+}, Cu^{2+}	Pd^{2+}, Pt^{2+}, Cu^+, Ag^+, $Au^{+,3+}$
Al^{3+}, Ga^{3+}, In^{3+}	Zn^{2+}	Cd^{2+}, Hg_2^{2+}, Hg^{2+}, $Tl^{+,3+}$
Si^{4+}, Sn^{2+}, Sn^{4+}, As^{3+}	Pb^{2+}	
metals in oxidation		metal in oxidation state 0 or lower
state 4+ or higher^a		
Ligands		
		H^-, CO, CN^-, R^-
		alkide ions, olefins
		aromatics
	N donors	P, As, Sb, Bi donors
F^-, O donors	Cl^-, Br^-	S, Se, Te donors, I^-
Solvents		
HF, H_2O, ROH	NH_3	$(CH_3)_2CO$, $(CH_3)_2SO$, $(CH_2)_4SO_2$,
		nitroparaffins, DMF

^{a} Exceptions exist if a large number of soft bases are coordinated to the metal.

low polarizability bases such as fluoride and the oxygen donors have been termed hard bases (8). The corresponding class **a** cations are called hard acids. The class **b** acids and the polarizable bases are termed soft acids and soft bases, respectively, with the general rule hard prefers hard and soft prefers soft. A classification is given in Table 4. The divisions are arbitrary; however, the trends are important. Stabilities generally increase for divalent 3*d* ions through copper and then decrease at zinc (9).

The chelate effect means that chelating ligands are usually more stable than the analogous unidentate ligands. This effect is particularly evident for five-membered rings, anionic ligands, and aromatic donors. Larger rings, unless rigid, lose appreciable entropy on coordination; anionic ligands displace large quantities of oriented solvent molecules; and aromatic ring donors are normally rigid. Detailed quantitative consideration between similar systems is often difficult because the ΔH and ΔS changes are normally a very small fraction of the enthalpies involved in the coordination bonds.

Steric Selectivity. In addition to the normal regularities that can be rationalized by electronic considerations, steric factors are important in coordination chemistry. To illustrate, whereas 8-hydroxyquinoline, or 8-quinolinol (Hq), at 100°C precipitates both Mg^{2+} and Al^{3+} from aqueous solution as hydrated Mg(q)$_2$ (formulated as Mg(q)$_2$(H$_2$O)$_2$ [56531-18-1]) and Al(q)$_3$ [2085-33-8], respectively, 2-methyl-8-hydroxyquinoline (3) precipitates only Mg^{2+} from aqueous solution. The 2-methyl group prevents three of the methyl-substituted ligands from coordinating in water. The crown ethers, such as 18-crown-6 or eicosahydrodibenzo[*b,k*][1,4,7,10,13,16]hexaoxacyclooctadecin (4) are highly selective for the potassium relative to the sodium ion (see Chelating agents). However, the cryptate 4,7,13,16,21,24-hexaoxa-1,10-diazabicyclo[8.8.8]hexacosane (5) is selective for the sodium ion (see Catalysis, phase-transfer).

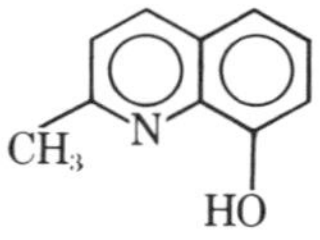

By varying ring sizes and donor atoms of the macrocyclic crown ethers and the fused ring cryptates (or cryptands) different metal ions can be selectively accommodated.

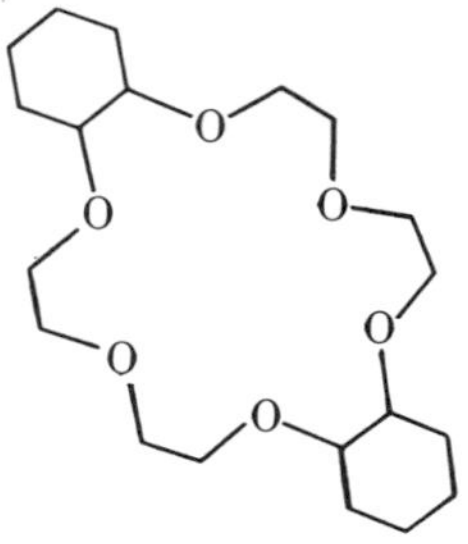

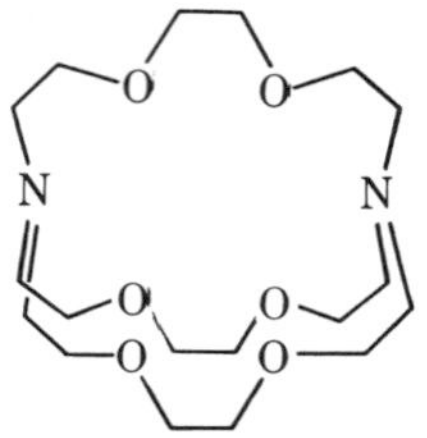

Another example of steric selectivity involves the homopoly and heteropoly ions of molybdenum, tungsten, etc. Each molybdenum(VI) and tungsten(VI) ion is octahedrally coordinated to six oxygen (oxo) ligands. Chromium(VI) is too small and forms only the well-known chromate-type species with four oxo ligands. The ability of other cations to participate in stable heteropoly ion formation is also size related.

Coordination stereochemistry (including various forms of isomerization) is a major academic area of interest. The sources should be consulted for details. This aspect of coordination is important for stereospecific catalytical applications.

Reactions

Substitution. Coordination species are often categorized in terms of the rate at which they undergo substitution reactions. Complexes that react with other ligands to give equilibrium conditions almost as fast as the reagents can be mixed by conventional techniques are termed labile. Included are most of the complexes of the alkali metals, the alkaline earths, the aluminum family, the lanthanides, the actinides, and some of the transition metal complexes. On the other hand, numerous transition metal complexes that resist substitution reactions are termed inert. Note that these terms imply substitution reactivity and not thermodynamic properties. To illustrate, the reaction rates of the aquated cations of some of the first-row transition elements with isotopically labeled water ($H_2^{18}O$) are listed in Table 5.

$$M(H_2O)_6^{n+} + H_2^{18}O \rightleftharpoons M(H_2O)_5(H_2^{18}O)^{n+} + H_2O$$

Table 5. Water Exchange Rates of Hexaaquo Complexes at 25°C

Ion	k_1, s^{-1}	E_a, kJ/mol[a]	Polyhydric structure	CAS Registry No.
V^{2+}	9×10^3		octahedron	[15696-18-1]
Cr^{2+}	7×10^9		distorted octahedron	[20574-26-9]
Mn^{2+}	3×10^7	36	octahedron	[15365-82-9]
Fe^{2+}	3×10^6	35	octahedron	[15365-81-8]
Co^{2+}	1×10^6	36	octahedron	[15276-47-8]
Ni^{2+}	3×10^4	51	octahedron	[15365-79-4]
Cu^{2+}	8×10^9	23	distorted octahedron	[14946-74-8]
Cr^{3+}	5×10^{-7}	112	octahedron	[14873-01-9]
Fe^{3+}	3×10^3	63	octahedron	[15377-81-8]

[a] To convert J to cal, divide by 4.184.

Although vanadium(II) and nickel(II) react more slowly than the other ions of oxidation state II listed, they are labile by the operational definition. On the other hand, chromium(III) and cobalt(III) complexes are inert. However, iron(III) is labile even though iron lies between chromium and cobalt in the periodic table.

In general, octahedral complexes of transition metal ions possessing 0, 1, or 2 electrons beyond the electronic configuration of the preceding inert gas (ie, d^0, d^1, d^2 configurations) are labile. The d^3 systems are usually inert; the relative lability of vanadium(II) may be charge and/or redox related. However, high-spin d^4, d^5, and d^6 species, which possess 4, 5, and 4 unpaired electrons, respectively, are labile as are d^7 through d^{10} octahedral complexes. In addition to the inert d^3 systems, low-spin d^4, d^5, and d^6 complexes are inert to rapid substitution. The d^8 species are the least labile of the configurations classed as labile.

Spin-paired octahedral d^6 ions and spin-free octahedral d^8 ions appear to react by largely dissociative (**D**) reactions

$$ML_n \rightarrow ML_{n-1} + L$$

$$ML_{n-1} + L' \rightarrow ML_{n-1}L'$$

although d^6 complexes which are just barely spin-paired, such as iron-(1,10-phenanthroline)$_3^{2+}$ [14708-99-7] and cobalt(ethylenediaminetetraacetato)$^-$ [15136-66-0], appear to undergo associative (**A**) reactions, with good nucleophiles.

$$ML_n + L' \rightarrow ML_nL' \rightarrow ML_{n-1}L' + L$$

Appreciable **A** character appears to be operative with the d^3 chromium(III) complexes and for the labile d^1 and d^2 ions. The four-coordinate planar molecules also show **A** behavior in their reactions. Because of solvent effects, the term interchange (**I**) is often used to indicate that the reactions show some concerted behavior. Species more dissociative than associative are termed dissociative interchange (**I$_d$**) reactions. For many species the effective atomic number (EAN) or 18-electron rule is helpful. Low-spin transition metal complexes with the EAN of the next inert gas (Table 6), which have 18 valence electrons (considering each normal donor to contribute two electrons and the remainder to be metal valence electrons), are usually inert and normally react by dissociation. The sixteen-electron complexes are often inert (if low spin), but can undergo associative substitution and oxidative–addition reactions.

Oxidation–Reduction. Redox or oxidation–reduction reactions are often governed by the hard–soft rule discussed above; eg, a metal in a low oxidation state (relatively

Table 6. Complexes with Effective Atomic Numbers of the Next Inert Gas [a]

Name	CAS Registry No.	Formula	Coordination no.	EAN
nickel tetracarbonyl	[13463-39-3]	Ni(CO)$_4$	4	36
iron pentacarbonyl	[13463-40-6]	Fe(CO)$_5$	5	36
chromium hexacarbonyl	[13007-92-6]	Cr(CO)$_6$	6	36
hexaammine platinum(IV)	[16893-12-2]	Pt(NH$_3$)$_6^{4+}$	6	86
tricarbonyldichlorobis-(triphenylphosphine)-molybdenum(II)	[17250-39-4]	Mo(CO)$_3$Cl$_2$[P(C$_6$H$_5$)$_3$]$_2$	7	54
tetrakis(8-quinolinolato)-tungsten(IV)	[17499-74-0]	W(C$_9$H$_6$NO)$_4$	8	86
nonahydridorhenate(VII)	[44863-47-0]	[ReH$_9$]$^{2-}$	9	86

[a] 18-Electron species.

soft) can be oxidized more easily if surrounded by hard ligands or a hard solvent. Metals tend toward hard-acid behavior on oxidation. Organometallic coordination species are usually between 16 and 18 electron systems (10). For example, the first step in hydrogenation by chlorotris(triphenylphosphine)rhodium(I) [14694-95-2] could be oxidative addition:

$$RhCl[P(C_6H_5)_3]_3 + H_2 \rightarrow Rh(H)_2Cl[P(C_6H_5)_3]_3$$

The 16-electron four-coordinate rhodium(I) complex becomes an eighteen-electron octahedral rhodium(III) species, which could lose a phosphine (18 → 16 electrons or dissociative substitution suggested above for d^6 octahedral complexes):

$$Rh(H)_2Cl[P(C_6H_5)_3]_3 \rightarrow Rh(H)_2Cl[P(C_6H_5)_3]_2 + P(C_6H_5)_3$$

An olefin could be added (16 → 18 electrons) to the Rh(III) bisphosphine followed by a tautomeric shift (18 → 16 electrons) to give a coordinated alkyl group:

$$Rh(H)_2Cl[P(C_6H_5)_3]_2 \xrightarrow{\text{olefin}} Rh(\eta^2\text{-}R)(H)_2(Cl)[P(C_6H_5)_3]_2 \rightarrow Rh(R')(H)(Cl)[P(C_6H_5)_3]_2$$

where R = olefin and R' = alkyl. These last two steps are often called olefin insertion. Addition of phosphine or a solvent could give an 18-electron complex:

$$Rh(R')(H)(Cl)[P(C_6H_5)_3]_2 + P(C_6H_5)_3 \rightarrow Rh(R')(H)(Cl)[P(C_6H_5)_3]_3$$

Reductive elimination produces the original catalyst or solvated equivalent:

$$Rh(R')(H)(Cl)[P(C_6H_5)_3]_3 \rightarrow RhCl[P(C_6H_5)_3]_3 + R'H$$

Redox rates are often limited by substitution rates of the reactant so that direct electron transfer can occur (11). If substitution is very slow, an outer-sphere or tunneling reaction may occur. The one-electron transfers are normally favored over multielectron processes, especially when three or more species must aggregate prior to reaction, although the oxidative addition and oxygen atom transfer are considered two-electron–two-electron transfers which are second only to one-electron–one-electron processes in general. The distinction between atom and electron transfer is academic when a bridging atom moves from one species to another during a redox reaction:

$$Co(NH_3)_5Cl^{2+} + Cr(H_2O)_6^{2+} \xrightarrow{H_2O} Cr(H_2O)_5Cl^{2+} + Co(H_2O)_6^{2+} + 5\,NH_3$$

[14970-14-0]   [20574-26-9]   [13820-87-6]   [7664-41-7]

In effect, the atom is transferred.

Photochemistry. Substitution rates of many complexes are enhanced by irradiation of the low energy d–d transitions ($t_{2g} \rightarrow e_g^*$ in octahedral coordination compounds). Quantum yields Φ, defined as the ratios of reacting molecules to incident photons, vary from very good (eg, for chromium(III) Φ ca 0.4) to poor (eg, for cobalt(III) $\Phi < 0.01$) for ligand substitution (12). The substituted ligand is normally the strongest ligand on the axis with the weakest net pair as determined by spectrochemical relationships; eg, $CN^- > NO_2^- > NH_3 > H_2O \approx F^- > Cl^-$. Exceptions do occur. Photochemical ligand dissociation is useful in the synthesis of multinuclear metal complexes:

$$Fe(CO)_5 \xrightarrow{h\nu} Fe_2(CO)_9 + CO$$

[13463-40-6]   [15321-51-4]

Or conversely, active radicals can be obtained by irradiating certain metal–metal bonded species:

$$Mn_2(CO)_{10} \xrightarrow{h\nu} 2\,Mn(CO)_5$$

[10170-69-1]　　[54832-42-7]

Irradiation of coordination compounds in the charge-transfer spectral region can often enhance redox reactions (12). Again, the quantum yields are variable. The use of photochemical redox for practical energy-transfer is being actively pursued (see Hydrogen; Hydrogen energy). For example, sufficient energy is available in visible photons to split water into hydrogen and oxygen.

Applications

Bactericides and Fungicides. Among the bactericides marketed today is the 2-pyridinethiol-1-oxide sodium salt [3811-73-2]; the milder zinc chelate [13463-41-7] (6) is used in shampoos (see Antibacterial agents; Hair preparations). The plant fungicides zineb [2122-67-7], $Zn(S_2CNHCH_2CH_2NHCS_2)$, and maneb [12427-38-2], $Mn(S_2CNHCH_2CH_2NHCS_2)$, are polymeric chelates (see Fungicides). A number of other bactericides, fungicides, and disinfectants have metal chelating capabilities, such as 8-hydroxyquinoline and its derivatives (see Disinfectants).

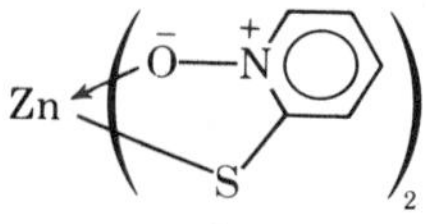

(6)

Catalysis. Hydrogenation reactions can be catalyzed by a wide variety (13) of coordination compounds (see Catalysis). For example, cobalt carbonyl [10210-68-1], $Co_2(CO)_8$, has been suggested to be a suitable catalyst for the hydrogenation of olefins to alkanes, aldehydes to alcohols, acid anhydrides to acids and aldehydes, as well as the selective hydrogenation of polyenes to monoenes and the hydrogenation and isomerization of unsaturated fats (see also Oxo process; Aldehydes). The list for chlorotris(triphenylphosphine)rhodium(I) is even longer (14). On the other hand, a number of chromium tricarbonyl arene derivatives such as benzenetricarbonylchromium [12082-08-5] appear to be quite selective for the hydrogenation of 1,3- and 1,4-dienes to form cis-monoenes by 1,4-addition.

The palladium chloride process for oxidizing olefins to aldehydes in aqueous solution (Wacker process) apparently involves an intermediate complex such as dichloro(ethylene)hydroxypalladate or a neutral aquo complex $PdCl_2(CH_2{=}CH_2)(H_2O)$ with uncertain mechanistic details (15) (see Acetaldehyde). The coordinated $PdCl_2$ is reduced to Pd during the olefin oxidation and reoxidized by the cupric–cuprous chloride couple, which in turn is reoxidized by oxygen, for a net reaction:

$$2\,RCH{=}CH_2 + O_2 \xrightarrow{PdCl_2{-}CuCl_2} 2\,RCH_2CHO$$

The oxo reaction or hydroformylation converts an olefin to an aldehyde containing one more carbon atom, using tetracarbonylhydrocobalt [16842-03-8]. Rhodium coordination species are also used for this purpose, and give a higher normal-to-isobutyraldehyde ratio in the hydroformylation of propene (16). Metal carbonyls and carbonyl clusters may catalyze the water–gas shift reaction (17) which should become more important as the price of petroleum fuels increases. Stereo irregular polymerization is catalyzed by coordination compounds (18) such as Ziegler-Natta catalysts (qv), which are heterogeneous $TiCl_3$–Al alkyl complexes (see also Organometallics). Cobalt carbonyl is a catalyst for the polymerization of monoepoxides, and several rhodium and iridium coordination compounds also catalyze certain polymerizations. Cyclooligomerization is promoted by nickel coordination compounds (19). The products depend on the sites available for coordination. For example, a complex with four octahedral sites available for coordination by acetylene produces cyclooctatetraene, three facial sites yield benzene, etc.

Olefin isomerization can be catalyzed by a number of catalysts such as molybdenum hexacarbonyl [13939-06-5], $Mo(CO)_6$. This compound has also been found to catalyze the photopolymerization of vinyl monomers, the cyclization of olefins, the epoxidation of alkenes and peroxo species, the conversion of isocyanates to carbodiimides, etc. Rhodium carbonylhydrotris(triphenylphosphine) [17185-29-4], $RhH(CO)[P(C_6H_5)_3]_3$, is a multifunctional catalyst which accelerates the isomerization and hydroformylation of alkenes. Many other examples could be given.

The applications discussed above are mostly examples of homogeneous catalysis. Coordination catalysts that are attached to polymers via phosphine, siloxy, or other side chains have also shown promise (20). The catalytic specificity is often modified by such immobilization. Metal enzymes are in effect anchored coordination catalysts immobilized by the protein chains. Even multistep syntheses are possible with alternating catalysts along a polymer chain (21). To obtain metal coordination catalysts closer in design to metal surfaces, metal cluster coordination compounds are being studied (22). Carbonyl clusters of rhodium are also employed in the production of ethylene glycol from carbon monoxide and hydrogen (23) (see Glycols). Other polynuclear coordination species such as the homopoly and heteropoly ions also have applications in reaction catalysis.

Coordination Polymers. In addition to catalysts, polymeric coordination compounds have applications in high temperature coatings utilizing poly(metal phosphinates), and also in the chelated fiber called Enkatherm, which is the zinc chelate of poly(terephthaloyl oxalic-bis-amidrazone), PTO (7). Enkatherm is said to resist a temperature of 1500°C, and be superior to any other flame-resistant fiber (24) (see

(7)

Flame-retardant textiles). Sensitivity to acid solutions, poor long-term stability, and photosensitivity have limited the use of such materials.

Dyes and Pigments. Over 4500 metric tons of metallated or metal coordinated phthalocyanine dyes (8) are sold annually in the United States (25). The partially oxidized metallated phthalocyanine dyes are good conductors and are called molecular metals (see Semiconductors, organic; Phthalocyanine compounds; Colorants for plastics). Azo dyes are also often metallated. The basic unit for a 2,2′-azobisphenol dye is shown as structure (9) (see Azo dyes). Sulfonic acid groups are used to provide solubility, and a wide variety of other substituents influence color and stability. Such complexes have also found applications as analytical indicators, pigments (qv), and paint additives.

(8)   (9)   (10)

Electroplating. Aluminum can be electroplated by the electrolytic reduction of cryolite, which is trisodium aluminum hexafluoride [13775-53-6] containing alumina. Brass can be electroplated from aqueous cyanide solutions which contain cyano complexes of zinc(II) and copper(I) with the soft CN^- stabilizing copper as copper(I). The two cyano complexes have comparable potentials. The potentials of aqueous zinc(II) and copper(II) without CN^- are over one volt apart, thus only copper precipitates. With careful control of concentration and pH, brass can be deposited also from solutions of citrate and tartrate. The noble metals are often plated from solutions in which coordination compounds help provide fine, even deposits (see Electroplating).

Petroleum Additives. Antiknock additives include lanthanide β-diketones, several mixed-ligand manganese carbonyl complexes and tetraethyl lead. Coordination compounds have been suggested as fuel oil additives. Metal deactivators for gasoline include Schiff base ligands that minimize oxidation state changes in the traces of copper dissolving from fuel lines. For example, the planar copper(II) chelate [14522-52-2] (10) resists reduction to copper(I) in gasoline (see Gasoline).

Therapeutic Chelates. Calcium ethylenediaminetetraacetate [38620-52-9] is used to treat lead poisoning; the free ligand causes calcium loss (see Lead compounds, toxicology). Platinum coordination compounds, such as platinum diammine dichloride [15663-27-1], cis-$Pt(NH_3)_2Cl_2$, are cancer therapeutics (1) (see Chemotherapeutics, antimitotic).

Technetium-99 coordination compounds are used very widely as noninvasive imaging tools (26) (see Radioactive drugs and tracers). Different coordination species are concentrated in different organs. Species of the $[Tc^VO(chelate)_2]^n$ type have been involved in certain complexes analogous to those used in imaging.

Miscellaneous. Tetrahedral cobalt(II) units are incorporated in cobalt blue glass. Tetrahedral to octahedral coordination and concurrent blue to pink indicator color change is obtained from moisture on cobalt(II) chloride. Nickel cyano clathrates are used for organic isomer separation, and the closely related zeolites as catalysts (27) and porous molecular sieves (qv). A number of dioxygen complexes are being considered for artificial blood applications; tetrakis(β-diketonato) anionic chelates of europium are employed in laser applications; the neutral tris(β-diketonato) lanthanide chelates are used for nmr shift reagents; and coordination compound membranes are under consideration for ion-selective electrodes (see Ion-specific electrodes).

BIBLIOGRAPHY

"Coordination Compounds" in *ECT* 1st ed., Vol. 4, pp. 379–391, by R. D. Johnson, The University of Pittsburgh; "Coordination Compounds" in *ECT* 2nd ed., Vol. 6, pp. 122–131, by N. Christian Nielsen, Central Methodist College.

1. D. R. Williams, ed., *An Introduction to Bio-inorganic Chemistry,* Charles C Thomas, Springfield, Ill., 1976, Chapt. 19.
2. R. L. Rawls, *Chem. Eng. News,* 24 (May 2, 1977).
3. *IUPAC Nomenclature of Inorganic Chemistry,* 2nd ed., Butterworths, London, Eng., 1971.
4. F. A. Cotton, *Chemical Applications of Group Theory,* 2nd ed., Wiley-Interscience, New York, 1971; G. Davidson, *Introductory Group Theory for Chemists,* Elsevier, Amsterdam, The Netherlands, 1971; M. Tinkham, *Group Theory and Quantum Mechanics,* McGraw-Hill Book Company, New York, 1964; J. P. Fackler, Jr., *Symmetry in Coordination Chemistry,* Academic Press, Inc., New York, 1971; W. E. Hatfield and W. E. Parks, *Symmetry in Chemical Bonding and Structure,* Charles E. Merrill, Columbus, Ohio, 1974; D. S. Urch, *Orbitals and Symmetry,* Penguin Books, Harmondsworth, Eng., 1970.
5. J. H. VanVleck, *J. Chem. Phys.* **3,** 803 (1935); H. B. Gray, *J. Chem. Educ.* **41,** 2 (1964); F. A. Cotton, *J. Chem. Educ.* **41,** 466 (1964); C. J. Ballhausen and H. B. Gray, *Molecular Orbital Theory,* W. Benjamin, New York, 1965; C. K. Jørgensen, *Absorption Spectra and Chemical Bonding in Complexes,* Pergamon Press, Oxford, Eng., 1962; C. K. Jørgensen, *Modern Aspects of Ligand Field Theory,* North Holland Press, Amsterdam, The Netherlands, 1971; L. E. Orgel, *An Introduction to Transition Metal Chemistry: Ligand Field Theory,* 2nd ed., John Wiley & Sons, Inc., New York, 1965; F. A. Cotton and G. Wilkinson, *Advanced Inorganic Chemistry,* 3rd ed., Wiley-Interscience, New York, 1972, Chapt. 20; K. F. Purcell and J. C. Kotz, *Inorganic Chemistry,* W. B. Saunders Company, Philadelphia, Pa., 1977, pp. 531–559; C. J. Ballhausen, *Introduction to Ligand Field Theory,* McGraw-Hill Book Company, Inc., New York, 1962; B. H. Figgis, *Introduction to Ligand Fields,* Interscience Publishers, New York, 1966; J. A. Griffith, *The Theory of Transition Metal Ions,* Cambridge University Press, Cambridge, Eng., 1961; H. L. Schlafer and G. Glieman, *Basic Principles of Ligand Field Theory,* John Wiley & Sons, Inc., New York, 1969; L. Pauling, *The Nature of the Chemical Bond,* 3rd ed., Cornell University Press, Ithaca, N.Y., 1960; L. Pauling, *The Chemical Bond,* Cornell University Press, Ithaca, N.Y., 1967.
6. L. G. Sillen and A. E. Martell, *Stability Constants,* Chemical Society, London, Eng., Special Publications No. 17 (1964) and 25 (1971); S. J. Ashcroft and C. T. Mortimer, *Thermochemistry of Transition Metal Complexes,* Academic Press, Inc., New York, 1970; F. J. C. Rossotti in J. Lewis and R. G. Wilkins, eds., *Modern Coordination Chemistry,* Interscience Publishers, New York, 1960, Chapt. 1; R. M. Smith and A. E. Martell, eds., *Critical Stability Constants,* Plenum Publishing Corporation, London, Eng.; K. B. Yatsimirskii and V. P. Vasil'ev, *Instability Constants of Complex Compounds,* Pergamon Press Ltd., London, Eng., 1960; F. J. C. Rossotti and H. Rossotti, *The Determination of Stability Constants,* McGraw-Hill Book Company, Inc., New York, 1961.
7. S. Ahrland, J. Chatt, and N. R. Davies, *Quart. Rev.* **11,** 265 (1958); S. Ahrland, *Struct. Bonding (Berlin)* **5,** 118 (1968).
8. R. G. Pearson, *Hard and Soft Acids and Bases,* Dowden, Hutchinson, and Ross, Stroudsburg, Pa., 1973; R. G. Pearson, *J. Chem. Educ.* **45,** 581, 643 (1968).
9. H. Irving and R. J. P. Williams, *Nature (London)* **162,** 746 (1948); H. Irving and R. J. P. Williams, *J. Chem. Soc.,* 3192 (1953).
10. C. A. Tolman, *Chem. Soc. Rev.* **1,** 337 (1972).
11. H. Taube, *Electron Transfer Reactions of Complex Ions in ⁻lution,* Academic Press, Inc., New York, 1970; A. Haim, *Acc. Chem. Res.* **8,** 265 (1975); N. Sutin, *Acc. Chem. Res.* **1,** 225 (1968); J. O. Edwards, *Inorganic Reaction Mechanisms,* W. A. Benjamin, Inc., Menlo Park, Calif.; K. F. Purcell and J. C. Kotz in ref. 5, pp. 654–693.
12. A. W. Adamson and P. D. Fleischauer, eds., *Concepts of Inorganic Photochemistry,* John Wiley & Sons, Inc., New York, 1975, 439 pp.; C. R. Bock and E. A. Koerner von Gustorf in J. N. Pitts, Jr., G. S. Hammond, and K. Gollnick, eds., *Advances in Photochemistry,* Vol. 10, John Wiley & Sons, Inc., New York, 1977, pp. 221–310; V. Balzani and V. Carossiti, *Photochemistry of Coordination Compounds,* Academic Press, Ltd., London, Eng., 1970, 432 pp.
13. D. Forster and J. R. Roth, *Homogeneous Catalysis-II, Advances in Chemical Series 132,* American Chemical Society, Washington, D.C., 1974, 344 pp.; G. Henrici-Olive and S. Olive, *Coordination and Catalysts,* Verlag Chemie, New York, 1976, 311 pp.; M. L. Bender, *Mechanisms of Homogeneous Catalysis from Protons to Proteins,* Wiley-Interscience, New York, 1971, Chapt. 8, Chapt. 17; F. A. Cotton and G. Wilkinson in ref. 5, Chapt. 24; K. Purcell and J. C. Kotz in ref. 5, Chapt. 17; R. B. King, ed., *Inorganic Compounds with Unusual Properties, Catalysis and Energy Transport Conversion, and*

Storage, American Chemical Society, Washington, D.C., in press; M. Strem, *The Strem Catalog, No. 8,* Strem Chemical, Newburyport, Mass., 1978, pp. 99–129.

14. M. Strem in ref. 13, p. 115.

15. J.-E. Bäckvall, B. Åkermark, and S. O. Ljunggren, *J. Chem. Soc. Chem. Commun.,* 264 (1977).

16. R. Fowler, H. Connor, and R. A. Baehl, *Chemtech* 772 (1976); *Chem. Eng. News* **54**(18), 25 (Apr. 26, 1976).

17. R. M. Laine, R. G. Rinker, and P. C. Ford, *J. Am. Chem. Soc.* **99,** 252 (1977).

18. J. C. W. Chien, ed., *Coordination Polymerization,* Academic Press, Inc., New York, 1975, 353 pp.

19. P. M. Maitlis, *Acct. Chem. Res.* **9,** 93 (1976).

20. J. C. Bailar, Jr., *Catal. Rev.* **10,** 17 (1974).

21. C. U. Pittman, Jr., L. R. Smith, and R. M. Hanes, *J. Am. Chem. Soc.* **97,** 1742 (1975).

22. A. L. Robinson, *Science* **194,** 1150, 1202 (1976); J. R. Shapeley, *Strem Chem.* **6,** 3 (1978).

23. U.S. Pats. 3,833,634 (Sept. 3, 1974); 3,878,214 (Apr. 15, 1975); 3,878,290 (Apr. 15, 1975); 3,878,292 (Apr. 15, 1975), (to Union Carbide).

24. D. W. Van Krevelen, *Chem. Ind.* **49,** 1396 (1971); F. C. A. A. Van Berkel and H. Grotjahn, *Appl. Polym. Symp.* **21,** 67 (1973); *Text. Prog.* **8,** 124, 156 (1976).

25. R. Price, *Chimia* **28,** 221 (1974); H. Zollinger, *Proc. Int. Wolltextil-Forschungscont. 5th* **1,** 167 (1976).

26. J. A. Siegel and E. Deutsch, *Ann. Rep. Inorg. Gen. Synth.* **4,** 311.

27. K. Seff, *Acc. Chem. Res.* **9,** 121 (1976).

General References

K. F. Purcell and J. C. Kotz, *Inorganic Chemistry,* W. B. Saunders Company, Philadelphia, Pa., 1977, pp. 514–979; this treatment spans coordination compounds from the classical Werner types through organo-metallics with a simplified molecular orbital approach.

F. A. Cotton and G. Wilkinson, *Advanced Inorganic Chemistry,* 3rd ed., Interscience Publishers, a division of John Wiley & Sons, Inc., New York, 1972, pp. 528–1115.

J. E. Huheey, *Inorganic Chemistry: Principles of Structure and Reactivity,* Harper & Row, New York, 1972, pp. 276–440; excellent bibliography (pp. 706–721).

J. C. Bailar, Jr., ed., *Chemistry of Coordination Compounds, ACS Monograph 131,* Reinhold Publishing Corporation, New York, 1956; somewhat dated but very good on the historical and classical aspects of coordination chemistry.

C. K. Jørgensen, *Absorption Spectra and Chemical Bonding in Complexes,* Pergamon Press, Oxford, Eng., 1962; this volume stresses a molecular orbital approach to complex bonding and structural interpretation.

J. Lewis and R. G. Wilkins, *Modern Coordination Chemistry,* Interscience Publishers, Inc., New York, 1960; emphasis on physical methods.

L. E. Orgel, *An Introduction to Transition Metal Chemistry,* John Wiley & Sons, Inc., New York, 1965.

F. P. Dwyer and D. P. Mellor, eds., *Chelating Agents and Metal Chelates,* Academic Press, Inc., New York, 1964.

F. Basolo and R. C. Johnson, *Coordination Chemistry,* W. A. Benjamin, Inc., New York, 1964; a fine, readable, small paperback, for the introductory level.

A. E. Martell, ed., *Coordination Chemistry,* Vol. 1 (1971), Vol. 2 (1978), *ACS Monograph 168, ACS Monograph 174,* ACS, Washington, D.C.

J. V. Quagliano and L. M. Vallarino, *Coordination Chemistry,* D. C. Heath and Company, Lexington, Mass., 1969; a recent paperback with a good discussion of the valence bond approach to complexes.

F. Basolo and R. G. Pearson, *Mechanisms of Inorganic Reactions,* 2nd ed., John Wiley & Sons, Inc., New York, 1967; an excellent volume which stresses the reactions of complexes in solution; a background and a detailed theory section is included which is largely crystal field theory, but the advantages and disadvantages of molecular orbital theory are included.

RONALD D. ARCHER
University of Massachusetts

COPOLYMERS

Synthetic polymeric materials have become some of the most useful commodities of the modern world. In the form of elastomers, plastics, and fibers they meet basic human needs (eg, shelter, clothing, transportation, etc), and improve the overall quality of human life. In recent years the emphasis in research, development, and production of synthetic macromolecules has been directed toward preparation of cost-effective multicomponent polymer systems (ie, copolymers, polyblends, and composites), rather than the preparation of new and frequently more expensive homopolymers.

This article reviews the preparation, properties, characterization, use, economic importance, and future of some of the important synthetic copolymers. Polyblends and composites are also mentioned (see also Polyblends; Composite materials). Biocopolymers are not included (see Biopolymers).

Homopolymers and Copolymer Structures

Homopolymers are high molecular weight molecules prepared by linking a large number of smaller molecules called monomers (eq. 1) (see also Polymers).

$$n\text{A} \rightarrow +\text{A}+_n \tag{1}$$

monomer   polymer

Macromolecules in which two or more different monomers (comonomers) are incorporated in the same polymer chain are copolymers (eq. 2).

$$\text{A} + \text{B} \rightarrow +\text{A-B}+_n \tag{2}$$

comonomers   copolymer

Copolymers can be further described by specifying the number and distribution of monomer units within the copolymer molecule. Thus a polymer with a statistical placement of monomer units, eg, ABBABABABAAB, is called a *random* copolymer.

An *alternating* copolymer consists of an alternating arrangement of the comonomers, ABABABABA.

On the other hand, a *block* copolymer has a long segment of one monomer followed by a long segment of another monomer, AAAABBBB.

A *graft* copolymer refers to a backbone polymer chain to which a second polymer is attached at intervals along the backbone.

AAAAA
B
B
B
B
B

A *network* copolymer is a cross-linked or three dimensional copolymer.

AAAABAAAA
C
C
C
C
AAAABAAAA

A *polyblend* is a physical or mechanical blend (alloy) of two or more homopolymers or copolymers. Although a polyblend is not a copolymer according to the above definition, it is mentioned here because of its commercial importance and the frequency with which polyblends are compared with chemically bonded copolymers. Another technologically significant relative to the copolymer is the composite, a physical or mechanical blend of a polymer with some unlike material, eg, fillers (qv) like SiO_2 and carbon black.

Obviously, these definitions represent only the most basic structures. They should apply to copolymers regardless of the mechanism of their formation, but their simplicity can be misleading, particularly in the case of step-growth copolymers. For example, an $+A-B+$ polyamide, prepared from a diacid and a diamine, could be considered either a homopolymer or a copolymer. In this review we have adopted the convention suggested by Odian (1), according to which step-growth copolymers are only those materials that have used more than the minimum number of monomers to effect homopolymerization (eg, a copolyamide from a diacid and two diamines), or that have used prepolymers as monomers. This greatly reduces the number of step-growth materials that qualify as copolymers. Consequently, this review emphasizes chain growth copolymers and copolymerization.

Nomenclature. There are myriad ways in which two or more monomers can be bonded to form copolymers. Consequently, a systematic nomenclature for copolymers is both a need and a challenge. Although there is presently no accepted IUPAC naming scheme for copolymers, the semisystematic system introduced by Ceresa (2) has received the most support (3-4). Ceresa's classification method is based upon a homopolymer nomenclature which uses *poly* plus the monomer, eg, poly(methylmethacrylate), polystyrene. Copolymers are distinguished from homopolymers and the various types of copolymers from each other by the use of prefixes. Thus a random copolymer of butadiene and styrene is named poly(butadiene-*co*-styrene); an alternating copolymer of propylene oxide and carbon dioxide becomes poly(propylene oxide-*alt*-carbon dioxide) [40081-35-4]. Block and graft copolymers are further distinguished by the prefixes -*b*- and -*g*-, respectively. The first polymer segment named corresponds to the homopolymer or copolymer that is prepared first, eg, as the backbone polymer of a graft copolymer.

Conventional diene microstructure prefixes can be inserted in their normal positions. Branching and cross-linking are indicated by the prefixes [br] and [c.l.]. The use of [c.1.] before poly indicates the whole structure is a three-dimensional network. Furthermore, tacticity is indicated by the use of [iso] for isotactic, [syndio] for syndiotactic, and [a] for atactic.

Chemical Abstracts Service (CAS) uses two types of nomenclature for copolymers. In one system a copolymer is simply identified as *monomer A—monomer B copolymer,* eg, butadiene–styrene copolymer. A more systematic scheme is used for indexing purposes. In the *Chemical Abstracts* index, copolymers are named as *monomer A, polymer with monomer B.* In turn, the respective monomers are named by the IUPAC nomenclature for organic compounds. Thus a butadiene–styrene copolymer is indexed as benzene, ethenyl-, polymer with 1,3-butadiene.

Copolymers extend the number and range of available materials, enabling the polymer scientist to achieve combinations of material properties (eg, tensile strength, solubility, solvent resistance, low temperature flexibility, etc) unattainable from the simple constituent homopolymers. As a result, a large number of copolymers have become commercially important. Table 1 lists some of them.

Table 1. Some Commercially Important Copolymers

Comonomers	Preferred semisystematic nomenclature	CAS Registry No.	CAS Indexing	Generic name, trade name and/or abbreviation
butadiene–styrene (random)	poly(butadiene-co-styrene)	[9003-55-8]	benzene, ethenyl-, polymer with 1,3-butadiene	synthetic rubber; GRS, SBR
ethylene–propylene (random)	poly(ethylene-co-propylene)	[9010-79-1]	ethene, polymer with 1-propene	EPM, EPR rubber
ethylene–propylene–diene (random)	poly(ethylene-co-propylene-co-5-ethylidene-2-norbornene)	[25038-36-2]	ethene, polymer with 1-propene and 5-ethylidene-2-norbornene	EPDM rubber
butadiene–acrylonitrile (random)	poly(butadiene-co-acrylonitrile)	[9003-18-3]	2-propenenitrile, polymer with 1,3-butadiene	NBR rubber
styrene–butadiene–styrene (triblock)	poly(styrene-b-butadiene-b-styrene)	[9003-55-8]	benzene, ethenyl- polymer with 1,3-butadiene	SBS thermoplastic rubber
isobutylene–isoprene (random)	poly(isobutylene-co-isoprene)	[29985-75-9]	1-butene, polymer with 1,3-butadiene, 2-methyl	butyl rubber, GR-1
vinyl chloride–vinyl acetate (random)	poly(vinyl chloride-co-vinyl acetate)	[9003-22-9]	chloroethene, polymer with acetic acid ethenyl ester	Vinylite flooring, Tygon tubing, coatings
vinyl chloride–acrylonitrile (random)	poly(vinyl chloride-co-acrylonitrile)	[9003-00-3]	chloroethane, polymer with 2-propenenitrile	Dynel fibers
vinyl chloride–vinylidene chloride (random)	poly(vinyl chloride-co-vinylidene chloride)	[9011-06-7]	chloroethene, polymer with ethene, 1,1-dichloro-	Saran packaging, fibers
acrylonitrile–vinyl acetate (random)	poly(acrylonitrile-co-vinyl acetate)	[24980-62-9]	2-propenenitrile, polymer with chloroethene	Orlon, Acrilon acrylic fibers
acid–glycol–diisocyanate (step-growth multiblock)	poly(acid-co-glycol-co-diisocyanate-co-diamine)		the acid, polymer with the glycol; the diisocyanate, and the diamine	Spandex fibers
acrylonitrile–butadiene–styrene (graft and polyblend)	poly(butadiene-g-styrene-co-acrylonitrile) + poly(styrene-co-acrylonitrile)	[9003-56-9]	2-propenenitrile, polymer with 1,3-butadiene and ethenylbenzene	ABS plastics
butadiene–styrene (graft and polyblend)	poly(butadiene-g-styrene) + polybutadiene + polystyrene	[9003-55-8]	benzene, ethenyl-, polymer with 1,3-butadiene	IPS impact styrene plastics
styrene–acrylonitrile (random)	poly(styrene-co-acrylonitrile)	[9003-54-7]	2-propenenitrile, polymer with ethenylbenzene	SAN plastics
styrene–maleic anhydride (alternating)	poly(styrene-alt-maleic anhydride)	[9011-13-6]	2,5-furandione, polymer with ethenylbenzene	SMA resins

800

Copolymerization Reactions

The mutual polymerization of two or more monomers is called copolymerization. This topic has been comprehensively reviewed by Ham (5). Monomers frequently show a different reactivity toward copolymerization than toward homopolymerization. In fact, some monomers that can be homopolymerized only with great difficulty, can be readily copolymerized. One such monomer is maleic anhydride (qv). It is rather inert to free radical homopolymerization, yet can be copolymerized conveniently with styrene under free radical conditions.

Chain-Growth Copolymerization Theory. The theory of chain-growth (eg, radical, anionic, etc) copolymerizations has received more attention than that of step-growth or other copolymerizations. Consequently, only the theory of chain-growth copolymerizations is discussed.

In the case of chain-growth copolymerization, growing polymer chains must choose between more than one monomer. Such a choice or relative reactivity has been quantitatively treated by the reactivity ratio (6–7) and Q–e schemes (8).

Reactivity Ratio Scheme. There are four possible propagation steps in the copolymerization of two monomers, M_1 and M_2, with two growing chain ends, M_1^* and M_2^*: (eqs. 3–6)

$$M_1^* + M_1 \xrightarrow{k_{11}} M_1M_1^* \tag{3}$$

$$M_1^* + M_2 \xrightarrow{k_{12}} M_1M_2^* \tag{4}$$

$$M_2^* + M_2 \xrightarrow{k_{22}} M_2M_2^* \tag{5}$$

$$M_2^* + M_1 \xrightarrow{k_{21}} M_2M_1^* \tag{6}$$

The reactivity ratios r_1 and r_2 are defined as follows:

$$r_1 = \frac{k_{11}}{k_{12}} = \text{relative reactivity of polymer chain } M_1^*$$

$$\text{toward monomer } M_1 \text{ and monomer } M_2 \tag{7}$$

$$r_2 = \frac{k_{22}}{k_{21}} = \text{relative reactivity of polymer chain } M_2^* \text{ toward } M_2 \text{ and } M_1 \tag{8}$$

From these propagation steps can be derived the so-called copolymerization composition equation:

$$\frac{d[M_1]}{d[M_2]} = \frac{[M_1]}{[M_2]} \cdot \frac{r_1[M_1] + [M_2]}{r_2[M_2] + [M_1]} \tag{9}$$

which relates the (instantaneous) copolymer composition with the monomer feed of M_1 and M_2. Values for r_1 and r_2 are usually determined by graphical methods (9–10).

There are five different types of reactivity ratio combinations (11).

$$\text{Type I: } r_1 \simeq r_2 \simeq 1 \ (r_1r_2 = 1) \tag{10}$$

Both r_1 and r_2 are approximately unity in an ideal copolymerization. In this case,

$k_{11} \simeq k_{12}$ and $k_{22} \simeq k_{21}$ and the growing chains show little preference for either monomer, M_1 or M_2. This copolymerization is essentially random, and the composition of the copolymer is approximately equal to that of the monomer feed at all feed compositions (Fig. 1).

$$\text{Type II: } r_1 \text{ and } r_2 < 1 \ (r_1 r_2 < 1) \tag{11}$$

In this situation, the copolymerization tends toward an alternating arrangement of monomer units. Figure 1 shows an example (II) of an alternating copolymer that has an azeotropic copolymer composition, ie, a copolymer composition equal to the monomer feed at a single monomer feed composition. This case is analogous to a constant boiling mixture in vapor–liquid equilibria (see Azeotropic and extractive distillation).

$$\text{Type III: } r_1 \text{ and } r_2 > 1 \ (r_1 r_2 > 1) \tag{12}$$

When both r_1 and r_2 are greater than 1, the two monomers become block copolymers or polyblends of homopolymers.

$$\text{Type IV: } r_1 \gg 1 \text{ and } r_2 \ll 1 \tag{13}$$

In this situation, copolymerization is very difficult to achieve, especially when $[M_1]$ is high (curve IV of Fig. 1). Monomers with such a combination of reactivity ratios and monomer charge tend to give homopolymer. However, when M_1 has been nearly exhausted, copolymerization becomes more favorable.

$$\text{Type V: } r_1 \simeq r_2 \simeq 0 \ (r_1 r_2 = 0) \tag{14}$$

Type V represents the case in which the copolymerization is strongly alternating (Fig. 1). Here, each growing chain reacts exclusively with the unlike monomer.

Table 2 shows characteristic reactivity ratios for some selected free radical, ionic,

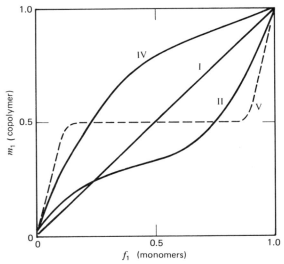

Figure 1.   Copolymer composition m_1 as a function of monomer feed ratio f_1 for various reactivity ratio combinations (11).

Table 2. Characteristic Reactivity Ratios [a]

M_1	M_2	r_1	r_2	Synthetic method
styrene	1,3-butadiene	0.78	1.38	free-radical, $T = 60°C$
styrene	vinyl acetate	55.0	0.01	free-radical, $T = 60°C$
styrene	maleic anhydride	0.01	0.0	free-radical, $T = 60°C$
acrylonitrile	vinyl acetate	4.2	0.05	free-radical, $T = 50°C$
methyl acrylate	vinylidene chloride	0.84	0.99	free-radical, $T = 60°C$
ethylene	propylene	17.95	0.06	coordination, $T = 25°C$
ethylene	vinyl acetate	1.2	1.1	free-radical, $T = 160°C$
styrene	1,3-butadiene	0.1	12.5	anionic, RLi, toluene solvent, $T = 25°C$
styrene	1,3-butadiene	0.74	1.03	anionic, RLi, THF solvent, $T = 30°C$
isobutylene	·isoprene	2.5	0.4	cationic, $T = -103°C$

[a] Ref. 12.

and coordination copolymerizations. It must be noted that the reactivity ratios predict only tendencies; some copolymerization, and hence some modification of physical properties, can occur even if r_1 and/or r_2 are somewhat unfavorable. For example, despite their dissimilar reactivity ratios, ethylene and propylene can be copolymerized to a useful elastomeric product by adjusting the monomer feed or by using a catalyst that increases the reactivity of propylene relative to ethylene.

The Q–e Scheme. The magnitude of r_1 and r_2 can frequently be correlated with structural effects, such as polar and resonance factors. For example, in the free-radical polymerization of vinyl acetate with styrene, both styrene and vinyl acetate radicals preferentially add styrene because of the formation of the resonance stabilized polystyrene radical.

Alfrey and Price (8) proposed a means of predicting monomer reactivity in copolymerization from two parameters, Q (a measure of resonance) and e (a measure of polar effects). These parameters have been related to the reactivity ratios by equations 15–17.

$$r_1 = \frac{Q_1}{Q_2} \exp[-e_1(e_1 - e_2)] \tag{15}$$

$$r_2 = \frac{Q_2}{Q_1} \exp[-e_2(e_2 - e_1)] \tag{16}$$

$$r_1 r_2 = \exp[-(e_1 - e_2)^2] \tag{17}$$

where Q_1 and e_1 are the monomer stability and polarity factors for M_1, and Q_2 and e_2 are the monomer stability and polarity factors for M_2. Negative e values are assigned to electron rich monomers, and positive e values to electron poor monomers. For conjugated monomers such as styrene, $Q \geq 0.2$; for nonconjugated monomers, $Q \leq 0.2$.

Synthetic Methods. The synthesis of random, alternating, and block (graft) copolymers involves mechanisms and processes (eg, bulk or mass, solution, emulsion, suspension) similar to those used in homopolymerization (see also Polymerization mechanisms and processes). Copolymerization mechanisms can be further divided into categories such as free-radical, anionic, cationic, coordination, ring opening, condensation, and post-polymerization reactions.

Free-radical copolymerizations have been performed in bulk (comonomers without solvent), solution (comonomers with solvent), suspension (comonomer droplets suspended in water), and emulsion (comonomer emulsified in water). On the other hand, most ionic and coordination copolymerizations have been carried out either in bulk or solution because water acts as a poison for many ionic and coordination catalysts. Similarly, few condensation copolymerizations involve emulsion or suspension processes. The following reactions exemplify the various copolymerization mechanisms.

Free-Radical. Examples of the types of copolymers formed by free-radical copolymerizations are shown in equations 18–20 (13–15).

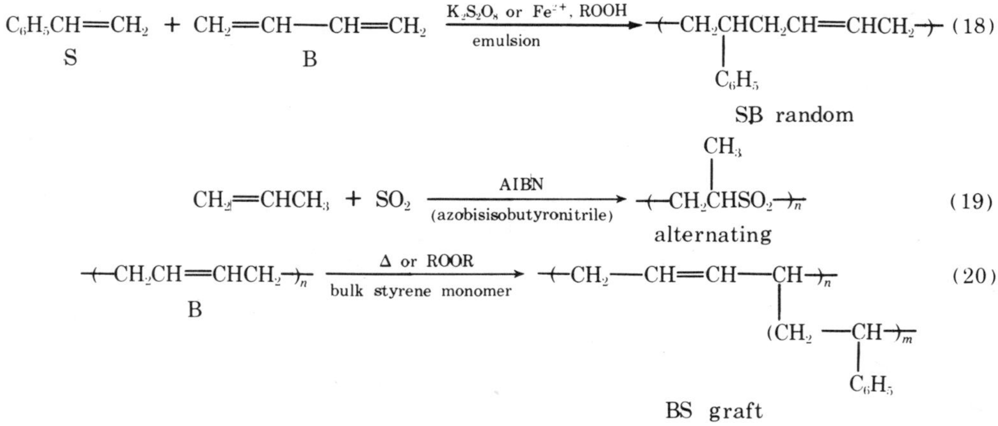

$$C_6H_5CH{=}CH_2 \;+\; CH_2{=}CH{-}CH{=}CH_2 \xrightarrow[\text{emulsion}]{K_2S_2O_8 \text{ or } Fe^{2+}, \text{ ROOH}} \;{+}{(}CH_2CHCH_2CH{=}CHCH_2{)}{+} \quad (18)$$
$$\underset{\text{S}}{} \qquad\qquad \underset{\text{B}}{}$$
$$\underset{C_6H_5}{|}$$
SB random

$$CH_2{=}CHCH_3 \;+\; SO_2 \xrightarrow[\text{(azobisisobutyronitrile)}]{AIBN} \;{+}{(}CH_2CHSO_2{)}{+}_n \quad (19)$$
$$\overset{CH_3}{|}$$
alternating

$${+}{(}CH_2CH{=}CHCH_2{)}{+}_n \xrightarrow[\text{bulk styrene monomer}]{\Delta \text{ or ROOR}} \;{+}{(}CH_2{-}CH{=}CH{-}CH{)}{+}_n \quad (20)$$
$$\underset{\text{B}}{} \qquad\qquad\qquad\qquad\qquad (CH_2{-}CH{)}{+}_m$$
$$\underset{C_6H_5}{|}$$
BS graft

Anionic. Typical copolymers prepared by anionic copolymerizations are shown in equations 21–23 (16–18).

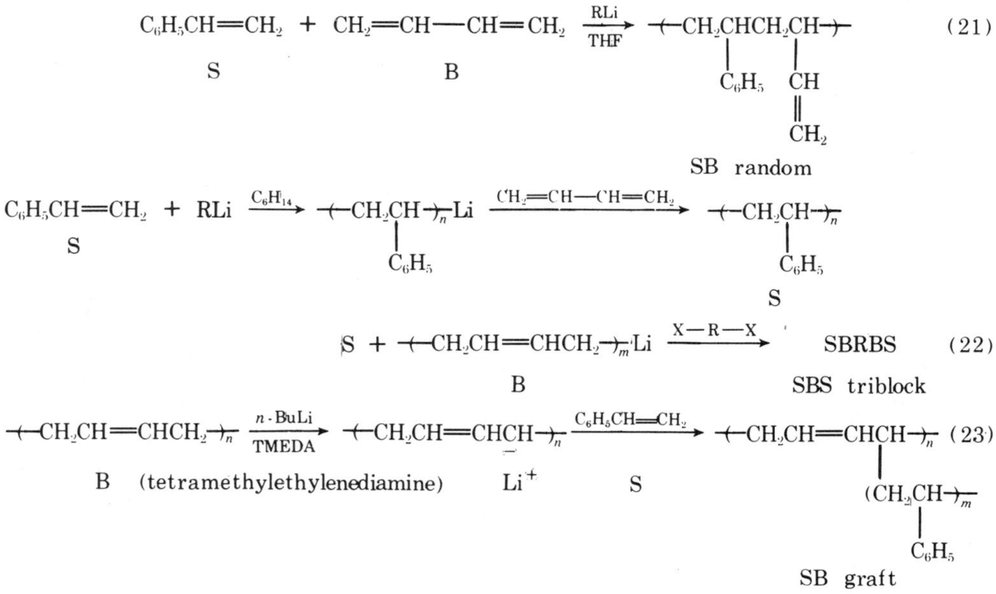

$$C_6H_5CH{=}CH_2 \;+\; CH_2{=}CH{-}CH{=}CH_2 \xrightarrow[\text{THF}]{RLi} \;{+}{(}CH_2CHCH_2CH{)}{+} \quad (21)$$
$$\underset{\text{S}}{} \qquad\qquad \underset{\text{B}}{} \qquad\qquad\qquad C_6H_5 \quad CH$$
$$\qquad\qquad\qquad\qquad\qquad \| \; CH_2$$
SB random

$$C_6H_5CH{=}CH_2 \;+\; RLi \xrightarrow{C_6H_{14}} \;{+}{(}CH_2CH{)}{+}_n Li \xrightarrow{CH_2{=}CH{-}CH{=}CH_2} \;{+}{(}CH_2CH{)}{+}_n$$
$$\underset{\text{S}}{} \qquad\qquad\qquad\qquad \underset{C_6H_5}{|} \qquad\qquad\qquad\qquad \underset{C_6H_5}{|}$$
$$\qquad\qquad\qquad\qquad\qquad\qquad\qquad\qquad\qquad \text{S}$$

$$S \;+\; {+}{(}CH_2CH{=}CHCH_2{)}{+}_m Li \xrightarrow{X{-}R{-}X} \quad \text{SBRBS} \quad (22)$$
$$\underset{\text{B}}{} \qquad\qquad\qquad\qquad \text{SBS triblock}$$

$${+}{(}CH_2CH{=}CHCH_2{)}{+}_n \xrightarrow[\text{TMEDA}]{n\text{-BuLi}} \;{+}{(}CH_2CH{=}CHCH{)}{+}_n \xrightarrow{C_6H_5CH{=}CH_2} \;{+}{(}CH_2CH{=}CHCH{)}{+}_n \quad (23)$$
$$\underset{\text{B}}{} \quad \text{(tetramethylethylenediamine)} \qquad \underset{Li^+}{} \qquad \underset{\text{S}}{} \qquad\qquad (CH_2CH{)}{+}_m$$
$$\qquad\qquad\qquad\qquad\qquad\qquad\qquad\qquad\qquad\qquad\qquad\qquad C_6H_5$$
SB graft

Cationic. Illustrative cationic copolymerizations are shown in equations 24–25 (19–20).

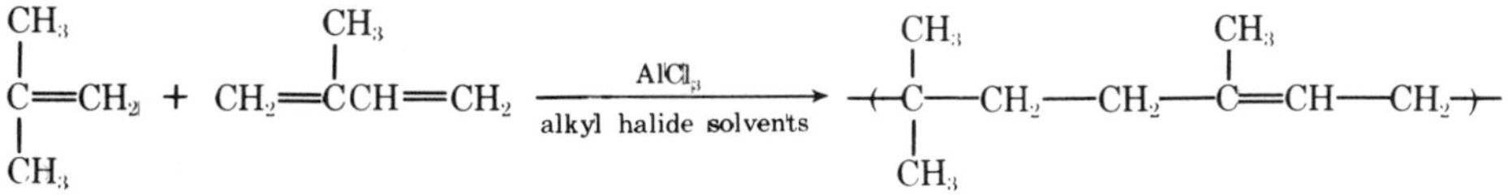

$$\text{(24)}$$

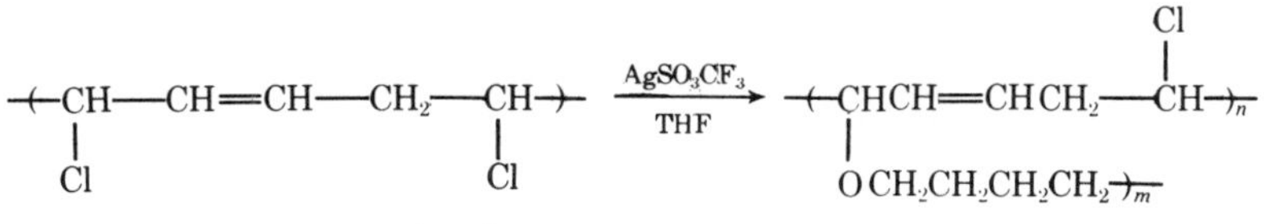

$$\text{(25)}$$

PVC with some allyl chloride sites   graft

Coordination. Examples of coordination copolymerization are shown in equations 26–27 (21).

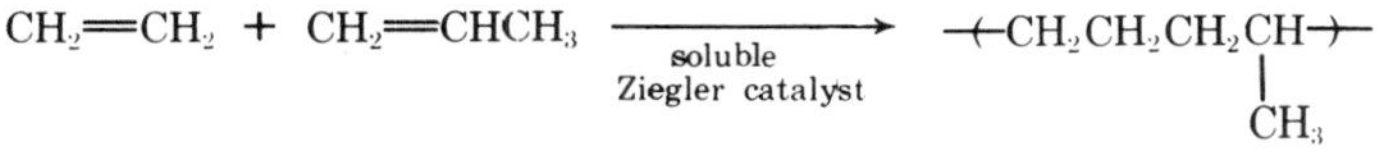

$$\text{(26)}$$

random

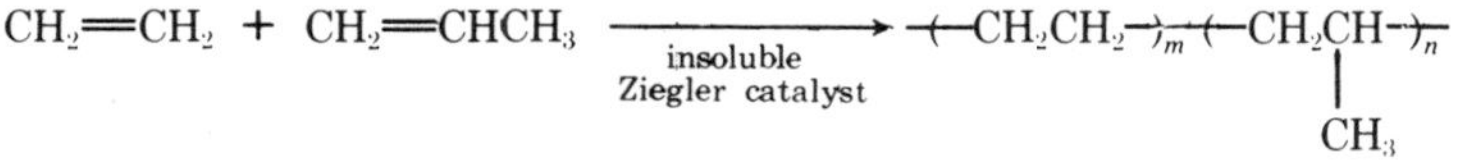

$$\text{(27)}$$

tapered block or homopolymer

A tapered block is an impure block of one monomer that has an increasing amount of a second monomer at one end.

Step Growth. A sample of a block copolymer prepared by condensation polymerization is shown in equation 28 (22). In this process, a prepolymer diol (HO-Z-OH) is capped with isocyanate end groups and chain-extended with a low molecular weight diol (HO-E-OH) to give a so-called segmented block copolymer, containing polyurethane hard blocks and O-Z-O soft blocks.

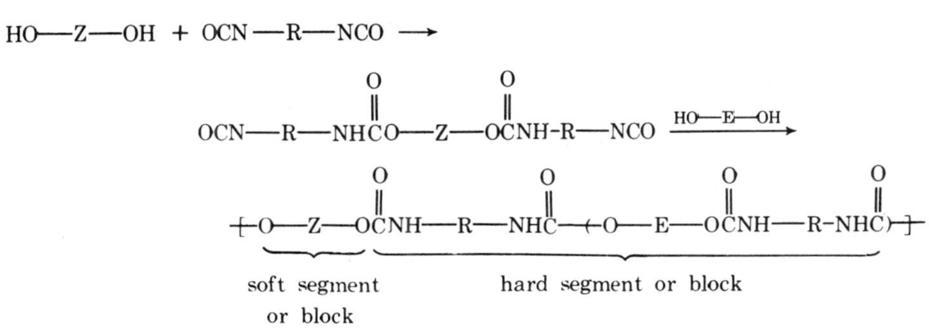

$$\text{HO—Z—OH} + \text{OCN—R—NCO} \longrightarrow$$

$$\text{(28)}$$

soft segment   hard segment or block
or block

segmented block copolymer

Random copolymers can be synthesized by step-growth copolymerization in an equimolar mixture of four monomers (1):

$$HOOC—R—COOH + HOOC—R'—COOH + H_2N—R—NH_2 + H_2N—R'''—NH_2 \rightarrow +CO—R—CONH—R''—NHCO—R'—CONH—R'''—NH+ \quad (29)$$

random

Ring Opening. Several examples of the formation of copolymers by ring opening reactions are shown in equations 30–31 (23–24).

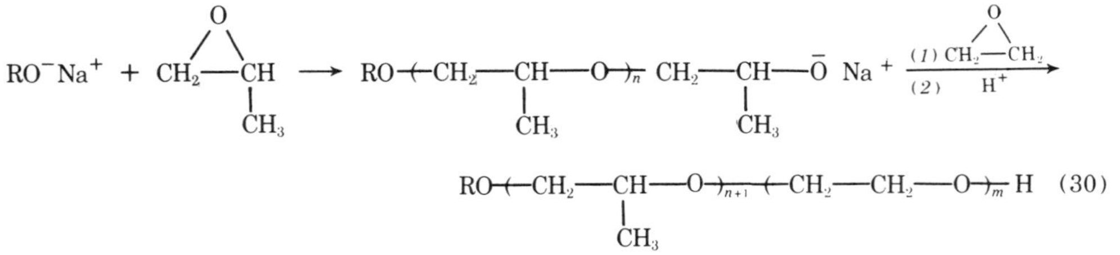

block

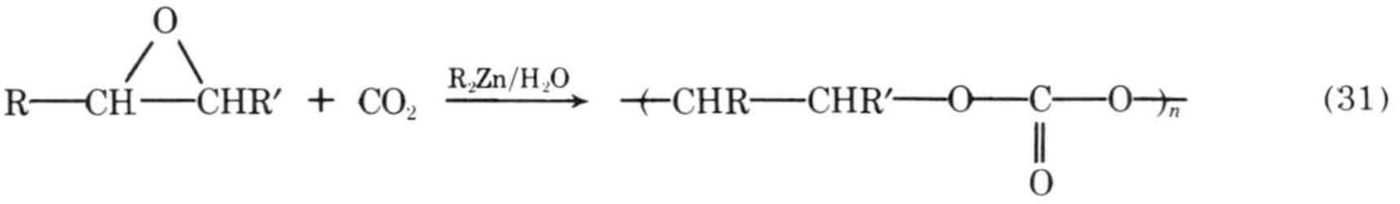

alternating

Post-Polymerization Reaction. A random copolymer can be formed by a post-polymerization reaction with a mixture of reagents (eq. 32) (25–26). This is the basis of Firestone's PNF semiinorganic rubbers (see Fluorine compounds, organic-fluoralkoxyphosphazenes).

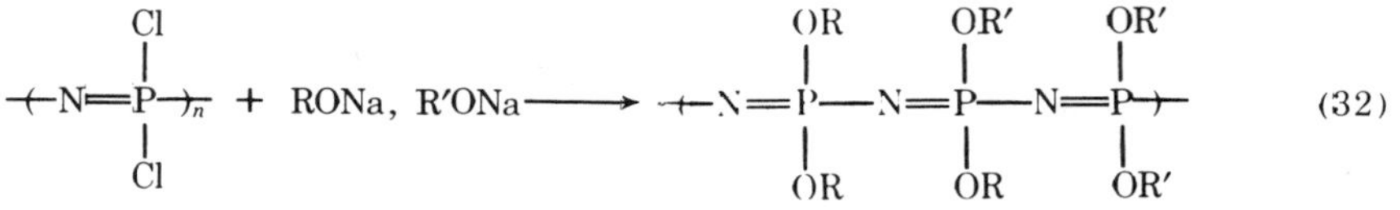

$R = —CH_2CF_3$ random

$R' = —CH_2(CF_2)_3CF_2H$

[*26085-02-9*] [*57579-81-4*]

Effects of Monomer Unit Arrangement on Physical Properties

Random Arrangement. The primary incentive for preparing copolymers is to attain certain properties in the products. The effect of random copolymerization on polymer properties is easily shown by differences in polymer crystallinity, melting point T_m, glass transition temperature T_g, and solubility between a copolymer and the corresponding homopolymers. Since random comonomer enchainment tends to reduce symmetry and modify intermolecular forces, it is not surprising that random copolymers have different melting behavior than the corresponding constituent crystalline homopolymers. Figure 2 shows the variation in T_m of two copolymers. The melting point of adipamide copolymer varies smoothly over the entire composition range. On the other hand, the sebacamide copolymer shows a eutectic point.

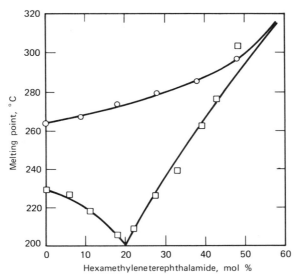

Figure 2. Mole percent hexamethyleneterephthalamide vs melting point (27). ○, adipamide; □, sebacamide.

The glass transition temperature in amorphous polymers is also sensitive to copolymerization. Generally, T_g of a random copolymer falls between the T_gs of the respective homopolymers. For example, T_g for solution polymerized polybutadiene is −96°C; that for solution-polymerized polystyrene is +100°C. Yet, a commercial random solution copolymer of butadiene and styrene (Firestone's Stereon) shows an intermediate T_g of −76°C (28). The glass transition temperature of the random copolymer can sometimes be related simply as follows:

$$\frac{1}{T_g} = \frac{W_1}{T_{g1}} + \frac{W_2}{T_{g2}} \tag{33}$$

T_{g1} and T_{g2} are the glass transition temperatures in K of the homopolymers; W_1 and W_2 are the weight fractions of the comonomers (29). Since glass transition temperature is directly related to many other material properties, changes in T_g by copolymerization cause changes in other properties too. Table 3 lists some of the properties of polymers that are related to the glass transition temperature.

Table 3.   Polymer Properties Dependent on the Glass Transition Temperature[a]

physical state (transition from brittle glass to rubber)
rate of thermal expansion
thermal properties
torsional modulus
refractive index
dissipation factor
impact resistance, brittle point
flow and heat distortion properties
minimum film-forming temperature of polymer latex

[a] Ref. 30. Courtesy of Academic Press, Inc.

The solubility of random copolymers of monomers whose homopolymers are noncrystalline also varies quite regularly as the relative amounts of the comonomers are changed. The solubility of random copolymers is often low in solvents for the respective homopolymers, yet high in solvent pairs (31).

Block (Graft) Arrangement. *Homogeneous.* The physical properties of homogeneous block or graft copolymers are similar to those of random copolymers. Homogeneous block polymers show considerable mixing of the blocks to yield a single-phase morphology. Thus the physical properties of these materials are usually intermediate between those of the respective homopolymers. Homogeneous block copolymers can result from low block molecular weights or from specific interactions, such as hydrogen bonding.

Heterogeneous. By far the most interesting block copolymers are those in which there is little or no mixing of the block phases. Such heterogeneous block copolymers tend to show the properties of the components rather than an averaging of homopolymer properties. Heterogeneous copolymers, which have domains or regions of one component dispersed within the matrix of the other component, show separate T_gs for each of the blocks. These copolymers also tend to be soluble in a wide variety of solvents. In fact, they themselves, may act like surfactants, if the respective blocks have widely different solubilities. Since the domain sizes in block copolymers are often much smaller than those of polyblends, the block copolymers tend to be more transparent than polyblends.

Not only do block copolymers have properties different from those of homopolymers, random copolymers, and polyblends, but the properties of block copolymers themselves differ depending upon the arrangement of the blocks. This is shown by the properties of high styrene resins of various block arrangements (Table 4) (32).

Some Important Random Copolymers

Many random copolymers have found commercial use as elastomers (qv) and plastics. For example, SBR (7,33), poly(butadiene-*co*-styrene), has become the most important synthetic rubber. It can be prepared in emulsion by use of free-radical initiators (qv), such as $K_2S_2O_8$ or Fe^{2+}/ROOH (eq. 18), or in solution by use of alkyl

Table 4.   Properties of High Styrene–Butadiene Block Copolymers[a]

Samples	S/B mole ratios (charged)	Elongation at break, %	Light transmittance	Charpy impact strength, J/m[b]	Heat distortion temperature, °C
S	100/0	7	90	20.6	97
SB	60/40	2	62	24.5	83
SBS	60/40	70	71	43.2	67
BSB	60/40	2	51	23.5	78
SBSB	60/40	23	70	30.4	55
commercial HIPS[c] (graft and polyblend)		30	0	68.6	75
polyblend	60/40	2	0	4.9	

[a] Ref. 32. Courtesy of IPC Business Press, Ltd.
[b] To convert J/m to ft·lb/in., divide by 53.38.
[c] HIPS = high impact strength.

lithium initiators. Emulsion SBR copolymers are produced under trade names by such companies as American Synthetic Rubber (ASPC), Firestone (FRS), B. F. Goodrich (Ameripol), Goodyear (Plioflex), Texas–U.S. Chemical (Synpol). Solution SBR is manufactured by such companies as Firestone (Stereon) and Phillips (Solprene). The total United States production of SBR in 1975 was 1.2 million metric tons (34).

SBR is a low T_g copolymer that is extendable with oil, reinforceable with carbon black, and vulcanizable with sulfur and an accelerator. Many of the physical properties of SBR vulcanizates (eg, tensile strength, resilience, hot tear strength, hysteresis, etc) are somewhat inferior to those of natural rubber (NR) (see Rubber). However, its low cost, cleanliness, aging properties, and abrasion resistance make it an attractive rubber for use in passenger tires. Indeed, most of the SBR produced in the United States is used in tires. Lesser quantities of SBR have been used in wire and cable materials, mechanical goods, footwear, and foam products.

Poly(butadiene-co-acrylonitrile), NBR (35), is another commercially significant random copolymer. This rubber is manufactured by free-radical emulsion polymerization. Important producers include Copolymer Rubber and Chemical (Nysyn), B. F. Goodrich Chemical (Hycar), Goodyear (Chemigum), and Uniroyal (Paracril). The total United States production of NBR (nitrile rubber) in 1975 was 55,000 t (34) (see Elastomers, synthetic). The most important property of NBR rubber is its oil resistance. It is used in oil well parts, fuel cell liners, fuel hose, and other applications requiring resistance to aromatic fuels, oils, and solvents (35).

EPM (ethylene–propylene) and EPDM (ethylene–propylene–diene) rubbers represent other commercial randomlike copolymers (36). Of these two, the EPDM terpolymer is the more important because it is sulfur-curable. These polymers are prepared by Ziegler-catalyzed coordination copolymerization (see Ziegler-Natta catalysts) and, in principle, should be normal alternating copolymers. However, the commercial copolymers do not usually follow an alternating sequence. Rather, they have polyethylene and polypropylene segments dispersed in longer segments of random copolymer. Typical United States producers of EPDM are DuPont (Nordel), Copolymer Rubber and Chemical (Epsyn), Exxon Chemical (Vistalon), B. F. Goodrich (Epcar), and Uniroyal (Royalene). The United States production of EPM and EPDM rubbers was 82,000 t in 1975 (34).

EPDM elastomers are low specific gravity rubbers with good vulcanizate properties (eg, tensile strength, and elongation) and good weathering resistance. As a consequence, EPDM has been used in tire sidewalls and coverstrips. In 1975, 72,500 t of EPDM was used in tires and related products in the United States. Another 72,500 t was consumed in other United States automotive applications. Lesser amounts were used in wire and cable (22,600 t), appliances, hose, belting, gaskets, and rolls (36).

SAN resins are random, noncrystalline copolymers of styrene and acrylonitrile (37). These materials are manufactured by emulsion, suspension, or continuous mass polymerization. Major producers are Dow (Tyril), BASF (Luran), and Monsanto (Lustran SANZI) (38). SAN copolymers are typically tougher and stronger than polystyrene. The presence of acrylonitrile imparts high heat deflection and chemical resistance to the copolymer. The important markets for SAN include appliances (knobs, refrigerator compartments), automotive uses (dashboard components), housewares (tumblers, ice buckets, bath accessories), furniture (chair backs, seats, lamps), medical disposables (parts for kidney dialysis equipment), packaging (containers, closures, dispensing parts), and alloys (blends with ABS and PVC) (37). In

1976 the total United States consumption of SAN was 47,000 t (39) (see Acrylonitrile polymers).

The family of random copolymers of vinyl chloride and vinyl acetate (40) is the most important class of vinyl chloride copolymers. Most poly(vinyl chloride-co-vinyl acetate) is manufactured by aqueous suspension copolymerization, using a free-radical initiator. Important producers are Air Products and Chemicals, Borden Chemical, Diamond Shamrock, B. F. Goodrich Chemical, Firestone Plastics, Stauffer Chemical, Tenneco Chemical, and Union Carbide Chemical and Plastics (see Vinyl polymers).

The copolymerization of vinyl chloride with vinyl acetate yields a material with improved processability as compared with vinyl chloride homopolymer. However, the physical and chemical properties of the copolymers are different from those of the homopolymer PVC. Generally, as the vinyl acetate content increases, the resin solubility in ketone and ester solvents and its susceptibility to chemical attack increases, the resin viscosity and heat distortion temperature decreases, and the tensile strength and flexibility increases slightly.

Poly(vinyl chloride-co-vinyl acetate) has found application in flooring, phonograph records, protective coatings, fibers, and some films and sheeting. Because of their low viscosity and good processability, such copolymers constitute the bulk of the vinyl tile market (80,000 t in 1976) and of the phonograph record market (60,000 t in 1976) (40–41).

Some Important Alternating Copolymers

Poly(styrene-*alt*-maleic anhydride) is a classical and commercial example of an alternating copolymer (42). This material is manufactured by free-radical bulk, solution, or emulsion copolymerization. Important producers are ARCO (SMA) and Monsanto (Lytron). Such copolymer resins are brittle and insoluble in most solvents. But they are soluble in alkaline solution and react with water to give acids, with alcohols to give esters, and amines to give amides. They can be converted to insoluble, infusible thermosets by heating with diamines or glycols. These resins and their derivatives are seldom used alone but are used as reactive additives for latex paints (to improve adhesion and gloss), pigment dispersants (to increase the pigment concentration), and floor polishes (to act as emulsifiers and protective colloids) (see Styrene plastics).

Poly(butadiene-*alt*-propylene), PBR, is a recently reported general purpose elastomer (43–44) prepared at the Maruzen Petrochemical Co. (Japan) by Ziegler coordination catalysis. PBR shows tack (self-adhesion) and green (unvulcanized) strength superior to that of SBR and BR. It also shows superior high temperature dynamic properties to those of BR and EPDM. Black-loaded vulcanizates can be compounded to give properties similar to those of SBR. PBR can also be covulcanized with SBR, BR, and EPDM (see also Olefin polymers).

Some Important Block Copolymers

Block copolymers have become commercially valuable commodities because of their unique structure–property relationships. They are best described in terms of their applications such as thermoplastic elastomers (TPEs), elastomeric fibers, toughened thermoplastic resins, and surfactants (see Elastomers, synthetic-thermoplastic).

Thermoplastic elastomers (45–47) result when block copolymers have an ABA, $(AB)_n X$, or $+AB+_n$ but not an AB diblock arrangement of A (thermoplastic) and B (rubbery) blocks. The hard A blocks may be glassy (eg, polystyrene) or crystalline (eg, polyester, polyurethane); the soft B blocks must be elastomeric (eg, polybutadiene, polyisoprene). When the hard segments are incompatible with the soft segments, the domains or regions of hard blocks act as reinforcing physical cross-links for the rubbery matrix. In contrast to chemically cross-linked rubbers, the physical network is thermally reversible. When the polymer is heated above the T_g (or T_m) of the hard block, the hard blocks soften and allow the rubber to flow and to be processed as a thermoplastic. Table 5 shows some commercially important TPE block copolymers and their United States consumption levels.

Table 5. Commercially Important Block Copolymer TPEs[a]

Comonomers	Block arrangement	Some important trade names (producers)	Estimated 1976 U.S. consumption,[a] 10^3 t
styrene–diene (hydrogenated styrene–diene)	ABA, $(AB)_n X$	Kraton (Shell) Solprene (Phillips)	24.1
urethane ester (ether)	$+AB+_n$	Estane (B.F. Goodrich) Texin (Mobay)	15.0
ester–ether	$+AB+_n$	Hytrel (DuPont)	2.7

[a] Ref. 48.

The manufacture of block copolymer TPEs depends upon the type and arrangement of the blocks. For example, butadiene–styrene AB, ABA, $(AB)_n X$ block copolymers are conveniently prepared by alkyllithium initiated anionic polymerization. Thermoplastic $+A-B+_n$ polyurethanes are synthesized by step-growth addition copolymerization of dihydroxy compounds such as polytetramethylene ether glycol and toluene diisocyanate. The copolyester–ether $+AB+_n$ copolymers are produced by the polycondensation of dicarboxylic acids (eg, terephthalic acid) with glycols or polyether glycols (see Urethane polymers; Polyesters).

The physical properties of block copolymer TPEs also depend upon the type and arrangement of the blocks. Table 6 compares the property advantages of various block copolymer thermoplastic elastomers.

The properties and prices of the various block copolymer TPEs greatly affect their markets. For example, the low cost butadiene–styrene block copolymers have found utility in footwear (sneakers, tennis shoes), injection-molded or extruded goods (automotive sight shields, fender extensions, toys, housewares), and adhesives (solvent cement and hot melt types) (47).

The polyurethane TPEs are used largely in fabric coatings (see Coated fabrics) and injection molded or extruded goods (exterior automotive parts, gears, gaskets, etc). In contrast, the copolyester–ether block copolymer TPEs are relatively expensive with high-performance characteristics. Their most important uses are in wire and cable

Table 6. Property Advantages of the Various Block Copolymer TPEs[a,b]

Property	Styrene–diene	Hydrogenated styrene–diene	Ester–ether	Urethane–ester
tensile			+	+
recovery	+	+		
upper use temperature			+	+
lower use temperature	+	+		
aging stability		+		
acid–base resistance	+	+		
oil resistance			+	+
electrical	+	+		
abrasion resistance				+
melt processability			+	
cost	+			

[a] Ref. 49.
[b] A designation of + indicates a performance strong point.

materials (eg, the coiled stretch telephone cords) (see Insulation, electric), injection molded articles (eg, small mechanical parts), and high pressure hoses (47).

Elastomeric (Spandex) fibers (50) represent another important application of block copolymers. Spandex fibers are segmented polyurethane copolymers with an $+AB+_n$ arrangement of A (hard urethane, or urethane-urea) and B (soft polyester, or polyether) segments. The production of such fibers in 1975 was estimated to be 4,200 t. Important producers for such fibers are DuPont (Lycra, Numa) and Globe (Glospan). The principal uses for such materials are in women's foundation garments, hosiery, swimwear, outerwear, and upholstery (see Fibers, elastomeric).

Toughened thermoplastic resins (46,51) can also be made from block copolymers. Styrene–diene block copolymers with a high volume fraction of styrene are such materials. They are priced between the low cost resins, polystyrene, polyethylene, etc, and the high cost resins, cellulosics, clear ABS, polycarbonates, etc. The family of K Resins manufactured by Phillips are such a class of high styrene block copolymer resins used in toys, housewares, storage units, lids, and a wide variety of packaging.

Certain block copolymers have also found application as surfactants (qv) (46). For example, A-B or ABA block copolymers in which one block is hydrophilic and one block is hydrophobic have proven useful for emulsifying aqueous and nonaqueous substances and for wetting the surface of materials. Examples of such surfactants are the poly(propylene oxide-b-ethylene oxide) materials, known as Pluronics (BASF Wyandotte Co.) (see Polyethers).

Some Important Graft Copolymers

Two commercially significant graft copolymers are ABS (acrylonitrile–butadiene–styrene) resins and IPS (impact polystyrene) plastics. Both of these families of materials were once just simple mechanical polyblends, but today such compositions are generally graft copolymers or blends of graft copolymers with homopolymers.

ABS is the sixth largest volume thermoplastic resin and the principal engineering (structural or load bearing) plastic (52). ABS is a terpolymer manufactured by copolymerizing acrylonitrile and styrene in the presence of polybutadiene rubber. Important producers of ABS plastics include Borg Warner (Cycolac), Monsanto (Lustran), Dow (Abtec), Rexene Styrenics (Rexene), and Carl Gordon Co. (see Acrylonitrile polymers; Engineering plastics).

The properties of ABS can be modified by varying the relative proportions of the basic components, the degree of grafting, and the molecular weight (53). For example, increasing the rubber content reduces tensile strength, modulus, and melt flow and increases impact strength.

A variety of ABS grades are available with Izod impact strengths ranging from 106.7 J/m (2 ft·lb/in.) to 640.5 J/m (12 ft lb/in.), heat distortion temperatures from 80–115°C, and tensile strengths from 27.6 MPa (4000 psi) to 55.2 MPa (8000 psi). Special ABS resins with high flow characteristics for the injection molding of intricate parts, controlled gloss, or deep-draw vacuum formability can also be made. In addition, certain ABS plastics are available in pellet or powder form for use in blending with other polymers such as PVC. ABS plastics can be processed by all the techniques common to thermoplastics.

The consumption of ABS in 1976 was estimated to be 422,000 t with projections to 674,000 t by 1981 (54). The principal markets for ABS are pipes and fittings (27% of total consumption), automotive parts (18%), appliances (11%), recreational equipment (9%), and business machines and telephones (4.5%).

Impact polystyrene (IPS) is one of a class of materials that contains rubber grafted with polystyrene. This composition is usually produced by polymerizing styrene (by mass or solution free-radical polymerization) in the presence of a small amount (ca 5%) of dissolved elastomer. Some important producers of impact resistant polystyrenes are: Shell (Kraton), BASF (Polystyrol), Dow (Styron), Monsanto (Lustrex), Cosden (Cosden), Foster Grant-American Hoechst (Tuf-Flex), Hammond/Carl Gordon (Gordon Superflex) (51).

In these rubber-modified polystyrene polymers, the rubbers should have low T_gs, large particle sizes (0.5–5 μm), graftable and cross-linkable sites, and should be compatible with styrene monomer (56). Polybutadiene, such as Firestone's Diene, meets all of these requirements and is used most frequently.

The small amount of dispersed and grafted rubber in the polystyrene glassy matrix results in tremendous improvements in IPS's resistance to brittle failure under high speed impact without a major sacrifice in mechanical or thermal properties. As the rubber content increases, impact strength increases, while rigidity, heat deflection temperature, and clarity gradually decrease. The exact mechanism for the improvement of polystyrene's impact strength by use of grafted rubbers is uncertain. A number of theories relating crack propagation, energy absorption, and shear band formation, have been advanced (57).

Impact polystyrenes have found use in the manufacture of refrigerator door liners, and in packaging. They are also commonly used for appliance housings, radio and television cabinets, sporting goods, toys, cameras, furniture, luggage, pipe and fittings, automotive parts, and women's shoe heels (58) (see Styrene plastics).

Characterization of Copolymers

The characterization of copolymers must distinguish copolymers from polyblends and the various types of copolymers from each other (59–60). In addition, the exact molecular structure, architecture, purity, and supermolecular structure must be determined.

Assessing whether a material is a copolymer or a mixture of homopolymers can sometimes be accomplished by extracting the prospective copolymers with solvents selective for the component homopolymers. However, this method is effective only when the copolymer segments differ significantly in solubility behavior. The situation is further complicated by the fact that block copolymers can themselves act as compatibilizing agents for the extraction solvents. Alternatively, solution fractionation has been used to test copolymers vs homopolymers or polyblends. Determining whether a copolymer is a block or random one can often be made by ^{1}H nmr, ^{13}C nmr, or from T_g measurements (differential scanning calorimetry (dsc) or thermal-mechanical tests).

An indication of whether block copolymer architecture is AB, ABA, or $+AB+_n$ can often be seen in its rheological behavior (see Rheological measurements). For example, ABA and $+AB+_n$ block copolymers exhibit higher melt viscosities than do AB diblock copolymers. The former two architectures partially preserve their physical network even in the melt. Gel permeation chromatography (gpc) is useful for determining the presence of homopolymer A and diblock AB impurities in anionically prepared ABA triblock copolymers (see Analytical methods).

The molecular structure of the copolymers is also important. Molecular weight measurements (osmometry, gpc) and functional group analyses are useful. Block copolymers require supermolecular (morphological) structural information as well. A listing of typical copolymer characterization tools and methods is shown in Table 7.

Table 7.   Copolymer Characterization[a]

Copolymer vs homopolymer	*Molecular structure*
solubility characteristics	osmometry
solid-phase extraction	solution light scattering
solution fractionation	ultracentrifugation
film clarity	gpc
solution compatibility	solution viscometry
molecular weight distribution	oligomer analysis
density gradient ultracentrifugation	selective degradation
gpc	[elemental and functional group analysis]
rheological characteristics	*Architecture and purity*
Block copolymer vs random	elastic recovery
copolymer	rheological characteristics
pmr	gpc
ir	density gradient ultracentrifugation
dynamic mechanical behavior	*Supermolecular structure*
dsc	dynamic mechanical behavior
electron microscopy	dsc
small-angle X-ray scattering	rheological characteristics
mechanical properties	electron microscopy and scanning
rheological properties	electron microscopy
crystallinity characteristics	wide-angle X-ray
solution light scattering	small-angle X-ray
thermomechanical analysis	birefringence
	small-angle light scattering

[a] Ref. 61. Courtesy of Academic Press, Inc.

Economic Importance of Copolymers

The economic importance of copolymers can be clearly illustrated by a comparison of United States production levels for various homopolymer and copolymer elastomers and resins (62–63). Figure 3 shows the relative contribution of elastomeric copolymers (SBR, ethylene–propylene, nitrile rubber) and elastomeric homopolymers (polybutadiene, polyisoprene) to the total production of synthetic elastomers. Clearly, SBR, a random copolymer, constitutes the bulk of the entire United States production. Copolymers of ethylene and propylene, and nitrile rubber (a random copolymer of butadiene and acrylonitrile) are manufactured in smaller quantities. Nevertheless, the latter copolymers approach or exceed the synthesis levels for the elastomeric homopolymers of butadiene or isoprene.

The relative United States production of styrene homopolymer and copolymer resins is also noteworthy (Fig. 4). The impact polystyrene (IPS) (graft and polyblend) copolymers are actually produced in slightly larger quantities than the styrene homopolymers. The ABS resins and styrene–butadiene latex resins are synthesized in lesser, yet significant, quantities (see Latex technology).

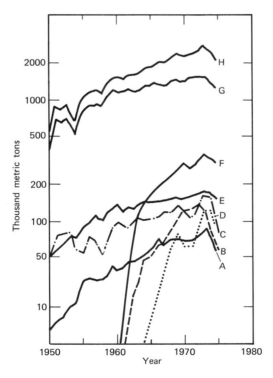

Figure 3. United States production of synthetic elastomers (62). A, nitrile; B, polyisoprene; C, butyl; D, ethylene–propylene; E, neoprene; F, polybutadiene; G, SBR; H, total.

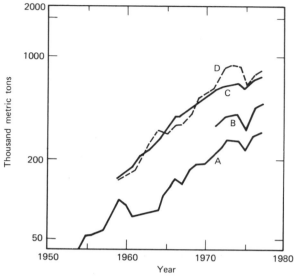

Figure 4.   United States production of polystyrene and styrene copolymers and resins (63). A, styrene–butadiene latex resins; B, ABS; C, straight polystyrene; D, impact polystyrene (rubber modified).

Future Trends

Prediction of the long range usage of copolymers is especially risky at this time. The copolymers of the future will be closely related to society's energy choices, which will greatly affect monomer and polymer availability and price.

Many of today's large volume copolymers owe their success to low cost petro-chemical fuels and monomers. However, future shifts to coal, nuclear, or solar energy, as well as to renewable resources (eg, agricultural and paper pulp chemicals, etc) may cause new materials to capture the general purpose market. High volume copolymers will tend to be those generated by energy (and cost) efficient processes. The total energy requirements of the monomers, synthetic methods, and processing will have to be considered.

Furthermore, increased governmental scrutiny of chemical substances will make it more difficult to bring a new product to market. The choice of comonomers and copolymers may be based partly upon EPA, FDA, OSHA, and TSCA rulings (see Regulatory agencies).

When new copolymers cannot be competitively introduced, new ways must be developed for improving old ones. This might involve copolymer blending or improved processing techniques. For example, the recent success of ABS–PVC alloys is an incentive for studying other copolymer blends (64). The importance of new processing methods for improved product economics has been verified by the introduction of reaction injection molding (RIM). This technique is an efficient method for fabricating thermoplastic polyurethane copolymers (65) (see Plastics technology).

There will also be a growing need for macromolecules for special end-use requirements. Such substances might be a copolymer with a wide service temperature range, such as Firestone's PNF elastomers (26,66), or one with a high resistance to flammability or degradation. Of special interest will be biomedical materials. Recent applications (67) of polyurethane copolymers to surgical devices, such as endotrachial tubes, synthetic blood vessels, heart valves, and even artificial hearts, suggest that specialty copolymers will play increasingly important roles in the future (see Prosthetic and biomedical devices).

BIBLIOGRAPHY

1. G. Odian, *Principles of Polymerization,* McGraw-Hill Book Co., New York, 1970, pp. 124–129.
2. R. J. Ceresa, *Block and Graft Copolymers,* Butterworths Co., Washington, D.C., 1962, Chapt. 1.
3. J. P. Kennedy in L. H. Sperling, ed., *Recent Advances in Polymer Blends, Grafts, and Blocks, Polymer Science and Technology,* Vol. 4, Plenum Press, New York, 1974, Chapt. 1.
4. R. B. Fox in N. Bikales, ed. *Encyclopedia of Polymer Science and Technology,* Vol. 9, Interscience Publishers, a division of John Wiley & Sons, Inc., New York, 1968, pp. 342–344.
5. G. Ham, ed., *Copolymerization, High Polymers,* Vol. 18, Interscience Publishers, a division of John Wiley & Sons, Inc., New York, 1964.
6. F. R. Mayo and F. M. Lewis, *J. Am. Chem. Soc.* **66,** 1594 (1944).
7. T. Alfrey, Jr., and G. Goldfinger, *J. Chem. Phys.* **12,** 205, 244 (1944).
8. T. Alfrey, Jr., and C. C. Price, *J. Polym. Sci.* **2,** 101 (1947).
9. M. Fineman and S. D. Ross, *J. Polym. Sci.* **5,** 259 (1950).
10. T. Kelen and F. Tudos, *J. Makromol. Sci. Chem.* **A9,** 1 (1975).
11. R. W. Lenz, *Organic Chemistry of Synthetic High Polymers,* Interscience Publishers, a division of John Wiley & Sons, Inc., New York 1967, pp. 382–383.
12. J. Brandrup and E. H. Immergut, eds., *Polymer Handbook,* 2nd ed., John Wiley & Sons, Inc., New York, 1975, Chapt. II-105.
13. W. Saltman in M. Morton, ed., *Rubber Technology,* 2nd ed., Van Nostrand Reinhold Co., New York, 1973, Chapt. 7.
14. W. G. Barb, *J. Am. Chem. Soc.* **75,** 224 (1953).
15. H. E. Frey and A. J. Wolfe in *Chemical Economics Handbook,* Stanford Research Institute, Menlo Park, Calif., 1977, p. 580.1502D.
16. L. E. Forman in J. P. Kennedy and E. G. Tornquist, eds., *Polymer Chemistry of Synthetic Elastomers II, High Polymers,* Vol. 23, Interscience Publishers, a division of John Wiley & Sons, Inc., New York, 1969.
17. L. J. Fetters, *J. Elastoplast.* **4,** 34 (1972).
18. A. F. Halasa in A. W. Langer, ed., *Polyamine-Chelated Alkali Metal Compounds,* ACS Advances in Chemistry, No. 130, American Chemical Society, Washington, D.C., 1974.
19. R. L. Zapp and P. Haus in ref. 13, Chapt. 10.
20. P. Dreyfus and J. P. Kennedy, *J. Polym. Sci. C,* **14,** 135 (1976).
21. T. G. Heggs in D. C. Allport and W. H. Jones, eds., *Block Copolymers,* John Wiley & Sons, Inc., New York, 1973, Chapt. 3.
22. D. C. Allport and A. A. Mohajer in ref. 21, Chapt. 5.
23. A. Noshay and J. E. McGrath, *Block Copolymers—Overview and Critical Survey,* Academic Press, Inc., New York, 1977: pp. 139–140.
24. S. Inone, *Chem. Technol.,* 588 (1976).
25. S. H. Rose, *J. Polym. Sci. B.* **6,** 837 (1968).
26. G. S. Kyker and T. A. Antkowiak, *Rubber Chem. Technol.* **47,** 32 (1974).
27. O. B. Edgar and R. Hill, *J. Polym. Sci.* **8,** 1 (1952).
28. *Firestone Product Specialized Elastomers, M-010-65,* Firestone Synthetic Rubber and Latex Co., Akron, Ohio, 1973.
29. W. R. Sorenson and T. W. Campbell, *Preparative Methods of Polymer Chemistry,* Interscience Publishers, a division of John Wiley & Sons, Inc., New York, 1968, p. 209.
30. S. R. Sandler and W. Karo, *Polymer Synthesis,* Vol 1, Academic Press, Inc., New York, 1974, p. 50.
31. F. W. Billmeyer, Jr., *Textbook of Polymer Science,* 2nd ed., Wiley-Interscience, a division of John Wiley & Sons, New York, 1971, p. 240.
32. M. Matsuo and co-workers, *Polymer* **9,** 425 (1968).
33. C. A. Harper in C. A. Harper, ed., *Handbook of Plastics and Elastomers,* McGraw-Hill Book Co., New York, 1975, Chapt. 1.

34. A. J. Wolfe in ref. 15, pp. 525.3230A–B.
35. J. P. Morrill in ref. 13, Chapt. 12.
36. E. L. Borg in ref. 13, Chapt. 9.
37. R. B. Otto in J. Agranoff, ed., *Modern Plastics Encyclopedia, 1976–1977,* Vol. 53 (IDA), McGraw Hill Co., New York, 1976, pp. 101–103.
38. M. J. Howard, ed., *The International Plastics Selector, 1977,* International Plastics Selector., Inc. of Cordura Publications, LaJolla, Calif., 1977, pp. 544–547.
39. *Mod. Plast.* **54**(1), 52 (1977).
40. M. W. Kline and E. Skiest in L. I. Nass, ed., *Encyclopedia of PVC,* Marcel Dekker, New York, 1976, Chapt. 4.
41. *Mod. Plast.* **54**(1), 51 (1977).
42. J. R. Stephens in H. Mark, ed., *Encyclopedia of Polymer Science and Technology,* Vol. 1, Interscience Publishers, a division of John Wiley & Sons, Inc., New York, 1964, pp. 80–84.
43. A. Kawaski and co-workers, *paper presented at 1975 International Rubber Conference, Tokyo, Japan, Oct. 14–17, 1975.*
44. U.S. Pat. 3,737,417 (June 5, 1973), K. Hayashi, A. Kawaski, and I. Maruyama (to the Maruzen Petro-chemical Co.).
45. H. L. Morris in J. Agranoff, ed., *Modern Plastics Encyclopedia, 1976–1977,* Vol. 53 (IDA), McGraw Hill, Inc., New York, 1976, pp. 103–108.
46. Ref. 23, pp. 69–75.
47. H. E. Frey in ref. 15, pp. 525.6622H–M.
48. H. E. Frey in ref. 15, p. 525.6621C.
49. Ref. 23, p. 71.
50. P. T. Wallace in ref. 15, pp. 543.4770A–G.
51. L. M. Fodor, A. G. Kitchen, and C. C. Baird in R. D. Deanin, ed., *New Industrial Polymers,* American Chemical Society, Washington, D.C., 1974, p. 37.
52. R. R. MacBride, *Mod. Plast.* **54**(5), 18 (1977).
53. G. A. Morneau in ref. 37, Vol. 53 (IDA), pp. 6–7.
54. R. R. Macbride, *Mod. Plast.* **54**(2), 16 (1977).
55. Ref. 38, pp. 443–450.
56. N. Platzer, *Chem. Technol.,* 634 (1977).
57. R. D. Deanin, A. A. Deanin, and T. Sjoblom in L. H. Sperling, ed., *Recent Advances in Polymer Blends, Grafts and Blocks,* Plenum Press, New York, 1974, Chapt. 2.
58. D. J. Griffin in C. A. Hamptel and G. G. Hawley, eds., *Encyclopedia of Chemistry,* 3rd ed., Van Nostrand Rheinhold, Co., New York, 1973, p. 902.
59. Ref. 23, pp. 48–60.
60. D. Braun, H. Cherdron and W. Kern, *Techniques of Polymer Synthesis and Characterization,* Wiley-Interscience, New York, 1971, pp. 88–89.
61. Ref. 23, p. 50.
62. A. J. Wolfe in ref. 15, pp. 525.3230A–B.
63. H. E. Frey and A. J. Wolfe in ref. 15, pp. 580.1502L–M.
64. R. L. Jalbert and J. P. Smejkal in ref. 37, pp. 108–109.
65. *Rubber and Plastics News* **VI**(26), 16 (July 25, 1977).
66. *PNF® Phosphonitrilic Fluoroelastomer,* technical publication of PNF Marketing Group, Firestone Tire & Rubber Co., Akron, Ohio, 1977.
67. D. J. Lyman in H. G. Elias, ed., *Trends in Makromolecular Science,* Gordon & Breach Science Publishers, Inc., New York, 1973, p. 55.

D. N. Schulz
D. P. Tate
Firestone Tire & Rubber Co.

COPPER

Copper [7440-50-8] has been of major importance in the development of civilization. Because of its unique physical and chemical properties and its tendency to concentrate in large ore bodies, copper has retained a position with iron and aluminum as one of the most important metallic elements.

Copper was critical in the development of civilization because it was the only metal found naturally in the metallic state suitable for the production of tools. Furthermore, the very factors that encourage the occurrence of copper metal in nature made it relatively easy for man to produce copper from naturally occurring minerals by reduction in a wood fire. The relative ease with which it could be reduced from the oxide form to the metal and its tendency to alloy to advantage with other metals naturally present in the ores promoted its broad use by emerging civilizations.

Copper is the first element of subgroup IB of the periodic table immediately above silver and gold. It is classed with silver and gold as a noble metal and like them it can be found in nature in the elemental form. Copper occurs as two natural isotopes, ^{63}Cu and ^{65}Cu (1).

Ancient man made use of copper's easy workability and beauty. Today, high electrical and thermal conductivities and corrosion resistance combine with these traditional attributes to give the metal its very wide range of commercial applications.

The earliest recorded use of copper by man was in northern Iraq about 8500 BC and then in Asia Minor and Egypt around 7000 BC. Copper items found on the Sinai Peninsula have been dated at about 3800 BC (2). The deposits on Cyprus were mined as early as 3000 BC. During this early period the Egyptians developed the metallurgical arts, and the use of bronze became moderately common. The mines on Cyprus were prized possessions of the empires that followed the Egyptians and became the chief source of metal for the Roman Empire. The metal was named aes cyprium and subsequently cuprum, from which is derived the English word copper and the symbol Cu.

Developments in the metallurgy of copper or its alloys are mentioned in 1556 in *De Re Metallica* where the processing of copper ore was described by Agricola. About that time, smelting operations commenced at Mansfeld, Germany, and at the Swansea smelter in Wales. Although there were differences in the processing at these two early smelters, both employed successive oxidations and reductions to eliminate sulfur. The process used in the Swansea smelter is similar to the current techniques.

Between 1869 and 1877 the Calumet and Hecla Company in Michigan became the largest individual copper producer in the world, although its annual production was less than 6200 metric tons of copper. By 1877, the mines of Rio Tinto in the Huelva province of southern Spain took a leading position with an annual output of slightly more than 24,000 t. This was surpassed in the 1890s by the Anaconda Copper Company (Montana) with an annual output of 34,000 t, which increased to more than 50,000 t/yr. The Anaconda mine maintained its position as the world's largest copper mine until 1920.

Large-scale mining of low-grade ores is a development of the twentieth century. The potential of the massive low-grade porphyry deposits was first realized when concentration methods were developed at the open-pit mine at Bingham Canyon, Utah (3).

Introduction of flotation for beneficiation of sulfide ores during the 1920s improved metal recovery and gave impetus to the exploitation of the low-grade porphyry deposits in Arizona which thus became the leading copper producing area in the United States. As a result of this development the United States became the major copper producer of the world.

The development of copper mining in Zaire and Zambia has moved Africa into second position in world production and the full potential of this area as yet has not been realized. Current estimates place the total production of the Union of Soviet Socialist Republics in third place; and exploitation of large ore bodies in Chile and Peru has made South America the fourth largest producer of copper.

Occurrence

Cosmically, copper is relatively abundant as compared to other heavy metals; 100–400 ppm are found in the metal phase of meteoric iron (see Planetary exploration). The high affinity of copper for sulfur is the major factor in determining the manner of occurrence in the earth's crust. Copper shows a strong tendency to combine with all available sulfur during the crystallization of rocks. Copper-iron sulfides are the last minerals to crystallize and fill the interstices between other minerals in igneous rocks which contain an average of about 60–70 ppm copper (4). Other copper compounds occurring in nature are oxides and silicates. The strong affinity of copper for sulfur is the prime factor in separating copper from iron in the pyrometallurgical reduction of copper from sulfide ore.

Copper ore minerals are classified as primary, secondary, oxidized, and native copper. Primary minerals are considered to have been concentrated in ore bodies by hydrothermal processes, whereas secondary minerals have been formed when copper leached from surface deposits by weathering and groundwater was reprecipitated near the water level (see Extractive metallurgy). The important copper minerals are classified and listed in Table 1. Of the sulfide ores, bornite, chalcopyrite, and tetrahedrite–tennantite are primary minerals, and covellite, chalcocite, and digenite were formed as secondary deposits. The oxide minerals such as chrysocolla, malachite, and azurite were formed by oxidation of the surface sulfides. Native copper is usually found in the oxidized zone. However, the major native copper deposits in Michigan are considered to be of a primary nature (5).

Most copper deposits are either (1) porphyry deposits and vein replacement deposits, (2) strata-bound deposits in sedimentary rocks, (3) massive sulfide deposits in volcanic rocks, (4) magmatic segregates of nickel copper in mafic intrusives, or (5) native copper as typified by the lava associated deposits of the Keweenaw Peninsula, Michigan.

The chief sources of copper are porphyry deposits. The term porphyry is generally applied to a type of disseminated copper deposit that is hydrothermal in origin and characterized by a large proportion of the minerals being rather uniformly distributed or in fractures and small veins. Copper contents are generally 1% or less. Figure 1 gives the worldwide distribution of copper porphyrys (7–8); these deposits contain almost two-thirds of the world's copper resources. In addition to the porphyrys, there are large bedded deposits in Germany, Poland, the USSR, and central Africa. The most extensive porphyry deposits are located in Canada, the southwestern United States, Mexico, and South America. In spite of their low copper contents, massive horizontal

Table 1. Important Copper Minerals[a]

Mineral	CAS Registry No.	Composition	Copper, %	Color	Crystal system	Luster	Mohs hardness	Specific gravity
sulfides								
bornite	[1308-82-3]	Cu_5FeS_4	63.3	between copper-red and brown	isometric	metallic	3	5.06–5.08
chalcopyrite	[1308-56-7]	$CuFeS_2$	34.5	brass-yellow	tetragonal	metallic	3.5–4	4.1–4.3
tetrahedrite[b]	[1317-91-5]	$Cu_{12}Sb_4S_{13}$	45.8	flint-gray to iron-black	isometric	metallic (splendent)	3–4.5	4.6
tennantite[b]	[12178-49-3]	$Cu_{12}As_4S_{13}$	51.6	blackish lead-gray to iron-black	isometric	metallic	3–4.5	4.37–4.49
chalcocite	[21112-20-4]	Cu_2S	79.8	blackish lead-gray	orthorhombic	metallic	2.5–3	5.5–5.8
covellite	[19138-68-2]	CuS	66.4	indigo-blue or darker	hexagonal	submetallic to resinous	1.5–2	4.6–4.76
oxides								
cuprite	[1308-76-5]	Cu_2O	88.8	red	isometric	adamantine to earthy	3.5–4	6.14
tenorite	[1317-92-6]	CuO	79.9	gray to black	monoclinic	metallic	3.5	5.8–6.4
malachite	[1319-53-5]	$CuCO_3 \cdot Cu(OH)_2$	57.3	bright green	monoclinic	adamantine to earthy	3.5–4	3.9–4.03
azurite	[1319-45-5]	$2CuCO_3 \cdot Cu(OH)_3$	55.1	azure-blue	monoclinic	vitreous, almost adamantine	3.5–4	3.77–3.89
brochantite	[12068-81-4]	$Cu_4SO_4(OH)_6$	56.2	green	orthorhombic	vitreous	3.5–4	3.9
atacamite	[1306-85-0]	$Cu_2Cl(OH)_3$	59.5	green	orthorhombic	adamantine to vitreous	3–3.5	3.76–3.78
chrysocolla	[26318-99-0]	$CuSiO_3 \cdot 2H_2O$	36.0	green to blue	orthorhombic	vitreous to earthy	2.4	2.0–2.4
native copper	[7440-50-8]	Cu	100	copper red	isometric	metallic	2.5–3	8.95

[a] Refs. 5–6.
[b] These are the limits of a series with all possible intergrades.

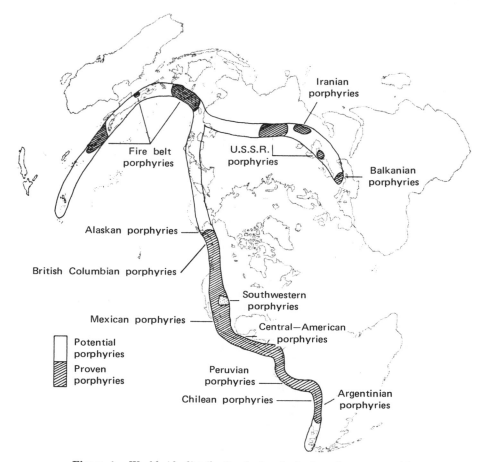

Figure 1.   Worldwide distribution (polar view) of porphyry copper (7).

development renders porphyry deposits amenable to large-scale production methods. Porphyry deposits are formed by mountain building processes so their occurrence might be expected in older mountains. The processes involved in mountain aging weather away all but the deepest deposits, however, leaving either oxidized or secondary copper in place of the porphyry.

The major copper mines throughout the world are listed in Table 2.

Exploration.   Despite an oversupply of copper on the world market from 1975 to 1978, significant new copper deposits are being explored or known reserves are being expanded. A high-grade deposit described as one of the top ten known massive sulfide deposits in North America was discovered in 1975 near Crandon, Wisconsin. In 1976 massive sulfide gradings up to 5.25% copper, 20.7% Zn, and 289 g Ag/t were discovered in Western Australia. The identification and drilling of three small to medium porphyry copper deposits in Pakistan were announced in 1976; and in Argentina, St. Joe Minerals Corporation expanded reserves at the Pachon porphyry copper deposit to approximately 725,000,000 t at 0.59% copper.

A less conventional area of exploration is of deep-sea ferromanganese nodules

Table 2. Locations of Major Copper Mines

Country	State, province, or region	Representative mineralogical geological type	Mining method	Mines producing more than 2,500,000 metric tons ore/year[a]
Australia	Mt. Isa, Q.	high-grade massive sulfide	underground	1
Bulgaria	Medet	disseminated porphyry copper	open-pit	1
Canada	British Columbia	disseminated porphyry copper	open-pit	6
		vein in volcanic	underground	1
	Manitoba		open pit	1
	Ontario	magmatic segregation in norite lopolith	underground	1
		replacement in volcanic and metamorphic	underground and open-pit	1
	Quebec		open-pit	1
Chile	Antofagasta	disseminated porphyry copper	open-pit	2
	Aconcagua, Atacama, O'Higgins	disseminated porphyry copper	underground	3
Mexico	Sonora	disseminated porphyry copper and limestone replacement	open-pit	1
Papau, New Guinea	Panguna P.N.G.	disseminated porphyry	open-pit	1
Peru	Moqueguo	disseminated porphyry copper	open-pit	1
Philippines	Marinduque, Negros Or.	probably disseminated	open-pit	2
	Mt. Prov.	porphyry copper	underground	1
	Cebu		underground and open-pit	1
South Africa	Transvaal	disseminated in a complex carbonatite intrusive	open-pit	1
Spain	Huelva	massive sulfide replacement in volcanics	open-pit	1
Sweden	Gällivare		open-pit	1
United States	Arizona	disseminated porphyry copper and contact metamorphic	underground	1
		disseminated porphyry copper	open-pit	15
	Michigan	bedded deposit chalcocite and native copper in Nonesuch shale formation	underground	1
	Montana	disseminated porphyry copper	underground and open-pit	1
	New Mexico	disseminated porphyry copper, bedded red bed deposit	open-pit	2
	Nevada	disseminated porphyry copper	open-pit	2
	Utah	disseminated porphyry copper	open-pit	1
Union of Soviet Socialist Republics	Urals	cupriferous pyrite and complex copper zinc ores	underground	
	Kazakhstan	disseminated porphyry copper and sedimentary sandstone	open-pit	2
	Uzbeckstan	disseminated porphyry copper	open-pit	1
	Armenia	disseminated porphyry copper, molybdenum	open-pit	1
	Siberia	cupriferous pentlandites probably magmatic segregation in basic intrusive	underground	

Table 2   (*continued*)

Country	State, province, or region	Representative mineralogical geological type	Mining method	Mines producing more than 2,500,000 metric tons ore/year[a]
Yugoslovia	Majdanpek	disseminated porphyry copper	open-pit	1
Zaire	Shaba	high-grade replacement in sedimentary formation	open-pit	2
Zambia	Mufilira Luanshya	high-grade replacement in sedimentary formation	underground	2
	Chingola, Kitwe	high-grade replacement in sedimentary formation	underground and open-pit	2

[a] Refs. 9–10.

that occur at many ocean sites. They consist primarily of manganese, with some deposits containing over 1% copper. The most valuable deposits are found in the Pacific Ocean. Copper resources from nodules must be considered tentative, although a number of companies are making active plans for recovering values from this source (11–12) (see Ocean raw materials).

A summary of estimated world primary copper resources is presented in Table 3 (13).

Properties

Like silver and gold, copper has an atomic structure that results in outstanding electrical and thermal conductivities and a high degree of malleability. Although the assigned electronic configuration of 2-8-18-1 implies a stable closed shell of 18 electrons, this shell is not inert. Rather, the underlying d orbitals appear to participate in metallic bonding by promotion of at least one d electron into a higher energy orbital of the outermost principal quantum level. There this electron is exceptionally available for participation in electrical and thermal conduction (14). A comparison of the properties of copper, silver, and gold is presented in Table 4. Disagreements between data obtained by different authors point to the difficulty in obtaining reproducible results because of variations in copper purity (14). Although commercial copper is of excellent purity, the differences between data for pure copper and electrolytic copper are very significant.

The unique nature of the electronic configuration of copper that contributes to its high electrical and heat conductivity provides chemical properties intermediate between transition and 18-shell-type elements even though the assigned electronic configuration is 18-shelled. The electron structure of the free copper atom is $1s^2 2s^2 2p^6 3s^2 3p^6 3d^{10} 4s^1$. It can give up the $4s$ electron to form the copper(I) ion or release an additional electron from the $3d$ subgroup to form the copper(II) ion.

The higher ionization energy and smaller ionic radius of copper contribute to its forming oxides much less polar, less stable, and less basic than the alkali metals (18). Because of the relative instability of its oxides, copper joins silver in occurring in nature in the metallic state.

Table 3.   World Copper Resources,[a] Million Metric Tons of Copper

Location	Reserves[b]	Other[c]	Total
North America			
United States	84	290	375
Canada	31	109	140
other	30	27	57
Total	*145*	*426*	*571*
South America			
Chile	84	118	202
Peru	32	36	68
other	20	63	83
Total	*136*	*217*	*353*
Europe, *Total*	6	*36*	*43*
Africa			
Zaire	25	27	53
Zambia	29	63	93
other	9	18	27
Total	*63*	*108*	*173*
Asia, *Total*	*27*	*63*	*91*
Oceania, *Total*	*18*	*54*	*73*
centrally planned economies	60	172	232
sea nodules[d]		689	689
World Total	*455*	*1765*	*2225*

[a] Ref. 13.

[b] Of the listed reserves, approximately one-third of the copper contained in the United States and market economy country totals is located in undeveloped deposits. These deposits can move between reserve and resource classifications depending on prevailing legal and economic conditions.

[c] Includes undiscovered (hypothetical and speculative) deposits.

[d] Estimate based on average of 1% copper per dry ton of nodules.

The standard oxidation potentials of copper(I) and copper(II) at 25°C are (16):

$$Cu^+ + e^- \rightleftarrows Cu \; E = 0.520 \tag{1}$$

$$Cu^{2+} + 2\,e^- \rightleftarrows Cu \; E = 0.337 \tag{2}$$

The simple copper(I) ion forms compounds with the anions of both strong and weak acids. Many of these compounds are insoluble in water and stable. There is a very strong tendency for disproportionation of copper(I) into copper(II) and metallic copper in aqueous solutions.

$$2\,Cu^+ \rightarrow Cu + Cu^{2+} \tag{3}$$

Compounds and complexes of copper(I) are almost colorless since the inner $3d$ orbital of the copper is completely filled.

The properties of the copper(II) ions are quite different. The ligands that form strong coordinate bonds complex the copper(II) ion readily to form complexes in which the copper has coordination numbers of 4 or 6, such as $[Cu(NH_3)_4]^{2+}$ and $[Cu(H_2O)_4]^{2+}$. Formation of copper(II) complexes in aqueous solution depends upon the ability of ligands to compete with water for coordination sites. Most copper(II) complexes are colored and paramagnetic due to the unpaired electron in tie $3d$ orbital (see Copper compounds).

Table 4.   Comparison of the Properties of Pure Copper with Those of Silver and Gold[a]

Property	Copper	Silver[b]	Gold[c]
atomic weight	63.54	107.87	196.97
atomic volume, cm^3/mol	7.11	10.27	10.22
mass numbers stable isotopes (relative abundance, %)	63 (69.1)	107 (51.35)	197 (100)
	65 (30.9)	109 (48.65)	
oxidation states	1, 2, 3	1, 2, 3	1, 2, 3
standard electrode potential,[d] 25°C, V	Cu/Cu$^+$ = 0.520	Ag/Ag$^+$ = 0.799	Au/Au$^+$ = 1.692
	Cu/Cu^{2+} = 0.337		
density, kg/m^3	8.96 × 10^3	10.49 × 10^3	19.32 × 10^3
crystal structure	fcc	fcc	fcc
metallic radius, nm[e]	0.1276	0.1442	0.1439
ionic radius, (M+), nm	0.096	0.126	0.137
covalent radius, nm[e]	0.138	0.153	0.150
electronegativity[e]	2.43	2.30	2.88
1st ionization energy[e], J/mol[f]	745 × 10^3	732 × 10^3	891 × 10^3
2nd ionization energy[e], J/mol[f]	1950 × 10^3	2070 × 10^3	
heat of atomization[e], J/mol[f]	339 × 10^3	286 × 10^3	354 × 10^3
thermal conductivity, W/(m·K)	394	427	289
electrical resistivity 20°C, $\mu\Omega$/cm	1.6730	1.59	2.35
temperature coef. of electrical resistivity at 0–100°C	0.0068	0.0041	0.004
melting point, °C	1083	960.8	1063
heat of fusion, J/kg[f]	212 × 10^3	102 × 10^3	67.4 × 10^3
boiling point, °C	2595	2212	2970
heat of vaporation, J/kg[f]	7369 × 10^3	2400 × 10^3	1860 × 10^3
specific heat at 20°C, J/(kg/°C)[f]	384	233	131 (18°C)
linear coef. of expansion × 10^6 per °C at 20°C	16.5	10.68	14.2
tensile strength, typical for annealed metal, kPa[g]	23 × 10^4	28 × 10^4	17 × 10^4
modulus of elasticity, for hard drawn metal, MPa[h]	10.2–12 × 10^4	7.75 × 10^4	7.85 × 10^4
modulus of rigidity, MPa[h]	44,000		
magnetic susceptibility at 18°C, cgs units/g	−0.086 × 10^{-6}	−0.20 × 10^{-6}	−0.15 × 10^{-6}
emissivity, unoxidized metal at 100°C	0.03	0.052	(0.02)?
viscosity at 1145°C, mPa·s(= cP)[i]	3.41		
surface tension at 1150°C, mN/m (= dyn/cm)[i]	1104		

[a] Ref. 15.
[b] See Silver.
[c] See Gold.
[d] Ref. 16.
[e] Ref. 14.
[f] To convert J to cal, divide by 4.184.
[g] To convert kPa to psi, multiply by 0.145.
[h] To convert MPa to psi, multiply by 145.
[i] Ref. 17.

Elemental copper is resistant to aerated alkaline solutions except in the presence of ammonia. It does not displace hydrogen from acid but dissolves readily in oxidizing acids such as nitric acid or in acid solutions that contain an oxidizing agent, ie, sulfuric acid solution containing ferric sulfate. Because of its corrosion resistance to salt solutions, it is used in marine applications and its resistance to oxidation by water vapor at high temperatures has made it a material of choice in cooling systems. Although the surface of copper oxidizes on exposure to the atmosphere, further corrosion is inhibited by the formation of a tightly adherent coating of corrosion products. In many instances this takes the form of a green patina that imparts a rich appearance to architectural and artistic uses. Studies of the chemistry of atmospheric corrosion have shown that the initial attack involves the formation of sulfides and oxides. With further oxidation and reaction with water, basic copper sulfates are formed such as $CuSO_4.Cu(OH)_2$ and $CuSO_4.3Cu(OH)_2$. The latter compound is found in nature as the mineral brochantite. A basic carbonate, $CuCO_3.Cu(OH)_2$, may also be present in some deposits (19).

Recovery and Processing

Most copper today is processed by mining, waste leaching and cementation, concentrating, smelting, and refining. Open-pit mining is more common than underground mining and the overburden, or waste, contains some copper. Frequently the waste is leached to extract the copper, which may be recovered by passing the leach solution through a bed of scrap iron, precipitating metallic copper, and dissolving the iron; the last operation is called cementation.

The copper ore from the mine, often containing less than 1% copper, is transported to the concentrator where it is first crushed and then ground with water. The ground ore slurry enters flotation cells, where copper concentrates are collected as a froth (see Flotation). Following dewatering, they enter the smelter. In the smelter the sulfide minerals react with oxygen and fluxes to produce impure copper metal, SO_2, and slag. Smelting occurs in two steps. In the reverberatory furnace, the copper concentrate is melted to produce matte, the mixed sulfides of copper and iron. Next, air is blown through the matte in the converters, producing impure copper plus a slag containing the iron. The impure copper is then cast into anodes and purified by plating onto pure copper in an electrolytic tankhouse. Figure 2 illustrates the major steps involved in this conventional approach to recovery and processing of copper.

Other hydrometallurgical processes include the direct leaching of ore followed by recovery of copper by cementation or electrowinning. Recently, however, hydrometallurgical treatment of concentrates in lieu of smelting is being developed in order to avoid the high cost of environmental control facilities required for new smelters.

Mining. The copper ore may be obtained from underground or surface mines. In underground mining, the ore is removed by tunneling or shafts, whereas in surface mining the entire overburden, or waste above the ore, is removed to expose the ore.

The operations in open-pit mining are drilling, blasting, loading, and hauling. Providing access for the mining and haulage equipment to remove the waste and ore leads to the characteristic inverted conical shape with terraces or benches of open-pit mines (see Fig. 3). The trend has been toward large open-pit mines (20–22) where the problem of declining ore grades is met by using very large capacity equipment such as 15–25-m^3 (500–900-ft^3) shovels and 100–250-t haulage trucks. Most of the copper mines in the southwestern United States mine 30,000–400,000 t of ore plus waste per day.

The economics of some of the newer mines frequently depend heavily on the revenues derived from leaching waste to recover additional copper (20–21). This waste material, which has to be removed to uncover the ore, may be hauled to specially constructed dumps for leaching where the sulfides, the most common form of copper,

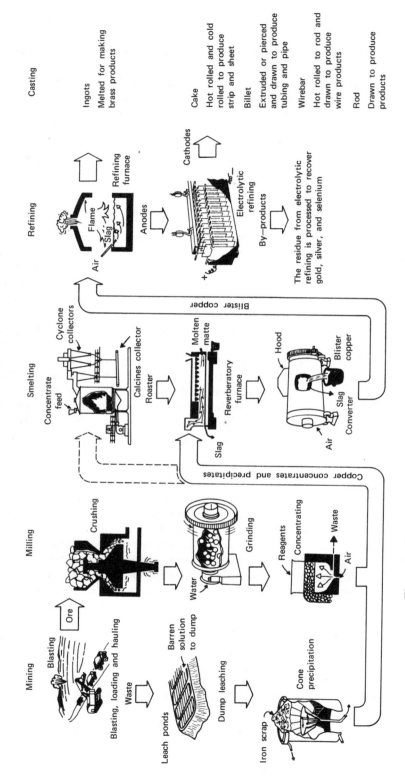

Figure 2. Recovery of copper from sulfide ore. Courtesy of Kennecott Copper Corporation.

Figure 3.   Kennecott open-pit mine, Bingham, Utah. Courtesy of Kennecott Copper Corporation, D. Green.

are oxidized and the leach solution can contact the waste material uniformly (see under Hydrometallurgical Processes). Unfortunately, the economics of mining sometimes preclude optimization of dumps for leaching operations, because of such factors as compaction, size distribution, and terrain.

Revenues from today's low-grade copper mines are also gained from such by-products as molybdenum, silver, gold, selenium, and tellurium.

Computer techniques are used for mine planning and design, integrating the effects of geology, cut-off grade, and by-product revenue, and for the optimum deployment of the mining and post-mining equipment.

Concentration.   A copper ore only rarely contains a sufficiently high percentage of copper to allow direct smelting. Ores with 1.0% copper or less are common and are upgraded by concentrating (23). This process consists of crushing and grinding to expose the copper mineralization, followed by flotation. In flotation, the copper minerals are separated and recovered as a froth concentrate wherein the solids typically contain about 20–30% copper (see Flotation). This concentrate is a suitable feed for smelting.

Figure 4 shows a simplified flow diagram for a typical copper concentrator. The term *mill* is used interchangeably with the term *concentrator* to describe the facility, although *milling* is more correctly applied to just the grinding step.

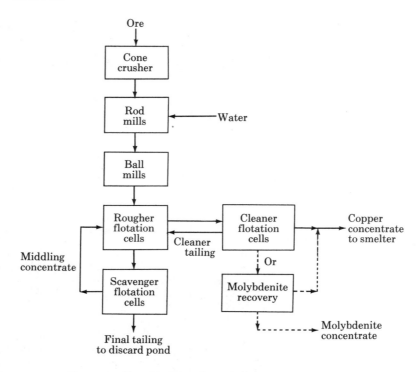

Figure 4.   Simplified flow diagram of a copper concentrater.

Crushing.   A typical copper concentrator in the southwestern United States processes ca 10,000–100,000 t/d of low-grade ore. Cone crushers are normally used for the first stages of size reduction, with two or three stages of cone crushers in series being common. A cone crusher consists of an eccentrically driven grinding cone that rotates in a fixed conical bowl sloped so that there is a wedge-shaped space between the grinding cone and the bowl. The eccentrically driven grinding cone and the bowl have a vertical axis, and as ore is introduced between them it is pinched and crushed. Typical product sizes for three crushing stages in series are <20 cm, <3 cm, and <1 cm. Dry screening is used between crushing stages so that fine ore bypasses those stages where it is finer than the crusher product.

Grinding.   Further size reduction (qv) is achieved by several stages of grinding with water to produce an ore slurry. Rod mills are frequently used for the first stage where the ore may be reduced to <3 mm. To reduce the size of the rod mill product so that about 75% is less than 0.25 mm in size, two stages of ball mills in series may be used.

Rod-and-ball mills are cylindrical vessels into which the ore, water, and steel balls or steel rods, called *grinding media,* are charged (see Fig. 5). The vessels rotate on horizontal axes to cascade the ore and grinding media, and thus grind the ore. Typical ball mill sizes range from 2.5 m dia and 3 m long, driven by a 187-kW motor, up to ca 5 m dia and 7 m long with a 3000-kW drive.

The product from each stage of grinding is size classified and the oversize material is recycled to be ground further. Wet cyclones are the most common type of classifier.

Figure 5. Ball mills grinding copper ore. Courtesy Kennecott Copper Corporation, D. Green.

Flotation. The slurry of ground ore leaving the grinding circuit may be separated from part of the water in thickeners or may go directly to the flotation cells. The latter are rectangular tanks into which air is injected or drawn in via impellers (see Fig. 6). The air adheres to the copper mineralization so that ore particles with exposed sulfide copper minerals float to the surface. The froth with the copper minerals overflows into collection launders. Reagents are added to adjust the pH, produce froth, and to collect some components and depress others.

Most copper flotation plants have three banks of flotation cells, ie, rougher cells, cleaner cells, and scavenger cells (see Fig. 4). The ore slurry enters the rougher cells where most of the copper mineralization is floated. The froth goes to cleaner cells where some of the gangue carried up with the sulfides is recycled for further processing while

Figure 6. Flotation cells concentrating copper ore. Courtesy of Kennecott Copper Corporation, D. Green.

the clean copper concentrate is dewatered and goes to the smelter. The material which does not float in the rougher cells goes to scavenger cells where additional copper is recovered. The froth from the scavenger cells may be reground to separate gangue and mineralization before it is returned to the rougher cells.

When the ore contains a large amount of clay slimes, the tailing from the scavenger cells can be cycloned to remove the slimes before the coarse material is floated in a tailings retreatment plant. The flotation product from the rougher cells of this plant can be reground and cleaned. This additional treatment of the tailings from the main copper flotation plant may improve the recovery of the metal values by 1–3%.

The pH of the pulp to the flotation cells is carefully controlled by the addition of lime, which optimizes the action of all reagents and is used to depress pyrite. A frother such as pine oil or another long-chain alcohol is added to produce the froth, an important part of the flotation process. Supported by the air bubbles, the desired minerals float while the gangue sinks. Typical collectors are xanthates, dithiophosphates, or xanthate derivatives, whereas typical depressants are calcium or sodium cyanide and lime.

Molybdenite normally floats with the copper sulfides. Therefore, the copper concentrate from the cleaner cells frequently has to be separated from molybdenite in a separate flotation circuit before the copper concentrate goes to the smelter. Gold, silver, selenium, and tellurium are valuable by-products separated with the copper concentrate.

Recent Developments.   A major development in concentrating has been the steady increase in equipment size as ore grades have declined and daily throughputs have increased. The newest flotation cells each have capacities of 28 m^3 (7400 gal) or more, whereas capacities in the 1950s rarely exceeded 3 m^3 (800 gal). The number of flotation cells in a concentrator can vary widely. Today, a concentrator for 90,000 t/d might require 100 flotation cells of 28-m^3 capacity; with the smaller cells about 1000 flotation cells would be necessary for the same service.

Recent developments have included automation and instrumentation such as automatic grinding control (24) and on-line analyses of process streams to continuously monitor metal value recovery (25).

Another innovation is the use of autogenous and semiautogenous grinding equipment. All or part of the ball charge in the ball mills is replaced by ore, so that the ore grinds itself. The mills are larger in diameter than conventional ball mills and perform part of the crushing operation in addition to grinding. Semiautogenous ball mills may use 10–20% more energy than conventional ball mills for the same service, and fully autogeneous ball mills use even more energy. Both provide operating cost savings relative to conventional grinding with full ball charges.

Smelting.   Chalcopyrite concentrate is a mixture of the sulfides of copper, copper–iron, and iron with smaller amounts of gangue minerals. It normally contains 20–30% copper.

Sulfur is removed as sulfur dioxide by heating the concentrate in the presence of air. Iron combines with silica either from the gangue or the added flux and is removed together with some other impurities in the resulting slag. Part of the small amount of impurities remaining in the copper is removed by fire refining. In this step, the impurities are rejected with slag and vapor by air oxidation and slagging. Gold, silver, selenium, and tellurium and some other impurities remain with the copper to be recovered as by-products in the final purification stage, namely electrorefining.

The term copper smelting designates the operations of melting the concentrate and extracting the copper by heat, flux and the addition of oxygen.

Reactions of Smelting (**26**).   When the concentrate is roasted, part of the sulfur is converted to sulfur dioxide in a concentration suitable for sulfuric acid production.

$$2\,CuFeS_2 + O_2 \rightarrow Cu_2S + 2\,FeS + SO_2 \tag{4}$$

$$FeS_2 \rightarrow FeS + S \tag{5}$$

$$S + O_2 \rightarrow SO_2 \tag{6}$$

$$2\,FeS + 3\,O_2 \rightarrow 2\,FeO + 2\,SO_2 \tag{7}$$

$$2\,SO_2 + O_2 \rightarrow 2\,SO_3 \tag{8}$$

$$FeO + SO_3 \rightarrow FeSO_4 \tag{9}$$

When melted, copper(I) sulfide and iron(II) sulfide are miscible, forming a matte. Molten iron silicate and iron oxide form a separate layer which floats on this matte and can be skimmed off in the smelting vessel to effect a partial separation of iron from copper. Typical reactions occurring in the smelting vessel are listed below. In addition, equation 4 occurs when the concentrate is not roasted before entering the smelting vessel.

$$FeS_2 + O_2 \rightarrow FeS + SO_2 \tag{10}$$

$$3\,FeS + 5\,O_2 \rightarrow Fe_3O_4 + 3\,SO_2 \tag{11}$$

$$Cu + CuFeS_2 \rightarrow Cu_2S + FeS \tag{12}$$

$$Cu_2O + FeS \rightarrow Cu_2S + FeO \tag{13}$$

$$3\,Fe_3O_4 + FeS + 5\,SiO_2 \rightarrow 10\,FeO.5SiO_2 + SO_2 \tag{14}$$

The copper and copper oxide in equations 12 and 13 are from copper production via leaching or from recycled converter slag.

Since iron(II) sulfide is more readily oxidized than copper(I) sulfide, iron can be separated from copper in the converters. Furthermore, sulfur combines with copper rather than with iron. Hence, copper sulfide remains in the converter after the iron has been oxidized and has combined with silica to be skimmed off as a slag. Typical converting reactions of iron sulfide are equations 11 and 15.

$$2\,FeS + 3\,O_2 + SiO_2 \rightarrow 2\,FeO.SiO_2 + 2\,SO_2 \tag{15}$$

Copper(I) sulfide combines with oxygen to yield copper(I) oxide and sulfur dioxide. Copper(I) sulfide and copper(I) oxide react to form metallic copper and gaseous sulfur dioxide in the converters:

$$2\,Cu_2S + 3\,O_2 \rightarrow 2\,Cu_2O + 2\,SO_2 \tag{16}$$

$$Cu_2S + 2\,Cu_2O \rightarrow 6\,Cu + SO_2 \tag{17}$$

$$3\,Cu_2S + 3\,O_2 \rightarrow 6\,Cu + 3\,SO_2 \tag{18}$$

Fire refining, the final smelting operation, removes further impurities and adjusts the oxygen level in the copper by air oxidation followed by reduction with green wood or hydrocarbons:

$$4\,Cu + O_2 \rightarrow 2\,Cu_2O \tag{19}$$

$$Cu_2O + H_2 \rightarrow 2\,Cu + H_2O \tag{20}$$

$$Cu_2O + CO \rightarrow 2\,Cu + CO_2 \tag{21}$$

Silver and gold stay mostly with the copper throughout the smelting operations rather than being lost with the slag.

Silica is added to the converters and sometimes to the reverberatory furnaces to form iron silicate slag by the reactions shown in equations 14 and 15. A second flux, limestone, is added to the reverberatory furnaces to lower the slag viscosity. The thermodynamics of copper smelting are discussed in refs. 27–28.

Roasting. The copper concentrates entering the smelter usually contain ca 10–15% moisture which may be reduced to 6–10% in a dryer. In some smelters the concentrates are roasted in multiple-hearth or fluid-bed roasters before smelting. Water is removed and part of the sulfur content is oxidized to sulfur dioxide. Off gases from roasters have a sulfur dioxide concentration suitable for conversion to sulfuric acid; reverberatory furnace gases are too dilute for this purpose. Hence plants that incorporate roasters have the potential for a higher sulfur capture than plants without roasters.

The multiple-hearth roaster is a brick-lined tower with horizontal brick hearths. The concentrate is introduced at the top hearth where rotating arms with rabble blades turn it over and move it to holes in the hearth. The concentrate is transferred successively to lower hearths and finally leaves the bottom hearth. Air for partial oxidation of the concentrates is blown into the bottom hearth and moves countercurrent to the concentrate, leaving the top hearth with a sulfur dioxide concentration of 2–6%.

In fluidized-bed roasters (29), the concentrate is suspended in an upward-moving air stream. The vessel is a refractory-lined steel shell with air entering through holes in a refractory-lined plate at the base. The sulfur dioxide concentration of the exit gas is 10–15%.

Reverberatory Smelting. Figure 7 is a simplified flow sheet for the reverberatory smelting process, showing typical material flows in metric tons per day. Concentrate, slag recycled from the converters, plus limestone flux and sometimes silica flux are charged to the reverberatory furnace. Fuel oil, natural gas, or coal burners supplement

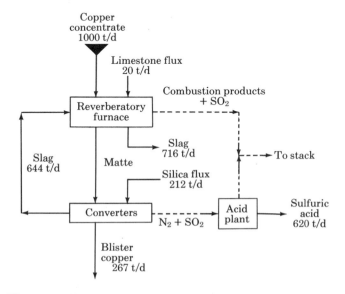

Figure 7.   Flow sheet for reverberatory furnace–converter smelting.

the energy obtained from the partial oxidation of the sulfides. Calcine feed is concentrate that has been roasted before entering the reverberatory furnace, whereas "green feed" had not been roasted.

Figure 8 shows a section of a rectangular reverberatory furnace typically 11 m wide and 37 m long, and constructed of refractory brick. Heat provided by burners at one end is reradiated from the furnace roof, hence the name. The hot gases leave through a waste heat boiler at the other end for heat recovery before discharging to a stack. The feed materials are added from charging hoppers along the sides of the furnace (31). The silica flux combines with iron(II) sulfide and iron(II) oxide to form slag. The fluidity of the slag in which unwanted impurities dissolve is controlled by the addition of limestone.

Converting. Figure 9 shows a typical converter separated from the reverberatory furnaces by a crane aisle. When the converter is rotated for charging and discharging, the air injection pipes, the tuyeres, are lifted above the molten bath so that they are not blocked when the air is turned off. Partial blockages are cleared during blowing by use of mechanical tuyere punchers (32).

Copper matte, the molten copper and iron sulfides from the reverberatory furnaces, is charged to the copper converters together with silica flux. Air is blown into the charge through the tuyeres in the sides of the converters. The sulfides are oxidized to sulfur dioxide which is usually sent to sulfuric acid plants (33). The iron sulfide component in the matte is oxidized by blowing with air in the slag blow and combined with silica to form iron silicate slag as shown in equation 15. This slag is returned to the reverberatory furnace to recover entrained and dissolved copper. After the slag has been poured off, further air in what is termed the *finish blow* provides oxygen to convert the copper sulfide to blister copper, containing about 99% copper plus some sulfur, oxygen, and impurities (eqs. 16–18).

Developments. The discharge of reverberatory furnace gases to the atmosphere has recently been challenged on environmental grounds. Copper producers have had to search for alternatives that would provide SO_2 concentrations of 5% or better in the off gas and thus allow sulfur recovery (34).

Electric smelters (34) and flash smelters (35–37) are designed as direct replacements for reverberatory furnaces and produce matte as feed for converters.

The Mitsubishi process combines smelting and converting (38–40) and is designed for continuous production of blister copper (Fig. 10).

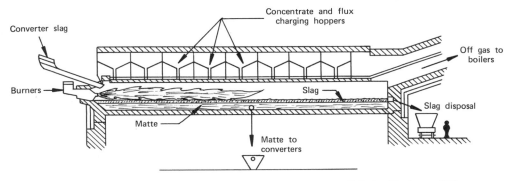

Figure 8. Reverberatory furnace. Courtesy of the Society of Mining Engineers (30).

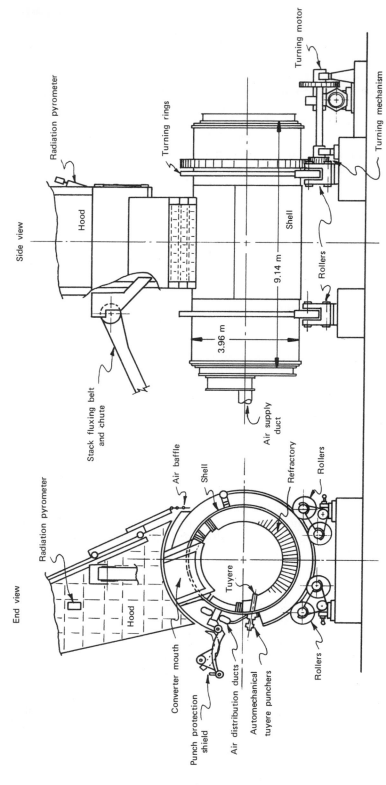

End view

Radiation pyrometer

Hood

Converter mouth

Punch protection shield

Air distribution ducts

Automechanical tuyere punchers

Air baffle

Shell

Tuyere

Refractory

Rollers

Rollers

Side view

Radiation pyrometer

Hood

Stack fluxing belt and chute

Turning rings

3.96 m

9.14 m

Shell

Rollers

Air supply duct

Turning motor

Turning mechanism

Figure 9. The Peirce-Smith converter.

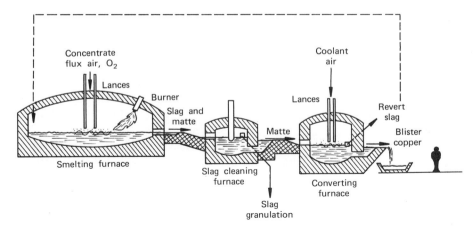

Figure 10. The Mitsubishi process. Courtesy of the Society of Mining Engineers (30).

The Noranda process (see Fig. 11) can be operated (41–42) to produce either a high-grade matte (Utah Copper Div., Kennecott Copper Corporation) or blister copper (Noranda Mines, Ltd.).

In the Outokumpu process (flash smelting) and the Noranda process, the exothermic converting reactions are carried out in part in the smelting vessel; their fuel requirements are below that for reverberatory converter smelting. The furnace slags from these processes, however, contain appreciable concentrations of copper which must be recovered.

Sulfur Removal. *Acid Plants.* Currently, sulfuric acid plants are the primary means for removing sulfur dioxide from smelter off gases. The gases from converters and roasters vary in sulfur dioxide concentration, but usually contain over 5% dioxide, which is a suitable strength for producing sulfuric acid. Reverberatory furnace gases from conventional equipment usually contain 0.5–2% sulfur dioxide, which is too dilute for sulfuric acid production; however, oxygen fuel burners mounted in the roof of a reverberatory furnace in the Chilean Caletones smelter (43) produce flue gases with a sulfur dioxide strength suitable for sulfuric acid production.

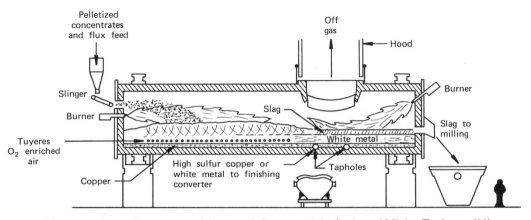

Figure 11. Noranda process smelting vessel. Courtesy of the Society of Mining Engineers (30).

Single-absorption sulfuric acid plants are commonly used for producing sulfuric acid from the off gases leaving copper smelters (see Sulfuric acid). Double-absorption plants permit a higher sulfur dioxide capture. However, they require more fuel for reheating the gas to maintain the conversion of sulfur dioxide to sulfur trioxide from the variable strength smelter gases. The pressure drop is higher than that for a single-absorption plant, and thus the blowers use more electrical energy. Furthermore, double-absorption plants are more expensive to construct and maintain than single-absorption acid plants.

Compression. The Inco flash smelter produces a very high strength sulfur dioxide gas by using oxygen for smelting; upon compression, liquid sulfur dioxide is obtained.

Scrubbers. Scrubbers for removing sulfur dioxide from smelter off gases have been under development for many years but few are used commercially.

The Smelter Control Research Association was founded by the United States copper industry to test and study scrubbing systems for smelters. Lime–limestone scrubbing and double alkali scrubbing have been piloted and the results are being evaluated. The latter is based on the following reactions:

$$(NH_4)_2SO_3 + H_2O + SO_2 \rightarrow 2\,NH_4HSO_3 \tag{22}$$

$$2\,NH_4HSO_3 + CaCO_3 \rightarrow (NH_4)_2SO_3 + CaSO_3 + H_2O + CO_2 \tag{23}$$

The calcium sulfite obtained from this process is suitable feed for wallboard production (see Sulfur recovery).

Fire Refining. The impurities in blister copper obtained from converters have to be reduced before it can be fabricated or cast into anodes to be electrolytically refined. High sulfur and oxygen levels result in excessive degassing during casting and uneven anode surfaces. Such anodes result in low current efficiencies and uneven cathode deposits with excessive impurity inclusions. Fire refining is essential whether the copper is to be marketed directly or electrorefined.

Fire refining adjusts the sulfur and oxygen levels in the blister copper and removes impurities as slag or volatile products. The fire-refined copper is sold for fabrication into end products, provided the chemistry permits product specifications to be met. Some impurities, such as selenium and nickel, remain, and are not removed sufficiently by the fire refining procedure. If these impurities are detrimental in fabrication or end use, the copper must be electrorefined. Other impurities such as gold, silver, selenium, and tellurium are valuable by-products that are only recovered via electrorefining.

Blister copper is fire-refined in reverberatory or rotary furnaces similar to converters. Both types have capacities of 100–400 t. Most plants fire-refine and cast the anodes or fire-refined ingots within the smelter building so that the fire-refining furnace is supplied with molten blister copper. Cold blister has to be melted first. Next, air is blown into the copper through iron pipes to oxidize some impurities and remove volatile impurities. Sodium carbonate flux may be added to remove arsenic and antimony, and finally the copper is reduced by poling with green wood poles or by feeding a hydrocarbon gas. The degrees of oxidation and reduction are determined by removing and casting small samples of the copper. An experienced operator knows whether the copper has the proper oxygen and sulfur content by the appearance of the samples.

The end product is either cast as anodes for electrolytic refining or as ingots for sale as fire-refined copper. A horizontal casting wheel with 12–24 horizontal molds is normally used for anode casting. A few plants use Hazelett continuous casting machines by which the copper is cast as a continuous strip to be cut to the required anode shape (44).

A fire-refining cycle with 8-h melting, 8-h refining, and 8-h casting is common.

Chemistry. Fire refining of blister copper is achieved by oxidation, fluxing, and reduction. The copper is partially oxidized by blowing air into the fire refining furnace where the copper oxide reacts with the impurities that oxidize preferentially (45).

The major impurities are eliminated in fire refining in the following sequence: slag, ie, oxides of iron, magnesium, aluminum, and silicon; flux–slag, ie, arsenic and antimony; and vapors, ie, sulfur dioxide, cadmium, and zinc.

Blowing the blister copper raises the copper oxide concentration to 6–10% which is too high for casting. The traditional reduction process is poling. In this operation, the end of a green tree trunk is inserted into the bath of molten metal in the fire-refining furnace. The resulting vigorous agitation of the bath combines with the reducing conditions from burning wood to adjust the oxygen concentration. Gaseous deoxidation with natural gas or propane, called *gaseous poling*, is now used in some plants. The gas may be reformed or used directly.

If the fire-refined copper is to be cast into anodes for electrorefining, the oxygen content of the copper is lowered to 0.05–0.2%. If the copper is to be sold directly for fabrication, the oxygen level is adjusted to 0.03–0.05%, which is the range for tough-pitch copper. The major reactions of fire refining, fluxing, and poling are shown below:

Fire refining:

$$4\,Cu + O_2 \rightarrow 2\,Cu_2O \tag{19}$$

$$2\,Fe + O_2 \rightarrow 2\,FeO \tag{24}$$

$$5\,Cu_2O + 2\,As \rightarrow As_2O_5 + 10\,Cu \tag{25}$$

$$5\,Cu_2O + 2\,Sb \rightarrow Sb_2O_5 + 10\,Cu \tag{26}$$

Fluxing:

$$2\,FeO + SiO_2 \rightarrow Fe_2SiO_4 \tag{27}$$

$$As_2O_5 + 3\,Na_2CO_3 \rightarrow 2\,Na_3AsO_4 + 3\,CO_2 \tag{28}$$

Poling or deoxidation:

$$2\,Cu_2O + C \rightarrow 4\,Cu + CO_2 \tag{29}$$

$$Cu_2O + CO \rightarrow 2\,Cu + CO_2 \tag{21}$$

$$Cu_2O + H_2 \rightarrow 2\,Cu + H_2O \tag{20}$$

Recent Developments. Fire-refined, electrolytic (electrowon) copper is becoming accepted as equivalent to electrorefined copper, and has entered the major copper market, the electrical industry. Kennecott Copper Corporation has patented a fire-refining process for removing bismuth, antimony, and arsenic impurities by treating molten blister copper with sulfur hexafluoride (46). This, or similar approaches, might eliminate the need for electrorefining (see Extractive metallurgy).

Electrorefining. Fire-refined copper is adequate for noncritical applications such as water tubing, bar stock, or ingots for alloying. Copper intended for electrical uses, however, is produced by electrorefining or sometimes electrowinning techniques.

In the electrorefinery, anodes produced from blister copper as described above, are dissolved electrolytically in acidic copper sulfate, and the copper is deposited on pure copper starting sheets to produce cathodes according to equation 2. The cathodes are sold directly or melted and cast into a number of forms.

Anode impurities either dissolve in the electrolyte or fall to the bottom of the electrolytic cell as anode slime. Since these slimes contain silver, gold, selenium, and tellurium, they represent a very significant value, and the recovery of by-products from the anode slime is an important operation.

The basic process for electrorefining copper was developed around the turn of the century, and all refineries are now using the same fundamental process. Table 5 presents operating data for six refineries.

Figure 12 is a simplified flow diagram of a modern electrorefinery, which consists of four stages: (1) the anode copper is dissolved in the tankhouse, and pure copper is deposited electrolytically on thin copper starting sheets in electrolytic cells to produce cathodes (see Fig. 12); (2) the electrolyte is treated by electrolyte purification to control the concentrations of copper and impurities (eg, stripping copper from solution in electrowinning cells); (3) the anode slime, ie, the residual material collected at the bottom of the electrolytic cells during refining, is processed for recovery of precious metals, selenium, and tellurium; and (4) the refined copper cathodes are melted down and cast into commercial shapes; in addition, anodes may be cast at the refinery.

Electrodeposition. Electrodeposition, the most important of the unit processes in electrorefining, is performed in lead or plastic-lined electrolytic cells of the type shown in Figure 13. A refinery with an annual production of 175,000 t might have as many as 1250 of these cells in its tankhouse. Nowadays, the cells are multiply connected with anodes and cathodes placed alternately and connected in parallel. Each cell is a separate unit and connected to adjacent cells by a bus bar.

Table 5.   Tankhouse Parameters of Six Refineries

Parameters	Copper Cliff Inco, Canada, 1928–1965[a]	James Bridge, U.K., 1967–1973[b]	Mulfulira, Zambia, 1947–1965[c]	Southwire, U.S.A. 1971[d]	Utah Copper, Kennecott, U.S.A., 1951	Tamano, Japan, 1972[e]
capacity, t/yr	165,500	56,000	180,000	60,000	180,000	93,400
source of copper	primary	secondary	primary	secondary	primary	primary
cell voltage, V	0.2			0.27	0.21	0.39
current density, Amps/m^2	172	350[f]	204	231	204	340[f]
current efficiency, %	94	85–90	91.5	96	88	95.8
anodes, weight, kg	363	331	308	363	317	350
cathodes, weight, kg	118	141	129	132	136	151
electrolyte	end-to-end	end-to-end	end-to-end	end-to-end	end-to-end	side-to-side

[a] Ref. 47.
[b] Ref. 48.
[c] Ref. 49.
[d] Ref. 50.
[e] Ref. 51.
[f] Periodic reverse current.

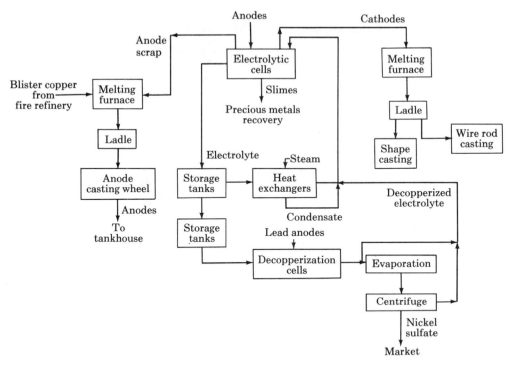

Figure 12. Flow diagram of electrorefinery.

The tankhouse is divided into the commercial and the stripper sections. In the latter, one-day deposits are prepared by electrorefining anode copper onto oiled copper or titanium blanks. Then these copper sheets are stripped from the blanks and fabricated into starter sheets for the commercial sections as starting cathodes. After 9–15 d, depending upon the specific tankhouse, full-term cathodes are pulled and washed and either sent to the casting department or sold directly.

The series system, which has been used in the past, is an alternative to the multiple system. In this system only anodes were charged and a potential was maintained between the ends of each cell so that copper dissolved from one anode was plated on the adjacent anode. After a sufficient period of time, all of the original anode copper was replaced by a cathodic deposit and the impurities were either in the form of anode slime or in solution. The series system demanded highly uniform anodes, a requirement that was difficult to meet with horizontal equipment. With the advent of continuous casting equipment of the Hazelett type, which can yield very smooth anodes of uniform thickness, the series system is reconsidered.

Composition, temperature, and flow rate of the electrolyte are of great importance to the quality of the cathode deposit and changes in any one of these parameters can have a serious effect. Storage and circulation of electrolyte are also of major importance. The total volume of electrolyte in a modern tankhouse is typically 6000 m³ (1.6 × 10⁶ gal) for a production level of ca 500 t Cu/d.

The copper contained in the electrolyte, anodes, and cathodes, and the normal circulating scrap load is equal to about 10% of the annual copper production of a typical

Figure 13.   Electrorefinery tankhouse. Courtesy of Kennecott Copper Corporation, D. Green.

electrorefinery. Some refineries (44) have changed from the traditional 25–30 d anode cycle to a 9–14 d cycle using smaller anodes to reduce the copper inventory in spite of the higher resulting scrap load.

Current densities and cell voltages have been increased in some refineries, but the majority of refineries are still operating at 175–230 A/m^2, copper concentrations of 30–50 g/L, electrolyte temperatures of 55–65°C, and circulation rates of 10–20 L/min to obtain good quality cathodes.

In electrorefining, copper is transferred according to the following reactions:

$$\text{Anode: Cu} \rightarrow \text{Cu}^{2+} + 2\,e^- \tag{30}$$

$$\text{Cathode: Cu}^{2+} + 2\,e^- \rightarrow \text{Cu} \tag{31}$$

The dendritic nature of the cathode deposits results in the occlusion of small but significant quantities of electrolyte and insoluble particulate matter; therefore, the dissolved impurities and suspended matter in the electrolyte are of great importance in determining the final impurity content of copper cast from cathodes. The quality of the cathode deposit is highly influenced by the use of organic additives which modify the crystalline deposit by influencing nucleation through surface adsorption or complexation. The purpose of the additive is to obtain a cathode deposit that (*1*) is free from deep striations or fissures that would entrap electrolyte impurities, (*2*) does not develop nodular growths that might cause short circuits as well as catch and entrap slimes, and (*3*) is not so hard as to preclude straightening bent sheets. Glue is most frequently used, and can be modified or extended by lignin derivatives (eg, Orzan or Goulac), sulfonated oils (Avitone), casein, and thiourea, as well as other proprietary reagents.

The total impurity content of anodes used in electrorefining is usually less than 1%, of which oxygen is the highest, ranging from ca 0.1–0.25%. This oxygen gives copper(I) oxide which then reacts with the acid of the electrolyte:

$$Cu_2O + 2\,H^+ \rightarrow Cu^{2+} + Cu + H_2O \tag{32}$$

The precipitated copper from this reaction is an important constituent of the slime that collects at the bottom of the electrolytic cells. Finely divided copper in the slime fraction can also result from sluffing of copper from the faces of the anodes. When a large amount of air is entrained in the electrolyte, a portion of the precipitated copper may be oxidized and dissolved. Chemical dissolution of copper depletes the sulfuric acid in the electrolyte and this must be replenished with fresh acid or recycled de-copperized electrolyte.

The accumulation of copper as well as of impurities such as nickel, arsenic, antimony, and bismuth is controlled by periodic bleed-off and treatment in the electrolyte purification section.

Although some changes occur in the melting furnace, cathode impurities are usually reflected directly in the final quality of electrorefined copper. It is commonly accepted that annealability of copper is unfavorably affected by tellurium, selenium, bismuth, antimony, and arsenic (in decreasing order of adverse effect). Silver in cathodes represents a nonrecoverable loss of silver to the refinery.

Rough deposits entrap electrolyte, resulting in high cathode impurity. If the copper content is maintained at the normal level of 40–50 g/L, arsenic, antimony, or bismuth do not codeposit. Under certain conditions, however, they form precipitates of variable compositions that can adhere to the cathode deposits at the solution line in association with calcium sulfate. Formation of these precipitates can be controlled partially by maintaining an appropriate ratio of arsenic to antimony and bismuth (52).

Cathode contamination with selenium and silver is due to one or more of the following factors: (1) precipitation of silver selenide from tankhouse electrolyte and the contamination of cathodes by particulate inclusions; (2) electrolytic codeposition of silver, which has been shown to be consistent with the limiting galvanic deposition rate; and (3) complexing of selenium with addition agents and its subsequent deposition at the cathode, particularly when there is insufficient silver in the anodes to precipitate selenium as Ag_2Se.

Electrolyte Purification. In electrolyte purification, copper is removed by electrowinning in cells that are similar to normal electrorefining cells, but with lead anodes in place of copper anodes. The electrolyte is cascaded through cells similar to those used for electrorefining. As the copper is depleted, the quality of the copper deposit is degraded and the impurity levels of the resulting liberator cathodes increase. The impurity level is low provided the copper concentration of the electrolyte is 12–18 g/L or higher. Refineries often charge the highest quality liberator cathodes to fire-refining furnaces or blend them with electrorefined cathodes as charges for arc or shaft furnaces. As copper is removed from the electrolyte by deposition on the cathodes, free acid is generated at the anode:

$$H_2O \rightarrow 2\,H^+ + \tfrac{1}{2}\,O_2 + 2\,e^- \tag{33}$$

The final cell product contains 250–300 g/L H_2SO_4. In the last stages of electrolyte purification, antimony and bismuth precipitate resulting in heavily contaminated cathodes that are recycled through the smelter. Arsenic reacts with hydrogen evolved at the cathodes at these later stages to form arsine, and hoods must be provided to collect the toxic gas.

In virtually all refineries, nickel is the major impurity in the electrolyte (up to 20 g/L). The nickel remains in the electrolyte as the copper is stripped out in the purification section and is recovered from the resulting acid solution by precipitation as the sulfate in evaporators. In some refineries dialyzers have been utilized to provide a partial separation of sulfuric acid from nickel and other impurities. The necessity for maintaining closed-circuit conditions in processing plants has increased efforts to recover by-products from electrolyte purification and to purify the acid to allow its use for makeup in the electrolyte.

By-Product Recovery.   The slime contains gold, silver, platinum, palladium, selenium, and tellurium. A chloride content of about 0.02 g/L modifies cathode deposits and precipitates any silver dissolving at the anode. The sulfur, selenium, and tellurium in the slimes combine with copper and silver to give precipitates (53).

As mentioned above, some arsenic, antimony, and bismuth can enter the slime

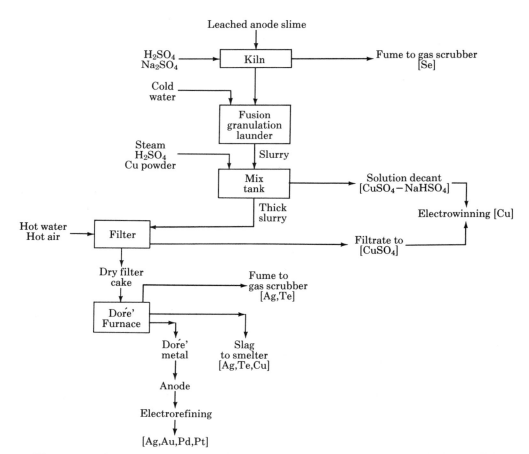

Figure 14.   Flow diagram of anode slime treatment. Courtesy of Kennecott Copper Corporation.

Table 6.   Doré Metal Analysis

Metal	Wt %
gold	8–9
silver	86–92
copper	0.5–1.0
palladium	0.16–0.18
platinum	0.005–0.009
lead	0.02
tellurium	0.003
selenium	0.00002

depending upon their concentrations in the electrolyte. Other elements that may precipitate in the electrolytic cells are lead and tin which form lead sulfate and $Sn(OH)_2SO_4$.

The mud or slime is collected from the bottom of the electrolytic cells and pumped to the silver refinery where it is processed for recovery of copper, precious metals, selenium, and in many cases tellurium. The anode slime contains 2–20% selenium as copper and silver selenides, whereas gold exists as the metal and in combination with tellurium. With direct smelting, excessive matte and slag formation causes heavy recirculation of precious metals, particularly silver, and poor recovery. In general the slime components are separated in silver refineries by employing hydrometallurgical methods prior to smelting.

Figure 14 illustrates the separation techniques used by the Utah Copper Div. of Kennecott Copper Corporation (54). Slime is water-leached, charged with sulfuric acid and sodium sulfate into a rotary kiln, and heated in the presence of air to 650°C. In the kiln, the following reactions occur:

$$Cu + 2\,H_2SO_4 \rightarrow CuSO_4 + SO_2 + 2\,H_2O \tag{34}$$

$$Ag_2Se + 2\,H_2SO_4 + O_2 \rightarrow Ag_2SO_4 + SeO_2 + SO_2 + 2\,H_2O \tag{35}$$

$$CuSe + 2\,H_2SO_4 + O_2 \rightarrow CuSO_4 + SeO_2 + SO_2 + 2\,H_2O \tag{36}$$

The fumes from the kiln are passed through a train of water-spray scrubbers and an electrostatic precipitator. In the scrubber, selenium dioxide reacts with sulfur dioxide [eq. (38)] to produce elemental selenium which is purified to provide a commercial product.

$$SeO_2 + 2\,SO_2 + 2\,H_2O \rightarrow Se + 4\,H^+ + 2\,SO_4^{2-} \tag{37}$$

The kiln product is leached and treated with elemental copper to precipitate dissolved silver. The settled solids are filtered, washed, dried, and charged to the Doré furnace where they are fire-refined to eliminate the arsenates, selenates, antimonates, tellurates, and residual copper.

When the copper content in the Doré metal has been reduced to less than 1% by fire refining, the metal is cast into anodes for electrolytic separation of silver. Typical analysis of Doré metal is presented in Table 6. The mud or slime from the silver separation is processed to form impure gold anodes. These anodes are then electrolyzed to yield purified gold and to separate platinum and palladium for subsequent recovery.

Casting. Casting operations in electrorefineries have changed considerably in the past few years. Copper formerly melted in reverberatory, arc, and induction furnaces is now being melted in shaft furnaces. Copper cathodes charged into the top of a shaft furnace pass through a gas-fired melting zone, and the molten copper collects in the bottom of the furnace. Shaft furnaces have the advantages of producing molten copper suitable for casting directly into rods for rolling and drawing into electrical wire, short start-up and shut-down times, and energy conservation compared with other melting furnaces. However, they are dependent on natural gas or propane for fuel.

In the past, wirebar, cake, and billets were cast in horizontal casting wheels which are being replaced by continuous casting (see Fig. 15).

Copper rod can be produced by three different processes without going through the intermediate wirebar stage. In the dip-forming process, developed at the General Electric Research Laboratory (55), copper is solidified onto a copper rod passed through a molten bath of copper. The belt-and-wheel process, based upon a continuous casting concept, was developed jointly by Western Electric Company and Southwire Company (56). The Contirod system uses a narrow Hazelett continuous casting machine having double belts with side dams rather than a belt and wheel, to cast a continuous bar which is then rolled into rod.

The rod resulting from the dip-forming process is substantially oxygen free because the oxygen content in the molten bath must be controlled at less than 20 ppm.

Recent Developments. Electrolytic refining requires a large capital investment, and labor costs per kilogram of copper produced are high. Most refineries have traditionally operated at current densities of about 240 A/m². Thus approximately 40

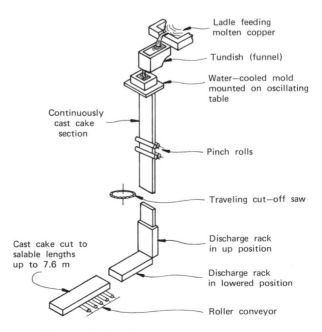

Figure 15. Continuous casting of copper cakes.

m^2 (430 ft^2) of tankhouse area is required per daily ton of copper produced. The use of higher current densities reduces capital requirements but may impair deposition efficiency and product quality.

Current density can be increased without impairing the quality of the copper by polishing the cathode surface by brief periodic current reversals (PCR). Reversed current electrolysis was first developed for electroplating and was tested in 1952 for copper refining with good results; however, no suitable electrical equipment for current reversal was available. The thyrister-controlled silicon rectifier, introduced in the 1960s, provided a means for industrial application of so-called periodic current reversal. Its first large-scale development was at the Bulgarian copper refinery in Peridot. Many copper refineries are now adopting reverse-current processes (48,51).

Current densities of 300–350 A/m^2 can be used for copper electrorefining by the PCR technique without decrease in the purity of the cathodes but at the expense of a noticeable though usually acceptable surface roughness. The technique efficiently expands the capacity of existing plants and provides flexibility of operation for new installations. It must be recognized, however, that doubling the current density essentially doubles the power cost and increases costs for power rectifiers and electrical distribution systems. Reversing the current also increases the energy consumption. This increased cost for energy must be balanced against savings in capital investment and possibly labor costs.

Other approaches to increase current density without impairing cathode quality include air sparging, high-velocity forced circulation of electrolyte, and the use of an abrasive belt or abrasive slurry for scrubbing the surface of the cathode. None of these approaches has as yet been developed to the point of economic feasibility on a commercial scale.

However, modernizations at the Onahama Smelting and Refining Company, Ltd., show more promise for cost savings. These include major reduction in labor requirement by abolition of inspection work, increased production per unit tankhouse area, and complete mechanization, resulting in a reduction in manpower from 0.91 h/t cathode to 0.17 h/t cathode. In addition, by going to a 9-day anode–cathode cycle using continuous cast anodes, the locked-up copper inventory is decreased by 60% (44).

Hydrometallurgical Processes. In hydrometallurgical processes, metal values are recovered in aqueous solution by chemical treatment of waste, ore, or concentrates (57–58). The metal and by-products are then recovered from solution by chemical or electrolytic processes. Leaching of copper ores in place by rain or natural streams and the recovery of copper from the runoff mine water as impure cement copper have been practiced for a long time. Until recently, most hydrometallurgical treatments were applied to ores or overburden in which the copper was present as the oxide, mixed oxide–sulfide, or native copper. Hydrometallurgy is now being studied extensively for treatment of sulfide concentrates because it offers an alternative to smelting which presents problems associated with sulfur dioxide recovery and environmental controls.

Hydrometallurgical processes can be categorized as follows: (1) acid extraction of copper from oxide ore; (2) oxidation and solution of sulfides in waste rock from mining, concentrator tailings, or in situ ore bodies; (3) dissolution of copper in concentrates to avoid conventional smelting; and (4) extraction of copper from deep-sea manganese nodules.

In recent years, in order to avoid air pollution problems posed by conventional smelting operations, increased emphasis has been placed on leaching techniques. These include the Anaconda oxygen–ammonia leaching process, the Lake Shore roast–leach–electrowin process, and ferric chloride leaching processes for the treatment of copper sulfides. The Duval ferric chloride (Clear process) plant is now producing over 25,000 t Cu/yr and is scheduled for expansion (59).

Hydrometallurgical recovery of copper from manganese sea nodules has been studied extensively in the past few years (11), as well as combined pyrometallurgical–hydrometallurgical processes (12). The timetable for actual commercial development of these processes depends upon political factors relating to recovery of manganese nodules from international waters.

Since 1965 increased application of leaching technology has had an important impact on copper production. In 1965 leaching yielded over 147,000 t or about 12% of United States annual new copper production (60). This increased in 1976 to about 240,000 t of copper or 17% of the annual new copper production (61).

Dump Leaching. As much as 10% of the United States copper production in 1977 was obtained by leaching waste rock and overburden at major open-pit copper mines. Production of large quantities of copper in solution has required the development of new techniques for recovering copper from solutions. The nature of copper porphyry ore bodies is such that no distinct boundary lies between barren country rock and the ore body itself. Consequently, in open-pit mining of porphyry ores, considerable quantities of copper are transported to waste dumps with the large quantities of overburden or waste rock which must be removed to reach ore-grade material suitable for treatment in flotation circuits. In 1970 ca 700,000 t/d of mine waste was accumulating in areas surrounding the open-pit copper mines of the Western United States (62). Since mine waste has been accumulating for many years, large quantities of low-grade copper material are in existing dumps.

Although mine waste has been leached for the recovery of copper for many years, systematic attempts to maximize copper recoveries from waste dumps did not commence until the early 1960s. For many years hydrometallurgical recovery of copper consisted simply of allowing normal underground water flow from mines, or water that had percolated through dumps, to pass over iron scrap in horizontal launders or troughs. Copper precipitated and the iron passed into solution. Recirculating the solution from which the copper had been removed was not systematic. It was then recognized that waste dumps represented a potential source of millions of tons of copper, and a number of major copper producers began to study methods to increase the copper recovery from waste dumps (63).

The following factors are important in dump leaching: (*1*) the role of bacteria; (*2*) the application of acid to prevent or delay precipitation of hydrated ferric sulfate; (*3*) oxidation to remove excess iron from mine water in settling pools, as shown in equations 38 and 39; (*4*) optimization of dump configuration for good solution distribution; and (*5*) availability of oxygen.

$$2\,Fe^{2+} + \tfrac{1}{2}\,O_2 + 2\,H^+ \rightarrow 2\,Fe^{3+} + H_2O \tag{38}$$

$$3\,Fe^{3+} + (n + 3)\,H_2O + 3\,SO_4{}^{2-} \xrightarrow{\text{pH 3.5}} Fe_2(SO_4)_3 \cdot Fe(OH)_3 \cdot n\,H_2O + 3\,H^+ \tag{39}$$

Many waste rock or overburden disposal systems result in compacted dumps with uncontrolled distribution of fines. In such dumps, solution distribution is poor, and there is little oxygen for reaction with the sulfides. Methods for managing these existing dumps to maximize copper recovery are being pursued actively by major copper producers.

Vat Leaching, Heap Leaching, and Agitated Leaching. Even more significant than dump leaching are systems utilizing heap or vat leaching to extract copper from oxidized ore. Major leaching installations in the Western United States are owned by Inspiration Consolidated, Kennecott, and Anamax in Arizona (64–65), and Anaconda at Weed Heights, Nevada (60). The Anamax plant is unique in that it treats 10,000 t/d of high-lime oxide ore and incorporates agitated acid leaching (58).

Vat leaching usually extracts copper from oxide or mixed oxide–sulfide ores containing more than 0.5% acid soluble copper. It is preferred to heap leaching if the ore is not porous and crushing is necessary to permit adequate contact between the leach solution and the copper minerals. Advantages of vat leaching compared with heap leaching include rapid copper recovery, with a higher extraction, reduced solution losses, and higher copper content in the effluent solution. Disadvantages are higher capital and operating costs, and shorter leach cycle times.

The Blue Bird Mine of Ranchers Exploration and Development Corporation is an important example of heap-leaching (66). The copper content of the pregnant liquor at the Blue Bird Mine is about 2 g/L which is a suitable feed to a liquid ion-exchange (solvent extraction) plant (see later). A smaller heap leaching operation was started in 1975 at Johnson Camp, Arizona (67). This plant is of particular interest because it uses ore heaps specially designed to ensure maximum copper extraction and because of the low capital cost ($1150 per annual t of copper produced).

The main differences between vat or heap leaching and dump leaching are: (1) in vat and heap leaching all of the material is leached instead of milling a high-grade component and leaching the waste; and (2) vat and heap leaching are usually performed with oxidized ores whereas dump leaching is applied to both oxide and sulfide mineralization. Vat and heap leaching are not applicable to ores containing large amounts of carbonate or other acid soluble, noncopper bearing minerals.

In-Place Leaching. Although some workers have defined *in-situ* or solution mining as the in-place extraction of metals from ores located within the confines of a mine, or in dumps, heaps, slag piles, or tailings ponds, in-place leaching is discussed here as the leaching of ore without its removal from the ground (68–70). Figure 16 illustrates three solution mining techniques according to the depth of deposit (68). A commercial application of solution mining is the in-place leaching of copper minerals in underground mines where block caving methods have been used, as for example at the Miami mine of Miami Copper Company in Arizona (106).

Cementation. Cementation is the precipitation of copper from copper leach solutions by replacement with iron. It is the most commonly used method of recovering the copper from leach solutions. The type of iron used is important, and the most widely used material at the present time is detinned light-gage shredded scrap iron. This operation can be performed, eg, by the scrap iron cone (Kennecott Copper Corporation) or a vibrating cementation mill that promises high copper precipitation efficiency with reduced iron consumption (71).

The fast copper precipitation rates obtainable with sponge or particulate iron (72) promise economic and processing advantages if particulate iron becomes cost-competitive with scrap iron. Another precipitant of potential importance is shredded automobile scrap using drum-type precipitators (73).

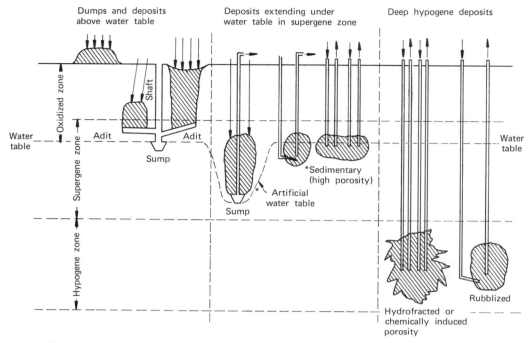

Figure 16.   Three types of solution mining; *-rubblized (explosives or mining), hydrofracted, or chemically induced porosity. Courtesy of the Society of Mining Engineers (68).

Solvent Extraction.   An alternative to cementation is the solvent extraction (liquid ion-exchange) technique (67–68,74) prior to copper recovery by electrowinning or other reduction processes. A variety of extractants are available (75–77), including Kelex 100 (Ashland Chemical Co.), SME 529 (Shell Chemical), Acorga (ICI, Ltd.), and LIX 64 (General Mills). The last is an α-hydroxyoxime designed specifically for the extraction of copper from aqueous solution (75). The general formula for α-hydroxyoxime extractants is $RR''C(OH)C(R'){=}NOH$ where R, R', and R'' may be any of a variety of aliphatic and alkylaryl radicals (R'' may also be hydrogen). The liquid ion-exchange resin is dissolved in the diluent, a hydrocarbon solvent. The extractants and their copper complexes are insoluble in water and are characterized by having a solubility of at least 2 wt % in the hydrocarbon solvent (generally kerosene) that constitutes the organic phase.

All commercial copper extractants selectively extract copper from weakly acidic aqueous leach solutions by the following general reaction:

$$(Cu^{2+} + SO_4{}^{2-}) \text{ aq} + (2\,RH)\text{ org} \rightarrow (R_2Cu)\text{ org} + (2\,H^+ + SO_4{}^{2-})\text{ aq} \qquad (40)$$

In the extraction reaction 2 protons are exchanged for the copper in the leach solution. Contacting the copper-bearing organic solution with a strong acid aqueous solution reverses equation (40) to allow stripping of the copper. Extractants have also been devised that function in basic ammonia systems (78).

A flow sheet of solvent extraction–electrowinning (SX–EW) is shown in Figure 17. The organic reagent transfers the copper from dilute impure leach solutions to concentrated higher-purity solutions that are appropriate feeds for electrowinning

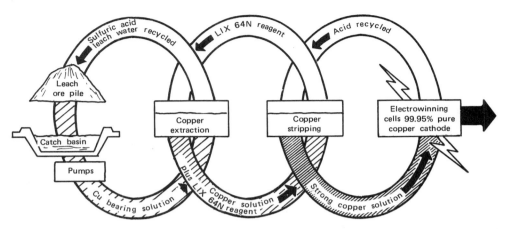

Figure 17. Conceptual flow sheet of solvent extraction–electrowinning. Courtesy of General Mills Chemical Incorporated.

plants. The SX–EW plant has three basic functions: extracting, stripping (both in mixer settlers), and electrowinning (in the electrowinning tankhouse).

The feasibility of using an extractant on a commercial scale depends on loading capacity, extraction rate, pH range, and the cost of the reagent ($1.00–1.50/kg) and the diluent. Loss of the extractant should be avoided because of its high cost. Organic losses to the aqueous phase are furthermore undesirable because of the deleterious effect on cathode deposits.

In summary, an organic reagent system for solvent extraction of copper should exhibit the following properties: (*1*) extremely low solubility in aqueous solutions; (*2*) high solubility in diluents such as kerosene; (*3*) low rates of biological or chemical degradation; (*4*) high copper-loading capacity; (*5*) rapid rate of copper extraction; (*6*) low cost; (*7*) good phase-separation characteristics; (*8*) ability to extract copper in the 1–7 pH range; (*9*) high selectivity; (*10*) low toxicity; and (*11*) low volatility at the operating temperature.

The largest solvent extraction–electrowinning facility is at Nchanga copper mines in Zambia (79) which has the capacity to treat over 60 m³/min (16,000 gal/min) of aqueous feed solution and produces approximately 90,000 t of cathode copper per year (80).

Electrowinning. *Applications.* Vat leaching often yields copper solutions with concentrations sufficiently high for direct electrowinning. However, high concentrations of cations other than copper and low copper concentrations make it more difficult to obtain high-purity electrolytic copper by direct electrolysis of leach solutions than by electrolysis of purified solutions obtained from solvent extraction.

Increasingly more copper is being produced by electrowinning because of the technical advances such as solvent extraction processes (81) and because of the increasingly stringent air pollution restraints on the more conventional concentrating, smelting, electrorefining processes. At present, copper obtained by electrowinning is not pure enough for wire drawing. However, certain brands obtained by SX–EW are traded as cathode quality.

Copper obtained by electrowinning tends to have a high lead content, partly due to the use of lead alloy anodes. Presence of cobalt in the electrolyte or substituting calcium for antimony in the lead anodes inhibits anode corrosion (80). The levels of other impurities such as arsenic, bismuth, selenium, tellurium, and silver are usually lower in copper obtained by electrowinning from solvent extraction solution than in copper from conventional electrorefining processes.

With continued improvement it is probable that copper from SX–EW will be suitable for most commercial uses (81).

Although the production of cathode-grade copper at the mine site is a strong incentive for applying SX–EW, the fundamental benefit of SX–EW is the substantially lower operating cost (in some cases 5¢/kg Cu). However, this advantage may be off-set by the significantly higher capital costs for SX–EW facilities.

Chemistry. In electrowinning, the cathode reaction is the same as for electrorefining:

$$Cu^{2+} + 2\,e^- \rightarrow Cu \tag{31}$$

However, because of the use of insoluble anodes, oxygen is released at the anode:

$$H_2O \rightarrow 2\,H^+ + \tfrac{1}{2}\,O_2 + 2\,e^- \tag{41}$$

The net reaction is:

$$Cu^{2+} + H_2O \rightarrow 2\,H^+ + \tfrac{1}{2}\,O_2 + Cu \tag{42}$$

In contrast to electrorefining, there is a minimum cell voltage of ca 1.67 V below which there is no appreciable current flow. Hence, the energy yield is only ca 0.3 kg of copper per kW·h as contrasted to about 3 kg/kW·h for electrorefining.

Iron, a common contaminant of electrowinning solutions, lowers the current efficiency by dissolving copper:

$$Cu + Fe_2(SO_4)_3 \rightarrow 2\,FeSO_4 + CuSO_4 \tag{43}$$

The ease with which ferrous ion can be oxidized to ferric ion in the electrowinning cell furthers this reaction. Attack on the copper is most apparent at the solution line where it results in corrosion of the loops supporting the cathodes, leading to dropped cathodes.

Secondary Recovery. Metal returning from the store of metal in use is referred to as old scrap in contrast to scrap generated within the copper fabrication process which is called new scrap. Scrap is a significant part of the United States copper supply (see Recycling). From 1965 through 1976 the percentage of the United States copper supply derived from old scrap was 14–19% of the total copper consumed. About half of the old scrap is used for producing refined copper; most of the remainder is used in the production of brass and bronze ingots. About two thirds of the new scrap is consumed by brass mills with most of the remainder used in the production of refined copper. Some estimates suggest that as much as 60% of the copper produced is ultimately recycled for reuse. Old scrap combined with new scrap from fabricating plants accounts for almost half the metallic input to domestic copper furnaces.

Effective recycling of copper scrap is an important factor in energy conservation. Processing scrap copper uses 4–40% of the energy required for primary copper production, depending on the purity of the scrap.

Grading of scrap follows standard definitions established by the National Association of Recycling Industries (82). The classifications are based upon the original source, ie, wire, turnings, or miscellaneous material, and the presence of contaminants.

Effluents from Production. *Gaseous.* Sulfur is eliminated when copper and iron sulfide minerals are oxidized, an essential step in copper production. Thus production of sulfur dioxide and trioxide gases is an inherent part of the commercial copper process. Although controversy continues as to the magnitude of the health hazard posed by sulfur oxides, stringent limitations have been placed on such emissions by many industralized countries (83). In the United States, the EPA sets rules for plant emissions whereas OSHA is concerned with the sulfur dioxide concentrations in working areas. The new source standards of the EPA specify that for plant expansions or new plants 95% of the sulfur in the plant feed must be captured. In locations where sulfur dioxide background levels are higher than specified, new smelters cannot be built. After all existing smelters in an area come into compliance, a new smelter can be built, provided it uses the best available technology and the other smelters reduce emissions by an amount equal to the emissions from the new smelter.

Formation of SO_2 in the smelting operation can be controlled by replacement of reverberatory furnaces with equipment that allows the formation of relatively high strength gases that can be treated to produce sulfuric acid in acid plants (30). Alternatively, the formation of SO_2 can be prevented by radical changes in copper processing methods employing hydrometallurgical techniques for treatment of copper sulfide ores or concentrates.

Solid and Liquid. Copper mining operations were always faced with major solid waste disposal problems caused by large tonnages of waste rock or overburden with open-pit mines, tailings from the concentration step, and slag and dust piles associated with smelters. Storage and disposal are well organized with a minimum of government regulations. In the United States, the Solid Waste Disposal Act, amended in 1976 by the Resources Conservation and Recovery Act, requires solid and hazardous wastes to be managed in accordance with regulations to be adopted by the EPA. Since specific methods for defining hazardous wastes are still being established, the full impact of this government regulation is not yet known.

Present and impending United States government regulations on water discharge require the installation of water treatment plants for smelter effluents carrying significant levels of heavy metals as well as discharges from the concentrator tailings ponds. It is anticipated that by 1985 all copper concentrators, smelters, and refineries in the United States will be practicing maximum water recycle (84). Such action will require new approaches to internal water treatment and possible modification of flotation reagent systems to compensate for build up in circulating loads of organic and inorganic constituents.

Since total recycle of water from dump-leaching operations is the current practice, actions should be restricted to whatever steps are necessary to keep leach solutions from entering ground-water systems. Uncontrolled runoff from mining operations requires impoundment with appropriate treatment before release into surface systems.

Energy. In 1977 energy represented ca 10% of the cost of copper production; yet in the total picture, the copper industry was not a major consumer of energy, using only ca 0.5% of the total energy used of the United States industrial sector.

Production of 1 t of copper consumes ca 130 GJ (123×10^6 Btu) distributed among mining (20%), concentrating (40%), smelting (30%), and refining (10%) (85).

Surface and underground mining differ in their energy requirements. The energy needed for open-pit mines is determined by the stripping ratios, mine profiles, and the distances between the mine, concentrator, and waste dumps. In many mines about half on the energy consumed is used for transportation.

Nearly all of the energy used in the concentrators is for electric motors driving ball mills, crushers, pumps, and the agitators of flotation cells. About half of the energy is used for grinding the ore to the proper size.

Smelting. *Reverberatory Furnaces.*   The fuel supplied to the furnace is in the range of 5–6 GJ/t ($4.7–5.7 \times 10^6$ Btu/t) concentrate. The heat balance for a reverberatory furnace based on smelting 700 t of concentrate per day is shown in Table 7.

The heat balance shows that steam produced in the waste heat boiler is equal to ca 60% of the energy supplied by the fuel. The additional heat recovered from the exit gases in the recouperator to preheat the combustion air is equal to ca 10% of the energy from the fuel. Hence, the heat recovered from the furnace is equal to ca 70% of the heat from the fuels.

The waste heat recovered in the boilers is in the form of steam and used either to generate electricity or to produce compressed air. The latter is used in the converter tuyeres. If electricity is generated, it is in turn used to produce compressed air for the converters. In practice, the energy made available in the waste heat boilers is almost equal to the energy required for producing the converter air.

The heat balance shows that the heat loss from the furnace walls is only ca 11% of the energy supplied by the fuel, and just slightly more than the sensible heat loss with the slag. The major heat loss is in the stack gases and is equivalent to ca 30% of the energy supplied by the fuel.

Table 7.   Heat Balance for Reverberatory Furnace[a] and Its Converter, GJ/d[b]

Heat	Reverberatory furnace	Converter
input		
fuel	4200	
preheated air	500	
heat of reaction	500	2300
recycled converter slag	600	
heat from reverberatory matte		600
Total	*5800*	*2900*
output		
recovered waste heat	2400	
matte to converter	600	
slag to dump	400	<100
wall losses	500	
preheated air (recycled)	500	
to stack	1400	
off gas		1100
blister		100
slag to reverberatory furnace		600
excess heat		1000
Total	*5800*	*2900*

[a] 700 t/d concentrate; 6 GJ/t (5.7×10^6 Btu/t) of concentrate.
[b] To convert GJ to Btu, multiply by 9.488×10^5.

Converters. A typical heat balance for a converter operating with a reverberatory furnace, also presented in Table 7, shows that converting is highly exothermic. Heat losses through the converter walls are small. Heat leaving with the slag and blister copper is substantially equal to that entering with the matte. About half of the heat from the converting reaction leaves with the converter off-gas and about half is excess heat. The latter is used for melting recycled copper, such as anode scrap, or frozen matte (skulls) from ladles used to transfer matte from the reverberatory furnace to the converters. When the excess heat exceeds the cooling capacity of recycled material, blister copper is cast and returned to the converters as a heat sink.

A few smelters have waste heat boilers to recover heat from converter off-gases (86). Most smelters do not have these waste heat boilers because the cyclic operation of converters results in unsteady production of steam.

Continuous Processes. Table 7 shows that in reverberatory furnace–converters the heat of reaction in the converters is equal to about half of the energy from the fuel to the reverberatory furnace. Hence, energy may be saved by partial conversion in the smelting vessel, as in continuous smelting processes such as the Noranda and Outokumpu processes. These save fuel by using part of the excess converting heat shown in the heat output of Table 7 (87–89). Furthermore, part of the energy is recovered in the off-gases passing through waste heat boilers.

Hydrometallurgical Processes. Most of the hydrometallurgical processes require slightly more total energy than is required for smelting processes.

Solution pumping uses most of the energy required in copper production via dump or heap leaching. Hence, the energy consumed depends on the height of the dump or heap and the distance from the copper extraction plant. Usually, slightly less energy is consumed for producing copper via dump and heap leaching than for copper production via smelting.

Scrap Copper. The energy required for production from scrap copper is appreciably less than that for copper production from ores. If the scrap is of low grade and has to be smelted and refined, the energy consumption is about 45 GJ/t (43×10^6 Btu/t) of copper (90). Scrap that only requires melting and casting uses about 5 GJ/t (4.7×10^6 Btu/t) of copper.

Outlook. The energy consumption per unit of copper production is expected to increase as copper grades decline. Furthermore, it will increase as facilities are added for environmental control. Therefore, a 40% increase from 1972 to 1980 in the energy consumption per unit of copper production in the United States smelters–electrorefineries has been projected (91).

A significant reduction in energy consumption could be achieved by installing new facilities. Ore haulage distances could be reduced by siting the concentrators near the mines. They could be designed to maximize gravity flow, and the use of large flotation cells would reduce the energy requirement. The new continuous smelting processes might reduce energy requirements by 50%, but new environmental control facilities would result in an increase. Electrolytic refineries could conserve energy by replacing electric melting furnaces by Asarco shaft furnaces and continuous casting. Although energy might be conserved by installing such new facilities, it is unlikely that facilities will be replaced merely to conserve energy since energy cost is a minor component of the total cost of copper production.

Economic Aspects

The United States is largely self-sufficient with respect to copper, meeting any shortfall by imports from Canada, Chile, and Peru. Australia and the Union of Soviet Socialist Republics consume most of their production on the domestic market. Japan and Western Europe, however, import substantial quantities of copper in the form of concentrates, blister, and refined copper.

Table 8 lists the copper production in the United States and elsewhere, and United States end uses from 1965–1976 (13). A more detailed outline for 1975 is presented in Figure 18.

The United States is the world's major copper producer and in 1976 accounted for approximately 20% of the total world production. In that year United States copper production was 1,450,000 metric tons with a sales revenue of 2.3×10^9 dollars, an increase of 14% in volume and 24% in value over 1975. More than 90% of this production came from the 25 major mines. The increase represented a return to the 1974 level following a severe slump in 1975. Imports of copper declined from 30% of consumption in 1950 to 15% in 1976.

The economics of the nationalized copper industries in Chile, Peru, Zambia, and Zaire are clearly different from the economics of the private producers in other parts of the world. Their production levels have been increased while the private producers have reduced output in order to bring supply in line with demand. In 1976 the output of these countries, combined with by-product copper, represented ca 60% of the copper production of the non-Communist world.

A detailed picture of the wide range and sources of the copper used in the United States and their relative importance for the period 1958–1977 can be obtained by a review of data presented by the Copper Development Association (92).

In spite of short-term overproduction and inroads into traditional copper usage by new developments, such as fiber optics (qv) in place of copper wiring for direct transmission of signals, there are strong indications that the future demand for copper

Table 8. United States Copper Production and Production Outside of United States, and United States End-Use Pattern, Thousand Metric Tons

Parameter	1965	1969	1972	1974	1976
production					
United States	1227	1402	1511	1449	1461
outside United States	3600	4246	5126	5909	5916
Total	*4827*	*5648*	*6637*	*7358*	*7377*
United States end-use pattern					
electrical	933	1083	1136	1134	1053
construction	377	309	392	386	302
machinery	272	230	272	309	263
transportation	206	180	206	233	209
ordnance	41	156	71	38	32
other	96	103	109	111	98
Total	*1930*	*2061*	*2186*	*2211*	*1956*
United States primary demand[a]	1464	1539	1770	1772	1602

[a] Industrial demand less scrap.

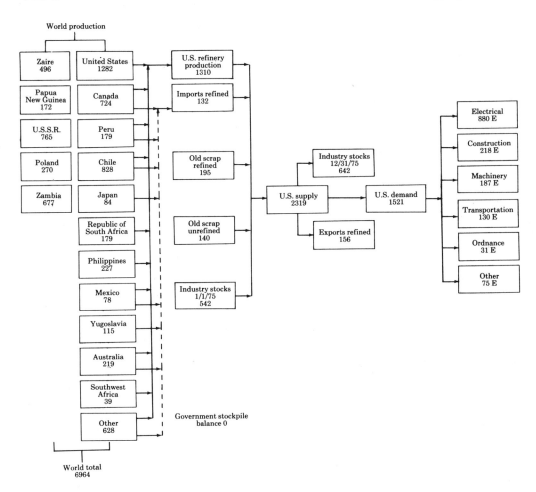

Figure 18. United States and world production of copper in 1975, including distribution of United States demand. Values in thousand metric tons; E, estimate; - - -, refined copper imports. Courtesy of the Bureau of Mines, U.S. Department of the Interior.

will exceed the production capacity. A contributing factor is that the mining industry in the United States is being subjected to ever-more stringent pollution regulations with attendant higher production costs. Estimates of 1977 production costs for United States copper companies indicated that at a time when the average selling price in the United States was held at $1.3/kg by excess supply, more than 60% of the copper produced was probably sold at a loss (93).

The United States demand for copper in the year 2000 is estimated to be (3.5–6.0) × 10⁶ t. As shown in Table 9, this represents an annual United States growth rate of 3.5% from 1975 to 2000 (13); a somewhat higher annual growth rate of 4.1% is projected outside of the United States. The annual growth rates in the use of secondary copper are projected at 5% in the United States and 4.5% outside of the United States.

Table 9.   Forecast of United States and World Demand, Thousand Metric Tons

Area	Year 1975	Year 1985	Year 2000	Projected average annual growth (1975–2000), %
United States				
primary	1,186	2,000	3,200	3.0
secondary	335	700	1,400	5.0
Total	*1,521*	*2,700*	*4,600*	*3.5*
World				
primary	6,456	10,200	17,700	3.7
secondary	1,735	3,100	6,000	4.7
Total	*8,191*	*13,300*	*23,700*	*3.9*

Specifications, Standards, and Quality Control

The ASTM designation B224-73 defines the standard classifications of copper according to method of refining and characteristics determined by method of casting or processing (94). Of late, definitions relating to methods of refining adopted by the International Standards Organization have been endorsed by the United States and are being introduced into United States domestic specifications (95). The accepted basic standard for electrolytic copper wire bars, cakes, slabs, billets, ingots, and ingot bars is ASTM B5-74. Commercial electrolytic tough-pitch (ETP) copper normally far surpasses the specifications of ASTM B5-74. A summary of the characteristics of a number of copper samples collected from worldwide sources is presented in Table 10. Although the copper contents are not specified in Table 10, it is noted that, by difference, the analyses would be significantly better than 99.90 as specified in ASTM

Table 10.   Average Impurity Levels and Physical Properties of Wirebar Copper Samples from Eight Worldwide Sources, 1976

Assay	Concentration, ppm	Standard deviation, ppm	No. of samples
antimony	3.41	0.55	14
arsenic	1.39	0.31	14
bismuth	0.36	0.14	15
iron	6.07	2.41	15
lead	4.08	3.06	14
nickel	3.41	1.54	14
oxygen	327.25	96.44	16
selenium	1.10	0.92	16
silver	11.19	2.93	16
sulfur	10.00	3.08	16
tellurium	1.16	1.16	16
tin	1.63	0.83	16
density, g/cm^3	8.48	0.15	16
conductivity, %	101.44	0.22	16
half-hardness temperature, °C	215.81	28.36	16

B5-74; and, in fact, ETP copper ranges from 99.94 to 99.96% copper. In addition, conductivity values are seen to be 101% or more in contrast to the 100% specified (100% = 0.15328 S/m).

Copper rod is usually produced in a continuous casting process rather than from wirebar; a significant proportion of the copper going into production of electrical wire is made by this process.

Purity. Electrolytic copper is one of the purest of the materials of commerce. The average copper content of ETP copper, for instance, is over 99.95% and even the highest level of impurities other than oxygen are found only to the extent of 15–30 ppm. Up to 0.05% oxygen is present in the form of copper(I) oxide. Even at these low impurity levels properties of interest to fabricators are affected in varying degree.

Electrical Conductivity. All dissolved impurities lower the conductivity (96). In ETP copper the presence of 100–500 ppm oxygen decreases the conductivity slightly because of its volume effect, but more importantly, it prevents most impurities from entering into solid solution in the metal by precipitation. Consequently their deleterious effect on conductivity is reduced. In the case of antimony and cadmium, the rate of precipitation is relatively slow but some effect is observed. Impurities such as silver, arsenic, nickel, selenium, tellurium, and sulfur do not form stable oxides and affect conductivity regardless of the oxygen content. The normal tendency of selenium and tellurium to precipitate from solution is inhibited by rapid quenching, and therefore, they have a greater effect than the other impurities discussed above.

Figure 19 contrasts the decrease in conductivity of ETP copper (**a**) with that for oxygen-free copper (**b**) as impurity contents are increased. The importance of oxygen in modifying the effect of impurities on conductivity is clearly illustrated. Phosphorus, which is often used as a deoxidizer, has a pronounced effect in lowering electrical conductivity in oxygen-free copper but little effect in the presence of excess oxygen.

Fabricability. Impurities in electrolytic copper are of such low level as to have little effect on hot- or cold-working operations. The concentration of cuprous oxide affects cold-working of copper more than do minor variations of other impurities. Even so in most applications, ETP copper with 0.02–0.05% oxygen can be cold-drawn to greater than 99% reduction in area without difficulty. Oxygen-free copper and high-purity copper (99.999+%) exhibit little or no apparent limit to cold-working. Key items of importance in determining fabricability characteristics of less pure copper are the individual impurities, their solubilities, and tendencies to form compounds.

Soluble impurities, such as silver, gold, nickel, and arsenic, do not affect hot-workability. Selenium, tellurium, sulfur, and oxygen form brittle compounds and decrease hot- and cold-workability. Sulfur contents above 0.0025% can cause problems in casting. Selenium and tellurium are sometimes added (0.5%) to improve free machining but there is a distinct loss in fabricating, especially cold-working properties. Bismuth and lead have limited solubility in copper and may separate as the temperature declines during hot-working of oxygen-free copper; an extremely brittle condition results. The practical limit for bismuth in oxygen-free copper is ca 0.002%. Because lead separates in globular form, ca 0.02% can be tolerated in oxygen-free copper. ETP coppers containing twice these concentrations of bismuth and lead can be rolled successfully. In the production of continuous cast rod for wire production, however, lead cannot be tolerated at more than about 5 ppm.

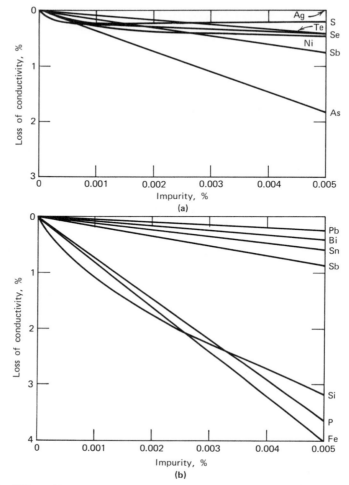

Figure 19. Effect of impurities on copper conductivity of copper: (**a**) decrease of conductivity of tough-pitch copper with content of impurities; (**b**) decrease of conductivity of oxygen-free copper with content of impurities. Courtesy of the American Chemical Society (96).

Annealability. Studies of the effect of impurities on the softening of copper have increased during the past decade. Magnet wire enamels that cure at low temperatures require copper that anneals during baking at these temperatures. The fil-coated wire must have a minimum springback after being wound on a form.

The first studies of impurity effects were made in conjunction with research on the production of spectrographically pure copper (97) and concerned iron and nickel in the 0.7–500 ppm range, and cobalt in the 20–500 ppm range. It was concluded that in oxygen-bearing copper, ie, oxygen in the ETP range (0.02–0.05%), these impurities did not change the recrystallization temperature. In subsequent studies the effects of silver, antimony, cadmium, tin, tellurium, phosphorus, arsenic, sulfur, and selenium were investigated. The results show that, in ETP copper, phosphorus has no effect on the softening temperature up to 200 ppm because of compound formation with oxygen. Arsenic (up to 5 ppm) and silver (up to 35 ppm) have no appreciable effect, and neither tin nor cadmium show any effect because they react with oxygen. Antimony, sulfur, tellurium, and selenium were found highly effective in increasing the softening temperature of copper subject, however, to heat treatment. The effect of antimony was only studied in the range of 18–600 ppm, an appreciably higher concentration than is normally encountered today in electrolytic copper.

Combinations of impurities were not studied (97). This was also the case in an investigation of the effect of phosphorus, cadmium, silver, arsenic, and tellurium on recrystallization of copper (98).

With regard to the influence of arsenic, bismuth, lead, antimony, and sulfur in the concentration range of 5–26 ppm, bismuth has the greatest unit effect (99). A decrease in the annealing temperature prior to cold deformation led to a decrease in the measured unit effectiveness, indicating that a low temperature bismuth is not in solid solution. Lead lowered the recrystallization temperature, provided the samples were annealed at 700°C or lower. This effect was explained on the basis of a precipitation reaction between lead and sulfur (100).

In a study of the springiness of copper wire, several commercial brands of copper were used (101). A strong effect of bismuth concentration in the range of 2–18 ppm and springiness was found. Selenium in the range of 1–5 ppm was also very effective.

Later the effects of various impurities on springiness was studied (102). The most deleterious impurities were found to be selenium and sulfur; silver, iron, nickel, cobalt, and oxygen had no effect. Bismuth, tellurium, antimony, and arsenic were harmful but not at the levels encountered in most electrolytic copper. The behavior of lead was erratic, sometimes appearing beneficial. The effects noted were based on the following ranges of impurities: selenium 0.2–3 ppm, sulfur 5–15 ppm, iron up to 40 ppm, silver 5–15 ppm, lead up to 20 ppm, nickel up to 45 ppm, and oxygen 120–1000 ppm.

The effects of trace quantities of silver, iron, sulfur, lead, selenium, and phosphorus on the conductivity and recrystallization behavior of copper were also investigated (103). To this purpose the ternary alloys Cu–Se–Ag, Cu–S–Ag, Cu–Se–Pb, Cu–Fe–S, Cu–Pb–S, and Cu–Fe–P were prepared by powder–metallurgical methods. A minimum recrystallization temperature occurs in the ternary systems Cu–Ag–Se and Cu–Pb–Se when Ag–Se or Pb–Se are present in stoichiometric ratios for the formation of selenides or double selenides.

Measurements of the annealing behavior of copper rod and wire made from wirebars containing various levels and combinations of tellurium, selenium, antimony, bismuth, lead, and silver showed tellurium to have the greatest unit effect. Selenium, antimony, and bismuth also raise the softening temperature. Silver decreases the recrystallization temperature (104). The effect of lead is dependent on the thermal history of the copper, decreasing the softening temperature under certain annealing conditions. Regression equations relating impurity concentrations and annealing characteristics were developed from the data.

Effect of Thermal History. Many of the impurities present in commercial copper are in concentrations above of the solid solubility at low temperatures, eg, 300°C. Other impurities oxidize in oxygen-bearing copper to form stable oxides at lower temperature. Hence, since the recrystallization kinetics are influenced primarily by solute atoms in the crystal lattice, the recrystallization temperature is extremely dependent on the thermal treatment prior to cold deformation.

Heating at temperatures below 600°C has the effect of precipitating certain impurities as metallic particles and others, as oxides whereas heating to about 800°C would ensure the solution of most of the impurities in copper with the exception of iron, cobalt, tin, phosphorus, calcium, and perhaps zinc. Therefore, when fabricating wire it is necessary to control mill conditions for rod to achieve the lowest annealing temperature. The most important parameter is the finishing temperature, ie, the temperature prior to quenching. Allowing the hot-rolled rod to cool slowly, renders many of the harmful impurities ineffectual.

Quality Control. The spectrometer is the most suitable instrument for determining most low-level residual impurities. The ASTM E414-71 is the standard method for the measurement of impurities in copper by the briquette d-c-arc technique (105). In this method, the sample in the form of chips, drillings, or powder is briquetted and excited in a d-c arc opposite a high-purity copper rod. Impurities can be measured in the ranges noted:

Impurity	Range, ppm	Impurity	Range, ppm
antimony	1–20	iron	5–25
arsenic	1–10	nickel	4–40
bismuth	0.1–5	tellurium	1–15
lead	3–50	tin	2–15

The specific resistance of the International Annealed Copper Standard copper is reported as 1.682×10^{-6} Ω at 20°C. This is defined as a conductivity of 100%. In performing conductivity tests, 0.48-cm (12-gage) wire is annealed at 500°C for 20 min, given a sulfuric acid pickle, quenched, dried, and compared against a standard wire. Most commercial electrolytic copper so measured gives conductivities over 101%. The average conductivity of the wirebar samples considered in Table 10 is 101.4% with values ranging from 101 to 101.8%.

Annealability, which is of importance in the workability of wire products, can be determined by the half-hardness temperature, residual hardness after cold-working and annealing, or spring elongation after work hardening and annealing.

The most important of these properties is the half-hardness temperature, defined as the temperature at which annealing for one hour will return the property measured (hardness, tensile strength, yield strength) to the midpoint between that for the fully worked metal and that for the fully annealed state. High-purity copper, after cold-working, reaches the half-hard stage when annealed for one hour at 140°C. Measurements on the ETP copper samples listed in Table 10 gave half-hardness temperatures of 176–265°C.

Analytical Methods

Microquantities of copper are readily identified by formation of (1) the deep blue cupramine complex ion (tetraammine–copper(II) ion) $[Cu(NH_3)_4]^{2+}$, (2) a red-brown precipitate with potassium ferrocyanide, and (3) a green precipitate with α-benzoin oxime. Substances containing copper when moistened with hydrochloric acid and heated on a platinum wire give a blue or green tinge to the flame. Numerous other reagents, chiefly organic, have been employed for the detection of microquantities of copper, usually by means of a spot test. Early methods (106) include iodometric and electrodeposition techniques; photometric and polarographic methods were used for measuring low levels of copper encountered in exploration or in organic materials.

Since 1960, new techniques, such as atomic absorption analysis, and x-ray fluo-rescence, are used for control of operations and environmental measurements. Iodo-metric techniques are still used in many industrial laboratories, particularly for measuring copper in concentrations ranging from 0.5 to 30%, but the newer techniques are more rapid and convenient. Table 11 is a compilation of methods and selected references for the analysis of the various copper-containing materials.

Health and Safety

Copper is required for all forms of aerobic life and most forms of anaerobic life. In man, the biological function of copper is related to the enzymatic action of specific essential copper proteins (132). Lack of these copper enzymes is considered a primary factor in cerebral degeneration, depigmentation, and arterial changes. Because of the abundance of copper in most human diets, chemically significant copper deficiency is extremely rare (133).

Accidental ingestion of large amounts of copper salts from food or beverages contaminated by copper released from copper vessels or pipes can cause gastrointes-tinal disturbances, and inhalation of copper fumes can cause metal fume fever. However, no chronic copper poisoning has been reported. The human metabolic system is very efficient in promoting a discriminating copper absorption. Copper is bound to albumin in blood plasma and large amounts can be stored in and eliminated through the liver. Therefore, although systemic effects, such as hemolysis, liver damage, and renal damage, have been reported after ingestion of large amounts of copper salts, recovery has usually been rapid upon treatment.

About 50% of copper in food is absorbed, usually under equilibrium conditions, and stored in liver and muscles. Excretion is mainly via the bile and only a few percent of the absorbed amount is found in urine. The excretion of copper from the human body is influenced by molybdenum. A low molybdenum concentration in the diet causes a low excretion of copper and a high intake results in a considerable increase in copper excretion (134). This copper–molybdenum relationship appears to correlate with copper deficiency symptoms in cattle. It has been suggested that, at the pH of the intestine, copper and molybdate ions react to form biologically unavailable copper molybdate (135).

Humans tolerate fairly large oral doses of copper without harmful effects, and medical uses range from a copper intrauterine device for contraceptive purposes (see Contraceptive drugs) to copper drugs in cancer therapy (132). Copper sulfate is a powerful emetic and has been used clinically as such in the treatment of intoxications. In adults, the ingestion of about 1 g of copper sulfate (ca 400 mg of copper) induces vomiting. Since systemic effects may be incurred from the absorption of copper sulfate, its use as an emetic has been decreasing during recent years.

Copper has been employed as a bactericide, molluscicide, and fungicide for a long time and is of importance in the control of schistosomiasis. In this case its addition to lake water acts as an efficient deterrent to transmittal of the disease by elimination of snails that act as host for the responsible parasite. It is commonly utilized at ca 0.1 mg/L as an algicide. In fresh water, acute toxicosis in fish is unusual if the concentration is below 0.025 mg/L (136) (see Poisons, economic; Fungicides).

The copper(II) ion appears to be toxic to marine invertebrates, and fish may be protected temporarily from copper toxicity by chelating agents such as ethylenedi-

Table 11. Selected References to Methods of Analysis for Copper in Various Materials

Sample type and conc. range	Atomic absorption	Colorimetric or photometric	Electrodeposition	Iodometric	Microprobe	Neutron activation	Electroanalytical	Spectrochemical	X-ray fluorescence
copper unrefined and refined, 95–100%			(107–108)	(108–109)					(110)
high copper alloys, <0.1–85%	(111)		(108–109)	(108–109, 112)				(113)	(109)
low copper alloys, <0.1–10%	(111)	(112)	(108–109)				(112)	(113)	(109)
copper compounds and concentrates, 15–30%			(106, 109)	(106, 109)					(110)
rocks, minerals, and ores, 0.1–5%	(114)	(106)		(106, 112)	(115)		(106, 112)	(116)	(110)
process solutions, 0.1–50 g/L		(112)		(106, 112, 117)			(118)		(110)
soils and fertilizers, 12–25 ppm	(119)	(106, 120)		(106)			(106)	(106)	
foods and beverages, 0.01–0.1 ppm	(121)	(106)					(106)		
vegetation and biological samples, 1–25 ppm	(122–123)	(106, 124)				(125)	(106)	(126)	(125)
water samples, 0.1–20 mg/L	(127–128)	(106, 124)					(128)	(106, 108)	
airborne particulates, 0.5–1000 $\mu g/m^3$	(129)							(130)	(110, 131)

864

aminetetracetic acid (EDTA) and nitrilotriacetic acid (NTA). Chelating agents could thus be used to detoxify copper as it passed through a critical section of contaminated river (136). In order to eliminate the possibility of damage to ecological systems, the EPA in the United States has placed a limit of 0.25 mg/L for copper levels in discharges from nonferrous operations to lakes and streams (137).

In the soil as little as 4 ppm of available copper provides the minimum requirement for most agricultural needs. Soils that are highly calcareous are likely to be deficient in copper as are newly reclaimed peaty soils, organic chalky soils, sandy soils with a high humus content, soils that are sandy and alkaline, and highly leached soils especially if acidic. High concentrations of nitrogen compounds can form complexes with soil copper rendering it less available (see Mineral nutrients).

The limits of toxicological tolerance for copper in man are so high that no problems arise in its use for parasite control on crops for which copper salts have been so used for more than 50 years (135).

Uses

The properties of copper and its alloys that make it a major metal of commerce may be summarized as follows: high electrical conductivity; high thermal conductivity; ease of casting, extrusion, rolling, and drawing to produce wire, tubing, and strip; low corrosion rate of copper when used for food preparation; excellent alloying characteristics; high esthetic appeal; and low toxicity to humans.

The occurrence of copper in nature in the metallic form led to its use since early times either as metallic copper, or alloyed with tin as bronze. It was used for tools, ornaments, pots for cooking, and coinage. Copper and brass, a copper–zinc alloy, continue to have appeal as ornaments.

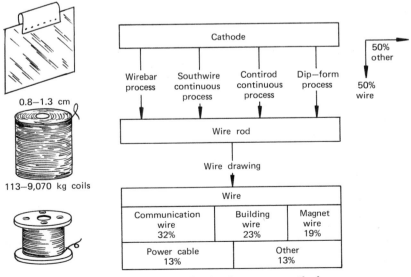

Figure 20. Wire production from copper cathodes.

The major use of copper in modern times has been as an electrical conductor, and about 50% of United States demand is for electrical uses. Copper has a very high electrical conductivity per unit volume. It can be drawn readily into wires, either single or multifilament, which can be bent readily and repeatedly without excessive work hardening. Copper wire is readily tinned, has excellent soldering characteristics, and resists corrosion at contact points. Figure 20 shows the flow of copper from cathode through to copper wire and the relative demand for the various wire types.

The resistance to salt water corrosion of admiralty brass, an alloy with 71% copper, 28% zinc, and 0.75–1.0% impurities, led to extensive use of this alloy in ships. This alloy has been largely replaced, in water applications with copper-nickel alloys which consume thousands of metric tons of copper annually. The resistance to corrosion of copper by food and the nontoxicity of copper in dilute concentrations has encouraged its use for food preparation equipment. Low corrosion rates coupled with ease of forming, bending, and of soldering resulted in extensive use of copper tubing for domestic water pipe. Copper radiators for automobiles utilize the high thermal conductivity of copper, and the ease of mechanical working and brazing.

Brass can be cast readily into intricate shapes and is used for many cast products having utilitarian or decorative applications. The ease of chrome plating brass has diversified the use of copper alloys where appearance and resistance to corrosion are major requirements.

BIBLIOGRAPHY

"Copper" in *ECT* 1st ed., Vol. 4, pp. 391–431, by J. L. Bray, Purdue University, and H. Freiser, University of Pittsburgh; "Copper" in *ECT* 2nd ed., Vol. 6, pp. 131–181, by H. Lanier, Kennecott Copper Corporation.

1. A. J. Wilson, *Eng. Min. J.* **178**(12), 68 (Dec. 1977).
2. J. Kronsbein and R. H. Ferminhac, "The Isotopes of Copper" in A. Butts, ed., *Copper—The Science and Technology of the Metal, Its Alloys and Compounds,* ACS Monograph 122, Reinhold Publishing Corp., New York, 1954, Chapt. 21.
3. A. Sutulov, *Copper Porphyries,* University of Utah Printing Services, Salt Lake City, Utah, 1974.
4. K. Rankama and T. G. Sahama, *Geochemistry,* University of Chicago Press, Chicago, Ill., 1950.
5. E. N. Pennebaker, "Copper Minerals, Ores, and Ore Deposits" in ref. 2, Chapt. 2.
6. E. S. Dana, revised by W. E. Ford, *A Textbook of Mineralogy,* 4th ed., John Wiley & Sons, Inc., New York, 1932.
7. Ref. 3, p. 28.
8. *World Mining Copper Map,* 2nd ed., Miller Freeman Publications Inc., San Francisco, Calif., 1976.
9. *Mining Mag.,* 211 (Sept. 1973).
10. Ref. 3, p. 92.
11. J. C. Agerwall and co-workers, *J. Met.* **28**(4), 24 (Apr. 1976).
12. R. Sridhar, W. E. Jones, and J. S. Warner, *J. Met.* **28**(4), 32 (Apr. 1976).
13. H. J. Schroeder, *Copper, Mineral Commodity Profiles, MCP-3,* U.S. Department of the Interior, Bureau of Mines, Pittsburgh, Pa., June 1977, p. 6.
14. R. T. Sanderson, *Chemical Periodicity,* Reinhold Publishing Corp., New York, 1960, p. 70–99.
15. J. W. Laist, "Copper Silver and Gold" in M. C. Sneed, J. L. Maynard, and R. C. Brasted, eds., *Comprehensive Inorganic Chemistry,* Vol. 2, D. Van Nostrand Company Inc., New York, 1954, p. 9.
16. M. Pourbaix, *Atlas of Electrochemical Equilibria in Aqueous Solutions,* Pergamon, New York, 1966.
17. *Metals Handbook,* American Society for Metals, Cleveland, Ohio, 1948, p. 906.
18. Ref. 14, p. 153.

19. P. T. Gilbert, "Chemical Properties and Corrosion Resistance of Copper and Copper Alloys" in ref 2, Chapt. 18.
20. R. E. Buckley, *Min. Congr. J.,* 30 (Feb. 1974).
21. S. D. Michaelson, *Min. Eng. Handbook,* Vol. 2, Society of Mining Engineers, Salt Lake City, Utah, 1973, pp. 17–89.
22. *Eng. Min. J.* **177**(6), 70 (June 1976).
23. C. O. Staples, *Denver Equipment Co., Denver, Colo., Bulletin No. N4,* B147.
24. B. J. Klee, *World Min.* 30 (Dec. 1976).
25. A. P. Langheinrich and W. M. Tuddenham, "Mining and Ore Processing" in H. K. Herglotz and L. S. Birks, eds. *X-ray Spectrometry,* Marcel Dekker, Inc., New York, 1978, p. 361.
26. J. L. Bray, *Nonferrous Production Metallurgy,* John Wiley & Sons, Inc., New York, London Chapman and Holt, Ltd., London, Eng.
27. A. Yazawa, *Can. Metall. Q.* **13**(3), (1974).
28. J. M. Toguri and co-workers, *Can. Metall. Q.* **3**(1), 197 (1964).
29. Mac Askill, *Eng. Min. J.* **174,** 82 (July 1973).
30. N. J. Themelis, *Min. Eng.* **28,** 43 (Jan. 1976).
31. R. J. Anderson in R. P. Erhlich, ed., *Copper Metallurgy—Proceedings of the Extractive Metallurgy Division Symposium on Copper Metallurgy, Denver, Col., Feb. 1970,* EMD–AIME, New York, 1970, pp. 146–172.
32. F. M. Aemone, *J. Met.* **20,** 33 (Sept. 1968).
33. K. M. Ogilvie, *paper presented at AIME Annual Meeting, Atlanta, Georgia, Mar. 1977.*
34. S. J. Salat, *Min. Mag.* **131,** 39 (July 1974).
35. T. Nagano and T. Suzuki in J. C. Yannopoulous and J. C. Agarwall, eds, *Extractive Metallurgy of Copper—International Symposium,* AIME, New York, 1976, pp. 488–507.
36. *CIM Bull.* **48,** 292 (May 1955).
37. S. Merla and co-workers, *paper presented at AIME Annual Meeting, San Francisco, Feb. 1972.*
38. *M. I. Process for Continuous Copper Smelting and Converting,* Mitsubishi Metal Corporation, Tokyo, Japan.
39. P. Rutledge, *Eng. Min. J.* **176**(12), 88 (Dec. 1975).
40. T. Nagano and T. Suzuki, "Commercial Operation of Mitsubishi Continuous Smelting and Converting Process" in J. C. Yannopoulous and J. C. Agarwall, eds., *Extractive Metallurgy of Copper—International Symposium,* Vol. 2, AIME, New York, 1976, p. 439.
41. N. J. Themelis and co-workers, *J. Met.* **24,** 24 (Apr. 1972).
42. Ref. 35, pp. 456–487.
43. J. H. Acuma, *paper presented at AIME Annual Meeting, Atlanta, Georgia, Feb. 1977.*
44. H. Ikeda and Y. Matsubara in ref. 40.
45. H. J. T. Ellingham, *J. Soc. Chem. Ind.* **63,** 125 (1944).
46. U.S. Pat. 4,010,030 (Mar. 1, 1977), R. O. French (to Kennecott Copper Corporation).
47. A. S. Gendron, R. R. Mathews, and W. C. Wilson, *CIM Bull.* **70,** 166 (Aug. 1977).
48. M. Owen and J. S. Jacobi, *J. Met.,* 10 (Apr. 1975).
49. I. S. Blair and L. R. Verney in ref. 31, pp. 275–313.
50. W. W. Brunson, and D. R. Stone, *Trans. Inst. Min. Metall. Sect. C* **85,** C150 (1976).
51. T. Kitamura and co-workers in ref. 40, pp. 525–538.
52. T. B. Braun, J. R. Rawling, and K. J. Richards in ref. 44, pp. 511–524.
53. M. Feller-Kniepmeier, and F. Pawlek, *Metall.* **1,** 36 (1963).
54. A. H. Leigh, "Precious Metal Refining Practice" in D. J. I. Evan and R. S. Shoemaker, eds., *International Symposium on Hydrometallurgy,* Chicago, Ill., 1973, AIME, New York, 1973.
55. R. P. Carreker, Jr., *J. Met.* **15,** 774 (1963).
56. K. J. Kinard, *Wire* **101,** 1 (1969).
57. R. R. Dimock, *Min. Eng.* **28,** 58 (1976).
58. C. Rampacek and J. T. Danham, *Min. Congr. J.* **62**(2), 43 (1976).
59. M. E. Wadsworth, *J. Met.* **29**(3), 8 (1977).
60. H. W. Sheffer and L. G. Evans, *U.S. Bur. Min. Inf. Cir.* **8341,** (1968).
61. R. J. Roman, H. Sheffer, and W. Stone, "Copper Leaching Practices in the Western U.S." *U.S. Bur. Min.* publication pending.
62. E. E. Malouf, E. Peters, and R. S. Shoemaker, *paper presented at Annual AIME Meeting, Denver, Col., Feb. 1970.*
63. E. E. Malouf and J. D. Prater, *Min. Congr. J.* **48**(1), 82 (1962).

64. D. L. Simpson, B. H. Ensign, and K. F. Marquardson, *paper presented at Operating Metallurgy Conference, AIME, Chicago, Ill., 1967.*

65. H. R. Moyer and R. F. Brindisi, *Min. Congr. J.* **62**,(10), 26 (1976).

66. K. L. Power in ref. 31.

67. *World Min.* **29,** 48 (Aug. 1976).

68. M. E. Wadsworth, *Min. Eng.* **29**(12), 30 (Dec. 1977).

69. I. R. Fletcher, *paper presented at Arizona Section Meeting, Milling Division, AIME, Apr. 1962.*

70. D. V. D'Andrea and S. M. Runke, "In Situ Copper Leaching Research at the Emerald Isle Mine," in A. Weiss, ed., *World Mining & Metals Technology,* AIME–MMIJ, Salt Lake City, Utah, 1976, pp. 409–419.

71. M. Esna-Ashari and co-workers, *Erzmetall* **30**(6), 262 (1977).

72. E. A. Back, *Trans. Soc. Min. Eng.* **238,** 12 (1967).

73. K. C. Dean, R. D. Groves, and S. L. May, *U.S. Bur. Min. Rep. Invest.* **7182,** (1968).

74. H. J. McGarr, *Chem. Eng.* **77,** 82 (1970).

75. D. W. Ayers and co-workers, *Trans. Soc. Min. Eng.* **235,** 191 (1966).

76. I. A. Hartlage, *paper presented at SME Fall Meeting, Salt Lake City, Utah, Sept. 1969.*

77. A. J. Van derZeeuw, *International Symposium—Copper Extraction and Refining,* AIME, New York, 1976, p. 1039.

78. G. M. Ritchey and A. W. Ashbrook, *Solvent Extraction in Process Metallurgy,* TMS–AIME, Denver, Col., 1978.

79. J. A. Holmes and co-workers in ref. 40, p. 907.

80. T. N. Andersen, D. L. Adamson, and K. J. Richards, *Met. Trans.* **5,** 1345 (1974).

81. J. L. Holman, J. F. McIlwaine, and L. A. Neumeir, *U.S. Bur. Min. Rep. Inv.* **8261,** (1978).

82. "Standard Classifications for Nonferrous Scrap Metals" *NARI Circular NF-77,* National Association of Recycling Industries, Inc., New York.

83. A. V. Colucci, *Sulfur Oxides: Current Status of Knowledge, EPRI EA-316, Final Report,* Electric Power Research Institute Palo Alto, Calif., Dec. 1976.

84. *Development Document for Interim Final Effuent Limitations Guidelines and Proposed New Source Performance Standards for the Primary Copper Smelting Subcategory and the Primary Copper Refining Subcategory of the Copper Segment of Nonferrous Metals Manufacturing, EPA-440/1-75/032-b.* EPA, Washington, D.C., Feb. 1975.

85. "Energy use Patterns in Metallurgical and Nonmetallic Mineral Processing," *Battelle, NTIS PB-245 759,* National Technical Information Service, Springfield, Va.

86. T. Nagano and co-workers, *J. Met.* **20,** 76 (July 1968).

87. D. A. Schultz, *J. Met.* **30,** 18 (Jan. 1978).

88. H. H. Kellogg and co-workers in ref. 40, pp. 373–415.

89. N. J. Themelis, *Min. Eng.* **27**(1), 42 (Jan. 1976).

90. H. H. Kellogg, *J. Met.* **28,** 29 (Dec. 1976).

91. *Fed. Reg.* **42**(111), Part II (June 9, 1977).

92. *Annual Data 1978, Copper Supply & Consumption, 1958–1977,* Copper Development Association, Inc., New York, 1978.

93. *Forbes* **121**(4), 49–50 (Feb. 20, 1978).

94. "Copper and Copper Alloys," *Annual Book of Standards,* Part 6, American Society for Testing and Materials, Philadelphia, Pa., 1976, pp. 393–397.

95. ISO Technical Reports, *ISO/TR 197/I-1976 (E), ISO/TR 197/II-1976 (E), ISO/TR 197/III-1976 (E), ISO/TR 197/IV-1976 (E),* International Organization for Standardization, Cologne, FRG, Oct. 1976.

96. J. S. Smart, Jr., "The Effect of Impurities in Copper" in ref. 2, Chapt. 19.

97. J. S. Smart, Jr., and A. A. Smith, Jr., *Trans. AIME* **166,** 144 (1946).

98. V. A. Phillips and A. Phillips, *J. Inst. Met.* **81,** 185 (1952).

99. S. Lundquist and S. Carlen, *Erzmetall* **9,** 145 (1956).

100. U.S. Pat. 2,897,107 (July 28, 1959), S. Carlen and S. Lundquist (to Bolidens Gruvaktiebolag).

101. S. Harper, A. R. Goreham, and J. Willmott, *J. Inst. Met.* **93,** 405 (1965).

102. K. E. Mackay and G. Armstrong-Smith, *J. Inst. Min. Metall.* **75C,** 269 (1966).

103. D. Muller, F. Pawlek, and H. Wever, *Z. Metallk.* **57,** 175 (1966).

104. D. A. Reese and L. W. Condra, *Wire J.* **2**(7), 42 (1969).

105. "Analytical Methods Spectroscopy, Chromatography," *Annual Book of Standards,* Part 42, American Society for Testing and Materials, Philadelphia, Pa., 1975, pp. 336–340.

106. W. C. Cooper, "Copper" in I. M. Koltloff and P. J. Elving, eds., *Treatise on Analytical Chemistry,* Part II, Sect. A, Vol. 3, Interscience, New York, 1961, pp. 1–37.

107. *Chemical Analysis of Metals and Metal Bearing Ores, Annual Standards,* Part 12, American Society of Testing and Materials, Philadelphia, Pa., 1976, p. 252.

108. W. T. Elwell and I. R. Scholes, *Analysis of Copper and Its Alloys,* Pergamon Press, New York, 1967.

109. R. S. Young, *Chemical Analysis in Extractive Metallurgy,* Chas. Griffin and Co., Ltd., London, Eng., 1971.

110. Ref. 25, Chapt. 13.

111. G. E. Peterson, *Atomic Absorption Newsletter (Perkin Elmer)* **16**(6), 133 (Nov.–Dec. 1977).

112. N. H. Furman, ed., *Scotts Standard Methods of Chemical Analysis,* 6th ed., Vol. 1.

113. *Analytical Methods—Spectroscopy; Chromatography, Annual Standards,* Part 42, American Society Testing and Materials, Philadelphia, Pa., 1975.

114. R. L. P. Butler and M. L. Kokot in R. E. Wainerdi and E. A. Uken, eds., *Modern Methods of Geochemical Analysis,* Plenum, New York, 1971, pp. 169–204.

115. H. G. Ansell, *Can. Min. Met. Bull.* **66**(736), 93 (Aug. 1973).

116. A. P. Langheinrich and D. B. Roberts, "Optical Emission Spectroscopy" in ref. 114, Chapt. 7.

117. T. H. Irvine, *The Chemical Analysis of Electroplating Solutions,* Chemical Publishing Co., Inc., New York, 1970.

118. S. Greenfield, *SPEX Speaker,* Vol. 22, Number 3, SPEX Industries, Metuchen, N.J., Sept. 1977, p. 1.

119. C. H. McBride, *Atomic Absorption News Letter (Perkin Elmer)* **3**(11), 144 (Dec. 1964).

120. E. R. Larsen, *Anal. Chem.* **46,** 1131 (1974).

121. G. K. Murthy, U. S. Rhea, and J. T. Pesler, *Environ. Sci. Technol.* **7**(11), 1042 (1975).

122. T. Grewling, *The Chemical Analyses of Plant Tissue,* Cornell University, Ithaca, New York, 1966, p. 50.

123. M. E. Tatro, W. L. Raynolds, and F. M. Costa, *Atomic Absorption Newsletter (Perkin Elmer)* **16**(6), 143 (Nov.–Dec. 1977).

124. H. D. Chapman and P. F. Pratt, *Methods of Analysis for Soils Plants and Waters,* University of California, Riverside, Calif., 1961, pp. 105–114.

125. *J. Water Poll. Control Fed.* **47,** 1711 (1975).

126. B. J. Mathis and T. F. Cummings, *J. Water Poll. Control Fed.* **45,** 1573 (1973).

127. *Methods for Chemical Analysis of Water and Wastes,* EPA, Cincinnati, Ohio, 1971, p. 106.

128. A. L. Wilson, *The Chemical Analysis of Water—General Principles and Techniques, Analytical Sciences Monograph No. 2,* The Society for Analytical Chemistry, London, Eng., 1974.

129. G. L. Hoffman and A. A. Duce, *Environ. Sci. Technol.* **5,** 1134 (1971).

130. D. W. Steiner and co-workers, *Appl. Spectrosc.* **25,** 270, 1971.

131. J. V. Gilfrich, P. G. Burkhalter, and L. S. Birks, *Anal. Chem.* **45,** 2002 (1973).

132. R. Osterberg, B. Sjolbert, and B. Branegard, *Final Report to INCRA,* Goteberg University, Goteberg, Sweden, Mar. 1975.

133. National Research Council Committee on Medical and Biological Effects of Enviornmental Pollutants, *Copper,* National Academy of Sciences, Washington, D.C., 1977.

134. *Toxicology of Metals—Volume II, EPA-600/1-77-022,* U.S. Environmental Protection Agency, Office of Research and Development, Health Effects Research Laboratory, Research Triangle Park, N.C., May 1977.

135. *Copper in Agriculture, Conciel International pour le Developpement du Cuivre, Geneva,* Aug. 1971.

136. Ref. 133, pp. 26–27.

137. *EPA, Effluent Guidelines and Standards for Nonferrous Metals,* 40 CFR 421, EPA, Washington, D.C., Oct. 1975.

W. M. TUDDENHAM
P. A. DOUGALL
Kennecott Copper Corporation